FOUNDATIONS *of* PALEOECOLOGY

FOUNDATIONS

Classic Papers with

OF PALEOECOLOGY

Commentaries

EDITORS

S. Kathleen Lyons,
Anna K. Behrensmeyer,
and Peter J. Wagner

THE UNIVERSITY OF CHICAGO PRESS
Chicago & London

The University of Chicago Press, Chicago 60637
The University of Chicago Press, Ltd., London

Published 2019
Printed in the United States of America

28 27 26 25 24 23 22 21 20 19   1 2 3 4 5

ISBN-13: 978-0-226-61817-3 (cloth)
ISBN-13: 978-0-226-61820-3 (paper)
ISBN-13: 978-0-226-61834-0 (e-book)
DOI: https://doi.org/10.7208/chicago/9780226618340.001.0001

Library of Congress Cataloging-in-Publication Data

Names: Lyons, S. Kathleen, editor. | Behrensmeyer, Anna K., editor. | Wagner, Peter J., editor.
Title: Foundations of paleoecology : classic papers with commentaries / editors, S. Kathleen Lyons, Anna K. Behrensmeyer, and Peter J. Wagner.
Description: Chicago ; London : The University of Chicago Press, 2019. | Includes index.
Identifiers: LCCN 2019007031 | ISBN 9780226618173 (cloth : alk. paper) | ISBN 9780226618203 (pbk. : alk. paper) | ISBN 9780226618340 (e-book)
Subjects: LCSH: Paleoecology.
Classification: LCC QE720 .F684 2019 | DDC 560/.45—dc23
LC record available at https://lccn.loc.gov/2019007031

⊗ This paper meets the requirements of ANSI/NISO Z39.48–1992 (Permanence of Paper).

Contents

PART TWO

PART FIVE

Preface

In many ways, the genesis of this book begins with the original formation of the Evolution of Terrestrial Ecosystems Program (ETE) at the National Museum of Natural History in 1987. At the time, Kay Behrensmeyer, John Damuth, Bill DiMichele, Rick Potts, and Scott Wing proposed that much more could be understood about evolution if there was a focus on its ecological context. They outlined an ambitious program to study the ecological context of evolution by evaluating changes in community and ecosystem structure and function through time. That initial five-year program is now in its 31st year and has spawned over 300 publications and numerous workshops and conferences.

The idea for this specific volume came about during discussions of the future directions of ETE following two major National Science Foundation–supported workshops involving ecologists and paleoecologists hosted by ETE at the National Museum of Natural History in 2007 and 2008. In thinking about how much we, as a scientific community, have learned about community and ecosystem dynamics over the past decades, we realized that the time was ripe to develop a list of classic papers and commentary for current and future graduate students. However, this book would have remained a nebulous idea were it not for several factors.

First, this book is truly a product of the Department of Paleobiology at the National Museum of Natural History, as all of the curators met multiple times and debated the papers included in this volume. This group includes Kay Behrensmeyer, Marty Buzas, Matt Carrano, Bill DiMichele, Doug Erwin, Brian Huber, Gene Hunt, Conrad Labandeira, Kate Lyons, Nick Pyenson, Hans Sues, Peter Wagner, and Scott Wing. We would like to thank, in particular, Doug Erwin and Brian Huber. For a variety of reasons, they did not provide a paper commentary, and so this is the only acknowledgment they will get for their efforts on behalf of this book.

Second, we'd like to thank the many paleoecologists around the globe who thoughtfully answered our emails listing the foundational papers in their subdisciplines and defending why they thought they were important. This helped us create the initial list of papers from which those included in this volume were derived. The commentary of those paleoecologists was particularly useful as we debated the significance of each of these papers.

Third, we'd like to thank the publishers and individuals who kindly allowed us to reprint their articles. Publishers include Wiley, *Nature*, *Science*, Cambridge, the Geological Society of America, Oxford University Press, *American Antiquity*, *American Journal of Science*, *American Scientist*, Missouri Botanical Garden, Springer, and the Linnean Society. We also thank Richard Bambach, David Jablonski, Dan Janzen, Leo Laporte, Larry Marshall, Christine Janis, Charles Thayer, Bruce Tiffney, Gary Upchurch, Blaire Van Valkenburgh, Geerat Vermeij, and John Warme for giving us permission to reprint their work.

Finally, we'd like to acknowledge and thank the current members of the National Science Foundation–supported ETE Research Coordination Network. These include Katie Amatangelo, Kay Behrensmeyer, Jessica Blois, René Bobe, Bill DiMichele, Andrew Du, Jussi Eronen, Tyler Faith, Nick Gotelli, Nathan Jud, Conrad Labandeira, Cindy Looy, Kate Lyons, Gary Graves, Brian McGill, David Patterson, Rick Potts, Brett Riddle, Rebecca Terry, Anikó Tóth, Hans Sues, Amelia Villaseñor, Jack Williams, and Scott Wing.

Funding for this project was provided by a National Museum of Natural History ETE

Program grant awarded to Kay Behrensmeyer, Kate Lyons, and Bill DiMichele and a National Science Foundation ETE Research Coordination Network grant awarded to Kate Lyons, Kay Behrensmeyer, and Nick Gotelli (DEB-1257625).

S. Kathleen Lyons,
Anna K. Behrensmeyer,
and Peter J. Wagner
Lincoln, Nebraska,
and Washington, DC,
January 2018

Introduction: Paleoecology as the Quintessence of Earth Studies

Peter J. Wagner, S. Kathleen Lyons, and Anna K. Behrensmeyer

In the late 1940s, a presidential address to the Paleontological Society by J. Brookes Knight touched on a question that had tacitly plagued paleontologists for years (Knight 1947): Was paleontology a geological science, a biological one, or its own science altogether? Some would have gone so far as to basically label paleontology a handmaiden of stratigraphy, as one of the primary uses of fossils was for correlating geographically distant rocks. The subsequent debate (e.g., Weller 1947, 1948; Moore 1948; Newell and Colbert 1948) might strike us as almost comical over a half century later. Nevertheless, one could revisit the issue and stand it on its head. "Geology" is, etymologically, the study of Earth. We typically think of this as the study of different basic types of rocks (sedimentary, igneous, and metamorphic) and the processes that create them, from the mineralogical ones to erosion, tectonics, chemistry, and so on. Indeed, we typically lump these together with atmospheric sciences and physical environmental sciences under rubrics such as "Earth sciences." However, nearly all of these processes and their resulting patterns could be studied on numerous planetary bodies in our solar system other than Earth! The one set of planet-shaping processes that are unique to our planet (at least within our solar system) are those relating to abundant life. And since at least the Cambrian, complex ecosystems among different elements of the biota have hugely affected basic marine and terrestrial environments. Moreover, only paleontologists can study how ecosystems have changed over time and the long-term relationships between ecosystems and the physical environment. Thus, in a slightly ironic, tongue-in-cheek sort of way that nobody would have considered in 1947, one could state that paleoecology is the truest form of geology.

It is therefore appropriate that, as paleontology began to ascend from being a mere handmaiden of stratigraphy to being head of the high table of macroevolutionary sciences, paleontologists began to aggressively address major ecological issues. This is in itself no small feat: although paleontologists in the 1940s might have lumped "ecology" within "biology," ecology itself encompasses a wide breadth of subdisciplines pertaining to every level of biotic and abiotic interactions. Moreover, these subdisciplines were rapidly evolving both methodologically and theoretically at the same time that paleontologists began to look to ecology for theory and (later) methods. To this end, we have organized the volume based on (very) general disciplines connected either with ecological theory or with ecological methods adapted to paleontological research. The first two sections address two aspects of community ecology. In some ways, these were the easiest issues for paleontologists to tackle with existing methods and techniques of the time. Their basic approaches to correlating different rock units (e.g., Shaw 1964) involved data and techniques similar to those ecologists used for ecological gradient analyses (e.g., Whittaker 1970), beta-diversity studies (e.g., Whittaker 1960, 1972), and attempts to characterize the exact structure of communities (e.g., Fisher et al. 1943; Preston 1948). Thus, paleontologists were fortuitously preadapted for a wide range of general ecological studies. Coupling this with the recent acceptance of plate tectonics also opened the door for historical biogeographical analyses in a theoretical context hitherto absent. And, of course, paleontologists were able to examine these patterns over a timescale that is orders of magnitude greater than anything ecologists could hope to study.

In addition to adapting ecological theory to issues concerning fossil data, paleontologists also began to aggressively scale up ecological theory to address the burgeoning field of macroevolution, which is addressed in the third part. From this arose assessments of the relative richness of species over time, culminating in the iconic historical diversity patterns documented by Sepkoski (e.g., Sepkoski 1997). However, paleontologists took this new effort well beyond "taxon counting" to basic issues involving the role of ecological mechanisms in higher taxa diversification, nonrandom associations between different taxonomic groups and environments, differential diversification of different ecologic guilds, et cetera, all of which became fair game for study. In this area in particular, one can see how paleontologists borrowed not just theory but also standard ecological methodology by scaling up population dynamic equations and modifying sampling-standardization techniques that ecologists also employed. Indeed, the issue of sampling is one of the most active areas of paleontological research, and although we are many steps removed from the original ecological papers on sampling standardization (e.g., Hurlbert 1971), the diversification of research programs centered on this attest to its importance (e.g., Marshall 1990; Alroy et al. 2001; Peters and Foote 2002; McGowan and Smith 2008). Perhaps even more important is the fact that the research program that we paleontologists hybridized from ecological research has spilled back into other biological sciences by inspiring molecular phylogeneticists to assess diversity patterns and attempt to accommodate extinction in ways that simple Yule models could never do (e.g., Nee et al. 1994; Pybus and Harvey 2000; Rabosky and Lovette 2008; Rabosky et al. 2012).

The fourth section emphasizes that a good sword cuts both ways. Using fossils to interpret physical environments was one of the handmaidens to the stratigraphers. It therefore is unsurprising that the oldest papers in this volume deal with that topic. In fact, these papers greatly predate the evolution of modern paleoecology. Just as with stratigraphic correlation, these papers helped establish general research programs preadapted to macroevolutionary paleobiology. We can also see yet another (sub)-discipline becoming increasingly important: geochemistry. Finally, we can see hybrid vigor at work: a side effect of the expansion of paleoecology and the application of contemporary ecological methods and theory to these issues was to greatly improve our abilities to use fossils as "classical" geological tools.

At one point or another in every paleontologist's career (and perhaps in most evolutionary biologist's careers), she or he is forced to assess the most important question about evolutionary history: Who would win in a fight between Dinosaur A and Dinosaur B? This is, of course, a very particular example of the more general issue of how species interact, and although this is much more mundane to the average five-year-old, it is no less important. In many ways, this is the heart of ecology, and the heart of what truly makes our planet unique: as we note above, even if it is not absurd that simple life has evolved elsewhere within our solar system, it is only here that complex interactions among them leave such obvious signs on a whole planet. Moreover, this topic feeds into another branch of macroevolution: the idea that particular traits affect the origination and extinction rates of large portions of phylogeny (e.g., Stanley 1975) has contributed to research programs involving molecular phylogenies that are (as of this writing) thriving (Sanderson and Donoghue 1994; Maddison et al. 2007; Alfaro et al. 2009; Pyron and Burbrink 2012; Rabosky et al. 2013). At the heart of this is the ecological theory intrinsic to things as basic as the Lotka-Volterra model: the success of any one species over any length of time depends at least in part on the traits and abilities of coexisting species. The papers in part 5 address this topic.

Our final section represents a branch of paleoecology that is a unique hybrid of paleontology and ecology: taphonomy—that is, the study of how organisms fossilize. It is not just the fossils themselves that tell us about environments: how organisms are fossilized provides information about the physical environment and organismal behavior that only recently has become fully appreciated. Once considered almost a trivial pursuit within paleoecology, taphonomy has burgeoned very nearly into its own discipline within paleontology. Correspondingly, this has spilled back into other aspects of paleontological research. For example, if we are worried about numbers of sampling opportunities, then it is important to have "taphonomic controls" (Bottjer and Jablonski 1988) in which we consider only those localities that preserve organisms of the sort in which we are interested (DiMichele and Aronson 1992; Wagner 1995; Droser et al. 1997; Kosnik et al. 2011).

Within each of these sections, we have assembled a variety of influential papers. Some of these works are *Meet the Beatles* analogs: that is, those that opened paleontology to new areas of (then) modern ecological theory or methods that had not been seriously considered by paleontologists before. However, others are *Sgt. Peppers Lonely Hearts Club Band* analogs or those that irrevocably shifted an existing paleoecological paradigm into some new direction. We also have included a brief introduction to each paper written by an active researcher familiar with the importance and impact of that paper. Some of these will seem almost personal, and that's because, in some cases, these are *the* papers that inspired the research programs of the people writing the introductions. To that end, we emphasize that paleoecology is still an evolving field. There are many papers that have appeared since 2000 that are pushing paleoecology in still new directions, such as complex modeling of sea-level patterns and environmental shifts, relative abundance distributions, and the extent to which patterns driving sedimentation also drive ecological and evolutionary processes. And, of course, it cannot be understated how important current paleoecological research probably will be for understanding and predicting the effects of climate change and for informing conservation biology efforts. If that means that some of the papers included here are "great-grandfathered" beyond the veil of acceptable numbers of references, then this is actually a good thing: it shows that the field is evolving and relevant. However, that also makes this an excellent time to stop and review how we have graduated from handmaiden to high table.

Literature Cited

Alfaro, M. E., F. Santini, C. Brock, H. Alamillo, A. Dornburg, D. L. Rabosky, G. Carnevale, and L. J. Harmon. 2009. Nine exceptional radiations plus high turnover explain species diversity in jawed vertebrates. *PNAS* 106:13410–14.

Alroy, J., C. R. Marshall, R. K. Bambach, K. Bezusko, M. Foote, F. T. Fürsich, T. A. Hansen, et al. 2001. Effects of sampling standardization on estimates of Phanerozoic marine diversity. *PNAS* 98:6261–66.

Bottjer, D. J., and D. Jablonski. 1988. Paleoenvironmental patterns in the evolution of post-Paleozoic benthic marine invertebrates. *Palaios* 3:540–60.

DiMichele, W. A., and R. B. Aronson. 1992. The Pennsylvanian-Permian vegetational transition: a terrestrial analogue to the onshore-offshore hypothesis. *Evolution* 46:807–24.

Droser, M. L., D. J. Bottjer, and P. M. Sheehan. 1997. Evaluating the ecological architecture of major events in the Phanerozoic history of marine invertebrate life. *Geology* 25:167–70.

Fisher, R. A., A. S. Corbet, and C. B. Williams. 1943. The relation between the number of species and the number of individuals in a random sample of an animal population. *Journal of Animal Ecology* 12:42–48.

Hurlbert, S. H. 1971. The nonconcept of species diversity: a critique and alternative parameters. *Ecology* 52: 577–86.

Knight, J. B. 1947. Palentologist or geologist. *Geological Society of America Bulletin* 58:281–86.

Kosnik, M. A., J. Alroy, A. K. Behrensmeyer, F. T. Fürsich, R. A. Gastaldo, S. M. Kidwell, M. Kowałewski, R. E. Plotnick, R. R. Rogers, and P. J. Wagner. 2011.

Changes in the shell durability of common marine taxa through the Phanerozoic: evidence for biological rather than taphonomic drivers. *Paleobiology* 37: 303–31.
Maddison, W. P., P. E. Midford, and S. P. Otto. 2007. Estimating a binary character's effect on speciation and extinction. *Systematic Biology* 56:701–10.
Marshall, C. R. 1990. Confidence intervals on stratigraphic ranges. *Paleobiology* 16:1–10.
McGowan, A. J., and A. B. Smith. 2008. Are global Phanerozoic marine diversity curves truly global? a study of the relationship between regional rock records and global Phanerozoic marine diversity. *Paleobiology* 34: 80–103.
Moore, R. C. 1948. Stratigraphical paleontology. *Geological Society of America Bulletin* 59:301–26.
Nee, S., E. C. Holmes, R. M. May, and P. H. Harvey. 1994. Extinction rates can be estimated from molecular phylogenies. *Philosophical Transactions of the Royal Society of London B* 344:77–82.
Newell, N. D., and E. H. Colbert. 1948. Paleontologist: biologist or geologist? *Journal of Paleontology* 22: 264–67.
Peters, S. E., and M. Foote. 2002. Determinants of extinction in the fossil record. *Nature* 416:420–24.
Preston, F. W. 1948. The commonness and rarity of species. *Ecology* 29:254–83.
Pybus, O. G., and P. H. Harvey. 2000. Testing macroevolutionary models using incomplete molecular phylogenies. *Philosophical Transactions of the Royal Society of London B* 267:2267–72.
Pyron, R. A., and F. T. Burbrink. 2012. Trait-dependent diversification and the impact of palaeontological data on evolutionary hypothesis testing in New World ratsnakes (tribe Lampropeltini). *Journal of Evolutionary Biology* 25:497–508.
Rabosky, D. L., and I. J. Lovette. 2008. Density-dependent diversification in North American wood warblers. *Philosophical Transactions of the Royal Society of London B* 275:2363–71.
Rabosky, D. L., F. Santini, J. Eastman, S. A. Smith, B. Sidlauskas, J. Chang, and M. E. Alfaro. 2013. Rates of speciation and morphological evolution are correlated across the largest vertebrate radiation. *Nature Communications* 4:nc2958.
Rabosky, D. L., G. J. Slater, and M. E. Alfaro. 2012. Clade age and species richness are decoupled across the eukaryotic tree of life. *PLoS Biology* 10:e1001381.
Sanderson, M. J., and M. J. Donoghue. 1994. Shifts in diversification rate with the origin of angiosperms. *Science* 264:1590–93.
Sepkoski, J. J., Jr. 1997. Biodiversity: past, present and future. *Journal of Paleontology* 71:533–39.
Shaw, A. B. 1964. *Time in Stratigraphy*. New York: McGraw-Hill.
Stanley, S. M. 1975. A theory of evolution above the species level. *PNAS* 276:56–76.
Wagner, P. J. 1995. Stratigraphic tests of cladistic hypotheses. *Paleobiology* 21:153–78.
Weller, J. M. 1947. Relations of the invertebrate paleontologist to geology. *Journal of Paleontology* 21: 570–75.
———. 1948. Paleontologist: biologist and geologist. *Journal of Paleontology* 22:268–69.
Whittaker, R. H. 1960. Vegetation of the Siskiyou Mountains, Oregon and California. *Ecological Monographs* 30:279–338.
———. 1970. *Communities and Ecosystems*. New York: Macmillon.
———. 1972. Evolution and measurement of species diversity. *Taxon* 21:213–51.

PART

1 Community and Ecosystem Dynamics

Edited by S. Kathleen Lyons, Cindy V. Looy, and Surangi Punyasena

Understanding the structure and function of ecological communities and how they change over time and space has long been of fundamental interest in modern ecological theory (e.g., Gleason 1926; Clements 1936). Yet the idea that you could study ecological communities from the fossil record and ask questions about how they have changed in response to perturbations, climate change, or simply long periods of time took much longer to make its way into the paleontological literature. Prior to the publication of papers like the ones highlighted in this section, much of paleontology was focused on tree of life–type research: finding and identifying new species and determining their likely evolutionary relationships, tempo, mode, and biostratigraphical usefulness.

The papers highlighted here provide us with a timeline of how the field of community and ecosystem dynamics within the paleobiological community evolved. Some papers address fundamental questions in ecological theory, including how communities respond to abiotic factors (e.g., Walker and Laporte, Davis, Scott and Jones) and the role of nutrient supply and energy flow through communities (Valentine). Other studies researched changes in functional ecology (Wolfe and Upchurch, Van Valkenburgh), changes in community structure beyond species richness (Buzas and Gibson, Davis), response of communities and ecosystems to major geological perturbations (Wolfe and Upchurch, Marshall et al.), and even the role of ecological processes such as competition on evolutionary dynamics (Marshall et al.).

Some of these studies helped create new research lines within their discipline. Wolfe and Upchurch's use of multiple plant proxies to reconstruct vegetational and environmental changes is now widely applied in paleobotany. Scott and Jones were forerunners in the relatively new field of fire history. Others introduced methods that are now the standard in their field; these include Davis's volumetric pollen counts and Van Valkenburgh's (paleo) guild studies. Walker and Laporte researched how Early Paleozoic shallow marine communities in different depositional environments were structured and how these communities evolved through time. It was studies like these that eventually lead to the large meta-analyses of marine communities through time that we are now so familiar with. The findings and concepts in the three remaining papers (Buzas and Gibson, Valentine, Marshall et al.) have become standard elements in our paleobiological discourse.

All these papers have in common that they were novel and unique when they were published. Their authors were among the first who attempted to study the ecology of communities and ecosystems over geologic timescales. They are all the more impressive when you consider that they were asking questions about ecological dynamics using data that were mostly collected for taxonomic purposes, rather than specifically to address the kinds of questions that they were asking. A further common thread among these papers is the clever and careful methods the authors developed to explore community and ecosystem function and dynamics using data fraught with potential flaws.

Unfortunately, the fundamental importance of these papers in introducing community and ecosystem dynamics to paleoecology is not always evident from their citation rates. Nevertheless, successive generations of scientists have profited from the compelling concepts introduced in these papers, which—often unwittingly—have contributed to, or even triggered, their current research. Indeed, each of these papers has inspired a host of similar studies, and it can be argued that there are whole subdisciplines of paleoecology that can be traced back to these original papers.

Although arguments can be made in favor of other papers that could easily have been included in this chapter and that may be cited more frequently, these papers span a wide range of ecological communities and taxonomic groups, and each represents a major step forward not only in our understanding of community and ecosystem dynamics but also in the kinds of questions that are answerable using the fossil record.

Literature Cited

Clements, F. E. 1936. Nature and structure of the climax. *Journal of Ecology* 24:252–84.

Gleason, H. A. 1926. The individualistic concept of plant association. *Bulletin of the Torrey Botany Club* 53: 7–26.

PAPER 1

Congruent Fossil Communities from Ordovician and Devonian Carbonates of New York (1970)

K. R. Walker and L. F. Laporte

Commentary

MARK E. PATZKOWSKY

In the late 1960s and early 1970s, marine paleoecologists began to recognize that associations of taxa with similar relative abundances recurred in space and time (communities or paleocommunities) and that these associations could also change through time (community evolution). Most studies emphasized either environmental patterns within a short time interval (Ziegler 1965) or long-term changes in community composition with only a generalized environmental framework (Bretsky 1968; Valentine 1968). The problem confronting marine paleoecologists was how to synthesize the wide range of spatial and temporal scales available in the fossil record into a model of community evolution. In this paper, Walker and Laporte provided a framework for linking community patterns, long-term taxonomic turnover, and global diversity trends.

Walker and Laporte compared the community structure of Late Ordovician (ca. 460 Ma) and Early Devonian (ca. 390 Ma) intertidal to shallow subtidal carbonate environments. They established the depositional environments based on detailed sedimentologic analyses to avoid the circular reasoning of determining depositional environments based on the biotic composition of assemblages. Notably, communities in these environments, separated by nearly 70 million years, were similar in taxonomic composition at high taxonomic levels. For example, each of the subtidal environments were characterized by similar taxa of calcareous algae, stromatoporoid sponges, colonial tabulate corals, solitary rugose corals, nautiloid cephalopods, dalmanellid brachiopods, high-spired snails, and bryozoans. Other aspects of the community structure were also similar between the two time periods, such as the onshore-offshore increase in taxonomic richness and the onshore-offshore decrease in eurytopes. Walker and Laporte also recognized some key differences between the two time intervals, such as the abundance of trilobites in the Ordovician communities, which are absent from the Devonian communities, and the abundance of tentaculitids in the Devonian, which are rare or absent in the Ordovician communities.

The main contribution by Walker and Laporte was to develop an approach to time-environment analysis that would be used subsequently by paleoecologists to study community change over evolutionary time (Jablonski et al. 1983; Sepkoski and Miller 1985). Detailed sedimentologic analyses establish an environmental framework, which is then compared through time. The recognition of long-term stability of community composition at higher taxonomic levels is a phenomenon later developed more fully with the recognition of Ecologic Evolutionary Units (Boucot 1983) and evolutionary faunas (Sepkoski 1981). Taxonomic differences between the Ordovician and Devonian marine communities generally reflect the global diversity trends of major groups, with trilobites waning in the Devonian and tentaculitids at peak diversity. Thus, local community abundance tends to reflect global diversity trends. Although Walker and Laporte did not make this connection, this study and others like it (e.g., West 1976) provided the basic framework for studying the link between local communities and environments and the waxing and waning of clades at the global scale.

Literature Cited

Boucot, A. J. 1983. Does evolution take place in an ecological vacuum? II. *Journal of Paleontology* 57:1–30.

Bretsky, P. 1968. Evolution of Paleozoic marine invertebrate communities. *Science* 159:1231–33.

Jablonski, D., J. J. Sepkoski Jr., D. J. Bottjer, and P. M. Sheehan. 1983. Onshore-offshore patterns in the evolution of Phanerozoic shelf communities. *Science* 222:1123–25.

Sepkoski, J. J., Jr. 1981. A factor analytic description of the Phanerozoic marine fossil record. *Paleobiology* 7 (1): 36–53.

Sepkoski, J. J., Jr., and A. I. Miller. 1985. Evolutionary faunas and the distribution of Paleozoic benthic communities in space and time. In *Phanerozoic Diversity Patterns*, edited by J. W. Valentine, 153–90. Princeton, NJ: Princeton University Press.

Valentine, J. W. 1968. Evolution of ecological units above the population level. *Journal of Paleontology* 42 (2): 253–67.

West, R. R. 1976. Comparison of seven Lingulid communities. In *Structure and Classification of Paleocommunities*, edited by R. W. Scott, and R. R. West, 171–92. Stroudsburg, PA: Dowden, Hutchinson, and Ross.

Ziegler, A. M. 1965. Silurian marine communities and their environmental significance. *Nature* 207:270–72.

JOURNAL OF PALEONTOLOGY, V. 44, NO. 5, P. 928–944, 10 TEXT-FIGS., SEPTEMBER 1970

CONGRUENT FOSSIL COMMUNITIES FROM ORDOVICIAN AND DEVONIAN CARBONATES OF NEW YORK

KENNETH R. WALKER AND LÉO F. LAPORTE
University of Tennessee, Knoxville, Tenn. and Brown University, Providence, R.I.

"Behind the history of every sedimentary rock there lurks an ecosystem, but what one sees first is an environment of deposition . . . The central problem of paleoecology, one that is approachable only through historical data, would appear to be the evolution of stable community structure."

—Edward S. Deevey

ABSTRACT—Comparison of the shallow water, marine carbonate rocks of the Black River Group (Ordovician) and Manlius Formation (Devonian) of New York shows strong similarities in lithofacies and biofacies. The lithologic resemblances record similar sedimentologic responses to slow epeiric transgression; the parallelism of the fossil assemblages at high taxonomic levels reflects equivalent ecologic control of major adaptive types.

Some of the more obvious facies similarities are seen in the supratidal and intertidal characterized by laminated, dolomitic limestones and dolomites with birdseye, mudcracks, algal mats, leperditiid ostracodes, and vertical U-shaped and straight burrowing suspension feeders. The subtidal, level bottom, facies which are massive, pelletal carbonate mudstones contain codiacean algae: *Hedstroemia* (Black River)/*Garwoodia* (Manlius); stromatoporoids: *Stromatocerium/Syringostroma;* solitary rugose corals: *Lambeophyllum/Spongophylloides;* tabulate corals: *Foerstephyllum/Favosites;* dalmanellid brachiopods: *Dalmanella/Dalejina;* nautiloid cephalopods: *Actinoceras/Anastomoceras;* as well as ramose ectoprocts, high-spired snails, and burrowing deposit feeders.

There are, however, possibly significant discrepancies. For example, the Black River has an abundant and diversified nautiloid fauna while the Manlius has only a single, uncommon species. Trilobites are common in the Black River but rare to absent in the Manlius, whereas tentaculitids are abundant in the Manlius but absent in the Black River. The tabulate coral *Tetradium* of the Black River has no obvious Manlius ecologic equivalent, but may be paralleled in function by a laminar growth form of the Devonian stromatoporoid *Syringostroma.* Despite these differences, however, the overall biotic aspects of comparable facies of the Black River and Manlius are quite close.

Our results demonstrate the feasibility of studying fossil assemblages in the geologic record as representatives of evolving communities rather than as isolated aggregations of organisms scattered in time and space.

INTRODUCTION

DETAILED paleoecological studies in New York State of the Middle Ordovician, Black River Group (Walker, 1969 and in press) and of the Lower Devonian, Manlius Formation (Laporte, 1967, 1969) show mutually similar carbonate lithofacies and biofacies of tidal flat and shallow subtidal origin. The resemblances between lithofacies—as indicated by constituent grain compositions, textures, and primary structures—record equivalent sedimentologic responses to relatively slow marine transgressions which occurred during the medial Ordovician and early Devonian in New York. The resemblances between biofacies—as indicated by relative organism abundances, dispersals, taxonomic affinities, and inferred modes of life—record similar ecologic controls upon the biotas exerted by the corresponding carbonate depositional environments that developed during each of these two marine transgressions.

Given the strong parallelism between the biotas of these two stratigraphic units, we might examine how the benthic communities of each are structured and arranged as well as document any replacement of taxa or changes in adaptive type occurring over the 60–70 million year interval separating Black River and Manlius deposition.

Most studies in the history of life have, for good and obvious reasons, documented and interpreted evolutionary phenomena expressed by diversification of taxa or by progressive trends within phyletic lineages. More recently, however, paleontologists have been turning their attention to evolution operating at the level of communities of taxonomically diverse organisms having various adaptive functions and ways of life. For example, Olson (1966) has defined three sorts of terrestrial vertebrate communities using food-chain relationships among community members and traced the changes in taxonomic composition and adaptive types from one community to another during the late Paleozoic and Mesozoic. He has described what is, in essence, a "phylogeny" of terrestrial vertebrate communities. Bretsky (1968) has done something similar for Paleozoic marine invertebrates within terrigenous clastic, onshore and offshore, shelf environments. He recognizes five major associations composed of stratigraphically delimited communities and notes their secular variations in generic composition and broad environmental setting. Valentine (1968) has provided a more general scheme which by implication includes both Olson's and Bretsky's approaches and provides a theoretical model for describing and interpreting community evolution. This model views ecological systems as a hierarchy from individual organisms to higher and increasingly more inclusive levels of populations, communities, provinces, and the biosphere. Evolution obviously takes place at all levels and involves changes in niches, niche occupants, and relative proportions and abundances of constituent taxa. Within this broader framework, this paper provides a rather detailed exmination of carbonate tidal flat and shallow subtidal marine communities at two widely separated points in time.

ACKNOWLEDGEMENTS

We acknowledge the assistance of a number of colleagues in developing the data and concepts of this paper. John Imbrie, K. M. Waage, D. C. Rhoads, and D. W. Fisher critically read certain earlier manuscripts related to this paper.

Financial support for part of the Manlius research was provided by NSF grants GA 407 and GA 1519 to Laporte. Walker's research was supported by the Museum and Science Service (Geological Survey) of New York state and by various grants from the Department of Geology, Yale University. Preparation of illustrations was supported by the Department of Geology, University of Tennessee.

STRATIGRAPHY AND FACIES

Black River Group.—The Black River Group in New York is a sequence of Middle Ordovician limestones and dolomites that crop out along the Black River valley and St. Lawrence-Lake Ontario lowlands on the southwestern side of the Adirondack Mountains (Text-figs. 1, 2). In the Black River Valley area studied by Walker the group varies in thickness from 60 to 150 feet and includes in ascending order the Pamela, Lowville, House Creek, and Chaumont formations (Walker, in press). These units consist of a number of different carbonate facies representing seven major environments. These environments existed contemporaneously along the medial Ordovician shoreline and arranged in onshore-offshore position include: 1) supratidal dolomitic mudflats, 2) high intertidal, pelletal mudflats, 3) shallow subtidal or low intertidal, quiet water muds, 4) shallow subtidal or low intertidal, channel bioclastic muds and sands, 5) subtidal wave baffles constructed by tabulate corals (*Tetradium*), 6) subtidal, wave baffle margin, bioclastic sands, and 7) subtidal, level-bottom, bioclastic muds. Although each stratigraphic formation is composed dominantly of the sediments of one or two of these environments, the overall aspect of the group is one of a complex facies mosaic that formed as these laterally adjacent environments migrated from place to place.

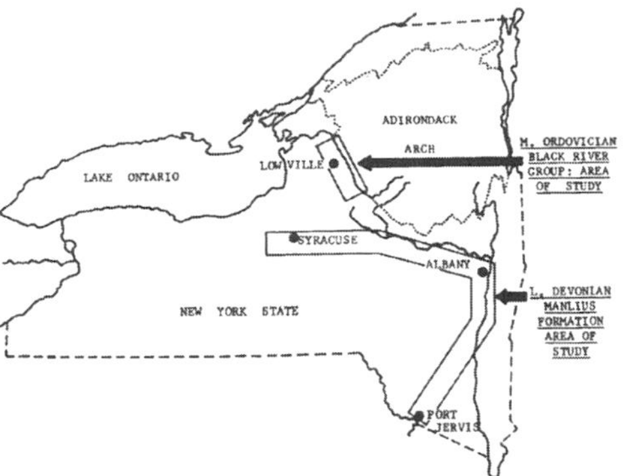

TEXT-FIG. *1*—Area of study of Middle Ordovician, Black River Group and Lower Devonian Manlius Formation in New York State.

Some 33 major fossil taxa have been recognized in the Black River Group of New York. These include not only the usual shelly invertebrates but also algal organo-sedimentary structures, burrows, and calcareous algae. These taxa occur in seven major communities which closely parallel the seven major depositional environments.

More complete and detailed discussion of the Black River Group can be found in Walker (1969, and in press).

Manlius Formation.—The Manlius Formation in New York is a Lower Devonian carbonate unit, 25 to 50 feet thick, that crops out from central to eastern New York and along the Hudson River Valley (Text-figs. 1, 2). The unit contains three major carbonate facies representing three sedimentary environments. Although the facies within the Manlius are more broadly defined than those in the Black River the same spectrum of environments exists, recording deposition along an early Devonian shoreline. In an onshore to offshore profile, these environments are : 1) supratidal dolomitic mudflats, 2) intertidal pelletal mudflats with oc-

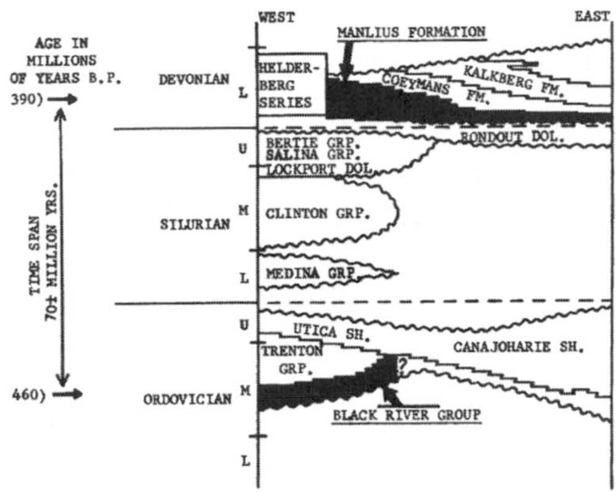

Tex-fig. 2—Simplified stratigraphic cross-section of Black River Group and Manlius Formation in New York. Note that both units are time-transgressive, becoming younger to east and west, respectively.

casional bioclastic sands, and 3) shallow subtidal, restricted lagoon muds, often with tabular accumulations of stromatoporoids. As in the Black River Group, certain horizons within the Manlius (usually designated as members) are dominantly composed of one or two facies. But again, the overall stratigraphic impression is one of complex lateral and vertical facies variations owing to secular changes in depositional regime.

Twenty-five to 30 major fossil taxa are reported from the Manlius, including algal structures, calcareous algae, and burrows as well as shelly invertebrates. As in the Black River Group, individual taxa occur characteristically in one of the three major sedimentary environments thereby forming distinct biofacies patterns which correspond to the lithofacies.

The Manlius is interpreted as having formed in a tidal flat–lagoon on the landward side of a broad epeiric sea within the central Appalachians (Laporte, 1969). Further discussion and interpretation of the Manlius Formation can be found in Laporte (1967).

Facies definition.—Both the Black River and Manlius carbonates record deposition in very shallow marine environments. Although there is clear evidence of periodic subaerial exposure in these environments, there is no direct way of telling how long any one subaerial or submarine "event" lasted. Fluctuations in local sea level can be caused by other phenomena than daily lunar tides. Thus, seasonal storms, periodic barometric changes, or shifting wind directions can significantly vary local sea level enough to inundate subaerially exposed sediment surfaces of low relief with several inches or feet of sea water. More likely than not these were the sorts of sea level fluctuations recorded in the Black River and Manlius units. The many scour surfaces, intraclasts, deep mudcracks, shell pavements, and early dolomitization suggest long intervals of subaerial exposure followed by innundation of marine waters, often with considerable turbulence.

Hence, the terms supratidal, intertidal, and subtidal are used here with broader connotations than usual. Supratidal is used to suggest that the carbonate sedimentary interface was only occasionally covered by sea water; intertidal, to suggest that approximately half the time the sediments were covered with sea water, the other half of the time subaerially exposed; and subtidal, to indicate that the sediment interface was virtually always within the marine environment. Moreover, these three environments are seen as somewhat idealized end-members of what was, of course, a complete intergrading depositional regime.

ENVIRONMENTAL CRITERIA

Recent studies of present day marine carbonate environments have provided reliable criteria for their identification in the ancient rock record. Many of these criteria, particularly for recognizing tidal flat carbonate deposits, are physical in nature. Others, while determined by organic activity, are sufficiently obvious that no refined or elaborate interpretations are required. It should be emphasized that any single criterion alone is not sufficient to indicate environment; it is the combination of several criteria which can be accurately interpreted. These criteria as listed in Table 1 have the combined virtue of not only being valuable in defining ancient tidal flat and shallow subtidal deposits but also preclude any assumptions or *a priori* interpretations about the ecology of the included organisms. Thus, by using these criteria, environmental interpretations of a stratigraphic unit can be made using lithofacies data alone. Having then recognized the environments, the organic components associated with the lithofacies can be interpreted independently thereby avoiding any circularity in reasoning about the organisms' ecology. This is especially important in studies of Paleozoic fossil assemblages, most of the members of which are extinct, without closely related living forms, and therefore had ecologic requirements as yet unknown or uncertain.

Because the interpretation of these criteria is so central to the environmental inferences and ecologic conclusions that follow, we will review them in some detail.

TABLE *1*.—Lithologic, paleontologic, and primary structural criteria for identifying tidal flat and shallow subtidal carbonate environments. X = features are common and typical; O = features are present. (*See* Roehl 1967, for integrated discussion of the environmental significance of many of these criteria as well as additional references.)

Criterion	Supra-tidal	High Inter-tidal	Low Inter-tidal	Sub-tidal	Main Reference Source
Birdseye	X				Shinn, 1968
Mudcracks	X	X	O		Ginsburg, 1957
Early Dolomitization	X	O			Pray and Murray, 1965
Intraclasts	O	X	X		Ginsburg, 1957
Massive Laminated Bedding	X				Logan and others, 1967
Massive Lumpy Bedding				X	Imbrie and Buchanan, 1965
Thin-Medium Bedding; Scour-and-fill		X	X		Ball and others, 1967
Algal Laminae	X	X	O		Logan and others, 1964
Algal Stromatolites		X	O		
Algal Oncolites			O	X	
Vertical Burrows	O	X	O		Rhoads, 1967
Horizontal Burrows			O	X	Rhoads, 1967
Rare Fossils	X				—
Few Taxa, Many Individuals		X	X		Laporte, 1968
Many Taxa, Many Individuals				X	Laporte, 1968

Birdseye.—Birdseye fabric is a petrographic term used in carbonate rock description for calcite or anhydrite filled voids or vugs several millimeters in diameter. (The term "birdseye" alludes to the glassy glint of these structures when seen on a freshly fractured rock surface.) Until recently the origin of birdseye fabric was in doubt and was usually attributed to formation of voids by algal activity, burrowers, gas bubbles, or internal desiccation; the voids later being infilled with secondarily precipitated calcite or anhydrite. Shinn (1968) in working with recent carbonate sediments has now shown, however, that these voids are formed by either gas bubbles (in the case of the more circular voids) or shrinkage resulting from desiccation (in the case of the more elliptical voids). Moreover, Shinn has established that both kinds of birdseye are common and typical in supratidal carbonate sediments, only sometimes present in intertidal sediments, and absent in subtidal sediments.

Mudcracks.—Polygonal desiccation cracks are often well developed in carbonate sediments that are exposed to the air. Although shape and depth of these mudracks depend on a number of variables such as thickness of the sedimentary layer being desiccated, grain and textural properties of the sediment, and local ardidity, it is usually the case that these mudcracks develop in the supratidal and intertidal zones. While some mudcracks have been reported from the shallow subtidal, these seem to be instances where intertidal sediments have been desiccated and at the time of the observation were covered by a few inches or feet of sea water owing to a temporary and ephemeral high water level. Ginsburg (1957) provides a brief but good discussion of mudcracks in the recent carbonates of South Florida.

Early dolomite.—Early replacement dolomite is a characteristic feature of supratidal and intertidal carbonate sediments in areas of high aridity (Pray and Murray, 1965). Although the kinetics of the process are not completely understood, it seems that sea water, either as pore water or as ponded surface water, increases in magnesium ion concentration sufficiently during evaporation to cause precipitation of, and replacement by, dolomite within the surrounding carbonate sediments. An alternate hypothesis for the origin of interlaminated limestone-dolomite carbonates laminated on a millimeter to centimeter scale (rather than massive dolomite) is that of Gebelein and Hoffman (1969) who note that present day algal mats in the intertidal and supratidal zones concentrate magnesium ions several fold. They suggest that it is this initial, algally induced magnesium concentration, rather than sea water evaporation, which determines the early dolomitization. Both hypotheses for early dolomite, however, involve mechanisms operating within the supratidal and intertidal environment.

Intraclasts.—Carbonate sediments exposed to the air will quickly dry out and become indurated. Further exposure results in desiccation cracking and weathering so that hardened carbonates will tend to break up into fragments or intraclasts of various sizes. Later flooding by high tides (from whatever cause) re-distributes these intraclasts across the sediment surface within the supratidal and intertidal zone. Carbonate intraclasts will also form along the banks of tidal flat channels or shores Ginsburg. (1957) discusses early lithification and induration of tidal flat carbonates with resulting formation of carbonate-pebble conglomerates.

Massive laminated bedding.—Irregularly to regularly laminated sediments, occurring in massive beds, are typical of tidal flat carbonate environments. These laminations are due to sediment trapping and binding by algal mats growing on sediment surfaces (Logan and others, 1964). While algal mats are also found in subtidal carbonate environments (Gebelein, 1969) they tend to be disrupted and non-continuous in normal marine situations by the abundant algal-grazing and sediment burrowing organisms found there. Hence, the preservation of abundant laminated bedding and lack of marine organisms is presumptive evidence of supratidal and intertidal environments. Subaerial laminated carbonates can also form in association with soil formation and weathering processes (Multer and Hoffmeister, 1968). Although these are often difficult to differentiate, the distinction should be made for accurate environmental interpretation.

Massive lumpy bedding.—Subtidal carbonates will have disrupted, mottled fabrics when the benthic biotas are so abundant that the accumulating sediments are constantly being biogenically re-worked. Of course, if processes of current re-working are great or salinities too high or low, some subtidal carbonates (e.g. oolite shoals, hypersaline lagoons) may retain primary bedding characteristics and lack the lumpy, mottled textures that result from organic re-working. Imbrie and Buchanan (1965) discuss the formation of primary structures in recent carbonate environments and their disruption by organisms. (See also Moore and Scruton, 1957; Rhoads, 1967.)

Thin-medium bedding, scour-and-fill.—In carbonate tidal flat environments water level changes are, by definition, frequent and characteristic. Related to these variations in water level are significant fluctuations in water turbulence, texture of sediment transported, and sediment induration. For example, carbonate muds which settled out of suspension on a previous tidal flooding will, upon subaerial exposure, become hardened and mudcracked. Later re-flooding of the flat will partially erode the carbonate mud layer and re-distribute the mudcracked clasts. Mixed with these carbonate intraclasts may be skeletal debris transported from the marginal subtidal environment. Finally, before the high waters recede, a layer of carbonate mud may settle out from the standing water which covers the tidal flat. After the waters recede, the flats are once again subaerially exposed and desiccated.

The physical stratigraphic result of these processes will be a sequence of sedimentation couplets composed of carbonate muds a few inches thick overlain "unconformably" by carbonate-pebble conglomerates and bioclastic sands of about the same thickness. Thus, thin to medium bedded, alternating carbonate mudstones with scour-and-fill surfaces, pebble-conglomerates, and skeletal lags are suggestive of tidal flat sedimentation. Ball and others (1967) discuss the geologic effects of storm tides on tidal flats and their analogues in the stratigraphic record.

Algal laminae, stromatolites, and oncolites.—Blue-green and green filamentous algae are today abundant in shallow seas. These algae form mats on wet sediment surfaces both in the shallow subtidal and periodically moistened intertidal and supratidal zones. The mats tend to bind sediment and inhibit their erosion (Neumann and others, 1969). Moreover, the blue-green algae secrete mucilaginous sheaths which trap sedimentary grains as they move across the mat surface. Successive algal mats will thus form structures consisting of alternating laminations of algal-rich and sediment-rich layers. Typically, the algal-rich layers soon decompose and the irregular void spaces that result are later infilled with secondarily precipitated calcium carbonate with no obvious evidence of the original algae being preserved (Logan and others, 1964). As these authors have further shown, the gross geometry of these "organo-sedimentary structures" is closely related to frequency and strength of water currents flowing across the algal matted, sediment surface. Although many individual exceptions exist, in general regular to irregular continuous laminae are typical of the higher part of the tidal flat environment, hemispherical heads attached to the substrate ("stromatolites") are typical of the middle portion, and free-lying unattached, algally laminated grains ("oncolites") are typical of the lower part. Other criteria, as discussed above, will usually be sufficient to corroborate just where within the tidal range these structures have formed in a given ancient situation.

Burrows.—Marine burrowers which live within the sediment substrate for either its sheltering effect or for its food content tend to increase in abundance with distance from the strandline. As Rhoads (1967) has shown, accompanying this increase in burrowing activity is a change in the style of burrowing. Burrowers in subtidal environments, where environmental conditions are relatively constant, burrow shallowly just below the sediment water interface. Moreover, those burrowers that are feeding on the sediment for its included organic matter will move laterally through the sediment leaving burrow mottles that have a well developed horizontal orientation. By contrast, burrowers in intertidal regions are forced to burrow more deeply in the sediment to avoid the ecologic stresses associated with intermittent subaerial exposure (heating or chilling, salinity changes, desiccation). In addition, many intertidal burrows are dwellings produced by suspension feeding organisms which require access to the filterable food resource in the overlying water mass. These burrows will tend, therefore, to have a well developed vertical orientation as contrasted with subtidal burrows. Consequently, burrow abundance and geometry can be important criteria for recognizing tidal flat and shallow subtidal carbonate environments.

Fossil abundance and diversity.—Marine organisms increase in abundance and diversity from tidal flat environments toward offshore, shallow subtidal marine environments. Conditions are less variable and more typically "normal" marine in the subtidal while the tidal flat environment represents, in fact, two different environments—the subaerial or terrestrial as well as the subtidal or marine. Usually, only a few organisms can tolerate the relatively frequent and drastic fluctuations in tidal flat environments. Even though diversity tends to be low in tidal flat environments, populations tend to be large because food resources are abundant and well distributed within the tidal flat regime, particularly due to the great productivity of benthic algae (Sanders and others, 1962) and the action of flood and ebb currents. Hence, trends in individual species abundance, overall abundance, and taxonomic diversity will, in conjunction with the other environmental criteria noted above, help define tidal flat and shallow subtidal environments (Laporte, 1968).

To summarize, then, these various criteria are valuable, when used together, in recognizing tidal flat and shallow subtidal environments in carbonate rocks. Moreover, they are relatively independent of any paleoecologic inferences about the associated organisms and therefore provide a reliable frame of reference for subsequent autecologic and synecologic interpretation.

FOSSIL COMMUNITY DEFINITION

We follow Johnson's (1964, p. 109) definition of "community" as an "assemblage of organisms inhabiting a specified space." Hedgpeth (1957, p. 40) employs a similar definition adding the phrase ". . . and presumed to have some relations with each other." As Thorson (1957, p. 467 ff.) indicates, once the co-occurrence of organisms is established, we may wish to know why these organisms are found in association with each other. It is at this point that further community definition and discussion loses much of its unanimity and divides into two major camps: one views communities as composed of species showing mutual interdependence and interaction, while another visualizes communities as ". . . associations of largely independent species that occur together because of similar responses to the physical environment" (Johnson, 1964). We will discuss Black River-Manlius communities as assemblages of organisms inhabiting a specified space: the "assemblage of organisms" being the fossil assemblage which includes trace fossils, algal structures, and calcareous algae as well as the shelly invertebrates, and the "space" being the inferred depositional environments. For the moment we will not attempt to explain the reasons for association.

Using the environmental criteria listed in Table 1 we have recognized four fossil communities within the Black River and Manlius carbonates. These communities are named in terms of the inferred depositional environment: supratidal, high intertidal, low intertidal, and subtidal. While ecologists studying marine benthic assemblages commonly identify the communities by one or more of the dominant constituent taxa, we have avoided this for the obvious reason that in dealing with temporally separated communities different taxa may be involved at different times. By contrast, the presumptive biotopes will show little change and therefore provide convenient labels for the communities.

All of the Manlius Formation can be assigned to one or the other of the four biotopes. In the Black River Group, however, there are several additional biotopes not represented in the Manlius (e.g. wave-baffle and wave-baffle margin). It may be noted here that the Black River Group wave-baffle biotope may be ecologically analogous to a laminar stromotoporoid biotope within the Manlius Formation subtidal facies. We have avoided making this analogy because it is much more uncertain than the comparable biotopes and communities which are discussed in detail. The Black River Group wave-baffle margin biotope existed because of certain peculiarities of growth habit of the major taxon of the wave-baffle *(Tedtradium cellulosum)*. Thus, the possible absence of comparable biotopes in the Manlius is not considered to detract significantly from the major thesis of our work. Although fully described by Walker (1969, and in press), these noncomparable biotopes are excluded from consideration here. Thus, we are only analysing comparable communities within these two stratigraphic units.

The rocks of the Black River-Manlius were assigned to one of these four biotopes using the environmental criteria discussed earlier to infer depositional environments. Next, we examined the fossils from these rocks to see which taxa occurred in which biotope. Virtually all taxa occurring with any regularity and in any significant abundance were consistently found in the same biotopes from sample to sample, and exposure to exposure. While this observation may initially seem too simple to be true, it agrees with distribution patterns delineated in Recent carbonate facies studies which show that there is nearly always a close correlation between carbonate lithofacies and biofacies patterns (Laporte, 1968). The reasons for the agreement between sediment and organism distribution are: 1) benthic organisms tend to be substrate specific; 2) carbonate sediments are, in large part, often composed of the accumulated remains of the local calcareous species; and 3) nonskeletal, calcareous sediments (oolites, mud, pellets, intraclasts) are sensitive to the same sorts of environmental parameters that determine the distribution and abundance of shelly benthic organisms (mainly water turbulence and turbidity, salinity, and circulation). Hence, strong correlation between shallow water carbonate lithofacies and biofacies should be expected. Moreover, the probability of appreciable postmortem transportation of skeletal remains is not especially great. For it has been shown in Recent carbonate environments that skeletal grains of sand-size and larger tend to accumulate in the same environments where produced (Laporte, 1968). We conclude, therefore, that fossils found in one carbonate biotope or another occurred there in life as well. This conclusion is confirmed by many occurrences in life position in both stratigraphic units. All Black River-Manlius taxa are assigned to one or more communities except for a few rare forms that have either unique occurrences or are found only a few times throughout several hundred samples. Table 2 lists Black River-Manlius taxa according to their community occurrences, and shows the major group to which each pair of taxa belongs.

AUTECOLOGY OF TAXA

In general, the ecology of the fossil taxa listed in Table 2 has been inferred from one or more of the following: 1) occurrence in life position; 2) analogy with the closest related living taxon; 3) adaptive morphology, when known, of fossil forms; and 4) consistency of organic structures with environmental parameters of the habitat as inferred independently from the enclosing rock.

Yonge (1954), Imbrie (1960), and McAlester (1968) have summarized feeding types and substrate mobility relations of major invertebrate groups. All agree that these habits remain fairly constant within higher taxonomic categories of recent invertebrates. We assume that these habits have also remained similar through time (unless there is evidence to the contrary), because a major shift in feeding type or substrate mobility would most likely be reflected in a corresponding shift in adaptive morphology resulting in different higher taxonomic assignment.

Ostracodes.—We consider the leperditiid ostracodes, *Leperditia* and *Herrmannina,* to have been benthonic rather than planktonic owing to the relatively high weight of the carapace as estimated from the thickness to length ratio. They were apparently not deep infaunal in habit as we can find no burrow traces of the appropriate size. If these ostracodes were benthic epifaunal forms, we further infer that they were surface deposit feeders or scavengers following the observations by Puri (1964) of modern benthic epifaunal ostracodes.

Berdan (1968) has concluded from her stud-

TABLE 2—Major taxa and their inferred modes of life in Black River and Manlius communities. For each pair of taxa, the first is from the Black River, the second from the Manlius. Large X in rectangle means taxon is typical of that environment, small 0 means taxon lived in that environment but only in small numbers. (See text for discussion.)

Major Taxon	Community Taxa Ordovician Black River Group/Devonian Manlius Formation	Community				Sal. Tol.		Relation to Substrate						Feeding Type				
									Epifaunal			Infaunal						
		Supratidal	High Intertidal	Low Intertidal	Subtidal	"Wide"	"Normal"	Nektobenthonic	Attached	Sessile	Vagile	Sessile	Vagile	Suspension Feeder	Deposit Feeder/ Scavenger	Browser	Predator	Primary Producer
Leperditiid Ostracoda	*Leperditia/Herrmannina*	X	X	X		X					X				X			
Filamentous Blue-Green Algae	Algal Mats/Algal Mats	X	X	X		X												X
???	Straight Vertical Burrower/Straight Vertical Burrower	O	X	O		X						X		X				
???	U-shaped Vertical Burrower/U-shaped Vertical Burrower		X			X						X		X				
Filamentous Green Algae	Algal Oncolites (Norm. Tube)/Algal Oncolites (Norm. Tube)			X	O	X												X
Strophomenid brachiopoda	*Strophomena/Mesodouvillina*			X	X	X				X				X				
Speriferid brachiopoda	*Zygospira/Howellella*			X		X			X					X				
???	Medium Burrower/Medium Burrower			X			?						X		X			
???	Large Burrower/Large Burrower			X			?						X		X			
Ramose Ectoprocta	*Stictopora*/Unidentified Trepostome			X			X		X					X				
Gastropoda	*Loxoplocus/"Loxonema"*				X		X				X					X		
Articulate Codiacean Algae	*Hedstroemia/Garwoodia*				X		X											X
Filamentous Green Algae	*Girvanella* Oncolites/*Girvanella* Oncolites				X		X											X
Stromatoporoidea	*Stromatocerium/Syringostroma*				X		X		X					X				
Solitary Rugose Anthozoa	*Lambeophyllum/Spongophylloides*				X		X		X								X	
Tabulate Anthozoa	*Foerstephyllum/Favosites*				X		X		X					X				
Nautiloid Cephalopoda	*Actinoceras/Anastomoceras*				X		X	X									X	
Dalmanellid Brachiopoda	*Dalmanella/Dalejina*				X		X		X					X				
Ramose Trepostome Ectoprocta	*Erydotrypa*/Unidentified Trepostome				X		X		X					X				
???	Small Horizontal Burrower/Small Horizontal Burrower				X		X						X		X			

ies of Paleozoic leperditiids that the smooth thick carapace, large adductor scars, and extensive ventral overlap of the valves, suggest adaptations to temporary subaerial exposure, and that subsidiary muscle scars indicate strong appendages capable of digging. Thus, she interprets these ostracodes as having "lived on tidal flats burrowing just below the surface when the tide went out, in the manner of modern horseshoe crabs and other intertidal arthropods."

Salinity tolerances of fossil ostracodes were rather wide, with smooth carapaces more common in hyposaline waters and more ornate forms increasing in abundance in more normal salinities (Benson, 1961). This supports our interpretation of Black River-Manlius ostracodes and is consistent with their abundance in the supratidal and intertidal communities.

Algal mats.—A number of recent publications (some of which we have already cited) demonstrate that blue-green algal mats are common and characteristic of tidal flat environments. They tolerate very wide salinity fluctuations, flourishing equally in fresh as well as hypersaline waters. They are quickly revived after long periods of desiccation and can tolerate very high temperatures. Besides being important sedimentologic agents as sediment trappers and binders, they also provide an important food source.

The morphology of algally produced laminae within the sediment is largely dependent upon water turbulence. For a discussion of this aspect of algal mats see above.

Burrowers.—Although the taxonomic affinities of Black River-Manlius burrowers are uncertain, their habits can be confidentally inferred. The vertical burrows, either straight or U-shaped, are inferred to have been made by suspension feeders as the burrows commonly terminate at bedding surfaces and usually lack back-fillings of pellets but instead are filled with secondarily precipitated calcium carbonate. The horizontally oriented burrows are interpreted as having been made by deposit feeders because the burrows are usually confined between bedding planes of one or two sedimentation units and are frequently back-filled with fecal pelletal sediment.

The much greater relative abundance of the horizontal burrows as compared to the vertical burrows further suggests the latter are made by suspension feeders as a more or less permanent residence while the former are continuously being made as the deposit feeding organism mines the sediment for palatable food. Thus, although one suspension feeder generally occupies only one or at most a few burrows in its lifetime, a single deposit feeder may leave a record of scores or perhaps even hundreds of burrow traces.

Lacking any data as to the taxonomic affinities of the burrowers, we cannot make any firm conclusions about their salinity tolerances except on the basis of evidence provided by associated fossil forms and inferred physical conditions of the environment of occurrence.

Brachiopods.—All Black River-Manlius brachiopods have been listed as suspension feeders based upon our knowledge of the feeding habits of modern articulate brachiopods. Taxa having a well developed pedicle opening in the adult stage are considered as attached epifaunal forms while those lacking pedicle openings as adults are interpreted as sessile, free-lying epifaunal types.

The small, pedically attached spiriferids, *Zygospira* and *Howellella,* and the somewhat larger, free-lying strophomenids, *Strophomena* and *Mesodouvillina,* belong to the low intertidal and subtidal communities. The only living intertidal brachiopods today are inarticulates which live in deep, vertical burrows and attached to the substrate with a long pedicle. No living articulates are found above the subtidal region. Apparently Black River-Manlius forms were able to withstand very brief subaerial exposure in the low intertidal areas.

Ectoproct bryozoans.—All living ectoprocts are considered to be attached epifaunal, colonial suspension feeders and presumably this was true of fossil forms as well. In the Black River-Manlius, ramose colonies of cryptostome and trepostome ectroprocts are found in both the low intertidal and subtidal communities and are usually preserved as unabraded but fragmented remains. The paleoecological significance of this growth form is not clear. While Schopf (1969) indicates that "the degree of dependence between environment and structural forms is still obscure" he presents data for living ectoprocts which suggest that erect and rigid colonies—similar to Black River-Manlius growth forms—are more characteristic of deeper, more offshore environments than shallower, inshore ones. Lagaaij and Gautier (1965), however, found the zoarial form of ectoprocts in the Rhône delta area depend more on sedimentation rate and water turbulence than simply water depth.

Gastropods.—Black River-Manlius snails, *Loxoplocus* and *"Loxonema"* are high spired forms with median labral sinuses. Both lack siphons and were probably therefore vagile epifaunal forms. *Loxoplocus* is an archaeogastropod and as such would be considered an herbivore by most workers. Recent revision of the Loxonematidae by Knight and others (1960) places "*Loxonema*" within the Caenogastropoda, a group of widely divergent feeding types. According to R. Batten (pers. comm.) the presence of median labral sinuses suggests adaptations for strong gill currents to clear the mantle cavity of mud. This interpretation is consistent with the occurrence of a muddy substrate in the subtidal biotope.

Calcareous algae.—The articulate calcareous algae, *Hedstroemia* and *Garwoodia,* occur in the subtidal communities. These fossil genera are judged by most workers (e.g. Johnson, 1961; Konishi, 1961) to be green, articulate codiacean algae and ecologically similar to the modern codiacean *Halimeda* which is abundant in shallow, well lit, warm seas. These algae, like their soft-bodied counterparts, provide an important food source within the subtidal community. As Cloud (1959) has observed, marine invertebrates and especially fish (and Paleozoic cephalopods?) will browse on calcareous algae. Forms transitional in morphology between *Hedstroemia* and *Garwoodia* have been recently discovered in the Middle Ordovician Trenton Group (see Fig. 2) of New York (Cameron, pers. comm., and Walker, pers. obs.) suggesting that these two algae may represent a phylogenetic sequence.

"*Girvanellid*" and normal-tube oncolites.—Some irregularly concentrically laminated carbonate grains are differentiated from other algal oncolites by the presence of an internal network of small tangled tubes. These oncolites are thus similar to those calcareous grains commonly found in Paleozoic limestones called *Girvanella,* which have an internal tubular microstructure and have been tentatively assigned to the blue-green algae (Johnson, 1961). Other oncolites of the Black River-Manlius show a different tube-morphology, one with straight tubes oriented normal to the coated surface. In both normal-tube and "girvanellid" oncolites individual tubes are generally greater than 50 microns in diameter. While living blue-green algae do, indeed, coat free-lying carbonate grains binding particles into irregular, multiple concentric laminations, the diameter of their filaments is but a few microns. The filamentous green algae, however, also form mats and coat carbonate grains but their filament diameters are significantly larger, ranging up to a millimeter or so. While the internal voids formed by the subsequent removal of the blue-green filaments are not usually visible in stromatolites or oncolites, those of the larger green filamentous algae are. There are also ecological differences between these two types of filamentous algae.

Unpublished studies by C. Gebelein, working with modern filamentous algae of South Florida and the Bahamas, indicate that while the blue-green algae range throughout the tidal flat and shallow subtidal environments, the green filamentous algae are essentially restricted to the continuously wet shallow subtidal and lowest intertidal. We conclude, therefore, that those Black River-Manlius oncolites having well defined, relatively large internal tubes were formed by green filamentous algae in the shallow subtidal and low intertidal environment. We are uncertain whether the two distinct types of tube orientations represent two adaptations within a single taxon or, alternatively, were formed by two (or more) separate algal taxa.

Corals.—Colonial hydrocorallines (stromatoporoids), tabulates, and solitary rugose corals occur in the Black River-Manlius. By analogy with living coelenterates these are interpreted as forms attached, at least initially, to the substrate, and feeding either as passive predators (the large polyp-bearing corals) or as suspension feeders (the small polyp-bearing stromatoporoids and tabulates). We make this distinction in feeding types because given the small size of the polyps in the latter groups it would be difficult in the case of these organisms to functionally separate passive predation of semi-microscopic food particles from filtering of the same particles, particularly when the process is being carried out by a colony of tiny polyps.

Cephalopods.—The orthoconic nautiloids, *Actinoceras* and *Anastomoceras,* are interpreted as swimming, bottom feeding predators. The greater relative abundance of nautiloids in the Black River as compared to the Manlius may be due to the post-Ordovician expansion of fishes which by the early Devonian caused significant replacement of their cephalopod competitors.

Our discussion of the inferred autecology of Black River-Manlius taxa has emphasized feeding type and substrate mobility relations. We believe the salinity tolerances of these taxa can be determined, but with less precision. As shown in Table 2, we judge individual taxa as having either "normal" or "wide" salinity tolerance. These interpretations are based somewhat loosely on the overall environmental framework for each stratigraphic unit as inferred from all our data. Although we cannot specify what these ancient salinities might have been, we wish to imply that "normal" salinity tolerance indicates adaptation to non-variable salinities of environments covered by normal, open-ocean

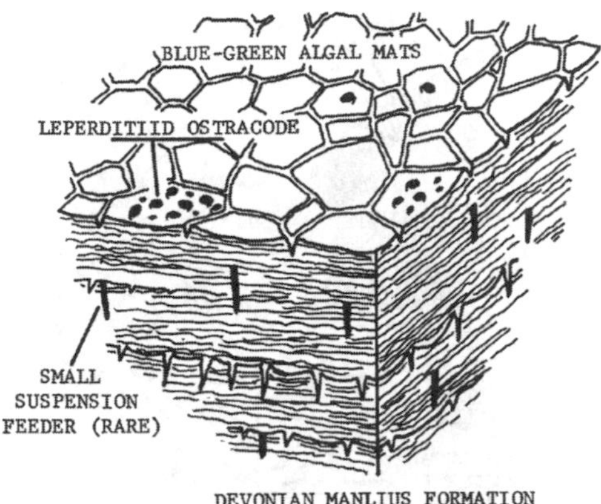

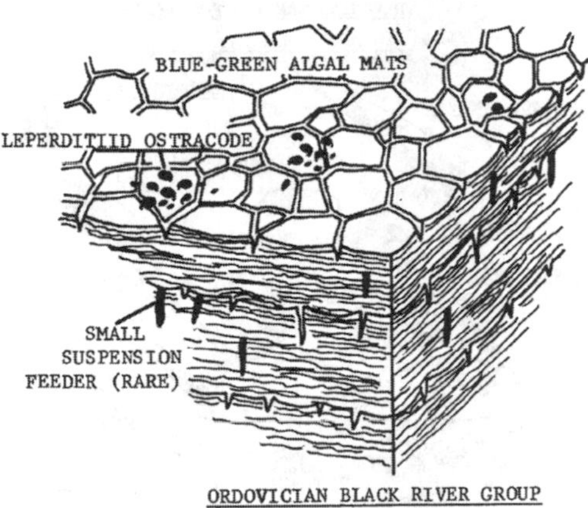

TEXT-FIG. 3—Reconstruction of *supratidal communities* from Black River (below) and Manlius (above).

seawater, while "wide" salinity tolerance means ability to withstand more variable salinities of higher and/or lower than open-ocean concentrations.

BLACK RIVER-MANLIUS COMMUNITIES

The four comparable communities of the Black River-Manlius are shown in Text-fig. 3, 4, 5, and 6. These schematic reconstructions illustrate the individual community members and suggest their relative abundance, dispersion, and inferred mode of life. A vertical view of the accumulating sediments is also given to show the texture, fabric, and structures associated with the depositional environment of each community. Such reconstructions can be misleading if viewed as "the way things really are (were)" rather than as idealized abstractions useful for summarizing large amounts of paleontologic and geologic data. Naturally, these community reconstructions are working hypotheses of our concepts of Black River-Manlius communities subject to future refinement.

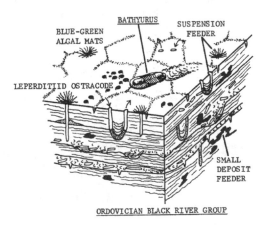

Supratidal community.—This community, which is the simplest of the four communities, is dominantly composed of epifaunal, vagile, deposit feeding/scavenging leperditiid ostracodes and algal mats. During unusually high water levels, the ostracodes moved up across the moist supratidal flats, feeding on organic detritus, later congregating in small irregular topographic lows where desiccation was less rapid as the flood waters receded. Scattered small suspension feeding burrowers were also rarely present, burrowing into the thin layers of soft,

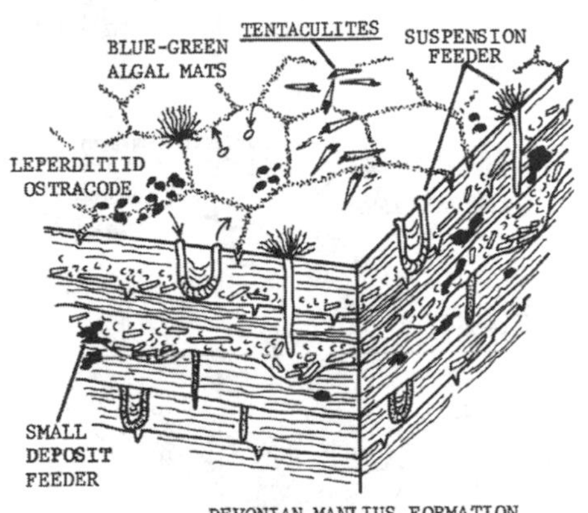

TEXT-FIG. 4—Reconstruction of *high intertidal communities* from Black River (below) and Manlius (above).

newly deposited sediment; these organisms, too, were left high and dry as the waters subsided.

The sediments associated with this community are algally laminated, carbonate mudstones with internal desiccation cracks (birdseye) and mudcracks. The sediment is characteristically dolomitized on a lamina by lamina scale.

High intertidal community.—Dominant community members are algal mats, leperditiid ostracodes, burrowing suspension and deposit feeders, and epifaunal, vagile, deposit feeders/scavengers: either the trilobite *Bathyurus* of the Black River or the tentaculitid *Tentaculites* of the Manlius. More frequent wetting in this environment kept the sediments sufficiently soft for burrowers to become established.

Obviously, as in the supratidal environment, some of the organisms of the high intertidal, for example the ostracodes, trilobites, and tentaculites, were not continuous inhabitants there, but moved up and down the flats with changing tides. Other organisms, especially infaunal forms, were capable of withstanding desiccation of the flats by retreating within the sediment.

The associated sediments are pelletal carbonate mudstones with algal laminations, some mudcracks, and frequent thin interbeds of carbonate pebble-conglomerates and skeletal calcarenites.

Low intertidal community.—Lower in the tidal flat environment, the biotas of the low intertidal community become more abundant and diverse. Except for the U-shaped burrowers which drop out, the organisms of the higher parts of the tidal flat remain with new elements being added, including strophomenid and spiriferid brachiopods, ramose ectoprocts, algal oncolites, and larger deposit feeding burrowers.

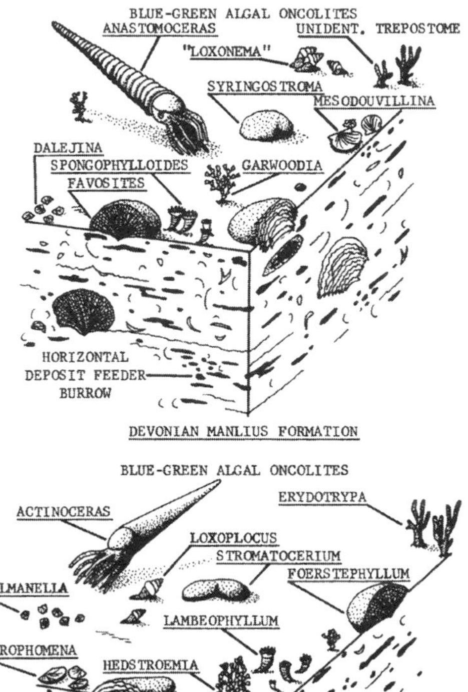

TEXT-FIG. 6—Reconstruction of *subtidal communities* from Black River (below) and Manlius (above).

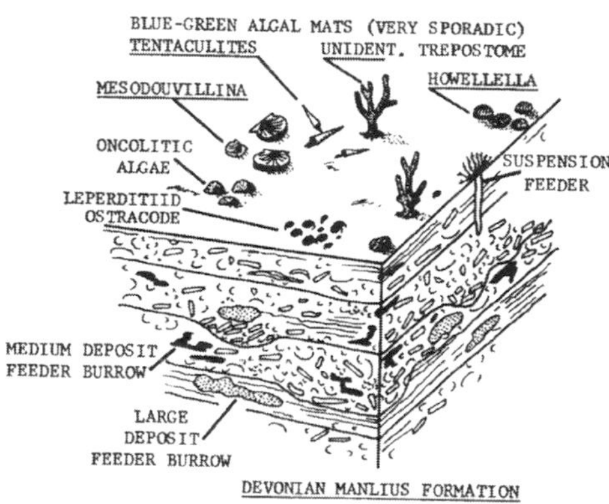

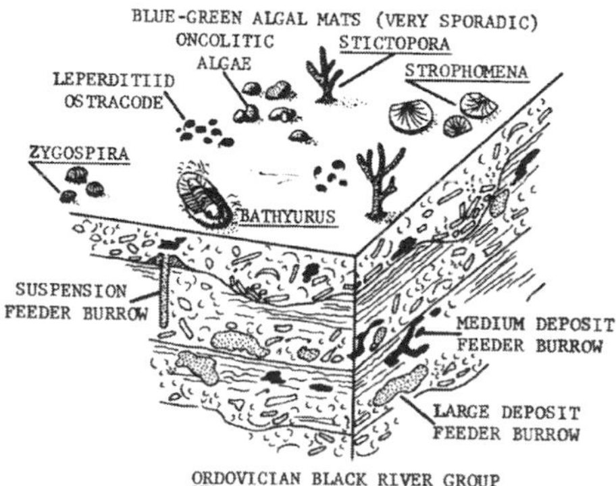

TEXT-FIG. 5—Reconstruction of *low intertidal communities* from Black River (below) and Manlius (above).

TABLE 3—List of significant discrepancies between equivalent Black River-Manlius communities. (See text for discussion.)

Noncomparable Taxa or Growth Habit		
Taxon	Environment	Unit of Occurrence
Bathyurus (Trilobite)	Hi & Lo Intertid.	O. Black River
Tentaculites	Hi & Lo Intertid.	D. Manlius Fm.
Spirorbis (Serpulid)	Lo Intertidal	D. Manlius Fm.
Tetradium (Tab. Coral?)	Subtidal	O. Black River
Endoceras (Nautiloid)	Subtidal	O. Black River
Michelinoceras (Naut.)	Subtidal	O. Black River
**Cyrtodonta* (Bivalvia)	Subtidal	O. Black River
Syringostroma (Stromatoporoid) of Manlius also occurs in laminar growth habit in addition to heads comparable to *Stromatocerium* of Black River.		

* Rare aviculoid or pterioid bivalves of Manlius may be comparable.

This environment was apparently wetted often enough that these marine organisms could populate it more or less continuously. Evidence of desiccation is relatively rare in the sediment of this environment.

The sediments are interbedded pelletal carbonate mudstones and calcarenites; layers of intraclasts and scour-and-fill structures are common. Algal laminations and mudcracks are occasionally present but are not as typical as in the supratidal and high intertidal sediments.

Subtidal community.—This community has the most abundant and diverse biota owing to continuous marine conditions in its environment of occurrence. The dominant members of this community are strophomenid and dalmanellid brachiopods; ramose ectoprocts; rugose, tabulate, and stromatoporoid corals; epifaunal snails; deposit feeding burrowers; girvanellid oncolites; and calcareous algae.

The sediments are lumpy thick-bedded, burrow-mottled, carbonate mudstones with abundant skeletal debris floating in the matrix. All of those features found in the sediments of the other environments which are typical of tidal flat sedimentation are absent here (e.g. mudcracks, intraclasts, erosional channels, algal laminations, etc.). There is also greater lithologic and paleontologic homogeneity.

While this community is the most marine of the communities seen in the Black River-Manlius, it should not be thought of as simply "the marine community" of the medial Ordovician or early Devonian, respectively. For, as our own studies have shown there are other marine communities that are essentially contemporaneous with the subtidal communities described here. Thus, for example, the subtidal communities of the Black River-Manlius are probably more restricted marine, shelf lagoon communities as compared with the more open marine, shelf communities recorded in the lower Trenton of the Ordovician or the Kalkberg-New Scotland of the Devonian.

Discrepancies between respective communities.—Despite the surprising similarities between respective communities of the Ordovician Black River and Devonian Manlius units, there are some significant and interesting differences (Table 3). Chief among these is the important abundance of the trilobite *Bathyurus* in the intertidal communities of the Black River and the virtual absence of any trilobites in the Manlius. We believe this is a real difference and not one due to variations in preservation or sampling, for trilobites have a very characteristic appearance in thin section and fragments as small as one millimeter can be confidentally identified. So while trilobites are found abundantly in other early Devonian carbonate environments of New York (Laporte, 1969) there are essentially none in the Manlius. Thus, the question which immediately arises is what Manlius organism filled a niche similar to that of the epifaunal, deposit feeding/scavenging trilobite, *Bathyurus.* It is tempting to suggest that the tentaculitids so abundant in the tidal flat communities of the Manlius (and absent in the Black River) are the ecologic equivalents of the tidal flat trilobite *Bathyurus.* Fisher (1962) postulated that the thick-shelled tentaculitids were nektobenthonic scavengers. However, the abundance of tentaculitids in the tidal flat facies of the Manlius, contrasted with their absence in the subtidal, argues for a benthic mode of life. Subtidal swimmers, even if bottom feeders, should occur in the subtidal facies; their presence in the tidal flat could then be explained by transportation of tests or individuals during times of tidal flat flooding. Of course it could be argued, alternatively, that a *Bathyurus*-like niche is filled (if filled at all) by an unpreserved, soft-bodied organism in the Devonian and the tentaculitid-like niche was filled by a soft-bodied Ordovician form. Although the evidence is by no means compelling, we suggest that Manlius tentaculitids had niches similar to the Black River *Bathyurus* niche in the high and low intertidal communities.

Another faunal element which is common in the Black River and which is without a comparable fossil form in the Manlius is the tabulate coral *Tetradium.* (The Manlius tabulate *Favosites* is presumably ecologically equivalent to the Black River tabulate *Foerstephyllum* which it closely resembles morphologically). We cannot choose among the various possibilities to explain this discrepancy: whether there was, in the Devonian, a soft-bodied ecologic descendant, an unfilled niche, or niche overlap by any one of a number of small, epifaunal suspension feeders assuming the role of *Tetradium.* It is also a distinct possibility as we have previously noted that the subtidal laminar form of the colonial *Syringostroma* (stromatoporoid) of the Manlius was ecologically similar, if not equivalent to the tiny colonial polyps of *Tetradium.*

There are several other differences between Black River and Manlius communities of less apparent significance. The encrusting calcareous worm *Spirorbis* is occasionally found in the low intertidal facies of the Manlius but absent in the Black River. The nautiloids, *Endoceras* and *Michelinoceras,* found sparsely in the Black River are absent in the Manlius, although as noted earlier, their community roles were probably filled by fish in the Devonian. *Cyrtodonta,* a byssally attached arcoid bivalve is common in the subtidal community of the Black River, occasionally in association with oolitic sediments. The rare aviculoid and pterioid bivalves of the Manlius may be the ecologic equivalents—curiously they, too, are found in oolitic horizons—and the difference in abundance may simply be due to variation in preservation.

There is a final difference between the Black River and Manlius and that is the presence of a laminar growth habit (besides a hemispherical one) of the Manlius stromatoporoid *Syringostroma* while the Black River stromatoporoids have only a hemispherical growth form. Differences in stromatoporoid shapes are usually attributed to variations in water agitation but this does not seem to be the case here as we believe that the range of water energy levels in the Black River and Manlius subtidal communities were fully comparable. As previously mentioned this laminar stromataporoid may be equivalent to *Tetradium* of the Black River Group.

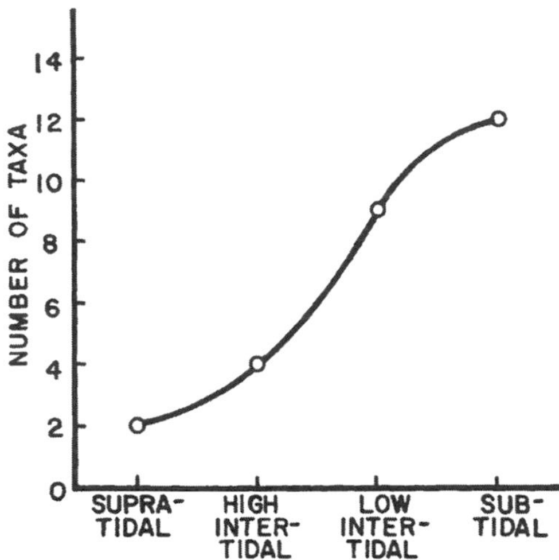

TEXT-FIG. 8—Increase in taxonomic diversity from onshore, supratidal community, to offshore, subtidal community within Black River-Manlius.

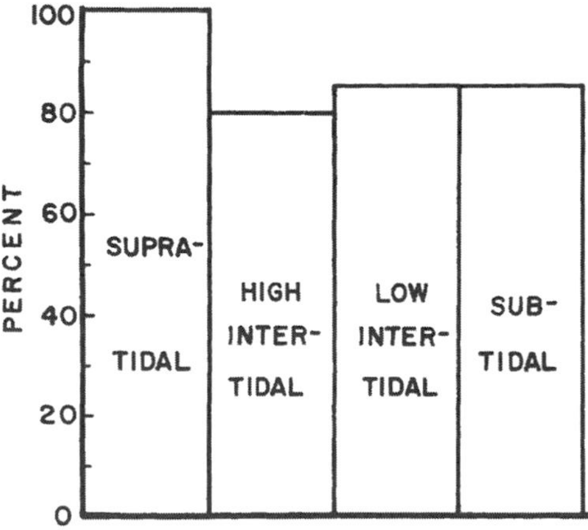

TEXT-FIG. 7—Percentage of similar taxa in equivalent Black River-Manlius communities. Despite discrepancies listed in Table 3, note high degree of taxonomic similarity between these two units.

Text-fig. 7 summarizes the percent of comparable taxa in corresponding Black River-Manlius communities. The percentage shown includes only those equivalent taxa about which we are confident. If more dubious equivalent pairs of taxa, such as *Bathyurus/Tentaculites* and *Tetradium*/laminar *Syringostroma,* were included, the percent of comparable taxa would rise to nearly 100% in all communities.

Inter-community trends.—Besides the similarities in taxonomic composition and ways of life of component organisms between corresponding communities of the Black River and Manlius, certain other interesting relationships can be seen when all four communities are viewed together.

The taxonomic diversity increases greatly from the onshore, supratidal, community to the offshore, subtidal community (Text-fig. 8). Such an increase is to be expected, of course, as the marine conditions are presumably both more normal and less variable in the subtidal environment. Temperature, salinity, and other widely varying physical factors excluded many taxa from the tidal flat environments.

The ratio of epifaunal to infaunal taxa also changes in a predictable way (Text-fig. 9). In

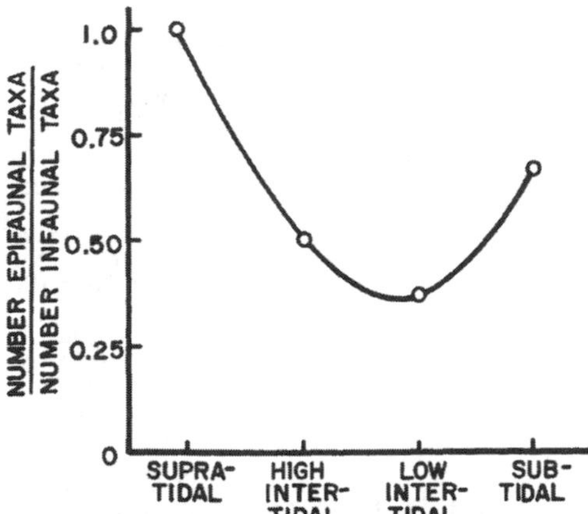

TEXT-FIG. 9—Changes in ratio of epifaunal to infaunal taxa within Black River-Manlius communities from onshore to offshore environments.

the supratidal community the few organisms present are all epifaunal, but the ratio of epifauna/infauna drops toward the subtidal as more infaunal burrowing forms appear. This increase in infauna can be explained by the softer (wetter) nature of the substrate permitting penetration by burrowers as well as the advantage of avoiding the high environmental stress conditions of the intertidal zone by living below the sediment interface. The overall increase in marine organisms in the subtidal, almost all of which are epifaunal (various corals, brachiopods, etc.), causes the ratio to rise once again offshore (Text-fig. 9).

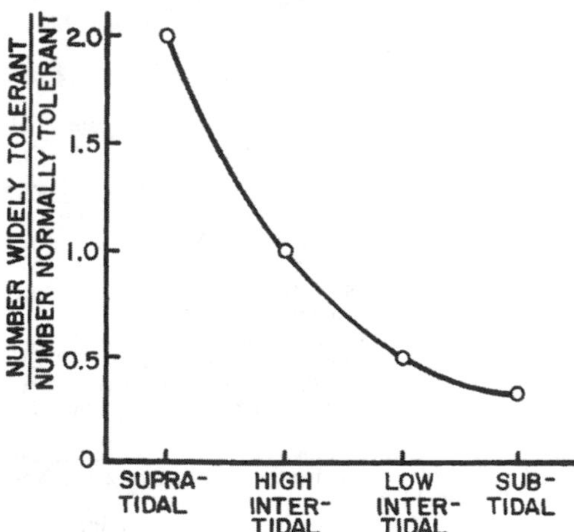

TEXT-FIG. 10—Systematic decrease in ratio of wide salinity tolerant to normal salinity tolerant taxa within Black River-Manlius communities from onshore to offshore environments.

Finally, the ratio of wide salinity tolerant forms to normal salinity tolerant taxa decreases sharply from the supratidal to the subtidal environments (Text-fig. 10). The data for calculating this ratio are not completely independant of our overall interpretations and so our reasoning if not circular is at least somewhat elliptical. For example, we can be fairly certain the spiriferid brachiopods *Zygospira* and *Howellella* were attached, epifaunal suspension feeders, but we are less certain that they had a "wide" salinity tolerance except that such an interpretation is consistent with everything else we know and infer about these environments. Nevertheless, we feel that the salinity tolerance trend is real and that it reflects the variable salinity conditions one expects in tidal flat environments where changes in salinity, owing to precipitation, runoff, or evaporation, are frequent and often drastic.

SUMMARY AND CONCLUSIONS

Marine fossils—including burrow traces, algal organo-sedimentary structures, calcareous algae, and shelly invertebrates—of the shallow water carbonate deposits from both the Middle Ordovician (Black River Group) and the Lower Devonian (Manlius Formation) of New York can be reliably subdivided into four major communities: supratidal, high intertidal, low intertidal, and subtidal. Each of the four corresponding communities from the Black River and Manlius has a similar taxonomic composition (usually at family or order level) reflecting similar adaptive morphology. Moreover, the inferred modes of life of Black River-Manlius paired taxa are essentially identical.

We consider the similarities between Black River and Manlius sedimentary rocks the result of equivalent carbonate depositional responses to similar shoreline environments which accompanied marine transgression during the medial Ordovician and early Devonian. Parallelism in the organic communities of the Black River-Manlius indicates that these shoreline environments made similar adaptive demands on the biotas inhabiting these environments.

We conclude from these observations that the ecosystems of the carbonate tidal flat and shallow subtidal environments of medial Ordovician and early Devonian time were virtually the same. By implication, then, these communities had evolved little or not at all in structure or arrangement during the 60 or 70 million years separating them. While the community roles remained the same, individual taxa did, of course, undergo considerable change.

Since the major expansion and diversification of marine invertebrates was completed by about the medial Ordovician, we should look to premiddle Ordovician marine carbonate rocks for significant variations in these tidal flat and shallow subtidal communities. Apparently, major evolutionary changes in community structure occurred during the Cambrian and early Ordovician and presumably on a scale similar to the changes in terrestrial vertebrate communities of the late Paleozoic and early Mesozoic (Olson, 1966).

Finally, we believe that we have demonstrated a useful approach for interpreting depositional environments thereby illuminating the ecosystems lurking behind the sedimentary rocks of the Black River and Manlius units. If our specific conclusions seem doubtful, we hope, at least, that our methods and points of view are nevertheless valid.

REFERENCES

Ball, M. M., E. A. Shinn, and K. W. Stockman. 1967. The geologic effects of hurricane Donna in south Florida. J. Geol. 75:583–597.

Benson, R. H. 1961. Ecology of ostracode assemblages, p. *Q*56–*Q*63. *In* Treatise invert. paleontol., Pt. Q, Arthropoda 3. Geol. Soc. Amer. and Kansas Univ. Press, Lawrence, Kansas.

Berdan, J. 1968. Possible paleoecological significance of Leperditiid ostracodes. Geol. Soc. Amer., Prog., Annu. Mtg., Northeastern Sect. Washington, D.C., p. 17. (Abstr.)

Bretsky, P. 1968. Evolution of Paleozoic marine invertebrate communities. Science 159:1231–1233; 161:491.

Cloud, P. E., Jr. 1959. Geology of Saipan, Mariana Islands: Pt. 4. Submarine topography and shoal-water ecology. U.S. Geol. Surv. Prof. Paper 280-K:361–445.

Deevey, E. S. 1965. Environments of the geologic past. Science 147:592–594. (Book Review)

Fisher, D. 1962. Small conoidal shells of uncertain affinities, p. *W*98–*W*143. *In* Treatise invert. paleontol., Pt. W. Miscellanea. Geol. Soc. Amer. and Kansas Univ. Press, Lawrence, Kansas.

Gebelein, C. D. 1969. Distribution, morphology, and accretion rate of Recent subtidal algal stromatolites, Bermuda. J. Sed. Petrol. 39:49–69.

—— and P. Hoffman. 1969. Algal origin of dolomite in interlaminated limestone-dolomite sedimentary rocks. Bermuda Biol. Sta. for Research, Spec. Pub. 3:226–235.

Ginsburg, R. N. 1957. Early diagenesis and lithification of shallow-water carbonate sediments in South Florida, p. 80–99. *In* Regional aspects of carbonate deposition. Soc. Econ. Paleontol. and Mineral. Spec. Pub. 5, Tulsa, Oklahoma.

Hedgpeth, J. W. 1957. Concepts of marine ecology, p. 29–52. *In* Treatise marine ecol. and paleoecol. Geol. Soc. Amer. Mem. 67(1).

Imbrie, J. 1959. Classification and evolution of major adaptive invertebrate types. Int. Oceanogr. Congr., Amer. Ass. Advance Sci. Preprints, p. 278.

—— and H. Buchanan. 1965. Sedimentary structures in modern carbonate sands of the Bahamas, p. 149–172. *In* Primary sedimentary structures and their hydrodynamic interpretation. Soc. Econ. Paleontol. and Mineral. Spec. Pub. 12, Tulsa, Oklahoma.

Johnson, J. H. 1961. Limestone-building algae and algal limestones. Colo. School of Mines, Boulder, Colo. 297 p.

Johnson, R. G. 1964. The community approach to paleoecology, p. 107–134. *In* J. Imbrie and N. D. Newell (ed.) Approaches to Paleoecology. John Wiley and Sons, New York.

Knight, J. B., R. L. Batten, E. L. Yochelson, and L. R. Cox. 1960. Supplement-Paleozoic and some Mesozoic Caenogastropoda and Opisthobranchia, p. *I* 310–*I*331. *In* Treatise invert. paleontol., Pt. I, Mollusca 1. Geol. Soc. Amer. and Kansas Univ. Press, Lawrence, Kansas.

Konishi, K. 1961. Studies of Paleozoic Codiaceae and allied algae. Pt. 1: Codiaceae (excluding systematic descriptions). Kanazawa Univ. Sci. Rept. 7: 159–261.

Lagaaij, R. and Y. V. Gautier. 1965. Bryozoa from the Rhône delta. Micropaleontology 11:39–58.

Laporte, L. F. 1967. Carbonate deposition near mean sea-level and resultant facies mosaic: Manlius Formation (Lower Devonian) of New York State. Amer. Ass. Petrol. Geol. Bull. 51:73–101.

——. 1968. Recent carbonate environments and their paleoecologic implications, p. 229–258. *In* E. T. Drake (ed.) Evolution and Environment. Yale Univ. Press, New Haven, Conn.

——. 1969. Recognition of a transgressive carbonate sequence within an epeiric sea: Helderberg Group (Lower Devonian) of New York State, p. 98–119. *In* Depositional environments in carbonate rocks. Soc. Econ. Paleontol. and Mineral. Spec. Pub. 14, Tulsa, Oklahoma.

Logan, B., R. Rezak, and R. N. Ginsberg. 1964. Classification and environmental significance of algal stromatolites. J. Geol. 72:68–83.

McAlester, A. L. 1968. The history of life. Prentice-Hall, Inc. Englewood Cliffs, N.J. 151 p.

Moore, D. G. and P. C. Scruton. 1957. Minor internal structures in some Recent unconsolidated sediments. Amer. Ass. Petrol. Geol. Bull. 41:2723–2751.

Multer, H. G. and J. E. Hoffmeister. 1968. Subaerial laminated crusts of the Florida Keys. Geol. Soc. Amer. Bull. 79:183–192.

Neumann, A. C., C. D. Gebelein, and T. P. Scoffin. 1969. Composition, structure, and erodability of subtidal mats, Abaco, Bahamas. Amer. Ass. Petrol. Geol. Bull. 53:734. (Abstr.)

Olson, E. C. 1966. Community evolution and the origin of mammals. Ecology 47:291–302.

Pray, L. and R. C. Murray. 1965. Dolomitization and limestone diagenesis—an introduction, p. 1–2. *In* Dolomitization and limestone diagenesis. Soc. Econ. Paleontol. and Mineral. Spec. Pub. 13, Tulsa, Oklahoma.

Puri, H. S. (*Editor*). 1964. Ostracodes as ecological and paleoecological indicators. Pub. Sta. Zool. Napoli 33 (suppl.) 612 P.

Rhoads D. C. 1967. Biogenic reworking of intertidal and subtidal sediments in Barnstable Harbor and Buzzards Bay, Mass. J. Geol. 75:461–475.

Roehl, P. O. 1967. Carbonate facies, Williston Basin and Bahamas. Amer. Ass. Petrol. Geol. Bull. 51: 1979–2032.

Sanders, H. L., E. M. Goudsmit, E. L. Mills, and G. E. Hampson. 1962. Study of the intertidal fauna of Barnstable Harbor, Mass. Limnol. and Oceanogr. 7:63–79.

Schopf, T. J. M. 1969. Paleoecology of ectoprocts (bryozoans). J. Paleontol. 43:234–244.

Shinn, E. A. 1968. Practical significance of birdseye structure in carbonate rocks. J. Sed. Petrol. 38: 215–223.

Thorson, G. 1957. Bottom communities (sublittoral or shallow shelf), p. 461–534. *In* Treatise mar. ecol. and paleoecol. Geol. Soc. Amer. Mem 67(1).

Valentine, J. W. 1968. Evolution of ecological units above the population level. J. Paleontol. 42:253–267.

Walker, K. R. 1969. Stratigraphy, environmental sedimentology, and organic communities of the Middle Ordovician Black River Group of New York State. Ph.D. Thesis. Yale Univ. 214 p.

—— in press. Stratigraphy and environmental sedimentology of the Middle Ordovician Black River Group of New York State. N.Y. State Mus. and Sci. Serv., Geol. Surv. Bull.

—— in preparation. Community ecology of Middle Ordovician Black River Group of New York State.

Yonge, C. M. 1954. Feeding mechanisms in the Invertebrata. Tabulae Biologicae 21(22):46–68.

Manuscript received November 3, 1969.

Brown University contributed $150.00 toward the publication of this paper.

PAPER 2

Mammalian Evolution and the Great American Interchange (1982)

L. G. Marshall, S. D. Webb, J. J. Sepkoski Jr., and D. M. Raup

Commentary

LARISA R. G. DESANTIS

Understanding the distribution of animals over space and time was the focus of many early naturalists, including Alfred Russel Wallace. Although he was the first to document the interchange of mammalian taxa between the Americas since the Miocene (Wallace 1876), now termed the Great American Biotic Interchange (Stehli and Webb 1985), Marshall and colleagues quantitatively clarified these dispersal patterns (accounting for potential sampling biases via rarefaction analysis) and tested biogeographic hypotheses. Contrary to prior work (e.g., Simpson 1950, 1980; Webb 1976) that suggested the replacement of southern genera by northern genera during the interchange, Marshall et al.'s paper was the first to demonstrate that relatively equal numbers of families and genera exchanged if pre-interchange diversities are considered (Jablonski et al. 1985). While more North American genera did immigrate south, this asymmetry is consistent with their greater generic diversity and lower turnover rates, as expected due to increased land area and consistent with MacArthur-Wilson species equilibrium theory (MacArthur and Wilson 1967). By considering origination and extinction rates of both northern and southern immigrants prior to and after the interchange, Marshall and colleagues further revealed that both higher origination rates and lower extinction rates of northern immigrants are in contrast to equilibrium theory and account for the imbalance in interchange fauna between these two continents. This work subsequently set the stage for determining potential reasons for this imbalance, leading to suggestions that northern immigrants were better able to fill narrow and/or more tropical niches due to their evolutionary history dealing with prior invasions and/or their ability to occupy environments similar to those occurring prior to or during the interchange (e.g., dry and open savanna-like conditions; Webb 1991, 2006; Woodburn 2010).

This study raised the bar for future paleoecological assessments of faunal dynamics, requiring quantitative analyses of fossil data, including accounting for potential sampling biases (similar to Raup 1975) and examining metrics of taxonomic evolution. Furthermore, Marshall and colleagues provided a classic example of long-term responses to taxonomic invasions with current conservation implications pertaining to invasive species (Malanson 2008; Hendry et al. 2010). This study continues to demonstrate the importance of considering equilibrium dynamics in evaluating changes in taxonomic diversity through time, particularly in response to major abiotic events.

From *Science* 215:1351–57. Reprinted with permission from AAAS.

Literature Cited

Hendry, A. P., L. G. Lohmann, E. Conti, J. Cracraft, K. A. Crandall, et al. 2010. Evolutionary biology in biodiversity sciences, conservation, and policy: a call to action. *Evolution* 64:1517–28.

Jablonski, D., K. W. Flessa, and J. W. Valentine. 1985. Biogeography and paleobiology. *Paleobiology* 11:75–90.

MacArthur, R. H., and E. O. Wilson. 1967. *The Theory of Island Biogeography*. Princeton, NJ: Princeton University Press.

Malanson, G. P. 2008. Extinction debt: origins, developments, and applications of a biogeographical trope. *Progress in Physical Geography* 32:277–91.

Raup, D. M. 1975. Taxonomic diversity estimation using rarefaction. *Paleobiology* 1:333–42.

Simpson, G. G. 1950. History of the fauna of Latin America. *American Scientist* 38:361–89.

———. 1980. *Splendid Isolation: The Curious History of South American Mammals*. New Haven, CT: Yale University Press.

Stehli, F. G., and S. D. Webb. 1985. *The Great American Biotic Interchange*. New York: Plenum Press.

Wallace, A. R. 1876. *The Geographical Distributions of Animals*. London: Macmillan.

Webb, S. D. 1976. Mammalian faunal dynamics of the great American interchange. *Paleobiology* 2:220–34.

———. 1991. Ecogeography and the Great American Interchange. *Paleobiology* 17:266–80.

———. 2006. The Great American Biotic Interchange: patterns and processes. *Annals of the Missouri Botanical Garden* 93: 245–57.

Woodburn, M. O. 2010. The Great American Biotic Interchange: dispersals, tectonics, climate, sea level and holding pens. *Journal of Mammalian Evolution* 17: 245–64.

12 March 1982, Volume 215, Number 4538 SCIENCE

Mammalian Evolution and the Great American Interchange

Larry G. Marshall, S. David Webb
J. John Sepkoski, Jr., David M. Raup

Summary. A reciprocal and apparently symmetrical interchange of land mammals between North and South America began about 3 million years ago, after the appearance of the Panamanian land bridge. The number of families of land mammals in South America rose from 32 before the interchange to 39 after it began, and then back to 35 at present. An equivalent number of families experienced a comparable rise and decline in North America during the same interval. These changes in diversity are predicted by the MacArthur-Wilson species equilibrium theory. The greater number of North American genera (24) initially entering South America than the reverse (12) is predicted by the proportions of reservoir genera on the two continents. However, a later imbalance caused by secondary immigrants (those which evolved from initial immigrants) is not expected from equilibrium theory.

Biogeographers have long recognized the late Cenozoic mingling of the previously separated American continental biotas as a monumental natural experiment, the Great American Interchange. Comparison of the "wonderful extinct fauna . . . discovered in North America, with what was previously known from South America" allowed Wallace to first recognize the existence of this event in 1876 (*1*). However, the direction in which representatives of the various animal groups had dispersed was not well understood by Wallace (*2*). It took another 15 years of intense paleontological exploration and study by Cope and Marsh in North America, and by Carlos and Florentino Ameghino in South America before sufficient data existed to permit clarification of many of the basic issues in this event (*3*). By 1891 a comprehensive and balanced overview of the interchange existed:

> . . . not only did North American taxa cross the newly opened land bridge, greatly expanding their ranges, but also South American autochthons began ranging into North America, and thus toward the end of the Pliocene epoch took place one of the most remarkable faunal exchanges that Geology has known [Karl A. von Zittel (*4*)].

Continued research in the present century has resulted in elaboration of the histories of the participant and nonparticipant taxa. This cumulative knowledge has been periodically summarized by Matthew (*5*), Scott (*6*), Simpson (*7*), Patterson and Pascual (*8*), and others (*9–12*).

Despite the wealth of accumulated knowledge, many finer details of the interchange have remained obscure. Recent improvements in paleontological sampling (especially screen-washing for taxa of small body size); refined taxonomic studies spanning both continents; and availability of an array of radioisotopic age determinations interpolated within the late Cenozoic land-mammal bearing strata on each continent permit clarification of unresolved earlier problems. There now exists sufficient knowledge of these aspects of the interchange to permit quantitative, rather than simply qualitative, examination of patterns of faunal dispersal and evolution. In this article we have compiled such quantitative data, and we use them to examine both empirical patterns of faunal interchange and correspondence to models of equilibrial diversities and biogeography.

Qualitative Aspects of the Interchange

South America was isolated from other continents during most of the Age of Mammals (*8, 13*). This isolation ended about 3 million years ago with the disappearance of the Bolivar Trough Marine Barrier in the area of northwestern Colombia and southern Panama, and the total emergence of the Panamanian land bridge (*8*). Thereafter the fossil record documents a reciprocal intermingling of the long-separated North and South American terrestrial biotas. Since the Bolivar Trough served as the final geographic barrier separating these biotas, the area to the north of it is here referred to as North America and the area south of it as South America. The area of the former Bolivar Trough is thus the "gateway" for the Great American Interchange.

On the basis of the timing and the means of dispersal, the participants in the Great American Interchange can be divided into two groups. The first group includes late Miocene waif immigrants, which are believed to have dispersed along island arcs before the final emergence of the land bridge (*7*). This group includes members of two families of North American origin: (i) procyonids (racoons and allies), which are first recorded in beds of late Miocene (Huayquerian) age in Argentina (*11, 14*), and (ii) cricetid rodents (New World rats and mice) of the tribe Sigmodontini (Hesper-

L. G. Marshall is an assistant curator of fossil mammals in the Department of Geology, Field Museum of Natural History, Chicago, Illinois 60605. S. D. Webb is curator of fossil vertebrates, Florida State Museum and professor of Zoology, University of Florida, Gainesville 32611. J. John Sepkoski, Jr., is an assistant professor of paleontology in the Department of the Geophysical Sciences, University of Chicago, Chicago, Illinois 60637. D. M. Raup is dean of science, Field Museum of Natural History, Chicago, Illinois 60605 and professor in the Department of the Geophysical Sciences, University of Chicago, Chicago, Illinois 60637.

0036-8075/82/0312-1351$01.00/0

omyini), which are first recorded in beds of early Pliocene (Montehermosan) age (*15*) in Argentina. It also includes members of the extinct South American ground sloth families Megalonychidae and Mylodontidae, which are first recorded in North America in beds of late Miocene (Hemphillian) age (*11*).

Included within the second group of participants are those taxa that walked across the bridge after its final emergence. The North American immigrants to South America include members of the families (i) Mustelidae (skunks and allies) and Tayassuidae (peccaries), which first appear in the late Pliocene (Chapadmalalan Age) (*8*); (ii) Canidae (dogs, wolves, foxes), Felidae (cats), Ursidae (bears), Camelidae (camels, llamas), Cervidae (deer), Equidae (horses), Tapiridae (tapirs), and Gomphotheriidae (mastodonts) which appear in the early Pleistocene (Uquian Age) (*8*); and (iii) Heteromyidae (kangaroo rats and allies), Sciuridae (squirrels), Soricidae (shrews), and Leporidae (rabbits) which are known only from Holocene or Recent (or both) (*8, 16, 17*). The South American immigrants to North America include members of the families (i) Dasypodidae (armadillos), Glyptodontidae (glyptodonts), Hydrochoeridae (capybaras), and Erethizontidae (porcupines), which appear in the late Pliocene (late Blancan Age) (*9, 11*); (ii) Didelphidae (opossums) and Megatheriidae (ground sloths), which appear in the early and middle Pleistocene (Irvingtonian Age) (*9, 11*); (iii) Toxodontidae (toxodonts), which are recorded in the late Pleistocene (Rancholabrean Age) (*8*); and (iv) Callitrichidae (marmosets and tamarins), Cebidae (New World monkeys), Choleopodidae (tree sloths), Bradypodidae (tree sloths), Cyclopidae (anteaters), Myrmecophagidae (anteaters), Dasyproctidae (agoutis, pacas), and Echimyidae (spiny rats), which are known only in Recent faunas (*17*).

The late Cenozoic record of fossil mammals in North and South America is relatively well documented. The one great deficiency of the South American record is that it is largely restricted to Argentina (*8*) and to a lesser extent Bolivia (*18*). Both the North and South American records are deficient for mammals from tropical latitudes. Nevertheless, inferences gleaned from these records yield generalities that are probably valid for the continents as a whole (*7*). Furthermore, the majority of faunas sampled appear to represent savanna-

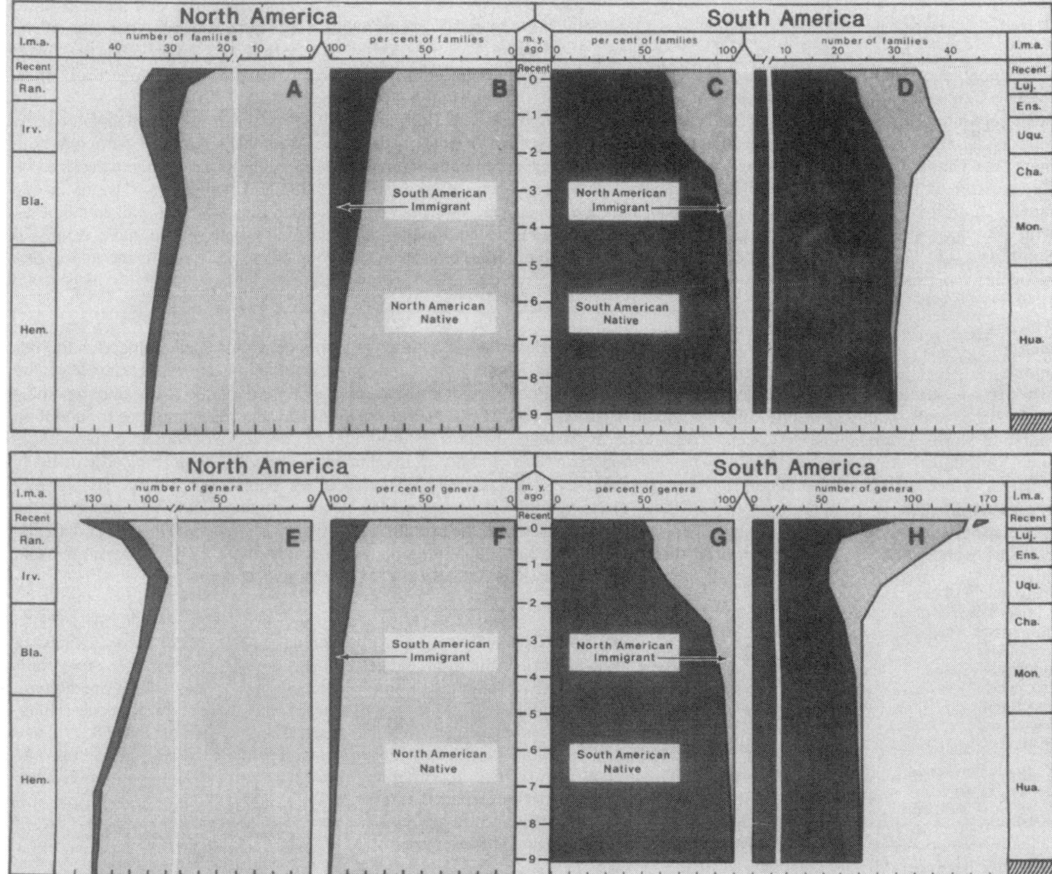

Fig. 1. Numbers (or "diversities") of known families (top) and genera (bottom) in successive late Cenozoic land mammal ages in North (left) and South (right) America. Graphs show total number of native and immigrant taxa and their percentage contribution to each land mammal age fauna. Note that scales are different for cumulative numbers of families and genera. The land mammal ages for North America are *Ran.*, Rancholabrean; *Irv.*, Irvingtonian; *Bla.*, Blancan; and *Hem.*, Hemphillian; for South America they are *Luj.*, Lujanian; *Ens.*, Ensenadan; *Uqu.*, Uquian; *Cha.*, Chapadmalalan; *Mon.*, Montehermosan; and *Hua.*, Huayquerian. Abbreviations: *l.m.a.*, land mammal age; *m.y.*, million years.

grassland habitats. We are thus dealing primarily with the evolutionary history of ecologically similar faunas during the interchange period, and this feature lessens the potential for sampling bias.

Aspects of Faunal Dynamics

The late Cenozoic stratigraphic ranges of families and genera of terrestrial mammals can be used to analyze simple aspects of taxonomic evolution [that is, measurements of changes in total numbers of taxa and changes of taxa within clades through time (*19–21*)]. For North America we analyze these data for the last 12 million years (divided into five standard land mammal ages) (*22*), and for South America we analyze these data for the last 9 million years (divided into six land mammal ages) (*18*) (Table 1). The perspective given by this temporal framework permits establishment of a pre–land bridge "basal metabolism" to which post–land bridge faunal dynamics can be compared.

Table 1 lists summations for each land mammal age of (i) familial and generic diversity, (ii) numbers of first and last fossil occurrences (that is, observed originations and extinctions) of genera of native and immigrant taxa, and (iii) aspects of faunal dynamics and taxonomic evolution (*21*) based on data in (*22*) for North America and (*18*) for South America. Numbers of families and genera in each land mammal age on each continent are shown in Fig. 1 (top and bottom, respectively), with shading indicating the continent of origin of the taxa (*23*) (Fig. 1). Below we consider aspects of the taxonomic evolution first of families and then of genera.

Families. The total number of known families remained relatively constant throughout the late Cenozoic on both continents; the average diversity from late Miocene to Recent in both North and South America was about 34. Today, the familial diversities remain similar, with 35 in South America and 33 in North America (*17*).

In South America the peak in familial diversity (39 families) followed the appearance of the land bridge and the arrival of members of eight North American families in the Uquian, raising the number of immigrant families to 12; the total number of families then dropped to 36 before Lujanian time (Table 1). For North America the record likewise indicates a sharp rise in South American immigrant families after emergence of the isthmus: representatives of a total of six new families appeared in the late

Table 1. Faunal dynamics (genera per million years) of late Cenozoic land mammal genera in South America (left) and North America (right). The number of families represented are listed in brackets. The land mammal ages for South America are (from oldest to youngest) H, Huayquerian; M, Montehermosan; C, Chapadmalalan; U, Uquian; E, Ensenadan; and L, Lujanian. For North America they are C, Clarendonian; H, Hemphillian; B, Blancan; I, Irvingtonian, and R. Rancholabrean.

Indices (*21*)	South American land mammal age						North American land mammal age				
	H	M	C	U	E	L	C	H	B	I	R
a. Durations (million years)	4.0	2.0	1.0	1.0	0.7	0.3	2.5	5.0	2.5	1.3	0.7
b. Number of genera											
North American	1[1]	4[2]	10[4]	29[12]	49[12]	61[12]	92[33]	128[33]	99[25]	90[26]	102[26]
South American	72[29]	68[30]	62[29]	55[27]	58[24]	59[24]	0[0]	3[2]	8[6]	11[8]	12[9]
Total	73[30]	72[32]	72[33]	84[39]	107[36]	120[36]	92[33]	131[35]	107[31]	101[34]	114[35]
c. Originations (No.)											
North American	1	3	6	21	26	13	43	75	54	28	22
South American	55	32	14	25	19	8	0	3	7	4	1
Total	56	35	20	46	45	21	43	78	61	32	23
d. Extinctions (No.)											
North American	0	0	2	6	1	20	37	81	40	9	23
South American	36	20	32	16	7	25	0	2	1	0	9
Total	36	20	34	22	8	45	37	83	41	9	32
e. Running means											
North American	0.5	2.5	6.0	15.5	35.5	44.5	52	50.0	65.0	71.5	79.5
South American	26.5	42.0	39.0	34.5	45.0	42.5		0.5	4.0	9.0	7.0
Total	27.0	44.5	45.0	50.0	80.5	87.0	52	50.5	69.0	80.5	86.5
f. Origination rates											
North American	0.3	1.5	6.0	21.0	37.0	43.3	17.2	15.0	21.6	21.5	31.4
South American	13.8	16.0	14.0	25.0	27.0	26.7		0.6	2.8	3.1	1.4
Total	14.1	17.5	20.0	46.0	64.0	70.0	17.2	15.6	24.4	24.6	32.9
g. Extinction rates											
North American	0	0	2.0	6.0	1.5	66.7	14.8	16.2	16.0	6.9	32.9
South American	9.0	10.0	32.0	16.0	10.0	83.3		0.4	0.4	0	12.9
Total	9.0	10.0	34.0	22.0	11.5	150.0	14.8	16.6	16.4	6.9	45.8
h. Turnover rates											
North American	0.1	0.8	4.0	13.5	19.3	55.0	16.0	15.6	18.8	14.2	32.2
South American	11.4	13.0	23.0	20.5	18.5	55.0		0.5	1.6	1.6	7.2
Total	11.5	13.8	27.0	34.0	37.8	110.0	16.0	16.1	20.4	15.8	39.4
i. Per-genus turnover	0.4	0.3	0.6	0.7	0.5	1.3	0.3	0.3	0.3	0.2	0.5
j. Breakdown estimate of immigrants											
Total number											
Primary	1	1	2	10	18	20	0	2	6	8	9
Secondary	0	3	8	19	31	41	0	1	2	3	3
Originations											
Primary	1	0	1	8	9	2	0	2	6	3	1
Secondary	0	3	5	13	17	11	0	1	1	1	0
Extinctions											
Primary	0	0	0	1	0	7	0	2	1	0	7
Secondary	0	0	2	5	1	13	0	0	0	0	2

Blancan (four families) and early Irvingtonian (two families). Both continents experienced a notable decline in familial diversity at the end of the Pleistocene (*24*).

A sharp rise in the number of known immigrant families between the late Pleistocene and Recent occurs on both continents. This rise is due to our ignorance of late Cenozoic tropical faunas (*17*). Today, members of 14 North American families occur in South America and contribute 40 percent to the familial diversity of that continent, whereas members of 12 South American families occur in North America and account for a nearly equivalent 36 percent of that continent's familial diversity.

In summary, the data show that at the family level the interchange was balanced (*12*). The fact that total familial diversity on both continents is virtually the same today as it was in pre–land bridge times might be construed as indicative of symmetrical replacement of native by immigrant taxa at high taxonomic levels (Fig. 1).

Genera. The known late Cenozoic diversity of fossil genera is, on the average, greater in North America than in South America (Table 1, row b); this is in contrast to the Recent fauna which is slightly more diverse in South America (170 genera) than in North America (141 genera) (*17*). In South America known diversity remained near 72 genera for at least 6 million years prior to the land bridge (Fig. 1, column H), suggesting that an equilibrium was established at about that level, at least in the environments sampled. After the appearance of the land bridge, generic diversity rose rapidly to 84 in the Uquian, 107 in the Ensenadan, and 120 in the Lujanian (Table 1, row b). During this time the North American immigrants increased sharply but steadily in number and progressively contributed a larger part of the South American land mammal fauna. Today, 85, or 50 percent, of the mammal genera in South America are derived from members of immigrant North American families (*17*).

In North America, observed numbers of genera drop from 131 in the Hemphillian to 101 in the Irvingtonian and then rise to 114 in the Rancholabrean (Table 1). The known South American immigrants rose from 3 to 12 during this period and came to contribute only 11 percent to the total North American land mammal fauna in the Rancholabrean. Today, 29 (21 percent) of the land mammal genera in North America are derived from immigrant South American families (*17, 25, 26*). Most of the genera not sampled as fossils in North America live in subtropical to tropical latitudes in the Neotropical Realm (*8*).

The generic diversities of Recent and pre–land bridge faunas in North America are similar. However, in South America a major increase in generic diversity followed the appearance of the land bridge, the result of adding immigrant taxa. At the same time, the number of native South American genera declined by 13 percent between pre–land bridge and Lujanian faunas, a percentage reduction comparable to the 11 percent decline among native genera in North America. Thus, as in the case of families, the percentage decline of native genera was virtually identical on both continents, and in this regard the interchange was balanced. But the increase in both numbers and percentages of immigrant taxa was much greater in South America, as is discussed further below.

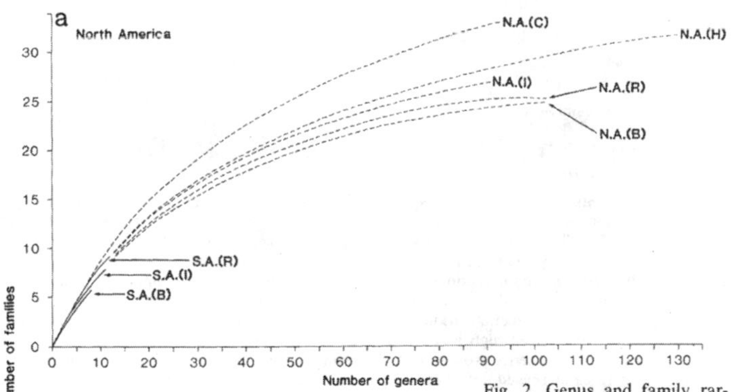

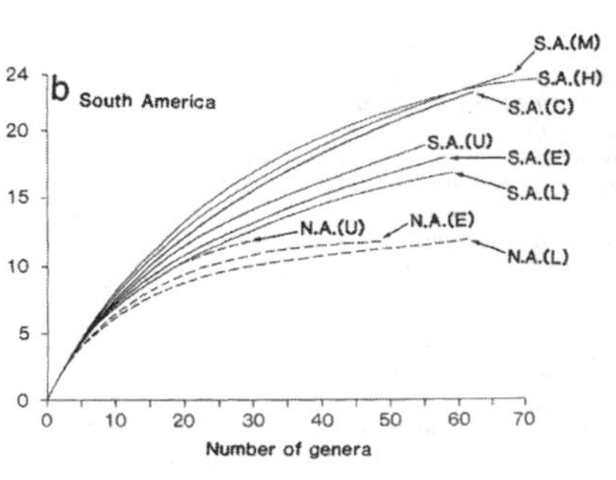

Fig. 2. Genus and family rarefaction curves for late Cenozoic land mammal age faunas in North (a) and South (b) America. (a) *S.A.*, genera that evolved from families of South American origin (solid lines); *N.A.*, genera that evolved from families of North American origin (dashed lines); (b) *S.A.*, genera that evolved from families of South American origin (solid lines); *N.A.*, genera that evolved from families of North American origin (dashed lines). For key to abbreviations of land mammal ages, see legend of Table 1.

Rarefaction Analysis

Before considering the dynamics of the Great American Interchange further, some aspects of the quality and robustness of the taxonomic data must be analyzed in more detail. The diversity values for land mammal age faunas shown in Table 1 and Fig. 1 are somewhat higher than the numbers of taxa actually recorded. Some taxa occur in preceeding and succeeding land mammal age faunas, and their presence in the intervening fauna or faunas is inferred. This usage of inferred ranges is conventional and legitimate for general analysis of diversity patterns, although an alternative analysis by rarefaction methods incorporates only the genera and families actually recorded in a particular land mammal age (*27*).

A taxonomic rarefaction curve is computed on the basis of the frequency distribution of genera within families in each land mammal age (Fig. 2). The distal end of each curve represents the number of genera and families actually recorded, and the curves provide estimates of the number of families that would have been recorded had fewer genera been found. Thus, rarefaction analysis provides answers to two basic questions: (i) Is the nature of sampling and taxonomic treatment consistent throughout the data set? and (ii) Did familial diversity differ significantly among the time intervals (land mammal ages) sampled?

All of the rarefaction curves in Fig. 2 have approximately the same shape, and crossing of curves is minimal, suggesting minimal overall differences in sampling both among land mammal ages and between continents. A sharp drop in North American native families (statistically significant, $P < .05$) occurs between the Clarendonian and Hemphillian (Fig. 2a), although no drop is seen in the number of actual or inferred families shown in Table 1. This drop could represent either a true decrease in diversity or a generic radiation among existing families. The diversity of native North American families within North America continues to drop after initiation of the interchange, but the changes are not statistically significant.

Diversity histories for South American natives in South America (Fig. 2b) reflect effects of the interchange. The post-interchange faunas (Uquian, Ensenadan, Lujanian) show significantly lower familial diversities than the pre–land bridge faunas, and the post-interchange curves for South American natives show decreasing diversity in chronostratigraphic order, the youngest being the lowest.

Thus, analysis of the data by rarefaction confirms the analysis of the raw family data presented above. The rarefaction work also lessens the possibility that the patterns observed in Fig. 1 are artifacts of sampling.

Patterns and Rates of Generic Evolution

The Great American Interchange has played a primary role in the development of basic biogeographic principles regarding tempo and mode of large-scale dispersal. Although these principles were formulated under a model of stationary continents, most can be applied to the dynamic paradigm with logical modifications and extensions (*28*). With the documentation and acceptance of plate tectonic theory, the Great American Interchange has become a classic example for studying the biological consequences of continental suturing.

Equilibrial biogeographic models have been suggested as applicable to the Great American Interchange (*9, 19, 29*) and have been used to test the extension of the island biogeographic theory of MacArthur and Wilson (*30*) to continental scales. This theory was first developed to explain biogeographic patterns on oceanic islands (*30, 31*) and only later was applied to continental and global systems (*19, 32–36*). The fundamental prediction of equilibrium models is that species diversity in any restricted area (that is, island, continent) will, under constant conditions, eventually attain a dynamic equilibrium maintained by balanced rates of origination (or immigration, or both) and extinction (Fig. 3). The resultant diversity and the time span required to attain equilibrium are largely dependent on the size of the area; thus continents will have a higher species diversity and lower per-species turnover rate and will require a longer time span to attain equilibrium, as compared to oceanic islands (*37*). Monte Carlo simulations and empirical studies of fossil data indicate that genera and families show patterns of diversification commensurate with diversification of their constituent species so long as large numbers of higher taxa are involved (*29, 34, 36*). This relationship permits study of patterns of diversification at higher taxonomic levels even though species are the real units of evolution.

As discussed above, North and South American land mammal faunas each appear to have attained equilibrial diversity prior to the Great American Interchange. These equilibria were dynamic, with diversity remaining steady despite continuous origination and extinction of taxa (Table 1). In South America, per-genus turnover rate averaged 0.4 genera per genus per million years from 9 to 2 million years ago, while in North America the per-genus turnover rate averaged only 0.3 genera per genus per million years over the same period. The greater generic diversity and lower overall turnover rates of mammals in North America prior to the interchange are consistent with its greater total area (24×10^6 square kilometers compared to 18×10^6 square kilometers for South America), as predicted by equilibrium models (*38*).

The emergence of the Panamanian land bridge ended the phase of simple equilibrium for both continents and made each a potential source of immi-

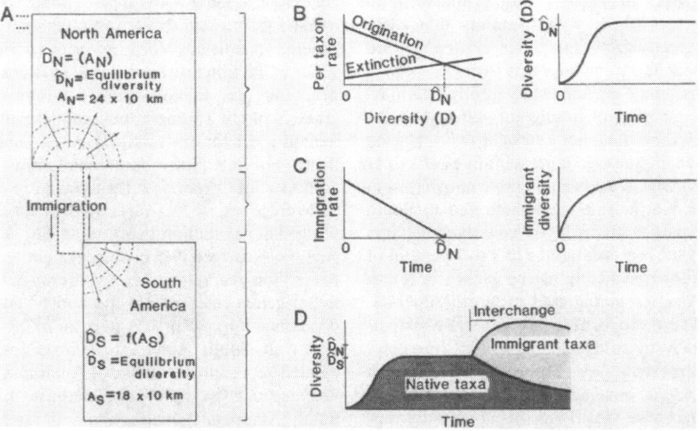

Fig. 3. Components of a hypothetical equilibrium model for the Great American Interchange. (A) Geographic constraints. North and especially South America were isolated continents through much of the Tertiary Period. As such, each should have supported a unique equilibrium diversity ($\hat{D}$), which may have been proportional, to a first approximation, to the area (A) of the continent (*29, 30*). Interconnection of the continents across the Panamanian land bridge terminated the equilibrium phase and permitted immigration of taxa between the two continents. (B) Closed-system diversification. Prior to interconnection, per taxon rates of origination (speciation) may have been high and rates of extinction low when diversity was low; with increasing diversity (and hence crowding of ecosystems), origination rates may have decreased, and extinction rates increased until becoming approximately equal at the equilibrium diversity $\hat{D}$ (*33–35*). This "diversity dependence" of evolutionary rates would result, in a simple deterministic system, in diversity increasing sigmoidally from some initial low to the equilibrium and then maintaining that equilibrium so long as the system remained closed (*43*). (C) Open-system immigration. Complete or partial interconnection of a large fauna at equilibrium with an area lacking fauna will initiate a flow of immigrants to the new area. The rate of immigration will be high at first and decline as the fauna in the newly colonized area becomes a larger and larger subset of the source fauna (*30, 37*). As a result, diversity in the new area should increase rapidly at first but later asymptotically approach an equilibrium determined by the equilibrium of the source area and by the local immigration and extinction rates. (D) Combined models. If taxa immigrate into a large area containing a native fauna, such as occurred in South America, the addition of immigrant taxa will, in essence, supersaturate the fauna of the new area. Extinction rates of both native and immigrant taxa will increase as diversity exceeds the equilibria of both faunal components. This will slow the increase in immigrant diversity and cause an exponential decline in native diversity. If the diversity of the source fauna is greater than the equilibrium of the native fauna, native diversity will eventually dwindle to zero in this simple model. More realistic constraints in the model (which could slow or prevent extinction of native taxa) would include backflow of immigrants north into North America and autochthonous evolution of taxa of immigrant ancestry, such as seen in the actual fossil records of North and South America.

grant taxa for the other. However, the tropical areas of North and South America seem to have acted as a barrier to dispersal of some representatives of genera and families. Only families with at least some constituent species distributed in tropical or subtropical areas took part in the interchange, whereas families with only temperate species did not disperse. Of all the families with part or all of their distribution in tropical areas, 17 South American families and 16 North American families dispersed; six South American and seven North American families did not. Thus, with regard to the potential family participants, the interchange was balanced (*10, 12, 17*).

Island biogeographic theory predicts that the effect of a source fauna on another fauna receiving immigrants should be proportional to the size (or diversity) of the source fauna (Fig. 3). In an attempt to apply this prediction to the interchange we divided the immigrant genera in Table 1 into primary immigrants (those with members that came directly from the other continent) and secondary immigrants (those whose founding species apparently evolved from primary immigrants after their arrival on the other continent) (*39*). During the interchange representatives of 1 to 11 percent of known available native genera in North America immigrated to South America in any given land mammal age, while representatives of 2 to 7 percent of known available native genera in South America immigrated to North America. These proportions are generally statistically indistinguishable (*40*). Ultimately, representatives of more North American genera immigrated to South America than vice versa, but this apparently represents a simple consequence of North America having a 60 percent greater average generic diversity than South America during the late Cenozoic. Thus, as predicted, the number of primary immigrants appears proportional to the size of the respective source faunas.

However, the subsequent evolutionary histories of the primary immigrants are significantly different. Various members of the 12 South American primary immigrants in North America gave rise to three secondary genera, whereas the 21 North American primary immigrants in South America gave rise to 49 secondary genera, derived subequally from members of five immigrant groups: cricetine rodents, carnivorans, proboscideans, perissodactyls, and artiodactyls (Table 1, row j). This difference in evolutionary histories represents nearly an order of magnitude difference in per-genus rates of origination between the respective primary immigrants (*26*). The Recent record further emphasizes this trend as demonstrated by the remarkable secondary diversity (more than 40 genera) of cricetine rodents.

The resulting faunal dynamics of native taxa on the two continents are predictable. North America, with a proportionately small "input" of primary immigrants, exhibits no detectable change in per-genus turnover rate; on the other hand, South America, where generic diversity eventually exceeded previous equilibrium levels by more than 50 percent, exhibits an increase of nearly 70 percent in per-genus extinction rates among native taxa (Table 1). The observed gradual decline in diversity of native South American genera subsequent to the land bridge (Fig. 1, column H and Table 1, row b) is consistent with patterns expected for a supersaturated biogeographic system (Fig. 3) [figure 6 in (*35*)]. North American immigrant genera in South America exhibit a marked increase in extinction rates over genera remaining in North America (per-genus extinction rate averages 0.3 genera per genus per million years for immigrants, and 0.2 genera per genus per million years for native North American taxa), but the North American immigrants in South America maintained lower average extinction rates (0.3 genera per genus per million years) than South American natives (0.5 genera per genus per million years) (*41*). These differences in per-genus rates reflect the continued diversification of North American immigrants in South America, their documented replacement of South American natives, and the significant increase in South American generic diversity and faunal enrichment on a continent-wide basis.

Conclusions

Some aspects of faunal dynamics of the Great American Interchange (that is, prior equilibrium, difference in turnover rates, importance of source faunas, and increased extinction with supersaturation) are predicted from elementary considerations of equilibrium theory. However, the significant and apparently rapid diversification of North American secondary immigrants within South America is not predicted by simple extrapolation of equilibrium models into evolutionary time frames. This radiation is thus the unique aspect of the interchange story; it alone seems to account for the long-observed asymmetry in interchange dynamics between the two continents and for the great change in taxonomic composition of the post–land bridge mammal fauna in South America (Fig. 1).

A possible but speculative explanation for the post–land bridge history of the South American fauna exists. During the late Cenozoic, a phase of orogeny beginning about 12 million years ago resulted in a significant elevation of the Andes Mountain range (*8*). A major phase of these orogenic movements occurred between 4.5 and 2.5 million years ago with a rise of from 2000 to 4000 meters (*42*). The newly elevated Andes served as a barrier to moisture-laden Pacific winds (*8*), and a rain shadow was created on the eastern (leeward) side. The southern South American habitats changed from primarily savanna-woodland to drier forests and pampas, and precocious pampas environments and desert and semidesert systems came into prominence at about that time. Many subtropical savanna-woodland animals retreated northward (*8*), and new opportunities favoring higher generic diversity arose for those animals able to adapt to these new ecologies.

The greater diversification of North American genera after they had reached South America is evident in such different groups as cricetid rodents, canid carnivores, gomphotheres, horses, llamas, and peccaries. If the relative success of northern groups is attributed to competitive displacement of equivalent southern groups, it becomes necessary to develop a number of complex scenarios with a great deal of uncertainty concerning which groups of species compete and on which adaptive bases (*10*). Perhaps it is more reasonable to attribute the success of the North American groups to some general ability inherent in their previous history to insinuate themselves into narrower niches (*8*). In any event, their success in South America is a clear pattern not predicted by simple equilibrium theory.

References and Notes

1. A. R. Wallace, *The Geographical Distribution of Animals* (Macmillan, London, 1876).
2. For example, Wallace (*1*) regarded *Galera* (a skunk), *Tapirus*, and *Lama* as moving from South America to North America, but the reverse is now known to be true.
3. It is somewhat ironic that these workers neither discussed, nor even believed in, the Great American Interchange as such (G. G. Simpson, personal communication, 11 June 1981); see also (*6*).
4. Translated from K. A. von Zittel, in *Handbuch der Palaeontologie*, vol. 4, *Band Vertebrata (Mammalia)* (Ouldenberg, Munich, 1891–1893), pp. 754–755. "Aber nicht nur nordamerikanische Typen benützen die neuröffnete Bahn, um ihr Verbreitungsgebiet zu vergrössen, sondern auch die südlichen Autochthonen begannen nach Norden zu wandern, und so vollzog sich am Schluss der Pliocaenzeit eine der merkwürdigsten Faunenüberschliebungen, welche die Geologie zu verzeichnen hat."

5. W. D. Matthew, *Ann. N.Y. Acad. Sci.* **24**, 171 (1915); *Spec. Publ. N.Y. Acad. Sci.* **1**, 1 (1939).
6. W. B. Scott, *A History of Land Mammals in the Western Hemisphere* (Hafner, New York, 1937).
7. G. G. Simpson, *J. Wash. Acad. Sci.* **30**, 137 (1940); *The Geography of Evolution* (Chilton, New York, 1965); in *Biogeography and Ecology in South America*, E. J. Fittkau *et al.*, Eds. (Junk, The Hague, 1969), p. 879; *Splendid Isolation: The Curious History of South American Mammals* (Yale Univ. Press, New Haven, Conn., 1980).
8. B. Patterson and R. Pascual, in *Evolution, Mammals and Southern Continents*, A. Keast, F. C. Erk, B. Glass, Eds. (State Univ. of New York Press, Albany, 1972), p. 274.
9. S. D. Webb, *Paleobiology* **2**, 216 (1976).
10. ——, *Annu. Rev. Ecol. Syst.* **8**, 355 (1977); *ibid.* **9**, 393 (1978); *Paleobiology* **4**, 206 (1978); L. G. Marshall and M. K. Hecht, *ibid.*, p. 203; L. G. Marshall, *ibid.* **5**, 126 (1979); in *Biotic Crises in Ecological and Evolutionary Time*, M. Nitecki, Ed. (Academic Press, New York, 1980), p. 133; I. Ferrusquía-Villafranca, *Univ. Nac. Auton. Mex. Inst. Geol. Bol.* **101**, 193 (1978); R. Hoffstetter, *Acta Geol. Hisp.* **16**, 71 (1981).
11. L. G. Marshall, R. F. Butler, R. E. Drake, G. H. Curtis, R. H. Tedford, *Science* **204**, 272 (1979).
12. J. M. Savage, *Nat. Hist. Mus. Los Angeles Cty. Contrib. Sci.* **260**, 1 (1974).
13. D. H. Tarling, in *Evolutionary Biology of the New World Monkeys and Continental Drift*, R. L. Ciochon and A. B. Chiarelli, Eds. (Plenum, New York, 1980), p. 1; M. C. McKenna, in *ibid.*, p. 43.
14. O. J. Linares, thesis, University of Bristol (1978).
15. O. A. Reig, *Publ. Mus. Munic. Cienc. Nat. Lorenzo Scaglia* **2**, 164 (1978).
16. P. Hershkovitz, in *Evolution, Mammals and Southern Continents*, A. Keast, F. C. Erk, B. Glass, Eds. (State Univ. of New York Press, Albany, 1972), p. 311.
17. S. D. Webb and L. G. Marshall, in *South American Mammalian Biology*, H. Genoways and M. Mares, Eds. [Special Publication Series of the Pymatuning Laboratory of Ecology, University of Pittsburgh (Univ. of Pittsburgh Press, Pittsburgh, Pa., in press)].
18. The range distributions of South American taxa are drawn from L. G. Marshall, R. Hoffstetter, R. Pascual, *Fieldiana Geol.*, in press; and L. G. Marshall *et al.*, *ibid.*, in press.
19. S. D. Webb, *Evolution* **23**, 688 (1969).
20. C. W. Harper, Jr., *J. Paleontol.* **49**, 752 (1975); H. R. Lasker, *Paleobiology* **4**, 135 (1978).
21. Indices used for measuring aspects of mammalian faunal dynamics in Table 1 are defined and computed as follows: (i) Durations of each land mammal age are given in millions of years (to the nearest tenth) and are based on all available radioisotopic, paleomagnetic, and biostratigraphic data, much of which is summarized in (*11*). (ii) The number of genera is the total number (or "diversity") of terrestrial mammal genera known for each land mammal age in each continent. We exclude *Homo*, and aquatic and volant mammals such as sea cows and bats. (iii) Originations are number of first appearances of a particular taxon (genus) in a given time interval (land mammal age) on each continent. This category combines three different kinds of originations: (a) New native autochthons (taxa whose members evolved in situ); (b) new immigrant allochthons (taxa with members that immigrated from outside the continent, or at least outside the area previously sampled); and (c) pseudo-originations produced by taxonomists when an evolving lineage changes enough to warrant a new name. Sufficient data are not available to consistently make distinctions among these alternatives. (iv) Extinctions are last appearances of a taxon on a given continent. These may not be "true extinctions" since the same taxon may continue to live on another continent, as with the Rancholabrean "extinctions" of *Tapirus* and *Equus* in North America, or the same population evolves to the point where it receives a new name, thus producing a taxonomic or pseudoextinction. As in the case of originations, the data do not consistently permit discrimination between these alternatives; however, pseudoextinctions appear to be of minor importance in this data set. (v) Running means (R_m) are expressions of the standing crop of a taxon (*19, 20*). This statistic compensates for time intervals of unequal duration by subtracting the average of originations (O_i) and extinctions (E_i) for a given age from the number of genera (S_i) for that age; thus, $R_m = S_i - (O_i + E_i)/2$. (vi and vii) Origination rates (O_r) are indices adjusted for time intervals of unequal length by dividing the total number of originations (O_i) of taxa occurring during a given time interval by the duration (d) of that interval; thus, $O_r = O_i/d$. By similar reasoning, the extinction rate, $E_r = E_i/d$. (viii) Turnover rates (T) are the average number of taxa of a given rank that either originate or go extinct during a given time interval (that is, rates of first and last appearances). Turnover rates represent the average of origination rates and extinction rates for a given time interval; thus, $T = (O_r + E_r)/2$. (ix) The per-genus turnover rate is the turnover rate adjusted for average diversity calculated by dividing the total turnover rate (T) by the total running mean (R_m).
22. The range distributions of North American genera are drawn from S. D. Webb, in *Pleistocene Extinctions, the Search for a Cause*, P. Martin and R. Klein, Eds. (Univ. of Arizona Press, Tucson, ed. 2, in press).
23. We have counted the trans-Beringian immigrants with the native North American genera, while freely recognizing the fact that North America was not a "closed system" as was South America. Three trans-Beringian genera (*Pseudocyon, Pseudoceras, Torynobelodon*) appear in the North American Clarendonian. After the Clarendonian, incursions into North America from the Old World grew steadily; trans-Beringian origins account for 18 genera in the Hemphillian, 20 genera in the Blancan and Irvingtonian, and 27 genera in the Rancholabrean. Most Hemphillian and Blancan immigrants were cricetid rodents and diverse Carnivora; by Irvingtonian and Rancholabrean time the principal taxa were arvicoline rodents and bovid and cervid ruminants. In fact, the balance of generic exchange between North America and Eurasia shifted strongly in favor of the latter during the late Cenozoic [C. A. Repenning, in *The Bering Land Bridge*, D. M. Hopkins, Ed. (Stanford Univ. Press, Stanford, Calif., 1967)], p. 288. Nonetheless, virtually all Irvingtonian and Rancholabrean immigrants from trans-Beringea were steppe-tundra grazers. Their ecological impact was surely concentrated in north temperate latitudes and considerably removed from major impact on the Great American Interchange in tropical North America.
24. Of the 36 families of mammals recorded in beds of Lujanian age in South America, 8 (22 percent) are now extinct. These include 6 of 24 (25 percent) native and 2 of 12 (17 percent) immigrant groups. All eight families were present in North America at about that time, and became extinct there as well. Of the 35 families of mammals recorded in beds of Rancholabrean age in North America, 11 (31 percent) are now extinct. These include 6 of 26 (23 percent) native families and 5 of 9 (56 percent) immigrant families.
25. Of the 120 genera known from the Lujanian of South America, 45 (40 percent) became extinct. Included were 25 of 59 (42 percent) native South American genera and 20 of 61 (33 percent) immigrant genera. These differences are not statistically significant (*26*). Of the 114 genera known from the Rancholabrean of North America, 32 (28 percent) became extinct. Included are 23 of 102 (23 percent) native North American genera, and 9 of 12 (75 percent) immigrant genera.
26. Tests for differences in proportions: $z = 1.08$, $P > .05$ for South American; $z = 3.82$, $P < .01$ for North American.
27. D. M. Raup, *Paleobiology* **1**, 333 (1975). The mathematical technique used here was that applied by Raup to higher taxa.
28. M. C. McKenna, in *Implications of Continental Drift to the Earth Sciences*, D. H. Tarling and S. K. Runcorn, Eds. (Academic Press, New York, 1973), p. 21; R. H. Tedford, in *Paleogeographic Provinces and Provinciality*, C. A. Ross, Ed. (Soc. Econ. Paleontol. Mineral. Spec. Publ. 21, 1974), p. 109.
29. K. W. Flessa, *Paleobiology* **1**, 189 (1975).
30. R. H. MacArthur and E. O. Wilson, *Evolution* **17**, 373 (1963); *The Theory of Island Biogeography* (Princeton Univ. Press, Princeton, N.J., 1967).
31. D. S. Simberloff, *Annu. Rev. Ecol. Syst.* **5**, 161 (1974); *Science* **194**, 572 (1976); in *Biotic Crises in Ecological and Evolutionary Time*, M. Nitecki, Ed. (Academic Press, New York, 1980).
32. R. H. MacArthur, *Biol. J. Linn. Soc.* **1**, 19 (1969).
33. M. L. Rosenzweig, in *Ecology and Evolution of Communities*, M. L. Cody and J. M. Diamond, Eds. (Belknap, Cambridge, Mass., 1975), p. 121.
34. J. J. Sepkoski, Jr., *Paleobiology* **4**, 223 (1978).
35. ——, *ibid.* **5**, 222 (1979).
36. D. S. Simberloff, *J. Geol.* **82**, 267 (1974).
37. ——, in *Models in Paleobiology*, T. J. M. Schopf, Ed. (Freeman, Cooper, San Francisco, 1972), p. 160.
38. Alternatively, the higher average turnover rates for South America may simply represent the need for a refined synthetic systematic review of these faunas [D. M. Raup and L. G. Marshall, *Paleobiology* **6**, 9 (1980)], and rate differences between the continental faunas may be due to different approaches to taxonomic treatment (lumpers versus splitters). We consider only potential mammal-mammal interactions yet recognize that nonmammalian groups may be involved as well [J. H. Brown and D. W. Davidson, *Science* **196**, 880 (1977)].
39. This subdivision of immigrants into primary and secondary groups has never before been formally attempted, although many workers have noted existence of these categories. Several caveats should accompany such an interpretive subdivision, the first being our wholly inadequate knowledge of late Cenozoic mammalian evolution in the American tropics, regarded as an undocumented source area for many of the immigrants (*8*). Nevertheless, our criteria for subdivision are simple and can be easily tested by further fossil discoveries. Primary immigrant genera are native genera (those that belong to native families) with members that occur on the other continent or are genera with members so closely related to known genera on the other continent that further taxonomic studies will probably show them to be congeneric (a criterion based on the observations of L.G.M. and S.D.W.). Secondary immigrant genera are those that belong to families native to the other continent but are unknown on that continent and apparently lack possible congeneric forms. By these criteria, the primary South American immigrants to North America are *Didelphis*, *Kraglievichia*, *Dasypus*, *Glyptotherium*, *Pliometanastes*, *Nothrotheriops*, *Eremotherium*, *Thinobadistes*, *Glossotherium*, *Hydrochoerus*, *Neochoerus*, and *Mixotoxodon*; and the primary North American immigrants to South America are *Calomys*, *Canis*, *Felis*, *Leo*, *Smilodon*, *Conepatus*, *Galera*, *Lutra*, *Mustela*, *Cyonasua*, *Nasua*, *Arctodus*, *Hemiauchenia*, *Odocoileus*, *Dicotyles*, *Platygonus*, *Equus*, *Hippidion*, *Tapirus*, *Cuvieronius*, and *Stegomastodon*. The remaining immigrant taxa listed in (*22*) for North America and in (*18*) for South America are regarded as secondary.
40. A z-test for differences in proportions [G. W. Snedecor and W. G. Cochron, *Statistical Methods* (Iowa State Univ. Press, Ames, ed. 6, 1967, p. 220)] applied to total numbers of primary immigrants moving north and south as proportions of the total size of their respective native faunas over the whole of the last 9 million years reveals no significant difference between North and South America ($z = 0.719$, $P > .05$). However, significant differences do occur within the Ensenadan-Lujanian interval (≈ Rancholabrean), when approximately 2 percent of available South American genera move north compared to 11 percent of available North American genera which move south ($z = 2.278$, $P < .05$).
41. It is possible that the comparatively low extinction rate for immigrant taxa in Table 1, row g may involve multiple immigrations rather than any special quality of the immigrants. If all members of a taxon endemic to continent A died out on that continent, it is for all practical purposes, regarded extinct. (Members of this taxon may have dispersed to continent B before their extinction on A, and then may have been able to reinvade A from B after the population on A died out, but the chance of this happening is regarded as having a very low probability.) If, on the other hand, an immigrant from continent B to continent A died out on A, its extinction would not be universal so long as it survived on continent B. This population from B could then reestablish itself on A by subsequent and repeated invasions from B and create the impression that the original population in A never really "became extinct." Such reestablishments of populations would result in lower extinction rates for immigrants relative to native taxa (that is, Table 1, row g) even if actual extinction rates in both groups were the same on each continent.
42. B. S. Vuilleumier, *Paleobiology* **1**, 273 (1975).
43. For applicability to mammalian evolution, see J. A. Lillegraven, *Taxon* **21**, 261 (1972).

44. We thank W. Burger, J. Cracraft, G. McGhee, K. Luchterhand, J. M. Savage, G. G. Simpson, S. Stanley, and W. D. Turnbull for reading the manuscript. Supported in part by NSF grant EAR 7909515 and DEB 7901976 (to L.G.M.), NSF grant DEB 7810672 (to S.D.W.), and NSF grant EAR 75-03870 (to D.M.R.).

PAPER 3

Resource Supply and Species Diversity Patterns (1971)

J. W. Valentine

Commentary

SETH FINNEGAN

Anyone who has thought much about paleoecology in the past five decades likely knows the experience of feeling that they have made a major theoretical or conceptual breakthrough only to go to the library (or Google Scholar) and discover that Jim Valentine thought of it first. This was a regular occurrence for me in graduate school, and despite my efforts to keep on top of his vast output, it still happens disturbingly frequently. It is, therefore, very difficult to pick a single one of Jim's many seminal papers to showcase in this volume.

What makes this paper stand out in such a crowded field is that, reasoning from first principles and a relative handful of relevant studies that were available at the time, Jim lays out a model of the relationship between nutrient availability and local (alpha) diversity in marine ecosystems that has since garnered considerable empirical support from studies of both modern and ancient communities. He begins by reviewing arguments for the idea that environments with relatively stable trophic resource supplies should support communities that are more diverse, more specialized, and have more even species-abundance distributions than those with unstable resource supplies. He then shifts to considering a question that had been much less thoroughly explored: What effect should total nutrient supply level (e.g., net primary productivity) have on diversity? By demonstrating that many of the same principles that pertain to stability of supply also pertain to productivity, he convincingly undermines the intuitive view that, regarding the relationship between productivity and diversity, "more is better." Finally, he applies this reasoning to considering the causes of mass extinctions in the geological record—a relatively new research topic in 1971. Arguing against the then-popular view that such climactic events may be caused by crises of food availability, Jim pointed out that in many cases just the opposite might be true.

From the vantage point of 2012, this argument appears particularly prescient. Long before the widespread application of stable isotopic and biomarker proxies, this paper anticipated the current view that some biotic crises in the fossil record are indeed related to oversupply, rather than undersupply, of nutrients. As a theoretical framework for considering the relationship between productivity and diversity trends through time, this paper remains remarkably relevant.

From *Lethaia* 4:51–61.

RESOURCE SUPPLY AND SPECIES DIVERSITY PATTERNS

JAMES W. VALENTINE

Valentine, J.W.: Resource supply and species diversity patterns. *Lethaia*, Vol. 4, pp. 51–61. Oslo, January 15th, 1971.

The adaptive strategies pursued by organisms faced with different trophic resource regimes may play a major role in diversity regulation. Selection for fitness in resource-rich or resource-unstable environments favors generalized, inefficient populations and results in low diversity; selection for fitness in resource-poor or resource-stable environments favors specialized, efficient populations and results in high diversity. Therefore, insofar as trophic resource levels influence diversifications and extinctions, diversification should be associated with lower, and extinctions with higher, resource levels. This conclusion is directly opposed to that of most recent investigators, but can be rationalized from theoretical considerations and deserves careful consideration.

James W. Valentine, Department of Geology and Institute of Ecology, University of California, Davis, California 95616, 19th May, 1970.

Paleontologists have long sought to explain the patterns of diversifications and extinctions of lineages in the fossil record. Since the control of species diversity within ecosystems has come to be recognized as an important ecological process, there have been attempts to link these ecological controls to an explanation of fossil diversity patterns (as Bretsky, 1969; Valentine, 1969a, 1969b; Lipps, 1970). By diversity is meant simply the number of species (or higher taxa when specified) that occur in the region or in the ecological unit under discussion. Opinions regarding the processes that regulate species diversity today have been rather diverse themselves (see for example, MacArthur, 1965; Pianka, 1966; Hessler & Sanders, 1967; Valentine, 1967, 1968; Stehli, Douglas & Newell, 1969). Nevertheless, continuing work is placing limits on the probable main factors, and, as a result, it is now possible to construct generalized models of ecological diversity control which satisfy many of the requirements of the data and which have theoretical bases. The theoretical predictions arise from considerations of the general principle governing energy flow in systems as inferred from information theory (Ashby, 1956). Some parameters of these models appear to operate in a fashion that is more or less the opposite of what might be expected at first thought; this is especially true of parameters that describe the levels of supply of nutrient and food resources. Since postulations concerning these resource supplies form important parts of some current hypotheses of diversification and extinction (Bramlette, 1965a, 1965b; Tappan, 1968; McCammon, 1969), the examination of resource regimes seems appropriate. Two major aspects of resource supplies have received special attention in connection with diversity controls: (1) the stability of the supply, and (2) its relative richness or level.

It is a pleasure to acknowledge the contributions of Dr. P.W. Bretsky and Mr. D.M. Lorenz, Northwestern University, Dr. J.H. Lipps, University of California, Davis, and Dr. A.L. McAlester, Yale University, all of whom read and criticized the manuscript.

Trophic resource regimes and species diversity

Resource stability is one aspect of general environmental stability which has been considered as a primary factor in regulating species diversity. The special significance of trophic resources is that they are density-dependent factors and this can regulate population densities and sizes, which in turn play important roles in the regulation of species diversity within ecosystems. Margalef (1968) treats this topic particularly well, while the accounts of MacArthur & Wilson (1967) and Levins (1968), though based upon widely different perspectives, tend towards similar conclusions. Bretsky & Lorenz (in press) have called attention to the significance of one aspect of the resource regime, stability, in interpreting the fossil record. It is only necessary to outline the salient points of this diversity regulation here, most of which are covered in these publications.

In brief, in environments wherein trophic resources fluctuate, populations must be adapted to utilize a resource that appears suddenly and to survive even though it disappears for a time. Populations with short life spans may correlate their life cycles closely to resource regimes, perhaps surviving low-resource periods as inactive zygotes. Populations with longer life spans may become relatively inactive during low-resource periods and commonly feed upon the more stable available food sources, such as detritus (Odum & De la Cruz, 1963). The more that environments fluctuate, the heavier the mortality is likely to be, and thus large populations are favored. Furthermore, because rapid population growth is possible during favorable periods, populations with high reproductive potentials are favored. In order to support as large a population as possible under such conditions, it is useful for species to have a broad tolerance for (1) food items, so that whatever prey happens to be present at any time may be utilized; (2) habitats, so that individuals may live at the greatest possible number of sites and thus maintain high population densities; and (3) all the density-independent factors that are fluctuating along with the trophic resources. These attributes add up to a generalized broad-niched species. An ecosystem composed of incompletely exploited populations that frequently suffer mass mortalities is quite inefficient, although it may be well adapted to the difficult problems of coping with fluctuating trophic resources. For a given environmental range, only a low number of inefficient populations with broad resource requirements can be accommodated.

Stable environments tend to favor selection for narrow functional ranges. In an environment which hardly fluctuates, populations need only narrow tolerances for temporally fluctuating parameters. Species occupying highly stable environments are rarely subjected to mass mortalities and are commonly not required to maintain large breeding populations. They may thus become quite specialized in food items, since the supply is reliable, and for habitats, for since high population density is not highly advantageous, it is not necessary to inhabit a wide range of habitat types. Thus, populations in stable environments tend to be specialized and rather small and stable in size and efficient in utilizing resources. For a given environmental range, a relatively high number of specialized and efficient populations can coexist.

Another aspect of the resource regime that may be of fundamental importance in regulating diversity is the level of resource supply, which has received much less attention than stability. Most writers have assumed that rich resource supplies would favor the evolution of diverse biotas, while low supplies would support only relatively impoverished ones (Connell & Orias, 1964; Tappan, 1968; McCammon, 1969). There are theoretical reasons, however, to expect that just the reverse is true in many situations (Margalef, 1963, 1965, 1968), and some recent studies have tended to support this view. The principle of adaptive strategy for resource level is much like that for resource stability. High food supply levels, for example, can obviously support larger populations of a given species than low levels. Therefore adaptive strategies that are inefficient in the utilization of energy sources can nevertheless be employed in a food-rich environment. Since ecologically flexible populations commonly contain heterogeneous arrays of individuals that have broad spectra of tolerances, they would probably be favored at high resource levels. On the other hand, in an environment in which resources are limited, only relatively small populations may be maintained, other things being equal. Conservation of population size may be promoted by optimizing the fitness throughout the population. Therefore specialization is favored.

As an example of population systems which support these inferred relations, Fryer & Iles (1969) have studied the relationship between species diversity, environmental stability, and resource levels among cichlid fishes of the African Great Lakes and associated aquatic environments. Some of the genera of this family have undergone spectacular diversification into great 'flocks' of species which co-inhabit the same lake; up to 200 species are present in some flocks. In general the species of the flocks are highly specialized in feeding and other habitat requirements, are restricted to a single lake, and have small and stable populations. The lake environments that support these flocks are relatively stable and have relatively low nutrient supplies (oligotrophic conditions). By contrast, other genera of cichlids (notably *Tilapia*) contain few species, of which only a very few co-exist in the same general habitat. They are broadly tolerant and relatively unspecialized species, vigorous, highly plastic phenotypically and tend to occur over wide geographical regions in a variety of habitats which include unstable river systems and such unusual environments as hot saline lakes. In environments where species of *Tilapia* are most usually found, nutrient supply is relatively high (eutrophic conditions).

As noted by Fryer & Iles (1969), this situation is precisely predicted by the somewhat theoretical suggestions of Margalef (1959, 1965). Selection in the case of the species flocks has been towards small stable populations with a low energy supply in a crowded environment which requires very efficient resource utilization; this is achieved by specialization. Selection of this sort has been called *K-selection* (MacArthur & Wilson, 1967); *K* represents the carrying capacity of the environment for a given population. Selection in the case of *Tilapia* has been in favor of high reproductive potential and adaptive flexibility and vigor under fluctuating but commonly rich resource supplies in an uncrowded environment. Resource utilization can be inefficient in this case. Selection of this sort has been called *r-selection* (MacArthur & Wilson, 1967); *r* represents the reproductive potential.

The large-scale latitudinal diversity gradient among the biota of the continental shelves can be interpreted in terms of these same factors (Margalef, 1968). Arctic molluscan faunas, for example, are of low diversity and tend to be relatively generalized functionally (Valentine, in preparation). The bivalves are chiefly deposit feeders or non-siphonate 'suspension' feeders that can live below the sediment surface and may well utilize interstitial detritus drawn in by their feeding currents which must commonly percolate between sedimentary grains (see Nicol, 1967, and Stanley, 1968). Most of the gastropods for which feeding habits are known are renowned for their broad carnivorous abilities as both predators and scavengers. Arctic populations tend to be large and polymorphic. Primary production fluctuates greatly, but during the peak season is extremely high. The low latitudes, by contrast, support highly diverse faunas that tend to be composed of highly specialized species with relatively small populations. There, the environment is relatively stable with a fairly constant food supply, but at a level well below the arctic peaks (Raymont, 1963; Dunbar, 1968).

Diversity patterns in the deep-sea are also amenable to interpretation by reference to resource stability (Hessler & Sanders, 1967) and supply. The deep-sea is an environment of high stability and low resource supply, wherein a high species diversity might be predicted. Hessler & Sanders (1967) have shown that deep-sea communities are indeed rich in species, essentially as rich as shallow-water counterparts from tropical shelves. The resource levels – especially of food – fall off rapidly in the deep-sea with distance from the continents and the density and biomass of the bottom fauna also decrease (Ekman, 1953; Zenkevitch, 1961; Sanders & Hessler, 1969). The deep-sea benthos is chiefly composed of small organisms that are widely dispersed. It would be expected that the diversity increases with lowering of resource levels into the basin centers. There are as yet few data on this point, but the data available suggest that this is the case. Hessler & Sanders (1967) report that the number of species in deep-sea samples decreases with depth less rapidly than the number of individuals. This suggests that the communities are becoming less dense but more diverse as the available food decreases. It is worth repeating that the data are inconclusive; nevertheless, they do not oppose the suggested inverse relation of diversity and resource level.

Another interesting feature of the distribution of deep-sea species is that while many range long distances they are rather restricted in their depth distributions (Sanders & Hessler, 1969). It is possible that this pattern is to be associated with food supply. If so, species may be adapted to certain levels of food supply and to the local features of density and diversity which result therefrom. In greater depths, a greater specialization, smaller population size, larger generation length and lowered reproductive potential would be expected of the average species, when compared with lesser depths. A population adapted to a specific resource regime would find dispersal much easier horizontally at similar depths where the resource regime is similar than vertically at different depths where the regime would change more rapidly over a given distance.

Finally, Sanders & Hessler (1969) believe that seasonal fluctuations in primary production that are notable at shallow depths are dissipated long before food particles reach abyssal bottoms. In this event, the increasing seasonality found in higher latitudes would not affect the deep-sea regime. However, the great seasonal bursts of productivity in high latitudes, which are inefficiently processed, probably result in a great food supply in abyssal waters at high latitudes (Foxton, 1956; Fish, 1954; Raymont, 1963). As abyssal waters move equatorward they may be progressively depleted in food, despite the rain of particles from overlying waters. Should this occur, then a longitudinal diversity gradient is expectable, resulting in a higher diversity in lower latitudes, and a poleward increase in species' depth ranges might be expected. Whether or not such patterns are present is not certain. In most of the examples reviewed here, it is obviously difficult to separate the effects of stability from those of resource supply level. It is possible that the pattern of diversity and of resources in the deep-sea will provide an invaluable test of the significance of these effects.

Community evolution in changing resource regimes

Fig. 1 summarizes the attributes expected of average populations in communities that have evolved in combinations of high and low resource stability with high and low resource supplies. We have already had examples of poor stable and rich unstable regimes. In a poor stable region (box 1, Fig. 1) populations need only narrow functional ranges and K-selection occurs. As resources are limited populations are small. The high specialization permits the presence of numerous species in a given community.

Community stability is probably simply a function of the environmental stability in these regions, but may also be related in some cases to a multiplicity of trophic interactions in the food webs (MacArthur, 1955). Actually, if the environment were absolutely stable, diversification might not proceed, but if slight fluctuations occur as in all real environments or if there is a gradation into less stable environments which can serve as a source of species (Valentine, 1967, 1968a), diversification may occur. In a rich unstable region (box 4, Fig. 1), populations are usually flexible and have high reproductive potentials which, when coupled with a periodic or episodic abundance of resources, permit the strategy previously described. Resource instability may reach levels where the resource supplies would prove inadequate at times to support any populations. Either extinction would then become general or populations could adapt by ceasing to employ resources at these times. Some forms of seasonal inactivity or migration would become more and more common features of the biota.

Now, consider the results of decreasing resource levels in a fluctuating environment (box 2, Fig. 1). So long as these resources act as density-dependent factors, either the average size of populations or the number of populations must decrease. Thus if the populations remain large and unspecialized, diversity must fall. However, can this condition be maintained? With a low resource level, the speed with which populations can respond to relatively favorable conditions is lessened, and adaptive strategies based upon reproductive prowess are undercut. Instead, strategies that minimize the fluctuations in a population's size, reaching a compromise between the maximum during favorable periods and the minimum fatality during unfavorable times, would probably be favored. This would involve raising the efficiency of the population and thus increasing the specialization.

Finally, conditions may be inferred in a rich stable environment (box 3, Fig. 1). If we compare it to a rich unstable environment, diversity clearly may rise with stability. However, the comparison with a poor stable environment is more difficult. The 'eutrophication' of a resource-poor, high-diversity environment would simply lead to increased diversity if the additional resources were utilized to support additional populations. If, on the other hand, the additional resources resulted in an expansion in population size of those lineages without internal checks to population growth (or without checks unrelated to resource levels), then the additional resource would result in the development of marked dominance. The dominant lineages would be able to afford the cost of exploration of evolutionary pathways that were closed previously, and thus would often evolve large niches, which would involve rising competition with species that controlled overlapping biospace. Probably some of the specialized, K-selected populations would be unable to respond to the pressure towards increased fecundity and would be eliminated. It is also probable that many of the narrowly-adapted populations could persist in the presence of abundant resources (MacArthur, 1965; Miller, 1967). In general, diversity would be somewhat

RESOURCES	STABLE	UNSTABLE
POOR	Box 1 K-selection Small stable population, highly specialized Highest diversity	Box 2 Compromise selection Moderate population size and specialization minimizing fluctuations Low diversity
RICH	Box 3 r-selection and K-selection Mixed population size and specialization High diversity	Box 4 r-selection Fluctuating population, often large, unspecialized Lowest diversity

Fig. 1. Comparison of population types and community diversities expected in four different resource regimes.

lower than in the poor, stable situation, and there would be a greater range of adaptive strategies represented by different populations in the same community.

Diversification and extinction patterns in the fossil record

The application of these principles to explain the patterns of diversity change and of turnover in taxa in the fossil record is necessarily highly speculative. Nevertheless, the attempt seems justified, for many current speculations run counter to those presented here and therefore a testing of the alternatives is highly desirable.

Temporally localized bursts of diversification are commonly ascribed to the operation of normal processes of evolution during times of unusual evolutionary opportunity. The opportunities are usually referred to two causes: the opening up of biospace (inhabitable environmental dimensions) through a preceding burst of extinction, or the evolutionary 'invention' of new adaptive modes which form the basis for major radiations, often by permitting the occupation of previously untenanted biospace. Another sort of opportunity may occur when the environment expands to create new biospace. This is essentially the process envisioned for much of late Mesozoic-Cenozoic diversification, when cooling on high-latitude shelves created new sorts of environments while old sorts were preserved in low latitudes (Valentine, 1967).

Many causes have been postulated to account for temporally localized bursts of extinction; good reviews are by Rhodes (1967) and Lipps (1970). Some of the suggested mechanisms have been rather catastrophic and while extinctions by poisoning of the oceans or by cosmic death ray are imaginable, perhaps the search for systematic ecological controls should not yet be abandoned. Some of the leading hypotheses that have ecological bases involve reductions in resource levels. For example, Bramlette (1965a, 1965b) has suggested that a reduction in nutrient supply to the oceans lowered the phytoplankton productivity, which so reduced the basic food supply as to cause marine extinctions across the Cretaceous-Tertiary boundary. McCammon (1969) has extended this reasoning to include the Permo-Triassic extinctions, especially of many Brachiopoda which she believes have relied heavily upon organic nutrients as a direct food source. Tappan (1968) has presented a more complex model that involves the assumption that a reduction in phytoplankton diversity is to be correlated with a reduction in phytoplankton productivity. The variations in productivity are inferred to account for depletion or enrichment of atmospheric oxygen and for other changes in the geochemical balance, which caused selective extinctions of animal taxa.

The view advocated here is that lowering of trophic recource levels may cause selective extinctions but favors processes of diversification in the long run. Major extinctions, on the contrary, are to be associated with times of increased (though fluctuating) resource levels.

Discussion

It is of course not certain to what extent variations in resource levels have contributed to Phanerozoic diversity patterns. If they played a role, however, the relations are likely to be those outlined here: rising resource levels would be associated with high productivity but lowered diversity; falling resource levels would be associated with lowered productivity but increased diversification. Changes in either direction would be likely to cause selective extinctions of less fit taxa and favor the diversification of pre-adapted taxa. The modest diversifications accompanying rising trophic resource levels would not fully replace the numbers of lineages that became extinct. The significant diversifications accompanying falling trophic resource levels would more than outstrip extinction and total diversity would rise. Judging from present patterns, changes in resource levels can be expected to produce less effect than changes in stability alone. Nevertheless, the resource patterns may control important secondary diversity trends and gradients, and may either reinforce or damp the effects of environmental stability levels.

The relations discussed here can be viewed as special cases of principles that have been developed in the field of cybernetics and general systems theory. Applications of the general theory to the special case of ecosystems have proven fruitful (for example, MacArthur, 1965; Margalef, 1968), and more attention should be given to their employment in interpreting biotic history. As nicely expounded by Margalef (1968), a principle in forming analogies between diversity regulation and the 'laws' of cybernetic systems is that the higher the energy flow within biotic systems, the lower the diversity. Margalef has taken as a measure of diversity the expression:

$$D = \Sigma p_i \log_2 p_i,$$

where p_i is the probability that an individual within the unit being measured belongs to species i. In this expression, diversity (D) is at a minimum if all individuals belong to the same species and at a maximum if each individual belongs to a different species. The expression is sensitive to the distribution of individuals into species; for example, if for a given number of species a few are abundant and the remainder rare, D is higher than if all the same species are present but are equally abundant. Using this definition of diversity, Margalef concludes that, insofar as energetics are concerned, high diversity is favored in communities that: (1) live in stable environments; (2) live in low-resource environments; (3) live in cold environments, as lower temperatures favor slow reaction rates and low energy flow; and (4) are exploited least, as cropping of populations requires an increase in reproduction among the prey, thus speeding energy flow. In this paper diversity has been measured simply as the number of species present, without regard to the relative sizes of their populations. In Margalef's account, the analogy between resource level and cybernetic principles was possible because when energy flow increases – when foods or nutrients are added to a system of populations – some populations benefit more than others. The relative sizes of populations therefore become more uneven and the diversity index D indicates falling diversity, although the number of species present need not change. The analogy in this paper is not based precisely upon this phenomenon, although it is related to it. Rather the analogy here is possible because the evolution of adaptive strategies for fitness in resource-rich (and/or unstable) and resource-poor (and/or stable) environments favors low species numbers in the former and high species numbers in the latter.

Unless the analogy is coincidental, there is an implication that the minimization of energy flow under ambient conditions contributes generally to fitness. It has been contended that selection tends to decrease the amount of energy required by an organism to maintain its niche (see Bock & Wahlert, 1965) and also tends to minimize the energy flow within ecosystems (Margalef, 1968). This more or less follows from the premise that the more efficient organism is better adapted, other things being equal.

REFERENCES

Ashby, W.R. 1956: *An Introduction to Cybernetics*. Chapman and Hall, London.

Bock, W.J. & von Wahlert, G. 1965: Adaptation and the form-function complex. *Evolution 19*, 269–299.

Bramlette, M.N. 1965a: Massive extinctions in biota at the end of Mesozoic time. *Science 148*, 1696–1699.

Bramlette, M.N. 1965b: Mass extinction of Mesozoic biota. *Science 150*, 1240.

Bretsky, P.W., Jr. 1969: Evolution of Paleozoic benthic marine invertebrate communities. *Palaeogeog., Palaeoclimat., Palaeoecol. 6*, 45–59.

Bretsky, P.W., Jr. & Lorenz, D.M. in press: Adaptive response to environmental stability. A unifying concept in paleoecology. In *Proc. N. Am. Paleontol. Conv.*, [Ed. E.L. Yochelson].

Connell, J.H. & Orias, E. 1964: The ecological regulation of species diversity. *Amer. Naturalist 98*, 399–414.

Dunbar, M.J. 1968: *Ecological Development in Polar Regions*. 119 pp. Prentice-Hall, Englewood Cliffs, N.J.

Ekman, S. 1953: *Zoogeography of the Sea*. 417 pp. Sedgwick and Jackson Ltd., London.

Fish, C.J. 1954: Preliminary observations on the biology of boreoarctic and subtropical oceanic zooplankton populations. *Symp. on Marine and Fresh-Water Plankton in the Indo-Pacific*, Bangkok, 3–9 [not seen].

Foxton, P. 1956: Standing crop of zooplankton in the Southern Ocean. *Discovery Repts. 28*, 193–235.

Fryer, G. & Iles T.D. 1969: Alternative routes to evolutionary success as exhibited by African cichlid fishes of the genus *Tilapia* and the species flocks of the Great Lakes. *Evolution 23*, 359–369.

Hessler, R.R. & Sanders, H.H. 1967: Faunal diversity in the deep-sea. *Deep-Sea Res. 14*, 65–79.

Levins, R. 1968: *Evolution in Changing Environments: Some Theoretical Explorations*. 120 pp Princeton University Press, Princeton, N.J.

Lipps, J.H. 1970: Plankton evolution. *Evolution 24*, 1–22.

MacArthur, R.H. 1955: Fluctuations of animal populations, and a measure of community stability. *Ecology 36*, 533–6.

MacArthur, R.H. 1965: Patterns of species diversity. *Biol. Rev. 40*, 510–33.

MacArthur, R.H. & Wilson, E.O. 1967: *The Theory of Island Biogeography*. 203 pp. Princeton University Press, Princeton, N.J.

Margalef, R. 1959: Mode of evolution of species in relation to their places in ecological succession. *XV Int. Congr. Zool., Sec. 10*, Paper 17.

Margalef, R. 1963: On certain unifying principles in ecology. *Amer. Naturalist 97*, 357–74.

Margalef, R. 1965: Ecological correlations and the relationship between productivity and community structure. *Mem. 1st Ital. Idrobiol. 18 (Suppl.)*, 355–364.

Margalef, R. 1968: *Perspectives in Ecological Theory*. 111 pp. Univ. Chicago Press, Chicago, London.

McCammon, H.M. 1969: The food of articulate brachiopods. *J. Paleontol. 43*, 976–985.

Miller, R.S. 1967: Pattern and process in competition. In *Adv. Ecol. Res.* Ed. J.B. Cragg, *4*, 1–74.

Nicol, D. 1967: Some characteristics of cold-water marine pelecypods. *J. Paleontol. 41*, 1330–1340.

Odum, E.P. & De la Cruz, A.A. 1963: Detritus as a major component of ecosystems. *Am. Inst. Biol. Sci. Bull. 13*, 39–40.

Pianka, E.R. 1966: Latitudinal gradients in species diversity. *Am. Naturalist 100*, 33–46.

Raymont, J.E.G. 1963: *Plankton and Productivity in the Oceans*. 660 pp. Pergamon Press, Oxford.

Rhodes, F.H.T. 1967: Permo-Triassic extinction. In *The Fossil Record. A Symposium with Documentation*. Eds. W.B. Harland et al., 57–76, Geol. Soc. London, London.

Sanders, H.C. & Hessler, R.R. 1969: Ecology of the deep-sea benthos. *Science 163*, 1419–1424.

Stanley, S.N. 1968: Relation of shell form to life habits in the Bivalvia (Mollusca). *Ph.D. Thesis*, Yale University, New Haven, Conn.

Stehli, F., Douglas, R. & Newell, N.D. 1969: Generation and maintenance of gradients in taxonomic diversity. *Science 164*, 947–949.

Tappan, H. 1968: Primary production, isotopes, extinctions and the atmosphere. *Palaeogr., Palaeoclimat., Palaeoecol. 4*, 187–210.

Valentine, J.W. 1967: Influence of climatic fluctuations on species diversity within the Tethyan provincial system. In *Aspects of Tethyan Biogeography*, Eds. C.G. Adams and D.V. Ager. *Syst. Ass. Pub. 7*, 153–166.

Valentine, J.W. 1968: Climatic regulation of species diversification and extinction. *Bull. Geol. Soc. Amer. 79*, 273–276.

Valentine, J.W. 1969a: Niche diversity and niche size patterns in marine fossils. *J. Paleontol. 43*, 905–915.

Valentine, J.W. 1969b: Patterns of taxonomic and ecological structure of the shelf benthos during Phanerozoic time. *Paleontology 12*, 684–709.

Zenkevitch, C.A. 1961: Certain quantitative characteristics of the pelagic and bottom life of the ocean. In *Oceanography*, Ed. M. Sears. *A.A.A.S. Public. 67*, 323–335. Washington, D.C.

PAPER 4

Leaf Assemblages across the Cretaceous-Tertiary Boundary in the Raton Basin, New Mexico and Colorado (1987)

J. A. Wolfe and G. R. Upchurch Jr.

Commentary

DENA M. SMITH

Recent and continuing climate change is predicted to result in dramatic changes to organisms and the ecosystems they inhabit (McCarthy et al. 2001). The only way to understand these changes and predict what their effects may be on Earth's biota, is to study the record of life preserved in the geologic past. Although previous researchers had already started to examine large-scale global patterns in taxonomic change over the Phanerozoic (Sepkoski 1982; Knoll 1984; Niklas 1985), few had taken an ecologic approach to examining major transitions in the history of life. Wolfe and Upchurch (1987) not only focused their work on a time interval in which dramatic change had been documented, the mass extinction event at the Cretaceous-Paleogene (K-Pg) boundary, but also used the physiognomy of leaf megafossils and dispersed cuticle from multiple sites to understand vegetation and climate change before, during, and after this event.

By using detailed stratigraphic sampling from numerous boundary localities, and an ecologic approach to looking at the K-Pg boundary, Wolfe and Upchurch were able to document an abrupt ecological disturbance (which they attributed to an asteroid impact) resulting in a major extinction in plant communities of the region. By relying on physiognomic characters and extant analogues, they were able to identify five primary phases in the vegetation that they believed corresponded to environmental change. The first phase, prior to the boundary, included a diverse Cretaceous flora, dominated by broadleaved evergreen taxa. The second phase, immediately above the boundary, consisted of an assemblage dominated by ferns (often referred to as the "fern spike"). The third through fifth phases showed low-diversity plant assemblages that started with the characteristics of modern early-succession forests and ended with assemblages that had more late-succession- to climax-forest components. The authors interpreted these phases to indicate a rapid event that was ecologically catastrophic, followed by a recovery pattern that was analogous to succession in modern vegetational systems, albeit over a much greater timescale.

Although later studies focused on improving sampling, stratigraphic resolution, and the inclusion of additional ecological variables, the paper by Wolfe and Upchurch served as a model for how to study vegetation and terrestrial ecosystems across important boundary events in the history of life. And while later workers found what amounted to very similar vegetation patterns in other basins, these works served mainly to elaborate and further corroborate the evidence of an impact at the K-Pg boundary (Johnson et al. 1989; Nichols 1990; Johnson 1992; Nichols et al. 1992). Perhaps most importantly, Wolfe and Upchurch changed the conversation from being focused solely on taxonomic changes across boundaries to a focus on the dramatic ecological changes that occur in vegetational communities across major events in the geological past (McElwain and Punyasena 2007). This conversation be-

From *Proceedings of the National Academy of Sciences USA* 84:5096–5100.

comes even more urgent as we strive to understand the processes that underlie ecosystem functioning during periods of dramatic and potentially catastrophic climate change.

Literature Cited

Johnson, K. R. 1992. Leaf-fossil evidence for extensive floral extinction at the Cretaceous-Tertiary boundary, North Dakota, USA. *Cretaceous Research* 13:91–117.

Johnson, K. R., D. J. Nichols, M. Attrep Jr., and C. J. Orth. 1989. High-resolution leaf-fossil record spanning the Cretaceous/Tertiary boundary. *Nature* 340:708–11.

Knoll, A. H. 1984. Patterns of extinction in teh fossil record of vascular plants. In *Extinctions*, edited by M. H. Nitecki, 21–68. Chicago: University of Chicago Press.

McCarthy, James, Osvaldo F. Canziani, Neil A. Leary, David J. Dokken, and Kasey S. White, eds. 2001. Climate Change 2001: Impacts, Adaptation and Vulnerability: Contribution of Working Group II to Intergovernmental Panel on Climate Change Third Assessment Report. Cambridge: Cambridge University Press.

McElwain, J. C., and S. W. Punyasena. 2007. Mass extinction events in the plant fossil record. *Trends in Ecology and Evolution* 22:548–57.

Nichols, D. J. 1990. Geologic and biostratigraphic framework of the nonmarine Cretaceous-Tertiary boundary interval in western North America. *Review of Palaeobotany and Palynology* 65:75–84.

Nichols, D. J., J. L. Brown, M. Attrep Jr., and C. J. Orth. 1992. A new Cretaceous-Tertiary boundary locality in the western Powder River basin, Wyoming: biological and geological implications. *Cretaceous Research* 13: 3–30.

Niklas, K. J. 1985. Patterns in vascular land plant diversification: an analysis at the species level. In *Phanerozoic Diversity Patterns: Profiles in Macroevolution*, edited by J. Valentine, 97–128. Princeton, NJ: Princeton University Press.

Sepkoski, J. J. 1982. Mass extinctions in the Phanerozoic oceans: a review. *Geological Society of America Special Paper* 191:283–89.

Proc. Natl. Acad. Sci. USA
Vol. 84, pp. 5096–5100, August 1987
Geology

Leaf assemblages across the Cretaceous–Tertiary boundary in the Raton Basin, New Mexico and Colorado

(paleobotany/quasisuccession/extinction/"impact winter")

JACK A. WOLFE AND GARLAND R. UPCHURCH, JR.*

Paleontology & Stratigraphy Branch, U. S. Geological Survey, MS-919, Federal Center, Denver, CO 80225

Communicated by Estella B. Leopold, May 6, 1987 (received for review September 4, 1986)

ABSTRACT **Analyses of leaf megafossil and dispersed leaf cuticle assemblages indicate that major ecologic disruption and high rates of extinction occurred in plant communities at the Cretaceous–Tertiary boundary in the Raton Basin. In diversity increase, the early Paleocene vegetational sequence mimics normal short-term ecologic succession, but on a far longer time scale. No difference can be detected between latest Cretaceous and early Paleocene temperatures, but precipitation markedly increased at the boundary. Higher survival rate of deciduous versus evergreen taxa supports occurrence of a brief cold interval (<1 year), as predicted in models of an "impact winter."**

Vegetational and floristic studies of pollen and leaf assemblages across the Cretaceous–Tertiary (K–T) boundary generally have presented evidence against major long-term effects of possible bolide impact(s). Although pollen data indicate major short-term disruption of the terrestrial ecosystem at the K–T boundary, low levels of extinction are reported (1, 2). In regard to leaf assemblages, vegetational change from the Lancian (late Maestrichtian) through the early Paleocene in the northern high plains of Wyoming and Montana was attributed to general Late Cretaceous cooling (3). Extinction levels across the K–T boundary were suggested to be comparable to those across the Paleocene–Eocene boundary in North America (3), although Krasilov (4) noted that extinction rates appeared to be much higher in northeastern Asia. These previous studies of extinction in leaf assemblages across the K–T boundary are based on assemblages (*i*) from sections that lack the iridium-rich boundary clay and (*ii*) that are stratigraphically widely spaced. Such studies cannot address rates of extinction or of climatic and ecologic changes.

That major ecologic disruption occurred throughout the western interior of North America, including the Raton Basin, cannot readily be disputed. Fern spores make up <25% of individual samples (including samples from swamp environments) during the latest Cretaceous in the Raton Basin, but fern spores reach abundances of 96–99% immediately above the iridium-rich boundary clay (1). This anomalous abundance of fern spores has been likened to mass-kill of land plants from known volcanic catastrophes such as the 1883 eruption of Krakatau (1), providing strong evidence for mass-kill of the angiosperm-dominated latest Cretaceous vegetation. However, mass-kill may not necessarily result in mass extinction.

This report is based on detailed sampling in the Raton Basin of Colorado and New Mexico (Fig. 1), where the iridium-rich boundary clay occurs in 15 local sections (5, 6). Almost all the fossil samples were obtained from sections that can be physically related to boundary clay. This clay also contains abundant shocked minerals (7), which are considered as strong evidence that the boundary clay represents fallout from a bolide impact. Palynological correlations (5, 6) indicate that the boundary clay in the Raton Basin is the same age as the iridium-rich clay in areas such as Hell Creek, Montana, where vertebrate data (8) also indicate that the clay approximates the K–T boundary. A previous report (9) of the Raton K–T boundary clay occurring in a zone of normal magnetic polarity is erroneous; the boundary occurs, as elsewhere, in a zone of reversed magnetic polarity (10).

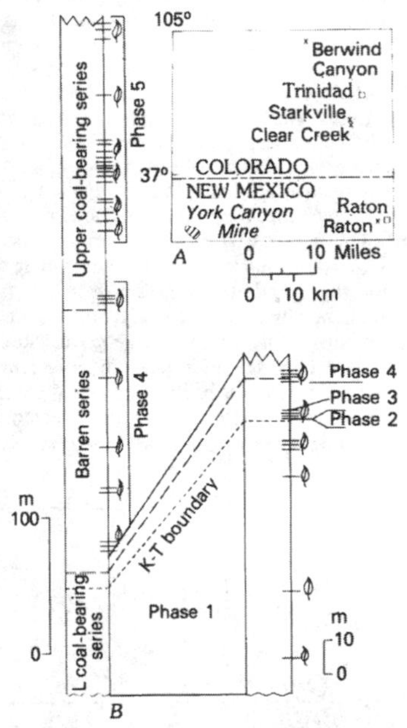

FIG. 1. (*A*) Map of part of Raton Basin and (*B*) generalized section of Raton Formation. Megafossil localities shown by leaf symbol have measured or estimated stratigraphic distances from the K–T boundary clay, which contains high amounts of iridium and shocked minerals.

METHODS OF ANALYSIS

High percentages of extinct plant genera and families in K–T boundary assemblages preclude accurately interpreting veg-

The publication costs of this article were defrayed in part by page charge payment. This article must therefore be hereby marked "*advertisement*" in accordance with 18 U.S.C. §1734 solely to indicate this fact.

Abbreviations: K–T, Cretaceous–Tertiary; aff., affinity.
*Present address: University Museum, Campus Box 315, University of Colorado, Boulder, CO 80309.

etation on the basis of ecologic tolerances of related extant taxa; general taxonomic affinities can yield only general concepts of ecologic and climatic change. Analyses of foliar physiognomy (e.g., leaf size, overall shape, apical shape, and characters of margin) offer a more precise method of inferring vegetational and climatic change because, in extant vegetation, changes in foliar physiognomy parallel overall changes in vegetation and thus climate (refs. 11–15; see particularly ref. 13, which discusses some of the prior critiques of foliar physiognomy). Various physiognomic characters of leaves, in conjunction with taphonomy, can also indicate probable ecologic preference within vegetation—e.g., whether streamside, understory, or early or late successional (14).

The analysis of cuticle—the waxy covering that preserves anatomical details of the leaf surface—adds a significant dimension to our data base. Individual cuticle morphology may be identified to leaf type on megafossils; subsequently, dispersed cuticle can be used to infer the presence of the whole-leaf taxon. Further, many additional groups can be recognized in the dispersed cuticle record by their diagnostic features. Like pollen, cuticle offers the potential of recovering many data from a small sample volume, including samples from stratigraphic intervals where megafossil assemblages are absent. This affords (*i*) a tighter stratigraphic sampling and control, (*ii*) increased accuracy in determining the ranges of whole-leaf taxa, and thus (*iii*) the potential to analyze vegetational change on a near-ecological time scale. Cuticles also have physiognomic characters (e.g., thickness, hairiness) that allow climatic inferences independent of their systematic affinities, thus affording additional physiognomic characters from which climate can be inferred (16, 17). Sampling in proximity to the K–T boundary was at intervals of 1–2 cm.

RATON LEAF SEQUENCE

Phase 1. Five vegetational–floristic phases are recognized (Fig. 2) in the Raton Basin. Phase 1, of Lancian age (about 66–68 million years), is represented by assemblages from the upper part of the Vermejo Formation (18) and from the base of the Raton Formation up to the K–T boundary clay. Leaf megafossil assemblages from the Lancian part of the Raton indicate high spatial heterogeneity as well as high specific diversity. The two main leaf localities in the Lancian Raton, each about 6 m below the boundary clay, produced 450–500 specimens and 36 species from excavations each <1 m^3; few species are common to the two localities. At least 75 species are known from the dispersed cuticle record.

The phase-1 flora contains palms and diverse dicots, including extinct relatives of Lauraceae (protolauraceans) and other Laurales (Fig. 2 *A, F, G, H, CC, DD, FF*), Illiciales (Fig. 2*EE*) and extinct relatives (protoillicialeans), Euphorbiaceae [extinct genera of Acalyphoideae such as "*Ficus*" *leei*, "*Cissites*" *panduratus* (Fig. 2*B*), and "*Zizyphus*" *fibrillosus* (Fig. 2*C*)], and Menispermaceae (Fig. 2*I*). One of the numerous extinct lauralean genera includes "*Artocarpus*" *dissecta* (Fig. 2 *G, FF*) and occurs as dispersed cuticle in the Mississippi Embayment, Atlantic Coastal Plain, and Western Interior during the late Campanian and Maestrichtian; this genus has a highly characteristic foliar morphology and cuticular anatomy and was obviously a major component of latest Cretaceous vegetation. Another common extinct lauralean group has lateral primary veins decurrent into the petiole [e.g., "*Ficus*" *praetrinervis* (Fig. 2*F*) and "*Cinnamomum*" *linifolium* (Fig. 2A)]. Many common Lancian dispersed-cuticle species occur up to the boundary but not above it; for example, immediately below (0–1 cm) the

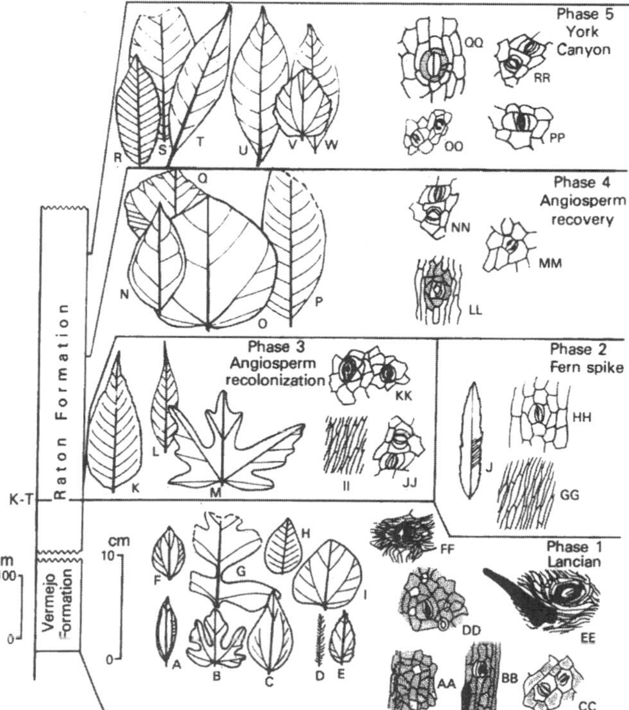

FIG. 2. Changes in leaf physiognomy from the latest Cretaceous to the early Paleocene in the Raton Basin. Phase 1 is characterized by a high-diversity flora and leaves (and leaf cuticles) indicative of warm subhumid climate. The iridium-rich clay interpreted as the K–T boundary occurs between phase 1 and phase 2. Immediately above the boundary (phase 2), the vegetation was dominated by a fern and probably represents vegetation following a mass-kill. Angiosperms again became dominant in phase 3 but were not diverse and were of early successional morphology; the leaves and cuticles indicate a warm, wet climate. Diversity gradually increased during phases 4 and 5, but even in phase 5 (about 1.5 million years after the K–T boundary), diversity was low compared to phase 1; leaves and cuticles indicate a warm rain-forest environment. *AA–RR*, cuticles (approximately ×120).

boundary clay at Berwind Canyon are the highest occurrences of dispersed cuticles of "*Cinnamomum*" *linifolium*, affinity (aff.) "*Ficus*" *praetrinervis*, "*Artocarpus*" *dissecta* and two closely related species, and two species groups of monocots.

Evergreen conifers are moderately diverse in the Vermejo. In the Lancian Raton, the highest megafossil assemblages contain *Geinitzia* (Fig. 2*D*) and a *Sequoia*-like plant; a dispersed-cuticle type allied to extant Australian–New Caledonian Cupressaceae (especially *Neocallitropsis* and *Callitris*) occurs 2 m below but not above the boundary (Fig. 2*AA*). Presumably conifers were emergents, analogous to extant vegetation of the Southern Hemisphere (15).

The plants of phase-1 megafossil assemblages were overwhelmingly broad-leaved evergreen, as indicated by thick, coriaceous leaves and robust cuticles. The high percentage of entire-margined leaves (Table 1) indicates that Lancian temperatures were megathermal (>20°C), as on lowland Taiwan or New Caledonia (15). Strong variation in taxonomic composition between coeval dispersed-cuticle assemblages from a given lithology indicates high spatial heterogeneity, characteristic of megathermal vegetation. Precise temperature assignment based on leaf-margin percentage is difficult and depends on whether analogies are made to extant east Asian or Southern Hemisphere vegetation (15). Occurrences of evergreen conifers and a low-diversity deciduous (thin-leaved) element suggest a Southern Hemisphere analog; if so, Lancian mean annual temperature was probably 21–22°C.

Elongated foliar apices (drip-tips) are uncommon and leaf size is small. Many taxa have dense hairs, some of which are cutinized (Fig. 2 *BB, EE*). Leaves of extant rain forest plants, if pubescent, typically have hairs only on the lower surface (11), but phase-1 leaves can have hairs on both surfaces, a condition that characterizes leaves of dry, sunny habitats (16, 17). In modern megathermal climates, such characters indicate only moderate precipitation. Such dry-adapted characters of both whole leaves and cuticles continue to the highest megafossil assemblages of the Lancian. Dispersed-cuticle assemblages immediately below the boundary are also characterized by adaptations to dryness, suggesting no significant change in precipitation during the Lancian.

Phase 2. Phase 2 occurs immediately above the iridium-rich boundary clay and includes the palynological "fern spike," in which fern spores are overwhelmingly dominant (1). Abundant fern fronds and rare specimens of a bryophyte occur <1 m above the clay in the Starkville and Clear Creek sections. The sterile fern fronds (Fig. 2*J*) represent an extinct genus related to the extant *Stenochlaena*, a primary colonizer in vegetation of Indomalaya and Africa. In open situations, *Stenochlaena* produces sterile fronds, while as a climber in closed forests, it produces fertile fronds. All cuticle assemblages are dominated by cuticles that have cells in files (Fig. 2*GG*), typical of herbaceous stems, monocot leaves, or fern rachises; in many samples, these cuticles are thin. Rare dicot cuticles also occur, all of which are smooth; we have been unable to identify most of these as any known Cretaceous leaf-species. Because of the limited assemblage, no temperatures are inferred for phase 2.

Phase 3. Phase 3 is characterized as the angiosperm recolonization phase. The first evidence of phase 3 occurs at the Starkville locality, where fragments of "*Cissites*" *panduratus* (Fig. 2*M*) occur directly above phase 2. This same species is well represented at Berwind Canyon, the best phase-3 locality. Here, >400 leaves were collected. Dominants are "*Cissites*" *panduratus* and a celastracean (Fig. 2*K*), with lesser abundances of a protolauracean (Fig. 2*L*), a fern similar to *Stenochlaena*, and a second fern. Only 8 megafossil and 10 dispersed-cuticle entities are known in phase 3. Although herbaceous-type cuticle is less dominant than in phase 2, all dispersed-cuticle assemblages are typically dominated by a single taxon. Dispersed-cuticle taxa include species of Laurales (Fig. 2*JJ*), a protoillicialean (Fig. 2*KK*), and a palm (Fig. 2*LL*); the last two have not been found below the boundary. Monocot, including palm, cuticle types in phases 3–5 typically represent different species than in phase 1.

The physiognomy and general affinities of the phase-3 flora indicate disturbed conditions. Phase 3 "*Cissites*" *panduratus* is large- and wide-leaved, palmately veined, and unlobed to deeply lobed; this morphology is found in extant megathermal colonizing species, including members of Acalyphoideae of the family Euphorbiaceae. The megafossil protolauracean and celastracean taxa are stenophyllous (narrow-leaved) as in extant stream-side plants. Although both taxa occur in carbonaceous shales during phase 3, both are restricted to fluviatile sandstones during phase 4, and the protolauracean occurs in fluviatile sandstones in phase 1. Stream-side plants are, of course, early successional, and the occurrence of such early successional plants in swamp habitats during phase 3 suggests ecologic disruption and extinction that allowed opportunistic species to invade stable habitats. All phase-3 taxa either indicate, or are consistent with, early successional

Table 1. Physiognomy of Vermejo and Raton leaf assemblages and lithologies sampled

	Megafossils							
Phase	Total localities	Total species	Total dicot species	Leaf-size index*	% entire-margined species	% drip-tips	Cuticle foliar hairs	Lithologies sampled for dispersed cuticle and/or megafossils
5 (York Canyon)	20	35	27	68	74	55	Noncutinized, lower surface	Coal, carbonaceous shale, overbank sandstone
4 (angiosperm recovery)	14	25	21	72	71	42	Noncutinized, lower surface (see text)	Coal, carbonaceous shale, overbank sandstone, channel sandstone
3 (angiosperm recolonization)	5	8	5	70	80	40	Absent	Coal, carbonaceous shale, overbank sandstone
2 (fern-spike)	3	2	0	—	—	—	Absent	Coal, carbonaceous shale
1 (Lancian)								
Raton	6	47	43	34	72	9	Cutinized and noncutinized, lower and upper surfaces	Coal, carbonaceous shale, overbank sandstone, channel sandstone
Vermejo	19	86	63	34	71	9	(Inadequately sampled)	Carbonaceous shale, overbank sandstone

*Leaf-size index is that of ref. 19; an assemblage composed entirely of mesophyllous and larger species has an index of >100, and an assemblage composed entirely of microphyllous and smaller species has an index of <0.

vegetation. The 2 m of coal and carbonaceous shales at Berwind Canyon that contain phase 3 have four rooted horizons, which we interpret as soil-forming stages; no lithologic evidence of flood events is present that would suggest periodic disruption of the environment. Although no precise estimates of duration of these soil-forming stages can be presently made, phase 3 apparently persisted for a considerably longer period than does extant early successional megathermal vegetation.

The low diversity of phase 3 yields a small statistical sample, but the dominance of large leaves in the flora, the abundance of large leaves at outcrops, and the highly attenuated apices probably indicate high precipitation. Species such as "*Cissites*" *panduratus* show a marked increase in leaf size and apical attenuation from the Lancian (Fig. 2*B*) into phase 3 (Fig. 2*M*). Dense, cutinized hairs are not known from phase-3 cuticle types, even when present on related taxa from phase 1; this condition characterizes leaves of wet megathermal vegetation (11).

Phase 4. The angiosperm recovery phase extends from just below the barren series (so named for the scarcity of coal) and into the lower part of the upper coal-bearing series, about 200 m above the K–T boundary (Fig. 1). Phase 4 contains leaf species in swamp environments that have physiognomy characteristic of late successional (or "climax") vegetation. Specimens of "*Cissites*" *panduratus* (Fig. 2*O*) are all unlobed, and these, along with stenophylls, are restricted to stream-side or near-stream depositional environments. Two deciduous dicots are also found in channel and overbank facies: the platanoid "*Cissus*" *marginatus* (Fig. 2*Q*) and the archaic rosid *Averrhoites affinis*. The leaves typically have drip-tips, and leaf-margin percentage and leaf size index are high (Table 1). Phase 4 is not diverse, although it is richer than phase 3 (Table 1). Dominants are palms (Fig. 2*LL*) and evergreen dicots: Lauraceae (Fig. 2*MM*), "*Ficus*" *praetrinervis*, aff. *Picramnia* (Fig. 2*N*), and Celastraceae. Cuticles from most samples are thick and smooth, as in megathermal rain forests. However, four samples from a 5-m interval in the upper part of phase 4 contain some cuticles that have cutinized hairs, suggesting a possible drier interval.

Phase 5. Phase 5 is distinguished by the first appearances in the Raton sequence of numerous taxa and is best represented in the upper coal-bearing series at York Canyon Mine in an interval 270–420 m above the K–T boundary (Fig. 1). Extensive commercial stripping over an area of about 6 km^2 allowed examination of many thousands of specimens at many sites in the field and large collections through a stratigraphic interval of about 50 m. The megafossil flora is of low diversity: if old collections from coal mine dumps elsewhere in the upper coal-bearing series are included (18), the phase-5 flora contains only 45–50 species from throughout the Raton Basin (>4000 km^2). Dispersed-cuticle assemblages show comparable low diversity. Palms (both fan and feather) are present, and evergreen dicots include protolauraceans, other lauraleans [including "*Magnolia*" *lesleyana* (Fig. 2*U*)], Euphorbiaceae, and Tiliaceae [aff. *Grewia* (Fig. 2*V*), aff. *Heliocarpus*, and aff. *Triumfetta*]; the "*Magnolia*" and Tiliaceae are not known lower in the Raton sequence. Phase 5 also contains the earliest Raton occurrence of the deciduous dicotyledons "*Carya*" *antiquorum* (Fig. 2*R*) and "*Eucommia*" *serrata* and a deciduous conifer (aff. *Glyptostrobus*, Fig. 2*QQ*).

The foliar (including cuticular) physiognomy of phase 5 is indicative of megathermal rain forest. Most leaf-megafossil species have marked drip-tips. The leaf species in phase 5 tend to have a similar gross morphology, unlike the morphologic diversity in phase 1.

CONCLUSIONS

Extinction and Floristic Change. Studies of leaf megafossils and dispersed cuticles from the Raton Basin corroborate palynological studies (1) in suggesting rapid extinction at the end of the Cretaceous. For example, archaic Laurales such as "*Artocarpus*" *dissecta*, "*Cinnamomum*" *linifolium*, and their allies disappear at the K–T boundary, as do other groups characteristic of the Lancian.

However, the amount of extinction in the Raton leaf record is apparently much greater than in the Raton pollen record, from which the regional extinctions of only three pollen genera and few pollen species were inferred (2). For example, acalyphoid Euphorbiaceae and archaic Laurales, which were diverse in the Lancian, are each represented by a single species in the early Paleocene. Although palms and other monocots occur both above and below the boundary, cuticular anatomy indicates that the latest Cretaceous taxa are distinct specifically, if not generically, from the early Paleocene taxa. Archaic conifers (e.g., *Geinitzia*) also disappear. Although some taxa that are rare below the boundary are common above the boundary and vice versa, about 75% of phase-1 leaf species are unknown in Paleocene floras either in the Raton Basin or elsewhere.

The inferred high level of extinction in the leaf flora contrasts with the low level based on pollen studies (1). Although the reasons for this discrepancy are not certain, the foliar studies indicate that many extinctions occurred in groups (*i*) that have pollen that rarely fossilizes (e.g., Laurales; see also ref. 20) or (*ii*) that have generalized pollen that is typically not diagnostic to specific or generic levels (e.g., Acalyphoideae and Palmae; see also refs. 2 and 20).

Most Raton phase-5 dominants are typically unknown in phase 1, but some have records prior to the K–T boundary. *Grewia*-type Tiliaceae have a Maestrichtian record in the Mississippi Embayment region (21), and the thin-leaved (deciduous) *Glyptostrobus*, "*Carya*" *antiquorum*, and "*Eucommia*" *serrata* have been reported from the Maestrichtian of Montana (22). The deciduous element in phases 4 and 5 appears to be derived largely from the north. Some broad-leaved evergreen dominants in phase 5, however, may have been derived from a southern source—e.g., "*Magnolia*" *lesleyana*, which is also abundant in the Paleocene of the Mississippi Embayment region (23).

Vegetational and Climatic Change. The dominance of ferns immediately above the K–T boundary in phase 2 is analogous to the abundance of ferns on Krakatau after the 1883 eruption (11), as has been emphasized in Raton pollen studies (1). Phase-3 foliar physiognomy is typical of early successional vegetation in modern megathermal climates but occurred over a much longer time period than in "normal" vegetational succession. Rate of recovery of diversity on Krakatau contrasts markedly with that following the K–T boundary. On Krakatau, angiosperm species numbered 15 after 3 years, 49 after 14 years, 73 after 23 years, and 219 after 51 years (11). Although the 45–50 species represented in phase-5 collections probably do not represent all taxa actually present during phase 5 in the Raton Basin, this factor is countered by (*i*) the variety of depositional environments sampled, (*ii*) the large number of localities, (*iii*) the large area sampled (>4000 km^2 in the Raton Basin versus 45 km^2 for Krakatau), (*iv*) the long time period represented by phase 5, and (*v*) the large number of specimens examined. Even assuming incomplete sampling of the phase-5 flora, recovery in diversity of early Paleocene vegetation was on an evolutionary, not ecological, time scale. Such long-term change (whether floristic or physiognomic) that mimics short-term ecologic succession is here termed *quasisuccession* (from Latin *quasi*, appearing as if or simulating).

Vegetation, except during phase 2, was primarily broad-leaved evergreen. The uniformity of leaf-margin percentages in the Vermejo–Raton sequence indicates no general temperature difference between the latest Maestrichtian and the early Paleocene. However, selective extinction could bias some aspects of physiognomic analyses. Nonentire margins are partially correlated with thin deciduous leaves (14). Selective extinction of evergreen versus deciduous taxa could lower the apparent paleotemperature derived from the leaf-margin analysis. Our data indicate that a higher percentage of evergreen taxa than deciduous taxa became extinct (see below), and thus the temperature for the early Paleocene could have been higher than indicated by the leaf-margin analysis.

Changes in foliar features such as leaf size, apical attenuation, and hairiness of cuticles indicate a major increase in precipitation at the K–T boundary in the Raton Basin. Because dispersed-cuticle samples 0–1 cm below the K–T boundary have physiognomy characteristic of other phase-1 samples, subhumid climate is inferred to have persisted up to the boundary. Samples immediately above the boundary and higher in the early Paleocene section have cuticular physiognomy characteristic of humid climate, thus suggesting that high precipitation had a rapid onset at the K–T boundary and continued through the early Paleocene.

Only 12 of 75 (16%) phase-1 evergreen dicots are known in the Paleocene, in contrast to 8 of 12 (67%) phase-1 deciduous dicots; most evergreen conifers also became extinct, with deciduous Taxodiaceae first appearing above the boundary. Higher survival rate of deciduous versus evergreen taxa suggests an event favoring plants that had dormancy mechanisms. Dormancy mechanisms would be triggered by drought, darkness, or low temperature. Given the observed major increase in precipitation at the boundary, which eliminates the drought factor, the event would probably have been a brief period (<1 year) of darkness, low temperature, or both, as predicted in models of an "impact winter" (24). The survival of ferns is also significant because rhizomes of ferns can withstand thermal shock (25), as shown on El Chichón. Moreover, the overall high rate of extinction indicates that both reproductive and vegetative organs of vascular plants were destroyed. As in the present tropics, latest Cretaceous megathermal plants that had never been exposed to low temperatures (19) would be particularly vulnerable to a low-temperature excursion (3, 26).

We thank C. L. Pillmore for assistance in stratigraphic problems. T. M. Bown, W. A. Clemens, L. J. Hickey, A. W. Knoll, E. B. Leopold, D. J. Nichols, C. B. Officer, P. H. Raven, R. A. Spicer, B. H. Tiffney, and S. L. Wing offered critiques of the manuscript.

1. Tschudy, R. H., Pillmore, C. L., Orth, C. J., Gilmore, J. S. & Knight, J. D. (1984) *Science* **225,** 1030–1032.
2. Tschudy, R. H. & Tschudy, B. D. (1986) *Geology* **14,** 667–670.
3. Hickey, L. J. (1981) *Nature (London)* **292,** 529–531.
4. Krasilov, V. A. (1983) *Paleontol. Zh.*, 93–95.
5. Orth, C. J., Gilmore, J. S., Knight, J. D., Pillmore, C. L. & Tschudy, R. H. (1982) *Geol. Soc. Am. Spec. Pap.* **190,** 423–433.
6. Pillmore, C. L., Tschudy, R. H., Orth, C. J., Gilmore, J. S. & Knight, J. D. (1984) *Science* **223,** 1180–1183.
7. Bohor, B. F., Foord, E. E., Modreski, P. J. & Triplehorn, D. M. (1984) *Science* **224,** 867–869.
8. Clemens, W. A. (1982) *Geol. Soc. Am. Spec. Pap.* **190,** 407–413.
9. Payne, W. A., Wolberg, D. L. & Hunt, A. (1983) *New Mexico Geol.* **5,** 41–44.
10. Shoemaker, E. M., Pillmore, C. L. & Peacock, E. W. (1987) *Geol. Soc. Am. Spec. Pap.* **209,** 131–150.
11. Richards, P. W. (1952) *The Tropical Rain Forest* (Cambridge Univ. Press, London).
12. Wolfe, J. A. (1971) *Palaeogeogr. Palaeoclimatol. Palaeoecol.* **9,** 27–57.
13. Wolfe, J. A. (1981) in *Paleobotany, Paleoecology, and Evolution*, ed. Niklas, K. J. (Praeger, New York), Vol. 2, pp. 79–101.
14. Givnish, T. (1979) in *Topics in Plant Population Biology*, eds. Solbrig, O. T., Jain, S., Johnson, G. B. & Raven, P. H. (Columbia Univ. Press, New York), pp. 375–407.
15. Wolfe, J. A. (1979) *U. S. Geol. Surv. Prof. Pap.* **1106.**
16. Coley, P. D. (1983) *Ecol. Monogr.* **53,** 209–233.
17. Fahn, A. (1967) *Plant Anatomy* (Pergamon, Oxford).
18. Knowlton, F. H. (1917) *U. S. Geol. Surv. Prof. Pap.* **101.**
19. Wolfe, J. A. & Upchurch, G. R. (1987) *Palaeogeogr. Palaeoclimatol. Palaeoecol.*, in press.
20. Muller, J. (1981) *Bot. Rev.* **47,** 1–142.
21. Berry, E. W. (1916) *U. S. Geol. Surv. Prof. Pap.* **91.**
22. Hickey, L. J. (1980) *Univ. Mich. Mus. Paleontol. Pap.* **24,** 33–49.
23. Berry, E. W. (1925) *U. S. Geol. Surv. Prof. Pap.* **136.**
24. Alvarez, W., Alvarez, L. W., Asaro, F. & Michel, H. V. (1982) *Geol. Soc. Am. Spec. Pap.* **190,** 305–315.
25. Spicer, R. A., Burnham, R. J., Grant, P. & Glicken, H. (1985) *Am. Fern J.* **75,** 1–5.
26. Ehrlich, P. R., Harte, J., Harwell, M. A., Raven, P. H., Sagan, C., Woodwell, G. M., Berry, J., Ayensu, E. S., Ehrlich, A. H., Eisner, T., Gould, S. J., Grover, M. P., Herrera, R., May, R. M., Mayr, E., McKay, C. P., Mooney, M. A., Myers, N., Pimentel, D. & Teal, J. M. (1983) *Science* **222,** 1293–1300.

PAPER 5

Climatic Changes in Southern Connecticut Recorded by Pollen Deposition at Rogers Lake (1969)

M. B. Davis

Commentary

ERIC C. GRIMM AND SHINYA SUGITA

During the 1960s, major advances were made in the use of pollen analysis to reconstruct past vegetation and climate, and a number of important innovations trace to Margaret Davis. In a 1963 paper, Davis developed the R-value model, which links percentages of tree pollen to forest abundances. However, she realized percentages have limitations, and in a series of papers (Davis 1964, 1967a, 1969), she pioneered the determination of absolute pollen accumulation rates or "pollen influx." The development of radiocarbon dating in the 1950s made it possible to calculate deposition times (yr/cm) of sediments. By simply dividing the pollen concentration (grains/cm^3) by the deposition time, one could calculate pollen influx (grains·cm^{-2}·yr^{-1}), which is related to species abundances on the landscape.

Davis's seminal 1969 paper showed that at Rogers Lake, Connecticut, the late-glacial herb zone, which has relatively high percentages of *Pinus* (>20%) and *Picea* (~5%) pollen, had very low values of pollen influx. So although the pollen percentages might suggest some kind of forest tundra, the influx values imply that this zone represented tundra "devoid of trees." Then about 12,000 ^{14}C yr BP, pollen influx increased dramatically, indicating the appearance of trees on the landscape, although the change in percentages was much less indicative of a major change. This paper demonstrated clearly the value of influx-based vegetation reconstruction.

Today, palynologists routinely calculate pollen concentrations and influx values. Because of the large within- and between-site variations of influx values owing to complex depositional processes, which Davis also studied (Davis 1967b, 1968), influx values have not been much used for broad-scale synoptic vegetation studies. However, with careful site selection and good chronological control, the number of influx-based studies has increased, especially in northern Europe (Giesecke and Fontana 2008). In particular, at the Nordic tundra-forest ecotones, influx values are invaluable for reconstruction of vegetation and land cover (Hicks 2001; Kuoppamaa et al. 2009). Although important theoretical and methodological issues remain unsolved (Sugita et al. 2010), Davis's pioneering work in the 1960s and 1970s still influence paleoecology in the twenty-first century.

Literature Cited

Davis, M. B. 1963. On the theory of pollen analysis. *American Journal of Science* 261:897–912.

———. 1964. Pollen accumulation rates: estimates from late-glacial sediment of Rogers Lake. *Science* 145: 1293–95.

———. 1967a. Pollen accumulation rates at Rogers Lake, Connecticut, during late- and postglacial time. *Review of Palaeobotany and Palynology* 2:219–30.

———. 1967b. Pollen deposition in lakes as measured by

From *Ecology* 50:409–22.

sediment traps. *Geological Society of America Bulletin* 78:849–58.

———. 1968. Pollen grains in lake sediments: redeposition caused by seasonal water circulation. *Science* 162: 796–99.

Hicks, S. 2001. The use of annual arboreal pollen deposition values for delimiting tree-lines in the landscape and exploring models of pollen dispersal. *Review of Palaeobotany and Palynology* 117:1–29.

Giesecke, T., and S. Fontana. 2008. Revisiting pollen accumulation rate estimates from lake sediments. *Holocene* 18:293–304.

Kuoppamaa, M., T. Goslar, and S. Hicks. 2009. Pollen accumulation rates as a tool for detecting land-use changes in sparsely settled boreal forest. *Vegetation History and Archaeobotany* 18:205–17.

Sugita, S., S. Hicks, and H. Sormunen. 2010. Absolute pollen productivity and pollen-vegtation relationships in northern Finland. *Journal of Quaternary Science* 25:724–36.

CLIMATIC CHANGES IN SOUTHERN CONNECTICUT RECORDED BY POLLEN DEPOSITION AT ROGERS LAKE

MARGARET B. DAVIS
Department of Zoology and Great Lakes Research Division
University of Michigan, Ann Arbor, Mich.

(Accepted for publication November 2, 1968)

Abstract. Rates of deposition of pollen grains throughout late- and postglacial time were determined from the pollen concentration in radiocarbon-dated sediment. Changes by a factor of 5 or more for all except rare pollen types from one level to the next were considered significant indication of changes in the pollen input to the lake, reflecting changes in the pollen productivity of the surrounding vegetation.

Low pollen deposition rates in the oldest sediments reflect the prevalence of tundra vegetation between 14,000 and 12,000 years ago. An increase in the rate for tree pollen occurred 12,000 years ago, when boreal woodland became established. The rates continued to increase until a sudden sharp rise for white pine, hemlock, poplar, oak, and maple pollen 9,000 years ago marked the establishment of forest, similar perhaps to modern forests of the northern Great Lakes region. Pine pollen rates declined 8,000 years ago, and deciduous tree pollen became dominant. Ragweed pollen was deposited at relatively high rates 8,000 years ago, reflecting changes in the vegetation associated with the "prairie period" recorded in the Great Lakes region at this time. Subsequent changes in pollen deposition rates reflect the immigration of beech (6,500 years B.P.), hickory (5,500 years B.P.), and chestnut (2,000 B.P.) to southern Connecticut. During the past few hundred years pollen deposition rates reflect changes in the vegetation caused by disturbance by European settlers. Throughout much of postglacial time the pollen assemblages deposited at Rogers Lake are different from assemblages known from modern sediment. This makes climatic interpretation difficult and suggests that the forest associations of the region as they are recognized now are of quite recent origin.

INTRODUCTION

Fossil pollen grains from terrestrial plants occur abundantly and almost ubiquitously in Quaternary sediments. They form a continuous fossil record in the organic muds accumulating in lakes and peat bogs, providing one of the richest sources of information about the terrestrial environment of the past. Analyses of fossil pollen, presented as pollen diagrams, show the percentages of different pollen types changing characteristically from one stratigraphic level to another, defining various stratigraphic horizons, and thereby aiding in correlation of sediments. Many changes resemble differences between modern pollen assemblages associated with different vegetation types. Because these changes in pollen are widespread, occurring in contemporaneously deposited sediment of all types, they must reflect changes in ancient vegetation similar to the geographical differences in modern vegetation. Thus pollen diagrams provide evidence for a sequence of vegetation changes, induced by a series of changes in regional climate (von Post 1967).

In this paper I have sought to augment the information contained in the traditional pollen percentage diagram by estimating rates of deposition of pollen grains in the accumulating sediment. Deposition rates at different times in the past can be compared meaningfully, since the yearly influx of each pollen type is independent of the influx of other pollen types. Changes in deposition rates infer changes in species abundances of plants that cannot be inferred with certainty from pollen percentages. Furthermore, many fossil assemblages are different from pollen assemblages known in modern sediment and cannot be compared with the present. Interpretation of these pollen sequences can now be made by comparing pollen deposition rates as they change from level to level (Tsukada and Stuiver 1966). As more information becomes available on present pollen deposition rates and their relationship to modern vegetation, it may become possible to read pollen deposition rate diagrams directly as a record of density of vegetation on the landscape (Davis and Deevey 1964). Interpretations offered here have been more cautious because little is known about factors that control pollen input to and distribution within lakes (Davis 1967*a*, 1968). Nevertheless, the results have enlarged upon and strengthened previous interpretations of vegetation based on pollen percentages, providing new insights into the climatic history of southern Connecticut.

LOCATION OF SITE

Rogers Lake (Fig. 1) is a large (107 ha) lake in southern Connecticut (41° 22′ N, 72° 7′ W), 6 km east of the Connecticut River and 8 km north

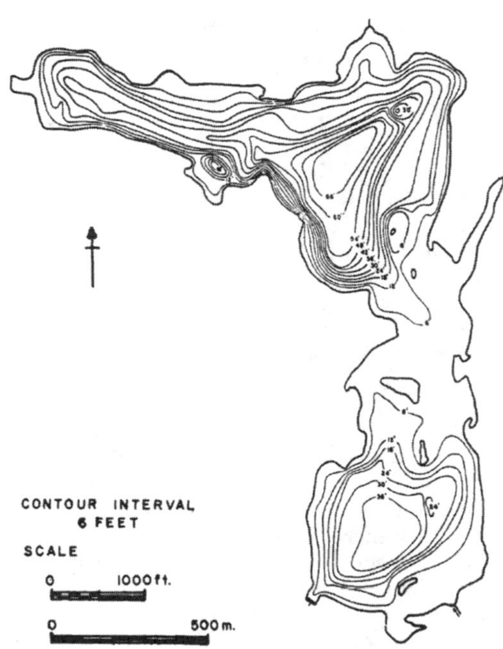

FIG. 1. Bathymetric map of Rogers Lake, Connecticut. Redrawn from Connecticut State Board of Fisheries and Game (1959).

of Long Island Sound. The lake has two basins, separated by a shoal only 2 m below the present water surface. The outlet at the south end of the lake has been dammed, raising the lake level about 1.3 m (Connecticut State Board of Fisheries and Game 1959).

Many cottages border Rogers Lake, but the surrounding landscape is heavily forested. The forest is almost entirely deciduous, with oaks making up almost half the basal area. Second to oak in abundance is red maple, followed in order by hickory, birch, and ash. Pine, beech, elm, and sugar maple are rare, and hemlock and alder still more infrequent (Davis 1965*b*, Davis and Goodlett, *unpublished data*). The forest is almost entirely second growth on abandoned farmland. Much of it has been further disturbed by cutting for cordwood and timber.

The topography of the region is irregular, with local relief of about 600 ft (ca. 200 m). Igneous bedrock is exposed at the surface on many hillsides, while other hills are covered with a mantle of glacial drift.

METHODS

Two cores of sediment, 5 cm in diameter, were collected from Rogers Lake by E. S. Deevey of Yale University and myself in September 1960, using a modified Livingstone piston corer (Deevey 1965) operated from a raft. An 11.5-m core was taken in 10.5-m water in the south basin. The upper 8 m of sediment was collected 1 or 2 m distant from the lower 4 m of sediment; correlation of the overlapping 1-m segment was confirmed by pollen analysis and radiocarbon dating. A second core was collected in 19.3-m water in the north basin of the lake. Here only 4.6 m of sediment was recovered before we encountered coarse sand.

The cores were stored in a cold room in the aluminum tubes in which they had been collected. Upon extrusion, samples were taken for pollen analysis by packing sediment into a 1-ml porcelain spatula. The remaining sediment was cut into segments, and 54 of these, varying in length from 5 to 20 cm, were submitted for age determination to Dr. Minze Stuiver of the Radiocarbon Laboratory at Yale University. Three 5-cm segments from the lowest meter of sediment collected in the north basin were also radiocarbon dated.

To assess the accuracy of the spatula method three series of five replicate samples were collected and weighed. The coefficients of variation of the mean wet weights were 11.5%, 3.4%, and 2.7%, respectively. Ninety-five per cent confidence intervals for mean wet weight per milliliter (assuming t-distribution) are $\pm$14.3%, $\pm$4.2%, and $\pm$3.4%, respectively. The 95% confidence interval of the pollen determinations was approximately $\pm$14% of the estimated number in the sample for counts of 200 grains; where 1,000 grains were counted, it was approximately $\pm$6% of the estimated number (Davis 1965*a*). The error in sample measurement is thus of the same magnitude as the error in pollen determination. If in future studies greater accuracy were desired, it would save time to measure the samples more accurately, perhaps by weighing them, rather than determining their pollen content more precisely by counting additional numbers of grains.

Samples were collected from the lowest meter of the Rogers Lake core by pushing a calibrated glass tube into the sediment. Samples collected with glass tubes were larger due to compaction than replicates collected with the spatula, and pollen counts from them were later corrected accordingly.

The 1-ml samples for pollen analysis were treated with KOH and acetolysis and, where necessary, with HF (Faegri and Iversen 1964). Assay of the pollen concentration was made by the aliquot slide method (Davis 1966). Control for loss of pollen during sample preparation was provided by adding a measured volume from a suspension of pure *Eucalyptus* pollen to the samples before preparation; the *Eucalyptus* pollen was counted together with the native pollen on the aliquot slides and the percentage recovery calculated. In only one case was the number of *Eucalyptus* pollen in prepared samples significantly different from the estimated number added. (The exception is the sample from 8.37-m depth.) The control was not in use at the time the late-glacial samples were prepared. In these samples the pollen concentration was determined using an older method (Davis 1965*a*) and cannot be accepted with equal confidence. From 200 to 1,000 grains total were counted on the aliquot slides; these counts were used both for estimation of the concentration of each type and for calculation of percentages.

Pollen from eastern white pine (*Pinus strobus*) was distinguished from pollen from the three other northeastern pine species (jack (*P. banksiana*), red (*P. resinosa*), and pitch (*P. rigida*), which were grouped as a single type) using the morphological criterion described by Ueno (1958). Microscope preparations from 11 samples were scanned under oil immersion and all pine grains were classified (Table 1); the results were extrapolated to the intervening levels. The mounting medium used was silicone fluid, so grains could be turned to improve visibility, but even so, almost one-third of the grains were torn or folded in such a way that critical identification could not be made. This proportion remained the same regardless of which type was dominant, however, indicating that failures to identify did not bias estimates of the relative abundances of the two types. Within the non-white pine species group, size will distinguish jack and/or red pine pollen from the much larger pollen of pitch pine (Whitehead 1964). Measurements were not made on the Rogers Lake material, but are available from stratigraphically equivalent levels at other sites.

TABLE 1. Percentage contribution of species to total pine pollen from selected levels in the Rogers Lake core (Values for other levels were obtained by interpolation)

Depth (m)	Estimated age of sample	*n*	White *P. strobus*	Other *P. banksiana*, *P. resinosa*, *P. rigida*	Unidentifiable
.075	+206	25	28	36	36
1.72	1,525	50	38	30	32
4.86	3,890	50	38	32	30
7.59	7,348	25	36	32	32
8.37	8,332	50	56	12	32
8.71	8,657	50	72	6	22
9.08	9,354	50	34	40	26
9.38	9,936	50	14	58	28
9.63	10,489	50	8	74	18
9.74	10,760	50	0	84	16
10.14	11,806	50	20	48	32

RESULTS AND DISCUSSION

Definition of terms

Pollen percentage is the percentage of any given pollen type among the total grains counted in a sample. Pollen percentages are plotted against depth or age of the sediment in the *pollen percentage diagram* (the traditional "pollen diagram").

Sediment matrix accumulation rate is the net thickness of sediment accumulated per unit time, after compaction and diagenesis. It may also be expressed as the amount of time per unit thickness of sediment; this has been termed *deposition time per centimeter* in a recent paper by Waddington (1969).

Pollen concentration is the number of grains per unit volume of wet sediment. Pollen concentration (per unit volume or weight of sediment) is often referred to in the literature as "absolute pollen frequency," or APF.

Pollen deposition rate, or *pollen accumulation rate,* is the net number of grains accumulated per unit area of sediment surface per unit time. It can also be called *pollen influx* per unit time. E. J. Cushing (*personal communication*) has suggested that this term is appropriate since we are measuring grains falling on an area. In the *pollen deposition rate diagram,* or *pollen influx diagram,* the number per square centimeter per year is plotted against the age of the sediment. This type of diagram has also been called an "absolute diagram," an undesirable term because it leaves unspecified whether the diagram shows pollen concentration or pollen deposition rate.

The general procedure in the study has been to multiply the pollen concentration at each level by the sediment matrix accumulation rate, in order to compute the pollen deposition rate.

Determination of sediment matrix accumulation rate

Matrix accumulation rate in the south basin of the lake was determined by plotting the radiocarbon ages of various levels in the core (Stuiver, Deevey, and Rouse 1963, Stuiver 1967) against depth. A smooth curve was fitted to the resulting points (Fig. 2); its slope represents the rate of accumulation. The deepest sediment, 11.5 m below the sediment surface, was 14,300 years old, indicating a fairly rapid average rate of accumulation of sediment during the lake's lifetime (about 0.08 cm per year). But marked fluctuations in accumulation rate of the sediment occurred from time to time (Fig. 2). Between 9,000 and 6,000 years ago, for example, the rate first increased to about 0.25 cm per year, then dropped to a rate

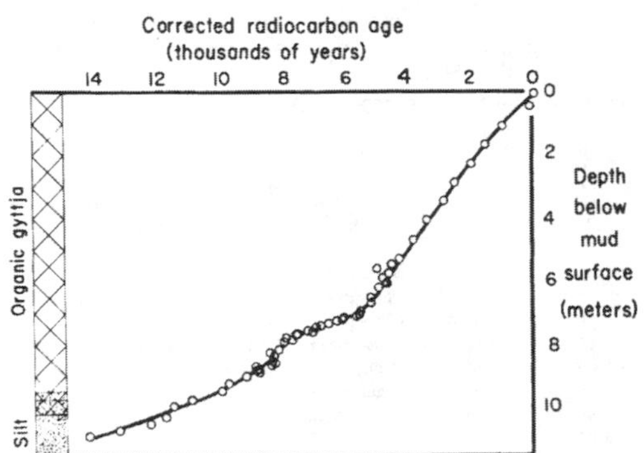

FIG. 2. Radiocarbon dates from the core from the south basin of Rogers Lake, plotted against the depth of the samples in the profile. The solid line is a composite curve drawn to express the general relationship between depth and age of the sediment.

one-fourth as rapid, and then increased again. There is no correlation between these changes and the type of sediment deposited (Fig. 2).

Matrix accumulation rate was also determined in the second core, from the north basin of the lake. The base of the core was 8,600 years old; the average rate of accumulation after that time was only 0.05 cm per year. Three radiocarbon dates (Y-1455, 1456, 1457) are available from the lowest 60-cm segment of the core, spanning the interval between 7,500 and 8,600 years B.P. (M. Stuiver, *personal communication*). Accumulation rate during this brief interval was about 0.06 cm per year. Rates in the south basin at that same time varied from 0.05 to 0.12 cm per year.

Two problems arise in evaluating estimates of matrix accumulation rates. One has its source in errors in the radiocarbon method itself. Because variations have occurred in the C^{14} concentration in the atmosphere, radiocarbon ages on material older than 2,500 years are slightly younger than the true ages of the samples, by an amount that increases proportionally with age (Damon, Long, and Greg 1966, Stuiver 1967). This anomaly causes the change in slope in the matrix accumulation curve at Rogers Lake at 2-m depth (2500 years B. P.) (Fig. 2) since samples below this level have C^{14} ages slightly younger than their true ages (Stuiver 1967). The difference in slope, however, is too small to be important in the present study.

Ancient carbon, apparently from ground water, is present in the water of Rogers Lake, giving an apparent age of 730 years (average of three samples) to the surface sediment. This amount has been subtracted from each age determination from the core, on the assumption that a fixed proportion of the carbon in the lake has always been C^{14}-deficient (Deevey et al. 1954, Broecker and Walton 1959, Stuiver et al. 1963). This assumption seems reasonable for the homogeneous postglacial sediment, although it might not hold for the older, silty sediment, deposited when the rate of carbon deposition was very low (Davis and Deevey 1964, Deevey 1964). Again, the change in slope resulting from a different correction would be too small to affect the results significantly.

The second problem in determining matrix accumulation rate is one of interpretation. The slope of the line fitted to the plot of radiocarbon dates vs. depth depends upon the shape of the curve used and on the closeness with which the curve has been fitted to the points. Some perspective on this problem is given by comparing the various curves that have been used for Rogers Lake, some using the less complete series of dates that were available earlier and others interpreting the existing data differently (Davis and Deevey 1964, Deevey 1964, 1968, Davis 1967*b*, Stuiver 1967). The differences between the resulting estimates of matrix accumulation rates, although larger than the differences mentioned above, are still relatively minor and should be considered a normal margin of error for this kind of study. Difficulties in interpretation arise because the size of the margin may not be predictable. The margin of error is influenced at any one site by the frequency and magnitude of fluctuations in accumulation rate and by the number of radiocarbon dates used to measure them.

Pollen concentration

The pollen concentration at various levels in the core has been plotted against radiocarbon age (not corrected for variation in atmospheric C^{14}, but corrected for C^{14} deficiency in the existing lake) (Fig. 3). Pollen concentrations vary from 20,000/ml for the oldest sediment to about 650,000/ml for sediment deposited 8,500 years ago. Samples from contiguous levels are generally similar in pollen concentration. An exception appears in sediment deposited about 10,200 years ago, where the pollen concentration suddenly falls to low levels, apparently because of dilution during an increase in matrix accumulation rate too brief to be detected by radiocarbon dating.

Deposition rate for total pollen

Pollen concentrations have been multiplied by the sediment accumulation rate (in centimeters per year) to obtain the pollen deposition rates (Fig. 3). The total influx of pollen to the sediment increased by a factor of 50 between 12,000 and 9,000 years ago, from 1,000 to 50,000 grains/cm^2 per year. Subsequently it fell to a rate of about

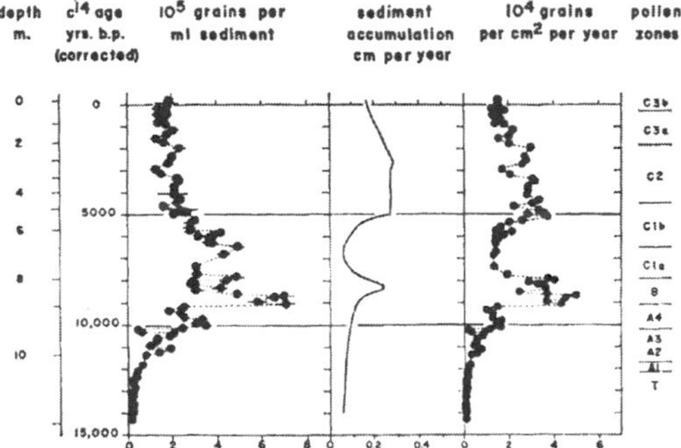

FIG. 3. To the left, concentrations of pollen grains per milliliter wet sediment plotted against the absolute age of the samples. Center column, accumulation rate for the sediment matrix, calculated from the data shown in Fig. 2. Right-hand column, deposition rates of total pollen from terrestrial plants.

TABLE 2. Pollen percentages not shown in Fig. 4 and 5 (Only samples in which rare types occur are included in the table)

Depth (m)	Thousands of radiocarbon years B.P.	Type and percentage
.08	+.21	Liguliflorae 0.2; *Zea* 0.7; Leguminosae 0.7
.32	0.17	Liguliflorae 0.4
.47	0.34	Gentianaceae 0.3
.58	0.45	*Plantago lanceolata* type 0.5; *Impatiens* 0.5
.79	0.67	Chenopodiaceae 0.2
.95	0.83	*Liriodendron* 0.4; Chenopodiaceae 0.4
1.00	0.88	*Plantago lanceolata* type 0.3
1.72	1.52	Leguminosae 0.3
1.98	1.76	Chenopodiaceae 0.3
2.22	1.95	*Impatiens* 0.5
2.70	2.32	Ranunculaceae 0.3
3.20	2.68	Betulaceae 0.2
3.84	3.14	*Plantago* sp. 0.4
4.24	3.43	Betulaceae 0.3
4.62	3.70	Scrophulariaceae 0.3; Leguminosae 0.3
4.86	3.89	Leguminosae 0.3
5.12	4.07	*Liriodendron* 0.6
5.66	4.47	Chenopodiaceae 0.2
6.06	4.76	*Liriodendron* 0.2
7.05	5.78	Chenopodiaceae 0.1
7.20	6.10	Betulaceae 0.3
7.24	6.22	Umbelliferae 0.2
7.25	6.23	*Impatiens* 0.1; Scrophulariaceae 0.2; Urticaceae 0.1; Leguminosae 0.1
7.34	6.65	*Smilax* 0.2; Chenopodiaceae 0.1
7.39	6.80	*Polygonum* 0.1; Chenopodiaceae 0.1
7.79	7.73	Chenopodiaceae 0.2; Umbelliferae 0.1
7.99	8.00	Chenopodiaceae 0.3
8.16	8.16	Chenopodiaceae 0.3
8.37	8.33	Chenopodiaceae 0.1
8.60	8.53	Leguminosae 0.2
8.66	8.59	Chenopodiaceae 0.2
8.70	8.66	*Eriocaulon* 0.1
8.87	9.00	Chenopodiaceae 0.2; Leguminosae 0.2
9.08	9.35	*Cornus* 0.2
9.15	9.49	Onagraceae 0.1
9.44	10.07	Chenopodiaceae 0.1
9.49	10.18	*Galium* 0.5
9.62	10.50	*Plantago* sp.
9.74	10.76	*Plantago* sp. 0.3
9.80	10.91	Umbelliferae 0.1
9.93	11.23	Liguliflorae 0.2; *Dryas* type 0.1; Betulaceae 0.2
9.96	11.32	*Plantago* sp. 0.4
10.33	12.36	Chenopodiaceae 0.1
10.38	12.50	Ranunculaceae 0.8
10.43	12.66	Liliaceae 0.3
10.49	12.84	*Polygonum* 0.2
10.53	12.97	*Galium* 0.6; Umbelliferae 0.3
10.63	13.29	Betulaceae 0.2; Leguminosae 0.2
10.78	13.79	Leguminosae 1.7
10.83	13.96	*Dryas* type 0.3

20,000/cm^2 per year, a rate maintained for several thousand years until a maximum again occurred 5,000 years ago, followed by a gradual decrease until the present.

Some of these changes in pollen deposition are undoubtedly related to changes in the pollen input to the lake, while others represent changes without significance, within an expected margin of error. For example, the drop in pollen deposition rate 2,500 years ago is an artifact of the error in estimating matrix accumulation prior to that time (see above). The minimum rates between 5,000 and 8,000 years ago are also closely correlated with matrix accumulation rate. There is no clear explanation for this relationship, although one can speculate that a factor such as a pattern of water circulation that decreased the rate of sediment accumulation also discouraged the accumulation of pollen grains at the sampling site. Pollen grains in themselves occupy too little volume to affect the sediment accumulation rate directly. Another explanation is systematic error in radiocarbon dates from this part of the profile, caused by a temporary change in the atmospheric concentration of radiocarbon. The evidence Stuiver (1967) has presented argues against this possibility for this portion of postglacial time.

The magnitude of changes in pollen deposition rates that can be considered significant will emerge when measurements of pollen deposition at several sites in the same region can be compared for consistency. For the present, only a limited comparison is available between the pollen influx measured in the short, radiocarbon-dated segment of core from the north basin and the equivalent segment of the south basin core. The pollen concentration is similar in both cores. The sediment accumulation rates are also similar, although the north core rate was determined merely by a straight-line fit to three radiocarbon dates from closely spaced samples, while the variable accumulation rate in the south basin was measured by a smooth fit to many samples. The pollen deposition rates, which varied between 12,000 and 40,000, are similar within a factor of two at any level. A threefold decrease in deposition seen in the south basin, however, is not apparent at the other site. On this basis, changes in deposition rate less than a factor of three cannot be considered significant. (The margin of error is greater for rare pollen types whose concentration has not been determined as accurately.) Taking into account the additional error introduced at certain levels by problems in radiocarbon dating, changes in pollen deposition by a factor of five are probably the smallest attributable to variations in pollen input to the lake. Future studies will indicate whether this opinion is too conservative or too optimistic.

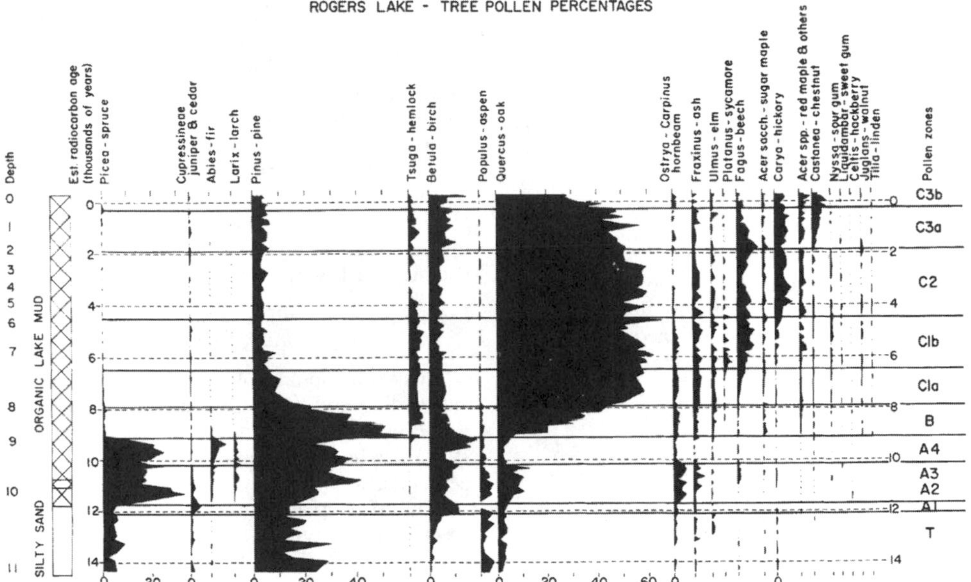

FIG. 4. Percentages for tree pollen types, calculated as percent total pollen from terrestrial plants, plotted against the absolute age of the sediment.

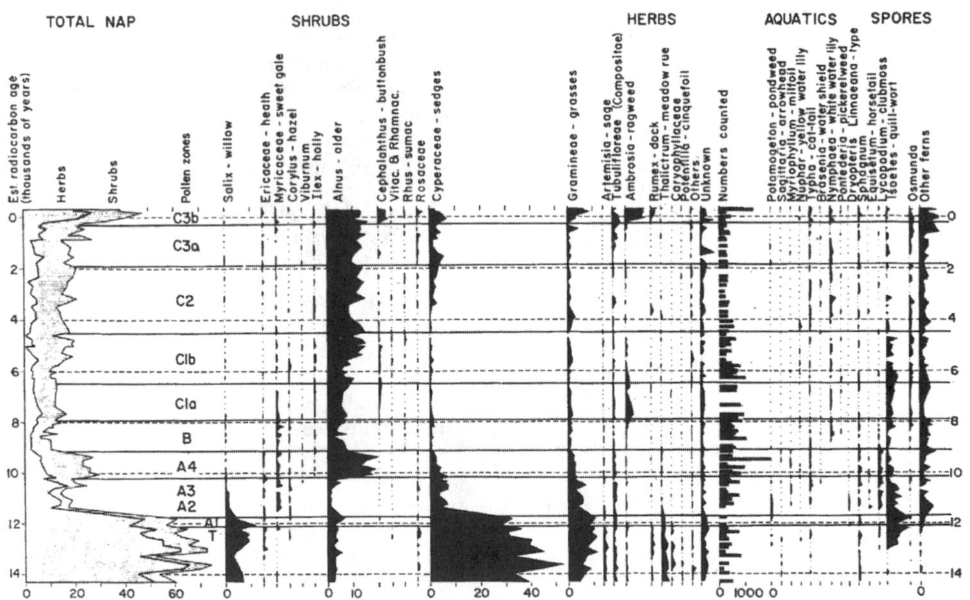

FIG. 5. Percentages for herb and shrub pollen types, calculated as percent total pollen from terrestrial plants, plotted against the absolute age of the sediment. Percentages for spores and aquatics, which are shown to the far right, are calculated as percentage of a sum equal to total terrestrial plant pollen.

Pollen stratigraphy and interpretation

The pollen percentage diagram from Rogers Lake (Fig. 4 and 5, Table 2) shows the stratigraphy described long ago for southern Connecticut (Deevey 1939, Leopold 1956, Beetham and Niering 1961). The new radiocarbon dates permit the use of absolute age, rather than depth, as the ordinate in the diagrams. The usual zonation of the pollen sequences for this region has been maintained here, except that zones C-1 and C-3 have been subdivided into rather different older (a) and younger (b) subzones. Zone B, however, is considered a single zone, rather than two subzones as at other sites (Davis 1958, 1960).

Deposition rate diagrams are given for all the major pollen types (Fig. 6 and 7). Several different scales have been used on the abscissae in order to magnify fluctuations in the rare pollen types. The taxa are arranged according to the stratigraphic level at which they reach maximum values.

Herb pollen zone (14,300–12,150 years ago)

The oldest, silty sediments are characterized by high percentages of pollen from herbaceous plants, especially sedges, grasses, *Artemisia*, *Rumex*, and *Thalictrum*. Willow is the most abundant shrub pollen, reaching 10% of the total, and pine, spruce, oak, and poplar are the most common tree pollen types.

The deposition rates (see especially Fig. 6C) show that all pollen types were in fact rare during this period. Even the herbs considered so characteristic of this pollen zone, such as grasses and *Artemisia*, are deposited at rates lower than those of the overlying spruce pollen zone. All trees are represented by very few pollen grains relative to the numbers deposited in sediment younger than 12,000 years; their prominence in the percentage diagram results from low pollen productivity by the herbaceous vegetation, which failed to dilute far-transported tree pollen in the sediment. Fluctuations in percentage values among the tree pollen types have been described from this zone at several sites in southern New England and have been attributed to climatic change (Leopold 1956, Deevey 1958, Leopold and Scott 1958, Ogden 1959, 1963). They involve small absolute numbers of grains, however, and are consequently of little or no significance as far as the local vegetation is concerned. Instead, they probably reflect chance fluctuations in the transport of pollen by wind from distant forests (Davis and Deevey 1964).

The vegetational and climatic meaning of the herb pollen zone in New England pollen diagrams has been the subject of much discussion. The presence of arctic or subarctic plant species in the region has been established from macrofossil evidence (Argus and Davis 1962). The percentages of non-arboreal pollen in the sediments resemble those in assemblages recovered from surface samples from arctic regions, very far from any forest, rather than from tundra along the forest margin (Davis 1967*c*). But it must be admitted that exact analogs to the fossil assemblages, which have appreciable percentages of pollen from oak and other temperate tree genera, have not yet been found in the modern Arctic. The very low deposition rates for all pollen types, however, are consistent with the low pollen productivity observed in modern Arctic vegetation (Aario 1940, Ritchie and Lichti-Federovich 1967).

Zone A-1, transition from herb to spruce zone (11,700–12,150 years ago)

This transitional pollen zone is recognizable in pollen diagrams from several sites in Connecticut and central Massachusetts (Leopold 1956, Davis 1958); it has a maximum of birch pollen, rising percentages of spruce pollen, and in some cases maximum poplar percentages. Earlier I had speculated (Davis 1958) that poplar might have formed the treeline of the advancing late-glacial forests. The deposition rates (Fig. 6B) show, however, that despite high percentage values in the lower part of the spruce zone, the deposition rates for poplar pollen are very low relative to rates later on, and probably do not represent an appreciable abundance of poplar trees. The maximum for birch pollen percentages in zone A-1 at Rogers Lake is also largely an artifact resulting from a decrease in deposition rates for sedge pollen (Fig. 6A and 6C).

Zone A-2-3, spruce-oak zone (10,200–11,700 years ago) and zone A-4, spruce-fir zone (9,100–10,200 years ago)

The pollen assemblage in zone A-2-3 is characterized by high percentages of spruce pollen and by a late-glacial percentage maximum (ca. 20%) for oak and other temperate deciduous trees such as hornbeam and ash. Pine percentages rise to a maximum near the upper boundary of the zone. Zone A-4 is characterized by decreased oak pollen percentages and maximal spruce, fir, and larch frequencies. Birch and alder pollen also increase.

Pollen deposition rates are much higher in the spruce zones than previously (Fig. 3 and 6B). Spruce rises steadily from fewer than 100 grains to maximum rates of 3,000 grains 10,000 years ago. Similar increases occur for fir, larch, and pine (Fig. 6B). The morphology of the pine

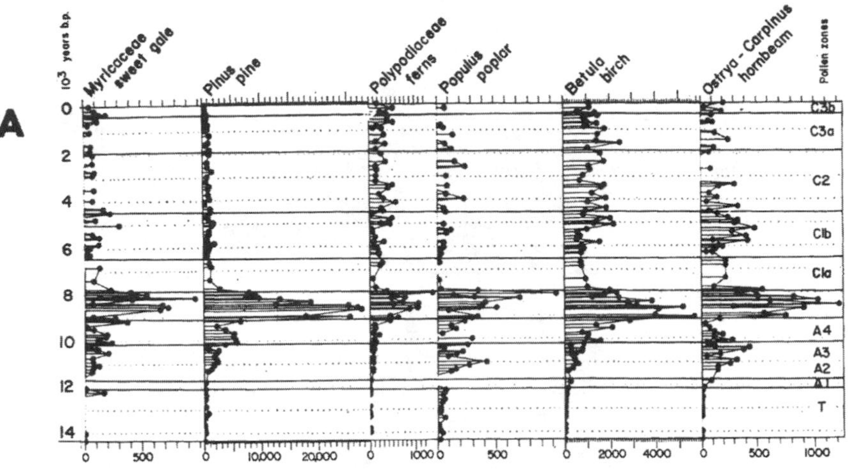
A
10^3 years b.p.
Myricaceae sweet gale
Pinus pine
Polypodiaceae ferns
Populus poplar
Betula birch
Ostrya - Carpinus hornbeam
Pollen zones
C3b
C3a
C2
Clb
Cla
A4
A3
A2
A1
T
0
2
4
6
8
10
12
14
0 500
0 10,000 20,000
0 1000
0 500
0 2000 4000
0 500 1000

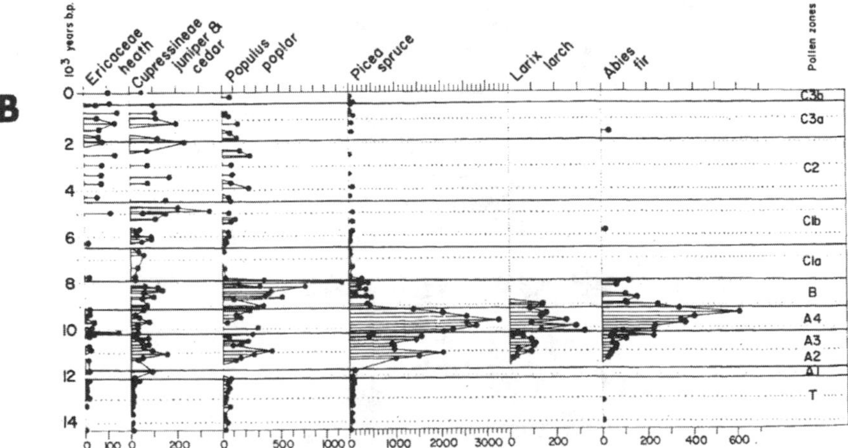
B
10^3 years b.p.
Ericaceae heath
Cupressineae juniper & cedar
Populus poplar
Picea spruce
Larix larch
Abies fir
Pollen zones
C3b
C3a
C2
Clb
Cla
B
A4
A3
A2
A1
T
0
2
4
6
8
10
12
14
0 100
0 200
0 500 1000
0 1000 2000 3000
0 200
0 200 400 600

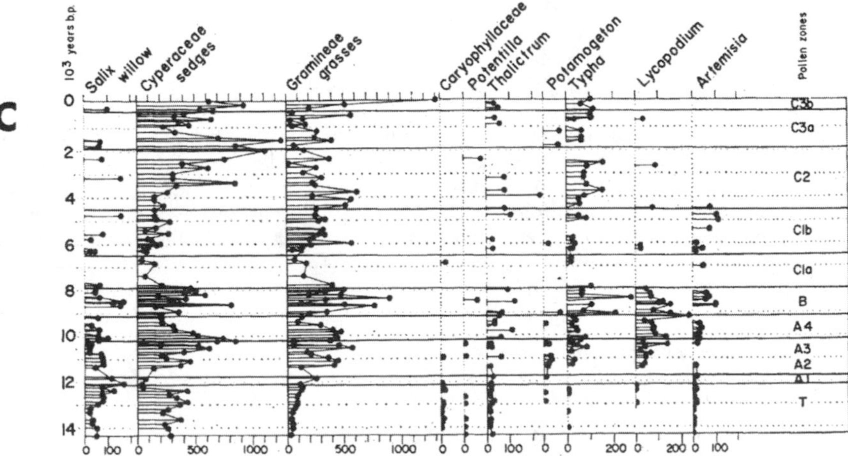
C
10^3 years b.p.
Salix willow
Cyperaceae sedges
Gramineae grasses
Caryophyllaceae
Potentilla
Thalictrum
Potamogeton
Typha
Lycopodium
Artemisia
Pollen zones
C3b
C3a
C2
Clb
Cla
B
A4
A3
A2
A1
T
0
2
4
6
8
10
12
14
0 100
0 500 1000
0 500 1000
0
0
0 100
0
0 200
0 200
0 100

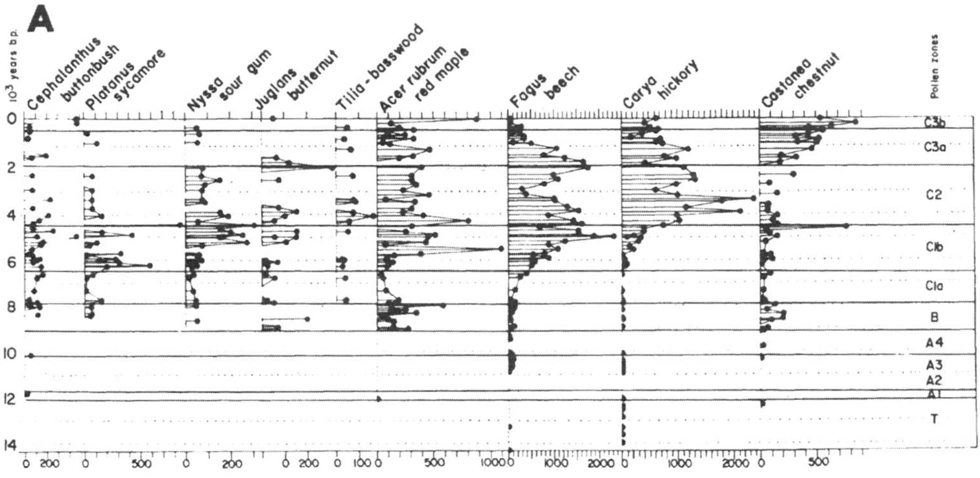

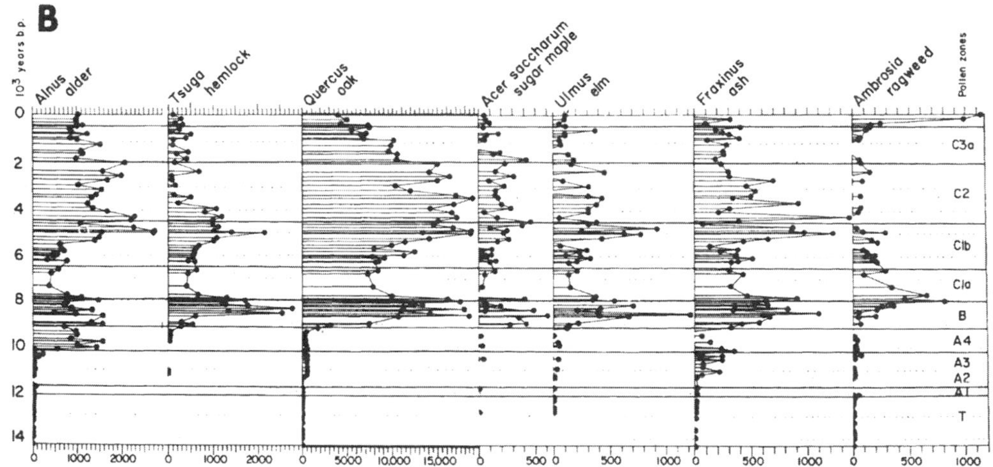

Fig. 7. A. Deposition rates for pollen types that display maximum rates in sediment younger than 6,000 years.

B. Deposition rates for pollen types that occur throughout the postglacial and for ragweed pollen.

pollen indicates that jack and/or red pine were the dominant species at this time (Fig. 8). (The small size of pollen measured at equivalent levels at other sites (Taunton, Mass., Davis, *unpublished data,* and Tom Swamp, Mass., Davis 1958) suggests that pitch pine was not present (Whitehead 1964)). Deciduous tree pollen deposition is significantly higher than in the preceding herb zone. The deposition rate for oak, however, is still very low (500) compared to the values attained in postglacial time (17,000) (Fig. 7B). Ash and hornbeam pollen influx (Fig. 6A, 7B), on the other hand, is not very different from the postglacial interval. Possibly the late-glacial vegetation of southern New England at the time of the spruce zone was similar in some respects to the contemporaneous vegetation of the Great Lakes region, which produced pollen assemblages dominated by spruce, in which ash pollen was an important component (Cushing 1965).

In zone A-4 the alder pollen deposition rate increased, while rates for the coniferous trees and

Fig. 6. A. Deposition rates for pollen types that reach maximum rates during the deposition of the pine zone (B). The different types of pine pollen are shown plotted separately in Fig. 8.

B. Deposition rates for pollen types that were abundant during the deposition of the spruce pollen zone (A).

C. Deposition rates for pollen types that occur in high percentages in the herb pollen zone (T).

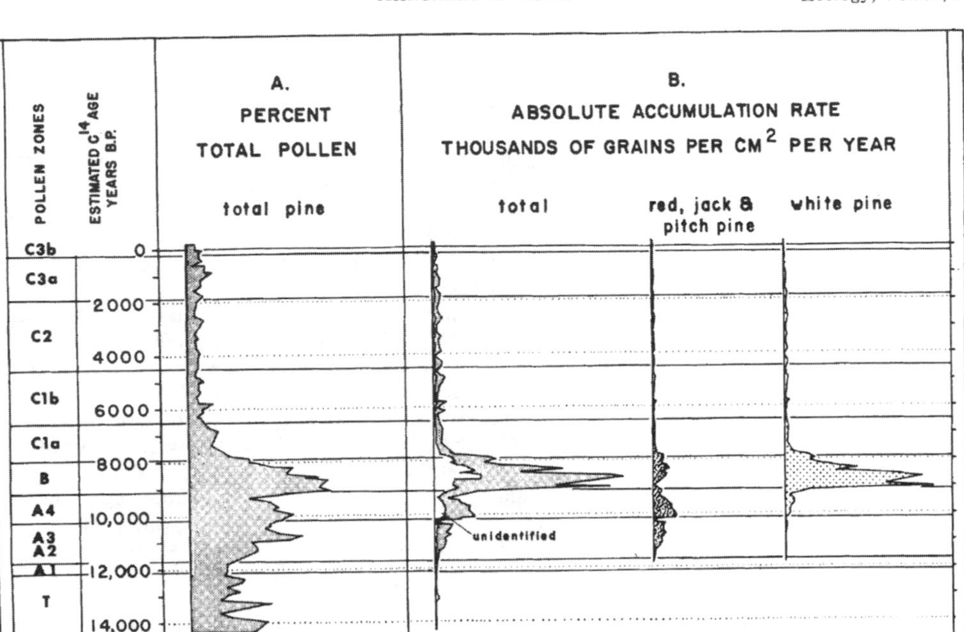

FIG. 8. Percentages and deposition rates for the various pine pollen types.

birch reached a maximum. Ash, hornbeam, and oak pollen deposition rates change slightly at this time, but the change is not quite large enough to be considered a significant indication of vegetation change.

The increased influx of tree pollen in the spruce zone, as well as the presence of macrofossils of trees at several other sites in the region (Davis 1965*b*) indicates that trees grew in southern New England at this time, increasing in abundance to maximum frequencies of boreal species in zone A-4. Whether the landscape was thickly forested, resembling the present-day boreal forest, remains in question, however. The pollen assemblages at most sites are unlike those deposited today within the boreal forest region. They resemble more strongly those from surface samples from a region north of the boreal forest in Quebec, where the prevailing vegetation type is open spruce woodland (Davis 1967*c*). At Rogers Lake, however, the comparison with Quebec is not satisfactory, as the ratio of pine to spruce pollen is higher than in the Quebec surface samples. High percentages for pine pollen are characteristic of zone A at all coastal sites in southern New England. Either spruce was less frequent at coastal sites and wind-blown pine pollen more common, or spruce was equally abundant at all sites while pine was more abundant near the coast. The pollen percentages in zone A-4 at Rogers are very similar to those of zone B-1 in central Massachusetts. Conceivably the two pollen zones could be the same age, although this is doubtful. The radiocarbon evidence is equivocal. (The only radiocarbon date from sites in Massachusetts is from the upper part of zone A-3. At 10,800 years, the date is a few hundred years older than equivalent levels at Rogers, but this is too small a difference to change the correlation for zone B-1, 30 cm higher in the section.) Percentages in zone B-1 in Massachusetts are similar to surface samples from southeastern Ontario, just south of the boreal forest. I am reluctant to interpret zone A-4 at Rogers as representing this kind of mixed coniferous-deciduous vegetation, however. White pine occurs in the vicinity of the surface sampling sites in Ontario (Rowe 1959), whereas its pollen is almost entirely absent from zone A-4 at Rogers.

Zone B, pine zone (7,900–8,100 years B.P.)

Pollen assemblages from zone B at Rogers are almost identical to those of zone B-2 in Massachusetts. White pine (*P. strobus*) dominates the assemblages, reaching percentages as high as 50% of total pollen. The deposition rate for white pine pollen increased rapidly at this time from a few hundred grains to 20,000 grains (Fig. 8). The increase was so rapid that pollen from several other genera, also increasing at this time, can hardly be noticed in the percentage diagram; the rate diagrams (Fig. 6A and 7B) show clearly, however, that pollen of the Myricaceae, poplar, birch, alder, hornbeam, oak, and hemlock, and fern spores reached a maximum. (The fern spores are from the Polypodiaceae, rather than *Pteridium*, which occurs characteristically in the pine zone in Minnesota (McAndrews 1966)). Also apparent (Fig. 6B) is the rapid decrease in deposition rates, as well as percentages, for boreal genera such as spruce, fir, and larch.

The climate must have changed dramatically at this time to produce such a striking change in the vegetation (Wright 1964, Ogden 1967). White pine, hemlock, and oak are distributed at present south of about 46° N lat in eastern Canada. Hemlock is the most restricted in range, occurring only to the south and east of Lake Superior. In the east white pine and hemlock range far southward along the Appalachian Mountain chain, but in the Midwest they drop out of the flora in southern Michigan, Wisconsin, and Minnesota (Fowell 1965). The climate of zone B time presumably approximated the modern climate somewhere within the geographical zone where these species ranges now coincide. Surface samples that precisely match the fossil percentages of zone B have not yet been found, but much of this area has not been sampled thoroughly. The high ratio of pine to birch pollen in zone B is similar to surface samples from central Canada northeast of the Great Lakes; it is in the Great Lakes region that analogs to the vegetation might occur.

Zones C-1, C-2, and C-3, oak zones (7,900 years ago to present)

The pollen diagrams show the postglacial stratigraphy described by Deevey (1939) many years ago: zone C-1, oak and hemlock, zone C-2, oak and hickory, and zone C-3, oak and chestnut. Additional details can now be seen: ragweed pollen occurs in the lower portion of zone C-1, designated here C-1a, while appreciable percentages of beech pollen characterize the upper portion (C-1b). In zone C-2 beech percentages (as well as hemlock) decline to a minimum, confirming the correlation of the beech minimum in Ohio with the hemlock minimum of zone C-2 in New England (Ogden 1966). In the upper portion of C-3 (C-3b) percentages of pollen from ragweed, *Rumex*, and other weeds increase.

The deposition rate for total pollen was relatively stable during this long period, and consequently the trends in rates for individual types are similar to the trends in percentages. Deposition rates for tree pollen seem to reach a minimum in zone C-1a, although the change is not large enough to be considered significant. At this time pollen from ragweed was deposited at increased rates. The maximum ragweed rate 8,000 years ago is approximately half as high as the modern rates, which record the response of this weedy species to forest clearance by European settlers. Further, the increased deposition of ragweed pollen in C-1a is a regional event: a similar maximum (of percentage) occurs in zone C-1 in a pollen diagram from Ohio (Ogden 1966). I believe that it indicates decreased density of forest, brought about by climatic changes associated with the expansion of prairie in the Great Lakes region 8,000 years ago (Cushing 1965). Wright (1968*b*), on the other hand, believes that the ragweed maximum in Northeastern profiles represents pollen blown from the expanded prairie in the Prairie Peninsula region of Illinois and Indiana. He points out that the warmest and driest part of postglacial time in the Northeast is generally considered to have occurred later, between 5,000 and 2,000 years ago. If the ragweed pollen were windblown, I would expect a ragweed maximum in postglacial sediment everywhere in New England. But although Vermont is no farther from the Prairie Peninsula than Connecticut, there is no ragweed maximum at Brownington Pond. There is, to the contrary, an increase in oak pollen percentages there at stratigraphically similar levels. In that region oak would increase in response to a drier climate (Davis 1965*b*). It is unfortunate that diagrams from Massachusetts (Davis 1958) are not detailed enough to show whether or not a ragweed maximum occurs in zone C-1.

An interesting feature in the Rogers Lake pollen sequence is the late arrival of several tree species (Fig. 7A, 7B). Red maple and hemlock first appeared 9,000 years ago (Figs. 7A, 7B), while hickory pollen was absent until 5,000 years ago, when its deposition rate suddenly increased to high levels. Chestnut pollen increased in abundance 2,000 years ago. Its pollen also appears sporadically throughout the profile. Presumably these late times of arrival and sudden increases to high frequencies are due to delayed migration from glacial refuges. This possibility has recently been emphasized by Wright (1964, 1968*a*). The absence of pollen from these important tree species in older sediment makes comparison with modern pollen assemblages and vegetation difficult. Surface samples from regions with a similar, although cooler, climate than southern Connecticut, for example central Massachusetts, contain about 20% pine pollen and 8% pollen from chestnut; none of the assemblages in Rogers Lake sediment is similar. In fact, only the uppermost levels can be compared with surface samples from anywhere in New England. The assemblages of C-1a might be compared with modern assemblages from the "Prairie Peninsula" region of the Midwest, since the high deposition rate for ragweed pollen is suggestive of forest with prairie openings. Furthermore, these regions are largely west of the present limit for beech, the pollen of which is rare in zone C-1a, while ash, hornbeam, and sycamore, the pollen of which is frequent in zone C-1a and C-1b, are characteristic trees of the prairie margin region in Indiana and Illinois. The comparison cannot be carried further, because surface sediments from that region rarely contain hemlock pollen and always contain hickory pollen, and thus they are very different from zone C-1a in Connecticut. This suggests that the forest in the vicinity of Rogers Lake during early postglacial time was different from any known today. Admittedly, much more intensive sampling of modern and presettlement sediments in the Midwest is necessary to prove that this view is correct. The soils and topography of southern New England are quite different from those of Michigan, Illinois, and Wisconsin. It seems probable that even in an identical climate, different abundances of species would prevail in the two regions.

The youngest pollen zones (C-3a and C-3b) date from the last two millenia; all the tree genera that are abundant in the modern forest surrounding Rogers Lake are present in the pollen assemblages. It should therefore be possible to make some comparisons between the fossil assemblages and the modern vegetation. In discussing the present distribution of tree species in southern Connecticut, Niering and Goodwin (1962) have emphasized the importance of fires set by prehistoric Indians. They believe that fire over a very long time interval has selected against fire-sensitive species, especially hemlock, while selecting for other more xerophytic types, such as oak, hickory, and chestnut. The pollen deposition rate diagram (Fig. 7B) shows that the frequency for hemlock has been low for the last 2,000 years; if fires were responsible, the effect must have started at around the time of the birth of Christ. Chestnut appears to have increased steadily in this interval; possibly its increase is related to disturbance, although climatic changes might also be responsible.

Later changes in the pollen content of the sediment (zone C-3b) are clearly related to forest clearance by European settlers. Ash and oak pollen percentages decline, and red maple, which can grow as a pioneer species in disturbed habitats, increases together with ragweed. *Rumex* and many other weedy herb pollen types increase in frequency. Buttonbush pollen increases (Fig. 5) perhaps as a result of changes in water level of the lake. The pollen changes are quite different from those that record agriculture in northwest Europe. In America ragweed is the prime indicator of agriculture, and plantain (*Plantago*) pollen, common in European deposits, is never abundant (Table 2). Cereal pollen is also extremely rare; only one *Zea* grain was found in the entire profile. In the uppermost samples (Fig. 4 and 5) a few very recent vegetation changes are recorded. A decline in total non-arboreal pollen corresponds with the very recent abandonment of many of the farms in the vicinity of Rogers Lake, and the two uppermost samples analyzed (which were not treated on an absolute basis) show decreased percentages for chestnut pollen, recording the chestnut blight, a disease that killed most of the trees of this species 30 to 45 years ago.

Conclusions

Application of a new method of pollen analysis to Rogers Lake sediments has confirmed and strengthened many of the previous interpretations of the pollen percentage diagrams from southern New England. New information has been obtained. Especially important is the sudden increase in pollen deposition rates for trees, from very low levels 12,000–14,000 years ago to high postglacial rates 9,000 years ago. The change in tree pollen fits well with the idea that a tundra vegetation, devoid of trees, was replaced first by woodland and then by forest as trees immigrated to the site and increased in frequency. Studies of modern pollen deposition have shown that although pollen productivity is much lower in tundra areas than in forest, the percentage composition of the pollen assemblages produced on either side of the northern tree line are very similar (Wright, Winter, and Patten 1963, Davis 1967*c*, Ritchie and Lichti-Federovich 1967). Consequently deposition rates are far more valuable than percentage diagrams as a pollen record of changes from tundra to ancient woodland or parkland.

Another interesting aspect of the late-glacial sequence at Rogers Lake is the sudden increase in the numbers of oak pollen grains deposited 11,500 years ago, a time just postdating the advance of the Valders ice sheet (Broecker and Farrand 1963). An increase in the abundance of oak somewhere within pollen dispersal distance is one possible explanation. Another is that the change reflects a change in prevailing wind direction from northwest to southwest correlated with the beginning of a time of very rapid glacial retreat.

The deposition rates throw new light on the nature of the vegetation 8,500–9,500 years ago, during the "pine period," when the pollen influx from white pine (*Pinus strobus*) increased very rapidly, implying a marked population increase for the species, followed by an equally rapid decline only a thousand years later. Many other pollen types, including poplar, maple, oak, and hemlock, also increase steeply 9,500 years ago, a change that is largely masked in the percentage diagram. Pollen influx for boreal species decreased, indicating that the change in the percentage pollen diagrams is not merely caused by dilution by increased pine pollen, but represents a real decline in numbers of boreal trees. The subsequent marked decline in deposition rate of pine pollen indicates a shift to almost purely deciduous forest. Perhaps the changes in species composition of the plankton in Rogers Lake at this time (Deevey 1968) are an additional reflection of changes in regional climate.

One of the new features of the Rogers Lake sequence is a maximum for ragweed pollen in zone C-1a, 7,000–8,000 years ago, a feature that has not been noticed in earlier analyses from the region, probably because emphasis was placed on tree pollen rather than upon herbs. Stratigraphically the ragweed maximum at Rogers Lake (Davis 1967*b*) is correlated with the Prairie Period of the Great Lakes region and reflects directly or indirectly vegetation changes associated with a xerothermic interval that occurred there between 5,000 and 8,000 years ago (Wright 1968*b*). The changes in pollen frequency that were originally thought to represent the zerothermic interval in southern New England (Deevey 1939) are in younger sediment, precisely dated now at 2,000–4,500 years B.P. These changes seem most easily interpreted now as the reflection of successive immigrations to Connecticut of first beech, then hickory, and then chestnut. Only during the last 2,000 years have the pollen assemblages in southern Connecticut been sufficiently similar to modern assemblages to permit comparison. This suggests that the forest communities there, as they are presently recognized, may be of very recent origin.

Acknowledgments

This paper is contribution 104 from the Great Lakes Research Division, University of Michigan. The work was initiated while the author was a research fellow at Yale University in 1960–61 and has been continued at the University of Michigan. Research support has been provided by the National Science Foundation, research grants G-19335, to Yale University, G-17830, GB-2377, and GB-5320 to the University of Michigan. I gratefully acknowledge the encouragement and cooperation of Edward S. Deevey, Jr. I also extend special thanks to Minze Stuiver, who has dated the sediment core from Rogers Lake.

Literature Cited

Aario, L. 1940. Waldgrenzen und subrezente Pollenspektren in Petsamo Lappland. Ann. Acad. Sci. Fenn., Ser. A, **54**(8) : 1–120.

Argust, G. W., and M. B. Davis. 1962. Macrofossils from a late-glacial deposit at Cambridge, Massachusetts. Amer. Midland Natur. **67**: 106–117.

Beetham, Nellie, and W. A. Niering. 1961. A pollen diagram from southeastern Connecticut. Amer. J. Sci. **259**: 69–75.

Broecker, W. S., and W. R. Farrand. 1963. Radiocarbon age of the Two Creeks forest bed. Geol. Soc. Amer. Bull. **74**: 795–802.

Broecker, W. S., and Alan Walton. 1959. The geochemistry of C^{14} in fresh-water systems. Geochem. Cosmochim. Acta **16**: 15–38.

Connecticut State Board of Fisheries and Game. 1959. A fisheries survey of the lakes and ponds of Connecticut. Lake and Pond Surv. Unit Rep. 1. 395 p.

Cushing, E. J. 1965. Problems in the Quaternary phytogeography of the Great Lakes region, p. 403–416. *In* H. E. Wright and D. G. Frey [ed.] The Quaternary of the United States. Princeton Univ. Press, Princeton, N. J. 922 p.

Damon, P. E., A. Long, and D. C. Greg. 1966. Fluctuation of atmospheric C^{14} during the last six millennia. J. Geophys. Res. **71**: 1055–1063.

Davis, M. B. 1958. Three pollen diagrams from central Massachusetts. Amer. J. Sci. **256**: 540–570.

———. 1960. A late-glacial pollen diagram from Taunton, Massachusetts. Bull. Torrey Bot. Club. **87**: 258–270.

———. 1965*a*. A method for determination of absolute pollen frequency, p. 674–686. *In* B. Kummel and D. Raup [ed.] Handbook of paleontological techniques. W. H. Freeman, San Francisco, Calif. 852 p.

———. 1965*b*. Phytogeography and palynology of northeastern United States, p. 377–401. *In* H. E. Wright, Jr., and D. G. Frey [ed.] The Quaternary of the United States. Princeton Univ. Press, Princeton, N. J. 922 p.

———. 1966. Determination of absolute pollen frequency. Ecology **47**: 310–311.

———. 1967*a*. Pollen deposition in lakes as measured in sediment traps. Geol. Soc. Amer. Bull. **849**: 858.

———. 1967*b*. Pollen accumulation rates at Rogers Lake, Connecticut, during late- and postglacial time. Rev. Palaeobot. **2**: 219–230.

———. 1967*c*. Late-glacial climate in northern United States: a comparison of New England and the Great Lakes region, p. 11–43. *In* E. J. Cushing and H. E. Wright, Jr. [ed.] Quaternary paleoecology. Yale Univ. Press, New Haven, Conn. 425 p.

———. 1968. Pollen grains in lake sediments: redeposition caused by seasonal water circulation. Science **162**: 796–799.

Davis, M. B., and E. S. Deevey, Jr. 1964. Pollen accumulation rates: estimates from late-glacial sediment of Rogers Lake. Science **145**: 1293–1295.

Deevey, E. S., Jr. 1939. Studies on Connecticut lake sediments. I. A postglacial climatic chronology for southern New England. Amer. J. Sci. **237**: 691–724.

———. 1958. Radiocarbon-dated pollen sequences in eastern North America. Veröff. Geol. Inst. Rübel (Zurich) **34**: 30–37.

PAPER 6

Species Diversity: Benthonic Foraminifera in Western North Atlantic (1969)

M. A. Buzas and T. G. Gibson

Commentary

ELLEN THOMAS

Deep-sea environments are the largest but most poorly known habitat on Earth: the Census of Marine Life (2010) concluded that the ~230,000 species in the World Register of Marine Species (WoRMS) are only 20–30% of living eukaryote marine species. The fact that the cold and eternally dark deep oceans are not azoic was established during the nineteenth century, with deep-sea expeditions in the 1840s–60s, culminating in the HMS *Challenger* Expedition (1872–76), which resulted in documentation of many inhabitants of the deep oceans in 83 beautifully illustrated volumes (e.g., Tizard 1885). It took almost a century more to establish that deep-sea ecosystems typically have a low density but very high species richness and diversity (Sanders 1968), comparable to tropical rain forests and coral reefs. Buzas and Gibson contributed, being the first to describe foraminiferal diversity in the deep oceans.

Their article was visionary and (in hindsight) far ahead of most papers on foraminifera at the time: they treated foraminifera as living organisms, with a population structure that could be evaluated using various measures of diversity rather than simple species richness. Foraminifera had been widely studied by micropaleontologists (many in the petroleum industry) rather than by biologists, who commonly used foraminifera as markers for geological age rather than organisms (Lipps 1981). Buzas and Gibson described species richness as well as diversity measures of incorporating evenness (now widely used), and linked their observations on living deep-sea benthic foraminifera to the geological history of deep-sea life. This could not have been done for other deep-sea organisms, most of which do not have fossilizable skeletons. Buzas and Gibson speculated that the last Pleistocene glaciation was not catastrophic for deep-sea faunas, at the time a novel argument because the widespread study of ancient deep-sea benthic foraminifera became feasible only with the beginning of the Deep-Sea Drilling Project (Edgar et al. 1973), the first volume of which was published in the same year as Buzas and Gibson's far-seeing paper. Only recently has evidence accumulated that an important but gradual extinction had occurred earlier during the Pleistocene (e.g., Hayward et al. 2010) and that some species described during the *Challenger* expedition were fossils and not alive in recent times.

Buzas and Gibson thus not only were among the first to treat foraminifera as the living beings they are, but they also set the stage for the study of deep-sea benthic foraminiferal assemblages in the geological past ("deep time"), which has led to recognition of past episodes of rapid global warming, combined with ocean acidification, deoxygenation, and extinction in the deep oceans, and to the use of the past to predict potential future worlds, with deep-sea benthic foraminiferal assemblages indicator species for deep-sea ecosystems.

From *Science* 163:72–75. Reprinted with permission from AAAS.

Literature Cited

Census of Marine Life. 2010. About the Census of Marine Life. Census of Marine Life. Accessed November 5, 2018. http://www.coml.org/about.html.

Edgar, N. T., J. B. Saunders, H. M. Bolli, R. E. Boyce, W. S. Broecker, T. W. Donnelly, J. M. Gieskes, et al. 1973. *Initial Reports of the Deep Sea Drilling Project*. Vol. 15. Washington, DC: U.S. Government Printing Office.

Hayward, B., K. Johnson, A. T. Sabaa, S. Kawagata, and E. Thomas. 2010. Cenozoic record of elongate, cylindrical deep-sea benthic Foraminifera in the North Atlantic and equatorial Pacific Oceans. *Marine Micropaleontology* 74:75–95.

Lipps, J. H. 1981. What, if anything, is micropaleontology? *Paleobiology* 7:167–99.

Sanders, H. L. 1968. Marine benthic diversity: a comparative study. *American Naturalist* 102:243–82.

Tizard, T. H., J. J. Wild, C. W. Thomson, and J. Murray. 1885. *Report on the Scientific Results of the Voyage of H.M.S. Challenger during the Years 1873–76*. Vol. 1, pt. 1, *Narrative*. London: Book LIV, 509 Seiten Illustrationen, Karten. https://www.biodiversitylibrary.org/item/72085#page/7/mode/1up.

fusion. The Prandtl number, the ratio of kinematic viscosity to thermal diffusivity, for liquid mercury is two orders of magnitude less than that of water. Linearized analysis of two-dimensional motions predicts that the ratio of the mean fluid velocity to the speed of the traveling thermal wave becomes large as the Prandtl number becomes small; thus the theory is not valid in the limit of Prandtl number approaching zero. Results from an extended analysis which includes nonlinear interactions between the perturbations and the mean flow are shown in Fig. 2, where the ratio of mean flow velocity to wave speed is plotted against Prandtl number. This velocity ratio, which is approximately proportional to the 15/4 power of the inverse Prandtl number for Prandtl numbers between 1 and 0.1, is of the order of unity for Prandtl numbers of the order 10^{-1}.

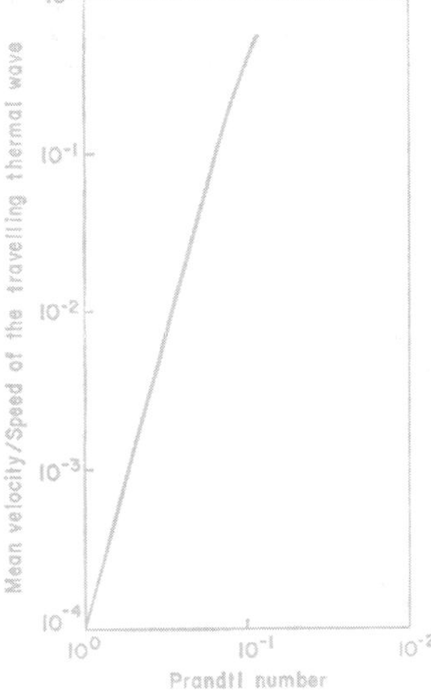

Fig. 2. Ratio of the mean velocity of the fluid to the speed of the forcing thermal wave plotted against Prandtl number. Fluid was confined within a two-dimensional channel, and temperature perturbations of the traveling wave were applied at the walls. The Boussinesq equations of motion were solved numerically for the following case: $\Delta T/T = 10^{-2}$, $\omega h^2/\kappa = 1$, $gh/U^2 = 10^4$, $kh = 10^{-2}$, where ω is the angular frequency, k is the wave number of the thermal wave, h is the channel height, κ is the thermal diffusivity, g is the acceleration of gravity, $U = \omega/k$, and $\Delta T/T$ is the relative magnitude of forced fluctuations of the wall temperature. The nonlinear interaction of the perturbations and the mean flow is included in the solution.

This experiment demonstrates that the periodic motion of a source of heat can lead to a mean fluid motion with speed several times faster than that of the source. This phenomenon may explain the relatively rapid displacements of clouds in the high atmosphere of Venus which have been observed in ultraviolet photographs (*5*). These observations suggest that at least the upper layers of the atmosphere of Venus are moving with speeds of 300 km/hour relative to the planet's surface. The overhead motion of the sun would provide a periodic traveling thermal source, and the zonal flow induced by this movement would be in the direction of the cloud motion, which is some 20 times faster than the overhead speed of the sun.

In the atmosphere of Venus, a near-infrared band of carbon dioxide absorbs a significant fraction of the incident solar radiation. At altitudes of tens of kilometers where pressures are of the order of an atmosphere or less (*6*), a kilometer of CO_2 absorbs several percent of the incident solar radiation (*7*). The radiative transfer would be characterized by an effective diffusion coefficient (*8*)

$$\kappa = 16\sigma T^3 l / 3\rho c_p$$

where σ is the Stefan-Boltzmann constant, T is the temperature, l is the mean free path of the radiation, ρ is the density, and c_p is the specific heat at constant pressure. At heights of tens of kilometers, $\kappa \approx 3l$ cm^2 sec^{-1}, and l is at least of the order of 10^5 cm (*7*). Momentum transport would at best be accomplished by turbulent mixing, for which the mixing coefficient is not likely to exceed 10^4 cm^2 sec^{-1} (*8*). Thus it is possible that in the high atmosphere of Venus periodic heating from above occurs in a medium that can transport heat more effectively than momentum. Under such circumstances, zonal motions at high velocity could be induced in the Venus atmosphere.

G. SCHUBERT

Department of Planetary and Space Science, University of California, Los Angeles

J. A. WHITEHEAD

Institute of Geophysics and Planetary Physics, University of California, Los Angeles

References and Notes

1. E. Halley, *Phil. Trans. Roy. Soc. London Ser. A Math. Phys. Sci.* 16, 153 (1686).
2. D. Fultz, *Meteorol. Monogr.* 4, 36 (1959).
3. M. E. Stern, *Tellus* 11, 175 (1959).
4. A. Davey, *J. Fluid Mech.* 29, 137 (1967).
5. B. A. Smith, *Science* 158, 114 (1967).
6. V. S. Avduevsky, M. Ya. Marov, M. K. Rozhdestvensky, *J. Atmos. Sci.* 25, 537 (1968).
7. P. Fabian, T. Sasamori, A. Kasahara, in preparation.
8. R. M. Goody and A. R. Robinson, *Astrophys. J.* 146 (1966).
9. We thank W. V. R. Malkus for providing the equipment and laboratory facilities for this experiment, and P. Cox for constructing the apparatus.

5 November 1968

Species Diversity: Benthonic Foraminifera in Western North Atlantic

Abstract. *Maximum species diversity occurs at abyssal depths of greater than 2500 meters. Other diversity peaks occur at depths of 35 to 45 meters and 100 to 200 meters. The peak at 35 to 45 meters is due to species equitability, whereas the other two peaks correspond to an increase in the number of species.*

Populations of benthonic Foraminifera exhibit species-diversity peaks at depths of 35 to 45 m, 100 to 200 m, and greater than 2500 m in the western North Atlantic. The peaks progressively increase in diversity as depth increases, the maximum diversity occurring in depths greater than 2500 m. The depths of the peaks correspond to effective wave base, the edge of the continental shelf, and the abyss. The peak in diversity at 35 to 45 m is due to species equitability rather than to an increase in the number of species, whereas the other two peaks correspond to an increase in the number of species. Data for this pattern are from 84 samples taken at depth ranges from 29 to 5001 m in the western North Atlantic (Fig. 1).

Many foraminiferal species have been recorded in abyssal depths in the Gulf of Mexico (*1*), off California (*2*), and off Panama in the eastern Pacific (*3*). The high foraminiferal diversities in abyssal environments closely reflect the diversity of the other groups of marine invertebrates including Mollusca, Arthropoda, and Echinodermata (*4*). The formerly erroneous viewpoint of very low diversity in the deep sea probably resulted from difficulty in obtaining enough individuals of larger invertebrates for an accurate estimate of the number of species (*4*).

The benthonic Foraminifera, being small, of high density and ubiquitous distribution, do not present many problems encountered in sampling larger organisms. Even in the abyss, hundreds of individuals can be obtained by one core or bucket sample of 100 ml of sediment.

That the patterns reported are not local phenomena, or artifacts of Pleistocene changes, or sedimentological processes is shown by the following. Although downslope displacement may occur, the large area of high diversities precludes the possibility of sampling shallow-water or mixed assemblages. Likewise, the extent of the abyssal plain, the distance of some stations from land, and the lack of shallow-water forms in the assemblages indicate that the fauna is endemic. Most important, although the total foraminiferal fauna (living and dead) was used, an examination of living assemblages gives the same pattern.

Numerous hypotheses (*5*), most not mutually exclusive, attempt to explain the increase in organic diversity toward the lower latitudes. The "stability" hypothesis, of relatively uniform environment over a long period, reasonably explains the progressive increase of diversity peaks observed here. Under such conditions, biologic components become more important than physical ones. Organisms no longer need to maintain a high degree of tolerance to survive in an ever-changing environment. Entire populations are not in danger of being destroyed by extremes of weather. This lack of physiological stress may permit a higher degree of specialization with complex interactions between species to use the energy in the system more efficiently, thus more species at all trophic levels.

Although diversity is usually measured as the number of species in some sample of specified size, diversity may also be measured as the number of species and their frequencies (*6, 7*). The information function (*7–9*) effectively measures the number of species and their proportions without making any assumptions as to an underlying distribution. The function is defined as

$$H(S) = \sum_{i=1}^{S} p_i \ln p_i$$

where p_i is the proportion of the *i*th species, and S is the number of species observed. Species with low frequencies contribute little to the value $H(S)$. For example, a species making up 30 percent of the assemblage contributes 0.36 to the value of $H(S)$, whereas a species with a frequency of 1 percent contributes only 0.05. Consequently, $H(S)$, is not as variable as S, which is more subject to fluctuation because of the presence or absence of a rare species. The maximum value of $H(S)$ (maximum diversity) occurs when all S species have equal frequencies, in which case $H(S)$ is ln S. When e, the base of the natural logarithms, is raised to the $H(S)$ power, the result is a measure of the equivalent number of equally distributed species (*10*). For equally distributed species $e^{H(S)} = S$, so that the ratio $e^{H(S)}/S$ is a measure of species equitability similar to some others proposed (*7, 9*). When the species are perfectly equally distributed the ratio = 1.

When $H(S)$ is plotted against depth (Fig. 2), the peaks in diversity occur at 35 to 45 m, 100 to 200 m, and greater than 2500 m. The highest peak occurs in waters deeper than 2500 m. Very shallow areas, less than 30 m, have values of $H(S)$ less than 1.5. In all traverses $H(S)$ increases to greater than 1.9 at about 35 to 45 m. This high is followed by a decrease to values with minima at about 70 m. Then $H(S)$ begins to increase, reaching a peak of 3.0 at 100 to 200 m. Below 200 m, the values slowly decrease down the continental slope to less than 2.0 at 2000 m. With increasing depth in the abyss, $H(S)$ rises sharply, and a maximum of 3.5 is reached at 4977 m.

Although the number of species per sample (S) is strongly correlated with $H(S)$ (Fig. 2), the $H(S)$ high at 35 to 45 m is not apparent in the species number; S usually is less than 20 in waters shallower than 70 m. Closer to the edge of the shelf, at depths of 70 to 200 m, the number of species increases to 30 to 49. As in $H(S)$, on the continental slope the number decreases to less than 30 species. At about 1000 m, the number begins to increase once more. At 2500 m, S increases to approximately 50 species, and at 4977 m (the second deepest station), 83 species, the maximum number in the study, occurs.

The ratio $e^{H(S)}/S$ has its highest set of values, 0.61 to 0.77, at the 35 to 45 m high of $H(S)$ (Fig. 2). This peak indicates that the most equitable distribution of species occurs here. A general decrease to a low of 0.22 occurs at

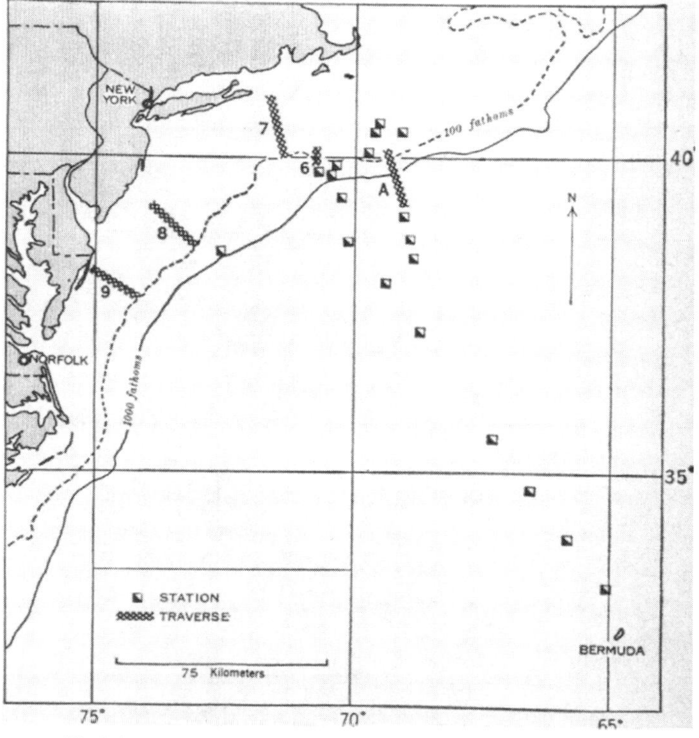

Fig. 1. Sample locations. Traverses 6 through 9 taken from Parker (*14*).

about 100 m. (The New Jersey traverse, however, shows a more complex pattern beyond the 35 to 45 m high, decreasing to 0.44 at 74 m, and then rising again to 0.68 at 90 m, after which the ratio decreases once more). All traverses have values usually between 0.30 and 0.50 at depths from 100 to 1000 m. Two stations at 393 and 469 m, however, have unusually high values of 0.66 and 0.68. Another low occurs at 1000 to 2000 m, with values ranging between 0.25 to 0.28. At 2000 to 5000 m the ratio again increases and ranges from 0.17 to 0.46. As Fig. 2 shows, the ratio is more variable than either *S* or *H*(*S*).

At the successive diversity-peaks, environmental variables fluctuate progressively less. The low diversities in shallow water can be explained by the extreme physical stress placed on organisms there. The low diversities on the continental slope may be due to environmental instability reflected in downslope movement of sediment on the slope. As mentioned earlier, the high at 35 to 45 m measured by *H*(*S*) is due to species equitability rather than an increase in the number of species. The significance of this high is not yet apparent.

To achieve the complex organization and thus high diversity would require a considerable period of environmental stability. Some authors (*11*) suggested that the abyssal and hadal faunas are a combination of relict species and species which have migrated into the abyss since Pleistocene glaciation. If the abyssal fauna was severely affected by the Pleistocene and is relatively recent in origin, than the organic diversity reported here cannot be explained by stability of environmental conditions over a long period.

Several lines of evidence, however, indicate that the Pleistocene glaciation was not a catastrophic event for the marine fauna. Shallow-water benthic foraminiferal faunas from the study area have undergone only slight change in the last 20 million years (*12*). Changes in sea level accompanying glaciation caused migration of shallow-water faunas, but with the rise in sea level, they simply repopulated the newly recovered shallow-water environments (*13*). If migration of shallow-water faunas into the abyss took place after the Pleistocene, species found in pre-Pleistocene shallow-water assemblages should now be found in living abyssal faunas. Comparison of Miocene and younger Tertiary shallow-water foraminiferal faunas with living abyssal faunas

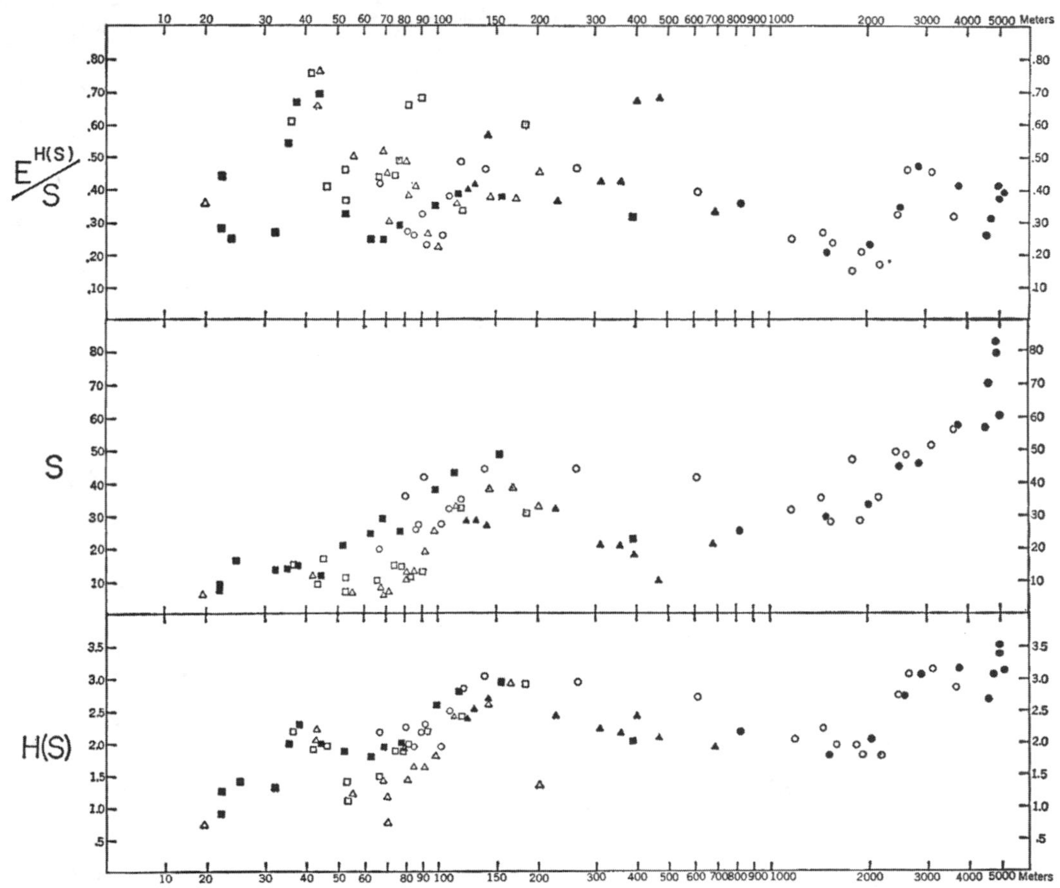

Fig. 2. Plots of species diversity and equitability with depth. Symbols: open triangle, Black Island (Parker); solid triangle, Martha's Vinyard (Parker); closed circle, Deep Sea (Gibson); open circle, Nantucket (Gibson); open square, New Jersey (Parker); closed square, Maryland (Parker).

shows this not to be the case; species found in shallow-water deposits in the Miocene and Pliocene are not presently found in abyssal depths although they still live in shallow waters. Likewise, one would expect deep-water assemblages in the Miocene and younger Tertiary to be largely or completely different from the post-Pleistocene. A sample of *Globigerina* ooze of early Miocene age was dredged at a depth of 847 m, and many of the species presently restricted to the abyssal fauna also occur only in the early Miocene deep-water assemblage. Thus, our evidence indicates that the abyssal fauna is not Recent in origin, and that time to develop the diverse faunas was available.

MARTIN A. BUZAS
Smithsonian Institution,
Washington, D.C. 20560

THOMAS G. GIBSON
U.S. Geological Survey,
Washington, D.C. 20242

References and Notes

1. F. L. Parker, *Mus. Comp. Zool. Bull.* **111**, 453 (1954).
2. O. L. Bandy, *J. Paleontol.* **27**, 161 (1953); ——— and R. E. Arnal, *Bull. Amer. Ass. Petrol. Geol.* **44**, 1921 (1960).
3. O. L. Bandy and R. E. Arnal, *ibid.* **41**, 2037 (1957).
4. R. R. Hessler and H. L. Sanders, *Deep-Sea Res.* **14**, 65 (1967).
5. Th. Dobzhansky, *Amer. Sci.* **38**, 209 (1950); A. G. Fischer, *Evolution* **14**, 64 (1960); P. H. Klopfer, *Amer. Natur.* **93**, 337 (1959); R. Margalef, *ibid.* **97**, 357 (1963); E. R. Pianka, *ibid.* **100**, 33 (1966); H. L. Sanders, *ibid.* **102**, 243 (1968).
6. R. A. Fisher, A. S. Corbet, C. B. Williams, *J. Anim. Ecol.* **12**, 42 (1943); F. W. Preston, *Ecology* **29**, 254 (1948); E. H. Simpson, *Nature* **163**, 688 (1949).
7. D. R. Margalef, *Gen. Syst.* **3**, 36 (1958).
8. R. H. MacArthur and J. W. MacArthur, *Ecology* **42**, 594 (1961).
9. B. B. Patten, *J. Mar. Res.* **20**, 57 (1962).
10. R. H. MacArthur, *Biol. Rev.* **40**, 510 (1965).
11. A. F. Bruun, *Nature* **177**, 1105 (1956); T. Wolff, *Deep-Sea Res.* **6**, 95 (1960).
12. T. G. Gibson, *Bull. Geol. Soc. Amer.* **78**, 631 (1967).
13. F. B. Phleger, *Mus. Comp. Zool. Bull.* **106**, 314 (1952).
14. F. L. Parker, *ibid.* **100**, 213 (1948).
15. We thank R. Beavers, R. Cifelli, J. F. Mello, J. Prioleau, and M. C. Taylor for their assistance.

10 September 1968; revised 4 November 1968 ■

Macroglobulin Structure: Homology of Mu and Gamma Heavy Chains of Human Immunoglobulins

Abstract. *The amino acid sequence of fragments obtained by cyanogen bromide cleavage of the mu-chain of a human γM-globulin is homologous to the NH_2-terminal sequences of the gamma-chain of human and rabbit γG-globulins and is related to that of human light chains. This supports the hypothesis that light and heavy chains evolved from a common ancestral gene.*

Immunoglobulins embrace antibodies and proteins related to antibodies in having a tetrachain structure composed of a pair of heavy and a pair of light polypeptide chains, and also include Bence Jones proteins (free light chains). The three major classes of immunoglobulins, γG, γA, and γM (or alternatively, IgG, IgA, and IgM), are demarcated by their heavy chains (γ, α, and μ, respectively); each class is divided into two antigenic types (K or L) based on the presence of a pair of kappa or of lambda light chains. Normal κ and λ light chains of all species are heterogeneous owing to variability in amino acid sequence of the NH_2-terminal half (about 110 residues); in contrast, the COOH-terminal half (105 to 107 residues) is invariant in sequence except for single amino acid substitutions, some of which are apparently allelic (*1*). This conclusion is based on complete amino acid sequence analysis of some ten individual human κ- and λ-type Bence Jones proteins excreted by patients with multiple myeloma or macroglobulinemia (*2–6*) and of the κ-type Bence Jones proteins from two mouse plasmacytoma strains (*3*). It is supported by partial sequence analysis of the NH_2-terminal and COOH-terminal peptides of the κ- and λ-type light chains of many species (*7*). A similar bipartite structure has been proposed for heavy chains (*1*); the latter are composed of somewhat more than 400 amino acid residues, and can be cleaved enzymatically (*8*) into an NH_2-terminal (Fd) and a COOH-terminal (Fc) fragment of approximately equal size. Although the Fd-fragment is believed to have the variable, and the Fc-fragment the constant sequence, this has not yet been established because the maximum length of sequence published comprises only the NH_2-terminal 84 residues of one human γ-chain from a myeloma patient designated Daw (*9*) and the COOH-terminal 216 residues of the γ-chain from pooled, normal γG-globulin of the rabbit (*10*).

We report here the complete sequence of three fragments (F1, F2, and F3) apparently comprising the first 105 residues of the NH_2-terminal portion of a μ-chain from a pathological human γM-macroglobulin designated Ou (Fig. 1), and also the sequence of two smaller fragments from the COOH-terminal (Fc) portion of the μ-chain (Fig. 2). When only one gap is placed in each chain, 61 residues in the first 84 (or 73 percent) are in identical positions in the human γ-chain Daw and the human μ-chain Ou, and the two chains are equally homologous to the NH_2-terminal F1 fragment comprising the first 35 residues of the normal rabbit γ-chain (*11*). Evolutionary relationships of heavy and light chains are suggested by these similarities in sequence of the three heavy chains and also by similarities to κ- and λ-chains.

The μ heavy chain of the κ antigenic type γM-globulin from patient Ou was prepared by mild reduction with dithiothreitol or β-mercaptoethanol to break the interchain disulfide bonds and subsequent alkylation with iodoacetamide. The κ light chains and the μ heavy chains were separated by gel filtration on Sephadex G-100 and three major peaks resulted. The heavy chain was cleaved by reaction with CNBr for 24 hours in 70 percent formic acid; this breaks the peptide bond on the carboxyl side of methionine and converts methionine to homoserine. The fragments formed by CNBr were fractionated with Sephadex G-100 and purified by repeated gel filtration or ion-exchange chromatography. The purified fragments were completely reduced and were aminoethylated with ethylenimine or were carboxymethylated with iodoacetamide to break intrachain disulfide bonds.

In two cases, reduction and aminoethylation yielded additional fragments. Nine fragments formed by CNBr were defined for the Ou heavy chain; altogether these accounted for the total amino acid content of the untreated μ-chain. The largest fragment was composed of about 130 amino acids (Sephadex peak 2); it accounted for the bulk of the carbohydrate in the original γM-globulin and had a COOH-terminal homoserine residue. Two homoserine containing units in peak 1, each composed of about 100 residues, appeared to be linked by an intrachain disulfide bond before reduction. The remaining fragments were smaller and varied in size from 4 to 50 amino acid residues. Their amino acid sequence, as determined by use of the Edman degradation method on the tryptic and chymotryptic peptides, is presented in this report. Identical fragments were also obtained by cleavage of the whole γM-globulin with CNBr followed by

PAPER 7

Trophic Diversity in Past and Present Guilds of Large Predatory Mammals (1988)

B. Van Valkenburgh

Commentary

NICHOLAS D. PYENSON

How does community structure evolve? The succession of fossil assemblages observed in the rock record provides the fundamental basis for answering this question, although many kinds of taphonomic filters obscure and hinder this investigation. Fortunately for vertebrate paleontologists, the skeletal parts of top predators that disproportionately compose their fossil record—skulls and jaws with teeth—are both taxonomically diagnostic and ecologically informative. In truly integrative fashion, Blaire Van Valkenburgh has devoted her career toward understanding the natural history of carnivores, using their ecomorphology as a basis for building data sets that meaningfully compare modern and fossil taxa. Her 1988 paper in *Paleobiology* capped a series of papers arising from her dissertation work on the paleoecology and evolution of late Cenozoic carnivores. This particular paper sought to examine guild structure—the aggregate tally of meat-eating carnivores, circumscribed broadly—as read by their functional dental features, along with estimates of body size. Her comparative scope was innovative because it included carnivorous guilds in two composite, well-sampled ones not only from the late Miocene and early Oligocene of North America but also from three modern environments (the Serengeti, Malaysia and Yellowstone), which provided a way to understand guild diversity and morphospace occupation through time and across a varied geography.

Van Valkenburgh's explicitly ecological and quantitative approach was distinct from previous descriptions of patterns in the fossil record of North America (e.g., Savage and Russell 1983), even though we would now use different ordination and disparity methods to analyze morphological diversity. Regardless, her results articulated a fundamentally novel view of the how ancient mammalian communities functioned: although carnivore guild diversity has changed through time (and differs by the same amount depending on geography in today's world), the morphological similarity among carnivore guild members has remained tight for the last ~31 million years. This finding is striking because carnivore lineages in the early Oligocene belong to entirely extinct groups, and, even in the latest Miocene, aberrant taxa (such as the extinct bear dogs or Amphicyonidae) occupied morphospaces with no modern analogs. A direct sequel to this paper followed, focusing on the iterative evolution of carnivores (Van Valkenburgh 1991), an idea that has continued to prove insightful (Van Valkenburgh et al. 2004), and it remains broadly relevant (e.g., in marine ecosystems, Velez-Juarbe et al. 2012). More importantly, the ecomorphological lens has persisted as a powerful heuristic for asking about the paleoecological dynamics of community structure, especially in its ability to reveal modes of occupancy in concert with changing climate (Janis et al. 2000; Jernvall and Fortelius 2004; Polly et al. 2011).

Literature Cited

Janis, C. M., J. Damuth, and J. M. Theodor. 2000. Miocene ungulates and terrestrial primary productivity: where have all the browsers gone? *Proceedings of the National Academy of Sciences* 97:7899–7904.

Jernvall, J., and M. Fortelius, 2004. Maintenance of trophic structure in fossil mammal communities: site occupancy and taxon resilience. *American Naturalist* 164: 614–24.

Polly, P. D., J. T. Eronen, M. Fred, G. P. Dietl, V. Mosbrugger, C. Scheidegger, D. C. Frank, J. Damuth, N. C. Stenseth and M. Fortelius. 2011. History matters: ecometrics and integrative climate change biology. *Proceedings of the Royal Society B* 278:1121–30.

Savage, D. E., and D. E. Russell. 1983. *Mammalian Paleofaunas of the World*. Reading, MA: Addison-Wesley.

Van Valkenburgh, B. 1991. Iterative evolution of hypercarnivory in canids (Mammalia: Carnivora): evolutionary interactions among sympatric predators. *Paleobiology* 17:340–62.

Van Valkenburgh, B., X. Wang, and J. Damuth. 2004. Cope's rule, hypercarnivory, and extinction in North American canids. *Science* 306:101–3.

Velez-Juarbe, J., D. P. Domning, and N. D. Pyenson. 2012. Iterative evolution of sympatric seacow (Dugongidae, Sirenia) assemblages during the past ~26 million years. *PLoS One* 7:e31294.

Paleobiology, 14(2), 1988, pp. 155–173

Trophic diversity in past and present guilds of large predatory mammals

Blaire Van Valkenburgh

Abstract.—Trophic diversity within guilds of terrestrial predators is explored in three modern and two ancient communities. The modern communities span a range of environments including savannah, rainforest, and temperate forest. The paleocommunities are North American, Orellan (31–29 Ma), and late Hemphillian (7–6 Ma), respectively. The predator guilds are compared in terms of: 1) species richness; 2) the array of feeding types; and 3) the extent of morphological divergence among sympatric species. Feeding type is determined from dental measurements that reflect the proportion of meat, bone, and non-vertebrate foods in the diet. Measurements include estimates of canine shape, tooth size, cutting blade length, and grinding molar area. Morphological divergence among sympatric predators is measured by calculating Euclidean distances among species in a six-dimensional morphospace. Results indicate that the number of predator and prey species are roughly correlated in both ancient and modern communities. Two of the predator guilds, the late Hemphillian and modern Yellowstone, contain relatively few species and appear to be the result of extinction without replacement. Despite differences in history, age, and environment, the extent of morphological divergence within guilds does not differ significantly for the sampled communities. It is clear that the basic pattern of adaptive diversity in dental morphology among coexisting carnivores was established at least 32 million years ago. It appears that interspecific competition for food has acted similarly to produce adaptive divergence among sympatric predators in communities that differ widely in time, space, and taxonomic composition.

Blaire Van Valkenburgh. Department of Biology, University of California, Los Angeles, California 90024-1606

Accepted: February 16, 1988.

Introduction

In the relatively recent past, numerous extinctions brought on by both Pleistocene and more recent events have radically altered the composition of many communities (cf. Olson and James 1984; McDonald 1984). Perhaps hardest hit were the large mammals; in North America, over seventy genera of mammals weighing more than 44 kg disappeared within the last 10,000 years (Kurtén and Anderson 1980; Martin 1984). Other continents suffered less, but in all cases, large species suffered relatively more extinctions than smaller species. As a consequence, many large mammals no longer coexist with taxa that could have profoundly affected their morphological and behavioral evolution (Guilday 1984; Graham and Lundelius 1984). For example, the wolf (*Canis lupus*) formerly coexisted with sabertooth cats (*Smilodon fatalis*) and dire wolves (*Canis dirus*), in addition to the coyotes, pumas, bears and bobcats with which it coexists today. Morphological and behavioral features observed in modern grey wolves may represent adaptations to previous conditions of competition and coexistence. Such possibilities create a frustrating problem for those who study living species and their communities: how can the effects of past ecological conditions be recognized apart from those of the present?

One way to approach this problem is through the study of communities that existed prior to the Pleistocene. By comparing the structure of fossil and living communities, the influence of Recent perturbations, such as extinction and human interference, should become apparent. In effect, the fossil communities are used as pre-perturbation controls against which their modern counterparts are compared.

An additional advantage of fossil communities is that they represent a long timespan. The dynamics of living systems can only be studied for a limited number of seasons or years, usually less than twenty. During such short intervals, the effects of major environ-

0094-8373/88/1402-0004/$1.00

mental catastrophes such as volcanic eruptions or large hurricanes are likely to be either missed or overestimated. (Woodley et al. 1981; Swanson 1987). By contrast, a fossil deposit will typically span a much greater time interval and records the community throughout a range of biotic and physical conditions. Admittedly, the resolution of paleontological studies is often less than that of neontological studies, but the fossil record can provide crucial evidence concerning ecological processes on a grand scale.

Of course, the fossils limit the kinds of data available. Due to the biases of preservation, it is difficult to estimate relative abundances of species (Behrensmeyer et al. 1979; Damuth 1982). However, the number of taxa and the morphological distance between taxa in a community can be quantified (Van Valkenburgh 1985). In the last 10 years, numerous studies of community structure have used these last two parameters to compare equivalent guilds in different localities, and the technique has been labeled the "ecomorphological approach" (e.g., Karr and James 1975; Findley 1976; Grant and Schluter 1984). The label reflects the use of morphology alone to infer significant aspects of a species niche, such as prey size from bill depth or jaw length.

In this paper, the ecomorphological approach is used to examine trophic diversity within the guild of large terrestrial predators in two ancient North American faunas of Orellan (Oligocene, 31.5–29 Ma) and late Hemphillian (Miocene, 7–6 Ma) age. The results are compared with those from a similar study of three modern guilds from different environments (Van Valkenburgh 1988). The Orellan paleoguild includes taxa that are phyletically distant from living Carnivora, early members of two extinct families of Carnivora (Amphicyonidae and Nimravidae) and terminal members of an extinct order of carnivorous mammals, the Creodonta. The late Hemphillian fauna includes representatives of the modern families of Carnivora, although several are from extinct subfamilies (e.g., Borophaginae, Machairodontinae).

The following questions are asked: 1) is species richness similar in modern and ancient guilds; 2) is the morphological distance between sympatric species equivalent in all guilds; and 3) what does the fossil evidence suggest about the impact of extinction on diversity within predator guilds? The previous study of three extant guilds indicated that species richness within predator guilds is greater when the species richness of prey is high and that the average morphological distance between sympatric species is similar among guilds. This suggests that competition for food is an important structuring force in guilds of large predators, an idea with considerable empirical and theoretical support (Hairston et al. 1960; Schoener 1974; Van Valkenburgh 1985 and references therein).

Materials and Methods

Fossil material of 17 extinct predators was measured in this study (Appendix 1). Numbers of specimens measured per species varied according to availability. When more than one specimen was measured, a species mean value was calculated.

Twelve functionally significant measures of cranial and dental anatomy were taken: skull length; length and width of the upper canine, largest lower premolar and all lower molars; and grinding area and cutting blade length of the lower molars (Fig. 1 and Appendix 2). As in the companion study of 47 living Carnivora (Van Valkenburgh 1988), these measurements were used to produce five morphometric ratios that describe canine shape, premolar shape and size, relative blade length, and grinding area of the lower molars (Appendix 2). Among living carnivores, these indices reflect the relative proportion of meat, bone, and non-vertebrate (e.g., fruits, insects) foods in the diet (Appendix 3).

Two of the indices, relative premolar size and relative blade length (RPS, RBL, Appendix 3), were able to clearly separate the extant Carnivora into four dietary groups on bivariate and multivariate plots. Fossil taxa were graphed on the same plots and placed in dietary categories according to their inclusion or proximity to one of the four dietary clusters. If the classification was uncertain, the values for other indices, such as canine shape and relative grinding area, were considered as well. The dietary categories are the same ones used in Van Valkenburgh (1988): 1) meat—greater than 70 percent meat in diet; 2) meat/bone—greater than 70 percent meat with the addition of large bones; 3) meat/non-vertebrate—50–70 percent meat with the balance fruit and/or insects and 4) non-vertebrate/meat—less than 50 percent meat with fruit and/or insects predominating (Table 1). Although the meat/bone category is occupied by only three species at present (hyaenas), it is included because of its presumed ecological significance. Bone-crushing abilities give bone/meat species access to calories that are unavailable to other meat specialists.

In addition to these indices, an estimate of body size was made for each fossil predator using linear regressions of weight against skull length and/or head-body length for 47 living Carnivora (Van Valkenburgh 1985). Table 1 lists the values of the morphometric indices and estimated body weight for each extinct species.

As in the earlier study of locomotor diversity within predator guilds (Van Valkenburgh 1985), two of the morphometric indices and body weight are used as axes of a three-dimensional morphospace (Fig. 2). The two indices, relative premolar size and cutting blade length (RPS, RBL; Appendices 2 and 3), are among the best predictors of dietary type. Meat specialists like felids tend to have relatively long cutting blades and average size premolars; bone eaters have both long cutting blades for slicing meat and enlarged premolars for crushing bone; and more omnivorous taxa have relatively shorter blades and narrow premolars (Ewer 1973; Jaslow 1987; Van Valkenburgh 1988). Body size is used as an axis because of its demonstrated importance in determining prey size, as well as life history traits, foraging radius, metabolic requirements, and interspecific dominance (Rosenzweig 1966, 1968; McNab 1971; Eaton 1979; Eisenberg 1981; Lamprecht 1981; Gittleman and Harvey 1982; Gittleman 1986).

The diversity and packing of feeding types within guilds can be visually compared by plotting the members of each guild separately. In addition, Fig. 2A presents a morphospace that contains the entire sample of living Carnivora used in the analysis of diet-dental

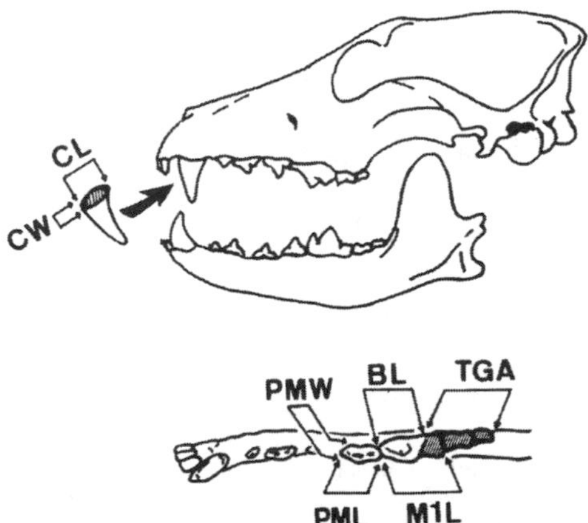

FIGURE 1. Dental measurements. CL, maximum anteroposterior length of canine. CW, maximum mediolateral width of canine. BL, blade length of carnassial tooth. PMW, maximum mediolateral width of premolar. PML, maximum anteroposterior length of premolar. M1L, maximum anteroposterior length of carnassial. TGA, total grinding area of lower molars (shaded area) measured from color slides by a polar planimeter. All other measurements were taken with dial calipers.

morphology correlations (Van Valkenburgh 1988). This graph displays the shape of the morphospace currently occupied by living Carnivora and serves as a framework for viewing each guild.

Because the 3-D plots can only display the morphological differences among sympatric predators in three characters, the Euclidean distances between pairs of guild members in six-dimensional space (one dimension per morphometric index plus body weight) are calculated. To compensate for differences in scale among the indices, the variables were transformed to standardized normal deviates (Sneath and Sokal 1973) before plotting or estimating distance values. The standardized deviates were calculated using data from predators of all the guilds.

Euclidean distance values are unaffected by correlations among variables and so can overlook the fact that some variables provide no new information (Van Valen 1974). To examine this possibility, correlations between the morphometric indices were tested with Spearman rank coefficients (Zar 1984). Of the 15 pairwise comparisons, only four differed significantly from zero ($P < 0.05$). In two of

TABLE 1. Values of the morphometric indices and diet group assignments for the examined fossil species. All values are means; see Appendix 1 for list of measured specimens. Morphometric indices: LBW, log body weight; CS, canine shape; RPS, relative premolar size; PMD, premolar shape; RBL, relative blade length; RGA, relative grinding area. Diet groups: M, meat; M/B, meat/bone; M/N-V, meat/non-vertebrate. For details, see text and Appendix 2.

Species	LBW	CS	RPS	PMD	RBL	RGA	Diet Group
Creodonta							
Hyaenodon horridus	1.97	0.68	2.28	0.56	1.00	0.00	M
H. crucians	1.30	0.75	2.23	0.54	1.00	0.00	M
Nimravidae							
Hoplophoneus occidentalis	1.87	0.54	1.76	0.47	1.00	0.00	M
H. primaevus	1.26	0.54	2.32	0.48	1.00	0.00	M
Dinictis felina	1.30	0.54	2.35	0.48	0.85	0.23	M
Eusmilus dakotensis	2.02	0.35	1.70	0.50	1.00	0.00	M
Felidae							
Machairodus coloradensis	2.21	0.42	2.14	0.42	1.00	0.00	M
Pseudaeleurus hibbardi	1.84	0.69	1.97	0.50	1.00	0.00	M
Adelphailurus kansensis	1.71	0.65	(2.00)[1]	(0.48)[1]	1.00	0.00	M
cf. *Felis proterolyncis*	1.19	0.77	2.05	0.49	1.00	0.00	M
Canidae							
Mesocyon sp.	1.11	0.64	1.91	0.48	0.68	0.88	M/N-V
Osteoborus cyonoides	1.62	0.70	3.11	0.66	0.74	0.73	M/B
Canis davisii	1.00	0.69	1.95	0.39	0.65	0.90	M/N-V
Vulpes stenognathus	0.84	(0.73)[2]	1.87	0.37	0.62	0.99	M/N-V
Amphicyonidae							
Daphoenus vetus	1.58	0.72	1.60	0.44	0.65	1.40	M/N-V
D. hartshornianus	1.41	0.72	1.56	0.46	0.67	1.24	M/N-V
Ursidae							
Agriotherium schneideri	2.54	0.73	2.02	0.55	0.60	1.39	M/N-V
Mustelidae							
Plesiogulo marshalli	1.53	0.77	2.50	0.61	0.64	0.57	M/B
Pliotaxidea nevadensis	0.36	(0.80)[3]	2.09	0.60	0.59	0.98	M/N-V

[1] There are no lower premolars known for *Adelphailurus kansensis*. These estimates are the mean value for PMS and PMD for six other sabertooth species (*Hoplophoneus occidentalis, H. primaevus, Dinictis felina, Eusmilus dakotensis, E. sicarius*, and *Machairodus coloradensis*).
[2] Because there are no upper canines known for *V. stenognathus*, their CD value was estimated by the mean value for 16 extant canids.
[3] Approximate (based on the alveolus of a single specimen figured in Hall [1946]).

these, the value of r fell below 0.5. The remaining two, CS versus RBL and RBL versus TGA, had r values of 0.63 and 0.84, respectively. Because the degree of independence among the variables was generally high, all six were retained in the distance analysis.

As in Van Valkenburgh (1985), three different estimates of dispersion were calculated based on the morphological distance between guild members: (1) the average link length of a minimum spanning tree connecting all guild members; (2) the average distance between each species and the guild centroid (determined as the mean value of each of the six characters); and (3) the average distance between any species and its nearest neighbor.

The species composition for each of the five guilds discussed in this paper is shown in Table 2. The guilds include all the non-aquatic species within the community that capture and consume prey. The guild is confined to predators weighing 7 kg or more (i.e., jackal size and larger), because the evidence for strong competitive interactions among these animals is substantial, whereas that for competition between them and smaller carnivores is relatively weak (see Van Valkenburgh 1985). The three extant guilds, Yellowstone, Malaysia, and Serengeti, and the Orellan paleoguild were used in two previous studies and detailed references on their physical, biotic, and in the case of the Orellan, paleoecological, characteristics are in Van Valkenburgh (1985).

The late Hemphillian paleoguild is a new addition. Although the North American Late Miocene record is generally good, this particular time was chosen because the taxonomy

TABLE 2. Predator guild composition. The guild includes all the terrestrial mammals larger than 7 kg that take prey. Some carnivore taxa have been excluded because they eat almost no meat: the sloth bear (*Melursus ursinus*) from Chitawan; the aardwolf (*Proteles cristatus*) from Serengeti; and the sun bear (*Helarctos malayanus*) from Malaysia. Within each community there are several species which might be considered borderline members according to this definition (e.g., pigs and some primates). However, such animals have remained omnivorous or tended to herbivory throughout their history and have probably had only minor competitive interactions with predatory mammals; thus, they are excluded with caution. By contrast, most of the living omnivorous members of the order Carnivora have descended relatively recently from more predacious ancestors (e.g., ursids and viverrids; see Petter 1969, Crusafont-Pairo and Truyols Santonja 1956, 1957). Their tendency to eat fruit and insects can be viewed as a possible response to competition within the guild.

Serengeti	Yellowstone
	Puma, *Puma concolor*
Lion, *Panthera leo*	Lynx, *Lynx canadensis*
Leopard, *P. pardus*	Bobcat, *Lynx rufus*
Cheetah, *Acinonyx jubatus*	Wolf, *Canis lupus*
Caracal, *Caracal caracal*	Coyote, *Canis latrans*
Serval, *Felis serval*	Red fox, *Vulpes vulpes*
Wild dog, *Lycaon pictus*	Wolverine, *Gulo gulo*
Blackbacked jackal, *Canis mesomelas*	Badger, *Taxidea taxus*
Golden jackal, *C. aureus*	Black bear, *Ursus americanus*
Sidestriped jackal, *C. adustus*	Grizzly bear, *Ursus arctos*
Spotted hyaena, *Crocuta crocuta*	
Striped hyaena, *Hyaena hyaena*	
Ratel, *Mellivora capensis*	
Civet, *Civettictis civetta*	

Malaysia
Tiger, *Panthera tigris*
Leopard, *P. pardus*
Clouded leopard, *Neofelis nebulosa*
Temminck's cat, *Felis temmincki*
Fishing cat, *F. viverrina*
Dhole, *Cuon alpinus*
Binturong, *Arctictis binturong*
Civet, *Viverra megaspila*

Hemphillian	Orellan
Machariodus coloradensis	
Pseudaeleurus hibbardi	*Hoplophoneus occidentalis*
Adelphailurus kansensis	*H. primaevus*
Felis proterolyncis	*Dinictis felina*
Canis davisii	*?Eusmilus dakotensis*
Vulpes stenognathus	*Daphoenus vetus*
Osteoborus cyonoides	*D. hartshornianus*
Plesiogulo marshalli	*Mesocyon* sp.
Agriotherium schneideri	*Hyaenodon horridus*
Arctonasua fricki	*H. crucians*

of the carnivores appeared stable relative to that of earlier faunas. Thus, the possibility of either missing or adding guild members due to taxonomic errors of splitting or lumping were minimized. The late Hemphillian guild represents a composite of three discrete localities, separated by 50 to 200 miles: The Coffee Ranch of Texas; the Optima (Guymon) of Oklahoma; and the Edson Quarry of Kansas (Schultz 1977). Two of these, Coffee Ranch and Edson Quarry, are lenticular deposits which may represent seasonal lakes or bogs (Dalquest 1969; Schultz 1977; Harrison 1983), whereas the Optima deposits are of stream channel origin (Savage 1941; Schultz 1977). Because fossil collections from each locality have numerous species in common, they are judged to have been roughly contemporaneous (Schultz 1977; Harrison 1983). However, no one locality has produced all ten species listed in Table 2, so the composite should be considered a probable maximum for late Hemphillian times. One omnivorous species, the badger (*Pliotaxidea nevadensis*), was excluded from the guild due to its small size. One additional large cat, *Nimravides catacopsis*, may be represented by a single tooth at Optima (Savage 1941; Martin and Schultz 1975), but its identity is questionable since the genus has been more recently listed as extinct by mid-Hemphillian times (Schultz 1977; Baskin 1981; Harrison 1983). It is excluded from the guild with caution.

For both paleoguilds, the possibility exists that species were present but not preserved or discovered, or species outside the order Carnivora that fulfilled the missing roles were overlooked. However, it appears unlikely that such artifacts of the record fully account for the great differences in species richness observed in the present paper (e.g., 9 versus 13 taxa). Taphonomic studies have shown that skeletons of species larger than 10 kg are subject to less post-mortem damage than those of smaller species and hence are more likely to be preserved in fossil deposits (Behrensmeyer et al. 1979). Skeletal and dental material from both the late Hemphillian and Orellan deposits described here is abundant, well-preserved, and probably provides a complete, or nearly complete, representation of the large predator community as it existed (cf. Dalquest 1969; Voorhies 1970; Damuth 1982; Harrison 1983).

The second possibility, that unexpected substitutes for the meat and meat/bone roles existed, such as large terrestrial reptiles or ground birds, can be largely ruled out since such potential surrogates are absent from Orellan and late Hemphillian faunas of North America (cf. Clark et al. 1967; Schultz 1977). However, as noted in the discussion of the Orellan guild below, there are some large artiodactyls (entelodonts) which may have scavenged occasionally.

Results

The distribution of the sample of 47 modern Carnivora within the morphospace defined by log body weight, relative premolar size, and cutting blade length is shown in Fig. 2A. Position on the blade length axis (RBL) reflects the relative proportions of meat and non-vertebrate foods in the diet. The strictest meat species, the cats, all have similarly long cutting blades (high RBL values) and cluster along the volume's left edge (RBL > 1, Fig. 2A). Slightly more omnivorous members of the meat group, such as the wolf and wild dog ("Cl," "Lp": Fig. 2A), have retained a talonid on their lower molar, have smaller blade length values than the felids, and are located near the center of the volume (RBL near 0). Adjacent to them, with still shorter shearing blades, are the members of the meat/non-vertebrate group (triangles). Members of the non-vertebrate/meat group have the relatively shortest cutting blades (low, negative RBL values) and form a clump near the right edge of the volume, although a few exist on the edge of the central cluster (squares: Fig. 2A). The premolar axis (RPS) distinguishes meat/bone specialists from all others and reveals species within other groups which tend toward bone-eating. Meat/bone eaters exhibit a combination of relatively wide premolars (high RPS) and long cutting blade (high RBL) and thus are positioned in the left, rear corner (stars: Fig. 2A).

Each of the four diet groups is characterized by a limited range of premolar size and shear length values, and each tends to occupy a separate region of the volume floor. In general, the species in the center of the volume are more generalized in their diet than those near the edges. Moreover, this central clump includes a greater taxonomic variety than do the three satellite clusters (for details, see Van Valkenburgh 1988). There are intriguing gaps between the four clumps, indicating that there are combinations of premolar size, shear length and body weight which are not exhibited by any of the 47 species of extant Carnivora. For example, there is no species of large size (LBW > 1) in the meat/non-vertebrate group, nor is there a species with average or narrow width premolars (RPS near 0) that has retained a small grinding basin on the carnassial (RBL = 0 to 1). This suggests that such species with such morphologies might not be viable at present, but, as will become apparent, does not exclude their past existence.

Results of guild comparisons.—The three Recent guilds differ in the total number of species and the array of dietary types (Figs. 2, 3). The

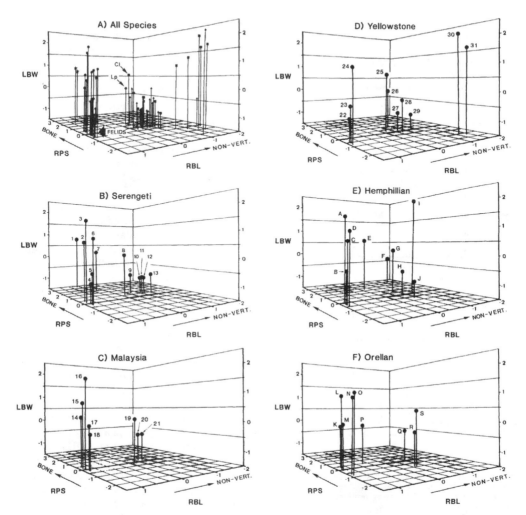

FIGURE 2. Guild morphospaces. For all plots, the axes are log body weight (LBW), relative premolar size (RPS), and relative blade length (RBL). Units are standardized normal deviates of values shown in Table 1. Shaded areas indicate the range of RPS and RBL values observed in (A), the sample of 47 Carnivora. (A) All species, includes 47 species described in Van Valkenburgh (1988). In (A), species are represented by symbols which indicate their dietary classification (Table 1). Meat, circles. Meat/bone, stars. Meat/non-vertebrate, triangles. Non-vertebrate/meat, squares. Abbreviations in (A) are as follows: Lp, *Lycaon pictus;* Cl, *Canis lupus.*

(B) the Serengeti predator guild, species as follows: 1, spotted hyaena; 2, leopard; 3, lion; 4, serval; 5, caracal; 6, cheetah; 7, striped hyaena; 8, wild dog; 9, ratel; 10, side-striped jackal; 11, blackbacked jackal; 12, golden jackal; 13, civet.

(C) the Malaysia guild, species as follows: 14, clouded leopard; 15, leopard; 16, tiger; 17, Temminck's cat; 18, fishing cat; 19, dhole; 20, civet; 21 binturong.

(D) Yellowstone guild, species as follows: 22, bobcat; 23, Canadian lynx; 24, puma; 25, wolf, 26, wolverine; 27, red fox; 28, coyote; 29, badger; 30, grizzly bear; 31, black bear.

(E) Late Hemphillian guild, species as follows: A, *Machairodus coloradensis;* B, *Felis proterolyncis;* C, *Pseudaeleurus hibbardi;* D, *Adelphailurus kansensis;* E, *Osteoborus cyonoides;* F, *Arctonasua fricki;* G, *Plesiogulo marshalli;* H, *Canis davisii;* I, *Agriotherium schneideri;* J, *Vulpes stenognathus.*

(F) Orellan guild, species as follows: K, *Hyaenodon crucians;* L, *Hyaenodon horridus;* M, *Hoplophoneus primaevus;* N, *Hoplophoneus occidentalis;* O, *Eusmilus dakotensis;* P, *Dinictis felina;* Q, *Mesocyon* sp.; R, *Daphoenus hartshornianus;* S, *Daphoenus vetus.*

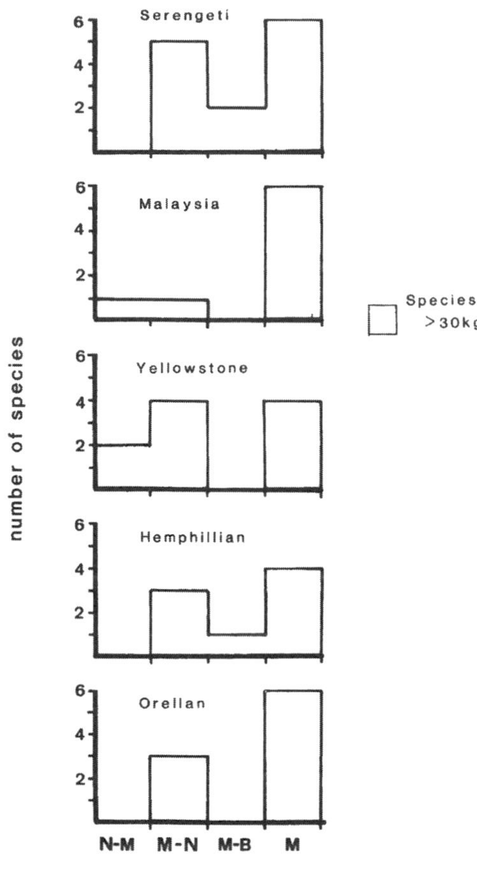

FIGURE 3. Species richness (n) within dietary categories for each guild as labeled. Diet categories: N-V/M, non-vertebrate/meat; M/N-V, meat/non-vertebrate; M/B, meat/bone; M, meat.

Serengeti is the most species-rich, with thirteen carnivores, drawn from each of the four diet categories except non-vertebrate/meat. Yellowstone has ten predators, representing each diet group except meat/bone. Malaysia has the smallest number of species, eight, and like Yellowstone, is without a meat/bone specialist. Both tropical guilds, Malaysia and the Serengeti, contrast with Yellowstone in their greater number of meat specialists, six as opposed to four, and their lack of non-vertebrate/meat species.

Despite these differences, the morphological distance between species is similar in the Serengeti and Malaysia, even when all six morphological characteristics are considered (Table 3). The greater average morphological distance between Yellowstone predators (average M.S.T. and D.F.C., Table 3) is due to the inclusion of bears in the guild. If the omnivorous bears are excluded, the dispersion among the remaining species is close to that shown in the two tropical guilds.

The three distance measures, M.S.T., N.N.D. and D.F.C., each contribute somewhat different information about the distribution of species within the volume. Of the three, the average link of the minimum spanning tree (M.S.T.) is probably the best indicator of overall morphological similarity among guild members. The mean distance from guild centroid (D.F.C.) increases with the addition of species, and nearest-neighbor distances (N.N.D.) are lowered by clumping within the volume. Unfortunately, neither M.S.T. or N.N.D. values can be compared statistically because their link lengths are not independent. Pairwise comparisons of D.F.C. values among the guilds (with or without bears included) showed no significant differences ($P > 0.10$, Mann-Whitney, two-tailed: Zar 1984).

The six-million-year-old late Hemphillian guild is similar to Malaysia and Yellowstone in low species-richness, but more like the Serengeti in the wide array of feeding types represented (Figs. 2E, 3). It differs from all the modern examples in the greater average morphological distance between species as measured by M.S.T. and N.N.D. (Table 3), even when the bear, *Agriotherium*, is removed. The values remain high even if the badger, *Pliotaxidea*, which was excluded from the analysis due to its small size, is added. The average D.F.C. is similar to that of the two larger guilds, Yellowstone and Serengeti, but greater than that of the similarly sized Malaysian guild (Table 3). However, the late Hemphillian D.F.C. value did not differ significantly from those of the other guilds ($P > 0.20$; Mann-Whitney, two-tailed). In spite of this, I doubt the packing of late Hemphillian predator species within the morphospace is equivalent to that of the other guilds, all of which are similar to one another in all three distance

TABLE 3. Morphological distance characteristics of each guild. The mean link length of the minimum spanning tree (M.S.T.), the mean distance to nearest neighbor (N.N.D.), and the average distance between each species and its guild centroid (D.F.C.). SD is the standard deviation of each mean for each guild. Values for the Yellowstone are shown both with ($N = 10$) and without ($N = 8$) the bears included. For the late Hemphillian, values are shown for the guild with *Pliotaxidea* and *Arctonasua* included ($N = 11$), excluded ($N = 9$), and with the bear, *Agriotherium*, excluded as well ($N = 8$). The fossil material of *Arctonasua fricki* is too incomplete for several of the dental measurements, so its values were estimated by those of its similarly sized relative, *A. eurybates*.

Guild	No. of species	Avg. link M.S.T.	SD	Avg. N.N.D.	SD	Avg. D.F.C.	SD
Serengeti	13	1.49	0.605	1.21	0.467	2.12	0.540
Malaysia	8	1.52	0.733	1.41	0.745	1.68	0.675
Yellowstone	10	1.67	1.06	1.11	0.695	2.33	0.784
	8	1.48	0.769	1.14	0.785	1.78	0.119
Late Hemphillian	11	1.95	0.934	1.97	0.930	2.61	0.796
	9	1.97	0.936	1.72	0.968	2.31	0.616
	8	1.84	0.922	1.57	0.917	2.20	0.684
Orellan	9	1.59	0.640	1.30	0.632	1.84	0.501

measures. Instead, it appears that D.F.C. values are relatively poor indicators of species packing when species tend to be distributed in clumps, as they are here. In such cases, the centroid lies far from either clump, and D.F.C. reflects the distance between clumps rather than between taxa. Thus, the distance between clumps is similar in the late Hemphillian and modern guilds, but that between taxa (as reflected by N.N.D. and M.S.T.) appears to be greater in the late Hemphillian.

There were ten predators in the late Hemphillian guild: four meat specialists (two sabertooth and two non-sabertooth felids), four meat/non-vertebrate species (an ursid, mustelid, and two canids), a meat/bone species (a canid), and a non-vertebrate/meat species (a procyonid). The meat/bone species, *Osteoborus cyonoides* (E in Fig. 2E), is a member of an extinct subfamily of canids, the Borophaginae, and represents a morphology (that is, combination of premolar size, shear length, and body weight) unknown within the sample of extant Carnivora (compare shaded area Figs. 2A, 2E). Note also that the meat/bone role was played by a canid in this instance, rather than a hyaenid as is true of the modern guilds. *Osteoborus* can be classified as a frequent consumer of bone with confidence, based on the measured characters (Table 1, Appendix 3) and previous descriptions of its dental morphology (Matthew and Stirton 1930; Dalquest 1969). The bear, *Agriotherium*, is unusual as well, both because it was much larger than any of the living members of the meat/non-vertebrate group and because it had a highly carnivorous dentition relative to modern ursids. These two important examples of convergent evolution suggest that phyletic constraints may be readily circumvented to produce adaptive divergence within guilds.

The Orellan paleoguild, which existed from 31.5 to 29 Ma, is considerably older than the other four guilds and is composed almost entirely of representatives from extinct families of mammals (Table 2). For example, the meat group is made up of creodonts and nimravid cats, both of which have no living descendants. Despite these important historical differences, the general configuration and dispersion of species within the guild volume is remarkably similar to modern examples (compare Fig. 2B–D with 2F; Table 3). Considering the examples of convergence discussed above, and those of the Orellan paleoguild, it cannot be argued that similarity among guilds is simply due to similarities in phylogenetic background.

In the Orellan paleoguild, there were nine species, six in the meat and three in the meat/non-vertebrate group; meat/bone and non-vertebrate/meat species are absent (Fig. 3). As in the Miocene paleoguild, there are species which occur outside the bounds of the morphological space occupied by Recent taxa. *Dinictis felina* (Fig. 2F, "P") is a sabertooth cat (Family Nimravidae) which retains a small heel on its carnassial, much like extant hyaenas. The two bear-dogs, *Daphoenus vetus* and

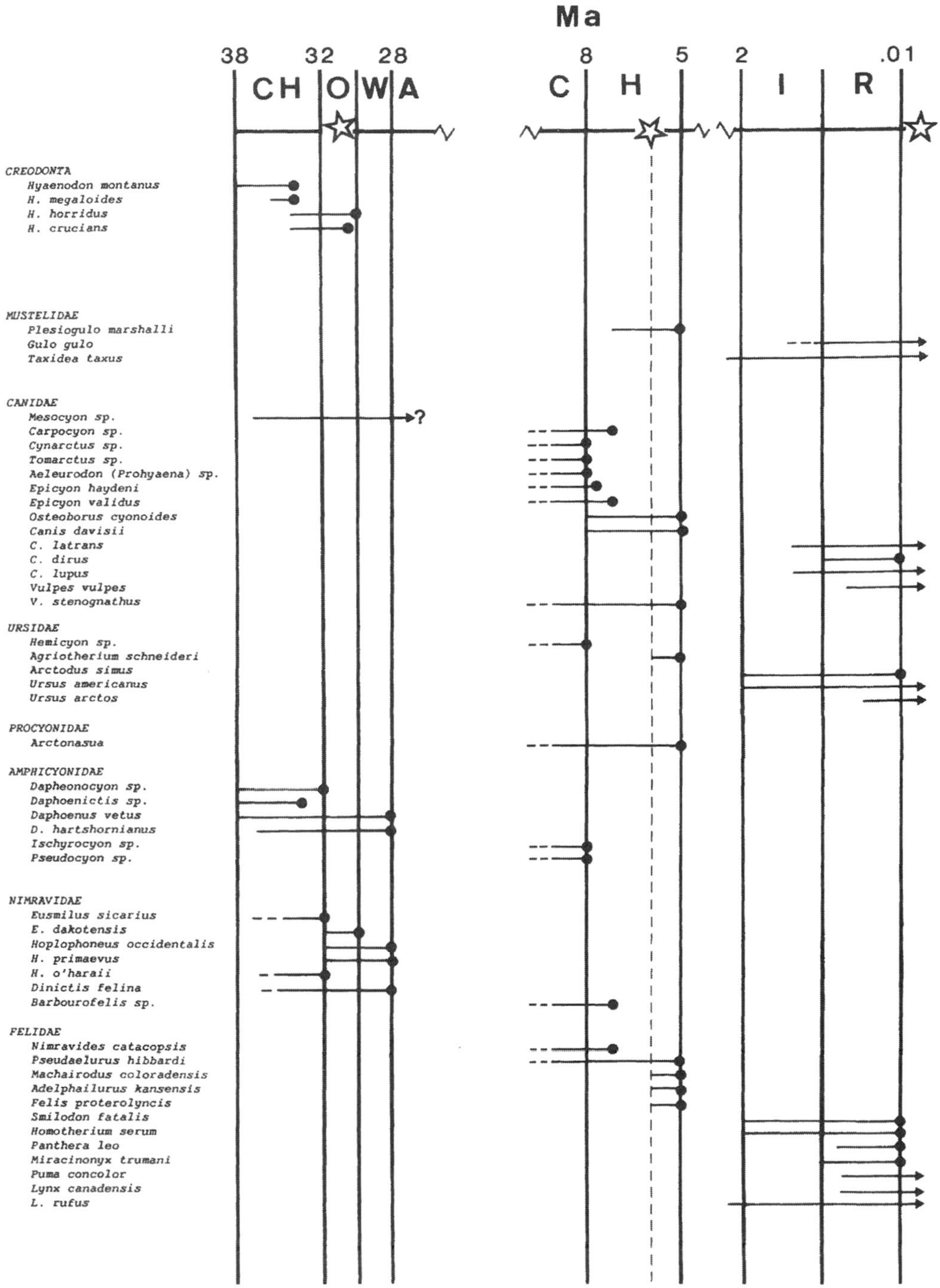

Ma

38 32 28 8 5 2 .01

CH O W A C H I R

CREODONTA
Hyaenodon montanus
H. megaloides
H. horridus
H. crucians

MUSTELIDAE
Plesiogulo marshalli
Gulo gulo
Taxidea taxus

CANIDAE
Mesocyon sp.
Carpocyon sp.
Cynarctus sp.
Tomarctus sp.
Aeleurodon (Prohyaena) sp.
Epicyon haydeni
Epicyon validus
Osteoborus cyonoides
Canis davisii
C. latrans
C. dirus
C. lupus
Vulpes vulpes
V. stenognathus

URSIDAE
Hemicyon sp.
Agriotherium schneideri
Arctodus simus
Ursus americanus
Ursus arctos

PROCYONIDAE
Arctonasua

AMPHICYONIDAE
Dapheonocyon sp.
Daphoenictis sp.
Daphoenus vetus
D. hartshornianus
Ischyrocyon sp.
Pseudocyon sp.

NIMRAVIDAE
Eusmilus sicarius
E. dakotensis
Hoplophoneus occidentalis
H. primaevus
H. o'haraii
Dinictis felina
Barbourofelis sp.

FELIDAE
Nimravides catacopsis
Pseudaelurus hibbardi
Machairodus coloradensis
Adelphailurus kansensis
Felis proterolyncis
Smilodon fatalis
Homotherium serum
Panthera leo
Miracinonyx trumani
Puma concolor
Lynx canadensis
L. rufus

FIGURE 4. Stratigraphic ranges of predators within the North American guilds and the predators which preceded the guild members. The "?" for *Mesocyon* sp. is due to its uncertain taxonomic identity. The time scale is shown at

D. hartshornianus (Fig. 2F; "R" and "S"), lie close to the area occupied by extant meat/non-vertebrate species and might be included if the sample of extant species were larger.

The important differences among the guilds can be summarized as follows:

(1) the Serengeti guild contains 13 species whereas all four others contain eight to ten;

(2) the two tropical guilds, the Serengeti and Malaysia, and the Orellan, include six meat species, whereas the Yellowstone and late Hemphillian include but four;

(3) meat/bone specialists are found only in the Serengeti and late Hemphillian guilds; and

(4) the average morphological distance between sympatric species is similar in all guilds, except the late Hemphillian, where it appears somewhat large.

Discussion

The results of the comparisons of the three Recent guilds are discussed in detail in Van Valkenburgh (1988) and are summarized here. In general, it appears that species richness within guilds, particularly of meat and meat/bone specialists, is greater in the environments where the biomass and species richness of prey are greater. The tropical guilds, Malaysia and Serengeti, have both a greater diversity and density of herbivore prey (arboreal and/or terrestrial) and meat eaters (including meat/bone specialists) than does the temperate environment of Yellowstone. In addition, the presence of scavengers (meat/bone species) in the Serengeti guild probably reflects the greater availability of carcasses in savannah as opposed to forest environments.

Although the distribution and biomass of prey in the tropical savannah and rainforest appear to reflect directly the abundance and availability of low-stature vegetation (Eisenberg and McKay 1974; Eisenberg and Seidensticker 1976; Coe et al. 1976), the same cannot be said of Yellowstone. There, history has probably played a more prominent role than environment in setting diversity levels. In North America, the impact of the Pleistocene on the large mammal fauna was severe, and six large predators were eliminated from the continent (Fig. 4 and references therein). Of the six, four were meat specialists (felids), one was a meat/bone species (the dire wolf, *Canis dirus*), and one was a relatively carnivorous bear (*Arctodus simus*). Thus, species dependent on large herbivores were hit hardest by Pleistocene events. Not surprisingly, the species richness of large herbivores in western North America declined sharply over the same period, from about 14 to six (Kurtén and Anderson 1980).

By contrast, the East African guild has undergone little change for close to two million years (Fig. 5 and references therein). Unfortunately, the fossil record of Malaysian carnivores is poor, but the evidence from the herbivores suggests that the rainforest community persisted throughout the Pleistocene, although its boundaries fluctuated (Verstappen 1975).

The similarity among the guilds in the average morphologic distance between sympatric predators is somewhat surprising. In the earlier study of locomotor diversity within predator guilds, predators were more similar to one another in the measured skeletal characters in the Serengeti, where terrestrial herbivore diversity and biomass levels are highest (Van Valkenburgh 1985). The difference in results between the locomotor and dietary analysis probably reflects the greater influence of environment on locomotor than dental anatomy. The Serengeti environment is predominantly grassland and the predators may have converged on an appropriate attack behavior (long-distance pursuit) and associ-

←

the top. The stars mark, from left to right, the times of the Orellan, Hemphillian and Yellowstone guilds, respectively. Abbreviations as follows: CH, Chadron; O, Orellan; W, Whitneyan; A, Arikareean; C, Clarendonian; H, Hemphillian; I, Irvingtonian; R, Rancholabrean. References: Jepsen 1933; Simpson 1941; Savage 1941; Scott and Jepsen 1941; Hough 1948, 1949; Mellett 1977; Hunt 1974; Munthe 1979; Kurtén and Anderson 1980; Baskin 1980, 1981; Harrison 1981, 1983; Savage and Russell 1983; Tedford et al. 1988.

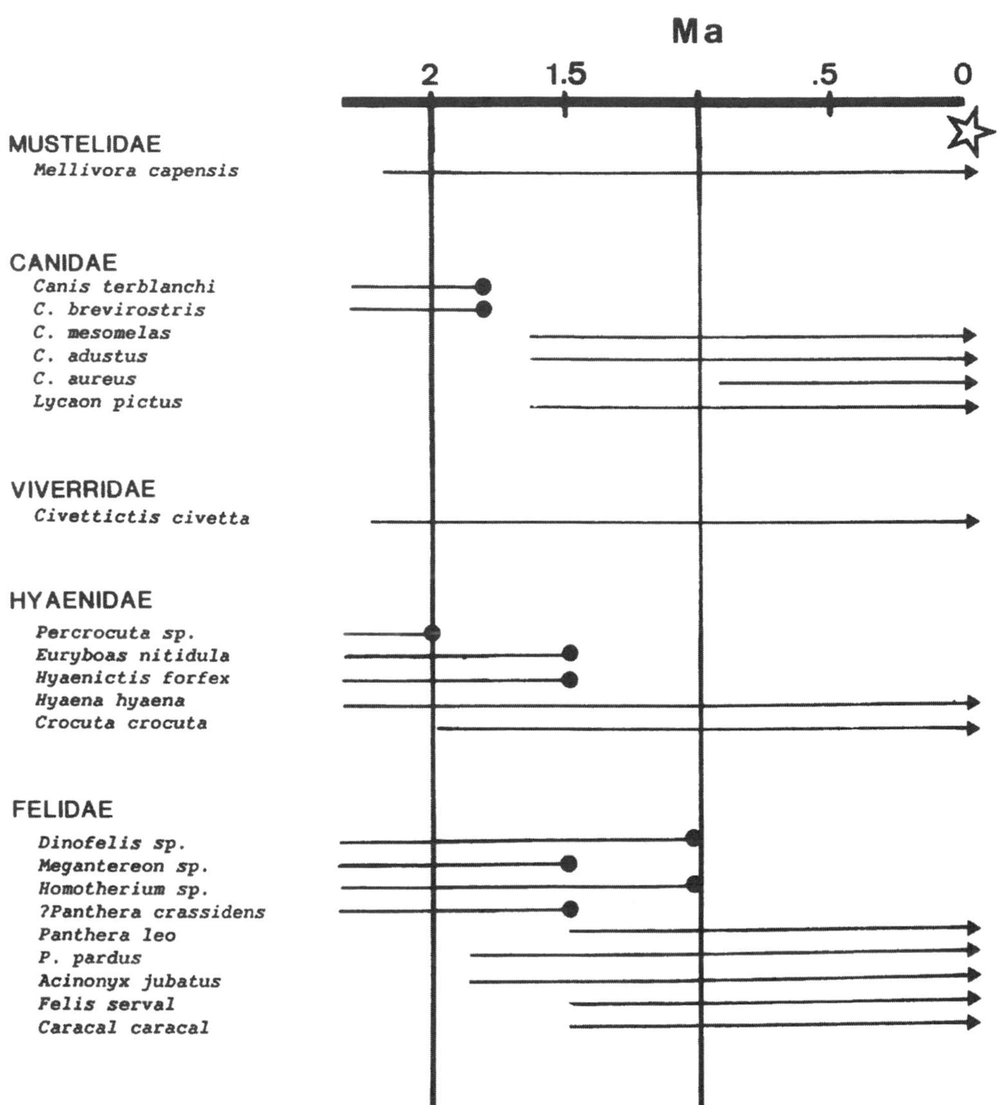

FIGURE 5. Stratigraphic ranges of the Serengeti predators and the predators that preceded them in Africa. The time scale is shown at the top; the star indicates the time of the studied guild (the present). References: Hendey 1974a, b, c; Savage 1978; Brain 1981; Walker 1984.

ated morphology. By contrast, the character of meat does not change with location and predators in very different environments are likely to have similarly adapted dental morphologies.

Nevertheless, it is remarkable that the impact of extinction on the Yellowstone guild is not apparent in the packing of species within the morphospace. The loss of six large predators might be expected to produce a guild that appears overdispersed relative to other, less disturbed guilds. Although this is not the case for the guilds examined here, it may be that the Yellowstone guild is overdispersed relative to the pre-Pleistocene paleoguild of the same area. I was unable to test this for the Wyoming area, but work in progress on the predators of southern California indicates that a 27 percent drop in the number of predators since the Rancholabrean was accompanied by nine percent and 25 percent rises in M.S.T. and N.N.D. values, respectively. Thus, it appears that extinction can result in greater morphological spacing within predator guilds.

The late Hemphillian paleoguild.—Paleobotanical evidence suggests that the dominant plant community in central North America during the Hemphillian was savannah-woodland mosaic with gallery forests (Leopold 1969; Wolfe 1978). Given this environment, the diversity of large herbivores is unexpectedly low, 13 species as opposed to the Serengeti's 24. As in the Recent examples, predator diversity is correspondingly low, ten species as opposed to 13. The presence of a bone-eater, *Osteoborus cyonoides,* suggest a relatively high predictability of carcasses, perhaps reflecting high ungulate biomass and/or the savannah environment.

As at Yellowstone, the low numbers of large herbivore and carnivore species in the late Hemphillian of North America appear to be the result of extinction without replacement. Six million years before, ungulate species richness was comparable to that of modern East Africa (Webb 1983). Two successive waves of extinction in the late Clarendonian (circa 9 Ma) and the early Hemphillian (circa 7 Ma) eliminated many herbivore and carnivore taxa. The causes of the decline are thought to be increased aridity and climatic cooling, both of which might lower plant productivity as savannah progressed toward steppe (Wolfe 1978; Webb 1983, 1984). As herbivore diversity dropped, competition for food among predators might have intensified, resulting in some extinctions.

The late Clarendonian guild included 14 predators, seven of which were lost by the early Hemphillian (Table 4). All seven were canids, amphicyonids or ursids; the sole nimravid and two felids persisted across the Mio-Pliocene boundary (Fig. 4). Two new canids appeared as replacements, *Osteoborus* and *Canis*

TABLE 4. Species richness of predators and prey in North America in the Late Miocene and Early Pliocene. For the late Clarendonian and early Hemphillian faunas, species richness is based on surveys of the collections of the American Museum of Natural History in New York and data in Tedford et al. (1988) and Webb (1983).

North American Land Mammal Age	Ma.	No. of large predators	No. of large herbivores
Lt. Clarendonian	9	14	18
E. Hemphillian	7	8	8
Lt. Hemphillian	6	10	13

davisii, bringing the total number of guild members to eight in the early Hemphillian. The mid-Hemphillian event eliminated three more species, two cats and a canid, but was followed by the immigration of five new predators, three felids, an ursid and a mustelid. Thus there was a net gain in species richness within the guild during Hemphillian time, from eight to ten (Table 4). This appears to have been associated with similar changes in the numbers of large herbivores throughout this interval, suggesting that predator species richness tracked that of the prey. This is to be expected if competition for food at all affects the structure of predator guilds. A smaller array of prey morphologies can support a more limited range of predator morphologies.

The role of competition in guild evolution is further supported by an examination of the sequence of extinctions and replacements during the Hemphillian. The loss of several dog-like predators in the late Clarendonian is partially filled by the subsequent appearance of two canids. Similarly, the extinction of two cat-like ambush predators precedes the immigration of three new felids.

The late Hemphillian paleoguild differs from the three modern examples in the greater average morphological distance between species, even when the bear, *Agriotherium,* is excluded (M.S.T., N.N.D.: Table 3). The apparent overdispersion of the late Hemphillian guild is probably due to the preceding extinction events. As mentioned above, extinctions might occur randomly with respect to morphological spacing within the guild and produce a relatively dispersed guild. Over time, given favorable conditions of prey richness and abundance, these gaps could be filled by new species and dispersion within the guild would decrease. Substantiation of this idea will require more comparisons of guilds before and after extinction events.

The Orellan paleoguild.—Unlike the late Hemphillian, the Orellan was not immediately preceded by a severe extinction event. There were extinctions of carnivore taxa at or near the Chadron–Orellan boundary, but all were replaced by similar species so that overall paleoguild diversity levels remained stable for 6–8 m.y. (Fig. 4).

Because the Carnivora (creodonts excluded) of the Orellan are closer, in a temporal sense, to the earliest representatives of their order than are extant carnivores, it might be expected that the Orellan carnivores would appear more similar to one another, and less dentally specialized, than extant sympatric predators. However, a glance at Fig. 2 shows this not to be true. Although only two of the four feeding categories are represented, the difference in dental morphology between the meat and meat/non-vertebrate species is almost always as great or greater than that observed between the two types today (compare Fig. 2F with 2B and 2D). The one exception is the sabertooth nimravid cat, *Dinictis felina* (Fig. 2F, "P"), which, unlike any extant and most extinct cats, did not devote its entire carnassial to a cutting blade.

These results contrast with those of the study of locomotor diversity within the same guilds (Van Valkenburgh 1985). In that study, all the predators of the Orellan appear to have been slower and more forceful than their modern counterparts. Apparently, there has been a tendency towards greater speed in both predator and prey over the Cenozoic (Bakker 1983), probably in response to a global increase in the amount of grassland and open woodland habitats (Webb 1977, 1978; Wolfe 1985). The texture of meat and bone is unlikely to have changed over the same timespan and so adaptations for consuming these substances evolved early and remained stable. This is not true of the dental adaptations of herbivores; a progressive increase in tooth height and resistance to wear has been documented for several groups (Scott 1913; Gregory 1971; Webb 1977). In this case, the quality of the food did appear to change over time as grasses became more resistant to grazing.

The strong divergence in dentition among the Orellan fissipeds seems not to have been accompanied by a similar divergence in cranial morphology. In a study of cranial shape characteristics of fissipeds, Radinsky (1982) showed that differences that distinguish modern families of carnivores were less pronounced in the Oligocene. The disparities in our results are surprising, in view of the close functional relationship between skull and dentition. However, the skull is undoubtedly subject to a wider variety of selection pressures than teeth since it serves many complex functions, such as housing the brain and sense organs, in addition to its role in feeding behavior. Consequently, teeth and skulls might be expected to evolve at different rates.

Given that the 31-million-year-old paleoguild is composed almost entirely of extinct families and orders of carnivores, it is remarkable that the array of feeding types is so similar to that of the modern guilds (Fig. 3). Like Malaysia, it includes six meat species (three over 30 kg), no meat/bone or non-vertebrate/meat species, and a small number of meat/non-vertebrate species. This suggests a rainforest environment with low ungulate diversity and density, but numerous folivores and frugivores. This is the environment first predicted by the locomotor adaptations of the Orellan predators, but contradicted by a lack of arboreal animals, the presence of a few open-habitat ungulates, and the paleobotanical data (Webb 1977; Retallack 1983; Van Valkenburgh 1985). A mixture of woodland, gallery forest, and some grassland seems most consistent with all the evidence.

In such an environment, terrestrial herbivore diversity should be lower than that of the Serengeti, since grasslands are less extensive. In addition, browsing, rather than grazing, ungulates should predominate. Bone/meat species might be present since carcass predictability should be higher than in a continuous forest, especially if the climate is dry or cool. The Orellan herbivore fauna is less diverse than the Serengeti (11 species versus 24), and browsing ungulates are more common than grazers (Gregory 1971; Webb 1977).

However, there are no bone/meat species, despite the fact that the Orellan appears to have been seasonally arid (Clark et al. 1967; Singler and Picard 1981; Retallack 1983).

It should be noted that there is an unusual group of Oligocene taxa, neither creodonts or carnivores, that may have been fulfilling the hyaena role in part: the entelodonts. The entelodonts were giant pig-like artiodactyls with relatively pointed premolars and large canines, two features which are reminiscent of carnivores (Matthew 1901; Scott and Jepsen 1941). As shown by the extant polar bear, meat-eating can be accomplished with teeth more typical of non-vertebrate/meat species (Van Valkenburgh 1988). A detailed examination of dental function and microwear in entelodonts might reveal if they were the hyaenas of the Orellan. If they were not, it appears that a potentially profitable mode of life remained unexploited within the Orellan community (see Stanley et al. 1983).

Despite its unique taxonomic composition, the Orellan paleoguild is remarkably similar in morphological diversity to the late Hemphillian and Recent guilds (Fig. 2). In part, this reflects the characters that were chosen to describe the morphology. For example, if I had used the shape of the lower second molar, then the creodonts, which have narrow, carnassial-shaped molars, would stand out as a unique morphology. However, the functional interpretation of the creodonts would not change; their diets would still be considered to have been 70–100 percent meat. So although the design and phylogeny of the Orellan animals differ markedly from their modern counterparts in some cases (e.g., creodonts, nimravid cats), the trophic structure within the paleoguild is similar to the Recent guilds. This tendency toward the evolution of convergent feeding adaptations in distantly related carnivores suggests that similarities in trophic structure among the guilds cannot be explained as a simple consequence of heritage constraints (i.e., all guilds look alike because they are composed of canids, felids, and hyaenids). Instead, the similarities in guild structure strongly suggest that guilds of large predators are shaped, at least loosely, by interspecific competition for food. As a result, predator guild structure ultimately reflects the resource base.

Summary and Conclusions

(1) Body weight and five dental indices are used to examine the diversity of dietary types within three Recent and two fossil guilds of large predators.

(2) Results show that guilds differ primarily in the number of member species and the relative representation of the four dietary types: meat, meat/bone, meat/non-vertebrate, and non-vertebrate/meat.

(3) The diversity of meat and meat/bone species within guilds is correlated with prey diversity in both ancient and modern guilds.

(4) Historical evidence suggests that the apparently depauperate Yellowstone and late Hemphillian guilds are the result of extinction without replacement.

(5) Predators have existed in the past with dental morphologies that are unique and possibly suboptimal relative to modern taxa, although these unique, extinct taxa coexisted with species of modern aspect.

(6) Despite the fact that the 31-million-year-old paleoguild is composed of members of extinct families and orders of predators, the array of dietary types is very similar to that of the Recent guilds, especially Malaysia.

(7) The basic pattern of adaptive divergence in dental morphology among sympatric carnivores was established at least 32 million years ago.

(8) The correlations between resource availability and dietary diversity in guilds, documented in the Recent and confirmed by the fossil communities, suggest that interspecific competition for food plays an important role in the evolution of predator guilds.

Acknowledgments

I am grateful to R. T. Bakker, A. R. Biknevicius, J. B. C. Jackson, S. M. Stanley, R. K. Wayne and S. D. Webb for comments on various drafts of the paper. I thank W. Dalquest, R. J. Emry, W. Harrison, R. M. Hunt, R. H. Tedford, W. Turnbull, and J. Wilson for access to the Oligocene and Miocene fossil material in their care. This paper represents work done in partial fulfillment of the Ph.D. degree at

The Johns Hopkins University. Funding was provided by the Geological Society of America, the American Association of University Women, and The Johns Hopkins University.

Literature Cited

BAKKER, R. T. 1983. The deer flees, the wolf pursues: incongruencies in predator-prey coevolution. Pp. 350–383. *In* Futuyma, D. J. and M. Slatkin (eds.), Coevolution. Sinauer Associates; Sunderland, Massachusetts.

BASKIN, J. A. 1980. The generic status of *Aeleurodon* and *Epicyon* (Carnivora, Canidae). Journal of Paleontology *54*:1349–1351.

BASKIN, J. A. 1981. *Barbourofelis* (Nimravidae) and *Nimravides* (Felidae) with a description of two new species from the late Miocene of Florida. Journal of Mammalogy *62*:122–139.

BASKIN, J. A. 1982. Tertiary Procyoninae (Mammalia:Carnivora) of North America. Journal of Vertebrate Paleontology *2*:71–93.

BEHRENSMEYER, A. K., D. WESTERN, AND D. D. BOAZ. 1979. New perspectives in vertebrate paleoecology from a recent bone assemblage. Paleobiology *5*:12–21.

BRAIN, C. K. 1981. The Hunters or the Hunted. University of Chicago Press; Chicago, Illinois. 365 pp.

BURT, W. H. 1931. *Machaerodus catocopsis* Cope from the Pliocene of Texas. University of California Publications in Geological Sciences *20*:261–293.

BUTTERWORTH, E. M. 1916. A new mustelid from the Thousand Creek Pliocene of Nevada. University of California Publications, Bulletin of the Department of the Geological Sciences *10*:21–24.

CLARK, J., J. R. BEERBOWER, AND K. K. KIETZKE. 1967. Oligocene sedimentation, stratigraphy, paleoecology and paleoclimatology in the Big Badlands of South Dakota. Fieldiana: Geological Memoirs *5*:1–158.

COE, M. J., D. H. CUMMINGS, AND J. PHILLIPSON. 1976. Biomass and production of large African herbivores in relation to rainfall and primary production. Oecologia *22*:341–354.

CRUSAFONT-PAIRO, M. AND J. TRUYOLS-SANTONJA. 1956. A biometric study of the evolution of fissiped carnivores. Evolution *10*:314–332.

CRUSAFONT-PAIRO, M. AND J. TRUYOLS-SANTONJA. 1957. Estudios masterometricos en la evolucion de los fissipedos. Boletin del Institute de Geologico y Minero España *68*:83–224.

DALQUEST, W. W. 1969. Pliocene carnivores of the Coffee Ranch (type Hemphill) local fauna. Bulletin of the Texas Memorial Museum *15*:1–44.

DAMUTH, J. 1982. Analysis of the preservation of community structure in assemblages of fossil mammals. Paleobiology *8*: 434–446.

EATON, R. L. 1979. Interference competition among carnivores: a model for the evolution of social behavior. Carnivore *2*:9–16.

EISENBERG, J. F. 1981. The Mammalian Radiations. University of Chicago Press; Chicago, Illinois. 610 pp.

EISENBERG, J. F. AND G. M. MCKAY. 1974. Comparison of ungulate adaptations in the New World and Old World tropical rainforests with special reference to Ceylon and the rainforests of Central America. Pp. 585–602. *In* Geist, V. and F. Walther (eds.), The Behavior of Ungulates and its Relation to Management. IUCN Publications, new series *24*; Morges, Switzerland.

EISENBERG, J. F. AND J. SEIDENSTICKER. 1976. Ungulates in Southern Asia: a consideration of biomass estimates for selected habitats. Biological Conservation *10*:293–308.

EWER, R. F. 1973. The Carnivores. Cornell University Press; Ithaca, New York. 494 pp.

FINDLEY, J. S. 1976. The structure of bat communities. American Naturalist *110*:129–139.

GITTLEMAN, J. L. AND P. H. HARVEY. 1982. Carnivore home-range size, metabolic needs and ecology. Behavior, Ecology and Sociobiology *10*:57–63.

GITTLEMAN, J. L. 1986. Carnivore life history patterns: allometric, phylogenetic and ecological associations. American Naturalist *127*:744–771.

GRAHAM, R. W. AND E. L. LUNDELIUS, JR. 1984. Coevolutionary disequilibrium and Pleistocene extinctions. Pp. 223–249. *In* Martin, P. S. and R. G. Klein (eds.), Quaternary Extinctions. University of Arizona Press; Tucson, Arizona.

GRANT, P. AND D. SCHLUTER. 1984. Interspecific competition inferred from patterns of guild structure. Pp. 201–233. *In* Strong, D. R., Jr., D. Simberloff, L. G. Abele, and A. B. Thistle (eds.), Ecological Communities: Conceptual Issues and the Evidence. Princeton University Press; Princeton, New Jersey.

GREGORY, J. T. 1971. Speculations on the significance of fossil vertebrates for the antiquity of the Great Plains of North America. Abhandlungen der Hessischen Landesamtes für Bodenforschung *60*:64–72.

GUILDAY, J. E. 1984. Pleistocene extinction and environmental change: case study of the Appalachians. Pp. 250–258. *In* Martin, P. S. and R. G. Klein (eds.), Quaternary Extinctions. University of Arizona Press; Tucson, Arizona.

HAIRSTON, N. G., F. E. SMITH, AND L. B. SLOBODKIN. 1960. Community structure, population control and competition. American Naturalist *94*:421–425.

HALL, E. R. 1946. A new genus of American Pliocene badger with remarks on the relationships of badgers of the Northern Hemisphere. Carnegie Institution of Washington Publications, Contributions to Paleontology *551*:9–23.

HARRISON, J. A. 1981. A review of the extinct wolverine *Plesiogulo* (Carnivora: Mustelidae) from North America. Smithsonian Contributions to Paleobiology *46*:1–27.

HARRISON, J. A. 1983. The Carnivora of the Edson local fauna (late Hemphillian), Kansas. Smithsonian Contributions to Paleobiology *54*:1–42.

HATCHER, J. B. 1895. Discovery in the Oligocene of South Dakota of *Eusmilus*, a genus of sabre-toothed cats new to North America. American Naturalist *29*:1091–1093.

HENDEY, Q. B. 1974a. Faunal dating of the late Cenozoic of Southern Africa, with special reference to the Carnivora. Quaternary Research *4*:149–161.

HENDEY, Q. B. 1974b. New fossil carnivores from the Swartkrans Australopithecine site (Mammalia: Carnivora). Annals of the Transvaal Museum *29*:27–47.

HENDEY, Q. B. 1974c. The late Cenozoic Carnivora of the southwestern Cape province. Annals of the South African Museum *63*:1–369.

HIBBARD, C. W. 1934. Two new genera of Felidae from the middle Pliocene of Kansas. Transactions of the Kansas Academy of Sciences *37*:239–255.

HOUGH, J. 1948. A systematic revision of *Daphoenus* and some allied genera. Journal of Paleontology *22*:573–600.

HOUGH, J. 1949. The subspecies of *Hoplophoneus*. Journal of Paleontology *23*:536–555.

HUNT, R. M. 1974. *Daphoenictis*, a cat-like carnivore (Mammalia, Amphicyonidae) from the Oligocene of North America. Journal of Paleontology *48*:1030–1047.

JASLOW, C. R. 1987. Morphology and digestive efficiency of red foxes (*Vulpes vulpes*) and grey foxes (*Urocyon cinereoargenteus*) in relation to diet. Canadian Journal of Zoology *65*:72–79.

JEPSEN, G. L. 1933. American eusmiloid sabre tooth cats of the Oligocene epoch. Proceedings of the American Philosophical Society *72*:355–369.

KARR, J. G. AND F. C. JAMES. 1975. Ecomorphological configurations and convergent evolution. Pp. 258–291. *In* Cody, M. L. and J. M. Diamond (eds.), Ecology and Evolution of Communities. Belknap Press; Cambridge, Massachusetts.

KAY, R. F. 1975. The functional adaptations of primate molar teeth. American Journal of Physical Anthropology *43*:195–216.

KITTS, D. B. 1958. *Nimravides*, a new genus of Felidae from the Pliocene of California, Texas and Oklahoma. Journal of Mammalogy *39*:368–375.

KURTÉN, B. AND E. ANDERSON. 1980. Pleistocene Mammals of North America. Columbia University Press; New York, New York. 442 pp.

LAMPRECHT, J. 1981. The function of social hunting in larger terrestrial carnivores. Mammal Reviews *11*:169–179.

LEOPOLD, E. B. 1969. Late Cenozoic palynology. Pp. 377–438. *In* Tschudy, R. H. and R. A. Scott (eds.), Aspects of Palynology. Wiley-Interscience; New York, New York.

MARTIN, L. D. AND C. B. SCHULTZ. 1975. Scimitar-toothed cats, *Machairodus* and *Nimravides*, from the Pliocene of Kansas and Nebraska. Bulletin of the University of Nebraska State Museum *10*:55–63.

MARTIN, P. S. 1984. Prehistoric overkill: the global model. Pp. 354–403. *In* Martin, P. S. and R. G. Klein (eds.), Quaternary Extinctions. University of Arizona; Tucson, Arizona.

MATTHEW, W. D. 1901. Tertiary mammals of northeastern Colorado. Memoirs of the American Museum of Natural History *1*:353–447.

MATTHEW, W. D. AND R. A. STIRTON. 1930. Osteology and affinities of *Borophagus*. University of California Publications in Geological Sciences *19*:171–217.

McDONALD, J. N. 1984. The reordered North American selection regime and late Quaternary megafaunal extinctions. Pp. 404–439. *In* Martin, P. S. and R. G. Klein (eds.), Quaternary Extinctions. University of Arizona; Tucson, Arizona.

McNAB, B. K. 1971. On the ecological significance of Bergmann's rule. Ecology *52*:845–854.

MELLETT, J. S. 1977. Paleobiology of North American *Hyaenodon* (Mammalia: Creodonta). Contributions to Vertebrate Evolution *1*:1–134. S. Karger; New York, New York.

MUNTHE, L. K. 1979. The skeleton of the Borophaginae: morphology and function. Unpublished Ph.D. thesis, University of California, Berkeley. 226 pp.

OLSON, S. L. AND H. F. JAMES. 1984. The role of Polynesians in the extinction of the avifauna of the Hawaiian Islands. Pp. 768–784. *In* Martin, P. S. and R. G. Klein (eds.), Quaternary Extinctions. University of Arizona; Tucson, Arizona.

PETTER, G. 1969. Interpretation evolutive caractères de la denture des viverrides Africains. Mammalia *33*:607–625.

RADINSKY, L. B. 1982. Evolution of skull shape in carnivores. 3. The origin and early radiation of the modern carnivore families. Paleobiology *8*:177–195.

RETALLACK, G. 1983. Late Eocene and Oligocene paleosols from Badlands National Park, South Dakota. Geological Society of America Special Papers *193*:1–82.

RIGGS, E. S. 1896. A new species of *Dinictis* from the White River Miocene of Wyoming. Kansas University Quarterly *4*:237–241.

ROSENZWEIG, M. L. 1966. Community structure in sympatric Carnivora. Journal of Mammalogy *47*:602–612.

ROSENZWEIG, M. L. 1968. The strategy of body size in mammalian carnivores. American Midland Naturalist *80*:299–315.

SAVAGE, D. E. 1941. Two new middle Pliocene carnivores from Oklahoma with notes on the Optima fauna. American Midland Naturalist *25*:692–710.

SAVAGE, D. E. AND D. E. RUSSELL. 1983. Mammalian Paleofaunas of the World. Addison-Wesley; Reading, Massachusetts. 432 pp.

SAVAGE, R. J. G. 1978. Carnivora. Pp. 249–267. *In* Maglio, V. J. and H. B. S. Cooke (eds.), Evolution of African Mammals. Harvard University Press; Cambridge, Massachusetts.

SCHOENER, T. W. 1974. Resource partitioning in ecological communities. Science *185*:27–39.

SCHULTZ, C. B. AND L. D. MARTIN. 1975. Bears (Ursidae) from the late Cenozoic of Nebraska. Bulletin of the University of Nebraska State Museum *10*:47–54.

SCHULTZ, G. B. 1977. The Ogallala formation and its vertebrate faunas in the Texas and Oklahoma panhandles. Pp. 5–105. *In* Schultz, G. E. (ed.), Guidebook, Field Conference on Late Cenozoic Biostratigraphy of the Texas Panhandle and Adjacent Oklahoma, August 4-6, 1977. Kilgore Research Center Special Publication *1*. West Texas State University; Canyon, Texas.

SCOTT, W. B. 1889. Notes on the osteology and systematic position of *Dinictis felina*, Leidy. Proceedings of the Academy of Natural Sciences in Philadelphia *41*:211–245.

SCOTT, W. B. 1913. A History of Land Mammals in the Western Hemisphere. Macmillan Press; New York. 786 pp.

SCOTT, W. B. AND G. JEPSEN. 1941. The mammalian fauna of the White River Oligocene. Transactions of the American Philosophical Society *28*:747–980.

SIMPSON, G. G. 1941. The species of *Hoplophoneus*. American Museum Novitates *1123*:1–21.

SINGLER, C. R. AND M. D. PICARD. 1981. Paleosols in the Oligocene of Northwest Nebraska. University of Wyoming Contributions to Geology *20*:57–68.

SNEATH, P. H. A. AND R. R. SOKAL. 1973. Numerical Taxomony. W.H. Freeman; San Francisco, California. 573 pp.

STANLEY, S. M., B. VAN VALKENBURGH, AND R. S. STENECK. 1983. Coevolution and the fossil record. Pp. 328–349. *In* Futuyma, D. J., and M. L. Slatkin (eds.), Coevolution: Sinauer Associates; Sunderland, Massachusetts.

SWANSON, F. J. 1987. Ecological effects of the eruption of Mount St. Helens. Pp. 1–2. *In* Bilderback, D. E. (ed.), Mount St. Helens 1980: Botanical Consequences of the Explosive Eruptions. University of California Press; Berkeley, California.

TEDFORD, R. H., T. GALUSHA, M. F. SKINNER, B. E. TAYLOR, R. W. FIELDS, J. R. MACDONALD, J. M. RENSBERGER, S. D. WEBB, AND D. P. WHISTLER. 1988. Faunal succession and biochronology of the Arikareean through Hemphillian interval (Late Oligocene through earliest Pliocene epochs), North America. Pp. 153–210. *In* Woodburne, M. O. (ed.), Cenozoic mammals of North America. University of California Press; Berkeley, California.

VAN VALEN, L. 1974. Multivariate structural statistics in natural history. Journal of Theoretical Biology *45*:235–247.

VAN VALKENBURGH, B. 1985. Locomotor diversity within past and present guilds of large predatory mammals. Paleobiology *11*:406–428.

VAN VALKENBURGH, B. 1988. Carnivore dental adaptations and diet: a study of trophic diversity within guilds. In press *In* Gittleman, J. L. (ed.), Carnivore Behavior, Ecology and Evolution. Cornell University Press; Ithaca, New York.

VAN VALKENBURGH, B. AND C. B. RUFF. 1987. Canine tooth strength and killing behaviour in large carnivores. Journal of Zoology *212*:1–19.

VERSTAPPEN, H. T. 1975. On palaeoclimates and landform development in Malesia. Pp. 3–36. *In* Bartstra, G. and W. A. Casparie (eds.), Modern Quaternary Research in Southeast Asia. A. A. Balkema; Rotterdam, Netherlands.

VOORHIES, M. R. 1970. Sampling difficulties in reconstructing late Tertiary mammalian communities. Proceedings of the North American Paleontology Convention *1*:454–468.

WALKER, A. 1984. Extinction in hominid evolution. Pp. 119–152. *In* Nitecki, M. H. (ed.), Extinction. University of Chicago Press; Chicago, Illinois.

WEBB, S. D. 1977. A history of savannah vertebrates in the New World. Part 1: North America. Annual Review of Ecology and Systematics *8*:355–380.

WEBB, S. D. 1978. A history of savannah vertebrates in the New World. Part. 2: South America and the Great Interchange. Annual Review of Ecology and Systematics *9*:393–426.

WEBB, S. D. 1983. The rise and fall of the late Miocene ungulate fauna in North America. Pp. 267–306. *In* Nitecki, M. H. (ed.), Coevolution. University of Chicago Press; Chicago, Illinois.

WEBB, S. D. 1984. Ten million years of mammal extinction in North America. Pp. 189–210. *In* Martin, P. S. and R. G. Klein (eds.), Quaternary Extinctions. University of Arizona Press; Tucson, Arizona.

WOLFE, J. A. 1978. A paleobotanical interpretation of Tertiary climates in the northern hemisphere. American Scientist *66*: 694–703.

WOLFE, J. A. 1985. Distribution of major vegetational types in the Tertiary. Geophysical Monographs *32*:357–375.

WOODLEY, J. D., E. A. CHORNESKY, P. A. CLIFFORD, J. B. C. JACKSON, L. S. KAUFMAN, N. B. KNOWLTON, J. C. LANG, M. P. PEARSON, J. W. PORTER, M. C. ROONEY, K. W. RYLAARSDAM, V. J. TUNNICLIFFE, C. M. WAHLE, J. L. WULFF, A. S. G. CURTIS, M. D. DALLMEYER, B. P. JUPP, M. A. R. KOEHL, J. NEIGEL, AND E. M. SIDES. 1981. Hurricane Allen's impact on Jamaican coral reefs. Science *214*:749–755.

ZAR, J. H. 1984. Biostatistical Analysis, Second Edition. Prentice-Hall; Englewood Cliffs, New Jersey. 718 pp.

Appendix 1: Fossil Specimens

Institution abbreviations: AM, American Museum of Natural History, New York; FAM, Frick Collection, The American Museum of Natural History; FM, Field Museum, Chicago; KU, University of Kansas, Lawrence; MSU, Midwestern State University; PU, Princeton University Museum, Princeton; TMM, Texas Memorial Museum, Austin; UC, University of California, Berkeley; UNSM, University of Nebraska State Museum, Lincoln; USNM, United States National Museum, Washington, D.C.; WT, West Texas State College, Panhandle Plains Museum, Canyon.

Creodonta

Hyaenodon horridus. FAM 7567, 75279, 75692, 75622, 75732; Mellett (1977).
H. crucians. FAM 75609, 75657, 75658, 75596, 75571, 75596; AM 647, 648, 1372; Mellett (1977).

Canidae

Mesocyon sp. FAM 63382, 63367, 63386, 63379, 27557, 54110.
Canis davisii. Harrison (1981).
Osteoborus cyonoides. FM UM310, 681, 682, 27161, 27162, 23626; WT 2175; and 37 specimens at MSU.
Vulpes stenognathus. Savage (1941).

Mustelidae

Plesiogulo marshalli. Harrison (1981).
Pliotaxidea nevadensis. MU 1334; Butterworth (1916); Hall (1946).

Procyonidae

Arctonasua fricki. Baskin (1982).

Ursidae

Agriotherium schneideri. WT 2137; MSU 41261-1, 41261-2; UNSM 76011; FM P27163; Savage (1941); Harrison (1983); Schultz and Martin (1975).

Amphicyonidae

Daphoenus vetus. KU 138, 8205, 8207, 9870, 2562, 5002; PU 11423, 13600, 13580, 13584, 63339, 12635; UNSM 26151, 25785; AM 38812, 39111, 63341, 12451, 11857, 39098, 9759; and 24 specimens from the Frick Collection.
D. hartshornianus. PU 10546; KU 165, 122, 8205; AM 6811; Scott and Jepsen (1941)

Nimravidae

Hoplophoneus occidentalis. AM 1407; FAM 62025, 62015.
H. primaevus. AM 38980, 38982, 38927, 62014, 62007, 69370, 38981, 62021, 62103, 62017, 62012; USNM 18191, 18189; Scott and Jepsen (1941).
Dinictis felina. FM 62031, 62032, 62051, 62040, 62053, 62058, 62030, 69375, 62055, 62049, 62045, 62052, 62035; AM 1393, 38805, SD230-4152, SD44-928, SD5-146; Riggs (1896); Scott (1889); Scott and Jepsen (1941).
Eusmilus dakotensis. USNM 12820; Jepsen (1933); Hatcher (1895).

Felidae

Pseudaeleurus hibbardi. MSU 11483; TMM 41261-3, 41261-5, 41261-6; UC 50664
Adelphailurus kansensis. Hibbard (1934); Harrison (1983).
Machairodus coloradensis. MSU 3527; TMM 41261-8, 41261-9, 41261-11, 5658; Dalquest (1969); Schultz and Martin (1975); Burt (1931).
Nimravides catacopsis. AM HIG227-1790, KAN41-7-271, 104044, HIG35-558; Kitts (1958).
cf. *Felis proterolyncus*. Savage (1941).

Appendix 2: Morphometric Indices of Dietary Type

Upper canine shape (CS).—The cross-sectional shape of the upper canine tooth was estimated as the ratio of its mediolateral width to its anteroposterior length at the dentine-enamel junction (CMW over CAL, Fig. 1). Canine shape appears to reflect prey-killing behavior as much or more than it reflects diet. Modern felids have relatively round canines and kill prey with a single strong bite By contrast, canids use their narrow canines to produce more shallow, slashing wounds. Most sabertooth cats had canines shaped more like living canids than felids, and thus their killing behavior probably differed from that of living felids (see Van Valkenburgh and Ruff [1987] for details).

Premolar shape and size (PMD, RPS).—The shape of the largest lower premolar (the fourth in all sampled species except the hyaenas where the third is largest) was measured as the ratio of maximum mediolateral width to maximum anteroposterior length (PMW over PML, Fig. 1). To gauge relative premolar size (RPS), the maximum width of the largest lower premolar (PMW, Fig. 1) was divided by the cube root of body weight. The premolars of meat/bone species tend to be relatively larger and more round in cross-section than those of species in other dietary groups (Appendix 3).

Relative blade length (RBL).—The relative proportion of the first lower molar devoted to slicing as opposed to grinding is estimated by the ratio of the anteroposterior length of the trigonid measured along the buccal margin (BL) divided by maximum M1 length (M1L, Fig. 1). This ratio proved to be a good indicator of the relative proportion of meat in the diet: both meat and meat/bone species have relatively longer blades than species within the two omnivorous groups (Appendix 3).

Relative grinding area (RGA).—The relative proportion of the molar area devoted to grinding as opposed to slicing is estimated by dividing the square root of the total grinding area of the molars (TGA, Fig. 1) by the total blade length of the carnassial (BL, Fig. 1). The entire occlusal area of the lower second and third (if present) molars, as well as that of the talonid of the lower M1 was measured with a polar planimeter. The area estimates were made from color transparencies of the lower molars, taken with the occlusal surface parallel to the plane of focus of the camera. This estimate of grinding area differs from that of Kay (1975), who measured individual wear facet areas, but is suitable for carnivores because they tend to wear the entire occlusal surface as a flat plane. Relative grinding area is greater among species in the two omnivorous groups as opposed to those in the meat and meat/bone groups (Appendix 3).

Appendix 3

Mean value ($\bar{x}$) and standard deviation (sd) of each morphometric variable for the four diet groups. The diet categories are defined in the text. Species included in each group are listed below the table. For references to behavioral data used to classify each species, see Van Valkenburgh (1988). LBW, log body weight; CS, canine shape; RPS, relative premolar size; PMD, premolar shape; RBL, relative blade length; RGA, relative grinding area. A superscript indicates that the mean is significantly different at the 0.05 level or better (Student's *t*, two-tailed test) from that of another group: 1, significantly different from the meat group; 2, meat/bone; 3, meat/non-vertebrate; 4, non-vertebrate/meat.

Group		LBW	CS	RPS	PMD	RBL	RGA
(1) Meat	$\bar{x}$	1.39	$73.9^{3,4}$	$2.14^{2,4}$	0.48^{2}	$0.94^{3,4}$	$0.07^{3,4}$
(*N* = 20)	SD	0.403	7.98	0.216	0.032	0.115	0.15
(2) Meat/Bone	$\bar{x}$	1.60	71.1	$3.79^{1,3,4}$	$0.65^{1,3}$	$0.85^{3,4}$	$0.0^{3,4}$
(*N* = 3)	SD	0 127	0.354	0 391	0.062	0.067	0.0
(3) Meat/Non-vert.	$\bar{x}$	1.09^{4}	69.6^{1}	$2.18^{2,4}$	$0.53^{2,4}$	$0.61^{1,2}$	$0.48^{1,2,4}$
(*N* = 17)	SD	0.409	10.3	0.430	0.076	0.080	0.237
(4) Non-vert./Meat	$\bar{x}$	1.65^{3}	66.2^{1}	$1.57^{1,2,3}$	0.60^{3}	$0.55^{1,2}$	$0.87^{1,2,3}$
(*N* = 7)	SD	0.666	4.32	0.540	0.071	0.064	0.247

Meat group: *Lynx rufus, L. canadensis, Felis yagouarundi, F. aurata, F. temmincki, F. viverrina, F. serval, Neofelis nebulosa, Caracal caracal, Uncia uncia, Puma concolor, Acinonyx jubatus, Panthera onca, P. pardus, P. leo, P. tigris, Canis lupus, Cuon alpinus, Lycaon pictus, Speothos venaticus.*

Meat/bone group: *Crocuta crocuta, Hyaena hyaena, H. brunnea.*

Meat/non-vertebrate group: *Viverra megaspila, Viverra zibetha, Civettictis civetta, Dusicyon culpaeus, Cerdocyon thous, Chrysocyon brachyurus, Canis aureus, C. adustus, C. mesomelas, C. latrans, Vulpes vulpes, Mellivora capensis, Gulo gulo, Taxidea taxus, Meles meles, Nasua nasua, Ursus maritimus.*

Non-vertebrate/meat group: *Arctictis binturong, Procyon lotor, Nyctereutes procyonoides, Selenarctos thibetanus, Tremarctos ornatus, Ursus americanus, Ursus arctos.*

PAPER 8

Fossil Charcoal: A Plant-Fossil Record Preserved by Fire (1991)

A. C. Scott and T. P. Jones

Commentary

CLAIRE M. BELCHER

Fire is an underappreciated process within the Earth system, and yet it is the most ubiquitous natural terrestrial disturbance. Fire consumes huge quantities of biomass in all ecosystems, ranging across all biomes from tundra to savanna and from boreal to tropical forests. Forty percent of our modern world is covered by fire-prone ecosystems, and fire promotes the expansion of flammable ecosystems in parts of the world that would otherwise be vegetated according to the physiognomic limits set by climate (Bond et al. 2005). Products of fire include chars, ashes, soots, and aromatic hydrocarbon species, which can be readily traced in modern sediments through to ancient ones. The recognition of these has enabled the interpretation of variations in, and the influence of, past fire activity throughout Earth history.

Identifying fire in the fossil record has been a subject of great scientific debate. Fusain was first formally described by Stopes in 1919 as one of the fundamental constituents of coal. Following this description, numerous authors believed fusain to be the result of wildfire (Harris 1958; Scott and Collinson 1978; Cope and Chaloner 1980), whereas others argued that it had a biochemical origin (Schopf 1975). Scott (1989) revealed that the physical and chemical characteristics of fusain and modern charcoal indicated that they share a similar origin. This earlier work by Scott also highlighted a wealth of evidence that supported the interpretation that fusain represents fossil charcoal, which enabled the idea to become accepted within the scientific community.

Scott and Jones reinforced this pioneering work. With the publication of their paper, the idea of studying the role of fire in ancient ecosystems was placed firmly in the minds of the paleoecology community. They highlighted the acceptance of fusain as fossil charcoal, provided renewed and vital creditability to previous suggestions that the record of fossil charcoal could be a useful indicator of ancient environments (cf. Cope and Chaloner 1980), and paved the way for studies of fire's role in Earth history. In this work, Scott and Jones draw attention to the palaeobotanical, palaeoenvironmental, and geological significance of studying fossil charcoal as a proxy for ancient wildfires. They highlight the excellent anatomical detail of preservation that charcoal provides, which as they note, has allowed the discovery of several important plant taxa. This is emphasized by the charcoalifed three-dimensional preservation of early angiosperm flowers from the Upper Cretaceous of Sweden (Friis and Skarby 1981), which has led to increased understanding of the evolution of the most diverse group of plants on our planet today.

The early synthesis of fire-related palaeoecological ideas by Scott and Jones was greatly expanded in Scott's 2000 paper "The Pre-Quaternary History of Fire," in which the impact of this earlier work can already be seen on the field of paleoecology. Ancient fire studies are now reaching to the forefront of palaeoecological and paleoenvironmental research and have, in the last decade, been used to assess a diversity of topics (cf. Bowman et al. 2009 for a re-

From *Geology Today* 7:214–16.

view), including the extent of wildfires ignited by the end Cretaceous extraterrestrial impact (Wolbach et al. 1985; Jones and Lim 2000; Scott et al. 2000; Belcher et al. 2003, 2005, 2009); the abundance of oxygen in the paleoatmosphere (Watson et al. 1978; Chaloner 1989; Scott and Glasspool 2006; Belcher and McElwain 2008; Glasspool and Scott 2010; Belcher et al. 2010b); and variations in fire activity in response to both pre-Quaternary and Quaternary climatic changes (Higuera et al. 2008; Marlon et al. 2009; Belcher et al. 2010a). It is without doubt that the fossil record of charcoal, as highlighted by Scott and Jones, will continue to provide a unique window into Earth's flammable past.

Literature Cited

Belcher, C. M., M. E. Collinson, A. R. Sweet, A. R. Hildebrand, and A. C. Scott. 2003. Fireball passes and nothing burns—the role of thermal radiation in the Cretaceous-Tertiary event: evidence from the charcoal record of North America. *Geology* 31:1061–64.

Belcher, C. M., M. E. Collinson, and Scott, A. C. 2005. Constraints on the Thermal Power Released from the Chicxulub Impactor: New Evidence from Multi-Method Charcoal Analysis. Journal of the Geological Society of London, 162; 591–602.

Belcher, C. M., and J. C. McElwain. 2008. Limits on combustion in low O_2 redefine palaeoatmospheric levels for the Mesozoic. *Science* 321:1197–1200.

Belcher, C. M., P. Finch, P., M. E. Collinson, A. C. Scott, and N. V. Grassineau. 2009. Geochemical evidence for combustion of hydrocarbons during the K-T impact event. *Proceedings of the National Academy of Sciences* 106:4112–17.

Belcher, C. M., L. Mander, G. Rein, F. X. Jervis, M. Haworth, I. J. Glasspool, S. P. Hesselbo, and J. C. McElwain, 2010a. Increased fire activity at the Triassic/Jurassic boundary in Greenland due to climate-driven floral change. *Nature Geoscience* 3:426–29.

Belcher, C. M., J. M. Yearsley, R. M. Hadden, J. C. McElwain, and G. Rein. 2010b. Baseline intrinsic flammability of Earth's ecosystems estimated from paleoatmospheric oxygen over the past 350 million years. *Proceedings of the National Academy of Sciences* 107:22448–53.

Bond, W. J., F. I. Woodward, and G. F. Midgley. 2005. The global distribution of ecosystems in a world without fire. *New Phytologist* 165:525–38.

Bowman, D. M. J. S., J. K. Balch, P. Artaxo, W. J. Bond, J. M. Carlson, M. A. Cochrane, C. M. D'Antonio, et al. (2009). Fire in the Earth system. *Science* 324:481–84.

Chaloner, W. G. 1989. Fossil charcoal as an indicator of palaeoatmospheric oxygen level. *Journal of the Geological Society London* 146:171–74.

Cope, M. J., and W. G. Chaloner. 1980. Fossil charcoal as evidence of past atmospheric composition. *Nature* 283:647–49.

Friis, E. M., and A. Skarby. 1981. Structurally preserved angiosperm flowers from the Upper Cretaceous of southern Sweden. *Nature* 291:485–86.

Glasspool, I. J., and A. C. Scott. 2010. Phanerozoic concentrations of atmospheric oxygen reconstructed from sedimentary charcoal. *Nature Geoscience* 3:627–30.

Harris, T. M. 1958. Forest fire in the Mesozoic. *Journal of Ecology* 46:447–53.

Higuera, P. E., L. B. Brubaker, P. M. Anderson, T. A. Brown, A. Kennedy, and F. S. Hu. 2008. Fires in ancient shrub tundra: implications of paleo-records for Arctic environmental change. *PLoS One* 3:e0001744. doi:10.1371/journal.pone.0001744.

Jones, T. P., and B. Lim. 2000. Extraterrestrial impacts and fire. *Palaeogeography, Palaeoclimatology, Palaeoecology* 164:57–66.

Marlon, J. R. and 22 others (2009). Wildfire responses to abrupt climate change in North America. *Proceedings of the National Academy of Sciences USA* 106: 2519–24.

Schopf, J. M. 1975. Modes of fossil preservation. *Review of Palaeobotany and Palynology* 20:27–53.

Scott, A. C. 1989. Observations on the nature and origin of fusain. *International Journal of Coal Geology* 12, 443–75.

———. 2000. The Pre-Quaternary history of fire. *Palaeogeography, Palaeoclimatology, Palaeoecology* 164: 281–329.

Scott, A. C., and M. E. Collinson. 1978. Organic sedimentary particles: results from scanning electron microscope studies of fragmentary plant material. In *Scanning Electron Microscopy in the Studies of Sediments,* edited by W. B. Whalley, 137–67. Norwich, UK: GeoAbstracts.

Scott, A. C., and I. J. Glasspool. 2006. The diversification of Palaeozoic fire systems and fluctuations in atmospheric oxygen. *Proceedings of the National Academy of Sciences* 103:10861–65.

Scott, A. C., B. H. Lomax, M. E. Collinson, G. R. Upchurch, and D. J. Beerling. 2000. Fire across the K-T boundary: initial results from the Sugarite Coal, New

Mexico, USA. *Palaeogeography, Palaeoclimatology, Palaeoecology* 164:381–95.

Stopes, M. C. 1919. On the four visible ingredients in banded bituminous coal: studies in the composition of coal, No. 1. *Proceedings of the Royal Society London B* 90:480–87.

Watson, A. J., J. E. Lovelock, and L. Marguilis. 1978. Methanogenesis, fires and the regulation of atmospheric oxygen. *Biosystems* 10:293–98.

Wolbach, W. S., R. S. Lewis, and E. Anders. 1985. Cretaceous extinctions: evidence for wildfires and search for meteoritic material. *Science* 230:167–70.

Fossil charcoal: a plant-fossil record preserved by fire

ANDREW C. SCOTT & TIMOTHY P. JONES

Small pieces of black organic material are common in many post-Devonian sequences, both sedimentary and volcanic. While many of these are coalified plant fragments, others are fossil charcoal, also known as fusain. Charcoalification preserves exquisite detail of the plant ultra-structure and is best viewed by scanning electron microscopy.

Lithotype: a macroscopic constituent of coal.

Maceral: individual microscopic constituents of coal.

Pyrolysis: chemical decomposition at high temperature.

Reflectance: the ratio of the light reflected by a material to that incident upon it.

Fulgurite: a tube of fused silica or silicates, formed by lightning strikes on sandy soil.

Middle lamella: ultra-structural feature seen in the middle of plant cell walls.

Fusain, or 'Mother of Charcoal', has been recognized since the early part of the last century. It is a common lithotype of coal, was formally described by Marie Stopes in 1919, and is composed mainly of the macerals fusinite and semifusinite. Fusain usually occurs as wedges or patches of black, fibrous, opaque material showing cellular structure, with the cell lumina being generally open. The majority of fusain fragments are charcoalified secondary wood, but many other plant organs are also preserved in this manner, including leaves, seeds and even flowers. It is the only lithotype of coal that gets your hands dirty. Fusain is almost pure carbon and is chemically highly inert. It therefore has superb fossilization potential. With its potential to provide new information on plants, as well as giving important geological information, it is now attracting increasing interest from geologists, palaeobotanists and palaeoecologists.

Fusain and charcoal

Charcoal is produced by the pyrolysis of plant material. The burning of plants by wildfire causes the complete combustion of a portion of the plant material, but some tissue can be starved of oxygen while still being subjected to sufficient heat for pyrolysis to occur, thus giving rise to charcoal. During pyrolysis, the cell walls become apparently homogenized, usually at around 250°C, depending upon the heating time. The plant tissue is converted to almost pure carbon, increasing in reflectance and becoming chemically highly inert. The resulting material has a high fossilization potential. It remains three-dimensional, except after physical crushing, when cell walls might shatter. Fusain has a very characteristic shape and size range, usually cuboid and up to 1 cm in size (Fig. 1). When embedded and polished, both modern charcoal and fusain have high reflectances, of 2–5%. Physical and chemical characteristics of fusain and modern charcoal indicate that they have a similar origin. The abundance of fusain points to the existence of extensive wildfires in the geological past.

There are several types of natural wildfire, most of which are started by lightning strikes. We even have a fossil record of this lightning in the occurrence of fulgurites. Fires may burn as crown fires in the tops of trees, ground fires with the burning of undergrowth, or litter fires, where dead plant litter is burned (Fig. 2). Even peat layers may burn during dry periods. It has been shown that the extent and intensity of the fires will depend not only on the vegetation but also on whether there have been long periods between fires. For example, the Yellowstone National Park (USA) fires of 1988 are thought to have been of such great intensity because of the park's earlier fire-suppression policy, allowing build-up of potential fuel.

Ground temperatures during wildfires may reach up to 900°C and, even several centimetres under the soil, temperatures may be up to 300°C. We have done experiments which show the physical and chemical changes that take place during pyrolysis. With our experimental conditions of heating times and burial depths (to restrict oxygen supply), temperatures around 230°C appear to be critical. Below this temperature the plant is charred and the middle lamella of the plant-cell walls remains visible. Above this temperature the plant-cell walls become apparently homogenized, and true charcoal is produced. With increasing charcoalification temperatures, reflectance increases proportionally, over the range 2–5%, and there are also significant chemical changes. The importance of the apparent homogenization of the cell walls for

Fig. 1. Fossil charcoal (fusain) in sandstone from the Middle Jurassic, Scalby Ness Formation, Scarborough, UK.

the identification of charcoal is self-evident. This is illustrated in the scanning electron micrographs shown in Fig. 3, which compare modern experimental charcoal, Jurassic fossil charcoal and Tertiary fossil wood.

Fusain occurs mostly in sedimentary rocks, but it is also found in volcanic rocks. It can be dissolved out of the rock using hydrochloric and hydrofluoric acids, subsequently picked under a binocular microscope, and mounted on a stub for examination under the scanning electron microscope (Fig. 4). It is also possible to impregnate fusain with different resins and prepare sections for reflectance microscopy, light microscopy and even transmission electron microscopy.

Palaeobotanical significance

Recent studies of fusain, from the Carboniferous and Cretaceous in particular, have led to the discovery of several important new plants. The earliest fossil conifer from the Upper Carboniferous of Yorkshire is preserved in this manner. In addition, a large number of early angiosperm flowers have also been found preserved as charcoal, from the Upper Cretaceous of Sweden (Fig. 5). Increasingly, it is being realized that many small plant parts may be preserved as fossil charcoal and that these can reveal excellent anatomical details of great importance to palaeobotanists (Fig. 6). We confidently believe that future investigations of fossil charcoal will yield many more new plants.

Palaeoenvironmental significance

The regular occurrence of fusain in the fossil record has wide-ranging palaeoenvironmental implications. Studies of the conditions under which charcoal is produced indicate that there are atmospheric constraints on its formation. Most importantly, the oxygen content of the atmosphere must reach in excess of 13% to be able to sustain any wildfires. Should the oxygen exceed 35%, wildfires would be so uncontrollable that sustained plant growth would become impossible. This hypothesis can be graphically demonstrated by dropping smouldering twigs into containers with reduced and elevated oxygen levels. In the former, any indications of combustion are instantly extinguished. In the latter, even damp twigs will flare up instantly and completely combust. Therefore, the continuous record of fossil charcoal from the Late Devonian to Holocene places constraints of 13% and 35% on atmospheric oxygen levels over this period.

The occurrence of a 'soot' layer at the Cretaceous/Tertiary boundary has led some authors to postulate a worldwide forest fire following an asteroid impact. Although the soot layer is an important facet of these bolide impact theories, it should be noted that fusain, as evidence of fires, is abundant in many marine sediments, often with particles under 10 μm. Conversely, supporters of a volcanic cause for the Cretaceous/Tertiary extinctions will point out that volcanic activity is a good way to start fires.

Fig. 2. The three main type of wildfire (from Scott, 1989).

(a)

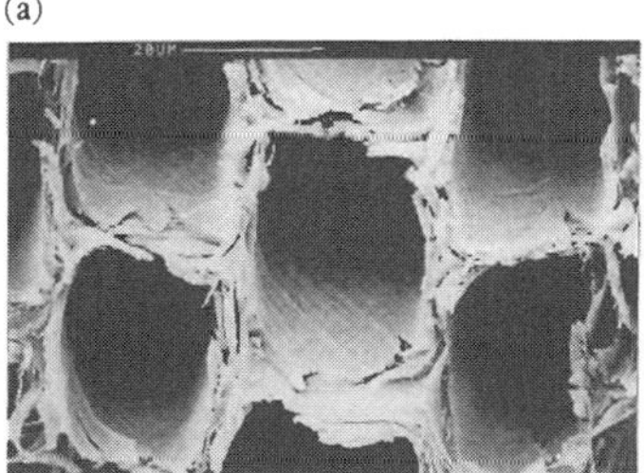

(b)

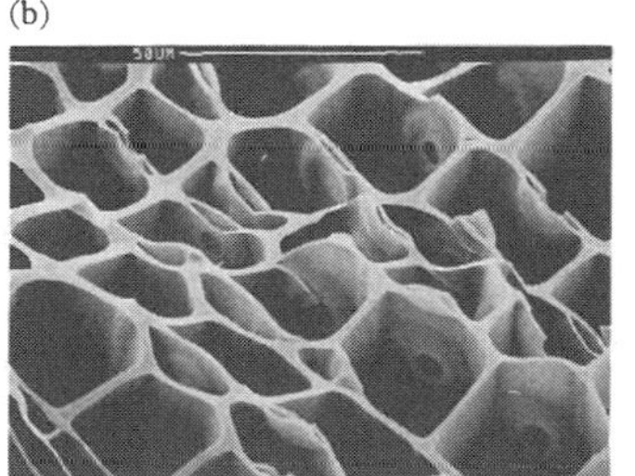

(c)

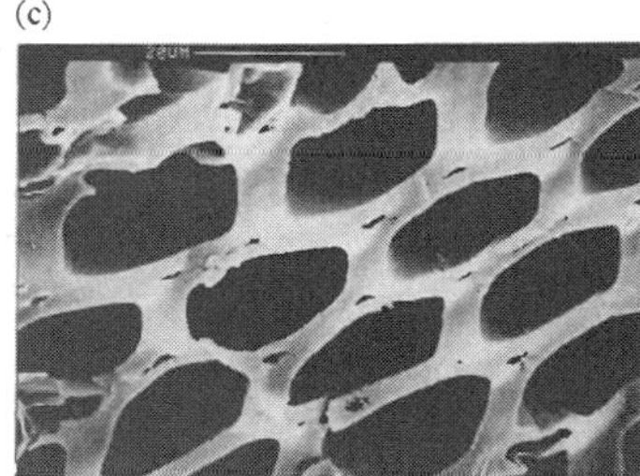

Fig. 3. Scanning electron micrographs comparing modern experimental charcoal, Jurassic fossil charcoal (fusain) and Tertiary fossil wood: (a) fossil wood from the Miocene, brown coals, Germany; (b) modern experimental charcoal, *P. sylvestris*, heated at 240°C for 1 hour, buried 20 mm in sand, homogenized cell walls; (c) Jurassic charcoal from the Scalby Ness Formation, homogenized cell walls.

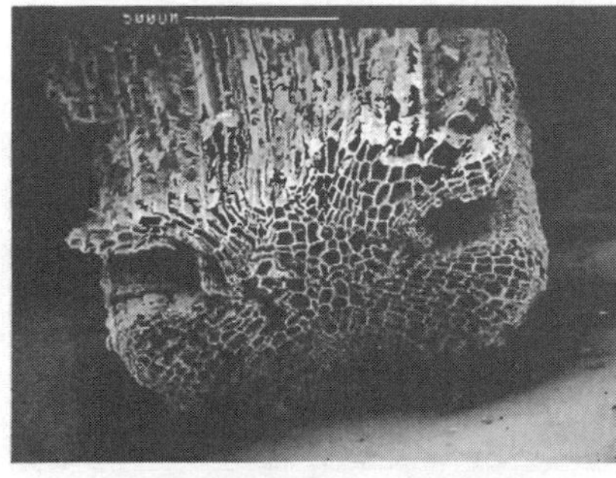

Fig. 4. Scanning electron micrograph of fusinized pteridosperm axis from the Lower Carboniferous of Scotland.

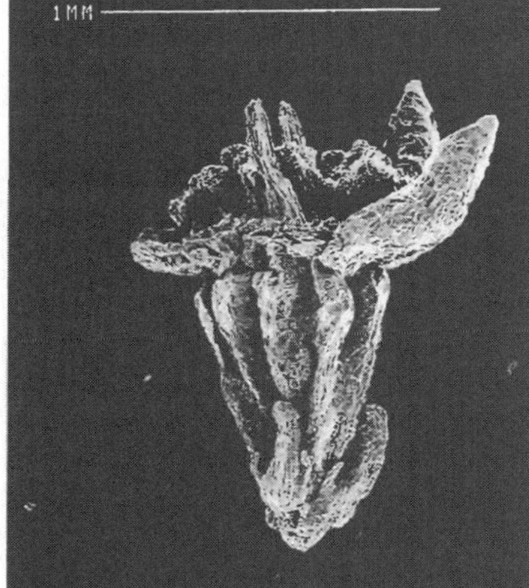

Fig. 5. Scanning electron micrograph of a flower preserved as fossil charcoal (fusain), from the Upper Cretaceous of Sweden.

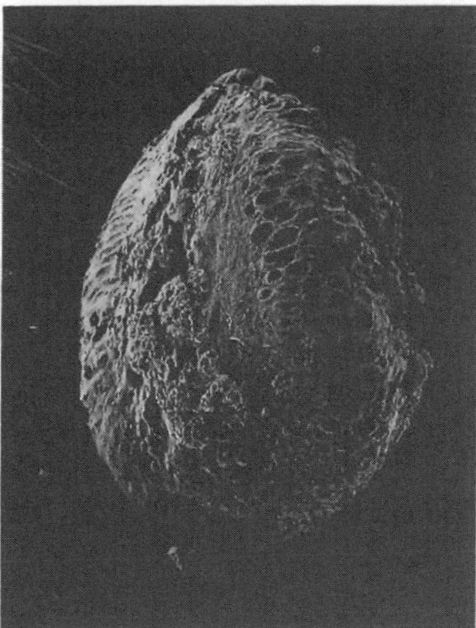

Fig. 6. Scanning electron micrograph of a fern sporangium preserved as fossil charcoal, from the Lower Carboniferous of Scotland.

Geological significance

Fusain is also found in volcanic rocks – not only volcanic ashes but also lavas. Studies of fusinized plants from the Lower Carboniferous basaltic ashes of the Midland Valley of Scotland have yielded many new plants, including several fern groups. Studies of modern wildfire-prone environments have shown that after fire has removed the vegetation helping to bind soil together, rates of erosion can increase up to 30 times the pre-fire levels until vegetation re-establishes itself. Whether or not an environment is prone to wildfire can have a direct correlation with the amount of material introduced into sedimentary systems. The Middle Jurassic sandstones of the North Sea often have numerous charcoal-rich laminations in otherwise characterless sequences, suggesting that wildfire-enhanced sediment supply could have contributed towards the formation of these deposits. This charcoal can be seen in sea-cliffs of the Scalby Formation on the Yorkshire coast (Figs 1 and 3). In peat-forming areas it has been shown that fires may burn the peat, creating depressions which accumulate clastic sediment, known as fire splays.

Modern wildfire

In many modern environments natural wildfires are common. Every year, news reports show fires running amok in some part of the globe. In Borneo during 1982–83, fires burned for 18 months, covering a region of some 37000 square kilometres, equivalent to the area of Belgium and Luxembourg combined. One study estimated that between 1940 and 1975 over 79131 fires were started by lightning strikes in the North America Rockies alone. Even in the wetland area of the Okefenokee swamp, USA, 318000 acres burned during the fires of 1954 and 1955. We must conclude, therefore, that natural wildfires have been common in the past and that they have left evidence of their passing in the fossil record.

Suggestions for further reading

Cope, M. J. & Chaloner, W. G. 1985. Wildfire, an interaction of biological and physical processes. In: B. H. Tiffney (ed.) *Geological Factors and the Evolution of Plants*. Yale University Press, Hartford, Connecticut, USA.

Scott, A. C. 1989. Observations on the nature and origin and fusain. *International Journal of Coal Geology*, v. 12, pp. 443–475.

Andrew C. Scott is senior lecturer and Timothy P. Jones is a research student in the Geology and Biology Departments, Royal Holloway and Bedford New College, University of London.

PART

2 Community Reconstruction

Edited by Scott L. Wing and Marty Buzas

The theme of this section is the reconstruction of paleocommunities. The word "reconstruction" might imply a static view—an attempt to reconstruct the spatial or geographic arrangement of organisms at a particular time in the past and to infer the environmental conditions under which they lived. Reconstruction can be thought of as a first step in paleoecology—documenting the species present in an ancient community, where and how they lived, and under what conditions. All of the papers reprinted here made methodological improvements to this basic paleoecological activity. They developed or applied techniques for quantifying the relative abundances of species in samples (see Johnson 1964; Ziegler 1965; Phillips et al. 1977; Cisne and Rabe 1978). They demonstrated that gradients in composition were related to observable and, in many cases, quantifiable variations in the enclosing sediments (see Newell 1957; Hickey and Doyle 1977). Some also introduced new methods of analysis to the field of paleoecology (Johnson 1964; Ziegler 1965). All of these papers used a combination of fossils and sedimentary features to understand the environmental gradients that controlled variation in the composition of ancient communities. They were also all inspired by close examination of modern environments and show the advantages of working the interface of paleontology and sedimentary geology with almost equal attention to both fields.

Even though these papers were selected because they are foundational to the practice of paleocommunity reconstruction, the authors clearly recognized that such reconstructions are important mostly because they improve our understanding of ecological and evolutionary change over long time periods. A static paleocommunity reconstruction is valuable, but the real advantage of working with the fossil record is realized when changes in environments and community structure are tracked over deep time. Those papers that remain most cited today did more than reconstruct a paleocommunity—they provided key insights into the environmental context for major ecological and evolutionary change. Paleoecology remains today most useful, and most important to sister disciplines, when it provides new ideas and hypotheses about why major evolutionary and ecological transitions have occurred.

PAPER 9

Coenocorrelation: Gradient Analysis of Fossil Communities and Its Applications in Stratigraphy (1978)

J. L. Cisne and B. D. Rabe

Commentary

MARK E. PATZKOWSKY

The growth of paleobiology as a research agenda in the 1970s led to a surge of interest in documenting long-term patterns of evolution and ecology from the fossil record. For evolution, this meant documenting and interpreting morphological changes within lineages through time (Eldredge and Gould 1972). For ecology, this meant documenting and interpreting the relationship between fossil taxa and their habitats through time (Ziegler 1965; Walker and Laporte 1970). Cisne and Rabe sought to combine these views in a study that applied gradient analysis to a suite of fossil assemblages to improve the resolution of stratigraphic correlations, which could then be used as a temporal framework to document microevolution within lineages. A specific concern of Cisne and Rabe was to identify environmental gradients through fossiliferous strata so that morphological change through time in lineages could be distinguished from clinal variation. This paper was far ahead of the field in its application of gradient analysis methods to the fossil record and it is one of the landmark papers in the area of stratigraphic paleobiology.

Cisne and Rabe analyzed the composition of fossil collections along a transect approximately 35 km long in the Middle Ordovician foreland basin of New York, where lithologic changes suggest a gradual deepening of the basin from south to north. Bulk fossil samples were collected from several stratigraphic sections along the depth transect. Abundances of individual fossil taxa in each sample were tallied, and these data were analyzed with several methods (percentage similarity, polar ordination, reciprocal averaging, principal components analysis) that determine similarity of taxonomic composition between two or more samples. These methods were borrowed from the growing ecological literature on gradient analysis of modern assemblages. The aim of this approach was to treat community composition as a continuous variable that changes along an environmental gradient. Gradient analysis of the Ordovician transect data produced curves of continuous change in water depth within sections, where the score for each sample represented a different depth. Parallel changes in water depth between sections permitted high-resolution correlation along the transect. The sections were independently correlated with numerous volcanic ash falls, interpreted to represent single, instantaneous events.

Cisne and Rabe demonstrated that gradient analysis performed in a detailed stratigraphic context could be used to solve a stratigraphic correlation problem. They then used this temporal framework to investigate evolution and speciation of Ordovician trilobites along a depth gradient (Cisne et al. 1980, 1982). This general approach was used later to investigate trilobite evolution from the Late Ordovician of Ohio by Webber and Hunda (2007), building on a correlation framework based on gradient analysis of fossil assemblages (Holland et al. 2001; Miller et al. 2001). More importantly, Cisne and Rabe demonstrated the interpretive

From *Lethaia* 11:341–64.

power of gradient analysis applied to the fossil record. Paleoecologists now use gradient analysis routinely to investigate a variety of problems of correlation, species evolution, and niche stability through time.

Literature Cited

Cisne, J. L., J. Molenock, and B. D. Rabe. 1980. Evolution in a cline: the trilobite Triarthrus along an Ordovician depth gradient. *Lethaia* 13:47–59.

Cisne, J. L., G. O. Chandlee, B. D. Rabe, and J. A. Cohen. 1982. Clinal variation, episodic evolution, and possible parapatric speciation: the trilobite *Flexicalymene senaria* along an Ordovician depth gradient. *Lethaia* 15:325–41.

Eldredge, N., and S. J. Gould. 1972. Punctuated equilibria: an alternative to phyletic gradualism. In *Models in Paleobiology*, edited by T. J. M. Schopf, 82–115. San Francisco: Freeman, Cooper.

Holland, S. M., A. I. Miller, B. F. Dattilo, and D. L. Meyers. 2001. The detection and importance of subtle biofacies in lithologically uniform strata: the Upper Ordovician Kope Formation of the Cincinnati, Ohio region. *Palaios* 16:205–17.

Miller, A. I., S. M. Holland, D. L. Meyers, and B. F. Dattilo. 2001. The use of faunal gradient analysis for high-resolution correlation and assessment of seafloor topography in the type Cincinnatian. *Journal of Geology* 109:603–13.

Walker, K. R., and L. F. Laporte. 1970. Congruent fossil communities from Ordovician and Devonian carbonates of New York. *Journal of Paleontology* 44: 928–44.

Webber, A. J., and B. J. Hunda. 2007. Quantitatively comparing morphological trends to environment in the fossil record (Cincinnatian Series; Upper Ordovician). *Evolution* 61:1455–65.

Ziegler, A. M. 1965. Silurian marine communities and their environmental significance. *Nature* 207:270–72.

Coenocorrelation: gradient analysis of fossil communities and its applications in stratigraphy

JOHN L. CISNE AND BRUCE D. RABE

LETHAIA

Cisne, John L. & Rabe, Bruce D. 1978 10 15: Coenocorrelation: gradient analysis of fossil communities and its applications in stratigraphy. *Lethaia*, Vol. 11, pp. 341–364. Oslo. ISSN 0024-1164.

Coenocorrelation is the correlation of positions in a stratigraphic sequence with corresponding positions along a paleoenvironmental gradient through gradient analysis of fossil communities. By ordination of community samples, the distribution and abundance of taxa along a depth gradient can be translated into a continuous scale that accurately measures the gradient and thus makes possible analysis of facies change on a continuum. The procedure is tested using data on marine benthic invertebrates along ten time-parallel transects down an Ordovician basin slope (Trenton Group, New York). Reversals between transgression and regression revealed in *coenocorrelation curves* served nearly as well as bentonite beds in time-correlation of sections. Ordinations gave estimates of depositional strike in close agreement with estimates from physical indicators. Walther's Law of Facies is evidently a more rigorous, quantifiable generalization than it has been revealed to be through classificatory approaches to communities and environments.

John L. Cisne, Department of Geological Sciences and Division of Biological Sciences, Cornell University, Ithaca, N.Y. 14853, U.S.A; Bruce D. Rabe, Department of Geological Sciences, Cornell University, Ithaca, N.Y. 14853, U.S.A; 7th October, 1977.

The key to interpreting very many stratigraphic sequences is finding an accurate means for relating positions in the sequence to positions along a paleoenvironmental gradient. The distribution and abundance of taxa provide a wealth of information useful for this purpose. Traditional approaches to study and analysis of fossil community phenomena have been concerned with classification of fossil assemblages in relation to abstract entities, biofacies or recurrent associations of taxa. The resulting discrete units are less than ideally suited to analyzing continuous paleoenvironmental change such as encountered along a basin slope or in a transgression. This paper considers an alternative approach, gradient analysis, as a means of obtaining accurate, quantitative information about paleoenvironmental gradients from related gradients in the composition of fossil assemblages, that is, directly from the distribution and abundance of taxa. Replicate samples from ten time-parallel transects down an Ordovician submarine slope are used to test the powers of gradient analytical methods in resolving relationships between the composition of fossil assemblages and their positions along a depth gradient. Results from certain methods seem promising.

This problem in stratigraphic analysis is very important in paleontological studies of evolution. In order to reliably trace microevolutionary patterns within a lineage, it is necessary to have a highly accurate means of time-correlation within the sequence studied. In order to distinguish temporal variation from clinal variation, if strictly possible, it is also necessary to have some means for measuring paleoenvironmental conditions represented in the sequence. The absence of such means has left considerable latitude for interpretation of results from microevolutionary studies (see Gould & Eldredge 1977). Gradient analytic methods promise to provide just such a scale, and, in certain instances, the means of time-correlation as well. It is this problem concerning microevolutionary studies that led us from an ongoing investigation of stratigraphic variation in lineages to consideration of the stratigraphic problem discussed here.

Practitioners of correlation by community have too frequently become bogged down in semantic problems of exactly what a community is. The word 'community' is given more than one meaning. This need not stand in the way of effective use of the concepts behind it. Whittaker's (1975) distinction between 'community' (the organisms or populations living together in the same place at the same time) and 'community-type' (recurrent association of taxa) is a useful one. 'Community' has been used in both senses,

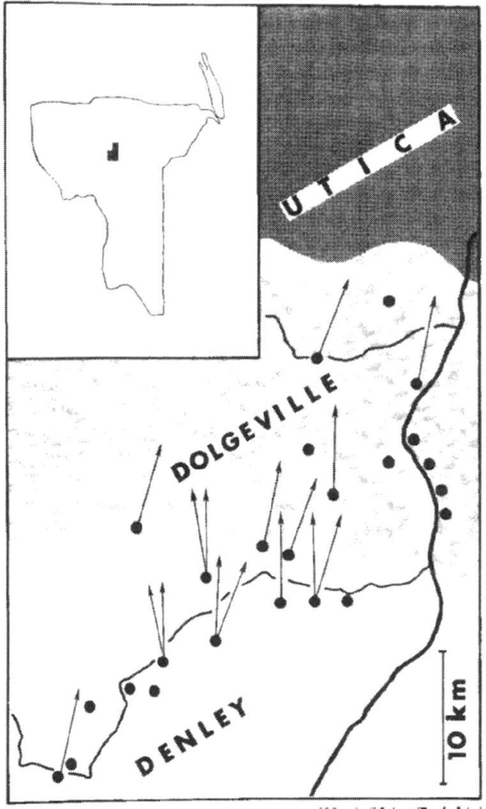

Fig. 1. Lithotope map of the study area at the time of deposition of the M 15 K-bentonite bed. Circles indicate localities studied. Arrows indicate downslope sediment flow directions determined from parallel-oriented fossils or cross bedding.

and even interchangeably, by paleoecologists. This has led to avoidable confusion. 'Fossil community', here used as a general term for phenomena, had been used more or less synonymously with others. Our definitions of the terms we use are as follows:

Paleocommunity – A representation of the organisms that lived together in about the same place at about the same time.
Paleocommunity sample – A collection of fossils believed to represent a paleocommunity.
Paleocommunity type – A recurrent association of taxa represented in paleocommunities.

Just as a community in Whittaker's (1975) sense may be indefinitely bounded in space at any one time, so a paleocommunity may be indefinitely bounded in both space and time. The nature and treatment of paleocommunity samples studied here are described below. We have taken the fossil record at face value: all specimens that we could identify were identified to the level of genus and counted, and data were in no way edited in the attempt to reconstruct original communities. Community reconstruction, an end in itself, is in a sense at cross purposes with use of paleocommunities in correlation. Isolating and eliminating taphonomic factors in community reconstruction may amount to subjectively throwing away information contained in the distribution and abundance of taxa that may be valuable in solving problems like the one used as an example here, telling where a sample falls along a slope.

Data for testing gradient analytic methods have been taken from a current study of marine benthic invertebrates in a certain portion of the Middle Ordovician Trenton Group that covers about two million years (Figs. 1–3). The sequence, described below, covers the transition from shelly facies to the deeper water graptolitic facies at the edge of a turbidite basin. Transects were laid out along altered volcanic ash beds (K-bentonites) that delineate time planes through the sequence, and parallel to the vector mean of physical slope indicators. It is unusual that circumstances provide such an opportunity for testing gradient analytical methods from transect studies within a well-known stratigraphic framework. If we seem to overemphasize our own data, it is because it is all we have available that would serve the purpose. The intent is simply to show what results can be obtained using gradient analytic methods, how they can be obtained, and how gradient analytic methods might be fruitfully applied in more usual sequences.

The test sequence and sample treatment

Middle Ordovician (higher Trenton Group) strata in the study area (Fig. 1) record the transition from a carbonate bank in the west to a deep basin in the east (Kay 1937, 1943, 1953; Titus & Cameron 1976). The transition can be traced laterally and vertically (Figs. 1–3). As the basin developed, bank sediments (thin bedded, shaly calcarenites of the Sugar River Limestone) were replaced upwards by upper slope deposits (thicker, interbedded gray to black calcarenites, calcisiltites, and shales of the Denley Lime-

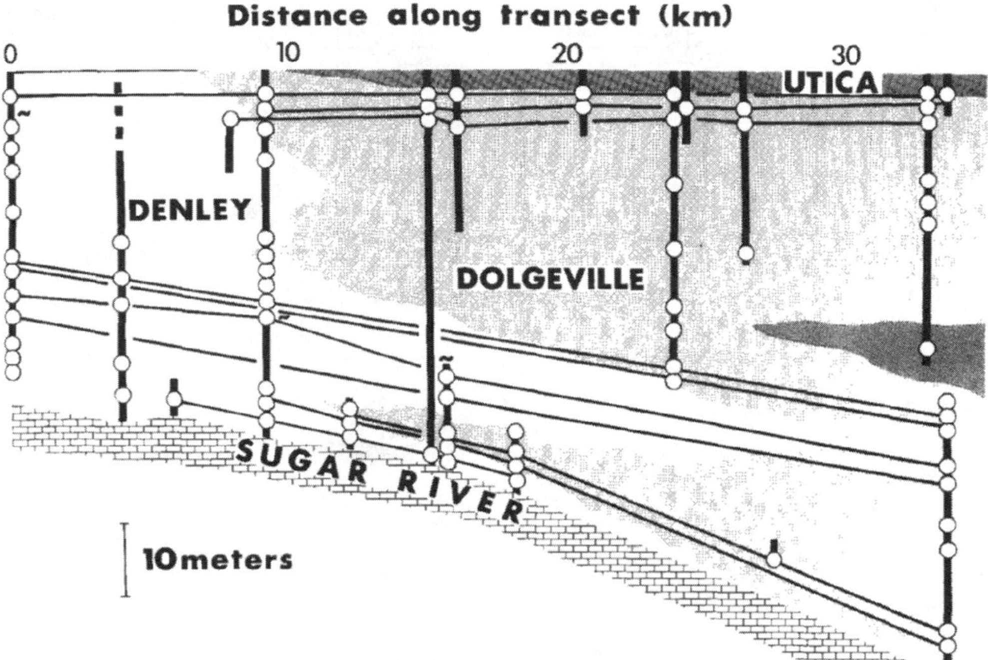

Fig. 2. Stratigraphic cross section through the Trenton Group sequence studied in the vector mean direction of slope indicators (N 83° E). The sequence thickens downslope into the turbidite basin in going from the Denley and Sugar River Limestones into the Dolgeville Formation and the Utica Shale. Vertical lines show the sections that can be correlated by bentonite beds, which are indicated by more or less horizontal lines. Circles indicate sampling stations. Three local unconformities evidently related to bentonite-floored slumps are indicated by the symbol (~).

stone), lower slope deposits (interbedded black calcisiltites and shales of the Dolgeville Formation, otherwise known as the Dolgeville Facies), and eventually basinal sediments (the black, graptolitic Utica Shale, this lower portion of which is also known as the Canajoharie Shale).

Following Kay (1953), correlation of sections (Fig. 2) was based on intertonguing relationships and altered volcanic ash beds, K-bentonites. Correlation of the long-mysterious Dolgeville Dam section with the Trenton Falls section exemplifies the successive approximation involved. Intertonguing relationships bracket the range of possible correlations of bentonite beds. Transgression and regressions marked by tongues in upper slope areas were found to be expressed as far from Trenton Falls (the standard section, $R = 0.0$ km) as Dolgeville Dam ($R = 33.6$ km). The regression marked by the Denley tongue near the middle of the sequence appears to be expressed at Dolgeville as one or two meters of rock on either side of the M 14 bentonite that, unlike the rest of the section, contain fossils characteristic of the upper slope rather commonly. The correspondence between sections suggests an hypothesis as to the correlation of bentonites that can then be tested using Shaw's (1964) graphical correlation method. Evidence strongly favors the hypothesis: positions of the four bentonites that would appear to be in common (M 8, M 11, M 14, M 15; bentonite beds are designated according to the meter in the composite standard section in which they occur, in this case, beds in the 8th, 11th, 14th, and 15th meters) are quite highly correlated statistically ($r > 0.999$, $p < 0.001$).

Ten bentonite beds usually of about 1 cm thickness were traced across 20–40 km distances. Kay (1953) noted five of these at at least one outcrop. Perhaps the most important beds for correlative purposes are the M 34–M 38 sequence. One bed about 10 cm thick (M 38) and 2–3 beds within a meter of it (one, no more than a few millimeters thick, cannot always be found)

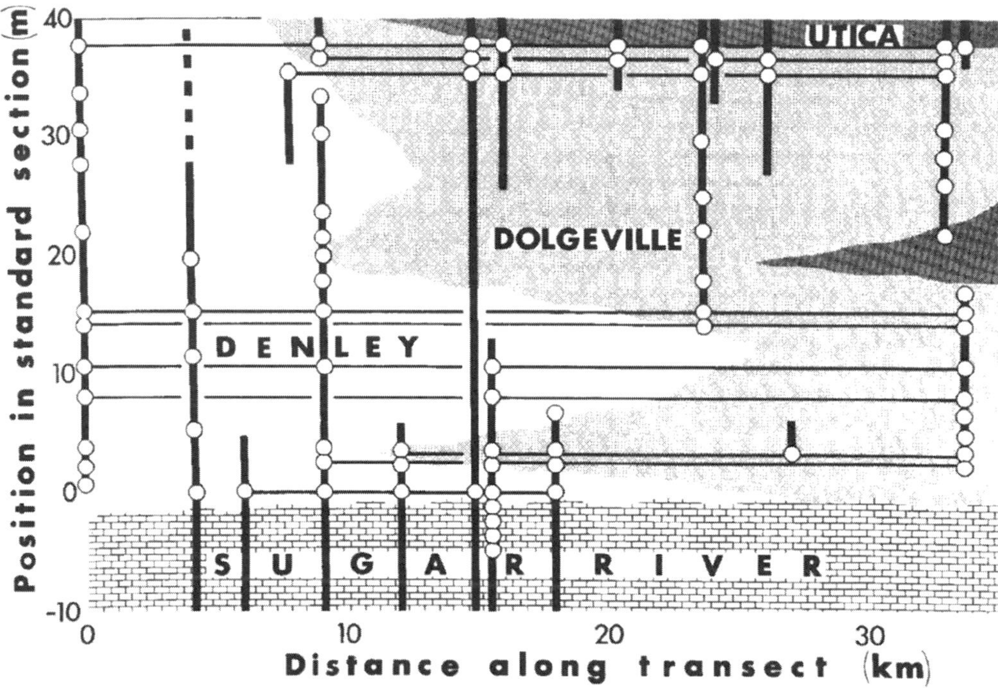

Fig. 3. Time-space cross section through the Trenton Group sequence based on Fig. 2. The composite standard section – the Trenton Falls section (transect position $R = 0.0$ km) except for the lowest few meters for which the Wolf Hollow Creek section ($R = 15.5$–15.7 km) is used – is taken as an approximate time-scale. Vertical lines show the sections that can be correlated by bentonite beds, which are indicated by horizontal lines. Circles indicate sampling stations.

can be traced from Trenton Falls (where one must stand more or less *in* the falls to see them) more than 30 km from outcrop to outcrop in the slope direction. Beds of usual thickness about two and four meters below the thickest one can be traced more than 20 km. As Kay (1953) noted, bentonites tend to thin and pinch out in the upslope direction. They commonly form floors for soft sediment folds in lower slope sediments, a consequence of their lower shear strength. They also appear to have formed floors for localized slumps, in which case their positions are occupied instead by brecciated beds. Kay (1953) noted apparent unconformities in positions corresponding closely to those expected from the M 14 and M 15 beds at Wolf Hollow Creek ('City Brook', $R = 15.5$–15.7 km) and the M 36 bed at Trenton Falls. The M 14 bed is similarly missing at Rathbun Brook ($R = 8.9$–9.3 km). An unconformity of this type is exposed spectacularly at the Dolgeville-Utica contact along the New York Thruway ($R = 24.1$–26.7 km), where in place of the M 38 bed is a thin, intermittent breccia. The M 34 and M 36 beds remain. All beds are present and conformably overlain by the Utica Shale at outcrops just a few km to the north along depositional strike.

The only notable difference between our interpretation of intertonguing relationships and Kay's (1953) concerns his correlation of 3–4 m of the Trenton Falls section with the Dolgeville beds. The resemblance goes as far as the gross form of limestone beds. Thick black shale interbeds characteristic of the Dolgeville are absent. We agree with Miller's (1909) conclusion that typical Dolgeville lithologies are not represented at this locality.

Because the lowest 2–3 m of Denley Limestone are only exposed underwater at Trenton Falls, the Wolf Hollow Creek section was used to fill out a composite standard section. The zero point in measuring stratigraphic thickness is the lowest (M 0) bentonite in this section. Net rates of rock accumulation were evidently very similar at the two locations during deposition of the lower Denley, as indicated by nearly identical spacing of the two bentonite beds in common (M 8, M 11) and otherwise by their correspondences with other sections in the correlation web (Kay 1953).

Crossbedding and parallel-oriented fossils (orthoconic cephalopod shells and graptoloid rhabdosomes oriented in sediment flows) give a remarkably consistent current direction (Fig. 1) believed to delineate the slope direction. Eighteen determinations from points throughout the sequence gave a vector mean direction of $97 \pm 17°$ (2σ) or S 83° E, which is essentially the same direction determined by Chenoweth (1952) from Trenton rocks over a larger region. The direction coincides very nearly with the slope direction reconstructed from regional stratigraphic and tectonic evidence by Bird & Dewey (1970).

The cross section (Figs. 2, 3) and transects along bentonite beds were laid out by projecting the locations of sections onto a line parallel to the vector mean current direction. Distances down the slope are measured along this line relative to the arbitrary zero point, the base of Trenton Lower High Falls.

The picture that emerges from new and more precise correlations of lower slope sections ($R = 14.8$–33.6 km) with already well-understood upper slope strata (Figs. 2, 3) shows the edge of an expanding and deepening turbidite basin very much as Kay (1937, pl. 4) reconstructed it.

Large (c. 10 kg) bulk samples were collected from 1–2 cm thick portions of shale beds at stations along transects paralleling bentonites and otherwise throughout the sequence to give more even coverage. Samples along transects were collected from shale beds 10–20 cm above and below the shale bed containing the respective bentonite. No consistent differences were found between samples above and below bentonite beds.

Rock pieces in a sample were cleaned ultrasonically and examined under a dissecting microscope. All identifiable specimens of benthic macroinvertebrates (ostracodes excluded) on bedding surfaces were identified to the level of genus and counted until 500 or more were recorded from the sample, if possible. Among more than 200 samples collected, just over 170 were worked up in this fashion. The 168 samples (80,000 + specimens) for which results are reported here exclude a few samples that contain no more than a few tens of specimens.

An effort was made to avoid beds that showed obvious signs of disturbance by current or sediment flow activity. We examined normal petrographic sections of limestone beds along the upper and lower interfaces of over 100 shale beds from which collections were made. We also examined tens of exceptionally large petrographic sections of limestone-shale couplets from the Trenton Falls section. Shale beds are evidently deposits of quiescent intervals, while the frequently crossbedded limestones are current-laid. Most every shale bed was buried by a limestone. Some transport of material no doubt took place. In general, shells transported any distance were evidently broken up. Shell pieces larger than sand-size are relatively rare in most current-laid limestone.

Ordination computations were made using ORDIFLEX (Cornell Ecology Programs 25A).

Coenocorrelation

Coenocorrelation, signifying 'community correlation', is the correlation of positions in a stratigraphic sequence with corresponding sets of paleoenvironmental conditions based on analysis of fossil communities. It is an essential part of ecostratigraphic correlation. Ecostratigraphy itself is more generally concerned with time-correlation of strata in relation to environmental events recorded in the stratigraphic distribution of taxa (Martinsson 1973, 1978). In approaching time-correlation, coenocorrelation is, then, the first stage in ecostratigraphic correlation.

Traditional approaches to fossil communities in coenocorrelation have been concerned in one way or another with classification of samples. Geologists were classifying paleontological samples in relation to Zones and biofacies long before it became popular to classify them into paleocommunity types. So far as classification itself is concerned, the process is not fundamentally different. Once the sequence of paleocommunity types along a paleoenvironmental gradient has been determined, it can be used as a scale for coenocorrelation. Ziegler's (1965; Ziegler et al. 1968) pioneering studies on Silurian marine invertebrate paleocommunity types are outstanding among numerous examples showing the good results attainable with a classificatory approach.

Classification of samples into paleocommunity types limits the resolution attainable in coenocorrelation. Paleoenvironmental information

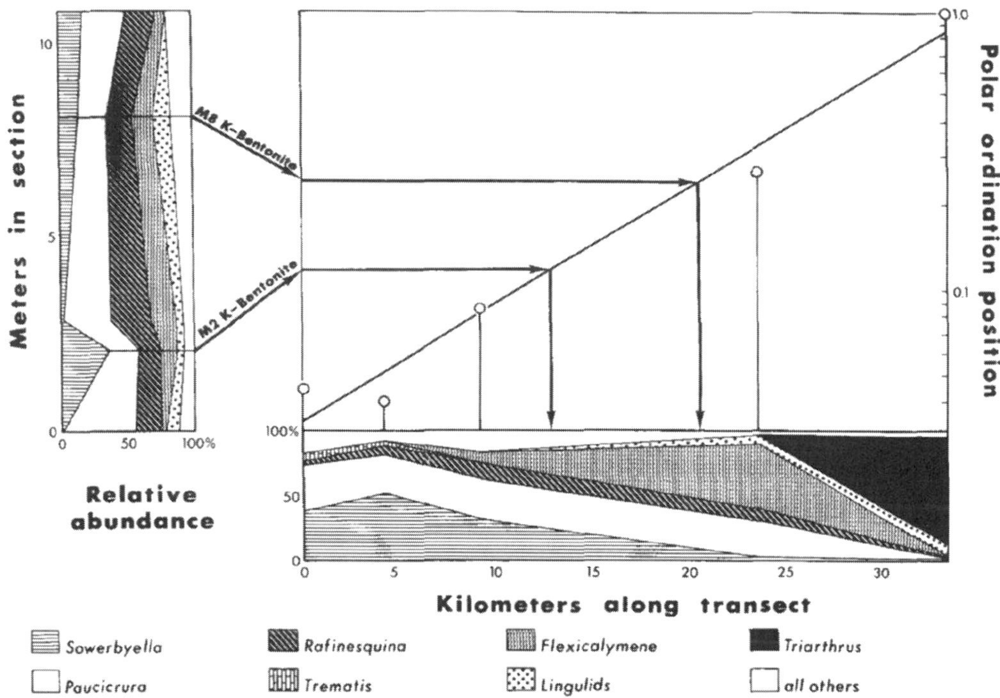

Fig. 4. Coenocorrelation of positions in a stratigraphic section (vertical) with positions along a time-parallel transect (M 15) (horizontal) down a Middle Ordovician (Trenton Group, central New York) slope based on gradient analysis of marine benthic paleocommunities. The approximately exponential ordination curve is based on direct polar ordination of paleocommunities from stations along the M 15 K-bentonite bed (horizontal scale) with respect to replicate samples at the upslope (transect position $R = 0.0$ km) and downslope ($R = 33.6$ km) ends of this transect. Its prime significance is to show that ordination position meaningfully reflects position along the depth gradient. Ordination of replicate samples above and below two other bentonite beds in the Wolf Hollow Creek section ($R = 15.5$–15.7 km) following the same procedure relates the samples to positions along the abstract axis that represents the paleocommunity gradient at one particular time, and thus to a position on a relative scale that measures the depth gradient.

Sowerbyella cf. *S. punctostriata*, *Paucicrura rogata*, and *Rafinesquina* spp. are articulate brachiopods. *Trematis terminalis* and lingulids are inarticulate brachipods. *Flexicalymene senaria* and *Triarthrus becki* are trilobites.

contained in the relative abundances of species may be lost or distorted in the classification process. Classification, as a human endeavor, is inevitably subjective to some extent. More important, creation of discrete units imposes a 'grain' on the resolution of coenocorrelation that can result in decreased accuracy. While abrupt, discontinuous changes in paleocommunities and paleoenvironments can be resolved in an appropriate fashion, continuous changes can only be resolved in terms of discrete steps.

Gradient analysis, an alternative approach, promises improvements in resolution attainable in coenocorrelation. It is concerned with analysis of the distribution and abundance of taxa relative to gradients in environmental factors. A methodology developed primarily in plant ecology, it has been used extensively in 'dissecting' community types in order to understand whatever ecological cohesiveness they may show (see Whittaker 1967, 1975 for review). Communities are treated in terms of continuous variables of composition, not the discrete variable of type. Ecological relationships between community compositions and environmental factors can be analyzed more exactly because the problem of 'grain' inherent in classification is circumvented. Applied to paleocommunities, this means coenocorrelation can be more precise. Continuous changes through a sequence can be treated as having been in fact continuous. Abrupt changes can be treated as well. The 'grain' of resolution is, in theory at least, infinitesimal.

Fig. 4 shows an actual example of coenocorrelation done according to methods considered at length in this paper. Two positions in a vertical stratigraphic sequence are related to corresponding positions along a time-parallel transect down a slope based on *ordination* of paleocommunity samples. Ordination is the process of ordering community samples in relation to environmental gradients. In *axial ordination,* as illustrated here, samples are ordinated along one or more abstract axes defined by taxa or taxonomic composition of samples. Positions of samples along an ordination axis may or may not meaningfully represent their positions along environmental gradients. The *ordination curve* (plot of ordination position against position along environmental gradient) in Fig. 4 depicts a good, useful ordination. That is, ordination position meaningfully describes the positions of samples along an environmental gradient, as the curve shows. The gradient in paleocommunities (Fig. 4, frequency diagram along horizontal scale) is closely, one might say immanently, related to the depth gradient. This same basic relationship has been found for other transects that bracket the positions along the gradient in paleocommunities by the same ordination method used on samples from the time-parallel transect. Their corresponding positions in the section can thus be related to positions along the downslope transect. Because ordination position and transect position are closely related measures of the depth gradient, either one could be used in coenocorrelation. However, use of ordination position is desirable because this less derived quantity involves lesser error than estimated position on transect for relating positions in the sequence to positions along the depth gradient.

In this simple case (Fig. 4), inspection of the distribution of taxa in the section and on the transect makes it clear that a transgressive change takes place between the two indicated points. Gradient analytic methods make possible an estimate of the amount of change. A decrease in the relative abundance of the brachiopod *Sowerbyella* is the principal effect measured by ordination position. The peak in abundance of *Sowerbyella* in the section is evidently not a matter of sampling error. It is found in replicate samples adjacent to the same bentonite in sections a few km upslope and downslope from the section diagrammed. Obviously, the correspondence in taxon distributions between the section and the transect is not perfect. For many reasons – sampling error, differing taphonomies of assemblages, change in paleocommunities over time – one would not expect it to be. Trying to find exactly corresponding points between two paleocommunity gradients, or in two community gradients for that matter, is the same problem as trying to touch a stream twice in the same place. Coenocorrelation is inherently not exact in a strict sense. The same argument can be made for populations and lineages as regards biostratigraphic correlation in general.

Gradient analysis

The essence of gradient analysis is the study of community ecology in terms of the interrelationships of gradients in community composition and in environmental factors. These interrelationships can be studied along transects laid out so as to represent spatial gradients in environmental factors. Because these factors may be so closely related as to be not strictly separable, the concept of *complex gradient,* a gradient in the complex of environmental factors actually encountered on a transect paralleling some important direction of environmental variation, has been developed. Numerous studies document continuous change in communities along continuous complex gradients. Whittaker (1967, 1973a) reviews the mass of data on vegetation. Gradient analytical studies on marine communities are rare. Studies by Sanders & Hessler (1969) and Johnson (1970, 1971) document continuous change in marine benthic communities along depth-related gradients. Results of gradient analyses are basic for the concept of *community gradient,* a gradient in community composition built up from the distributions of species along the corresponding complex gradient. A community gradient and complex gradient together make up an *ecocline* (see Whittaker 1975, for a thorough account of concepts and terminology).

The concepts of *paleocommunity gradient, paleoenvironmental complex gradient* and *paleoecoline* carry over from community ecology to community paleoecology without fundamental modification. Just as paleocommunity is a biased representation of an original community, a paleocommunity gradient is a biased representation of an original community gradient. The composition of paleocommunities is determined in part by the composition of the communities

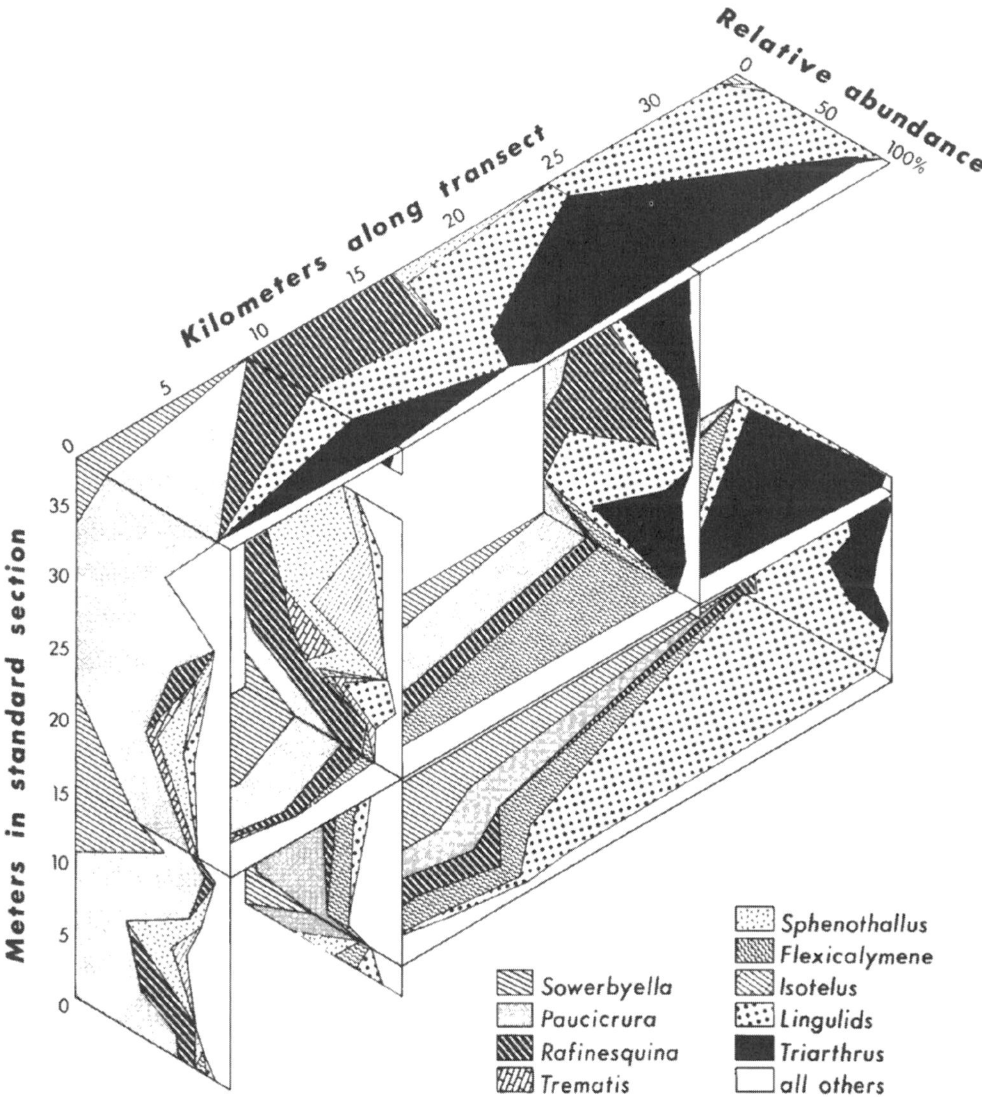

Fig. 5. Distribution of taxa along representative sections (vertical) and transects (horizontal) in the Trenton Group sequence studied. Transgressive and regressive changes in paleocommunities seen in sections parallel changes seen along time-parallel transects.

The sections are: Trenton Falls at the upslope end (transect position $R = 0.0$ km), Rathbun Brook ($R = 8.9$–9.3 km), North Creek ($R = 23.4$–24.5 km), and Dolgeville Dam ($R = 33.6$ km). The transects parallel the M 2, M 15, and M 38 K-bentonite beds. *Sowerbyella* cf. *S. punctostriata*, *Paucicrura rogata*, and *Rafinesquina* spp. are articulate brachiopods. *Trematis terminalis* and lingulids collectively are inarticulate brachiopods. *Flexicalymene senaria*, *Isotelus gigas*, and *Triarthrus becki* are trilobites. *Sphenothallus* sp. is a conularid.

from which they are formed, and in part by environmental factors that affect the transformation from community to paleocommunity. The original complex gradient is reflected not only in the composition of the original community but also in the community's taphonomy in becoming a paleocommunity. Taphonomic factors have come to be regarded as something of a nuisance in attempts to reconstruct communities from paleocommunities and to use paleocommunities in coenocorrelation. In fact, taphonomic factors, immanently related to an original complex gradient, may represent simply additional ways in which paleoenvironmental information is recorded in paleocommunities. The information content of the paleocommunity gradient might actually be enriched over that of the original community gradient. Warme et al. (1975) discuss the matter and give an example of this phenomenon. They found the composition of death assemblages to be more highly correlated with ambient environmental conditions than the composition of life assemblages.

Distribution of fossils along a complex gradient – an example

Like extant communities along uninterrupted complex gradients, benthic paleocommunities from the Trenton Group change more or less continuously along a depth gradient (Fig. 5). Articulate brachipods (*Sowerbyella* cf. *punctostriata* and *Paucicrura rogata*) numerically dominate in the shelly facies of the upper slope. Both articulate brachiopods (*Paucicrura* and *Rafinesquina* spp.) and the trilobite *Flexicalymene senaria* dominate in the shelly facies of the lower slope. The latter association is the *Geisonoceras* Paleocommunity Type of Titus & Cameron (1976), *Geisonoceras* spp. being very obvious but numerically unimportant cephalopods. Inarticulate brachiopods (several taxonomically difficult lingulids including *Pseudolingula* spp. that are all but indistinguishable in their most frequently encountered younger stages) and/or the nektobenthic trilobite *Triarthrus becki* dominate in the graptolitic facies. *Triarthrus* dominates over a relatively narrow zone along the slope, as Titus & Cameron (1976) noted (hence their *Triarthrus* Paleocommunity Type). The lingulids that dominate on either side of this zone probably belong to different species, though this would be quite difficult to ascertain quantitatively owing to combined problems of preservation and taxonomy. Ostracodes are also very abundant in the region of facies transition. No effort has been made to treat them here.

Examination of more than 80,000 specimens in 168 collections reveals continuities in distributions of taxa along the complex gradient, even across the transition from shelly to graptolitic facies. Downslope transport of remains is no doubt a factor behind continuity of the paleocommunity gradient, but certain evidence leads us to believe that it was not a very important factor. Distributions of species show no preferred downslope orientation. Fossils associated with the shelly facies are present, though generally rare, in graptolitic shales; and fossils associated with the graptolitic facies are present, though likewise generally rare, in rocks of the shelly facies. In samples, fossils that can be said to have a life position are not infrequently in it (e.g. the ectoproct *Prasopora* spp.). In certain collections from burrowed, black, graptolitic shales, complete, articulated specimens of *Paucicrura* and *Sowerbyella* have been found most frequently in what would appear to be positions occupied in life. Many samples representing the *Triarthrus* Paleocommunity Type are similar in composition to the Beecher's trilobite bed fauna (Cisne 1973), a natural census of a benthic paleocommunity, despite the fact that they are not themselves such census assemblages. We share Johnson's (1972) opinion that transportation of remains has been overemphasized. When it was popular among paleoecologists to believe that communities were organized as discrete units, just as it once was among ecologists, it was convenient to invoke transportation as a device for tidying up the edges. The main idea to come from gradient analyses is that communities intergrade continuously except across truly sharp discontinuities in environmental conditions. While transportation may enhance intergradation of paleocommunities, intergradation itself is to be expected among the communities from which they formed.

The shape of a paleocommunity gradient can be analyzed by comparing samples along a transect using any of a number of sample similarity measures. These fall into two basic types: (1) those based on the abundance, or more generally, measures of importance (e.g. biomass), of species, and (2) those based on species composition (see Goodall 1973, and Sepkoski 1975, for good reviews).

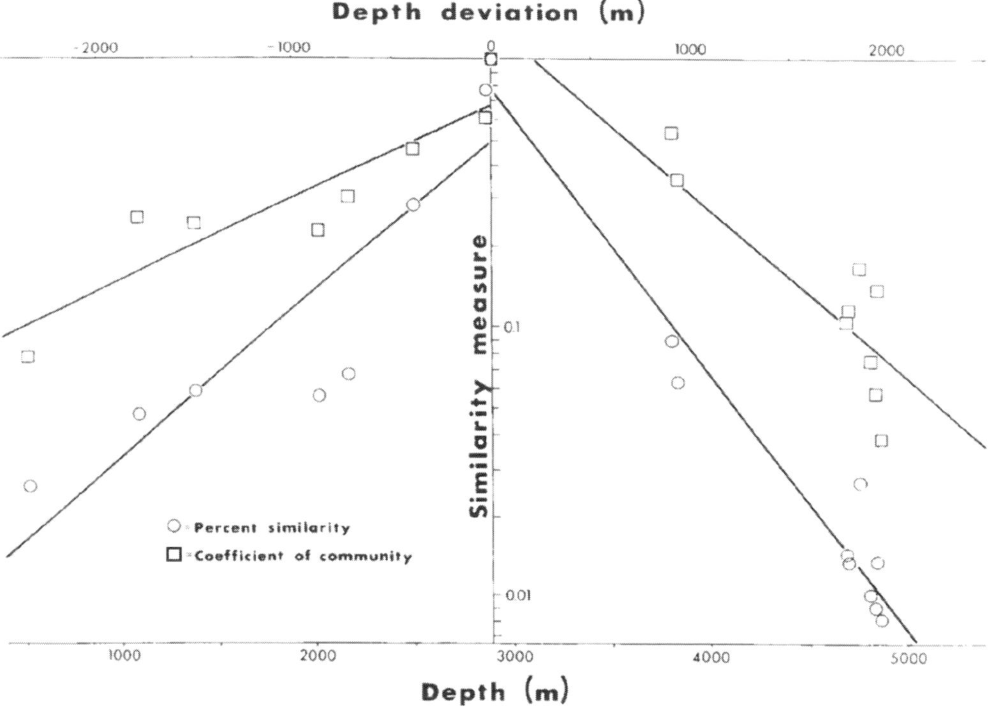

Fig. 6. Relationship between Percentage Similarity and Coefficient of Community relative to a sample from the lower continental slope for modern marine benthic communities along the Gay Head-Bermuda transect. Both similarity measures decrease approximately exponentially with increasing separation along the depth gradient, as consistent with the Gauch-Whittaker (1972a) community gradient model. Data are from Sanders & Hessler (1969).

Percentage Similarity (PS) is perhaps the most frequently used of the first type:

$$PS = 1 - 1/2 \; \Sigma_i \, |p_i - q_i| \qquad (1)$$

where p_i and q_i are the importance values for the ith species in the two samples compared. Coefficient of Community (CC) is perhaps the most popular measure of the second type:

$$CC = 2 \, S_{AB}/(S_A + S_B) \qquad (2)$$

where S_{AB} is the number of species in common between samples A and B and S_A and S_B are the number of species in each one. Although the conceptual bases for the two measures would appear to be quite different, they are closely related mathematically, being two ends of spectrum as regards weighting dominance or diversity in assessing similarity (Sepkoski 1975). Percentage Similarity emphasizes dominance and gives little weight to rare species. Coefficient of Community emphasizes species composition and diversity. It is less sensitive to errors in measuring the abundance of dominant species, and more sensitive to rare species. It is quite sensitive to differences in sample size. In the paleontological examples to follow, PS is used with relative abundance as the measure of species importance. The choice was in large part dictated by the nature of the samples. Bulk samples at approximately the same size (c. 10 kg) contained highly variable numbers of specimens (roughly 10–10,000). Low fossil density is characteristic of samples representing the graptolitic facies. Otherwise, changes in dominance, not diversity, appear to be the main ecological effect along downslope transects (see Figs. 4, 5).

Both PS and CC tend to decrease exponentially with distance along a transect paralleling a continuous community gradient. Whittaker (1960, 1967) and Terborgh (1971) demonstrated the pattern from field data on woody plants and birds, respectively. Gauch & Whittaker (1972)

Table 1. Paleocommunity statistics on ten transects down a slope in the Middle Ordovician Trenton Group. Transects, laid out parallel to bentonites, are designated by the meter in the standard section in which the bentonite occurs (e.g. 'M 0' means 'meter zero'). The extent is given as the positions *R* in km down the slope direction from endpoint stations in the extreme upslope position. A half-change unit is the distance in km corresponding to a decrease in Percentage Similarity (PS) by a factor of one-half. It is estimated from the regression of log (PS) on *R* for each transect. Percentage Similarities relative to replicate samples at the extreme upslope station are used when samples along a transect more frequently represent the shelly facies (S). Values relative to samples at the extreme downslope station are used when samples more frequently represent the graptolitic facies (G). Stations are indicated in Figs. 2 and 3.

Transect	Number of samples	Dominant facies	Half-change unit (km)	Half-changes
M 0	17	S	[1,2]53.4	[1,2]0.2
M 2	10	S	7.9	3.1
M 3	10	S	7.3	3.0
M 8	6	S	10.4	3.2
M 11	9	S	[2]5.3	[2]6.3
M 14	8	S	15.3	2.2
M 15	14	S	[2]6.8	[2]5.0
M 34	15	G	6.8	3.2
M 36	8	G	6.9	2.6
M 38	18	G	[2]14.8	[2]2.3

[1]log (PS) does not show significant regression on *R* at 0.01 level.

[2]log (PS) shows significant departures from exact linear dependence on *R* at 0.05 level.

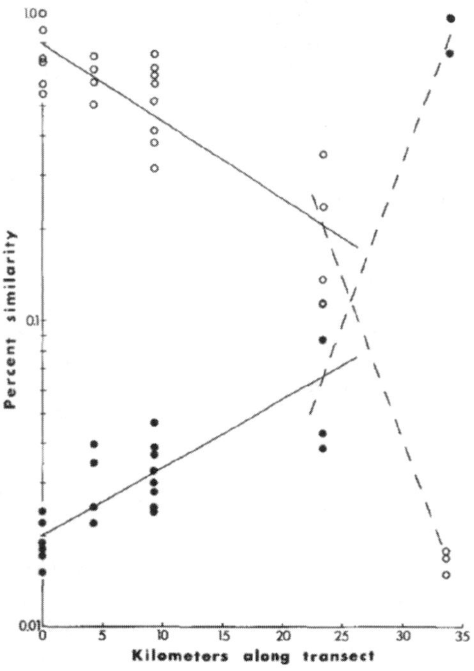

Fig. 7. Relationship between Percentage Similarity relative to replicate samples at upslope (transect position $R = 0.0$ km) and downslope ($R = 33.6$ km) ends of the M 15 transect for paleocommunity samples along this transect through the Trenton Group. The similarity measure tends to decrease exponentially with separation along the depth gradient within the shelly facies (solid lines, $R = 0.0$–23.4 km) as consistent with the Gauch-Whittaker (1972a) model of a continuous community gradient. Taxonomic turnover along the gradient is relatively more rapid across the transition between shelly and graptolitic facies, as indicated by the steepness of the dashed lines. There is evidently an inflexion in the paleocommunity gradient between facies, but, as indicated by distributions of taxa (Figs. 4, 5), not a sharp discontinuity.

modeled a continuous community gradient as built up from Gaussian distributions of species abundance with modes distributed at random along the transect – a pattern suggested by data on vegetation. Gauch (1973) found that, according to this model, PS should decrease as the error function complement of distance between samples along the transect. He derived a similar relationship for CC inductively. Over a certain interval, the complemented error function approximates an exponential decay curve (Gauch 1973).

Transect data can be tested against this pattern for consistency with the Gauch-Whittaker community gradient model. One need not rely entirely on inspection of distribution patterns for judging degrees of continuity. Data on the distribution and abundance on a transect from the New England continental shelf into the deep-sea (Sanders & Hessler 1969) show the same exponential decrease in PS and CC with depth difference between samples (Fig. 6). Depth relative to the standard sample accounts for about 90% of the variance in log (PS) and about 80% of the variance in log (CC). These highly significant results ($p < 0.001$ in each of four cases) are quite consistent with Sanders' and Hessler's conclusion that communities intergrade continuously between continental shelf and deep sea.

Within a facies, paleocommunities in the Trenton Group show the same exponential decrease in PS value with separation of samples along a depth gradient according to

$$PS = a\, e^{-bR} \qquad (3)$$

where *R* is the distance in km measured from an endpoint station in the direction of the slope, and

$a \sim 1$ and b are constants. Nine of ten transects show highly significant ($p < 0.001$) regressions of log (PS) on R for PS values relative to samples at the upslope ends of transects primarily covering the shelly facies (M 2–M 15) and at the downslope ends of transects primarily covering the graptolitic facies (M 34–M 36) (see Table 1). The one exception, the lowest and shortest transect (M 0) shows no significant regression of PS value on transect position nor on two-dimensional geographic coordinates (i.e. $b \sim 0$ in Eq. 3). At this particular time, the slope was just beginning to form and there very well may have been no appreciable paleocommunity gradient over this transect through the shelly facies. Position on transect accounts for 65–90% of the total variance in log (PS) for the six other transects through the lower part of the sequence. Only two of these, M 11 and M 15 (Fig. 7) show significant ($p < 0.05$, by F-test) deviations from exponential decrease of PS value with distance over the entire range. When the *Triarthrus*-dominated samples from the extreme downslope ends of these transects are excluded so that only shelly facies samples are considered in the regression analysis, PS value is found to decrease exponentially (see Fig. 7). Among the three higher transects, only one (M 38) shows significant deviation from exponential decrease in PS value with distance. Again, the deviation is due to occurrence of *Triarthrus*-dominated samples, but in the middle of the transect, not at the end. Otherwise, position on transect accounts for 38–80% of the variance in log (PS).

Deviations from exponential decrease in PS value with distance can indicate an interruption in a paleocommunity gradient, or simply a breakdown in the exponential approximation to the exact prediction from the Gauch-Whittaker community gradient model for PS values around 0.01 or below. In the present instance, deviations from the exponential pattern as seen in Fig. 7 evidently indicate an inflexion in the paleocommunity gradient, not a true discontinuity in the distributions of taxa but rather an interval of accelerated change in paleocommunities along the gradient in which abundances of taxa change more rapidly than they do within facies. The inflexion coincides with what has long been recognized as a marked change in oxygenation of waters analogous with oxygen sills in the Black Sea and some continental borderland basins off southern California (Ruedemann 1926, 1935; Bretsky 1969, 1970; Cisne 1973; Titus & Cameron 1976). The pronounced variation in paleocommunities across the transition between shell and graptolitic facies may be due to fluctuations in this sill.

The particular expression of the irregularity in the paleocommunity gradient is in large part due to the peculiar abundance of *Triarthrus* over a few km along transects. Where the trilobite occurs, it tends to be represented by a profusion of disarticulated parts. While these are probably molted parts, by and large, it is impossible to definitely identify them as such. Specimens preserving the gut and muscles often show the beginnings of disarticulation at what is obviously a very early stage of decomposition (see Cisne 1975). Histological studies on trilobite cuticle in general that may make possible distinction of molted parts from remains of shell in life are just now being carried out by several workers. Our policy of not editing data, combined with the taphonomy of this trilobite, no doubt leads to an exaggeration of its abundance relative to non-arthropods. However, as noted, the trilobite's apparent importance is consistent with its importance in a natural census (Cisne 1973). The association of *Triarthrus* with some sort of boundary layer is reminiscent of the association of modern lophogastrid mysid crustaceans with the oceanic oxygen minimum.

When a similarity measure decreases approximately exponentially with distance along a transect, as PS does (Eq. 3), it provides a unit for measuring the ecological length of the corresponding section of community gradient, *the half-change*, the analogue of the half-life of a radioisotope. From Equation 3, the half-change unit (HC, in km) for PS along a transect is given by

$$HC = \ln 2/|b| \quad (4)$$

The half-change units and half-change lengths for the ten Trenton transects are given in Table 1. These have been estimated from regressions of log (PS) on R. A consequence of using replicate samples from a station is a reduction in the mean PS value at the endpoint position. Because this initial PS value is not unity, half-change lengths tend to be underestimated by no more than half a half-change. The overall effect of deviations from the exponential pattern and of 'noise' in general, is in this case to increase the estimated half-change unit.

Ordination in coenocorrelation

The paleoecological relationships applied in coenocorrelation can be resolved using a variety of ordination methods, that is, methods for arringing species or samples in relation to paleoenvironmental gradients or gradients in sample composition that may reflect them. Of concern here are sample ordination methods; in particular, axial methods in which samples are related to positions along axes. Whittaker (1973) reviewed the numerous methods applied in plant ecology, the area in which ordination techniques have been developed and applied most extensively. He recognized two basic modes of ordering species or samples: *direct ordination*, in which entities are arranged with respect to environmental gradients or samples chosen to represent them, and *indirect ordination*, in which entities are arranged with respect to directions of maximum variance within a body of data when plotted in species-defined space (R-mode analysis) or sample-defined space (Q-mode analysis). Indirect ordination methods are of special interest in coenocorrelation because they do not require independent, and possibly unavailable, evidence on the orientation of paleoenvironmental gradients.

Axial ordination methods are simply mechanisms for resolving patterns in data structure that may or may not have ecological meaning. When a method's powers are understood through repeated applications, it can be a useful tool for translating ecological patterns into numerical form for further analysis. Different ordination techniques may give much different ordination curves for the same pattern. The curve in Fig. 4, for instance, the combined product of the paleocommunity gradient and the sampling and ordination·scheme, just happens to conform to an exponential pattern. The ideal ordination method would represent a gradient in communities as a precisely defined ordination curve conforming to a simple, tractable form such as this (Fig. 4), best of all perhaps, as the straight line that is very rarely found in practice. Some ordination curves resemble pretzels more closely. Power and utility in resolving ecologically meaningful patterns are the prime criteria for judging a method. Testing a method is a matter of curve-fitting. In order to carry out an accurate analysis of the regression of ordination position on transect position or some other measure of environmental conditions, the form of the relationship must be known. In general it is not, though data may suggest a particular pattern that can itself be tested as to 'fit'.

Gauch and Whittaker have devoted much effort to testing the wide range of ordination methods (see Gauch 1973b; Gauch & Chase 1974; Gauch, Chase & Whittaker 1974; Gauch & Wentworth 1976; Gauch & Whittaker 1972a, b, 1976; Gauch, Whittaker & Wentworth 1977; Noy-Meir & Whittaker 1977; Whittaker 1973). This paper considers the axial methods they consider best among those not requiring assignment of species weights (i.e. weighted averages) nor knowledge of data dimensionality (i.e. non-metric multidimensional scaling), namely, polar ordination (PO), a direct or indirect technique, and principal components analysis (PCA) and reciprocal averaging (RA), two indirect techniques. These multivariate techniques have been rarely used in any connection with gradient analytical coenocorrelation (but see Park 1968; Gevirtz *et al.* 1971). Principal components analysis and its relative principal coordinates analysis, however, have been used extensively in classifying samples in relation to biofacies (see Hazel 1977 for review) as well as in numerical taxonomy of organisms.

All ordination methods have a definite range over which they can provide meaningful resolution of community gradients. Ecologists have explored limitations as to the 'length' of community gradient in half-changes that can be treated satisfactorily by various methods (see references above). The half-change lengths of transects studied (Table 1) fall within the useful ranges of methods considered here. The problem of time range remains to be explored. With originations and extinctions of taxa, the basic structure of a paleocommunity gradient will change over time. The accuracy of ordination methods in coenocorrelation will correspondingly decrease with increasing duration of the time interval considered. One might liken the falling off in accuracy with time to the falling off in accuracy with distance.

Direct ordination

The simplest form of direct ordination is the plotting of measures of species importance against position along a complex gradient (e.g. horizontal transects in Figs. 4, 5). Represented in this fashion, species distributions delineate the general patterns in paleocommunities. Inspection of direct ordination results can be valuable

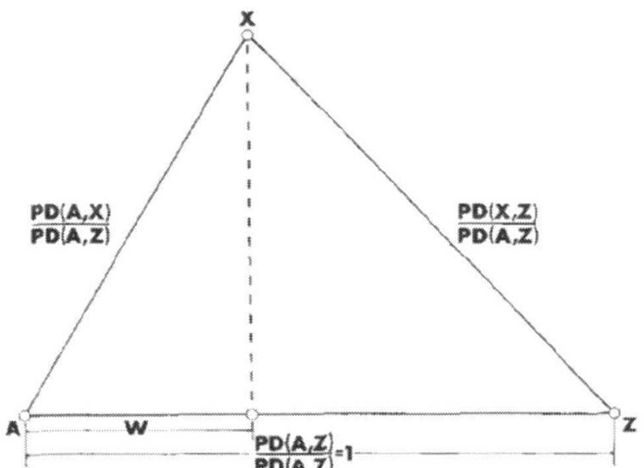

Fig. 8. The geometry of polar ordination as applied in this paper. Sample *X* is ordinated with respect to samples *A* and *Z* based on Percentage Dissimilarity (PD) values to determine ordination position *w* on the ordination axis, which is standardized to unit length.

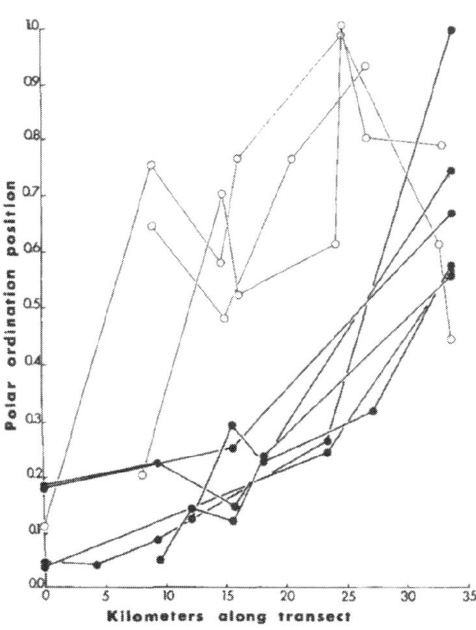

Fig. 9. Paleocommunity ordination curves for nine time-parallel transects down the Trenton Group slope from direct polar ordination with respect to replicate samples from upslope (transect position $R = 0.0$ km) and downslope ($R = 33.6$ km) ends of the M 15 transect. Ordination scores for samples from six transects across the lower half of the sequence (M 2, M 3, M 8, M 11, M 14, M 15; solid circles) increase approximately exponentially with distance down the slope (see Figs. 4, 5). Ordination scores for samples from three transects near the top of the sequence (M 34, M 36, M 38; open circles) increase to a maximum, then decrease. The non-monotonic trend is a distortion effect due to very large Percentage Dissimilarity values between samples at the extreme downslope end of the higher transects and those samples used as endpoints in ordination (see Fig. 8). Such distortion effects limit the usefulness of ordination methods to a certain range along a community gradient.

in choosing appropriate axial ordination methods for further analysis of patterns. There are numerous examples of what could be called direct ordinations in the paleontological literature, plots of the characteristic distributions of taxa, if not their abundances, in relation to lithotope or, in a few instances, to an onshore-offshore gradient (e.g. Elias 1937; Elles 1939). Stratigraphers encountered and dealt with problems of ordination well before ecologists did, though it remained for ecologists to develop quantitative methods for treating them in greater detail. As applied to biofacies, Walther's Law of Facies (Walther 1890–1893; see discussion in 'Conclusion') is a generalization about the distribution of fossil organisms in relation to environmental gradients. Using paleocommunity types rather than biofacies as the classified entities, paleoecologists have found more precise ways of applying Walther's Law in stratigraphic analysis (e.g. Ziegler *et al.* 1968). Coenocorrelation in this case is based on a direct ordination of paleocommunity types in relation to generalized onshore–offshore transect.

Ordination by similarity value. – This is the simplest type of axial ordination. Samples are arranged not in a plane or space defined by coordinate axes but along a half-line or line segment according to similarity value relative to a sample or samples at a particular station. This sort of ordination has been illustrated for PS value (Fig. 7). Its principal importance would seem to be in measuring the shape and length of paleocommunity gradients in transect studies. Ordination results can be important in choosing other ordination methods appropriate to the situation, in checking transects for irregularities not connected with facies change which may indicate miscorrelation of sections, or in determining whether a particular distribution of paleocommunities would be better treated through classification or gradient analysis.

Polar ordination. – This is an axial method for ordinating samples according to their similarity values relative to two endpoint samples (samples A and Z in Fig. 8). In direct ordination, endpoint samples are chosen to represent an environmental gradient. More than one gradient can be represented by defining more than one axis in this way. The similarity value used in following examples is Percentage Dissimilarity (PD):

$$PD = 1 - PS \quad (5)$$

(see Eq. 1). For each ordination, the axis is standardized to unit length so that the expression for the ordination position *w* for some Sample *X* becomes

$$w = \frac{PD^2(A, Z) + PD^2(A, X) - PD^2(X, Z)}{2\,PD^2(A, Z)} \quad (6)$$

(see Fig. 8). The reason for standardization in the present case is the use of replicate samples from certain stations as endpoint samples. All possible combinations between samples at one endpoint station and those at the other are used in defining different axes assumed to represent one and the same complex gradient. For example, pairs at either station are used to define four axes. The assumption is expressed in the standardization of each axis to unit length.

Over a certain range along the paleocommunity gradient, PO gives good ordination curves useful in coenocorrelation (Fig. 9). Endpoint stations, at either end of the M 15 transect, were selected to represent paleocommunities in the range between *Sowerbyella*-dominated ones of extreme upslope areas and *Triarthrus*-dominated ones of regions just downslope from the transition to the graptolitic facies. The choice amounts to selection of the range over which PO could serve in measuring the community gradient in a meaningful way. Ordination curves for the six lower transects (M 2–M 15; M 0 shows no significant gradient) show an approximately exponential increase in ordination position with position on transect (Fig. 9). By design, these samples fall in the useful range in this particular ordination. The curvature is evidently a reflection of accelerating change along the paleocommunity gradient as the facies transition is approached. The curve in Fig. 4 (also shown in Fig. 9) gives an idea of how good the exponential approximation can be. However, data from only two transects show no significant deviations from the exponential pattern (M 14, M 15; at the 0.05 level). Position along the depth gradient accounts for highly significant amounts of the variance in log (ordination position) (65–77%; $p < 0.001$), and in ordination position (59–95%; $p < 0.001$). All ordination curves show significant deviations from linear form. Results from regression analyses according to both models indicate that position on transect is the principal determining factor behind variation in ordination position within the prescribed range along the paleocommunity gradient.

Results for the three highest transects (M 34–M 38) illustrate distortions in ordination curves that result from ordination of samples from parts of a paleocommunity gradient outside the section between endpoint samples (Fig. 9). The ordination curves first increase to a maximum, as seen in curves for lower transects, in going downslope and then decrease toward a *w* value near 0.5. The reason is that lingulid-dominated samples from the extreme downslope ends of the higher transects give PD values near unity with respect to samples used to define either endpoint of the ordination axis. As a result, they are ordinated to positions very near the middle of the axis (see Fig. 8). Thus *w* values near 0.5 can have at least two meanings: in one case, it would correspond to a sample of composition intermediate with compositions of endpoint samples and, in the other, it would correspond to all samples having little or nothing in common with endpoint samples. Good judgment (and the differing heights of ordination triangles, Fig. 8) serves to make what would seem to be an obvious distinction.

Indirect ordination

Polar ordination. – In indirect polar ordination the two endpoint samples are selected automatically as the pair showing maximum dissimilarity according to the similarity measure used. Otherwise, the procedure is the same as in direct polar ordination. Use of only two endpoint samples, as called for in the original recipe (Bray & Curtis 1957), is a vulnerability, particularly if samples having little in common with others are represented in the data set. Samples that happen to show maximum dissimilarity may define an ecologically meaningless axis. This problem could manifest itself in an interesting way in paleontological data sets. If the samples cover too great a time range, the axis could reflect temporal variation more than ecological variation – an interesting way to get at time-correlation but a potential source of inaccuracy in coenocorrelation.

Ordination curves from the indirect technique coincide almost exactly with the direct ordina-

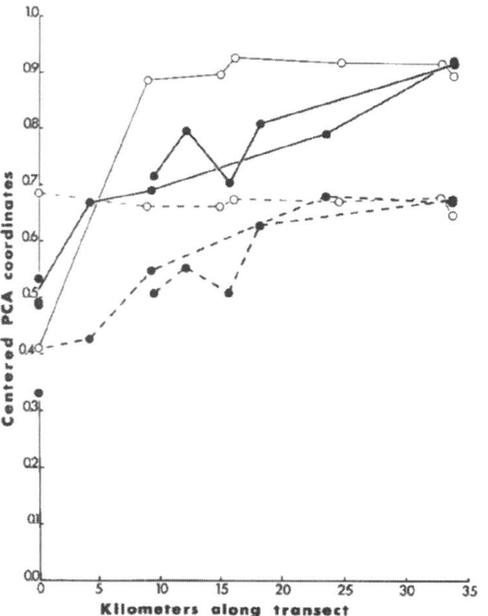

Fig. 10. Paleocommunity ordination curves for three time-parallel transects down the Trenton Group paleoslope from 'centered' principal components analysis. Curves for the first principal component (solid lines) show a more or less linear trend for lower transects (M 2, range $R = 9.3–33.6$ km, and M 15, $R = 0.0–33.6$ km; solid circles) and a markedly curvilinear trend for the transect at the top of the sequence (M 38; open circles). While the first principal component curves show strong sensitivity to the difference between shelly facies (lower R values) and graptolitic facies, curves for the second principal component (dashed line; transects designated as above) do not. The fact that neither set of curves covers more than half of the unit length axis indicates that the ordination is less sensitive to the main paleocological effect, the depth-related paleocommunity gradient, than polar and reciprocal averaging ordinations (see Figs. 9, 13).

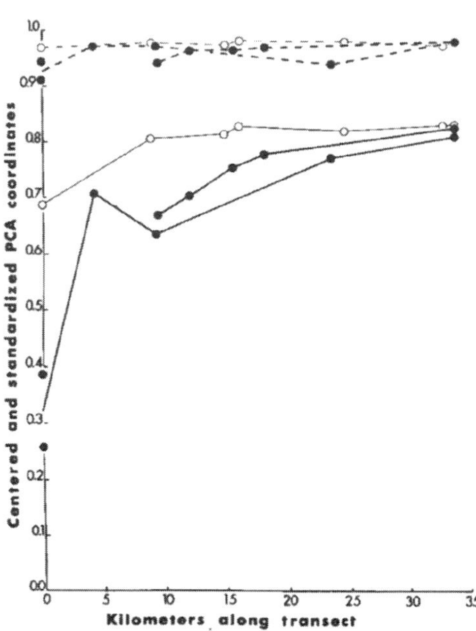

Fig. 11. Paleocommunity ordination curves from 'centered and standardized' principal components analysis for the same three transects diagrammed in Fig. 10. Aside from the first principal component's (solid lines) sensitivity to relatively small differences among shelly facies paleocommunities, both sets of curves indicate the method's insensitivity to the paleocommunity gradient.

tion curves (Fig. 9). Ordination positions are very highly correlated (e.g., $r = 0.97$, $N = 56$ for data from the M 15 transect). The reason for the similarity is selection of endpoint samples different from, but very similar to, endpoint samples chosen in direct ordination. While the latter choice was arbitrary, it is none the less reassuring that the same useful results can be obtained by indirect methods not requiring knowledge of precise time relationships nor of paleoenvironmental gradients. Despite the fact that automatically selected endpoint samples come from different transects (M 15, M 38), ordination is not affected appreciably.

Principal components analysis. – In R-mode PCA, samples are represented as points in species-defined space. Each species is taken to define an axis calibrated in some measure of species importance. Some number N samples in which S species are represented would be depicted as a cloud of N points in S-dimensional space. The shape of this cloud is characterized using basic methods of linear regression analysis applied in S dimensions. The directions of maximum variance are extracted by translation and rotation of mutually perpendicular axes. These are ranked in order of the proportion of the total variance for which they account, beginning with the first, which accounts for the most. In the basic method, here called 'Centered PCA' (PCA-CENT), the origin is translated from the zero-point of all axes to the centroid of points, as in regression analysis, and the first axis is rotated to the position in which it coincides with the maximum dimensions of the cloud, again as in regression analysis (see Pielou 1969). The second and successive axes are extracted in the same manner subject to the constraint that they be perpendicular to all previously extracted

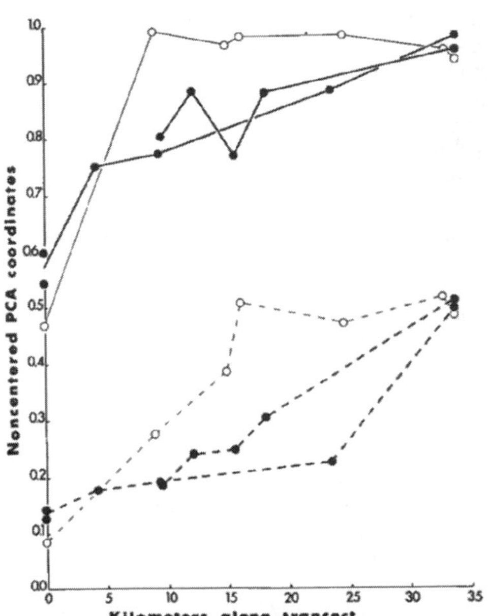

Fig. 12. Paleocommunity ordination curves from 'non-centered' principal components analysis for the same three transects diagrammed in Fig. 10. Both the first (solid lines) and second (dashed lines) principal components curves show a marked sensitivity to the paleocommunity gradient, though they express it in variably linear to curvilinear form very much as in 'centered' principal components analysis (Fig. 10).

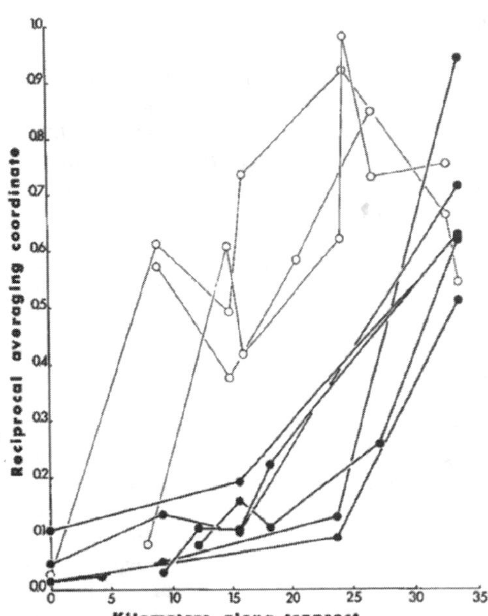

Fig. 13. Paleocommunity ordination curves from reciprocal averaging for the same nine transects diagrammed in Fig. 9. Curves for transects across the lower portion of the sequence (M 2–M 15; solid circles) show an exponential increase in ordination position down the slope. Curves for transects near the top of the sequence (M 34–M 38; open circles) show much the same distortion as polar ordination curves.

ones. A fundamental problem with PCA is that relationships among species importance values are assumed to be of linear form where in reality they are not (e.g. Figs. 4, 5). Sometimes logarithmic, square root, or other transformations are used in the attempt to improve ordinations. In the present instances simple relative abundance is used. In the attempt to standardize possibly non-comparable measures, or to give greater weighting to rarer species, axes can be standardized to unit variance (see Pielou 1969). This procedure leads to 'centered and standardized PCA' (PCA – C + ST). Noy-Meir (1971) suggested a technique known as 'non-centered PCA' (PCA–NONC) in which the origin is retained at the zero point of all axes, a procedure that can be useful in analyzing transect data covering many half-changes or showing discontinuities.

Each of the three PCA variants gives ordination curves of varying form and utility (Figs. 10–12). Curves for at least the first principal component in each case express the main paleoecological effect, the depth-related paleocommunity gradient. Position on transect accounts for highly significant amounts of the total variance in first principal component ordination position on the three transects diagrammed (M 2, M 15, M 38) – 44–85% for PCA–CENT (Fig. 10), 42–58% for PCA–C+ST (Fig. 11), 36–80% for PCA–NONC (Fig. 12). The first axis accounts for respectively 46%, 8% and 51% of the total variance in samples compositions as plotted in species-defined space. In other words, ordination positions are roughly as sensitive to regularities within the body of data that reflect the paleocommunity gradient as they are to regularities that do not. This is apparent in the scaling of points on the ordination curves. Axes are of unit length, yet first and second principal component ordination curves cover no more than half this length. Second axis ordination curves from PCA–CENT and PCA–C+ST show at best weak relationships between ordination position and transect position. Second axis curves from PCA–NONC, however, show relationships comparable with first axis ordinations. One of them in fact comes very close to the ideal, linear ordination curve.

As compared with other indirect ordination methods studied, the one interesting, potentially important feature of PCA is its ability to detect longer range trends along the paleocommunity gradient. While PCA–C+ST ordination curves (Fig. 11) serve only to distinguish extreme upslope paleocommunities from everything else, the basically monotonic PCA–CENT and PCA–NONC curves (Figs. 10, 12) indicate the regularity of the paleocommunity gradient in a way that non-monotonic PO and RA curves do not.

Reciprocal averaging. – RA can be approached through eigenanalysis, as PCA is, or through iterative treatment of a matrix of species importance values in obtaining solutions in which the weight assigned a species (i.e. a multiplier of importance values) equals the average score of samples in which the species occurs. Hill (1973) lucidly explains the method as applied to presence-absence data. Continuous species importance data are treated in the same way (Hill 1974). To begin with, species are assigned arbitrary weights that may or may not express their positions along a community gradient. Random weights are used in computer calculations. Samples are then scored according to the sum of weighted species importance values. Species are again weighted according to the average score for samples in which they occur. As back and forth iteration proceeds, a unique ordering of species and samples is approached by successive approximation. There are also 'trivial' solutions with the reciprocal averaging property. These correspond to ordinations along axes other than the first principal component in PCA.

First axis ordination curves (Fig. 13) are very similar to PO curves (Fig. 9), which is surprising in view of their much different mathematical derivation. Except for greater curvature of RA ordination curves for lower transects (M 2–M 15), RA and PO represent the paleocommunity gradient in essentially the same way, even to the point of showing similar distortions for lingulid-dominated samples from downslope ends of the higher transects (M 34–M 38). The fact that the RA axis is derived by reaching a sort of consensus among all samples and taxa, rather than by simple selection of the two most dissimilar samples as in indirect PO, tends to give one greater confidence in RA ordinations. Thus the correspondence between RA curves and direct PO curves is specially reassuring as regards the time range over which the two methods would provide meaningful ordinations, about two million years in the present case. Very accurate time-correlation is evidently not necessary for obtaining good ordinations, direct or indirect.

Apart from distortion outside a certain range along the paleocommunity gradient, RA gives the ordination curves that conform most closely to one simple pattern. Ordination curves for the six lower transects (Fig. 13) suggest an exponential pattern. Regression analyses show no significant departures from this pattern in each case (by F-test, at 0.05 level). Position on transect accounts for 71–92% of the log (ordination position). These results suggest that RA would be particularly useful in coenocorrelation. The exponential pattern in the present cases is in part a consequence of accelerated change along the paleocommunity gradient in going from shelly to graptolitic facies. Gauch *et al.* (1977) found first axis ordination curves for simulated continuous community gradients to be more nearly linear. The main point is that the basic form is simple.

As Gauch *et al.* (1977) noted for analyses for field data, second and higher axes may or may not have ecological meaning and, in particular, may only serve to separate one very distinctive sample from all others. A sample dominated by an otherwise rare species makes for a 'trivial' solution, in which great weight is assigned to this species and negligible weights to all others. This is quite obvious when it occurs, as it did for the second axis in the present case. Another distinctive sample served in this way to concentrate other samples toward the opposite end of the third through fifth ordination axes. While only a few samples dominated by otherwise rare species were represented in the data set, they had a profound effect insofar as dispersing the cloud of points representing samples in species-defined space. This is reflected in some potentially misleading statistics: according to a linear model, the first ordination axis accounts for 19% of the total variance, the second for 14%, and the first five for 61%. From this evidence of internal consistency, which is often all there is on which to judge an ordination, one would surmise that PCA–CENT and PCA–NONC were much superior to RA. In actuality, RA resolves the paleocommunity gradient more meaningfully, provided 'trivial' axes can be distinguished from the principal one.

Discussion

Axial ordination methods served to resolve the Trenton Group paleocommunity gradient with varying clarity. Among indirect methods, PO and RA gave comparably good ordination curves, better curves than PCA. Gauch *et al.* (1977) reached basically the same conclusion from tests using field data and simulated community gradients. However, they found lesser degrees of distortion in PO and RA ordination curves covering the same range in half-changes along a community gradient. For several reasons, distortions may be relatively more pronounced in present ordinations. There is an inflexion in the paleocommunity gradient across the facies transition. Use of replicate samples leads to possible underestimation of half-change length along transects, as do deviations from exponential decay of similarity value along the transect. Perhaps the most important factors are the taphonomy of *Triarthrus* and lack of precision in identifications of lingulid brachiopods. Taken at face value PO and RA ordination curves for the higher transects (M 34–M 38, Figs. 9, 13) appear to indicate a reversal in the paleocommunity gradient, as if the transects proceeded up a slope on the other side of the basin. One can even construct a geological interpretation that explains the pattern while ignoring a mass of evidence to the contrary: the Adirondack Arch (see Kay 1937), a topographic high in earlier Ordovician time, formed the basin's opposite wall. The point is that erroneous interpretations can arise from problems with data and from limitations in the useful range of ordination methods.

Results from gradient analysis

As has been shown, appropriately chosen ordination methods can translate the distributions of taxa along transects, or the mutual associations of taxa in samples, into a numerically calibrated scale that measures paleoenvironmental conditions in a meaningful way. Using scales from various ordination methods, it is possible to obtain good estimates of the orientation of paleoenvironmental gradients even when other evidence is unavailable. It is further possible to trace transgressive and regressive changes through a single stratigraphic section, and to correlate patterns of change from section to section.

In the Trenton examples, direct and indirect ordination give fairly accurate estimates of the orientation of the slope as otherwise determined from physical evidence. The two-dimensional ordination curves (Figs. 7, 9–13) represent three-dimensional *ordination surfaces* that are, so to speak, viewed on edge, as if one were looking along depositional strike. After getting some idea as to the form of the relationship, the attitude of the ordination surface can be determined by multiple regression methods. One can in effect measure the strike and dip of a paleocommunity gradient. On a slope, the strike of the gradient is likely to coincide with depositional strike. The slope's direction of dip is otherwise a matter of interpretation.

Given a rough idea of time relations and depositional strike from stratigraphic and paleontological evidence, the estimate of strike can be improved upon using direct ordination. A scenario can be constructed for the Trenton example in which depositional strike is measured almost as accurately without knowledge of bentonites and slope indicators; a scenario that may correspond to a more typical field situation. Field observations make clear the facies transition from east to west (see Cushing 1905). Intertonguing relationships through the lower half of the sequence establish which rocks represent the extremes in depositional environment, and provide a rough time-correlation among sections (see Figs. 2, 3; and Kay 1953). Thus, without necessarily resorting to biostratigraphic correlation, a certain time-constrained sampling interval can be set off and a knowledgeable choice of endpoint samples for polar ordination can be made. To judge from present results, choice of one or another more or less representative sample is not critical. Samples from the M 15 transect used in direct PO are quite similar to others from their surrounding bodies of rock, and the selection of more or less contemporaneous samples such as these could have been based on intertonguing relationships rather than bentonites. Indirect PO provides an objective choice of endpoint samples; and in the present instance, the choice was quite similar to the one made in direct PO. Despite the fact that the time-correlation, based on transgressive and regressive change, is bought at the expense of not having a stationary paleocommunity gradient, the ordination surface defined collectively by samples along the M2–M 15 transects gives a remarkably accurate estimate of depositional strike: $-13°$ (N 13° W) versus $7 \pm 17°$ (2σ) (N 7° E) as determined from eighteen measurements

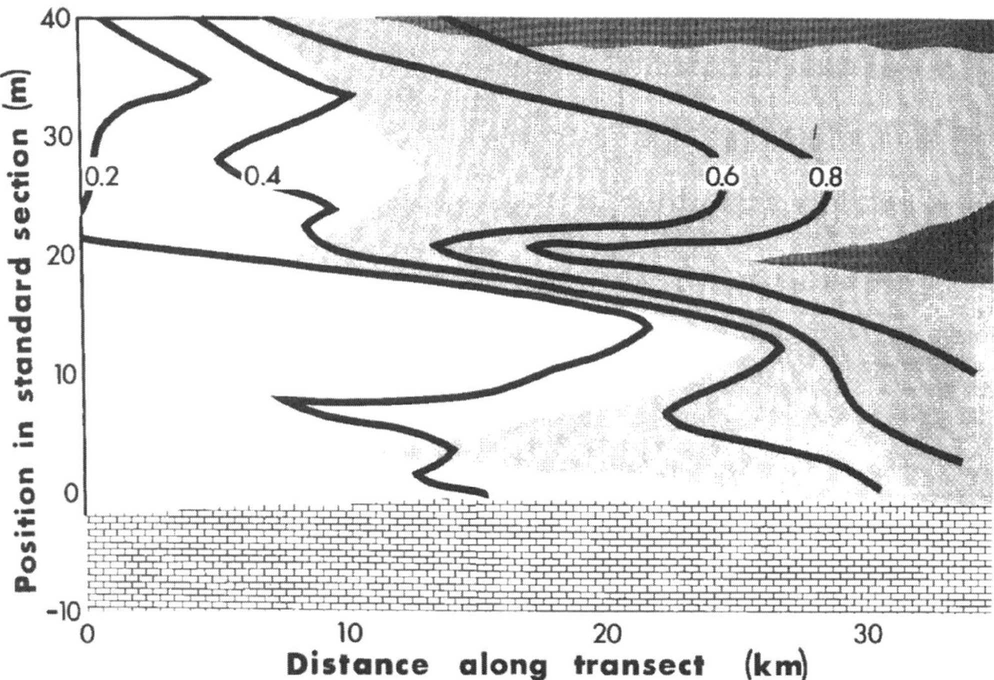

Fig. 14. Isocoene ('equal commonality') diagram for the Trenton Group sequence studied – a contour diagram of ordination position on a time-space diagram (names of stratigraphic units and locations of sampling stations are shown in Fig. 3, the base diagram) based on direct polar ordination of 168 samples with respect to replicate samples at the extreme upslope (transect position $R = 0.0$ km) and downslope ($R = 33.6$ km) ends of the M 15 transect (see Figs. 4, 5). The curves are lines of equal ordination position as determined by averages of ordination positions of samples within five meter by five kilometer squares on the grid (see Fig. 9 for the relationship between ordination position and position along the slope). Contours beyond 0.8 are 'off scale' owing to distortion effects (see Fig. 9). Biotopes, as measured through ordination, and lithotopes have basically parallel distributions. Ordination of paleocommunities reveals transgressions and regressions that would otherwise be inferred from lithostratigraphic evidence. Moreover, it measures facies change in such a way that patterns and relative rates of facies change can be appraised. The overall pattern is a transgression, as indicated by change in ordination position such as would accompany shoreward movement of coastline (see Fig. 9), that proceeded through a series of smaller transgressions and relatively slower regressions.

on paleocurrent indicators. Estimates from individual transects are likewise nearly coincident with estimates from physical evidence, the vector mean being $-20 \pm 20°$ (2σ) (N 20° W). Geographic position accounts for 67–97% of the variance in ordination position for the individual transects, or not much more than position on transect does in ordination curves, and 68% of the variance for the transects collectively.

The RA ordination, which could have been obtained without knowing the orientation of the paleoenvironmental gradient, gives very similar results. The nearly planar ordination surface for log (RA position) of samples on transects M 2–M 15 estimates strike as $-3°$ (N 3° W). Planes for individual transects give a vector mean direction of $-13 \pm 61°$ (2σ) (N 13° W). The reason for the large error of estimate is that results from the M 2 transect differ substantially from others. Geographic position accounts for 71–94% of the variance in log (RA position) for individual transects, scarcely more than position on transect does in ordination curves, and for 68% of the variance for the transects collectively.

The general pattern can be detected by other means. Similarity value ordination using PS (see Fig. 7 and Eq. 3) gave similar though generally less precise strike estimates than PO and RA. Porter & Park (1969) detected a general east-west trend in paleocommunities from lower Trenton rocks using R-mode cluster analysis.

Coenocorrelation diagrams, in which ordination positions are related to positions in a stratigraphic sequence, can be useful means for summarizing and presenting information on a paleoecocline for examination and further analysis. The *isocoene diagram* (a contour map of ordination position on a stratigraphic cross section or time-space diagram) ('isocoene' signifies 'equal commonality') in Fig. 14 shows the correspondence between contoured ordination position and lithology in the Trenton case. Ordination position reflects lithology rather closely. Transgressions and regressions one would infer from ordination data by and large correspond to transgressions and regressions one would infer from lithologic evidence. The words 'transgression' and 'regression' take on specific meanings in connection with coenocorrelation patterns. A transgression or regression in this sense is indicated by a change in ordination position such as would accompany the appropriate shift in the location of a paleocommunity gradient with change in water depth. Biostratigraphic and lithostratigraphic indications need not coincide. Ordination data can be misleading in this regard if a method's peculiarities and limitations are not well enough understood that distortion effects can be recognized for what they are.

The basic pattern in Fig. 14 is a transgression (overall increase in ordination position upwards in the sequence on the particular direct PO scale). Two episodes of regression climaxed at times corresponding approximately to the 15 and 30 meter points in the composite section. These separate three episodes of transgression. The isocoenes bear a certain resemblance to sawtooth global sea-level curves deduced from seismic data (Vail, Mitchum & Thompson 1977). However, just the reverse of the sea-level curves, the isocoenes appear to indicate generally more rapid transgression than regression. The effect is somewhat enhanced by increased net rate of rock accumulation in downslope areas. (Note the thickening of the sequence down dip in Fig. 2.) The composite section does not provide a strictly linear time scale. The difference in pattern is likely due to differences in geographic scale and tectonic setting. Coenocorrelation patterns probably reflect the tectonics of a growing basin more than they do global sea-level. Perhaps gradient analytic methods will prove useful in further resolving the still mysterious sea-level patterns and in getting at the dynamics behind them.

Transgressive and regressive changes within

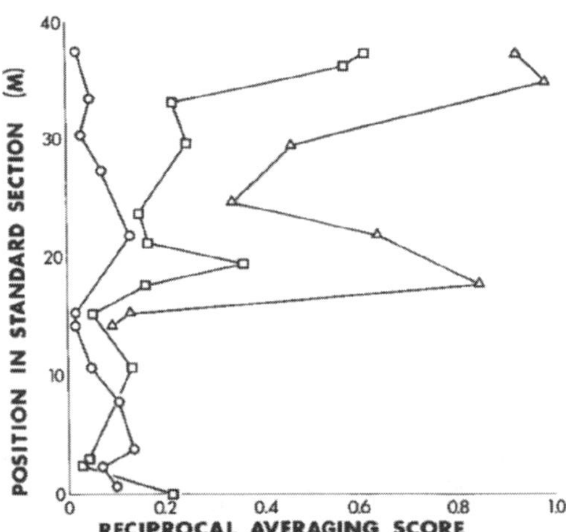

Fig. 15. Coenocorrelation curves for Trenton Group sections at transect positions $R = 0.0$ km (left), 8.9–9.3 km (middle), and 23.4–24.5 km (right). Each curve is a plot of reciprocal averaging ordination position (averaged over samples from each station) versus position of samples in their respective stratigraphic section. Section thicknesses have been standardized with respect to the composite section so that time lines are horizontal (see Fig. 3). All three sections record the pronounced regression (change in ordination position such as would accompany a seaward shift of coastline; see Fig. 13) near the 15 meter point. Sections from the downslope areas (right hand curves) record a pronounced pulse of transgression just above this point that coincides so well that, like the regressive pulse, it would serve to time-correlate the sections nearly as accurately as the bentonite beds actually used to correlate them. The weak reciprocal relationship between seeming transgression in extreme upslope areas (left curve) and regression in lower slope areas near the five and twenty meter points may indicate a tectonic hinge at the edge of the turbidite basin between transect positions $R = 0.0$ km and $R = 8.9$–9.3 km (see text).

an individual stratigraphic section can be revealed in a *coenocorrelation curve,* a plot of ordination positions against positions of samples in the section. In coenocorrelation curves for the three longest sections studied (Fig. 15), note the correspondence in peaks for transgressions (near the 20 and 40 meter marks) and regressions (near the 15 and 30 meter marks). Maxima and minima in the curves parallel bentonites quite closely. The patterns for Rathbun Brook (squares) and North Creek (triangles), downslope sections that cover the transition from shelly to graptolitic facies over time, are more pronounced owing to the curvilinear dependence of RA ordination position on distance down the slope (see Fig. 13). The Trenton Falls section (circles) at the extreme upslope end shows small but interesting discrepancies from the patterns for the two others.

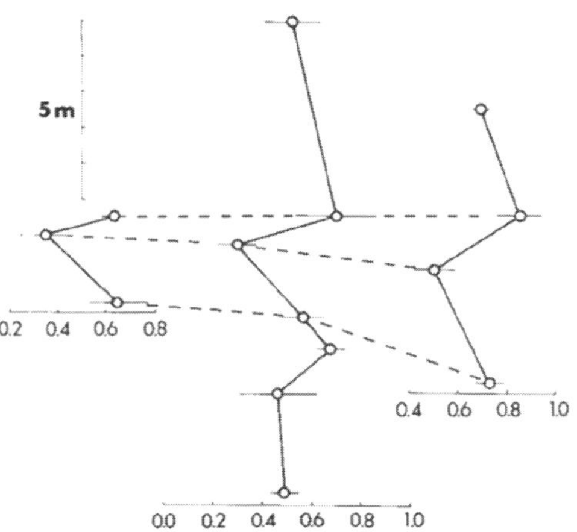

Fig. 16. Coenocorrelation curves for three relatively short sections at transect positions $R = 12.2$ km (left), 15.6 km (middle), and 18.1 km (right) along the downslope transect through the Trenton Group sequence studied. Each curve is a plot of direct polar ordination position (relative to extremes of shelly facies samples from the M 15 transect at positions 0.0 km and 23.4 km) versus position of samples in their respective sections. Curves show a regressive pulse (change in ordination position such as would accompany a seaward movement of coastline) that correlates well from section to section relative to three bentonite beds (dashed lines; M 0, M 2, M 3). This relatively small change has been detected by shortening the ordination axis (see Fig. 8) to make the method more sensitive, hence the relatively long one standard deviation error bars. The pulse marks a small event of potential importance for time-correlation.

Near the 5 and 25 meter points, the Trenton Falls curve indicates change in the opposite sense of downslope sections, as if transgression were taking place there while regression was taking place in downslope areas. While the discrepancies may represent no more than noise, they could also be taken no indicate a tectonic hinge at the edge of the basin located somewhere around the five km point on transects. There is independent evidence of a small dome, the possible hinge, at just this point, and clear evidence of its activity in times before and after deposition of the rocks in question (Kay 1953, Fig. 32) but otherwise no indication of its activity in intervening times. The point is that coenocorrelation curves can perhaps be used to detect rather subtle features in the local paleogeography and tectonics.

Coenocorrelation curves can be used to detect transgressive and regressive changes on a smaller scale. Coenocorrelation curves for three short sections adjacent to one another along the transect show a pulse of regressive change that can be correlated from section to section relative to three bentonite beds (Fig. 16). The event this pulse records – a marked increase in the abundance of the brachiopod *Sowerbyella* within a few tens of cm above and below the M 2 bentonite that can be traced over about 10 km from section to section down the slope (Figs. 4, 5) – probably took place over a geologically very short period of time, possibly no more than a few decades. While increased abundance of this brachiopod is otherwise associated with regression (as in the Denley Limestone tongue near the 15 m point in the standard section) the pulse in coenocorrelation curves may simply represent a large fluctuation in the *Sowerbyella* population or a change in taphonomic factors affecting its shells, not a marine regression in the usual sense. For practical purposes, the possibly very complicated reasons for the increase in abundance are beside the point. The coenocorrelation curves merely resolve an event, the nature of which is a matter of interpretation, that appears to have practical significance in time-correlation.

Conclusion

As applied to biofacies, Walther's Law of Facies (Walther 1890–1893) is a far more precisely quantifiable generalization than it has been revealed to be through traditional classificatory approaches to paleocommunity phenomena. Evidence presented suggests that it can be restated, perhaps in a less broadly applicable manner, in terms of gradient analysis of paleocommunities: the paleocommunity gradient found at a given time along a depth-related paleoenvironmental complex gradient is repeated vertically in sections through a transgressive or regressive stratigraphic sequence. Walther's Law is evident from the distribution of taxa in the sequence studied (Figs. 4, 5). While such obvious transgressive and regressive patterns can be described adequately in terms of biofacies and paleocommunity types, they can be described more precisely through axial ordination of samples. The paleocommunity gradient in short time intervals can be characterized, sometimes quite precisely, by ordination curves (Figs. 7, 9–13). Coenocorrelation diagrams (Figs. 14, 15) show repeated transgressive and regressive fluctuations in ordination positions that parallel transgressive and regressive changes in the distribution of rock units.

Present gradient analyses suggest that the fossil record contains more and better information on paleoenvironments than has been commonly believed, quantifiable information contained in the distribution and abundance of taxa that can be recovered by appropriate quantitative sampling and analysis. While one cannot help but be struck by the variety of fossil assemblages from surficial collecting at an outcrop, and by the overwhelming number of possibly random factors that could have contributed to this variety, differences tend to average out on a larger geographic scale, and surprisingly well-defined patterns can emerge. In present examples, ordinations gave apparently quite accurate estimates to depositional strike. Geographic position along the depth gradient accounted for the greatest amount of variance in ordination position from better methods. Other factors such as transportation and differential preservation are evidently not so important, at least as random factors, on this larger scale.

These analyses further suggest that Walther's Law of Facies can be applied more rigorously than has been done in quantitative facies analysis, and that such application can lead to results of importance in time-correlation. The fact that certain ordination methods can resolve transgressive and regressive changes precisely enough that reversals could be used to time-correlate sections practically as accurately as bentonites illustrates at once the precision of the information available to be recovered and the precision with which it actually can be. This finding also illustrates the big problem with biostratigraphic correlation as time-correlation, sensitive paleoecological and taphonomic control of distribution and abundance of taxa. This problem can be turned to an advantage. Through coenocorrelation, fossils can be used to detect paleoenvironmental events of importance in time-correlation that may be far more tightly constrained in time than originations and extinctions of taxa. Much remains to be done in developing gradient analytical methods for this purpose as well as for exploring the 'evolution' of paleocommunity gradients.

Acknowledgements. – R. H. Whittaker and H. G. Gauch helped us very much with gradient analysis. J. J. Sepkoski, K. M. Waage, and R. H. Whittaker reviewed earlier versions of this paper. We specially benefited from conversations with R. K. Bambach, B. Cameron, D. W. Fisher, M. Kay, and C. Mehrtens (who also kindly lent us large thin sections of Trenton Falls strata). Many people in the Mohawk Valley vicinity kindly cooperated with our investigations. We thank these friends for their help. Errors are our own. Work was supported by the National Science Foundation Grants DEB 75-09651 A01 and DEB 78-03179.

Whatever worth is found in this paper is dedicated to Ralph Gordon Johnson (1927–1976). To have studied with him was a privilege and a pleasure we both cherish. His memory is a continuing inspiration.

References

Bird, J. M. & Dewey, J. F. 1970: Lithosphere plate-continental margin tectonics and the evolution of the Appalachian orogen. *Geol. Soc. Am. Bull. 81,* 1031–1060.

Bray, J. R. & Curtis, J. T. 1957: An ordination of the upland forest communities of southern Wisconsin. *Ecol. Monogr. 27,* 325–349.

Bretsky, P. W. 1969: Central Appalachian Late Ordovician communities. *Geol. Soc. Am. Bull. 80,* 193–212.

Bretsky, P. W. 1970: Upper Ordovician ecology of the central Appalachians. *Peabody Mus. Nat. Hist. (Yale Univ.) Bull. 34.* 150 pp.

Chenoweth, P. A. 1952: Statistical methods applied to Trentonian stratigraphy in New York. *Geol. Soc. Am. Bull. 63,* 521–560.

Cisne, J. L. 1973: Beecher's trilobite bed revisited: ecology of an Ordovician deepwater fauna. *Postilla 160.* 25 pp.

Cisne, J. L. 1975: Anatomy of *Triarthrus* and the relationships of Trilobita. *Fossils and Strata 4,* 45–63.

Cushing, H. P. 1905: Geology of the vicinity of Little Falls, Herkimer County. *N.Y. St. Mus. Bull. 77.* 95 pp.

Elias, M. 1937: Depth distribution of the Big Blue sediments in Kansas. *Geol. Soc. Am. Bull. 48,* 403–432.

Elles, G. L. 1939: Factors controlling graptolite succession and assemblages. *Geol. Mag. 76,* 181–188.

Gauch, H. G., Jr. 1973a. The relationship between sample similarity and ecological distance. *Ecology 54,* 618–622.

Gauch, H. G., Jr. 1973b: A quantitative evaluation of the Bray-Curtis ordination. *Ecology 54,* 829–836.

Gauch, H. G., Jr. & Chase, G. B. 1974: Fitting the Gaussian curve to ecological data. *Ecology 55,* 1377–1381.

Gauch, H. G., Jr., Chase, G. B. & Whittaker, R. H. 1974: Ordination of vegetation samples by Gaussian species distributions. *Ecology 55,* 1382–1390.

Gauch, H. G., Jr. & Wentworth, T. R. 1976: Canonical correlation analysis as an ordination technique. *Vegetatio 33,* 17–22.

Gauch, H. G., Jr. & Whittaker, R. H. 1972a: Coenocline simulation. *Ecology 53,* 446–451.

Gauch, H. G., Jr & Whittaker, R. H. 1972b: Comparison of ordination techniques. *Ecology 53,* 868–875.

Gauch, H. G., Jr. & Whittaker, R. H. 1976: Simulation of community patterns. *Vegetatio 33,* 13–16.

Gauch, H. G., Jr., Whittaker, R. H. & Wentworth, T. R. 1977: A comparative study of reciprocal averaging and other ordination techniques. *J. Ecol. 65,* 157–174.

Gevirtz, J. L., Park, R. A. & Friedman, G. M. 1971: Paraecology of benthic foraminifera and associated micro-organisms of the continental shelf off Long Island, New York. *J. Paleontol. 45,* 154–177.

Goodall, D. W. 1973: Sample similarity and species correlation. *In:* Whittaker, R. H. [ed.]: *Handbook of Vegetation Science V,* 105–156. Junk, The Hague.

Gould, S. J. & Eldredge, N. 1977: Punctuated equilibria: the tempo and mode of evolution reconsidered. *Paleobiology 3*, 115–150.

Hazel, J. E. 1977: Use of certain multivariate and other techniques in assemblage zonal biostratigraphy: examples utilizing Cambrian, Cretaceous, and Tertiary benthic invertebrates. *In:* Kauffman, E. G. & Hazel, J. E. [eds.]: *Concepts and Methods of Biostratigraphy*, 187–212. Dowden, Hutchinson & Ross, Stroudsburg, Pennsylvania.

Hill, M. O. 1974: Correspondence analysis: a neglected multivariate method. *J. Roy. St. Soc., Ser. C, 23*, 340–354.

Hill, M. O. 1973: Reciprocal averaging: an eigenvector method of ordination. *J. Ecol. 61*, 237–249.

Johnson, R. G. 1970: Variations in diversity within benthic marine communities. *Am. Nat. 104*, 285–300.

Johnson, R. G. 1971: Animal-sediment relations in shallow water benthic communities. *Mar. Geol. 11*, 93–104.

Johnson, R. G. 1972: Conceptual models of benthic marine communities. *In:* Schopf, T. J. M. [ed.]: *Models in Paleobiology*, 148–159. Freeman, Cooper & Co., San Francisco, Calif.

Kay, G. M. 1937: Stratigraphy of the Trenton Group. *Geol. Soc. Am. Bull. 48*, 233–302.

Kay, G. M. 1943: Mohawkian Series on West Canada Creek, New York. *Am. J. Sci. 241*, 597–606.

Kay, M. 1953: Geology of the Utica Quadrangle, New York. *N.Y. St. Mus. Bull. 347*. 126 pp.

Martinsson, A. 1973: Editor's column: Ecostratigraphy. *Lethaia 6*, 441–443.

Martinsson, A. 1978: Project Ecostratigraphy. *Lethaia 11*, 84.

Miller, W. J. 1909: Geology of the Remsen Quadrangle. *N.Y. St. Mus. Bull. 126*. 51 pp.

Noy-Meir, I. 1971: Multivariate analysis of the semi-arid vegetation of southeastern Australia: nodal ordination by component analysis. *Proc. Ecol. Soc. Aust. 6*, 159–193.

Noy-Meir, I. & Whittaker, R. H. 1977: Continuous multivariate methods in community analysis: some problems and developments. *Vegetatio 33*, 79–98.

Park, R. A. 1968: Paleoecology of *Venericardia sensu lato* (Pelecypoda) in the Atlantic and Gulf Coastal Province: an application of paleosynecologic methods. *J. Paleontol. 42*, 955–986.

Pielou, E. C. 1969: *An Introduction to Mathematical Ecology*. 286 pp. Wiley-Interscience, New York, N.Y.

Porter, L. A. & Park, R. A. 1969: Biofacies analysis of the Middle Trenton (Ordovician) of New York State. *Geol. Soc. Am. Abstr. Progr. N.E. Reg. Mtg. 49*.

Ruedemann, R. 1926: Faunal and facies differences of the Utica and Lorraine shales. *N.Y. St. Mus. Bull. 267*, 61–78.

Ruedemann, R. 1935: The ecology of black mud shales of eastern New York. *J. Paleontol. 9*, 79–91.

Sanders, H. L. & Hessler, R. R. 1969: Ecology of the deep-sea benthos. *Science 163*, 1419–1424.

Sepkoski, J. J., Jr. 1975: Quantified coefficients of association and measurement of similarity. *Math. Geol. 6*, 135–152.

Shaw, A. B. 1964: *Time in Stratigraphy*. 365 pp. McGraw-Hill, New York, N.Y.

Terborgh, J. 1971: Distribution on environmental gradients: theory and a preliminary interpretation of distributional patterns in the avifauna of the Cordillera Vilcabamba, Peru. *Ecology 52*, 23–40.

Titus, R. & Cameron, B. 1976: Fossil communities of the lower Trenton Group (Middle Ordovician) of central and northwestern New York State. *J. Paleontol. 50*, 1209–1225.

Vail, P. R., Mitchum, R. M., Jr. & Thompson, S. III. 1977: Global cycles of relative changes of sea level. *In:* Vail, P. R. & Mitchum, R. M., Jr. [eds.]: Seismic Stratigraphy – Applications to Hydrocarbon Exploration. *Assoc. Petrol. Geol. Mem. 26*, 83–97.

Walther, J. 1890–1893: *Einleitung in die Geologie als historisch Wissenschaft 1–3*. 1055 pp. G. Fisher, Jena.

Warme, J. E., Ekdale, A. A., Ekdale, S. F. & Peterson, C. H. 1976: Raw material of the fossil record. *In:* Scott, R. W. & West, R. R. [eds.]: *Structure and Classification of Paleocommunities*, 143–169. Dowden, Hutchinson & Ross, Inc., Stroudsburg, Pennsylvania.

Whittaker, R. H. 1960: Vegetation of the Siskiyou Mountains, Oregon and California. *Ecol. Monogr. 30*, 279–338.

Whittaker, R. H. 1967: Gradient analysis of vegetation. *Biol. Rev. 42*, 207–264.

Whittaker, R. H. 1973a: Direct gradient analysis: results. *In:* Whittaker, R. H. [ed.]: *Handbook of Vegetation Science* V, 35–51. Junk, The Hague.

Whittaker, R. H. 1973b: Evaluation of Ordination techniques. *In:* Whittaker, R. H. [ed.]: *Handbook of Vegetation Science* V, 289–321. Junk, The Hague.

Whittaker, R. H. 1975: *Communities and Ecosystems*. 2nd Ed. 385 pp. Macmillan & Co., New York, N.Y.

Ziegler, A. M. 1965: Silurian marine communities and their environmental significance. *Nature 207*, 270–272.

Ziegler, A. M., Cocks, L. R. M. & Bambach, R. K. 1968: The composition and stucture of Lower Silurian marine communities. *Lethaia 1*, 2–27.

PAPER 10

Silurian Marine Communities and Their Environmental Significance (1965)

A. M. Ziegler

Commentary

THOMAS D. OLSZEWSKI

At first glance, Ziegler's paper using fossil assemblages to recognize a marine depth gradient in the Silurian rocks of Wales seems like a solid, but hardly innovative, piece of paleoecological research. The idea of depth-related benthic faunal zones in modern seas and their significance for interpreting ancient sedimentary rocks was well established by the beginning of the twentieth century (Forbes 1844; Petersen 1924). In 1937, Elias applied this concept to the fossil record to infer water depth changes in a cyclic stratigraphic succession from the Permian of Kansas, anticipating many aspects of Ziegler's work by several decades. To a paleontologist in the twenty-first century, Ziegler's approach seems familiar, a straightforward application of basic paleoecological principles.

However, the importance of Ziegler's paper has to be seen in the context of its time. Prior to the 1960s, the major research focus of paleoecology was to document the environmental and geographic distribution of fossils and to use knowledge of the natural history of fossil organisms to interpret sedimentary facies (e.g., virtually all the papers under the heading "Selected Analyses from the Geologic Record" in Ladd's monumental 1957 *Treatise on Marine Ecology and Paleoecology*). Although the main question that Ziegler tackled—identifying an ancient marine depth gradient—was clearly rooted in the older tradition of paleoecology, by recognizing and mapping distinct assemblages he demonstrated that communities were distributed *independently* of lithofacies. Communities were not merely aspects of sedimentary facies, they were distinct entities that presumably could only be fully understood in terms of the ecology of their component organisms.

Ziegler's work "signaled a major line of research" (Schopf 1972) in paleoecology: a community-based approach focused on ecological processes (Fagerstrom 1964; Johnson 1964) that allowed paleoecologists to explore the history of life in a novel ecological context (e.g., Bretsky 1968; Valentine 1973). What makes this a foundational paper is how it redefined what could be done with fossil assemblages. But perhaps what is most remarkable is the fact that the ideas that had such an impact in 1965 have been so thoroughly absorbed by paleontologists that it hardly seems possible that there was a time when they were not at the core of paleoecology.

Literature Cited

Bretsky, P. W. 1968. Evolution of Paleozoic invertebrate communities. *Science* 159:1231–33.

Elias, M. K. 1937. Depth of deposition of the Big Blue (Late Paleozoic) sediments in Kansas. *Geological Society of America Bulletin* 48:403–32.

Fagerstrom, J. A. 1964. Fossil communities in paleo-

Reprinted by permission from Macmillan Publishers Ltd: *Nature* 207:270–72, copyright 1965. https://doi .org/10.1038/207270a0.

ecology: their recognition and significance. *Geological Society of America Bulletin* 75:1197–16.

Forbes, E. 1844. Report on the Mollusca and Radiata of the Aegean Sea, and on their distribution, considered as bearing on geology. In *Report of the 13th Meeting of the British Association for the Advancement of Science*, 130–93. London: John Murray.

Johnson, R. G. 1964. The community approach to paleoecology. In *Approaches to Paleoecology*, edited by J. Imbrie and N. D. Newell, 107–34. New York: Wiley.

Ladd, H. S., ed. 1957. *Treatise on Marine Ecology and Paleoecology*. Vol. 2, *Paleoecology*. Geological Society of American Memoir, vol. 67. [New York:] National Research Council (U.S.), Committee on a Treatise on Marine Ecology and Paleoecology.

Petersen, C. G. J. 1924. A brief survey of the animal communities in Danish waters, based upon quantitative samples taken with the bottom sampler. *American Journal of Science* 7:343–54.

Schopf, T. J. M. 1972. Editorial introduction to chapter 8, Conceptual models of benthic marine communities, by R. G. Johnson. In *Models in Paleobiology*, edited by T. J. M. Schopf, 148. San Francisco: Freeman, Cooper.

Valentine, J. W. 1973. *Evolutionary Paleoecology of the Marine Biosphere*. Englewood Cliffs, NJ: Prentice-Hall.

270 NATURE July 17, 1965 VOL. 207

timing of the replication cycle or differential sensitizing, protective or additive effects related to phases of replication.

[1] Deschner, E. E., and Gray, L. H., *Radiat. Res.*, 11, 115 (1959).
[2] Gray, L. H., *Radiation Biology*, edit. by Martin, J. H., 76 (Butterworths Scientific Publications, London, 1959).
[3] Jamieson, D., and van den Brenk, H. A. S., *Int. J. Rad. Biol.*, 6, 529 (1963).
[4] Hornsey, S., Howard-Flanders, P., and Moore, D., *Int. J. Rad. Biol.*, 2, 37 (1960).
[5] Gray, L. H., Howard, A., and Hawes, C. A., *Fortieth Annual Report B.E.C.C.*, Part 2, 238 (1962).
[6] van den Brenk, H. A. S., Elliott, K., and Hutchings, H., *Brit. J. Cancer*, 16, 518 (1962).
[7] Terasima, T., and Tolmach, L. J., *Biophys. J.*, 3, 11 (1963). Dewey, W. C., and Humphrey, R. M., *Radiat. Res.*, 16, 503 (1961). Sinclair, W. K., and Morton, R. A., *Nature*, 199, 1158 (1963).
[8] Hornsey, S., and Howard, A., *Ann. N.Y. Acad. Sci.*, 6, 915 (1956). Basergs, R., *Arch. Path.*, 75, 156 (1963).
[9] Adams, A., *Biochem. J.*, 6, 297 (1912). de Almeida, A. O., *C.R. Soc. Biol. (Paris)*, 116, 1225 (1934). MacAllister, T. A., Stark, J. M., Norman, J. N., and Ross, R. M., *Lancet*, 1040 (1963).
[10] Gelfant, S., *Intern. Rev. Cytol.*, 14, 1 (1963).
[11] Jamieson, D., Ladner, K., and van den Brenk, H. A. S., *Austral. J. Exp. Med. Sci.*, 41, 491 (1963).

SILURIAN MARINE COMMUNITIES AND THEIR ENVIRONMENTAL SIGNIFICANCE

By A. M. ZIEGLER

Division of the Geological Sciences, California Institute of Technology, Pasadena, California

PETERSEN[1], in a classic investigation in marine ecology, demonstrated that the benthonic animals of Danish waters tend to occur in several distinctive communities, and he produced maps showing the distribution of these communities. Other workers[2,3] have followed Petersen's lead in treating the animal community as the basic ecological unit, and the general implication is that communities are ultimately controlled by environmental factors. If this is true, then maps showing the distribution of co-existing communities in remote geological times should reflect the different environments of those times; furthermore, the relationships of the communities to such geologically verifiable features as shorelines should provide a clue to the type of environmental factors responsible. Although some palaeontologists[4,5] have defined specific fossil communities in Palaeozoic rocks, few have been able to map coexisting communities, probably because of the difficulties of stratigraphical control.

Table 1

Community name	Characteristic species	Associated species
1. *Lingula* community	*Lingula pseudoparallela* '*Camarotoechia*' *decemplicata* '*Nucula*' *eastnori*	'*Hormotoma*' sp. '*Pterinia*' sp. *Cornulites* sp.
2. *Eocoelia* community	*Eocoelia* spp.* '*Leptostrophia*' *compressa* *Dalmanites weaveri*	*Howellella crispa* *Salopina* sp. '*Pterinia*' sp.
3. *Pentamerus* community (*Pentameroides* community)*	*Pentamerus* spp.* *Atrypa reticularis* *Dalejina* sp.	*Eocoelia* spp.* *Howellella crispa*
4. *Stricklandia* community (*Costistricklandia* community)*	*Stricklandia* spp.* *Eospirifer radiatus* *Atrypa reticularis*	*Resserella* sp.
5. *Clorinda* community	*Clorinda* spp.* *Dicoelosia biloba* *Cyrtia exporrecta* *Skenidioides lewisi*	*Plectodonta millinensis* *Coolinia applanata* *Plectatrypa marginalis*

* Chronological species or genera succeed one another in the same community.

The early Silurian deposits of Wales and the Welsh Borderland contain fossil assemblages which fall naturally into several groups. The distinctions between some of the assemblages were noticed by earlier workers who invoked proximity to volcanoes[6] or submarine ridges[7] to account for the differences. At least five assemblages or communities may now be recognized, and these are briefly defined in Table 1. In the column headed 'Characteristic Species' are the animals which normally occur abundantly in the community; the 'Associated Species' usually occur in the community, but are not as abundant or regular in their appearance. The communities are probably completely intergrading, and it should be emphasized that their recognition depends as much on relative abundances of species as it does on occurrences of particular species; thus *Eocoelia* is very abundant in its own community, but it frequently occurs in the *Pentamerus* community and is found occasionally in the other communities. The communities are widely distributed and, in fact, the same associations occur in collections known to me from as far afield as Alabama and Hudson's Bay.

The five communities: (1) *Lingula*; (2) *Eocoelia*; (3) *Pentamerus*; (4) *Stricklandia*; (5) *Clorinda*; are related to each other in a linear fashion; adjacent communities, for example 1 and 2, or 4 and 5, have many species in common, while 1 and 4, or 2 and 5, are mutually exclusive. The communities defined, with the exception of the *Lingula* community, are dominated by articulate brachiopods which typically constitute more than 85 per cent of the fossil remains, so the distinctions between the communities are based largely on the brachiopods. In the case of the *Lingula* community, only one species of

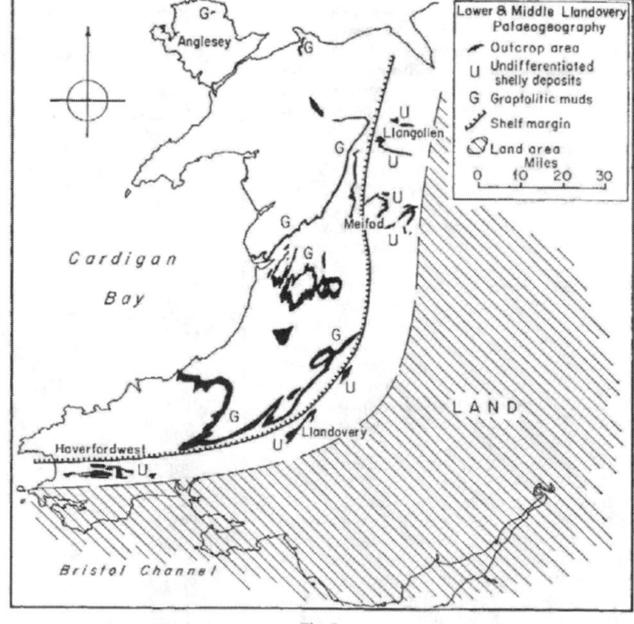

Fig. 1

No. 4994 July 17, 1965 NATURE 271

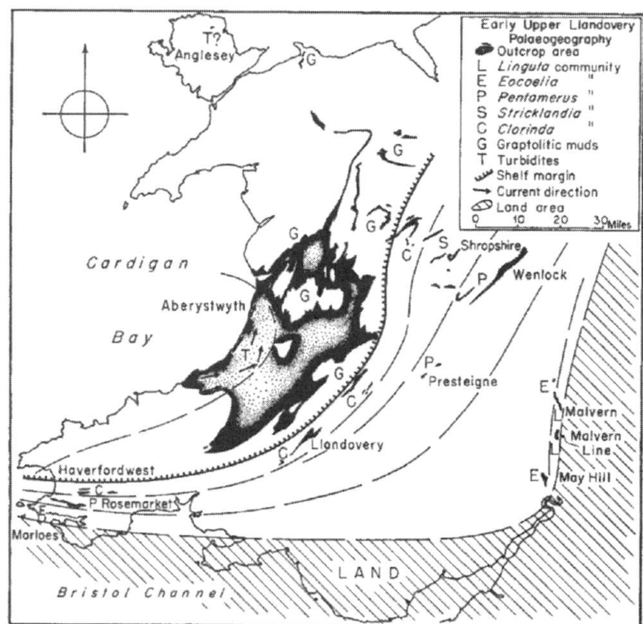

Fig. 2

The off-shore sequence of the communities may be clearly demonstrated with palaeogeographical maps showing two phases of the Upper Llandovery transgression (Figs. 2 and 3). The communities of the Lower and Middle Llandovery have not yet been studied in detail, but a map of this time period has been included (Fig. 1) to show the extent of the land area later flooded by the Upper Llandovery transgression, and also to contrast the widths of the shelf area during different times. The accuracy of these maps depends largely on the correlation of the various sections, and this has been accomplished by studying evolutionary trends in various brachiopod lineages[9]. The greywackes and the graptolitic muds are included on the maps; these deposits typically do not contain benthonic animals. The graptolites are largely restricted to these deposits, only because the environment of deposition was such that their delicate remains were preserved; they probably lived in the surface waters[10] and drifted wherever seas existed.

A first approximation of the depths at which the various communities existed may be derived from volcanic flows which occur both at Marloes, Pembrokeshire, and Tortworth, Gloucestershire[12]. At Renny Slip, near Marloes, a 20-ft. thick pillow basalt flowed out on deposits containing an *Eocoelia* community. The water was apparently shallowed by an amount equal to the thickness of the flow and this was enough to displace the *Eocoelia* community by its neighbour, the *Lingula* community which occurs in the beds just above the flow. However, the duration of the *Lingula* community at this locality was short as the *Eocoelia* community occurs about 40 ft. higher in the succession, its return being due to the

articulate brachiopod, a rhynchonellid, is present, and many groups such as the corals, crinoids, and trilobites are completely unrepresented. Because of the restricted nature of this community and because of its occurrence at the base of transgressive sequences, such as the Kenley Grit of Shropshire, or the Cowleigh Park Beds of the Malvern Hills, a coastal environment is postulated, similar perhaps to the environment inhabited by many modern lingulids[8].

Sediment type does not seem to have been a controlling factor in community distribution, as each community has been found in a wide range of sedimentary rocks. But there is a general tendency for communities 1 and 2 to occur in sandstones, communities 3 and 4 in sequences of varying proportions of sandstone and shale beds, and community 5 in shale.

It is evident that a transgressing sea, such as the Upper Llandovery sea of the Welsh Borderland, would deposit stratigraphical sequences representing environments progressively farther from the shore. The Damery Beds of Tortworth show the stratigraphic sequence of communities 2, 3, and 4, whereas at May Hill the community sequence is 2 (Huntley Hill Beds), 3, 4 and 5 (Yartleton Beds). In the Malvern District the community sequence is at Eastnor 1 (Cowleigh Park Beds), 3, and 4 (Wyche Beds); at Gulley quarry it is 3 and 4 (Wyche Beds); and at Old Storridge Common it is 2 (Cowleigh Park Beds), 3, 4, and 5 (Wyche Beds). In Shropshire the general sequence is 1, 3 and 5, representing the Kenley Grit, *Pentamerus* Beds and Purple Shales respectively. Thus the communities, as numbered, represent progressively offshore environments, though the complete sequence of communities is not present in any one area.

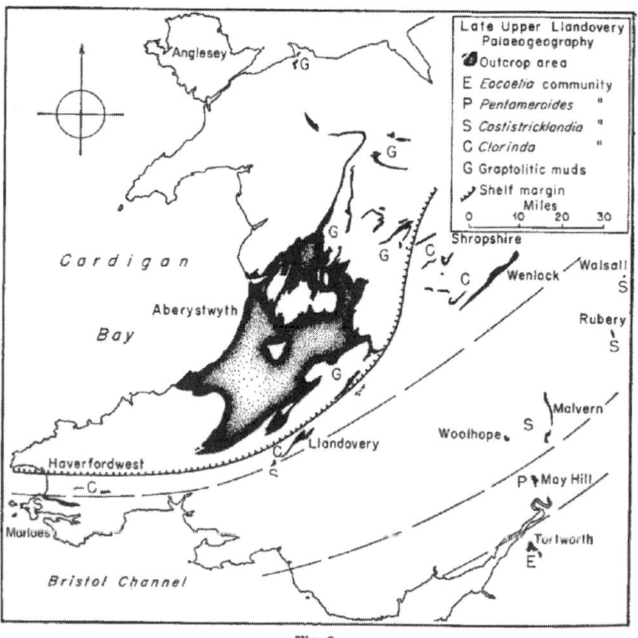

Fig. 3

272 NATURE July 17, 1965 VOL. 207

generally transgressive nature of the sea at this time. Two miles away, however, at Marloes Bay, the basalt is much thicker (130 ft.) and therefore had a greater effect. Two flows occurred and the lower flow has a reddened surface[11], suggesting that deposition of the lower flow was at least in part subaerial. The *Lingula* community did not return until several hundred feet of conglomerates and sandstones had been deposited on top of the basalts. The relationships of these volcanic flows suggest that the depth ranges of these communities were of the order of tens of feet rather than hundreds of feet.

In the Tortworth inlier, an andesite flow of variable thickness occurs on top of deposits containing the *Costistricklandia* community. At Woodford, its thickness is small, perhaps 15 ft., and the *Costistricklandia* community is developed on top of the flow. However, less than two miles to the south, the flow is 150 ft. thick and the *Eocoelia* community is developed in the sediments above the flow; here, both the *Costistricklandia* and *Pentameroides* communities were displaced. Relief of 135 ft., then, seems likely between the levels of the *Eocoelia* and *Costistricklandia* communities, and the *Pentameroides* community must have occupied some intermediate range.

In summary, the various lines of evidence on the community environments are: (1) comparison with modern communities in the case of *Lingula*; (2) sediment type; (3) stratigraphical succession; (4) areal relationships; (5) volcanic displacement. All these considerations point to an off-shore sequence of the communities, the *Lingula* community being the closest to the strand and the *Clorinda* community furthest from the strand. If this interpretation is correct, then several conclusions regarding the geological history follow.

The transgression of the Upper Llandovery sea occurred in two pulses. In early Upper Llandovery times the sea extended to cover much of South Wales and most of the Welsh Borderland. The record of its relatively rapid transgression is preserved in the *Eocoelia* community low in the succession of Presteigne and the *Lingula* community of the Kenley Grits of Shropshire. This was just a transient phase, however, and Fig. 2 shows the relationships during much of early Upper Llandovery time, with the shoreline existing at the 'Malvern Line'[13]. About this time, turbidity currents began to deposit the Aberystwyth Grits in the geosyncline[14]. These turbidites are known to have extended further to the east and north with time, eventually covering much of central Wales[15]. It is a singular fact that there is a gap in the sequences of the shelf area which corresponds quite well with the deposition of the turbidities in the geosyncline; for example, north-west of Malvern at Old Storridge Common, Wyche Beds of C_5 age rest directly on Cowleigh Park Beds of C_1 or C_2 age. Possibly there was a slight regression causing much of the sediment that had accumulated on the shelf to be transported to the shelf margin where it was carried to the depths of the geosyncline by turbidity currents. Confirmation of this regression is found at Llandovery, about the only place where the sequence is continuous, where the *Stricklandia* community of C_4 beds succeeds the *Clorinda* community of C_2–C_3 beds.

The second pulse of the transgression occurred about C_5 time. It is perhaps significant that a transgression occurred on the north side of the geosyncline as well, in County Galway, Ireland[16], suggesting that the change of sea-level was eustatic. Much of south-east England became submerged, to judge by bore-hole information[17], turbidite deposition stopped abruptly, and fine-grained shales accumulated over much of the shelf and deeper area. By the end of Upper Llandovery times, the *Clorinda* community existed as far east as May Hill and Old Storridge Common, showing that subsidence continued at a faster pace than the accumulation of sediment.

To conclude, it is clear that the animal community technique of the ecologists is applicable to fossil assemblages and can provide the basis for interpreting past environments. This article reports some preliminary results and is intended only as an announcement of a much more complete treatise, which would define the communities quantitatively, describe the various stratigraphical successions, and present a detailed geological history of the Llandovery of the British Isles.

[1] Petersen, C. G. J., *Amer. J. Sci.*, Ser. 5, 7, 343 (1924).
[2] Thorson, G., *Mem. Geol. Soc. Amer.*, 67, 461 (1957).
[3] Parker, R. H., *Vid. Medd. fra Dansk Naturh. Foren.*, 126 (1964).
[4] Allan, R. S., *N.Z. Dept. Sci. Indust. Res., Palaeont. Bull.*, 14 (1935).
[5] Fagerstrom, J. A., *Bull. Geol. Soc. Amer.*, 75, 1197 (1964).
[6] Jones, O. T., *Quart. J. Geol. Soc.*, 77, 144 (1921).
[7] Whittard, W. F., *Quart. J. Geol. Soc.*, 83, 737 (1928).
[8] Craig, G. Y., *Trans. Edin. Geol. Soc.*, 15, 110 (1952).
[9] Ziegler, A. M., *Palaeontology* (in the press).
[10] Bulman, O. M. B., *Quart. J. Geol. Soc.*, 120, 455 (1964).
[11] Cantrill, T. C., Dixon, E. E., Thomas, H. H., and Jones, O. T., *Mem. Geol. Surv. Gt. Brit.*, Sheet 227 (1916).
[12] Curtis, M. L. K., in *Bristol and its Adjoining Counties, Brit. Assoc. Adv. Sci.*, 3 (1955).
[13] Ziegler, A. M., *Geol. Mag.*, 101, 467 (1964).
[14] Wood, A., and Smith, A. J., *Quart. J. Geol. Soc.*, 114, 163 (1959).
[15] Bassett, D. A., *Quart. J. Geol. Soc.*, 111, 239 (1955).
[16] McKerrow, W. S., and Campbell, C. J., *Sci. Proc. Roy. Dublin Soc.*, A, 1, 27 (1960).
[17] Bullard, E. C., Gaskell, T. F., Harland, W. B., and Kerr-Grant, C., *Phil. Trans. Roy. Soc.*, A, 239, 29 (1940).

PARTICIPATION OF ADRENOCORTICAL HYPERACTIVITY IN THE SUPPRESSIVE EFFECT OF SYSTEMIC ACTINOMYCIN *D* ON UTERINE STIMULATION BY OESTROGEN

By BARBARA M. LIPPE* and PROF. CLARA M. SZEGO

Department of Zoology, University of California, Los Angeles

INVESTIGATIONS in several laboratories have been directed toward localization of the oestrogen-sensitive step(s)[1-3] in uterine metabolism. In a recent series of reports[4-7] reviewed elsewhere[8], evidence has been provided that augmentation of RNA synthesis is an early correlate of oestrogen action in the uterus. Thus, actinomycin *D*, which inhibits DNA-dependent RNA synthesis in isolated systems[9,10] and depresses *in vivo* the incorporation of labelled orthophosphate into all forms of RNA in rat liver[11], also inhibits some characteristic early actions of oestrogen on the rat uterus including imbibition of water and accentuation of phospholipid and protein synthesis[4,12]. Moreover, augmentation of uterine RNA polymerase activity has been noted within 2 h of oestrogen pretreatment[5,13]. Puromycin administration abolished this response[5], as well as the anabolic effects of the hormone[3,14].

Enhancement of RNA synthesis is undoubtedly an early indicator of the uterine stimulatory action of oestrogen. This would be anticipated from the generalized accentuation by oestrogen of anabolic responses including that of protein elaboration[1]. What is less clear, however, is whether the hormone influences RNA synthesis directly as a primary step on which the metabolic stimulatory pattern depends[4,12,15,16], or whether the accentuated rate

* Supported by medical student research training grant *5T5 GM* 42-03, U.S. Public Health Service, 1-SO11-*FR* 05355-01.

PAPER 11

Early Cretaceous Fossil Evidence for Angiosperm Evolution (1977)

L. J. Hickey and J. A. Doyle

Commentary

SCOTT L. WING AND NATHAN JUD

This paper, following on Doyle and Hickey (1976), showed that the stratigraphic distribution of angiosperm pollen and leaves in the Potomac Group represents the primary diversification of extant flowering plants, an idea that until that time had been controversial. More important from a paleoecological perspective, this paper reviewed current interpretations of the functional morphology of early angiosperm leaf fossils and combined that information with new data on their sedimentary context to build a consistent hypothesis about the environments in which this major evolutionary radiation took place. The resulting "riparian weed" hypothesis of angiosperm radiation has remained influential ever since. The paper has necessarily been abstracted here to less than a quarter of its original length, emphasizing the conclusions and ideas rather than the data.

The two primary lines of evidence for early angiosperm ecology came from the size and form of Early Cretaceous angiosperm leaves and from their distribution in fluvial facies. Leaf fossils of the oldest angiosperms were small and poorly organized, suggesting that they were borne by small, short-lived plants. They also were rare and occurred exclusively in disturbed environments such as sandy fluvial channels and poorly sorted mudstones deposited as levees and crevasse splays. Through the Early Cretaceous, angiosperm leaves became more abundant, larger, better organized, and more diverse in form and were found in a greater variety of fluvial subenvironments, including ponds and floodplains. Hickey and Doyle concluded that flowering plants initially diversified as fast-growing, short-lived plants of disturbed habitats and that their ecological scope expanded into more stable and competitive habitats as the group diversified.

The influence of this paper is seen in recent work on the functional biology of angiosperm leaves, including papers on vein density (Boyce et al. 2009; Brodribb and Feild 2009) and leaf mass per area (Royer et al. 2010). Also, it is now routine to infer the habitat preferences of extinct plants from the sedimentary facies in which they are found, with this approach being used even to characterize the land plant radiation of the Silurian-Devonian (Hotton et al. 2001) and the seed plant radiation of Carboniferous-Permian (DiMichele and Aronson 1992).

Literature Cited

Boyce, C. K., T. J. Brodribb, T. S. Feild, and M. A. Zwieniecki. 2009. Angiosperm leaf vein evolution was physiologically and environmentally transformative. *Proceedings of the Royal Society B* 276:1771–76.

Brodribb, T. J., and T. S. Feild. 2010. Leaf hydraulic evolution led a surge in leaf photosynthetic capacity during early angiosperm diversification. *Ecology Letters* 13: 175–83.

From *Botanical Review* 43:2–104.

DiMichele, W. A., and R. B. Aronson. 1992 The Pennsylvanian-Permian vegetational transition: a terrestrial analogue to the onshore offshore hypothesis. *Evolution* 46:807–24.

Doyle, J. A., and L. J. Hickey. 1976. Pollen and leaves from the mid-Cretaceous Potomac Group and their bearing on early angiosperm evolution. In *Original and Early Evolution of Angiosperms*, edited by C. B. Beck, 139–206. New York: Columbia University Press.

Hotton, C. L., F. M. Huber, D. H. Griffing, and J. S. Bridge. 2001. Early Terrestrial Plant Environments: an Example from the Emsian of Gaspé, Canada. In *Plants Invade the Land: Evolutionary and Environmental Perspectives*, edited by P. G. Gensel and D. Edwards, 174–212. New York: Columbia University Press.

Royer, D. L., I. M. Miller, D. J. Peppe, and L. J. Hickey. 2010. Leaf economic traits from fossils support a weedy habit for early angiosperms. *American Journal of Botany* 97 (3): 438–45.

EARLY CRETACEOUS FOSSIL EVIDENCE FOR ANGIOSPERM EVOLUTION

LEO J. HICKEY

Division of Paleobotany
Smithsonian Institution
Washington, D.C. 20560, USA

AND

JAMES A. DOYLE

Museum of Paleontology
and
Department of Ecology and Evolutionary Biology
The University of Michigan
Ann Arbor, Michigan 48109, USA

ABSTRACT

Morphological, stratigraphic, and sedimentological analyses of Early Cretaceous pollen and leaf sequences, especially from the Potomac Group of the eastern United States, support the concept of a Cretaceous adaptive radiation of the angiosperms and suggest pathways of their initial ecological and systematic diversification. The oldest acceptable records of angiosperms are rare monosulcate pollen grains with columellar exine structure from probable Barremian strata of England, equatorial Africa, and the Potomac Group, and small, simple, pinnately veined leaves with several orders of reticulate venation from the Neocomian of Siberia and the basal Potomac Group. The relatively low diversity and generalized character of these fossils and the subsequent coherent pattern of morphological diversification are consistent with a monophyletic origin of the angiosperms not long before the Barremian. Patuxent-Arundel floras (Barremian-early Albian?) of the Potomac Group include some pollen and leaves with monocotyledonous features as well as dicotyledonous forms. Patuxent angiosperm pollen is strictly monosulcate and has exine sculpture indicative of insect pollination. Rare Patuxent-Arundel angiosperm leaves are generally small, have disorganized venation, and are largely restricted to sandy stream margin lithofacies; the largest are comparable to and may include ancestors of woody Magnoliidae adapted to understory conditions. Patapsco floras (middle to late Albian?) contain rapidly diversifying tricolpate pollen and several new complexes of locally abundant angiosperm leaves. Ovate-cordate and peltate leaves in clayey pond lithofacies may include ancestors of aquatic Nymphaeales and Nelumbonales. Pinnatifid and later pinnately compound leaves with increasingly regular venation which are abundant just above rapid changes in sedimentation are interpreted as early successional "weed trees" transitional to but more primitive than the modern subclass Rosidae. Apparently related palmately lobed, palinactinodromous leaves which develop rigidly percurrent tertiary venation and become abundant in uppermost Potomac stream margin deposits (latest Albian-early Cenomanian?) are interpreted as riparian trees ancestral to the order Hamamelidales. Comparisons of dated pollen floras of other regions indicate that one major subgroup of angiosperms, tricolpate-producing dicots (i.e., excluding Magnoliidae of Takhtajan) originated in the Aptian of Africa-South America at a time of increasing aridity and migrated poleward into Laurasia and Australasia. However, the earlier (Barremian) monosulcate phase of the angiosperm record is represented equally in Africa-South America and Laurasia before marked climatic differentiation between the two areas. These trends are considered consistent with the hypothesis that the angiosperms originated as small-leafed shrubs of seasonally arid environments, and underwent secondary expansion of leaf area and radiated into consecutively later successional stages and aquatic habitats after entering mesic regions as riparian "weeds," as opposed to the concept that they arose as trees of mesic forest environments.

fragmented former continent of Gondwana, consisting of Africa-South America (West Gondwana), interrupted by a rift valley system destined to become the South Atlantic Ocean, the minor continents of Madagascar and India in uncertain spatial relation to each other and to the other blocks, and Antarctica-Australasia (East Gondwana), which remained united until the Early Tertiary. Depositional settings for sedimentary sequences containing early angiosperm floras included rift valleys soon to be invaded by marine waters, as in Brazil, Gabon, and the Congo (Bahia and Cocobeach sequences); coastal plains on the subsiding margins of earlier Mesozoic rift oceans, such as the Atlantic Coastal Plain (Potomac Group) and Gulf Coastal Plain (Fredericksburg and Tuscaloosa Groups) of United States and the coastal plain of Portugal; interior continental basins (the Wealden Group of England); margins of shallow epicontinental seas, as in the Western Interior of the United States (Cheyenne Sandstone and Dakota Group of Kansas; Lakota and Fall River Formations of the Black Hills) and in Kazakhstan; and older continental borders, such as the underthrust Pacific margin of North America (Shasta Group of California) and the Ussuri River region of maritime Siberia.

The Potomac Group, which serves as the focus for the present discussion, consists of a wedge of terrestrial sediments deposited on river flood plains and upper deltas, dipping seaward at approximately 1° ESE. It is exposed in a discontinuous belt up to 30 km wide extending from the Nottoway, Appomattox, and James Rivers south of Richmond, Virginia, through Fredericksburg, Virginia, Washington, D.C., Baltimore, Maryland, northeastern Maryland and adjacent Delaware, and an uncertain distance into New Jersey (Fig. 2; Glaser, 1969; Hansen, 1969; Owens and Sohl, 1969; Wolfe and Pakiser, 1971; Schluger and Roberson, 1975).

In its type area between Baltimore and Washington, the Potomac Group has been customarily divided into three or four formations (following Clark and Bibbins, 1897). The predominantly sandy Patuxent Formation at the base is overlain by the highly organic Arundel Clay, followed by the more heterogeneous Patapsco Formation. The generally sandy beds which overlie typical Patapsco strata in Maryland and Delaware were for many years misidentified

→

Fig. 2. Outcrop map of the Lower Cretaceous Potomac Group of Virginia, Maryland, and Delaware, not including Elk Neck beds. Localities mentioned in the text are identified by numbers in approximate stratigraphic order from oldest to youngest: 1, Dutch Gap Canal; 2, Trent's Reach; 3, Potomac Run; 4, Vinegar Hill; 5, Federal Hill in Baltimore; 6, Drewrys Bluff; 7, Fredericksburg; 8, Bank near Brooke; 9, Deep Bottom; 10, Wellhams; 11, Mount Vernon; 12, White House Bluff; 13, Aquia Creek; 14, Quantico; 15, West Brothers clay pit; 16, Red Point; 17, Stump Neck; 18, Widewater; 19, White Point; 20, Bull Mountain; 21, Brightseat; 22, Cedar Point. An additional locality, the Hylton pit, is located on Pennsauken Creek in northern Camden County, New Jersey. (Base map modified after Glaser, 1969.)

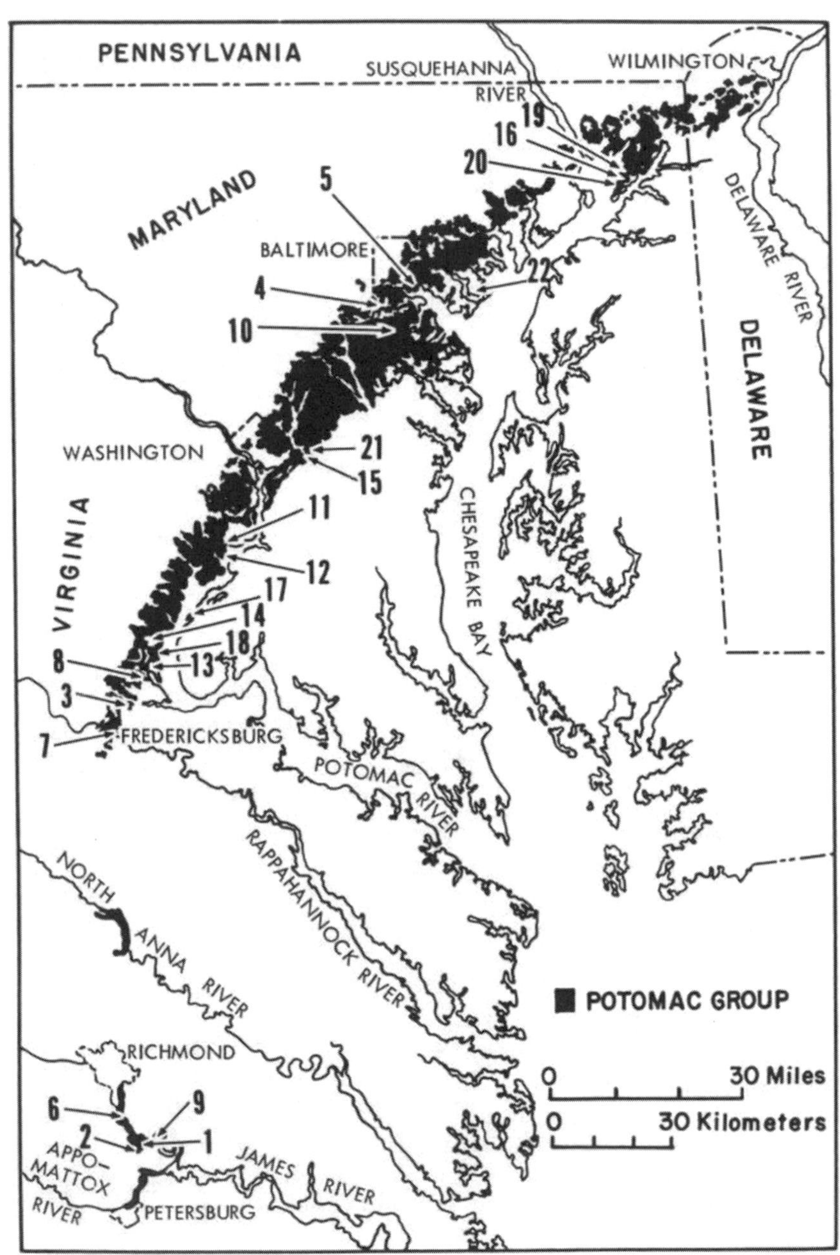
PENNSYLVANIA
SUSQUEHANNA RIVER
WILMINGTON
MARYLAND
BALTIMORE
DELAWARE RIVER
DELAWARE
WASHINGTON
VIRGINIA
CHESAPEAKE BAY
FREDERICKSBURG
POTOMAC RIVER
RAPPAHANNOCK RIVER
NORTH ANNA RIVER
POTOMAC GROUP
RICHMOND
0 30 Miles
0 30 Kilometers
APPOMATTOX RIVER
JAMES RIVER
PETERSBURG
1
2
3
4
5
6
7
8
9
10
11
12
13
14
15
16
17
18
19
20
21
22

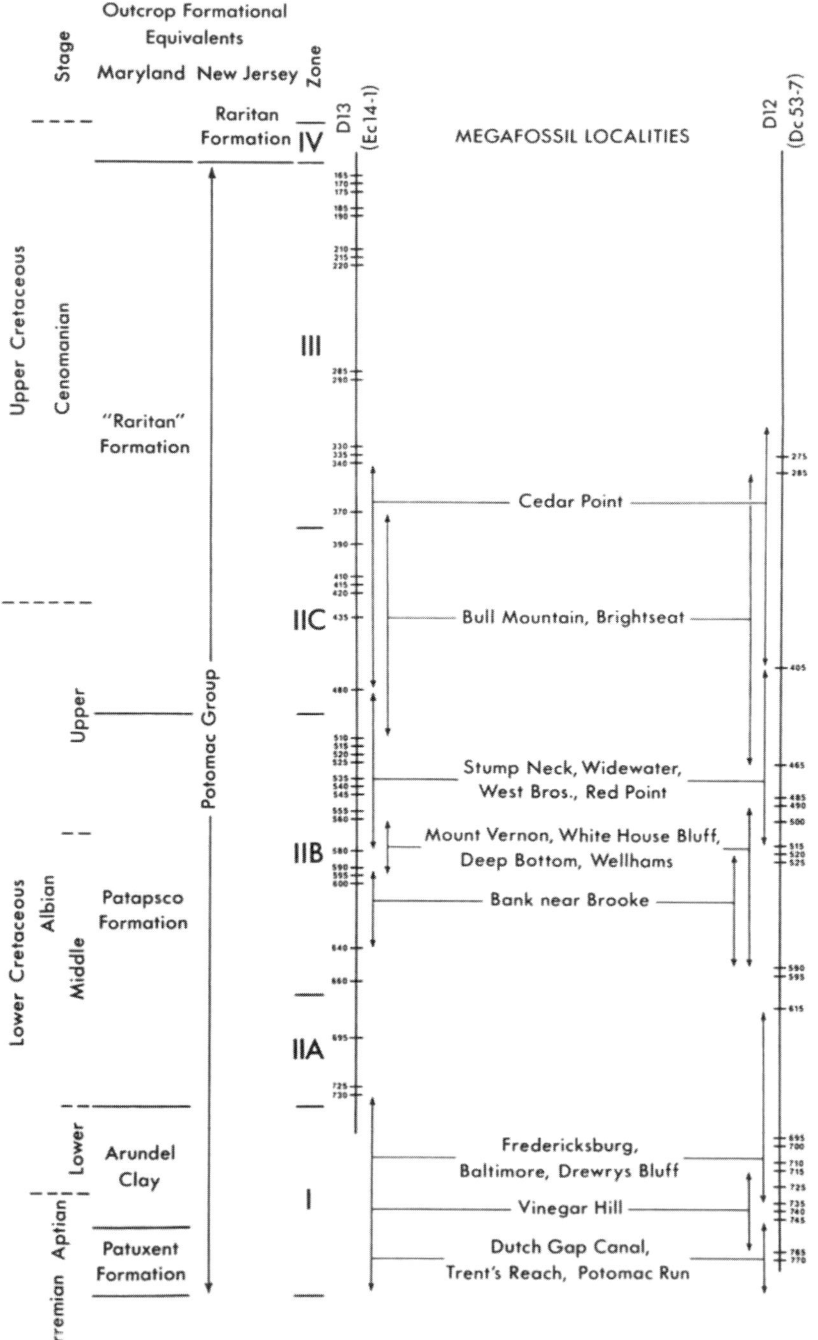
Stage
Outcrop Formational Equivalents
Maryland
New Jersey
Zone
Raritan Formation
IV
D13 (Ec14-1)
MEGAFOSSIL LOCALITIES
D12 (Dc53-7)
Upper Cretaceous
Cenomanian
III
"Raritan" Formation
Cedar Point
IIC
Bull Mountain, Brightseat
Upper
Potomac Group
Stump Neck, Widewater, West Bros., Red Point
Mount Vernon, White House Bluff, Deep Bottom, Wellhams
IIB
Bank near Brooke
Lower Cretaceous
Albian
Middle
Patapsco Formation
IIA
Lower
Arundel Clay
I
Fredericksburg, Baltimore, Drewrys Bluff
Vinegar Hill
Aptian
Patuxent Formation
Dutch Gap Canal, Trent's Reach, Potomac Run
Barremian

with the younger and lithologically distinct Raritan Formation of New Jersey (e.g., McGee, 1888; Clark, 1897; Berry, 1916). As a result of sedimentological and palynological studies, these "Maryland Raritan" beds are now considered part of the Potomac Group (Weaver et al., 1968; Owens and Sohl, 1969; Wolfe and Pakiser, 1971). Wolfe and Pakiser (1971) have treated them simply as part of the Patapsco Formation, but we prefer to distinguish them informally as the Elk Neck beds (cf. Doyle and Robbins, in press). Outside the Baltimore-Washington area, where the Arundel Clay is not recognizable, there is no consistent way to subdivide the Potomac Group lithologically, and it is therefore referred to as the Potomac Formation.

Direct dating of the Potomac sequence in terms of the standard European stages of the Cretaceous (Table I) is precluded by the absence of marine fossils below the Woodbridge Clay Member of the Raritan Formation, which is dated by mollusks as mid-Cenomanian (N. F. Sohl, unpublished, cited by Doyle, 1969; Wolfe and Pakiser, 1971). In addition, the rapid lithological changes characteristic of such fluvial strata, extremely limited outcrops, and the sporadic distribution of plant megafossils have long frustrated attempts at internal stratigraphic subdivision and correlation. This situation has improved markedly in the past two decades as a result of discovery and investigation of abundant pollen and spores in the Potomac Group. Thanks to published studies on remarkably similar successions and associations of pollen and spore types in faunally dated sediments elsewhere in North America and Europe, palynology has clarified age relations not only within the Potomac-Raritan sequence, but also with the European stages (Groot and Penny, 1960; Groot et al., 1961; Brenner, 1963, 1967; Doyle, 1969, 1973; Wolfe and Pakiser, 1971; Doyle and Robbins, in press).

An important aspect of these studies has been establishment of an informal palynological zonation, beginning with Brenner's (1963) proposal of Zone I and Subzones II-A and II-B as a result of work on well sections of the Patuxent through Patapsco Formations of Maryland. This scheme has been extended

---

←

Fig. 3. Subdivisions of the Potomac Group and palynological correlations of its megafossil localities. Correlations are expressed (right) by comparison with two subsurface Potomac Group sequences near Delaware City, Delaware: well D13 (Ec14-1) and well D12 (Dc53-7). Double-headed arrows terminate above and below the depths (in feet) of well samples that are detectably older and younger than the outcrop sample (e.g., the age of Mount Vernon is bracketed as post-595', pre-560' in well D13). Overlaps of arrows (especially next to unsampled well intervals) do not necessarily indicate possible age equivalence, but rather reflect the fact that exact age limits cannot be defined where there is insufficient sampling in the reference section (e.g., Mount Vernon is considered definitely older than Stump Neck). Inferred correlations with the Potomac-Raritan palynological zonation, formational units, and the standard European series, stages, and substages are indicated to the left. (Modified after Doyle and Hickey, 1976.)

to younger beds by recognition of Subzone II-C and Zones III and IV, based on wells penetrating Potomac and Raritan strata in Delaware and southern New Jersey (Doyle, 1973; Wolfe et al., 1975; Doyle and Hickey, 1976; Doyle and Robbins, in press). Strictly speaking, these palynostratigraphic units are defined primarily on the stratigraphic distribution and associations of spore and pollen species, and secondarily on regional changes in relative abundance of major elements of the flora (Brenner, 1963; Doyle and Robbins, in press). However, for the present purposes they and the external correlations derived from them may be roughly characterized in terms of the first rare appearances and subsequent diversification of major angiosperm pollen types. Thus reticulate, columellar monosulcates comparable to those from the Barremian and Aptian of England (Couper, 1958; Kemp, 1968; Hughes, this symposium) occur throughout Zone I (Patuxent Formation and Arundel Clay and their equivalents) and younger units. Very rare reticulate tricolpates enter in the upper part of Zone I (Arundel Clay and equivalents), as in the early Albian of England (Kemp, 1968; Laing, 1975, 1976), and become regularly present and more diverse in Subzones II-A and II-B (Patapsco Formation), as in the middle and late Albian elsewhere (e.g., Hedlund and Norris, 1968). Tricolporoidates, with rudimentary ora in the centers of their colpi, appear and become more abundant in Subzone II-B; in Subzone II-C and Zone III (Elk Neck beds) they include increasing numbers of very small, smooth, and often triangular types similar to those first seen elsewhere near the Albian-Cenomanian boundary (Norris, 1967; Pacltová, 1971; Singh, 1971, 1975; Laing, 1975, 1976). Larger, definitely tricolporate forms are first recognizable in Zone III and become more common in Zone IV (lower members of the Raritan Formation of New Jersey), where they are joined by the first triangular triporates of the Normapolles complex, comparable to species which appear in the middle Cenomanian of Europe (Góczán et al., 1967; Pacltová, 1971;

←

Fig. 4. Representative angiosperm pollen from Zone I of the Potomac Group (Barremian-early Albian?), light micrographs, × 1000. Coordinates: UMMP Zeiss RA microscope 4767359.

a, b: cf. *Clavatipollenites hughesii* Couper, Trent's Reach (71-8-1d, 17.2 × 101.1).

c, d: *Clavatipollenites* sp. A, Potomac Run (71-14-1b, 20.0 × 96.0).

e, f: *Retimonocolpites* sp. aff. "*Clavatipollenites*" *minutus* Brenner, Dutch Gap Canal (71-15-1b, 6.0 × 95.8).

g, h: *Liliacidites* sp. A, two grains, Trent's Reach (71-8-1d, 8.0 × 89.1).

i, j: *Stellatopollis* sp., Dutch Gap Canal (71-15-1b, 4.7 × 83.9).

k, l: *Liliacidites* sp. B, Trent's Reach (71-8-1d, 13.2 × 87.8).

m, n: aff. *Retimonocolpites peroreticulatus* (Brenner) Doyle, Dutch Gap Canal (71-15-1a, 8.6 × 87.3).

o, p: aff. *Tricolpites crassimurus* (Groot and Penny) Singh, Baltimore (71-6-1b, 10.1 × 83.1).

q, r: aff. *Retimonocolpites dividuus* Pierce, Fredericksburg (71-21-1c, 12.7 × 84.9).

Laing, 1975, 1976). The inferred ages of Zones I through III—from the Barremian or Aptian through the early Cenomanian—allow us to estimate the duration represented by the Potomac Group as approximately 15–20 million years.

MATERIALS AND METHODS

Our first step in integrating the Potomac Group angiosperm pollen and leaf records (Doyle and Hickey, 1972, 1976) was to correlate Potomac megafossil localities with the zonation scheme outlined above. This was accomplished by comparing the pollen and spore assemblages obtained from matrix samples of the classic Potomac megafossil collections of Fontaine (1889), Ward (1895, 1905), and Berry (1911a, 1916), as well as from several new localities, with reference sections consisting of two wells drilled through the Potomac Formation near Delaware City, Delaware (Brenner, 1967; Doyle and Robbins, in press). These correlations, documented in greater detail in our previous paper (Doyle and Hickey, 1976), are summarized in Figure 3.

This analysis shows that Potomac megafossil localities fall at five main stratigraphic levels, with an unfortunate gap in sampling in Subzone II-A and the base of Subzone II-B. Because pollen grains are so much more transportable than leaves, there are more conspicuous environmentally controlled, or facies, differences among megafossil assemblages from a single presumed horizon than among pollen assemblages. In some cases, notably that of Fredericksburg and Baltimore, these differences were incorrectly ascribed by earlier workers to age effects. Although each locality yields only a partial representation of the total leaf flora, the greater ease of palynological correlation between facies has allowed us to reconstruct what appears to be a fairly complete picture of the range of leaf morphology and ecological adaptations in the angiosperm flora for at least three horizons, namely upper Zone I and the middle and upper parts of Subzone II-B. In some cases, intensive col-

→

Figs. 5–10. Angiosperm leaves from Zone I of the Potomac Group. These and all other line drawings of leaves are reproduced from Doyle and Hickey (1976) by permission of Columbia University Press. All figures × 1.

Fig. 5. *Rogersia angustifolia* Fontaine (USNM 192339), from Fredericksburg, Virginia.

Fig. 6. *Vitiphyllum multifidum* Fontaine (USNM 31824) from Baltimore, Maryland. Note the highly irregular venation and lack of venational bracing of the sinuses.

Fig. 7. *Acaciaephyllum spatulatum* Fontaine (USNM 175802A), bent axis with attached obovate, acrodromous leaves from Dutch Gap Canal near Richmond, Virginia.

Fig. 8. Leaf fragment belonging to the *Celastrophyllum latifolium* Fontaine complex from Baltimore, Maryland. Originally described and illustrated as *Celastrophyllum obovatum* Fontaine (in Ward, 1905, p. 560, Pl. 117, Fig. 2; USNM 31814).

Fig. 9. *Proteaephyllum reniforme* Fontaine (USNM 3915) from Fredericksburg, Virginia.

Fig. 10. *Proteaephyllum dentatum* Fontaine (USNM 31820) from Baltimore, Maryland.

(Carlquist, personal communication). In any case, the ability of early angiosperms to survive in particular environments must be evaluated in terms of their ability to compete with Early Cretaceous gymnosperms and ferns, not with higher dicot types which had not yet evolved.

In our previous paper, we also proposed a model for the subsequent vegetative adaptive radiation of mid-Cretaceous angiosperms, based in part on sedimentary associations and in part on functional-morphological analysis of trends in the Potomac leaf record (Fig. 66). We stressed mechanical considerations and relations between light gathering strategies and ecological succession, following arguments presented by Horn (1971, 1975). Many of Horn's inferences should now be refined and reformulated in terms of a more comprehensive theory of the adaptive trade-offs involved in leaf architecture developed by Givnish (1976, in press) and Givnish and Vermeij (1976).

One of the main conclusions of our analysis is that the general pattern of the Potomac leaf record is more consistent with our secondary expansion hypothesis and Stebbins' (1965, 1974) concept that the first angiosperms were shrubs of semixerophytic origin which entered mesic areas as colonizers of unstable habitats—the "weeds" of the Early Cretaceous—than it is with the more conventional postulate that they were broad-leafed, mesic forest trees comparable to living Magnoliales (e.g., Bews, 1927; Cronquist, 1968; Takhtajan, 1969; Thorne, 1976). Sedimentological evidence that small-leafed Zone I angiosperms were adapted to disturbed habitats, specifically stream margins, is presented below. Here we will only remark that their small size, poor petiole differentiation, and disorganized venation are more consistent with low, highly branched shrubs growing in sunny habitats, where small leaf size is favored (cf. Givnish and Vermeij, 1976), than with arborescent stature. The occurrence of apparently herbaceous monocot-type leaves from the beginning of the Potomac record is also easier to explain if one postulates that the first angiosperms were low shrubs rather than trees.

In functional terms, the most notable variants from the norm among Zone I angiosperms are the large, broad, undissected, low rank leaves of the *Ficophyllum* type. Such leaves suggest Horn's (1971) monolayer light gathering strategy, where photosynthetic tissue is arranged in a single layer to catch all available light. Since such a strategy is advantageous in poorly lit situations, we proposed (Doyle and Hickey, 1976) that these leaves belonged to some of the first angiosperms to enter conifer-dominated Early Cretaceous forests, not as direct competitors of the conifers but as understory trees. This would have been a relatively minor step from our hypothetical shrubby prototype. Although Givnish and Vermeij (1976) question some of Horn's assumptions regarding light gathering strategies, their analysis of trade-offs between transpirational losses and photosynthetic gains also leads to the conclusion that large, undissected leaves should be favored at intermediate levels of the understory. Horn

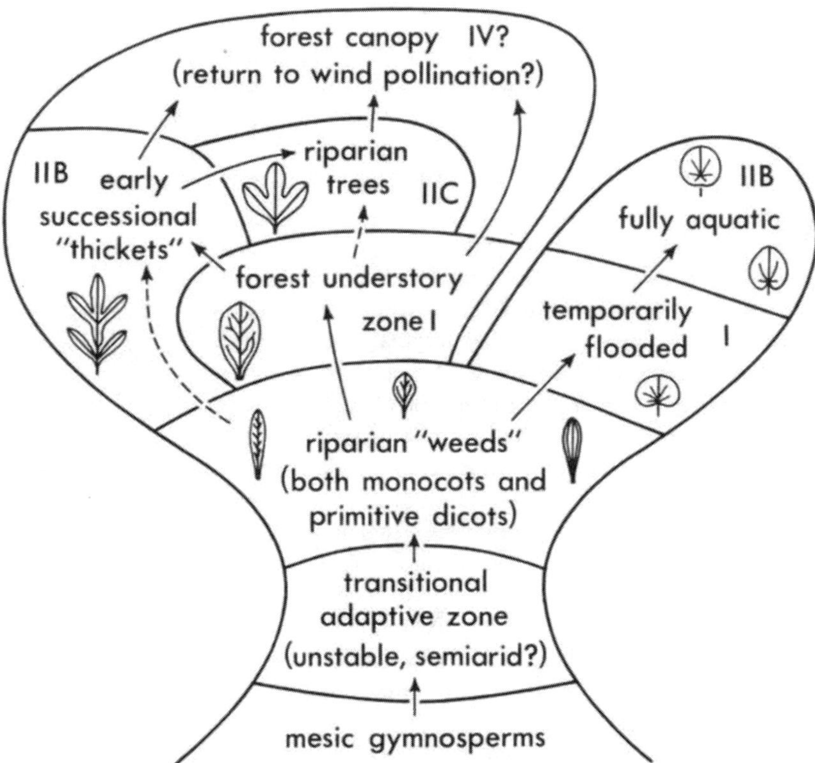

Fig. 66. Model for early ecological-adaptive evolution of the angiosperms (reprinted from Doyle and Hickey, 1976, Fig. 28, p. 186, by permission of Columbia University Press). We argue that trends in the Early Cretaceous leaf and pollen records are consistent with Stebbins' (1965, 1974) hypothesis that angiosperms arose from mesic gymnospermous ancestors (seed ferns?) under pressures for efficient reproduction in an unstable, semiarid environment, as opposed to the concept that they arose as trees of mesic forest habitats. We infer that the primary adaptive radiation of the angiosperms followed perfection of their distinctive reproductive features and coincided with their reinvasion of mesic regions such as the Potomac basin, first as "weedy" shrubs of disturbed stream margin habitats. The course of ecological evolution from "riparian weeds" onward is based on functional interpretations of observed trends and sedimentological associations of Potomac Group leaves. Alternative interpretations for certain trends, such as the possibility that the pinnately and palmately lobed groups originated as dry-season deciduous plants of tropical regions before entering mesic areas as early successional "weed trees", are discussed in the text. Roman numerals (I, IIB, IIC, and IV) refer to Atlantic Coastal Plain pollen zones by which particular adaptive types are inferred to have evolved.

(1971, 1975) argues that a monolayer strategy is also advantageous for climax forest canopy trees; however, observations by Hickey (unpublished) show that virtually no modern canopy species match the low degree of vein regularity or petiole differentiation displayed by Zone I leaves. This is probably related to the mechanical disadvantages of irregular vein networks in the high wind and rain stresses experienced at the canopy level. Also consistent with an understory habit for *Ficophyllum* is the fact that it is most abundant in the rich fern-gymnosperm flora from Fredericksburg, but even there it makes up only a small proportion of the total flora.

In our previous discussion, we suggested that the appearance of pinnatifid, pinnately compound, and palmately lobed leaves in Subzone II-B represents evolution of early successional "weed trees" from understory or shrubby ancestors. Following Horn (1971), we argued that lobation, by decreasing internal leaf diameter and hence the length of full shadows cast by leaves, allows lower leaves to be stacked in the partial shadows of higher leaves in a multilayer configuration that makes more efficient use of full sunlight. Again, although Givnish (1976, in press) points out that Horn does not explain why lobate or compound leaves are superior to branches bearing collections of small, narrow leaves, he too concludes that lobed and compound leaves are advantageous for early successional trees. Givnish argues that in pioneer tree species competing to fill a space as rapidly as possible, natural selection should tend to maximize the amount of photosynthetic laminar tissue and minimize the amount of slower-growing, non-productive support tissue.

One solution to the problem of obtaining the most laminar tissue using the least support tissue is exemplified by Subzone II-B and II-C "platanoids"—a large, lobate leaf in which thin laminar tissue is stretched like fabric on the ribs of an umbrella between the primaries radiating from the leaf base, rather than supported by numerous pinnate secondary veins. Considering the resultant tensional stresses on the laminar tissue between the major veins, it is not surprising that the platanoids are the first group to develop rigid percurrent tertiary venation running perpendicularly from one secondary to the next.

Another solution to the same problem is the pinnately compound or pinnatifid *Sapindopsis*-type leaf, which escapes the necessity for extensive internal bracing by developing a large amount of foliar tissue supported mainly by the rachis and leaflet midveins, while keeping the size of the individual laminae relatively small. The smaller size of contiguous laminar areas may well be the reason why the leaf ranking trend in the *Sapindopsis* complex lags behind that in the platanoids. The pinnately compound leaf, which can be regarded functionally as a cheap "throwaway" branch system, represents an especially advantageous solution for an early successional tree, where rapid upward growth is more important than the capacity for true lateral branching.

Givnish (1976, in press) argues that pinnately compound leaves are also

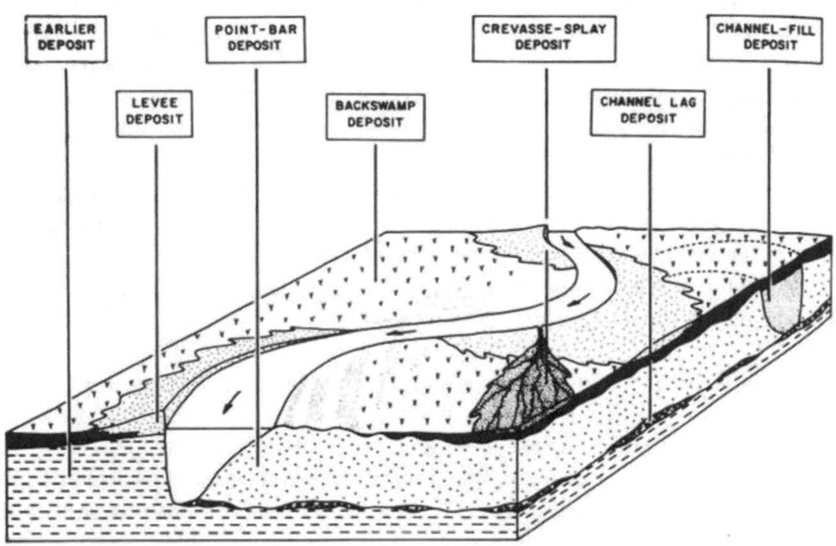

Fig. 67. Lithofacies relationships in the floodplain deposits of a meandering river illustrated by an idealized block diagram. (Reprinted from Allen, 1964, Fig. 4, p. 168, by permission of Elsevier Publishing Company.)

advantageous for dry-season deciduous trees of tropical areas, since the ability to shed their ultimate support framework along with the leaflets allows them to eliminate a major source of transpirational water loss. This explanation is not directly applicable to leaves from the Potomac Group, where the climate appears to have been moist and subtropical and there is sedimentological evidence in favor of the early successional interpretation. However, it is possible that the pinnatifid-leafed *Sapindopsis* complex originated in the tropics in response to seasonal drought, and was preadapted to early successional habitats upon later entering mesic areas, rather than *vice versa* (cf. below).

A final aspect of Givnish's (1976) synthesis which applies to our analysis of the early angiosperm record is his prediction that in leaves where there is no requirement for the veins to fill a support function, the blade should tend to become circular about the supply entry point (petiole attachment), and the veins should assume a roughly radial configuration. A somewhat similar solution for an optimal transport network with major trunk lines radiating from a central but elongate supply line was derived by Sen (1971). In addition, Givnish and Vermeij (1976) argue that whereas a cordate shape best resolves mechanical stresses in liana leaves with petioles held at an angle, a peltate form is favored when the petiole is held vertically. These observations further support the interpretation of actinodromously veined Subzone II-B peltate leaves as floating- or emergent-leafed aquatics, though they do not resolve the question of the habit of the related cordate forms. Both groups, however, show a tendency for extensive dichotomy and looping of the radiating primary veins well within the margin, a feature which is typical of aquatic leaves of both Nymphaeales and unrelated groups such as the genus *Nymphoides*, but which contrasts with the tendency for primary veins or their branches to run directly to the margin or an intramarginal vein in lianas such as Menispermaceae. It may be noted that the early appearance of aquatic angiosperms is easier to explain if the first angiosperms were weedy shrubs, especially if they occupied stream margin habitats, rather than mesic trees.

Sedimentological Evidence on Ecology of Early Angiosperms

In addition to providing an integrated suite of micro- and megafloral remains, the Potomac Group supplies important sedimentological insights into the paleoecology of early angiosperms. The unit consists of lenses of sandstone interbedded with claystone and mudstone and is comparable to ancient and modern sedimentary sequences laid down by meandering river systems on flood plains. Analysis of such fluvial deposits has advanced rapidly during the past 20 years, allowing the development of a rather detailed picture of sedimentary environments and the genesis of constituent rock types (Moore and Scruton, 1957; Allen, 1964, 1965, 1970; Beerbower, 1965; Laming, 1966; Cotter, 1971; Rigby and Hamblin, 1972; Blatt et al., 1972). Because of the importance of the fluvial environment as our window on early angiosperm evolution, we feel that a brief description of its sedimentology will be of value in understanding the fossil record.

The deposits of a meandering river are coarsest near the channel and become finer with increasing distance from the stream (Fig. 67). On the inner, or convex, side of the channel-meander the river lays down point-bar deposits. These consist of lenses of cross-bedded sand becoming finer-grained and horizontally stratified upward. Episodes of lateral accretion as the river channel shifts toward the outside of the meander alternate with quiet water intervals, leaving the upper surface of the point-bar deposit marked by a series of concentric ridges separated by swales. The swales tend to be occupied by standing water where accumulation of parallel-laminated, organic-rich beds of clay or mud takes place, especially on the portion of the point bar away from the stream. Natural levees, made up of cross-bedded and cross-laminated silt and fine-grained sand, are laid down on the stream margins as the river overflows its banks during times of flood, especially on the outer, or concave, portion of the meander-loop. The low-lying back-swamp areas behind the point bars and levees receive much finer sediment which settles from suspension as flood

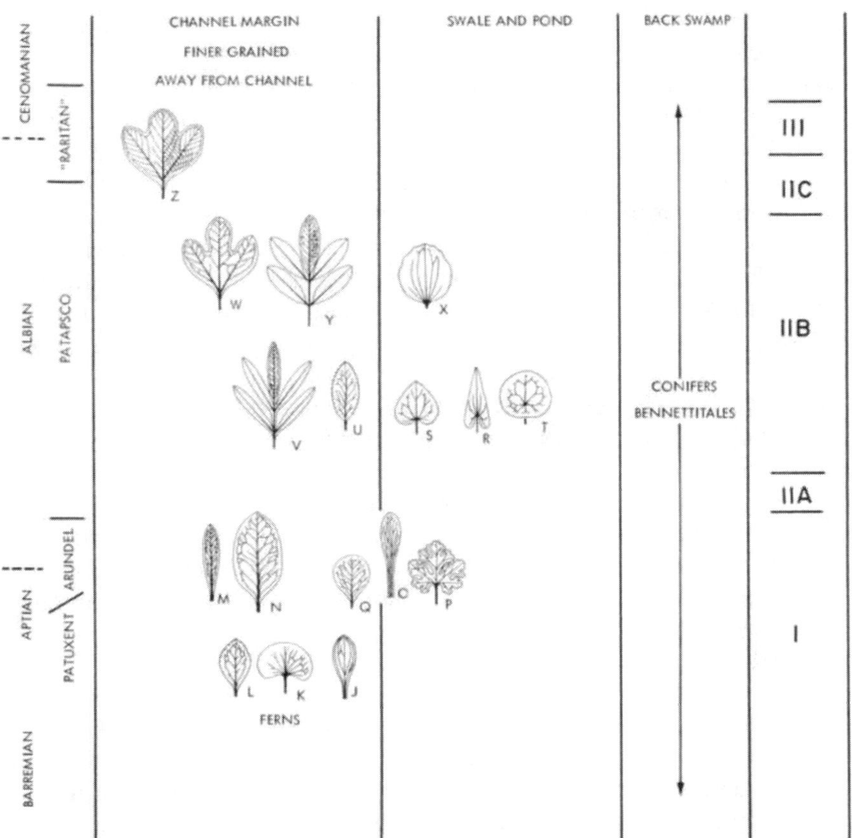

Fig. 68. Relationship of certain leaf types to the sedimentary lithofacies of the Potomac Group. Distance from the left-hand margin of the diagram is roughly correlated with decrease in grain size of the enclosing sediments and increase in features indicating lower energy deposition. The channel-margin lithofacies includes point-bar, levee, and crevasse-splay deposits. The letters identifying the various leaf types are the same as in Fig. 60 except for **W**, indicating "*Sassafras*" *potomacensis*.

waters recede; they are characterized by massive to parallel-laminated mud and clay with a high content of organic matter. During times of flood, the levee may be breached and fine-grained, cross-laminated sand or silt carried into the back-swamp to form what are called crevasse-splay deposits.

During our megafossil collecting in the Potomac Group, we noted a strong correlation between sedimentary lithofacies and the types of plants encountered. In fact, these observations often allowed us to predict whether a bed was likely to yield fossil leaves, to specify the assemblage we would encounter, and to assign previously collected material to a general facies. Figure 68 shows the principal Potomac angiosperm leaf types placed within one of three major fluvial lithofacies based on the characteristics of the enclosing sediments. Relative distance from the stream channel is indicated by increasing distance of the leaf sketches from the left-hand side of the diagram. The channel-margin lithofacies includes point-bar, levee, and crevasse-splay deposits. Preliminary observations on lithofacies associations of early angiosperm leaves in Kazakhstan and the Western Interior of the United States lead us to believe that the correlations illustrated are typical of the early angiosperm record in at least the Southern Laurasian area.

First, this analysis demonstrates the striking localization of Zone I angiosperm leaf remains to relatively coarse, channel-margin sediments. In any specific case, it cannot be determined whether these leaves represent plants that were actually growing on the stream margin site where they were preserved, or plants whose leaves were transported from some distance upstream. However, this distance could not have been great if they were to escape mechanical degradation. Furthermore, their almost complete absence from finer-grained back-swamp deposits even where these are extremely rich in ferns, cycadopsids, and conifers supports the notion that Zone I angiosperms occupied only areas relatively close to stream courses.

In middle Subzone II-B, the majority of angiosperm leaves, including *Sapindopsis* and the platanoids, continue to occur in channel-margin deposits, but in contrast to the situation in Zone I they are locally abundant and tend to extend into finer-grained sediments. The presence of a nearly pure mat of *Sapindopsis magnifolia* just above the contact between a horizontally bedded sandstone and a mudstone in the inferred crevasse-splay deposit at Brooke (Fig. 69), associated with lignitized and charred fragments of conifer wood, is dramatic evidence of both its ability to form dense stands and the instability of the local environment in which it grew. This agrees well with our functional-morphological arguments that *Sapindopsis* was a "weed tree" adapted to early successional conditions. Later members of the *Sapindopsis* complex with pinnately compound leaves are often found in the same sort of setting (e.g., just above a clay-pebble conglomerate with abundant charcoalized wood at the West Brothers pit). However, by the top of Subzone II-B and in Subzone II-C *Sapindopsis* and the platanoids have begun to show an imperfect segregation into different lithofacies, with the platanoids especially abundant in the coarser channel-margin facies (e.g., White Point), recalling the riparian habitat of modern *Platanus*, and *Sapindopsis* sometimes occurring in quite fine, clayey sediments (e.g., Red Point, Severn Clay Mine).

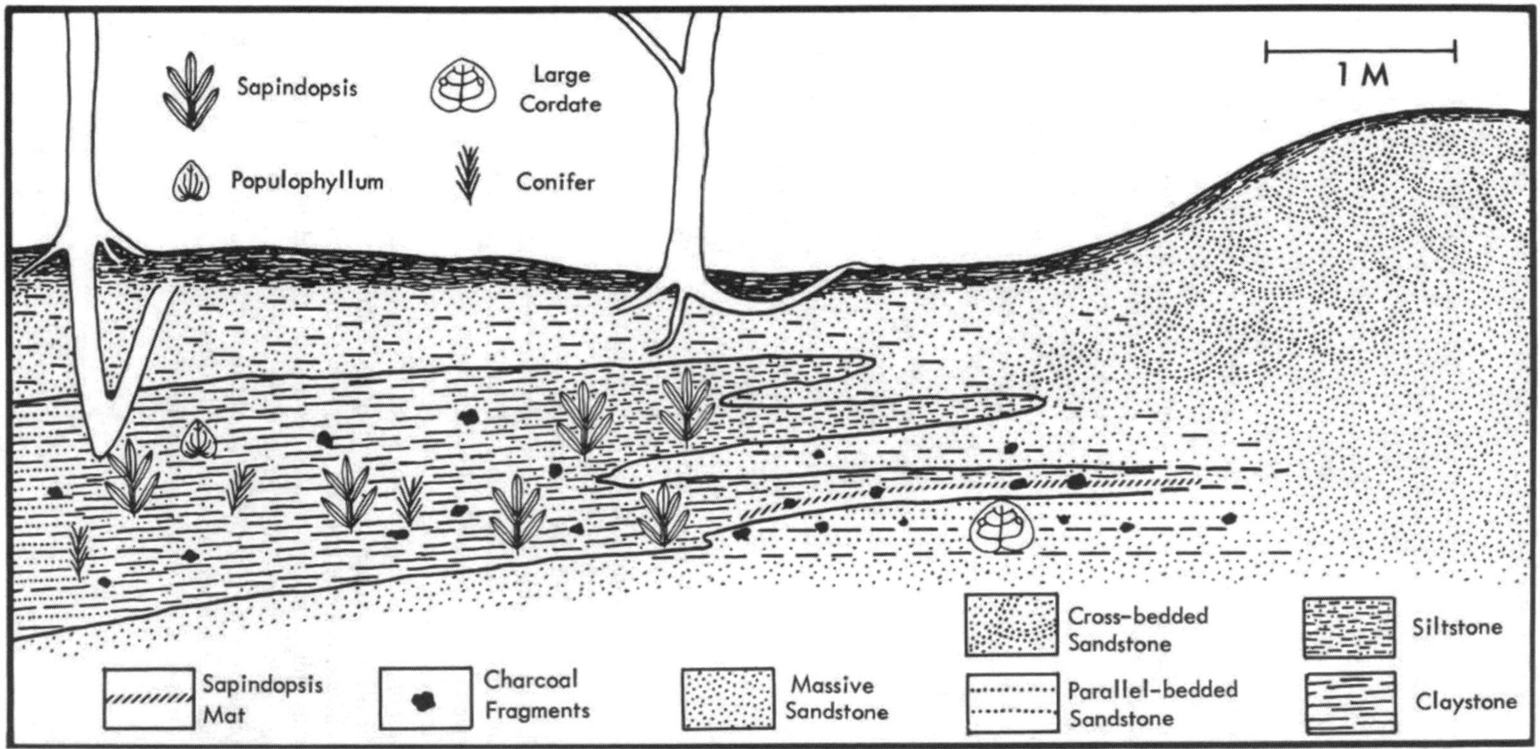

Fig. 69. Sketch of sedimentary facies at the Brooke, Virginia locality showing the major leaf types found in them. Note the symbol indicating the mat of *Sapindopsis magnifolia* leaves near the base of the lowest claystone unit at the right, and the general decrease in grain size and the increase in conifer material toward the left.

Another important development seen in the middle of Subzone II-B is the appearance of cordate and peltate actinodromous leaves in fine-grained beds that are often thin and interbedded with sandstone. The inference that at least the peltate forms were growing in place in the standing water of swale or pond environments is supported by the generally parallel lamination of their enclosing sediments, as well as by the leaf architectural features mentioned above. In one such deposit, at Quantico, the peltate leaves occur in clusters spaced about a meter apart along the outcrop, suggesting a rooted aquatic habit.

Even in upper Subzone II-B and in Zone III, we have found fine-grained deposits containing mostly ferns and gymnosperms, suggesting that many environments well removed from the stream channels remained dominated by non-angiospermous groups throughout deposition of the Potomac Group. Such observations agree with pollen evidence, noted as early as 1961 by Pierce, that even in the Cenomanian, gymnosperms continued to play a greater role in the regional vegetation than had been inferred from the abundance of angiosperm leaves in some contemporaneous fluvial facies, such as the sandstones of the Dakota Group of Kansas.

The results of this sedimentological analysis reinforce our functional-morphological inferences concerning early evolution of the angiosperm vegetative body. In particular, the gradual spread of angiosperms from coarse-grained channel-margin deposits agrees with Stebbins' (1965, 1974) hypothesis that the first angiosperms were "weedy" semixerophytic shrubs preadapted to disturbed habitats, rather than mesic trees comparable to modern Magnoliales. Stream margins are the least stable sites in a flood plain environment, where plants are subjected to periodic inundation, flood-training, and rapid sedimentation, and the areas where closed-canopy conifer forests would have the least chance to develop. In addition, when viewed in terms of the time-scale and dynamics of ecological succession, the evolutionary trends inferred in the subsequent angiosperm leaf record make sense as responses to selective pressures operating in a stream margin situation. On the one hand, the repeated cycles of lateral migration of the channel and return to more stable back-swamp conditions which are typical of a fluvial system would place plants growing on stream margins under selective pressure to adapt to later successional stages. *Sapindopsis* and the platanoids are interpreted as examples of such a trend. On the other hand, periodic flooding and shifting of the borders between stream margin and swale or pond habitats would favor adaptations for partially and eventually fully aquatic habit, as inferred for the peltates. If correct, this picture constitutes a remarkable example of Margalef's (1968) concept that in most groups evolutionary trends should tend to parallel changes realized in succession, except when unusual evolutionary events (e.g., neoteny) allow groups to "drop back" to earlier successional stages.

LITERATURE CITED

Allen, J. R. L. 1964. Studies in fluviatile sedimentation: Six cyclothems from the lower Old Red Sandstone, Anglo-Welsh Basin. Sedimentology **3**: 163–198.

________. 1965. A review of the origin and characteristics of Recent alluvial sediments. Sedimentology **5**: 89–91.

________. 1970. Studies in fluviatile sedimentation: A comparison of fining-upward cyclothems, with special reference to coarse-member composition and interpretation. J. Sediment. Petrol. **40**: 298–323.

Andrews, H. N. 1963. Early seed plants. Science **142**: 925–931.

Archangelsky, S. and J. C. Gamerro. 1967. Spore and pollen types of the Lower Cretaceous in Patagonia (Argentina). Rev. Palaeobot. Palynol. **1**: 211–217.

Arkhangel'skiy, D. B. 1966. Zvezdchataya skul'ptura ekziny pyl'tsevykh zeren, *in* M. I. Neyshtadt (editor), Znachenie palinologicheskogo analiza dlya stratigrafii i paleofloristiki, pp. 22–26. Nauka, Moscow.

Axelrod, D. I. 1952. A theory of angiosperm evolution. Evolution **6**: 29–60.

________. 1959. Poleward migration of early angiosperm flora. Science **130**: 203–207.

________. 1960. The evolution of flowering plants, *in* S. Tax (editor), The evolution of life, pp. 227–305. Univ. Chicago Press, Chicago.

________. 1970. Mesozoic paleogeography and early angiosperm history. Bot. Rev. **36**: 277–319.

________. 1972. Edaphic aridity as a factor in angiosperm evolution. Amer. Naturalist **106**: 311–320.

Bailey, I. W. 1949. Origin of the angiosperms: Need for a broadened outlook. J. Arnold Arbor. **30**: 64–70.

Banks, H. P. 1968. The early history of land plants, *in* E. T. Drake (editor), Evolution and environment, pp. 73–107. Yale Univ. Press, New Haven.

________. 1970. Evolution and plants of the past. Wadsworth, Belmont, California.

Barghoorn, E. S. 1971. The oldest fossils. Sci. Amer. **224**(5): 30–42.

Beck, C. B. 1960. The identity of *Archaeopteris* and *Callixylon*. Brittonia **12**: 351–368.

________. 1970. The appearance of gymnospermous structure. Biol. Rev. Cambridge Philos. Soc. **45**: 379–400.

________. 1976. Current status of the Progymnospermopsida. Rev. Palaeobot. Palynol. **21**: 5–23.

Beerbower, J. R. 1965. Cyclothems and cyclic depositional mechanisms in alluvial plain sedimentation. Kansas State Geol. Surv. Bull. **169**: 31–42.

Bell, W. A. 1956. Lower Cretaceous floras of Western Canada. Geol. Surv. Canada Mem. **285**.

Berry, E. W. 1911a. Systematic paleontology, Lower Cretaceous: Fossil plants, *in* W. B. Clark (editor), Lower Cretaceous, pp. 214–508. Maryland Geol. Surv., Baltimore.

———. 1911b. The flora of the Raritan Formation. New Jersey Geol. Surv. Bull. **3**.

———. 1916. Systematic paleontology, Upper Cretaceous: Fossil plants, *in* W. B. Clark (editor), Upper Cretaceous, pp. 757–901. Maryland Geol. Surv., Baltimore.

———. 1922. The flora of the Cheyenne Sandstone of Kansas. U. S. Geol. Surv. Profess. Pap. **127-I**: 199–225.

Bews, J. W. 1927. Studies in the ecological evolution of angiosperms. New Phytol. **26**: 1–21, 65–84, 129–148, 209–248, 273–294.

Bierhorst, D. W. 1971. Morphology of vascular plants. Macmillan, New York.

Blatt, H., R. Middleton and R. Murray. 1972. Origin of sedimentary rocks. Prentice-Hall, Englwood Cliffs, New Jersey.

Brenner, G. J. 1963. The spores and pollen of the Potomac Group of Maryland. Maryland Dept. Geol., Mines and Water Resources Bull. **27**.

———. 1967. Early angiosperm pollen differentiation in the Albian to Cenomanian deposits of Delaware (U.S.A.). Rev. Palaeobot. Palynol. **1**: 219–227.

———. 1968. Middle Cretaceous spores and pollen from northeastern Peru. Pollen & Spores **10**: 341–383.

———. 1976. Middle Cretaceous floral provinces and early migrations of angiosperms, *in* C. B. Beck (editor), Origin and early evolution of angiosperms, pp. 23–47. Columbia Univ. Press, New York.

Brideaux, W. W. and D. J. McIntyre. 1975. Miospores and microplankton from Aptian-Albian rocks along Horton River, District of Mackenzie. Geol. Surv. Canada Bull. **252**: 1–85.

Brown, R. W. 1956. Palmlike plants from the Dolores Formation (Triassic) in southwestern Colorado. U. S. Geol. Surv. Profess. Pap. **274-H**: 205–209.

Burger, D. 1970. Early Cretaceous angiospermous pollen grains from Queensland. Bur. Mineral Resources, Geol. and Geophys., Canberra, Bull. **116**: 1–10.

———. 1973. Palynological observations in the Carpentaria Basin, Queensland. Bur. Mineral Resources, Geol. and Geophys., Canberra, Bull. **140**: 27–44.

Carlquist, S. 1975. Ecological strategies of xylem evolution. Univ. California Press, Berkeley and Los Angeles.

Casey, R. 1964. The Cretaceous Period, *in* The Phanerozoic time-scale. Quart. J. Geol. Soc. London **120s**: 193–202.

Chaloner, W. G. 1967. Spores and land-plant evolution. Rev. Palaeobot. Palynol. **1**: 83–93.

———. 1970. The rise of the first land plants. Biol. Rev. Cambridge Philos. Soc. **45**: 353–377.

Chesters, K. I. M., F. R. Gnauck and N. F. Hughes. 1967. Angiospermae, *in* W. B. Harland and others (editors), The fossil record (Geol. Soc. London Special Publ.), pp. 269–288.

Clark, W. B. 1897. Outline of present knowledge of the physical features of Maryland, *in* Maryland Geological Survey, vol. 1, pp. 139–228. Baltimore.

Clark, W. B. and A. B. Bibbins. 1897. The stratigraphy of the Potomac Group in Maryland. J. Geol. **5**: 479–506.

Cloud, P. E. 1948. Some problems and patterns of evolution exemplified by fossil invertebrates. Evolution **2**: 322–350.

_______. 1976. Beginnings of biospheric evolution and their biochemical consequences. Paleobiology **2**: 351–387.

Cotter, E. 1971. Paleoflow characteristics of a Late Cretaceous river in Utah from analysis of sedimentary structures in the Ferron Sandstone. J. Sediment. Petrol. **41**: 129–138.

Couper, R. A. 1953. Upper Mesozoic and Cainozoic spores and pollen grains from New Zealand. New Zealand Geol. Surv. Paleontol. Bull **22**: 1–77.

_______. 1958. British Mesozoic microspores and pollen grains. Palaeontographica, Abt. B, **103**: 75–179.

_______. 1960. New Zealand Mesozoic and Cainozoic plant microfossils. New Zealand Geol. Surv. Paleont. Bull. **32**: 5–82.

Cracraft, J. 1974. Continental drift and vertebrate distribution. Annual Rev. Ecol. Syst. **5**: 215–261.

Cronquist, A. 1968. The evolution and classification of flowering plants. Houghton Mifflin, Boston.

Davis, P. N. 1963. Palynology and stratigraphy of the Lower Cretaceous rocks of northern Wyoming. Ph.D. thesis, Univ. Oklahoma, Norman.

Delevoryas, T. and C. P. Person. 1975. *Mexiglossa varia* gen. et sp. nov., a new genus of glossopteroid leaves from the Jurassic of Oaxaca, Mexico. Palaeontographica, Abt. B, **154**: 114–120.

Dettmann, M. E. 1973. Angiospermous pollen from Albian to Turonian sediments of eastern Australia. Geol. Soc. Australia Special Publ. **4**: 3–34.

Dilcher, D. L. 1974. Approaches to the identification of angiosperm leaf remains. Bot. Rev. **4**: 1–157.

Dilcher, D. L., W. L. Crepet, C. D. Beeker and H. C. Reynolds. 1976. Reproductive and vegetative morphology of a Cretaceous angiosperm. Science **191**: 854–856.

Doyle, J. A. 1969. Cretaceous angiosperm pollen of the Atlantic Coastal Plain and its evolutionary significance. J. Arnold Arbor. **50**: 1–35.

_______. 1973. Fossil evidence on early evolution of the monocotyledons. Quart. Rev. Biol. **48**: 399–413.

_______. (in press). Patterns of evolution in early angiosperms, *in* A. Hallam (editor), Patterns of evolution. Elsevier, Amsterdam.

Doyle, J. A. and L. J. Hickey. 1972. Coordinated evolution in Potomac Group angiosperm pollen and leaves. Amer. J. Bot. **59**: 660 (abstract).

_______ **and** _______. 1976. Pollen and leaves from the mid-Cretaceous Potomac Group and their bearing on early angiosperm evolution, *in* C. B. Beck (editor), Origin and early evolution of angiosperms, pp. 139–206. Columbia Univ. Press, New York.

Doyle, J. A. and E. I. Robbins. (in press). Angiosperm pollen zonation of the continental Cretaceous of the Atlantic Coastal Plain and its application to deep wells in the Salisbury Embayment. Palynology **1**.

Doyle, J. A., P. Biens, A. Doerenkamp and S. Jardiné. (in press). Angiosperm pollen from the pre-Albian Lower Cretaceous of equatorial Africa. Proc. 7th West African Micropaleontol. Colloquium (Ile Ife, 1976).

Doyle, J. A., S. Jardiné and A. Doerenkamp. 1976. Evolution of angiosperm pollen in the Lower Cretaceous of equatorial Africa. Bot. Soc. Amer., Abstr. of Papers, p. 25 (abstract).

Doyle, J. A., M. Van Campo and B. Lugardon. 1975. Observations on exine structure of *Eucommiidites* and Lower Cretaceous angiosperm pollen. Pollen & Spores **17**: 429–486.

Erdtman, G. 1948. Did dicotyledonous plants exist in early Jurassic time? Geol. Fören. Stockholm Förh. **70**: 265–271.

———. 1952. Pollen morphology and plant taxonomy. Part I. Angiosperms. Chronica Botanica, Waltham, Massachusetts.

Esau, K. 1953. Plant anatomy. Wiley, New York.

Faegri, K. and L. van der Pijl. 1966. The principles of pollination ecology. Pergamon, Oxford.

Florin, R. 1951. Evolution in cordaites and conifers. Acta Horti Berg. **15**: 285–388.

Fontaine, W. M. 1889. The Potomac or younger Mesozoic flora. U. S. Geol. Surv. Monogr. **15**.

Givnish, T. J. 1976. Leaf form in relation to environment: A theoretical study. Ph.D. Thesis, Princeton Univ.

———. (in press). The adaptive significance of compound leaves, with particular reference to tropical trees, *in* P. B. Tomlinson and M. H. Zimmermann (editors), Tropical trees as living systems. Cambridge Univ. Press.

Givnish, T. J. and G. J. Vermeij. 1976. Sizes and shapes of liane leaves. Amer. Naturalist **110**: 743–778.

Glaser, J. D. 1969. Petrology and origin of Potomac and Magothy (Cretaceous) sediments, Middle Atlantic Coastal Plain. Maryland Geol. Surv. Rep. Invest. **11**.

Glen, W. 1975. Continental drift and plate tectonics. Merrill, Columbus, Ohio.

Góczán, F., J. J. Groot, W. Krutzsch and B. Pacltová. 1967. Die Gattungen des "Stemma Normapolles Pflug 1953b" (Angiospermae). Paläontol. Abh., Abt. B, **2**: 429–539.

Groot, J. J. and C. R. Groot. 1962. Plant microfossils from Aptian, Albian and Cenomanian deposits of Portugal. Comun. Serv. Geol. Portugal **46**: 133–176.

Groot, J. J. and J. S. Penny. 1960. Plant microfossils and age of nonmarine Cretaceous sediments of Maryland and Delaware. Micropaleontology **6**: 225–236.

Groot, J. J., J. S. Penny and C. R. Groot. 1961. Plant microfossils and age of the Raritan, Tuscaloosa, and Magothy Formations of the eastern United States. Palaeontographica, Abt. B, **108**: 121–140.

Grow, J. A., J. S. Schlee, R. E. Mattick and J. C. Behrendt. 1976. Recent marine geophysical studies along the Atlantic continental margin. Geol. Soc. Amer. Abstr. with Programs **8**: 186 (abstract).

Hansen, H. J. 1969. Depositional environments of subsurface Potomac Group in southern Maryland. Bull. Amer. Assoc. Petrol. Geol. **53**: 1923–1937.

Hara, N. 1964. Ontogeny of the reticulate venation in the pinna of *Onoclea sensibilis*. Bot. Mag. (Tokyo) **77**: 381–387.

Harris, T. M. 1932. The fossil flora of Scoresby Sound, East Greenland. 2. Description of seed plants *incertae sedis* together with a discussion of certain cycadophyte cuticles. Meddel. Grønland **85**: 1–112.

Hedlund, R. W. and G. Norris. 1968. Spores and pollen grains from Fredericksburgian (Albian) strata, Marshall County, Oklahoma. Pollen & Spores **10**: 129–159.

Helal, A. H. 1966. Jurassic plant microfossils from the subsurface of the Kharga Oasis, Western Desert, Egypt. Palaeontographica, Abt. B, **117**: 83–98.

Herngreen, G. F. W. 1973. Palynology of Albian-Cenomanian strata of borehole 1-QS-1-MA, State of Maranhao, Brazil. Pollen & Spores **15**: 515–555.

———. 1974. Middle Cretaceous palynomorphs from northeastern Brazil. Sci. Géol., Bull., Strasbourg, **27**: 101–116.

Heslop-Harrison, J. 1971. Sporopollenin in the biological context, *in* J. Brooks, P. R. Grant, M. Muir, P. van Gijzel and G. Shaw (editors), Sporopollenin, pp. 1–30. Academic Press, London.

———. 1976. The adaptive significance of the exine, *in* I. K. Ferguson and J. Muller (editors), The evolutionary significance of the exine (Linn. Soc. London Symp. Series No. 1), pp. 27–37. Academic Press, London.

Hickey, L. J. 1971. Evolutionary significance of leaf architectural features in the woody dicots. Amer. J. Bot. **58**: 469 (abstract).

———. 1973. Classification of the architecture of dicotyledonous leaves. Amer. J. Bot. **60**: 17–33.

Hickey, L. J. and J. A. Wolfe. 1975. The bases of angiosperm phylogeny: Vegetative morphology. Ann. Missouri Bot. Gard. **62**: 538–589.

Hideux, M. J. and I. K. Ferguson. 1976. The stereostructure of the exine and its evolutionary significance in Saxifragaceae *sensu lato*, *in* I. K. Ferguson and J. Muller (editors), The evolutionary significance of the exine (Linn. Soc. London Symp. Series No. 1), pp. 327–378. Academic Press, London.

Hollick, A. 1906. The Cretaceous flora of southern New York and New England. U. S. Geol. Surv. Monogr. **50**.

Horn, H. S. 1971. The adaptive geometry of trees. Princeton Univ. Press, Princeton, New Jersey.

———. 1975. Forest succession. Sci. Amer. **232**(5): 90–98.

Hughes, N. F. 1961a. Fossil evidence and angiosperm ancestry. Sci. Progr. **49**: 84–102.

———. 1961b. Further interpretation of *Eucommiidites* Erdtman, 1948. Palaeontology **4**: 292–299.

———. 1976. Palaeobiology of angiosperm origins. Cambridge Univ. Press.

———. 1977. Palaeo-succession of earliest angiosperm evolution. Bot. Rev. **43**: 105–127.

Jardiné, S. and L. Magloire. 1965. Palynologie et stratigraphie du Crétacé des bassins du Sénégal et de Côte d'Ivoire. Mém. Bur. Rech. Géol. Minières **32**: 187–245.

Jardiné, S., A. Doerenkamp and P. Biens. 1974a. *Dicheiropollis etruscus*, un pollen caractéristique du Crétacé inférieur afro-sudaméricain. Conséquences pour l'évaluation des unités climatiques et implications dans la dérive des continents. Sci. Géol., Bull., Strasbourg, **27**: 87–100.

Jardiné, S., G. Kieser and Y. Reyre. 1974b. L'individualisation progressive du continent africain vue à travers les données palynologiques de l'ère secondaire. Sci. Géol., Bull., Strasbourg, **27**: 69–85.

Jarzen, D. M. and G. Norris. 1975. Evolutionary significance and botanical relationships of Cretaceous angiosperm pollen in the western Canadian interior. Geosci. & Man **11**: 47–60.

Kaplan, D. R. 1973. The problem of leaf morphology and evolution in the monocotyledons. Quart. Rev. Biol. **48**: 437–457.

Kemp, E. M. 1968. Probable angiosperm pollen from British Barremian to Albian strata. Palaeontology **11**: 421–434.

———. 1970. Aptian and Albian miospores from southern England. Palaeontographica, Abt. B, **131**: 73–143.

Krassilov, V. A. 1967. Rannemelovaya flora Yuzhnogo Primorya i yeye znacheniye dlya stratigrafii. Nauka, Moscow.

———. 1973. Mesozoic plants and the problem of angiosperm ancestry. Lethaia **6**: 163–178.

———. 1975. Dirhopalostachyaceae—a new family of proangiosperms and its bearing on the problem of angiosperm ancestry. Palaeontographica, Abt. B, **153**: 100–110.

Krutzsch, W. 1970. Atlas der mittel- und jungtertiären dispersen Sporen- und Pollen- sowie der Mikroplanktonformen des nördlichen Mitteleuropas. Lieferung VII. Monoporate, monocolpate, longicolpate, dicolpate und ephedroide (polyplicate) Pollenformen. Gustav Fischer Verlag, Jena.

Kuyl, O. S., J. Muller and H. T. Waterbolk. 1955. The application of palynology to oil geology, with special reference to western Venezuela. Geol. & Mijnb., n.s., **17**: 49–76.

Laming, D. J. C. 1966. Imbrication, paleocurrents, and other sedimentary features in the lower New Red Sandstone, Devonshire, England. J. Sediment. Petrol. **36**: 940–959.

Laing, J. F. 1975. Mid-Cretaceous angiosperm pollen from southern England and northern France. Palaeontology **18**: 775–808.

———. 1976. The stratigraphic setting of early angiosperm pollen, *in* I. K. Ferguson and J. Muller (editors), The evolutionary significance of the exine (Linn. Soc. London Symp. Series No. 1), pp. 15–26. Academic Press, London.

Lesquereux, L. 1892. The flora of the Dakota group. U. S. Geol. Surv. Monogr. **17**.

Luyendyk, B. P., D. Forsyth, and J. D. Phillips. 1972. Experimental approach to the paleocirculation of the oceanic surface waters. Bull. Geol. Soc. Amer. **83**: 2649–2664.

Mamay, S. H. 1976. Paleozoic origin of the cycads. U. S. Geol. Surv. Profess. Pap. **934**.

Margalef, R. 1968. Perspectives in ecological theory. Univ. Chicago Press.

McElhinny, M. W. 1973. Palaeomagnetism and plate tectonics. Cambridge Univ. Press.

McGee, W. J. 1888. Three formations of the Middle Atlantic Slope. Amer. J. Sci., Ser. 3, **35**: 120–143.

Meeuse, A. D. J. 1966. Fundamentals of phytomorphology. Ronald Press, New York.

———. 1970. The descent of the flowering plants in the light of new evidence from phytochemistry and from other sources. Acta Bot. Neerl. **19**: 61–72, 133–140.

———. 1975. Floral evolution in the Hamamelididae. Acta Bot. Neerl. **24**: 155–179.

Melville, R. 1962. A new theory of the angiosperm flower. I. The gynoecium. Kew Bull. **16**: 1–50.

———. 1963. A new theory of the angiosperm flower. II. The androecium. Kew Bull. **17**: 1–63.

———. 1969. Leaf venation patterns and the origin of the angiosperms. Nature **224**: 121–125.

Mersky, M. L. 1973. Lower Cretaceous (Potomac Group) angiosperm cuticles. Amer. J. Bot. **60**(4, suppl.): 17–18 (abstract).

Millioud, M. E. 1967. Palynological studies of the type localities at Valangin and Hauterive. Rev. Palaeobot. Palynol. **5**: 155–167.

Moore, D. G. and P. C. Scruton. 1957. Minor internal sedimentary structures of some Recent unconsolidated sediments. Bull. Amer. Assoc. Petrol. Geol. **41**: 2723–2751.

Mouton, J. A. 1970. Architecture de la nervation foliaire. Compt. Rend. 92e Congr. Natl. Soc. Savantes (Strasbourg et Colmar, 1967) **3**: 165–176.

Müller, H. 1966. Palynological investigations of Cretaceous sediments in northeastern Brazil, *in* J. E. van Hinte (editor), Proc. 2nd West African Micropaleontol. Colloquium (Ibadan), pp. 123–136. Brill, Leiden.

Muller, J. 1959. Palynology of Recent Orinoco delta and shelf sediments. Micropaleontology **5**: 1–32.

———. 1970. Palynological evidence on early differentiation of angiosperms. Biol. Rev. Cambridge Philos. Soc. **45**: 417–450.

Němejc, F. 1956. On the problem of the origin and phylogenetic development of the angiosperms. Sborn. Nár. Mus. v Praze, Řada B, Přír. Vědy, **12**: 59–143.

Newberry, J. S. 1895. The flora of the Amboy Clays. U. S. Geol. Surv. Monogr. **26**.

Norris, G. 1967. Spores and pollen from the lower Colorado Group (Albian-?Cenomanian) of central Alberta. Palaeontographica, Abt. B, **120**: 72–115.

Norris, G., D. M. Jarzen and B. V. Awai-Thorne. 1975. Evolution of the Cretaceous terrestrial palynoflora in western Canada, *in* W. G. E. Caldwell (editor), The Cretaceous System in the Western Interior of North America. Geol. Assoc. Canada Special Pap. **13**: 333–364.

Owens, J. P. and N. F. Sohl. 1969. Shelf and deltaic paleoenvironments in the Cretaceous-Tertiary formations of the New Jersey Coastal Plain, *in* S. Subitzky (editor), Geology of selected areas in New Jersey and eastern Pennsylvania and guidebook of excursions, pp. 235–278. Rutgers Univ. Press, New Brunswick, New Jersey.

Pacltová, B. 1961. Zur Frage der Gattung *Eucalyptus* in der böhmischen Kreideformation. Preslia **33**: 113–129.

———. 1971. Palynological study of Angiospermae from the Peruc Formation (?Albian-Lower Cenomanian) of Bohemia. Ústřední Ústav Geol., Sborn. Geol. Věd, Paleontol., Řada P, **13**: 105–141.

Pannella, G. 1966. Palynology of the Dakota Group and Graneros Shale of the Denver Basin. Ph.D. thesis, Univ. Colorado, Boulder.

Pettitt, J. M. and C. B. Beck. 1968. *Archaeosperma arnoldii*—a cupulate seed from the Upper Devonian of North America. Contr. Univ. Michigan Mus. Paleontol. **22**: 139–154.

Pierce, R. L. 1961. Lower Upper Cretaceous plant microfossils from Minnesota. Minnesota Geol. Surv. Bull. **42**: 1–86.

Playford, G. 1971. Palynology of Lower Cretaceous (Swan River) strata of Saskatchewan and Manitoba. Palaeontology **14**: 533–565.

Pocock, S. A. J. 1962. Microfloral analysis and age determination of strata at the Jurassic-Cretaceous boundary in the western Canada plains. Palaeontographica, Abt. B, **111**: 1–95.

Pray, T. R. 1955. Foliar venation of angiosperms. II. Histogenesis of the venation of *Liriodendron*. Amer. J. Bot. **42**: 18–27.

———. 1960. Ontogeny of the open dichotomous venation in the pinna of the fern *Nephrolepis*. Amer. J. Bot. **47**: 319–328.

———. 1962. Ontogeny of the closed dichotomous venation of *Regnellidium*. Amer. J. Bot. **49**: 464–472.

———. 1963. Origin of vein endings in angiosperm leaves. Phytomorphology **13**: 60–81.

Raven, P. R. and D. I. Axelrod. 1974. Angiosperm biogeography and past continental movements. Ann. Missouri Bot. Gard. **61**: 539–673.

Read, R. W. and L. J. Hickey. 1972. A revised classification of fossil palm and palm-like leaves. Taxon **21**: 129–137.

Regali, M. S., N. Uesugui and A. S. Santos. 1974. Palinologia dos sedimentos meso-cenozóicos do Brasil. Bol. Técn. Petrobrás **17**: 177–191, 263–301.

Reymanówna, M. 1968. On seeds containing *Eucommiidites troedssonii* pollen from the Jurassic of Grojec, Poland. J. Linn. Soc., Bot., **61**: 147–152.

Reyment, R. A. and E. A. Tait. 1972. Biostratigraphical dating of the early history of the South Atlantic Ocean. Philos. Trans., Ser. B, **264**: 55–95.

Rigby, J. K. and W. K. Hamblin (editors). 1972. Recognition of ancient sedimentary environments. Soc. Econ. Paleontol. Mineral. Special Publ. **16**.

Sampson, F. B. 1976. Aperture orientation in *Laurelia* pollen (Atherospermataceae syn. subfamily Atherospermoideae of Monimiaceae). Grana **15**: 153–157.

Samylina, V. A. 1960. Pokrytosemennye rasteniya iz nizhnemelovykh otlozheniy Kolymy. Bot. Žurn., SSSR, **45**: 335-352.

________. 1968. Early Cretaceous angiosperms of the Soviet Union based on leaf and fruit remains. J. Linn. Soc., Bot., **61**: 207-218.

Scheckler, S. E. and H. P. Banks. 1971. Anatomy and relationships of some Devonian progymnosperms from New York. Amer. J. Bot. **58**: 737-751.

Schluger, P. R. and H. E. Roberson. 1975. Mineralogy and chemistry of the Patapsco Formation, Maryland, related to the ground-water geochemistry and flow system: A contribution to the origin of red beds. Bull. Geol. Soc. Amer. **86**: 153-158.

Schopf, J. W. 1970. Precambrian micro-organisms and evolutionary events prior to the origin of vascular plants. Biol. Rev. Cambridge Philos. Soc. **45**: 319-352.

________. 1975. The age of microscopic life. Endeavour **34**: 51-58.

Schulz, E. 1967. Sporenpaläontologische Untersuchungen rätoliassischer Schichten im Zentralteil des germanischen Beckens. Paläontol. Abh., Abt. B, **2**: 542-633.

Schuster, R. M. 1976. Plate tectonics and its bearing on the geographical origin and dispersal of angiosperms, *in* C. B. Beck (editor), Origin and early evolution of angiosperms, pp. 48-138. Columbia Univ. Press, New York.

Scientific Party for Leg 43 of the Deep Sea Drilling Project. 1975. Glomar Challenger drills in the North Atlantic. Geotimes **20**(12): 18-21.

Scott, R. A., E. S. Barghoorn and E. B. Leopold. 1960. How old are the angiosperms? Amer. J. Sci. **258 A** (Bradley vol.): 284-299.

Scott, R. A., P. L. Williams, L. C. Craig, E. S. Barghoorn, L. J. Hickey and H. D. MacGinitie. 1972. "Pre-Cretaceous" angiosperms from Utah: Evidence for Tertiary age of the palm wood and roots. Amer. J. Bot. **59**: 886-896.

Sen, L. 1971. The geometric structure of an optimal transport network in a limited city-hinterland case. Geogr. Analysis **3**: 1-14.

Seward, A. C. 1931. Plant life through the ages. Cambridge Univ. Press.

Simpson, G. G. 1953. The major features of evolution. Columbia Univ. Press, New York.

Singh, C. 1971. Lower Cretaceous microfloras of the Peace River area, northwestern Alberta. Bull Res. Council Alberta **28**.

________. 1975. Stratigraphic significance of early angiosperm pollen in the mid-Cretaceous strata of Alberta. Geol. Assoc. Canada Special Pap. **13**: 365-389.

Slade, B. F. 1957. Leaf development in relation to venation as shown in *Cercis siliquastrum* L., *Prunus serrulata* Lindl., and *Acer pseudoplatanus* L. New Phytol. **56**: 281-300

Sporne, K. R. 1972. Some observations on the evolution of pollen types in dicotyledons. New Phytol. **71**: 181-185.

Stanley, E. A. 1967. Cretaceous pollen and spore assemblages from northern Alaska. Rev. Palaeobot. Palynol. **1**: 229-234.

Stanley, S. M. 1976. Ideas on the timing of metazoan diversification. Paleobiology **2**: 209–219.

Stebbins, G. L. 1965. The probable growth habit of the earliest flowering plants. Ann. Missouri Bot. Gard. **52**: 457–468.

———. 1974. Flowering plants: Evolution above the species level. Harvard Univ. Press, Cambridge, Massachusetts.

Takhtajan, A. L. 1969. Flowering plants: Origin and dispersal. Oliver and Boyd, Edinburgh.

———. 1976. Neoteny and the origin of flowering plants, *in* C. B. Beck (editor), Origin and early evolution of angiosperms, pp. 207–219. Columbia Univ. Press, New York.

Teixeira, C. 1948. Flora mesozóica portuguesa. Part I. Serv. Geol. Portugal, Lisbon.

Thorne, R. F. 1976. A phylogenetic classification of the Angiospermae, *in* M. K. Hecht, W. C. Steere and B. Wallace (editors), Evol. Biol. **9**: 35–106.

Tidwell, W. D., S. R. Rushforth, J. L. Reveal and H. Behunin. 1970a. *Palmoxylon simperi* and *Palmoxylon pristina*: Two pre-Cretaceous angiosperms from Utah. Science **168**: 835–840.

Tidwell, W. D., S. R. Rushforth and A. D. Simper. 1970b. Pre-Cretaceous flowering plants: Further evidence from Utah. Science **170**: 547–548.

Tralau, H. 1968. Botanical investigations into the fossil flora of Eriksdal in Fyledalen, Scania. II. The Middle Jurassic microflora. Sveriges Geol. Undersökning, Ser. C, Årsbok **62**(4): 1–185.

Vagvolgyi, A. and L. V. Hills. 1969. Microflora of the Lower Cretaceous McMurray Formation, northeast Alberta. Bull. Canad. Petrol. Geol. **17**: 154–181.

Vakhrameev, V. A. 1952. Stratigrafiya i iskopaemaya flora melovykh otlozheniy Zapadnogo Kazakhstana. Regional'naya Stratigrafiya SSSR **1**.

———. 1973. Pokrytosemennye i granitsa nizhnego i verkhnego mela, *in* A. F. Chlonova (editor), Palinologiya Mezofita (Trudy III Mezhdunarodnoy Palinologicheskoy Konferentsii), pp. 131–137. Nauka, Moscow.

Vakhrameev, V. A., I. A. Dobruskina, E. D. Zaklinskaya and S. V. Meyen (editors). 1970. Paleozoyskie i mezozoyskie flory Yevrazii i fitogeografiya etogo vremeni. Nauka, Moscow.

Valentine, J. W. and C. A. Campbell. 1975. Genetic regulation and the fossil record. Amer. Sci. **63**: 673–680.

Van Campo, M. 1971. Précisions nouvelles sur les structures comparées des pollens de Gymnospermes et d'Angiospermes. Compt. Rend. Hebd. Séances Acad. Sci., Sér. D, **272**: 2071–2074.

———. 1976. Patterns of pollen morphological variation within taxa, *in* I. K. Ferguson and J. Muller (editors), The evolutionary significance of the exine (Linn. Soc. London Symp. Series No. 1), pp. 125–137. Academic Press, London.

Van Campo, M. and B. Lugardon. 1973. Structure grenue infratectale de l'ectexine des pollens de quelques Gymnospermes et Angiospermes. Pollen & Spores **15**: 171–187.

Van Konijnenburg-van Cittert, J. H. A. 1971. *In situ* gymnosperm pollen from the Middle Jurassic of Yorkshire. Acta Bot. Neerl. **20**: 1–97.

Wagner, W. H., Jr. 1964. The evolutionary patterns of living ferns. Mem. Torrey Bot. Club **21**: 86–95.

Walker, J. W. 1971. Pollen morphology, phytogeography, and phylogeny of the Annonaceae. Contr. Gray Herb. **202**: 1–131.

________. 1974. Aperture evolution in the pollen of primitive angiosperms. Amer. J. Bot. **60**: 1112–1137.

________. 1976. Evolutionary significance of the exine in the pollen of primitive angiosperms, *in* I. K. Ferguson and J. Muller (editors), The evolutionary significance of the exine (Linn. Soc. London Symp. Series No. 1), pp. 251–308. Academic Press, London.

Walker, J. W. and J. A. Doyle. 1975. The bases of angiosperm phylogeny: Palynology. Ann. Missouri Bot. Gard. **62**: 664–723.

Ward, L. F. 1888. Evidence of the fossil plants as to the age of the Potomac formation. Amer. J. Sci., Ser. 3, **36**: 119–131.

________. 1895. The Potomac Formation. U. S. Geol. Surv., 15th Annual Rep., pp. 307–397.

________. 1905. Status of the Mesozoic floràs of the United States. U. S. Geol. Surv. Monogr. **48**.

Weaver, K. N., E. T. Cleaves, J. Edwards and J. D. Glaser. 1968. Geologic map of Maryland. Maryland Geol. Surv., Baltimore.

Whitehead, D. R. 1969. Wind pollination in the angiosperms: Evolutionary and environmental considerations. Evolution **23**: 28–35.

Wolfe, J. A. 1972a. Significance of comparative foliar morphology to paleobotany and neobotany. Amer. J. Bot. **59**: 664 (abstract).

________. 1972b. Phyletic significance of Lower Cretaceous dicotyledonous leaves from the Patuxent Formation, Virginia. Amer. J. Bot. **59**: 664 (abstract).

________. 1973. Fossil forms of Amentiferae. Brittonia **25**: 334–355.

Wolfe, J. A. and H. M. Pakiser. 1971. Stratigraphic interpretations of some Cretaceous microfossil floras of the Middle Atlantic states. U. S. Geol. Surv. Profess. Pap. **750-B**: B35–47.

Wolfe, J. A., J. A. Doyle and V. M. Page. 1975. The bases of angiosperm phylogeny: Paleobotany. Ann. Missouri Bot. Gard. **62**: 801–824.

Young, K. 1966. Texas Mojsisovicziinae (Ammonoidea) and the zonation of the Fredericksburg. Mem. Geol. Soc. Amer. **100**.

PAPER 12

The Community Approach to Paleoecology (1964)

R. G. Johnson

Commentary

MARTY BUZAS

Ralph Gordon Johnson brilliantly analyzed the community concept in paleoecology and forecasted the direction of research for the next half century. His analysis is based on the modern macro-fauna, mostly mollusks, inhabiting the "level bottom community" of the continental shelf (shallow water to 200 m). While great effort and attention is afforded to "coral reef communities," the level bottom communities of the world's continental shelves that Johnson examines are a vastly larger biological entity.

A "concrete" definition of community is given simply as "the assemblage of organisms inhabiting a specified space." The reality of communities is established through field observations. Not all combinations of species are observed, but rather a finite number of recognizable specific assemblages recur. These recurring assemblages give rise to the definition of community type. In a similar fashion, paleoecologists also recognize in the field that similar assemblages recur in time. In the Neogene, before extinction had taken its toll, the same species associations are recognizable again and again. Johnson believed that the evidence from modern communities indicated interspecific neutralism. That is, changes in composition are associated with physical rather than biological aspects of the environment. While a foray into a belief of functional groups or a biotic component of communities gained popularity for a while among some paleoecologists, today, most researchers agree with the Johnsonian view.

The description and analysis of the distribution of organisms in space and time is inherently a quantitative undertaking. In the early 1960s, with the advent of computers, quantitative paleoecology was in its infancy. Johnson recognized the importance of a quantitative approach to introduce objectivity. While examining features of modern communities such as mode of reproduction, composition, spatial relations, and changes in abundance and composition, he outlined the areas for future quantitative investigations. At the same time, he cautioned that quantitative approaches will require great effort for data collection, reduction, and analysis. And so they have. They have provided us, however, with a much more detailed view of ecology and paleoecology. Detailed quantitative analysis indicates that, often, phenomena regarded as simple or homogenous are, on closer examination, more complex or heterogeneous than previously imagined. For example, the shelf can often be zoned into depth assemblages and is not a single entity as Johnson imagined.

Johnson gives examples from his own work as well as from those of others on the relationship between living and dead populations. He sets the stage for what has become an intense area of investigation to this day. The great contribution of Johnson's paper is that the principles learned from modern community ecology unequivocally lead to an understanding of communities in the past. This outlook encouraged many paleoecologists to study living populations in order to understand the past. In so doing, they became ecologists in their own right, although with a larger perspective. These ecologist-paleoecologists have contributed greatly to our understanding of modern communities as well as those of the past.

From *Approaches to Paleoecology*, 107–34, ed. J. Imbrie and N. D. Newell (New York: Wiley).

The Community Approach to Paleoecology

by Ralph Gordon Johnson *Department of the Geophysical Sciences, University of Chicago, Chicago, Illinois*

ABSTRACT

The application of the principles of community ecology to studies of ancient benthic marine communities should yield information of both geological and biological significance. The potential value of this approach is offset by imposing practical and theoretical difficulties. The community approach is evaluated by reviewing the reproduction, composition, spatial relations, and changes in modern marine communities as these features relate to paleoecological problems.

Features of larval ecology bear upon interspecific correlation, shifts in habitat, local extinctions, and substrate relations. Shallow water communities can comprise several or several hundred species. About 30% of the species and individuals in modern shallow-water communities are preservable by virtue of possessing resistant hard parts. Sediments that are most favorable for the preservation of animals *in situ* usually contain the least number of preservable animals. Almost all of the fluctuations in animal numbers in benthic communities that have been studied are associated with changes in the physical rather than the biological environment. In the context of geologic time, recovery from anomalous fluctuations of the environment and invasions of new areas is virtually instantaneous. Evidence is cited in support of the view that benthic communities are associations of largely independent species that occur together because of similar responses to the physical environment.

The author is indebted to J. W. Hedgpeth and E. C. Olson for stimulating discussions on the problems considered here. Data from Tomales Bay, California were obtained with the aid of a grant from the National Science Foundation.

Introduction

The application of the principles of community ecology to studies of ancient benthic marine communities should yield information of both geological and biological significance. Such studies may reveal features of the distribution and organization of ancient communities that can be used in the interpretation of evolutionary events. The stratigraphic and environmental implications of a natural assemblage of species should be far more specific than those of a single species or taxonomic group. The potential value of this approach is offset by imposing practical and theoretical difficulties. Chief among these difficulties are the inherent limitations of the fossil record, problems of the recognition of life associations, data gathering and analysis, and inadequate knowledge of modern benthic communities.

Statistical methods for recognizing and analyzing life assemblages in the fossil record have been developed by several workers. The labor involved in obtaining appropriate data for statistical analysis raises the question of whether or not the community approach can be used to solve real problems. Only a few paleontologists have actually stressed the community approach in their work (e.g., E. C. Olson, D. I. Axelrod, J. A. Shotwell, J. R. Beerbower, R. G. Johnson). Since the literature is sparse, we should explore the subject by reviewing major aspects of the structure of modern benthic communities, since these features bear upon paleoecological problems. This review is based mainly on the literature since Thorson's 1957 summary and on the author's studies of modern marine communities.

The Community Concept

While most biologists regard communities as real units of biological integration, a few have denied the reality of communities (e.g., Muller, 1958). Critics of the community concept have observed that the species composition of associations belonging to a single recognized community type may vary considerably from place to place. Biologists have not been able to define a community in such a way as to encompass the varied viewpoints of all ecologists, partially because communities do not exhibit a fixed composition and because a community may broadly intergrade with neighboring communities. One of the striking features of the biosphere, however, is that all species living today do not occur in all possible combinations in natural assemblages. Rather, we observe around us a finite number of recurring associations of species. The high recurrence of species combinations would appear to be a phenomenon of considerable biological significance. Our failure to arrive at a rigorous description of this phenomenon is not proof against its reality but is probably a reflection of our ignorance and propensity for preconceived notions.

Both concrete and abstract definitions have been proposed for the term "community." The concrete definitions specify a particular assemblage of organisms at a particular place. The abstract definitions are based on some assumed general attribute of communities such as the recurrence of suites of species, the degree of interspecific interaction, or economic independence. For this discussion, a concrete definition will be used for the term "community": the assemblage of organisms inhabiting a specified space. The recurrence concept will be used for a definition of community type: an assemblage of organisms which often occur together. These definitions seem useful inasmuch as they involve attributes that can be demonstrated from field observations.

In order to apply the concepts of community ecology to the fossil record we must be able to recognize the components of particular communities in fossil assemblages. Paleontologists have used three kinds of evidence for recognizing ancient communities: (1) field evidence of burial *in situ,* (2) taxonomic analogy with a modern community, and (3) recognition of recurring suites of species analogous to the recurrence observed among modern communities.

Rarely is there unequivocal field evidence that the fossil assemblage has been buried in place without the introduction of foreign elements. Taxonomic analogues are particularly useful in studies of late Cenozoic assemblages but their utility diminishes with greater geologic age. The recognition of community types seems to be the most generally useful approach and has been employed, at least subconsciously, by most paleontologists. The field paleontologist, after some experience in an area and stratigraphic section, comes to recognize recurring assemblages of species and hence occasional irregular associations. Recently, several workers have attempted to quantify this experience and thereby introduce objectivity (Beerbower and McDowell, 1960; Johnson, 1962). Measures of interspecific association are used to recognize pairs of species which frequently occur together. These pairs are assembled into groups of highly associated species. The ecologic aspect, stratigraphic, and lithologic distribution of the groups may then be studied for evidence that the groups are composed of members of the same community type. These techniques appear to offer considerable promise but represent a great effort in acquiring suitable collections, data reduction, and analysis. This approach cannot be evaluated until we have tried it a number of times. Meanwhile we can judge whether or not all the effort is justified by examining modern benthic communities to find features that could illuminate paleontological problems. The following discussion is limited to level bottom communities as these are the most commonly encountered types in the marine fossil record.[1] It is one of the theses of this paper that the level bottom community types differ in certain important ways from rocky intertidal, coral reef, and terrestrial community types.

Modern Marine Communities of the Level Bottom

Reproduction and Larval Ecology

The mode of reproduction or recruitment of benthic marine communities is one of their most singular features. As Thorson has emphasized in several papers, 70 to 90% of modern bottom dwelling invertebrates possess pelagic larval stages in their life histories. Curiously enough, the presence or absence of pelagic larvae in the ontogeny of a marine organism cannot be directly correlated with the extent of geographic range or any other obvious single feature. A number of cosmopolitan and circumpolar species exhibit direct development.

Pelagic larvae of benthic animals may spend from a few hours to several months in the overlying water mass. A benthic species really represents two kinds of animals inhabiting grossly different kinds of environments, possessing separate sets of adaptions, and presumably undergoing more-or-less separate evolutionary histories. In recent years, evidence has been accumulated indicating that the larvae of many species actively select a suitable environment for settlement.

In Table 1, the environmental factors associated with the settlement of pelagic larvae are shown for 39 species. Experimental evidence indicates that the larvae of many species actively search for a suitable environment and are able to put off metamorphosis considerably beyond the normal period if necessary (e.g., see Wilson, 1952). As the larval period is so extended, discrimination diminishes, and eventually the larva settles regardless of the suitability of the environment.

Marine biologists have long recognized a high degree of association between the distribution of species and substrate (e.g., see Pratt,

[1] The term "level bottom communities" is used in the sense of Thorson (1957) to apply to the benthic communities developed on or in essentially flat deposits and to exclude most intertidal and all coral reef communities.

1953; Sanders, 1958; Jones, 1952; Stickney and Stringer, 1957; Southward, 1957). Most of the species for which the larval ecology has been studied appear to respond actively to some feature of the substrate (Table 1). Settlement may be associated with particle size, particle shape, organic content, surface roughness or with some cryptic substance that can be isolated from preferred substrates. The fact that the nature of the substrate is involved in the settling response of many species in several phyla is of considerable interest to the paleontologist. The deposit type and many of its properties are environmental features that can be studied in the fossil record. Thorson and others have pointed out that while the substrate may act as the key factor in inducing larval settlement, this does not necessarily mean that the properties of the substrate are themselves the significant factors of the effective environment of the adult organism. The sediment at a particular site is related to many other features of the environment, such as distance from land, depth, and water movement. Properties of the sediment are often important ecological factors affecting the growth and mortality of a benthic species (see, e.g., Loosanoff and Tommers, 1948; Pratt, 1935). As size, sorting, and mineralogy are readily measured, these properties have received the most attention from paleontologists. Evidence is accumulating, however, indicating that other properties of the sediment, such as organic content, porosity, proportion of clay, surface roughness, and the firmness or cohesiveness of the deposit, may be even more directly related to the mode of life of many benthic animals.

Larvae exhibiting selective behavior must still possess tolerance for conditions grading away from those for optimal development. Larvae must react to substrata possessing the minimal conditions for settlement, as they cannot have the forsight to refuse a substrate in anticipation of a more favorable site downcurrent. Thus chance will be an important factor in the distribution of a species even though the animal may exhibit considerable selectivity.

Most of the species whose larval development has been studied in detail have restricted environmental distributions. Animals that are able to flourish in a variety of environments probably settle at random to survive in favorable circumstances or die if they chance upon an unfavorable environment. A high degree of selectivity could be expected among deposit feeders and forms that must attach themselves to stable surfaces. Lower selectivity could be expected in predators, scavengers, or filter feeders. Only a few observations have been made, in this regard, on species having broad environmental tolerances. Kristensen (1957) suggests that the larvae of the filter feeder,

TABLE 1. *Environmental factors found to be associated with the settlement of pelagic larvae in 39 modern species*

Species	Factors Associated with Settlement	Author
Annelida		
Polychaeta		
Hydroides norvegica	Temperature, light, orientation of substrate	Wisely, 1959
Notomastus latericeus	Sediment size	Wilson, 1937
Ophelia bicornis	Sediment size, shape, organic content, Microbenthos	Wilson; 1948, 1953, 1954, 1955
Owenia fusiformis	Sediment size	Wilson, 1932
Pygospia elegans	Unknown feature of substratum	Smidt, 1951
Scolecolepis fuliginosa	Sediment size	Day & Wilson, 1934
Spirorbis borealis	Algal filmed surfaces, presence of own species	Knight-Jones, 1951; 1953 Wisely, 1960
Spirorbis pagenstecheri	" "	Crisp & Ayland, 1960
Archiannelida		
Protodrilus rubropharyngeus	Sediment size, water soluble inorganic substance in substrate	Jägersten, 1940
Bryozoa		
Alcyonidium hirsutum	Species of macro-algae, surface texture, mucous-free surfaces	Ryland, 1959
Alcyonidium plyoum	" "	
Celleporella hyalina	" "	
Flustrellidra hispida	" "	
Bugula avicularia	Temperature	Wisely, 1959
Bugula neritina	"	
Bugula flagellata	Clean surfaces	Crisp & Ryland, 1960
Watersipora cucullata	Light, temperature	Wisely, 1959
Acanthodesia tenuis	Horizontal surfaces " "	Pomerat & Reiner, 1942
Electra hastingseae	" "	
Phoronida		
Phoronis mulleri	Sediment size, clay content	Silen, 1954
Phoronis pallida	" " " "	

TABLE 1. *Environmental factors found to be associated with the settlement of pelagic larvae in 39 modern species (Continued)*

Species	Factors Associated with Settlement	Author
Mollusca		
Pelecypoda		
Cardium edule	Sediment size, water movement	Orton, 1937
Crassostrea commercialis	Temperature	Wisely, 1959
Mytilus edulis	Depth, orientation of substrate	Chippenfield, 1953
Ostrea edulis	Water movement, orientation of substrate, surface texture	Cole & Knight-Jones, 1939 Erdmann, 1934 Korringa, 1940
Ostrea gigas	Surface orientation	Schaefer, 1937
Ostrea lurida	" "	Hopkins, 1935
Ostrea virginica	Copper concentration in water, salinity, temperature, surface texture	Prytherch, 1934
Teredo norvegica	Organic extract of wood	Harrington, 1921
Venerupis pullastra	Surface orientation	Quayle, 1952
Gastropoda		
Nassarius obsoletus	Water soluble substance in substratum	Scheltema, 1961
Arthropoda		
Cirripedia		
Balanus, 6 species	Presence of adults, temperature, surface texture, light, algal film	Pomerat & Reiner, 1942 Gregg, 1945 Weiss, 1947 Crisp & Barnes, 1954 Knight-Jones, 1955 Wisely, 1959 Connell, 1961
Eliminius modestus	Presence of adults, surface texture	Knight-Jones, 1950; 1955 Crisp & Barnes, 1954
Crustacea, Cumacea		
Cumella vulgaris	Sediment size, organic content	Wieser, 1956

Cardium edule, show little or no selective behavior. Holme (1961) argues that widely distributed species probably do not exhibit selective behavior in settling. The experiments of Scheltema (1956, 1961) indicate an interesting difference in the larval behavior of two species of the gastropod genus *Nassarius.* Scheltema demonstrated that *Nassarius obsoletus* exhibits a complex behavior pattern leading to preferential settling on a particular substratum. He obtained from the preferred substratum a water soluble substance capable of inducing settlement in sea water not in contact with a favorable substratum. *Nassarius vibex,* which lives in the same region as *Nassarius obsoletus,* does not metamorphose in response to bottom sediment. Since *Nassarius obsoletus* is a deposit feeder, its selective behavior would seem to have considerable adaptive value. *Nassarius vibex,* on the other hand, is a scavenger and not as directly dependent on sediment properties.

In his 1957 review Thorson stated, "we have evidence enough to abandon the 'raining down at random' theory hitherto accepted by most marine biologists" (Thorson, 1957, p. 482). It now seems more probable, however, that benthic invertebrates exhibit a spectrum of selectivity, some species showing highly specific behavior while others approach random fallout.

Thorson has discussed the possible influence of adult populations on the settlement of larvae. There is evidence that the larvae of *Ostrea, Balanus,* and *Spirorbis* are attracted to sites occupied by adults of the same species (Thorson, 1957; Table 1). It also seems possible that the effect of the adult population on the substratum, through fecal accumulation, for example, may render a particular site unattractive to larvae. Some field observations tend to support this possibility, but no direct evidence is yet available. Certainly the presence of a dense population of adults may directly affect settlement by mechanical obstruction or other means. Kristensen describes experiments in which larvae were inhaled by adult clams and subsequently smothered by the coating of mucus they received in the process (Kristensen, 1957). Such deterring effects by the adult population could be responsible for the populations of animals of nearly the same size that are occasionally encountered.

If the presence of adults at a certain density does repel larvae of the same species, certain important results are theoretically possible. Since the life span of most benthic species is seldom more than a few years, the species composition at some site might change over a relatively short period of time. The composition of the entire community might remain the same, while the spatial distribution of species within the community continually changes. Such a model as this might explain the lack of correlation between the numbers of individuals of different species and the high incidence of sympatric species in these kinds of communities. There would be insufficient time for subtle interactions between neighboring sedentary individuals to affect the natural regulation of numbers. As the carrying capacity of the environment is approached, interaction might become quite important. The wide fluctuations in animal numbers observed in some level bottom communities and the variation from place to place suggests, however, that the full potential of most environments is seldom utilized in nature. If neutral relations between species are common in level bottom communities, this fact would have important implications bearing on the evolution of benthic organisms.

The reproductive ecology of benthic invertebrates is important to paleoecology in several ways. The dual nature of most benthic species makes the task of interpreting change in the fossil record more difficult. It has often been pointed out that the continuity of any species at a particular site is dependent on the weakest link in its life cycle, which almost invariably involves reproductive phases. Since pelagic larvae are often instrumental in selecting the bottom environment of the adults, shifts in adult habitat may partly be due to changes in larval response. Failure of a species to maintain itself at some locale may be caused either by these changes in larval response or by the failure of larvae to reach the site due to some chance combination of current, temperature, biological activity, etc. Offsite conditions could thus influence the permanency of a population irrespective of the stability of local environmental conditions. The disappearance of a species in a local stratigraphic sequence does not always, therefore, imply a change in the local benthic environment.

While the larva represents a complex of adaptions to an environment quite different from that of the adult, the evolutionary history of both phases must be closely linked. Failure of an appreciable number of larvae to reach an environment suitable for adult growth and reproduction will ultimately result in the local extermination of the species. Changes in pelagic ecosystems in geological time might lead to the world-wide extinction of many species without leaving the paleontologist any trace of the causal factors. The duality of benthic species may have other evolutionary implications. It is conceivable, for example, that genes having selective value for larval life could exert influence on adult morphology by pleiotropy in a fashion analagous to the mechanism of the theory of senescence proposed by Williams (1957).

Not all of the sediment attributes involved in settling behavior of different species may be preserved in the fossil record. For this reason we cannot consistently hope to obtain a high association between the occurrence of a fossil and lithology even when the species may be highly responsive to some property of the substratum in life. Undoubtedly there are sediment properties of ecological significance that are preserved but are not yet recognized. The lithologic distribution of a fossil assemblage derived from a single community type may reveal such properties to us. As imperfectly preserved as it may be in the fossil record, the substrate is one complex of environmental factors for which we can have direct evidence. It is important to note here that a variety of processes may produce very similar sedimentary rocks. Diagenetic changes can further mask subtle differences between products of different processes. The fossils can provide another criterion for selecting a unique set of processes but only after we are able to demonstrate that the fossils were derived from a single community type.

Composition

The composition of a benthic community can be described in terms of species, numbers of individuals, biomass, and the proportion of various ecologic roles represented.

The number of species at any station within a community may vary from a few to several hundred per square meter. The entire community can comprise several species or hundreds of species. These ranges are comparable to those encountered within stratigraphic units in the fossil record. Evidence from several extensive surveys (e.g., see Barnard, Hartman, and Jones; 1959) suggests that on the level sea bottom (in shallow waters to 200 meters) there is no marked trend in the number of species present in a community with respect only to depth or distance from shore. The finer silts and coarser sands support fewer species than intermediate-sized sediments, but this is not invariably the case. In very shallow water the presence of plants is often associated with a larger number of species than the same substrate without plants. In modern seas to 200 meters depth, there appears to be no regular relation between the proportion of various taxonomic groups and depth. Frequently mollusks are more abundant and varied in waters from 1 to 50 meters than in deeper water, but not invariably. Paleontological speculation on the depth distribution of organisms probably has been unduely influenced by the results of

TABLE 2. *Number of animals/m² reported in several recent shallow water investigations. Various sampling methods and mesh sizes employed*

Number of Animals/m^2	Place	Author
10–292	English Channel	Holme, 1953
266–1290	Japanese Bays	Miyadi, 1940, 1941
2356	English Channel	Mare, 1942
3500 (average)	Southern California Shelf	Barnard, Hartman, Jones, 1959
1064–12,576	Buzzard's Bay, Mass.	Sanders, 1958
4554–23,014	Scottish Loch	Raymont, 1949
5556–46,404	Long Island Sound	Sanders, 1956
40–63,600	Dutch Waddensea	Smidt, 1951

the major oceanographic expeditions which encompass great ranges in depth.

Table 2 contains tabulated data on numbers of individuals per square meter as obtained in eight recent ecological surveys. These data are for organisms larger than 1 or 1.5 mm. Such data are difficult to compare since different collecting gear and screen sizes were used. At any rate, as few as 10 and as many as 63,000 individuals per square meter have been reported. Only a few studies have been made of the microbenthos (e.g., see Mare, 1942). The numbers of individuals reported range from 10^5 to 10^9 individuals per square meter. All environments in shallow seas, studied to date, support life, and usually abundant life. In the fossil record, if we identify a particular sediment as being of marine origin, we may safely assume it supported life.

Table 3 shows some of the preliminary results of Barnard, Hartman, and Jones (1959) from an intensive survey of the communities on the continental shelf off southern California. In this area the gray sands and muds support greater numbers of individuals than the coarser red sands and rock and rubble. There are a greater number of species on the gray sands but not much greater than on the other substrates. The lower biomass together with the large numbers of individuals on the gray sands reflect the fact that this substrate supports large numbers of small species. The nature of the bottom deposits in this area is not independent of distance from shore, as the coarser sediments occur nearer shore. When the numbers of species and individuals are arranged by depth, no consistent differences can be rec-

TABLE 3. *Number of specimens, species, and biomass reported in an extensive survey of the continental shelf area of Southern California (Barnard, Hartman, Jones, 1959). Number of stations shown in parentheses*

Bottom Deposit	Average Number of Specimens/m^2	Average Number of Species/0.25 m^2	Average Weight in gm/m^2
Black mud	3711 (10)	79 (10)	252 (28)
Green mud	4540 (61)	87 (61)	294 (193)
Gray sand	5047 (18)	96 (18)	157 (53)
Red sand	2536 (11)	67 (11)	200 (14)
Rock or rubble	3210 (4)	73 (4)	291 (16)

ognized from 4 to 300 meters. The biomass does drop off, implying smaller species with depth. If the sediments were uniform on the shelf, we would probably observe a decrease in biomass and numbers with distance from shore as related to the transportation of food materials.

A striking characteristic of shallow water benthic communities is the marked dominance of one or a few species. Almost invariably one to a dozen species account for 95% of the individuals and biomass of the community. This feature can be illustrated from the data of Sanders (1960). In a soft-bottom community in Buzzards Bay, Massachusetts, he found that one species made up 59% of the individuals and another species accounted for 29% of the biomass of a community consisting of 79 species. Eleven species accounted for 95% of the individuals and 8 species for 95% of the biomass. A similar circumstance is found in many fossil assemblages but does not necessarily reflect the species abundance relations of a living community. The fossil assemblage, as an accumulation, more closely approaches the total production of preservable individuals of the community rather than a standing crop. Interpretation of species abundance is further complicated by the fact that the number of individual fossils present per unit space is influenced by the rates of mortality and deposition. Since the dominant species in the community may not be preserved, we may obtain a false impression of the importance of a species in an ancient community. It also follows that the number of individuals does not necessarily reflect the suitability of an environment for a particular species or group of species. Thus, a high fossil density may not indicate near optimum conditions for settlement or growth but high mortality and/or low rates of deposition. Conversely, a relatively low fossil density may actually represent near optimum conditions for survival and growth but a high rate of deposition. Reworking can further modify density. For these reasons, it is difficult to biologically interpret differences in fossil density from place to place. Further, it is difficult to interpret measures of association based on the numbers of specimens as related to lithology or the numbers of other species.

The composition of modern benthic communities provides some insight into the preservability of these kinds of life assemblages. We have reviewed data from modern communities of the level sea bottom and determined the percentage of the macrobenthos possessing resistant hard parts. These are animals which we have guessed to be preservable under the ordinary circumstances of the preservation of death assemblages; they are chiefly mollusks and some echinoderms. No attempt was made to evaluate other factors affecting preservation (e.g., rate of deposition, permanency of deposit, selective solution, etc.). In Table 4, data from 9 modern quantitative surveys, involving 534 samples, are shown. As before, different collecting methods were employed. If an average value could be computed for these data, it would probably fall around 30% for both the percentage of species and percentage of individuals. Such a value is as high as it is due to the abundance of mollusks today and is as low as it is primarily because of the predominance of polychaetes in recent shallow seas. The higher values were obtained in communities dominated by one or a few

TABLE 4. *Range of percentage of animals possessing resistant hard parts, in shallow water benthic communities*

Bottom Deposit	Number of Samples	Species, %	Individuals, %
Mud	142	10–55	8–70
Muddy sand	319	7–35	1–72
Clean sand	47	21–50	18–79
Gravel	26	7–67	4–87
	534		

Data based upon 9 modern surveys and 534 samples. Data from Smith, 1932; Thorson, 1933; Jones, 1952, 1956; Holme, 1953; Sanders, 1956, 1960; Southward, 1957; Jones, 1961.

TABLE 5. *Average percentage of animals possessing resistant hard parts in some benthic communities in the Irish Sea. Depths between 9 and 26 meters (based on data from Jones, 1952)*

Bottom Deposit	Number of Samples	Species, %	Individuals, %
Mud	7	10	8
Muddy sand	93	30	68
Fine sand	37	35	79
	137		

preservable species. Thus, while 87% of the individuals of the community may be preservable by virtue of possessing resistant skeletons, the preservable component of the community may be extremely biased with regard to the total species present. Nevertheless, these figures seem encouraging although we cannot apply them directly to ancient faunas. Certainly, the benthic marine community today is the most preservable community on earth from the viewpoint of preservable hard parts and environments favoring rapid burial.

As shown in Table 4, the number of preservable forms relative to the type of bottom deposit is quite variable. However, there is some evidence of a relationship to deposit type when a single geographic region is considered, as shown in Tables 5 and 6. Table 5 shows the percentage of the macrobenthos possessing resistant hard parts in some communities in the Irish Sea. Low percentages are found associated with muds. In Table 6 similar data are shown for fewer stations in the English Channel. Here again, the finest sediment is associated with the least number of preservable animals. This result is confirmed in other surveys. The sediments that in some respects are the most favorable for the preservation of animals *in situ* usually contain the least number of preservable animals. This circumstance probably reflects the dominance of polychaetes in the finer sediments and the feature that organisms inhabiting high energy environments (strong waves and/or currents) generally possess more substantial supporting structures.

Data from modern communities can also be used to examine the degree of correspondence between life and death assemblages. In the summer of 1959, an ecological survey of Tomales Bay, California was conducted by the author. Eighty-two stations were occupied throughout the bay.[2] Mollusks are the only common group that possess resistant hard parts in these waters. At ten stations, on various substrates, no preservable species were found either alive or dead. At the remaining stations an average of 20% of the living species of mollusks was also represented in the local death assemblage. At 31 stations the living molluscan assemblage was totally different from the dead molluscan assemblage. A clear relationship was observed between the correspondence of life and death assemblages and the type of bottom deposit. Muds and sands were about equally represented at the 82 stations. One third of the stations in mud showed no correspondence between life and death assemblages, while two thirds of the stations in coarser sediments lacked correspondence. The ten samples showing the highest agreement between life and death assemblages were all in clays, silts, and the finer muddy sands. These results agree with the expectation for the relation of correspondence and water movement. The proportion (39%) of stations showing completely dissimilar life and death assemblages seems high but probably is representative of the upper limits in shallow seas, since Tomales Bay is quite shallow and is subjected to considerable current and wave action. In deeper and hence quieter waters (below 15 or 20 meters) correspondence between life and death assemblages is probably considerably higher.

It was also noted in the Tomales Bay study that the areal distribution of the remains of a particular species was nearly always greater than its living representatives. If was found, for example, that the small pelecypod, *Transenella tantilla*, found at 37 stations, occurred

[2] Analysis of these data was made possible by a grant from the Atomic Energy Commission, supporting several projects pertaining to the marine geology and biology of Tomales Bay.

TABLE 6. *Average percentage of animals possessing resistant hard parts in some benthic communities in the English Channel. Depths between 40 and 70 meters (data from Holme, 1953)*

Bottom Deposit	Number of Samples	Species, %	Individuals, %
Muddy sand	3	11	9
Clean sand	10	32	33
Gravel	2	25	23
	15		

both alive and dead at 16 stations, dead only at 16 stations, and alive only at 5 stations. Such results are also to be expected under shallow bay conditions. The zones of gradation or intermingling between these death assemblages are generally broader than the zones of gradation or the ecotones between neighboring communities. This would suggest that abrupt lateral changes in the composition of fossil assemblages may be indicative of quiet waters and little or no transportation of remains.

Another important aspect of the composition of level bottom benthic communities in shallow water is the dominance of detritus feeders. About 70 to 80% of benthic invertebrates feed upon detritus (Sanders, 1956, 1960; Smith, 1932; Mare, 1942). In the fine sediments, the majority of the species are deposit feeders subsisting upon the detritus either on or in the sediment. In coarser sediments, suspension feeders may dominate. Under most circumstances, there is little difference between the kinds of organic matter being utilized by these feeding types (Holme, 1961). The important feature is that the food web and hence the biologic organization of these communities is simpler than that of the rocky intertidal, coral, and terrestrial communities. The overwhelming predominance of detritus feeders would seem to imply that many species may have more or less neutral relations with one another. This circumstance has important implications for the evolution of benthic invertebrates and will be discussed later.

Spatial Relations

Most living benthic communities exhibit vertical stratification as recognized in the distinction between epifauna and infauna (Thorson, 1957). Horizontal stratification is often pronounced in the intertidal zone, but in the absence of plants it is seldom evident below the low tide mark.

Most species and individuals of the infauna occur in the upper 15 cm of the sediment (Holme, 1953, reviews the literature on vertical distribution). The few individuals found below 15 cm tend to be large and may contribute most of the biomass of the infauna. Large pelecypods may occur as deep as a meter below the surface of the substrate (e.g., *Schizothaerus* of the western coast of North America). Bacteria are found in appreciable numbers well below 1 meter (ZoBell, 1957). Thus, while the highest concentration of organisms is near the surface of the substrate, biological activity may extend considerably below and still affect the alteration of organic remains buried quite deeply in the sediment.

One consequence of the vertical stratification of benthic communities is that a few animals will be found living amidst the relics of the preceding population. If a site is continually occupied for an appreciable length of time, a resulting fossil record will consist of mixtures of succeeding populations and overlapping life spans. Low rates of deposition and reworking may greatly exaggerate the disparity in age of remains within the same stratum. While mixing of this type limits our temporal resolution, it can be neglected as trivial in the context of geologic time for the purposes of most paleontological studies. Perhaps it is just as well that seasonal and annual variations will tend to be obscured, since they might distract our attention from the long term variations. This type of temporal mixing has long been recognized by paleontologists. The important implication of these circumstances is that abrupt vertical changes in the fossil record must, more often than not, indicate a major ecological event.

In the presence of plants, a diverse epifauna, itself strongly stratified, may be developed as on *Macrocystis*. In the absence of plants, the epifauna on a soft bottom may be virtually absent or limited to solid objects on the surface of the deposit such as stones or shells. A diverse epifauna in a fossil assemblage would strongly suggest the presence of plants or a firm substrate. The predominance of such epifaunal forms as the articulate brachiopods and crinoids in many pure, fine-grained limestones may mean that the ancient substrate was firm either as bare rock or stiff mud. The presence of a diverse gastropod assemblage, on the other hand, would suggest the presence of plants.

Thorson has pointed out that there is considerable difference in the environment of the epifauna and the infauna (Thorson, 1957). The infaunal environment of a particular deposit type tends to be more uniform in conditions for life from place to place and less subject to the effects of variations in the hydroclimate. In association with the uniformity of the infaunal environment, Thorson has argued that similar substrates throughout the world are inhabited by ecologically similar and phylogenetically related organisms. Thorson has termed such similar assemblages in similar substrates as parallel communities. As a generalization the concept suffers, as do all ecological generalizations, from the fact that deviations are common. Nevertheless, there remains a remarkable similarity between the infaunas of similar deposits throughout the world in sharp contrast to the distinctiveness of epifaunas. This circumstance has several paleoecological implications. We might expect infaunal and epifaunal organisms to differ in their rates of evolution. Since the composition of infaunas is less subject to local hydroclimates than the epifauna, infaunal organisms should make better index fossils, except for the fact that they may evolve more slowly and have greater stratigraphic ranges. These inferences might be tested with the fossil record.

Within the photic zone, the horizontal stratification of plants may confer horizontal stratification upon the animal populations. This is evident on hard substrates at least (Aleem, 1956). In the absence of plants and below the low tide mark there is little evidence of horizontal stratification in benthic communities. Environmental heterogeneity or interactions between organisms may result in distinct patterns of dispersion. Commonly individuals of a species are found to occur in clumps or colonies presumably brought about by the patchiness of the environment, the settlement of larvae in swarms, or the development of clones. Such patchiness is also a feature of the distribution of small algae and protozoa (Lackey, 1961). Within clumps, the individuals may occur clumped, evenly or randomly distributed (Holme, 1950; Connell, 1956; Johnson, 1959). Jones (1961) studied the pattern of dispersion of 19 species in shallow water. Fifteen of these species were found to be randomly dispersed in the study areas. Also, many sedentary organisms are quite capable of moving about. Some infaunal pelecypods (e.g., *Macoma, Solen, Cardium*) move about actively but gradually lose their ability with age.

The pattern of dispersion of fossils in an outcrop can be studied readily but is very difficult to interpret because of the strong likelihood that any pattern in life will be lost (Johnson, 1960). Positive evidence of patchiness in the distribution of fossils in a single stratum can perhaps be interpreted ecologically.

Changes in Abundance and Composition

Most of the studies of change in benthic marine communities involve periods of only a few years. Much of the information is from comparisons of recent and earlier surveys of the same region in connection with some disastrous event. Nearly always the compared surveys were made by different workers using different collecting methods at different times of the year. Data on fluctuations in benthic communities over a period of years are scarce.

Seasonal fluctuations appear to be far more pronounced in very shallow waters (e.g., Allee, 1919; Blegvad, 1951). Below the low tide mark there are tremendous regional differences in the extent of fluctuations in numbers. In some areas there appears to be considerable fluctuation (Poulsen, 1951; Segerstråle, 1960; Jones, 1961), while elsewhere variation seems slight (Sanders, 1956; Holme, 1953). Most of the communities that are characterized by marked fluctuations in numbers are found in very shallow waters and are exposed to frequent and considerable fluctuations in salinity or temperature. As discussed earlier, the normal temporal mixing in the fossil record would obliterate any trace of short term fluctuations in abundance under ordinary circumstances of preservation.

Occasionally a drastic environmental fluctuation has been observed to cause major changes in the local fauna. In northern regions, particularly hard winters may produce profound changes in the shallow water fauna. Kristensen (1957) found that the severe winter of 1946–1947 killed nearly all the shallow water cockles in the Dutch Wadden Sea. Allee (1919) described the effects of a hard winter on the benthic communities in the vicinity of Woods Hole, Massachusetts. He found that species in very shallow water and species at the northern extent of their ranges were the most affected. Raymont (1949) has observed the deleterious effects of lowered oxygen tensions in a sea loch. Hoese (1960) and Goodbody (1961) have recently described the biotic changes associated with salinity fluctuations. The detrimental effects of domestic and industrial wastes have been studied by numerous authors.

One of the major biological disasters of recent times was that associated with the eel grass epidemic of the early 1930's. Eel grass (*Zostera marina*) is a flowering plant of major importance in the cool, shallow (0–10 feet) waters of the northern hemisphere. In 1931–1932 vast beds of eel grass disappeared along the Atlantic Coast of North America and it declined markedly along the European Coast. There is still no agreement among plant pathologists as to the cause of the disease. The fungus-like mycetozoan *Labyrinthula* has been shown to be associated with the disease but it is not yet clear why the pathogen suddenly became active or why only particular strains of eel grass were affected (Cottam and Munro, 1954). At any rate, the biotic consequences of the disappearance of the eel grass were followed by many workers and reported in over 50 papers.

The early consequences of the disappearance of the eel grass were similar everywhere. The first animals to become rare were those whose habitat was the leaves of the plant (Wilson, 1949; Dexter, 1944; Poulsen, 1951). Some species managed to survive locally on algae (Dexter, 1944). Another common result of the epidemic was drastic changes in the bottom deposits previously covered by beds of eel grass. The plant effectively anchors organic muds and diminishes current and wave action. In many places the loss of the plant resulted in the fine sediments being swept away, leaving the coarser fractions behind

(Wilson, 1949; Cottam and Munro, 1954). An infauna more characteristic of a coarser deposit commonly invaded the area. Local changes in depth and current patterns also resulted.

The secondary, biological effects of the disappearance of eel grass were by no means general but instead were controlled by local circumstances. On the eastern coast of North America the Eastern Brant declined rapidly in numbers. Eel grass had constituted 80% of the bird's winter diet. In several years the Brant population had partially recovered, switched to a new diet of *Ulva, Enteromorpha,* and other algae, and had altered its migration routes.

Another species which appeared to undergo a rapid decline associated with the eel grass epidemic was the bay scallop, *Aequipecten irradians.* At the time it was thought that the decline of the scallop was due to its dependence on eel grass as a substrate for the attachment of spat. In contrast to the general decline of this species, in most places, Marshall (1947) observed that the scallop increased in such proportions in the Niantic estuary as to constitute a new fishery. Subsequent investigations (Marshall, 1960) indicate that the loss of the eel grass so improved circulation adjacent to the bottom that more nutrients were available in the area than before. Adequate sites for attachment of young were available on algae and other surfaces. It now seems likely that the decline of the scallop elsewhere was due to the changes in the substrate resulting from the loss of the eel grass. Conditions in the Niantic estuary apparently led to a neat balance between circulation of nutrients and bottom stability.

The loss of eel grass as a source of detritus was thought by some observers to be responsible for the observed decline of several important infaunal species. *Mya* and *Ensis* declined on the eastern coast of North America. These changes may have been due, however, to the alteration of the substrate or the fact that spat formerly held in place by eel grass were now being washed into less favorable environments. The observations made by the Danish Biological Station in the Limfjord of Jutland, which connects the North Sea with the Kattegat, are particularly pertinent to the ecological role of eel grass (Blegvad, 1951; Poulsen, 1951). Annual quantitative surveys, involving over 100 samples per survey, have been made in this area since the work of C. G. J. Petersen in 1909. In his classical studies of benthic communities Petersen had placed great emphasis on eel grass as a primary source of food for benthic organisms. The data accumulated by the Danish Biological Station over a long period, however, indicated no reduction in animal numbers that could be associated with the disappearance of eel grass, except for 3 or 4 species that lived directly upon or among the plants (Poulsen, 1951). A major species in these waters had been destroyed by a general epidemic with only very minor changes in the local biota! The fluctuations in numbers over the years since 1909 in the Limfjord appears to be more directly associated with the effects of severe winters (Blegvad, 1951).

The biological effects of the eel grass epidemic have interesting implications bearing on the effects of a major disaster and the organization of benthic communities. Except for a few instances, the general effect of this disaster on animal populations was an indirect one. The primary consequence of the disappearance of the eel grass appears to have been the freeing of the soil and the improvement of bottom circulation. The biological consequences of these changes were largely controlled by very local conditions. In some places the fauna associated with the eel grass was replaced by another; in other places, nothing happened. Frequently the animals directly involved with the plants switched to new substrates for attachment or new sources of food. In the Limfjord a dominant plant species disappeared without greatly affecting the fauna. This seems quite contrary to our expectations for a community as a biocoenosis in the original context of that term. Almost all of the fluctuations in animal numbers in benthic communities studied to date appear to be associated with changes in the physical rather than the biological environment. Even the artificial introduction of new species in a marine community is not accompanied by such drastic changes in the communities involved as have been observed with the introduction of species in terrestrial communities (see Elton, 1958). Paleontologists often ascribe sudden faunal changes to local disasters. The evidence cited here shows how complex and as yet unpredictable the biological consequences of such events are. The reactions of the population may be due to factors far removed from the primary environmental fluctuation.

Data on the recovery of populations following anomalous environmental fluctuations in the environment, and on succession, demonstrate the enormous colonizing capacity of marine invertebrates. The animals of Lagos Harbor, a lagoon system 160 miles long, are annually destroyed by the seasonal influx of fresh water (Sandison and Hill, 1959). The area is repopulated from very small populations surviving in protected places. A large population of cockles in shallow water of the Waddensee was destroyed by a particularly severe winter, but replenished in one year (Kristensen, 1957). Many other similar cases could be cited. Reish (1961) and many others have studied succession on newly available surfaces. It has been found that usually the first stage in colonization is the development of an algal film. Thereafter the course of events is swift, and a stable biota is achieved very rapidly. In the context of geologic time, recovery from anomalous fluctuations of the environment and the invasion of newly available and suitable areas would be instantaneous.

We conclude from these considerations that the widespread disappearance of a species or a fauna at some horizon in the fossil record cannot commonly be ascribed to anomalous fluctuations of some environmental factor. Instead, some major environmental change must be involved in most cases. The evidence available suggests that biological factors alone probably could not produce such an effect.

Discussion

Individuals of two or more species might live together because they interact with one another and/or have a similar response to features of the local environment. They may occur together by chance. Organisms may interact with one another in several ways (see Burkholder, 1952, for some possible kinds of relations). Our present knowledge of level bottom communities suggests that direct interactions between individuals is of less importance in determining the character of level bottom communities than in others. Many, if not most, of the species in level bottom communities appear to be quite independent of one another. These communities are dominated by detritus feeding organisms and hence are characterized by a comparatively simple food web. Fluctuations of numbers of individuals of different species are commonly associated with fluctuations of the physical environment and show little or no interspecific correlation. The loss of a numerically important species, such as eel grass, may have little or no effect on the remaining species. It is common to find two or more closely related members of the same genus living in the same area. All of these features suggest interspecific neutralism. The importance of predation, parasitism, and competition for space as ecological factors in level bottom communities cannot be denied, but these factors appear to be less important in such communities than in intertidal, coral reef, and terrestrial communities.

Species may occur in a particular community by chance. The distribution of species with broad environmental tolerances could be partially determined by the vagaries of currents. Transient species are often found in level bottom communities. A population of such species may persist locally for only one or two generations, disappear for a few years, and reappear again. The high recurrence of particular suites of species observed in modern environments, however, suggests that communities are not largely chance associations.

High recurrence of particular combinations of species could occur if such species had similar environmental responses and suitable environments were common. It seems likely that most of the species in level bottom communities occur together because of similar responses to the physical features of their environments. This seems evident from settling reactions of larvae, from the effects of environmental fluctuations, and from the modes of life represented in level bottom communities.

The independence of benthic species and their direct dependence on the physical environment suggests a simple organization. Such independence would seem highly adaptive from the standpoint of individual species. It is conceivable that this degree of independence and hence the relatively low order of organization of level bottom communities is the product of a long evolutionary history that might even have proceeded from a higher level of complexity.

The viewpoint that is developed here, and perhaps overstated, is that benthic communities are commonly associations of largely independent species occurring together because of similar responses to the physical environment. Such a concept is far removed from that of a community as a biocoenosis as originally defined by Möbius (Möbius, 1877, quoted in Allee et al., 1949, p. 35). If this is the case, then the biological organization of a level bottom community may not be an important influence on the evolutionary history of the members of the community. Rather, we should expect that the complexes of environmental factors represented by the hydroclimate and the sediment are of more direct influence.

Probably most adaptations to the substrate and to the hydroclimate involve physiological changes. We cannot directly observe such features but physiological adaptations to environment frequently, if not always, alter the spatial distribution of a species. Change in environmental distribution must commonly be associated with physiological changes affecting the settling behavior of pelagic larvae. By utilizing all species as a complex analyzer of environment—the community approach—we should be able to detect the results of adaptations we cannot directly perceive. Permanent changes in the association between species and between species and deposit type, with or without morphological change, must constitute the first signs of such adaptive shifts. Subtle shifts in association should, theoretically, be evident first in marginal environments, the ecotone in space, and its gradational equivalent in time.

To recognize such shifts in the fossil record we must utilize the evidence of high recurrence of the event at many localities. This means that large amounts of data must be collected in appropriate ways, reduced to some manageable form, and analyzed using the conceptual framework of marine biology. Statistical methods, such as the several measures of interspecific association that are now available, can facilitate data reduction and analysis. Concurrently, study of modern marine communities from the viewpoint of the paleontologist will provide more accurate analogues and the context for the interpretation of results.

Studies of shifts in interspecific association may guide us to other features of the environment of ancient communities that have left traces in the physical record. New insights into the evolution of species and communities may result from the combination of the quantitative studies of morphologic and community integration. These possibilities do seem to be worth the effort involved in the application of the community approach to paleoecology.

Invertebrate paleontology, by necessity, has been and is now largely a descriptive science. This work must continue, but at the same time we are moving into stages of synthesis and generalization. We have lauded the objectives of the paleoecological approach in what seems to be an endless chain of symposia, presidential addresses, and papers. Now the time has come to utilize the quantitative methods and the biological concepts that are available to exploit our tremendous accumulation of stratigraphic and systematic data.

REFERENCES

Aleem, A. B., 1956, Quantitative underwater study of benthic communities inhabiting kelp beds off California: *Science,* v. 123, p. 183.

Allee, W. C., 1919, Note on animal distribution following a hard winter: *Biol. Bull.,* v. 36, pp. 96–104.

Allee, W. C., O. Park, A. E. Emerson, T. Park, and K. P. Schmidt, 1949, *Principles of Animal Ecology:* Philadelphia, W. B. Saunders, 837 pp.

Axelrod, D. I., 1950, Evolution of desert vegetation in Western North America: *Pub. Carnegie Inst. Wash.,* n. 590, pp. 217–306.

Barnard, J. L., O. Hartman, and G. F. Jones, 1959, Oceanographic survey of the continental shelf area of Southern California; Benthic biology of the mainland shelf of Southern California: *Calif. State Water Pollution Control Board,* pub. n. 20, pp. 265–429.

Beerbower, J. R., and F. W. McDowell, 1960, The Centerfield biostrome; an approach to a paleoecological problem: *Proc. Penn. Acad. Sci.,* v. 34, pp. 84–91.

Blegvad, H., 1951, Fluctuations in the amounts of food animals of the bottom of the Limfjord in 1928–1950: *Rept. Danish Biol. Station,* n. 53, pp. 3–16.

Burkholder, P. R., 1952, Cooperation and conflict among primitive organisms: *Am. Scientist,* v. 40, pp. 601–631.

Chipperfield, P. N. J., 1953, Observations of the breeding and settlement of *Mytilus edulis* (L.) in British waters: *J. Marine Biol. Assoc. U. K.,* v. 32, pp. 449–476.

Cole, H. A., and E. W. Knight-Jones, 1939, Some observations and experiments on the settling behavior of larvae of *Ostrea edulis: Journal du Conseil,* v. 14, pp. 85–105.

Connell, J. H., 1956, Spatial distribution of two species of clams *Mya arenaria* L. and *Petricola pholadiformis* Lamarck in an intertidal area: *Invest. Shellfisheries Mass.,* rept. n. 8, pp. 15–28.

________ 1961, Effects of competition, predation by *Thais lapillus,* and other factors on natural populations of the barnacle *Balanus balanoides: Ecol. Monog.,* v. 31, pp. 61–104.

Cottam, C., and D. A. Munro, 1954, Eel grass status and environmental relations: *J. Wildlife Management,* v. 18, pp. 449–460.

Crisp, D. J., and H. Barnes, 1954, The orientation and distribution of barnacles at settlement with particular reference to surface contour: *J. Animal Ecol.,* v. 23, pp. 142–162.

________ and J. S. Ryland, 1960, Influence of filming and of surface texture on the settlement of marine organisms: *Nature,* v. 185, p. 119.

Day, J. H., and D. P. Wilson, 1934, On the relation of the substratum to the metamorphosis of *Scolecolepis fuliginosa* (Claparide): *J. Marine Biol. Assoc. U. K.,* v. 19, pp. 655–662.

Dexter, R. W., 1944, Ecological significance of the disappearance of eel-grass at Cape Ann, Mass.: *J. Wildlife Management,* v. 8, pp. 173–176.

Elton, C. S., 1958, *The Ecology of Invasions by Animals and Plants:* New York, John Wiley and Sons, 181 pp.

Erdman, W., 1934, Untersuchungen uber die Lebensgeschicte der Auster, n. 5, Uber die Entwicklung und die Anatomie der "ansatzreifer" Larven von *Ostrea edulis* mit Bemerkungen uber die Lebensgeschichte der Auster: *Wiss. Meeres. Abt. Helgoland,* v. 19, pp. 1–25.

Goodbody, I., 1961, Mass mortality of a marine fauna following tropical rains: *Ecology,* v. 42, pp. 150–155.

Gregg, J. H., 1945, Background illumination as a factor in the attachment of barnacle cyprids: *Biol. Bull.,* v. 88, pp. 44–49.

Harrington, C. R., 1921, A note on the physiology of the ship-worm (*Teredo norvegica*): *Biochem. J.,* v. 15, pp. 736–741.

Hoese, H. D., 1960, Biotic changes in a bay associated with the end of a drought: *Limnol. and Ocean.,* v. 5, pp. 326–336.

Holme, N. A., 1950, Population dispersion in *Tellina tenuis* da Costa: *J. Marine Biol. Assoc. U. K.,* v. 29, pp. 267–280.

________ 1953, The biomass of the bottom fauna in the English Channel off Plymouth: *J. Marine Biol. Assoc. U. K.,* v. 32, pp. 1–49.

________ 1961, The bottom fauna of the English Channel: *J. Marine Biol. Assoc. U. K.,* v. 41, pp. 397–461.

Hopkins, A. E., 1931, Factors influencing the spawning and settling of oysters in Galveston Bay, Texas: *Bull. U. S. Bureau Fish.,* v. 47, pp. 57–83.

Jagersten, G., 1940, Die Abhangigkeit der Metamorphose vom Substrat des Biotopes bei *Protodrilus: Arkiv. F. Zool.,* v. 32 A, pp. 1–12.

Johnson, R. G., 1959, Spatial distribution of *Phoronopsis viridis* Hilton: *Science,* v. 129, p. 1221.

________ 1960, Models and methods for the analysis of the mode of formation of fossil assemblages: *Bull. Geol. Soc. America,* v. 71, pp. 1075–1086.

________ 1962, Interspecific association in Pennsylvanian fossil assemblages: *J. Geology,* v. 70, pp. 32–55.

Jones, N. S., 1952, The bottom fauna and the food of flatfish off the Cumberland Coast: *J. Animal Ecol.,* v. 21, pp. 182–205.

——— 1956, The fauna and biomass of a muddy sand deposit off Port Erin, Isle of Man with an appendix on methods used for the analysis of deposits: *J. Animal Ecol.,* v. 25, pp. 217–252.

Jones, M. I., 1961, A quantitative evaluation of the benthic fauna off Point Richmond, California: *Univ. Calif. Pub. Zool.,* v. 67, pp. 219–320.

Knight-Jones, E. W., 1951, Gregariousness and some other aspects of the settling behavior of a *Spirorbis: J. Marine Biol. Assoc. U. K.,* v. 30, pp. 201–222.

——— 1953, Decreased discrimination during settling after prolonged planktonic life in larvae of *Spirorbis borealis* (Serpulidae): *J. Marine Biol. Assoc. U. K.,* v. 32, pp. 337–345.

——— 1955, The gregarious setting reaction of barnacles as a measure of systematic affinity: *Nature,* v. 173, p. 266.

——— and J. P. Stevenson, 1950, Gregariousness during settlement in the barnacle *Eliminius modestus* Darwin: *J. Marine Biol. Assoc. U. K.,* v. 29, pp. 281–287.

Korringa, P., 1940, Experiments and observations on swarming pelagic life and setting in the European flat oyster, *Ostrea edulis* L.: *Arch. Neerland. Zool.,* v. 5, pp. 1–249.

Kristensen, I., 1957, Differences in density and growth in a cockle population in the Dutch Wadden Sea: *Arch. Neerland Zool.,* v. 12, pp. 351–454.

Lackey, J. B., 1961, Bottom sampling and environmental niches: *Limnol. and Ocean.,* v. 6, pp. 271–279.

Loosanoff, V. L., and F. D. Tommers, 1948, Effect of suspended silt and other substances on rate of feeding of oysters: *Science,* v. 107, pp. 69–70.

Mare, M. F., 1942, A study of a marine benthic community with special reference to the microorganisms: *J. Marine Biol. Assoc. U. K.,* v. 25, pp. 517–554.

Marshall, N., 1947, An abundance of bay scallops in the absence of eelgrass: *Ecology,* v. 28, pp. 321–322.

———, 1960, Studies of the Niantic River, Connecticut, with special reference to the bay scallop, *Aequipecten irradians: Limnol. and Ocean.,* v. 5, pp. 86–105.

Miyadi, D., 1940, Marine benthic communities of the Tanabe-wan: *Annot. Zool. Japon.,* v. 19, pp. 136–148.

——— 1941, Ecological survey of the benthos of the Ago-wan: *Annot. Zool. Japon.,* v. 20, pp. 169–180.

Muller, C. H., 1958, Science and philosophy of the community concept: *Am. Scientist,* v. 46, pp. 294–308.

Olson, E. C., 1952, The evolution of a Permian vertebrate chronofauna: *Evolution,* v. 6, pp. 181–196.

Orton, J. H., 1937, Some interrelations between bivalve spatfalls, hydrography, and fisheries: *Nature,* v. 140, p. 505.

Pomerat, C. M., and E. R. Reiner, 1942, The influence of surface angle and light on the attachment of barnacles and other sedentary organisms: *Biol. Bull.,* v. 82, pp. 14–25.

Poulsen, E. M., 1951, Changes in the frequency of larger bottom invertebrates in the Limfjord in 1927–1950: *Rept. Danish Biol. Station,* n. 53, pp. 17–34.

Pratt, D. M., 1953, Abundance and growth of *Venus mercenaria* and *Callocardia morrhuana* in relation to the character of the bottom sediments: *J. Marine Res.,* v. 12, pp. 60–74.

Prytherch, H. F., 1934, The role of copper in the setting, metamorphosis and distribution of the American oyster, *Ostrea virginica: Ecol. Monographs,* v. 4, pp. 49–101.

Quayle, D. B., 1952, Structure and biology of the larvae and spat of *Venerupis pullastra* (Montagu): *Trans. Roy. Soc. Edinburgh,* v. 62, pp. 255–297.

Raymont, J. E. G., 1949, Further observations on changes in the bottom fauna of a fertilized sea loch: *J. Marine Biol. Assoc. U. K.,* v. 28, pp. 9–19.

Reish, D. J., 1961, A study of the benthic fauna in a recently constructed boat harbor in Southern California: *Ecology,* v. 42, pp. 84–91.

Ryland, J. S., 1959, Experiments on the selection of algal substrates by polyzoan larvae: *J. Exp. Biol.,* v. 36, pp. 613–631.

Sanders, H. L., 1956, Oceanography of Long Island Sound 1952–1954, X; the biology of marine bottom communities: *Bull. Bingham Ocean. Coll.,* v. 15, pp. 345–414.

——— 1958, Benthic studies in Buzzards Bay, I; Animal-sediment relationships: *Limnol. and Ocean.,* v. 3, pp. 245–258.

——— 1960, Benthic studies in Buzzards Bay, III; The structure of the soft bottom community: *Limnol. and Ocean.,* v. 5, pp. 138–153.

Sandison, E. E., and M. B. Hill, 1959. The annual repopulation of Lagos Harbor by sedentary marine animals: *Intern. Ocean. Congress Preprints,* pp. 584–585.

Schaefer, M. B., 1937, Attachment of the larvae of *Ostrea gigas,* the Japanese oyster, to plane surfaces: *Ecology,* v. 18, pp. 523–527.

Scheltema, R. S., 1956, The effect of the substrate on the length of planktonic existence in *Nassarius obsoletus: Biol. Bull.,* v. 111, p. 312.

——— 1961, Metamorphosis of the veliger larvae of *Nassarius obsoletus* (Gastropoda) in response to bottom sediment: *Biol. Bull.,* v. 120, pp. 92–109.

Segerstråle, S. G., 1960, Fluctuations in the abundance of benthic animals in the Baltic area: *Soc. Sci. Fennica Commentationes Biol.,* v. 23, pp. 1–19.

Shotwell, J. A., 1958, Intercommunity relationships in Hemphillian (Mid-Pliocene) mammals: *Ecology,* v. 39, pp. 271–382.

Silen, L., 1954, Developmental biology of Phoronidae of the Bullmar Fiord area (west coast of Sweden): *Acta Zoologica,* v. 35, pp. 215–257.

Smidt, E. L. B., 1951, Animal production in the Danish Waddensea: *Medd. Komm. Danmarks Fisk Havinders Ser. Fiskeri:* v. 11, pp. 1–151.

Smith, J. E., 1932, The shell gravel deposits and the infauna of the Eddystone grounds: *J. Marine Biol. Assoc. U. K.,* v. 18, pp. 243–278.

Southward, E. C., 1957, The distribution of Polychaeta in offshore deposits in the Irish sea: *J. Marine Biol. Assoc. U. K.,* v. 36, pp. 49–75.

Stickney, A. P., and L. D. Stringer, 1957, A study of the invertebrate bottom fauna of Greenwich Bay, Rhode Island: *Ecology,* v. 38, pp. 11–122.

Thorson, G., 1933, Investigations on shallow water animal communities in the Franz Joseph Fjord (East Greenland) and adjacent waters: *Medd. om Grønland,* v. 100, pp. 1–68.

——— 1957, Bottom communities, pp. 461–534, *in Treatise on Marine Ecology and Paleoecology* (J. W. Hedgpeth, ed.), v. 1, Ecology: Geol. Soc. Am. Mem. 67, 1296 pages.

Weiss, C. M., 1947, The effect of illumination and stage of tide on the attachment of barnacle cyprids: *Biol. Bull.,* v. 93, pp. 240–249.

Wieser, W., 1956, Factors influencing the choice of substratum in *Cuenella vulgaris* Hart (Crustacea Cumacea): *Limnol. and Ocean.,* v. 1, pp. 274–285.

Williams, G. C., 1957, Pleiotropy, natural selection, and the evolution of senescence: *Evolution*, v. 11, pp. 398–411.

Wilson, D. P., 1932, On the Mitraria larvae of *Owenis fusiformis* Delle Chiaje: *Phil. Trans. Roy. Soc. Lond.*, ser. B, v. 221, pp. 231–334.

——— 1937, The influence of the substratum on the metamorphosis of *Notomastus*-larvae: *J. Marine Biol. Assoc. U. K.*, v. 22, pp. 227–243.

——— 1948, The larval development of *Ophelia bicornis* Savigny: *J. Marine Biol. Assoc. U. K.*, v. 27, pp. 540–553.

——— 1949, The decline of *Zostera marina* L. at Salcombe and its effects on the shore: *J. Marine Biol. Assoc. U. K.*, v. 28, pp. 395–412.

——— 1952, The influence of the nature of the substratum on the metamorphosis of the larvae of marine animals, especially the larvae of *Ophelia bicornis* Savigny: *Ann. Inst. Oceanog.*, v. 27, pp. 49–156.

——— 1953, The settlement of *Ophelia bicornis* Savigny larvae: *J. Marine Biol. Assoc. U. K.*, v. 33, pp. 361–380.

——— 1954, The attractive factor in the settlement of *Ophelia bicornis: J. Marine Biol. Assoc. U. K.*, v. 33, pp. 361–380.

——— 1955, The role of microorganisms in the settlement of *Ophelia bicornis: J. Marine Biol. Assoc. U. K.*, v. 34, pp. 531–543.

Wisely, B., 1959, Factors influencing the settling of the principal marine fouling organisms in Sydney Harbour: *Australian J. Marine Freshwater Res.*, v. 10, pp. 30–44.

——— 1960, Observations on the settling behavior of larvae of the tubeworm *Spirorbis borealis* (Daudin Polychaeta): *Australian J. Marine Freshwater Res.*, v. 11, pp. 55–73.

ZoBell, C. E., 1957, Marine bacteria, pp. 1035–1040, in *Treatise on Marine Ecology and Paleoecology* (J. W. Hedgpeth, ed.), v. 1, Ecology: Geol. Soc. America, Mem. 67, 1296 pp.

PAPER 13

Paleobotany of Permineralized Peat (Coal Balls) from the Herrin (No. 6) Coal Member of the Illinois Basin (1977)

T. L. Phillips, A. B. Kunz, and D. J. Mickish

Commentary

IAN GLASSPOOL

Floral anatomy preserved in coal balls has been studied since the mid-nineteenth century, leading to the description of hundreds of new species. These studies have facilitated taxonomic and systematic treatments and allowed an improved understanding of plant biology (Scott and Rex 1985). Coal balls also provide unparalleled evidence of peat-swamp ecology. Phillips et al. set out to investigate the paleoecology and elemental composition of the Middle Pennsylvanian Herrin Coal at four localities across the southern Illinois Basin using a cm^2 grid system to quantify the percentage volume and number of organs and taxa in coal ball peels. While the study demonstrated that arborescent lycopods dominated the seam volumetrically, more importantly it showed that the floral composition changed over time and that the seam was divisible into discrete zones. These "stratified compositional data" were not just taxonomic but also included plant organ associations. The statistical approach adopted by the authors proved that the taxonomic composition of peats was relatively consistent across widespread areas of similar age and environment. Going further, the authors offered deductions about the early onset of coal ball formation and the sequential occurrence of these events as the seam accumulated and they demonstrated that the seam's chemistry was influenced more by ground water chemistry than by floral composition.

Additionally, the study demonstrated the limitations of palynology in investigating peat-swamp paleoecology and provided insights into the role of fire through the quantification of fusain. These data directly demonstrate that the abundance of charcoal (fusain) in coal balls and inertinite in the same bituminous coal are volumetrically comparable. This supports recent arguments that inertinite is charcoal and hence produced by fire (Scott and Glasspool 2007) and shows the importance of this methodology to those studying the origins of coal formation. However, it is for providing a workable, quantitative methodology to explore the paleoecological data contained within coal balls (Pryor 1988) that this paper is most important. Once published, this methodology rapidly became fundamental in the study of permineralized peat paleoecology. While initially used to examine seam floral successions, the methodology has been applied in much broader-scale spatial and temporal studies, and these have had profound importance for our understanding of climatically driven ecological change during the late Paleozoic (e.g., Phillips and Peppers 1984).

From *Interdisciplinary Studies of Peat and Coal Origins*, 18–49, ed. P. H. Given and A. D. Cohen, Geological Society of America Microform Publications, vol. 7 (Boulder, CO: Geological Society of America).

Literature Cited

Phillips, T. L., and R. A. Peppers. 1984. Changing patterns of Pennsylvanian coal-swamp vegetation and implications of climate control on coal occurrence. *International Journal of Coal Geology* 3:205–55.

Pryor, J. S. 1988. Sampling methods for quantitative analysis of coal-ball plants. *Palaeogeography, Palaeoclimatology, Palaeoecology* 63:313–26.

Scott, A. C., and I. J. Glasspool. 2007. Observations and experiments on the origin and formation of inertinite group macerals. *International Journal of Coal Geology* 70:55–66.

Scott, A. C., and G. Rex. 1985. The formation and significance of Carboniferous coal balls. *Philosophical Transactions of the Royal Society London B* 311:123–37.

The Geological Society of America, Inc.
Microform Publication 7

Interdisciplinary Studies of Peat and Coal Origins

Edited by
P. H. Given
and
A. D. Cohen

Published by
The Geological Society of America, Inc.
3300 Penrose Place
Boulder, Colorado 80301

18

PALEOBOTANY OF PERMINERALIZED PEAT (COAL BALLS) FROM THE HERRIN (NO. 6) COAL MEMBER OF THE ILLINOIS BASIN

Tom L. Phillips
Botany Department
University of Illinois
Urbana, Illinois 61801

A. Barry Kunz and Daniel J. Mickish*
Physics Department
University of Illinois
Urbana, Illinois 61801

ABSTRACT

Study of the permineralized peat in coal-ball zones permits reconstruction of the plant communities inhabiting Carboniferous coal-forming swamps. Quantitative data on the taxa and tissue-organ composition of eight or more coal-ball zones are compared from four localities along a 142 km east-west traverse in the Herrin Coal Member, Carbondale Fm., Desmoinesian Series, in the southern part of the Illinois Basin. Vertical sections of coal-ball zones at collecting sites account for 54-95 percent of the thickness of coal seams at the sites. These thicknesses were two to three times the average thickness of adjacent areas. Botanical identifications were quantified using a cm^2 grid system, with the aid of which the percent volume and percent number of taxa and organs in the permineralized peat and coal-ball zones could be established, from the study of areas of peel preparations from transverse sections.

Comparisons of similar peat zones based on percent volume of distinctive plant types established four to six good matches among the peat profiles of the four sites, using a persistent clay parting ("the blue band") as a datum line for correlating zones. Of the identified botanical constituents, lycopods were dominant (72-75 percent by volume); ferns are usually second (11-17 percent); pteridosperms were the most variable (6-14 percent); sphenopsids accounted for 1-4.5 percent and cordaites ≤0.5 percent. With the major exception of *Psaronius* roots, the identified aerial portions of plants formed 53-73 percent of the volume; fructifications made up 2-4 percent, leaves 13.5-14.5 percent, and stems 35-57 percent.

Mass spectrographic analyses of 17 trace and 13 major elements are given for coal-ball zones along with determinations of boron and selenium by other techniques. Correlation coefficients of ≥0.95 were found between iron and whole sulfur, and between F, Ce, Ga, La, Mn, Nd, Pr, Sc and Sr, taken in pairs. Comparison of these data with δC^{13} values for the carbonates in the coal balls indicate that the abundances of these elements are dependent on the degree of marine influence on the original peat swamps.

*Present Address: E. I. duPont de Nemours and Company
Wilmington, Delaware 19898

19

INTRODUCTION

The quantitative analyses of Pennsylvanian coal swamp vegetation and major trace elements from the coal balls of the Herrin Coal Member provide the first such profiles of composition and change in a bituminous coal seam derived from estimates of the permineralized peat. It has long been supposed that different kinds of coal result from differences in plant associations which contributed to the peat stage, and the anatomically preserved peats from multiple zones of coal balls are the most direct evidence of the botanical constituents of the coal. The stratigraphic patterns of composition and change in Pennsylvanian coal swamp vegetation (Kosanke, 1947; Peppers and Phillips, 1972; Phillips and others, 1974) in the Illinois Basin, and concurrent differences in parts of the Western Interior Basin, seem to point geologically and geographically to potentially broad predictable trends in the botanical constituents of the extensive bituminous coal reserves of vascular plant origin.

The establishment of sampling techniques and means of quantifying the types of plants, organs, and tissues in profiles of coal balls, are, we feel, important, because they show how information on the botanical constituents of coal seams can be used as a primary frame of reference in basic studies of coals. In particular, some insight into the nature of ancient peat swamps can be obtained. The quantitative analysis of Pennsylvanian coal swamp vegetation is a means of getting at stratified compositional data, such as plant associations, root and aerial (or combined) zones, and patterns that reflect various events in the swamps that are botanically detectable. The major and trace element analysis is one of several means necessary for determinations of environmental and other geological factors which relate to the peat, coal balls, and adjacent coal. Summary data from coal palynology and study of the permineralized peat from coal balls from the Herrin Coal are reasonably corroborative (Phillips and others, 1974), but precise quantitative comparisons have not been made for profiles of each in the same coal.

Pennsylvanian and equivalent age coal swamp vegetation from coal balls in North America, western Europe, and the U.S.S.R. consists of five major taxa of vascular plants: lycopods, ferns, pteridosperms, sphenopsids, and cordaites. Each plant group is represented largely or exclusively by tree forms, and any given coal is dominated in percent volume by a single genus or only a few genera. This is the initial basis for groupings of taxa, plant types, organs, and tissues under major categories for comparsons and summaries of results.

In the coal balls studied thus far from the Herrin Coal, lycopods dominate (72-75 percent by volume). These are represented by *Lepidophlois* assemblages and to a lesser extent by *Lepidodendron* along with their root system, *Stigmaria*; the two stem genera were usually not distinguishable within sections of coal-ball profiles. The ferns are dominated by *Psaronius*, pteridosperms by *Medullosa* (and *Myeloxylon*), and the sphenopsids by *Calamites* (*Arthropitys*). Cordaites contributes <0.5 percent volume and is associated with the ovule, *Cardiocarpus oviformis*. Small plants, including the coenopterid ferns, *Selaginella*, *Sphenophyllum* and pteridosperm assemblages other than those of *Medullosa* and *Sutcliffia*, accounted for <3 percent of the total volume at all sites and averaged 1.6 percent.

The Herrin Coal is the most important coal economically in Illinois and has yielded one-fifth of 75 known coal-ball occurrences in the Illinois Basin. Eight or more zones of coal balls were sampled at four localities along a 142 km traverse of the southern part of the basin (Fig. 1, next frame). From one of the earliest coal-ball collections in the Herrin Coal near Nashville, Illinois, Schopf (1938) first made quantitative use of the permineralized peat as an index to the constitution of coal.

Stratigraphic, locality, and outcrop data for the five sources of coal balls are given in Appendix 1. and the sampling and quantification techniques are described in Appendix 2.

VERTICAL ANALYSIS OF PERMINERALIZED PEAT IN COAL BALLS

The four study sites in the Herrin Coal were low areas within the coal swamp and subject to repetitive introductions of carbonate-bearing waters. These waters probably saturated root zones and peat accumulations at numerous times during the existence of the coal swamp. The permineralized peat occurs as stratified sheetlike masses or, more often, as one or more layers of rounded to lenticular coal balls. The coal-ball zones were formed from the bottom zone upward. Various characteristics of the coal ball assemblies support the view that the successive permineralized peat zones were formed independently during the temporal span of the Herrin coal swamp. These characteristics are: preservational quality, and compactional and collapse features, of plant material in the various zones; the topographic relationships of successive coal-ball zones and the blue band (concerning the latter, see Appendix 1); and the correlation of certain major and trace elements with differing (freshwater-marine) carbonate sources (M. E. Brownlee, unpub. data). Subsequent coal swamp forests developed on top of successive permineralized peat zones until the marine transgression wiped out the coal swamp forests.

Plant tissues and organs extend throughout coal balls, although there are some areas of a square centimeter or more without discernible plant material. Such areas are usually calcite fillings in the fractured or fissured peat or are heavily pyritized zones, often contoured along the outer boundary of the coal ball. In these areas plant material may be obliterated beyond recognition. The actual organic matter in coal balls represents about 2 percent by weight (C. P. Rao, unpub. data). Unidentified tissues, organs, and taxa, as well as humic-like matter and zones of calcite and pyrite have, for convenience in summaries, been lumped together as the botanically *unidentified* portion. This is 13-19 percent of the profiles analyzed. With the exclusion of calcite and unidentifiable pyritic areas, about 90 percent of the plant material has been identified, indicating that permineralization of the peat occurred usually during relatively early stages of accumulation. Most of the shallow root systems or basal supportive root mantles were preserved *in situ* and most of the stratified aerial portions of plants accumulated essentially where they fell.

TAXONOMIC COMPOSITION

Trees account for >97 percent of the identified plant volume of the four coal-ball profiles. Lycopods dominate (72-75 percent of identified material; see Fig. 2, next frame), or 60-62 percent if *unidentified* matter is included

21

Figure 1. Coal-ball sources in the Herrin Coal.

	LYCOPODS	FERNS	PTERIDOSPERMS	SPHENOPSIDS	CORDAITALES
DELTA	72.1	11.1	14.2	2.6	<.1
SAHARA	72.2	16.8	6.0	4.5	.5
SHAWNEETOWN	74.9	14.8	8.6	1.7	<.1
PARADISE	75.1	13.0	10.7	1.3	<.1

Figure 2. Summary of percent volume of major taxa in permineralized peat profiles from coal balls from the Herrin Coal.

in the total. The percent data given hereafter on taxa and organ-tissues have been normalized with deletion of the *unidentified* (for discussion of the nature of the *unidentified* material, see below). Ferns are usually second in abundance (11-17 percent by volume); pteridosperms are the most variable with 6-14 percent, sphenopsids are 1-4.5 percent and cordaites ≤0.5 percent. Cordaites are not considered further because of their dearth. Small plants range in total volume 0.8-1.4 percent at three localities with a maximum at Sahara of 2.7 percent. Small pteridosperms are more numerous at all localities than are coenopterid ferns; these two groups comprise three-fourths of the small plants.

Profiles

The most extensive taxonomic profile of the permineralized peat zones (Fig. 3-6, next frames) is the Shawneetown core, in which stigmarian root systems are found in the basal zone. With the finer divisions of zones from the core sample (see Appendix 1) there is evidence of a strongly cyclic alternation of root zones and deposits heavy in aerial plant parts through the lower 130 cm up to the blue band (Fig. 3). Common to the profiles of Shawneetown, Paradise, and Delta are the marked developments of tree ferns and/or pteridosperms immediately above the blue band, the maximum development of pteridosperms being in a zone or two higher. In the Shawneetown core this pteridosperm zone is numbered 22 (Fig. 3), the only major root zone of pteridosperms; greater abundances of pteridosperms occur still higher at and near the top of the seam - a pattern not seen in other profiles. The outer roots of *Psaronius* are plotted simply as roots because of our inability to distinguish supportive aerial roots from those anchored in the substrate. *Psaronius* shows a maximum development at Delta. It is found in a somewhat higher zone in the upper half of the sections at Shawneetown and Paradise. A similar set of pteridosperm and fern abundances in the Sahara profile is somewhat blurred by the thicknesses of intervening coal and the thick coal-ball zones.

A best match comparison (Fig. 7, next frame) of the four taxonomic profiles of coal-ball zones utilized abundances of distinctive plant groups, that is, ferns and pteridosperms, with the blue band (see Appendix 1) as a datum line except in the case of Sahara. The middle point position of matched zones was plotted on a linear scale of location within each seam from the bottom of that seam along the abscissa, against the thinnest profile (Delta) along the ordinate. A plot of the Delta profile along the abscissa with its ordinate plot would, of course, bisect the ordinate-abscissa angle. No compaction ratio was introduced. Four to six good matches were obtained for all profiles with essentially straight line plots. The two key permineralized zones, which are apparently persistent and widespread in the profiles, are the pteridosperm and upper fern zones.

FUSAIN

The total volume of fusinized plant material in the four profiles is 5.0-7.2 percent (mean for unweighted zones = 5.4-8.0 percent). Maximum fusain composition of zones ranged from one-tenth to one-half. Coal-ball zones fairly high in fusinized material occur immediately above and below the blue band,

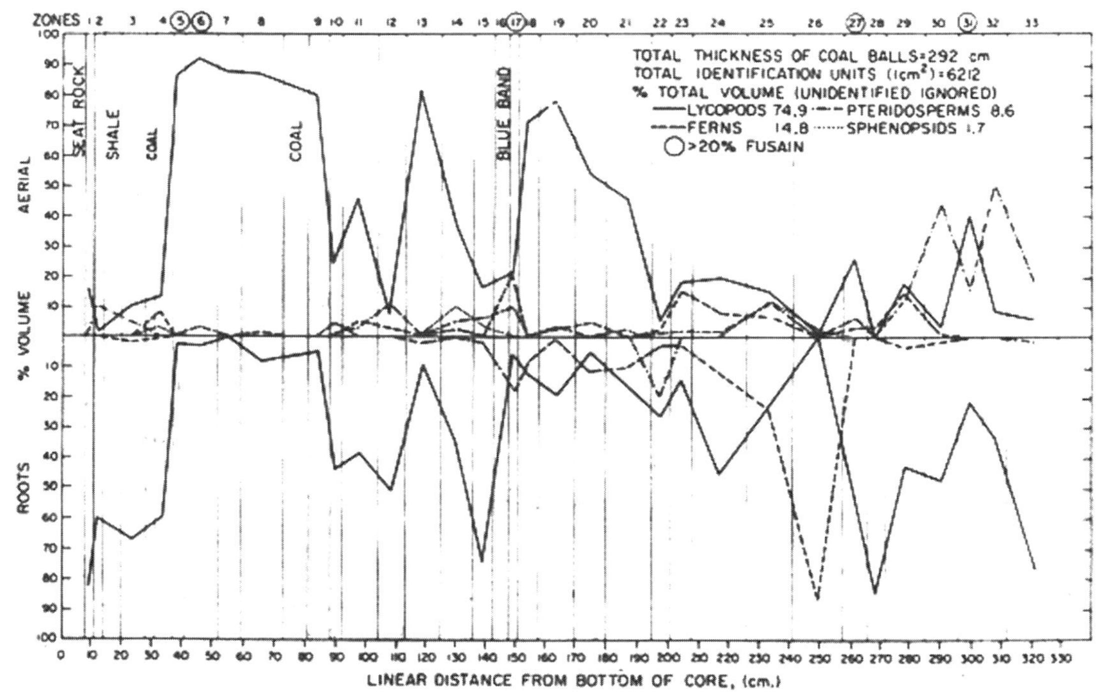

Figure 3. Percent volume of major taxa in profile of Shawneetown core. Middle horizontal line separates aerial and root portions.

25

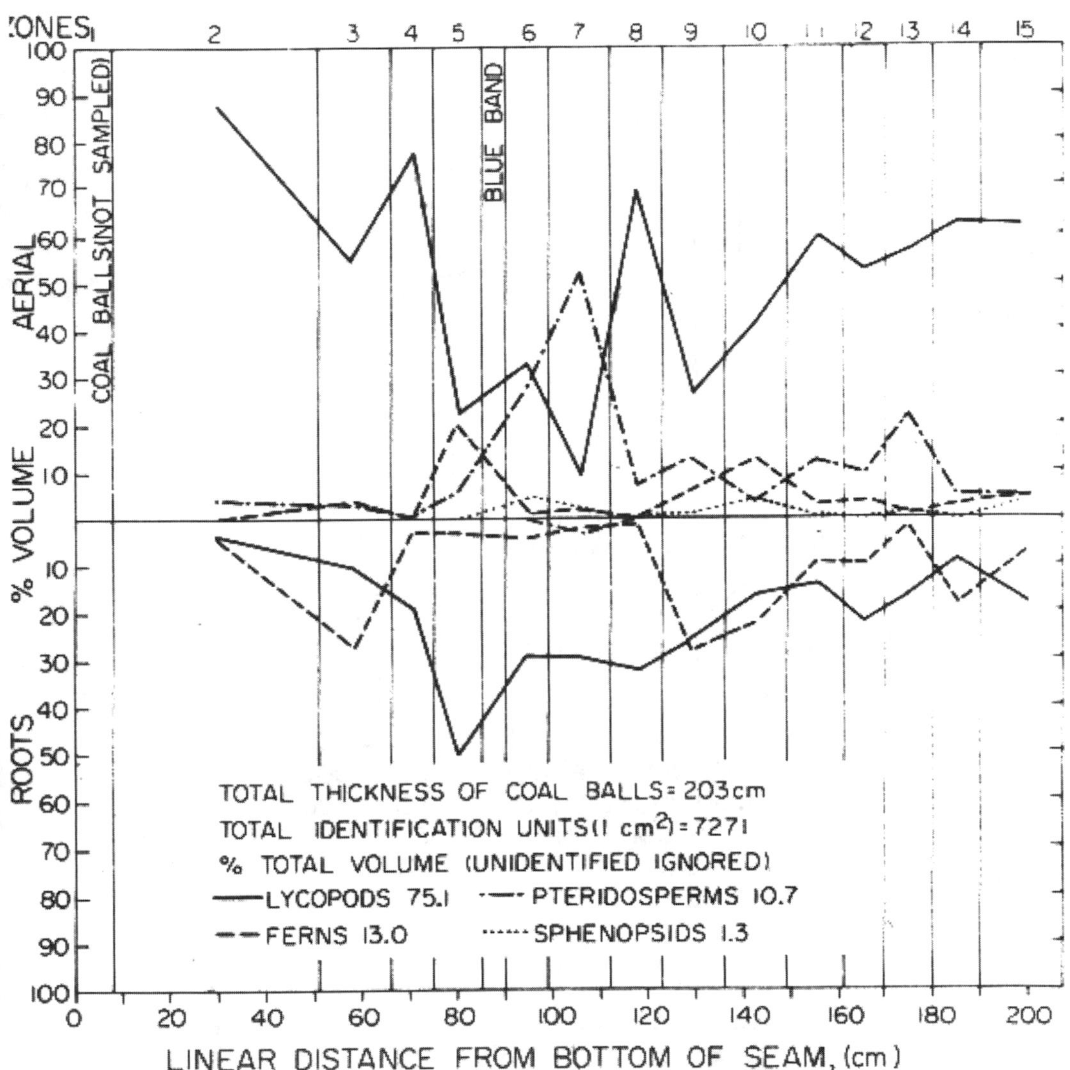

Figure 4. Percent volume of major taxa in profile of coal-ball zones at Paradise.

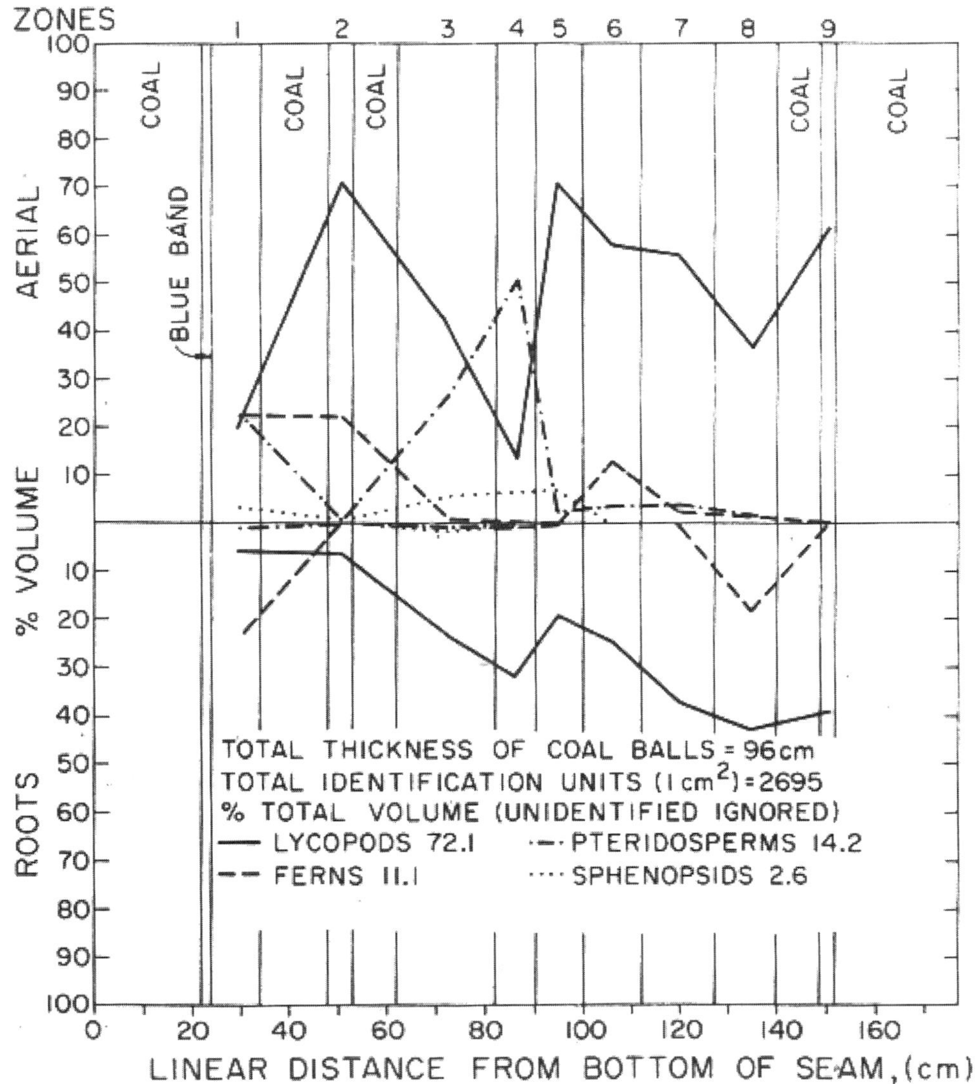

Figure 5. Percent volume of major taxa in profile of coal-ball zones at Crab Orchard (Delta).

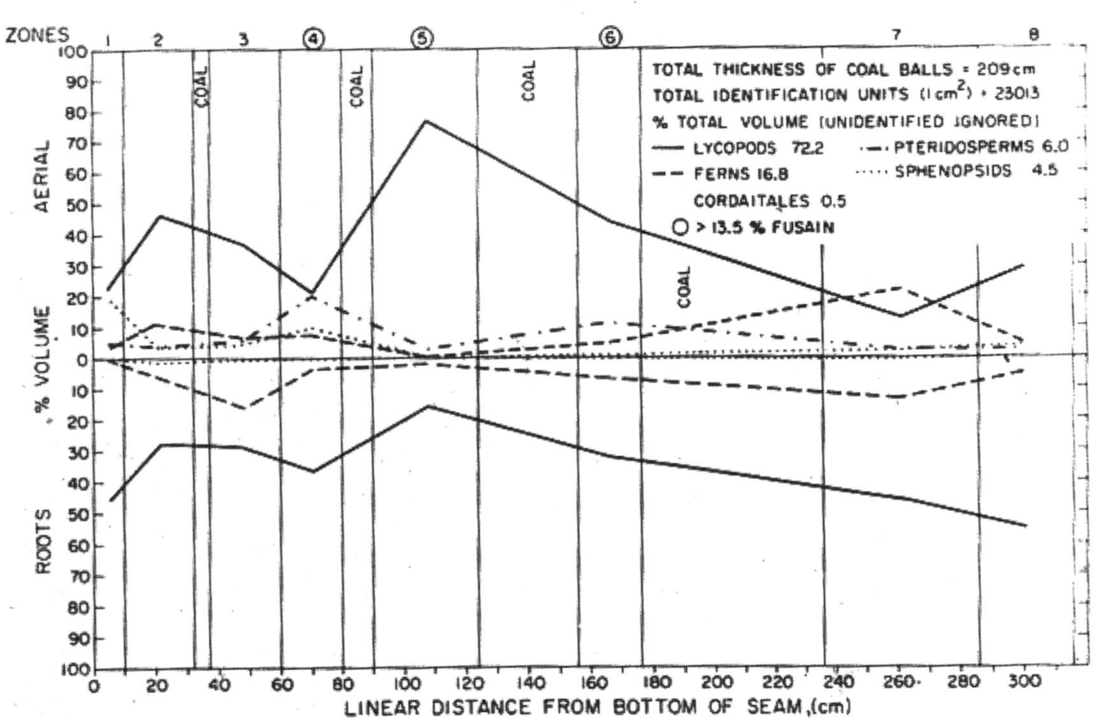

Figure 6. Percent volume of major taxa in profile of coal-ball zones at Carrier Mills (Sahara).

28

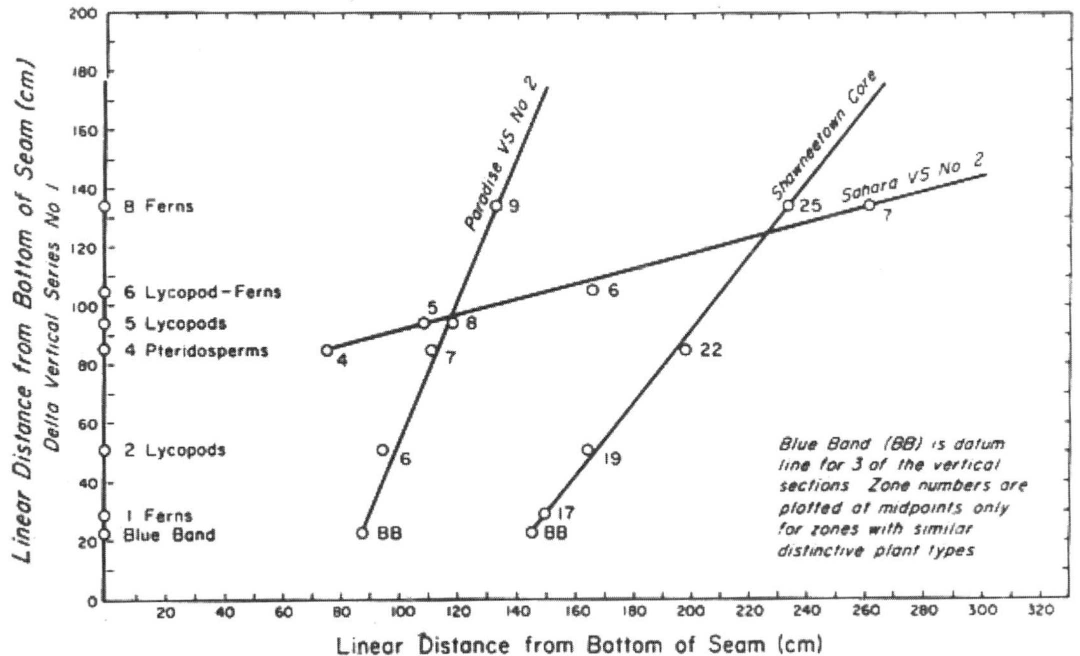

Figure 7. Comparison of similar petrified peat zones based on percent volume abundance of distinctive plant types in the vertical sections of coal-ball material from Delta (Crab Orchard), Sahara (Carrier Mills), Shawneetown and Paradise. Delta data are plotted on the ordinate.

and high fusain zones often coincide with higher pteridosperm content. Zones high in fusain occur at three successive coal-ball zones at Sahara (Fig. 6), three at Paradise (two zones below blue band, 7.5-22 percent, and zone 6 with 11 percent, Fig. 4), and at Delta (zones 3-5, 7-35 percent, Fig. 5). The remaining fusain zone at Delta has only 5 percent and is immediately above the blue band. At Shawneetown are five zones containing >20 percent; above the blue band fusain is 21 percent and below 9 percent.

With the exception of the Shawneetown core (one-eighth identifiable fusain) 57-82 percent of fusinized tissue was identified taxonomically or histologically. Of the identifiable fusain 74-95 percent (mean = 82 percent) is lycopod tissue and 3.5-20 percent (mean = 11 percent) is pteridosperm. Although the fraction of total fusain recognizable as derived from sphenopsids was the lowest, fusinized sphenopsid tissue usually represented a relatively large fraction of total sphenopsid material. Of the fusinized material 77-97 percent is stem tissue and, with the exception of the Shawneetown core, less than one-eighth of stem tissue is fusinized; 1-18 percent is foliage, mostly pteridosperm. Less than 1 percent fusain was root material, primarily the inner root zone of *Psaronius*.

ORGAN-TISSUE COMPOSITION

With ranges of 53-73 percent, aerial portions of plants are more abundant in identifiable peat than roots at all localities (outer roots of *Psaronius* are plotted as roots). About three-fourths of the aerial organs are stems. Leaves are the most consistent in volume from site to site with 13.4-14.5 percent and fructifications (cones, fertile foliage, sporangia and seeds) represent 2.3-4.4 percent (Fig. 8, see next frame).

Stigmaria (60-77 percent) and *Psaronius* (19-39 percent) account for 96-99 percent of the roots. Root systems of lycopods form 16-36 percent of the identifiable peat with free outer roots of *Psaronius* contributing 6-11 percent. Of the stem tissue, 83-96 percent is from lycopod trees with one noteworthy exception occurring at Sahara, where the 11 percent calamites stem material accounted for twice as much of total stem tissue as did stem from plants of any non-lycopod group. Sterile foliage is dominated by pteridosperms (34-80 percent), which with leaves of lycopods constitute 75-91 percent of the sterile foliage. Ferns contribute 10-12 percent of the sterile foliage except at Sahara, where they contribute 24 percent. Lycopods and ferns contribute 86-94 percent of the fructification material in about equal amounts; fern fructifications included pinnules as well as sporangia. Pteridosperm fructifications, largely medullosan, are 4.5-11 percent of total fructifications.

Roots constitute 54-82 percent of fern assemblages, 22-50 percent of lycopod assemblages and only 1.2-3.8 percent of pteridosperms. Stems comprise 85-91 percent of sphenopsid material, 41-73 percent of lycopods, 11-18 percent of pteridosperms and <5 percent of ferns. Sterile foliage forms 73-84 percent of pteridosperm assemblages, 10-19 percent of ferns and 2-8 percent of lycopod assemblages. The lowest percentage of assemblages and the smallest range for fructifications are in the lycopods with 1.7-2.3 percent of their total volume.

Wood as chunks of area one square centimeter or more constitutes only 2.5-5.5 percent of the identifiable peat, although total xylem tissues may contribute more than twice those amounts. The ratios of periderm to wood

	ROOTS	AERIAL	STEMS	LEAVES	FRUCTIFICATIONS
DELTA	34.1	65.9	48.2	14.5	3.2
SAHARA	46.9	53.1	35.3	13.4	4.4
SHAWNEETOWN	43.1	56.9	38.8	14.0	4.1
PARADISE	27.3	72.7	56.8	13.6	2.3

Figure 8. Summary of percent volume of organs from the coal-ball profiles at Delta, Sahara, Shawneetown and Paradise.

vary from 3:1 at Sahara, where lycopod periderm is lowest at 17 percent, to 21:1 at Paradise, where periderm constitues one-half of identified tissues. An average of about one-third of the identified wood is lycopod, pteridosperm and calamites, respectively, with any of the three contributing as much as 46 percent at a given locality. Except for Sahara, lycopod periderm constitutes 87-97 percent of the lycopod stem tissue preserved.

Profiles

Profiles or organs preserved in coal-ball zones above the blue band (Fig. 9, next frame) tend to be markedly richer in either aerial or root accumulations, depending on the locality. At all four localities the bulk of the leaves and fructifications are found preserved in the upper half of the coal, or above the blue band where present, and the coal-ball zone immediately above the blué band contains abundant foliage.

BOTANICALLY UNIDENTIFIED CATEGORIES

A detailed breakdown of botanically *unidentified* material is shown in Fig. 10 (next frame). The major components are plant materials that could not be assigned to taxonomic or tissue-organ categories. Calcite fillings were most abundant at Paradise, and were largely associated with chunks of periderm,as elsewhere. Large local concentrations of pyrite were rare in the Shawneetown and Paradise samples, but were more abundant at the Sahara and Crab Orchard sites.

As relative quantitative indicators of the distribution of different types of pyrite occurrences, areas of concentrated pyrite were recorded, as were heavily pyritized plant tissues. Comparisons of the data with pyrite analysis by weight of the same coal-ball zones at Sahara (C. P. Rao, unpub. data) indicate that these approximations do delineate major pyrite zones. The pyrite tends to be concentrated often in the outer boundaries of the coal balls and is often high in stigmarian root zones. The main pyritic coal-ball zones at Crab Orchardand Sahara include the bottom and topmost zones; at Sahara, zones 1, 3 and 8; at Crab Orchard, zones 2, 7 and 9, the last of which was mostly pyritized lycopod periderm and stigmarian roots. The total volume of pyrite and pyritized plant material is 4.6 percent at Sahara, which is about four times that found at Crab Orchard. The iron and sulfur contents of an adjacent profile of coal-ball zones at Crab Orchard (Delta Vertical Series 2), given in the chemical analyses, corroborate the high pyrite content observed in the bottom and top coal-ball zones.

TAXONOMIC COMPOSITION OF SELECT AND RANDOM SAMPLES

The first and only published quantitative data on the botanical contents of coal balls (Schopf, 1938) were obtained from seven coal balls found in an upper bench of the Herrin Coal at the Clarkson Mine near Nashville, Illinois. The following data were obtained for similar coal-ball collections from Nashville (Fig. 1) and Paradise (Appendix 1) for comparison with Schopf's results, with the profiles just described, and with the coal palynology. Random samples from Sahara and Shawneetown were also analyzed for comparisons with select zone samples, as above, and with summary data from taxonomic profiles of the mulitiple

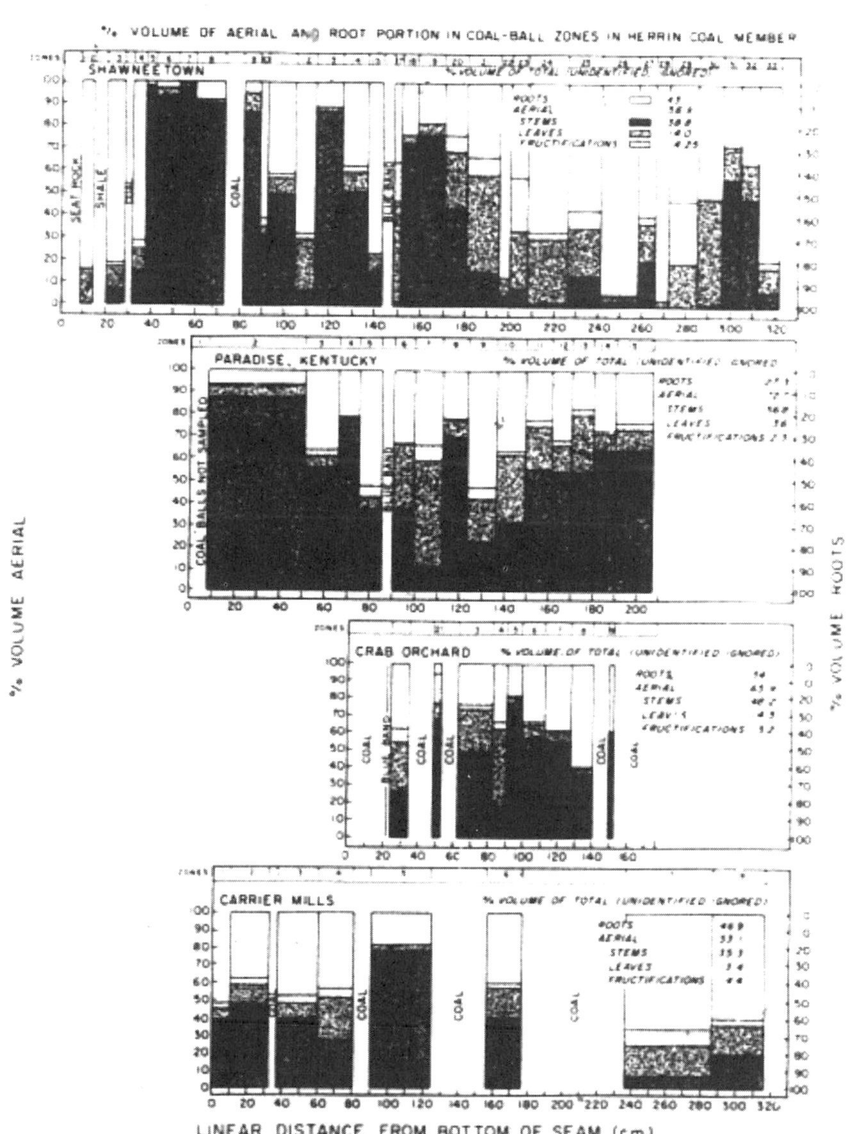

Figure 9. Histograms of percent volume of identified plant material in organ categories for the coal-ball profiles at Shawneetown, Paradise, Crab Orchard (Delta) and Carrier Mills (Sahara).

33

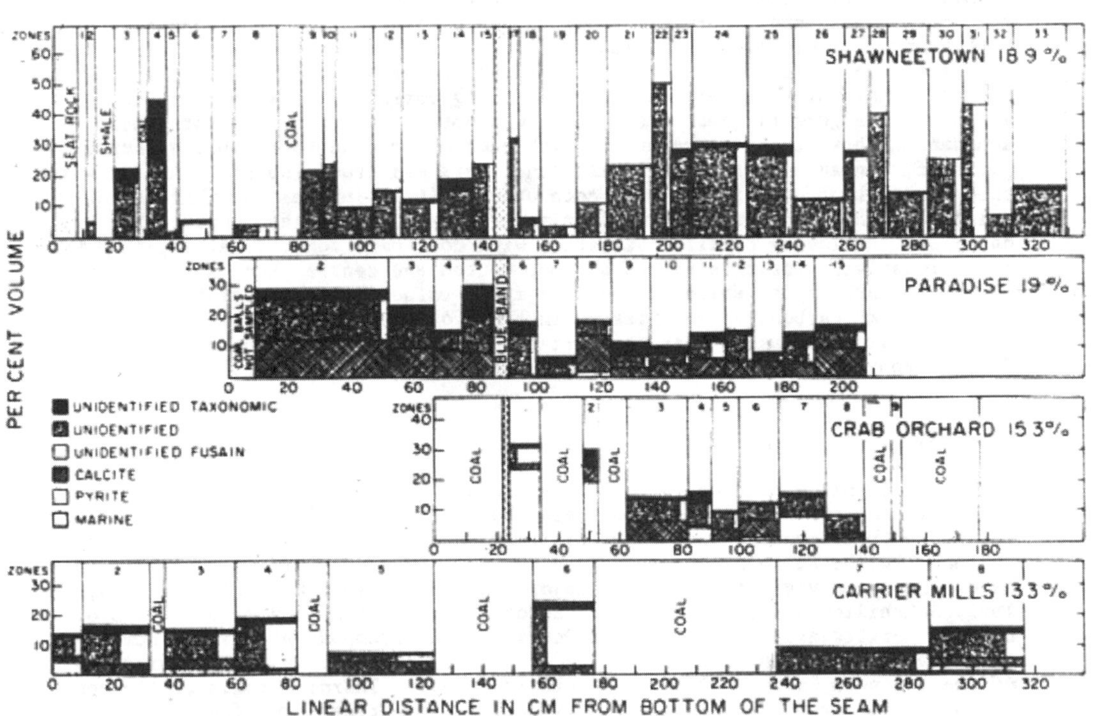

Figure 10. Histograms of botanically unidentified portions of coal-ball zones in profiles at Shawneetown, Paradise, Crab Orchard (Delta) and Carrier Mills (Sahara).

zones of coal balls from the Herrin Coal. Our results and the modified grouping of Schopf's data are given in Table 1 (next frame).

MAJOR AND TRACE ELEMENTS

The objective of these analyses was to determine whether the patterns of abundance and changes in elements through profiles correlated with vegetational changes, terrestrial or marine sources of carbonates and profiles of the adjacent coal. Mass spectrographic analyses of 30 major and trace elements along, with abundances of boron and selenium determined by other techniques were made for seven coal-ball zones at the Delta (Crab Orchard) locality (Table 2, next frame). Many points of elements showed strong correlations with each other, as shown in Figure 11 (next frame). It was noted that those elements whose abundances show strong correlations with each other tend to exhibit W-shaped plots of abundance against coal-ball zone; this was true also of Fe and S, which show a very high correlation with each other. It is interesting that the δC^{13} values for the carbonates in the same samples (M. E. Brownlee, unpublished data) show a closely parallel variation with coal-ball zone. That is, δC^{13} shows relatively positive values at top, bottom and centre, such as are found in marine carbonates, while the more negative values in intermediate zones correspond to carbonate precipitated in fresh or brackish waters. There is thus a clear implication that the variations in abundance of the elements indicated above are associated with the extent of marine influence.

DISCUSSION

Not all bituminous coals of Pennsylvanian age and originating from vascular plants in the northern hemisphere yield abundant coal balls. Those that do, fall into three general categories distinguished by the dominant type of vegetation. In the Western Interior Basin of North America, cordaites either dominates the swamp vegetation or is a major contributor (Darrah, 1939; Leisman, 1961; Schabilion and others, 1974). Secondly, in Westphalian A to D coals of Europe, particularly those from the Donetz Basin, lycopods are either dominant or quite abundant (Snigirevskaya, 1972), as they are also in the Desmoinesian of the Illinois Basin (Phillips and others, 1974). Thirdly, coals of the Upper Pennsylvanian or Stephanian are mostly derived from tree ferns (*Psaronius*).

The first explored sampling of coal-ball zones was in the great coal-ball horizon of western Europe, in the Bouxharmont seam of the lower Westphalian A of Belgium (Leclercq, 1930). From three layers distinguished in a 95 cm thick coal, 196 coal balls were examined but no major changes were detected vertically in the repetitive lycopod-dominated floras. No quantitative data were given, and sampling techniques were not defined. In the Western Interior Basin an ecological-floristic comparison of coal-ball plants between Kansas Coal Members (Weir-Pittsburg, Mineral, Fleming and Bevier), and the Herrin Coal at Sahara, was initiated several years ago by Gilbert A. Leisman. Although his techniques differ from ours, the data on abundance of plant groups in the Herrin Coal are in general agreement with our own (Leisman, unpub. data).

Schopf's study (1938) illustrated what could be done quantitatively with coal balls as an index to the botanical constitution of coal. He had over 3 m of linear transects and presented data for the composition of plants in seven coal balls (Table 1). While our sampling tests (Appendix 2) indicate

TABLE 1. PERCENT VOLUME OF TAXA IN HERRIN COAL BALLS

Sample no.	1	2	3	4	5
Lycopods	31.2	45.8	53.3	71.2	64.4
Ferns	63.9	38.6	28.8	18.4	14.0
Pteridosperms	3.9	14.4	16.0	8.3	17.0
Sphenopsids	1.0	1.2	1.9	1.9	2.2
Cordaites	0	0	0	0.06	0.08

1. Normalized percentages from Schopf (1938), 7 coal balls from Nashville.
2. 20 coal balls from upper bench at Nashville, 1,443 cm^2.
3. 10 coal balls from upper bench at Paradise, 1,467 cm^2.
4. 10 coal balls from Sahara, random sample, 2,000 cm^2.
5. 10 coal balls from Shawneetown, random sample, 1,330 cm^2.

36

TABLE 2. MAJOR AND TRACE ELEMENTS OF COAL BALLS BY ZONE

Zone	1	2	3	4	5	6	8(=top)	Mean
				Wt. Percent				
Fe	6	3	1	3	0.8	1	9	3.4
S	7	3	0.4	3	0.4	1	10	3.5
Si	0.5	0.4	0.3	0.5	0.7	1	0.2	0.5
Mg	3	0.2	0.8	3	3	7	10	3.9
				Mean Concentration <10,000 ppm				
Al	1,000	400	≤70	300	≤200	1,000	1,000	567
Ba	20	40	60	30	90	100	200	77
Cl	80	500	<80	100	300	300	1,000	337
F	30	30	20	30	30	70	200	59
P	100	70	30	100	600	100	700	243
K	80	200	<90	50	200	400	50	153
Mn	900	2,000	900	1,000	1,000	1,000	9,000	2,257
Na	50	60	100	100	500	100	200	159
Sr	1,000	700	500	600	700	1,000	2,000	929
				Mean Concentration <30 ppm				
B	13.9	4.5	9.3	3.0	1.6	17.0	13.0	8.9
Co	0.5	0	0	0.4	0	10	9	2.8
V	2	0.8	0	0.4	≤0.6	30	20	7.7
Sb	0	≤1	≤4	0	≤3	0	0	1.1
As	3	10	4	2	2	6	10	5.3
Cu	0	0	0	0	9	0	10	2.7
Sn	0	≤0.3	≤6	0	≤10	0	0	2.3
Ti	20	70	≤10	≤3	60	10	≤9	26.0
Zn	≤9	≤6	20	≤9	50	≤30	≤40	23.4
Ce	≤20	≤10	≤4	≤8	≤3	10	90	20.7
Cr	≤4	0	0	0	0	0	20	3.4
Ga	0	0	0	0	0	4	9	1.9
La	10	10	3	7	4	8	50	13.1
Nd	≤5	5	≤3	≤5	≤4	10	50	11.7
Pr	1	2	≤3	2	≤1	0	10	2.7
Rh	0	0	0	0	0	7	20	3.9
Se	1.07	0.82	0.52	1.9	0.63	0.7	4.54	1.45
Sc	1	0.7	0	0.6	0	1	7	1.5
Y	8	4	2	6	2	0	50	10.3

Note on Table 2: All elements except selenium and boron determined by mass spectrographic analysis. Matrix is calcium. Zone 7 was excluded from analysis because of small coal-ball sample size. Each zone analysis was based on a mixture of 2.5 cm diameter, centered vertical cores (coal free) from each of five coal balls. Coal-ball material included <3 percent weight of plant material. Mass spectrographic analyses by R. J. Blattner; photographic plates 521-22, 539-40 and 598-600 are on file at Materials Research Laboratory, University of Illinois, Urbana. Neutron activation and radiochemical separation of selenium and colorimetric determination of boron (Bohor and Gluskoter, 1973) by Illinois State Geological Survey.

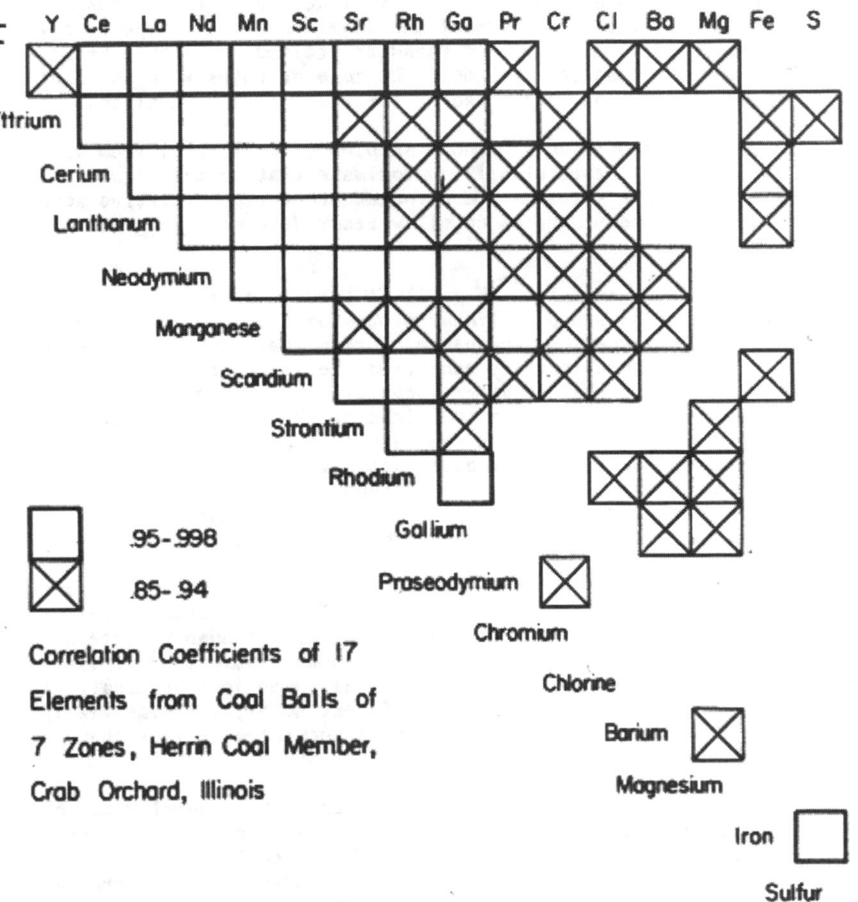

Figure 11. Diagram showing the trace elements between which the correlation coefficient for abundance is greater than 0.85. The correlations are for 17 elements in 7 coal-ball zones at the Crab Orchard (Delta) locality; see Figure 2.

that his sample size was too small to characterize with adequate statistical significance the botanical constituents of the upper bench of the Herrin Coal, his data do emphasize the abundance of root material in the peat and establish *Psaronius* as very abundant in part of the uppermost layers of the coal. Examination of a larger sample from the same collections show that lycopods are more abundant than ferns and, as in Schopf's (1938) data, that ferns are particularly abundant in the upper part of the seam. A similar type of collection from the uppermost coal-ball zones of Paradise yielded comparable results (Table 1) but 10 percent less fern volume. The zone or zones with abundant ferns near the top of Herrin Coal were noted in the four profiles analyzed.

The botanical composition of random samples of coal balls from sites with multiple zones (Table 1) more closely approximate that of taxa from profiles (Fig. 2), but the sample sizes used are inadequate for quantitative accuracy. However, such small random samples do allow ready determination of the dominant vegetation type.

In the best match comparison (Fig. 7) of taxonomic profiles above the blue band, the inverse of the slope of linear plots for each profile was normalized for comparisons of thicknesses accumulated during what was presumably the same time interval. The Sahara locality exhibited the greatest amount of preserved accumulation, 4-10 times that of other localities, and there was as much coal in the Sahara section as in adjacent portions of the seam. A downward extrapolation of linear plots for Sahara is consistent with the lack of a blue band, that is, the Sahara locality was probably not a site of peat accumulation when the blue band was deposited. The level of identifiable plant material was highest at Sahara, and the coal-ball profiles contained the most root material, ferns including coenopterids, sphenopsids, and fructifications.

The strongly cyclic alternation of root zones and aerial plant parts in the lower half of the Shawneetown profile (Fig. 3) may be interpreted as a sequence of 4-5 successive forest stands. If a similar number contributed to the upper half of the section, there were probably some ten successive forest stands during the span of the Herrin Coal swamp in that area. With estimations of 50-100 years for each major forest cover, it seems likely that the plant constituents of the coal were accumulated in less than 1,000 years.

Miospore palynology of the Herrin Coal seems to be generally corroborative in the limited comparisons that can be made with coal-ball vegetation, except for pteridosperms. Lycopod spores dominate (64-79 percent of the whole; *Lycospora* genus, 67-94 percent of lycopod spores), while ferns account for up to 15 percent of spores (R. A. Peppers, unpub. data). In the "megafossil" fraction *Monoletes* is about 15 percent (Winslow, 1959). Coal palynology of the upper 30 cm from Nashville indicates lycopod spores (*Lycospora* and *Cappasporites*) accounting for 61.5 percent of the whole and fern spores 30.5 percent; analysis of the upper 20 cm (above upper shale parting) from near Shawneetown yielded 64 percent lycopod spores and 31 percent for ferns (R. A. Peppers, unpub. data). Among the trends of spore frequency in the Herrin Coal (Kosanke, 1950, p. 76) is a general decrease in *Lycospora* near the top of the coal (above upper shale parting) except in western Illinois.

Fusinized plant material (here treated as equivalent to the coal maceral group, inertinite) in coal balls represents 5.0-7.2 percent of the total volume. An average of 82 percent of the identifiable fusain and semifusain is derived from lycopod tissue, with 11 percent from pteridosperms. The inertinite content (mineral-free basis) of coals from two of the sites studied are in the range 4.1-7.1 (N. H. Bostick, unpub. data).

The major minerals in coal balls from the Illinois Basin are calcite, ferroan dolomite and pyrite (Rao and Pfefferkorn, 1971); 8 percent of the total carbonate fraction is ferroan dolomite. Pyrite in coal balls from the Delta locality was previously noted to increase toward coal-ball margins and was regarded as probably secondary (Eggert and Cohen, 1973); framboidal pyrite occurred in most coal balls and was especially common in fusinized tissues. Further significant observations on pyritization of plant tissues in concretions in coals are given by Kizil'shtein (1974). Chemical composition of American coal balls, including major and some trace elements, has been reported by Mamay and Yochelson (1962) who also cite other reports; none of these include coal balls from the Herrin Coal. A comprehensive study of the mineralogy of coal balls from the Illinois Basin, however, has been completed (C. P. Rao, unpub. data).

It has been noted above that the vertical distribution of certain trace elements, as well as iron and sulfur, in the coal-ball zones is W-shaped and correlates strongly with changes in $\delta^{13}C$ values in the carbonates in the same zones. It was inferred that the variations in abundances of the elements reflect fluctuations in the salinity of the ground waters during formation of the permineralized material. It is perhaps not surprising that trace element signatures characterizing early changes in environment should be retained within a calcitic/pyritic matrix for so long. Nevertheless, that it is so adds a further interest to the study of coal balls as a key to the understanding of the formation of coals in peat swamps. At the same time, it should be noted that no strong correlation between element distributions and taxonomic abundances were established.

The average total sulfur content of the Herrin Coal is 3.4 percent compared to 3.5 percent for the Crab Orchard coal-ball profile. With the exception of calcium and magnesium, all of the other elements determined occur in lower concentration in coal balls (mean concentration in all zones) than in the coal itself at the same mine (Ruch and others, 1974); however, the mean fluorine value was almost identical.

CONCLUSIONS

Techniques have been devised and implemented for the quantitative botanical analysis in profile of permineralized peat as indices of the plant constituents of humic bituminous coals.

Determinations of peat composition from coal balls from the Herrin Coal show lycopod trees dominated the plant communities. This supplemented our preliminary studies of other coals with different vegetational types. Such vegetational patterns should have predictive value for characterizing bituminous coal reserves of Pennsylvanian age; sampling control should be possible by using random samples of coal balls and coal palynology. Valid use of random samples of coal balls is, of course, dependent on their distribution within a seam. The usefulness of palynological data, either as summaries of channel samples or profiles delineating vertical changes, is dependent on our understanding the quantitative relationships of palynological data to the plant composition of the original peat. Without such quantitative relationships, the relative contributions of various plants to peats cannot be inferred from spore counts, because of the natural wide variability of productivity of spores displayed by different genera and species.

Regional correlation of distinctive assemblages of plants at sites along a 142 km traverse of the southern part of the Illinois Basin can be made. Determination of the fraction by volume of material representing the dominant and subdominant types of vegetation demonstrates the widespread relatively consistent taxonomic composition of peats of comparable age in similar environments. However, efforts are continuing to determine trends of change in the Herrin Coal in order to understand better the ancient environments of deposition. Fluctuations in the contents of various macerals (Bostick and Foster, 1974) and palynological analyses (Kosanke, 1950) have been detected most strongly in western Illinois.

Multiple coal-ball zones and zones that locally constitute most of the seam thickness provide entombed sources of inorganic data, less subject to changes than those relating to the coal itself. Permineralization has combined elements of the original swamp with the secondary carbonates and associated minerals of the coal balls. The association of inter-element correlations for a number of elements with the character of the carbonate sources (relatively marine or freshwater) allows some separation of the elements into groups. These associations, and the variations of $^{13}C/^{12}C$ in carbonates, demonstrate the importance of environmental influences in at least certain areas of the swamp and indicate possible source areas for some major and trace elements.

ACKNOWLEDGMENTS

This study was carried out in cooperation with the Coal and Stratigraphy Sections of the Illinois State Geological Survey; we wish to especially thank Russel A. Peppers, Lindell H. Van Dyke, Matthew J. Avcin, Kenneth R. Cope, Harold J. Gluskoter, Mary E. Brownlee, Neely H. Bostick and William G. Miller. We appreciate the field assistance of Allen Williamson, Kentucky Geological Survey, Donald L. Eggert, Indiana Geological Survey and James F. Mahaffy, Botany Department, University of Illinois.

APPENDIX 1. STRATIGRAPHY - LOCALITY AND OUTCROP DESCRIPTIONS

Coal balls were obtained from the Herrin Coal Member in Illinois, Brereton Cyclothem, Carbondale Fm, Kewanee Group, Desmoinesian Series (Kosanke and others, 1960) or the equivalent in western Kentucky, the Kentucky No. 11 Coal Member, Carbondale Fm, Allegheny Series (Mullins and others, 1965). The coal is middle Pennsylvanian in age and is Westphalian D or Moscovian C_2^m in European and Soviet time-stratigraphy (Havlena, 1967).

Vertical sections of coal-ball zones came from the following four localities where the Herrin Coal in adjacent areas was 90-110 cm thick: (1) *Paradise* - Peabody Coal Co. Ken Mine, Ohio Co., Paradise 7 1/2' Quad. 37° 16' 33" N, 86° 57' 42" W, 2 km E-NE of Paradise, Ky. Seam thickness was 220 cm with coal balls constituting 208 cm; coal occurred in thin layers of variable thickness among the 15 coal-ball zones. The coal-ball zone on the underclay and to a lesser extent that on top of the blue band* had apparently been leached by ground water. In vertical series #2, specimens (183) are numbered from top to bottom, 12,180-12,362. Coal balls from the upper 30 cm, collected at random, are 11,628-11,637. (2) *Shawneetown* - Peabody Coal Co. Eagle Mine No. 2, Gallatin Co., Shawneetown 7 1/2' Quad. Sec. 4, T 10 S, R 9 E, 5 km SE of "New" Shawneetown, Ill. Core of seam, 5 cm in diam, 322 cm thick with 292 cm of coal-ball material; seam was described in the compaction study by Kosanke and others (1958). Peel specimens are designated 11,433-11,465 for 32 coal-ball zones and the blue band, numbered from the bottom. A random sample of coal balls was made from Sec. 9 about 1.7 km SE of the core section and is numbered 11,669-11,678. (3) *Sahara* or *Carrier Mills* - Sahara Coal Co. Mine No. 6, Saline Co., Harrisburg 7 1/2' Quad. Sec. 24, T 9 S, R 5 E, 3 km NW of Ledford, Ill. Vertical series #2 was collected from benches over an area of about 100 m^2. Estimated thickness of the bed is 316 cm with 209 cm of coal balls in eight zones. Specimens (189) are numbered from bottom, 9,529-9,717. Two other samples from the

*The blue band is a persistent clay or clay shale parting 1.5-10 cm thick within the Herrin Coal, usually 30-60 cm (somewhat higher in thicker beds) above the bottom of the bed (Kay, 1922; Giles, 1934; Cady, 1952) although it is not present in all areas of the Herrin Coal. The blue band was shown to be directly related to a main channel ("Walshville channel") in the 2,700 square mile area in southwestern Illinois studied by computer plotted mapping of lithologic units from the Herrin Coal to the Piasa Limestone (Johnson, 1972). The blue band thickens to more than 20 cm adjacent to the channel, particularly on the outsides of meanders. Johnson (1972) concluded that the blue band probably represented an overbank deposit during a time of major flooding as earlier proposed by E. S. Moore *in* Giles, 1934). Johnson (1972) stated, "Inasmuch as the peat swamp contained numerous channels, and because the depositional slope was probably less than one degree, flooding would have affected a wide area - probably several thousand square miles."

area include ten coal balls from six zones in Sec. 23, 60 m W of vertical series #2, numbered 9,519-9,528, and ten random samples from the immediate area, 11,679-11,688. No blue band was present at the outcrops examined. (4) *Crab Orchard* - AMAX Delta Mine, Pit 5 1/2, Williamson Co., Marion 15' Quad. Sec. 28, T 9 S, R 4 E, 4 km E-SE Crab Orchard, Ill. Vertical series #1 and 2 were obtained from a seam 177 cm thick with 96 cm of coal balls in 8-9 zones; specimens numbered from the bottom are 11,466-11,562 (96) and 12,547-12,840, 13,141-13,228 (382). Vertical series #1 is a channel sample 20-40 cm wide and #2 was collected along a 10 m face adjacent to #1. All coal balls occurred above the blue band.

The total of intervening coal layers at the above collection sites equalled 12 percent (Paradise), 30 percent (Shawneetown), 81 percent (Crab Orchard) and an estimated 100 percent (Sahara) of the average seam thickness near the sites.

A random sample of 20 coal balls from the Clarkson Shaft Mine, Washington Co. at Nashville, Ill.was obtained from collections of the Illinois State Geological Survey for comparison with Schopf's (1938) data. The location is given as Nashville 15' Quad. SW, SE, Sec. 13, T 2 S, R 3 W. With the exception of Illinois Geological Survey collections, specimens are filed in the Paleobotanical Collections (Morrill Hall), Botany Department, University of Illinois, Urbana.

APPENDIX 2. SAMPLING AND QUANTIFICATION TECHNIQUES

The basic sample unit is a coal ball although some samples were fragments of larger balls or of a continuous layer of permineralized peat. The layers of coal-ball material from outcrops at Carrier Mills, Crab Orchard and Paradise were manually excavated from each zone and color-coded with spray paint on top surfaces. Saw cuts about 2.5 cm apart were made at right angles to the top-bottom surfaces to expose maximum cross-sectional area of stratified peat inside the coal ball. Sizes and shapes dictated whether all cuts of a coal ball were parallel or at different angles; diagrams of coal-ball orientation, shape and slice positions were made to assist in sampling procedures. Peel preparations were made for all cut surfaces.

Peels were examined with a boom-type stereomicroscope primarily at 8X with use of higher magnifications when necessary. An overlay of cellulose acetate was used, ruled at 1 cm intervals with a grid system (26 × 20 cm), the intersections of the rules being numbered (1-520); this allowed identification maps of major contents of the peel to be made on similarly grid-marked sheets with a simple code system for taxa, type of plant, organ or tissue, and preservational or mineral aspects such as fusain, pyrite, calcite or unidentified. Particular care was taken to avoid identifications that were doubtful; categories were established with a flexible language for unidentified organs or tissues as opposed to unidentifiable organic matter or mineral areas lacking observable plant material.

The coded identifications were transcribed to data sheets and then to IBM computer cards with the grid numbers serving as locations of the recorded contents in each peel and the interconnections of plant tissues larger than one cm^2 for purposes of counting occurrences. A one cm^2 grid was adopted as a workable area for an overlay system although many organs or fragments may occur within such an area; the numerical count of plant parts based on area occupied on the peel is significant for plant structures equal to or larger than one cm^2 in area, and consistently recorded at each occurrence. Data sorting and calculations were carried out on Xerox Sigma 5 and IBM 360/75 computers.

The minimum amount of data necessary to adequately characterize the botanical constituents of a coal ball was determined by recording identifications from totals of all peels of a coal ball sliced as indicated above and comparing the data to that from one, center-most, peel. The peel from the center-most slice seemed to combine the features of maximum cross-sectional area and representativeness of content; therefore, data comparisons from totals of various peel combinations from coal balls were made with center peels to establish the reliability of one middle peel.

Area of peel occupied by plant material is treated as volume in percentage calculations based on comparative sampling data of all peels of 10 coal balls and selected multiple peels (based on size and shape of 189 coal balls), compared in both cases with data from only the center-most peel of each coal ball. Ten intact coal balls were taken from six different zones to provide variety of plant material, and the data from 110 faces were computed and plotted from all the peels, every second peel, third and so on of each coal ball, in each case including the center-most peel, until only one peel per coal ball was used in determination of percent volume (Fig. 12, next frame). With the diminution of data from about 6,000 cm^2 to about 700 cm^2 there is a slight fluctuation in results. The 189 balls from a second outcrop were used to test the *one middle peel* approach to sampling. Peel selections were made for each coal ball based on size and shape, including the center-most peel. Comparisons (Table 3, next frame) indicate that the middle peel is adequate sampling with a maximum deviation of 0.7 percent for any taxonomic category.

Similar procedures were used to determine adequate coal-ball samples for zones at the Sahara mine (Fig. 13, next frame). Determinations of the results for all coal balls for each zone were plotted against data from every other coal ball, every third, and so on, and determinations made of the level of sampling at which results become erratic. For the Herrin Coal the coal-ball sample size per zone at a given locality should include 10-15 coal balls representing more than 1,500 cm^2 based on a middle peel from each. Zones 7 and 8 (Fig. 13) contained a less heterogeneous mixture of vegetation in the permineralized peat.

Cores 5 cm in diameter were broken along coal boundaries and the coal-ball portions were sliced longitudinally. Peels were prepared from both diameter faces and both semi-circumferences. All four resultant faces were used in identifications. For summary data of botanical constituents analysis of any of the four faces of the core is adequate (Table 4, next frame), but all the data were used for plots of vertical composition because of limited surface area.

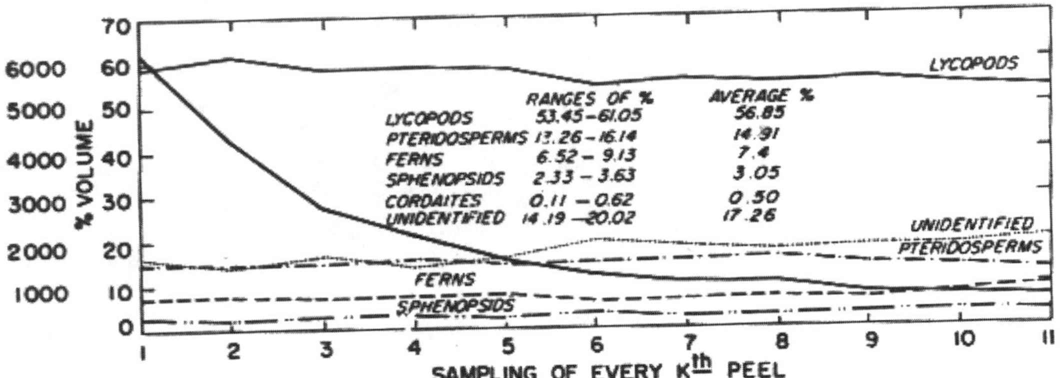

Figure 12. Percent volume of major taxa and unidentified based on all peels from 10 coal balls, every other peel, every third and so on to only one peel (centermost) per coal ball.

TABLE 3. PERCENT VOLUME COMPARISON OF SAMPLING 189 COAL BALLS, HERRIN COAL

	Middle Peel	Selected Peels and Middle Peel
Total cm^2	13,604	23,013
Unidentified	14.1	13.3
Cordaitales	0.6	0.4
Ferns	13.8	14.5
Lycopods	62.6	62.2
Pteridosperms	4.7	5.1
Sphenopsids	4.2	4.5

TABLE 4. COMPARISON OF PERCENT VOLUME OF TAXA IN SHAWNEETOWN CORE

	4 faces	face 1	face 2	face 3	face 4
Unidentified (18.0-19.9)	18.9	18.7	19.9	18.0	19.2
Lycopods (60.4-61.8)	60.7	60.1	61.8	60.6	60.4
Pteridosperms (6.4-7.7)	7.1	7.7	6.4	7.2	7.2
Ferns (10.1-13.1)	12.0	11.8	10.1	13.1	12.4
Sphenopsids (0.8-1.9)	1.3	1.7	1.9	1.2	0.8
Cordaitales (0.0)	0.0	0.0	0.0	0.0	0.0

47

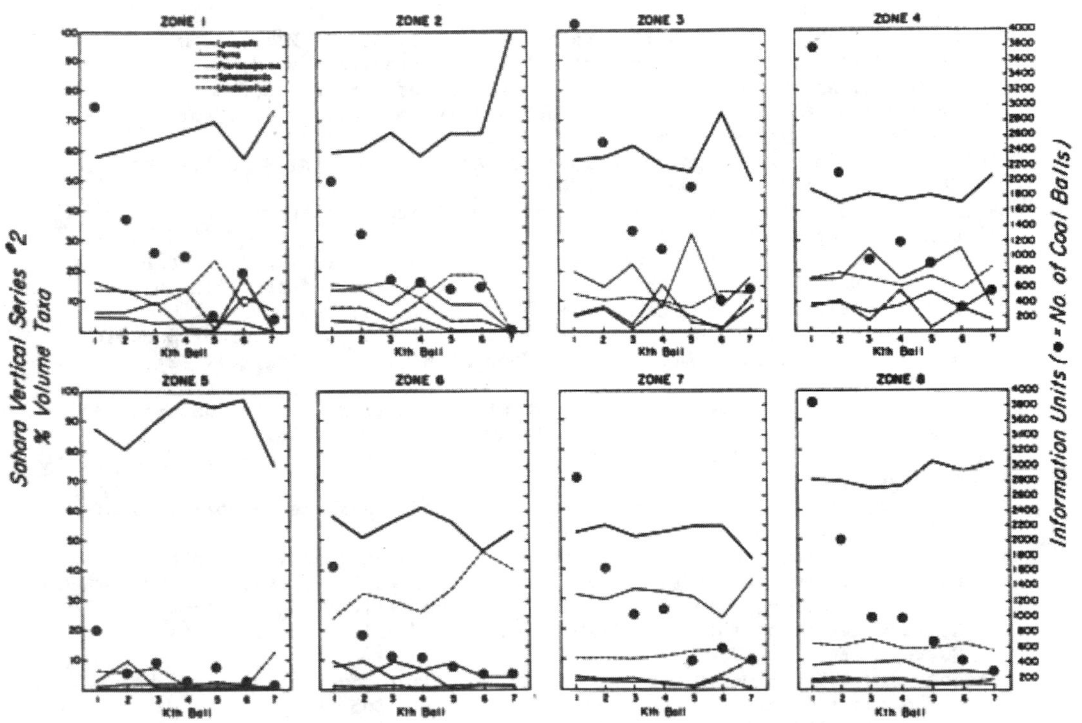

Figure 13. Percent volume of major taxa and unidentified for the 8 zones at Carrier Mills (Sahara; see Fig. 6). Number of coal balls in black circles is plotted by total information units (cm^2). Successive plots of 1-7 include data from every coal ball; 2, every other coal ball; 3, every third coal ball and so on. For further explanation, see text.

48

REFERENCES CITED

Bohor, B. F., and Gluskoter, H. J., 1973, Boron in illite as an indicator of paleosalinity of Illinois coals: Jour. Sed. Petrology, v. 43, p. 945-956.

Bostick, N. H., and Foster, J. N., 1974, Regional distribution of macerals in the Herrin (No. 6) Coal, Illinois: Geol. Soc. America Abs. with Programs, v. 6, p. 662.

Cady, G. H., 1952, Minable coal reserves of Illinois: Illinois Geol. Survey Bull. 78, 138 pp.

Darrah, W. C., 1939, The fossil flora of Iowa coal balls. Part 1. Discovery and occurrence: Harvard Univ. Bot. Mus. Leaflet, v. 7, p. 125-136.

Eggert, D. L., and Cohen, A. D., 1973, Petrology of coal balls from the Herrin (No. 6) Coal, Carbondale Fm., Williamson Co., Ill.: Geol. Soc. America Abs. with Programs, v. 5, p. 312-313.

Giles, A. W., 1934, Partings in coal beds: Am. Inst. Mining Metallurgical Engineers Trans., v. 108, p. 31-40.

Havlena, V., 1967, Stratigraphische Tabelle des Karbon von Perm: Fachschaft Geologie Münster, 1 pp.

Johnson, D. O., 1972, Stratigraphic analysis of the interval between the Herrin (No. 6) Coal and the Piasa Limestone in southwestern Illinois [Ph.D. thesis]: Urbana-Champaign, Univ. Illinois, 105 pp.

Kay, F. H., 1922, Coal resources of District VII (Coal No. 6 west of Duquoin anticline): Illinois Geol. Survey Coop. Min. Inv. Series Bull. 11, 233 pp.

Kizil'shtein, L. Y., 1974, The formation of iron sulfide concretions in coal seams: Litologiya i Poleznye Iskopaemye, No. 2, p. 58-65.

Kosanke, R. M., 1947, Plant microfossils in correlations of coal beds: Jour. Geology, v. 55, p. 280-284.

——— 1950, Pennsylvanian spores of Illinois and their use in correlation: Illinois Geol. Survey Bull. 74, 128 pp.

Kosanke, R. M., Simon, J. A., and Smith W. H., 1958, Compaction of plant debris-forming coal beds: Geol. Soc. America Bull., v. 69, p. 1599-1600.

Kosanke, R. M., Simon, J. A., Wanless, H. R., and Willman, H. B., 1960, Classification of the Pennsylvanian strata of Illinois: Illinois Geol. Survey Rept. Inv. 214, 84 pp.

Leclercq, S., 1930, Etude d'une coupe verticale dans une couche à coal-balls du houiller de Liége: Soc. Géol. Belgique, p. B63-B67.

Leisman, G. A., 1961, A new species of *Cardiocarpus* in Kansas coal balls: Kansas Acad. Sci. Trans., v. 64, p. 117-122.

Mamay, S. H., and Yochelson, E. L., 1962, Occurrence and significance of marine animal remains in American coal balls: U. S. Geol. Survey Prof. Paper 354-I, p. 193-224.

Mullins, A. T., Lounsbury, R. E., and Hodgson, D. L., 1965, Coal reserves of northwestern Kentucky: Tennessee Valley Authority, 28 pp.

Peppers, R. A., and Phillips, T. L., 1972, Pennsylvanian coal swamp floras in the Illinois Basin: Geol. Soc. America Abs. with Programs, v. 4, p. 624-625.

Phillips, T. L., Peppers, R. A., Avcin, M. J., and Laughnan, P. F., 1974, Fossil plants and coal: patterns of change in Pennsylvanian coal swamps of the Illinois Basin: Science, v. 184, p. 1367-1369.

Rao, C. P., and Pfefferkorn, H. W., 1971, Occurrence and mineralogy of coal balls in the Illinois Basin: Geol. Soc. America Abs. with Programs, v. 3, p. 678.
Ruch, R. R., Gluskoter, H. J., and Shimp, N. F., 1974, Occurrence and distribution of potentially volatile trace elements in coal: A final report: Ill. Geol. Survey Environmental Geology Notes, No. 72, 96 pp.
Schabilion, J., Brotzman, N., and Phillips, T. L., 1974, Two coal-ball floras from Iowa: Am. Jour. Bot., v. 61, p. 19.
Schopf, J. M., 1938, Coal balls as an index to the constitution of coal: Illinois Acad. Sci. Trans., v. 31, p. 187-189.
Snigirevskaya, N. S., 1972, Studies of coal balls of the Donetz Basin: Rev. Palaeobotany, Palynology, v. 14, p. 197-204.
Winslow, M. R., 1959, Upper Mississippian and Pennsylvanian megaspores and other plant fossils from Illinois: Illinois Geol. Survey Bull. 86, 135 pp.

PAPER 14

Paleoecology of Permian Reefs in the Guadalupe Mountains Area (1957)

N. D. Newell

Commentary

RICHARD K. BAMBACH

Norman Newell was a central intellectual inspiration in the maturation of traditional paleontology into the modern discipline of paleobiology. The paper reprinted here is a summary of Newell et al. (1953), probably the most influential pioneering effort in modern paleoecology. That study and the massive two-volume *Treatise on Marine Ecology and Paleoecology* edited by Hedgpeth and Ladd and published in 1957 (in which this summary of Newell et al. [1953] appeared) are the founding documents of modern paleoecology, which received full identity with the publication of Ager (1963) and Imbrie and Newell (1964). Along with the work on Permian reefs, Newell and colleagues also began a series of studies on modern carbonates (Newell et al. 1951, 1959) that led, in the early 1960s, to a revolution in the study of carbonate sedimentology as well.

Although the Middle and Upper Permian rocks in the Guadalupe Mountains of Texas and New Mexico were recognized as a basin reef–back reef complex long before Newell's study was published (Adams and Frenzel 1950 and references therein), the synthesis done by Newell and colleagues reviewed in the paper reprinted here brings to life the reef environment and its communities in Permian times. Newell's sensitivity to modern organisms and carbonate environments informs the geological study throughout. Over the years, some doubts have been raised about some of the interpretations both from definitional concerns and because of issues related to the common recrystallization and diagenetic alteration of the sediments. However, recent studies still support the basic conclusions of Newell's work.

Six relatively recent works summarize current understanding of this classic paleoecological setting: Wood, Dickson, and Kirkland-George (1994) and Wood, Dickson, and Kirkland (1996) differentiate reef-framework and cryptic ecological assemblages; Fagerstrom and Weidlich (1999, 2005) examine the structure of the reef-building assemblages in detail and the effects of different environmental stresses in producing varied patterns (with analogy to modern systems); and Olszewski and Erwin (2009) and Fall and Olszewski (2010) document environmental patterns in brachiopod communities from the Permian reef complex, including data from the extensive collecting of exceptionally preserved silicified faunas by G. A. Cooper and R. E. Grant.

Literature Cited

Adams, J. E., and H. N. Frenzel. 1950. Capitan barrier reef, Texas and New Mexico. *Journal of Geology* 58: 289–312.

Ager, D. V. 1963. *Principles of Paleoecology: An Introduction to How and Where Animals and Plants Lived in the Past*. New York: McGraw-Hill.

From *Geological Society of America Memoir* 67:407–36.

Fagerstrom, J. A., and O. Weidlich. 1999. Strengths and weaknesses of the reef guild concept and quantitative data: application to the Upper Capitan massive community (Permian), Guadalupe Mountains, New Mexico–Texas. *Facies* 40:131–56.

———. 2005. Biologic response to environmental stress in tropical reefs: lessons from modern Polynesian coralgal atolls and Middle Permian sponge and Shamovella-microbe reefs (Capitan Limestone USA). *Facies* 51:501–15.

Fall, L. M., and T. D. Olszewski. 2010. Environmental disruptions influence taxonomic composition of brachiopod paleocommunities in the Middle Permian Bell Canyon Formation (Delaware Basin, West Texas). *Palaios* 25:247–59.

Hedgpeth, J. W., and H. S. Ladd, eds. 1957. *Treatise on Marine Ecology and Paleoecology*. Geological Society of America Memoir, vol. 67. [New York:] National Research Council (U.S.), Committee on a Treatise on Marine Ecology and Paleoecology.

Imbrie, J., and N. D. Newell, eds. 1964. *Approaches to Paleoecology*. New York: Wiley.

Newell, N. D., J. Imbrie, E. G. Purdy, and D. L. Thurber. 1959. Organism communities and bottom facies, Great Bahama Bank. *Bulletin of the American Museum of Natural History*, vol. 117, article 4, 181–228.

Newell, N. D., J. K. Rigby, A. J. Whiteman, and J. S. Bradley. 1951. Shoal-water geology and environments, eastern Andros Island, Bahamas. *Bulletin of the American Museum of Natural History*, vol. 97, article 1, 1–29.

Newell, N. D., J. K. Rigby, A. G. Fischer, A. J. Whitman, J. E. Hickox, and J. S. Bradley. 1953. *The Permian Reef Complex of the Guadalupian Mountains Region, Texas and New Mexico*. San Francisco: W. H. Freeman and Company.

Olszewski, T. D., and D. H. Erwin. 2009. Change and stability in Permian brachiopod communities from western Texas. *Palaios* 24:27–40.

Wood, R., J. A. D. Dickson, and B. Kirkland-George. 1994. Turning the Capitan Reef upside down: a new appraisal of the ecology of the Permian Capitan Reef, Guadalupe Mountains, Texas and New Mexico. *Palaios* 9:422–27.

Wood, R., J. A. D. Dickson, and B. L. Kirkland. 1996. New Observations on the ecology of the Permian Capitan Reef, Texas and New Mexico. *Palaeontology* 39 (no.3): 733–62.

Geol. Soc. America Memoir 67, 1957
p. 407–436, 11 figs. Made in U. S. A.

Chapter 15

Paleoecology of Permian Reefs in the Guadalupe Mountains Area[1]

NORMAN D. NEWELL

Geologist, The American Museum of Natural History and Columbia University, New York, N. Y.

ABSTRACT

Fossil reefs of Permian age in West Texas and southern New Mexico are remarkably well developed and are ideally exposed. This study is an attempt to interpret the environmental conditions under which some of these interesting structures were formed.

Three adjacent geologic provinces are characterized by stratigraphically equivalent rocks and fossils of strongly contrasting facies: (1) the Delaware basin with laminated detrital drab-to-black limestones and quartz sandstones and a pelagic fauna; (2) the basin margin, occupied by light-colored very fossiliferous massive reefs and by banks of detrital limestone and dolomite; and (3) a shelf area covered by comparatively thin-bedded unfossiliferous, light-colored rocks, evaporites, dolomite, and quartz sandstone. Regional analysis leads to the following paleoecologic conclusions:

The land around the Permian seaway was very low after earliest Permian time, and the climate was warm and dry. The marine faunas are most similar to contemporaneous faunas at low latitudes in the Eastern Hemisphere (Tethys). This probably is a reflection of circumequatorial conditions. The lithologic, paleontologic and structural characteristics of the Delaware basin suggest deposition in quiet waters which at times were at least 1800 feet deep. On the other hand reefs and banks at the basin rim were formed near the surface where wave attack and occasional collapse resulted in a succession of wedges of detrital limesand and talus seaward and lagoonward from the basin rim. These marginal deposits, unlike those of the basin and shelf, are mainly composed of skeletal material reflecting relatively greater organic productivity here of calcareous deposits. The lagoonal, or shelf, deposits are relatively poor in skeletal material. This and other characteristics of the rocks show that environmental conditions on the shelf were unfavorable for many kinds of organisms. Waters of the shelf area probably were generally deeper than at the rim, but it is unlikely that they exceeded a few tens of feet.

The regional relationships suggest a shelfward flow of surface waters over evaporating pans where hypersaline waters were trapped behind the low barrier of slightly elevated banks and reefs. Basin waters of nearly normal salinity evidently were continuously renewed through one or more shallow inlets, probably at the south side of the Delaware basin. The depths of the Delaware basin below inlet threshold were little disturbed by this flow, and because of mild winter temperatures there was only limited seasonal turnover. Hence, the deeper waters were generally stagnant.

INTRODUCTION

Fossil organic reefs, formed under narrowly limiting conditions by intense metabolism of plants and animals in agitated clear and shallow waters, offer exceptional

[1] This is a summary of a longer work by Norman D. Newell, J. Keith Rigby, Alfred G. Fischer, A. J. Whiteman, John E. Hickox, and John S. Bradley (1953).

opportunities for the development and demonstration of principles and techniques of paleoecology. Communities of reef organisms are among the most complex and specialized found anywhere in the sea, and their geologic results are impressive. Characteristically, they occupy circumscribed areas surrounded and isolated by contrasting environments. For this reason fossil reefs contrast strikingly with the adjacent rocks in both lithologic and paleontologic characters. These contrasts are helpful in recognition and classification of facies, but they also result in complex stratigraphic relationships in which there are few stratigraphic datum horizons. Ordinarily there is a high order of correlation between assemblages of fossils and lithologic facies of the containing rock because both lithofacies and biofacies are commonly dependent on the same or related environmental factors. Distinctive assemblages of fossils commonly characterize particular lithologic facies, but the importance of these fossils as selective indicators of environment has not been sufficiently stressed. They are geologically as important as are the sought-for stratigraphic guide fossils which may be less closely related to the lithologic facies.

PERMIAN REEFS OF THE GUADALUPE MOUNTAINS AREA

Scattered over western Texas and southern New Mexico are extensive and excellent exposures of marine Permian rocks, most of which lie around and within the Delaware

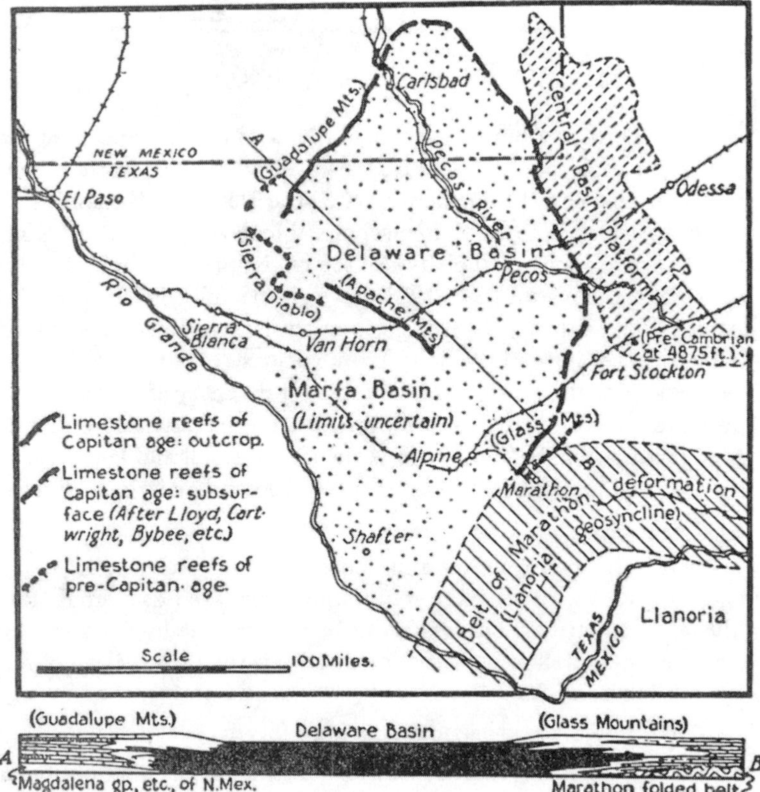

FIGURE 1.—*Permian structural provinces in Western Texas and adjoining parts of New Mexico* After P. B. King (1934, p. 704)

FIGURE 2.—*Structural features of Permian age in West Texas and ajoining part of New Mexico*
After P. B. King (1942, p. 665)

basin (Figs. 1, 2). This contribution is devoted to one of the principal outcrop belts at the northwestern margin of the Delaware basin, including the southern part of the Guadalupe Mountains and the northern part of the Delaware Mountains. The area has been recently described in detail in a scholarly monograph by P. B. King (1948). A general outline of the geological history of the region has been published by Adams and Frenzel (1950), and Newell *et al.* (1953) have given a fully documented description of the area together with a paleoecological analysis; it is thus unnecessary here to review the physical setting and previous investigations.

The Guadalupe Mountains area is remarkable for diversity of lithologic and faunal facies. There are widely distributed evaporites, lagoonal deposits, and bituminous rocks of a stagnant basin. The basin rim is the site of small and large reefs, some of which form the most extensive and best-displayed fossil barrier reefs known.

FOSSILS

Well-preserved fossil invertebrates and deposits of lime-secreting algae locally are abundant, but not all the rocks of the contrasting facies are fossiliferous. Progressive stages in diagenetic destruction of fossils can be traced laterally from areas in which the fossils are generally well preserved. Noteworthy is the preservation of fine details of structure in silicified shells.

DEPOSITIONAL ENVIRONMENTS

Three major depositional environments are indicated by the distribution and character of the Permian rocks: (1) comparatively deep and at times stagnant waters of the Delaware basin (basin phase), (2) bank and reef environment restricted to a narrow belt around the rim of the basin, and (3) shallow, generally hypersaline waters of the shelf (lagoonal, backreef phase) or stable interior of the continent extending over thousands of square miles around the basin. Possibly the only communication with the open ocean was through a strait (Hovey channel) which extended southward toward Mexico. Another strait connected the Delaware basin with the similar Midland basin to the east in which Permian rocks are covered by later deposits (Fig. 2).

An abundance of probably stenohaline forms in rocks of the basin (ammonoids) and in the reefs (fusulines, fenestelloid bryozoans, echinoderms, calcareous sponges, and many brachiopods) indicates that at least the surface waters of these areas may have been similar to those of the Permian oceans. The carbonate rocks of the Delaware basin and the reefs, with a few exceptions, are dominantly calcitic limestones. Permian rocks of the shelf areas on the other hand are dolomitic; and, some distance away from the basin, dolomitic limestones and sandstones are replaced by anhydrite and red beds.

Fossils generally are uncommon in the shelf deposits except near the reef. They have been generally obliterated or modified by dolomitization. In Leonardian and Guadalupian times the shelf waters were only sparsely populated, and the youngest faunas of the shelf area are limited to several kinds of gastropods, only a few of which are similar to those found in and around the reefs, and a few species of pelecypods, brachiopods, nautiloids, and scaphopods. Bryozoans, corals, ammonoids, sponges, and echinoderms are extremely rare or unknown in most of the backreef formations. Fusulines and dasycladacean algae are abundant near the edge of the shelf. Evidently many of the forms found in the shelf deposits were euryhaline and were adapted to life in hypersaline waters. The faunas of this phase are not well known, but representative faunules have been described by Girty (1909), Newell *et al.* (1940), and Clifton (1942).

DISTRIBUTION OF THE PERMIAN REEFS

It may thus be concluded that the general setting and distribution of environments were somewhat unlike those of any modern reef. The reef tract around the Delaware basin was roughly elliptical and was broken at the south by the Hovey and Sheffield channels (Fig. 2). Obviously it cannot be likened to the structure of an atoll, since the relative positions of sea and lagoon are reversed. The development of reefs in the area, as shown by P. B. King (*e.g.*, 1934, p. 788–790), was limited very closely to the rim of the Delaware basin. The basin rim, however, antedated the earliest reefs of the area and was produced primarily by relative subsidence in the basin area.

Modern reefs, which require sunlight, grow most rapidly in shallow waters within the agitated zone of surface waves. Commonly, the adjacent waters are relatively deep, and the bottom may drop steeply to oceanic depths. Reef animals benefit from the oxygen released by the algae. The carnivorous forms obtain most of their food at night when animal plankton rise to the surface from intermediate depths. Optimum conditions for the community of reef organisms occur along the seaward margins of shallow submarine platforms overlooking waters some hundreds or thousands of feet in depth. All the great modern reefs are so situated.

During the earliest Permian (Wolfcampian), scattered reef knolls or mounds grew around the basin. Several of these are exposed along the front of the Sierra Diablo along the west side of the basin. Probably few of them reached into very shallow waters; at least, they are not flanked by conspicuous talus fans.

At subsequent epochs the inland seas became progressively more and more restricted while organic production of carbonate sediments was greatly increased at and near the basin rim. During later Permian time (Leonardian, Guadalupian) the rim of the Delaware basin was built up by organisms well within the zone of wave activity, as recorded clearly by fans of coarse limestone detritus that were successively built out into the basin. Sedimentation in the reef tract was appreciably more rapid here than in the basin and on the shelf. Probably during much of the later history of the basin the reef crest between shallow channels stood above the level of low spring tides. In this condition the reef formed an effective barrier between the deep waters of the Delaware basin and the shallow evaporation pan of the shelf sea, inhibiting the return of heavy shelf waters into the basin. In a descriptive sense it was a barrier reef, but in many respects it was quite unlike modern barrier reefs, most of which face the open ocean.

REEF FORMING ORGANISMS

There is a general consensus that scleractinian corals and coralline algae are basically responsible for most marine reef development at the present time and that deposition of calcium carbonate by hydrocorallines, foraminifers, bryozoans, tube-forming worms, etc., is only accessory. These organisms find optimum life conditions in the reef habitat, but they do not alone produce extensive reefs. Today, stony corals are among the most conspicuous reef organisms. Therefore, it is not inappropriate to refer to coral-covered reefs as coral reefs. Some students of coral reefs believe, however, that the coralline algae and other reef dwellers would form reefs where coral reefs now occur, even if there were no stony corals (Howe, 1932).

Corals did not play an important part in formation of reefs in the Permian basin; they are represented by inconspicuous tabulates and small, solitary rugose forms. Paleozoic corals generally were unimportant sediment formers after the Mississipian period although a few *Chaetetes* were locally important in the early half of the Pennsylvanian. It is doubtful that there are Permian coral reefs anywhere.

The Texas Permian reefs were wave resistant, hence are true reefs in the preferred sense of the term (Ladd, 1944). They could not have been formed mainly, therefore, of inorganic precipitates of carbonate as suggested by C. L. Baker (1929). The evidence, summarized below, indicates that they resulted mainly from the activities of several kinds of low algae aided by contributions from calcareous sponges, hydrocorallines, and bryozoans.

MAJOR PROVINCES

Three sedimentary provinces are represented in the Permian rocks of the Guadalupe Mountains area: (1) the Delaware basin, (2) the margin of the shelf at the rim of the Delaware basin, and (3) the shelf area (Fig. 2). Each province is characterized by several lithologic types, each of which is considered here to form a separate lithofacies. There are many distinctive associations of fossils and these are regarded as biofacies.

Table 1.—*Rock formations, Guadalupe Mountains*

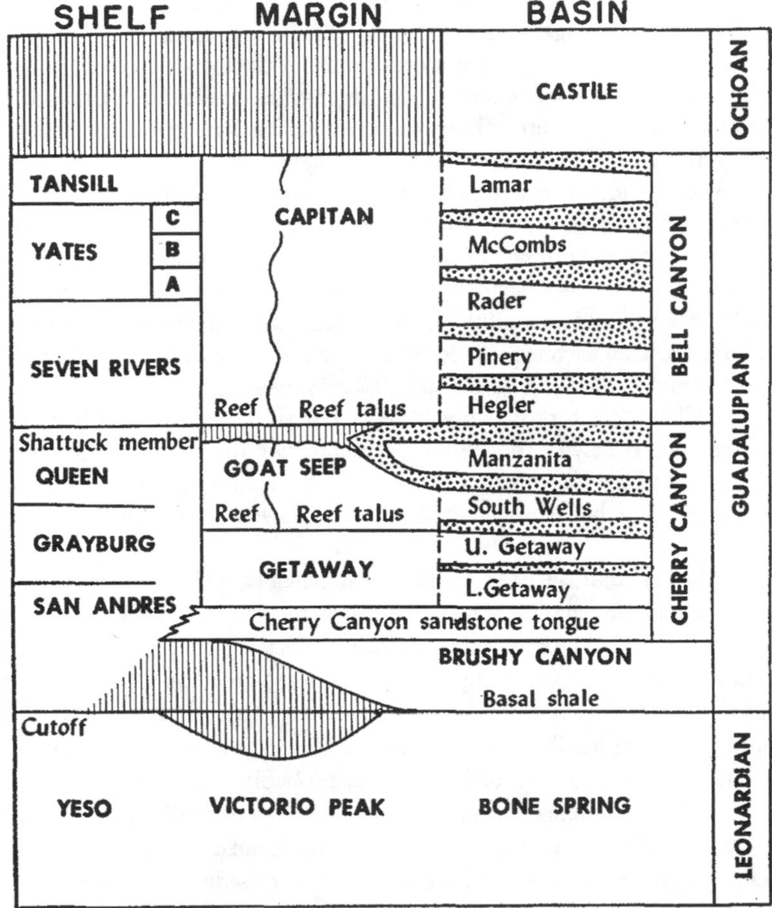

Vertically ruled areas represent hiatus; stippled areas, unnamed sandstone tongues. The Mc-Combs member, named from flagstone quarries on the Mc-Combs ranch, replaces King's Flaggy limestone bed.

The oldest Permian (Wolfcampian) rocks are buried in the Guadalupe Mountains; therefore they are not considered here. The youngest Permian rocks of the area do not contain reefs, and they are unfossiliferous. Accordingly, the following discussion is limited to rocks of Leonardian and Guadalupian age (Table 1).

BASIN PHASE

Rocks of Leonardian and Guadalupian age in the Delaware basin comprise more than 7000 feet of fine-grained quartz sandstones, black bituminous limestones, and, in the deeper parts of the basin, black bituminous "shales".[2] The limestones contain little magnesium in contrast to equivalent rocks outside the basin. Almost all investigators regard these rocks as the deposits of a comparatively deep and stagnant basin.

[2] The shales contain almost no clay. They are very platy calcareous sandstones.

Over most of the Delaware basin the rocks are stratified in very thin and uniform laminations characterized by differences in composition, texture, and color. It is difficult to estimate the average thickness of the laminae because they range from a fraction of a millimeter to more than a centimeter in thickness. In most places there are well-defined groupings of fine laminae in larger units. It is evident that sedimentation was rhythmic, and it is not unlikely that the laminations are seasonal deposits as suggested by Udden (1924), Lang (1935), and others.

Paleontological evidence suggests that the Leonardian and Guadalupian rocks represent much of Permian time, perhaps 10 to 15 million years. Unconformities and other indications of interruptions of sedimentation are not recognized in the deeper parts of the basin. There is a fairly well-defined faunal change at the boundary between the Leonardian and Guadalupian rocks, but this change is not attributed to a hiatus. Average annual increments required to deposit 7000 feet of sediments in 10 million years would measure approximately 0.2 mm. Average thickness of laminations in the basin sediments is certainly much greater than 0.2 mm., possibly as great as several millimeters. It may be concluded that deposition was very slow, and probably a large proportion of the planes between laminae represent long periods of nondeposition. It is also possible that these deposits were laid down in much less than 10 million years.

In spite of the fact that the deep waters of the basin provided conditions favorable for relatively rapid accumulation of sediments, organic deposition of limestone around the margin of the basin at times exceeded rates of deposition in the basin. The marginal reef limestones of late Guadalupian age (Capitan) are about 75 per cent thicker than equivalent beds (Bell Canyon) in the Delaware basin. It is evident that sedimentation was extraordinarily slow in the basin. Most of the sediments show features of deposition in quiet, stagnant, and therefore, probably deep waters. The primary structure of reef-derived talus deposits around the edge of the basin indicates great local relief in front of the Capitan barrier reef. Hence it must be concluded that subsidence was not equalled or exceeded by accumulation of sediments, as ordinarily has been the case on the continent during most of geologic history. An extraordinarily slow rate of terrigenous sedimentation probably was the result of aridity and subdued surface topography over much of the North American continent.

The rocks of the stagnant basin phase characteristically are bituminous and pyritiferous. Rocks of fine grain are black or dark gray, and coarser rocks range from tan to light gray. Quantitatively most important is remarkably uniform arkosic sandstone of "very-fine" to "fine" texture in the Wentworth scale. The sand grains characteristically are angular and bright. Moderately fresh microcline and plagioclase are minor accessories. Trails and burrows of benthonic animals occur sparsely in the sandstones at a few places near the margin of the basin, but generally the thin laminations are unmarked and unmodified by organisms. Fossil shells and shell detritus in the sandstone likewise are most abundant near the edge of the basin. There is every indication that the bottom sands of deeper quiet waters were barren wastes devoid of benthonic forms. Sandstones of this type occur in a few thin beds in the Bone Spring formation. They make up nearly all of the Brushy Canyon formation and perhaps three-quarters of the higher Guadalupian beds. Sandstones of the Brushy Canyon formation are coarser, with grains up to a maximum of about 0.5 mm. in diameter, and there are numerous massive beds, some of which are rippled. The Brushy Canyon formation is confined to the Delaware and Midland basins and is evidently equivalent to an obscure hiatus on the shelf (Fig. 3). Structural relationships and the characteristics of the formation indicate that the basin was more or less filled with sand during Brushy Canyon time (Fig. 4B). Deposition began in comparatively deep waters and ended in shallow waters.

Horizontal textural gradients have not been recognized in any of these sandstones, and they are sufficiently fine to have been carried by the wind. The grains generally are angular, but this may not be significant since they are very small. The ultimate source or sources of the sand are uncertain.

Limestones of the Delaware basin vary somewhat in detail, but they have many characteristics in common. Several miles from the rim of the basin they are very fine-grained to aphanitic, and dark gray to black. Small voids locally contain free petroleum, and when broken the rock generally emits an odor of hydrogen sulfide and petroleum. Stratification is in beds a few inches thick, and the beds are delicately laminated. The rock consists of very fine grains of calcite, probably mainly detrital, and many scattered fine detrital grains of quartz.

Some horizons are replete with entire shells of ammonoids representing complete growth series. Microscopic examination shows that the rock is crowded with juvenile shells ranging upward in size from protoconchs. Scattered *in situ* shells of the brachiopod *Leiorhynchus* occur in these rocks, but they are uncommon. Thin sections of the fine-grained black limestones show that they contain a large proportion, up to 40 per cent or more by volume, of small spicules of siliceous sponges. Monaxons are prevalent, and all are characterized by an axial canal. Specimens from some localities can be freed readily from the calcareous matrix by dilute acid, but commonly the spicules, presumably composed originally of opal, have been replaced by calcite, so that the rocks have become less siliceous than at time of deposition. Radiolarians and algal spheres are also common constituents, but teeth and bones of fishes are exceedingly rare. In older rocks (Bone Spring) this phase also contains scattered *in situ* nuculoid pelecypods and small gastropods.

Not all the black limestone beds are delicately laminated. Probably the bottom was at times populated by soft-bodied scavengers, but if so they have left almost no record in burrows and trails. Scattered brachiopods (*Leiorhynchus*) and nuculoid pelecypods retaining both valves indicate that the bottom was habitable at times. The evidence of the ubiquitous sponge spicules is equivocal, however, because the spicules are dissociated. They have undergone sorting and some transportation.

Black, sandy limestones form most of the Bone Spring formation. The Brushy Canyon formation does not contain beds of this type, but the higher Cherry Canyon and Bell Canyon formations contain several limestone members which extend as thin tongues from the basin rim far out into the basin.

Toward the margin of the basin the limestones gradually become coarser-grained and lighter-colored. The transition is not uniform. At some horizons tongues of coarse shell detritus and limestone fragments extend several miles into the basin, but at

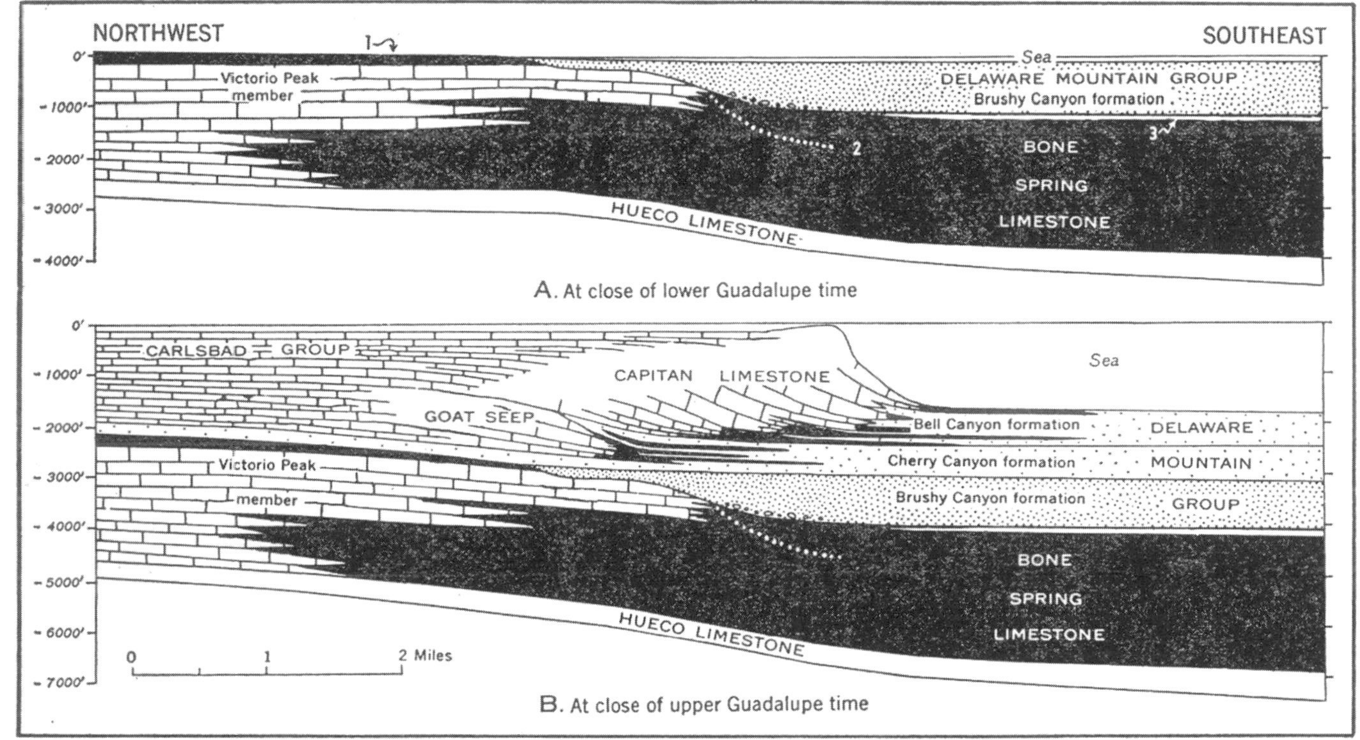

FIGURE 3.—*Reconstructed profiles through El Capitan showing inferred structural and stratigraphic relations following deposition of Brushy Canyon formation (A), and following deposition of the Capitan reef (B)*

Vertical exaggeration ×2. 1, Cutoff member of Bone Spring limestone; 2, Base of Cutoff member in Delaware basin; 3, Shale with *Waagenoceras* forming basal member of Brushy Canyon formation. Modified from P. B. King (1948)

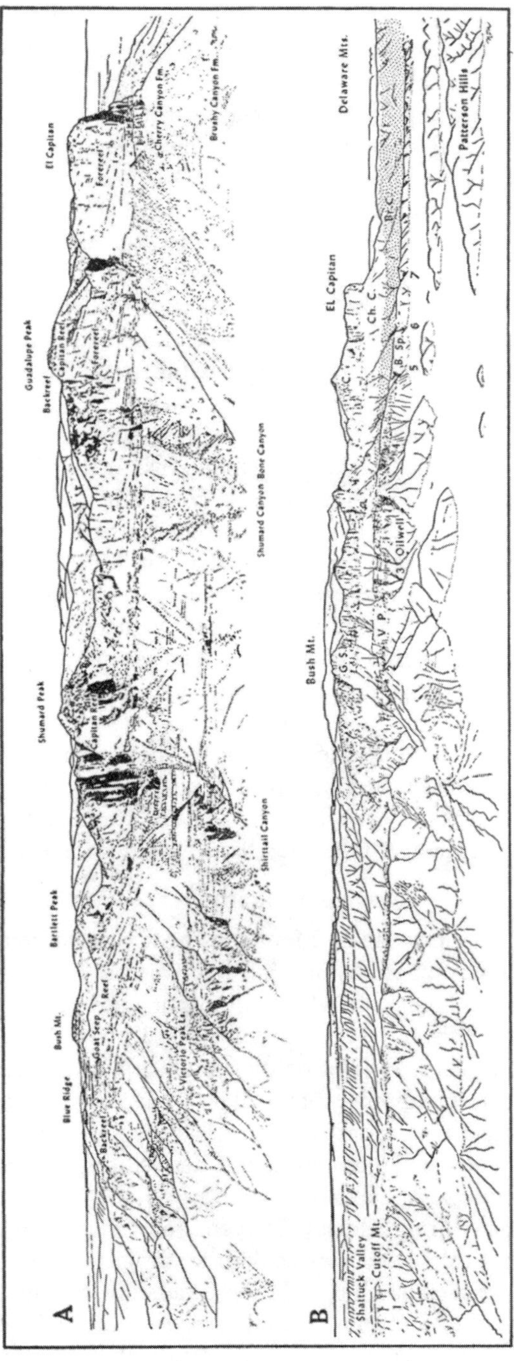

FIGURE 4.—*Panoramas of west side of the Guadalupe Mountains*

Drawn on aerial photographs; taken by Newell (A) and Muldrow (B). Stippled area in B corresponds to the outcrop of the Brushy Canyon formation.

others aphanitic black limestones reach within a quarter of a mile or less of the reef (Fig. 3). Nevertheless, there is a general coarsening of the rock toward the basin rim.

The fossils of these detrital beds show indications of transportation. The smallest and lightest shells have been transported the greatest distance with the least damage, and there has been a great deal of selective sorting. This observation accords well with experimental conclusions of Menard and Boucot (1951).

Fusulines have been transported relatively far basinward. They are associated with detrital rocks and show signs of extensive wear. Toward the margin of the basin unbroken fossils are relatively numerous, but cemented forms, such as *Prorichthofenia* and *Leptodus*, rarely are represented in the detrital fans. Attached forms are abundant in the reefs but rarely are found elsewhere. Lime-secreting organisms were abundant and varied at the shelf margin, and much of the detritus spread from the rim into the basin consists of shells of shallow-water forms.

Submarine talus fans flank the barrier reefs around the basin and form a large part of the massive limestones of the Guadalupe Mountains. Inclined massive beds of limestone dipping at 35° or more toward the basin have long attracted the attention of geologists (Fig. 4A). Lloyd (1929, p. 645–658) recognized that the inclined bedding planes are the detrital slopes of the seaward side of a reef. Most investigators apparently have regarded the talus deposits as an essential part of the reef. Indeed these rocks are frequently considered to represent the most characteristic expression of the reef limestone. However, the detrital deposits shed by a reef hardly can be considered part of the reef any more than arkose fans are part of the granite masses from which they are derived.

The forereef talus beds were formed of loose blocks, calcareous sand, and mud derived mainly from wave erosion and slump of the advancing reef front. Possibly lagoonal sediments were at times brought into the basin through reef channels, and some of the finer matrix of the reef talus may have had this origin.

REEF AND BANK PHASE

The Permian reefs of the Guadalupe Mountains area were at time of deposition solid masses of limestone surrounded by unlithified sediments. Thus they are not comparable to bars and channel fillings of shell detritus sometimes classed as "bioherms." Structures resulting from differential compaction are well shown in rocks which surround the smaller reefs (Figs. 5, 6). Not only have underlying strata been squeezed aside and truncated but steeply dipping marginal beds have been somewhat deformed by lateral shift as the rigid masses settled into relatively soft strata below. All the reefs are associated with lenses and aprons of reef-derived clastic detritus composed of angular fragments and reworked fossils apparently derived entirely from the reefs.

Small reefs in the Bone Spring, Brushy Canyon, and Cherry Canyon formations are only a few tens of feet thick, and they could have projected very little above the sea bottom (Fig. 6). Most such patch reefs in the Guadalupe Mountains area are associated with relatively small and thin detrital fans. In some examples (Fig. 6) the reef-derived detritus is limited to local lenses of reef-derived angular pebbles which blanket the surface of the reef. These patch reefs probably developed at depths reached only rarely by storm waves. The barrier reefs, exemplified by the Goat Seep

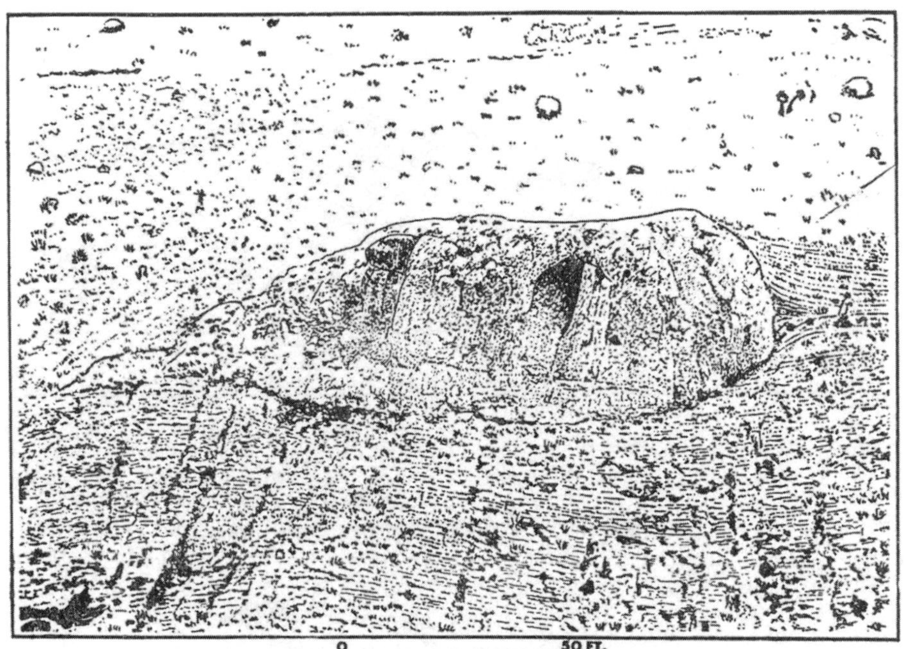

FIGURE 5.—*Small reef at the mouth of Bone Canyon*

The grassy slope in the upper half of the view is underlain by Cutoff shale. Truncation of limestone beds below the reef and the steep inclination of beds at the right are results of differential compaction and lateral shift of the reef before lithification of the surrounding rocks.

and Capitan reefs, shed enormous talus fans on the seaward flank. Forward growth near sea level carried the migrating reef front many miles into the Delaware basin across older talus accumulations (Figs. 3, 4).

The Capitan reef, at the outcrop in the Guadalupe Mountains, is approximately 2½ miles wide. The massive limestones that form the Capitan formation are 1500–2000 feet thick measured vertically (Fig. 3). But the lower two-thirds to half of this interval is composed of inclined beds of reef talus. There is evidently a similar relationship between reef and reef talus in the Goat Seep formation, but primary features in this unit are poorly preserved and have been modified greatly by dolomitization. Henson (1950) has shown that carbonate deposits genetically related to reef deposits form a distinctive association of unlike facies which he has termed the "reef complex". Organic deposits which expand against wave erosion to form marine barriers were referred to as "reef-walls" (p. 227). According to this terminology the Capitan formation consists of reef talus below and reef wall above (Adams and Frenzel, 1950). "Reef wall" is objectionable in this sense because it implies topographic features that commonly are not visible in fossil reefs. This is the organic reef from which the talus deposits were formed by wave erosion. Most students of living reefs agree that basically they are *in situ* accumulations of calcareous organic debris which characteristically are cemented in place as they are deposited. According to the interpretation herein

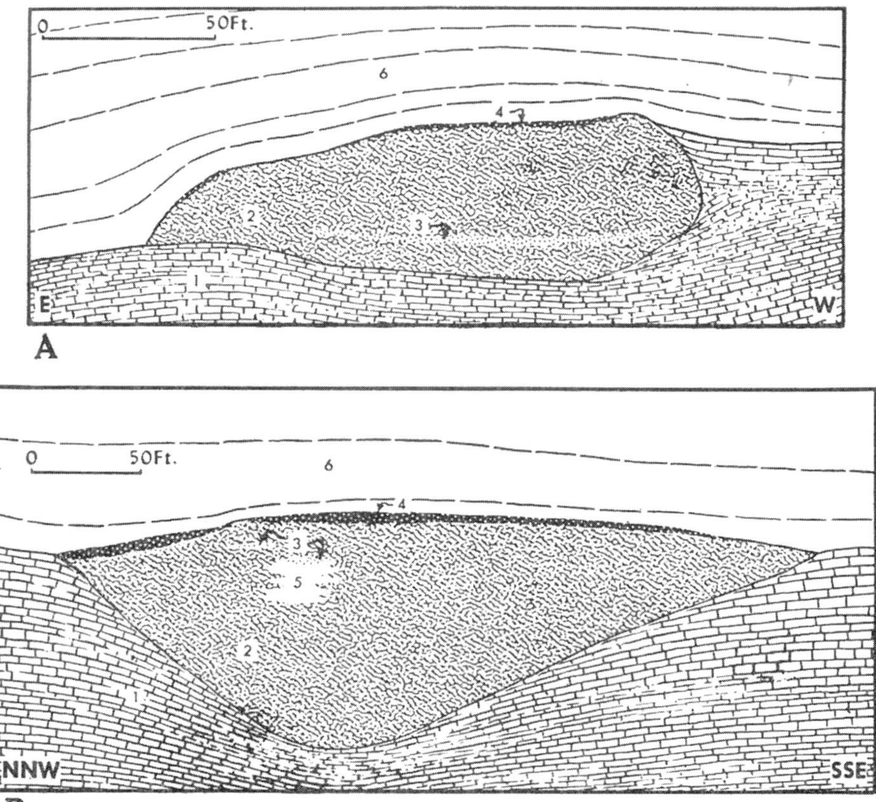

FIGURE 6.—*Diagrammatic cross sections through small Bone Spring reef at the mouth of Bone Canyon*

1, Black thin-bedded limestone; 2, Unbedded granular gray limestone containing *in situ* fossils; 3, Calcareous quartz sandstone; 4, Brecciated limestone containing fusulines; 5, Semilithographic, light-gray limestone containing *in situ* fossils; 6, Bituminous black shale and even-bedded black limestone of Cutoff member. The two outcrops probably represent a single elongate reef of very low relief. Pinching and warping of beds have been caused by differential compaction.

followed the Capitan formation consists of reef-derived talus in the lower part and reef limestone above.

The reef limestone of the Capitan, unlike the talus and backreef deposits, is composed mainly of calcitic limestone and is unstratified. The reef talus characteristically contains an appreciable percentage of magnesium, and the backreef carbonate beds are dolomitic throughout. In hand specimens much of the reef consists of semilithographic, light-gray, cream-colored or white limestone containing well-preserved fossils and innumerable openings, most of which have been partly or wholly filled by concentric layers of fibrous calcite which closely resemble cavity fillings described by Cullis in the Funafuti cores (Cullis, 1904). Many of the cavities were lined with layers of fibrous calcite and were later filled with fine quartz sand like that of the shelf and basin deposits. From this it is evident that deposition of the fibrous calcite (perhaps aragonite originally) occurred while in contact with the sea. Quartz sand is fairly common throughout the reef in small pockets and sandstone dikes ranging in thickness from a fraction of an inch to 1 foot. There is no chert in the reef, and fossils there have not been silicified. Pockets, lenses, and fissure fillings of detrital calcium carbonate sand are abundant, but much of the detrital material has undergone recrystallization. Many of the most characteristic features of the reef limestone result from diagenetic changes before and after burial.

Diagenetic modification and consequent loss of primary textures is highly characteristic of fossil reefs and the old parts of living reefs. Very probably the porous, open texture inherent in the mode of deposition in the reefs is responsible for the lack of stratification and provides avenues after burial for circulation of waters. Primary voids in reefs occupy the space between the skeletal structures of incrusting reef organisms. Commonly those crevices and pits serve as traps for detrital sand and mud which become incorporated in the substance of the reef; but openings free from loose sediment also are sealed by the lateral spread of incrusting organisms. Furthermore, many skeletal types are themselves quite porous, so that when formed the rock of reefs is characteristically spongy or cavernous. Living reefs tend to grow upward until they reach the effective ceiling of growth at mean sea level, and further growth is principally at the seaward front of the reef where optimum conditions of temperature, aeration, and nutrition are found. A flat reef surface is thus formed which periodically is exposed to atmospheric agencies. Boring and burrowing organisms find refuge in the rock of modern reefs in such great numbers that probably more organisms live within the limestone than upon the depositional surface. Worms, mollusks, filamentous algae, and even bacteria bore deeply into the rock, and all these increase the permeability.

Well-defined animal burrows are rare in the Capitan reef and are limited to scattered microscopic borings in shells. There is no evidence that the mollusks, echinoids, and larger worms played an important role as they do today in perforating the rocks at the surface of the reef. Smaller organisms, however, may have been active in limestone destruction.

Semilithographic limestone is a characteristic rock type in fossil reefs in many parts of the world. Dolomitization of limestones commonly results in partial or complete destruction of primary textures. Characteristically such rocks are of uniform crystal size. In some cases faint outlines of fossils demonstrate that the rock texture is considerably finer than the coarser elements of the original sediment. Hadding (1950, p. 403) described semilithographic reef limestone from the Silurian of Gotland. He considered this material to represent lithified pure calcium carbonate mud until he discovered examples that show traces of delicate lamination which he considers to be of organic origin. The primary organic structures generally are obliterated by recrystallization, and the secondary texture is exceedingly fine. It has been shown by Cullis (1904, p. 398) that recrystallization of calcium carbonate mud does not necessarily result in a coarsening of the texture.

Walther (1885) was interested in this problem and concluded that through recrystallization under certain conditions algal limestones may become much finer in texture than the original sediments. He described Tertiary algal limestones, at Syracuse, similar to living *Lithothamnion* banks (*seccas*) in the Bay of Naples. The Tertiary algal nodules are loosely bound in a friable matrix, and they are well preserved in microscopic structure. In near-by areas the algal nodules lose the structure, and the containing rocks pass horizontally into semilithographic structureless limestones. The lateral modification was mainly a result of recrystallization of the limestone. The Triassic Dachstein of the Alps is a semilithographic dolomite containing fossil forms which Walther believed could not have lived in a soft mud environment. He suggests that organic matter contained in the sediment may have been responsible for recrystallization and consequent destruction of some of the original sedimentary textures, but dolomitization may also have destroyed all primary textures. Crystal elements of invertebrate and algal structures are very small. Recrystallization of these elements with or without moderate increase in crystal dimensions may result in textures much finer than those of the original sediment.

Not all the Capitan reef limestone is semilithographic. Extensive areas of outcrop are characterized by slightly coarser textures. Thin sections of this rock reveal detrital texture or a crystalline mosaic produced by recrystallization.

Probably the most essential organisms in the building of reefs are the incrusting forms which bind and cement loose material into a solid reef frame. Among living forms two groups, red coralline algae (nullipores) and stony corals, play leading roles. Corals, however, played a very minor role in the Permian. Red algae (*Solenopora*) are represented as far back as the lower Ordovician, but they were not primarily responsible for formation of reefs during the Paleozoic. It is clear that reef organisms of the Paleozoic were very unlike those of the present. Yet the formation of the reefs undoubtedly has always been governed by those conditions which permit high rates of precipitation of calcium carbonate from sea water by organisms.

The Silurian reefs of Gotland contain abundant stromatoporoids, corals, and bryozoans, but the algae probably were the most important as frame builders (Hadding, 1950). Of the latter a laminar incrusting deposit similar to *Spongiostroma* is quantitatively most important. This type usually is regarded as deposits of low forms of blue-green algae.

Stromatoporoids, tabulate corals, and schizocorals are important frame builders and detritus traps in Silurian reefs in Illinois (Lowenstam, 1950, p. 444). The most important frame builder according to Lowenstam (p. 439) is a problematical structure *Stromatactis*, originally described from the Devonian of Belgium. He observes that this looks more like deposits of algae than stromatoporoids, although calcareous algae have not been established definitely in these reefs (p. 444).

Many well-defined structures of calcareous algac have been recognized by Johnson (1942, 1951) in Permian limestones of West Texas. These include red algae (*Solenopora*) distantly related to living nullipores, green algae of the families Dasycladaceae (*Athracoporella*, *Diplopora*, *Macroporella*, *Mizzia*) and Codiaceae (*Gymnocodium*); blue-green or possibly green algae of the Porostromata (*Girvanella*, *Ortonella*), and blue-green algae of the Spongiostroma (*Colenella*). Probably not all the reef algae have been recognized because most of Johnson's material was collected in the back-reef facies. We have observed *Solenopora* only in the reef facies where it is invariably associated with *Acanthocladia* (a bryozoan), but it is not quantitatively important. The colonies are small, and probably they were not important frame builders. *Cole-*

nella forms large masses in the reef limestone. It was at least locally important as rigid frame builder and sediment trap. This fossil is commonly more recrystallized than associated brachiopods and sponges and in places is barely distinguishable from the matrix. Other thin-laminated deposits are abundant through the reef. They are not related to primary cavities and they are not fibrous. These structures probably were laid down as rigid incrustations formed by blue-green algae.

Other genera, mainly siphonous algae, are abundant in the backreef rocks immediately behind the Capitan reef, but they are rare in the reef, where possibly they were trapped occasionally only as detrital fragments.

In addition to algae, there were three supplementary groups of frame building organisms: sycon (calcareous) sponges, fistuliporoid bryozoans, and a problematical incrusting and stoloniferous hydrocoralline.

The calcareous sponges are exceedingly abundant in most reefs. Commonly they are found crowded in upright position surrounded by detrital material. They were rigid and effective traps for loose sand. Representative sponge genera are *Monarchopemmatites*, *Cystothalamia*, *Cystauletes*, *Polyphymaspongia*, *Guadalupia*, *Steinmannia*, *Amblysiphonella*, and *Girtyocoelia*. Less conspicuous, but nevertheless very abundant, is the hydrocoralline referred to above. Superficially it has somewhat the general expression of *Solenopora*, forming lamellar expansions, rounded tubercles, and cylindrical stolons. Internally the fossil is nearly structureless except for erratic tubular perforations, the work probably of some boring organism. It is easily distinguished in thin section from *Solenopora* by greater opacity, a feature that makes this problematical form conspicuous even as detrital fragments.

Small, unbranched colonies of the bryozoan *Fistulipora* are abundant in some of the patch reefs. They are of unusual external form. They are conical to subcylindrical, half an inch to an inch in diameter and 1–2 inches in height. The base and sides of the zoaria are covered by an extensive epitheca.

Bryozoa in growth position referred provisionally by Girty to *Domopora* form a tangled mat in parts of the Capitan reef. Representatives of *Acanthocladia* and fenestrate forms are abundant but somewhat less important as rock formers. Unquestionably the bryozoans were efficient traps for calcium carbonate sand and they most commonly occur in a matrix of calcarenite. It is doubtful, however, that these forms were resistant to strong waves, and they must have been very fragile. As a matter of fact they characteristically are rare where the sponges are common. Bryozoan remains are extremely rare as detrital fragments in the backreef deposits, but they are scattered abundantly throughout the reef-talus deposits. These facts suggest that fragile types of bryozoans were mainly limited to the deeper, less-agitated waters. Probably optimum environmental conditions were limited to relatively quiet waters some tens of fathoms deep at the base of the reef front. Probably the range of the bryozoans and some of the other reef dwellers extended over the upper part of the talus slope. Cementation by bryozoans and other organisms contemporaneously with deposition might have been responsible for the steepness of the upper slopes of the talus beds.

The reef limestones are rich in brachiopods, but mollusks are subordinate. Corals and echinoderm plates are common but are not important rock formers. Fusulines of the genera *Codonofusiella*, *Schubertella*, *Parafusulina*, and *Polydiexodina* are abundantly represented. In some cases (Fig. 6) they are profuse in the reef facies and lacking in neighboring contemporaneous basin facies. Presumably the fusulines led a free benthonic existence, and probably they were readily rolled about by currents and waves. Concentrations of fusulines at the top of some of the small reefs (Fig. 6) suggest deposition in quiet waters. However, the Capitan reef front, which evidently was formed in a surf zone, contains scattered concentrations of *Polydiexodina*. It is difficult to understand how these giant foraminifers were able to cling to the reef surface. Perhaps they were trapped in pits on the reef flat. In any case this may not have been the preferred environment for these animals because their tests are much more abundant in the proximal part of the backreef beds.

Bank deposits of carbonate rocks are well illustrated by the Victorio Peak member of the Bone Spring formation (Fig. 3). They consist mainly of granular dolomites and subordinate amounts of limestone in very thick even beds. Reworked crinoid fragments occur in vast numbers, constituting a large proportion of some beds. Fusulines (*Parafusulina*) are only slightly less abundant at a few horizons. This bank deposit grades imperceptibly into characteristicly thinner-bedded and finer-grained shelf deposits 5 miles or so behind the rim of the Delaware basin. Although the bank deposits commonly are thoroughly dolomitized so that primary textures are largely obliterated, it is clear that a large proportion of these rocks consisted of detrital shell sand.

At the edge of the Delaware basin the thick-bedded bank dolomites are abruptly replaced by thin-bedded black limestones of the basin phase. There are no talus aprons extending into the basin and no other indications that the bank deposits were lithified, wave-resistant masses at the time of deposition. Localization of the Victorio Peak bank around the rim of the Delaware basin presumably resulted from the same factors that made this the site of extensive reef growth at later stages. The bank deposits are essentially the result of acceleration of shelf-type deposition in the favorable belt at the edge of the Delaware basin. They are not reefs.

SHELF PHASE

Small patch reefs in the Bone Spring formation lie in a narrow belt at the rim of the Delaware basin, but these reefs do not clearly separate bank deposits from basin deposits. The reefs generally are surrounded by black limestones of the basin phase, and the latter interfinger with the bank phase. Thin beds of quartz sandstone wedge out toward the rim of the basin marking the location of a structural arch (Bone Spring flexure) along the rim. This arch was partly responsible for the abrupt and striking changes in facies and the localization of reefs and banks at the rim of the Delaware basin during late Bone Spring time.

Barrier development began with the formation of the Goat Seep reef in about the same position as the earlier Victorio Peak bank. Circulation between basin and shelf was inhibited and restricted probably to reef channels. Barrier-reef growth probably was interrupted briefly at the end of deposition of the Goat Seep reef (end of South Wells time), as there is a hiatus representing latest Cherry Canyon time between this and the Capitan reef (Fig. 3).

Reef limestones pass abruptly into backreef dolomitic limestones (Fig. 7). Within

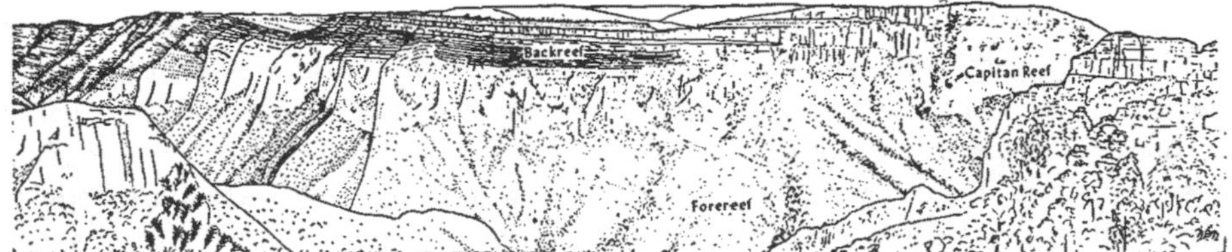

FIGURE 7.—*View northeastward across Big Canyon*
Showing relationships of backreef and forereef facies to the reef

a few tens of yards striking changes occur in mineralogical composition, texture, primary structure, and fossil content. The rocks near the reef are composed dominantly of reef-derived detritus intermingled with enormous quantities of siphonous algae (Dasycladaceae and Codiaceae) and fusulines of many genera. This facies ranges in breadth from a few feet to about 1 mile. The majority of fossils of the backreef phase occur in these rocks. A few species of reef brachiopods and a calcareous sponge have been identified a few feet behind the reef. Probably these were washed into the lagoon from the reef. Echinoderms, bryozoans, and corals are rare in this facies and unknown at most horizons. A few species of gastropods (*Naticopsis* and *Bellerophon*) and a species or so of scaphopods are characteristic elements of the fauna. The rock of this facies is distinguished from the reef by the detrital texture, even stratification in thick beds, and dolomitic composition. Replacing this facies lagoonward are thin beds of pisolite. Commonly they are not fossiliferous, although fossil fragments form the nuclei in some examples. Pisolite beds a few inches thick extend almost to the reef, but rocks of this facies are most abundant about half a mile to a mile from the reef where they constitute an appreciable part of the entire sequence. The pisolites are considered by Johnson to have been formed through the activities of primitive blue-green algae (Johnson, 1942), but Pia (1940) regarded them as inorganic concretions. Evidence for organic origin is the nonfibrous structure of the lamellae and included layers of quartz sand. The sand may have been entrapped in organic fibers and slime at the growing surface. It is difficult to understand how quartz sand grains might become incorporated in inorganic pisoliths. The pisoliths commonly contain both calcite and dolomite.

Beyond the calcarenites and pisolites the carbonate rocks are composed dominantly of even, relatively thin beds of fine-grained, homogeneous dolomite. Thin tongues of this facies extend almost to the Capitan reef, especially in the younger beds, but this rock type becomes prevalent about 1–2 miles behind the reef. The fine-grained dolomites contain very few recognizable fossils. Poorly preserved and rare pelecypods and gastropods indicate that the general lack of fossils may be a result of complete recrystallization. On the other hand, circumstantial evidence of the facies relationships suggests that the rock may have been formed of calcium carbonate mud precipitated inorganically in waters of elevated salinity, somewhat comparable to conditions west of Andros Island on the Great Bahama Bank (Newell *et al.*, 1951). Original textures, however, have no doubt been completely altered by diagenetic dolomitization. It is doubtful that these rocks are a result of primary precipitation of dolomite from hypersaline sea water.

Thin wedges of quartz sandstone are intercalated within the bedded carbonate rocks of the backreef phase, and many extend nearly to the Capitan reef. Like the sandstone tongues of the Delaware basin, they must indicate times when the contribution of terrigenous sediments greatly exceeded deposition of detrital calcium carbonate. Probably some of the sandstone beds of the backreef area correspond in age with those of the marginal area of the Delaware basin; there are more beds of sandstone in the backreef area than in the basin, therefore simple matching of sandstone beds is not practicable. Each bed of sandstone may correspond to a time of reduced rate of reef growth, or these beds may record times of accelerated rate of supply of quartz sand into the region. The former possibly is favored here because (1) The sediments in the central part of the Delaware basin are composed chiefly of quartz sand, and all the carbonate rocks contain an appreciable proportion of the same material. A fairly constant supply of sand is indicated. (2) The sandstone beds, although very rich in carbonate near the reef, rarely contain conspicuous reef fragments and shells. Apparently the reef contributed very little detrital material, most of it very fine, during deposition of the most extensive basin sandstone units.

At distances behind the reef ranging from 6 to 20 miles, many of the fine-grained dolomite beds are replaced by thin tongues of anhydrite and fine red sandstone.

PHYSICAL CONDITIONS DURING PERMIAN SEDIMENTATION

CLIMATE

Maritime climates at the latitude of the Delaware basin (31°–32° N. Lat.) are now warm-temperate to subtropical and are generally mild, depending mainly on the character of oceanic circulation. The inland seas of southeastern Europe enjoy a mild climate at even higher latitudes. The mean temperature of southwestern United States may have been somewhat higher during the Permian than at present. Shallow epicontinental seas covered much of the continent of North America, and the land areas evidently were much lower, on the average, than now.

There are, however, more direct indications of prevailingly high mean temperatures during the Permian. Extensive evaporite deposits are scattered throughout much of the Permian sequence from Kansas to New Mexico and West Texas. Lang (1937) has shown that quantitatively important deposition of evaporites began in the early Leonardian (Wellington) in Kansas, and during successive stages the center of greatest accumulation of evaporites migrated toward the southwest until the final stages of withdrawal of the sea resulted in deposition of evaporites over the area of the Delaware basin (Fig. 8). The quantity of normal sea water which must be evaporated to

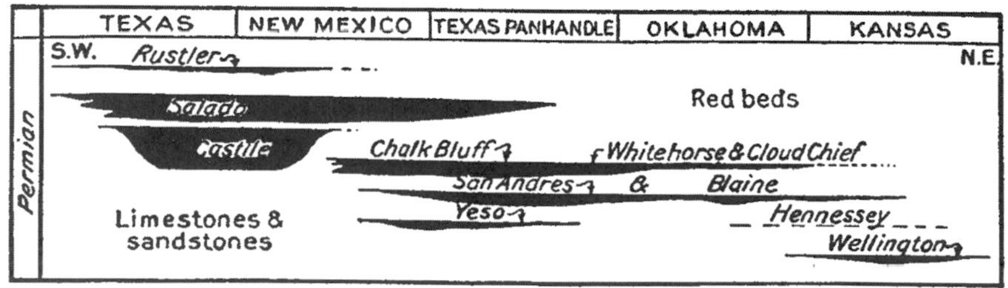

FIGURE 8.—*Schematic diagram of the Permian basin from Kansas to West Texas*

Showing the shift southwestward of the basin of saline deposition during Permian time. Evaporite deposits shown by black. After W. B. Lang (1937, p. 885)

produce these products would be very great. Anhydrite is common throughout the Permian succession in central New Mexico, and gypsiferous beds are found at several horizons in the Permian of Arizona (McKee, 1938) and southeastern Wyoming (Thomas, 1934).

The extensive distribution of Permian evaporites indicates regional rather than local causes. Evidently prevailing rates of evaporation exceeded rates of replenishment. Excessive evaporation over so large an area suggests low annual precipitation and relatively high temperatures, hence aridity over much of the region. On the other hand there is some evidence that the eastern shore of the Permian seaway in central Texas received more moisture; or perhaps westward-flowing streams built well-watered deltas in that area. (See paleogeographic maps by Hills, 1942.) Locally abundant fossil amphibians there in Leonardian (Wichita and Clear Fork) red beds indicate availability of ample fresh waters.

Some of the invertebrate fossil faunas of West Texas are more closely related to Permian faunas of the Mediterranean region than to those, for example, of Idaho. Permian faunas of West Texas, the Mediterranean region, Asia Minor, south Asia, Indonesia, and Japan have much in common. They are characterized by rich and varied faunas, including especially large fusulines and distinctive brachiopods. Most of the aberrant brachiopod types, such as leptodids and richthofenids, are limited to this belt although some examples range as far north as the southern Urals. This includes the great Permian Tethyan province, the known range of which is between about 45° N. Lat. and the equator. On the other hand the Permian faunas of northern Europe and the Arctic islands form quite a different, boreal facies.

The unlike geographic distributions of these faunal types in the northern hemisphere suggest climatic zonation roughly parallel to the modern equator. The Permian of West Texas and Mexico lies in what may be termed a Permian tropical zone. Lack of well-defined latitudinal zonation in the boreal faunas of high latitudes, on the other hand, suggests prevailing mild climates well into arctic regions. Permian faunas of the southern hemisphere are not particularly illuminating with respect to climatic zonation, perhaps because they are not well known.

Converging lines of evidence, incomplete as they are, suggest that the Permian climate of the West Texas region was arid, more equable, and probably warmer than at present.

PERMIAN TOPOGRAPHY

Prevailing characteristics of the Permian rocks over much of the southwestern United States suggest that most of the terrigenous sediments were derived from distant sources. Local uplands, such as the Amarillo Mountains, Llanoria, and the Ancestral Rockies, were nearly or completely buried in sediments before the close of the Permian, and at no time did they appreciably affect the quantity or character of terrigenous sediments entering the region. These sediments consist for the most part of very fine- to medium-grained angular arkosic sand of fresh appearance. There is an astonishingly small quantity of clay minerals in the West Texas area. Accessory minerals in the sands are microcline, plagioclase, and minor quantities of zircon, tourmaline, and apatite, all of which suggest weathering under arid conditions.

The fine sand was deposited evenly over shelf and basin areas, and only during certain times was it concentrated especially in the basins.

In the Delaware basin area the maximum dimension of terrigenous sand grains rarely exceeds about 0.5 mm and more commonly is less than 0.2 mm. Very little of the sand is finer than "very fine" sand in the Wentworth scale (1/8 to 1/16 mm). The sand is so uniform, without recognizable textural or mineralogical gradients, as to suggest a single mode of sorting and constant supply. Lang (1937, p. 888), who was unable to distinguish stratigraphic horizons by means of accessory minerals in the sandstones, suggested (p. 889) a source in the eastern part of the continent.

The Permian rocks to the north and northeast of the Delaware basin are dominantly red beds. Presumably the oxidation of iron responsible for the red color was acquired under more humid conditions before or during transit. Sediments deposited above sea level over broad fluviatile plains to the north and northeast have retained the red color. The terrigenous sediments that reached the reducing submarine environments of the Delaware basin and vicinity lost the red colors and are now shades of tan, gray, and black. Tongues of reddish sandstone thin out and change color toward the rim of the basin, but 25 miles or so away nearly all the shelf sandstones are reddish.

These lithologic characteristics indicate that the subaerial topography of the entire region was exceedingly flat. Hills of pre-Permian rocks undoubtedly rose here and there above the general depositional surface (Hills, 1942), they had only a local effect on sedimentation.

Submarine topography was pronounced. Nearly all students of the region are in essential agreement that the Delaware basin contained relatively deep waters which overflowed the surrounding flat shelf.

By analogy with modern examples it is judged that the reefs grew in shallow waters at or near sea level. Estimates of the maximum depths of the Delaware basin range from about 500 feet by P. B. King (1934, p. 788) to 2400 feet by Adams (1936, p. 789). Kroenlein (1939, p. 1684) thought that the waters near the reef might have been 1800–2000 feet deep. King (1948, p. 85) later revised his estimate to a depth of 1000 feet near the basin margin during the last stages of reef growth, but Adams and Frenzel (1950, p. 310) think that 1800 feet is a more accurate estimate of the depth.

Estimates of the depth of the sea in the Delaware basin are made by tracing an inclined bedding plane from the reef front to a point where stratification is nearly horizontal in front of the reef escarpment. This local structural relief along the reef front measures approximately 1800–2000 feet at the top of the Capitanian beds in the Guadalupe Mountains. Adams and Frenzel (1950, p. 310) think that the observed structure is essentially the same as the original Permian topography and has not been been significantly modified by differential subsidence or compaction subsequent to deposition.

These beds of reef talus dip basinward as much as 35°–40° at the top, steeper than would be expected for the repose angle of *unconsolidated* subaqueous sediments. Lagoonal rocks for a few hundred feet immediately behind the reef tend to dip toward the Delaware basin at angles up to about 10°. The lithologic character of these back-reef beds indicates that they were formed largely of detrital material swept off the top of the reef, and they surely were deposited nearly horizontally, probably sloping gently away from the reef source. Basinward tilting must have progressively affected the near-reef part of the lagoon during reef growth. Strata of the lagoonal area a few hundred yards behind the reef are not bent downward toward the basin. It is obvious that the downwarp is a local phenomenon related to the reef itself and not due to differential subsidence of the basin. During reef growth massive limestone deposits were built horizontally over water-laden basin sediments (Fig. 3). Increments of load at the reef front probably produced a narrow migrating belt of subsidence as unstable talus deposits beneath the newly formed part of the reef were undergoing adjustment by slump and loss of water. There has been appreciable progressive basinward tilting along the reef during deposition, and this must be considered in estimating the original relief of the submarine topography of the area as King has done. Estimates made in this way, by assuming that the backreef beds were essentially horizontal at time of deposition, indicate that the Capitan reef during final stages of growth stood approximately 2000 feet above the neighboring floor of the basin.

Probably the basin floor sloped gently southeast since well records indicate slight thickening of nearly all units toward the deeper part of the basin (R. E. King, 1942, Pl. 2). The structural relief and inclination of beds of reef talus increase at successive horizons in the reef system; therefore, it may be concluded that the waters of the Delaware basin progressively deepened during reef growth. The characteristics of a large proportion of the basin sediments suggest deposition below wave base in quiet waters.

Depths of the shelf seas are conjectural. Individual bedding planes at some horizons can be traced for several miles shoreward from the reef. Between bedding planes the rocks are unstratified, suggesting deposition in agitated waters. Beds of dolomitic limestone thicken rapidly before they pass into reef facies, and interbedded quartz sandstones thin out before reaching the reef. These facts considered with the petrology of the rocks indicate that the carbonate sediments near the reef were detrital aprons derived mainly from the reef. A unit near the middle of the Yates formation thins rapidly in the first mile from the reef beyond which the bedding planes are approximately parallel. On the assumption that the bottom of the unit was horizontal during the final stages of deposition, it would appear reasonable to conclude that at a distance of 1 mile from the reef the sea was 100 feet deeper than it was over the reef. The differential basinward tilt of beds near the reef, however, discussed above, complicates this problem and casts doubts on any estimate. The sediments and contained fossils indicate deposition in well-aerated waters. Presumably the waters of the shelf seas were commonly much less than 100 feet deep, perhaps in places no more than 20 feet deep.

HYDROGRAPHY OF THE DELAWARE BASIN

Deep lakes and marine basins develop their own characteristic stratification and circulation at various depths. Marked fluctuations in annual temperatures such as are characteristic of intermediate latitudes result in convection in waters of different densities. Persistent winds and occasional storms produce surface effects which in turn may result in circulation at depths.

Hydrographic conditions in the Mediterranean Sea are instructive. Summer loss by evaporation greatly exceeds addition by precipitation so that the salinity of surface waters becomes abnormally high. During the winter this hypersaline surface water is cooled until it becomes heavier than the waters immediately below. In unusually cold winters the heavy surface water sinks into the deepest basins of the Mediterranean, some of which are more than 2000 fathoms below sea level. Usually, however displacement of lighter waters causes a vertical mixing so that a homogeneous stratum of warm water of high salinity is formed down to about 350 fathoms. In the summer a new layer of light water forms on the surface by the continuous influx of oceanic waters through the Straits of Gibraltar, and this layer is warmed by the summer sun. The heavy waters that sink to the bottom gradually escape across the threshold at the Straits of Gibraltar. Thus, a westward-flowing current of heavy water passes into the Atlantic along the bottom of the straits. The volume of this water is about 16 times the flow of the Mississippi River, and it is exceeded slightly by the surface inflow of light water through the straits.

The enormous quantity of evaporites in the middle and upper part of Permian deposits from Kansas to Texas has suggested to many that very high rates of evaporation over the shallow interior seas depressed the water surface appreciably below the level of the ocean, thus causing a continuous inflow of normal sea water through one or more narrow straits in the southwest. The Gulf of Karabugaz on the Caspian, cited by Ochsenius to illustrate his bar theory of the origin of evaporite deposits, affords a useful analogy with the seas of the Permian basin. The Gulf of Karabugaz on the east side of the Caspian is about 80 miles wide and 95 miles long. It is very shallow, not more than 45 feet deep, so that the water is heated throughout by the sun, and waves prevent marked density stratification. It is separated from the Caspian by two sand spits. The salinity is 164 parts per thousand as compared with 12–13 parts per thousand in the Caspian Sea. Evaporation in the gulf depresses the surface appreciably below that of the Caspian, so that a strong current constantly flows into the Gulf of Karabugaz. A delta has been built at the entrance, and gypsum is deposited in large quantities for some distance into the gulf. During the winter glauberite is deposited over a great area on the bottom and around the shores, but the salt is partly dissolved during the summer. Higher concentrations are required for precipitation of halite.

Observed geologic and paleontologic conditions in the Delaware basin are understandable by analogy with conditions in the Mediterranean and the Gulf of Karabugaz. The rocks and fossils of the deeper parts of the basin indicate generally stagnant conditions. Winter cooling may have resulted in effective mixing in the early history of the region when the difference in depth between the threshold and the basin was small (Fig. 9A). Before the barrier was formed around the basin rim heavy waters of the shelf no doubt flowed toward deeper waters and were replaced above by lighter waters.

Progressive deepening of the basin accompanied by reef development and restriction of circulation accentuated the environmental differences between the basin and shelf area. Influx of hypersaline waters of the shelf area into the basin was impeded and limited, probably, to times of protracted storms. Heavy waters would accumulate in the basin during stagnation until the thermocline rose to a level higher than that of the threshold (Fig. 9B) at which stage they would overflow the sill of the basin forming a counter current beneath the influx of normal waters. Salinities did not become

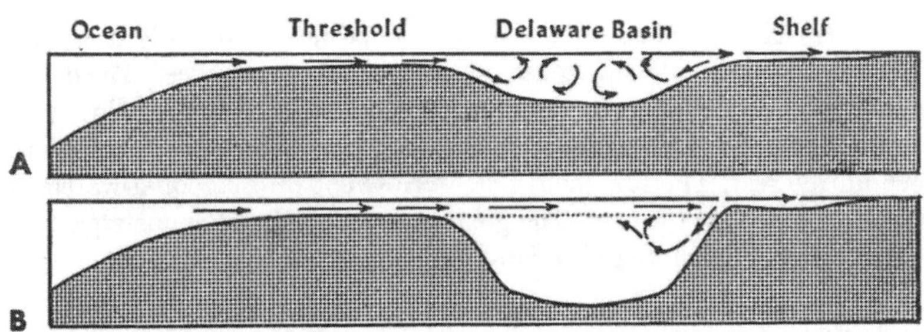

FIGURE 9.—*Theoretical hydrographic relations in the Delaware basin area during (A) early Permian (Wolfcampian, Leonardian) time and (B) during late Guadalupian time*

During early history (A) surface sea water sinks, made relatively heavy by evaporation and winter cooling. This results in convective overturn and intermittent ventilation of deep waters, and conditions favorable for partial stagnation and accumulation of organic sediments. During later history gradual accumulation of hypersaline waters below the level of the threshold favors complete stagnation in deeper waters. Storm turbulence and turbidity currents bring hydrogen sulfide into shallower waters causing mass mortality of ammonoids.

excessive in the basin, however, until withdrawal of the sea from the shelf area. The Delaware basin then abruptly became a center of precipitation of enormous deposits of anhydrite (Castile formation).

The thermocline may have been periodically disturbed by the annual turnover of shallow waters and by influx of heavy shelf waters through reef channels during storms. In this manner nutrient-charged waters would be made available to the phytoplankton in the sunlit waters near the surface. All life processes, and the deposition of organic sediments, and probably the formation of source materials of petroleum, depended directly on the productivity of these unicellular plants. It is not at all improbable that the characteristic lamination or banding of the basin sediments resulted directly from seasonal fluctuations in the quantity of organic matter which settled to the bottom of the basin.

MAJOR BIOTOPES

Interpretation of the paleoecology of the Permian marine deposits in the Guadalupe Mountains area must take into account the fact that many of the fossils occur in detrital deposits under circumstances that clearly indicate transportation from the original sites. The talus deposits which flank the Capitan reef on the seaward side extend as thin tongues for some miles from the reef source. Fossils in these deposits with few exceptions belong to species that also are found in the reef. Many of the fossils studied by Girty (1908; and *in* King, 1948) were collected in this detrital facies. The evidence indicates that most species recognized in these beds were reef dwellers. They did not dwell on the lower slopes of the talus.

There is progressive attrition of shells basinward, and this is accompanied by selective elimination of the more fragile and relatively heavy forms. The most securely attached reef forms such as the incrusting algae, sycon sponges, *Leptodus*, *Prorichthofenia*, and massive bryozoans are poorly represented in the detrital waste. Shells have been transported much farther basinward than lagoonward. This is in part a function of the relative relief on the two sides of the barrier reef, but probably it is related also to marked ecologic and biotic differences on the two sides of the reef. The detrital fans lagoonward from the reef contain very few species. Evidently wave velocity was generally insufficient to wash the shells of the larger animals across the reef flat into quiet water behind the reef. The few shells that were swept into lagoonal waters came to rest immediately behind the reef. The great quantities of siphonous algae and fusulines immediately back of the reef must be near the sites where they grew most abundantly. Related living algae are most common in quiet waters, and they can endure rather marked fluctuations in salinity. Some of the fusulines show effects of abrasion and fracture, but most are undamaged. Locally they are aligned roughly at right angles to the reef axis. Perhaps this was caused by wind-formed currents parallel to the reef. The cylindrical form of these fusulines probably rendered them highly mobile, and the distinctive shape indeed may be an adaptation for transportation by currents. Orientation of these forms in the reef is random, an indication that the reef dwellers were protected from the effects of currents; no doubt they were lodged with detrital shell sand in pits. Fusulines in rocks of the Delaware basin are in places rudely aligned, and the mean orientation commonly is at right angles to the dominant trend of the basin rim as shown by King. These facts suggest that bottom currents capable of rolling the fusulines were dominantly parallel with the basin rim. It is concluded that optimum conditions for the fusulines during Leonardian and Guadalupian time were limited to the shallow waters at and possibly just beyond the rim of the Delaware basin. Fusulines of the basin rocks characteristically are broken and abraded. Probably all were transported after death from shallow waters at the margin of the basin (Fig. 10).

Many of the fossils in the reef limestone are unbroken, and some are clearly in position of growth. Many of the sponges are erect in the rock. A large proportion of the brachiopods retain both valves, and the longest and most nearly complete bryozoan zoaria occur here. It must not be inferred, however, that complete fossils are the rule. Probably more than half the limestone of the reefs consists of material that was originally loose shell sand with minor proportions of calcium carbonate mud. The rigid frame of the reef, composed mainly of incrusting algae, sponges, and bryozoans, forms most of the remainder. The majority of brachiopods and mollusks are damaged, and only a few that may have lived in protected places are entire.

Quartz sandstones in the region generally are barren of fossils at all horizons. Worn fusulines are common at some horizons, and trails of wormlike animals are common at others, but they are exceptional. It is concluded that the sandy bottoms generally were not inhabited by forms capable of leaving fossil records. General conclusions regarding the major habitats are summarized in Figure 11.

BASIN PHASE

During Leonardian and Guadalupian times the deeper waters of the Delaware basin were poorly ventilated. At times the bottom must have been nearly devoid of organisms capable of fossilization. Nevertheless, benthonic brachiopods (*Leiorhynchus, Avonia*), pelecypods (nuculoids), and small gastropods did inhabit at least the marginal deep waters of the basin. The waters generally were too deep for bottom-

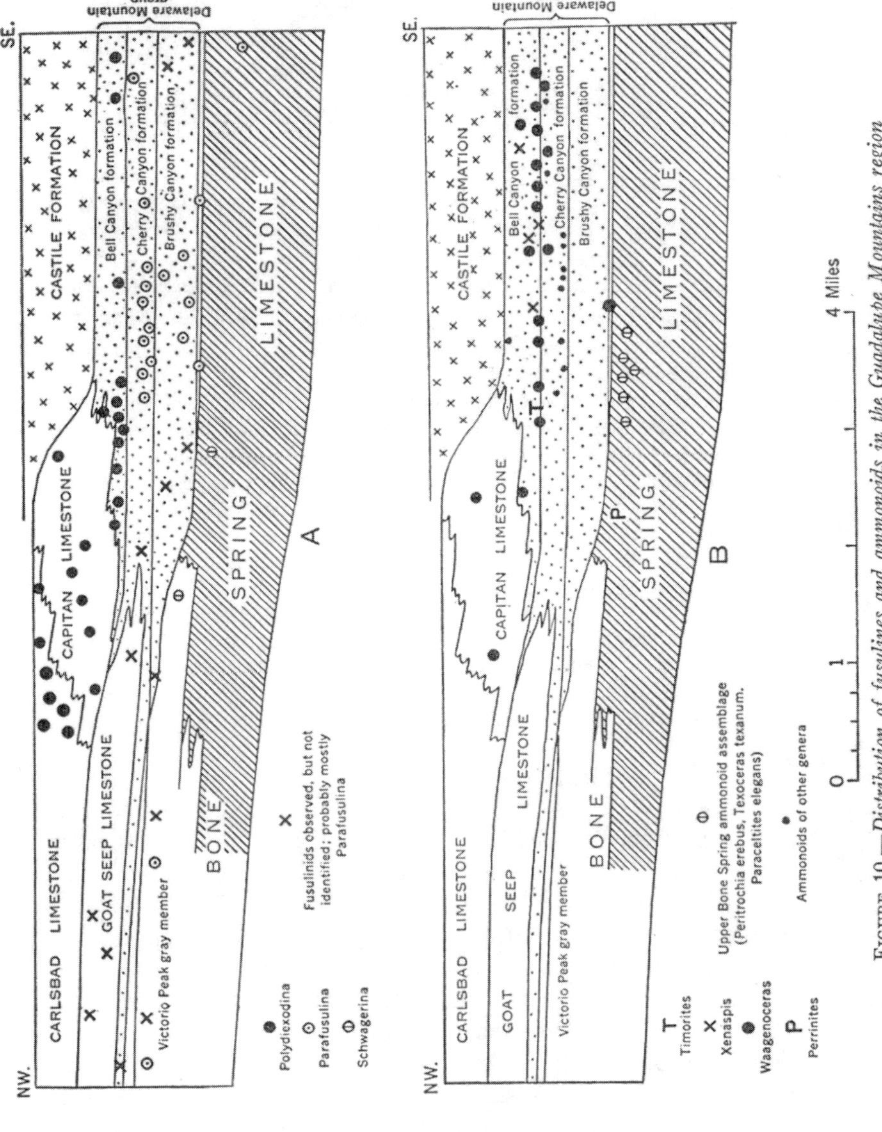

FIGURE 10.—*Distribution of fusulines and ammonoids in the Guadalupe Mountains region*

The fusulines of the formations in the Delaware basin have been transported after death from the reef and bank areas. Ammonoids are common in the basin facies but rare elsewhere. Fusulines and ammonoids indicate that the shale below the Brushy Canyon formation should be included in the Delaware Mountain group. From P. B. King (1948)

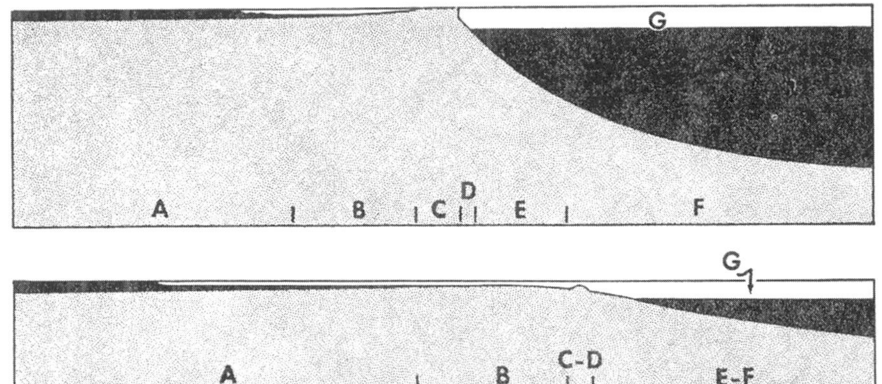

FIGURE 11.—*Chief inferred ecological zones (biotopes) of Permian time in the Guadalupe Mountains area*

Delaware basin at the right. Black areas represent heavy waters which tend to become stagnant where deep. White areas represent aerated waters of salinity similar to that of the Permian oceans.

Lower represents late Bone Spring (late Leonardian) time; *Upper* represents Guadalupian time.

A. Hypersaline waters, characterized in early stages by impoverished fauna, essentially barren later. *B*. Calcarenite bank, in early history populated mainly by crinoids, fusulines, and dasycladaceans; populated after development of barrier reef mainly by pisolite-forming algae, dasycladaceans, and fusulines. Becomes differentiated later into (*C*) reef top and (*D*) reef front. Reef top populated mainly by calcareous sponges, coralline and filamentous algae, cemented brachiopods, massive bryozoans, and fusulines (in pockets); reef front occupied mainly by pedunculate brachiopods, fenestrate and pinnulate bryozoans, and various types of algae.

E–F. Facies of black, sandy mud dominated by monactinellid sponges, nuculacean pelecypods, spired gastropods, and *Leiorhynchus*, later becoming differentiated into zones. *E*. Intermittently stagnant waters of upper talus slope populated by monactinellid sponges, *Posidonia*, *Euomphalus*, and echinoids. *F*. Generally sterile bottom.

G. Pelagic fauna rich in ammonoids and radiolarians. Probably very rich in phytoplankton. Possibly fishes were abundant.

dwelling green plants. At times the lower slopes of the reef talus were inhabited by a pelecypod, *Posidonia*, a gastropod, *Euomphalus*, and echinoids. *In situ* glass sponges (Lyssacina) have been found in black limestones of the Bone Spring, but they are not common.

The life distribution of the monactinellid sponges responsible for enormous quantities of scattered spicules throughout the black limestones of the Delaware basin is completely conjectural. The spicules are coextensive with the basin phase, and the sponges must have extended into the shallowest depths at which these deposits were formed. The spicules of monactinellids typically become separated on decay of the supporting tissues so that it is not surprising that the fossil spicules are scattered. Probably these sponges grew most abundantly on the lower slopes at the sides of the Delaware basin, and the spicules were winnowed to lower levels by gentle bottom currents. Possibly, however, the sponges ranged farther into deep stagnant waters than other organisms preserved as fossils. These conclusions are summarized in Figure 11 E, F.

Pelagic forms are well represented by Radiolaria, hollow calcareous algal spheres, and abundant ammonoids. Seasonal fluctuations in production of phytoplankton may be responsible for alternations in shades of brown and black that characterize some of the laminated limestones.

Ammonoids are scattered throughout the basin limestones (Fig. 10). They are most abundant in the deeper part of the Delaware basin and become progressively less abundant toward the margin. They are quite rare in the talus, reef, and shelf deposits. Some of the shells are crushed and otherwise poorly preserved, but generally they are entire. At several localities in the Cherry Canyon and Bell Canyon formations the rocks are crowded with ammonoid shells ranging from a fraction of a millimeter to 8 or 10 cm in diameter. These represent complete ontogenetic series of many generations of ammonoids that evidently lived and died together. There is no indication of current sorting or secondary concentration of shells after death. Probably these accumulations were a result of the simultaneous and catastrophic death of entire communities. Probably these animals normally lived well above the bottom. Occasionally, bottom waters charged with poisonous H_2S may have been brought to the surface during persistent or violent wind storms, or as a result of occasional sediment slides at the basin margin with consequent destruction of the pelagic forms of shallower aerated waters. Probably fishes were destroyed occasionally in these mass mortalities. If so, their remains generally were decomposed at the bottom before fossilization. These conclusions are indicated in Figure 11 G.

REEF AND BANK PHASES

Optimum supplies of oxygen and nutrients needed by lime-secreting organisms evidently were limited to an area along the edge of the shelf bordering the deep waters of the Delaware basin, and this area became progressively more restricted during reef development. Presumably a constant supply of nutrients was provided by upwelling waters from the relatively stagnant depths. Equally favorable conditions did not extend far from the basin because of depletion of nutrients and increasing salinity in the shallow, warm shelf seas.

An extensive bank of shell sand (Victorio Peak member) was formed around the margin of the Delaware basin late in Leonardian time. Some of the carbonate rocks of the Victorio Peak bank are completely dolomitized. The most conspicuous fossils are robust fusulines (*Parafusulina* spp.), but small fusulines (*Schubertella*) and other Foraminifera are common where primary textures are not obliterated. Ramose, pinnulate, and fenestrate Bryozoa are abundant in some beds. Robust brachiopods (*Dictyoclostus bassi*, *Neospirifer pseudocameratus*) and dasycladaceans are scattered throughout the facies. The most abundant fossils, however, are detrital fragments of crinoids.

During Cherry Canyon and Bell Canyon times the area occupied previously by bank conditions was differentiated into a barrier reef and an area of deposition of calcium carbonate sand immediately behind the reef. Neither of these two environments was very like the bank conditions of Victorio Peak deposition, but the backreef biota is essentially an impoverished bank fauna from which elements most sensitive to high salinities have been eliminated.

The earliest reefs in the area are localized structures a few hundred feet basinward from the Victorio Peak bank. These patch reefs (Figs. 5, 6) are remarkable for the small proportion of associated reef-derived detritus and for the rarity of calcareous sponges. Dominant fossils are algal deposits (siphonous, red, and blue green), hydrocorallines, cylindrical fistuliporoid bryozoans, fusulines, and various brachiopods. A few ammonoids (*Perrinites?*) were found here, but they are exceptional in this facies.

Small patch reefs in the Brushy Canyon formation south of Bone Canyon are completely dolomitized, and there are no recognizable fossils. All younger reefs of the area are characterized by abundant sycon sponges, fenestrate, pinnulate, and ramose bryozoans, in addition to various kinds of algae and fusulines. The restricted distribution of the calcareous (sycon) sponges suggests that they occupied a very limited, specialized habitat. Probably they occupied only the highest part of the reefs in very shallow water. Their rarity in the Bone Spring reefs may be related to slightly greater depth at which those reefs were formed. The reef brachiopods are for the most part cemented forms, but pedunculate types were very abundant at the front of the reef and possibly were most abundant below the surf zone where they were associated with the fragile types of bryozoans. The near absence of brachiopods and bryozoans in the sediments immediately behind the reef shows that they did not live in close proximity to the lagoon. Figure 11 C, E shows the inferred distribution of the reef biota.

SHELF PHASE

The shelf environments precluded the development of normal and balanced communities like those of the Delaware sea. A number of characteristically stenohaline groups such as the corals, bryozoans, echinoderms, and ammonoids are very poorly represented and are absent in the younger beds. Brachiopods of the shelf environment were limited to a few hardy types that tended to be ubiquitous. Presumably these forms were euryhaline. All the evidence points to hypersaline waters in the Leonardian and Guadalupian shelf seas. A few species of pelecypods and gastropods are almost ubiquitous and are found in almost all faunas collected in the Yeso, San Andres, and Carlsbad (Whitehorse) faunas. Many of these species are also found around the margin of the Delaware basin; the wide distributions suggest they were euryhaline. Most organisms of these seas probably were adapted to mud and sand bottom or were pelagic. There are no records of rock-bottom communities in the shelf facies.

Fusulines and siphonous algae are most abundant in the bank phase and the zone immediately behind the barrier reefs. Possibly the fusulines were tolerant of or preferred slightly higher salinity than the reef organisms. The relatively quiet shallow waters of the bank and lagoon may have been more suitable for these forms than the turbulent waters in front of the reef. The inferred life distributions are shown in Figure 11.

REFERENCES

ADAMS, J. E. with discussion by W. B. Wilson. Oil pool of open reservoir type: Bull. Amer. Assoc. Petrol. Geol., vol. 20, pp. 780–796, incl. index map, 1936.

—— AND FRENZEL, HUGH N. Capitan barrier reef, Texas and New Mexico: Jour. Geol. vol. 58, pp. 289–312, 1950.

BAKER, C. L. Depositional history of the red beds and saline residues of the Texas Permian: Univ. Texas Bull. 2901, pp. 9–72, 1929.

CLIFTON, ROLAND LEROY. Invertebrate faunas from the Blaine and the Dog Creek formations of the Permian Leonard series: Jour. Paleon., vol. 16, pp. 685–699, 1942.

CULLIS, C. GILBERT. The mineralogical changes observed in the cores of the Funafuti borings in The atoll of Funafuti: Royal Soc. London, Sec. XIV, pp. 392–420, 1904.

GIRTY, G. H. The Guadalupian fauna: U. S. Geol. Survey Prof. Paper 58, 651, pp. 1908.

Girty, G. H. Paleontology of the Manzano group of the Rio Grande valley, New Mexico: U. S. Geol. Survey Bull. 389, pp. 41–136, 1909.

Hadding, Assar. Silurian reefs of Gotland, Jour. Geol., vol. 58, pp. 402–409, 1950.

Henson, F. R. S. Cretaceous and Tertiary reef formations and associated sediments in Middle East: Bull. Amer. Assoc. Petrol. Geol., vol. 34, pp. 215–238, 1950.

Hills, J. M. Rhythm of Permian seas, a paleogeographic study: Bull. Amer. Assoc. Petrol. Geol., vol. 26, pp. 217–255, 1942.

Howe, M. A. The geologic importance of the lime-secreting algae: U. S. Geol. Survey Prof. Paper 170, pp. 57–64, 1932.

Johnson, J. H. Permian lime-secreting algae from the Guadalupe Mountains, New Mexico: Bull. Geol. Soc. America, vol. 53, pp. 195–226, 1942.

——. Permian calcareous algae from the Apache Mountains, Texas: Jour. Paleon., vol. 25, pp. 21–30, 1951.

King, P. B. Permian stratigraphy of trans-Pecos Texas: Bull. Geol. Soc. America, vol. 45, pp. 697–798, 1934.

——. Permian of West Texas and Southeastern New Mexico: Bull. Amer. Assoc. Petrol. Geol., vol. 26, no. 4, pp. 535–763, 1942.

——. Geology of the southern Guadalupe Mountains, Texas: U. S. Geol. Survey Prof. Paper 215, 183 pp. 1948.

King, R. E., et al. Résumé of geology of the South Permian basin, Texas and New Mexico: Bull. Geol. Soc. America, vol. 53, pp. 539–560, 1942.

Kroenlein, G. A. Salt, potash, and anhydrite in Castile formation of southeast New Mexico: Bull. Amer. Assoc. Petrol. Geol. vol. 23, pp. 1682–1693, 1939.

Ladd, H. S. Reefs and other bioherms: Rept. Comm. Marine Ecology as related to Paleontology, 1943–44, Nat. Research Council, no. 4, pp. 26–29, 1944.

Lang, W. B. Upper Permian formations of Delaware basin of Texas and New Mexico: Bull. Amer. Assoc. Petrol. Geol., vol. 19, pp. 262–270, 1935.

——. The Permian formations of the Pecos valley of New Mexico and Texas: Bull. Amer. Assoc. Petrol. Geol., vol. 21, pp. 833–898, 1937.

Lloyd, E. R. Capitan limestone and associated formations of New Mexico and Texas: Bull. Amer. Assoc. Petrol. Geol., vol. 13, pp. 645–658, 1929.

Lowenstam, Heinz A. Niagaran reefs of the Great Lakes area: Jour. Geol., vol. 58, pp. 430–487, 1950.

McKee, E. D. The environment and history of the Toroweap and Kaibab formations of northern Arizona and southern Utah: Carnegie Inst. Washington Pub. 492, VIII, 268 pp., 1938.

Menard, H. W. and Boucot, A. J. Experiments on the movement of shells by water: Amer. Jour. Sci., vol. 249, pp. 131–151, 1951.

Newell, N. D. *et al.* Invertebrate fauna of the late Permian Whitehorse sandstone: Bull. Geol. Soc. America, vol. 51, pp. 261–336, 1940.

Newell, N. D., Rigby, J. Keith, Whiteman, A. J., and Bradley, J. S. Shoal-water geology and environments, eastern Andros Island, Bahamas: Bull. Amer. Mus. Nat. History, vol. 97, Art. 1, pp. 1–30, 1951.

Newell, N. D., Rigby, J. Keith, Fischer, Alfred G., Whitman, A. J., Hickox, John E. and Bradley, John S. The Permian Reef Complex of the Guadalupe Mountains Region, Texas and New Mexico: 236 pp., San Francisco, California, W. H. Freeman and Company, Inc., 1953.

Pia, J. V. Vorläufige Übersicht der Kalkalgen des Perms von Nordamerika: Akad. Wiss. Wien, Math.-Naturwiss. Kl., Anz. 9, preprint, June 13, 1940.

Thomas, H. D. Phosphoria and Dinwoody tongues in lower Chugwater of central and southeastern Wyoming: Bull. Amer. Assoc. Petrol. Geol., vol. 18, pp. 1655–1697, 1934.

Udden, J. A. Laminated anhydrite in Texas: Bull. Geol. Soc. America, vol. 35, pp. 347–354, 1924.

Walther, Johannes. Die gesteinsbildenden Kalkalgen des Golfes von Neapel und die Entstehung strukturloser Kalke: Zeitschr. d. deutsch. Geol. Ges. Bd. XXXVII, pp. 329–357, 1885.

Manuscript Received by the Editor June 7, 1953.

PART

3 Diversity Dynamics

Edited by Peter J. Wagner and Gene Hunt

In a review of a Festschrift for George Gaylord Simpson, the late Leigh Van Valen (1973) famously stated that "a plausible argument could be made that evolution is the control of development by ecology." Van Valen goes on to lament that neither development nor ecology had figured prominently in evolutionary biology since Darwin. However, Van Valen wrote this just as macroevolutionary research was beginning to embrace both developmental and ecological theory in attempts to document and explain how evolution generated both the range of anatomical forms and the richness of organisms possessing those forms that we see in the fossil record. This volume is dedicated to only half of Van Valen's lamentation, although the two issues are not independent: the "loose genes vs. loose ecosystems" debate about the Cambrian explosion and other major radiations (e.g., Valentine 1980) fundamentally was an argument about whether ecology or development was more important for the generation of major evolutionary innovations. At lower scales, paleontologists borrowed population dynamics theory and methods to model major richness patterns over time. Paleontologists also paralleled another branch of ecology, conservation biology, not just in examining extinctions and their effects but also by examining the effects of long-term shifts in habitat availability and even how sampling affects our perceptions of diversity.

The papers presented in this section fueled much of the major research that filled the pages of *Paleobiology* and that provided paleontological contributions to other evolutionary biology journals in the 1980s, the 1990s, and the first decade of the twenty-first century. All sought to understand the factors that determined the diversity that paleontologists encountered in local, regional, or global compilations of fossil taxa. Computational tools and data sources such as the Paleobiology Database (http://paleobiodb.org) have now progressed to the point that current methods and the data far outstrip what was available for these studies. Nevertheless, these papers continue to be cited for the important concepts that they introduce. Indeed, several of these papers have seen a marked increase in numbers of citations in recent years. With one exception, each of these papers garnered the greatest number of citations in the last decade. Now, to some extent this likely reflects an increase in total numbers of papers published. However, clearly these papers are still influencing the hypotheses, methods, and discussions covered in other papers: workers still choose to cite these "first generation" papers in addition to (or instead of) the "second" or "third" generation papers of the subsequent decades influenced by these works.

We do not think that Van Valen's lament could be made today: much of what we now call macroevolutionary theory has invoked both developmental and ecological theory and methods. From the paleontological perspective, these papers were in the vanguard for the integration of ecological ideas into macroevolutionary thought. Their heavy citation rates, even to the present day, indicate that evolutionary biologists in general and paleontologists in particular are still finding new ways to do this and that this will continue to be a fruitful source of questions, explanations, and methods for paleontologists.

Literature Cited

Valentine, J. W. 1980. Determinants of diversity in higher taxonomic categories. *Paleobiology* 6:444–50.

Van Valen, L. 1973. Festschrift. Review of *Evolutionary Biology*, vol. 6, edited by Theodosius Dobzhansky, Max K. Hecht, William C. Steere. *Science* 180:488.

PAPER 15

Taxonomic Diversity during the Phanerozoic (1972)

D. M. Raup

Commentary

SHANAN E. PETERS

In the *Origin of Species*, Charles Darwin closed the chapter not subtly titled "On the Imperfection of the Geological Record" with an analogy: "I look at the natural geological record, as a history of the world imperfectly kept . . . of this history we possess the last volume alone. . . . Of this volume, only here and there a short chapter has been preserved; and of each page, only here and there a few lines." Here, for the first time, David Raup quantitatively addresses the relationships between the geologic record, fossil sampling, and our understanding of the history of life on Earth. Raup focuses on the long-term trajectory of marine animal taxonomic diversity, which was then gaining attention by geologists in the emerging field of plate tectonics and by evolutionary biologists trying to understand and model evolutionary process. Does evolution lead to an indefinite increase in diversity or is there an equilibrium or steady-state biodiversity on Earth? Raup's dispassionate and quantitative dissection of this fundamental question, and his clear account of most of the difficulties in answering it, defined an agenda for several decades of research in paleobiology, some of which is ongoing.

There are many insights in this paper, but two stand out as formative. Foremost is the recognition that large-scale paleobiological and geological patterns should be interrogated quantitatively and in parallel, if for no other reason than our understanding of the history of life derives from fossils that are preserved in sedimentary rocks. Raup's novel comparison of global compilations of rock quantity to diversity estimates based on the stratigraphic ranges of fossil taxa yielded compelling correlations that did "justify further investigation." The second insight is that null models based on sampling theory should at least be considered as alternatives to a face-value interpretation of the fossil record. Although recent work has shown that some of the results interpreted as bias by Raup, such as the correlation between diversity and rock quantity, may instead reflect shared responses to Earth systems changes, such as global sea level, Raup's call to quantitative action makes this a foundational contribution to paleoecology.

From *Science* 177:1065–71. Reprinted with permission from AAAS.

22 September 1972, Volume 177, Number 4054 **SCIENCE**

Taxonomic Diversity during the Phanerozoic

The increase in the number of marine species since the Paleozoic may be more apparent than real.

David M. Raup

The evolution of taxonomic diversity is receiving increasing attention among geologists. The immediate reason for this is that diversity data may have a direct bearing on problems of plate tectonics and continental drift. The tantalizing possibility exists that diversity may be a good indicator of past arrangements of continents or climatic belts, or both. Valentine (*1*, *2*) has related temporal changes in fossil diversity to changes in climate and to the evolutionary consequences of continental drift. Stehli (*3*) and others have used spatial differences in diversity to interpret paleoclimates and paleolatitudes for single intervals of time.

Diversity information from the fossil record is also important because of its bearing on general models of organic evolution. Is the evolutionary process one that leads to an equilibrium or steady-state number of taxa, or should diversification be expected to continue almost indefinitely? Has equilibrium (or saturation) been attained in any habitats in the geologic past? If mass extinction has led to a significant reduction in diversity, what are the nature and rate of recovery? The answers to these and comparable questions depend in part on theoretical arguments, but their documentation must come ultimately from the fossil record itself.

The large-scale analysis of taxonomic diversity has been facilitated in the past few years by several important publications. The American *Treatise on Invertebrate Paleontology* (*4*) and the Russian *Osnovy Paleontologii* (*5*) are particularly valuable in having brought together vast amounts of taxonomic data with a minimum of inconsistency. Also, the British publication *The Fossil Record* (*6*) provides a useful synthesis of the geologic ranges of the higher taxa. This new literature, plus advances in data-processing technology, makes possible a more sophisticated study of diversity problems than has been possible heretofore (*7*).

Valentine (*1*, *2*) used the newly published data to estimate temporal changes in diversity during the Phanerozoic, the geologic time since the end of the Precambrian. His conclusions were not dramatically different from those of earlier workers, but the breadth of documentation was far greater.

My purpose in this article is to investigate the nature of the diversity data to determine if more can be learned from it. In particular, I will examine the proposition that systematic biases exist in the raw data such that the actual diversity picture may be quite different from that afforded by a direct reading of the raw data. My study will be limited to the major groups of readily fossilizable marine invertebrates (as was Valentine's) and to changes in their worldwide diversity through time.

The author is professor of geology at the University of Rochester, Rochester, New York 14627.

Traditional View

Figure 1 shows three histograms of taxonomic diversity for the Phanerozoic. The three sets of data differ somewhat in scope. Those of Valentine (*1*) and Newell (*8*) are principally tied to the family level, whereas Müller's (*9*) are numbers of genera. All three are limited mostly to the major groups of fossilizable marine invertebrates: Protozoa, Archaeocyatha, Porifera, Coelenterata, Bryozoa, Brachiopoda, Arthropoda, Mollusca, and Echinodermata, but Newell's data also include vertebrates. All three sets of data inevitably include some nonmarine and terrestrial taxa, but in none is this influence numerically significant.

The important fact is that all three show essentially the same picture and the one that has constituted the consensus for many years. The overall pattern is one of (i) a rapid rise in the number of taxa during the Cambrian and Early Ordovician, (ii) a maximum at about the Devonian, (iii) a slight but persistent decline to a minimum in the Early Triassic, and (iv) a rapid increase to an all-time high in diversity at the end of the Tertiary. Valentine (*1*, *2*) has suggested that the rise in diversity at the species level in Mesozoic to Tertiary time was an exponential one, with the late Tertiary having up to 20 times more species than the average for the mid-Paleozoic. This rise would appear even greater if insects, land plants, and terrestrial vertebrates were considered. These are particularly "noticeable" groups, important to man, and the history of their diversity has influenced thinking on the general subject.

It should be emphasized that the Phanerozoic diversity pattern yielded by the published taxonomic data depends on the choice of taxonomic level. As Valentine has pointed out, diversities at the levels of phylum, class, and order have behaved very differently from those at the lower levels. The number of phyla has been essentially constant since the Ordovician, for example.

Sedimentary Record and Diversity

It has been established that the general quality of the sedimentary rock record improves with proximity to the Recent (*10, 11*). That is, the younger parts of the record are represented by larger volumes of rock (per unit of time), and the amount of metamorphism, deformation, and cover by overlying rocks is generally less. This is usually interpreted as resulting from the fact that the younger rocks are closer to "the top of the stack" and that, being younger, they have had less chance to be destroyed by erosion, metamorphism, and the like.

Figure 1 includes a graphic display of Gregor's estimate (*12*) of change in the sedimentary record through the Phanerozoic. The vertical coordinate in the lower graph is what Gregor calls the "survival rate" and is expressed as cubic kilometers of sediment per year now known and dated stratigraphically. This shows, for example, that the Devonian is represented by about twice the volume of sediments as the Cambrian (after adjustment for the relative durations of the periods). Gregor's survival data are comparable to estimates made on quite different bases by others (*10, 13*).

There is unquestionably a strong similarity between the patterns of taxonomic diversity at the genus and family levels and the pattern of sediment survival rate. This similarity suggests that changes in the quantity of the sedimentary record may cause changes in apparent diversity by introducing a sampling bias.

In spite of the fact that the patterns in Fig. 1 are correlated, a causal relationship is by no means demonstrated. Furthermore, the correspondence is not perfect, and both the diversity and sedimentary data are subject to many errors and uncertainties. The remainder of this article is devoted to a more detailed assessment of these relationships.

Gregor's data (Fig. 1) are estimates of survival rate for all sedimentary rocks, without distinction between marine and nonmarine. This detracts from the comparison with diversity because the biologic data are nearly free of nonmarine elements. Also, with the exception of the interval from the Devonian through the Jurassic, Gregor's numbers are derived from estimates of maximum sediment thickness (*14*). This part of the data is suspect because of the logical problems involved in going from the maximum known thickness (in a local section) for a geologic system to the total volume of rock in that system (*11*). Furthermore, Gregor's rates are all sensitive to errors in estimates of the absolute time durations of the periods.

Thus, although there is little doubt about the general validity of Gregor's pattern, the inherent weaknesses prevent its use in more rigorous analysis.

By far the best data for sediment volumes are those published by Ronov (*15*). They are based on the results of a 10-year project of compiling lithological-paleogeographic maps and must be considered the most comprehensive data available. They are limited, however, to the Devonian-Jurassic interval. Ronov's data were used by Gregor where possible, but were modified by his calculation of survival rates. Ronov carefully distinguished between continental clastics, marine clastics, evaporites, marine and lagoonal carbonate rocks, and volcanics.

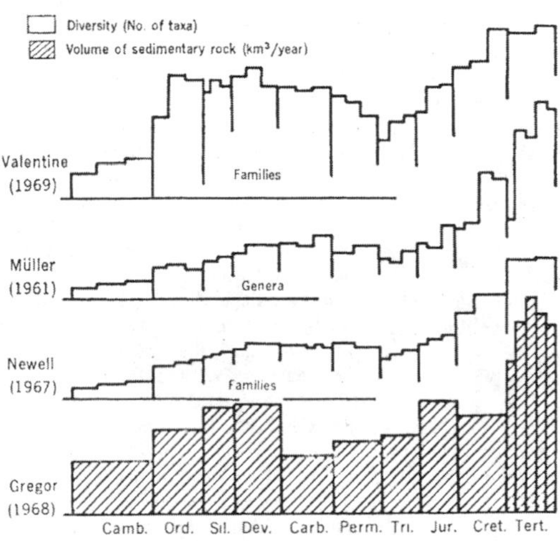

Fig. 1. Comparison of the number of taxa and the volume of sedimentary rock during the Phanerozoic. The diversity data are based mainly on well-skeletonized marine invertebrates (*1, 8, 9, 12*).

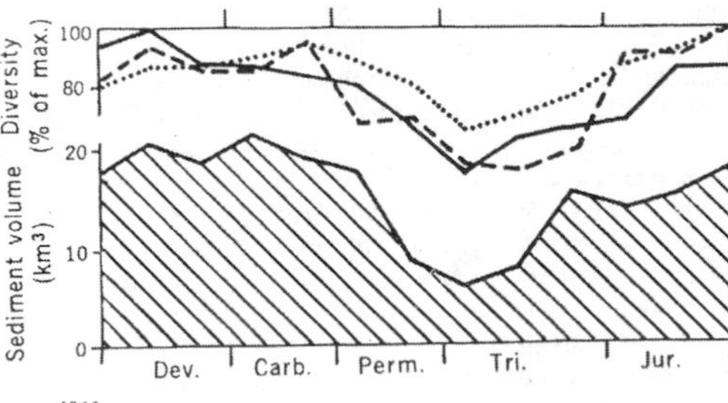

Fig. 2. Apparent taxonomic diversity compared with estimated volume of marine and lagoonal clastic and carbonate sediments. The diversity data are from Fig. 1. (Solid line) Valentine (*1*), (dotted line) Müller (*9*), (dashed line) Newell (*8*).

In Fig. 2, the taxonomic diversity data of Newell, Müller, and Valentine are compared with Ronov's estimates for the total volume of marine and lagoonal clastics and carbonates. Absolute time does not enter in because for each stratigraphic series total number of taxa and total sediment volumes are used. The diagram is thus free of most of the effects of errors in radiometric dating.

The correspondence between diversity and quantity of sediments is much stronger than indicated in Fig. 1. In particular, it should be noted that the Early Triassic diversity minimum coincides with a sediment minimum, which was not the case when Gregor's data were used. This is primarily because Gregor used Ronov's data for all sedimentary facies and because of the effect of Gregor's rate calculation.

It could be argued that the similarity between the patterns in Fig. 1 is due simply to a broad but independent increase in both sediment volume and diversity from the Cambrian through the Tertiary and that similarity in detail is quite accidental. Figure 2 largely denies this interpretation because the Carboniferous-Permian interval shows the reverse trend in both measures. Thus, although a causal relation is not proved, the empirical relation appears to be strong enough to justify further investigation.

There is no disagreement on the proposition that the number of taxa known from the fossil record is less than the number that actually lived. This stems simply from the fact that some taxa (particularly at the species level) are rarely or never preserved. The effect is most striking when late Tertiary diversity is compared with the diversity of living organisms. There is no evidence for widespread extinction in the late Tertiary yet most groups have much smaller Tertiary records than would be predicted from neontological data. Furthermore, it is agreed that some biologic groups show fossil diversities closer to their actual diversities than do other groups because of inherent differences in preservability. Crustaceans, for example, are clearly underrepresented as fossils when compared with brachiopods or bivalves. The real problem, however, in the present context, is to evaluate relative changes in diversity over time, using the fossil record as the only available measure.

Sampling Problems

Many fossil taxa remain to be discovered. At the species level, this number probably exceeds even the number that have been described, although this would vary greatly from group to group. The diversity problem is thus in the realm of sampling theory and can be attacked from a mathematical viewpoint.

Exploration for fossils is analogous to problems in probability theory known variously as cell occupancy and urn problems. Consider a wooden tray which is divided into small compartments or cells, and assume that small balls are thrown randomly at the tray in such a way that each ball falls into a cell, without being influenced by the position of the cell or whether it is already occupied. The first ball thrown will inevitably result in the occupancy of one cell. The second ball may fall in the same cell and thus not add to the number of cells occupied: The probability of this event will be greatest if the total number of cells is small. At some point, all the cells will be occupied by at least one ball, and the waiting time necessary to accomplish

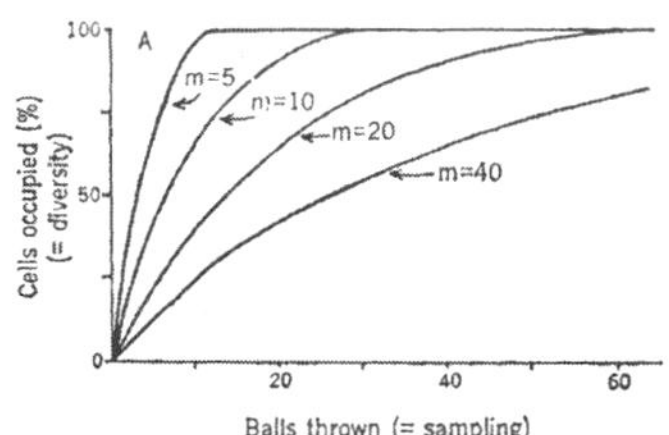

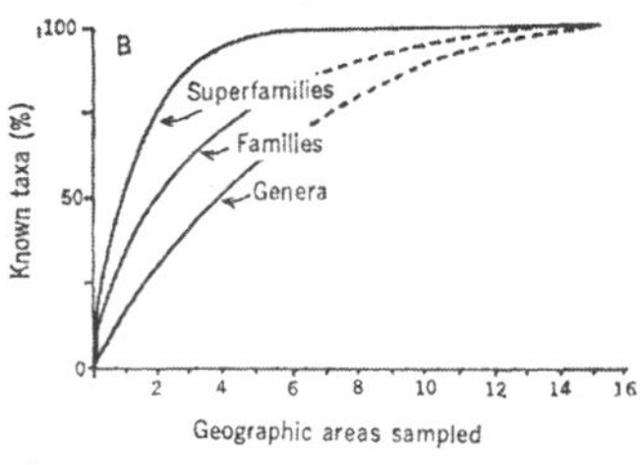

Fig. 3. Diversity as a function of sampling. (A) Illustration of cell occupancy problem. The average waiting time for cell occupancy varies with the number of cells to be occupied (m). (B) Effect of sampling on apparent diversity in fossil ammonoids of the *Meekoceras* zone (Triassic).

this (measured in number of balls thrown) will depend only on the number of cells in the tray.

As noted above, the waiting time for occupancy of one cell is equal to 1 (one ball thrown). It can be shown (*16*) that the average additional waiting time for occupancy of a second cell is:

$$\frac{m}{m-1}$$

where m is the total number of cells in the tray. The additional waiting time for the third occupancy is:

$$\frac{m}{m-2}$$

and so on. The total waiting time for complete cell occupancy then becomes:

$$m\left(\frac{1}{m}+\frac{1}{m-1}+\frac{1}{m-2}+\ldots+\frac{1}{2}+1\right)$$

Calculated curves for the expected waiting time for various values of m are shown in Fig. 3A.

The appropriate paleontological analogy is as follows: Let m be the total number of taxa available for discovery (thus, one cell equals one taxon), and let the balls thrown be the number of fossils found and identified or described. The first fossil discovered inevitably means recognition of one taxon. The second fossil may be the same or it may be from a second taxon (second cell occupied). Groups with fewer subgroups will require less sampling to be completely discovered.

This reasoning can be applied directly to the influence of taxonomic level on observed diversity. In any fossiliferous rock unit, the number of families represented is inevitably equal to or greater than the number of phyla, the number of genera is equal to or greater than the number of families, and so on. Thus, much less sampling is required to find all or nearly all the phyla (low m) than the families or genera (higher m values). At any point in the sampling process, a larger percentage of the phyla will be known than of the lower taxa. In Fig. 3, the curves of low m are what would be expected for discovery of high taxa, and the curves of high m would be representative of lower taxa. It should be noted that as sampling progresses, the ratio of the numbers of lower to higher taxa (genera per family, for example) steadily increases.

Figure 3 also shows a paleontological analog of the calculated curves. It is based on published data for ammonoids of the *Meekoceras* zone (Lower Triassic) (*17*). The data include the known occurrences of 58 genera in 15 geographic assemblages around the world. The ammonoid data (Fig. 3B) show the relationship between sampling and apparent diversity. Sampling is in this case expressed as the number of sites or areas sampled and is analogous to the number of balls thrown in the cell occupancy problem. The number of taxa found at one site in the *Meekoceras* case depends, of course, on which site is used. China, for example, yields well over half the genera and about three-quarters of the families; at the other extreme, the assemblage from the Caucasus has only two of the genera. The curves in Fig. 3 are therefore based on average expectations. For each taxonomic level, some of the values could be calculated directly; other values were determined by simulation based on a random selection of the published distributional data. The remainder (dashed lines) were extrapolated.

The ammonoid example demonstrates that apparent diversity is severely controlled by (i) the extent of sampling and (ii) the taxonomic level. At the order level (Ammonoidea) any one of the 15 sites is sufficient to yield 100 percent of the known diversity. At the generic level it requires (in this case) an average of 5 sites to exceed 50 percent. It should be emphasized that the leveling off of the curves at 100 percent does not mean that the 15 sites yield all of the ammonoid diversity in the *Meekoceras* zone: New genera and new localities are still being found.

As noted above, an increase in sampling is accompanied by predictable increases in the apparent number of genera per family, and so forth, and the effect is seen in the *Meekoceras* zone data. That this is a general phenomenon was noted by Simpson (*18*) as follows: "Sampling at few, restricted localities certainly reveals a much higher percentage of the genera than of the species that existed at any one time."

The sampling problem need not be analyzed only in the context of geographic extent of collecting. The sampling axes of Fig. 3 could be replaced by various measures of the intensity of collecting or study (such as number of paleontologists or years of study) or by measures of the quality of the fossil or rock record (extent of outcrops, type of preservation, and even accessibility of outcrops). The fact that new taxa are constantly being defined or discovered means that the fossil record is still in a relatively early stage of sampling and thus may be represented by the steeper parts of the curves in Fig. 3.

Sources of Error in Diversity Data

In the following numbered sections I consider seven major sources of error that may affect any set of diversity data. All of them certainly have influenced published diversity data of the type shown in Fig. 1.

1) *Range charts.* When the objective of a diversity study is to estimate how many taxa lived during a given interval of geologic time, the primary source of information is usually a range chart drawn at the appropriate taxonomic level. If a family has a range from the base of the Silurian to the top of the Lower Devonian, for example, it is assumed that the family lived throughout the entire range. Thus, the family is registered for the Upper Silurian even though the Upper Silurian fossil record may not actually contain species of the family.

This procedure is valid biologically as a means of estimating actual diversity, but it does have the effect of overestimating "observed" diversity for relatively unfossiliferous intervals. In fact, an interval can be completely unfossiliferous yet still be credited with having considerable fossil diversity. This source of error becomes important when one is assessing the biasing effect of low sediment volume, as in the Permian-Triassic of Fig. 2. In this instance, the drop in fossil content of Permian rocks may be greater than it appears from the range chart data.

More important, the use of range charts introduces a systematic, time-related bias, as follows. Many (or most) range charts are incomplete in that the true first and last occurrences have not yet been found. In fact, the fossil record may not even contain the first or last occurrences (due to nonpreservation). Ranges of taxa may be truncated at either end, but truncation at the older end (first occurrence) has a higher probability because the older rocks have a greater chance of nonexposure or destruction by erosion and metamorphism. This means that the Phanerozoic diversity data are inevitably biased toward an increase in observed diversity through time.

2) *Influence of "extant" records.* Cutbill and Funnel (*7*) have already noted the biasing effect of the fact that ranges of fossil taxa are generally said to include the Recent if the taxa have living representatives. A not uncommon example would be a living group which has only one fossil occurrence, let us say in the Jurassic. Its range would be listed as Jurassic-Recent which, again, is valid for many purposes but causes problems in the present context. If the group had the same sparse fossil record but had not survived to the Recent, its range would be given as simply Jurassic. Cutbill and Funnel concluded that truncation at the "last occurrence" end of a range through nondiscovery is less likely if the group has living representatives, and since younger rocks contain more extant forms, the late Mesozoic and Cenozoic diversity data are consistently biased toward larger diversity and fewer extinctions than older parts of the column.

3) *Durations of geologic time units.* Consider the effects of the durations of periods and epochs on the diversity data in Fig. 1. The horizontal axis in the diagram is roughly adjusted for relative durations—albeit with little justification in many cases—but the vertical axes showing numbers of taxa are not. The height of each bar on the histograms indicates the total taxa which are found anywhere in the system or series or which have ranges that include those rocks. All things being equal, a long time interval will show a higher diversity than a short one. The effect of the bias is probably to overestimate diversity in the early Paleozoic, where period and epoch durations are generally greater (*7*). This bias thus operates in a direction opposite to that of the two discussed above.

Furthermore, the bias is not easily corrected. The calculation of a simple ratio, such as families per million years, is valid when working with, for example, extinction rates, but only makes matters worse in the present context, where "standing crop" is the objective.

4) *Monographic effects.* The effects of the quality and quantity of taxonomic activity on apparent diversity are well known. It has been noted, for example, that the peak number of brachiopod genera shifted from Devonian to Ordovician largely as a result of the publication of one monograph (*19*). It is interesting to note that the generic peak has since shifted back to the Devonian.

Some of the monographic effects stem from the stratigraphic distribution of taxonomic specialists and taxonomic and phylogenetic philosophy, and perhaps even from the geographic distribution of taxonomists. Fossiliferous rocks in western Europe and eastern North America are more likely to be fully studied and thus to show higher diversity than rocks in other parts of the world.

If monographic effects are randomly distributed among the major phyla and throughout the stratigraphic column, then the consequences for overall trends in Phanerozoic diversity are minimal. Whether this lack of systematic bias exists is difficult to prove. If more families and superfamilies have been defined in the lower Paleozoic than in other parts of the geologic time table, it is impossible to say whether the difference reflects a tendency of lower Paleozoic paleontologists to be quick to erect such taxa, or whether it results from different kinds of diversity and states of preservation. At the very least, the monographic factors make highly precise studies of diversity impossible.

One special type of monographic effect is surely time dependent. If a group of organisms has many living representatives, and if biologists have subdivided it into many higher taxa, fossil representatives of these higher taxa are more likely to be recognized than if living forms are absent. This says in effect that it is easier to recognize a fossil taxon as distinct if the classification has already been established on the basis of the more complete morphological information afforded by living species. This bias has the effect that diversity is underestimated in extinct groups relative to nonextinct groups. For example, the discovery in Japan of a bivalved gastropod led to the reassignment of its Eocene counterpart from the Bivalvia to the Gastropoda. This greatly extended the stratigraphic range of the gastropod order Sacoglossa and thus increased the apparent gastropod diversity of the Tertiary (*20*).

5) *Lagerstätten*. Our knowledge of the history of life would be very different were it not for the occasional instances of spectacular preservation of large assemblages (Lagerstätten). Individual formations such as the Solnhofen, the Burgess shale, and the Baltic Amber as well as unusually fossiliferous groups of rocks such as in Timor and Madagascar have significant effects on diversity curves. In some cases, the lack of Lagerstätten is also significant. For example, the observed diversity of insects during the Cretaceous is essentially zero, but this is presumably only an artifact resulting from the lack of the special conditions required for good insect preservation during that period.

The distribution of Lagerstätten through time does not appear to be systematic although they are probably more common in younger rocks. To the extent that this is true, there will be a bias toward high diversity in younger rocks. The greatest effect, however, is to add "noise" to the diversity data in much the same way that monographic bursts produce irregularity in diversity trends in the affected groups.

6) *Area-diversity relationships*. When a new geographic region is opened to exploration, new taxa are almost inevitably discovered. This is due in part to increased sampling, but it also results from the fact that taxa tend to be geographically restricted because of either climatic factors or barriers to dispersal. Also, diversity has been shown empirically to be area dependent (*21*).

Many instances of geographic effects could be cited. One example comes from Mortensen's tabulation of distributions of living cidarid echinoids, which shows that the 148 species and subspecies of the 27 genera are distributed among 18 geographic regions (*22*). Only one genus, *Eucidaris*, is found in as many as half the 18 regions, and 63 percent of the genera are confined to fewer than four regions. No single region contains even one-third of the species. This is in spite of the fact that most cidarids have a free-swimming larval stage.

If the cidarid distribution is looked at in terms of the probable fossil record it will leave, the potential effect of geographic restriction becomes greater. The biogeography of living echinoids is based on a reasonably good sampling of three-quarters of the earth's surface—that is, the oceanic areas. In the fossil record, sampling is limited for all intents and purposes to one-quarter of the earth's surface (the continents and islands), and a significant part of that quarter has remained out of the marine realm by being emergent during most of the Phanerozoic. Thus, the paleontologists can examine only a small fraction of the ocean area for any point in the geologic past. If one were to look at only 5 percent of the present ocean area (or even 5 percent of the present continental shelf area), the apparent diversity in groups such as echinoids would be greatly reduced at all taxonomic levels. This is particularly true since, in most geologic systems, the bulk of the record is usually concentrated in a few areas—rather than being randomly scattered over the world.

The effect of biogeography on diversity is greatest at the species level and decreases upward in the taxonomic hierarchy. Most modern phyla have worldwide distributions but even so are missing in some large regions, mainly due to climatic factors. At the family level, endemism becomes much more common, although this varies greatly from group to group.

The net effect of the biogeographic factor in the present context is to make the observed fossil diversity dependent not only on the area of rock exposure but also on the nature of the world distribution of exposures. Relatively small exposures on several continents are likely to yield a higher overall diversity than the same total exposure concentrated on one continent.

Although Gilluly has demonstrated a clear increase in area of exposure through the Phanerozoic column (*10*), no studies have been made on the manner in which these rocks are distributed spatially. However, because the probability of finding older Phanerozoic rocks is less than that of finding younger ones (assuming equal time durations) it would seem reasonable that geographic coverage improves toward the Recent. This should produce higher observed diversities in younger rocks.

7) *Sediment volume*. This article started with the empirical correlation between sediment volume per unit of time and diversity of major marine groups. It is clear from sampling considerations that more sedimentary record should produce more diversity. The correlation shown in Fig. 2 is thus quite plausibly a causal one. But the strength of the resulting bias depends on (i) the taxonomic level and (ii) the kinds of differences in sediment volume from one part of the column to another. A figure for sediment volume for one geologic system [such as used by Ronov (*15*)], may be higher than the figure for another geologic system for many reasons. Discontinuous sedimentation may mean that many short-lived taxa are not preserved, but the fossil record of longer-lived taxa, characteristic of families and orders, may not be much affected. Thus, for example, if the Paris Basin had twice the volume of sediments, species diversity would be higher but family diversity little if any different. If sediment volume figures are influenced by differences in area of sedimentation, then the biogeographic relationships discussed above become significant, even at high taxonomic levels.

Postdepositional destruction or covering of sediments is the most widely accepted explanation for the temporal trends in sediment volume. Such losses of record are likely to have a spotty geographic distribution. That is, loss of the sedimentary record from one or more whole regions is more likely than small-scale reductions in all areas. This suggests that loss of biogeographic coverage is the important factor for diversity and that the sediment volume bias is closely tied to the geographic bias discussed earlier.

Models for Phanerozoic Diversity

Figure 4 shows in generalized form Phanerozoic diversity patterns at several taxonomic levels for shelf invertebrates with well-developed skeletons. The illustration is a composite of several from Valentine (*1*) and one from Müller (*9*). Minor irregularities were removed in making the composite, and vertical scales were adjusted. Valentine based the species curve on inference, but all the others were drawn directly from observed diversities.

Valentine concluded that the patterns are a plausible result of a combination of the evolutionary process of diversification and certain events in the physical history of the Phanerozoic. The basic biologic process envisioned requires that diversification take place first at high taxonomic levels (phylum, class, order) and later at successively lower taxonomic levels. The number of phyla (not shown) reached a maximum during or before the Early Ordovician, classes and orders later in the Ordovician, families in the Devonian, and genera and species in the Carboniferous or earliest Permian. According to Valentine, the diversity of the higher taxa (except phyla) declined after the initial peaks because as high taxa became extinct, they were replaced not by equally distinct groups but rather by specialized lower taxa (genera and species) within the surviving groups.

Still following Valentine's interpretation, the Permian-Triassic mass extinctions sharply reduced the diversity at all levels, and this was followed by a dramatic rise in diversity at the family, genus, and species levels, leading to the present-day array. Valentine argues that the driving forces behind this Mesozoic-Cenozoic rediversification were (i) continental drift and (ii) an increase in latitudinal temperature gradients. The diversity increase would presumably have taken place anyway—but to a lesser degree—as a continuation of the trend to specialization that was interrupted by the Permian-Triassic extinctions.

Figure 4 and its interpretation represent, therefore, one model for Phanerozoic diversity. It is an appealing one in that it is based largely on a "face value" use of empirical data and because it is biologically and ecologically plausible.

The foregoing interpretations are subject to several problems. The patterns in Fig. 4 contain elements that are qualitatively those which would be predicted from the biases discussed in this paper, as follows:

1) If the quality or quantity of sampling increases through time, it is inevitable that the ratios of species to genera, genera to families, and so on, will also increase.

2) Time-dependent biases should produce a rise in diversity at the lower taxonomic levels as the Recent is approached. The post-Paleozoic increase in numbers of families, genera, and species seen in Fig. 4 may be due to this factor.

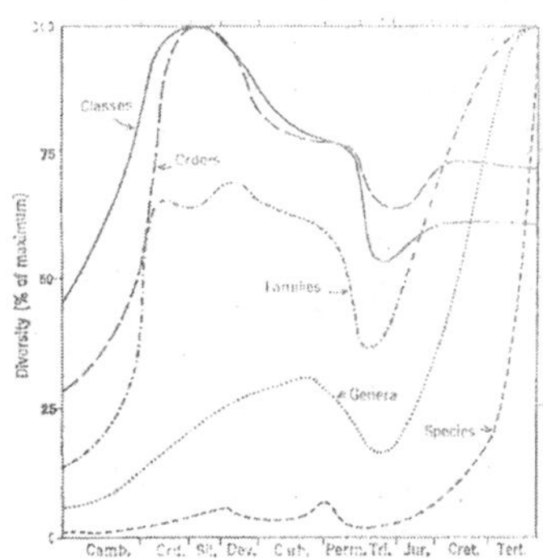

Fig. 4. Variation in apparent taxonomic diversity for several taxonomic levels of well-skeletonized marine invertebrates during the Phanerozoic.

3) Time-dependent biases should also shift any diversity peak toward the Recent (to the right in Fig. 4), and the amount of shift should be greatest at the lowest taxonomic levels. The fact, noted by Valentine (*1*), that diversities at lower taxonomic levels appear to have peaked after those at higher levels may actually be due to the effects of biases.

The last point deserves more consideration. From an evolutionary viewpoint, it is certainly plausible that diversity maxima for species and genera should occur after those for higher taxa in the same group. The question is whether the time lag is large enough and sufficiently universal to produce distinct offsets when diversities of several major animal groups are plotted together as in Fig. 4. If this were the case, periods of widespread extinction should be followed by recognizable intervals of low diversity, during which rediversification takes place. But the fact is that most major extinctions are not followed by periods of low diversity. Lowered diversity must have occurred at such times, but it evidently did not last long enough to be noticeable on the time scale used here. Valentine points out that the Permian-Triassic extinction is the only one which is followed by a diversity drop. Figure 2 indicates that in that interval the diversity drop may be an artifact of sampling.

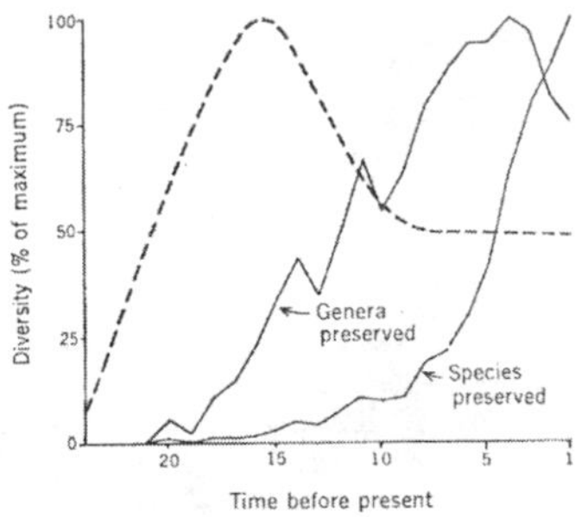

Fig. 5. Computer simulation of taxonomic diversity. The dashed line is a hypothetical diversity distribution before fossilization and is based on simulated ranges of 2000 species constituting 100 genera. The solid lines indicate the diversity trends after biases are applied to the range data.

An alternate model for Phanerozoic diversity is suggested by the dashed line in Fig. 5 and consists of a diversity maximum followed by a decline to an equilibrium level. The time scale in Fig. 5 is arbitrary, but a mid-Paleozoic position for the maximum is implied: The curve was suggested by the curves in Fig. 4 for classes and orders (where effects of biases should be least). The alternative model makes no distinction between taxonomic levels and thus is meant to apply to all levels below phylum. Thus, the assumption has been made that the offset of diversity peaks caused by gradual diversification either is not large enough to be observed at this scale or is masked by noise resulting from the fact that many animal groups with different evolutionary histories are plotted together. The proposed model is, of course, valid only if the biases described in this article are quantitatively significant.

The plausibility of the alternative model was checked by a computer simulation. By using random numbers, hypothetical first and last occurrences were generated, and a range chart was constructed showing the distributions in time of 2000 hypothetical species (segregated into 100 genera). The dashed diversity curve in Fig. 5 was computed from the simulated range chart. The curve thus represents a hypothetical diversity pattern before biasing factors are applied.

Next, information was removed from the range chart by a random process designed to simulate the biasing factors. For each species, portions of the record were "destroyed," with the probability of destruction increasing back in time. Record losses occurring only inside a range had no effect. If, however, a loss included the beginning or end of the range, the range was shortened accordingly. In many cases, species were completely removed by this process. The Recent was made immune from these information losses to stimulate the biasing effect of "extant" records.

Finally, a new range chart was constructed from what was left after the information removals. The diversity curves computed from this are also shown in Fig. 5. Species diversity increases sharply toward the Recent whereas generic diversity shows a maximum, offset to the right of the original maximum. When genera are grouped into hypothetical families (not shown), the diversity maximum is offset to the right but not as far.

The simulation demonstrates that diversity patterns such as are observed in the fossil record can be produced by the application of known biases to quite different diversity data. The simulation does not, of course, prove the alternative model for Phanerozoic diversity because of our present ignorance of the actual impact of the biases. The simulation does suggest, however, that the model proposed in Fig. 5 is a plausible one for the Phanerozoic record of marine invertebrates.

The alternative model cannot be applied literally to land-dwelling forms because the exploitation of terrestrial habitats started much later in geologic time and may be still going on. The fossil record of terrestrial organisms is subject to the same biases, however, and so should be read with caution.

Summary

Apparent taxonomic diversity in the fossil record is influenced by several time-dependent biases. The effects of the biases are most significant at low taxonomic levels and in the younger rocks. It is likely that the apparent rise in numbers of families, genera, and species after the Paleozoic is due to these biases. For well-skeletonized marine invertebrates as a group, the observed diversity patterns are compatible with the proposition that taxonomic diversity was highest in the Paleozoic. There are undoubtedly other plausible models as well, depending on the weight given to each of the biases. Future research should therefore be concentrated on a quantitative assessment of the biases so that a corrected diversity pattern can be calculated from the fossil data. In the meantime, it would seem prudent to attach considerable uncertainty to the traditional view of Phanerozoic diversity.

References and Notes

1. J. W. Valentine, *Paleontology* **12**, 684 (1969).
2. ———, *Bull. Geol. Soc. Amer.* **79**, 273 (1968); *J. Paleontol.* **44**, 410 (1970).
3. F. G. Stehli, in *Evolution and Environment*, E. T. Drake, Ed. (Yale Univ. Press, New Haven, Conn., 1968), p. 163.
4. R. C. Moore *et al.*, Eds., *Treatise on Invertebrate Paleontology* (Geological Society of America and Univ. of Kansas Press, Lawrence, 1953–1972).
5. Y. A. Orlov, *Osnovy Paleontologii* (Akademiia Nauk SSSR, Moscow, 1958–1964).
6. W. B. Harland *et al.*, Eds., *The Fossil Record* (Geological Society of London, London, 1967).
7. J. L. Cutbill and B. M. Funnel, *ibid.*, p. 791.
8. N. D. Newell, in *Uniformity and Simplicity*, C. C. Albritton, Jr., Ed. (Geological Society of America, New York, 1967), p. 63.
9. A. H. Müller, *Grossabläufe der Stammesgeschichte* (G. Fischer, Jena, Germany, 1961).
10. J. Gilluly, *Bull. Geol. Soc. Amer.* **60**, 561 (1949).
11. J. D. Hudson, in *The Phanerozoic Time-Scale*, W. B. Harland *et al.*, Eds. (Geological Society of London, London, 1964), p. 37.
12. C. B. Gregor, *Proc. Kon. Ned. Akad. Wetensch.* **71**, 22 (1968).
13. See also the general discussion by R. M. Garrels and F. T. Mackenzie, *Evolution of Sedimentary Rocks* (Norton, New York, 1971).
14. The data on maximum thickness are mostly from A. Holmes, *Trans. Edinburgh Geol. Soc.* **17**, 117 (1959); M. Kay, in *Crust of the Earth*, A. Poldervaart, Ed. (Geological Society of America, New York, 1955), p. 665.
15. A. B. Ronov, *Geokhimiya* **1959** 397 (1959), translated in *Geochemistry USSR* **1959**, 493 (1959).
16. M. Dwass, *Probability* (Benjamin, New York, 1970).
17. B. Kummel and G. Steele, *J. Paleontol.* **36**, 638 (1962). A few of the ammonoid occurrences were designated as doubtful, these were eliminated for the present purpose with the effect that one of the original 16 localities was eliminated.
18. G. G. Simpson, *The Major Features of Evolution* (Columbia Univ. Press, New York, 1953), p. 31. See also the excellent discussions of sampling problems and biases by J. W. Durham, *J. Paleontol.* **41**, 559 (1967) and by G. G. Simpson, in *Evolution After Darwin*, S. Tax, Ed. (Univ. of Chicago Press, Chicago, 1960), vol. 1, pp. 117–180.
19. G. A. Cooper, *J. Paleontol.* **32**, 1010 (1958); see also the general discussion of monographic effects in A. Williams, *Geol. Mag.* **94**, 201 (1957).
20. L. R. Cox and W. J. Rees, *Nature* **185**, 749 (1960).
21. F. E. Preston, *Ecology* **43**, 185 (1962); N. D. Newell, *Amer. Mus. Nov. 2465* (1971).
22. T. Mortensen, *A Monograph of the Echinoidea* (Reitzel, Copenhagen, 1928), vol. 1.

Maize and Its Wild Relatives

Teosinte and Tripsacum, wild relatives of maize, figured prominently in the origin of maize.

H. Garrison Wilkes

The close relatives of maize, teosinte and the genus Tripsacum, have assumed increasing importance in the understanding of the evolution under domestication of the New World's most important plant food. Tripsacum hybridizes with maize under experimental conditions, and teosinte crosses with maize in its native habitat, Mexico and Central America. Much of the heterotic vigor of maize is attributed to introgressive hybridization from its closest relative, teosinte. Today, the maize crop is the single largest harvest in the United States and is the staple food for most of the inhabitants of Latin America.

Considering the importance of the hybridization of maize (*Zea mays* L.) with its wild relatives (Fig. 1) teosinte [*Z. mexicana* (Schrad.) O. Ktze] (1), an annual grass looking very much like maize, and *Tripsacum* (2), perennial grasses quite distinct from maize in appearance, it is startling to realize how little is known about this phenomenon in the wild. Maize and teosinte are genetically compatible and hybridize freely with each other in places where the isolating mechanisms between the two have broken down, as in the Sierra Madre Occidental of northern Mexico, the Central Plateau and Valley of Mexico in central Mexico, and in Huehuetenango of northern Guatemala. *Tripsacum* does not hybridize readily with maize in the field, but hybrids can be produced under experimental conditions. There is reason to be alarmed by the rapid extinction of these wild relatives in and around maize fields where teosinte is known to have hybridized with maize for at least three millennia. This extinction of the native populations of teosinte is disastrous from the standpoint of future introgression, since it

The author is a botanist in the department of biology at the University of Massachusetts, Boston 02116.

PAPER 16

Onshore-Offshore Patterns in the Evolution of Phanerozoic Shelf Communities (1983)

D. Jablonski, J. J. Sepkoski Jr., D. J. Bottjer, and P. M. Sheehan

Commentary

RICHARD B. ARONSON

The 1980s were heady days for paleobiology. Paleobiologists grappled with biotic catastrophes that were all but inconceivable in their scope and intensity, pondered diversification over the broad wash of geologic time, and argued excitedly over the scale dependence or scale independence of evolutionary pattern and process. Paleoecology slunk into this period of intellectual ferment as the poor cousin of mainline evolutionary paleobiology. Paleoecology received scant respect because it struck many in the field as an uncritical, tautological application of ecological principles on inappropriate scales. By publishing this groundbreaking paper in *Science*, Jablonski and colleagues got paleoecology back on track. The paper unified the ecological and evolutionary flavors of paleobiology by providing an ecological context for macroevolutionary change in the marine biosphere.

The central idea of the paper is simple and appealing. Living marine-benthic communities vary along an environmental gradient from nearshore, shallow-water habitats, across shelf habitats, and into the bathyal zone and the deep sea. Marine-benthic assemblages in the fossil record—interpreted as paleocommunities—are likewise distributed along that environmental gradient. Jablonski et al. reported that novel types of paleocommunities, which are assemblages containing novel higher taxa with novel ecological roles, appeared preferentially in nearshore environments. The new assemblages then expanded offshore, displacing the previously incumbent paleocommunities in those environments. Offshore expansion was no surprise because there was nowhere else for marine biotas to go from coastal environments, but the onshore origination of novel paleocommunities required explanation. The authors suggested two nonexclusive possibilities: (1) greater extinction rates of clades offshore and (2) greater origination rates of higher taxa and novel ecologies onshore.

In one bold stroke, this first version of what became known as the onshore-offshore theory crystallized a large body of broadly overlapping observations of the marine-fossil record. The onshore-offshore theory explained how Sepkoski's three evolutionary faunas (1981) succeeded each other through the Phanerozoic; showed that Vermeij's Mesozoic marine revolution (1977) and Stanley's related Mesozoic infaunalization of bivalves (1968) were part of a larger macroevolutionary pattern; and clarified Bambach's data on niche occupancy through time (1993), as well as parallel data from Ausich and Bottjer (1982), Meyer (1979), Thayer (1979), and others on niche diversification of suspension feeders and bioturbators. Furthermore, it lent scientific credence to the popular conception that the deep sea is the refuge of living fossils and retrograde communities. DiMichele and Aronson (1992) adapted the idea to explain the environmental bias of evolutionary and ecological novelty in terrestrial floras: there is a good reason that swamps look primeval, and it is the same reason as for the deep sea. Jablonski and colleagues subsequently refined the onshore-offshore theory with pa-

From *Science* 222:1123–25. Reprinted with permission from AAAS.

pers showing that novel communities did not expand from the nearshore as integrated, Clementsian units; rather, individual taxa shifted their distributions and the new paleocommunity types reassembled progressively farther offshore.

Decades ago, when I was an undergraduate, my chemistry professor quipped that most theories are either too good to be true or too true to be good. Either they are so general and simplistic that they explain very little about a lot of things or they are so constrained and detailed that they explain a lot about very few things. The onshore-offshore theory hits the sweet spot of goodness and truth, for it is conceptually simple and it explains a lot. I realized from it that humans are an onshore evolutionary novelty, because fishing pressure and its strong, cascading, trophic impacts begin nearshore and expand offshore. Fishing from onshore to offshore progressively transforms Sepkoski's modern communities into "postmodern" communities, with humans as the apex predators.

Because *Homo sapiens* is an onshore evolutionary novelty, this paper will be a theoretical cornerstone of the emerging field of conservation paleobiology. The anthropogenically driven, hemorrhagic loss of marine diversity and the global-scale ecological homogenization that is its ineluctable consequence are part of a biological rhythm in deep time, rolling from the coast out to sea. That may be a depressing thought, but as Paul Valéry famously pointed out, trend is not destiny.

Literature Cited

Ausich, W. I., and D. J. Bottjer. 1982. Tiering in suspension-feeding communities on soft substrata throughout the Phanerozoic. *Science* 216:173–74.

Bambach, R. K. 1993. Seafood through time: changes in biomass, energetics, and productivity in the marine ecosystem. *Paleobiology* 19:372–97.

DiMichele, W. A., and R. B. Aronson. 1992. The Pennsylvanian-Permian vegetation transition: a terrestrial analogue to the onshore-offshore hypothesis. *Evolution* 46:807–24.

Meyer, D. L. 1979. Length and spacing of the tube feet in crinoids (echinodermata) and their role in suspension feeding. *Marine Biology* 51:361–69.

Sepkoski, J. J., Jr. 1981. A factor analytic description of the Phanerozoic marine fossil record. *Paleobiology* 7: 36–53.

Stanley, S. M. 1968. Post-Paleozoic adaptive radiation of infaunal bivalve molluscs: a consequence of mantle fusion and siphon formation. *Journal of Paleontology* 42:214–29.

Thayer, C. W. 1979. Biological bulldozers and the evolution of marine benthic communities. *Science* 203: 458–61.

Vermeij, G. J. 1977. The Mesozoic marine revolution: evidence from snails, predators, and grazers. *Paleobiology* 3:245–58.

pearance of eastward flow at the surface. This has been observed previously during April through September, when the trade winds weaken [K. Wyrtki *et al.*, *Science* 211, 22 (1981)].
4. Observed fluctuations in sea level are superimposed on the mean dynamic topography relative to 500 decibars [K. Wyrtki, *J. Phys. Oceanogr.* 5, 450 (1975)].
5. ______, *ibid.*, p. 572; J. McCreary, *ibid.* 6, 632 (1976).
6. J. McCreary, *Philos. Trans. R. Soc. London* 298, 603 (1981); S. G. H. Philander and R. C. Pacanowski, *J. Geophys. Res.* 85, 1123 (1980).
7. S. G. H. Philander and R. C. Pacanowski, *Tellus* 33, 201 (1981).
8. S. G. H. Philander [*J. Phys. Oceanogr.* 11, 176 (1981)] predicted the disappearance of the undercurrent following relaxation of the trade winds. This is a linear feature of his nonlinear model.
9. We thank W. Austin, the crew of the R.V. *Machias*, and the technicians who made possible the longtime series of current measurements. The work was funded primarily by the National Science Foundation, with additional support from the National Oceanic and Atmospheric Administration and the State of Hawaii. This is contribution No. 1435 from the Hawaii Institute of Geophysics.

11 July 1983; accepted 28 September 1983

Onshore-Offshore Patterns in the Evolution of Phanerozoic Shelf Communities

Abstract. *Cluster analysis of Cambrian-Ordovician marine benthic communities and community-trophic analysis of Late Cretaceous shelf faunas indicate that major ecological innovations appeared in nearshore environments and then expanded outward across the shelf at the expense of older community types. This onshore-innovation, offshore-archaic evolutionary pattern is surprising in light of the generally higher species turnover rates of offshore clades. This pattern probably results from differential extinction rates of onshore as compared to offshore clades, or from differential origination rates of new ecological associations or evolutionary novelties in nearshore environments.*

The broad outlines of the Phanerozoic history of skeletonized marine animals are now reasonably well known (*1*). However, comparatively little is known about how changes in global diversity relate to local environments (*2*). In light of this situation, we have analyzed marine faunal changes within environmental gradients for two pivotal intervals of the Phanerozoic: the Cambrian and Ordovician periods in the early Paleozoic and the Late Cretaceous Epoch at the end of the Mesozoic. The Cambro-Ordovician interval encompassed the origination of all three of the great "Evolutionary Faunas" that compose the Phanerozoic marine fossil record (*3*), with the appearance of the first shelly fauna (the "Cambrian Fauna") in the Early Cambrian, the rapid expansion of the more complex and diverse "Paleozoic Fauna" in the early and middle Ordovician, and, finally, the rise of early members of the modern fauna in the mid-to-late Ordovician. The Late Cretaceous interval, some 340 million years later, included a major reorganization of communities within the Modern Fauna, with the Mesozoic marine revolution bringing diversification of durophagous predators and infaunal bioturbators, decline of epifaunal suspension feeders, and increase in both global and local species richness (*1, 4*). Our analyses of the faunal changes within an environmental framework during both intervals indicate that the major new community types appeared first in nearshore settings and then expanded into offshore settings, despite higher rates of species-level evolution in the offshore habitats.

For the Cambro-Ordovician interval, a Q-mode cluster analysis was performed on 102 animal communities with well-documented macrofaunas [tabulated in (*5*)]. These communities, as illustrated in Fig. 1, were selected to give broad coverage of all marine environments from nearshore to continental slope and deep basin over the whole of the 140-million-year interval (*6*). The communities were clustered (*7*) in order to see what environments had similar ordinal–level faunas and therefore to determine where and when major faunal changes were occurring along the shelf-slope gradient. The analysis revealed four primary clusters of communities, represented by the patterned boxes in Fig. 1. The oldest two clusters correspond to the Cambrian Fauna (*3*) and are differentiated only by the dominant trilobite orders. The last appearance of the second Cambrian Fauna cluster is markedly time-transgressive, so that this grouping encompasses all shelf and slope localities in the Middle and Upper Cambrian, but then becomes restricted to progressively more offshore environments through the Ordovician—a pattern of faunal replacement first recognized by Berry (*8*). The third cluster corresponds to the Paleozoic Fauna, extending across the shelf after its initial diversification near the shoreline in the early Ordovician. Finally, a fourth primary cluster occurs in nearshore environments in the late Ordovician; this group represents the first appearance of the Modern Fauna in a distinct environmental association (*2*, p. 11; *9*). Thus, by the end of the Ordovician, the three major evolutionary faunas of the Phanerozoic oceans were arrayed in distinct community associations across the continental shelf and slope: the remnants of the Cambrian Fauna on the slope, the Paleozoic Fauna on the mid- to outer shelf, and the early members of the Modern Fauna on the inner shelf.

A parallel onshore-offshore pattern of faunal change was found in the distribution of adaptive types (as opposed to higher taxonomic groups) within the Modern Fauna over the course of the post-Paleozoic (Fig. 2). Late in the Cretaceous (Santonian-Maestrichtian) of the Gulf and Atlantic Coastal Plain and the Western Interior Provinces of North America, nearshore assemblages were dominated largely by infaunal suspension-feeders, whereas more fine-grained or midshelf assemblages (or both) were trophically mixed, containing a large complement of deposit feeders; these results correspond well to environmental patterns in modern marine benthos (*4, 10, 11*). Unlike their modern counterparts, however, Late Cretaceous mid-to-outer shelf and slope assemblages were still dominated numerically by immobile epifaunal suspension feeders (*10, 12*), the prevalent adaptive type across much of the Paleozoic and earlier Mesozoic shelf (*10, 13, 14*). These Late Cretaceous epifaunal dominants on soft substrata include pycnodont and exogyrine oysters, inoceramid bivalves, articulate brachiopods, and cyclostome and cheilostome bryozoans (*10*). In contrast, in offshore settings today immobile epifauna occur almost exclusively on hard substrata or on firm, coarse sediments that are either relict or maintained by current action (*10, 11*). Soft, offshore muds today are dominated by deposit feeders and carnivores, and lack the epifaunal suspension-feeding mode of life so prevalent in the Paleozoic and in certain Mesozoic habitats.

Several alternative evolutionary dynamics could have given rise to the onshore-offshore patterns of faunal change documented here. The patterns cannot be driven simply by differential speciation rates, because origination rates at low taxonomic levels actually tend to be higher offshore than onshore in both the Paleozoic (*15*) and post-Paleozoic (*16*). Two alternative mechanisms are:

1) *Differential extinction.* The greater extinction-resistance of nearshore clades (*16*) increases both the probability that nearshore innovations persist long

enough to diversify and the total number of speciation events within a clade over its lifetime (*17, 18*).

2) *Differential origination.* Although speciation rates are lower onshore, the temporal and spatial heterogeneity of nearshore environments may be conducive to the production of evolutionary novelties or new ecological associations (*19*); new community types could then expand across the shelf in the wake of attritional extinction of offshore taxa (*20*).

Both of these hypotheses are testable with detailed data on rates of origination and extinction of clades within their paleoenvironmental context. Whatever the underlying mechanism, the data summarized here indicate that major evolutionary ecologic changes were not accomplished entirely by random species replacement throughout the marine environment. Rather, they occurred by outward expansion of evolutionary innovations and new community types from a nearshore evolutionary crucible into more conservative offshore habitats.

DAVID JABLONSKI
Department of Ecology and Evolutionary Biology, University of Arizona, Tucson 85721
J. JOHN SEPKOSKI, JR.
Department of Geophysical Sciences, University of Chicago, Chicago, Illinois 60637
DAVID J. BOTTJER
Department of Geological Sciences, University of Southern California, Los Angeles 90007
PETER M. SHEEHAN
Department of Geology, Milwaukee Public Museum, Milwaukee, Wisconsin 53233

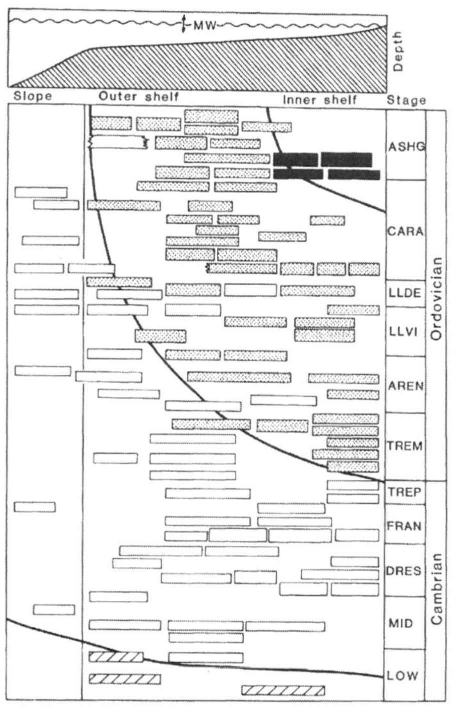

Fig. 1. Time-environment diagram showing the distribution of four primary clusters of Cambro-Ordovician fossil communities. Each box in the diagram represents a single community; the vertical position shows the age of the community; the horizontal position shows its approximate environmental range. Cluster membership of each community is indicated by patterning (diagonal ruling, Lower Cambrian shelf cluster unified by the joint possession of redlichiid trilobites, hyolithids, and inarticulate brachiopods; blank, Middle Cambrian to lower Ordovician shelf and Ordovician slope cluster unified by the joint possession of diverse ptychopariid trilobites and lingulid and acrotretid brachiopods; stippling, Ordovician shelf cluster unified by diverse orthid brachiopods, archeogastropods, trepostome and cryptostome bryozoans, crinoids, and some ptychopariid trilobites; solid black, Upper Ordovician inner shelf cluster distinguished by the dominance of bivalves, especially modiomorphoids, nuculoids, and pterioids). Cluster boundaries are strongly time-transgressive, indicating that major faunal associations originate in the nearshore environments and spread across the shelf. Stages from bottom to top are as follows: Lower, Middle, Dresbachian, Franconian, Trempealeauan, Tremadocian, Arenigian, Llanvirnian, Llandeilian, Caradocian, and Ashgillian.

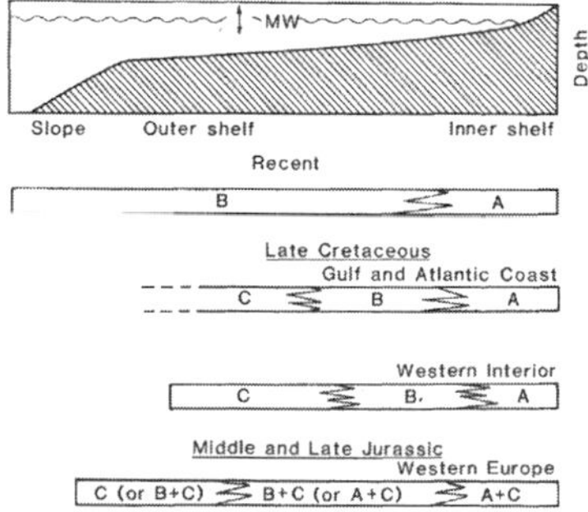

Fig. 2. Generalized macrofaunal adaptive types in generalized shelf transects in two Late Cretaceous provinces, and for comparative purposes, Recent and Middle-Late Jurassic. For the Cretaceous transects, note prevalence of immobile epifaunal suspension-feeders in soft-bottom offshore settings, reminiscent of Paleozoic and earlier Mesozoic shelf communities; in modern seas such habitats are occupied predominantly by small-bodied deposit feeders, again suggesting an onshore-new, offshore-archaic pattern of replacement for major benthic ecologic groupings. Predominant Late Cretaceous epifaunal taxa in the Gulf and Atlantic Province include gryphaeid oysters and, locally, articulate brachiopods and inoceramids, and in the Western Interior Province, inoceramid bivalves and an assortment of their epibionts.

References and Notes

1. N. D. Newell, *Geol. Soc. Am. Spec. Pap.* **89**, 63 (1967); J. W. Valentine, *Palaeontology* **12**, 684 (1969); *Evolutionary Paleoecology of the Marine Biosphere* (Prentice-Hall, Englewood Cliffs, N.J., 1973); J. J. Sepkoski, Jr., R. K. Bambach, D. M. Raup, J. W. Valentine, *Nature (London)* **293**, 435 (1981); D. M. Raup and J. J. Sepkoski, Jr., *Science* **215**, 1501 (1982).
2. A. J. Boucot, *J. Paleontol.* **57**, 1 (1983).
3. J. J. Sepkoski, Jr., *Paleobiology* **7**, 36 (1981).
4. G. J. Vermeij, *ibid.* **3**, 245 (1977); R. K. Bambach, *ibid.*, p. 152; M. LaBarbera, *ibid.* **7**, 510 (1981); C. W. Thayer, *Science* **203**, 458 (1979); W. I. Ausich and D. J. Bottjer, *ibid.* **216**, 173 (1982).
5. J. J. Sepkoski, Jr., and P. M. Sheehan, in *Biotic Interactions in Recent and Fossil Benthic Communities*, M. J. Tevesz and P. L. McCall, Eds. (Plenum, New York, 1983), p. 673.
6. Analyzed communities were confined to North America in order to avoid confounding environmental variation with biogeographic differentiation.
7. The analysis summarized in Fig. 1 was performed as follows. The frequency of species within each of 76 orders in each of the 102 communities was tabulated. On the basis of these data, similarities between communities were then measured with the index of proportional similarity [J. Imbrie and E. G. Purdy, *Am. Assoc. Petrol. Geol. Mem.* **1**, 253 (1962)]. This matrix was clustered by the unweighted pair-group method with arithmetic averages [P. H. A. Sneath and R. R. Sokal, *Numerical Taxonomy* (Freeman, San Francisco, 1973)]. Other similarity indices and statistical procedures, including Q-mode factor analysis, gave similar results, which suggests that the observed patterns are statistically robust.
8. W. B. N. Berry, *Lethaia* **5**, 69 (1972); *J. Geol.* **82**, 371 (1974) [see also R. Ludvigsen, *Geol. Assoc. Can. Spec. Pap.* **18**, 1 (1978); T. P. Crimes, *Sedimentology* **20**, 105 (1973); *Nature (London)* **248**, 328 (1974)].
9. See also P. W. Bretsky, *Science* **159**, 1231 (1968); *Palaeogeogr. Palaeoclimatol. Palaeoecol.* **6**, 45 (1969); K. W. Flessa, *Geol. Soc. Am. Abstr.* **5**, 160 (1973); N. J. Morris, *Philos. Trans. R. Soc. London Ser. B* **284**, 259 (1978).
10. D. Jablonski and D. J. Bottjer, in *Biotic Interactions in Recent and Fossil Benthic Communities*, M. J. Tevesz and P. L. McCall, Eds. (Plenum, New York, 1983), p. 747.
11. For example; D. C. Rhoads, *Oceanogr. Mar. Biol. Annu. Rev.* **12**, 263 (1974); A. A. Neyman, in *Marine Production Mechanisms*, M. J. Dunbar, Ed. (Cambridge Univ. Press, New York, 1979), p. 269; J. S. Levinton, *Marine Ecology* (Prentice-Hall, Englewood Cliffs, N.J., 1982).
12. Since many of the dominant taxa in offshore sediments secrete calcitic skeletons that are resistant to postdepositional dissolution, the observed faunal trends could be artifactural. However, preservation of aragonitic ammonites in

these rocks, preservation of aragonitic fossils in coeval deep-sea carbonates, and similar epifauna-rich community composition in rare occurrences of well-preserved offshore faunas, all indicate that diagenesis is only emphasizing a true ecologic pattern; see F. Surlyk, *Biol. Skr. Dan. Vidensk. Selsk.* **19** (No. 2), 1 (1972); E. G. Kauffman, *Treatise on Invertebrate Paleontology* (1979), p. A418; D. E. Hattin, *Kans. State Geol. Surv. Bull.* **225**, 78 (1982).

13. For example, A. Hallam, *Lethaia* **9**, 245 (1976); K. L. Duff, *Palaeontology* **18**, 443 (1975); F. T. Fürsich, *ibid.* **20**, 357 (1977).
14. See also B. S. Morton [*Malacologia* **21**, 35 (1981)] on onshore-offshore displacement of infaunal pholadomyoid bivalves by infaunal veneroid bivalves in the Cenozoic.
15. P. W. Bretsky and D. M. Lorenz, *Proceedings of the North American Paleontology Convention, Section E* (Allen Press, Lawrence, Kans., 1970), p. 522; R. A. Fortey, *Paleobiology* **6**, 24 (1980); J. F. Pachut, *J. Paleontol.* **56**, 703 (1982).
16. J. B. C. Jackson, *Am. Nat.* **108**, 541 (1974); D. Jablonski, *Paleobiology* **6**, 397 (1980); ——— and J. W. Valentine, *Evolution Today, Proceedings of the 2nd International Congress of Systematic and Evolutionary Biology* (1981), p. 441.
17. However, high extinction rates are generally balanced in offshore clades by high speciation rates that can also confer extinction-resistance [S. J. Gould and N. Eldredge, *Paleobiology* **3**, 115 (1977); T. A. Hansen, *ibid.* **6**, 193 (1980); S. M. Stanley, *Macroevolution* (Freeman, San Francisco, 1979)], so that probabilistic replacement of one clade by another due to differential species-extinction rates may not be sufficient to generate the pattern.
18. In addition to background extinctions, mass extinctions could also contribute to the pattern, especially if marine habitats are differentially affected.
19. The incessant local extinctions and recolonizations in frequently disturbed nearshore habitats may promote the origin of major new community types through repeated sorting and recombining of new and established species (*5*). Alternatively, the evolutionary novelties themselves may arise preferentially nearshore because new isolates in those habitats are commonly small and drawn from panmictic populations, and are thus more likely to undergo genetic revolutions or transiliences that could produce rapid shifts in morphology or physiology than the more frequent speciation events in offshore environments [see D. Jablonski and R. A. Lutz, *Biol. Rev. Phil. Soc.* **58**, 21 (1983); J. W. Valentine and D. Jablonski, in *Evolution, Time and Space: The Emergence of the Biosphere*, R. W. Sims, J. H. Price, P. E. S. Whalley, Eds. (Academic Press, New York, 1983); A. R. Templeton, *Evolution* **34**, 719 (1980); E. Mayr, *ibid.* **36**, 1119 (1982); but see B. Charlesworth, R. Lande, M. Slatkin, *ibid.* p. 474].
20. G. J. Vermeij, *Biogeography and Adaptation* (Harvard Univ. Press, Cambridge, Mass., 1978) for a different sort of combined physiological-competitive mechanism.
21. We thank A. J. Boucot, K. W. Flessa, F. T. Fürsich, A. Hallam, E. G. Kauffman, S. M. Kidwell, and G. J. Vermeij for critical comments. This study was partially supported by NSF grants DEB 81-21212 (D.J.), DEB 81-08890 (J.J.S.), and EAR 82-13202 (D.J.B.).

13 June 1983; accepted 22 October 1983

Productive Infection and Cell-Free Transmission of Human T-Cell Leukemia Virus in a Nonlymphoid Cell Line

Abstract. *Human T-cell leukemia virus (HTLV), American PL isolate, was transmitted by cocultivation and by cell-free filtrates to a nonlymphoid human osteogenic sarcoma (HOS) cell line, designated HOS/PL, but not to nine other lines bearing receptors for HTLV. HOS and HOS/PL cells are not dependent on interleukin-2 and do not express interleukin-2 receptors that are recognized by anti-Tac monoclonal antibody. HTLV released by the Japanese MT2 cell line was also transmitted to HOS cells. The infected HOS cells release substantial titers of progeny HTLV which is antigenically indistinguishable from parental virus and is able to transform T cells.*

Human T-cell leukemia virus (HTLV) is a C-type RNA tumor virus associated with a mature form of adult T-cell leukemia-lymphoma (ATLL). HTLV was first isolated and characterized from patients in the United States (*1*) and later in Japan (*2*), and in patients of West Indian origin (*3*) and in Israel (*4*). Human umbilical cord lymphocytes and peripheral blood lymphocytes cocultivated with HTLV-releasing lymphoma cells become infected and transformed in vitro (*4, 5*). Transformation of simian and rabbit peripheral blood T cells by HTLV has also been reported (*6*). Several of the T-cell lines transformed in vitro produce larger quantities of HTLV particles than the original tumor lines.

We have recently demonstrated that cocultivation of HTLV-producing cells with a variety of human and animal nonlymphoid cell types induces cell fusion, leading to the formation of large, multinucleated syncytia as a result of HTLV expression (*7*). These observations indicate that HTLV interacts with the surface of a number of cell types. Further studies with vesicular stomatitis virus (VSV) pseudotypes bearing the envelope glycoproteins of HTLV showed that there is a broad range of cells susceptible to pseudotype infection (*8*). Thus the expression of HTLV receptors is not restricted to lymphoid cells, because many cell types derived from diverse mammalian species are permissive for HTLV adsorption and penetration.

In this report we describe the productive infection of a nonlymphoid human cell line by American and Japanese strains of HTLV. Furthermore, we show that cell-free transmission of HTLV is achieved in this line.

Permissivity of HOS cells to HTLV replication. Five human and five animal cell lines known to have receptors for HTLV (*8*) were cocultivated with HTLV-producing C91/PL T cells. The human cells were 7605L embryonic lung fibroblasts, HOS osteogenic sarcoma cells, RD rhabdomyosarcoma cells, HeLa cervical carcinoma cells, and EJ bladder carcinoma cells. Animal cells were Vero African green monkey kidney cells, Fcf2th canine thymus murine sarcoma virus (MSV)-transformed S+L− cells, feline CCC MSV-transformed S+L− cells, CCL64 mink lung cells, and XC Rous sarcoma virus (RSV)-induced rat sarcoma cells. In the first set of experiments HTLV-producing cells were not x-irradiated but during serial passage the lymphoma cells were soon lost from the adherent cultures. The cells were maintained in Dulbecco-modified Eagle's medium with 5 to 10 percent fetal calf serum and were passaged for 5 months.

Although each of the ten cell types cocultivated with HTLV-producing cells was susceptible to HTLV penetration and eight were susceptible to HTLV-induced cell fusion, only one cell type, the HOS cell line (*9*), was permissive for HTLV replication. During the first 2 weeks of cocultivation, cell fusion occurred among the HOS cells, but with the loss of C91/PL cells on passage, the

Table I. Virus production, syncytium induction, pseudotype formation, and antigen expression by HOS/PL cells.

Cell line	Reverse transcriptase*	Syncytium induction†	VSV (HTLV) pseudotype titer‡	Percentage of cells immunofluorescent: HTLV antigens§ ATLL	HTLV antigens§ p19	T-cell marker¶	IL-2 receptor‖
C91/PL	16361	+++	3 × 10^8	87	89	72	85
HOS/PL	18890	++++	5 × 10^4	98	82	0	0
HOS	926	−	< 10^1	0	0	0	0

*Assay of viral RNA-directed DNA polymerase, expressed as the counts per minute of [^{3}H]TMP incorporated during incubation for 60 minutes at 37°C (*7*). †XC indicator cells were cocultivated with test cells for 18 hours and examined for syncytia (*7*). The results are expressed as the percentage of nuclei contained within syncytia: −, no syncytia; +++, 30 to 50 percent; ++++, > 50 percent. ‡Plaque-forming units per milliliter of vesicular stomatitis virus (VSV) with envelope antigens specific to HTLV (*8*). §Indirect immunofluorescence on fixed cells using serum from an antibody-positive ATLL patient (*7*) and monoclonal antibody to p19 (*10*). ¶Indirect immunofluorescence with UCHT1 monoclonal antibody (*12*) on live cells. ‖Indirect immunofluorescence with anti-Tac monoclonal antibody (*14*) on live and fixed cells.

PAPER 17

Patterns of Taxonomic and Ecological Structure of the Shelf Benthos during Phanerozoic Time (1969)

J. W. Valentine

Commentary

MICHAEL FOOTE

The many shades of ink and graphite in the margins of my >25-year-old, dog-eared copy of Jim Valentine's 1969 paper reflect the numerous times I have returned to this work to learn anew from his insights. Recently, Valentine (2009) cited several people who influenced *his* thinking, but clearly his 1969 paper portrayed his own strikingly original worldview, a synthesis of ecological and evolutionary theory with biogeographical, geological, and paleontological data.

What strikes me is that the questions he framed, the approaches he developed, and the claims he made remain with us today, representing a broad sampling of the major themes in paleobiology. It can be a parlor game to mine modern ideas out of classic works, but in Valentine's case no reading between the lines is necessary, and the pedigree of his ideas can easily be traced forward to the present day. These include: the relationship between ecological and taxonomic hierarchies; the evolution of ecosystems in addition to taxa; the relationship between alpha and global diversity; energy flow (not just genes) as ecological currency; contributions of provinciality and climatic gradients to global diversity; the need to model species richness rather than tabulate it directly, given incomplete sampling; the relative importance of ecospace expansion, niche partitioning, and niche overlap in accommodating increased taxonomic diversity; temporal changes in the size of evolutionary transitions; and systematic differences among taxonomic levels in diversity trajectories and rates of origination.

In my latest read, I was struck by the discussion on page 694 related to diversity dependence of diversification. Valentine contributed to this area in later years, touching on the degree of diversity dependence of speciation rates (high) versus species extinction rates (low to absent; Walker and Valentine 1984). A point I had not fully appreciated before, however, was his 1969 argument that familial extinction within one higher taxon is often followed by enhanced diversification of families in a *different* higher taxon. Whether this will ultimately be borne out by more up-to-date treatment is unclear, but it adds interesting texture to discussions of biotic interaction in the generation of global diversity at different hierarchical levels.

Literature Cited

Valentine, J. W. 2009. The infusion of biology into paleontological research. In *The Paleobiological Revolution*, edited by D. Sepkoski and M. Ruse, 385–97. Chicago: University of Chicago Press.

Walker, T. D., and J. W. Valentine. 1984. Equilibrium models of evolutionary species diversity and the number of empty niches. *American Naturalist* 124:887–99.

From *Palaeontology* 12:684–709.

PATTERNS OF TAXONOMIC AND ECOLOGICAL STRUCTURE OF THE SHELF BENTHOS DURING PHANEROZOIC TIME

by JAMES W. VALENTINE

ABSTRACT. The taxonomic and ecological structure of the shelf biota are intimately related at the species-population levels. Early Paleozoic faunas contained relatively few species representing relatively many higher taxa, and ecosystems were relatively generalized. Medial and late Paleozoic faunas contained more species representing fewer higher taxa, and ecosystems were relatively specialized. This suggests that, as higher taxa became extinct, they were not replaced except at lower taxonomic levels; diversification was proceeding through increasing specialization. After Permo-Triassic extinctions, rediversification was chiefly confined to low taxonomic levels. Late Mesozoic and Cenozoic diversification at lower taxonomic levels has been remarkably great, resulting not only from increasing specialization at the population level but from a marked increase in provinciality due to rising latitudinal temperature gradients on the shelves and to the fragmentation and isolation of shelf environments by continental drift.

THIS paper examines the historical relationships between the ecological and taxonomic structures of the marine biosphere, and attempts to account in a general way for the patterns of their evolution. Each of these structures is hierarchic. The units composing the levels of the ecological hierarchy include individuals, populations, communities, and provinces, while the units composing the levels of the taxonomic hierarchy are such categories as species, genera, and families.

There has been relatively little theoretical discussion of the evolution of these structures for marine invertebrates, yet the geological record of skeletonized taxa of the shallow marine invertebrate benthos is longer and more complete than for any comparable group of organisms. This paper therefore deals with the rich and lengthy record of shallow marine environments.

This restriction to a specific group of communities has some special advantages. The diversity pattern for the world at large is obviously very much influenced by the deployment of organisms into new environments, such as the invasion of the terrestrial habitat by vertebrates. By restricting the data to a limited group of communities it may be possible to investigate the patterns of diversity changes within ecosystems.

Much of the structural evolution which the taxonomic and ecological hierarchies have undergone is a product of the diversification and extinction of species. There has been much discussion of the patterns of taxonomic diversifications and extinctions through geologic time, especially of higher taxa, and ecological relations are commonly invoked to account for these patterns, particularly for extinctions. The processes of diversification assumed herein are those of the synthetic theory of evolution based on Darwinian selection and upon modern genetic concepts. Speciation and the origin of higher taxa have been discussed from this viewpoint in a number of larger works (for example Huxley 1942; Mayr 1963; Rensch 1947; and Simpson 1953). As diversity rises, there must be a mechanism of accommodation of the new forms in ecological systems; such mechanisms are discussed by Klopfer (1962), MacArthur and Wilson (1967), and

[Palaeontology, Vol. 12, Part 4, 1969, pp. 684–709.]

Miller (1967), among others. Possible causes of extinction, and hypotheses of the processes of extinction that have operated to create the Permo-Triassic faunal change, have been reviewed by Rhodes (1967).

THE ECOLOGICAL STRUCTURE OF THE BIOSPHERE

The ecological hierarchy is regarded as being composed of the levels that are depicted in text-fig. 1. This paper is chiefly concerned with the functional aspects of the hierarchy,

GENETIC HIERARCHY		ECOLOGICAL HIERARCHY		
UNIT	*COLLECTIVE*	*DESCRIPTIVE UNITS*	*FUNCTIONAL UNITS*	*LEVEL*
$*^2$	$*^3$ →	Marine Shelf Biota	Shelf Realm of the Biosphere	High
$*^1$	$*^2$ →	Province	Provincial System	
Gene Pool	$*^1$ →	Community	Community System (Ecosystem)	
Genotype	Gene Pool →	Population (Deme, Species)	Population System (Niche)	
Functional Genetic Unit	Genotype →	Individual	Ontogenetic System	Low

$*^1$ *Collection of gene pools*
$*^2$ *Collection of gene pool collections*
$*^3$ *Collection of collected gene pool collections*

TEXT-FIG. 1. Some levels of organization in the ecological hierarchy employed in this paper (after Valentine 1968*b*).

that is, with the interacting systems of organisms and environments. From the highest level down, each functional system is composed of subsystems representing the systems of the next lower level. The lowest functional level in the figure, that of the individual, is certainly capable of further subdivision into sorts of functions of 'unit characters', each underpinned by a system of genes and its regulators. For the most part, however, the present discussion concerns population and higher levels.

It is convenient to consider the ecological units in terms of the environment with which they interact. Hutchinson (1957, 1967) has developed a formal conceptual model that treats the environment as a multi-dimensional region (see also Simpson 1944, 1953). Only an informal treatment, based on Hutchinson's model, is required here. If each separate environmental parameter is visualized as a single geometric dimension of this region, then all possible environments are represented by the resulting multi-dimensional space or hyperspace, which contains as many dimensions as there are possible environmental parameters. The space extends along each dimension to the physical limits of each parameter. It is assumed that this multi-dimensional environment model is standardized by having each axis allotted an arbitrary but permanent direction to form an

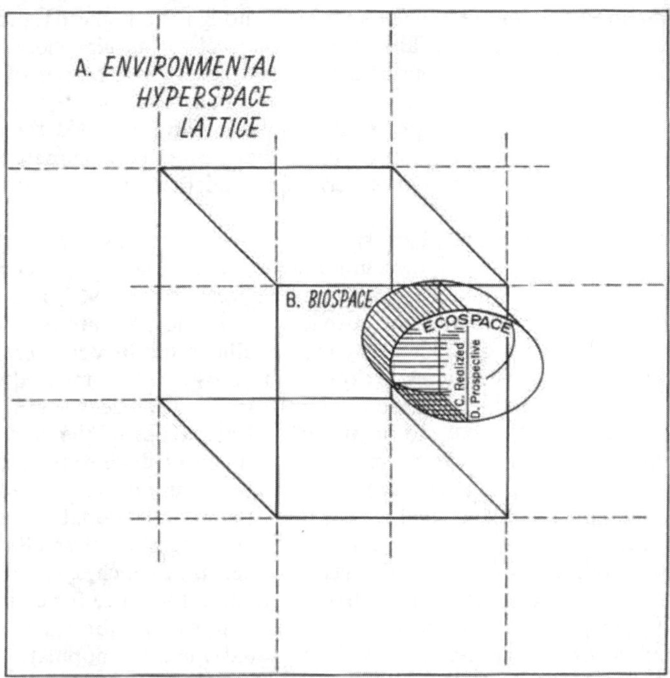

TEXT-FIG. 2. Highly diagrammatic representation of some aspects of environment–organism relations, visualized as a multi-dimensional space, of which each dimension is some environmental factor, physical or biotic. Each point within the lattice represents a unique combination of factors. Only three of the many dimensions are depicted. A, the total possible range of all environmental factors represented as a multi-dimensional lattice. B, the portion of the environment that actually exists on earth, the *biospace*; it is available for occupation by organisms. C and D, the region of environmental space that coincides with factors tolerated by an organism and that is bounded by its limits of tolerance—the *ecospace* of the organism. Only a portion of the ecospace is realized (C); the remainder is prospective ecospace (D) that may be inhabited if the environment fluctuates so as to include more of that portion of the lattice. The ecospace concept may be expanded to population, community, province or biosphere levels, and to species, genus, family, and higher taxonomic levels.

environmental hyperspace lattice, hereinafter called simply a lattice for brevity. Only a certain portion of the total possible lattice (the prospective lattice) actually represents conditions of the environment. This 'realized ecological hyperspace' may be called *biospace* (text-fig. 2), a term employed in a similar but less generalized sense by Doty (1957).

For any organism there is some more or less small volume (actually a hypervolume) within the lattice corresponding to the range of environmental conditions under which it may live. This functional hypervolume will be called the *ecospace* of that organism (text-fig. 2). Each population also has its own ecospace, which is the hypervolume of its niche within the lattice. Indeed, the ecological units at all levels have ecospaces. A community ecospace is the multi-dimensional model of its ecosystem, and a provincial ecospace is the model of the provincial system. Although the highest functional level standing above that of the provincial system is the level of the biosphere, the system of the shallow marine realm is being used here in its place as a matter of simplicity, and this realm has its own ecospace. The total ecospace that an organism or other ecological unit may utilize if it is physically available may be called the *prospective ecospace*, while the portion of the ecospace that actually overlaps with realized biospace may be called the *realized ecospace*. These terms are modelled on the discussion of Parr (1926) and Simpson (1944, 1953).

Dimensions of the lattice which have special properties are those that represent the *real* dimensions of space, in which discontinuities occur that permit the occupation of similar functional regions in different geographic regions (Miller 1967) and of time, in which the changing shapes and sizes of ecospaces and of biospace are perceived.

The structure of the ecological hierarchy may be illustrated by considering just one level, for example the community level. Community ecospace is composed of the ecospaces of all the niches of the component populations, and includes some dimensions that are not niche properties but are organizational properties of the ecosystem. The size of the community ecospace, measured by the number of dimensions occupied and the extent of occupation along each dimension, depends upon the sizes of the component niche ecospaces and to a small extent upon the organizational properties. Into a community ecospace of a given size, a relatively large number of small niches or a relatively small number of large niches may be packed. All niches in a community ecospace overlap to some degree, for all share a common tolerance for certain salinity ranges, for example, and for certain oxygen concentrations, and for other parameters. The more that niches overlap, other things being equal, the more populations that can be packed into a community ecospace of a given size (Klopfer 1962; Miller 1967).

Consider, then, a community (*A*) composed of relatively few populations that have very large niches that overlap only narrowly on the whole. The animals tend to be rather generalized feeders, so that energy flows in relatively broad streams through the trophic levels. This community, though of low diversity, may displace a large biospace in the lattice, that is, may have a large ecospace. Consider another community (*B*) composed of many populations of different species that tend to have very small niches which overlap broadly on the average. The animals are highly specialized with relatively narrow ranges of food sources, so that energy flows through the trophic levels in relatively discrete paths along chains of organisms that tend to be rather isolated owing to their high specialization. Energy flow is not like a stream but more like a shower that breaks up into numerous jets. A community of this sort, though rich in species, may displace no more biospace in the lattice than community (*A*) and may displace considerably less.

These communities have vastly different structures in the lattice, and yet it seems possible for one to evolve from the other. They may thus represent relatively early (*A*) and advanced (*B*) stages in the evolution of a community 'lineage' that has inhabited a similar biotope through its history. In this event the ecospaces of the two communities will approximately coincide, although the way in which each community biospace is occupied by niches is different. The community structure has evolved.

Structural states of ecological systems at other levels may be described in an analogous way. All the systems evolve by changes in the quality, relative proportions, and diversity of their subsystems (Valentine 1968*b*). Thus evolution of ecological systems need not involve organic evolution, but may result merely from the readjustment of existing populations in new patterns of association. However in the present discussion the chief interest lies in changes that *are* based upon organic evolution, upon changes in gene frequencies within populations that produce changes in niches, and upon the accommodation of the changed niches in ecosystem structures. Enough is now known of these processes to permit the construction of a provisional model of the diversification of ecosystems. But before proceeding to the model, it is appropriate to examine the main patterns of taxonomic structure during the Phanerozoic.

THE TAXONOMIC STRUCTURE OF THE BIOSPHERE

The taxonomic hierarchy is too well known to require any general remarks. For purposes of this paper only a few levels need be considered: phylum, class, and order, which will be called 'higher' taxonomic categories; and family, genus, and species, which will be called 'lower' categories. It is possible to visualize the ecospace of any genus as composed of the ecospaces displaced by all its component species, and the ecospace of a family as composed of all the generic ecospaces, and so on. Thus defined, the ecospace of a higher taxon displaces the actual regions of the lattice that have been occupied by the members of that taxon. Thus the taxonomic hierarchy possesses a precise structure at any time. This structure changes through time in well-defined patterns.

The main trends of evolution of the taxonomic structure may be characterized by considering the trends of diversity among higher and lower taxa through geologic time. The fossil record of diversity, however, is certainly biased. An important source of bias is the differential preservation of taxa. It seems possible to use the skeletonized taxa that are best represented as a sample, from which to attempt to generalize to the entire biota. The basic data from which generalizations will be attempted are the records of easily fossilized shallow benthonic taxa of nine phyla: Protozoa, Porifera, Archaeocyatha, Coelenterata, Ectoprocta, Brachiopoda, Mollusca, Arthropoda, and Echinodermata. The ranges of these phyla and of their taxa are taken chiefly from the *Treatise on Invertebrate Paleontology* (ed. Moore 1953–67), the *Fossil Record* (ed. Harland *et al.* 1967), and the Russian *Osnovy Paleontologii* (Orlov 1958–64).

As the assignment of groups of organisms to taxonomic categories involves a large element of subjectivity, it is fair to ask to what extent the trends in taxonomic diversity are real. In the first place, if one constructs a hierarchical classification of fossils that appear at different times, the average time of appearance of higher taxa will be earlier than that of lower, simply because some of the lower taxa appeared later than others, but none appeared earlier than the higher taxa to which they belong. The mode of first appearance should shift progressively towards the present at lower and lower taxonomic levels (Simpson 1953, pp. 237–9). Similarly, the mode of highest diversity will tend to shift towards the recent at progressively lower levels provided that the earlier taxa at each level persist or are replaced. These considerations account for such shifts in the mode of appearance and diversity in text-fig. 3.

Secondly, there is no doubt that the present data contain monographic artifacts (for

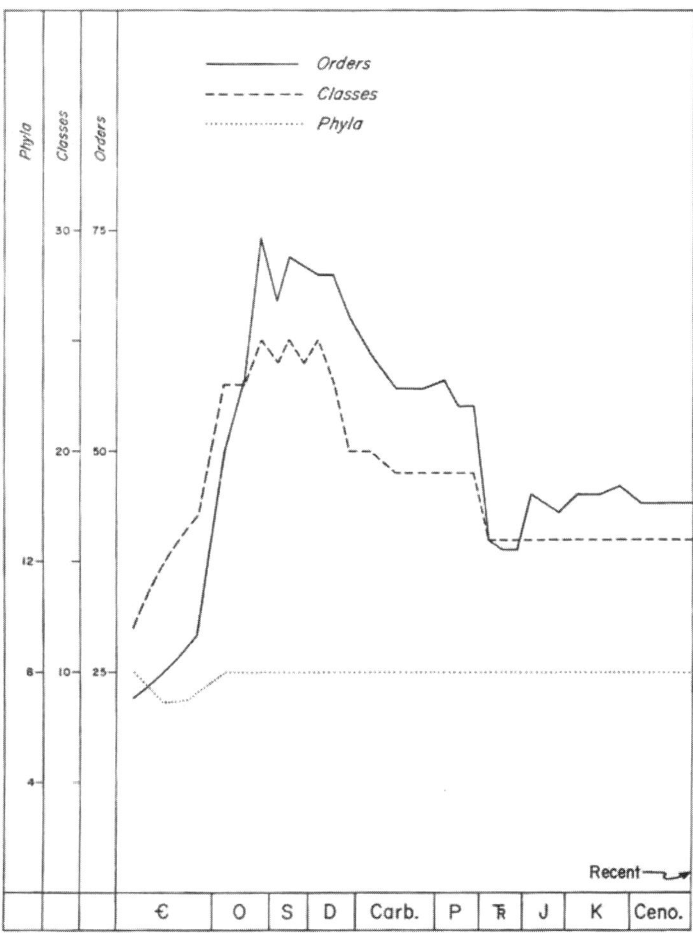

TEXT-FIG. 3. Stratigraphic variation in diversity of higher taxa of well-skeletonized marine shelf invertebrates. Data chiefly from Harland *et al.* (1967) and Moore (1953–7).

an instructive example see Williams 1957). However some of the main points to be discussed here concern relative diversities among the several taxonomic levels. Presumably, monographic artifacts would tend to appear at all levels, and relative diversities would be much less affected than absolute diversities. It is unlikely that there is a consistent monographic bias in the same direction among a majority of the taxa, and I therefore believe that the major trends are real.

Another important consideration is the extent to which trends among skeletonized taxa represent the biota as a whole. At present the non-skeletonized Invertebrata have the same biogeographic and synecological patterns as skeletonized groups (Lipps, *in press*), and there is no reason to expect that patterns of diversity of non-skeletonized taxa would follow different trends than the skeletonized ones. Furthermore Lipps has pointed out (pers. comm.) that the preserved groups are morphologically diverse and unrelated, yet they often exhibit similar patterns. Therefore it is assumed that major trends among skeletonized and non-skeletonized groups tend to be in phase.

Experience has shown clearly that the chances of preservation of an organism that does not possess a well-mineralized skeleton are exceedingly small. Indeed, the lack of a record of a taxon that does not have a relatively high probability of preservation can hardly be taken as proof that the taxon was not living at the time. And the probabilities of preservation cannot yet be specified even for taxa with highly mineralized skeletons, under many of the stratigraphic situations common in the geologic record. It is therefore difficult to assess the significance of negative records.

Most of the known phyla had appeared in the record by Cambrian time, although even among our sample of nine, one (Ectoprocta) does not appear until the Lower Ordovician. The phyla are well-differentiated and some contain relatively complex organisms when they first appear, so that a fairly long period of evolution can be assumed to have preceded their appearance in the record. However, it is possible to argue for many phyla that their final organization into the ground-plans that are now considered as characteristic may have only narrowly preceded their appearance in the record (Cloud 1949, 1968). It has been suggested that such a great evolutionary event may have been permitted by an increase in atmospheric oxygen past a critical level (Berkner and Marshall 1965). At any rate, it is likely that nearly all of the invertebrate phyla had become established before the Cambrian, and the relative timing of their appearance in the record may partly indicate the order in which they acquired hard parts or the chance occurrences of unusual preservations. Nicol (1966) and others have suggested that the acquisition of hard parts may have ensued as a result of widespread phyletic body-size increases.

Many of the nine phyla in the sample contain taxa that are not members of the shelf benthos or that have relatively low probabilities of preservation. Examples are the planktonic Scyphozoa and the soft-bodied Keratosa. Such taxa are excluded from the tallies. A few other taxa probably participated only partly in the benthonic ecosystems. An important example is the Ammonoidea. The effects of such taxa on the diversity curves are considered separately. Diversity graphs for phyla, classes and orders that seem to be chiefly members of the benthos are presented in text-fig. 3; diversities are classed by geological epochs, and therefore do not exactly represent the standing diversities at any given time. The diversity levels depicted in text-fig. 3 represent a balance between diversification and extinction, but the amount of taxonomic turnover that has occurred in any epoch cannot be inferred from the diversity levels. Text-fig. 4 depicts the numbers of appearances and disappearances (presumed to be extinctions) among the higher taxa per epoch.

From text-figs. 3 and 4 the following history of higher taxonomic diversity can be inferred. The higher the category the earlier it tends to reach its maximum diversity. If the phyla were not all present throughout the Cambrian, at least they all appear by early Ordovician time, but the highest diversity is recorded in the Middle and Upper Cambrian. New classes continue to appear until the Lower Carboniferous, but the highest class diversity is recorded in the Middle and Upper Ordovician. Orders have

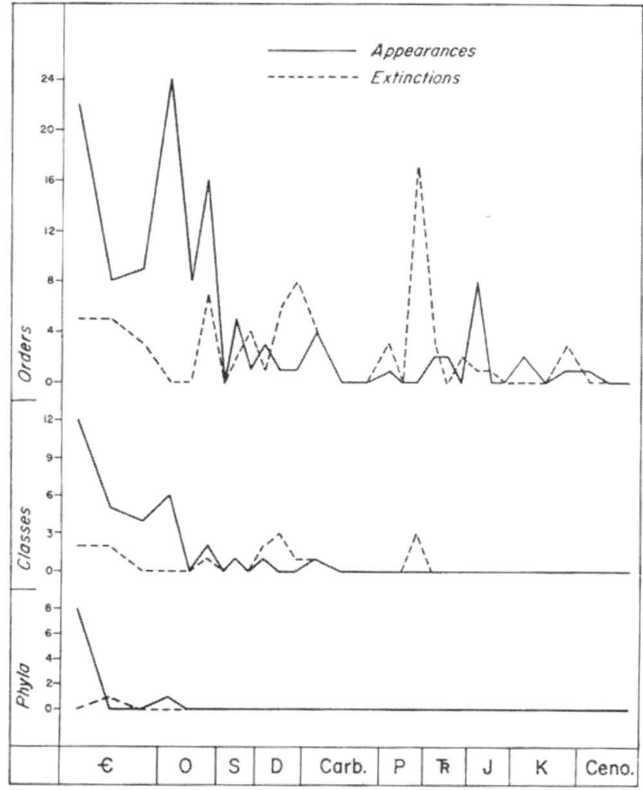

TEXT-FIG. 4. Appearances and extinctions of phyla, classes and orders of shelf benthos in the sample, classed by Epochs. Data as in text-fig. 3.

continued to appear until the end of the Cretaceous, but achieved their highest recorded diversity in medial Ordovician time. Ordinal diversification, however, was great during early Ordovician time, whereas the greatest class diversification was during the Cambrian (text-fig. 4).

It also appears that the higher the taxonomic category the less it has been ravaged by extinction, and the earlier extinction has stopped. Only one phylum (Archaeocyatha), which comprises 11% of the sample, disappears, although the sample is so small that this figure cannot be taken as very precise. However, 16 classes comprising 50%, and

75 orders comprising 64%, disappear. These extinct taxa are never fully replaced by other higher taxa, at least not from among the taxa that we are considering, so that the diversity of each higher taxonomic category has decreased, rather markedly in the cases of classes and orders, since the early Paleozoic. For the taxa in the sample, the extinction of phyla is complete by the end of Cambrian time, of classes by the end of Permian time, and of orders by the end of the Cretaceous.

TEXT-FIG. 5. Stratigraphic variation in diversity of skeletonized families of shelf benthos belonging to phyla included in text-fig. 3. Data chiefly from Harland *et al.* (1967), Moore (1953–7), and Orlov (1958–64)

The lower taxa present a somewhat different pattern. Text-fig. 5 depicts the geological record of family diversity in the sample, again excluding unsuitable taxa such as the planktonic foraminiferal families, and text-fig. 6 depicts the record of appearances and disappearances. About 100 families are recorded by the end of the Cambrian, and 300 occur in the Upper Ordovician; diversity remains near 300 until the later Paleozoic, when it gradually falls off. A marked drop occurs in late Permian and early Triassic times. The pattern has until this point been not too unlike that of the higher taxa, with the mode of major diversity shifted towards the present. The Jurassic rise is even anticipated on the ordinal level (text-fig. 3). However it is in the great diversity rise of the Cretaceous and Cenozoic that the pattern of the families departs in a fundamental way

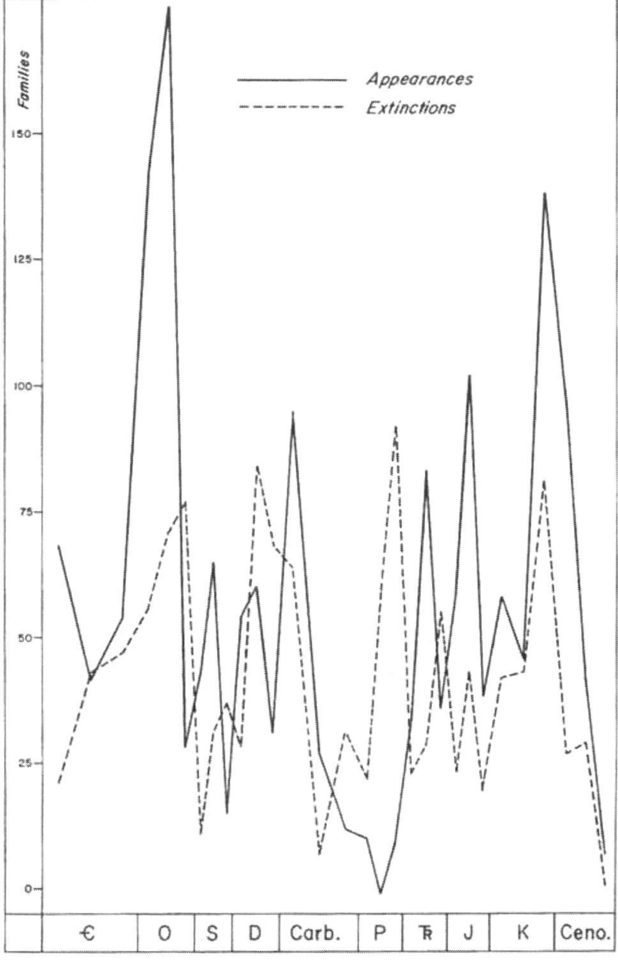

TEXT-FIG. 6. Appearances and extinctions of the families of shelf benthos in the sample, including Nautiloidea and Ammonoidea, classed by Epochs. Data as in text-fig. 5.

from the pattern of the higher taxa, for the higher taxa reach and maintain rather steady levels of diversity from Ordovician onwards for phyla, Lower Triassic onwards for classes, and lower Jurassic onwards for orders. The Cretaceous and Cenozoic diversification of families marks a major change in the evolutionary trends of taxonomic structure.

It can be seen from text-fig. 6 that the times of diversification of families tend to alternate with times of extinction. This pattern is well shown in a figure by Newell (1967, fig. 7) that is based upon a different taxonomic sample. The pattern certainly suggests that times favourable for diversification and those favourable for extinction were distinct, and that there is therefore a complementary relation between these processes. Newell suggests in effect that there have been extinctions to provide unoccupied biospace before there are major diversifications, and of course there must be many taxa resulting from diversification before there can be major extinctions. However, the data in text-figs. 5 and 6 do not entirely bear out this thesis. The very high extinction peaks in the Ordovician, Devonian, and Cretaceous are not accompanied by massive reductions in standing diversity. In fact, the early Ordovician and Cretaceous extinction highs are nearly hidden in text-fig. 5 owing to the great contemporary diversifications. Late Ordovician and Devonian extinction peaks reduce the diversity level somewhat, but only by about 6 and 13% respectively, because diversification is fairly high at these times. Only near the Permo-Triassic boundary, when diversification is exceedingly low (text-fig. 5), does the diversity level suffer a major decline of about 50%. The unusually low level of Permo-Triassic diversity is not unique because of the extinction peak alone, but because of the lack of a corresponding peak of diversification.

It is interesting to examine the patterns of family diversification and extinction within each of the higher taxa. Newell (1967) has presented graphs of family diversification among a number of higher taxa; I have prepared similar charts for diversification as well as for extinction of the families of higher taxa in the present sample which confirm the patterns he has presented. In general, however, the high rates of family diversification and of extinction within a higher taxon do not alternate in time but are highly correlated. For example, the Brachiopoda diversify strongly during the Ordovician and Devonian, but extinction peaks are found at these times also. Secondary levels of diversification during the Silurian and Lower Carboniferous correlate with secondary peaks of extinction. Only in the Permian is there a lack of correlation; the great extinction is not accompanied (nor is it followed) by diversification at the family level, but is accompanied by the lowest diversification rate known for brachiopods during the Phanerozoic.

Trilobites display a similar correlation. Cambrian diversification rises to a peak in the late Cambrian and falls off progressively during Ordovician epochs; extinction levels do precisely the same, except that they rise a bit in latest Ordovician. There is no following rise in diversity, however, although there is a secondary peak of extinction in medial and late Devonian time. Certainly there is no alternation of diversification and extinction. The same may be said of diversity patterns of the Porifera, the Echinoidea, and several of the Paleozoic echinoderm groups. The Foraminiferida, Anthozoa, and Gastropoda have more complex patterns, but there is no suggestion of alternating extinction and diversification except at the Permo-Triassic boundary. In the Ostracoda there is some indication that extinction follows diversification and not the reverse, and similar trends are found during parts of the record of other taxa. Even in the Ammonoidea and Nautiloidea peaks of diversification and extinction tend to correlate and certainly do not alternate.

The alternation of peaks of diversification and extinction of all families in the sample (text-fig. 6), then, are chiefly due to the alternation of high rates of family extinction of some higher taxa with high rates of diversification of different higher taxa (which is usually accompanied by a rise in extinction among these different taxa also). Except at the Permo-Triassic boundary, all this tends to be accomplished while diversity as a whole remains surprisingly stable considering the magnitude of extinction and diversification peaks. The major events that alter standing diversity on the family level are the diversification in Cambro-Ordovician times, the Permo-Triassic diversity low, and the Cretaceous–Cenozoic diversification (text-fig. 5).

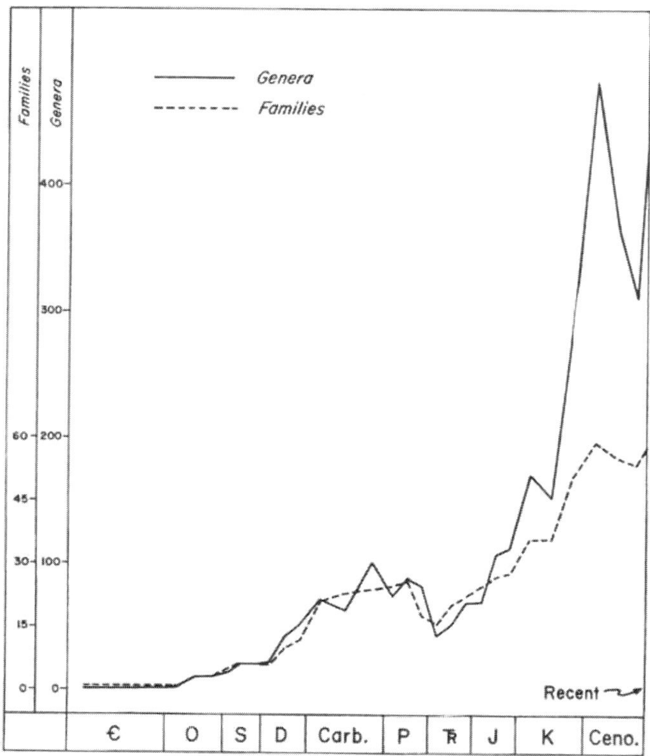

TEXT-FIG. 7. Stratigraphic variation in the diversity of benthonic shelf families (dashed line) and genera (solid line) of Foraminiferida, excepting the poorly skeletonized Allogromiina. Data from Loeblich and Tappan (1964).

It is possible to demonstrate that with the genera as with the families, there is a striking rise in numbers in the Cretaceous and Cenozoic. Text-fig. 7 depicts the diversity of families and of genera of benthonic Foraminiferida through geological time. This is one of the taxa that contributes strongly to the late Cretaceous and Cenozoic rise in family diversity. Throughout the Paleozoic there are, on the average, about 3–4 genera per family described. Across the Permo-Triassic boundary this ratio drops and then rises again in the Jurassic and early Cretaceous. In the late Cretaceous the genus/family ratio climbs to nearly 6, and in the early Cenozoic to over 8, where it stands at present after a late Cenozoic decline. The size of families is somewhat a matter of opinion. There is no special reason, however, to believe that the disproportionate Cretaceous–Cenozoic rise is a taxonomic artifact. The data are certainly subject to monographic and other biases, but have all been reviewed by the same team of authorities (Loeblich and Tappan 1964). It is interesting in this regard that the same trend can be inferred from the data charted by Henbest (1952) based chiefly on the work of Cushman (1948). If the trend is an artifact it is an enduring one. Incidentally, peaks of extinction of genera of foraminiferida correlate rather than alternate with peaks of diversification, just as is common among invertebrate families.

Another taxon that contributes heavily to the Cretaceous rise in family diversity is the Gastropoda, among which the same pattern of disproportionate generic diversification is present. There is no satisfactory recent review of all marine gastropod genera,

ORDERS	GEOLOGIC RANGE	FAMILIES	Genera & Subgenera	Genera-Subgenera / Family
Archaeogastropoda[1]	U. Cambrian-Recent[2]	82	1314	16.0
Mesogastropoda	U. Ord. (Caradocian)-Recent	75	1467	19.6
Opisthobranchia	L. Carb. (Visean)-Recent	9	240	26.6
Neogastropoda	L. Cretaceous (Albian)-Recent	20	1119	56.0

[1] Including Bellerophontacea.

[2] Possibly from L. Cambrian, depending upon ordinal assignment of early groups.

TEXT-FIG. 8. Families, genera, and subgenera of orders of shallow marine gastropods. The most advanced orders contain higher numbers of genera and subgenera, on the average, than primitive orders. Data from Taylor and Sohl (1962).

but Taylor and Sohl (1962) have published a census of gastropod genera and subgenera combined by family and higher taxa. It is possible to show that the more recently an order has appeared in the record, the more genera and subgenera per family it contains on the average (text-fig. 8). The Neogastropoda, which appear in the Lower Cretaceous (Albian or possibly earlier) and which diversify chiefly in the Upper Cretaceous and later, have 56 genera and subgenera per family on the average. This is 3½ times as many as the average of the Archaeogastropoda. Furthermore, even the groups of archaeogastropods with the largest living representatives, such as the trochaceans, tend to have differentiated strongly at the generic level in the Upper Cretaceous and Cenozoic. For the Trochacea, for example, there are 14 genera and subgenera recorded from the Lower Cretaceous, 32 from the Upper Cretaceous, and 66 from the Upper Cenozoic (data from Moore 1953–7). The other orders of Gastropoda are intermediate, both in time of appearance and in genus–subgenus/family ratios, between the Archaeogastropoda and the Neogastropoda. It follows from this situation that, for the shallow marine shelled Gastropoda, the late Cretaceous–Cenozoic rise in family diversity (from 37 families in the Lower Cretaceous to 66 in the Upper Cretaceous and to 83 or 84 in the Cenozoic) is disproportionately exaggerated on the generic level. This agrees with the data for Foraminiferida. Two other groups that contribute especially strongly to the rise in Upper Cretaceous and Cenozoic family diversity, the Ectoprocta and the

Echinoidea, appear to display similar trends (see Newell 1952 and the appropriate *Treatise* volumes).

In contrast to the preceding groups, the phylum Brachiopoda displayed its greatest familial diversity during the Paleozoic. It has not diversified during the Cretaceous–Cenozoic but maintained an average of about 12 or 13 families and about 50–65 genera

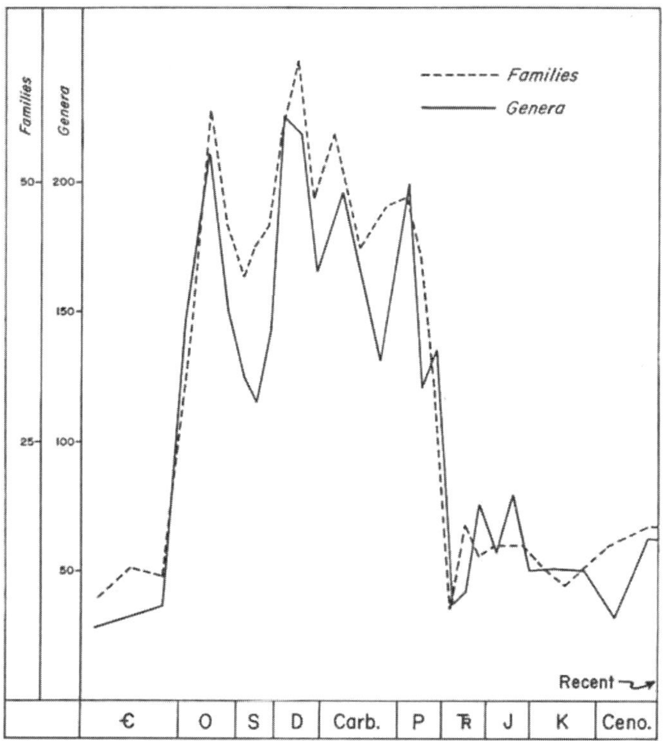

TEXT-FIG. 9. Stratigraphic variation in brachiopod diversity by families (broken line) and by genera (solid line). The genera/families ratio remains near 4. Genera after Williams (1965), and family data from Williams *et al.* (1965).

during this time (text-fig. 9). It is interesting, therefore, that the brachiopods have about the same genus/family ratio recorded for the Paleozoic as in the Mesozoic and Cenozoic —about 4 times as many genera as families throughout the Phanerozoic. Even during the greatest periods of diversification, the numbers of genera per family did not rise disproportionately, although generic turnover was significantly greater than family turnover. Judging from a comparison of the graphs of evolutionary rates among trilobite genera (Newell 1952) with family diversity trends, disproportionate generic diversity is not found among trilobites during their time of greatest family diversity either.

In summary, diversification on the family level during the Paleozoic and early Mesozoic seems to be accompanied chiefly by simple proportionate diversification on the generic level. Diversification on the family level during the late Cretaceous and Cenozoic seems to be accompanied by a disproportionately high diversification on the generic level.

It is impracticable to attempt a census on the species level for even a few higher taxa, and there are strong reasons for doubting the significance of fossil species counts in any event. It is necessary to approach species diversity at least in part from a theoretical point of view. Species evolve at greater frequencies than genera or families or higher taxa, and their standing diversities are therefore more volatile. The appearance of isolated habitats can produce swarms of closely related species, and the development of specialized communities may permit the development of a great number of species, not necessarily closely related, but endemic to the community. Reef communities, for example, appear to contribute large opportunities for both these types of speciation. A great number of specializations are possible on reefs, and thus large numbers of relatively specialized species from various phylogenetic backgrounds may appear. Reef tracts are also characterized by patchy and discontinuous distributions of reefs, and the isolation of outlying patches might often serve as a basis for speciation. Communities such as those on reefs that appear, endure long enough for a highly specialized biota to develop, and then disappear or become greatly reduced, can produce temporally localized but significantly large fluctuations in standing species diversity. If species diversification is great within such communities, generic diversity would also be enhanced. Since the species endemic to such communities are normally specialized, the average niche size of the shelf biota would be decreased while they flourish and increased when they wane.

At times the middle and upper Paleozoic record contains numerous reef associations and at these times it is likely that species diversity reaches disproportionately high levels, relative to families. It is expected that generic diversity might also rise disproportionately at these times. Although early and middle Permian reefs are widespread and contain probably the most specialized Brachiopoda recorded (Rudwick and Cowen 1968), a disproportionate generic diversity peak does not appear at that time in the available data (text-fig. 9). The description and evaluation of Permian reef biota is far from finished, however.

Another important way in which specific (and generic) diversity may be disproportionately multiplied is through a rise in provinciality. For example, theoretical considerations suggest that there are many more shelf species today than in the past, owing to the high degree of shallow-water provinciality at present (Valentine 1967, 1968*a*). This provinciality is both latitudinal, correlating with the great latitudinal temperature gradients at present, and longitudinal, owing to the presence of efficient biogeographic barriers of continents and ocean deeps. In the early Jurassic provinciality was not strongly developed. A Middle and Upper Jurassic Boreal fauna that contains endemic forms has been widely recognized (Neumayr 1883; Arkell 1956). Evidence has now been advanced to suggest that the Boreal fauna of the Jurassic signifies a low-salinity facies in a region of rather stable palaeogeography rather than a climatic province (Hallam 1969). Whatever its environmental basis, the appearance of the widespread fauna marks an increase in environmental heterogeneity on a sub-continental scale and a rise in species diversity. The general trend towards increasing provinciality in late Cretaceous and Cenozoic times must have greatly enhanced the numbers of species on the shelves (Valentine 1967, 1968*a*). Many genera which contain several species in a given province are now represented in different provinces by separate suites of species, so that the total numbers of their species are immense. Such is the case with species of *Nucula*, *Macoma*, and *Mactra* among the Bivalvia and *Conus*, *Calliostoma*, and *Fissurella* among the Gastropoda, to choose a few of the many examples. This situation must have been much less marked during times of low provinciality. For these reasons alone it is contended that the number of species in the shelf environment has increased disproportionately relative to the genera, especially during the Cretaceous and Cenozoic. Thus a species diversity curve would have about the same pattern as a generic diversity curve, but the peaks would be exaggerated (and the curve would be offset slightly towards the present). There are still other reasons, discussed below, for believing this pattern to be correct.

The major diversity trends in time among the fossil taxa are assumed to reflect real diversity trends among the ancient shelf biota, and they can be described in terms of the structure of the taxonomic hierarchy (text-fig. 10). In the earliest Paleozoic each phylum was represented by only a few classes, each class by relatively few orders, and so on down the hierarchy. By the close of Ordovician time, however, the average phylum was well differentiated into classes, and the average class into orders. Some phyla became extinct, but were not replaced by other phyla. After the Middle Ordovician the diversity of classes and of orders declined but the diversity of families rose, so that the structure became relatively more diversified among the lower taxa (text-fig. 10). It is possible that the average generic and specific diversity of the Upper Carboniferous indicated in text-fig. 10 is too low, owing to a disproportionate diversification at these lower levels that culminated during the early Permian. At about the Permo-Triassic boundary, both the numbers of classes and of orders were reduced by just less than half relative to their Middle Ordovician peaks, as were the families. The Permo-Triassic diversity low was most marked at lower taxonomic levels (contrast the familial and ordinal diversity decreases).

After the Permian the only gains in diversity registered among higher taxa are on the ordinal level. The numerous lost classes are not replaced, and even the rise in ordinal diversity is relatively small. The great climb in diversity at the familial level returns the hierarchy to a pattern commensurate with the early Upper Paleozoic pattern by Middle Jurassic time. Thereafter the number of families per higher taxon increases, especially during the Upper Cretaceous, and the diversities of lower taxa follow suit (text-fig. 10).

The so-called nekto-benthonic cephalopod groups Nautiloidea and Ammonoidea have not been included in the basic sample because of uncertainty as to the degree to which they participated in benthonic ecosytems. Nevertheless, some of them were surely regular members of a benthonic food chain. In text-fig. 5, the families of these cephalopod taxa are added to the families of the sample. The pattern of diversity is not much altered thereby; there was a slight increase in the steepness of the diversity rise from the Triassic to the Lower Cretaceous and the appearance of a rough plateau during the late Cretaceous and Cenozoic. The effect on the ordinal level is to emphasize the Ordovician rise in diversity and the mid- and upper-Paleozoic decline. Therefore, the exclusion of these Cephalopoda from the sample has a conservative effect insofar as the major trends are concerned, and in no way contributes to special conclusions.

Another major group which might have merited some representation in the figures is the fishes. At the family level their inclusion would raise the curve in text-fig. 5 to an even higher Devonian peak, and somewhat steepen the Mesozoic trend of rediversification (as

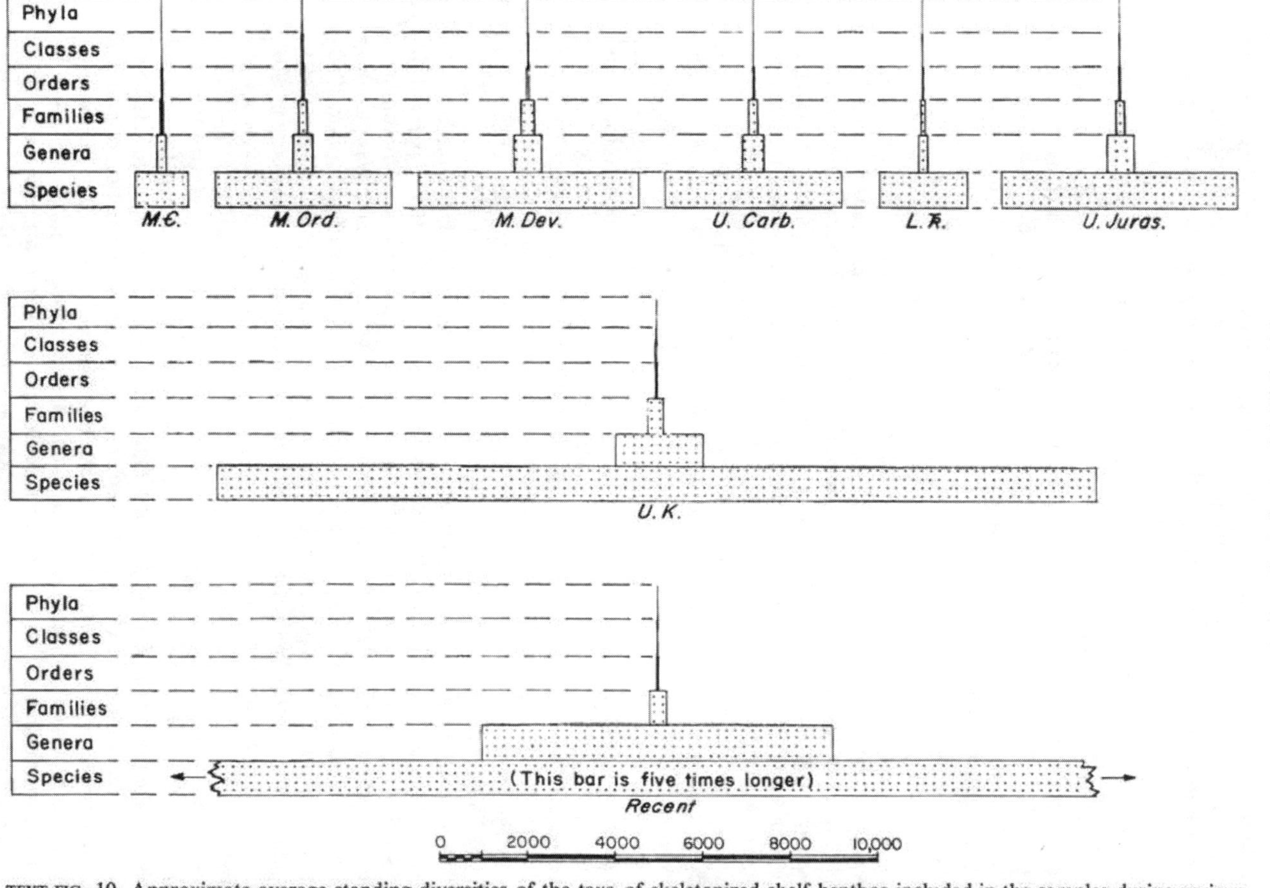

TEXT-FIG. 10. Approximate average standing diversities of the taxa of skeletonized shelf benthos included in the samples during various Series and at present. Diversities of familial and higher levels are based upon actual tallies. Diversities at generic and specific levels are calculated from scattered data on their proportions to families, except for the Recent. Recent generic and specific diversities are conservative estimates based on the literature.

would almost any additional taxon). At higher levels, they would raise ordinal diversity in medial Paleozoic and class diversity in early Paleozoic times, emphasizing graph patterns. Clearly, their exclusion does not affect the major diversity trends.

COADAPTATION OF ECOLOGICAL AND TAXONOMIC HIERARCHIES

The correlation of the structures of the two hierarchies under consideration is best approached at the species level. The fluctuations in species diversity, which have changed through a whole order of magnitude and more at times (Valentine 1967), naturally cause wide fluctuations in the numbers of populations in the ecological hierarchy, which must be accommodated in some manner.

In general there are three ways in which new species populations might be accommodated in ecological units (Klopfer 1962): (1) they may colonize parts of the environmental lattice that were previously unoccupied, in which case their niches represent an extension of ecospace in the ecological unit but average niche size is little affected; (2) space for their niches may be created in the lattice by shrinking one or several of the pre-existing niches, in which case the existing ecospace becomes more crowded by partitioning and average niche size decreases; or (3) parts of their niches may overlap with one or several pre-existing niches and thus crowd the lattice by overlap rather than by partitioning. The last sort of accommodation, by niche overlap, occurs hand in hand with one of the other sorts. The fitting of a new niche into the lattice may commonly involve all these sorts of accommodation. Theoretical or practical aspects of niche partitioning, which were touched on by Darwin (1859), have been considered in a modern perspective by a number of workers (for example, Bray 1958; Brown and Wilson 1956; Klopfer and MacArthur 1960, 1961; Klopfer 1962; McLaren 1963; MacArthur and Levins 1967; MacArthur and Wilson 1967; Hutchinson 1967; and Miller 1967). Yet data on variations in niche size and its relation to species diversity among marine shelf invertebrates is scanty. Marine research includes the work of Kohn (1959, 1966) on the gastropod *Conus* and Connell (1961) on some intertidal barnacles.

The most recent major rise inferred in species diversity from the late Cretaceous to the present seems to have involved an extension of ecospace, an invasion of parts of the lattice which were becoming newly realized, thus expanding the available environment. This increase in environmental heterogeneity in the shallow marine realm was evidently due partly to the cooling of shelf waters in high latitudes (Smith 1919; Durham 1950; Valentine 1967, 1968*a*), which permitted the rise of new biogeographic provinces in separate chains along north–south-trending coastlines. New provinces may also have been created by the drifting apart of some continents, progressively isolating, from about medial Cretaceous time, shelf regions that had previously been connected. This permitted an increase in endemism.

Thus this expansion of ecospace is envisaged as due primarily to two factors: (1) extension of the thermal factors and of numerous other parameters that are related to temperature, creating one set of biogeographic barriers; and (2) the creation of another set of barriers through the breaking up of formerly continuous or nearly continuous epicontinental seaways and continental shelves through continental drift (which in effect multiplies biospaces along dimensions of real space). The relative timing of these events is not yet clear, although there is a suggestion that longitudinal provinciality was strengthening in the late Cretaceous (Sohl 1961) while latitudinal provinciality was still weak. The Cretaceous–Cenozoic expansion of ecospace was fundamentally on the level of the biosphere and involved the rise of provinciality, which in turn permitted new communities based upon endemic populations to appear in each province. The increase in isolation among the populations living in separate provinces or separate communities permitted the formation of many new species.

The increase in generic diversity follows from the multiplication of species and their isolation in separate provinces or communities. Different but related species with similar morphological adaptations would arise in similar habitats in different provinces, forming associations such as Thorson's (1957) parallel bottom communities or becoming 'geminate' or twin species such as occur on opposite sides of the Isthmus of Panama (Ekman 1953). The increase in generic diversity would be proportionately less than in species diversity, owing to the multiplication of similar morphological types in distinct provinces that would be grouped as genera by taxonomic practice. The same principle seems to apply to the family level. A marked increase in the diversity and provinciality of genera would lead to a more or less modest increase in family diversity. That the present biogeographic pattern of diversity is similar at the familial, generic, and specific levels has been well documented for the Bivalvia by Stehli, McAlester, and Helsley (1967). Kurtén (1967) suggested that mammalian diversity is high at the ordinal level because of endemism arising through the isolating effects of continental drift.

At higher taxonomic levels on the marine shelves a different factor must be operating to control diversity, since higher taxa do not much participate in the Mesozoic–Cenozoic diversification. Perhaps this diversification has occurred too recently for evolution to have proceeded to the level of higher taxa. The pattern of ordinal diversity suggests that this may be the case. Yet class diversity has declined since the Ordovician. A better explanation may be that the available biospace of the epicontinental seas and shelves was nearly fully occupied since at least very early in the Phanerozoic, and most increases in taxonomic diversity had to be accomplished by ecospace partitioning and overlap (see Rhodes 1962, pp. 270–2; Nicol 1966). Thus only in times of unusual expansion of the marine shelf biospace, when the realized parts of the environmental lattice increased persistently, would significant diversity increases result simply from the invasion of new biospace. Ecospace partitioning involves a decrease in the niche sizes of populations, at least along dimensions where competition may occur (Miller 1967), and this is not a process that lends itself to the appearance of organisms with wholly new ground-plans or with major modifications thereof, such as are required for the development of higher taxa. It is instead a process suited more to the modification in detail of pre-existing morphological types, so as to accommodate to smaller ecospaces—in other words, a process suited to the increase of specialization.

By Cambrian time or shortly thereafter the ground-plans that are the hallmarks of the major invertebrate phyla had been established. Most of the species were by present standards primitive and functionally generalized; modal niche size was no doubt far larger than at present, though there certainly may have been some highly specialized forms. Evidently, diversification in the Cambrian and Ordovician led to the presence in late Ordovician time of a large number of higher taxa. This may have been partly due to an expansion of biospace and should have included an increase in resources. As much of the former marine shelf biospace may have been occupied by soft-bodied organisms, the expansion of skeletonized forms may have involved the appropriation of some resources that had formerly been utilized by soft-bodied groups. On the other hand, on the assumption that soft-bodied taxa should have responded to the same opportunities as skeletonized taxa, it seems even more likely that there was a concomitant diversification of soft-bodied lineages. These are debatable points, but it is clear that it cannot be assumed that Cambro–Ordovician radiation was occurring in vacant biospace. Vast regions of the lattice must have been occupied, and this may have served to channel the evolutionary pathways of diversifying lineages.

Certainly there was unusual extinction among higher taxa during this time. Higher taxa in the sample that appear in the Cambrian, but which are not known to have survived the Ordovician, include the phylum Archaeocyatha, the echinoderm classes Homostelea and Helicoplacoidea, the inarticulate brachiopod orders Obolellida, Paterinida, and Kutorginida, the monoplacophoran order Cambridoida, and the trilobite orders Redlichiida and Corynexochida. Numbers of other taxa that are poorly known also disappeared early, and may have represented higher taxonomic levels. They were not included in the sample because of their questionable status. These include some trilobitoids, some early echinoderm stocks, and early gastropod-like molluscan stocks. Perhaps most of these are functionally generalized forms, the morphological architecture of which proved unsuitable to the demands for specialization.

Biospace formerly occupied by populations of extinct lineages would soon be recolonized by populations of the lineages that remained, if such colonization did not actually precede and contribute to the extinction. This easily leads to the diversification of extant lower taxa, but not to the creation of taxa on a comparably high level. In the early Phanerozoic the large average size of the former ecospaces of extinct taxa provided ecological room, so to speak, which allowed the lineages that reoccupied vacated biospace a certain leeway for progressive morphological modification that could still lead to the establishment of a higher taxon. Later, when vacated biospace was to be in smaller parcels, opportunities for morphological modifications became limited, as during medial Paleozoic time, when the diversity of classes declined markedly (text-fig. 3). Although diversity of orders (text-fig. 3) and families (text-fig. 4) also declined during that time, the decline was less marked, and order/class and family/order ratios were both rising. This suggests that the extinction of a higher taxon was not usually accompanied by replacement at the level of the higher taxon but at a lower level. It also suggests that biospace was decreasing. In the Lower and Middle Permian the numbers of species and genera may well have been disproportionately higher than is suggested by the number of families, owing to the development of reef associations.

Towards the close of the Paleozoic the change in the taxonomic structure suggests a great reduction in the heterogeneity of the shelf environment. Provinciality was already low. It is not certain, therefore, whether a significant further reduction occurred or indeed was even possible in late Permian time. However the numbers of communities certainly decreased (for example, the late Paleozoic reefs disappear) and, although there are not well-documented field studies, it appears from faunal lists that the numbers of populations in the remaining communities also declined. Thus the ecological structure shrank at all, or nearly all, levels, suggesting a general decline in biospace. There are too few data on the ecological hierarchy of Permo-Triassic times, however, to support speculation on the precise causes of the extinctions on this basis. Rhodes (1967) has reviewed the major hypotheses of extinction and remarks that probably none of them alone would cause the sorts of changes found at the Permo-Triassic boundary.

Rediversification evidently began by medial Triassic time, unless the increase in family diversity then is an artifact. If it is real (and it includes the beginnings of Scleractinian radiation as well as the expansion of gastropod families), it suggests that generic and specific diversification was proceeding at even higher rates. Presumably biospace had expanded (or was expanding) once more and the newly realized parts of the environmental lattice were being recolonized. However, compared with Cambro-Ordovician lineages, Triassic lineages were rather specialized, with smaller modal niche sizes, so that the average colonizing lineage must have occupied a relatively smaller part of the lattice. The opportunities now presented for the formation of higher taxa could not be much exploited by the relatively specialized populations. There may be exceptions; the Scleractinia may have taken advantage of biospace vacated by Paleozoic coelenterate lineages to become skeletonized and reoccupy some of the same biospace. Finally, at some time in the Mesozoic a diversification involving the marked rise in provinciality discussed previously began to occur as well. It seems likely that the Jurassic and early Cretaceous diversity increases, which are not inconsiderable and which take place at a high rate (text-fig. 5), are partly owing to an increase of latitudinal provinciality due to cooling poles and to an increase of longitudinal provinciality due to the separation of some continental masses in Jurassic time. A thorough analysis of the biogeographic patterns of Mesozoic diversification is badly needed.

The Cretaceous-Cenozoic boundary is marked by extinctions of some benthonic marine groups (see Hancock 1967), but if there was any appreciable alteration in the taxonomic diversity structure at the family and higher levels it was of so short a duration that it does not appear in the present data. The structure of communities, provinces and of the entire shelf realm must have undergone qualitative changes, but this is a more or less continuous process on the broad scale we are considering. The Late Cretaceous rise in diversity extends unabated across the Cretaceous-Cenozoic boundary. Surely there was no large-scale reduction in biospace.

Two major modes of taxonomic diversification have been described, both proceeding at progressively lower taxonomic levels through time. The first involves a biospace that fluctuates about some size that does not vary much in time. Diversification at the population level at first proceeds by colonization of untenanted biospace, but soon must be accompanied by a progressive decrease in average niche size. Communities therefore become increasingly packed with more and more specialized populations and begin to fragment into portions, each of which has an energy flow that is partially independent of the others. The isolation and independence of these portions will increase with further specialization until they form ecosystems that are as independent as the original one from which they fragmented. Slighter and slighter environmental discontinuities will form community boundaries until the environmental mosaic of a given primitive community, relatively heterogeneous but occupied by primitive populations with large niches, is broken up into a number of smaller environments, each more homogeneous than the original and each occupied by more specialized populations. Provinces become packed with more and more communities that have progressively smaller ecospaces. The diversity of provinces is not so sensitive to this progressive specialization, although it may eventually be affected if the trend continues long enough. Any widespread partitioning of temperature or temperature-correlated parameters would lead to increased provinciality even in the absence of progressive changes in the latitudinal temperature gradient. If partitioning were to continue, smaller and smaller changes in thermal regimes would act to localize range end-points, and thus form provincial boundaries (Valentine 1966). In sum, in this mode the ecological structure evolves from lower levels towards higher.

The other mode of diversification involves a biospace that expands to create new environments or to add new dimensions (or at least to extend old dimensions) to old environments. New environments may be created, for example, through climatic changes, and old environments may be extended through the improvement of limiting factors, that is through the amelioration of conditions which tended to inhibit diversification.

From what is known and can be inferred of the hierarchies of the Paleozoic, diversification (at least after an initial radiation of skeletonized taxa and probably before) was proceeding chiefly in the first mode, from the bottom of the ecological hierarchy upwards, implying a relatively stable biospace. Rediversification following the Permo-Triassic extinction was probably in the second mode, involving an amelioration of factors that had inhibited diversification, and the Upper Cretaceous and Cenozoic diversification seems to have also been in the second mode, but involved the creation of new environments. Nevertheless there must have been continuing specialization and thus the first mode was also active in the taxonomic and ecological evolution, Processes that bring species that appeared during biospace expansion into sympatry with older lineages, such as 'species pumps' of various kinds (Valentine 1967, 1968*a*), may link these two modes into a single system of diversification. Finally, the more lineages that exist the greater the opportunity for large-scale diversification under appropriate circumstances in either mode. This factor is certainly at work in the disproportionate multiplication of lower taxa during Cretaceous and Cenozoic times.

A number of authors have suggested that extensive changes in sea level may control some of the diversification and extinction patterns (Newell 1952, 1956, 1963; Moore 1954; see Rhodes 1967 for other references). Widespread epicontinental seas, it is asserted, provide more inhabitable area for shelf invertebrates and therefore more opportunity for diversification, while regressions reduce the inhabitable area and thus the diversity. There is some theoretical support for this position in the species-area work of Preston (1962), Williams (1964), MacArthur and Wilson (1967), and others. Little work has been done in marine environments; the areas of ancient shelf seas, especially during regressive phases, are difficult to estimate (though see Ronov 1968); and the effects that could be expected in a biosphere of vastly different ecological and taxonomic structure and composition are largely uncertain. The problem is further complicated by facies differences between epicontinental seas and shelves bordered by open oceans. Although uncertainties in calculations must be great, preliminary estimates suggest that the species-area effect would have been far too small to account for major diversifications and extinctions by itself. Moreover, we live at present in a time of great continental emergence yet the shelves are richly diverse in lower taxa and in ecological units at all levels, precisely the opposite of the pattern of Permo-Triassic extinction. Indeed, a relative lowering of sea level must commonly result in the emergence of land barriers which isolate regions formerly connected and permit the rise of an endemic biota in each region. This would have the effect of increasing the total number of species in these regions.

Nevertheless, the elimination or rise of species resulting from shelf-area fluctuations would certainly contribute to diversity patterns, and further evaluation of this subject is clearly merited.

CONCLUSIONS: THE PROGRESSIVE CANALIZATION OF ECOSPACE

It is concluded that a major Phanerozoic trend among the invertebrate biota of the world's shelf and epicontinental seas has been towards more and more numerous units at all levels of the ecological hierarchy. This has been achieved partly by the progressive partitioning of ecospace into smaller functional regions, and partly by the invasion of previously unoccupied biospace. At the same time, the expansion and contraction of available environments has controlled strong but secondary trends of diversity. Present marine biospace is in fact unusually extensive, and the world's shelf seas are therefore unusually heterogeneous and support a large number of ecological units today. The relations of these trends to trends within the taxonomic structure of the benthonic invertebrates are intimate.

Assuming for the moment that evolutionary trends among marine benthonic invertebrates will continue and that biospace does not change much (a dim prospect in view of rising pollution), what might be predicted of the future structures of the ecological and taxonomic hierarchies? Speculation on this point may be of some value to underline the sort of process that is postulated to have gone before. Clearly, the trend towards specialization would further reduce the average niche sizes of species. It would be increasingly difficult for evolving lineages to depart much from their modal functions and morphologies, as biospace would become available only in increasingly smaller compartments. The amount of change necessary to produce a new family would be increasingly difficult to attain, and eventually no new families could appear. In fact, some families would become extinct so that familial diversity would decrease, and lineages from other families would fill any vacated biospace. After some time genera could no longer appear, for biospace would be packed too tightly to permit morphological variation even at that level, and generic diversity would decline for a while as some extinction, inevitably, occurred. Eventually, all the biospace would become filled with evolving lineages with an incredibly small modal niche size, each lineage constrained by the presence of all the others to evolve in only a narrow pathway directed by the trends of evolution of the entire biota, and of changes in the entire biosphere. Ecological units are now exceedingly small by today's standards, with virtually every few food-chains forming a separate community and every moderate topographic irregularity forming a provincial boundary. Canalization of ecospace is complete. The biosphere has become a splitter's paradise.

Although this extrapolation cannot be taken too seriously, it does point to some important consequences of ecospace partitioning. First, average species of the early Paleozoic, with their broad niches, may have had different patterns of morphological variation than the specialized species of today. Secondly, the occurrence in the early Paleozoic of numbers of unusual 'aberrant' higher taxa that contain few lower taxa is not necessarily due to a poor fossil record but is probably the natural consequence of adaptive strategies that prevailed in primitive ecosystems of low diversity. Finally, extinction of taxa of high diversity is less likely than extinction of taxa of low diversity, other things being equal (Simpson 1953), simply because so many more lineages must disappear. Similarly, the markedly rising provinciality of the late Cretaceous and Cenozoic will tend to make the extinction of the newly diverse taxa that have representation in many provinces—a common situation even on the generic level—more difficult.

Acknowledgements. Gratitude for extensive discussion of the ideas presented herein is expressed to Dr. A. Hallam (Oxford University) and Professor A. L. McAlester (Yale University). The manuscript was carefully reviewed by Dr. W. S. McKerrow (Oxford University) and Professor R. Cowen, Professor J. H. Lipps, and Robert Rowland (University of California, Davis). All this attention resulted in much improvement. The manuscript was written during a Guggenheim Fellowship spent at the Department of Geology and Mineralogy, Oxford University, and at the Department of Geology and Geophysics, Yale University. The generosity of the Guggenheim Foundation and the hospitality of these departments is gratefully acknowledged.

REFERENCES

ARKELL, W. J. 1956. *Jurassic geology of the world.* Edinburgh.

BERKNER, C. V. and MARSHALL, C. C. 1965. Oxygen and evolution. *New Scientist*, **28**, 415–19.

BRAY, J. R. 1958. Notes towards an ecologic theory. *Ecology*, **39**, 770–6.

BROWN, W. L., JR. and WILSON, E. O. 1956. Character displacement. *Syst. Zool.* **5**, 49–64.

CLOUD, P. E. 1949. Some problems and patterns of evolution exemplified by fossil invertebrates. *Evolution, Lancaster, Pa.* **2**, 322–50.

—— 1968. Pre-metazoan evolution and the origin of the metazoa. *In* DRAKE, E. T. (ed.), *Evolution and environment.* 1–72. New Haven, Conn.

CONNELL, J. H. 1961. The influence of interspecific competition and other factors on the distribution of the barnacle *Chthamalus stellatus*. *Ecology*, **42**, 710–23.

CUSHMAN, J. A. 1948. *Foraminifera, their classification and economic use.* Cambridge, Mass.

DARWIN, C. R. 1859. *On the origin of species by means of natural selection.* London.

DOTY, M. S. 1957. Rocky intertidal surfaces. *In* HEDGPETH, J. W. (ed.), Treatise on marine ecology and paleoecology. *Mem. geol. Soc. Am.* **67**, 1, 535–85.

DURHAM, J. W. 1950. Cenozoic marine climates of the Pacific Coast. *Bull. geol. Soc. Am.* **61**, 1243–64.

EKMAN, S. 1953. *Zoogeography of the sea.* London.

HALLAM, A. 1969. Faunal realms and facies in the Jurassic. *Palaeontology*, **12**, 1–18.

HANCOCK, J. M. 1967. Some Cretaceous–Tertiary marine faunal changes. *In* HARLAND, W.B. *et al.* (eds.), *The fossil record*, 91–104. London (Geological Society).

HARLAND, W. B. *et al.* (eds.). 1967. *The fossil record.* London (Geological Society).

HENBEST, L. G. 1952. Significance of evolutionary explosions for diastrophic division of earth history—introduction to the symposium. *J. Paleont.* **52**, 299–318.

HUTCHINSON, G. E. 1957. Concluding remarks. *Cold Spring Harbor Symp. quant. Biol.* **22**, 415–27.

—— 1967. *A treatise on limnology, Volume 2. Introduction to lake biology and the limnoplankton.* New York.

HUXLEY, J. S. 1942. *Evolution, the modern synthesis.* London.

KLOPFER, P. M. 1962. *Behavioral aspects of ecology.* Englewood Cliffs, N.J.

—— and MACARTHUR, R. H. 1960. Niche size and faunal diversity. *Am. Nat.* **94**, 293–300.

—— —— 1961. On the causes of tropical species diversity: niche overlap. *Am. Nat.* **95**, 223–6.

KOHN, A. J. 1959. The ecology of *Conus* in Hawaii. *Ecol. Monogr.* **29**, 47–90.

—— 1966. Food specialization in *Conus* in Hawaii and California. *Ecology*, **47**, 1041–3.

KURTÉN, B. 1967. Continental drift and the palaeogeography of reptiles and mammals. *Soc. Scient. Fennica*, **31** (1), 1–8.

LEVINS, R. 1962. Theory of fitness in a heterogeneous environment. II. Developmental flexibility and niche selection. *Am. Nat.* **97**, 75–90.

LIPPS, J. H. (in press). Plankton evolution. *Evolution.*

LOEBLICH, A. R. and TAPPAN, HELEN. 1964. Protista 2, Sarcodina, chiefly 'Thecamoebians' and Foraminiferida. *In* MOORE, R. C. (ed.), *Treatise on invertebrate paleontology, Part C*. Geol. Soc. Amer. and Univ. Kansas Press.

MACARTHUR, R. H. and LEVINS, R. 1964. Competition, habitat selection and character displacement in a patchy environment. *Proc. nat. Acad. Sci. U.S.* **51**, 1207–10.

—— —— 1967. The limiting similarity, convergence, and divergence of coexisting species. *Am. Nat.* **101**, 377–85.

—— and WILSON, E. O. 1967. *The theory of island biogeography*. Princeton, N.J.

MAYR, E. 1963. *Animal species and evolution*. Cambridge, Mass.

MCLAREN, I. A. 1963. Effects of temperature on growth of zooplankton, and the adaptive value of vertical migration. *J. Fish res. Bd. Canada*, **20**, 685–727.

MILLER, R. S. 1967. Pattern and process in competition. *In* CRAGG, J. B. (ed.), *Adv. Ecol. Res.* **4**, 1–74.

MOORE, R. C. (ed). 1953–7. *Treatise on invertebrate paleontology*. Geol. Soc. Amer. and Univ. Kansas Press.

NEUMAYR, M. 1883. Ueber klimatische Zonen während der Jura- und Kreidezeit. *Denkschr. Akad. Wiss., Wien, Math.-nat. Kl.* **18**, 277–310.

NEWELL, N. D. 1952. Periodicity in invertebrate evolution. *J. Paleont.* **26**, 371–85.

—— 1956. Catastrophism and the fossil record. *Evolution, Lancaster, Pa.* **10**, 97–101.

—— 1963. Crises in the history of life. *Scient. Am.* **208**, 76–92.

—— 1967. Revolutions in the history of life. *Spec. Pap. Geol. Soc. Am.* **89**, 63–91.

NICOL, D. 1966. Cope's rule and Precambrian and Cambrian invertebrates. *J. Paleont.* **40**, 1397–9.

ORLOV, Y. A. 1958–64. *Osnovy Paleontologii*, Akad. nauk SSSR., Moscow (in Russian).

PARR, A. E. 1926. Adaptiogenese und Phylogenese; zur Analyse der Anpassungserscheinungen und ihre Entstehung. *Abh. Theor. org. Entw.* **1**, 1–60.

PRESTON, F. W. 1962. The canonical distribution of commonness and rarity. *Ecology*, **43**, 185–215, 410–32.

RENSCH, B. 1947. *Neuere Probleme der Abstammungslehre*. Stuttgart.

RHODES, F. H. T. 1962. *The evolution of life*. Baltimore, Md.

—— 1967. Permo-Triassic extinction. *In* HARLAND, W. B., *et al.* (eds.), *The fossil record*, 57–76. London (Geological Society).

RONOV, A. B. 1968. Probable changes in the composition of sea water during the course of geological time. *Sedimentology*, **10**, 25–43.

RUDWICK, M. J. S. and COWEN, R. 1968. The functional morphology of some aberrant strophomenide brachiopods from the Permian of Sicily. *Bol. Soc. Paleont. Ital.* **6**, 113–76.

SIMPSON, G. G. 1944. *Tempo and mode in evolution*. New York.

—— 1953. *The major features of evolution*. New York.

SMITH, J. P. 1919. Climatic relations of the Tertiary and Quaternary faunas of the California region. *Proc. Calif. Acad. Sci.* (4) **9**, 123–73.

SOHL, N. F. 1961. Archaeogastropods, Mesogastropods, and stratigraphy of the Ripley, Owl Creek and Prairie Bluff Formations. *Prof. pap. U.S. geol. Surv.* **331A**, 151 pp.

STEHLI, F. G., MCALESTER, A. L., and HELSLEY, C. E. 1967. Taxonomic diversity of Recent bivalves and some implications for geology. *Bull. geol. Soc. Am.* **78**, 455–66.

TAYLOR, D. W. and SOHL, N. F. 1962. An outline of gastropod classification. *Malacologia*, **1**, 7–32.

THORSON, G. 1957. Bottom communities (sublittoral or shallow shelf). *In* HEDGPETH, J. W. (ed.), Treatise on marine ecology and paleoecology. *Mem. geol. Soc. Am.* **67** (1), 461–534.

VALENTINE, J. W. 1966. Numerical analysis of marine molluscan ranges on the extratropical northeastern Pacific shelf. *Limnol. Oceanogr.* **11**, 198–211.

—— 1967. Influence of climatic fluctuations on species diversity within the Tethyan Provincial System. *In* ADAMS, C. G. and AGER, D. V. (eds.), *Syst. Ass. Pub.* **7**, 153–66.

—— 1968*a*. Climatic regulation of species diversification and extinction. *Bull. geol. Soc. Am.* **79**, 273–76.

—— 1968*b*. The evolution of ecological units above the population level. *J. Paleont.* **42**, 253–67.

—— 1969. Niche diversity and niche size patterns in marine fossils. Ibid. **43**, 905–15.

WILLIAMS, A. 1957. Evolutionary rates in brachiopods. *Geol. Mag.* **94**, 201–11.

WILLIAMS, A. 1965. Stratigraphic distribution. *In* MOORE, R. C. (ed.), *Treatise on invertebrate paleontology, Part H*. H237–50. Geol. Soc. Am. and Univ. Kansas Press.

—— *et al.* 1965. *In* MOORE, R. C. (ed.), *Treatise on invertebrate paleontology, Part H*. Geol. Soc. Am. and Univ. Kansas Press.

WILLIAMS, C. B. 1964. *Patterns in the balance of nature and related problems in quantitative ecology*. New York.

J. W. VALENTINE
Department of Geology
University of California
Davis, California, 95616
U.S.A.

Typescript received 8 April 1969

PAPER 18

A Kinetic Model of Phanerozoic Taxonomic Diversity. I. Analysis of Marine Orders (1978)

J. J. Sepkoski Jr.

Commentary

ARNOLD I. MILLER

As luck would have it, I was an undergraduate during the 1970s at the University of Rochester, where Jack Sepkoski had his first faculty appointment. At the time, the paleontology program had an amazing cadre of faculty and graduate students, and I enjoyed the atmosphere of departmental colloquia and rathskeller gatherings, even though I was a marginal student and understood only a limited percentage of what I was hearing. But one thing that I recall distinctly were attempts by several students at appropriately inflected renditions of a mantra attributed to Jack: "I can explain all of Phanerozoic diversification with the logistic equation!"

Of course, I didn't entirely understand this bold declaration at the time, and I don't think that any of the students fully appreciated what Jack was up to behind the scenes. Jack was about to indelibly change the paleontological landscape with the paper reprinted here—known as "Kinetic I"—in which he treated the world's oceans as a kind of global island and adapted the logistic equation, previously used in island biogeography, to produce a diversity trajectory that closely approximated that of Phanerozoic marine orders, initiating a debate that continues to this day about whether global biodiversity has unfolded under equilibrial constraints. But, even before "Kinetic I" was published in 1978, Jack was compiling data and producing graphs (Miller 2009) that were included in a remarkable series of follow-up papers published over the next seven years that presented a taxonomically comprehensive global marine diversity trajectory at the family level; delineated evolutionary faunas based on factor analysis and conceptualized them as interactive entities with their own characteristic equilibria in a *coupled* logistic system; and sought to document an underlying, environmental cohesiveness for each of the evolutionary faunas in an onshore-offshore framework.

"Kinetic I" was compelling not only because it adapted a set of seminal ecological principles to global patterns through deep time but also because it contained important subtexts, including the suggestion that the Cambrian explosion was the inevitable outcome of an exponential diversification initiated during the Vendian. More broadly, following on Jim Valentine's pioneering efforts, "Kinetic I" propelled the full infusion of the geological time dimension and macroevolution into paleoecology.

Literature Cited

Miller, A. I. 2009. The consensus that changed the paleobiological world. In *The Paleobiological Revolution: Essays on the Growth of Modern Paleontology*, edited by D. Sepkoski and M. Ruse, 365–82. Chicago: University of Chicago Press.

Paleobiology, 4(3), 1978, pp. 223–251

A kinetic model of Phanerozoic taxonomic diversity I. Analysis of marine orders

J. John Sepkoski, Jr.

Abstract.—A simple equilibrial model for the growth and maintenance of Phanerozoic global marine taxonomic diversity can be constructed from considerations of the behavior of origination and extinction rates with respect to diversity. An initial postulate that total rate of diversification is proportional to number of taxa extant leads to an exponential model for early phases of diversification. This model appears to describe adequately the "explosive" diversification of known metazoan orders across the Precambrian-Cambrian Boundary, suggesting that no special event, other than the initial appearance of Metazoa, is necessary to explain this phenomenon. As numbers of taxa increase, the rate of diversification should become "diversity dependent." Ecological factors should cause the per taxon rate of origination to decline and the per taxon rate of extinction to increase. If these relationships are modeled as simple linear functions, a logistic description of the behavior of taxonomic diversity through time results. This model appears remarkably consistent with the known pattern of Phanerozoic marine ordinal diversity as a whole. Analysis of observed rates of ordinal origination also indicates these are to a large extent diversity dependent; however, diversity dependence is not immediately evident in rates of ordinal extinction. Possible explanations for this pattern are derived from considerations of the size of higher taxa and from simulations of their diversification. These suggest that both the standing diversity and the pattern of origination of orders may adequately reflect the behavior of species diversity through time; however, correspondence between rates of ordinal and species extinction may deteriorate with progressive loss of information resulting from incomplete sampling of the fossil record.

J. John Sepkoski, Jr. Department of the Geophysical Sciences, University of Chicago; 5734 South Ellis Ave., Chicago, Illinois 60637

Accepted: February 2, 1978

Introduction

This paper presents a simple mathematical model for the growth and maintenance of taxonomic diversity through geologic time. The model attempts to describe interrelationships among a small number of variables, specifically origination rate, extinction rate, and number (or "diversity") of taxa, and show how these should vary with respect to one another and to time. The major result of the inferred interrelationships is the prediction that the diversification of taxa within any large, relatively discrete ecological system should be fundamentally a logistic process. This prediction is tested with data on numbers of marine metazoan orders in the world ocean through geologic time, examining first patterns of diversification across the Precambrian-Cambrian Boundary and then expanding to consider patterns throughout the whole of the Phanerozoic.

Like most previous quantitative models of large-scale diversification over evolutionary time, the one presented here is an equilibrium model. Models of this type contend that numbers of taxa (and especially species) within well-defined areas should in time reach steady states sustained by balanced rates of origination (i.e. speciation) and extinction. This genre of models is an outgrowth of the seminal work of MacArthur and Wilson (1963, 1967) on equilibrial island biogeography and their separate suggestions (MacArthur 1969; Wilson 1969) that equilibrial situations may extend beyond ecological time into evolutionary time (see also May 1973; Rosenzweig 1975). The paleontological consequences of such evolutionary equilibria have been discussed and analyzed by a number of workers, including Simpson (1960, 1969), Webb (1969, 1976), Lillegraven (1972), Raup (1972), Raup et al. (1973), Cisne (1974), Schopf (1974), Simberloff (1974), Flessa (1975), Sepkoski (1976), Bakker (1977), Bambach (1977), Gould et al. (1977), and Mark and Flessa (1977). These authors have found various patterns within

 0094–8873/78/0403–0001/$1.00

the fossil record that are consistent with (although not necessarily compelling evidence for) an equilibrial view of taxonomic diversity through all or portions of the Phanerozoic.

The primary purpose of this paper is to provide a more generalized and precise formulation of the concept of evolutionary equilibria as it relates to paleontology. This formulation, although greatly simplified, predicts specific quantitative patterns in taxonomic diversity and in relative magnitudes of origination and extinction rates that can be tested statistically with paleontological data. It is hoped that this will provide a more direct basis for assessing the applicability of the concept of evolutionary equilibria to the fossil record. Still, the model in its simplicity is limited. It can predict neither the timings of events nor the absolute magnitudes of rates; it simply describes the behavior or relative "motion" of diversity with respect to time once such parameters are fixed. For this reason I choose to call it a "kinetic" model.

The preliminary tests of the kinetic model discussed herein are conducted using data on the diversity of fossil orders. These taxa have not been chosen because they are superior to other taxonomic categories. Indeed, lower taxa, and particularly species, might provide a stronger test. Yet, orders do possess certain characteristics that make them more immediately tractable for this type of paleontological analysis:

1. The total number of orders known from the fossil record is small (about 500), making compilation of complete data on their geologic ranges relatively easy.
2. Orders are comparatively large taxonomic units, leaving them less susceptible to the vagaries of sampling than lower taxonomic units (see Raup 1972, 1975).
3. As large entities, orders tend to smooth out, much like running averages, many relatively minor temporal fluctuations in diversity, thus making the fundamental patterns more readily discernible.

The last portion of this paper is devoted to examining by means of computer simulations of hierarchial taxonomy and paleontological sampling how well higher taxa might be expected to reflect species diversity, thereby testing the assumption that orders are appropriate for analyzing patterns of Phanerozoic diversity.

Diversification as a Growth Process

Models of paleontological diversity can be constructed with forms analogous to those of classical population ecology. Eldredge and Gould (1972), Stanley (1975), and Gould and Eldredge (1977) have forcefully argued that the majority of species, like individuals, are discrete and fairly static entities with constant or mildly varying morphologies and ecologies throughout most of their durations. Origination and extinction probably involve periods of time that are relatively short compared to the total durations of the species (see also Van Valen 1973a). If we assume with proper caution that these properties make species analogous to individuals, an entire suite of conceptual and mathematical models of population dynamics becomes available for prudent use in studies of cladogenesis. Van Valen (1973a) certainly made such an assumption, at least implicity, in his use of life-table techniques to analyze taxonomic extinction, and Stanley (1975) very explicitly analogized species and individuals in his extension of natural selection to "species selection" (see also Gould and Eldredge 1977; Gould 1976).

Discrete species should have phylogenies which when viewed against time have the appearance of simple branching diagrams (Raup et al. 1973), not unlike the genealogies of individuals in asexual populations. Higher taxa, as discrete packages of species, should also have this phylogenetic pattern, as illustrated by the example in Fig. 1. Any phylogeny begins with a single ancestor that produces one or more descendents, and these more, and so on, causing the phylogeny to expand through time. Such multiplication of descendents suggests that at any given time a phylogeny's rate of growth, or "diversification," should be roughly proportional to its diversity, that is, number of constituent taxa (cf. Stanley 1975, 1977a). This relationship can be expressed as a simple difference equation, describing the change in number of taxa or diversity D per interval of time t as

$$\Delta D/\Delta t = r_d D \qquad (1)$$

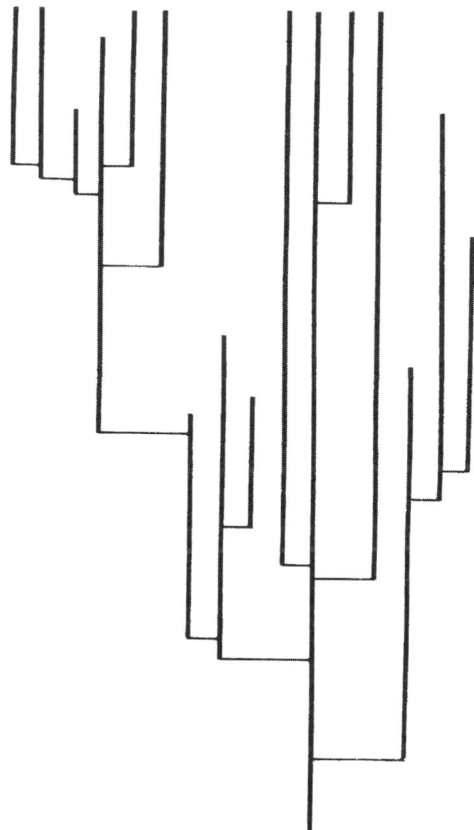

FIGURE 1. An example of a phylogeny drawn as a simple orthogonal branching diagram with time in the vertical dimension. New branches (i.e. taxa) in a phylogeny can be derived only from pre-existing branches. Thus, if probabilities of branching are approximately constant throughout the phylogeny, the number of new branches appearing during any interval of time should be roughly proportional to the number of pre-existing branches in that interval. The illustrated phylogeny represents families of ammonites descended from the Desmoceratidae during the Albian through Turonian Stages of the Cretaceous; it is redrawn from Arkell (1957) against the time scale of van Hinte (1976b).

The rate of diversification r_d is the difference between the rate (or, more properly, probability) of origination (i.e. speciation) r_s and the rate of extinction r_e. These latter two terms are measured as

$$r_s = \frac{1}{D} \cdot \frac{S}{\Delta t} \tag{2a}$$

and

$$r_e = \frac{1}{D} \cdot \frac{E}{\Delta t}, \tag{2b}$$

where S is the number of originations (or speciations) observed during an interval of time Δt, and E is the number of extinctions. The rates defined above are *per taxon* rates of origination and extinction, as indicated by the reciprocal of D in the righthand sides of the equations. These rates represent average "probabilities" of individual taxa branching or becoming extinct during a unit of time. The more familiar *total* rates of origination, R_s, and extinction, R_e, which describe the cladogenetic behavior of a phylogeny as a whole, are defined as

$$R_s = S/\Delta t \tag{3a}$$

and

$$R_e = E/\Delta t. \tag{3b}$$

Without the normalization term for diversity, these two rates might be expected to vary directly with the size of a phylogeny, whereas per taxon rates, if unaffected by species interactions, might remain more constant.

If, for the purposes of argumentation, we assume that diversification is a continuous-time process with a constant rate of diversification, Equation 1 can be rewritten as a differential and integrated to yield the familiar exponential growth equation

$$D = D_o \exp(r_d t). \tag{4}$$

Here, D_o is the initial diversity at $t = 0$. For analysis it is usually best to write the exponential growth equation in its logarithmic form

$$\ln D = \ln D_o + r_d t. \tag{5}$$

This form yields a straight line when graphed with a logarithmic ordinate.

This simple exponential model of diversification is probably applicable to a variety of evolutionary radiations, particularly those in which major taxonomic groups radiate into new adaptive zones or previously "unoccupied" ecospace (see Stanley 1975, 1977a; also Cailleux 1950, 1954). In fact, this model may adequately describe the greatest radiation of all time, the so-called "explosion" of marine invertebrates near the beginning of

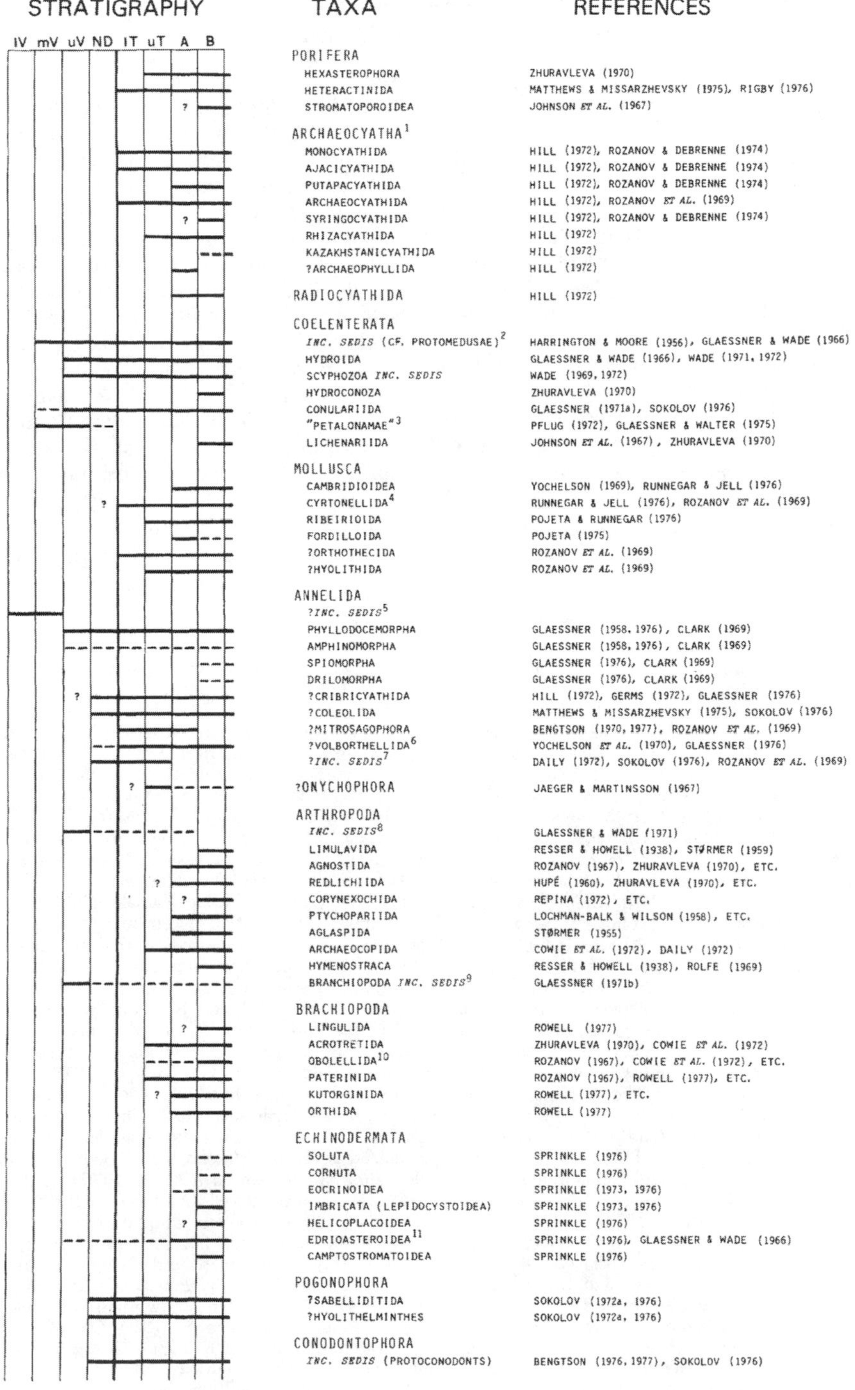
STRATIGRAPHY
TAXA
REFERENCES
lV mV uV ND lT uT A B
PORIFERA
HEXASTEROPHORA ZHURAVLEVA (1970)
HETERACTINIDA MATTHEWS & MISSARZHEVSKY (1975), RIGBY (1976)
STROMATOPOROIDEA JOHNSON ET AL. (1967)
ARCHAEOCYATHA[1]
MONOCYATHIDA HILL (1972), ROZANOV & DEBRENNE (1974)
AJACICYATHIDA HILL (1972), ROZANOV & DEBRENNE (1974)
PUTAPACYATHIDA HILL (1972), ROZANOV & DEBRENNE (1974)
ARCHAEOCYATHIDA HILL (1972), ROZANOV ET AL. (1969)
SYRINGOCYATHIDA HILL (1972), ROZANOV & DEBRENNE (1974)
RHIZACYATHIDA HILL (1972)
KAZAKHSTANICYATHIDA HILL (1972)
?ARCHAEOPHYLLIDA HILL (1972)
RADIOCYATHIDA HILL (1972)
COELENTERATA
INC. SEDIS (CF. PROTOMEDUSAE)[2] HARRINGTON & MOORE (1956), GLAESSNER & WADE (1966)
HYDROIDA GLAESSNER & WADE (1966), WADE (1971, 1972)
SCYPHOZOA INC. SEDIS WADE (1969, 1972)
HYDROCONOZA ZHURAVLEVA (1970)
CONULARIIDA GLAESSNER (1971a), SOKOLOV (1976)
"PETALONAMAE"[3] PFLUG (1972), GLAESSNER & WALTER (1975)
LICHENARIIDA JOHNSON ET AL. (1967), ZHURAVLEVA (1970)
MOLLUSCA
CAMBRIDIOIDEA YOCHELSON (1969), RUNNEGAR & JELL (1976)
CYRTONELLIDA[4] RUNNEGAR & JELL (1976), ROZANOV ET AL. (1969)
RIBEIRIOIDA POJETA & RUNNEGAR (1976)
FORDILLOIDA POJETA (1975)
?ORTHOTHECIDA ROZANOV ET AL. (1969)
?HYOLITHIDA ROZANOV ET AL. (1969)
ANNELIDA
?INC. SEDIS[5]
PHYLLODOCEMORPHA GLAESSNER (1958, 1976), CLARK (1969)
AMPHINOMORPHA GLAESSNER (1958, 1976), CLARK (1969)
SPIOMORPHA GLAESSNER (1976), CLARK (1969)
DRILOMORPHA GLAESSNER (1976), CLARK (1969)
?CRIBRICYATHIDA HILL (1972), GERMS (1972), GLAESSNER (1976)
?COLEOLIDA MATTHEWS & MISSARZHEVSKY (1975), SOKOLOV (1976)
?MITROSAGOPHORA BENGTSON (1970, 1977), ROZANOV ET AL. (1969)
?VOLBORTHELLIDA[6] YOCHELSON ET AL. (1970), GLAESSNER (1976)
?INC. SEDIS[7] DAILY (1972), SOKOLOV (1976), ROZANOV ET AL. (1969)
?ONYCHOPHORA JAEGER & MARTINSSON (1967)
ARTHROPODA
INC. SEDIS[8] GLAESSNER & WADE (1971)
LIMULAVIDA RESSER & HOWELL (1938), STØRMER (1959)
AGNOSTIDA ROZANOV (1967), ZHURAVLEVA (1970), ETC.
REDLICHIIDA HUPÉ (1960), ZHURAVLEVA (1970), ETC.
CORYNEXOCHIDA REPINA (1972), ETC.
PTYCHOPARIIDA LOCHMAN-BALK & WILSON (1958), ETC.
AGLASPIDA STØRMER (1955)
ARCHAEOCOPIDA COWIE ET AL. (1972), DAILY (1972)
HYMENOSTRACA RESSER & HOWELL (1938), ROLFE (1969)
BRANCHIOPODA INC. SEDIS[9] GLAESSNER (1971b)
BRACHIOPODA
LINGULIDA ROWELL (1977)
ACROTRETIDA ZHURAVLEVA (1970), COWIE ET AL. (1972)
OBOLELLIDA[10] ROZANOV (1967), COWIE ET AL. (1972), ETC.
PATERINIDA ROZANOV (1967), ROWELL (1977), ETC.
KUTORGINIDA ROWELL (1977), ETC.
ORTHIDA ROWELL (1977)
ECHINODERMATA
SOLUTA SPRINKLE (1976)
CORNUTA SPRINKLE (1976)
EOCRINOIDEA SPRINKLE (1973, 1976)
IMBRICATA (LEPIDOCYSTOIDEA) SPRINKLE (1973, 1976)
HELICOPLACOIDEA SPRINKLE (1976)
EDRIOASTEROIDEA[11] SPRINKLE (1976), GLAESSNER & WADE (1966)
CAMPTOSTROMATOIDEA SPRINKLE (1976)
POGONOPHORA
?SABELLIDITIDA SOKOLOV (1972a, 1976)
?HYOLITHELMINTHES SOKOLOV (1972a, 1976)
CONODONTOPHORA
INC. SEDIS (PROTOCONODONTS) BENGTSON (1976, 1977), SOKOLOV (1976)

the Cambrian (Cloud 1948, 1968), as discussed below.

Exponential Diversification and the Cambrian "Explosion"

Figure 2 illustrates the stratigraphic ranges of all fossil orders of Metazoa (including Parazoa) known from the Lower Cambrian and uppermost Precambrian, or "Vendian." This range chart, which represents an expansion and refinement of those presented by Zhuravleva (1970) and Stanley (1976a), was compiled from a number of sources, listed in the caption and right hand column of Fig. 2. The stratigraphic units in the chart, representing four for the Vendian (left) and four for the Lower Cambrian (right), are based upon divisions of the Precambrian and Cambrian in the Soviet Union and their tentative worldwide correlation. A discussion of the definitions of these units along with estimates of their absolute ages appears in the appendix to this paper.

The number of fossil orders inferred as present in each stratigraphic unit is plotted logarithmically against absolute time in Fig. 3. As evident, the locus of points in this graph approximates a straight line remarkably well, especially considering the many uncertainities in the data (see discussion below and in Appendix). A fit of the exponential diversification model in Equation 5 to the data from this limited time interval produces an excellent goodness-of-fit (i.e. squared correlation) of 0.994. This is considerably better than fits of other simple models such as a linear function ($r^2 = 0.747$) or power function ($r^2 = 0.772$). Thus, the better fit as well as the a priori application of the exponential model suggests that this simple function is an appropriate description of the Precambrian-Cambrian radiations.

This result is particularly significant when

←

FIGURE 2. The known stratigraphic ranges of Vendian (uppermost Precambrian) and Lower Cambrian metazoan orders. The sixty or so orders listed by phylum in the central column are based primarily on the taxonomy of the *Treatise on Invertebrate Paleontology* (Moore and Teichert 1953–1975) and *Fossil Record* (Harland et al. 1967); this taxonomy has been modified and augmented in various places with works cited in the righthand column. Question marks preceding taxonomic names indicate orders questionably assigned to the given phyla. Assignments of various problematic taxa to orders follow the judgments of workers considered major authorities on the groups; this is particularly true for the Vendian taxa. Stratigraphic ranges are indicated by the bars in the chart on the left, with major references for this information also cited in the righthand column ("etc." indicates numerous other works in addition to those cited). Solid bars indicate fairly definite ranges; broken bars indicate probable ranges; question marks indicate possible ranges. Stratigraphic abbreviations are as follows: *lV* = "lower" Vendian; *mV* = "middle" Vendian; *uV* = "lower upper" Vendian; *ND* = "upper upper" Vendian (including Nemakit-Daldyn Horizon); *lT* = "lower" Tommotian (Sunnagin Horizon); *uT* = "upper" Tommotian (Kenyadan Horizon); *A* = Atdabanian; *B* = Botomian. The basis for this stratigraphic division is discussed in the Appendix to this paper. Footnotes: 1) Archaeocyathids may be calcareous algae rather than metazoans (Öpik 1975; D. C. Fisher, pers. comm.). 2) This represents an artificial grouping of Ediacaran and younger medusoids of uncertain class affinities; one or more unique orders may be represented here. 3) This includes the Ediacaran taxa that Glaessner assigned to the Pennatulacea in his earlier paper (see Glaessner 1966, 1972b; Glaessner and Wade 1966; etc.). 4) Reports of Lower Cambrian archaeogastropods are discounted, following the taxonomy of Runnegar and Jell (1976) and the considerations of Yochelson (1975) and Bockelie and Yochelson (1976). The presence of possible mollusc-like trace fossils in the uppermost Vendian (*"Bilobites"*; see Sokilov 1976) may indicate molluscs or related organisms extend into the Precambrian. 5) This represents lower Vendian trace fossils of probable coelomate origin (see Appendix); the worm-like fossils described by Cloud et al. (1976) might also be placed here. 6) *Wyattia* is included here. This fossil occurs below the lowest trilobitomorphid trace fossils (*Rusophycus*, etc.; see Alpert 1976) in the White-Inyo Mountains and therefore may be uppermost Precambrian, as suggested by Taylor (1966) (but see also Bergström 1970; Daily 1972). 7) This includes various Tommotian and uppermost Vendian problematica, including *Anabarites, Cambrotubulus, Sunnaginia, Tumulduria*, etc. Several authors have included some or all of these along with the coleolids in the questionably monophyletic order Angustiochreida (see Rozanov et al. 1969; Matthews and Missarzhevsky 1975; Bengtson 1977). 8) *Praecambridium* and *Vendia* are included here. Trilobitomorphid trace fossils in pre-trilobite, Tommotian strata might also be considered evidence of "trilobitoid *incertae sedis*." 9) This represents *Parvancorina*. 10) Obolellid-like brachiopods in pre-Botomian strata are included here. 11) *Tribrachidium* is included (with uncertainty) here, based on its gross morphologic similarities to early ontogenetic stages of edrioasteroids (see Bell 1976a) as well as affinities suggested by other authors (e.g. Glaessner and Wade 1966; Paul 1977).

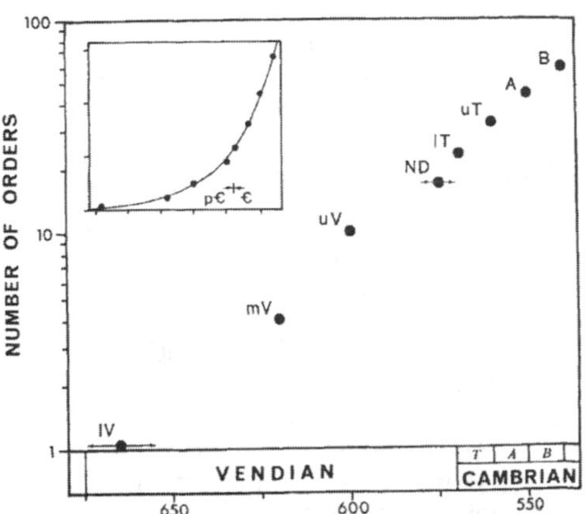

FIGURE 3. Graphs of known ordinal diversity in the Vendian and Lower Cambrian (see Fig. 2) plotted against absolute time. The points in the large semi-logarithmic graph coincide very closely with a straight line, as expected in exponential diversification. The inset graph in the upper left illustrates the same data on linear axes; the solid line is a least-squares fit of an exponential function ($r^2 = .994$), illustrating the interpreted continuously accelerating diversification across the Precambrian-Cambrian Boundary. Letters above points are abbreviations for stratigraphic units listed in the caption to Fig. 2. Points are plotted at the upper boundaries of these units since each represents the cumulative originations but not extinctions of that time interval. Arrows extending from points indicate possible intervals of time for points whose absolute ages are particularly uncertain (see Appendix).

considered in terms of the paleontological controversies surrounding the Precambrian-Cambrian biotic transition. Most previous explanations for the appearance of the Cambrian fauna have involved changes in extrinsic factors leading to an acceleration of evolutionary tempos. Hypotheses have included changes in the chemistry of the world ocean (Lane 1917; Vinogradov 1959; Chilingar and Bissell 1963; and others), increase in concentration of atmospheric oxygen (Nurshall 1959; Berkner and Marshall 1964, 1965; Towe 1970; Rhoads and Morse 1971; and others), changes in the stability of the world climate (Rudwick 1964; Fischer 1965; Valentine and Moores 1972; and others), initial appearance of predators (Schuchert and Dunbar 1933; Raymond 1935; Hutchinson 1961; Bengtson 1977; and others), and so on. All of these hypotheses have been predicated on the assumption that rates of diversification were low in the late Precambrian and abruptly much higher in the Early Cambrian. However, none have adequately differentiated total rates from per taxon rates. The fit of the exponential model indicates that total rate of diversification certainly was accelerating across the Precambrian-Cambrian Boundary, as implied by the inset in Fig. 3. However, this acceleration resulted simply from the multiplicative effect of the continuous addition of taxa. *The per taxon rate of diversification seems to have remained constant.* Therefore, in the absence of change in this fundamental rate of diversification (i.e. r_d), there seems to be no reason to invoke any extrinsic trigger specifically at the Precambrian-Cambrian Boundary.

This conclusion, of course, does not deny the necessity of special events in explaining the early diversification of metazoans. Rather, it shifts attention away from the Precambrian-Cambrian (or Vendian-Cambrian) Boundary and focuses it toward explaining events and processes surrounding the initial appearance of metazoans in the fossil record near the beginning of the Vendian, more than 100 Myr earlier. Although the factors influencing the occurrence and timing of this event have not yet been determined definitively, intrinsic processes, such as the slow evolution of eukaryotic diploid sexuality (J. W. Schopf et al. 1973) or prior appearance of cropping protists (Stanley 1973, 1976a), may have been important. Even the extrinsic mechanisms listed above may have played a role, although the analysis here suggests that once metazoans appeared changes in such mechanisms became insignificant in limiting their diversification, at least until the end of the Early Cambrian.

Under the present hypothesis, the "explosion" of taxa in the Cambrian becomes simply a rapid growth phase in a continuous process of exponential diversification, as illustrated by the inserted linear graph in Fig. 3. Obviously, this is a non-mechanistic explanation for this event, stating simply that taxa (specifically orders) could originate in a rapid though regular fashion; it does not specify the particular mechanisms by which this could have occurred. However, considerations of evolutionary mechanisms such as neoteny and paedomorphosis (Termier and Termier 1976; Gould 1977) and mutations within genetic regulatory systems (Valentine and Campbell 1975; Valentine 1977) suggest at least in theory that the rapid evolution of both taxonomic and morphologic diversity was possible if not likely following the initial appearance of metazoans. The appearance of skeletons among a number of distinct taxa during the rapid growth phase rather than earlier may represent simply the multiple attainment of a highly adaptive grade of organization over a few tens of millions of years during a time when numerous other lineages were also evolving, as argued by Zhuravleva (1970) and Stanley (1976b).

This hypothesis still requires considerably more documentation. The data presented in Figs. 2 and 3 are by no means definitive and objections can be raised to both the taxonomy and stratigraphy. Perhaps the most serious problem is the uncertainty in the estimates of absolute ages of the Vendian and Cambrian subdivisions. Small changes in the estimated durations of these subdivisions can move points in Fig. 3 substantially with respect to one another, disrupting the apparent exponential curve. Taxonomic problems may be less serious. Although orders represent rather coarse taxonomic units that are subject to continuous revision and stratigraphic extension, they still may parallel the diversification of species fairly closely, as I shall argue in the last section of this paper. More serious may be non-preservation of many soft-bodied organisms that must have existed during the Vendian. Valentine (1969) has argued that soft-bodied organisms as a whole do not differ ecologically from skeletonized organisms and thus should be subject to the same evolutionary controls of diversification (see also T. J. M. Schopf 1977). If soft-bodied organisms did in fact undergo exponential diversification, even at a rate different from fossilized organisms, their addition to Fig. 3 would not substantially reduce the fit of the model, so long as their diversification had approximately the same beginning point as the data already illustrated. Some evidence of a more universal exponential diversification is afforded by trace fossils. Recent studies of these fossils (Glaessner 1969; Banks 1970; Cowie and Spencer 1970; Webby 1970; Young 1972) suggest they were indeed diversifying at an accelerating rate from the late Vendian into the Early Cambrian. However, further refinement of the systematics and stratigraphic distribution of trace fossils is necessary before the exact pattern of this diversification can be determined.

Diversity-Dependent Diversification

If we now step forward from the Precambrian-Cambrian Boundary and examine some more of the early history of metazoan orders, as illustrated in Fig. 4, we find that the initial rate of diversification quickly tapered following the Early Cambrian. Indeed, if the Early Cambrian rate had continued, extrapolation of the curve in the inset of Fig. 3 indicates that the seas would now be teeming with more than half a billion metazoan orders! This number is clearly absurd since it exceeds the number of modern marine *species* by more than two orders of magnitude (cf. Cisne 1974). In reality, only about 248 metazoan orders now inhabit the world ocean.

Taxonomic diversity in the seas is probably limited because the rate of diversification is itself "diversity dependent" (Valentine 1973), such that it declines through negative feedback as numbers of taxa increase. MacArthur (1969) and Rosenzweig (1975) have argued that this decline should result from extinction rates increasing more rapidly with diversity than speciation rates. Reasons why the per taxon rate of extinction might increase as species are added into a region include

1. additional species may create new directed competition and intensify diffuse competition, causing exclusion and thus possible extinction of some species (see Mayr 1963; MacArthur and Levins 1967; MacArthur 1969; Rosenzweig 1975);
2. increased species packing may cause average local population sizes to be reduced, leaving some rarer species more prone to extinction by adverse environmental perturbations (e.g. severe storms, unusual cold spells, etc.), overpredation, epidemics, etc. (see Mayr 1965; MacArthur and Wilson 1967; MacArthur 1972; Ricklefs and Cox 1972; Simberloff 1972, 1974; May 1973; Rosenzweig 1975);
3. intensification of directed and diffuse competition may make colonization of

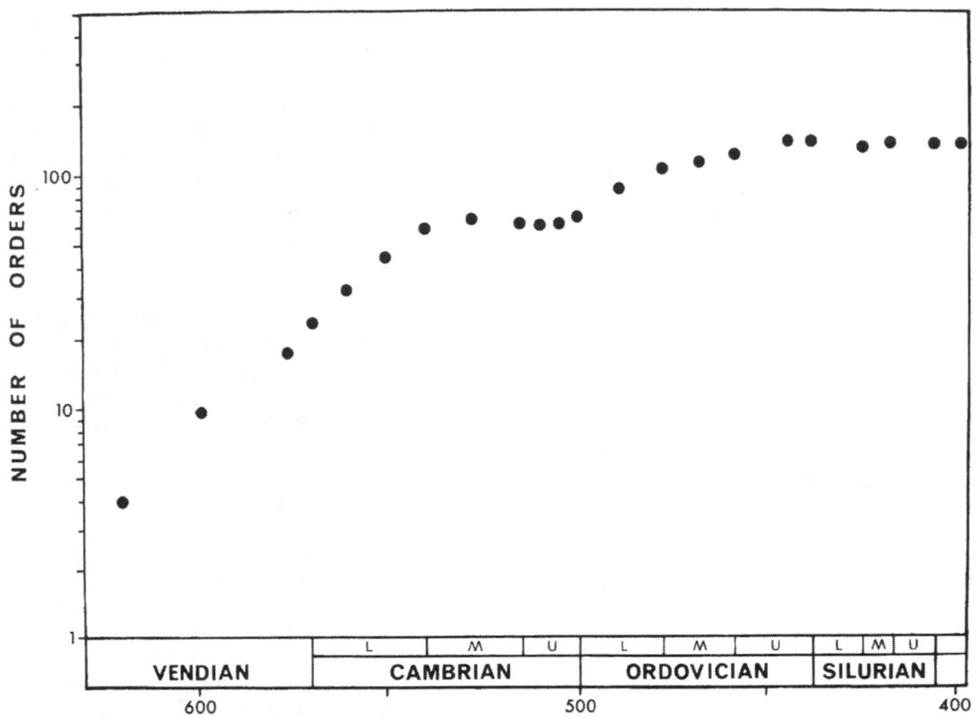

Figure 4. A further segment of the early diversification of marine metazoan orders, plotted as a semilogarithmic graph. The locus of points indicates that the initial exponential diversification, appearing in the lefthand third of this graph, was not long lived but quickly tapered following the Early Cambrian. Diversity appears to "plateau" in the Middle and Late Cambrian and then rise again but with a decelerating rate, reaching an approximate steady state in the Late Ordovician.

new areas and repopulation of areas recently abandoned (by local extinction or emigration) more difficult, thereby reducing the average size of the species' geographic ranges (see MacArthur 1972; Cody 1975; Rosenzweig 1975); reduction of range will reduce total population size and thus possibly enhance chances of extinction, as discussed above (see also Hancock 1967; Bretsky 1973; Van Valen 1973a; Thayer 1974; Boucot 1975, 1976).

These three ecological factors should also influence speciation. If most species originate from small, isolated populations (Mayr 1963), intensified competition and predation associated with increasing diversity should reduce the probability of any species shedding a daughter because

1. contraction of species' ranges and reduction of the general ease of colonization may leave species less likely to cross barriers and produce potentially diverging isolates (see MacArthur and Wilson 1967; Rosenzweig 1975);
2. increased species packing and resultant reduction in local population sizes may cause isolated populations to become extinct more frequently so that fewer attain the status of independent species (see Mayr 1965; Stanley 1977b);
3. reduction of the ease of colonization also may prevent more incipient species from expanding their ranges beyond their area of origin, dooming them to an early extinction before they have any effect on either the regional diversity or

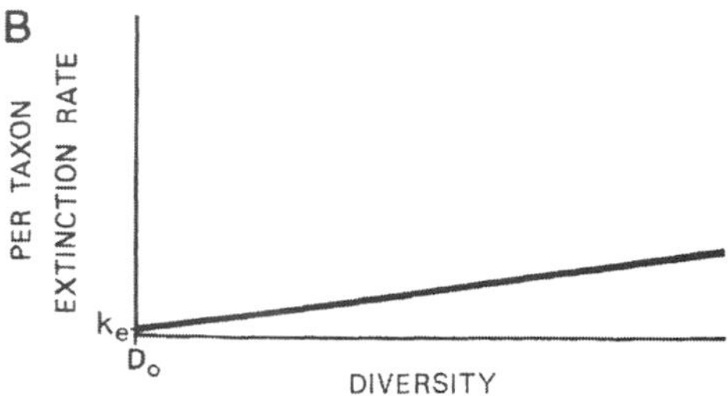

FIGURE 5. Linear models of diversity-dependent *per taxon* rates of origination (A) and extinction (B). The effects of diffuse competition and shrinking population sizes are inferred to cause the per taxon rate of origination to decline from some maximum rate k_s at an initial diversity D_o as diversity (i.e. number of taxa) increases. Conversely, the same factors are inferred to cause the per taxon rate of extinction to increase away from some minimum k_e at D_o.

the phylogenetic history of their taxocene (see Stanley 1977a).

In addition to these ecological arguments, Rosenzweig (1975) discusses several other genetic and biogeographic factors that also may cause speciation rate to decline with diversity.

The effects of variable rates of origination and extinction upon temporal patterns of diversity can be ascertained by proceeding beyond prosaic formulations and constructing models of diversity dependence. Presently, ecological theory is not sufficiently advanced to permit exact specification of diversity-dependent functions. However, first (and probably crude) approximations can be constructed from the directions of relationships suggested in the simple ecological arguments above.

Based on the assumption that diversity dependence causes the *per species* rate of origination to decrease monotonically with diversity, a simple model can be constructed as a linear function with a negative slope:

$$r_s = k_s - aD. \tag{6}$$

Here, k_s is the initial per species rate of origination at some small diversity D_o and a is the slope of the function; a graph of this function is presented in Fig. 5A. Using the assumption of monotonicity again, the per species rate of extinction also can be approximated by a linear function but now with a positive slope:

$$r_e = k_e + bD. \tag{7}$$

As before, k_e is the initial per species rate of extinction at the same small diversity, and b is the slope. This function is graphed in Fig. 5B. These two linear models are analogous to those used by MacArthur and Wilson (1963, 1967) to describe *total* rates of immigration and extinction. Like those, they are oversimplified, particularly since the number of potential interactions among species increases with the second rather than first power of species number (Terborgh 1973; Simberloff 1974). Still, the simpler linear approximations permit greater ease of derivation and analysis of diversity dependence.

Total rates of origination and extinction, which here are second-order functions, can be derived from per taxon rates by multiplying through by D. Thus, the relationship for total rate of origination becomes

$$R_s = k_s D - aD^2, \tag{8}$$

and for total rate of extinction becomes

$$R_e = k_e D + bD^2. \tag{9}$$

These two functions, which relate the total number of originations and extinctions in a time interval to the standing diversity, define oppositely facing parabolas, as illustrated in Fig. 6.

These parabolic functions have the immediate implication that the common paleontologic practice of plotting total numbers or rates of origination and extinction against geologic time may not provide the most interesting information about the history of a taxonomic group. Although such plots will give an impression of the times of high and low origination and extinction, most of this information should be obtainable from a simple graph of standing diversity (assuming the group already has been tested for diversity dependence). What may be more interesting

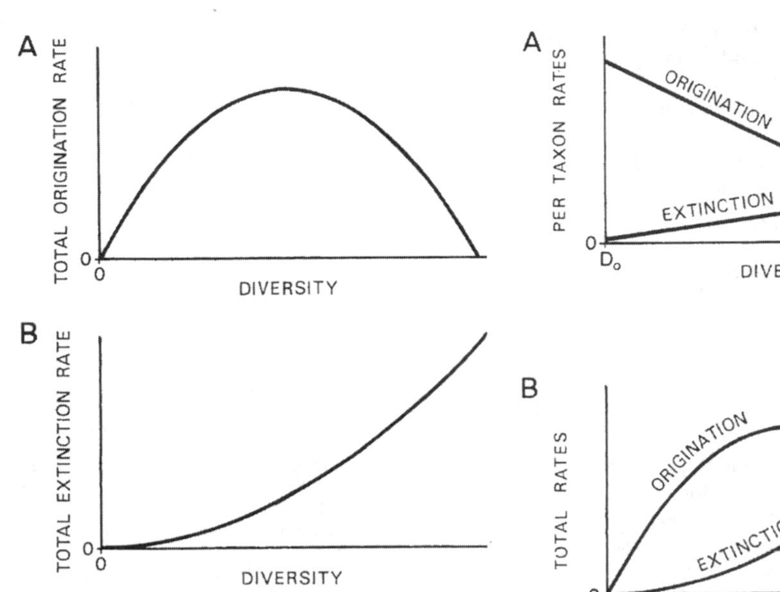

FIGURE 6. Parabolic models of diversity-dependent *total* rates of origination (*A*) and extinction (*B*), derived from the linear models by multiplying through by diversity (see text Equations 8 and 9). These models indicate that the total number of originations per interval of time should first increase with diversity and then decline while the total number of extinctions should increase continuously with an accelerating rate.

FIGURE 7. Combined models of origination and extinction for both per taxon rates (*A*) and total rates (*B*), illustrating their intersections at some diversity $\hat{D}$. This is an equilibrium diversity, with originations exceeding extinctions at lower diversities and extinctions exceeding originations at higher diversities.

in terms of paleobiologic history are plots of deviations from Equations 8 and 9, since these may identify periods of *unusually* high and low rates of origination and extinction.

The functions for both per species and total rates of origination and extinction have opposite signs for their second terms and are thus constrained to intersect at some diversity, assuming, of course, that k_s is greater than k_e. This relationship is shown graphically in Fig. 7. The point of intersection, labelled $\hat{D}$ in this figure, is an equilibrium diversity. At $\hat{D}$, rate of origination by definition equals rate of extinction so that their difference, the rate of diversification, is zero. Below $\hat{D}$, rate of origination exceeds extinction, causing diversity to climb. Above, rate of extinction exceeds origination, causing diversity to fall back toward $\hat{D}$.

The equilibrium diversity can be defined explicitly in terms of the diversity-dependent per species rates. The definition that origination rate equals extinction rate at equilibrium implies that the righthand sides of Equations 6 and 7 are equal when D equals $\hat{D}$. Manipulation of this equation yields

$$\hat{D} = \frac{k_s - k_e}{a + b}. \qquad (10)$$

However, the value of $\hat{D}$, or even its existence, may be difficult to determine from analysis of rates alone, as attempted by Mark and Flessa (1977). It is probably more appropriate to analyze an entire curve of standing diversity with respect to time to determine if its general shape is consistent with an equilibrial model (Rosenzweig 1977). The diversity-dependent model constructed thus far predicts that this curve should be logistic with an initial sigmoidal phase of diversification followed by an indefinite period of steady-state diversity (see Fig. 8). An algebraic description of this curve can be derived by combining the definitions of diversity-dependent rates in Equations 6 and 7 with the initial model of proportional diversification in Equation 1. The rate of diversification is the difference between the rates of origination and extinction. Substituting this difference into Equation 1 and treating it as a differential gives

$$dD/dt = [(k_s - aD) - (k_e + bD)] \cdot D .$$

Rearrangement and insertion of the definition of equilibrium diversity yields

$$dD/dt = r_o D(1 - D/\hat{D}) , \qquad (11)$$

where,

$$r_o = k_s - k_e . \qquad (12)$$

This last term is the initial per species rate of diversification and is the same rate as appears in the exponential model of Equation 4. Note that this simple model is still applicable to the early phase of diversity-dependent diversification, such as seen with metazoan orders, since Equation 11 reduces to Equation 1 when D is much smaller than $\hat{D}$.

The solution of Equation 11 with respect to time is

$$D = \frac{D_o \hat{D}}{D_o + (\hat{D} - D_o) \cdot \exp(-r_o t)} . \qquad (13)$$

Equations 11 and 13 are, of course, the familiar logistic growth equations commonly used by ecologists to describe the temporal growth of single-species populations. The model presented here now extends their use to describe the diversification of multi-species systems.

Logistic Diversification and Phanerozoic Orders

Two solutions of the logistic model of diversification are presented in Fig. 8. The upper graph represents a deterministic solution, showing the ideal shape of the curve. Diversity increases sigmoidally in time from some low initial value D_o and then asymptotes toward the equilibrium diversity $\hat{D}$. As previously noted, the earliest portion of the sigmoidal rise approximates an exponential curve, as can be seen by comparison with the inset in Fig. 3 (see also Whittaker 1975, p. 18).

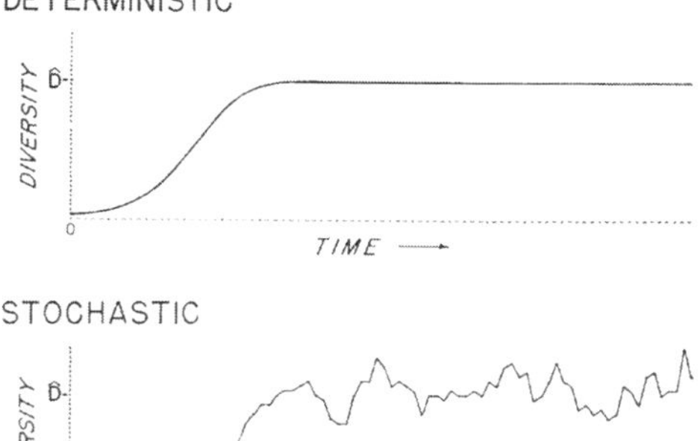

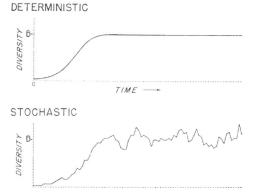

FIGURE 8. Logistic models of the behavior of diversity with respect to time, derived from the linear models of diversity-dependent per taxon rates of origination and extinction. The upper graph represents a deterministic solution, illustrating the ideal form of diversification with an initial sigmoidal rise followed by a constant equilibrium. The lower graph shows a simulated stochastic solution, probably more typical of natural systems; this solution was derived by adding small random variation to the rates of origination and extinction.

The deterministic solution is, of course, overconstrained and can never be expected to appear in a natural system. The lower curve in Fig. 8 represents a more realistic stochastic solution of the logistic model. This was obtained through simulation by permitting the per species rates of origination and extinction to vary with slight random components, representing perturbations in the environment, peculiarities of the taxa involved, etc. (cf. May 1973). The resultant curve still exhibits the elements of logistic diversification. Diversity increases with a perceptibly sigmoidal pattern and then fluctuates about the equilibrium in an irregular although limited fashion.

The entire Phanerozoic history of marine ordinal diversity is eminently comparable to this stochastic curve, as illustrated by the complete stage-by-stage record of ordinal diversity in Fig. 9. The logistic pattern is so striking that most specific historical events, represented by knicks and bumps in the curve, appear no more important than simple stochastic fluctuations. The curve increases approximately exponentially through the Vendian and Early Cambrian and then

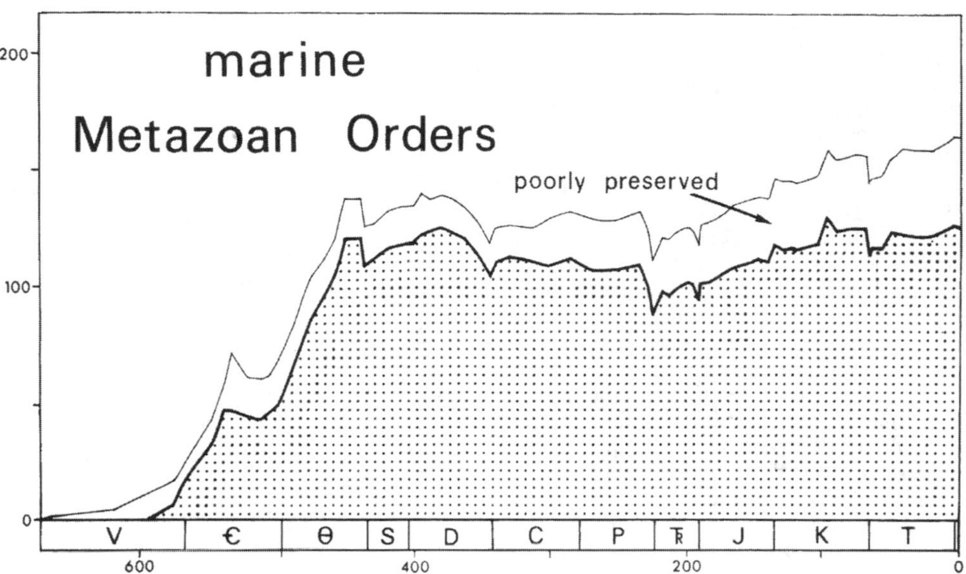

FIGURE 9. The complete Phanerozoic record of the diversity of marine metazoan orders, drawn with linear axes (number of orders on the ordinate, geologic time with systems indicated on the abscissa). This graph illustrates the basic consistency of a logistic model (Fig. 8) with the known pattern of Phanerozoic ordinal diversity. This diversity curve differs from previous curves for orders, such as presented by Simpson (1969), Valentine (1969), and others, largely as a result of inclusion of more taxa and incorporation of stratigraphic extensions based on recent discoveries. The data represented in this figure, as well as in Fig. 4, were derived primarily from the compilation of Kukalová-Peck (1973) which, in turn, was based largely upon the taxonomy and stratigraphy of the *Treatise, Fossil Record,* and *Vertebrate Paleontology* (Romer 1966). Sources used to modify these data, in addition to those cited in Fig. 2, include Conway Morris (1978), Jell and Jell (1976), Johnson and Richardson (1968), and Ossian (1973) for coelenterates; Jeletzky (1965) for mollusks; Cisne (1974), Jell (1974), and Schram (1969) for arthropods; Bell (1976b) and Moore and Strimple (1973) for echinoderms; Lindström (1970) for conodontophorids; Conway Morris (1977a,b,c), Jones and Thompson (1977), Richardson (1966), Schram (1973), and Welch (1976) for annelids and miscellanea; and Bardack and Zangerl (1968) and Greenwood (1975) for vertebrates. Ecological discussions in the *Treatise* and in Romer (1966) were used to separate marine from exclusively continental taxa. The field labelled "poorly preserved" in the graph represents those orders of soft-bodied metazoans whose fossil record is controlled solely by exceptional deposits, such as the Pound Quartzite, Burgess Shale, Hunsrück Schiefen, Solenhofen, etc. Orders represented only in these deposits, such as medusoid coelenterates, nematodes, chaetagnaths, etc., exhibit a net increase in number through time that seems to have resulted simply from chance preservation coupled with the complete modern record. The time scale on the abscissa of the graph is composite, based on Berggren (1972) for the Cenozoic, van Hinte (1976a,b) for the Cretaceous and Jurassic, Boucot (1975) for the Devonian, Silurian, and Upper Ordovician, and Harland et al. (1964) for the remainder, excluding the Lower Cambrian and Vendian which are discussed in the Appendix.

"plateaus" temporarily for the remainder of the Cambrian. (The peak at 75 orders in the total curve in the Middle Cambrian results largely from the inclusion of the Burgess Shale trilobitoids.) Rapid diversification recommences in the Ordovician but stops rather suddenly at 136 orders in the Late Ordovician. This termination appears somewhat more abrupt than might be expected in a perfectly logistic curve, although this is not uncommon when such curves have stochastic components. Following the terminal Ordovician extinctions, ordinal diversity attains a second peak at 140 orders in the Early Devonian and then slightly declines to a nearly level plateau at 125 to 132 orders through the remainder of the Paleozoic. This plateau makes the middle Paleozoic peaks appear almost as overshoots of a slightly lower equilibrium diversity. The Late Permian extinctions are quite visible on the ordinal curve but still appear much less severe than in curves for lower taxa (see, for example, Valentine 1969); the mild response of orders to these great extinctions is a property of their composite nature, as will be discussed later. Following the Permo-Triassic minimum, orders slowly rediversify with a decelerating rate characteristic of the upper portions of logistic curves (compare to Fig. 8). Equilibrium seems to be attained once again in the Cretaceous at a level slightly higher than that of the Paleozoic; this may result from sampling biases inherent in the fossil record (Raup 1972) or, perhaps, from the existence of slightly different multiple equilibria in the history of life (Bambach 1977). Note, however, that at the present, approximately 248 metazoan orders are recognized in the oceans; this represents an increase of over 45% from the inferred fossil diversity of the Cenozoic.

Although this pattern of standing diversity for marine metazoan orders appears largely consistent with the kinetic model, it does not represent a complete test. A variety of functions for per taxon rates of origination and extinction can produce approximately logistic curves so long as these functions intersect (and, of course, k_s is greater than k_e); the functions need not have opposite slopes. For example, a model could be conceived in which originations decline with diversity but so do extinctions, perhaps as a result of stability gained through increasing complexity of communities (cf. Valentine 1969). Thus, it is imperative to analyze the patterns of diversification as well as that of diversity.

Statistical analysis, however, is not without complications. Because diversity D appears in both the calculation of per taxon rates (Equations 2a and b) and the model for diversity dependence, Equations 6 and 7 should not be fit directly to data. All measured variables (S, E, D, and Δt) in these equations are subject to error, but S and E as point counts are probably subject to proportionally more error than D, which is a cumulative variable. This might cause plots of per taxon rates of origination and extinction versus diversity to behave as plots of $1/D$ against D, thus possibly effecting spurious negative correlations. This statistical artifact can be avoided by analyzing only total rates of origination and extinction (Equations 8 and 9).

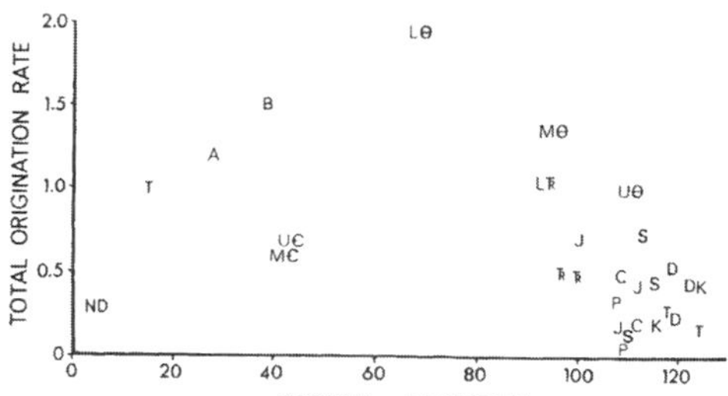

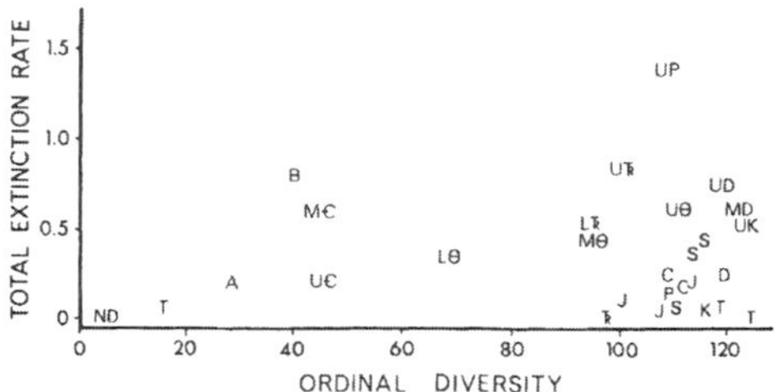

FIGURE 10. Graphs of total rates of ordinal origination and extinction plotted against ordinal diversity. Rates were calculated as the number of well-preserved orders appearing or disappearing in a series divided by the estimated duration of that series. Series rather than stages were used (except for the Lower Cambrian) in order to reduce sampling effects (ranges of a fair number of orders are documented only to series) and to minimize the proportional error in estimates of geologic time. Systems were rejected as too coarse and too few for sufficient analytic resolution. The time scale employed is the same as in Fig. 9. Letters, plotted in place of points in the graphs, are systemic abbreviations for the series and are the same as those used in Figs. 2 and 9. Lower, middle, and upper ("L," "M," and "U") series are indicated for lower Paleozoic systems and for series with high rates of origination (above) and extinction (below). The locus of points in the upper graph for total origination rates has a basically downward-expanding parabolic form (not drawn) as expected from the diversity-dependent model (compare to Fig. 6); a least-squares fit of this model accounts for 42% of the variance in the calculated rates. The graph for the extinction rates below, however, does not exhibit the expected upward-expanding parabolic form but rather includes two clusters of high and low rates, one containing Cambrian points and the other points at higher diversities; a fit of the diversity-dependent model to these data accounts for less than 1% of the variance in the calculated rates.

Figure 10 presents two graphs of total origination and extinction rates for well-preserved orders plotted against diversity. (Poorly preserved orders are excluded in order to avoid effects of the sporadic temporal distribution of exceptional deposits.)

The graph of origination rates at the top of Fig. 10 appears to be fairly consistent with the model of diversity dependence. A least-squares fit of Equation 8 accounts for 42% of the variance in calculated rates. Much of the residual variation is contributed by lower Paleozoic points at low to intermediate ordinal diversities. Elimination of the Middle and Upper Cambrian points, which seem to deviate particularly far from the expected pattern (perhaps for deterministic reasons; see Sepkoski 1977), adds more than 10% to the goodness-of-fit. The large cluster of points at high diversities exhibits a fairly strong negative trend with numbers of orders, as predicted. Note that this cluster includes two replicates: from the Ordovician through Devonian (and beyond) origination rates decline with diversity, and then again from the Triassic through Cenozoic rates also decline, although apparently beginning at a slightly lower level.

The graph of total rates of ordinal extinction against diversity in the lower part of Fig. 10 is far less encouraging. The expect upward-expanding parabola illustrated in Fig. 6B is not evident in the scatter of points. Instead, the data seem to be concentrated into two clusters. The larger corresponds to higher ordinal diversities and includes both the highest and lowest calculated extinction rates in no obvious trend with diversity. The smaller cluster includes Cambrian points at comparatively low diversities and again contains both high and low rates.

Possible causes of the model's poor fit to ordinal extinctions are numerous. One explanation is that the parameters in the extinction models of Equations 7 and 9 as applied to orders are so small that the curve for total extinctions barely leaves the ordinate, permitting statistical variation to swamp the basic pattern. Indeed, the proximity of the outer limb of the origination curve to the abscissa might suggest that the total extinction curve for orders is very shallow. The variation about this curve might be augmented substantially by the existence of more than one mode of extinction. As suggested by Terborgh (1973), normal, diversity-dependent extinction might be sporadically punctuated by periods of intensified, possibly diversity-*independent* extinction caused by changes in climate, regressions of epicontinental seas, etc. (cf. Valentine 1973).

The poor fit might also result in part from substantial errors in the data. For example, the possibly erroneous inclusion of archaeocyathids among the Metazoa (see Fig. 2, footnote 1) may be partially responsible for the high Cambrian extinction rates. On the other hand, the "pull of the Recent" (Raup 1972, 1977a, 1978) may be responsible for some low extinction rates in the Mesozoic and especially Cenozoic. Inaccuracies in the estimates of series' durations also may be quite serious since these appear in the denominators of calculated rates. A small error of several million years can substantially change a calculated rate, especially if the series is short or the number of extinctions or originations is large. Since the accuracy of measurement of geologic time decreases with age, this source of error might be expected to be greatest in the older series. Indeed, the correlation between absolute values of residuals from a parabolic fit to the origination rates in Fig. 10 and the age of the series is a statistically significant 0.486 ($p < .01$). Still, errors in estimated durations cannot account for all outlying points. Some calculated origination and extinction rates, such as for the Middle Cambrian, deviate from expectation in opposite directions, so that adjustment of the series' duration would reduce one deviation while increasing the other.

A final, more subtle source of error may be contributed by the composite nature of orders. Up to this point I have assumed without explicit justification that higher taxa can be analyzed in terms of a model derived from considerations of species. Yet, orders, as relatively large ensembles of species, may not have histories that coincide exactly with their component species. Two hypotheses incorporating this idea can be generated to explain some of the apparent lack to fit among extinctions in Fig. 10:

1. The disappearance of a polytypic order is contingent upon a series of events, namely the extinctions of all its constituent species. If such events are essentially independent, they could contribute considerable variance to the timing and pattern of ordinal extinctions, particularly since orders vary greatly in size.

2. As a composite entity, each order has its own history involving internal diversification and decline. Early in its history when it contains few species, an order's extinction is contingent upon only a small number of species' extinctions and thus may have a relatively high probability (see Stanley 1977a; Raup 1978). But if the order survives and enlarges, its probability of extinction may decline considerably. This property may produce higher than expected rates of ordinal extinction during and just after periods with high rates of ordinal origination, when young, small orders are numerous, such as seen in the Cambrian. Incomplete sampling, which serves to further shorten observed stratigraphic ranges of orders (Sepkoski 1975), may augment this effect.

These two hypotheses will be analyzed further in the simulations of the next section, which considers the broader problem of whether logistic patterns among higher taxa reflect similar patterns among species, the fundamental units of diversity.

Species, Higher Taxa and Diversity

A model for species diversity ideally should be tested with data on species themselves. However, as is well known, this is rarely possible in paleontology. Knowledge of fossil species is far from complete with at best only a few percent discovered and described (see Raup 1972, 1976a,b; Sepkoski 1975, and references therein). As a result, one is usually forced to use higher taxa, whose records are far better documented, to analyze the nature of paleontological diversity. This practice, however, is vulnerable to criticism, since higher taxa often are regarded as large and arbitrary units subject to the idiosyncrasies of evolution and the whims of taxonomists. Yet, as I shall argue in this section, these properties may not necessarily prevent higher taxa from paralleling the patterns and kinetics of species' diversification, even when information loss resulting from incomplete sampling of the fossil record masks these patterns in the species themselves.

The few comparative studies of diversity patterns in recent species and higher taxa have shown that several of the fundamental patterns of species diversity are adequately reflected in higher taxa. For example, Stehli et al. (1967) demonstrated that families of bivalves exhibit the same equator-to-pole diversity gradient that is seen in species, although with slightly more gentle slope (see also Cook 1969; Fischer 1960). More recently, Flessa (1975) presented data which suggest that the well-known species-area relationship may be reflected throughout much of the taxonomic hierarchy. His data, pertaining to mammals inhabiting areas ranging from oceanic islands to continents, show a strong positive correlation between area and number of taxa, from species to orders. Again, the slope of the relational function decreases with increasing taxonomic rank. Simberloff (1970, 1974) considered theoretical aspects of this relationship and also concluded that species-area patterns should be reflected adequately in higher taxa.

From a strictly empirical standpoint, these parallels should not be surprising. Most higher taxa are actually rather small, as documented by Anderson (1974) and Anderson and Anderson (1975) (see also Lillegraven 1972; Van Valen 1973b). Their data indicate, for example, that more than half of all mammalian families contain fewer than 10 species and half of mammalian orders contain fewer than 17 species. In fact, monotypic taxa, regardless of group or rank, are generally far more common than any other single type, producing a "hollow curve" in taxonomic frequency distributions (see review by Anderson 1974). Figure 11 illustrates this pattern for fossil taxa. The two frequency distributions exhibit hollow curves for numbers of fossil genera within invertebrate families and numbers of families within marine orders. These indicate that the median sizes of both taxonomic groups are rather small. More than half of fossil families contain three or fewer genera, and over half of all orders similarly contain three or fewer families. This, of course, is countered by the fact that half the genera are concentrated in only 15% of the families, and half the families are concentrated in only 9% of the orders. Still, so long as the small and large families are more or less randomly intermixed in space and time, variations in numbers of small taxa should track

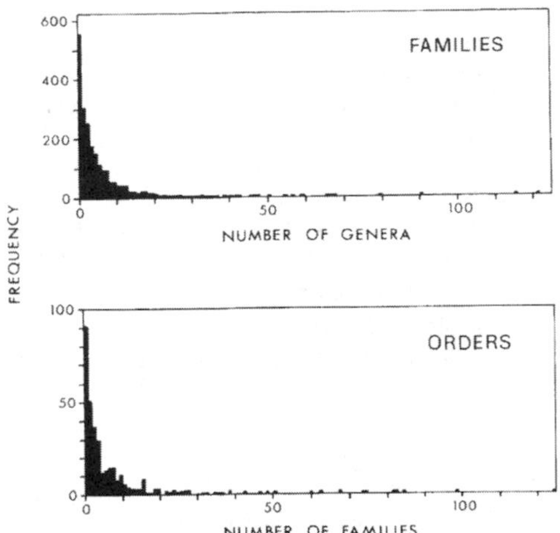

FIGURE 11. Taxonomic frequency distributions of fossil genera within invertebrate families (above) and families within Phanerozoic marine metazoan orders (below). The ordinates in these graphs represent the frequencies of families or orders containing the number of genera or families, respectively, indicated on the abscissae. The resulting frequency distributions appear as "hollow curves," indicating most higher taxa contain comparatively few constituent lower taxa. Frequencies of genera within invertebrate families (excluding protists) were tabulated from a computer compilation of stratigraphic data in the *Treatise on Invertebrate Paleontology* made by D. M. Raup; frequencies of families within all marine metazoan orders were compiled from a variety of sources, including those used for compilation of ordinal diversity in Figs. 3, 4, and 9.

the patterns of species diversity, despite "background noise" contributed by a few enormous taxa.

This argument can be made more concrete with simulations of the diversification of higher taxa and component species. I have used a modification of the "MBL" computer program developed by Raup et al. (1973) for this purpose. The MBL program uses Monte Carlo techniques and discrete time to simulate cladogenesis as a simple "birth-and-death" branching process with the number of branches (i.e. "lineages") extant at any time constrained in a dynamic equilibrium. Raup et al. (1973) and Raup and Gould (1974) originally used this program to explore the applicability of random processes to the interpretation of the fossil record. Their initial findings were fairly favorable, suggesting that evolutionary processes are so numerous and varied that over long periods of time they interact to produce phylogenetic patterns that cannot be easily differentiated from random (see also Raup 1977b,c). Thus, because stochastic patterns do mimic many of the essential features of the fossil record, similar simulations can now be used not just for simple exploration but also for tests of specific models in order to determine their more subtle effects on phylogenetic topologies (cf. Gould et al. 1977).

I have made two major modifications of the original MBL program for the simulations presented here. First, I have substituted the model of diversity-dependent origination and extinction rates in Equations 6 and 7 for the somewhat similar "dampening function" of Raup et al. (1973) (described explicitly by Simberloff 1974). Secondly, I have used a new routine for splitting completed phylogenies into monophyletic (or, more properly, paraphyletic) clades (i.e. higher taxa). In their comparison of real and simulated clades, Gould et al. (1977) found that the original size-dependent taxonomy of Raup et al. (1973) produced clades that are less variable than those observed in the fossil record whereas a random taxonomy produced clades that are more variable. To avoid these extremes, I have developed an intermediate "biased random" taxonomy, performed in three steps:

1. A single lineage is selected at random to become the originator of a potential, monophyletic clade. (This step alone constitutes the entire algorithm for the random taxonomy.)

2. The numbers of lineages in both the potential clade and the remaining ancestral clade are counted and the smaller of the two numbers is designated m.

3. A rejection probability p is calculated as a function of m and then compared to a random number between 0 and 1. If this number is larger than p, the clade is accepted; otherwise, it is rejected and the three steps are performed again.

These three steps are repeated until a predetermined number of clades has been created. In most simulations this procedure was run twice, first defining "superclades," each containing an average of 30 to 40 lineages, and then further subdividing these into "clades" with an average of 10 lineages each. The rejection function in Step 3 was developed through experimentation and is defined as

$$p = \exp(-m/10)\,. \qquad (14)$$

Although absolutely arbitrary, this function creates clades with "clade statistics" that fall within the range measured for the fossil record by Gould et al. (1977). It also produces hollow taxonomic frequency distributions similar to those in Fig. 11.

Thus, clades simulated in this fashion are arbitrary in definition and quite variable in size, similar to the real clades with which they share many formal properties. Yet, they still behave similarly to lineages in the growth and maintenance of their diversity. This was observed in a number of simulations using different parameters in the functions for lineage origination and extinction rates. The basic results of these various simulations were all similar although a few scale-dependent patterns did emerge, as will be noted in the discussion below. For this discussion, I have chosen a single, more or less typical simulation with intermediate parameters to exemplify the results. Selected patterns of interest from this simulation are illustrated in Fig. 12.

The upper portion of Fig. 12 illustrates the temporal patterns of diversification in lineages, clades, and superclades from the selected simulation. The logistic nature of lineage diversification was, of course, directly determined by Equations 6 and 7; the stochastic variation in the curve, comparable to that in Fig. 8, reflects the Monte Carlo methods employed. The pattern of diversification for clades and superclades was, however, only indirectly controlled through formation from lineages. Yet, the diversification of these "higher taxa" is remarkably logistic, closely paralleling the diversification of lineages. Systematic differences that do occur are relatively minor, as noted below:

1. Clades attain apparent equilibrium with a slight lag behind lineages; this lag tends to increase in proportion to average clade size.
2. Clades initially "overshoot" their apparent equilibrium and then gently decline in diversity, as best seen in clade diversity in Fig. 12A. This overshoot results from the "hindsight" nature of the simulated clades, since older potential clades tend to be slightly larger and therefore less often rejected than younger potential clades (cf. Raup et al. 1973); strictly random taxonomy lacking any hindsight does not produce this pattern.
3. Upon attainment of equilibrium, clades exhibit less short-term stochastic variation in diversity than do lineages. This variation decreases as clades become larger and as turnover rate among lineages declines.
4. Clades and superclades occasionally exhibit long-term departures from apparent equilibria, particularly when their numbers are small and the turnover rate among lineages is high; such departures are temporary and previous equilibria are eventually reattained.

These patterns seem to have a number of parallels in the observed diversity of Phanerozoic higher taxa. Ordinal diversity, as previously noted, seems to exhibit a slight overshoot of equilibrium in the mid-Paleozoic followed by a mild decline up through the Permian. Although patterns such as this conventionally have been interpreted as elimination of intermediate or less-adapted groups at the close of an adaptive radiation (Simpson 1953; Gould 1970), the simulations suggest that early departures from apparent equilibria may simply reflect the small numbers and large average sizes of higher taxa. The sequence with which higher categories in the fossil record attain their maximum diversities is opposite that seen in the simulations (see Simpson 1953; Valentine 1969, 1973, 1977); however, as argued by Raup (1972) and further illustrated below, this pattern may result from incomplete sampling of the fossil record. Following attainment of apparent equilibrium, orders like simulated clades exhibit only mild fluctuations in their diversity; this probably represents a dampening of much greater fluctuations in species diversity. This is certainly true for the Permo-Triassic interval in which ordinal diversity underwent only mild decline whereas species numbers must have fallen radically. However, because this interval was relatively short

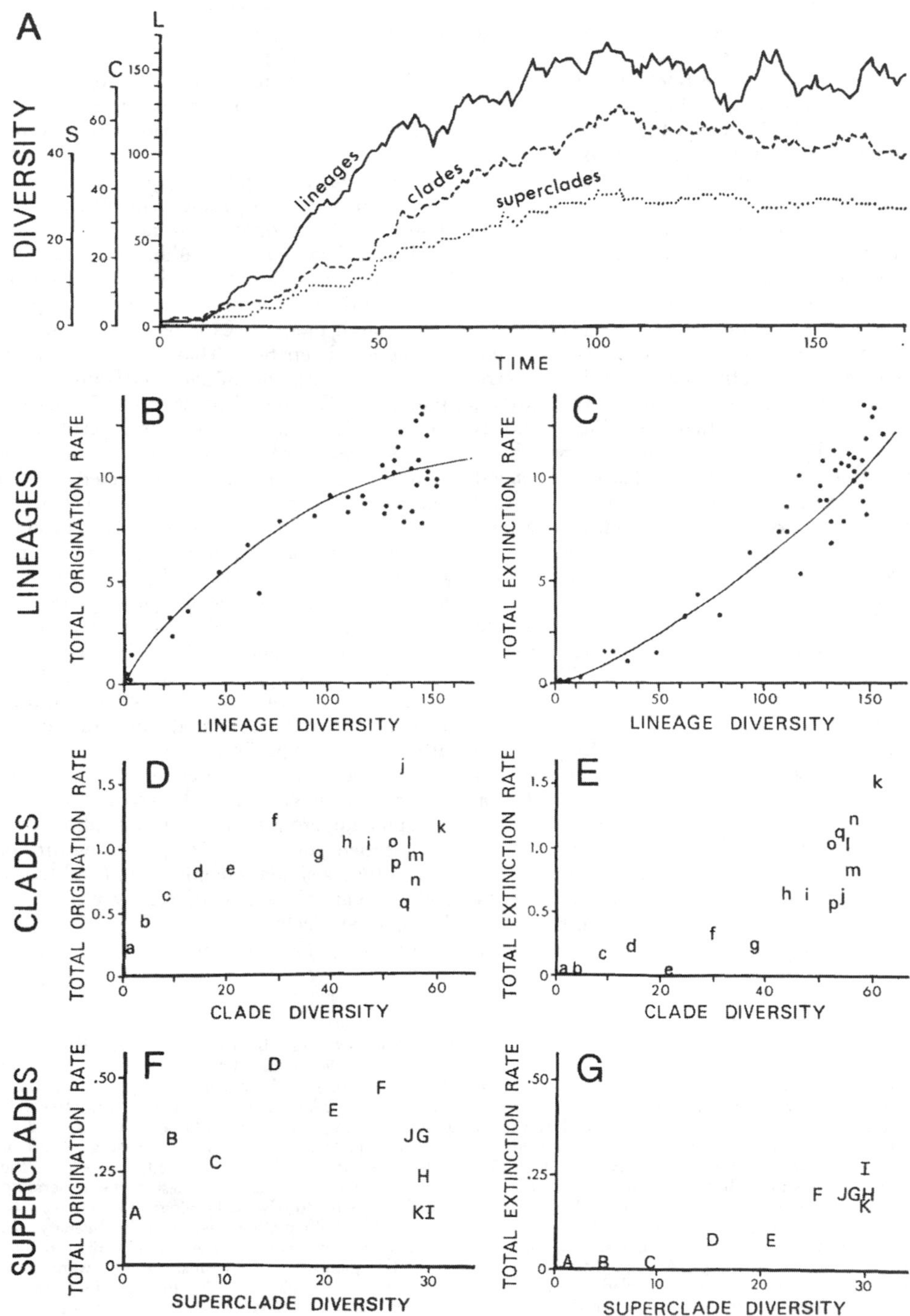
A
DIVERSITY
S
C
L
lineages
clades
superclades
TIME
B
C
LINEAGES
TOTAL ORIGINATION RATE
TOTAL EXTINCTION RATE
LINEAGE DIVERSITY
D
E
CLADES
CLADE DIVERSITY
F
G
SUPERCLADES
SUPERCLADE DIVERSITY

(compared to the total duration of the Phanerozoic), orders seem to have been able to withstand a decline in average size without suffering extensive extinctions. (This phenomenon is analogous in form, if not cause, to Flessa's [1975] demonstration that numbers of mammalian orders vary much more slowly with respect to habitat area than do lower taxa.)

These initial results suggest that higher taxa may very well reflect the standing diversity of species so long as sufficient total numbers are involved. This conclusion is actually more general than the analysis of the logistic model presented here. Other models, such as continuous diversification or patternless diversification with long-term major fluctuations, also produce clades that track lineage diversity with slight lags and with dampened short-term stochastic variation. This more general result strengthens the argument, suggesting that the observed logistic pattern of ordinal diversity does indeed indicate a similar underlying pattern in Phanerozoic species diversity.

This conclusion can be extended to the kinetics of diversification, although the results here are not so transparent. The lower portion of Fig. 12 illustrates total rates of origination and extinction for lineages, clades, and superclades plotted against their respective diversities in the same manner as ordinal rates in Fig. 10. Graphs of rates for lineages, plotted along with their generating functions (solid curves), are presented in Fig. 12B and C in order to illustrate some of the "noise" inherent in a stochastic system. As evident, this noise is not sufficient to mask the underlying parabolic patterns, and fits of the diversity-dependent models (Equations 8 and 9) still account for nearly 90% of the variation in total rates.

The plots of total rates for clades and superclades appear much like those for the lineages although with some minor differences. Total extinction rates in Fig. 12E and G have upward-expanding parabolic trends only slightly more concave than that seen for lineages. Total origination rates (Fig. 12D and F) show convex parabolic trends like the lineages, although those for the "higher taxa" tend to be considerably more complete with rough outer limbs extending downward at higher diversities. These outer limbs are associated with the initial overshoots of apparent equilibrium diversities discussed previously. As equilibria are approached, origination rates of clades and superclades remain slightly higher than expected from a simple logistic model, causing the overshoots. In time, however, these rates decline, producing the downward "dribbles" of points in the graphs as well as the mild drops in the standing diversities; the temporal pattern of this decline can be seen most clearly in the sequence of letters in Fig. 12F. A similar pattern is also apparent in the graph of ordinal origination rates in Fig. 10; middle Paleozoic points associated with the initial attainment of equilibrium tend to be slightly higher than points for later series. The simulation thus suggests that the nearly complete parabola apparent for ordinal origination rates may reflect more the composite nature of orders than a similar completeness (and therefore similarly low rate of turnover at equilibrium) in species.

However, the simulations still do not in-

←

FIGURE 12. Results of simulations of the diversification of lineages and "higher taxa" (i.e. clades and superclades; see text). *A.* Patterns of standing diversity (ordinate) in simulated time (abscissa); note that different scales are used on the ordinate for lineages (L), clades (C), and superclades (S). The patterns of diversification among the hierarchial categories are remarkably similar with only minor, quantitative differences, as described in the text. *B.* Total rate of lineage origination plotted against lineage diversity and *C.* total rate of lineage extinction plotted against diversity. Points on these two graphs represent averages of simulated rates in consecutive intervals of four time units; thus, the total noise originally present in the stochastic system has been somewhat modulated. The solid curves in the graphs represent the generating functions for the points, showing that they correspond fairly closely to the deterministic functions. *D.* Total rate of clade origination and *E.* total rate of clade extinction, both plotted against clade diversity. *F.* Total rate of superclade origination and *G.* total rate of superclade extinction. Letters, rather than points, are plotted in these two pairs of graphs to show the temporal sequences of the points. (Points were averaged over 10 time units for the clades and 15 time units for the superclades.) The four graphs illustrate that the patterns of diversification for these "higher taxa" are similar to those for the lineages from which they were derived.

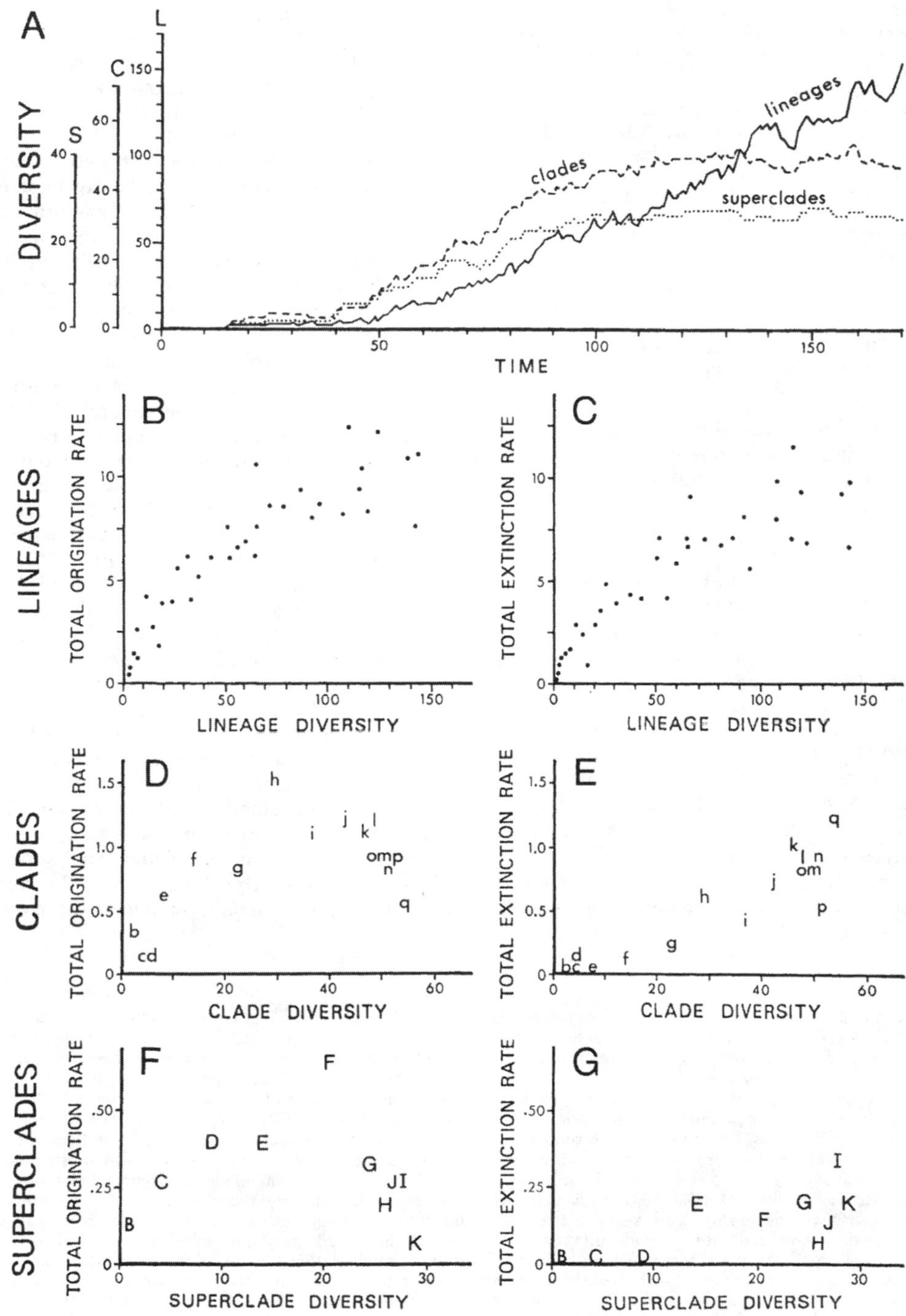
A
DIVERSITY
S
C
L
lineages
clades
superclades
TIME
B
C
LINEAGES
TOTAL ORIGINATION RATE
TOTAL EXTINCTION RATE
LINEAGE DIVERSITY
D
E
CLADES
CLADE DIVERSITY
F
G
SUPERCLADES
SUPERCLADE DIVERSITY

TABLE 1. Values of goodness-of-fit for a simulation of the diversification of higher taxa. Each number is the squared correlation measured for a fit of the parabolic diversity-dependent model in text Equations 8 and 9 to the total rates of origination ("Orig") or extinction ("Ext") observed for clades and superclades in the complete or partially sampled system. The fits indicate that the model should tend to fit total origination rates as well as or better than total extinction rates of higher taxa only when the numbers and temporal ranges of their constituent species are imperfectly known as a result of incomplete sampling.

	Complete		Sampled	
	Orig	Ext	Orig	Ext
CLADES	.41	.84	.83	.85
SUPERCLADES	.59	.95	.69	.53

dicate why the model of diversity dependence fits ordinal originations so much better than extinctions. In fact, contrary to expectations, parabolic curves fit the total extinction rates of simulated clades and superclades better than the total origination rates, as documented in Table 1. This last result, however, is a property only of systems that are completely known. If systems are only partially known, a pattern more similar to that observed for real orders emerges.

Figure 13 illustrates the simulated system of Fig. 12 after having been subjected to "partial sampling" using the model discussed by Raup (1972) and Sepkoski (1975). Lineages were recovered (i.e. sampled) only from some of the time units, with the probability of recovery decaying exponentially backward through simulated time. Diversities were calculated following this sampling by extending lineages' ranges between their oldest and youngest points of recovery.

The effects of partial sampling on simulated standing diversities are illustrated in Fig. 13A. As might be expected, sampling most severely affects lineage diversity. Most indications of the original equilibrium are lost, and lineage diversity now appears to rise nearly linearly toward the "Recent." Diversity patterns of clades and superclades, on the other hand, still retain early sigmoidal rises followed by mild fluctuations about apparent equilibria. Differences from their original patterns in Fig. 12A are largely quantitative: the period of sigmoidal diversification is shifted toward the "Recent"; the tendency to overshoot equilibrium initially is suppressed; and the apparent equilibrial diversity is slightly lowered. There is a reversal in the sequence of equilibration with superclades now attaining apparent equilibrium first, followed by clade and finally lineages, mimicking the pattern observed in the fossil record (cf. Raup 1972).

Rates of diversification in the sampled system are illustrated as before in the lower portion of Fig. 13. Again, lineages have been most severely affected, with their total extinction rate (Fig. 13C) actually reversing its apparent functional relationship to diversity and appearing now as a downward-expanding parabola. This result further suggests that the true kinetics of diversification as well as patterns of standing diversity may be difficult to ascertain from lower taxa in the fossil record. Simulated clades and superclades are less affected by sampling. In fact, partial sampling actually enhances the para-

←

FIGURE 13. Results of simulations of the diversification of lineages and "higher taxa" after subjection to partial sampling (see text for explanation of sampling procedure). All plotting conventions are the same as in Fig. 12. Only 65% of the lineages from Fig. 12 are represented here, and many of these have had their temporal ranges severely shortened. Clades and superclades are better represented with 89% and 96%, respectively, present; these higher percentages reflect the necessity of only a single lineage to document the presence of a multi-lineage clade (cf. Simpson 1953; Cisne 1974; Raup 1975). The simulated sampling efficiencies for the various taxonomic units used here are undoubtedly much higher than those for the actual fossil record but are probably sufficient for qualitative conclusions. *A*. Patterns of standing diversity show that lineages are severely affected by partial sampling whereas clades and superclades retain most aspects of their original logistic diversity. *B*. Total rate of lineage origination retains much of its original functional relationship to diversity whereas *C*. total rate of lineage extinction reverses its relationship to diversity. *D*. Total rate of clade origination exhibits an improved parabolic relationship to diversity, whereas *E*. total rate of clade extinction exhibits about the same relationship as in the completely known system (Fig. 12E) although with slightly less correlation at the higher diversities. *F*. Total rate of superclade origination also exhibits a slightly enhanced parabolic relationship to diversity, whereas *G*. total rate of superclade extinction shows a deteriorated relationship with greater variance in rates at intermediate to high diversities.

bolic form of their total origination rates, as evident in Fig. 13D and F; this is further reflected in the better fits of the diversity-dependent model to origination rates in the sampled as compared to complete system, as indicated in Table 1. The better fits result from suppression of overshoots and from slight upward creeps in clade diversities during "equilibrium" phases. Together, these cause the outer limbs in the graphs of origination rates to stretch out slightly, producing stronger negative trends between total rates and higher diversities. This suggests that the strong negative trend between total origination rate and higher diversities evident for orders in Fig. 10 may reflect not only the biology of the system but also the effects of incomplete sampling of the fossil record.

While patterns of origination artificially improve after sampling, patterns of extinction for clades and especially superclades begin to deteriorate, as also seen for lineages. The variation in extinction rates through time increases, tending to produce some comparatively low rates at high diversities as well as occassional higher rates at intermediate diversities (see especially Fig. 13G). As documented in Table 1, this increased variation does not affect the parabolic fit for clade extinctions but does substantially reduce the fit for the larger superclades; in fact, superclade extinctions are now less accurately described by the simple diversity-dependent model than are their originations. Thus, information loss tends to reverse the trend seen previously in the complete system and produce a pattern approaching that observed for orders in the fossil record.

This final result completes the suggestive evidence that the patterns of diversification observed for Phanerozoic orders may indeed reflect underlying patterns for species, despite variance imparted by the composite nature of orders and amplified by information loss. Specifically, the simulations suggest that

1. a logistic diversification of species should be reflected among higher taxa even when information loss masks this pattern in the species themselves;
2. a parabolic functional relationship between total origination rate and diversity in species should be preserved in higher taxa;
3. a parabolic relationship between total extinction rate and diversity in species may be disrupted in higher taxa as a result of information loss.

The simulations indicate, however, that the tendency toward disruption of extinction patterns will probably not be complete, since fits of the diversity-dependent model for extinction rates even of sampled superclades are still significant (Table 1). Therefore, it must be concluded that other factors, particularly periods of extraordinary extinction, must also have been operative in producing the chaotic pattern of ordinal extinctions in Fig. 10. But it is also entirely possible that the kinetic model does not afford an appropriate description of Phanerozoic extinctions and that other, perhaps less obvious models could predict the patterns more precisely. Thus, testing of the kinetic model is far from complete. Definitive analysis must await greater sophistication of our concepts and further refinement of our knowledge of the history of life.

Summary and Conclusions

The major points discussed in this paper are summarized below:

1. Phylogenies, like populations, expand through proportional growth, suggesting that under constant conditions their constituent species should diversify exponentially.
2. An exponential model seems to describe adequately the diversification of known metazoan orders across the Precambrian-Cambrian Boundary. This suggests that the "explosive" appearance of the Early Cambrian fauna was simply one phase of a continuously accelerating diversification and that no special event, other than the much earlier initial appearance of the Metazoa, was necessary for the occurrence of this diversification.
3. Diversification is probably limited at some point by "crowding" effects among species. Considerations of diversity-dependent processes suggests that per species rates of origination should fall and per species rates of extinction should rise as diversity increases.

4. These diversity-dependent rates can be modeled as linear functions and manipulated to construct a simple logistic model of diversity which suggests, in its stochastic form, that numbers of taxa should rise sigmoidally in time and then fluctuate about some constant equilibrium.
5. The complete known history of the diversity of marine metazoan orders is remarkably consistent with a logistic model. Ordinal diversity rises roughly sigmoidally from the Venedian into the Ordovician and then fluctuates mildly about an apparent equilibrium. This logistic pattern is also seen in the variation in total origination rates as a function of diversity but not in total extinction rates.
6. Considerations of the size of higher taxa and simulations of their diversification suggest that higher taxa should adequately track both the standing diversity and the patterns of diversification of species, providing additional credence to tests involving ordinal diversity.
7. Further simulations suggest that observed patterns of standing diversity and calculated rates of diversification for higher taxa should be much more robust to the effects of partial sampling than species although patterns of extinction may be partially masked.

The model and analysis presented here are admittedly oversimplified. However, what I have been seeking is not a complete causative account of the history of life but rather a description of the fundamental patterns in the temporal behavior of taxonomic diversity. The kinetic model is a first approximation of this behavior but one that can be modified and expanded to describe secondary patterns of diversification and also to test various causal hypotheses. Rates of diversification can be manipulated and "equilibrium" can be made to vary as a function of habitat area, continental dispersion, climatic change, or even time itself. Further refinements of the mathematics of the model and of the data from the fossil record should make such possibilities eminently feasible.

Acknowledgements

I especially acknowledge the numerous systematists whose works are cited herein because without their careful observations and meticulous papers, theoretical work such as this would not be possible. This work was completed while I was in the Department of Geological Sciences and Center for Evolution and Paleobiology, University of Rochester. The writing of this paper greatly benefited from discussions with a number of colleagues, particularly David M. Raup, to whom I am also grateful for supplying data for a portion of Figure 12. I thank Daniel C. Fisher for discussing problems of archaeocyathid affinities and S. J. Gould, D. M. Raup, T. J. M. Schopf, P. Signor, J. C. Tipper, and two anonymous reviewers for their critiques and suggestions for the manuscript. The research received partial support from the Earth Sciences Section, National Science Foundation, NSF Grant DES75-03870.

Literature Cited

Alpert, S. P. 1976. Trilobite and star-like trace fossils from the White-Inyo Mountains, California. J. Paleontol. *50*:226–239.

Anderson, S. 1974. Patterns of faunal evolution. Q. Rev. Biol. *49*:311–332.

Anderson, S. and C. S. Anderson. 1975. Three Monte Carlo models of faunal evolution. Am. Mus. Novit. *2563*:1–6.

Arkell, W. J. 1957. Introduction to Mesozoic Ammonoidea. Pp. L81–L129. In: Moore, R. C., ed. Treatise on Invertebrate Paleontology, Pt. L. 490 pp. Geol. Soc. Am. and Univ. Kans. Press; Lawrence, Kansas.

Bakker, R. T. 1977. Tetrapod mass extinctions—a model of the regulation of speciation rates and immigration by cycles of topographic diversity. Pp. 439–468. In: Hallam, A., ed. Patterns of Evolution. 591 pp. Elsevier, Amsterdam.

Bambach, R. K. 1977. Species richness in marine benthic habitats through the Phanerozoic. Paleobiology. *3*:152–167.

Banks, N. L. 1970. Trace fossils from the late Precambrian and Lower Cambrian of Finmark, Norway. Pp. 19–34. In: Crimes, T. P. and J. C. Harper. Trace Fossils. 547 pp. Geol. J. Spec. Issue 3. Seel House Press; Liverpool.

Bardack, D. and R. Zangerl. 1968. First fossil lamprey—a record from the Pennsylvanian of Illinois. Science. *162*:1265–1267.

Bell, B. M. 1976a. Phylogenetic implications of ontogenetic development in the Class Edrioasteroidea (Echinodermata). J. Paleontol. *50*:1001–1019.

Bell, B. M. 1976b. A study of North American Edrioasteroidea. N.Y. State Mus. Mem. 21, 446 pp.

Bengtson, S. 1970. The Lower Cambrian fossil *Tommotia*. Lethaia. *3*:363–392.

Bengtson, S. 1976. The structure of some Middle Cambrian conodonts and the early evolution of conodont structure and function. Lethaia. *9*:185–206.

Bengtson, S. 1977. Aspects of problematic fossils in the early Palaeozoic. Acta Universitatis Upsaliensis, No. 415, 71 pp.

Berggren, W. A. 1972. A Cenozoic time-scale—some implications for regional geology and paleobiogeography. Lethaia. *5*:195–215.

Bergström, J. 1970. *Rusophycus* as an indication of Early Cambrian age. Pp. 35–42. In: Crimes, T. P. and J. C. Harper, eds. Trace Fossils. 547 pp. Geol. J. Spec. Issue 3. Seel House Press; Liverpool.

Berkner, L. V. and L. C. Marshall. 1964. The history of oxygenic concentration in the earth's atmosphere. Discuss. Faraday Soc. *37*:577–595.

Berkner, L. V. and L. C. Marshall. 1965. History of major atmospheric components. Proc. Nat. Acad. Sci. USA. *53*:1215–1225.

Bockelie, T. G. and E. L. Yochelson. 1976. "Worm" tubes from the Valhallfonna Formation, Spitzbergen. Geol. Soc. Am. Abstr. with Program. *8*:139.

Boucot, A. J. 1975. Evolution and Extinction Rate Controls. 427 pp. Elsevier; Amsterdam.

Boucot, A. J. 1976. Standing diversity of fossil groups in successive intervals of geologic time viewed in the light of changing levels of provincialism. J. Paleontol. *49*:1105–1111.

Bretsky, P. W. 1973. Evolutionary patterns in the Paleozoic Bivalvia: Documentation and some theoretical considerations. Geol. Soc. Am. Bull. *84*: 2079–2096.

Cailleux, A. 1950. Progression géométrique du nombre des espécies et vie en expansion. C.R.S. de la Soc. Geol. de France. *13*:222–224.

Cailleux, A. 1954. How many species? Evolution. *8*:83–84.

Chilingar, G. V. and H. J. Bissell. 1963. Note on possible reason for scarcity of calcareous skeletons of invertebrates in Precambrian formations. J. Paleontol. *37*:942–943.

Cisne, J. L. 1974. Evolution of the world fauna of aquatic free-living arthropods. Evolution. *28*:337–366.

Clark, R. B. 1964. Dynamics in Metazoan Evolution. 313 pp. Clarendon Press; Oxford.

Clark, R. B. 1969. Systematics and phylogeny: Annelida, Echiura, Sipuncula. Pp. 1–68. In: Florkin, M. and B. T. Scheer, eds. Chemical Zoology. V. 4. 548 pp. Academic Press; New York.

Cloud, P. E. 1948. Some problems and patterns of evolution exemplified by fossil invertebrates. Evolution. *2*:322–350.

Cloud, P. E., Jr. 1968. Pre-metazoan evolution and the origins of the Metazoa. Pp. 1–72. In: Drake, E. T., ed. Evolution and Environment. 470 pp. Yale Univ. Press; New Haven, Conecticut.

Cloud, P. 1973. Pseudofossils: A plea for caution. Geology. *1*:123–127.

Cloud, P., J. Wright, and L. Glover III. 1976. Traces of animal life from 620 m.y. old rocks in North Carolina. Am. Sci. *64*: 396–406.

Cody, M. L. 1975. Towards a theory of continental species diversities. Pp. 214–257. In: Cody, M. L. and J. M. Diamond, eds. Ecology and Evolution of Communities. 544 pp. Belknap Press; Cambridge, Massachusetts.

Conway Morris, S. 1977a. Aspects of the Burgess Shale fauna, with particular reference to the non-arthropod component. J. Paleontol. *51*(Suppl.): 7–8.

Conway Morris, S. 1977b. A redescription of the Middle Cambrian worm *Amiskwia saggitiformis* Walcott from the Burgess Shale of British Columbia. Paläontol. Z. *51*:271–287.

Conway Morris, S. 1977c. Fossil priapulid worms. Palaeontol. Assoc. London, Spec. Pap. Palaeontol., No. 20, 95 pp.

Conway Morris, S. 1978. *Laggania cambria* Walcott: A composite fossil. J. Paleontol. *52*:126–131.

Cook, R. E. 1969. Variation in species density of North American birds. Syst. Zool. *18*: 63–84.

Cowie, J. W. 1961. Contributions to the geology of North Greenland. Medd. Grønland. *164*:1–47.

Cowie, J. W. and M. F. Glaessner. 1975. The Precambrian-Cambrian Boundary: A symposium. Earth-Sci. Rev. *11*:209–251.

Cowie, J. W. and A. Yu. Rozanov. 1974. I.U.G.S. Precambrian-Cambrian Boundary working group in Siberia, 1973. Geol. Mag. *111*:237–252.

Cowie, J. W., A. W. A. Rushton, and C. J. Stubblefield. 1972. A correlation of Cambrian rocks in the British Isles. Geol. Soc. London, Spec. Rep. No. 2, 42 pp.

Cowie, J. W. and A. M. Spencer. 1970. Trace fossils from the late Precambrian/Lower Cambrian of East Greenland. Pp. 91–100. In: Crimes, T. P. and J. C. Harper, eds. Trace Fossils. 547 pp. Geol. J. Spec. Issue 3. Seel House Press; Liverpool.

Crawford, A. R. and B. Daily. 1971. Probable non-synchroneity of late Precambrian glaciations. Nature. *230*:111–112.

Daily, B. 1972. The base of the Cambrian and the first Cambrian faunas. Univ. Adelaide Centre for Precambrian Res., Spec. Pap. No. 1, pp. 13–42.

Eldredge, N. and S. J. Gould. 1972. Punctuated equilibria: An alternative to phyletic gradualism. Pp. 82–115. In: Schopf, T. J. M., ed. Models in Paleobiology. 250 pp. Freeman, Cooper & Co.; San Francisco, California.

Fischer, A. G. 1960. Latitudinal variations in organic diversity. Evolution. *14*:64–81.

Fischer, A. G. 1965. Fossils, early life, and atmospheric history. Proc. Nat. Acad. Sci., U.S.A. *53*: 1205–1215.

Flessa, K. W. 1975. Area, continental drift, and mammalian diversity. Paleobiology. *1*:189–194.

Fritz, W. H. 1972. Lower Cambrian trilobites from the Sekwi Formation type section, MacKenzie Mountains, northwestern Canada. Geol. Surv. Can. Bull., No. 212, 58 pp.

Germs, G. B. 1972. New shelly fossils from the Nama Group, South West Africa. Am. J. Sci. *272*: 752–761.

Glaessner, M. F. 1958. New fossils from the base of the Cambrian in South Australia. Trans. R. Soc. South Aust. *81*:185–188.

GLAESSNER, M. F. 1966. Precambrian paleontology. Earth-Sci. Rev. *1*:29–50.

GLAESSNER, M. F. 1968. Biological events and the Precambrian time scale. Can. J. Earth Sci. *5*:585–590.

GLAESSNER, M. F. 1969. Trace fossils from the Precambrian and basal Cambrian. Lethaia. *2*:369–393.

GLAESSNER, M. F. 1971a. The genus *Conomedusites* Glaessner & Wade and the diversification of the Cnidaria. Paläontol. Z. *45*:7–17.

GLAESSNER, M. F. 1971b. Geographic distribution and time range of the Ediacara Precambrian fauna. Geol. Soc. Am. Bull. *82*:509–514.

GLAESSNER, M. F. 1972a. Precambrian paleozoology. Univ. Adelaide Centre for Precambrian Res., Spec. Pap. No. 1, pp. 43–52.

GLAESSNER, M. F. 1972b. Precambrian fossils—a progress report. 1968 Proc. IPU, XXIII Int. Geol. Congr. Pp. 377–384.

GLAESSNER, M. F. 1976. Early Phanerozoic annelid worms and their geological and biological significance. J. Geol. Soc. London. *132*:259–275.

GLAESSNER, M. F. AND M. WADE. 1966. The late Precambrian fossils from Ediacara, South Australia. Palaeontology. *9*:599–628.

GLAESSNER, M. F. AND M. WADE. 1971. *Praecambridium*—a primitive arthropod. Lethaia. *4*:71–77.

GLAESSNER, M. F. AND M. R. WALTER. 1975. New Precambrian fossils from the Arumbera Sandstone, Northern Territory, Australia. Alcheringa. *1*:59–69.

GOULD, S. J. 1970. Evolutionary paleontology and the science of form. Earth-Sci. Rev. *6*:77–119.

GOULD, S. J. 1976. The interpretation of diagrams. Nat. Hist. *85*(7):18–28.

GOULD, S. J. 1977. Ontogeny and Phylogeny. 501 pp. Belknap Press; Cambridge, Massachusetts.

GOULD, S. J. AND N. ELDREDGE. 1977. Punctuated equilibria: The tempo and mode of evolution reconsidered. Paleobiology. 3:115–151.

GOULD, S. J., D. M. RAUP, J. J. SEPKOSKI, JR., T. J. M. SCHOPF, AND D. S. SIMBERLOFF. 1977. The shape of evolution: A comparison of real and random clades. Paleobiology. *3*:23–40.

GREENWOOD, P. H. 1975. A History of Fishes. 467 pp. Ernest Benn Ltd; London.

HANCOCK, J. M. 1967. Some Cretaceous-Tertiary marine faunal changes. Pp. 91–104. In: Harland, W. B. et al., eds. The Fossil Record. 828 pp. Geol. Soc. London; London.

HARLAND, W. B. 1964a. Evidence of late Precambrian glaciation and its significance. Pp. 119–149. In: Nairn, A. E. M., ed. Problems in Palaeoclimatology. 705 pp. Wiley; New York.

HARLAND, W. B. 1964b. Critical evidence for a great Infra-Cambrian glaciation. Geol. Rundsch. *54*:45–61.

HARLAND, W. B. 1974. The Pre-Cambrian-Cambrian Boundary. Pp. 15–42. In: Holland, C. H., ed. Cambrian of the British Isles, Norden, and Spitsbergen. 300 pp. Wiley; New York.

HARLAND, W. B. ET AL., EDS. 1967. The Fossil Record. 828 pp. Geol. Soc. London; London.

HARLAND, W. B., A. G. SMITH, AND B. WILCOCK, EDS. 1964. The Phanerozoic Time-Scale: A Symposium. 458 pp. Geol. Soc. London; London.

HARRINGTON, H. J. AND R. C. MOORE. 1956. Protomedusae. Pp. F21–F23. In: Moore, R. C., ed. Treatise on Invertebrate Paleontology, Pt. F. 498 pp. Geol. Soc. Am. and Univ. Kans. Press; Lawrence, Kansas.

HILL, D. 1972. Archaeocyatha. Pp. E1–E158. In: Teichert, C., ed. Treatise on Invertebrate Paleontology, Pt. E (revised). V. 1, 158 pp. Geol. Soc. Am. and Univ. Kans. Press; Lawrence, Kansas.

HOFMANN, H. J. 1971. Precambrian fossils, pseudofossils and problematica in Canada. Geol. Surv. Can. Bull. No. 189. 146 pp.

HUPÉ, P. 1960. Sur le cambrien inférieur du Maroc. 21st Int. Geol. Cong., Copenhagen. Rep., Pt. 8, pp. 75–85.

HUTCHINSON, G. E. 1961. The biologist poses some problems. Pp. 85–94. In: Sears, M., ed. Oceanography. Am. Assoc. Adv. Sci., Publ. 67. 654 pp.

JAEGER, W. AND A. MARTINSSON. 1967. Remarks on the problematic fossil *Xenusion auerswaldae*. Geol. fören. Stockholm förh. *88*:435–452.

JELETSKY, J. A. 1965. Taxonomy and phylogeny of fossil Coleoidea (= Dibranchiata). Geol. Surv. Can. Pap. *65-2*:72–76.

JELL, P. A. 1974. Faunal provinces and planetary reconstruction of the Middle Cambrian. J. Geol. *82*: 319–350.

JELL, P. A. AND J. S. JELL. 1976. Early Middle Cambrian corals from western South Wales. Alcheringa. *1*:181–195.

JOHNSON, G. A. L., I. D. SUTTON, F. M. TAYLOR, AND G. THOMAS. 1967. Coelenterata. Pp. 347–378. In: Harland, W. B. et al., eds. The Fossil Record. 828 pp. Geol. Soc. London; London.

JOHNSON, R. G. AND E. S. RICHARDSON, JR. 1968. The Essex Fauna and Medusae. Fieldiana: Geol. *12*:109–115.

JONES, D. AND I. THOMPSON. 1977. Echiura from the Pennsylvanian of northern Illinois. Lethaia. *10*: 317–325.

KUKALOVÁ-PECK, J. 1973. A Phylogenetic Tree of the Animal Kingdom (Including Orders and Higher Categories). 78 pp. Nat. Mus. Can. Publ. Zool. No. 8. Ottawa, Canada.

LANE, A. C. 1917. Lawson's correlation of the pre-Cambrian Era. Am. J. Sci. *43*:42–48.

LILLEGRAVEN, J. A. 1972. Ordinal and familial diversity of Cenozoic mammals. Taxon. *21*:261–274.

LINDSTRÖM, M. 1970. A suprageneric taxonomy of the conodonts. Lethaia. *3*:427–445.

LOCHMAN-BALK, C. AND J. L. WILSON. 1958. Cambrian biostratigraphy in North America. J. Paleontol. *32*:312–350.

MACARTHUR, R. H. 1969. Patterns of communities in the tropics. Biol. J. Linn. Soc. *1*:19–30.

MACARTHUR, R. H. 1972. Geographical Ecology. 269 pp. Harper & Row; New York.

MACARTHUR, R. H. AND R. LEVINS. 1967. The limiting similarity, convergence, and divergence of coexisting species. Am. Nat. *101*:377–385.

MACARTHUR, R. H. AND E. O. WILSON. 1963. An equilibrium theory of insular zoogeography. Evolution. *17*:373–387.

MACARTHUR, R. H. AND E. O. WILSON. 1967. The Theory of Island Biogeography. Monogr. Population Biol., No. 1. 203 pp. Princeton Univ. Press; Princeton, New Jersey.

MARK, G. A. AND K. W. FLESSA. 1977. A test for evolutionary equilibria: Phanerozoic brachiopods and Cenozoic mammals. Paleobiology. *3*:17–22.

MATTHEWS, S. C. AND V. V. MISSARZHEVSKY. 1975. Small shelly fossils of late Precambrian and Early Cambrian age: A review of recent work. J. Geol. Soc. London. *131*:289–304.

MAY, R. M. 1973. Stability and Complexity in Model Ecosystems. Monogr. Population Biol., No. 6. 236 pp. Princeton Univ. Press; Princeton, New Jersey.

MAYR, E. 1963. Animal Species and Evolution. 767 pp. Belknap Press; Cambridge, Massachusetts.

MAYR, E. 1965. Avifauna: Turnover on islands. Science. *150*:1587–1588.

MOORE, R. C. AND H. L. STRIMPLE. 1973. Lower Pennsylvanian (Morrowan) crinoids from Arkansas, Oklahoma, and Texas. Univ. Kans. Paleontol. Contrib., Art. 60, 84 pp.

MOORE, R. C. AND C. TEICHERT, EDS. 1953–1975. Treatise on Invertebrate Paleontology. Geol. Soc. Am. and Univ. Kans. Press; Lawrence, Kansas.

NORTH, F. K. 1971. The Cambrian of Canada and Alaska. Pp. 219–324. In: Holland, C. H., ed. Cambrian of the New World. 456 pp. Wiley; New York.

NURSHALL, J. R. 1959. Oxygen as a prerequisite to the origin of the Metazoa. Nature. *183*:1170–1172.

ÖPIK, A. A. 1967. The Ordian Stage of the Cambrian and its Australian Metadoxidae. Aust. Bur. Mineral Res., Geol. and Geophy., Bull. *92*:133–169.

ÖPIK, A. A. 1975. Cymbric Vale fauna of New South Wales and Early Cambrian biostratigraphy. Aust. Bur. Min. Res., Geol. and Geophys., Bull. 159, 74 pp.

OSSIAN, C. R. 1973. New Pennsylvanian scyphomedusan from western Iowa. J. Paleontol. *47*:990–995.

PALMER, A. R. 1977. Biostratigraphy of the Cambrian System—a progress report. Annu. Rev. Earth and Planet. Sci. *5*:13–33.

PAUL, C. R. C. 1977. Evolution of primitive echinoderms. Pp. 123–158. In: Hallam, A., ed. Patterns of Evolution. 591 pp. Elsevier; Amsterdam.

PFLUG, H. D. 1972. The Phanerozoic-Cryptozoic Boundary and the origin of Metazoa. Proc. 24th Int. Geol. Congr., Montreal, Sec. 1, pp. 58–67.

POJETA, J., JR. 1975. *Fordilla troyensis* Barrande and early pelecypod phylogeny. Bull. Am. Paleontol. *67*:363–384.

POJETA, J. AND B. RUNNEGAR. 1976. The paleontology of rostroconch mollusks and the early history of the phylum Mollusca. U.S. Geol. Surv. Prof. Pap. 968. 88 pp.

RAUP, D. M. 1972. Taxonomic diversity during the Phanerozoic. Science. *177*:1065–1071.

RAUP, D. M. 1975. Taxonomic diversity estimation using rarefaction. Paleobiology. *1*:333–342.

RAUP, D. M. 1976a. Species diversity in the Phanerozoic: A tabulation. Paleobiology. *2*:279–288.

RAUP, D. M. 1976b. Species diversity in the Phanerozoic: An interpretation. Paleobiology. *2*:289–297.

RAUP, D. M. 1977a. Removing sampling biases from taxonomic diversity data. J. Paleontol. *51* (Suppl.):21.

RAUP, D. M. 1977b. Probabilistic models in evolutionary paleobiology. Am. Sci. *65*:50–57.

RAUP, D. M. 1977c. Stochastic models in evolutionary palaeontology. Pp. 59–78. In: Hallam, A., ed. Patterns of Evolution. 591 pp. Elsevier; Amsterdam.

RAUP, D. M. 1978. Cohort analysis of generic survivorship. Paleobiology. *4*:1–15.

RAUP, D. M. AND S. J. GOULD. 1974. Stochastic simulation and evolution of morphology—towards a nomothetic paleontology. Syst. Zool. *23*:305–322.

RAUP, D. M., S. J. GOULD, T. J. M. SCHOPF, AND D. S. SIMBERLOFF. 1973. Stochastic models of phylogeny and the evolution of diversity. J. Geol. *81*:525–542.

RAYMOND, P. E. 1935. Pre-Cambrian life. Geol. Soc. Am. Bull. *46*:375–392.

REPINA, L. N. 1972. Biogeography of Early Cambrian of Siberia according to trilobites. 1968 Proc. IPU, XXIII Int. Geol. Congr., pp. 289–300.

RESSER, C. E. AND B. F. HOWELL. 1938. Lower Cambrian *Olenellus* Zone of the Appalachians. Geol. Soc. Am. Bull. *49*:195–248.

RHOADES, D. C. AND J. W. MORSE. 1971. Evolutionary and ecologic significance of oxygen-deficient marine basins. Lethaia. *4*:413–428.

RICHARDSON, E. S., JR. 1966. Wormlike fossil from the Pennsylvanian of Illinois. Science. *151*:75–76.

RICKLEFS, R. AND G. COX. 1972. Taxon cycles in West Indian avifauna. Am. Nat. *106*:195–219.

RIGBY, J. K. 1976. Some observations on occurrences of Cambrian Porifera in western North America and their evolution. Pp. 51–60. In: Robison, R. A. and A. J. Rowell, eds. Paleontology and Depositional Environments: Cambrian of Western North America. 227 pp. Brigham Young Univ. Geol. Studies. *23*(2).

ROBISON, R. A., A. V. ROSOVA, A. J. ROWELL, AND T. P. FLETCHER. 1977. Cambrian boundaries and divisions. Lethaia. *10*:257–262.

ROLFE, W. D. I. 1969. Phyllocarida. Pp. R291–R331. In: Moore, R. C., ed. Treatise on Invertebrate Paleontology, Pt. R. V. 1. 650 pp. Geol. Soc. Am. and Univ. Kans. Press; Lawrence, Kansas.

ROMER, A. S. 1966. Vertebrate Paleontology. 3rd ed. 468 pp. Univ. Chicago Press; Chicago, Illinois.

ROSENZWEIG, M. L. 1975. On continental steady states of species diversity. Pp. 121–140. In: Cody, M. L. and J. M. Diamond, eds. Ecology and Evolution of Communities. 544 pp. Belknap Press; Cambridge, Massachusetts.

ROSENZWEIG, M. L. 1977. Does the fossil record provide for natural experiments? Paleobiology. *3*: 322–324.

ROWELL, A. J. 1977. Early Cambrian brachiopods from the southwestern Great Basin of California and Nevada. J. Paleontol. *51*:68–85.

ROZANOV, A. YU. 1967. The Cambrian lower boundary problem. Geol. Mag. *104*:415–434.

ROZANOV, A. YU. AND F. DEBRENNE. 1974. Age of archaeocyathid assemblages. Am. J. Sci. *274*:833–848.

ROZANOV, A. YU., V. V. MISSARZHEVSKY, N. A. VOLKOVA, L. G. VORONOVA, I. N. KRYLOV, B. M. KEL-

LER, I. K. KOROLYUK, K. LENDZION, P. MIKHNIAK, N. G. PYKHOVA, AND A. D. SIDOROV. 1969. Tommotsky yarus i problema nizhney granitsy Kembria [The Tommotian Stage and the problem of the Cambrian lower boundary]. Trans. Geol. Inst. Acad. Sci. USSR. *206*:5–380.

RUDWICK, M. J. S. 1964. The Infra-Cambrian glaciation and the origin of the Cambrian fauna. Pp. 150–155. In: Nairn, A. E. M., ed. Problems in Palaeoclimatology. 705 pp. Wiley; New York.

RUNNEGAR, B. AND P. A. JELL. 1976. Australian Middle Cambrian molluscs and their bearing on early molluscan evolution. Alcheringa. *1*:109–138.

RUSHTON, A. W. A. 1974. The Cambrian of Wales and England. Pp. 43–121. In: Holland, C. H., ed. Cambrian of the British Isles, Norden, and Spitsbergen. 300 pp. Wiley; New York.

SCHOPF, J. W. 1975. Precambrian paleobiology: Problems and perspectives. Annu. Rev. Earth and Planet. Sci. *3*:213–249.

SCHOPF, J. W., B. N. HAUGH, R. E. MOLNAR, AND D. F. SATTERTHWAIT. 1973. On the development of metaphytes and metazoans. J. Paleontol. *47*:1–9.

SCHOPF, T. J. M. 1974. Permo-Triassic extinctions: Relation to sea-floor spreading. J. Geol. *82*:129–143.

SCHOPF, T. J. M. 1977. Patterns and themes of evolution among the Bryozoa. Pp. 159–207. In: Hallam, A., ed. Patterns of Evolution. 591 pp. Elsevier; Amsterdam.

SCHRAM, F. R. 1969. The stratigraphic distribution of Paleozoic Eumalacostraca. Fieldiana: Geol. *12*: 213–234.

SCHRAM, F. R. 1973. Pseudocoelomates and a nemertine from the Illinois Pennsylvanian. J. Paleontol. *47*:985–989.

SCHUCHERT, C. AND C. O. DUNBAR. 1933. Historical Geology. 241 pp. Wiley; New York.

SEPKOSKI, J. J., JR. 1975. Stratigraphic biases in the analysis of taxonomic survivorship. Paleobiology. *1*:343–355.

SEPKOSKI, J. J., JR. 1976. Species diversity in the Phanerozoic: Species-area effects. Paleobiology. *2*: 298–303.

SEPKOSKI, J. J., JR. 1977. The enigma of the Cambrian diversification. Geol. Soc. Am. Abstr. with Program. *9*:1168.

SIMBERLOFF, D. S. 1970. Taxonomic diversity of island biotas. Evolution. *24*:23–47.

SIMBERLOFF, D. S. 1972. Models in biogeography. Pp. 160–191. In: Schopf, T. J. M., ed. Models in Paleobiology. 250 pp. Freeman, Cooper & Co.; San Francisco, California.

SIMBERLOFF, D. S. 1974. Permo-Triassic extinctions: Effects of area on biotic equilibrium. J. Geol. *82*:267–274.

SIMPSON, G. G. 1953. The Major Features of Evolution. 434 pp. Columbia Univ. Press; New York.

SIMPSON, G. G. 1960. The history of life. Pp. 117–180. In: Tax, S., ed. Evolution after Darwin. v. 1. 629 pp. Univ. Chicago Press; Chicago, Illinois.

SIMPSON, G. G. 1969. The first three billion years of community evolution. Pp. 162–177. In: Woodwell, G. M. and H. H. Smith, eds. Diversity and Stability in Ecological Systems. Brookhaven Symp. Biol. No. 22, 264 pp.

SOKOLOV, B. S. 1972a. Vendian and Early Cambrian Sabelliditida (Pogonophora) of the USSR. 1968 Proc. IPU, XXIII Int. Geol. Congr. pp. 79–86.

SOKOLOV, B. S. 1972b. The Vendian Stage in earth history. Proc. 24th Int. Geol. Congr., Montreal, Sec. 1, pp. 78–84.

SOKOLOV, B. S. 1973. Vendian of northern Eurasia. Pp. 204–218. In: Pitcher, M. G., ed. Arctic Geology. 74 pp. Am. Assoc. Pet. Geol.; Tulsa, Oklahoma.

SOKOLOV, B. S. 1976. Precambrian Metazoa and the Vendian-Cambrian Boundary. Paleontol. J. *10*: 1–13.

SPRINKLE, J. 1973. Morphology and Evolution of Blastozoan Echinoderms. Spec. Publ. Mus. Comp. Zool., Harvard Univ. 283 pp.

SPRINKLE, J. 1976. Biostratigraphy and paleoecology of Cambrian echinoderms from the Rocky Mountains. Pp. 61–73. In: Robison, R. A. and A. J. Rowell, eds. Paleontology and Depositional Environments: Cambrian of Western North America. 227 pp. Brigham Young Univ. Geol. Studies. *23*(2).

STANLEY, S. M. 1973. An ecological theory for the sudden origin of multicellular life in the late Precambrian. Proc. Nat. Acad. Sci., USA. *70*:1486–1489.

STANLEY, S. M. 1975. A theory of evolution above the species level. Proc. Nat. Acad. Sci., USA. *72*: 646–650.

STANLEY, S. M. 1976a. Ideas on the timing of metazoan diversification. Paleobiology. *2*:209–219.

STANLEY, S. M. 1976b. Fossil data and the Precambrian-Cambrian evolutionary transition. Am. J. Sci. *276*:56–76.

STANLEY, S. M. 1977a. Trends, rates, and patterns of evolution in the Bivalvia. Pp. 209–250. In: Hallam, A., ed. Patterns of Evolution. 591 pp. Elsevier; Amsterdam.

STANLEY, S. M. 1977b. Influence of rates of speciation and extinction on the diversity and evolutionary stability of higher taxa. J. Paleontol. *51* (Suppl.):26–27.

STEHLI, F. G., A. L. MCALESTER, AND E. C. HELSLEY. 1967. Taxonomic diversity of Recent bivalves and some implications for geology. Geol. Soc. Am. Bull. *78*: 455–466.

STØRMER, L. 1955. Merostomata. Pp. P4–P41. In: Moore, R. C., ed. Treatise on Invertebrate Paleontology, Pt. P. 181 pp. Geol. Soc. Am. and Univ. Kans. Press; Lawrence, Kansas.

STØRMER, L. 1959. Trilobitoidea. Pp. O23–O37. In: Moore, R. C., ed. Treatise on Invertebrate Paleontology, Pt. O. 560 pp. Geol. Soc. Am. and Univ. Kans. Press; Lawrence, Kansas.

TAYLOR, M. E. 1966. Precambrian mollusc-like fossils from Inyo County, California. Science. *153*: 198–201.

TCHERNYCHEVA, N. E. 1959. Système Cambrien. Pp. 185–244. In: Markovsky, A. P., ed. Structure Géologique de l'U.R.S.S. Tome 1. Stratigraphic (P. de Saint-Aubin and J. Roger, trans.). 803 pp. Centre National de la Recherche Scientifique; Paris.

TERBORGH, J. 1973. On the notion of favorableness in plant ecology. Am. Nat. *107*:481–501.

TERMIER, H. AND G. TERMIER. 1976. The Ediacaran fauna and animal evolution. Paleontol. J. *10*: 264–270.

Thayer, C. W. 1974. Environmental and evolutionary stability in bivalve mollusks. Science. *186*: 828–830.

Towe, K. M. 1970. Oxygen-collagen priority and the early metazoan fossil record. Proc. Nat. Acad. Sci., USA. *65*:781–788.

Valentine, J. W. 1969. Patterns of taxonomic and ecological structure of the shelf benthos during Phanerozoic time. Paleontology. *12*:684–709.

Valentine, J. W. 1973. Evolutionary Paleoecology of the Marine Biosphere. 511 pp. Prentice-Hall; Englewood Cliffs, New Jersey.

Valentine, J. W. 1977. General paterns of metazoan evolution. Pp. 27–57. In: Hallam, A., ed. Patterns of Evolution. 591 pp. Elsevier; Amsterdam.

Valentine, J. W. and C. A. Campbell. 1975. Genetic regulation and the fossil record. Am. Sci. *63*:673–680.

Valentine, J. W. and E. M. Moores. 1972. Global tectonics and the fossil record. J. Geol. *80*: 167–184.

van Hinte, J. E. 1976a. A Jurassic time scale. Am. Assoc. Pet. Geol. Bull. *60*:489–497.

van Hinte, J. E. 1976b. A Cretaceous time scale. Am. Assoc. Petrol. Geol. Bull. *60*:498–516.

Van Valen, L. 1973a. A new evolutionary law. Evol. Theory. *1*:1–30.

Van Valen, L. 1973b. Are categories in different phyla comparable? Taxon. *22*:333–373.

Vinogradov, P. O. 1959. The origin of the biosphere. Pp. 23–38. In: Oparin, A. I. et al., eds. The Origins of Life on the Earth. V. 1. Clark, F. and R. L. M. Synge, trans. 961 pp. Pergamon Press; New York.

Wade, M. 1969. Medusae from uppermost Precambrian or Cambrian sandstone, central Australia. Palaeontology. *12*:351–365.

Wade, M. 1970. The stratigraphic distribution of the Ediacara fauna in Australia. Trans. R. Soc. S. Aust. *94*:87–104.

Wade, M. 1971. Bilateral Precambrian chondrophores from the Ediacaran fauna, South Australia. Proc. R. Soc. Victoria. *84*:183–188.

Wade, M. 1972. Hydrozoa and Scyphozoa and other medusoids from the Precambrian Ediacara fauna, South Australia. Palaeontology. *15*:97–225.

Webb, S. D. 1969. Extinction-origination equilibria in late Cenozoic land mammals of North America. Evolution. 23:688–702.

Webb, S. D. 1976. Mammalian faunal dynamics of the great American interchange. Paleobiology. *2*: 220–234.

Webby, B. D. 1970. Late Precambrian trace fossils from New South Wales. Lethaia. *3*:79–109.

Welch, J. R. 1976. *Phosphannulus* on Paleozoic crinoid stems. J. Paleontol. *50*:218–225.

Whittaker, R. H. 1975. Communities and Ecosystems, 2nd ed. 385 pp. MacMillan Publ. Co., Inc.; New York.

Wilson, E. O. 1969. The species equilibrium. Pp. 38–47. In: Woodwell, G. M. and H. H. Smith, eds. Diversity and Stability in Ecological Systems. Brookhaven Symp. Biol., No. 22, 264 pp.

Yochelson, E. L. 1969. Stenothecoida, a proposed new class of Cambrian Mollusca. Lethaia. *2*:49–62.

Yochelson, E. L. 1975. Discussion of Early Cambrian "molluscs." J. Geol. Soc. London. *131*:661–662.

Yochelson, E. L., J. W. Pierce, and M. E. Taylor. 1970. *Salterella* from the Cambrian of central Nevada. U.S. Geol. Surv. Prof. Pap. 643-H: H1–H7.

Young, F. G. 1972. Early Cambrian and older trace fossils from the southern Cordillera of Canada. Can. J. Earth Sci. *9*:1–17.

Zhuravleva, I. T. 1970. Marine faunas and Lower Cambrian stratigraphy. Am. J. Sci. *269*:417–445.

APPENDIX: VENDIAN-CAMBRIAN STRATIGRAPHY

The eight stratigraphic units employed in the compilation of ordinal ranges in Fig. 2 are based upon the subdivision of the uppermost Precambrian and Lower Cambrian in the Soviet Union. Although the biostratigraphic zonation of these units is not fully documented and their worldwide correlations are certainly not complete (Palmer 1977; Robison et al. 1977), they are now sufficiently known to permit at least approximate separation of the times of origin and extinction of the oldest metazoan groups. The definitions of these units and the rationale behind correlation and assignment of absolute ages are outlined below. (Letters in parenthesis refer to stratigraphic abbreviations used in Figs. 2 and 3.)

Lower Cambrian: Three of the four now-familiar stages of the Siberian Lower Cambrian are represented in Fig. 2:

7. Botomian (*B*). This unit appears to correlate with much of the upper Lower Cambrian of Europe (Cowie et al. 1972) and with the upper *Nevadella* zone and *Bonnia-Olenellus* Zone of North America. Rosanov and Debrenne (1974) and other recent authors have applied the name "Lenian" to this unit.
6. Atdabanian (*A*). This unit includes the first trilobite faunas throughout most of the world; it appears to correlate with the *Fallotaspis* and lower *Nevadella* Zones in North America (see Fritz 1972; Rozanov and Debrenne 1974).
5. Tommotian. This includes the common pre-trilobite skeletal faunas of the world. In Siberia, it is sub-divided into two major units (Rozanov et al. 1969; Cowie and Glaessner 1975):
   b. upper Tommotian (*uT*) or Kenyadan Horizon;
   a. lower Tommotian (*lT*) or Sunnagin Horizon;

Correlations of these Siberian units with Lower Cambrian sections in other parts of the world (which must be emphasized as tentative) were based largely upon the chart presented by Cowie et al. (1972, Plate 5) and augmented by the discussions of Rozanov (1967), Zhuravleva (1970), Daily (1972), Hill (1972), Repina (1972), Cowie and Rozanov (1974), Rozanov and Debrenne (1974), and Cowie and Glaessner (1975). The Lenian (or "Elankian") Stage was not used in this compilation since it overlaps with the lower Middle Cambrian of much of the world and especially of North America and Australia (Öpik 1967; Jell 1974; Rozanov and Debrenne 1974). Absolute ages of the boundaries within the Lower Cambrian are difficult to determine, and those presented here can best be considered guesses. The standard age estimates for the base of the Cambrian (here taken to be the base of the Tommotian) and top of the Lower Cambrian (here taken to be near the top of the Botomian) are 570 and 540 myr, respectively (see Harland et al. 1964; Cowie and Glaessner 1975; Robison et al. 1977). The three stages and their correlated units vary greatly in thickness from area to area around the world but no single unit seems to be greatly thicker or thinner than the others (see stratigraphic sections presented by Alpert 1976; Cowie et al. 1972; North 1971; Rozanov and Derenne 1974; Rushton 1974; Tchernycheva 1959). This suggests that as a first approximation the durations of the three stages can be considered roughly equal, providing approximate ages of 560 and 550 myr for the tops of Tommotian and Atdabanian, respectively (see also Robison et al. 1977). The Sunnagin Horizon is relatively thin, on the average encompassing slightly more than 10% of the thickness of the entire Tommotian in Siberia (Rozanov et al. 1969); if thickness is considered proportional to time, this leads to a rough estimate of about 569 myr for the age of top of this horizon.

Vendian: The Vendian, as described in English summaries by Sokolov (1972b, 1973, 1976), is the uppermost division of the Precambrian on the Russian Platform, spanning the interval from the base of the Baltic Series (which is probably in part pre-Tommotian) down to near the base of the highest Precambrian (upper "Varangian" or "Laplandian") tilloids. These tilloids may be roughly correlatable around the world (see Harland 1964a,b, 1974; but also Cowie 1961; Crawford and Daily 1971; and others), giving the Vendian potential global significance. The "Ediacaran" faunas always lie well above the upper tilloids, where present (see Glaessner 1966, 1968, 1971b, 1972a; Harland 1974; Sokolov 1972b, 1976), and thus are referable to the Vendian. Although biostratigraphic control does not yet permit accurate correlations within the Vendian, the sequence of Ediacaran faunas in the Soviet Union does suggest a very tentative division of major biotic units, as outlined below:

4. Valdai Series and lower Baltic "Series" (Rovno Horizon) of the Russian Platform and Nemakit-Daldyn Horizon (*ND*) of Siberia. Precambrian skeletal fossils, including sabelliditids, hyolithelminthids, etc., are present in the upper 80 m (Nemakit-Daldyn Horizon) of the calcareous Yudoma "Complex" in Siberia (Rozanov et al. 1969; Zhuravleva 1970) and in the lower portion (Rovno Horizon) of the terrigenous Baltic "series" in Russia, which Sokolov (1976) argues is pre-Tommotian. These fossils are separated by a considerable gap, corresponding to all or most of the Valdai Series, from the uppermost occurrences of Ediacaran-type fossils in the Soviet Union. This barren interval is present in most other parts of the world and appears to correspond to a major diversification of trace fossils (Glaessner 1969; Webby 1970; Sokolov 1976; see also Banks 1970; Cowie and Spencer 1970; Young 1972). According to Sokolov (1976), the Valdai Series has a duration of about 30 Myr, giving ages of 600 and 570 Myr to the boundaries of the uppermost informal unit of the Precambrian. However, this unit is separated from the Tommotian in Siberia by a slight although easily perceptible disconformity (Rozanov et al. 1969; Cowie and Rozanov 1974) and from the upper Baltic "Series" in Russia by a thin unfossiliferous interval (Sokolov 1976), making the effective upper age of the Precambrian skeletal faunas several million years older than the Cambrian.
3. "Lower upper" Vendian (*uV*). Relatively diverse Ediacaran assemblages are present in the lower upper Vendian or Redkino Series in the Soviet Union (Sokolov 1972b, 1973). The "type" Ediacaran fauna of the Pound Quartzite in Australia, which occurs much closer to the base of the Cambrian than to the uppermost Precambrian tilloid (Wade 1970; Webby 1970; Daily 1972), shares a number of elements in common with the Soviet faunas, suggesting possible correlation. Sokolov (1972b, 1973, 1976) gives an estimated age of 620 Myr for the base of the Redkino Series.
2. "Middle" Vendian (*mV*). The oldest representatives of the Ediacaran fauna in the Soviet Union occur in relatively depauperate assemblages within the upper portion of the Volhyn Series, which includes all of the lower Vendian in Russia (Sokolov 1972b, 1973, 1976). Although the lower age of these occurrences is not known, it does not appear to be considerably older than the 620 Myr age of the top of the Volhyn Series.
1. "Lower" Vendian (*lV*). The lowermost portion of the Vendian contains no body fossils but does contain the first unquestioned trace fossils, appearing just above the Varangian tilloids (Glaessner 1968, 1972b, Stanley 1976b). These traces, which probably record the first appearance of coelomate metazoans (cf. Clark 1964), are assumed to represent one order, very questionably referred to "Annelida *incertae sedis*" in Fig. 2. [Numerous examples of older trace fossils have been reported, but most have been invalidated (e.g. Cloud 1968, 1973; Hofmann 1971). Remaining examples are still doubtful, both because of their rarity (since once burrowers appeared, even if monospecific, there is no reason why their population sizes in certain environments should have remained substantially smaller than in the later Phanerozoic) and because of the apparent absence of advanced intermediate evolutionary grades among Middle and Upper Riphean microbiotas (see J. W. Schopf 1975).] The age of the lower boundary of the Vendian is estimated to be 675 to 680 Myr (Sokolov 1972b, 1973, 1976; Harland 1974; Cowie and Glaessner 1975), making the first definite appearance of metazoans just slightly younger.

PAPER 19

Plate-Tectonic Regulation of Faunal Diversity and Sea Level: A Model (1970)

J. W. Valentine and E. M. Moores

Commentary

MATTHEW G. POWELL

In this short work, Valentine and Moores suggested that secular changes in diversity were largely a response to plate motions—at a time when the concept of plate tectonics was in its infancy and the faunal diversity a new line of inquiry. The birth of mountains through continent-continent collision (Moores's area of expertise) was only newly recognized, and the publication of the *Treatise on Invertebrate Paleontology* had just made possible a new compilation of Phanerozoic marine diversity. Drawing on these advances, Valentine and Moores were able to suggest convincingly that long-term changes in diversity were coincident with continental collision and break-up and were causally linked via the accompanying changes in climate, habitat fragmentation, and sea-level change. In their view, continental amalgamation depleted diversity as a result of greater seasonality (which increased nutrients in the shallow ocean due to vertical mixing), fewer barriers to migration, and the loss of habitable area as epicontinental seas regressed. Continental break-up increased diversity as a result of a more equable climate, more geographically isolated faunas, and increased habitable area during transgression.

Their work made several particularly significant contributions. Most fundamentally, they were among the first to recognize that a supercontinent—"Pangaea I," which was later named Rodinia—was present in the late Proterozoic Eon. Their work was also among the first to explore controls on diversification and extinction, almost a decade before the seminal work of Dave Raup, Jack Sepkoski, and others. Finally and most basically, their work integrated paleontology, tectonics, and climate in a way that anticipated a much later recognition of Earth as a system of interacting components.

Significant details in our understanding have certainly changed in the more than forty years since their paper. For example, the break-up of Rodinia probably occurred ~800–700 Ma, much earlier than the Cambrian diversification (although, intriguingly, another supercontinent may have existed in the intervening time, and its break-up may have triggered the Cambrian diversification), and it is now recognized that the end-Permian mass extinction happened much too rapidly to be accounted for by plate tectonics. Yet their contribution remains a model of the interdisciplinary, whole-system approach to understanding paleoecological patterns that enlivens our discipline today.

Reprinted by permission from Macmillan Publishers Ltd: *Nature* 228:657–59, copyright 1970. https://doi.org/10.1038/228657a0.

NATURE VOL. 228 NOVEMBER 14 1970 657

The logarithmic stress–strain relationship for igneous rocks is

$$\varepsilon = \frac{P}{\mu}[1 + q\log(1 + at)] \quad (1)$$

where ε is the strain, P the stress, μ the rigidity, q and a are experimentally determined constants, a being of the order of $10^3\ s^{-1}$. The wave equation based on (1) (ref. 2) gives a constant value of Q in the frequency range $10^{-8} \leqslant f \leqslant 1$Hz, provided that Q (which is proportional to q^{-1}) is greater than 600.

It has been established experimentally[4] that the values of Q for most rocks are constant over the frequency range $10^{-4} \leqslant f \leqslant 10^6$ Hz, the absolute values being about 200. Thus Q is constant over a range of frequencies which is not explicable in terms of the logarithmic stress–strain relationship and it must therefore be presumed that there are other mechanisms that can give rise to this behaviour.

Futterman has shown[7] that the constant Q behaviour cannot extend to zero frequency if the causative mechanism is linear. Experimental evidence for this linearity has been provided for metals by Savage and Hasegawa[8] and for sediments by myself[9] and shows that the superposition theorem is valid for stress waves showing constant Q propagation in these materials. Hence the constant Q mechanism cannot manifest itself at zero frequency as a particular stress–strain relationship and its presence cannot be taken as evidence for or against such a relationship.

An analogous situation arises in the study of the damping of electromagnetic waves in dielectrics. The values of Q of many dielectrics are independent of frequency[10] and cannot be derived from the d.c. conductivities. But the presence of the constant Q mechanism cannot be taken as evidence that a steady state, unidirectional flow of charge cannot occur through dielectrics, because it is known experimentally that they have measurable d.c. conductivities.

If the material of the mantle can undergo steady state creep it would be expected that a "viscous" attenuation mechanism with a characteristic relaxation time would be observed. McConnell has estimated[11] the viscosity of the upper mantle as 10^{20} to 10^{22} poise from the rate of uplift of Fennoscandia. The implied relaxation time is between 5×10^7 and 5×10^9 s. Bhatia shows[12] that stress waves with periods considerably less than the relaxation time are not attenuated or dispersed by the viscous mechanism. Thus there will be no evidence for this damping mechanism from stress waves of "ordinary" seismic periods. Jeffreys indicates[1] that the free nutation of the Earth's axis (an oscillation with a period of about 4×10^7 s) shows a viscous damping with a time constant of about 10^8 s. The agreement between the two relaxation times lends support to the idea that the upper mantle material can exhibit steady state creep.

C. McCANN

Department of Geology,
University of Reading,
Reading RG6 2AB.

Received June 5, 1970.

1 Jeffreys, H., *Nature*, 225, 1007 (1970).
2 Lomnitz, C., *J. Appl. Phys.*, 28, 201 (1957).
3 Press, F., *Science*, 124, 1204 (1956).
4 Anderson, D. L., and Kovach, R. L., *Proc. US Nat. Acad. Sci.*, 51, 168 (1964).
5 Anderson, D. L., in *The Earth's Mantle* (edit. by Gaskell, T. F.) (Academic Press, London, 1967).
6 Attewell, P. B., and Ramana, Y. V., *Geophysics*, 31, 1049 (1966).
7 Futterman, W. I., *J. Geophys. Res.*, 67, 5279 (1962).
8 Savage, J. C., and Hasegawa, H. S., *Geophysics*, 32, 1003 (1967).
9 McCann, C., thesis, Univ. Wales (1968).
10 Gross, B., *Theories of Viscoelasticity* (Hermann, Paris, 1953).
11 McConnell, R. K., *J. Geophys. Res.*, 70, 517 (1965).
12 Bhatia, A. B., *Ultrasonic Absorption* (Oxford Univ. Press, London, 1967).

Plate-tectonic Regulation of Faunal Diversity and Sea Level: a Model

A GROWING body of evidence and theory supports the idea that processes of plate tectonics have been operating for the past 3×10^9 years[1]. The hypothesis of ocean-driven plates implies that certain mountain systems represent the remains of former large ocean basins. An important corollary of this notion is that the history of continental assembly and fragmentation can be inferred from the geological record of the various mountain systems. We show here how the diversity of marine fauna and fluctuations in sea level can be related to patterns of continental fragmentation and reassembly. It seems that changes in the diversity of fauna predicted from these patterns can account for a great deal of the variation in the diversification and extinction of marine biota that has long puzzled palaeontologists. The timing of major transgressions and regressions predicted from these patterns also agrees well with the observed record.

Among the factors that regulate species diversity are environmental stability[2,3], food supply[2,4,5] and provinciality[6,7]. Consider a large land mass composed of an assembly of all the presently dispersed continents. Such a continental state favours great seasonality of marine climate, partly as a result of excess summer heating and winter cooling of the interior compared with the surrounding oceans. Off-shore winter and spring winds cause upwelling which results in relatively low surface temperatures and high supplies of nutrients in coastal waters; on-shore summer and autumn winds favour the on-shore drift of relatively warm surface waters and low levels of nutrients in coastal waters. Barriers to migration are few, and although some endemism would occur in epicontinental oceanic arms and along topographically distinctive coastal stretches, species would tend to be widespread, especially if climatic differences around the continental margin were not too extreme.

If such a supercontinent were to fragment, the marine climates around the small daughter continents would have a different character. In general, the climates of these smaller land masses would be more equable, with less distinct seasonality arising from monsoon effects and with less vertical mixing and therefore fewer nutrients in shelf waters. In this more stable, more "oligotrophic" environment, increasing species diversity within communities is favoured[2,5,8]. Furthermore, there is now growing isolation of the biotas of the daughter continents, leading to growing endemism and the eventual appearance of distinctive biotas on each continent, increasing the total biotic diversity. Total species diversity has thus increased as a result of the increased stabilization of the shelf environment and the accompanying oligotrophic tendency, and the increased provincialization.

Subsequent reassembly of smaller continents into larger ones leads to an increase in seasonality in shelf waters and to increased supplies of nutrients, and also eventually brings endemic biotas into association. A significant reduction in diversity would thus occur. Extinction would be greatest among groups that were least pre-adapted to fluctuating conditions (usually the deeper shelf forms[9,10] and very specialized forms in general), and among groups that were unable to meet the competition or to withstand the predation pressures from invading members of the other biotas. Diversity would tend to be lowest when all continental fragments were reassembled into a single supercontinent, and highest when continents are most fragmented. Climatic zonation is, however, another major cause of provincialization; if continental configurations are such that latitudinal temperature gradients are high, then the biota of each north–south continental coast will be subdivided into climatic provinces and diversity levels will be multiplied by this factor. This situation partly accounts for the unusually high number of species at present[11,12].

658 NATURE VOL. 228 NOVEMBER 14 1970

A model of continental breakup and reassembly can be inferred from existing mountain chains[1]. Phanerozoic events were probably directly affected by the breakup of an early supercontinent (Pangaea I, Fig. 1, event *A*) which by the time of the upper Cambrian had given rise to a pre-Appalachian and pre-Hercynian and pre-Uralian ocean[13-17] (Fig. 1, event *B*). Judging from the emplacement of ophiolites[18,19], the pre-Caledonide–Acadian ocean began closing by the mid-Ordovician and was sutured in the late Silurian to late Devonian (Fig. 1, event *C*). The pre-Hercynian and pre-Appalachian oceans closed in the late Pennsylvanian or early Permian (Fig. 1, event *D*). Finally, the closing of the pre-Uralian ocean near the end of Permian time completed the reconstruction of a supercontinent (Pangaea II, Fig. 1, event *E*).

seem to correlate well with the events described. The correspondence seems close enough to support one of our principal theses, that a major part of the regulation of species diversity on continental shelves has been accomplished through changes in environmental stability, nutrient supply and provinciality which act in concert and are correlated with changes in continental configurations. The continuing rediversification during the Mesozoic and Cainozoic is partly the result of the widening latitudinal spread of continents and of the development of latitudinal chains of provinces as shelf waters have cooled in high latitudes[9,10].

The patterns of continental assembly and fragmentation inferred earlier also suggest a model which may explain a long standing enigma: the timing and correla-

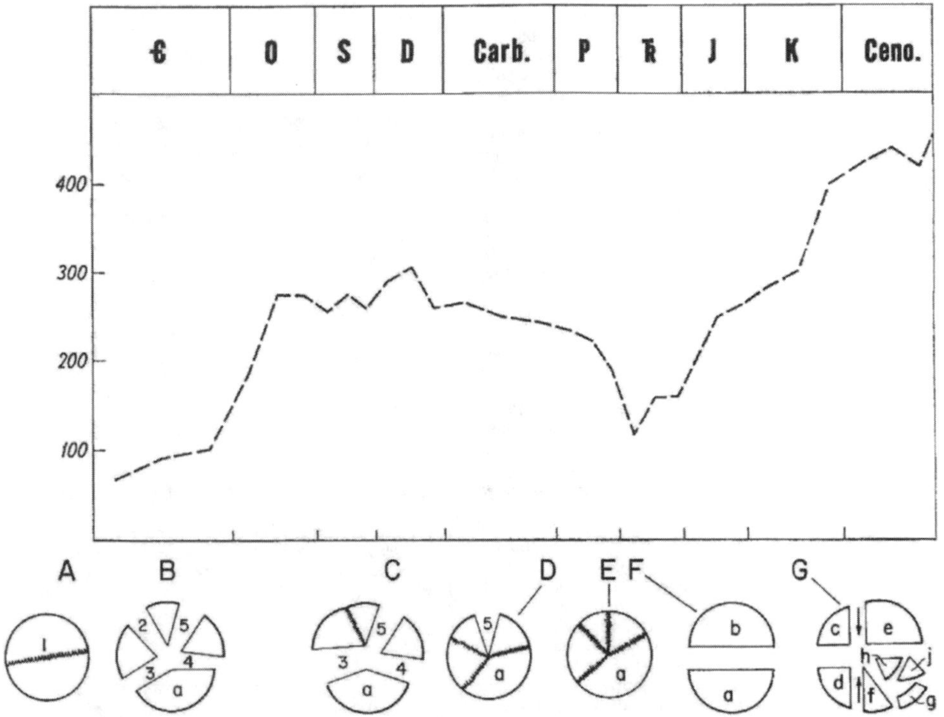

Fig. 1. Correlation of standing levels of diversity with patterns of continental assembly and fragmentation. Broken line is number of families. Events: *A*, Eocambrian suturing of Pan-African-Baikalian (1) system—formation of Pangaea 1; *B*, fragmentation of *A* in Cambro-Ordovician time to give pre-Caledonian–Acadian (2), pre-Appalachian (3), pre-Hercynian (4), and pre-Uralian (5) oceans; *C*, Siluro-Devonian suturing of Caledonian–Acadian orogenic system; *D*, Pennsylvanian–Permian suturing of Appalachian–Hercynian orogenic system; *E*, Permian-Triassic suturing of Urals—formation of Pangaea II; *F*, Triassic–lower Jurassic extension of Tethyan seaway; *G*, Cretaceous–Recent closing of Tethys, opening of Atlantic, fragmentation of Gondwana. *a*, Gondwana; *b*, Laurasia; *c*, North America; *d*, South America; *e*, Eurasia; *f*, Africa; *g*, Antarctica; *h*, India; *j*, Australia.

This Permo-Triassic supercontinent was short-lived, however, and separated into Laurasia and Gondwana as a result of the extension of the Tethyan ocean (Fig. 1, event *F*). Subsequent breakup of both these continental masses has been accompanied by a partial closure of Tethys, probably commencing in the late Jurassic to early Cretaceous (judging by the emplacement of ophiolites[20-23]) and may have been accelerated by the opening of the North Atlantic and fragmentation of Gondwana in mid-Cretaceous times (Fig. 1, event *G*).

The diversity of skeletonized benthonic shelf families of nine major invertebrate phyla is shown in Fig. 1 [11,12]. The number of families present (family diversity) is represented by epoch; the major trends in standing diversity

tion of major transgressions and regressions on presently separated continents.

Our model involves an expansion and application of a principle suggested by Russell[24]. Continental fragmentation occurs at ridges where ocean floor spreading is in progress. These ridges stand above the general abyssal plains to form broad topographic highs which, even on the scale of the world ocean, contain significant volume. A rough calculation based on published data[25-28] indicates that the volume of present ridges is about $2{\cdot}5 \times 10^8$ km^3, which would be sufficient to lower the ocean level by about 0·5 km across the entire surface of the Earth, or by about 0·65 km over the present ocean basins, were they to subside.

NATURE VOL. 228 NOVEMBER 14 1970 659

The lateral displacement of continents involves at least partial destruction in subduction zones[23] of either the plate on which they are borne or of another plate; these zones are expressed as topographic lows. As the continents move apart they displace ocean water from deeper regions bordering their outer edges, while the sea floor appearing between them, partly composed as it is of a ridge, stands relatively high. The total volume of the ocean basins is therefore reduced and water tends to rise relative to the continental platforms, which undergo transgressions.

If, on the other hand, continents are being joined, their suturing prevents further relative motion between the plates on which they ride. In this event, the component sea floor spreading of the world rift system which accounted for the relative motion between the now-sutured plates must eventually decay or new subduction zones must develop. There is evidence that if the spreading stops, the ridges subside[30-33], perhaps resulting from cooling and downward growth of the lithosphere. If new subduction zones develop, they would be associated with trenches. In either case, there would be an increase in volume of the ocean basins, and therefore an epicontinental regression.

According to this model, assembly of continents should be accompanied by regressions, and fragmentation by transgressions. In so far as the record is amenable to interpretation in these terms, these relations seem to hold. The emergent Eocambrian supercontinent (or continents; Fig. 1, event *A*) fragmented and was then transgressed during Cambrian time (Fig. 1, event *B*). Caledonian-Acadian, Hercynian and Appalachian suturings (Fig. 1, events *C*, *D*) were accompanied by regressions. Permo-Triassic continental assembly (Fig. 1, event *E*) was accompanied by a profound regression. Subsequent transgressions may mark episodes of fragmentation (Fig. 1, events *F*, *G*), as is suggested by the correlation of major continental breakup near the mid-Cretaceous with a widespread transgression. Post-Cretaceous sea level changes may have involved such events as movements on the "Darwin Rise"[27], suturing and continental underthrusting in the Alpine-Himalayan belt, and the well known eustatic effect of ice and snow storage.

The expected effects of transgressions and regressions tend to reinforce the effects of the ecological regulators of diversity. Widespread transgressions would tend to moderate the seasonal climates, while widespread regressions would tend to enhance continentality. Furthermore, the species-area effects of expanding and shrinking of the epicontinental seas, which may have less influence on marine biotic diversity than these other factors[5], would nevertheless be of some significance. According to the present hypothesis, all these factors operate in concert.

J. W. Valentine
E. M. Moores

Department of Geology,
University of California,
Davis, California 95616.

Received April 8; revised July 16, 1970.

1 Dewey, J. R., and Horsfield, B., *Nature*, **225**, 521 (1970).
2 Margalef, R., *Perspectives in Ecological Theory* (1968).
3 Bretsky, P. W., and Lorenz, D. M., *Proc. N. Amer. Paleont. Conv., Soc. Syst. Zool.* (1970).
4 Valentine, J. W., and Conoboy, J. T., *Proc. Geol. Soc. Amer. Ab. with Prog.*, **2**, 156 (1970).
5 Fryer, G., and Iles, T. D., *Evolution*, **23**, 359 (1969).
6 Valentine, J. W., *Publs. Syst. Ass.*, 7, 153 (1967).
7 Valentine, J. W., *Palaeontology*, **12**, 684 (1970).
8 MacArthur, R. G., and Wilson, E. O., *The Theory of Island Biogeography* (1967).
9 Bretsky, P. W., *Science*, **159**, 1231 (1968).
10 Bretsky, P. W., *Palaeogeog., Palaeoclimatol., Palaeoecol.*, **6**, 45 (1969).
11 Valentine, J. W., *Palaeontology*, **12**, 684 (1970).
12 Valentine, J. W., *J. Paleontology*, **44** (in the press).
13 Bird, J. M., and Dewey, J. F., *Bull. Geol. Soc. Amer.*, **81**, 1031 (1970).
14 Wilson, J. T., *Nature*, **211**, 676 (1968).
15 Dorn, P., *Geologie von Mitteleuropa* (1960).
16 Rodgers, J., *Amer. J. Sci.*, **265**, 408 (1967).
17 King, L. C., *Morphology of the Earth* (1962).
18 Smith, C. H., *Mem. Geol. Surv. Canad.*, 290.
19 Hess, H. H., *Spec. Pap. Geol. Soc. Amer.*, No. 62, 391 (1955).
20 Moores, E. M., *Spec. Pap. Geol. Soc. Amer.*, No. 118 (1969).
21 Moores, E. M., and Vine, F. J., *Trans. Roy. Soc.* (in the press).
22 Moores, E. M., *Nature* (in the press).
23 Trumpy, R., *Bull. Geol. Soc. Amer.*, **71**, 843 (1960).
24 Russell, K. L., *Nature*, **218**, 861 (1969).
25 Von Arx, W. S., *An Introduction to Physical Oceanography* (1962).
26 Heezen, B. C., *Sci. Amer.* **203**, 98 (1961).
27 Heezen, B. C., and Tharp, M., *Phil. Trans. Roy. Soc.*, A, **258**, 90 (1965).
28 Menard, H. W., *Marine Geology of the Pacific* (1964).
29 Nelson, T. H., and Temple, P. G., *Trans. Amer. Geophys. Un.*, **50**, 634 (1969).
30 Schneider, E. D., and Vogt, P. R., *Nature*, **217**, 1212 (1968).
31 Menard, H. W., *Earth Planet. Sci. Lett.*, **6**, 275 (1969).
32 Van Andel, Tj. H., *Earth Planet. Sci. Lett.*, **7**, 228 (1969).
33 Hsu, K. J., and Schlanger, S. O., *Proc. Twenty-third Intern. Geol. Cong.*, **1**, 91 (1968).

Molecular and Crystal Structure of Nonachloroclovononaborane, B_9Cl_9

In a recent publication[1] we described the isolation and characterization of a yellow-orange boron subchloride, B_9Cl_9. The structure of this compound has now been determined from three-dimensional X-ray data collected by counter methods.

Attempts to grow single crystals by sublimation were unsuccessful, but well formed orange rhombs were obtained by crystallization from methylene chloride in high vacuum conditions. The compound crystallizes with four molecular units in the space group *A*2/*m* of the monoclinic system. Crystal data: $a = 9·25$ Å, $b = 12·66$ Å, $c = 13·50$ Å; $\beta = 95·1°$. Volume of unit cell = 1549 $Å^3$. Measured density (flotation in mixtures of organic solvents) was 1·7 g/cm^3, calculated density 1·78 g/cm^3.

Diffraction data to $2\theta = 50°$ (Mo-K_α radiation) were collected on a manual General Electric XRD-6 diffractometer. The structure was solved by Patterson, Fourier and least-squares refinement techniques. The coordinates and anisotropic thermal parameters of all atoms were refined to the present discrepancy factor of $R = 6·5$ per cent using 1,305 independent reflexions above background. The relevant details of the parameters of the molecule are summarized in Table 1.

Table 1. INTERATOMIC DISTANCES IN (Å)

Atoms		Distances	Standard deviation
B	B	1·729	0·009
B	B	1·733	0·010
B	B	1·826	0·009
B	B	1·751	0·009
B	B	1·812	0·009
B	B	0·739	0·009
B	B	1·735	0·009
B	B	1·765	0·009
B	Cl	1·729	0·008
B	Cl	1·748	0·008
B	Cl	1·738	0·008
B	Cl	1·741	0·006
B	Cl	1·746	0·006
B	Cl	1·721	0·007

As shown in Fig. 1, B_9Cl_9 has a boron-cage structure with terminal chlorine atoms on each boron. The cage has the shape of a tri-capped trigonal prism with three boron atoms and their associated chlorines sitting on the crystallographic mirror plane; within experimental error the molecule has a three-fold axis although this is not a space group requirement. The B–B and B–Cl bond lengths are in close agreement with those found by Lipscomb for the related compound[2,3], B_8Cl_8. Both B_9Cl_9 (G. F.

PAPER 20

Species Richness in Marine Benthic Habitats through the Phanerozoic (1977)

R. K. Bambach

Commentary

ANDREW M. BUSH

Fossil assemblages—the time-averaged remains of biological communities—hold much information on the relationships between ancient organisms and environments. However, Bambach showed that local assemblages can be critical to interpreting global-scale paleobiological patterns as well. In the 1970s, competing theories held that global marine species richness increased, decreased, or remained constant from the Paleozoic to the Cenozoic. Unfortunately, global-scale analyses of biodiversity were complicated by preservational biases, making it difficult to distinguish between these hypotheses.

Drawing on the work of modern ecologists, Bambach divided global diversity into a within-habitat (alpha) component and a between-habitat (beta) component. Alpha diversity was relatively easy to study compared with global diversity, and it was not afflicted by many of the same biases. Bambach found that the average alpha diversity in the Cenozoic was about twice the maximum observed in previous eras. Because between-habitat components of diversity were not believed to have decreased, this result supported an increase in global diversity, albeit a modest one. This finding was a key component of the "consensus paper" (Sepkoski et al. 1981), which showed that several data sets subject to different biases all showed a moderate increase in biodiversity through geological time.

Following Bambach's work, analyses of the spatial structure of diversity have become central to paleontological diversity analysis. These analyses have become more sophisticated, particularly in dealing with heterogeneity in sampling and preservation (Powell and Kowalewski 2002; Bush and Bambach 2004; Kowalewski et al. 2006; Alroy et al. 2008). In fact, it is easy to undervalue a classic study such as Bambach's (conducted before the advent of easily accessible computing power) because it did not quantitatively account for potential biases. However, Bambach displayed an intimate knowledge of and concern for these biases, which he addressed with arguments that were practical and informed.

Bambach's paper stands as a powerful and influential statement of robust, first-order patterns; refining these patterns will continue to provide fruitful opportunities for research as data and methods are developed further.

Literature Cited

Alroy, J., M. Aberhan, D. J. Bottjer, M. Foote, F. T. Fürsich, P. J. Harries, A. J. W. Hendy, et al. 2008. Phanerozoic trends in the global diversity of marine invertebrates. *Science* 321:97–100

Bush, A. M., and R. K. Bambach. 2004. Did alpha diversity increase during the Phanerozoic? lifting the veil of taphonomic, latitudinal, and environmental biases in the study of paleocommunities. *Journal of Geology* 112:625–42.

Powell, M. G., and M. Kowalewski. 2002. Increase in evenness and sampled alpha diversity through the Phanerozoic: comparison of early Paleozoic and Cenozoic marine fossil assemblages. *Geology* 30: 331–34.

Kowalewski, M., W. Kiessling, M. Aberhan, F. T. Fürsich, D. Scarponi, S. L. Barbour Wood, and A. P. Hoffmeister. 2006. Ecological, taxonomic, and taphonomic components of the post-Paleozoic increase in sample-level species diversity of marine benthos. *Paleobiology* 32: 533–61.

Sepkoski, J. J., Jr., R. K. Bambach, D. M. Raup, and J. W. Valentine. 1981. Phanerozoic marine diversity: a strong signal from the fossil record. *Nature* 293: 435–37.

Paleobiology. 1977. vol. 3, pp. 152–167.

Species richness in marine benthic habitats through the Phanerozoic

Richard K. Bambach

Abstract.—The distribution of numbers of species and the median number of species from 386 selected fossil communities are tabulated for high stress, variable nearshore, and open marine environments during the Lower, Middle, and Upper Paleozoic, the Mesozoic and the Cenozoic. The number of species always increases from high stress to variable nearshore to open marine environments. Within-habitat variation in number of species is small for long intervals of the Phanerozoic. The median number of species in communities from high stress environments remains fixed at about 8 from the Cambrian to the Pleistocene. In open marine environments, the median is near 30 for the Middle and Upper Paleozoic and almost the same for the Mesozoic. Increases of 50% in median number of species between the Lower and Middle Paleozoic and 2 times between the Mesozoic and Cenozoic occur in open marine environments with parallel, but less pronounced, increases in variable nearshore environments. Conditions controlling overall within-habitat species richness changed at those times. These changes do not correlate directly with evolution of new major taxa, change in physical conditions, predation, space availability or oxygen supply. They may be related to changes in resource availability influenced by factors such as the developing terrestrial flora, to lag-time inherent in the evolutionary process of diversification, or to as yet undetermined factors. Although provinciality determines total species richness for the biosphere, the within-habitat data suggest that the number of marine invertebrate species in the world has increased since the Middle Paleozoic, contrary to Raup's (1976b) contention, but possibly only by about 4 times, not the order of magnitude or more suggested by Valentine (1970).

Richard K. Bambach. Department of Geological Sciences, Virginia Polytechnic Institute and State University, Blacksburg, Virginia 24061

Accepted: January 13, 1977

"Why are there so many kinds of animals?" G. E. Hutchinson (1959)

Introduction

If we are to understand the history of life we must determine the course of metazoan species diversification between the faunally barren Precambrian and the biotically rich modern world (Fischer 1965; Cloud 1968, 1976; Stanley 1973). The empirical model of Valentine (1969, 1970, 1973a, 1973b), the bias simulation model of Raup (1972), and the equilibrium model of Raup and others (Raup, Gould, Schopf, and Simberloff 1973; Gould 1975; Gould and Raup 1975; Raup 1976b) are attempts to describe the history of species diversification for marine invertebrates. The models differ in their conclusions about the timing of increase in species richness and in their estimates of the number of marine species in the Paleozoic. Figure 1 illustrates these basic differences.

Using the empirical model, which is derived from a tabulation of higher taxa (genera to phyla), Valentine (1969, 1970) concluded that the number of species increased initially to a plateau in the Middle and Late Paleozoic, decreased sharply in the Late Permian and Triassic, and then increased continuously through the remainder of the Mesozoic and Cenozoic to a unique high level in the Recent. Valentine asserts that modern species richness may be an order of magnitude greater than the Paleozoic maximum (Valentine 1969, p. 699–700; 1970, p. 413; 1973a, p. 1,078).

In the bias simulation model, Raup (1972, p. 1,070) hypothesized that the maximum number of species in the Paleozoic may have been twice the Recent number of species. He argued that preservational and sampling biases could reduce the number of preserved species recognized to proportions similar to those in the empirical model. Raup (1972, 1976b) questioned the assumption of the empirical model that the variations observed in

US ISSN 0094-8373

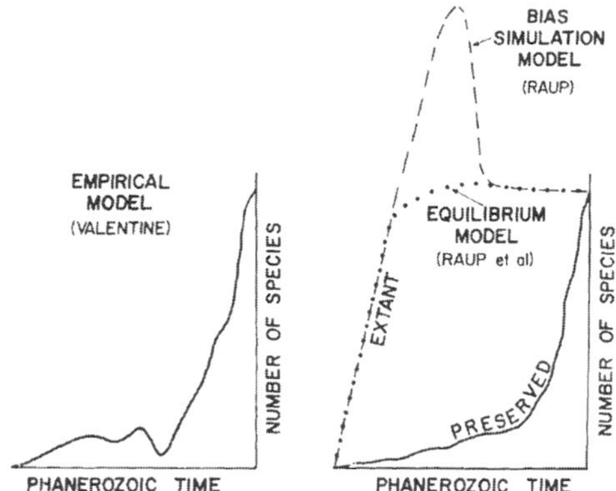

FIGURE 1. Comparison of the empirical model (order of magnitude increase in number of species between the Paleozoic and Late Cenozoic), bias simulation model (decrease from a Paleozoic high in extant species but with a biased fossil record resembling the empirical model), and equilibrium model (Early Paleozoic rise in number of species and no significant change since). Derived from Valentine (1970), Raup (1972), and Raup (1976b).

the number of species in the fossil record directly reflect changes in the total number of species that existed in the past.

Using an equilibrium model in an analysis of the pattern of evolutionary growth and extinction of clades, Gould and Raup (1975, p. 1,088) reached the preliminary conclusion that significant increase in the number of species was restricted to the Cambrian and Ordovician. In this application of the equilibrium model, the number of species would probably have remained close to the modern value since diversification in the Early Paleozoic.

Gould (1975) commented that Valentine (1970) and Raup (1972) have probably defined only the end members of the range of possibilities for the course of species diversification. The real sequence lies somewhere in between. This paper generates a new analytic approach to species diversification and provides new information on the course of species diversification within marine ecosystems. It is an attempt to narrow the range of possibilities.

Species diversity can be considered at three levels. Whittaker has called them alpha diversity (within-habitat), beta diversity (between habitat, expressed as a rate of ecologic change), and gamma diversity (whole landscape diversity, a composite of alpha and beta diversities) (Whittaker 1965, 1970; Peet 1974). The studies by Valentine and Raup and his colleagues have been directed toward the gamma diversity of the marine realm, whereas the analysis attempted in this paper directs attention to alpha diversity through geologic time.

The biosphere is the most all-encompassing category of the nested hierarchy of ecological units (population, community, province, biosphere) (Valentine 1968, 1973b). Because it is the sum of its component parts, one can, in theory, examine the whole unit without resolving it into all its components (Valentine 1968). This can be done accurately only when reliable data for the whole unit are known. Both Valentine and Raup have attempted to model species richness of the marine portion of the biosphere directly. However, because of biases inherent in the geologic record (Darwin 1859; Raup 1972, 1976b), it is not possible to directly determine all the necessary data about numbers of species on a worldwide basis for any part of the biosphere in the past. That is the heart of Raup's concern about Valentine's model. It is also a problem for the alternate models Raup and others have proposed (Raup 1976b). We need to know the numbers of species in ecological subdivisions of the biosphere for which reliable paleontologic data are available. Such data are available for benthic marine communities throughout the Phanerozoic. Study of numbers of species within communities provides a different approach to analyzing the growth of species richness through time. It will only be after we know the course of species diversification within communities and provinces that we will be able to extrapolate to the biosphere level with any confidence.

The question of the biological relevance of numbers of species to the organisms themselves can be examined only at ecological levels below that of the biosphere. Because biosphere species richness (gamma diversity) is the sum of all the different species assemblages (communities and provinces), it does not reflect the conditions which directly control the number of species within ecosystems (alpha diversity). Changes in world species richness can result simply from changes in the number of zoogeographic provinces or communities. These may be the result of shifts in the geographic distribution of environmental conditions (Valentine and Moores 1972; Schopf 1974) which are of fundamental interest in historical geology but have only indirect influence on the interactions of organisms within communities. By examining the history of species richness within communities, changes in the number of species within habitats can be detected. Change in the numbers of species coexisting within habitats would indicate that the basic structure of units in the ecological hierarchy may be altered through time. Study of species richness within habitats through time can provide biological insight not available when considering worldwide distributions.

Use of Fossil Communities

The use of data from well preserved fossil assemblages alleviates or avoids most of the problems of bias raised by Raup (1972, 1976b). Reliability of data derived directly from fossil assemblages is greater than that extrapolated to a worldwide level from the incompletely known fossil record. Variation in detail of data (monographic biases) is circumvented by using data only from thoroughly studied faunas. Because samples from particular environmental settings are available from most parts of the geologic column, and because local communities may be adequately sampled even from a restricted outcrop, there is no lack of information due to inadequate geographic sampling.

The major loss of information about the fauna of a local area occurs between the death of the individual animals and their burial (Lawrence 1968; Schäfer 1972). Because part of the record is not preserved and because proportions of skeletal (more readily preserved) and non-skeletal components of the fauna may vary from one environment or time to another, there is no exact method for determining how much of any particular fauna is missing in the fossil record. Comparison of fossil assemblages with the range of conditions in the Recent and a general evaluation of the structure of the fossil record both indicate that there has been no systematic increase in the fossilizeable portion of faunas with time. The record is imperfect but does not appear to be skewed by any time dependent factors.

The number of well skeletonized major taxa of marine invertebrates is constant at the phylum level through the Phanerozoic. The number of skeletonized taxa at the class and order levels present at any one time actually declines from Ordovician high values to the Cenozoic (Valentine 1969). The decrease has been 32% for classes (from 25 in the Ordovician and Silurian down to 17 in the Mesozoic and Cenozoic) and 40% for orders (from 75 in the Ordovician down to 45 in the Cenozoic). Valentine (1969) has demonstrated that the course of evolution is not characterized by the addition of new skeletonized higher taxa. Instead, major adaptive breakthroughs, replacement of extinct taxa, and diversification of species occur primarily within skeletonized groups already extant.

A survey of the literature on trace fossils (Crimes and Harper 1970; Frey 1975; Häntzschel 1975) gives no indication that the overall variety or abundance of trace fossils has declined. Nor are there any major changes in trace fossil time ranges which correspond with the shifts in numbers of skeletonized species. Many changes in the trace fossil record do parallel the taxonomic changes noted for skeletonized forms. This may simply be because many traces are made by skeletonized organisms. Nonetheless, there is no evidence from the fossil record that the diversity of the soft bodied fauna has declined. This suggestion is reinforced by observations on the degree of bioturbation in the geologic record. Most Paleozoic sections are relatively well bedded. Thin sections of Paleozoic sediments often reveal some trace of primary lamination, even in bioturbated beds (Raup and Stanley 1971, p. 320). This contrasts with the frequent complete mottling or biogenic homogenization seen in many Cenozoic and Recent deposits (Moore and Scruton 1957). Although these are qualitative observations, the conclusion seems most reasonable that the soft bodied fauna in the Cenozoic is fully equivalent to that in the Paleozoic.

Therefore, although final proof will probably always be lacking, I take the constancy of representation of higher skeletal taxa and the lack of a systematic change in the record of bioturbation in the trace fossil record as support for the hypothesis that there has not been a significant change in the average proportion of preservable skeletonized species through the Phanerozoic. For Recent benthic communities the proportion of the local fauna actually preserved varies from community to community but the average proportion of the species likely to be fossilized is about 30% (Johnson 1964; Craig and Jones 1964; Lawrence 1968). The assumption is made that a similar relationship holds for the Phanerozoic.

Well preserved fossil assemblages are equivalent and directly comparable regardless of their age. Most fossil assemblages are derived from ecologically controlled species associations (communities) (Walker and Bambach 1971; Bowen, Rhoads, and McAlester 1974). Those fossil assemblages which were mixed or transported before deposition and burial can usually be recognized as such by distinctive features in the enclosing sediments or from the condition of the fossils themselves (Johnson 1960). Data on fossil communities are less biased by differential preservation than is commonly realized. After a fossil deposit is first formed, there is little change in paleontologic information content unless later intense diagenetic alteration or actual metamorphism occurs. When such destructive events do occur, their presence is almost always detectable through telltale lithologic features such as grain fabrics or modes of preservation (for example, McAlester 1962; Klement and Toomey 1967; Boyd and Newell 1972). Remarkable amounts of biological data are recorded not only in stable hardparts but also in molds and biogenic structures, the redundant aspects of preservation (Lawrence 1968; Häntzschel 1975; Frey 1975). For instance, although aragonite is much less stable than calcite, there is a superb record of originally aragonitic fossils from the Paleozoic, such as cephalopods and bivalved molluscs, which rivals that of calcitic forms, such as brachiopods. Preservation of bivalves is primarily as composite molds (McAlester 1962; Bambach 1969, 1973) and cephalopods often are preserved as internal casts of the chambers. Nonetheless, diagenesis has not eliminated them from the record. Impressions (traces, casts, molds) and pseudomorphs (replacements, permineralization) retain information in the fossil record, especially, for well lithified rocks where biogenic sedimentary structures and delicate molds are readily recovered without special techniques to prevent disintegration. When evaluating the number of species rather than the relative abundance of individuals, partly preserved but identifiable remains are even of use. Differential rate of preservation is not a handicap if even a few specimens of each taxon get preserved. Well preserved Cambrian fossil assemblages retain quite as much taxonomic detail as is preserved in Pleistocene deposits.

Data Compilation

The data summarized in Figs. 2 and 3 are taken from studies of 386 fossil communities or large collections representing single communities. The data are grouped both by environmental category and by age. The sources are listed in the Appendix.

Only fossil assemblages are considered. In this way no arbitrary selection or elimination of species was necessary, as must be done when comparing Recent faunas with fossil assemblages. The species tabulated are exclusively benthic marine invertebrate megafossils. No regional or whole formation faunal lists are used. Each fauna is from a single large collection, from a single bed or from a set of collections made in sequence through a restricted part of a single section. In all cases where more than one collection was used in compiling a species list, faunal homogeneity and contiguous occurrence were necessary criteria before the combined list was accepted as representing a single community.

One of the criteria for use in selecting the data was that the species list be based on a large number of specimens. Two sorts of sample type are included: single collections and pooled collections. The range in sample size for single collections is from about 100 specimens to over 1000. The majority of these samples include 200 to 400 specimens. For example, the five Silurian collections tabulated in Ziegler, Cocks and Bambach (1967) are 373 individuals (16 species), 1,337 individuals (31 species), 198 individuals (18 species), 325 individuals (20 species), and 416 individuals (31 species). Ager (1963) tabulated 850 and 1,000 specimens for collections containing 27 and 30 species. The pooled collection data are for generally larger numbers of specimens. For example Bayer (1967) lists 3,325 specimens (18 species) and 2,200 specimens (14 species) in tables of collections for two Ordovician communities. The seven

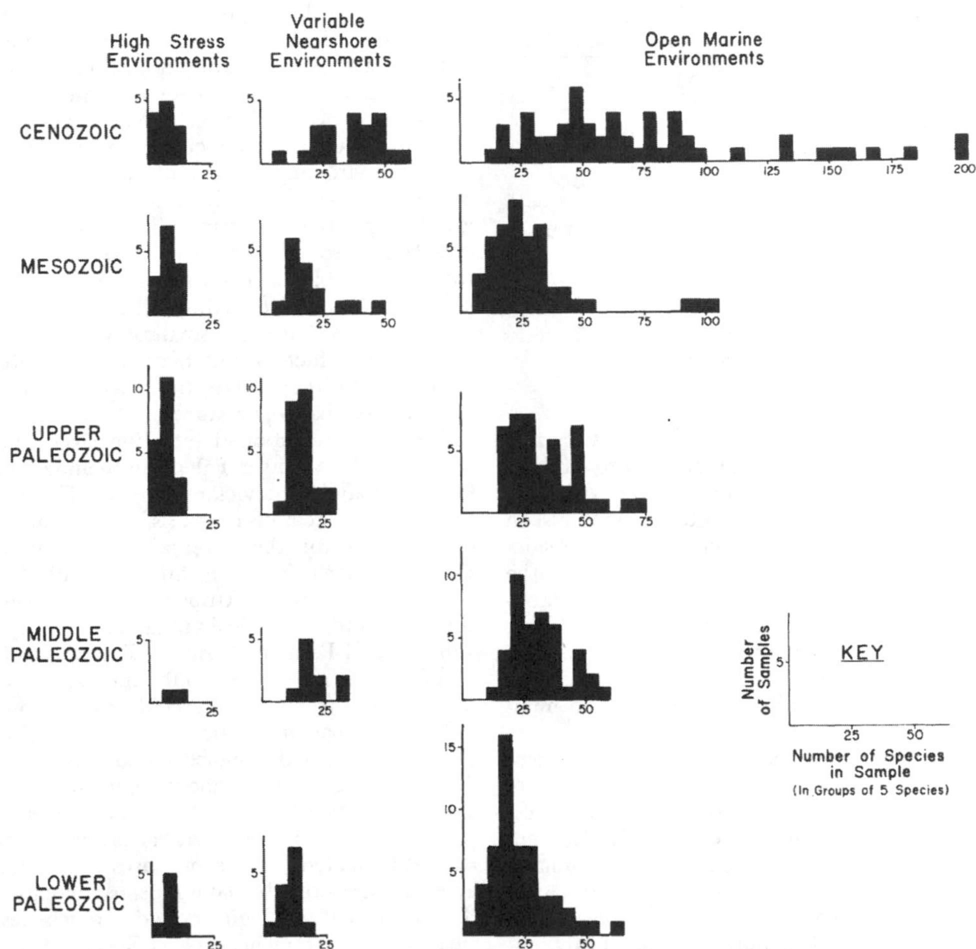

FIGURE 2. Histograms of data on numbers of species in selected fossil communities. Grouping by environmental category and time segment explained in text. Vertical dimension is number of communities, horizontal dimension is number of species in the community. Number of species in groups of 5 species. Data sources are listed in the Appendix.

Silurian faunal lists compiled from McLearn (1924) and Bambach (1969) are based on 14,000 bivalve specimens and comparable numbers of other taxa. The sample sizes for pooled collection community lists range from about 500 to over 10,000 specimens. Although exact abundance data are not available for all the lists used, it is apparent from the authors' discussions that sample sizes are generally in these ranges for all time intervals considered, with the possible exception of the Cambrian. There are no systematic increases in sample size from the Ordovician to the Cenozoic for the data included in this study.

The communities used are all level bottom communities (Thorsen 1957). Reefs and banks were excluded because the numbers of species in such habitats differs systematically from level bottom communities (see discussion below, p. 159). Faunal lists were used for Paleozoic communities only when they included species listings for all major taxa. For the Cenozoic many of the lists were only mollusks. This causes little bias because mollusks domi-

Median Number of Species	High Stress Environments	Variable Nearshore Environments	Open Marine Environments
CENOZOIC	8.5	39	61.5
MESOZOIC	7.5	17	25
UPPER PALEOZOIC	8	16	30
MIDDLE PALEOZOIC	9.5	19.5	30.5
LOWER PALEOZOIC	7	12.5	19

FIGURE 3. The median for the numbers of species for the data in each histogram of Fig. 2.

nate the skeletal portions of most Cenozoic faunas. Few of the complete Cenozoic faunal lists for assemblages from clastic sediments contained more than six to ten non-molluskan megafossil species. Nonetheless, the reader should be aware of this inadequacy of the data. Cenozoic numbers of species as tabulated are somewhat less than they should be. However, this bias turns out to have a conservative effect and actually enhances the basic conclusions about changes in the number of species within habitats.

The environmental assignments for the communities were determined by evaluation of facies characteristics and relationships discussed by the authors of each study. The environmental classification used is intentionally broad, with only three categories, to avoid erroneous placement of communities which often are described only in general environmental terms. Although generalized, these categories are adequate to discriminate differences in the number of species between three different basic types of habitat: (1) high physiological stress environments, (2) variable nearshore marine environments, and (3) open marine environments. High physiological stress environments are tidal flats, estuaries and similar settings as described by authors or determined by stratigraphic position in which temperature, salinity, availability of water or other conditions frequently change rapidly to extreme values. Variable nearshore marine environments are those listed by authors as deltaic, shoreline or nearshore or those which occur by stratigraphic position between high stress and open marine settings. The range and rate of change of physiologically important factors fluctuate less in them than in high stress environments, but they are more variable in both physiological factors and the frequency of high energy disturbance of the bottom from waves and storms than open marine environments. Open marine environments are those listed as such by authors or those which occur by stratigraphic position seaward of environments specified as shoreline or nearshore. They are the most stable or even environments considered. Because of mixing with the oceanic reservoir they have the smallest variation in physiological factors and because of greater water depth they have the least frequent rate of disturbance by storms.

The data are grouped into five large time segments. The Lower Paleozoic includes the Cambrian and Ordovician Periods. The Eocambrian Ediacaran fauna is also included in the data for this interval. The Middle Paleozoic comprises the Silurian and Devonian Periods and the Upper Paleozoic covers the Carboniferous (Mississippian and Pennsylvanian) and Permian Periods. The Mesozoic and Cenozoic are defined in the standard manner. In this report no Triassic faunas and only six faunas from the Paleogene (Paleocene, Eocene, and Oligocene) are included. The five large time segments were used because they clearly show the general trends, whereas adequate numbers of samples are not yet tabulated (or in some cases even available) to make reliable comparisons among all the individual geologic periods, much less smaller time increments such as Epochs. Comparisons of Periods and Epochs is made in the discussion below only for those for which adequate numbers of samples have been obtained. More data for this evaluation are being compiled. A later report, detailing the course of within-habitat fluctuations in numbers of species, especially through critical parts of the geologic record, will be forthcoming.

General Trends

Figure 2 is a set of histograms showing the distribution of the data for each of the three environmental categories in each of the five time segments. Figure 3 is a table of the median value for the number of species corresponding to each histogram of Fig. 2. The open spaces between boxes in Fig. 3 mark differences between environment types or time segments which are considered significant in average number of species. The data distribution has been analyzed statistically by J. Sepkoski, and all the differences between environmental categories and time segments discussed here are significant at the 0.05 level or more (Sepkoski, unpublished manuscript).

Three general trends are apparent in the data (Figs. 2 and 3).

(1) Numbers of species increase from high stress to open marine environments. This is true throughout the Phanerozoic. This is documented by the narrow range and consistently low values in the numbers of species in high stress environments, and the increasingly greater range in numbers of species plus the displacement of the mode of each histogram for the variable nearshore and open marine environments toward higher numbers of species (Fig. 2). This is paralleled by the tabulated increase in the median for number of species in the sequence of environmental categories (Fig. 3) in each time segment.

(2) The numbers of species in high physiological stress environments remains low through the entire Phanerozoic. The narrow range of numbers of species, the consistently low numbers of species (no sample contained more than 15 fossil species) and the constant mode of the histograms for communities from high stress environments (Fig. 2) emphasize the invariance of numbers of species in high physiological stress habitats through time.

(3) The numbers of species in more stable marine environments increases significantly twice, once in the transition from the Lower to Middle Paleozoic and a second time between the Mesozoic and Cenozoic. Differences in both shape and mode distinguish the histogram for Lower Paleozoic open marine environments from those for the Middle Paleozoic, Upper Paleozoic, and Mesozoic. The similarities in shapes, ranges of number of species and modes for the histograms (Fig. 2) and the similarity in the median numbers of species (Fig. 3) for communities from open marine environments for the Middle and Upper Paleozoic and Mesozoic indicate that there was no significant increase in the number of species within marine habitats over a time span of 350 million yr. The corresponding histogram for Cenozoic open marine environments is totally unlike any of the four preceding ones. These features, coupled with the large number of samples tabulated for open marine environments from each time segment (53 for the Lower Paleozoic, 42 for the Middle Paleozoic, 47 for the Upper Paleozoic, 49 for the Mesozoic, and 56 for the Cenozoic) establish that the changes in median number of species between the Lower and Middle Paleozoic and between the Mesozoic and Cenozoic (Fig. 3) reflect real increases in the number of preserved species occurring in communities from open marine benthic environments at these times. The shapes of the histograms for variable nearshore environments are less regular (Fig. 2). This is probably related to sampling biases which cause variation in representation of the widely different local habitat conditions included in this category from different time intervals. Nonetheless, the Lower Paleozoic variable nearshore environment has a lower median value and a smaller range of values than any later time segment. The histogram for Cenozoic variable nearshore environments is so unlike those for earlier time segments in both shape and mode that the Cenozoic median number of species for this category undoubtedly also represents a real increase from the past (Fig. 3).

Discussion

Comparison between habitat types.—In every time segment from the Lower Paleozoic through the Cenozoic the range in number of species, median number of species, and maximum number of species increase from high stress to open marine environments (Figs. 2 and 3). This increase is entirely independent of possible time dependent factors such as taxonomic change (evolution) and possible biases in the record (differential preservation, collection sizes). The increase in number of species matches the gradient for increasing stability or evenness in environmental conditions from unstable estuarine and intertidal through shallow, frequently disturbed nearshore environments to deeper, quieter, more even offshore marine conditions. It also matches a generalized sequence of increasing habitat area for each environmental category. High stress environments are always circumscribed geographically, compared to more open marine conditions. Variable nearshore environments include communities which inhabited smaller

CORRELATIVE NON-BIOHERM (SHELF) ENVIROMENTS	BIOHERMS
8.34 (N=248)	32.50 (N=77)

MEAN NUMBER OF BRACHIOPOD SPECIES - PERMIAN, WEST TEXAS

FIGURE 4. The average (mean) number of brachiopod species for bioherm and correlative non-bioherm samples in the Permian of West Texas. Data derived from Grant (1971).

areas and narrower belts than those from open marine shelf environments. Although these data do not distinguish between species-area effects (MacArthur and Wilson 1967; Simberloff 1972, 1976) and the effect of increased stability (Sanders 1968, 1969; Slobodkin and Sanders 1969), they suggest that the controls which produce a similar gradient in marine benthic species richness in the Recent have operated throughout the Phanerozoic.

Secondary influences in open marine environments.—The three broadly defined environmental categories do not represent "pure" habitats. Further subdivision would be possible if adequate environmental and paleogeographic data were reliably available. Detailed analysis of the community data for open marine environments reveals two effects which influence species richness in these environments. One is habitat complexity and the other is an "epeiric sea" effect.

Increased habitat complexity or heterogeneity increases species richness. This is seen in the species richness of tropical forests contrasted to prairies or coral reefs compared to sand or mud bottoms. The influence of habitat complexity is observed in the fossil record as well. For example, Fig. 4 compares the average number of brachiopod species from non-bioherm level bottom samples with samples from bioherms in Grant's study of the Permian of West Texas (Grant 1971). There is an average of four times as many brachiopod species from each bioherm sample as from each non-bioherm sample. Because of the systematic influence of habitat complexity on numbers of species, a separate environmental category would be necessary to evaluate species richness for the reefs, banks and similar complex habitats which have come and gone through time (Laporte 1974). Unfortunately, there are not enough species level data available on whole faunas from reefs and bioherms to permit comparisons for them over the length of the Phanerozoic.

Even though only level bottom communities were included in this study some "noise" in the species numbers is apparently created by variations in habitat complexity. The community with the highest number of species in the entire Paleozoic (73) is a crinoid dominated community (Lane 1973). Many of the other Paleozoic communities with higher numbers of species have either crinoids, bryozoa, or corals as conspicuous elements. For instance, in the Lower Paleozoic, bryozoa are abundant in communities with 63, 49, and 44 species, all among the longest of those species lists. The higher species richness in these communities may reflect increased habitat complexity and niche availability created by upright colonies and crinoid stems. These communities may have been partly stratified, rather than strictly level-bottom community types. Detailed niche-group or guild analysis (Bambach 1974, 1977) supports this interpretation.

One important difference in the geographic setting for a community is whether it is from an ocean-facing shelf or an epeiric sea. Differences in environmental stability and resource supplies may characterize these two settings. Those differences could influence community species richness. A more detailed time breakdown of the data for open marine environments (Fig. 5) reveals a definite "epeiric sea" effect of lower species richness. Nineteen of the twenty species lists from the Silurian are from what were ocean-facing shelves. The Silurian median value is 34.5 species. Twenty of the twenty-two Devonian lists are from epeiric sea environments. The Devonian median is 25 species. The sixteen lists for the ocean-facing shelf communities of the Carboniferous of California (Watkins 1974) have a median value of 39 species, whereas the median value for the twenty-two lists from the epeiric sea of the midcontinent in the Pennsylvanian is 25 species. If the four ocean-facing shelf faunas for the Ordovician are removed, the median value for the remaining thirty-one Ordovician epeiric sea communities becomes 19 species. Epeiric seas were apparently less species rich than ocean-facing shelves.

Some of the variation in species richness may be the result of grouping communities from different climatic regimes. Climate-related gradients in species richness are well-known in the Recent. The detailed influence of past geographic position must await the delineation of ancient crustal plate boundaries and the course of their change through time. The data used in this paper is exclusively for tropical and subtropical climates for the Paleozoic, subtropical to temperate for the Mesozoic and subtropical to cool temperate for the Cenozoic. The four highest Cenozoic values are for subtropical communities. As more precisely located paleogeographic positions can be made in the future, the effect of climatic zonation on species richness should be detectable.

Trends in within-habitat species richness (alpha diversity) through time.—Although species richness in high stress environments remained low and unvarying throughout the Phanerozoic, it has increased in more open marine habitats (Figs. 2 and 3). This increase was not continuous or gradual but episodic, occurring over three short intervals. The initial metazoan diversification occurred in the Eocambrian, the second increase in within-habitat species richness was at the transition from the Lower to Middle Paleozoic, and the third took place in the Paleogene (Lower Cenozoic). The intervening intervals were times of relative stability for within-habitat species richness.

In the Lower Paleozoic both the Cambrian and Ordovician are characterized by a large number of communities with low species richness. This is apparent from the narrow, high peak in the Lower Paleozoic open marine histogram (Fig. 2) as well as the low median values for both periods (Fig. 5). The fortunate preservation of two unique soft bodied faunas, the Ediacaran fauna of 27 species in the Eocambrian and the Burgess Shale fauna (*Pagetia bootes* faunule) of 32 species from the Middle Cambrian suggest that metazoan diversification had already produced single fauna species richness typical of the Lower

Open Marine Medians Periods and Epochs	
62	PLEISTOCENE
60	MIOCENE
34	CRETACEOUS
21.5	JURASSIC
25	PENNSYLVANIAN
39	CALIF. CARBONIF.
25	DEVONIAN
34.5	SILURIAN
23	ORDOVICIAN
17.5	CAMBRIAN

FIGURE 5. The median for the numbers of species in communities from open marine environments for Periods and Epochs for which 16 or more faunas are available.

Paleozoic at the very start of the metazoan record and that soft bodied species richness was not extremely different from the species richness observed for skeletal faunas in the Lower Paleozoic. The median number of species for Ordovician communities from epeiric sea habitats was only 19. This is close to the Cambrian median of 17.5 and considerably less than that for similar environments later in the Paleozoic (a median of 25 for both Devonian and Pennsylvanian epeiric sea communities (Fig. 5)). Despite the taxonomic turnover between the Cambrian and Ordovician there was little increase in species richness through the Lower Paleozoic.

The median number of species increases by 50% between the Lower and Middle Paleozoic, but the number of species in both the Middle and Upper Paleozoic is strikingly similar (Figs. 2 and 3). These similarities imply a stability for within-habitat species richness over this long interval. This idea is supported by the more detailed data in Fig. 5. Similar median values for ocean facing shelf communities in the Silurian (34.5) and the Carboniferous of California (39) and identical medians for epeiric sea communities in the Devonian and Pennsylvanian (25) suggest that each habitat type developed its own species richness characteristics and that they did not change through the Middle and Upper Paleozoic.

In the Mesozoic, the Jurassic has a low average species richness (median 21.5), suggestive of Lower Paleozoic figures, whereas the Cretaceous has a median of 34, similar to Middle and Upper Paleozoic numbers (Fig. 5). The data for the Jurassic come from broad shelves and small epeiric seas at various distances from narrow ocean basins. The Cretaceous data come primarily from ocean facing shelves (median 35.5 species for 14 communities). The Jurassic low numbers of species may represent more than just an "epeiric sea effect." If the Permian extinction affected within-habitat species richness (alpha diversity) as well as world species richness, then the Jurassic low numbers of species might still be reflecting this, with within-habitat species richness matching Paleozoic levels again only in the Cretaceous. As a result of the taxonomic changes which occurred in the Triassic, Cretaceous faunas are mollusk dominated and resemble Cenozoic, rather than Paleozoic faunas in their overall composition. It is striking that Cretaceous species richness is so similar to that of the Middle and Upper Paleozoic. This may indicate pre-Cenozoic limits to species richness which were in effect from the Middle Paleozoic through the Mesozoic.

Cenozoic communities have a large range in species richness but on average are twice as species rich as those of preceding times. The Cenozoic median of 61.5 is close to the maximum values for any earlier time (Fig. 2). This increase in species richness for the Cenozoic is seen in variable nearshore as well as open marine environments. The similarity in the medians for the Miocene (60) and Pleistocene (62) suggests that increase in species richness within habitats for the Cenozoic occurred in the Paleogene and that stability in species richness for particular habitats may have become re-established by the Neogene.

The increases in numbers of species through time shown in Figs. 2 and 3 represent real increases in average species richness in communities, not systematic bias in the data from the fossil record. As discussed above (p. 154), analysis at the community level avoids most of the biases which hamper evaluation of the fossil record (Raup 1976b). If soft bodied taxa were replaced by skeletal forms at the times when fossil species richness increases, we would expect to see changes in the proportional representation of particular skeletal groups as they diversified to replace the soft bodied organisms. The major times of taxonomic "reorganization" in the preserved fossil record do not match the times of increase in species richness. Raup (1976a, p. 284, Table 3 and p. 287, Figure 2) shows that the major times of change in the taxonomic structure of the fossil record are in the Ordovician with the replacement of the trilobite dominated Cambrian faunas by a variety of skeletal taxa and in the Triassic with the rise of mollusk dominated faunas. Very little change in the proportional representation of different taxa occurs between the Ordovician and Silurian or between the Cretaceous and Cenozoic, the times of increase in species richness. Apparently these increases in species richness are not related to the replacement of particular soft bodied, or even skeletal, groups. As discussed on p. 154 no evidence indicates that there has been a decrease in the diversity of non-skeletal organisms with time.

Speculations on controls for species richness within-habitats.—The very long intervals of stability in average numbers of species imply some sort of equilibrium in species richness, possibly like species-area equilibria seen in the study of island biogeography (MacArthur and Wilson 1963, 1967; Wilson 1969; Simberloff and Wilson 1969, 1970; Simberloff 1976) with taxonomic turnover (evolution and extinction) balanced through time for particular community types. Raup (1972); Raup, Gould, Schopf and Simberloff (1973) and Gould and Raup (1975) have used the equilibrium concept in generating their stochastic models of evolutionary phenomena. At the level of communities, the idea of species packing to saturation is controversial. Theoretical arguments can be advanced to demonstrate that species create niches into which more species can be fit and there should be no end to potential increase in the number of species. From a sampling standpoint alone, as sample sizes are increased more species are collected. Nevertheless, the data presented here indicate no change in species richness has occurred in high stress environments for the last 570 million yr and that very little change in species richness occurs in open marine benthic communities for most of the Lower Paleozoic and again for the entire interval from the Silurian to the end of the Cretaceous, intervals of 150 million and 350 million yr. Different habitats have different levels of species richness. As these habitats are defined more closely the similarities in species richness for them over time become more apparent. The "epeiric sea effect" described above is an example. Some property of the ecosystem must operate to control species richness within habitats and limit or prevent its increase, despite evolutionary turnover, during long periods of time.

Along with the phenomenon of stability in species richness within habitats, there is the counter phenomenon of increase in species richness. If some property or properties of ecosystems can limit species richness, what changes are responsible for permitting episodic increases? Why should these increases occur in some environmental settings and not in others?

Features such as environmental stability, habitat complexity and geographic area are known to be involved in controlling species richness. They may be important in differentiating the levels of species richness shown by the three environmental categories examined in this paper, as well as other effects such as the epeiric sea effect. But they do not change systematically with time to permit the regular increase of species richness observed.

The times of increase in numbers of species within communities do not correlate with the times when major new taxa evolve (Raup 1976a, Table 3). Although physical conditions such as climate have fluctuated markedly through geologic time and may have had local effects on species richness, no physical changes have operated in a progressive fashion which could act as a general influence on species richness within habitats. Cropping and predation are related to high species richness in some Recent communities (Paine 1966; Dayton 1971). All major taxa of marine predators first appear in the fossil record during times when little change occurs in the numbers of species (Carter 1968; Sohl 1969). The decrease in area of epeiric seas may have had a major role in the Permian extinction (Valentine and Moores, 1972; Schopf, 1974; Simberloff, 1974), but systematic changes in space or habitat area corresponding to the increase in species richness within communities have not occurred. Oxygen supply has been suggested as a major influence in the course of evolution (Fischer 1965; Tappen 1968, 1971; Rhoads and Morse 1971; Cloud 1976), but the proposed times of change in oxygen supply since the Eocambrian do not correlate with the times of change in species richness in communities.

Two possibilities suggest themselves as potential influences on increase in within-habitat species richness: systematic expansion of ecospace through the development of new or increased resource supplies, and possible inherent lag time in species diversification. Increased resource supplies can increase the potential for resource subdivision, niche partitioning and an increase in species richness within a habitat. Calef and Bambach (1973 and in preparation) have advanced the hypothesis that Lower Paleozoic oceans had very low nutrient and food supplies. The development of a terrestrial flora first with the origin of land plants in the Silurian (Gray and Boucot 1971; Pratt, Phillips, and Dennison 1975) and then the evolutionary blossoming of the angiosperms in the Late Cretaceous and Cenozoic (Hughes 1976) coincides with increases in marine benthic species richness. Nutrient supply to the oceans, both organic and chemical, would have been affected by these events. Oceanic nutrient recycling also may have been influenced by changes in thermohaline circulation caused by climatic changes, such as the onset of Paleozoic glaciations in Africa in the Late Ordovician (Fairbridge 1971; Bennacef et al. 1971; Crowell and Frakes 1970) and the Cenozoic formation of the deep cold oceanic water mass in the Late Eocene or Early Oligocene (Benson 1975). These changes in oceanic mixing could change food resources in the shelf seas. These changes may be associated with the increases in marine habitat species richness. On the other hand, the timing of these changes may simply be a result of a lag time inherent in the process of species diversification. This has been suggested by Simberloff (1974) for the Early Mesozoic and for both the Early Paleozoic and Mesozoic by Sepkoski (1976a, 1976b).

World species richness.—Because the data

in this paper speaks only to community conditions, the interpretation of world species richness requires extrapolation. World species richness is the sum of all the local (province and community) numbers of species (Valentine 1970, 1973a). Valentine correctly asserts that biosphere species richness depends on the degree of provinciality. However, when we try to determine the provinciality and community distributions in the geologic past, we run afoul of bias and inadequate data, which have fueled the questions over what to believe about worldwide interpretation of species richness data (Raup 1972, 1976b; Valentine 1973a). Nevertheless, the history of species richness within habitats does permit two general conclusions.

(1) The within-habitat increases in numbers of fossil species represent real increases in species richness. At the biosphere level the number of marine invertebrate species must have increased at the times when numbers of species within habitats increased. This would fail to occur only if provinciality had declined. There is consensus in the literature that it has increased, especially in the Neogene. Therefore, the conservative conclusion of Raup (1976b, p. 289) that there is "no compelling evidence for a general increase in the number of invertebrate species from Paleozoic to Recent" can not be accepted.

(2) The average number of species within open marine environments doubles between the Middle Paleozoic through Mesozoic interval and the Cenozoic. Neither Raup's figures (1976a) nor those in Figs. 2 and 3 support Valentine's contention that the Late Cenozoic may have ten times or more the number of species than the Paleozoic peak (Valentine 1969, 1970, 1973a). World marine species richness probably has increased less dramatically. Raup (1976a) calculates the Neogene number of species as 2.4 times the Devonian number. The Devonian is the time of greatest provinciality in the Paleozoic (Boucot 1975). If Devonian provinciality was comparable to Cenozoic provinciality, then the correspondence between the Middle Paleozoic to Cenozoic ratio in numbers of species within communities (2) and the ratio observed by Raup for his Devonian–Cenozoic comparison (2.4) could actually be measuring the proportional increase in the number of species in the biosphere. To create an order of magnitude increase in species richness would require a four to five fold difference in provinciality between the Paleozoic and Cenozoic. This seems too great, given our developing knowledge of ancient plate configurations and past climatic zonation (Ziegler et al. 1977). If we accept a doubling of provinciality for the Neogene over the Devonian then an estimate of four fold increase in the number of species in the marine realm since the Paleozoic seems reasonable.

"It is certainly not the least charm of a theory that it is refutable."

F. W. Nietzsche (1885–6)

Acknowledgments

Discussions with James Valentine and David Raup started my thinking on this project. David Raup kindly supplied copies of his two recent (1976) papers in advance of publication. J. John Sepkoski served as a sounding board, suggested ways of looking at the data and has assumed the task of a separate statistical analysis. I have had stimulating and helpful discussions on various facets of this study with Steven Stanley (Johns Hopkins), Karl Turekian (Yale), Ida Thompson (Princeton), and Jack Webster, Tom Jensen, Duncan Porter and Richard Connor (V.P.I. and S.U.). The critics for Paleobiology did an exceptionally helpful job with the manuscript. My wife edited each version of the paper. Gail Walbridge, Mary Ellen Saunders and Carol Eiss typed the initial manuscript and Judy Baker prepared the final version, both under great time pressure.

My interest in larger problems in paleontology was stimulated many times by the active mind of a good friend, the late Ralph Gordon Johnson. The profession misses him. If there is any value in this paper I dedicate it to his memory.

Literature Cited

BAMBACH, R. K. 1973. Tectonic deformation of composite-mold fossil Bivalvia (Mollusca). Am. J. Sci., Cooper Vol. 273-A:409–430.

BAMBACH, R. K. 1974. Resource partitioning in Paleozoic benthic communities. Geol. Soc. Am. Abstr. with Programs. 6:643.

BAMBACH, R. K. 1977. Patterns of niche-partitioning in the history of life. In preparation for symposium on the Control of Diversity in Ecological and Evolutionary Time (S. J. Gould, convener) at North Am. Paleontol. Conv. II.

BENNACEF, A., S. BEUF, B. BIJU-DUVAL, O. DE CHARPAL, O. GARIEL, AND P. ROGNON. 1971. Example of cratonic sedimentation: Lower Paleozoic of Algerian Sahara. Am. Assoc. Pet. Geol. Bull. 55:2225–2245.

BOUCOT, A. J. 1975. Evolution and Extinction Rate Controls. 427 pp. Elsevier; New York.

BOWEN, Z. P., D. C. RHOADS, AND A. L. MCALESTER. 1974. Marine benthic communities in the Upper Devonian of New York. Lethaia. 7:93–120.

BOYD, D. W. AND N. D. NEWELL. 1972. Taphonomy and diagenesis of a Permian fossil assemblage from Wyoming. J. Paleontol. *46*:1–14.

CALEF, C. E. AND R. K. BAMBACH. 1973. Low nutrient levels in Lower Paleozoic (Cambrian-Silurian) oceans. Geol. Soc. Am. Abstr. with Programs. 5:565.

CALEF, C. E. AND R. K. BAMBACH. In preparation. The nutrient poor Lower Paleozoic marine ecosystem.

CARTER, R. M. 1968. On the biology and palaeontology of some predators of bivalved Mollusca. Palaeogeog., Palaeoclimatol., Palaeoecol. *4*:29–65.

CLOUD, P. E. 1968. Pre-metazoan evolution and the origins of the metazoa. pp. 1–72. In: Drake, E. T., ed. Evolution and Environment. Yale Univ. Press; New Haven, Connecticut.

CLOUD, P. E. 1976. Beginnings of biospheric evolution and their biochemical consequences. Paleobiology. *2*:351–387.

CRAIG, G. Y. AND N. S. JONES. 1966. Marine benthos, substrate, and palaeoecology. Palaeontology. 9:30–39.

CRIMES, T. P. AND J. C. HARPER, eds. 1970. Trace Fossils. 547 pp. Geol. J. Spec. Issue 3. Seel House Press; Liverpool.

CROWELL, J. C. AND L. A. FRAKES. 1970. Phanerozoic glaciation and the causes of ice ages. Am. J. Sci. *268*:193–224.

DARWIN, C. 1859. The origin of species by means of natural selection. 386 pp. The Modern Library (Random House); New York.

DAYTON, P. K. 1971. Competition, disturbance, and community organization: the provision and subsequent utilization of space in a rocky intertidal community. Ecol. Mon. *41*:351–389.

FAIRBRIDGE, R. W. 1971. Upper Ordovician glaciation in northwest Africa? Reply. Geol. Soc. Am. Bull. *82*:269–274.

FISCHER, A. G. 1965. Fossils, early life, and atmospheric history. Proc. Natl. Acad. Sci. 53:1205–1213.

FREY, R. W., ed. 1975. The study of Trace Fossils. 562 pp. Springer-Verlag; New York.

GOULD, S. J. 1975. Diversity through time. Nat. Hist. *84*(8):24–32.

GOULD, S. J. AND D. M. RAUP. 1975. The shape of evolution: a comparison of real and random clades. Geol. Soc. Am. Abstr. with Programs. 7: 1088.

GRANT, R. E. 1971. Brachiopods in the Permian reef environment of West Texas. Proc. North Am. Paleontol. Conv., Chicago, 1969. Part J:1444–1481.

GRAY, J. AND A. J. BOUCOT. 1971. Early Silurian spore tetrads from New York: earliest New World evidence for vascular plants? Science. *173*:918–921.

HÄNTZSCHEL, W. 1975. Trace fossils and problematica (Part W Miscellanea) Supplement 1 of Treatise on Invertebrate Paleontology. 269 pp. Univ. Kansas; Lawrence.

HUGHES, N. F. 1976. Palaeobiology of Angiosperm Origins. 242 pp. Cambridge Univ. Press; Cambridge.

HUTCHINSON, G. E. 1959. Homage to Santa Rosalia or why are there so many kinds of animals? Am. Nat. *93*:145–159.

JOHNSON, R. G. 1960. Models and methods for analysis of the mode of formation of fossil assemblages. Geol. Soc. Am. Bull. *71*:1075–1086.

JOHNSON, R. G. 1964. The community approach to paleoecology. pp. 107–134. In: Imbrie, J. and N. D. Newell, eds. Approaches to Paleoecology. 432 pp. John Wiley; New York.

LAPORTE, L. F. 1974. Reefs in Time and Space. 256 pp. Soc. Econ. Pal. and Min. Sp. Pub. No. 18.

LAWRENCE, D. 1968. Taphonomy and information losses in fossil communities. Geol. Soc. Am. Bull. *79*:1315–1330.

KLEMENT, K. W. AND D. F. TOOMEY. 1967. Role of the blue-green algae *Girvanella* in skeletal grain destruction and lime-mud formation in the Lower Ordovician of West Texas. J. Sediment. Pet. *37*: 1045–1051.

MACARTHUR, R. AND E. O. WILSON. 1963. An equilibrium theory of insular geography. Evolution. *17*:373–387.

MACARTHUR, R. H. AND E. O. WILSON. 1967. The Theory of Island Biogeography. 203 pp. Princeton Univ. Press; Princeton, N.J.

MCALESTER, A. LEE. 1962. Mode of preservation in Early Paleozoic pelecypods and its morphologic and ecologic significance. J. Paleontol. *36*:69–73.

MOORE, D. G. AND P. C. SCRUTON. 1957. Minor internal structures of some recent unconsolidated sediments. Bull. Am. Assoc. Petrol. Geol. *41*:2723–2751.

NIETZSCHE, F. W. 1885–1886. Beyond Good and Evil. *1*:part 18.

PAINE, R. T. 1966. Food web complexity and species diversity. Am. Nat. *100*:65–75.

PEET, R. K. 1974. The measurement of species diversity. pp. 285–307. In: Johnston, R. F., P. W. Frank, and C. D. Michener, eds. Annual Review of Ecology and Systematics. Vol. 5. Annu. Rev. Inc.; Palo Alto, California.

PRATT, L. M., T. L. PHILLIPS, AND J. M. DENNISON. 1975. Nematophytes from Early Silurian (Llandoverian) of Virginia provide oldest record of probable land plants in Americas. Geol. Soc. Am. Abstr. with Programs. 7:1233–1234.

RAUP, D. M. 1972. Taxonomic diversity during the Phanerozoic. Science. *177*:1065–1071.

RAUP, D. M. 1976a. Species diversity in the Phanerozoic: a tabulation. Paleobiology. *3*:279–288.

RAUP, D. M. 1976b. Species diversity in the Phanerozoic: an interpretation. Paleobiology. *3*:289–297.

Raup, D. M., S. J. Gould, T. J. M. Schopf, and D. Simberloff. 1973. Stochastic models of phylogeny and the evolution of diversity. J. Geol. *81*: 525–542.

Raup, D. M. and S. M. Stanley. 1971. Principles of paleontology. 388 pp. Freeman; San Francisco.

Rhoads, D. C. and J. W. Morse. 1971. Evolutionary and ecologic significance of oxygen-deficient marine basins. Lethaia. *4*:413–428.

Sanders, H. L. 1968. Marine benthic diversity: a comparative study. Am. Nat. *102*:243–282.

Sanders, H. L. 1969. Benthic marine diversity and the stability-time hypothesis. pp. 71–81. In: Woodwell, G. M. and H. H. Smith, eds. Diversity and Stability in Ecological Systems. Brookhaven Symp. in Biol. No. 22.

Schäfer, W. 1972. Ecology and Paleoecology of of Marine Environments. 568 pp. Univ. of Chicago Press; Chicago.

Schopf, T. J. M. 1974. Permo-Triassic extinctions: relation to seafloor spreading. J. Geol. 82:129–143.

Sepkoski, J. J., Jr. 1976a. A kinetic model of Phanerozoic diversity. Geol. Soc. Am. Abstr. with Programs. *8*:1098–1099.

Sepkoski, J. J., Jr. 1976b. Species diversity in the Phanerozoic: species-area effects. Paleobiology. *2*: 298–303.

Simberloff, D. 1972. Models in biogeography. pp. 160–191. In: Schopf, T. J. M., ed. Models in Paleobiology. 250 pp. Freeman, Cooper and Co.; San Francisco.

Simberloff, D. 1974. Permo-Triassic extinctions: effects of area on biotic equilibrium. J. Geol. *82*: 267–274.

Simberloff, D. 1976. Experimental zoogeography of islands: effects of island size. Ecology. *57*: 629–648.

Simberloff, D. and E. O. Wilson. 1969. Experimental zoogeography of islands. The colonization of empty islands. Ecology. *50*:278–296.

Simberloff, D. and E. O. Wilson. 1970. Experimental zoogeography of islands. A two-year record of colonization. Ecology. *51*:934–937.

Slobodkin, L. B. and H. L. Sanders. 1969. On the contribution of environmental predictability to species diversity. pp. 82–95. In: Woodwell, G. M. and H. H. Smith, eds. Diversity and Stability in Ecological Systems. Brookhaven Symp. in Biol. No. 22.

Sohl, N. F. 1969. The fossil record of shell boring by snails. Am. Zool. *9*:725–734.

Stanley, S. M. 1973. An ecological theory for the sudden origin of multicellular life in the Late Precambrian. Proc. Natl. Acad. Sci. *70*:1486–1489.

Stanley, S. M. 1976. Ideas on the timing of metazoan diversification. Paleobiology. *2*:209–219.

Tappan, H. 1968. Primary production, isotopes, extinctions and the atmosphere. Palaeogeog., Palaeoclimatol., Palaeoecol *4*:187–210.

Tappan, H. 1971. Microplankton, ecological succession and evolution. Proc. North Am. Paleontol. Conv. Chicago, 1969. Part H:1058–1103.

Thorsen, G. 1957. Bottom communities. pp. 461–534. In: Hedgepeth, J. W., ed. Treatise on Marine Ecology and Paleoecology. Vol. 1. Ecology. Geol. Soc. Am. Mem. 67.

Valentine, J. W. 1968. The evolution of ecological units above the population level. J. Paleontol. *42*:253–267.

Valentine, J. W. 1969. Patterns of taxonomic and ecological structure of the shelf benthos during Phanerozoic time. Palaeontology. *12*:684–709.

Valentine, J. W. 1970. How many marine invertebrate fossil species? A new approximation. J. Paleontol. *44*:410–415.

Valentine, J. W. 1973a. Phanerozoic taxonomic diversity: a test of alternate models. Science. *180*:1078–1079.

Valentine, J. W. 1973b. Evolutionary Paleoecology of the Marine Biosphere. 511 pp. Prentice-Hall; Englewood Cliffs, N.J.

Valentine, J. and E. M. Moores. 1972. Global tectonics and the fossil record. J. Geol. *80*:167–184.

Walker, K. R. and R. K. Bambach. 1971. The significance of fossil assemblages from fine-grained sediments: time-averaged communities. Geol. Soc. Am. Abstr. with Programs. *3*:783–784.

Whittaker, R. H. 1965. Dominance and diversity in land plant communities. Science. *147*:250–260.

Whittaker, R. H. 1970. Communities and ecosystems. 162 pp. Macmillan; New York.

Wilson, E. O. 1969. The species equilibrium. pp. 38–47. In: Woodwell, G. M. and H. H. Smith, eds. Diversity and Stability in Ecological Systems. Brookhaven Symp. in Biol. No. 22.

Ziegler, A. M., K. S. Hanson, M. E. Johnson, M. A. Kelly, C. R. Scotese, R. Van der Voo. 1977. Silurian continental distributions, paleogeography, climatology, and biogeography. Tectonophysics.

APPENDIX

Sources for the data presented in Figures 2 and 3.

LOWER PALEOZOIC

Bayer, T. N. 1967. Repetitive benthonic community in the Maquoketa Formation (Ordovician) of Minnesota. J. Paleontol. *41*:417–422.

Bretsky, P. W. 1970a. Upper Ordovician Ecology of the Central Applachians. 150 pp. 44 pl. Peabody Mus. Nat. Hist., Yale Univ. Bull. 34.

Bretsky, P. W. 1970b. Late Ordovician Benthic Marine Communities in North-Central New York. 34 pp. N. Y. State Mus. and Sci. Serv. Bull. 414.

Bretsky, P. W. and S. S. Bretsky. 1975. Succession and repetition of Late Ordovician fossil assemblages from the Nicolet River Valley, Quebec. Paleobiology. *1*:225–237.

Cisne, J. L. 1973. Beecher's Trilobite Bed Revisited: Ecology of an Ordovician Deepwater Fauna. 25 pp. Peabody Mus. Nat. Hist., Yale Univ. Postilla No. 160.

Dixon, J. 1975. Ordovician and Silurian fossils from the Lang River and Allen Bay Formations of Prince of Wales and Somerset Islands, Northwest Territories. Bull. Can. Petrol. Geol. 32:172–184.

Fritz, W. H. 1971. Geological setting of the Burgess

Shale. Proc. North Am. Paleontol. Conv., Chicago, 1969. Part I, pp. 1155–1170.

GLAESSNER, M. F. AND M. WADE. 1966. The Late Precambrian fossils from Ediacara, South Australia. Palaeontology. 9:599–628.

LOCHMAN, C. AND C.-H. HU. 1962. Upper Cambrian faunas from the northwest Wind River Mountains, Wyoming. Part III. J. Paleontol. 36:1–28.

NELSON, S. J. 1963. Ordovician Paleontology of the Northern Hudson Bay Lowland. 152 pp. Geol. Soc. Am. Mem. 90.

PESTANA, H. R. 1960. Fossils from the Johnson Spring Formation, Middle Ordovician, Independence Quadrangle, California. J. Paleontology 34:862–873.

PICKERILL, R. K. 1973. *Lingulasma tenuigranulata*—palaeoecology of large Ordovician linguloid-trilobite community. Palaeogeogr., Palaeoclimatol., Palaeoecol. 13:143–156.

PLANTS, F. 1977. Communities in the Martinsburg Formation (Upper Ordovician), Catawba Mountain, Virginia. Ms. Thesis in preparation. Virginia Polytechnic Inst. and State Univ., Blacksburg, Va.

ROBISON, R. A. 1964. Late Middle Cambrian faunas from western Utah. J. Paleontol. 38:510–566.

STEELE, H. M., AND G. W. SINCLAIR. 1971. A Middle Ordovician Fauna from Braeside, Ottawa Valley, Ontario. 97 pp. Geol. Surv Can. Bull. 211.

TITIES, R. AND B. CAMERON. 1976. Fossil communities of the Lower Trenton Group (Middle Ordovician) of central and northwestern New York State. J. Paleontol. 50:1209–1225.

WALKER, K. R. 1972. Community Ecology of the Middle Ordovician Black River Group of New York State. Geol. Soc. Am. Bull. 83:2499–2524.

WALKER, K. R. AND L. P. ALBERSTADT. 1975. Ecological succession as an aspect of structure in fossil communities. Paleobiology. 1:238–257.

WILLOUGHBY, R. 1975. Unpublished data—M.S. thesis in preparation at Virginia Polytechnic Inst. and State Univ., Blacksburg, Va.

MIDDLE PALEOZOIC

AGER, D. V. 1963. Principles of Paleocology. 371 pp. McGraw-Hill; New York.

BAMBACH, R. K. 1969. Bivalvia of the Siluro-Devonian Arisaig Group, Nova Scotia. 376 pp. 85 pl. Unpubl. Ph.D. diss. Yale Univ., New Haven.

BRAY, R. B. 1972. The paleoecology of some Ludlowville brachiopod clusters (Middle Devonian), Erie County, New York. 24 Int. Geol. Cong., Sect. 7. pp. 66–73.

CALEF, C. E. AND N. J. HANCOCK. 1974. Wenlock and Ludlow marine communities in Wales and the Welsh borderland. Palaeontology. 17:779–810.

ERDTMANN, B.-D. AND D. R. PREZBINDOWSKI. 1973. Niagaran (Middle Silurian) interreef fossil burial environments in Indiana. N. Jb. Geol. Paläont. Mh. 1973. pp. 624–640.

FISHER, J. N. 1970. The paleoecology of the Hamilton Group (Middle Devonian) in southeastern New York State. 145 pp. Unpubl. Honors thesis (summa cum laude). Smith Coll. Northampton, Mass.

HURST, J. M. 1975. Wenlock carbonate, level bottom, brachiopod-dominated communities from Wales and the Welsh borderland. Palaeogeog., Palaeoclimatol., Palaeoecol. 17:227–255.

LINSLEY, R. M. 1968. Gastropods of the Middle Devonian Anderdon Limestone. Bull. Am. Paleontol. 54:333–465.

MAZZULLO, S. J. 1973. Deltaic depositional environments in the Hamilton Group (Middle Devonian), southeastern New York State. J. Sed. Petrol. 43: 1061–1071.

MCLEARN, F. H. 1924. Palaeontology of the Silurian rocks of Arisaig, Nova Scotia. 180 pp. 30 pl. Geol. Surv. Canada Mem. 137.

VOPNI, L. K. AND J. F. LERBEKMO. 1972. The Horn Plateau Formation: a Middle Devonian coral reef, Northwest Territories, Canada. Bull. Can. Pet. Geol. 20:498–548.

WALKER, K. R. AND L. P ALBERSTADT. 1975. Ecological succession as an aspect of structure in fossil communities. Paleobiology. 1:238–257.

WALLACE, P. 1969. Specific frequency and environmental indicators in two horizons of the Calcaire de Ferques (Upper Devonian), northern France. Palaeontology. 12:366–381.

WILLIAMS, H. S. 1913. Recurrent *Tropidoleptus* zones of the Upper Devonian in New York. 103 pp. U.S. Geol. Surv. Prof. Paper 79.

ZIEGLER, A. M., L. R. M. COCKS, AND R. K. BAMBACH. 1968. The composition and structure of Lower Silurian marine communities. Lethaia. 1:1–27.

UPPER PALEOZOIC

BIRD, S. O. 1968. A pelecypod fauna from the Gaptank Formation (Pennsylvanian), West Texas. Bull. Am. Paleontol. 54:111–185.

CRAIG, G. Y. 1954. The palaeoecology of the Top Hosie Shale (Lower Carboniferous) at a locality near Kilsyth. Q. J. Geol. Soc. London. 110:103–119.

DONAHUE, I., H. B. ROLLINS, AND G. D. SHAAK. 1972. A symmetrical community succession in a transgressive-regressive sequence. 24th Int. Geol. Cong., Sect. 7. pp. 74–81.

EDWARDS, W. AND C. J. STUBBLEFIELD. 1948. Marine bands and other faunal marker—horizons in relation to sedimentary cycles of the middle coal measures of Nottinghamshire and Derbyshire. Q. J. Geol. Soc. London. 103:209–256.

FERGUSON, L. 1962. The paleoecology of a Lower carboniferous marine transgression. J. Paleontol. 36:1090–1107.

JOHNSON, R. G. 1962. Interspecific associations in Pennsylvanian fossil assemblages. J. Geol. 70:32–55.

LANE, N. G. 1973. Paleontology and Paleoecology of the Crawfordsville Fossil Site (Upper Osagian: Indiana). 141 pp. 20 pl. Univ. Cal. Publ. Geol. Sci. 99. Univ. Cal. Press; Berkeley, Cal.

MCCRONE, A. W. 1963. Paleoecology and biostratigraphy of the Red Eagle Cyclothem (Lower Permian) in Kansas. 114 pp. Geol. Surv. Kansas Bull. 164.

MUDGE, M. R. AND E. P. YOCHELSON. 1962. Stratigraphy and paleontology of the Uppermost

Pennsylvanian and Lowermost Permian rocks in Kansas. 213 pp. 17 pl. U.S. Geol. Surv. Prof. Paper 323.

NICOL, D. 1965. An ecological analysis of four Permian molluscan faunas. Nautilus. *78*:86–95.

RUNNEGAR, B. AND N. D. NEWELL. 1971. Caspian-like relict molluscan fauna in the South American Permian. Bull. Am. Mus. Nat. Hist. *146*:1–66.

SHAAK, G. D. 1975. Diversity and community structure of the Brush Creek marine interval (Conemaugh Group, Upper Pennsylvanian) in the Appalachian Basin of Western Pennsylvania. Bull. Fla. State Mus. Biol. Sci. *19*:69–133.

WATKINS, R. 1974. Carboniferous brachiopods from northern California. J. Paleontol. *48*:304–325.

MESOZOIC

DUFF, K. L. 1975. Palaeoecology of a bituminous shale—the Lower Oxford Clay of central England. Palaeontology. *18*:443–482.

HALLAM, A. 1960. A sedimentary and faunal study of the blue lias of Dorset and Glamorgan. Phil. Trans. R. Soc. London. Series B. *243*:1–44.

HALLAM, A. 1971. Facies analysis of the Lias in west central Portugal. N. Jb. Geol. Paläont. Abh. *139*:226–265.

HUDSON, J. D. 1963. The ecology and stratigraphical distribution of the invertebrate fauna of the Great Estuarine Series. Palaeontology. *6*:327–348.

KAUFFMAN, E. G. 1973. A brackish water biota from the Upper Cretaceous Harebell Formation of northwestern Wyoming. J. Paleontol. *47*:436–446.

MCKERROW, W. S., R. T. JOHNSON, AND M. E. JAKOBSON. 1969. Palaeoecological studies in the Great Oolite at Kirtlington, Oxfordshire. Palaeontology. *12*:56–83.

PERKINS, B. F. 1960. Biostratigraphic studies in the Comanche (Cretaceous) Series of northern Mexico and Texas. Geol. Soc. Amer. Mem. 83. 138 pp.

RHOADS, D. C., I. G. SPEDEN, AND K. M. WAAGE. 1972. Trophic group analysis of Upper Cretaceous (Maestrichtian) bivalve assemblages from South Dakota. Am. Assoc. Pet. Geol. Bull. *56*:1100–1113.

SCOTT, R. W. 1970. Paleoecology and paleontology of the Lower Cretaceous Kiowa Formation, Kansas. 94 pp. Univ. Kansas Paleontological Cont. Article 52.

SCOTT, R. W. 1974. Bay and shoreface benthic communities in the Lower Cretaceous. Lethaia. 7: 315–330.

STEPHENSON, L. W. AND W. H. MONROE. 1940. The Upper Cretaceous Deposits. 296 pp. Miss. State Geol. Surv. Bull. 40.

CENOZOIC

DALEY, B. 1972. Macroinvertebrate assemblages from the Bembridge Marls (Oligocene) of the Isle of Wight, England, and their environmental significance. Palaeogeog., Palaeoclimatol, Palaeoecol. *11*: 11–32.

DODD, J. R. AND R. J. STANTON, JR. 1975. Paleosalinities within a Pliocene bay, Kettleman Hills California: a study of the resolving power of isotopic and faunal techniques. Geol. Soc. Am. Bull. *86*:51–64.

DUBAR, J. R. 1958. Stratigraphy and paleontology of the Late Neogene strata of the Caloosahatchee River area of Southern Florida. 267 pp. Fla. Geol. Surv. Bull. No. 40.

DUBAR, J. R. AND D. W. BEARDSLEY. 1961. Paleoecology of the Choctawatchee deposits (Late Miocene) at Alum Bluff, Florida. Southeast. Geol. *2*:155–189.

DUBAR, J. R. AND J. R. CHAPLIN. 1963. Paleoecology of the Pamlico Formation (Late pleistocene); Nixonville Quadrangle, Horry County, South Carolina. Southeast. Geol. *4*:127–165.

DUBAR, J. R. AND D. S. TAYLOR. 1962. Paleoecology of the Choctawhatchee deposits, Jackson Bluff, Florida. Gulf Coast Assoc. Geol. Soc. Trans. *12*: 349–376.

CERNANT. R. E. 1970. Paleoecology of the Choptank Formation (Miocene) of Maryland and Virginia. 63 pp. Md. Geol. Surv. Rept. Inv. No. 12.

GOLDRING, W. 1922. The Champlain Sea. N.Y. State Mus. Bull. *239–240*:153–194.

JUNG, P. 1969. Miocene and Pliocene mollusks from Trinidad. Bull. Am. Paleontol. *55*:293–657.

JUNG, P. 1971. Fossil mollusks from Carriacou, West Indies. Bull. Am. Paleontol. *61*:147–262.

KERN, J. P. 1971. Paleoenvironmental analysis of a Late Pleistocene estuary in southern California. J. Paleontol. *45*:810–823.

LIPPS, J., J. W. VALENTINE, AND E. MITCHELL. 1968. Pleistocene paleoecology and biostratigraphy, Santa Barbara Island, California. J. Paleontol. *42*:291–307.

SMITH, A. B. 1959. Paleoecology of a molluscan fauna from the Trent Formation. J. Paleontol. *33*: 855–871.

STUMP, T. E. 1975. Pleistocene molluscan paleoecology and community structure of the Puerto Libertad region, Sonora, Mexico. Palaeogeog., Palaeoclimatol., Palaeoecol. *17*:177–226.

WEISBROD, N. E. 1969. Some Late Cenozoic Echinoidea from Cabo Blanco, Venezuela. Bull. Am. Paleontol. *56*:277–371.

PAPER 21

Seed Size, Dispersal Syndromes, and the Rise of the Angiosperms: Evidence and Hypothesis (1984)

B. H. Tiffney

Commentary

HALLIE J. SIMS

The role of the seed in plant ecology has fascinated scientists and laypeople alike since Charles Darwin reported that a cup of seemingly sterile soil placed on his windowsill had sprouted seedlings without any assistance on his part. (On hearing of this, Henry David Thoreau repeated the endeavor and was inspired by the metaphysical implications of a largely invisible seed bank to write his last manuscript, eventually published as "Faith in a Seed.") Subsequent workers have marveled at the ability of some propagules to remain dormant for decades or even centuries, as well as at the intricate architectures that facilitate seed dispersal by wind, water, and animal vectors. Studies of seed ecology flourished during the twentieth century in both the field and the lab, greatly expanding our understanding of development, genetics, and phylogeny. However, although seeds have an excellent fossil record, discussions of seed evolution in deep time remained largely anecdotal until Bruce Tiffney published his thoughtful, semiquantitative paper in 1984.

In order to explore the roles of dispersal syndrome and environmental variation in the evolution of angiosperms, Tiffney assembled a data set of Cretaceous (K) and Tertiary (T) fossil seed assemblages. Perhaps the most thought-provoking aspect of his paper was the resulting plot (fig. 2 in his paper), which suggested a striking increase in within-community mean seed size at or around the K-T boundary. Tiffney's discussion of the relative impacts of the extinction of nonavian dinosaurs, the environmental perturbations associated with the K-T mass extinction, and the radiation of mammals and birds on the evolution of angiosperm seed ecology is exquisitely holistic. Although data sets of greater magnitude and precision have been gathered in the years since Tiffney was published, his results are supported more often than not and his cogitations continue to provide fodder for present and future macroecologists.

From *Annals of the Missouri Botanical Garden* 71:551–76.

SEED SIZE, DISPERSAL SYNDROMES, AND THE RISE OF THE ANGIOSPERMS: EVIDENCE AND HYPOTHESIS[1]

Bruce H. Tiffney[2]

The seeds and fruits of angiosperms serve the functions of nurturing, protecting, and dispersing the embryonic plant, and thus form an evolutionarily sensitive portion of the life cycle of the whole organism. Two of these functions also enhance the probability of fossilization of these disseminules. Protection is often achieved through lignification of the fruit or seed wall, predisposing the organ to preservation. Dispersal increases the probability of a propagule arriving at a fossilizing environment. It is, therefore, not surprising that fruits and seeds are a major source of information on the fossil record of the angiosperms, particularly from the Tertiary (Tiffney, 1977a).

This information has generally appeared in descriptive reports of fossil floras and their composition [e.g., the Eocene London Clay Flora (Reid & Chandler, 1933) and the middle Tertiary floras of central Europe (Mai, 1964)]. These floristic studies have formed the basis for synthetic undertakings such as the elucidation of biogeographic patterns (Wood, 1972; Wolfe, 1975; Tiffney, 1980; Mazer & Tiffney, 1982) and the inference of climatic history (Leopold, 1967; Mai, 1970; Friis, 1975; Gregor, 1980a; Collinson et al., 1981). Consideration of evolutionary questions has been largely restricted to the demonstration of species sequences within single genera (e.g., *Stratiotes* L., Chandler, 1923; *Aldrovanda* L., Dorofeev, 1968; *Toddalia* Juss., Gregor, 1979) and families (e.g., Juglandaceae, Manchester & Dilcher, 1981, unpubl. data). However, fruits and seeds additionally offer an excellent starting point for paleobiological inquiry based on modern ecological studies. Of particular note are two considerations: (1) the relation of seed size to the habit and habitat of the parent plant, and (2) dispersal syndromes.

Harper et al. (1970) [after Salisbury (1942)] have demonstrated a strong correlation between seed weight and the stature and successional status of the parent plant. Herbaceous plants, and those of early successional stages, tend to have small propagules, while dominant forest trees and plants of late successional status tend to have large propagules. Some shrubs and "weedy" trees tend to have propagules of intermediate sizes. The mode of dispersal of a living plant may often be inferred from the morphology of the fruit or seed, together with the mode of its presentation to the dispersal agent (Ridley, 1930; van der Pijl, 1969). While fossilization precludes knowledge of the mode of presentation, many of the morphological characters of the fossils permit inference of the mode of dispersal in at least a broad sense. These two features, propagule size and dispersal, have been examined only in modern plants and generally have been treated separately. In the present paper I extend observations on propagule size and dispersal type through the fossil record and propose that these (1) have been related throughout the history of the angiosperms and (2) underwent an intensive period of change in the latest Cretaceous and early Tertiary. My emphasis will be on propagule size; the subject of dispersal syndromes through time warrants a separate study and is not treated in detail here.

Methods

In the following discussion, I will use the general term "diaspore" to indicate the reproductive unit that is dispersed or sown. Thus, in the case of a capsule, which releases its contents, the term will apply to the morphological seed. In the case of a drupe or berry, the term will encompass fruit tissue. However, in cases in which reference is

[1] I thank Leo J. Hickey (Yale University) for his devil's advocacy, which has clarified my thinking; Karl J. Niklas (Cornell University) for his helpful suggestions; Paul Olsen (Yale University) for advice on reptiles; Robin Gowen Tiffney for drafting the figures, and Leo J. Hickey, Karl J. Niklas, Daniel Axelrod, and Maureen Stanton (University of California, Davis), Steven N. Handel (Yale University), Steven Manchester and David L. Dilcher (Indiana University), and Else Marie Friis (Aarhus University) for a critical reading of the manuscript. Research partially supported by NSF grant DEB 79-05082.

[2] Peabody Museum and Department of Biology, Yale University, P.O. Box 6666, New Haven, Connecticut 06511.

Ann. Missouri Bot. Gard. 71: 551–576. 1984.

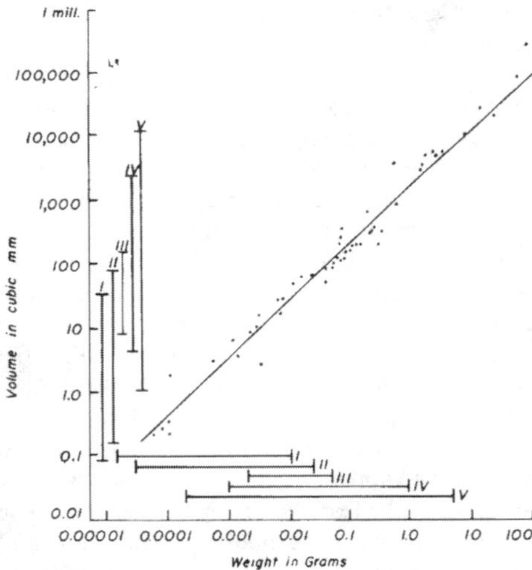

FIGURE 1. Log-log plot of weight versus volume for the propagules of 52 modern angiosperms. The five categories of seed weight (after Harper et al., 1970) on the horizontal axis (weight) are transposed to the vertical axis (volume) by use of the regression line. Categories of plants: I. open habitat; II. woodland margin; III. woodland ground; IV. woodland shrubs; V. woodland trees.

to a specific morphological structure, and particularly when discussing the nutrient reserves of a dispersed seed, I will use the appropriate morphological term.

Salisbury (1942), Harper et al. (1970), and other workers have quantified diaspore size using weight. This approach cannot be applied to a comparative study of fossil seeds because they may be preserved as original organic matter or by replacement with minerals; while lignin has a specific gravity of about 1.2, silicon dioxide has one of 2.65, and pyrite of 5.01. Linear measurements (e.g., length) are also inappropriate because they do not account for variation in three-dimensional shape. I have, therefore, chosen to estimate size from volume. This also permits the calculation of diaspore size from published reports as well as from actual specimens. The use of volume involves two assumptions: (1) that weight and volume are related in fruits and seeds, and (2) that the volumes may be calculated in an accurate and repeatable manner. To test these assumptions, the diaspores of 52 modern species were weighed to the nearest one thousandth of a gram and measured to the nearest tenth of a millimeter. The results are plotted in Figure 1. A regression of weight versus volume yields $r = 0.928$, indicating a significant correlation between the two. This correlation further suggests that the measurement of volume was sufficiently accurate for the purposes of this study.

Diaspore volumes of seven Cretaceous and 20 representative Tertiary and Quaternary floras were then calculated from specimens and the literature (Table 2). In order to obtain accurate identification and measurement of the individual diaspores, only floras with three-dimensional, well-preserved fossils were used. Volumes were obtained only for those fossils that represented diaspores as defined above. Calculations were based on average width, length, and thickness of the specimens as described. In cases where one or more dimensions were not cited, the missing value(s) was estimated from illustrations. In cases of extreme compaction, thickness was assumed to be a value equal to 0.66 × the width. This value was arrived at empirically, and is an outgrowth of the $\frac{2}{3}$ power law governing the relation of surface area to volume. The volume of spherical diaspores was estimated at $\frac{4}{3}\pi r^3$. On those occasions where spines or other projections seriously hampered accurate measurement, estimates were made of the volume. The stratigraphic ages of the deposits are those provided by the authors, with modification in light of recent data (Gregor, 1980b) as appropriate. The conversion of stratigraphic age to absolute age is made from van Eysinga (1975) and Gregor (1980b). Assignment of absolute age is necessary to permit calculation of regression values and aids in the relative location of the floras. However, the ages are *approximate* and should be recognized as such. Regression values were calculated using a pre-programmed Texas Instruments TI-55 calculator. Readers are cautioned that the use of numbers with regard to these fossils may convey a false sense of precision. While the numbers used are certainly valid within the relative framework of the present discussion, they often involve subjective judgements and should be regarded as educated approximations, not as absolutes.

DATA

CRETACEOUS FRUITS AND SEEDS

The consideration of Cretaceous fruits and seeds falls into two sections, since floras of three-dimensional fruits and seeds have, to date, only been found in the Late Cretaceous. Before this time, the record of angiosperm reproductive structures involves isolated fossils.

A summary of the better documented fruits and seeds of Early Cretaceous and Cenomanian age (115–95 million years ago, henceforth Ma) is presented in Table 1. The majority of these are preserved as casts or impressions; of the compressions, only a few can be or have been studied in anatomical detail. As a result, many of the earliest reported forms cannot be clearly assigned to the angiosperms and may well be gymnospermous. This has been suggested in the case of *Onoana californica* Chandl. & Axelr. (Chandler & Axelrod, 1961), and by inference *Onoana nicanica* Krass. (Krassilov, 1967), by Wolfe et al. (1975). The same arguments apply to several other Early Cretaceous endocarp-like forms including "*Carpolithus*" (Chandler, 1958), *Nyssidium* Saml. (Samylina, 1961), *Prototrapa* Vas. (Vasil'yev, 1967), *Araliaecarpum* Saml. and *Caricopsis* Saml. (Samylina, 1960) and *Knella* Saml. (Samylina, 1968). Retallack and Dilcher (1981) have similarly viewed many of the above reports as potentially non-angiospermous. These reports will not be considered further.

The remaining reports tabulated in Table 1 fall into two categories. The first includes several structures reported by Fontaine (1889) under the genus "*Carpolithus*" and interpreted by Dilcher (1979) and Retallack and Dilcher (1981) to represent multifollicles. I have not personally examined these specimens and accept the judgement of these authors. The second category includes well-preserved fruits, often containing seeds. These involve clearly angiospermous material such as *Caspiocarpus paniculiger* Vach. & Krass. (Vachrameev & Krassilov, 1979), *Ranunculaecarpus quinquiecarpellatus* Saml. (Samylina, 1960), *Carpites liriophylli* Lesq. (Dilcher et al., 1976) and a host of forms from the Dakota Formation of central North America.

The majority of these Early Cretaceous angiosperm fruits are small, individual carpels ranging from 1 to 15 mm in length and from 0.5 to 8 mm in width, or are capsules of from 10 to 12 mm in diameter. In the five cases where seeds are known from these fruits, the seeds are small, ranging from 0.2 mm^3 (*Caspiocarpus paniculiger*) to approximately 7.5 mm^3 (estimated for the "unpublished five-carpellate fruit" from the Dakota Group; Dilcher, 1979). The one exception to this tendency to small size is *Carpolithus curvatus* Font., which is a carpel about 40 mm long and 15 mm wide. This specimen is not well preserved, and there is no indication as to the size of the included seeds.

The most common fruit morphology is a dehiscent follicle, borne on a central axis, although capsules are also frequently observed. This is in keeping with the classic hypothesis that the conduplicate carpel, and dispersal by morphological seeds, are the primitive conditions in the group (Cronquist, 1968; Takhtajan, 1969). The one potential exception to this pattern is the report of a fleshy fruit from the Cenomanian (98 Ma; Dilcher, 1979). However, the status of this fossil is not clear because Retallack and Dilcher (1981: 49) imply that no fleshy fruits are known from Cenomanian and older sediments. The reported seeds are all apparently thin-walled and without any distinctive features related to dispersal. The capsular-follicular morphology of the fruits and the small, unspecialized, nature of the seeds are characters indicative of a general adaptation to abiotic dispersal mechanisms, a conclusion also reached by Retallack and Dilcher (1981).

Individual fruits are also reported in the Late Cretaceous, often as constituents of compression or impression leaf floras, and several reports exist of isolated occurrences of seeds or seed-like objects (Miner, 1935; Schemel, 1950; Hall, 1963, 1967; Binda, 1968; Colin, 1973; Knobloch, 1981). However, of greater importance to the present work are several fairly diverse (10–50 species) floras of three dimensionally preserved fruits and seeds from fluvial and lacustrine sediments. The most important of these are listed in rows 1–7 of Table 2; several others of lower diversity have not been included but are of a similar nature (Knobloch, 1971, 1977). Although some of these seeds have been assigned to extant families (Caryophyllaceae, Cyperaceae, Menispermaceae, Myricaceae, Theaceae, Urticaceae: Knobloch, 1977; Jung et al., 1978) and orders (Juglandales: Friis, 1984), the majority have been placed in the organ genus *Microcarpolithes* Vangerow erected for seeds or one-seeded fruits of angiospermous affinities. [This genus requires renaming. The type species, *M. hexagonalis* Vangerow (Hall, 1963) has been shown to be an insect coprolite (Knobloch, 1977).]

The average size of the seeds in these floras is approximately 1.7 mm^3 (see Table 2, column $\bar{x}$ and Fig. 2, floras 1–7). This small size does not appear to be a function of mechanical sorting, or of ecological separation, for a variety of reasons. The Santonian-Campanian floras (about 77 Ma) reported by Friis (1984) from Åsen, Sweden, and

TABLE 1. Summary of individually reported fruits and seeds of presumed angiospermous affinities from Early Cretaceous and Cenomanian localities. Judgement of angiospermous affinities in the "comment" column is by the present author unless otherwise noted. "No distinguishing angiospermous features" only implies that the specimen is not *clearly* angiospermous.

Age[a]	Name	Locality	Type	Size	Reference	Comment
Tithonian-Berrasian (134)	"*Tyrmocarpus*"	Tyrma R., Siberia	"capsule-like fruit"	ca. 6 mm diam.	Krassilov (1973) Hughes (1976)	No distinguishing angiospermous features.
Valangian (127)	*Carpolithus*	Vaucluse, France	unclear	22 mm × 12 mm	Chandler (1958)	The original (now lost) was a sandstone cast with no distinctively angiospermous features.
Barremian (117)[b]	*Nyssidium orientale* Sam.	Siberia, U.S.S.R.	unclear—inferred as endocarp	10 mm × 6 mm	Samylina (1961)	No distinguishing angiospermous features. Illustrated specimens show little relation to *Nyssa*. Found in a totally gymnospermous flora.[a]
Barremian (117)[b]	*Nyssidium* Sp.	Siberia, U.S.S.R.	unclear—inferred as endocarp	10 mm × 5 mm	Samylina (1961)	
Barremian (117)[b]	*Onoana californica* Chand. & Axelr.	California, U.S.A.	unclear—inferred as endocarp	20 mm × 15 mm	Chandler and Axelrod (1961)	No distinguishing angiospermous features (Wolfe et al., 1975).[c]
Late Barremian-Early Aptian (112)[d]	*Carpolithus geminatus* Font.	Virginia, U.S.A.	multifollicle	6 mm × 9 mm[e]	Fontaine (1889)	Footnote f.
Late Barremian-Early Aptian (112)[d]	*Carpolithus sessilis* Font.	Virginia, U.S.A.	multifollicle	12 mm × 4 mm	Fontaine (1889)	Footnote f.
Late Barremian-Early Aptian (112)[d]	*Carpolithus virginiensis* Font.	Virginia, U.S.A.	multifollicle	7–10 mm × 4–6.5 mm	Fontaine (1889)	Footnote f.
Aptian (110)	*Onoana nicanica* Krass.	Primorye, U.S.S.R.	unclear—inferred as endocarp	8–10 mm × 5.5–7.5 mm	Krassilov (1967)	No distinguishing angiospermous features (cf. Wolfe et al., 1975). Preservation poor.[c]
Aptian-Albian (107.5)	*Prototrapa douglasi* Vass.	Victoria, Australia	endocarp	1–3 mm × 0.5–1.5 mm	Vasil'yev (1967)	Resemblance to *Trapa* is superficial; angiospermous affinities unclear. Impression.

TABLE 1. (Continued).

Age[a]	Name	Locality	Type	Size	Reference	Comment
Aptian-Albian (107.5)	*Prototrapa praepomelii* Vass.	Victoria, Australia	endocarp	2 mm × 1 mm	Vasil'yev (1967)	Resemblance to *Trapa* is superficial; angiospermous affinities unclear. Impression.
Aptian-Albian (107.5)	*Prototrapa tenuirostrata* Vass.	Victoria, Australia	endocarp	1.2 mm × 0.5 mm	Vasil'yev (1967)	Resemblance to *Trapa* is superficial; angiospermous affinities unclear. Impression.
Albian (105)[b]	*Araliaecarpum kolymensis* Sam.	Siberia, U.S.S.R.	unclear; possibly a winged endocarp?	6 mm × 6 mm	Samylina (1960)	Affinities unclear to present author.
Albian (105)[b]	*Caricopsis compacta* Sam.	Siberia, U.S.S.R.	unclear	3–5 mm × 2 mm	Samylina (1960)	Affinities unclear to present author.
Albian (105)[d]	*Carpolithus conjugatus* Font.	Virginia, U.S.A.	multifollicle	7.5 mm × 3.6 mm[e]	Fontaine (1889)	Footnote f.
Albian (105)[d]	*Carpolithus curvatus* Font.	Virginia, U.S.A.	multifollicle	42 mm × 14.2 mm[e]	Fontaine (1889)	Footnote f.
Albian (105)[d]	*Carpolithus fasiculatus* Font.	Virginia, U.S.A.	multifollicle	15 mm × 8 mm[e]	Fontaine (1889)	Footnote f.
Albian (105)[d]	*Carpolithus ternatus* Font.	Virginia, U.S.A.	multifollicle	8–11 mm × 4–7 mm[e]	Fontaine (1889)	Footnote f.
Albian (105)[g]	*Carpolithus karatscheensis* Vachr.	Kazakhstan, U.S.S.R.	multifollicle	—	Vachrameev (1952)	Original article not seen; data from Retallack and Dilcher (1981).
Albian (105)	*Caspiocarpus paniculiger* Vachr. & Krass.	Kazakhstan, U.S.S.R.	dehiscent follicle	Fruit—1 mm × 0.5 mm Seed—0.8 × 0.5 mm	Vachrameev and Krassilov (1979)	Clear angiospermous affinities.
Albian (105)[b]	*Knelia harrisiana* Sam.	Kolyma R., U.S.S.R.	unclear—endocarp?	16 mm × 5 mm	Samylina (1968)	Poor preservation, angiospermous affinities not demonstrated. See also Hughes (1976).[c]

TABLE 1. (Continued).

Age[a]	Name	Locality	Type	Size	Reference	Comment
Albian (105)[b]	*Ranunculaecarpus quinquiecarpellatus* Sam.	Kolyma R., U.S.S.R.	Dehiscent follicle	Fruit—10 mm × 5 mm × 2 mm Seed—1.5 mm × 0.6 mm	Samylina (1960)	Angiospermous affinities fairly certain.
Cenomanian (97.5)	*Carpites liriophylli* Lesq.	Dakota Group, U.S.A.	Dehiscent follicle	Fruit—15–20 mm × 3–4 mm Seed—1.4 mm × 0.6 mm	Dilcher et al. (1976); Dilcher (1979)	Clear angiospermous affinities.
Cenomanian (97.5)	*Carpites tiliaceus* Lesq.	Dakota Group, U.S.A.	Five-valved capsule	ca. 10 mm diam.[h]	Lesquereux (1892); Dilcher (1979)	Clear angiospermous affinities.
Cenomanian (97.5)	*Laurus macrocarpa* Lesq.	Dakota Group, U.S.A.	Syncarpous fruit	12 mm × 8.3 mm[h]	Lesquereux (1874); Dilcher (1979)	Clear angiospermous affinities.
Cenomanian (97.5)	*Platanus primaeva* Lesq.	Dakota Group, U.S.A.	Spherical mass of individual fruits	head ca. 3–4 mm diam.[h]	Lesquereux (1892); Dilcher (1979)	Clear angiospermous affinities.
Cenomanian (97.5)	"*Salix*"	Dakota Group, U.S.A.	Dehiscent follicle	3.3 mm × 1.5 mm[h]	Lesquereux (1892); Dilcher (1979)	Clear angiospermous affinities.
Cenomanian (97.5)	un-named	Dakota Group, U.S.A.	Fleshy fruit	5–6 mm diam.[h]	Dilcher (1979)	Dilcher (1979) interprets as angiospermous.

TABLE 1. (Continued).

Age[a]	Name	Locality	Type	Size	Reference	Comment
Cenomanian (97.5)	"unpublished 5-carpellate fruit"[i]	Dakota Group, U.S.A.	Five-valved capsule	Fruit—10 mm diam. Seeds—3.3 mm × 2.5 mm × 1.2 mm[i]	Dilcher (1979)	Clear angiospermous affinities.
Cenomanian (97.5)	"un-named follicular axis associated with *Magnoliaephyllum*"	Dakota Group, U.S.A.	Dehiscent follicle	Fruit—2.7 mm × 2.0 mm Seed—0.8 mm × 0.5 mm[h]	Dilcher (1979)	Clear angiospermous affinities.
Cenomanian (97.5)	"un-named globose heads"	Dakota Group, U.S.A.	Globose mass of individual fruits	Head ca. 10 mm × 7.5 mm diam.[h]	Dilcher (1979)	Fairly clear angiospermous affinities.
Cenomanian (94)	Platanaceae	Dakota Group, U.S.A.	Spherical mass of individual fruits	Not given	Schwarzwalder and Dilcher (1981)	Infructescences and leaves demonstrably related to Platanaceae.

[a] Absolute ages after van Eysinga (1975).
[b] Age after Hughes (1976).
[c] Retallack and Dilcher (1981) consider the angiospermous affinities of this species to be unproven.
[d] Age after Doyle and Hickey (1976).
[e] Measurements are approximate; made from Fontaine's (1889) illustrations.
[f] The identification of this species as a multifollicle is provided by Retallack and Dilcher (1981), although Fontaine (1889) placed it as a seed of a gymnosperm. I have not viewed this material.
[g] Age after Vachrameev and Krassilov (1979).
[h] Measurements are approximate; made from Dilcher's (1979) illustrations.
[i] This unpublished specimen may be the same as *Carpites tiliaceus* Lesq. (Dilcher, 1979). Since the measurements given for both seed and fruit are made from Dilcher's illustrations, they must be viewed as approximate. The measurements for the seed are taken from the presumed seed-cavity cast.

TABLE 2. Data on individual Late Cretaceous and Tertiary fruit and seed floras. N = total number of seeds measured in flora, $\bar{x}$ = average value, s.d. = standard deviation, CV = covariance.

Stratigraphic Age	Estimated Numeric Age[a]	Locality	Reference	N	$\bar{x}$ (mm^3)	s.d.	Co. var. (s.d./$\bar{x}$)	Largest (mm^3)	Smallest (mm^3)
1. Santonian-Campanian	77	Åsen, Sweden	Friis (1984)	>50	—	—	—	27[b]	0.02[b]
2. Santonian-Campanian	77	Gay Head, Massachusetts, U.S.A.	Tiffney (unpubl. data)	41	5.7	10.5	1.82	55	0.03
3. Santonian-lower Campanian	77	Staré Hamry 1, Czechoslovakia	Knobloch (1977)	19	1.4	0.7	0.5	3	0.41
4. Senonian	75[c]	Aachen, West Germany	Vangerow (1954)	11	0.2	0.17	0.83	0.5	0.03
5. Senonian	75[c]	Petrovice, Czechoslovakia	Knobloch (1964)	11	0.3	0.2	0.625	0.73	0.11
6. Campanian-Early Paleocene	73–63?[d]	Horní, Bečva, Czechoslovakia	Knobloch (1977)	29	1.5	1.4	0.91	6.4	0.15
7. Late Senonian	69	Kössen, Austria	Knobloch (1975)	9	1.3	0.9	0.69	3	0.15
8. Late Senonian	69	Kössen, Austria	Jung et al. (1978)	13	1.5	1.2	0.82	3	0.15
9. Maastrichtian-Middle Paleocene	67–61?[d]	Rusava, Czechoslovakia	Knobloch (1977)	20	1.2	0.8	0.66	3	0.06
10. Late Paleocene	55	Woolwich and Reading Beds, England	Chandler (1961)	18	129	234	1.82	731	1.2
11. Ypresian (Early Eocene)	52	London Clay, England	Reid and Chandler (1933)	202	1,957	5,932	3.03	61,318	0.25
12. Lutetian (Middle Eocene)	45	Geiseltal, East Germany	Mai (1976)	25	308	643	2.1	3,182	2.1
13. Auversian (Late Eocene)	42	Clarno, Oregon, U.S.A.	Scott (1954); Bones (1979)	33	3,729	10,626	2.8	59,150	0.25
14. Middle Oligocene	32	Haselbach, East Germany	Mai and Walther (1978)	79	268	1,220	4.5	9,294	0.07
15. Middle Oligocene	30	Bovey Tracey, England	Chandler (1957)	33	68	225	3.3	1,300	0.35
16. Late Oligocene	25	Tomsk, Siberia U.S.S.R.	Nikitin (1965)	95	19	123	6.4	1,200	0.07
17. Lower to Middle Ottnangian (Lower Miocene)	18.5	Chomutov-Most-Teplice Basin, Czechoslovakia	Bůžek and Holý (1964)	22	19	38.6	2	180	1.27

TABLE 2. (Continued).

Stratigraphic Age	Estimated Numeric Age[a]	Locality	Reference	N	$\bar{x}$ (mm³)	s.d.	Co. var. (s.d./$\bar{x}$)	Largest (mm³)	Smallest (mm³)
18. Lower Miocene	18	Rusinga, Kenya	Chesters (1957)	29	1,319	2,044	1.6	10,935	63
19. Middle Ottnangian (Lower Miocene)	18	Wiesa, East Germany	Mai (1964)	71	1,410	3,263	2.3	15,611	0.25
20. Middle to Upper Ottnangian (Lower Miocene)	18	Hartau, East Germany	Mai (1964)	51	774	2,303	3	11,600	1
21. Upper Ottnangian (Lower Miocene)	17.5	Turów, Poland	Czeczott and Skirgiello (1959, 1961a, 1961b, 1967, 1975, 1980a, 1980b)	47	2,501	4,340	1.7	15,096	2
22. Carpathian (Lower Miocene)	17	Nowy Sącz Basin, Poland	Łancucka-Środoniowa (1979)	79	7	22.8	3.3	179	0.014
23. Badenian (Middle Miocene)	14.5	"Gdów Bay," Poland	Łancucka-Środoniowa (1966)	52	95	366	3.9	2,125	0.04
24. Pliocene	3.5	Kranichfeld, East Germany	Mai (1965)	35	13	63	5	1,215	0.1
25. Pliocene	3.5	Bergheim, West Germany	van der Burgh (1978)	83	100	245.5	2.5	1,400	0.9
26. Plio-Pleistocene	1.8	Rippersroda, East Germany	Mai et al. (1963)	66	68.1	301	4.4	2,125[e]	0.014
27. Holocene	0.035 0.068	New Haven, Connecticut, U.S.A.	Pierce and Tiffney (unpubl. data)	43	1,077	4,116	3.8	25,000[f]	1.1

[a] Numerical ages after van Eysinga (1975), millions of years (Ma).
[b] Estimated from absolute possible smallest and largest sizes of angiosperm fruiting remains presented in Friis (1984). The largest fruiting remain is a seed-bearing fruit (Friis, pers. comm.) and is likely not the unit of dispersal.
[c] Deposit cited as "Senonian;" 75 Ma is taken as the midpoint of the Senonian.
[d] Exact stratigraphic position of deposit not determined; possible range indicated.
[e] This value is for a nut of *Corylus*.
[f] Value for *Juglans cinera*; the next largest value is 8,000 mm³ for *Carya*.

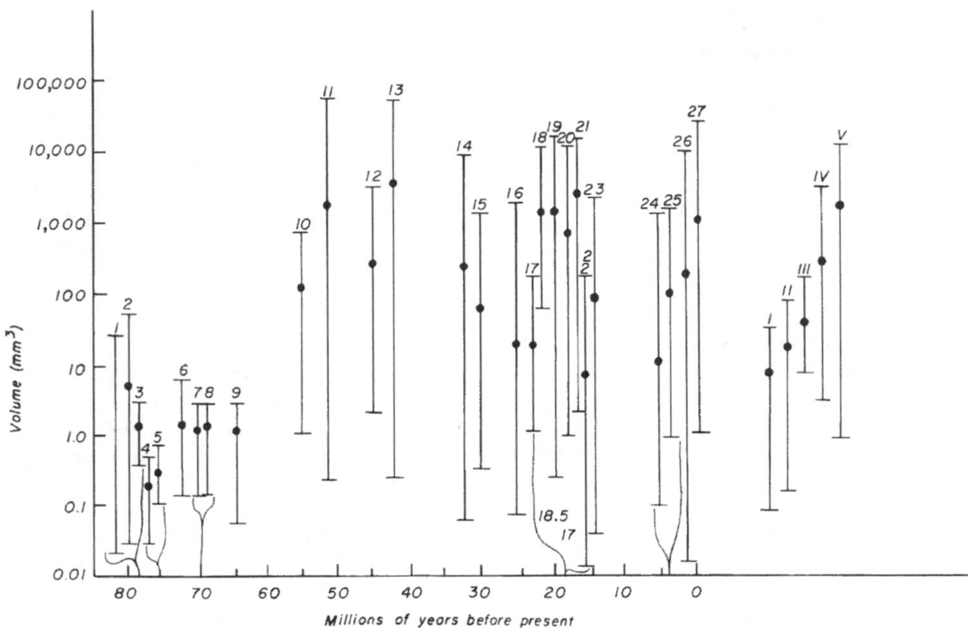

FIGURE 2. Plot of seed volume (vertical axis, logarithmic) versus time (horizontal axis) for 27 Cretaceous and Tertiary fruit and seed floras. The vertical line for each flora indicates the range of diaspore volume; the central dot, the average diaspore volume. The lines I–V at the far right correspond to the volume equivalents of the ecological classes of Harper et al. (1970) as derived in Figure 1. (I. Open habitat; II. woodland margins; III. woodland ground; IV. woodland shrubs; V. woodland trees.) Note that the average volume is not available for flora #1, and that in several cases (floras 1–3, 7, 8, 17–22, 24, 25), more than one flora occurs at a single time.—1. Åsen, Sweden (Friis, 1984).—2. Gay Head, Massachusetts, U.S.A. (Tiffney, unpubl. data).—3. Staré Hamry 1, Czechoslovakia (Knobloch, 1977).—4. Aachen, West Germany (Vangerow, 1954).—5. Petrovice, Czechoslovakia (Knobloch, 1964).—6. Horní Bečva, Czechoslovakia (Knobloch, 1977).—7. Kössen, Austria (Knobloch, 1975).—8. Kössen, Austria (Jung et al., 1978).—9. Rusava, Czechoslovakia (Knobloch, 1977).—10. Woolwich and Reading beds, England (Chandler, 1961).—11. London Clay, England (Reid & Chandler, 1933).—12. Geiseltal, East Germany (Mai, 1976).—13. Clarno, Oregon, U.S.A. (Scott, 1954; Bones, 1979).—14. Haselbach, East Germany (Mai & Walther, 1978).—15. Bovey Tracey, England (Chandler, 1957).—16. Tomsk, Siberia, U.S.S.R. (Nikitin, 1965).—17. Chomutov-Most-Teplice Basin, Czechoslovakia (Bůžek & Holý, 1964).—18. Rusinga, Kenya (Chesters, 1957).—19. Wiesa, East Germany (Mai, 1964).—20. Hartau, East Germany (Mai, 1964).—21. Turów, Poland (Czeczott & Skirgiello, 1959, 1961a, 1961b, 1967, 1975, 1980a, 1980b).—22. Nowy Sącz Basin, Poland (Łancucka-Środoniowa, 1979).—23. "Gdów Bay," Poland (Łancucka-Środoniowa, 1966).—24. Kranichfeld, East Germany (Mai, 1965).—25. Bergheim, West Germany (van der Burgh, 1978).—26. Rippersroda, West Germany (Mai et al., 1963).—27. New Haven, Connecticut, U.S.A. (Pierce & Tiffney, unpubl. data).

by Tiffney (unpubl. data, see Tiffney, 1977b) from Massachusetts, U.S.A., are both in fluvial deposits containing large pieces of wood and, in the case of Tiffney's material, conifer cones (Miller & Robison, 1975). The same situation exists in lagoonal sediments of a similar age from Cliffwood Beach, New Jersey, U.S.A. (Tiffney, unpubl. data). Thus, the size of the angiosperm reproductive remains from these deposits cannot be explained by mechanical sorting.

It is possible that a uniformly small seed size could result from the derivation of the fossils from a single, ecologically-specialized aquatic community, or a combination of aquatic and river-floodplain communities. At the outset, it is noteworthy that in one case (Petrovice, Czechoslovakia, flora #4) Knobloch (1964) described a seed flora of 11 species of *Microcarpolithes* comparable in all respects with other central European Cretaceous fruit and seed floras, but in the same sample as a flora of large leaves including *Araliophyllum* sp. Ett., *Debeya bohemica* Knob., *Pseudoprotophyllum senonense* Knob., *Quercophyllum triangulodentatum* Knob., *Laurophyl-*

lum elegans Holl., *Proteophyllum* sp., *Platanophyllum* sp., *Cinnamomophyllum* sp., and three species of *Dicotylophyllum*. This clearly suggests that at least some of these seeds were borne by woody, nonaquatic vegetation. The question remains, do these floras sample only unstable floodplain forests? The answer lies in an examination of Tertiary and modern deposits, where we have a better idea of the community affinity of the fossils through their taxonomy. The fluvial deposits of the Tertiary (e.g., Bergheim, or those described by Gregor, 1978, 1980b) often include a wide range of fruit and seed sizes that are presumed on taxonomic grounds to be derived from several separate communities, including upland mesic ones. This is further supported by the New Haven, Connecticut flora (#27), which is of Holocene age and includes many upland taxa. Even those Tertiary floras least affected by transport include occasional samples of plants growing in the mesic sites surrounding the water. For example, the Oligocene (30 Ma) lacustrine Bovey Tracey flora of England (Chandler, 1957) includes larger seeds of such trees as *Magnolia* L., *Fagus* L., and *Nyssa* Gronov. ex L. In conclusion, it seems unlikely that these Cretaceous floras are solely records of aquatic or floodplain vegetation.

The morphology of these Late Cretaceous fruits and seeds is somewhat more diverse than that of the Early Cretaceous and Cenomanian remains. Follicles and capsules are still common (Massachusetts and New Jersey deposits), but some evidence is at hand for nuts and drupes (Friis, 1984; Tiffney, unpubl. data). While many of the seeds in these deposits have thin or fragile walls, others have rather thick walls and well-developed surficial sculpture. The dehiscent fruits and thin-walled small seeds suggest the continuing importance of abiotic dispersal mechanisms. However, the presence of drupes and seeds with thick walls is circumstantial evidence for at least the potential for adaptation to animal dispersal, if not for its presence in a limited degree.

Exceptions exist to the general rule of small Cretaceous diaspores. Monteillet and Lappartient (1981) reported a Late Campanian to Maastrichtian (70–66 Ma) flora from Senegal, including seven species of angiosperm fruits with an average volume of 51,950 mm³. The fossils are poorly preserved, and in the cases of "*Annona*" L., "*Cola*" Schott, and perhaps "*Cordyla*" Lour. and "*Trichilia*" P.Br., I am not convinced that the specimens are of plants. The illustrated specimens of *Borassus* L., *Meliacea* (? new form genus), and perhaps *Nauclea* L. are more convincing. Chesters (1955) has also reported large fruits of Annonaceae, Icacinaceae, and possible other angiosperms from the Maastrichtian (68 Ma) of Nigeria. Although large fruit size is no guarantee of large seed size (*Nauclea* has tiny seeds, Willis, 1973); it appears that larger diaspores were becoming more common in the late Cretaceous.

TERTIARY FRUITS AND SEEDS

Individual large diaspores are known from the Paleocene (Brown, 1962; Koch, 1972a, 1972b), but the only published fruit and seed floras are those of the Woolwich and Reading beds of southern England and possibly Horní Bečva and Rusava, Czechoslovakia. The Rusava flora (Knobloch, 1977) is between latest Cretaceous (67 Ma) and middle Paleocene (60 Ma) in age, while Horní Bečva is inferred from Table 1 of Knobloch (1977) as being between Campanian (73 Ma) and Early Paleocene (63 Ma) in age. Both floras are quite similar to those of Staré Hamry 1 and Petrovice, to which they are geographically close.

By contrast, the flora of the Late Paleocene (55 Ma) Woolwich and Reading beds (Chandler, 1961) includes a diverse array of large and small diaspores with an average volume of 129 mm³. This sets the pattern for the remaining Tertiary floras, which vary in percentage composition of larger and smaller diaspores, but which always include both. The basic trends can be discerned from columns "$\bar{x}$," "largest diaspore," and "smallest diaspore" in Table 2, and from floras 10–27 of Figure 2. Very large diaspores first appear in numbers in the Early Eocene (52 Ma) London Clay flora and dominate the Middle Eocene (45–42 Ma) Geiseltal and Clarno floras and the mid-Oligocene (32 Ma) Haselbach flora, resulting in high average diaspore sizes for these floras. From the Late Oligocene, there is a general tendency for the average diaspore size of a flora to decrease through Pliocene/Pleistocene time, although this trend is not statistically significant. This decrease in average diaspore size is not due to a decrease in the size of the largest diaspores in each flora (by regression of largest seed size versus age ($P = 0.1$; $r = 0.40$, $N = 18$), but to the occurrence of fewer large diaspores in each flora.

Herbaceous angiosperms began to diversify dramatically in the latest Paleogene and the early Neogene (Tiffney, 1981). Herbs normally have small seeds (Harper et al., 1970), and their increasing importance during this time is reflected in the reduced average diaspore volumes of the floras commencing with the Lagernogo Sad deposit (Tomsk, Siberia; age from Dorofeev, 1963) and carrying through the later Tertiary floras from the Chumotov-Most-Teplice basin (Czechoslovakia), Nowy Sącz basin (Poland), and Kranichfeld (East Germany). While the average diaspore volume (ADV) of these floras is small relative to that of other Tertiary floras, it is markedly larger than the ADV of the Cretaceous floras. The large average volumes for the Miocene Gdów Bay (Poland) and Pliocene Rippersroda (East Germany) floras are due to the presence of a few large diaspores. Deletion of *Corylus* L., *Fagus* L., and *Carya* Nutt. from the former flora brings the average volume down to 9.2 mm^3, and deletion of *Trapa* L. and *Corylus* L. from the Rippersroda flora brings the average volume down to 21.5 mm^3. The large average volume for the Pliocene Bergheim flora (Mine Fortuna-Garsdorf 1) of West Germany results from the river sands of this deposit having a large allochthonous component derived from upland forest trees (e.g., *Magnolia* L., *Persea* Mill., *Corylus* L., *Castanea* Mill., *Quercus* L., *Halesia* J. Ellis ex L., *Styrax* L.). In spite of these individual differences in ecological and taphonomic setting, it is interesting that each of these floras shows a greater range of diaspore size and a higher average diaspore volume than the Cretaceous floras.

The dispersal mechanisms and syndromes of Tertiary angiosperm fruits and seeds may be inferred from their morphology and from their living relatives. Neither source is totally satisfactory; many morphological features are not preserved, and present dispersal adaptations of a genus or family are no guarantee of past mechanisms. However, both lines of evidence suggest that a wide range of fruits and seeds adapted to animal dispersal were present by the Eocene and Oligocene. This included a variety of sizes from the smaller berries of the Vitaceae to the larger aggregate fruits of the Annonaceae or the drupes of the Mastixiaceae. This diversity of fruit types and sizes offered opportunities to a range of dispersal agents, but particularly to those able to deal with larger disseminules. With the Miocene diversification of herbaceous angiosperms (Retallack, 1981; Tiffney, 1981), an array of small seeds and fruits became available, which was probably important to ground-dwelling rodents and granivorous birds. Thus, the Tertiary appears to be the time in which major dispersal patterns (diaspore morphologies, relations with particular agents) first achieved their modern form and diversified among angiosperms.

SUMMARY

Cretaceous diaspores are generally small. Cretaceous seed floras are marked by a small average diaspore volume (ADV) and a limited range of diaspore volume (RDV). Early Tertiary floras exhibit a major increase in ADV. Succeeding floras show a broad trend of decreasing ADV, but with no decrease in RDV. The change in ADV through the Tertiary is a result of changes in the relative proportions of large and small diaspores in each flora. Small diaspores show no trend in size change through the Cretaceous and Tertiary, and after their appearance in the Tertiary, large diaspores also show no trend in size change. The increase in diaspore size is paralleled by an apparent change from the dominance of abiotic dispersal mechanisms in the Cretaceous to the increasing importance of biotic dispersal agents commencing in the earliest Tertiary.

This pattern of change in size and mode of dispersal cannot be ascribed to taphonomic or ecologic factors because similar depositional environments are sampled in both Cretaceous and Tertiary deposits. Certainly the deposits of Åsen and Massachusetts demonstrate that larger fruits and seeds could have been carried into the deposit and preserved if present. Many of the Tertiary deposits (e.g., Nowy Sącz, Kranichfeld) demonstrate that even deposits dominated by aquatic vegetation may be expected to include some elements of mesic communities.

Climate could have influenced the composition of the vegetation or the presence of dispersal agents. However, Cretaceous climates from the first appearance of the angiosperms through to the latest Cretaceous are generally felt to have been as warm as those of the early Tertiary (Savin, 1977; Barron et al., 1981; Thompson & Barron, 1981), although there is good evidence for a latest Cretaceous–Paleocene cool phase (Hickey, 1981). In addition, the European Pliocene and Quaternary include climates that were cooler than those of any period of angiosperm history, but floras from these epochs have at least a few large diaspores. Perhaps the only unanswerable bias is that all the fruit and seed data are derived from northern hemisphere, primarily European, localities. It is possible that different patterns in the evolution of diaspore size could have taken place in other portions of the world, but this cannot be evaluated from existing paleontological data. However, extant tropical (Levin, 1974) and temperate (Salisbury, 1942; Harper et al., 1970) angiosperm fruits and seeds apparently exhibit the same range of sizes.

The timing of this transition from small to large diaspores, and from the dominance of abiotic dispersal to the increased importance of biotic dispersal, is not clear from present knowledge. Since the mid- to Late Cretaceous and the early to mid-Tertiary of Europe both possessed warm climates, they presumably had a similar potential to host tropical plants. If plants with large diaspores were present in the Cretaceous tropics, they should have been seen in the European Cretaceous, much as they were in the Tertiary. However, a Cretaceous-Tertiary boundary cooling at higher latitudes (Hickey, 1981) could have masked the evolution of angiosperms with large diaspores in the tropics in the latest Cretaceous. These could have then appeared in northerly latitudes with the return of subtropical climates in the early Tertiary.

INTERPRETATIONS

The observed pattern in seed size can be explained most simply as a response to one or both of two ecological factors. The first is the relation between seed size and the habit or ecological site of the parent plant. The second is the importance of dispersal agents, which exert pressure on the morphology and size of fruits and seeds, as evidenced by the existence of distinct "dispersal syndromes" in the angiosperms (van der Pijl, 1969). Each of these factors will be treated in turn.

SEED SIZE AND PARENT PLANT HABIT/HABITAT

Harper et al. (1970) and Silvertown (1981), following on the classic work by Salisbury (1942), have demonstrated a correlation between the habit, habitat, and diaspore weight of individual plants. Short-lived or weedy plants of open or unstable habitats generally have many small diaspores that may be dispersed widely, often by abiotic mechanisms (wind, water). These seeds provide very little nutrient reserve to the germinating seedling, so that seedlings generally survive only in open, sunny habitats. However, the large numbers and wide dispersal of these diaspores increase the likelihood that a few seedlings will germinate in suitable habitats. At the other end of the scale, dominant, long-lived, forest trees of large stature tend to bear fewer, larger diaspores, often involving large seeds. Because of their mass, such diaspores are often dispersed by biotic vectors, although less frequently they may be transported by gravity or water. A large seed provides a massive reserve of nutrients to the young seedling and enables it to become established in the shade of the deep forest. Between the two extremes are groups of plants with intermediate habit, habitat, diaspore, and seed size including (in order of decreasing stature and seed size) woodland shrubs, woodland herbs, and herbs of woodland margins.

It should be noted that this is a general tendency, rather than an invariant rule. Habit and habitat adaptation may interact in a complicated manner and influence seed weight. Several early seral (weedy) trees have seeds as small as those of herbs, but possess a tree habit. However, such species (e.g., *Populus* L., *Betula* L., and *Fraxinus* L. in temperate forests; *Cecropia* Loefl. in the New World tropics) are often fast-growing and short-lived, and tend not to form time-stable, closed communities. Further, other features, including water availability and degree of seasonality, may influence diaspore and seed size (Baker, 1972; Levin & Kerster, 1974), and seed sizes in each ecological class appear to be slightly larger in tropical communities than in temperate ones (Levin, 1974). However, an overview of this variation suggests that the basic pattern of correlation of seed size with the habit and ecology of the parent plant holds as a broad principle in a wide range of environments.

A graphic summary of the average diaspore volume and range of diaspore volume for several modern ecological groups is presented in Figure 2 (cf. Harper et al., 1970). The values for each category were originally calculated by weight (Salisbury, 1942; Levin, 1974), but I have converted this to cubic millimeters by use of the graph presented in Figure 1.

Comparison of the values for the average and range of diaspore volumes (ADV, RDV) for each of these ecological groups with the ADV and RDV for the fossil floras reveals a clear pattern. Cretaceous floras (#1–7) have ADVs equal to or less than that for modern plants of open communities. Further, only in the case of the flora from Massachusetts (#2) does the RDV exceed that seen in modern plants of open habitats. The sedimentary context (Doyle & Hickey, 1976;

Hickey & Doyle, 1977), and the small seeds of the earliest angiosperms, support the contention that they were "weedy" plants of unstable or transient habitats outside of the climax gymnosperm forest (Takhtajan, 1976; Hickey & Doyle, 1977; Doyle, 1978; Niklas et al., 1980; Tiffney, 1981). Note that this statement does not exclude the possibility that the Cretaceous floras included trees, for the ADVs of several Cretaceous floras overlap with the lower end of the range of tree diaspore volumes. However, although trees and shrubs may have been present in these communities, the small sizes of the diaspores involved imply that these were likely early successional plants rather than canopy dominants. An example may be provided by *Platanus* L., which is probably present as early as the Cenomanian (97 Ma) (Dilcher, 1979; Schwarzwalder & Dilcher, 1981). *Platanus* is an early successional tree in modern floras (Braun, 1950) and has been demonstrated to occupy unstable, stream-side habitats in the Eocene (Wing, 1981).

By contrast, the Tertiary floras possess a wide range of diaspore volumes embracing all five ecological categories of diaspore size. This suggests that each flora has the potential to contain plants of any and all habits and habitats. It is not possible to be certain that any one flora was dominated by plants of a particular habit or habitat from diaspore size for three reasons. First, the diaspore sizes for the five modern ecological categories do overlap. Second, taphonomic factors have resulted in the mixing of disseminules from different communities in the fossil record. Third, the average diaspore volume (ADV) for each fossil flora is not a fully trustworthy indicator of the dominant physiognomy of the community; one or two large fruits can drastically affect the ADV of a flora. For example, the elimination of the three largest diaspores (two species of *Carya* and one of *Juglans*) from the New Haven flora (Fig. 2, #27; N = 43 species) drops the ADV from 1,077 mm^3 to 133 mm^3. The degree of influence of large specimens on the ADV may be approximated by the coefficient of variation (s.d./$\bar{x}$, see Table 2). Large values of the coefficient of variation indicate that the mean is not that of a randomly distributed population but is an artifact of a polymodal distribution. The value of this coefficient is high through the Tertiary and shows no significant directional change during this time (commencing with the Woolwich and Reading beds, a regression of the coefficient of variation with time yields $P > 0.20$; $r = -0.27$, N = 18).

In a broad manner, the ADV decreases through the Tertiary, although not in a statistically-significant manner (regression of ADV versus time yields $r = 0.27$, N = 18). The floras (Fig. 2) of the London Clay and Clarno have ADVs very close to that for trees in the modern day. This is not unexpected, as both floras are presumed on taxonomic bases to be related to the modern forests of Indomalaysia (Chandler, 1964). From this high, the ADV falls off through the Tertiary to values close to those for modern plants of open environments (note flora #22, Nowy Sącz, and #24, Kranichfeld). This trend indicates an increasing dominance of smaller-seeded plants and parallels the climatic deterioration and increase in climatic variability that occurred in temperate regions in the later Tertiary (Mai, 1970; Buchardt, 1978). These cooler and more variable temperate climates could be expected to result in the evolution of new, open, unstable communities, populated by plants with a rapid life cycle. This is what is observed in the taxonomic composition of late Tertiary communities, which show a diversification of herbaceous angiosperms (Niklas et al., 1980; Tiffney, 1981). However, it is of great importance to note that while small-seeded forms dominated in the later Tertiary, large-seeded trees remained as part of the flora, although diminished in importance.

In summary, the diaspore size data suggest that Cretaceous angiosperms were small and/or opportunistic plants, and that only in the latest Cretaceous or early Tertiary did the group clearly evolve to include physiognomically-dominant trees of stable, climax forests. This does not exclude angiosperms from forming forests in the Cretaceous, but the diaspore size data suggest that such forests would be restricted to unstable environments, while gymnosperms would be the physiognomic dominants in stable environments. These predictions can be tested in part by examining the sedimentological settings of Cretaceous angiosperm and gymnosperm floras. Angiosperm floras should be more commonly associated with sediments indicative of unstable environments (e.g., river margins), while gymnosperms should be associated with sediments representative of more stable environments (e.g., back swamps or uplands).

The foregoing interpretation rests on at least two assumptions that require consideration. First, does the correlation of diaspore size and habit/habitat witnessed in the modern day hold with respect to Cretaceous angiosperms, for which there are few modern homologues? I cannot answer this question directly, but suggest that Cretaceous angiosperms do indeed follow the same pattern as modern ones since the relation of diaspore size and habit/habitat holds across a wide range of taxonomic groups. Chaloner and Sheerin (1981) have successfully applied this concept to an explanation of early land plant reproductive strategies, in which the evolution of larger plant size is directly correlated with an increase in disseminule size. Also, while the fact that these Carboniferous plants are extinct makes the inference of successional status tenuous, it appears that the dominants of the relatively more stable lowland swamp communities (e.g., medullosans and certain arborescent lycopods) had larger disseminules than plants of less stable habitats (e.g., calamitaleans, cordaitaleans, and conifers). The Mesozoic flora was dominated by the gymnosperms, members of which had a wide range of diaspore sizes. Some (cycads, araucarian conifers) had quite large seeds that appear to have necessitated biotic dispersal (van der Pijl, 1969), while others had small seeds (some seed ferns, cycadeoids, and taxodioid conifers) morphologically adapted to abiotic dispersal. Retallack and Dilcher (1981) suggested that cycadeoids may have been restricted to unstable stream margins in the Cretaceous, while conifers dominated the more stable upland communities. With the advent of large-seeded angiosperms in the latest Cretaceous and Tertiary, gymnosperms declined in importance (Niklas et al., 1980; Tiffney, 1981). It is unclear if this was an unrelated event or a direct result of the expansion of the angiosperms. The latter possibility deserves consideration, because the dominant modern group of gymnosperms in the Northern Hemisphere are the Pinaceae, which have relatively small seeds and are generally restricted to early successional positions or to sites from which angiosperms are excluded by physiological factors. These elements of circumstantial evidence suggest that the correlation of diaspore size and the habit and/or habitat of the parent plant generally holds for land plants and may be assumed to have done so for early angiosperms.

Second, the association between diaspore size and habit/habitat was demonstrated using temperate plants (Salisbury, 1942; Harper et al., 1970). Does it hold with warm-temperate to often tropical taxa, which commonly occur in the Tertiary? Again, the answer is circumstantial but positive. The relationship of seed size to habit/habitat of the parent plant seems to be general among land plants, and anecdotal evidence suggests that it holds in the modern tropics (van der Pijl, 1969; Stebbins, 1971; Opler et al., 1980; Janzen & Martin, 1982). In one case where seeds of plants of the five ecological categories were measured (Levin, 1974), the average seed size in each category was a bit larger (two to five times) than observed in the temperate flora. While interesting, the magnitude of this variation is too small to affect the hypothesis of Cretaceous and Tertiary angiosperm seed size presented here.

SEED SIZE AND MODE OF DISPERSAL

Seed (diaspore) dispersal is an important element in the life cycle of seed plants (cf. Levin & Kerster, 1974). Abiotic dispersal (wind, water) is successfully employed by a wide range of angiosperms, including many trees. However, there is little question that biotic dispersal is of greater importance, if not dominant, among angiosperms in the modern day. Biotic dispersal agents exert a strong selective pressure on angiosperm fruit and seed size and morphology, as evidenced by the evolution of a wide range of adaptations for animal dispersal (cf. Ridley, 1930; van der Pijl, 1969). I am unaware of any estimate of the absolute proportion of the world's angiosperm flora that is animal-dispersed, but in the few reports of individual communities, the proportion of biotically-dispersed species is often high (Jones, 1956; Smythe, 1970; Stiles, 1980; Handel et al., 1981) and reaches 90% in some Central American examples (Frankie et al., 1974; Janzen, 1977). This may be affected by edaphic factors, however, as suggested by Janzen's (1977) observation that a low degree of biotic dispersal occurs in some Indomalaysian forests growing on nutrient-poor soils.

There are five animal groups commonly involved in the dispersal of angiosperm fruits and seeds; ants, fish, reptiles, birds, and mammals (including bats). All have, to greater or lesser degrees, affected the size and shape of angiosperm disseminules. The history and general influence of each group is considered in turn.

Ants. Ant dispersal (myrmecochory) is primarily known among forest floor herbs, particularly in the temperate zone (van der Pijl, 1969; Handel et al., 1981) although it is also reported from other areas (e.g., Berg, 1975). Morphological adaptations to ants usually involve small diaspore size and the presence of an oil body or elaiosome as a food source on the exterior of the diaspore. Although ants are known from the Cretaceous (Burnham, 1978), they would affect only small seeds.

Fish. Fish are generally assumed to have a minor role in the dispersal of angiosperms (Ridley, 1930; van der Pijl, 1969), although a recent study of Amazonian plant communities (Golding, 1980) suggested that fish may disperse diaspores, particularly in time of high water. The degree to which this dispersal syndrome involves adaptations in diaspore morphology and size, and its importance outside the Amazon basin, are not clear. It may not be so much a "coevolved syndrome" as a glorified case of scavenging. Fish have been around since the Paleozoic (Romer, 1966) and may well have served as generalist dispersal agents in swamps and rivers since the Carboniferous.

Reptiles. Reptilian dispersal (saurochory) is a recognized syndrome, often involving brightly colored and odoriferous seeds or fruits borne near the ground and of a wide range of sizes (van der Pijl, 1969). The important modern representatives include turtles and tortoises, which first spread as a group in the Triassic (Romer, 1966), and lizards, particularly iguanas. The latter group appears in the Eocene, although its forerunners may go back to the Upper Jurassic (Romer, 1966). Perhaps the decline of the reptiles at the end of the Cretaceous, just as the angiosperms were undergoing a major expansion, explains the relative lack of dispersal syndromes involving the two groups in the modern day. The possibility must also be entertained that the primary dispersal vectors of the large seeds of the physiognomically-dominant Mesozoic gymnosperms were reptiles, and that the decline of the reptiles may have influenced the demise of some gymnosperm groups in the late Cretaceous (Krassilov, 1978). If so, this would also imply that Late Cretaceous and early Tertiary plant communities were in a state of flux, and open to angiosperm invasion.

Birds. Birds are among the most important of angiosperm dispersal agents, affecting very small to very large diaspores in temperate and tropical communities (Ridley, 1930; van der Pijl, 1969). There are many morphological adaptations of angiosperm disseminules to bird dispersal (ornithochory) because transport may be internal or external. The most common syndrome involves odorless, brightly colored, edible, fleshy fruits with hard, resistant, inner seeds. These are often clearly displayed; and, in dehiscent fruits, the seeds often dangle from the fruit at maturity.

The fossil record of the birds has been reviewed at the family level by Brodkorb (1971), to which I have added data provided by Kurochkin (1976) (Fig. 3). Bird families often contain organisms of diverse dietary habits, but, based on information provided by van Tyne and Berger (1976), individual families can be described as "carnivorous" (no plant material consumed), "omnivorous" (some plant material consumed), or "vegetarian" (dominantly plant material consumed). Figure 3 presents a summary of the diversity of bird families from the Early Cretaceous to the present, broken into these three dietary groups. No fossil families are included; they are few in number, and it would be difficult to ascertain their dietary affinities. The family level is used for ease of tabulation. A generic or specific level summary is beyond the scope of this paper and would not greatly alter the trends seen in Figure 3, although the family level does mask the effect of the late Tertiary diversification of the species-rich, dominantly omnivorous or vegetarian, passerines (Brodkorb, 1971).

Cretaceous families for which diets may be surmised from modern relatives are predominantly carnivorous and marine; one omnivorous family is present (Cracraft, 1973; Brodkorb, 1976). This conclusion is generally supported by the beak morphology of the fossils. The fossil record of the birds diversifies greatly in the Eocene, and the number of modern families continues to increase throughout the Tertiary. The proportion of omnivores remains steady during this time, but the proportion of vegetarian families (which first appear in the Eocene) rises consistently, while that of the carnivorous families falls.

As displayed in Figure 3, the avian fossil record would suggest that birds were not important in angiosperm dispersal until the early Tertiary, whereupon they became increasingly important and continued to mount in significance to the present day. However, the available fossil record of any group may be seen either as a real reflection of evolutionary events, or as too biased to be trusted directly. Cracraft (1973) considered the fossil record to be less important than the information that can be drawn from a comparison of the timing of Cretaceous and Cenozoic continental movements with the modern distri-

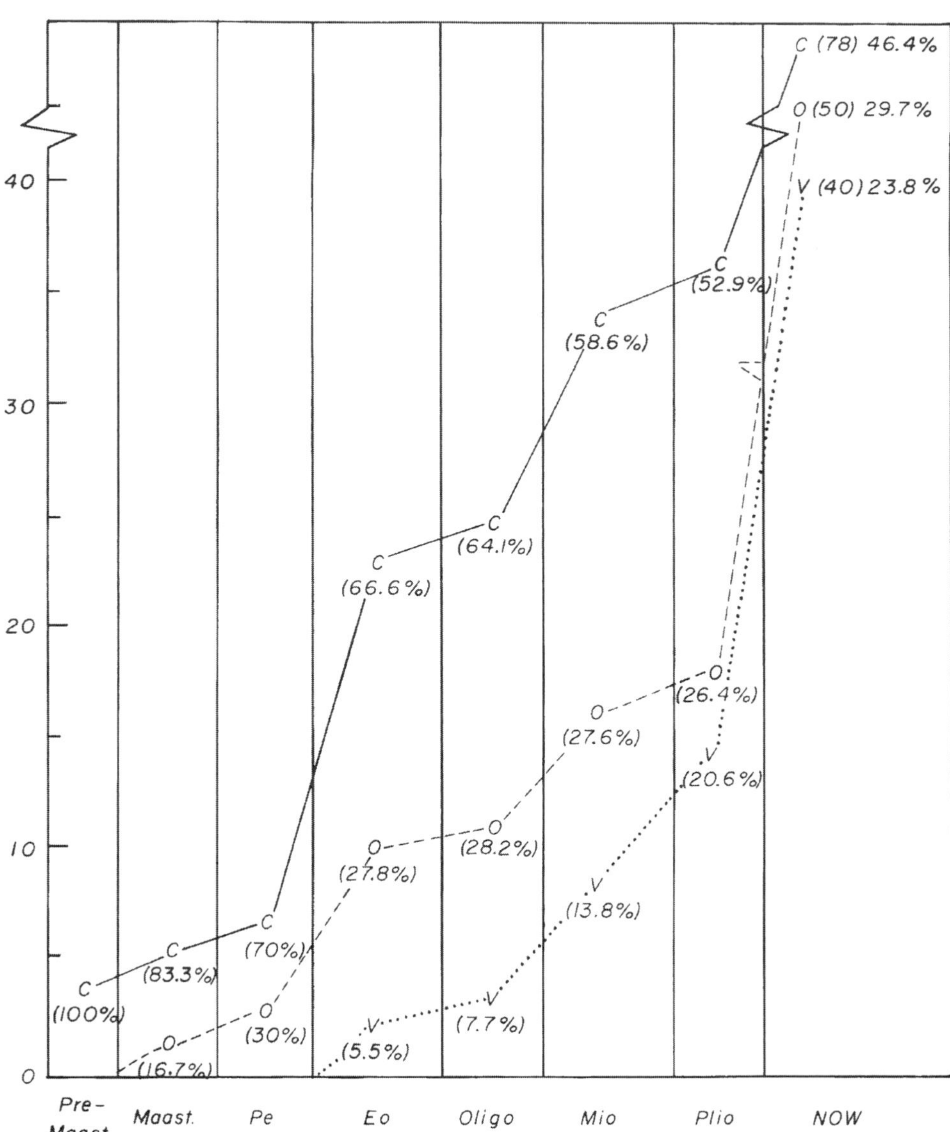

FIGURE 3. Cumulative diversity of modern families of birds through the Cretaceous and Tertiary broken into three dietary groups: carnivorous (c), omnivorous (o), and vegetarian (v). Percentages are of the total number of families in each epoch. Data from Brodkorb (1971, 1976); Kurochkin (1976); van Tyne and Berger (1976). Pre-Maast = pre-Maastrichtian, Maas = Maastrichtian, Pe = Paleocene, Eo = Eocene, Oligo = Oligocene, Mio = Miocene, Plio = Pliocene.

butional patterns of bird families. On this basis, he inferred that the immediate ancestors, if not the actual families, of modern birds were present by the mid- and Late Cretaceous. Brodkorb (1971) also suggested that modern families were present in the mid- and Late Cretaceous. He reasoned that the modern appearance, specialization, and diversity of early Tertiary birds argue for considerable antecedent evolution of the group.

Neither author accepts the fossil record at face value. This logic is reminiscent of the school of paleobotany that sought the origins of the angiosperms in the latest Paleozoic and early Mesozoic on the basis of their "diverse" Cretaceous record (Axelrod, 1960, 1970). Subsequent study of the angiosperms (Doyle & Hickey, 1976; Hickey & Doyle, 1977; Doyle, 1978) has demonstrated that the initial radiation of the angiosperms was confined to the Cretaceous and matches the known fossil record of the group. I choose to accept the fossil record of the birds as indicative of a major radiation in the Tertiary, perhaps with its beginnings in the latest Cretaceous, paralleling that of mammals (Colbert, 1969) and of modern angiosperm families (Niklas et al., 1980; Tiffney, 1981; Muller, 1981). A recent consideration of avian evolution (Cracraft, 1982) does not alter this view, because it emphasizes the aquatic and presumably carnivorous nature of Mesozoic birds as they are presently understood.

Mammals (excluding bats). The dispersal of angiosperm seeds by mammals is probably only a little less important than that by birds (Ridley, 1930; Martin et al., 1951; van der Pijl, 1969; Halls, 1977). Van der Pijl gives the impression that it is of greater importance in the tropics than in the temperate regions. The fruit and seed characters associated with mammal dispersal are quite varied and may involve internal or external transport. Mammals often possess a good sense of smell and thus many mammal-dispersed fruits have a distinctive odor (van der Pijl, 1969). The thickness of the seed wall tends to vary with seed size. Small seeds that would probably be passed through the digestive tract do not require hard coats, whereas large ones with edible contents require protection against direct predation. One important aspect of mammal dispersal is that mammals can move seeds of quite large sizes, often from within the forest canopy (Ridley, 1930; van der Pijl, 1969). Mammal-dispersed fruits of more open habitats are often adapted to external transport (van der Pijl, 1969).

The Tertiary is the "age of mammals," and is the time of radiation of a host of important dispersalist groups (Romer, 1966; Colbert, 1969). Rodents, lagomorphs, primates, artiodactyls, and perissodactyls all commenced radiations in the Eocene. Other, more specialized omnivorous lineages appeared in the mid-Tertiary. While modern terrestrial mammals are almost entirely products of Tertiary radiations, the possibility remains that some Cretaceous lineages of mammals could have been important dispersal agents. Lillegraven (1979) suggested that perhaps mammals and flowering plants were establishing the basic features of their coevolutionary relationship in the Cretaceous. Certainly, several major groups of Cretaceous mammals are presumed to be omnivorous and a few herbivorous (Kron, 1979; Clemens & Kielan-Jaworowska, 1979; Clemens, 1979), although all are generally of small size (Lillegraven, 1979). While I do not doubt that Cretaceous mammals had some interactions with angiosperms, it seems unlikely to me that such interactions were significant and widespread. The known Cretaceous thick-walled seeds are not especially common, and they show no trends in morphology or size to suggest their specialized adaptation to biotic dispersal. Further, if such interactions were established to a significant degree in the Cretaceous, it is surprising that they should not have carried over into the Tertiary and the present. Instead, today the fruits of primitive families or of primitive lineages in families are more commonly abiotically-dispersed, while derived families and lineages are dispersed by organisms of Tertiary origin (Schuster, 1976). Parallel to this pattern is a second one, demonstrated by the Rosaceae, for the geographic co-occurrence of fleshy-fruited angiosperm lineages and "advanced" (products of Tertiary radiations) mammalian lineages (Schuster, 1976). This information, while circumstantial, suggests that mammals became important as dispersal agents only in the Tertiary.

Bats. Bats are important in the dispersal of angiosperm propagules in the warmer regions of the world where fruiting occurs throughout the year. This is borne out by a wide range of anecdotal and scientific observations (Constantine, 1970; Smith, 1976). Bat-dispersed disseminules are normally large, odoriferous, fleshy fruits with a hard stone, or a similar seed with a sarcotesta or aril (van der Pijl, 1957, 1969).

The fossil record of bats (Smith, 1976) is limited and often fragmentary. Bats first appear in the Early Eocene and possess dentition indicative of an insectivorous diet. The first frugivorous bat was thought to be *Archaeopteropus transiens* Meschin. of the Italian Oligocene (Jepsen, 1970), but more recent investigation reveals that the dentition of this specimen is that of an insectivore (Smith, 1976). At present, the earliest record of a frugivore is of a phyllostomatid bat from the late Pleistocene (Smith, 1976).

While some authors feel that bats may have been present in the Late Cretaceous (Jepsen, 1970; Smith, 1976), the absence of pre-Eocene bats, and the rapid diversification of the group in the Eocene and Oligocene, suggests that bats were unimportant, if extant, in pre-Eocene time. Further, the earliest bats were insectivores, and frugivory appears to be a derived condition (Smith, 1976). It is thus likely that the morphological characters of the bat-dispersed fruit evolved as a dispersal syndrome at the earliest in the Eocene.

In summary, pre-Tertiary angiosperm dispersal agents probably included ants, fish, reptiles, and certain groups of archaic mammals. Of these, the last group was probably the most significant, but probably was generalized in its adaptations and of restricted influence on the angiosperms. It appears that the important groups of modern biotic dispersal vectors all underwent their most important period of radiation in the early Tertiary. The sudden appearance of such a wide variety of dispersal agents would be expected to lead to clear changes in seed morphology and size. Such changes are seen in the early Tertiary.

Conclusions

Two separate features have been explored in an attempt to explain the observed pattern of diaspore size change. The first involves the relationship between seed size and the habit and habitat of the parent plant. This leads to the conclusion that the observed increase in diaspore size is related to the angiosperms becoming physiognomically-dominant plants of stable forest communities. The second approach turns to the evolution of dispersal strategies as inferred from the fossil record of the dispersal agents and from the morphology of the disseminules themselves. This suggests that biotic dispersal vectors influenced angiosperm fruit and seed morphology only in the latest Cretaceous or early Tertiary. Each hypothesis provides an adequate explanation of the observed pattern, although the first does not explain its timing. However, the two are not mutually exclusive and may be synthesized to provide a new perspective on angiosperm evolution. As a preamble to this synthetic interpretation, it is necessary to explore briefly the assumption that canopy dominance generally requires large seeds, which, in turn, normally require biotic dispersal agents.

For an angiosperm to achieve dominance (in a physiognomic, not a numerical, sense) of the community, its seedlings must be able to grow in the shade of the parent (or other) trees, thus insuring the continuity of the population. Any increase in seed size in an abiotically-dispersed plant would provide the seedlings with more nutrients and the capability of growing in a more shaded habitat. Logically, one could envision a slow increase in the seed size and shade tolerance of seedlings in one population, ultimately leading to in situ dominance of the community. However, larger seeds generally have a reduced radius of dispersal, particularly in closed communities or in the absence of means of water transport. Thus, in an abiotically-dispersed plant, increased seed size would result in smaller population sizes and perhaps increased endemism and a higher potential for extinction. All this implies that angiosperms required the presence of biotic dispersal agents in order to attain canopy dominance in a closed community. This conclusion appears correct in light of examples of forest trees that have lost their dispersal agents. *Ginkgo* L. attained a wide distribution in the Mesozoic, and maintained it into the Tertiary (Tralau, 1968), but is now a highly restricted endemic in its natural state. Its fleshy and odoriferous seed seems well-adapted to attracting reptiles, and perhaps these were the vectors by which it spread in the Mesozoic. The loss of these vectors in the early Tertiary would be of little immediate importance, because the genus had a wide distribution in this time of warm climate. However, with the climatic changes of the later Tertiary, its range became severely restricted, and, in the absence of a dispersal vector, it was unable to re-expand its range in periods of favorable climate, thus coming to the brink of extinction in the Holocene. Similar, although less dramatic, examples of the effect of the loss of a dispersal agent on range and population structure have been described by Janzen and Martin (1982) for angiosperms in Central America. These circumstantial examples support the tentative conclusion that attainment of canopy dominance and the maintenance of a stable population structure in closed forests are generally linked to the establishment of biotically-mediated dispersal.

The scenario involving seed size and dispersal in the evolution of the angiosperms is simple. Our understanding of Early Cretaceous land plant communities is limited but conveys the impression that there were few existing seed plants of an early successional nature. The angiosperms first appeared about 120 Ma as an "r" strategy (weedy generalist) group with small stature, rapid life cycle, and small, abiotically-dispersed seeds. They spread to occupy a wide variety of early-successional sites (Doyle & Hickey, 1976; Hickey & Doyle, 1977; Doyle, 1978). This diversification was probably paralleled by the appearance of a variety of adaptive vegetative morphologies and may have resulted in the "blocking out" of the general character complexes of several modern suprageneric groups. Evidence from leaf architecture (Doyle & Hickey, 1976; Hickey & Doyle, 1977) indicates that the angiosperms had attained shrub and tree stature by the late Albian or early Cenomanian (about 100 Ma). These plants may have formed extensive angiosperm-dominated communities in consistently unstable environments (e.g., aggradational river bottoms) but probably were displaced by the large-seeded and dominant gymnosperms in more stable habitats. Some angiosperms may have formed an understory of shrubs and small trees in open-canopied gymnosperm forests such as those of the uplands of New Caledonia in the present day (L. J. Hickey, pers. comm.).

Therefore, while Cretaceous angiosperms were probably diverse in certain habitats and perhaps numerically-dominant over the gymnosperms, they did not dominate the world vegetation in the physiognomic sense that they do in the modern day. Rather, they were a more specialized "weedy" group that initially radiated to fill a specific aspect of the community, but that did not continue their radiation at the same rate throughout the Cretaceous. This implies that angiosperm diversity should have risen slowly, rather than dramatically, during the later Cretaceous, which is what is observed (Krassilov, 1977; Niklas et al., 1980; Tiffney, 1981). That the Cretaceous angiosperms included few if any large canopy trees capable of sustaining a closed climax forest is inferred strictly from the paucity of large angiosperm diaspores. While large seeds would have permitted the angiosperms to achieve physiognomic dominance of the community, their size would necessitate association with biotic dispersal agents. Both the record of diaspore size and of animals in the Cretaceous suggest that the appropriate dispersal agents for angiosperms were few, and that the advantages of large seeds were outweighed by the disadvantage of their poor dispersal.

This impasse was broken by the radiation of birds and mammals (including bats) in the early Tertiary or perhaps latest Cretaceous, leading to the swift development of many biotic dispersal syndromes, which in turn influenced the evolution of the angiosperms in the Tertiary.

The total diversity of angiosperms increased dramatically in the early Tertiary (Niklas et al., 1980; Tiffney, 1981). This is to be expected as a result of the development of plant-animal dispersal interactions for three reasons. First, coevolution favored increasing specialization and speciation (cf. Regal, 1977). Second, greater distance of dispersal favored allopatric speciation, an ultimate example of which is the modern eastern North America–eastern Asia pattern of disjunction (Wood, 1972; Wolfe, 1975). Finally, the establishment of dispersal syndromes with animals opened the way for angiosperms to explore all possible habits and habitats within the community. This is partially reflected in the high value of the coefficients of correlation (Table 2) in the Tertiary, indicating the increased polymodality of diaspore size in Tertiary floras. Increased seed size led to an arborescent community with decreased light penetration, diversified biotic competition, and presumably tighter species packing, all features favoring increased diversity. Further, all the changes in the rate of diversification initiated by the interaction of dispersal agents and angiosperms would both accentuate and be accentuated by parallel coevolutionary interactions with insect pollinators (Crepet, 1979, 1984).

The latest Cretaceous and early Tertiary is also remarkable as a time of rapid "modernization" of the world's flora. This involves two aspects; the sudden appearance of large numbers of modern families and genera (Niklas et al., 1980; Muller, 1981; Tiffney, 1981) and their swift spread over the Northern Hemisphere (Wolfe, 1975). Both follow logically from the establishment of dispersal relationships. The rapid appearance of modern taxa results from the increased rate of speciation occasioned by the initiation of the coevolutionary spiral that extends to the present day. The rapid spread of these taxa follows from their association with the effective dispersal agents, although three other factors were of importance. First, the Late Paleocene and the Eocene were periods of warm climate in the northern hemisphere (Buchardt, 1978; Wolfe, 1978). Second, apparently both North Atlantic and North Pacific land bridges were available to terrestrial organisms in the Early Tertiary (Lehmann, 1973; McKenna, 1975; Tiffney, 1980). Finally, and perhaps most importantly, the newly-evolved angiosperm taxa probably included many adaptations to previously unfilled "niches" and were spreading at a time when several gymnosperm groups had recently declined or gone extinct (Krassilov, 1978; Niklas et al., 1980; Tiffney, 1981; Vachrameev, 1982).

The question remains: what initiated the increased level of interaction between plants and dispersal agents in the latest Cretaceous and early Tertiary? Did external factors (e.g., the decline of the reptiles) lead to the evolution of new groups of birds and mammals, which in turn spurred angiosperm evolution? Alternatively, did angiosperms (perhaps responding to the appearance of new groups of insect pollinators, Crepet, 1984) begin to diversify first and thereby stimulate the evolution of potential dispersal agents? I do not believe that the data given here will support interpretations of cause and effect, if indeed, such considerations are not rendered irrelevant by the synergistic nature of coevolutionary relationships. Further, while I have emphasized the historical importance of the development of angiosperm dispersal syndromes in this paper, it is only one of three coevolutionary features that must have had a strong influence on the course of angiosperm evolution. Pollination syndromes have perhaps had an even greater influence in view of the vast array of morphological and ethological permutations involved. Indeed, the interactions of angiosperms with modern pollinators may have been established at a slightly earlier date than the interactions with modern dispersal agents (cf. Crepet, 1984). Additionally, interactions between herbivores and plants have not been explored in the fossil record but must also have been of significance (cf. Niklas, 1978).

Finally, I would like to explore briefly three ancillary points.

First, the evidence presented here suggests that the "Durian Theory" (Corner, 1949, 1964) is untenable. The "Durian Theory" assumes that the primitive angiosperm seed was arillate, of moderate size, probably animal dispersed, and was contained in a dehiscent fruit borne of a large, pachycaulous tree. According to the fossil record of angiosperm seeds, such a combination of characters could have evolved only in the Tertiary, following the establishment of widespread biotic dispersal syndromes. Certainly no evidence is seen of "moderate-sized" angiosperm seeds in the Cretaceous: they are all small. Further, no evidence is seen of widespread animal dispersal in the Cretaceous, or of pachycauly, although the scarcity of Cretaceous angiosperm wood (Wolfe et al., 1975) renders the last a statement based on negative evidence.

Second, the minimum and maximum seed sizes did not appear to have undergone any significant directional change of size during the time period measured. Two regressions were run: (1) the size of the smallest seed of each of the floras examined (Table 2) against time, and (2) the size of the largest seed from each of the Tertiary floras against time. The first regression was not significant ($0.20 > P > 0.10$; $r = 0.29$, $N = 26$); the Rusinga flora was excluded on account of its anomalously large "smallest seed"—probably a function of the collection of the flora from surficial lag deposits. In the second case ($N = 18$, $r = 0.40$), the r value is marginally insignificant ($P = 0.10$), however, this figure may be influenced by the temperate adaptations of the source plants of the later Tertiary. This suggests that the two seed classes have achieved some form of balance between the selective features that affect size. The time stability of small seed size implies that the appearance of larger seeds in the early Tertiary did not alter the selective advantage of small seed size in certain environments. This could be further extended to imply that the basic habitats available to smaller-seeded plants have not greatly altered during the history of the angiosperms, although in some times habitats favoring opportunistic forms are less widespread (early Tertiary) than in others (late Tertiary, Pleistocene). With respect to the larger seeds, it appears that there has been no distinct trend of size increase through the Tertiary. This could imply that there is an optimal upper limit for seed size, one that strikes a balance between available endosperm and efficiency of dispersal, and that was achieved by the early Tertiary. However, this observation may be influenced by climate. The fossils are primarily from Europe and sample a tropical vegetation in the early Tertiary, but an increasingly more temperate one through the later Tertiary. Limited data (Levin, 1974) suggest that modern tropical lowland communities have slightly larger seeds than modern temperate ones. If so, then possibly the size of the largest seeds did increase slightly through the Tertiary.

Finally, both Harper (1961) and Margelef (1968) have suggested that the evolutionary history of a group should tend to parallel its successional history, and that an evolving group should "climb its own seral tree." This is what is seen in the fossil record, with the angiosperms initially appearing as weedy plants and in due time evolving to become dominant members of the climax community. The fossil record suggests that this transition required the appearance of dispersal vectors to permit the dispersal of large seeds of the plants of later seral stages. This implies that the unique characters of the angiosperms (rapidity of life cycle, potential for insect pollination, specialized conducting tissue, etc.) were not sufficient separately or jointly to directly ensure the final dominance of the group. However, the developmental plasticity of the angiosperms did permit them to evolve a diversity of fruit and seed dispersal adaptations in response to the appearance of dispersal agents. This observation raises interesting questions about the structure and function of pre-angiosperm communities. Were dispersal agents involved in previous climax communities? Do climax communities in which dispersal agents are not available have a different, perhaps more open, canopy structure than those in which dispersal agents are present? [For example, could the seeming diversity of lowland Carboniferous coal swamps as contrasted to the upland Carboniferous vegetation be influenced by the availability of mechanisms permitting the dispersal of large seeds in the lowland community (water, fish) and their absence in the upland communities?]

SUMMARY

Analysis of Cretaceous and Tertiary fruit and seed floras from the Northern Hemisphere reveals a change in the average size and range of size of angiosperm diaspores through time. Cretaceous floras are composed almost entirely of small diaspores. Early Tertiary floras are dominated by large diaspores but include many as small as those of the Cretaceous. Later Tertiary floras are primarily composed of smaller diaspores but consistently include a few very large ones. Analyses suggest that the minimum diaspore size for angiosperms has not changed since their appearance in the Cretaceous, and that their maximum size has not increased greatly, if at all, from the time of appearance of large diaspores in the earliest Tertiary to the present. There are two major features that influence diaspore size: (1) the relation between seed size and the ecological characteristics of the parent plant, and (2) dispersal mechanisms. The observed pattern in angiosperm diaspore size through time may be interpreted in light of these two selective forces. Cretaceous angiosperms were primarily small-seeded, abiotically dispersed shrubs or opportunistic trees, perhaps occupying marginal or open habitats in the gymnosperm-dominated vegetation, but probably not forming a closed-canopy climax community. The relative paucity of dispersal agents in the Cretaceous limited the success of large angiosperm diaspores and the closed-canopy forest that they could be expected to give rise to. The latest Cretaceous or early Tertiary radiation of birds, bats, and terrestrial mammals reversed this situation, permitting a biotic interaction favoring large, animal-dispersed propagules. This in turn allowed the establishment of angiosperm seedlings in areas of low light intensity and led to the development of stable, closed-canopy, climax communities, physiognomically- as well as numerically-dominated by angiosperms, and similar in structure for the first time to those of the modern day. This interaction of seed size and dispersal agents may have occurred with, or slightly later than, the establishment of interactions between angiosperms and modern pollinators. Regardless of sequence, the establishment of biological interactions between angiosperms and their pollinators and dispersers was reflected in the rapid appearance of modern families and genera, and of their swift spread around the northern hemisphere, in the latest Cretaceous and early Tertiary.

LITERATURE CITED

AXELROD, D. I. 1960. The evolution of flowering plants. Pp. 227–305 *in* S. Tax (editor), The Evolution of Life. Univ. Chicago Press, Chicago.

———. 1970. Mesozoic paleogeography and early angiosperm history. Bot. Rev. (Lancaster) 36: 277–319.

BAKER, H. G. 1972. Seed weight in relation to environmental conditions in California. Ecology 53: 997–1010.

BARRON, E. J., S. L. THOMPSON & S. H. SCHNEIDER. 1981. An ice-free Cretaceous? Results from climatic model simulations. Science 212: 501–508.

BERG, R. Y. 1975. Myrmecochorous plants in Australia and their dispersal by ants. Austral. J. Bot. 23: 475–508.

BINDA, P. L. 1968. New species of *Spermatites* from the Upper Cretaceous of southern Alberta. Rev. Micropaléontol. 11: 137–142.

BONES, T. J. 1979. Atlas of fossil fruits & seeds from North Central Oregon. Oregon Mus. Sci. Industr. Occas. Pap. 1: 1–23.

BRAUN, E. L. 1950. Deciduous Forests of Eastern North America. Blakiston Co., Philadelphia.

BRODKORB, P. 1971. Origin and evolution of birds. Pp. 19–55 *in* D. S. Farner & J. R. King (editors), Avian Biology, Volume I. Academic Press, New York.

———. 1976. Discovery of a Cretaceous bird, apparently ancestral to the orders Coraciiformes and Piciformes (Aves: Carinatae). Smithsonian Contr. Paleobiol. 27: 67–73.

BROWN, R. W. 1962. Paleocene flora of the Rocky Mountains and great Plains. Profess. Pap. U.S. Geol. Surv. 375: 1–119.

BUCHARDT, B. 1978. Oxygen isotope palaeotemperatures from the Tertiary period in the North Sea area. Nature 275: 121–123.

BURNHAM, L. 1978. Survey of social insects in the fossil record. Psyche 85: 85–133.

BŮŽEK, C. & F. HOLÝ. 1964. Small-sized plant remains from the coal formation of the Chomutov-Most-Teplice Basin. Sborn. Geol. Věd Paleontol. 4: 105–138.

CHALONER, W. G. & A. SHEERIN. 1981. The evolution of reproductive strategies in early land plants. Pp. 93–100 *in* G. G. E. Scudder & J. L. Reveal (editors), Evolution Today, Proceedings of the Second International Congress of Systematic and Evolutionary Biology. Hunt Institute, Pittsburgh.

CHANDLER, M. E. J. 1923. Geological history of the genus *Stratiotes*: an account of the evolutionary changes which have occurred within the genus during Tertiary and Quaternary times. Quart. J. Geol. Soc. London 79: 117–138.

———. 1957. The Oligocene flora of the Bovey Tracey Lake Basin, Devonshire. Bull. Brit. Mus. (Nat. Hist.), Geol. 3: 71–123.

———. 1958. Angiosperm fruits from the Lower Cretaceous of France and the Lower Eocene (London Clay) of Germany. Ann. Mag. Nat. Hist., Ser. 13, 1: 354–358.

———. 1961. The Lower Tertiary Floras of Southern England. I. Palaeocene Floras. London Clay Flora (Supplement). British Museum (Natural History), London.

———. 1964. The Lower Tertiary Floras of Southern England. IV. A Summary and Survey of Findings in the Light of Recent Botanical Observations. British Museum (Natural History), London.

——— & D. I. AXELROD. 1961. An Early Cretaceous (Hauterivian) angiosperm fruit from California. Amer. J. Sci. 259: 441–446.

CHESTERS, K. I. M. 1955. Some plant remains from the Upper Cretaceous and Tertiary of West Africa. Ann. Mag. Nat. Hist., Ser. 12, 8: 498–504.

———. 1957. The Miocene flora of Rusinga Island, Lake Victoria, Kenya. Palaeontographica, Abt. B, Paläophytol. 101: 30–71.

CLEMENS, W. A. 1979. Marsupialia. Pp. 192–220 *in* J. A. Lillegraven, Z. Kielan-Jaworowska & W. A. Clemens (editors), Mesozoic Mammals. Univ. California Press, Berkeley.

——— & Z. KIELAN-JAWOROWSKA. 1979. Multituberculata. Pp. 99–149 *in* J. A. Lillegraven, Z. Kielan-Jaworowska & W. A. Clemens (editors), Mesozoic Mammals. Univ. California Press, Berkeley.

COLBERT, E. H. 1969. Evolution of the Vertebrates, a History of the Backboned Animals Through Time, 2nd edition. J. Wiley & Sons, New York.

COLIN, J.-P. 1973. Microfossiles vegetaux dans le Cenomanien et le Turonien de Dordogne (S. O. France). Palaeontographica, Abt. B, Paläophytol. 143: 106–119.

COLLINSON, M. E., K. FOWLER & M. C. BOULTER. 1981. Floristic changes indicate a cooling climate in the Eocene of southern England. Nature 291: 315–317.

CONSTANTINE, D. G. 1970. Bats in relation to the health, welfare and economy of man. Pp. 319–449 *in* W. A. Wimsatt (editor), Biology of Bats, Volume II. Academic Press, New York.

CORNER, E. J. H. 1949. The durian theory or the origin of the modern tree. Ann. Bot. (London) n.s. 13: 367–414.

———. 1964. The Life of Plants. World Publ. Co., Cleveland.

CRACRAFT, J. 1973. Continental drift, paleoclimatology, and the evolution and biogeography of birds. J. Zool. London 169: 455–545.

———. 1982. Phylogenetic relationships and monophyly of Loons, Grebes and Hesperornithiform birds, with comments on the early history of birds. Syst. Zool. 31: 35–56.

CREPET, W. L. 1979. Insect pollination: a paleontological perspective. BioScience 29: 102–108.

———. 1984 [1985]. Advanced (constant) insect pollination mechanisms: pattern of evolution and implications vis-à-vis angiosperm diversity. Ann. Missouri Bot. Gard. 71: 607–630.

CRONQUIST, A. 1968. The Evolution and Classification of Flowering Plants. Houghton Mifflin Co., Boston.

CZECZOTT, H. & A. SKIRGIELLO. 1959. The fossil flora of Turów near Bogatynia. Systematic description of plant remains. (1). Dicotyledones: Hamamelidaceae, Nymphaeaceae, Sabiaceae, Vitaceae, Nyssaceae. Prace Muz. Ziemi 3: 93–112, 121–128.

——— & ———. 1961a. The fossil flora of Turów near Bogatynia. Systematic description of plant remains. (2). Juglandaceae. Prace Muz. Ziemi 4: 51–73, 103–113.

——— & ———. 1961b. The fossil flora of Turów near Bogatynia. Systematic description of plant remains. (2). Aceraceae. Prace Muz. Ziemi 4: 78–81, 116–117.

——— & ———. 1967. The fossil flora of Turów near Bogatynia. Systematic description of plant remains. (3). Araceae, Betulaceae, Menispermaceae, Meliaceae, Sterculiaceae, Passifloraceae, Combretaceae, Trapaceae, Symplocaceae, Styracaceae. Prace Muz. Ziemi 10: 97–141, 143–166.

——— & ———. 1975. The fossil flora of Turów near Bogatynia. Systematic description of the plant remains. (4). Magnoliaceae, Celastraceae, Cornaceae, Sapotaceae. Prace Muz. Ziemi 24: 25–56.

——— & ———. 1980a. The fossil flora of Turów

near Bogatynia. Systematic description of the plant remains. (5). Illiciaceae, Lauraceae, Rosaceae, Rutaceae, Staphyleaceae, Buxaceae. Prace Muz. Ziemi 33: 5–15.

——— & ———. 1980b. The fossil flora of Turów near Bogatynia. Systematic description of the plant remains. (5). Sparganiaceae, Zingiberaceae. Prace Muz. Ziemi 33: 17–21.

DILCHER, D. L. 1979. Early angiosperm reproduction: an introductory report. Rev. Palaeobot. Palynol. 27: 291–328.

———, W. L. CREPET, C. D. BEEKER & H. C. REYNOLDS. 1976. Reproductive and vegetative morphology of a Cretaceous angiosperm. Science 191: 854–856.

DOROFEEV, P. I. 1963. Tretichnye Flory Zapadnoĭ Sibiri. Izdat. Akad. Nauk S.S.S.R., Moscow.

———. 1968. Oligocene flora of Transuralia. Paleontol. J. (Transl.: Amer. Geol. Inst.) 1968: 248–255.

DOYLE, J. A. 1978. Origin of angiosperms. Annual Rev. Ecol. Syst. 9: 365–392.

——— & L. J. HICKEY. 1976. Pollen and leaves from the mid-Cretaceous Potomac Group and their bearing on early angiosperm evolution. Pp. 139–206 *in* C. B. Beck (editor), Origin and Early Evolution of Angiosperms. Columbia Univ. Press, New York.

FONTAINE, W.M. 1889. The Potomac or younger Mesozoic flora. Monogr. U.S. Geol. Surv. 15: 1–377.

FRANKIE, G. W., H. G. BAKER & P. A. OPLER. 1974. Comparative phenological studies of trees in tropical wet and dry forests in the lowlands of Costa Rica. J. Ecol. 62: 881–919.

FRIIS, E. M. 1975. Climatic implications of microcarpological analyses of the Miocene Fasterholt Flora, Denmark. Bull. Geol. Soc. Denmark 24: 179–191.

———. 1984 [1985]. Preliminary report of Upper Cretaceous angiosperm reproductive organs from Sweden and their level of organization. Ann. Missouri Bot. Gard. 71: 403–418.

GOLDING, M. 1980. The Fishes and the Forest, Explorations in Amazonian Natural History. Univ. California Press, Berkeley.

GREGOR, H.-J. 1978. Die Miozänen Frücht- und Samen-floren der Oberpfälzer Braunkohle I. Funde aus den Sandigen Zwischenmitteln. Palaeontographica, Abt. B, Paläophytol. 167: 8–103.

———. 1979. Systematics, biostratigraphy and paleoecology of the genus *Toddalia* Jussieu (Rutaceae) in the European Tertiary. Rev. Palaeobot. Palynol. 28: 311–363.

———. 1980a. Eine neues Klima- und Vegetations-Modell für das untere Sarmat (Mittelmiozän) Mitteleuropas unter spezieller Berücksichtigung floristischer Gegebenheiten. Verh. Geol. B.-A. 1979: 337–353.

———. 1980b. Die Miozänen Frücht- und Samenfloren der Oberpfälzer Braunkohle II. Funde aus den Kohlen und Tonigen Zwischenmitteln. Palaeontographica, Abt. B, Paläophytol. 174: 7–94.

HALL, J. W. 1963. Megaspores and other fossils in the Dakota Formation (Cenomanian) of Iowa, (U.S.A.). Pollen & Spores 5: 425–443.

———. 1967. Invalidity of the name *Chrysotheca* Miner for microfossils. J. Paleontol. 41: 1298–1299.

HALLS, L. K. 1977. Southern fruit-producing woody plants used by wildlife. U.S.D.A. Forest Serv. Gen. Techn. Rep. SO-16: 1–235.

HANDEL, S. N., S. B. FISCH & G. E. SCHATZ. 1981. Ants disperse a majority of herbs in a mesic forest community in New York State. Bull. Torrey Bot. Club 108: 430–437.

HARPER, J. L. 1961. Approaches to the study of competition. Symp. Soc. Exp. Biol. 15: 1–39.

———, P. H. LOVELL & K. G. MOORE. 1970. The shapes and sizes of seeds. Annual Rev. Ecol. Syst. 1: 327–356.

HICKEY, L. J. 1981. Land plant evidence compatible with gradual, not catastrophic, change at the end of the Cretaceous. Nature 292: 529–531.

——— & J. A. DOYLE. 1977. Early Cretaceous fossil evidence for angiosperm evolution. Bot. Rev. (Lancaster) 43: 3–104.

HUGHES, N. F. 1976. Palaeobiology of Angiosperm Origins. Cambridge Univ. Press, Cambridge.

JANZEN, D. H. 1977. Promising directions of study in tropical animal-plant interactions. Ann. Missouri Bot. Gard. 64: 706–736.

——— & P. S. MARTIN. 1982. Neotropical anachronisms: the fruits the Gomphotheres ate. Science 215: 19–27.

JEPSEN, G. L. 1970. Bat origins and evolution. Pp. 1–64 *in* W. A. Wimsatt (editor), Biology of Bats, Volume I. Academic Press, New York.

JONES, E. W. 1956. Ecological studies on the rainforest of southern Nigeria. II. J. Ecol. 44: 83–117.

JUNG, W., H. H. SCHLEICH & B. KÄSTLE. 1978. Eine neue, stratigraphisch gesicherte Fundstelle für Angiospermen-Früchte und -Samen in der oberen Gosau Tirols. Mitt. Bayer. Staatssamml. Paläontol. Hist. Geol. 18: 131–142.

KNOBLOCH, E. 1964. Neue Pflanzenfunde aus dem südböhmischen Senon. Jahrb. Staatl. Mus. Mineral. Dresden 1964: 133–201.

———. 1971. Fossile Früchte und Samen aus der Flyschzone der mährischen Karpaten. Sborn. Geol. Věd Paleontol. 13: 7–46.

———. 1975. Früchte und Samen aus der Gosauformation von Kössen in Österreich. Věstn. Ústřed. Ústavu Geol. 50: 83–92.

———. 1977. Paläokarpologische Characteristik der Flyschzone der mährischen Karpaten. Sborn. Geol. Věd Paleontol. 19: 79–137.

———. 1981. Die gattung *Costatheca* Hall in der mitteleuropäischen Kreide. Sborn. Geol. Věd Paleontol. 24: 95–114.

KOCH, B. E. 1972a. Fossil Picrodendroid fruit from the Upper Danian of Nûgssuaq, West Greenland. Meddel. Grønland 193(3): 1–33.

———. 1972b. Coryphoid palm fruits and seeds from the Danian of Nûgssuaq, West Greenland. Meddel. Grønland 193(4): 1–38.

KRASSILOV, V. A. 1967. Rannemelovaya Flora Yuzhnogo Primor'ya i ee Znachenie dlya Stratigrafii. Izdat Akad. Nauk S.S.S.R., Moscow.

———. 1973. Mesozoic plants and the problem of angiosperm ancestry. Lethaia 6: 163–178.

———. 1977. The origin of angiosperms. Bot. Rev. (Lancaster) 43: 143–176.

———. 1978. Late Cretaceous gymnosperms from Sakhalin and the terminal Cretaceous event. Palaeontology 21: 893–905.

KRON, D. G. 1979. Docodonta. Pp. 91–98 *in* J. A. Lillegraven, Z. Kielan-Jaworowska & W. A. Clemens (editors), Mesozoic Mammals. Univ. California Press, Berkeley.

KUROCHKIN, E. N. 1976. A survey of the Paleogene birds of Asia. Smithsonian Contr. Paleobiol. 27: 75–86.

ŁANCUCKA-ŚRODONIOWA, M. 1966. Tortonian flora from the "Gdów Bay" in the south of Poland. Acta Palaeobot. 7: 3–135.

———. 1979. Macroscopic plant remains from the freshwater Miocene of the Nowy Sącz Basin (West Carpathians, Poland). Acta Palaeobot. 20: 3–117.

LEHMANN, U. 1973. Zur Paläogeographie des Nordatlantiks im Tertiär. Mitt. Geol.-Paläontol. Inst. Univ. Hamburg 42: 57–69.

LEOPOLD, E. B. 1967. Late-Cenozoic patterns of plant extinction. Pp. 203–246 *in* P. S. Martin & H. E. Wright, Jr. (editors), Pleistocene Extinctions, The Search for a Cause. Yale Univ. Press, New Haven.

LESQUEREUX, L. 1874. Contribution to the fossil flora of the western territories. Part I. The Cretaceous flora. Rep. U.S. Geol. Surv. 6: 1–136.

———. 1892. The flora of the Dakota Group. Monogr. U.S. Geol. Surv. 17: 1–400.

LEVIN, D. A. 1974. The oil content of seeds: an ecological perspective. Amer. Naturalist 108: 193–206.

——— & H. W. KERSTER. 1974. Gene flow in seed plants. Evol. Biol. 7: 139–220.

LILLEGRAVEN, J. A. 1979. Introduction. Pp. 1–6 *in* J. A. Lillegraven, Z. Kielan-Jaworowska & W. A. Clemens (editors), Mesozoic Mammals. Univ. California Press, Berkeley.

———, M. J. KRAUS & T. M. BOWN. 1979. Paleogeography of the world of the Mesozoic. Pp. 277–308 *in* J. A. Lillegraven, Z. Kielan-Jaworowska & W. A. Clemens (editors), Mesozoic Mammals. Univ. California Press, Berkeley.

MCKENNA, M. C. 1975. Fossil mammals and Early Eocene North Atlantic land continuity. Ann. Missouri Bot. Gard. 62: 335–353.

MAI, D. H. 1964. Die Mastixioideen-Floren im Tertiär der Oberlausitz. Paläontol. Abh., Abt. B, Paläobot. 2: 1–192.

———. 1965. Eine Pliozäne Flora von Kranichfeld in Thüringen. Mitt. Zent. Geol. Inst. 1965: 37–64.

———. 1970. Change of climate and biostratigraphy in the continental younger Tertiary of boreal province. Giorn. Geol. (2)35: 85–90.

———. 1976. Eozäne Floren des Geiseltales. Abh. Zent. Geol. Inst. 26: 93–149.

——— & H. WALTHER. 1978. Die Floren der Haselbacher Serie im Weisselster-Becken (Bezirk Leipzig, DDR). Abh. Staatl. Mus. Mineral. Dresden 28: 1–200.

———, J. MAJEWSKI & K. P. UNGER. 1963. Pliozän und Altpleistozän von Rippersroda in Thüringen. Geologie 12: 765–815.

MANCHESTER, S. R. & D. L. DILCHER. 1981. [Abstract:] Fossil fruits and the history of the walnut family. Bot. Soc. Amer. Misc. Ser. Publ. 160: 45–46.

MARGELEF, R. 1968. Perspectives in Ecological Theory. Univ. Chicago Press, Chicago.

MARTIN, A. C., H. S. ZIM & A. L. NELSON. 1951. American Wildlife and Plants. McGraw-Hill Book Co., Inc., New York.

MAZER, S. J. & B. H. TIFFNEY. 1982. Fruits of *Wetherellia* and *Palaeowetherellia* (?Euphorbiaceae) from Eocene sediments in Virginia and Maryland. Brittonia 34: 300–333.

MILLER, C. N., JR. & C. R. ROBISON. 1975. Two new species of structurally preserved pinaceous cones from the Late Cretaceous of Martha's Vineyard Island, Massachusetts. J. Paleontol. 49: 138–150.

MINER, E. L. 1935. Paleobotanical examinations of Cretaceous and Tertiary coals. Amer. Midl. Naturalist 16: 585–615.

MONTEILLET, J. & J.-R. LAPPARTIENT. 1981. Fruits et graines du Crétacé Supérier des carrières de Paki (Sénégal). Rev. Palaeobot. Palynol. 34: 331–344.

MULLER, J. 1981. Fossil pollen records of extant angiosperms. Bot. Rev. (Lancaster) 47: 1–142.

NIKITIN, P. I. 1965. Akvitanskaya Semennaya Flora Lagernogo Sada (Tomsk). Izdat Tomskogo Univ., Tomsk.

NIKLAS, K. J. 1978. Coupled evolutionary rates and the fossil record. Brittonia 30: 373–394.

———, B. H. TIFFNEY & A. H. KNOLL. 1980. Apparent changes in the diversity of fossil plants. Evol. Biol. 12: 1–89.

OPLER, P. A., H. G. BAKER & G. W. FRANKIE. 1980. Plant reproductive characteristics during secondary succession in Neotropical lowland forest ecosystems. Biotropica, Suppl. 12: 40–46.

REGAL, P. J. 1977. Ecology and evolution of flowering plant dominance. Science 196: 622–629.

REID, E. M. & M. E. J. CHANDLER. 1933. The London Clay Flora. British Museum (Natural History), London.

RETALLACK, G. 1981. Fossil soils: indicators of ancient terrestrial environments. Pp. 55–102 *in* K. J. Niklas (editor), Paleobotany, Paleoecology and Evolution, Volume I. Praeger Publishers, New York.

——— & D. L. Dilcher. 1981. A coastal hypothesis for the dispersal and rise to dominance of the flowering plants. Pp. 27–77 *in* K. J. Niklas (editor), Paleobotany, Paleoecology and Evolution, Volume II. Praeger Publishers, New York.

RIDLEY, H. N. 1930. The Dispersal of Plants Throughout the World. L. Reeve & Co., Kent.

ROMER, A. S. 1966. Vertebrate Paleontology, 3rd edition. Univ. Chicago Press, Chicago.

SALISBURY, E. J. 1942. The Reproductive Capacity of Plants. Studies in Quantitative Biology. G. Bell & Sons, London.

SAMYLINA, V. A. 1960. Pokrytosemennye rasteniya iz nizhnemelovykh otlozhenii Kolymy. Bot. Žurn. (Moscow & Leningrad) 45: 335–352.

———. 1961. Novye dannye o nizhnemelovoĭ flore yuzhnogo Primor'ya. Bot. Žurn. (Moscow & Leningrad) 46: 634–645.

———. 1968. Early Cretaceous angiosperms of the Soviet Union based on leaf and fruit remains. J. Linn. Soc., Bot. 61: 207–218.

SAVIN, S. M. 1977. The history of the earth's surface

temperature during the last 100 million years. Annual Rev. Earth Planet. Sci. 5: 319–355.

SCHEMEL, M. P. 1950. Cretaceous plant microfossils from Iowa. Amer. J. Bot. 37: 750–754.

SCHUSTER, R. M. 1976. Plate tectonics and its bearing on the geographical origin and dispersal of angiosperms. Pp. 48–138 *in* C. B. Beck (editor), Origin and Early Evolution of Angiosperms. Columbia Univ. Press, New York.

SCHWARZWALDER, R., JR. & D. L. DILCHER. 1981. [Abstract:] Platanoid leaves and infructescences from the Cenomanian of Kansas. Bot. Soc. Amer. Misc. Ser. Publ. 160: 47.

SCOTT, R. A. 1954. Fossil fruits and seeds from the Eocene Clarno Formation of Oregon. Palaeontographica, Abt. B, Paläophytol. 96: 66–97.

SILVERTOWN, J. W. 1981. Seed size, life span, and germination date as coadapted features of plant life history. Amer. Naturalist 118: 860–864.

SMITH, J. D. 1976. Chiropteran evolution. *In* R. J. Baker, J. K. Jones, Jr. & D. C. Carter (editors), Biology of Bats of the New World Family Phyllostomatidae, Part I. Special Publ. Mus. Texas Tech Univ. 10: 49–69.

SMYTHE, N. 1970. Relationships between fruiting seasons and seed dispersal methods in a neotropical forest. Amer. Naturalist 104: 25–35.

STEBBINS, G. L. 1971. Adaptive radiation of reproductive characteristics in angiosperms, II: seeds and seedlings. Annual Rev. Ecol. Syst. 2: 237–260.

STILES, E. W. 1980. Patterns of fruit presentation and seed dispersal in bird-disseminated woody plants in the eastern deciduous forest. Amer. Naturalist 116: 670–688.

TAKHTAJAN, A. 1969. Flowering Plants, Origin and Dispersal. Smithsonian Inst. Press, Washington, D.C.

———. 1976. Neoteny and the origin of flowering plants. Pp. 207–219 *in* C. B. Beck (editor), Origin and Early Evolution of Angiosperms. Columbia Univ. Press, New York.

THOMPSON, S. L. & E. J. BARRON. 1981. Comparison of Cretaceous and present earth albedos: implications for the causes of paleoclimates. J. Geol. 89: 143–167.

TIFFNEY, B. H. 1977a. Fossil angiosperm fruits and seeds. J. Seed Technol. 2: 54–71.

———. 1977b. Dicotyledonous angiosperm flower from the Upper Cretaceous of Martha's Vineyard, Massachusetts. Nature 265: 136–137.

———. 1980. [Abstract:] The Tertiary flora of eastern North America and the North Atlantic land bridge. Second Int. Congr. Syst. Evol. Biol. 373.

———. 1981. Diversity and major events in the evolution of land plants. Pp. 193–230 *in* K. J. Niklas (editor), Paleobotany, Paleoecology and Evolution, Volume II. Praeger Publishers, New York.

TRALAU, H. 1968. Evolutionary trends in the genus *Ginkgo*. Lethaia 1: 63–101.

VACHRAMEEV, V. A. 1952. Regionalinay Stratigrafii S.S.S.R. Tom. 1. Stratigrafii i Iskopaemay Flora Meloviih Otlozhenii Zapadnogo Kazakhstana. Izdat. Akad. Nauk S.S.S.R., Moscow.

———. 1982. Ancient angiosperms and the evolution of the flora in the middle of the Cretaceous period. Paleontol. J. (Transl.: Amer. Geol. Inst.) 1982: 1–11.

——— & V. A. KRASSILOV. 1979. Reproductive organs of flowering plants from the Albian of Kazakhstan. Paleontol. J. (Transl.: Amer. Geol. Inst.) 1979: 112–118.

VAN DER BURGH, J. 1978. The Pliocene flora of Fortuna-Garsdorf I. Fruits and seeds of angiosperms. Rev. Palaeobot. Palynol. 26: 173–211.

VAN DER PIJL, L. 1957. The dispersal of plants by bats (Chiropterochory). Acta Bot. Neerl. 6: 291–315.

———. 1969. Principles of Dispersal in Higher Plants. Springer-Verlag, Berlin.

VAN EYSINGA, F. W. B. 1975. Geologic Time Table, 3rd edition. Elsevier Publ. Co., Amsterdam.

VANGEROW, E. F. 1954. Megasporen und andere pflanzliche Mikrofossilien aus der Aachner Kreide. Palaeontographica, Abt. B, Paläophytol. 96: 24–38.

VAN TYNE, J. & A. J. BERGER. 1976. Fundamentals of Ornithology, 2nd edition. J. Wiley & Sons, New York.

VASIL'YEV, V. N. 1967. A new genus of Trapaceae. Paleontol. J. (Transl.: Amer. Geol. Inst.) 1967: 92–97.

WILLIS, J. C. 1973. A Dictionary of the Flowering Plants and Ferns, 8th edition. Revised by H. K. A. Shaw. Cambridge Univ. Press, Cambridge.

WING, S. L. 1981. A Study of Paleoecology and Paleobotany in the Willwood Formation (Early Eocene, Wyoming). Ph.D. dissertation, Yale Univ., New Haven.

WOLFE, J. A. 1975. Some aspects of plant geography of the northern hemisphere during the Late Cretaceous and Tertiary. Ann. Missouri Bot. Gard. 62: 264–279.

———. 1978. A paleobotanical interpretation of Tertiary climates in the northern hemisphere. Amer. Sci. 66: 694–703.

———, J. A. DOYLE & V. M. PAGE. 1975. The bases of angiosperm phylogeny: paleobotany. Ann. Missouri Bot. Gard. 62: 801–824.

WOOD, C. E. 1972. Morphology and phytogeography: the classical approach to the study of disjunctions. Ann. Missouri Bot. Gard. 59: 107–124.

PAPER 22

Patterns of Ecological Diversity in Fossil and Modern Mammalian Faunas (1979)

P. Andrews, J. M. Lord, and E. M. Nesbit Evans

Commentary

CATHERINE BADGLEY

Ecological diversity, in contrast to taxonomic diversity, refers to ecological properties of species in modern or ancient biotas. Just as taxonomic diversity of an ecosystem is measured in terms of species or generic richness, ecological diversity is measured by the frequency of taxa in classes of ecological properties, such as body-weight classes. This approach captures information from many species rather one or two "indicator species" for habitat reconstruction, provides a snapshot of ecological interactions and three-dimensional substrate use, and permits faunas (biotas) that differ strikingly in taxonomic composition to be compared in terms of ecological structure.

Andrews et al. documented the ecological diversity of 23 modern and five Neogene mammalian faunas from tropical Africa in terms of taxonomic affiliation (mammalian order), body-weight class, locomotor adaptation, and major feeding habit of species in each assemblage. These ecological attributes are feasibly estimated for fossil mammals when appropriate skeletal elements are preserved. The authors noted that ecological diversity in modern faunas is diagnostic of habitat (vegetation) and proceeded to infer the paleohabitat for five hominoid-bearing fossil sites in East Africa. Ecological-diversity profiles for three fossil faunas corresponded to modern vegetation categories, one fauna had attributes of two categories and demonstrated strong preservational bias, and one fauna did not correspond to any of the modern habitats represented. Despite some methodological weaknesses, the paper presented both strong rationale and evidence for the significance of ecological diversity and stimulated vertebrate paleoecologists to take similar approaches to a wide range of modern and ancient faunas.

In addition to habitat inference, ecological-diversity analysis has broader implications for paleoecology. Biogeographic reconstructions of past intervals in Earth history require information about geographic variation in the ecological character of ecosystems across continents and oceans. Hypotheses of top-down or bottom-up regulation of local diversity require evaluation of changes in trophic diversity. The fossil record provides ecological-diversity profiles to test the concept of community convergence. Finally, the continuing controversy about ecological limits to diversification and diversity at different scales requires information about not only numbers of taxa but also their ecological attributes for mechanistic understanding. Although this paper was modest in its goals, it was influential in enlarging the scope of diversity analyses to include ecological diversity.

From *Biological Journal of the Linnean Society* 11:177–205.

Biological Journal of the Linnean Society, 11: 177–205. With 7 figures
March 1979

Patterns of ecological diversity in fossil and modern mammalian faunas

PETER ANDREWS, J. M. LORD

British Museum (Natural History), London

AND

ELIZABETH M. NESBIT EVANS

Department of Zoology, University of Cambridge, Cambridge

Accepted for publication February 1978

Ecological diversity provides a means of analysing the community structure of fossil mammalian faunas in order to obtain information on the habitat of the fauna. As a basis for the analysis, 23 modern mammalian communities from distinct habitats have been used to establish patterns of community structure for tropical African habitats according to their species diversity by taxonomic group, size, locomotor zonal adaptation, and feeding adaptation. All the communities tested were in tropical Africa, but additional analyses on tropical forest communities in Australia, Malaya, and Panama have shown that these communities, which all have completely different species composition, nevertheless have community structures very similar to each other and to those of the African forest communities.

Analysis of the ecological diversity patterns of African mammalian communities has made it possible to differentiate statistically between five habitat types, lowland forest, montane forest, woodland–bushland, flood plain and short grass plains, and to identify these habitats by the pattern of their faunas alone. When five fossil faunas, ranging in age from the early Miocene to the Pleistocene, were analysed by this method, considerable differences were detected between the faunas, and three of the faunas had ecological diveristy patterns which paralleled those from modern habitats. It is suggested that the three fossil faunas which most resemble the modern ones, Songhor, Fort Ternan and Olduvai, were probably derived from habitats like those of their modern counterparts, namely lowland forest, woodland–bushland, and woodland–grassland respectively. This is partly supported by earlier investigations of indicator species. The ecological diversity patterns for two of the fossil faunas, Rusinga and Karungu did not show such strong resemblance to modern patterns. The Rusinga fauna was evidently a mixed fauna, derived from at least two distinct habitat types with taphonomical biases altering it further, and the fauna from Karungu may have come from an ecosystem which was unlike any of the modern faunas which we have analysed in this study.

KEY WORDS: — ecological diversity — mammal communities — habitats — palaeoenvironments.

CONTENTS

0024-4066/79/020177-29/$02.00/0

INTRODUCTION

Palaeoecology is the study of fossilized plants and animals in relation to their environments. It is far less precise a discipline than the ecology of living organisms because the material available for study is always incomplete. As a result, more reliance is placed on hypothesis and interpretation than is acceptable in biological sciences concerned with present day phenomena. Moreover, the interpretations of palaeoecology are often limited to only one aspect, the palaeoenvironment, using evidence from the conditions of deposition and the nature of the organisms present in the fossil assemblage (Andrews & Van Couvering, 1975). The purpose of this paper is to suggest a new approach to the interpretation of palaeoecology through the analysis of the ecological diversity of fossil mammal assemblages.

Ecological diversity (Fleming, 1973) is a recent concept in modern biology. It was developed from earlier studies on species diversity in avian and mammalian communities (Karr, 1971; Simpson, 1964). These have shown that there are changes in numbers (or diversity) of animal and plant species following a north-south gradient on different continents, the numbers increasing with decrease in latitude. Superimposed on this gradient is a minor topographical gradient (Simpson, 1964) with numbers of species increasing with greater variety of topography. In addition to these changes in the numbers of species, the ecological adaptations of the species also change from one community to another in such things as the distribution of feeding habits, locomotor zonal adaptations and body sizes of the animals. This results in structural changes within the communities, and this variation is termed ecological diversity. It has been demonstrated that the ecological diversity of mammal communities changes significantly with latitude (Fleming, 1973), and we show here that the ecological diversity of mammal communities within one latitudinal zone varies significantly with habitat type. When both latitude and habitat type are similar, however, the ecological diversity of different mammal communities is remarkably constant.

This consistency of ecological diversity extends even to mammal communities from different continents, in the habitats so far examined, despite the fact that they may have few or no species in common. The ecological diversity patterns for tropical forest communities from four continents are compared in Fig. 1, and it is immediately apparent that despite minor differences their basic pattern is very similar. All the faunas are dominated by the bats, but there are higher proportions of insectivorous bats in the Australian and Malayan forests (Harrison, 1962) and of frugivorous bats in the Central American Forest

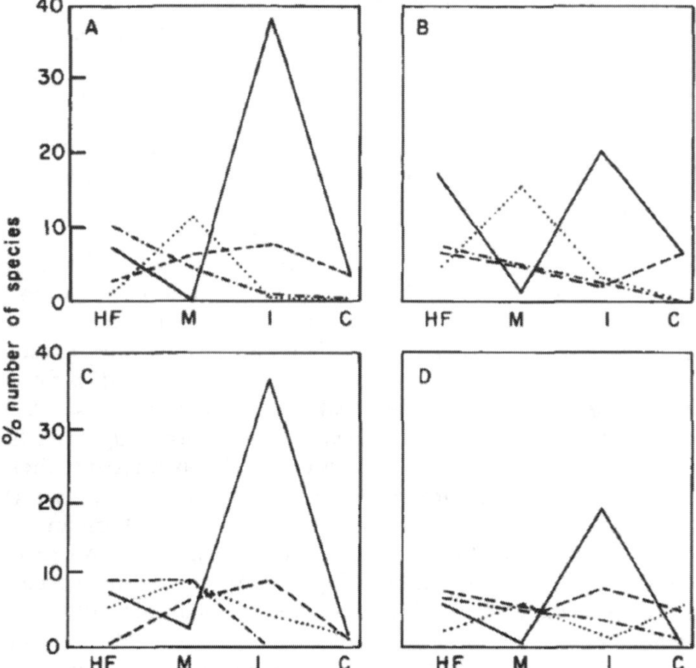

Figure 1. Proportions of forest mammals in communities in Malaya, Australia, Panama, and Zaire by locomotor zonal adaptations and feeding habits. The locomotor adaptations shown are Aerial, which includes bats and flying squirrels; Arboreal, which covers upper canopy small branch adapted animals like primates; Scansorial, which includes clawed large branch adapted animals such as squirrels; and SGM, the small ground mammals that come a little way into the lowest canopy or on to fallen trees, but which live mainly on the ground. The larger exclusively terrestrial mammals have not been included because they are relatively uncommon. They always follow identical patterns in the four diagrams. The feeding classes are: HF, plant eaters which includes herbivores and frugivores; I, insectivores; C, carnivores; M, mixed feeders.

A. Malaya, all forested areas, $N=161$. B. Panama, dry forest type, $N=70$. C. Australia, Cape York forest, $N=50$. D. Zaire, Irangi forest, $N=96$. —— Aerial; —— SGM; · · · · scansorial; —·—·— arboreal.

(Fleming, 1973). Compared with the other communities, the latter also has a higher proportion of scansorial animals, mostly represented by squirrels. There is a particularly striking similarity in ecological diversity between the Australian marsupial fauna and the mammal fauna of Malaya, even though the taxonomic composition and numbers of species are so different (Fig. 1), and this is taken to show that the available ecological niches are filled in parallel fashion by the different faunas. Thus it can be concluded that the habitat of a mammal community can be determined on the basis of its ecological diversity, even though the species composition of the fauna may be partly or completely unique to that community.

The predictive value of ecological diversity analysis, irrespective of species composition of the faunas, is of direct application to palaeoecology. By establishing patterns of community structure for a series of modern mammal communities, and then calculating corresponding patterns for fossil communities, it is possible to compare the communities directly even though the species composition of the latter is completely different. Their similarities and differences can then be analysed so that inferences can be made about the habitats from which the fossil faunas may have come. This approach contributes towards the solution of the three major problems which are encountered in palaeoecology. These are that fossil faunas are largely composed of extinct species that have unknown habitat preferences; that various biases may alter the composition of the faunas; and that evolution of ecosystems may have changed the community structure in certain habitats.

In the first case, the majority of fossil faunas consist of species which are extinct, so that there is no direct information as to either the ecological niches of these species or the significance of the association of several species together (Andrews & Van Couvering, 1975). The palaeoecology of fossil localities has usually been discussed in terms of so-called indicator species present in the fauna. These are species which are considered to have well defined habitat requirements and their presence in the fossil fauna is held to indicate the presence of that habitat type. The preferred habitats of these fossil animals had to be inferred from those of living animals that are either closely related to them or have convergent or parallel adaptations related to their ecology, such as occur in the dentition (a guide to the feeding behaviour, Kay, 1977) or in locomotor adaptations. It must be recognized, however, that even if such relationships are well established, there is no way of knowing whether the extinct species actually lived in its inferred habitat or not. The analysis of the ecological diversity of fossil faunal assemblages that is attempted here makes it possible to examine the structure of the community as a whole and to by-pass the reliance on predictions of the ecology of individual species.

The second problem in the interpretation of fossil faunas is more complex. Fossil organisms occur together in what is described as a "death assemblage", which may bear no relation to the community in which the organisms once lived but which merely reflects the fact that the remains of animals and plants were together at the time that they died. Workers in the field of taphonomy, the study of the transition of animal remains from the biosphere into the lithosphere (Efremov, 1940), have found that the material in a death assemblage can be further biased during the process of fossilization by the operation of selective factors which lead to non-random fossil aggregations through the concentration of some types of material and loss of others. The causes of these biases have been found to be due to such agents as differential bone transport and selective bone preservation and fossilization, but no reliable correction factors have yet been produced to allow for these (Behrensmeyer, 1975). Analysis of ecological diversity is not concerned with the correction of such biases, but it may be useful in identifying the presence of discrepancies in fossil assemblages so that estimates can be made of the extent to which they are biased and so diverge from living communities. Special aspects, such as the derivation of the fauna from more than one habitat, or rarity of carnivore species in the fossil record, can also be highlighted.

A third problem that must be briefly considered is the possibility of evolutionary change in the structures of ecosystems themselves. It has been suggested by Olson (1966) that ecosystems evolve over long periods of time. This may be caused by one of the constituent members of a community evolving in such a way as to affect its inter-relations with other members, e.g. predator–prey interactions, but there is little evidence to show how great a change may have occurred since the Angiosperms diversified to their present extent. The existence of such changes, if they have occurred, can be expected to be apparent in their efforts on community structure, so that the analysis of ecological diversity of the communities should show whether or not the changes have occurred.

METHODS

Distribution of localities

The faunas from five fossil localities in the equatorial zone of Africa have been analysed. The sites were Karungu, Rusinga, Songhor and Fort Ternan, four Miocene sites for which we have information on the sedimentology and taphonomy and from which hominoid fossils have come. An analysis of the Pleistocene hominid site of Olduvai is also included because this site shows interesting intermediate patterns of ecological diversity, and all other available evidence suggests that the palaeoenvironment closely resembled the habitat found in that area today.

Faunal associations from single fossiliferous horizons were analysed from Karungu (Andrews, 1974) and Olduvai (Leakey, 1971). The Rusinga fauna was from one level but two discrete faunal associations were combined (Andrews & Van Couvering, 1975) as the sample size was otherwise too small. The Songhor and Fort Ternan faunas (Andrews & Walker, 1976) were both composite, but in the course of the analysis discrete associations, from bed 5 at Songhor (Andrews & Pickford, 1973) and from level 2 at Fort Ternan (Churcher, 1970), were analysed separately and found to have comparable patterns to the composite faunas. The latter are illustrated here because they had larger sample sizes. The association of faunal elements indicated by their presence together in single fossiliferous levels ensures that they constitute discrete time assemblages, although there are of course many other potential biases, some of which will be discussed later. The references to the fauna and the details of the sites are given in Table 1.

For comparison with the fossil faunas, the ecological diversity of 23 modern mammalian communities was analysed. Like the fossil sites, the modern localities were restricted to the tropical zone to minimize any latitudinal variation, and particular care was taken to pick localities which represent uniform topographical zones and vegetation types. Habitats large enough to give an acceptable sample size are never totally uniform, due to the existence, for instance, of rivers or roads with their associated vegetation, but collections which spanned several major habitat types or lacked sound ecological data were not used. The main variable being compared, therefore, is the structure of the mammal communities living at one time in certain well defined habitats. The modern localities selected are representative of the following vegetation types: lowland evergreen forest, two localities; lowland semi-deciduous forest, three localities; intermediate semi-deciduous forest, two localities; lowland deciduous forest, one locality; montane forest, three localities; woodland, four localities; bushland, four localities; grassland, one locality; and flood plain, three localities (Michelmore, 1939; Richards, 1952, 1973; Pratt, Greenway & Gwynne, 1967; Andrews, 1973; Andrews, Groves & Horne, 1975).

Table 1. Fossil localities

Locality	Country	Latitude	Longitude	Altitude	Age	Horizon	References*
Songhor	Kenya	0°2'S	35°13'E	1480	Early Miocene	Redbed Member 1972 Collection	Shackleton, 1951; Andrews & Pickford, 1973
Rusinga	Kenya	0°33'S	34°19'E	1160	Early Miocene	Kaswanga Point Below marker bed I 1971 Collection	Van Couvering & Miller, 1969; Andrews & Van Couvering, 1975
Karungu	Kenya	0°52'S	34°12'E	1184	Early Miocene	Bed 16, 1973 Collection	Oswald, 1914; Andrews, 1974
Fort Ternan	Kenya	0°14'S	35°20'E	1740	Middle Miocene	All levels	Gentry, 1970; Churcher, 1970; Andrews & Walker, 1976
Olduvai	Tanzania	2°59'S	35°20'E	1530	Early Pleistocene	Bed 1, FLKN I L/3	Leakey, 1971

* References to faunal descriptions from the Miocene localities are as follows: (MacInnes, 1936, 1942, 1943, 1951, 1953, 1956, 1957; Leakey, 1943, 1967; Le Gros Clark & Leakey, 1951; Le Gros Clark & Thomas, 1951; Le Gros Clark, 1952; Whitworth, 1954, 1958; Butler, 1956, 1965, 1969; Butler & Hopwood, 1958; Napier & Davis, 1959; Lavocat, 1964, 1973; Savage, 1965, 1973; Patterson, 1965; Hooijer, 1966, 1968; Churcher, 1970; Gentry, 1970; Andrews, 1973, 1974a; Pickford, 1975, 1975a; Wilkinson, 1976).

Faunal lists were compiled for each locality, based as far as possible on single major collections. With two exceptions mentioned below, the faunal lists were not altered by adding animals which had not been collected even if they occurred in similar habitats or in localities nearby. Similarly, if ecologically unexpected species had been recorded, they were not rejected from the faunal list on this account, and, as a result, the faunal lists show some species occurring in habitats in which they are not typically found. This is an inevitable consequence of compiling the faunal lists in this manner, but it does make the lists more directly comparable with the fossil faunas since both consist of associations of animals present at one time in one place. Examples of the kinds of distortion that occur are grassland rodent species entering forest areas along elephant tracks and the grass verges of roads, and predominantly forest species living in thick riverine vegetation along watercourses in otherwise pure grass or bushland.

Two adjustments have been made to the faunal lists which we used. Bats were omitted from the lists to enable comparison to be made more easily between living and fossil faunas, for although they provide much useful information, and increase the discrimination between modern faunas from different habitats, they are very rarely found as fossils. We also took account of local extinctions of large mammals arising from human activity. Some animals, particularly elephants, rhinoceroses and large predators, have been recorded in the past from several of the localities used in this study, and these have been included in the faunal lists even though the more recent collections on which this work is based have shown that they are no longer present. Details of the 23 modern communities are given in Table 2.

Ecological diversity analysis

The fossil and modern mammalian communities have been broken down into four categories: systematic position, body size, locomotor zonal adaptation and feeding habits. The systematic divisions are by Order because at this taxonomic level there has been little change during the Neogene. We have also analysed the faunas at lower taxonomic levels, for example by genus and family, but allowance then has to be made for extinct taxa and the results are not as directly comparable as the ones given here.

The size divisions of living mammals are based on weights given by the following authors: Walker (1964), Dorst & Dandelot (1970), Kingdon (1971, 1974), Meester & Setzer (1971), Schaller (1972), Sachs (1967), Delany (1975). Weights of fossil species were estimated by comparisons of comparable body parts with related living animals, based for the most part on the descriptions of the fossil species given in Table 1. Eight weight subdivisions were recognized initially, but some of these were later combined as they tended to obscure differences between the faunas.

The locomotor categories are fairly broad, and an attempt has been made to follow a zoning classification (Harrison, 1962) whereby the animals in a community are divided into zones corresponding to physical layers in the environment. Starting from the highest part of the ecosystem there is the air above the vegetation to which aerial mammals such as bats are restricted; in the upper canopy or small branch zone the arboreal mammals are found; in the middle and lower vegetation strata the scansorial mammals like squirrels range

Table 2. Present-day localities

Locality	Country	Latitude	Longitude	Altitude (m)	Vegetation	References
1 Irangi	Zaire	1°55′S	28°28′E	1000	Lowland evergreen forest; primary, secondary-riverine cultivation types	Rahm, 1965, 1966
2 Kafue	Zambia	15°40′S	26–28°E	1000	Flood plain grassland with bordering woodland, many shallow lakes, raised levees	Sheppe & Osborne, 1971
3 Mt Kenya	Kenya	0°15′S	37°12′E	2–2,700	Montane evergreen forest, primary, secondary with glades	Moreau, 1944, 1944a; Coe, 1967; Coe & Foster, 1972; Duncan & Wrangham, 1971
4 Legaja	Tanzania	3°0′S	35°02′E	1530	Lake in short grass plains with fringing woodland	Swynnerton & Hayman, 1950; Swynnerton, 1958; Grzimek & Grzimek, 1960; Talbot & Talbot, 1963; Anderson & Talbot, 1965; Verschuren, 1965; Misonne & Verschuren, 1966; Dorst & Dandelot, 1970; Laurie, 1971; Kingdon, 1971, 1974; Schaller, 1972; Kruuk, 1972
5 Serengeti	Tanzania	2°55′S	35°10′E	1530	Short grass plains, small patches of bush, rocky valleys	
6 Banagi	Tanzania	2°15′S	34°49′E	1480	Deciduous woodland and bushland with rocky hillsides	
7 Amani	Tanzania	5°6′S	38°38′E	880	Intermediate semi-deciduous forest	Allen & Loveridge, 1927; Moreau, 1935; Swynnerton & Hayman, 1950; Kingdon, 1971, 1974
8 Rukwa	Tanzania	7°55′S	31°58′E	850	Flood plain grassland with bordering woodland and lake swamps	Vesey-Fitzgerald, 1963, 1964; Kingdon, pers. comm., 1976
9 Kakamega	Kenya	0°20′N	34°52′E	1520	Intermediate semi-deciduous forest, mainly secondary, with glades, bush covered hills: forest is impoverished Congo type	Stewart & Stewart, 1963; Williams, 1967; Kingdon, 1971, 1974; Dorst & Dandelot, 1970
10 Tana	Kenya	2°15′S	40°12′E	40	Flood plain grassland with patches of deciduous forest, shallow lakes and sandy levees	Stewart & Stewart, 1963; Dorst & Dandelot, 1970; Kingdon, 1974; Andrews, Groves & Horne, 1975

11 Karamoja	Uganda	3°00′S	34°20′E	1000	Arid bushland with rocky hills	Watson, 1948, 1949, 1950, 1951, 1952; Ross, 1968; Harrington & Ross, 1974; Kingdon, 1976, pers. comm.
12 Seredou	Guinea	7°35′N	9°20′W	500–1000	Lowland evergreen forest with much cultivation and plantation	Roche, 1971; Dorst & Dandelot, 1970; Kingdon, 1971, 1974
13 Gabiro	Kagera, Rwanda	1°33′S	30°23′E	1500	Wooded grassland on hillside connecting with lake flats	Curry-Lindahl, 1956; Verschuren, 1965; Misonne, 1965
14 Kapiti	Kenya	1°35′S	36°54′E	1600	Grassland–bushland mosaic; fire induced; narrow riverine woodland	Williams, 1967; Hartman, 1967; Foster & Coe, 1968
15 Jebel Mara	Sudan	13°10′N	24°22′E	1–2000	Woodland with rocky outcrops, many glades	Happold, 1966
16 Tsavo	Kenya	3°02′S	38°20′E	700	Bushland and arid bush, rocky hills, strip of riverine woodland	Dollman, 1914; Williams, 1967
17 Sokoke-Gedi	Kenya	3°18′S	39°56′E	800	Lowland deciduous forest and woodland	Stewart & Stewart, 1963; Kingdon, 1971, 1974; Dorst & Dandelot, 1970
18 Budongo	Uganda	1°45′N	31°26′E	1100	Lowland semi-deciduous forest, mainly secondary and monotypic types	Williams, 1967; Kingdon, 1971, 1974, pers. comm., 1976; Delany, 1975
19 Rwenzori	Uganda	0°30′S	29°55′E	1000	Lowland semi-deciduous forest, Budongo type	Delany, 1964a, b, 1975; Neal, 1967; Williams, 1967; Dorst & Dandelot, 1970; Field & Laws, 1970; Kingdon, 1971, 1974
20 Rwenzori	Uganda	0°5′S	29°50′E	1000	Short grass plains–bush–woodland mosaic	
21 Semliki	Uganda	0°4′N	30°0′E	750	Lowland semi-deciduous forest, Budongo type	Kingdon, 1971, 1974, pers. comm., 1976; Dorst & Dandelot, 1970; Delany, 1975
22 Semliki	Uganda	0°4′N	30°0′E	1800	Montane evergreen forest continuous with the lowland	
23 Lemera	Zaire	2°08′S	28°50′E	2150	Montane evergreen forest in river valley	Rahm & Christiaensen, 1963; Rahm, 1967

up and down on tree trunks; on the ground, but not wholly restricted to it, are the small terrestrial mammals that also frequent lower branches of bushes and fallen trees; restricted to the ground are the large terrestrial mammals; below the ground are the fossorial mammals and in water are the aquatic mammals. The emphasis of this classification is therefore based on ecological niche structure rather than behavioural criteria which are not known for fossils. In both fossil and living animals adaptations to certain parts of the physical habitat are clearly recognizable in the postcranial anatomy, for instance adaptations for flying, swimming or digging, but other adaptations are not as clear. The difference between arboreal and scansorial animals is essentially one between animals with grasping hands and mobile limbs and clawed animals with more restricted limb movements. Similarly the difference between the two types of terrestrial animals is one of limb mobility, animals with restricted limb movements being confined to the ground while others with more mobile limbs are able also to scramble on to fallen trees or bushes.

The feeding classes are also broad and they were chosen mainly on the basis of primary dietary adaptation as indicated by tooth morphology, this being the evidence available from fossils. This classification also gives rise to certain difficulties. For example the diets of insectivores and carnivores often overlap to a considerable extent, smaller carnivores eating insects and larger insectivores eating small mammals. Despite this overlap in actual diet, however, the dental adaptations of the two groups are quite distinctive, and the positions they occupy in the ecological food chain are also for the most part separate. For these reasons we have distinguished between them, as we have also between grazers and browsers in the herbivore class, even though such simplistic divisions are no longer recognized by modern ecologists. The divisions used are carnivore, insectivore, herbivore, divided into grazing and browsing subdivisions (the latter including folivores), frugivores, and omnivores. The omnivorous category was kept as small as possible by assigning dual feeding classes to mammals with less specific feeding adaptations: combinations such as HI, herbivore–insectivore, or HF, herbivore–frugivore, were found to be more precise than the mixed feeding category of Harrison (1962) and the omnivorous category of Fleming (1973) and yet still capable of extension to fossil animals.

The data on the dental and postcranial morphology of the fossil animals, on which their dietary and adaptive zone classes were based, were taken from the references listed in Table 1. These classes are defined in Figs 2–6. The data for living mammals on food habits and locomotor adaptations were derived from Walker (1964), Dorst & Dandelot (1970), Kingdon (1971, 1974), Delany (1975) and Hanney (1965).

The data for both present-day and fossil communities were tested for association using 2 × 3 and 5 × 3 contingency tables. Individual fossil communities were tested against present day forest bushland and grassland communities in 2 × 3 contingency tables and the probabilities of deviation from independence calculated (Simpson, Roe & Lewontin, 1960). Probabilities of less than 0.05 were taken to indicate that the data were not independent, or in other words that a non-random trend was apparent in the data. Similarly, the five fossil communities were compared between themselves and with the means of the five types of modern community to test the significance of the trends apparent in the diversity histograms. In every case, several of the ecological and taxonomic

Table 3. Results of the 5 × 3 contingency tests (8 degrees of freedom). *P* shows the level of probability of the null hypothesis: that the data are independent and show no trend. Values of *P* less than 0.05 are taken to show that the null hypothesis is not true

		Fossil localities	Means of present-day localities
Taxonomic categories			
1, rodents and insectivores;	χ^2	19.31	16.85
2, primates;			
3, artiodactyls and carnivores	P	0.025	0.05
Size categories			
1, less than 1 kg;	χ^2	13.54	14.15
2, 1–45 kg;			
3, more than 45 kg	P	0.1	0.1
Locomotor categories			
1, large ground mammals;	χ^2	25.91	46.97
2, small ground mammals;			
3, arboreal, scansorial, aerial	P	0.005	0.001
Feeding categories			
1, insectivores and frugivores;	χ^2	34.25	30.67
2, herbivorous browsers;			
3, herbivorous grazers and carnivores	P	0.001	0.001

classes were combined to produce sets of three, which are defined in Table 3, in order to raise the expected values in the individual cells to above one.

To test the significance of the diversity distributions another way, the correlations were calculated between the fossil and modern faunas for their distribution of classes within each category. The diversity distributions of the fossil faunas were first ranked within each of the four taxonomic and ecological categories, and the correlations between them and the same distribution of classes for the means of the five modern community types were then calculated. The correlation coefficients were tested for significance, and significant correlations were taken to indicate a high level of similarity in class distributions between the faunas being compared, or, in other words, similar faunal constitution. Regressions were also calculated and in some cases plotted out, but they contributed little and the results are not given here.

Ecological diversity indices

The differences in ecological diversity between animal communities can be expressed numerically by an index of diversity using the Shannon–Wiener information function:

$$H' = p_i \log_e P_i$$

Where p_i is the proportion of species present in each class. This indicates the extent to which the classes of each ecological category are uniformly filled, so that a community having a high degree of diversity in one ecological category will have all classes filled and therefore a high value of H'. The same community may have a lower degree of diversity in another ecological category, with greater disproportion in the classes of that category, and it would therefore have a lower value of H' for that category. The index is given by the relation $E=H'/H$ where H is the natural logarithm of the number of classes.

RESULTS

The results of the ecological diversity analyses are shown in histogram form for the fossil localities (Fig. 2) and the modern communities (Figs 3–6). These show how the species proportions are divided between taxonomic, size, locomotor and feeding categories. For the fossil faunas the four categories are shown together, but for the larger comparative sample of present-day communities each category is shown separately in Figs 3–6. In these figures there is a results part, where individual histograms are given for each one of the 23 mammalian faunas used in this study, and a summary part, where four enlarged histograms give the means of the data from four community groups, namely lowland forest, montane forest, flood plain and bushland. The single grassland fauna, which forms a fifth community type, is shown separately at the bottom of each figure. These data are summarized in Fig. 7, in which the ecological diversity indices in the size, locomotor and feeding categories are shown on a three-dimensional model for all of the fossil and living faunas. Finally the results of some of the statistical tests are summarized in Table 3.

Taxonomy

The results of the taxonomic analysis of the fossil faunas are shown in Fig. 2. The mammalian fauna has been divided into six groups, the Rodentia, Insectivora, Primates, Artiodactyla, and Carnivora, and the remainder were grouped in a single category. The species proportions belonging to these six groups are shown in histogram form in Fig. 2.

The fossil faunas differ from the modern communities in having low proportions of carnivore species, but apart from that, some of them show distinctive similarities with the modern communities. The pattern for Songhor (Fig. 2) is closest to the lowland forest communities (Fig. 3), and while there is no significant difference between it and the modern forest faunal mean, it differs significantly ($P=0.01$) from both the bushland and grassland communities. The particularly poor representation of carnivores and artiodactyls suggests the presence of bias in the fossil record (Andrews & Pickford, 1973). The Rusinga fauna (Fig. 2) also resembles forest communities, but its additional similarities to the flood plain communities suggests that the habitat may have been mixed. The Karungu fauna (Fig. 2) most closely resembles that of the woodland–bushland faunas except for the low number of carnivore species which has already been commented on. The Fort Ternan fauna (Fig. 2) appears to be mixed, both the assemblages from single levels (levels 2 and 4) (Churcher, 1970) and that from all levels which is shown here. Artiodactyls are the most abundant group, both in species proportions and the numbers of individuals in each species, and this combination outweighs the relatively high proportion of primate species, which are represented by low numbers of specimens and individuals; the strongest

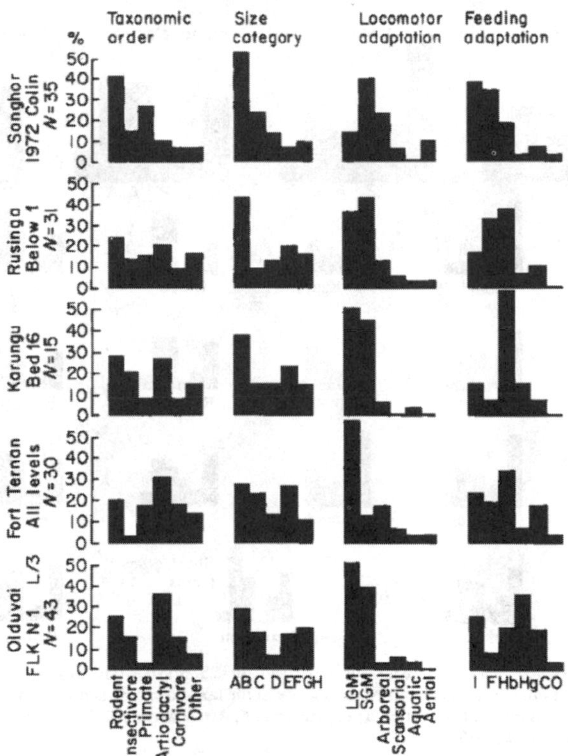

Figure 2. Ecological diversity histograms for five fossil localities. The site details are given on the left with N signifying the number of species available in the sample. The vertical scale represents the percentage number of species for each of the same four categories used in Figs 3–6. The fossil localities are listed in Table 1.

affinities, therefore, are seen to be with the woodland–bushland communities (Fig. 3). In the tests for association no significant difference was found between the Fort Ternan faunas and either the forest community or the bushland community means. Finally, the Olduvai fauna (Fig. 2) is close to both the Serengeti short grass plains community and, to a lesser extent, the woodland–bushland communities, except again for the low number of carnivores (Fig. 3). No significant difference was found between any of these three faunas, but the Olduvai fauna is significantly different from the forest faunas.

In addition to testing the differences between individual fossil faunas and the means of the modern communities, the variation within the five fossil faunas and the five modern community means was also tested on 5 × 3 contingency tables (Table 3). This confirms that the fossil faunas display a statistically significant ($P=0.025$) trend within themselves, but, more importantly, it also confirms ($P=0.05$) that in the modern community means the observed trend from high proportions of rodents and primates in forest faunas to high proportions of artiodactyls and carnivores in bushland and grassland faunas is statistically

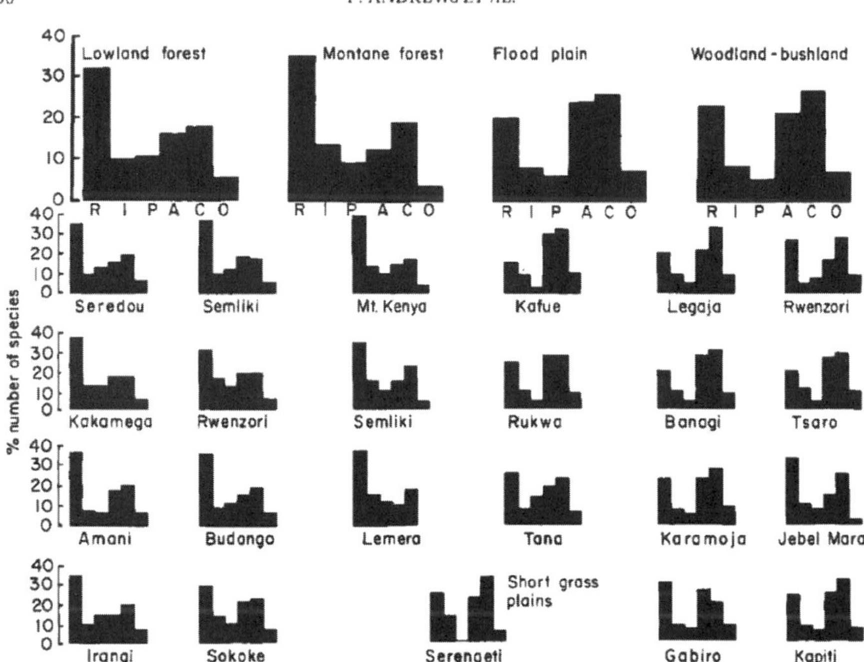

Figure 3. Ecological diversity (taxonomic category) for 23 present-day localities in Africa. The vertical scale gives the percentage number of species in the taxonomic division based on Orders. Bats are excluded. R, Rodentia; I, Insectivora; P, primates; A, Artiodactyla; C, Carnivora; O, others. The localities are listed in Table 2.

significant. There was no significant difference between lowland and montane forest, or between bushland and grassland.

Body size

The results of the body size analysis are shown in Fig. 2. The size categories are based on body weight and have been grouped in certain cases to simplify the data. In this analysis, the Songhor fauna shows a very strong resemblance to the modern forest patterns, particularly in this case to the montane forest communities with their high proportions of small mammals (Fig. 4). There is no significant difference from the modern forest faunas. The lowland forest communities are dominated by small mammals less than 10 kg in weight (Fig. 4), which include all of the rodents and insectivores, the small carnivores that prey on them, and some of the primates that live in the tree canopies and small artiodactyls that live on the forest floor. Large mammals are comparatively rare, both in terms of numbers of species, illustrated here, and in terms of species density. This pattern is seen in even more exaggerated form in the montane forest communities, where, as in the Songhor fauna, the small mammals below 1 kg make up a higher proportion of the fauna (Fig. 4). Both patterns contrast with the non-forest communities, which show greater diversity or more even distribution of size classes, and the Songhor faunal pattern is significantly different ($P = 0.05$) from them.

The Rusinga fauna does not resemble any of the 23 modern communities so

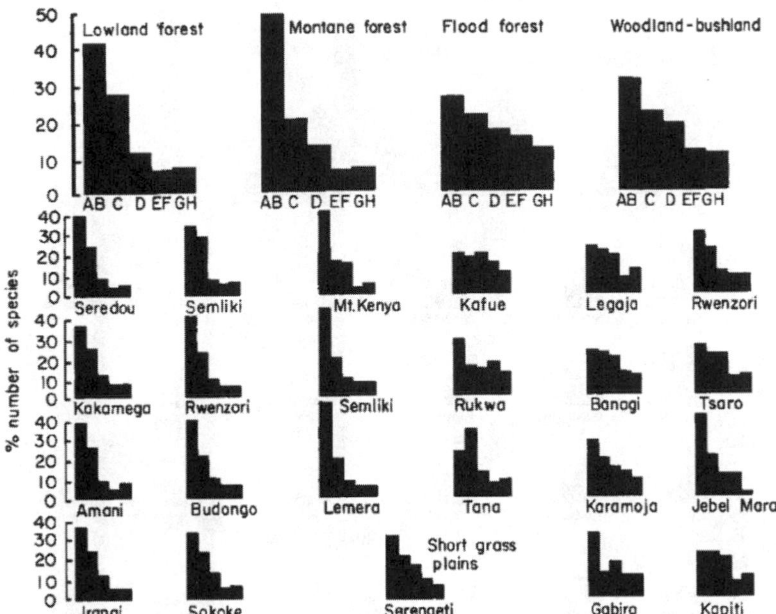

Figure 4. Ecological diversity by size for 23 present-day localities in Africa. The vertical scale gives the percentage numbers of species within the size categories, which are divided as follows; AB, less than 1 kg; C, 1–10 kg; D, 10–45 kg; EF, 45–180 kg. GH more than 180 kg. The localities are listed in Table 2.

far analysed. The smallest size categories predominate in the faunas of both Karungu and Rusinga (Fig. 2), but apart from this, the communities are distributed more evenly between the remaining classes with relatively high proportions of large animals. This could indicate that the faunas are either mixed or heavily biased. The Fort Ternan and Olduvai faunas are both alike and the nearest modern equivalent is seen in the bushland communities. The low proportion of small mammals in the Fort Ternan fauna may be due to a size bias during the accumulation of the bones in the deposits, but it may also indicate the presence of local flooding. Present day flood plain environments are notable for the comparative rarity of small mammals (Sheppe & Osborne, 1971; Andrews, Groves & Horne, 1975) which are unable to survive the periodic flooding, and the Fort Ternan faunal pattern is particularly close to that of the Kafue flood plain fauna of the present time (Fig. 4). The low proportions also of category D (10–45 kg) animals at both Fort Ternan and Olduvai are linked with the poor representation of carnivores, many of which fall into this category in living faunas. Despite these features, no significant difference was found between either of the fossil faunas and the woodland and grassland faunas. A significant difference ($P = 0.01$) was found, however, between both the Fort Ternan and Olduvai faunas and the modern forest faunas.

Locomotor zonal adaptation

The results of the locomotor zonal analysis are shown in Fig. 2. Of the fossil communities the Songhor fauna most resembles the present day forest

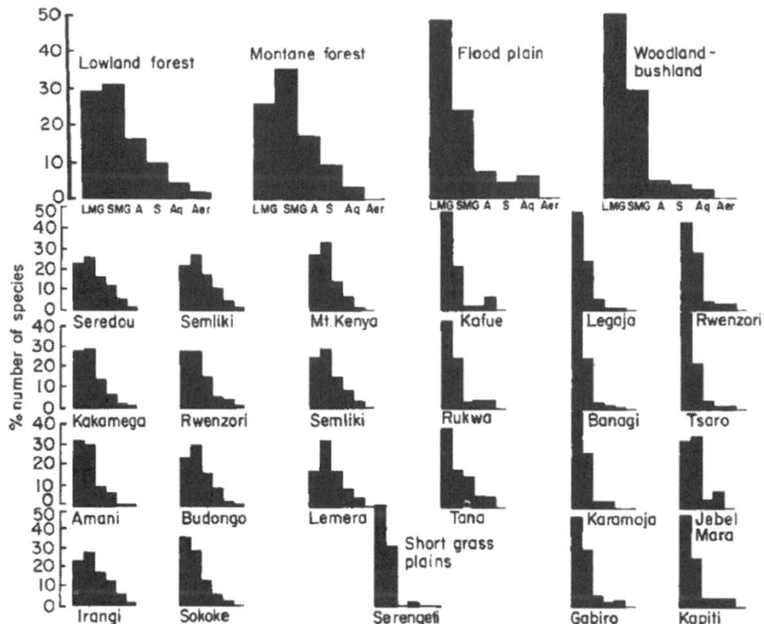

Figure 5. Ecological diversity by locomotor zonal adaptation for 23 present-day localities in Africa. The vertical scale gives the percentage numbers of species in the categories of locomotor adaptation. SGM signifies small ground mammals which range upwards on to fallen trees and bushes and are not therefore strictly confined to the ground. LGM signifies completely terrestrial large mammals. A, Arboreal; S, scansorial; Aq, aquatic; Aer, aerial. The localities are listed in Table 2.

communities (Fig. 5). The low proportion of large terrestrial mammals (LGM) is probably a bias linked with the poor representation of artiodactyls and carnivores already mentioned, but even with this bias the Songhor fauna is not significantly different from the present day forest communities, and it is significantly correlated with the montane forest community mean ($P=0.05$). It is different ($P=0.001$) from the woodland–bushland and grassland communities. The Rusinga fauna also resembles the montane forest communities most closely. The present day montane forest communities differ from the lowland ones in having higher proportions of SGM compared with LGM and in the absence of any aerial forms, although in both communities the small ground and bush scrambling animals (SGM) are the most common; large terrestrial animals (LGM) are less common and they are closely followed in frequency by arboreal and scansorial animals. When bats are included they augment the discriminatory power of this analysis very greatly, as they increase the small difference seen here in the aerial class (Fig. 5), but for reasons already explained the bats have been omitted from this presentation.

The pattern observed for the Karungu fauna (Fig. 2) is unlike that of any of the modern communities (Fig. 5). It is more similar to the non-forest communities, and it is significantly correlated with the woodland–bushland faunas ($P=0.05$) but in the tests of association no significant results were obtained. The similarity with the non-forest communities is based on the fact that they are all composed almost entirely of terrestrial species, with LGM the most common. There is no essential difference in this category between the floodplain bushland and grassland communities, although the single grassland community stands out as by far the most extreme of this type (Fig. 5), but when they are compared with the forest communities, the contingency test on all five types of modern community gives a probability of much less than 0.001 that the observed differences are due to chance (Table 3). It can be concluded therefore that the trend towards increasing terrestrial adaptation in bushland and grassland communities is statistically significant but that the Karungu fauna does not fit the pattern. The Fort Ternan fauna has an exaggerated bushland pattern because of its large number of terrestrial species (Fig. 2). It has an unusually low proportion of species in the SGM category, due to the paucity of small mammals, and the relatively high proportion of arboreal mammals is due to the primates; both of these have already been commented on. The Olduvai fauna is very similar indeed to the bushland communities. Both the Olduvai and Fort Ternan faunas are significantly different from the forest communities ($P=0.001$) but do not differ from woodland–bushland or grassland communities. They also differ from each other ($P=0.05$). They are significantly correlated with the non-forest communities, Fort Ternan most highly correlated with the flood plain communities ($P=0.05$) and Olduvai with the grassland community ($P=0.01$).

Feeding adaptation

In the feeding categories (Fig. 2) the Songhor fauna once again resembles the forest communities, particularly those of lowland forest, and it differs from non-forest faunas at a high level of significance ($P=0.001$). There was a significant correlation ($P=0.05$) between the Songhor and present-day lowland forest faunas in their feeding class distribution. There is little difference in this category between lowland and montane forest faunas, although there is a suggestion (Fig. 6) that the proportion of insectivorous mammals is higher and that of frugivores is lower in montane forest than in lowland forest. Species of browsing herbivores are much more common than grazers in forest habitats, whereas all the woodland and bushland communities have approximately equal proportions of grazing and browsing herbivore and carnivore species, and on this basis the Rusinga pattern also is similar to the forest pattern. The Karungu pattern is very distinctive and is unlike that of any modern community. The Fort Ternan pattern is closest to that of the woodland–bushland communities, except for the low proportion of carnivores, and the Olduvai pattern is closest to the grassland pattern, also with the exception of the carnivores. These similarities have been confirmed by statistical analysis, although the levels of differences between the Olduvai and Fort Ternan faunas and the forest faunas ($P=0.01$) are not quite as great as found for the locomotor categories. None of the correlation coefficients are significant except for the one between the Songhor fauna and the lowland forest mean, but it is interesting that all four of the Miocene faunas have negative correlations with the grassland community, and only the Olduvai fauna has a positive correlation. The Olduvai fauna is most highly correlated, although well below the 0.05 level of significance, with the flood plain and woodland–bushland communities.

The patterns of association that emerge from these results can be summarized as follows: the Songhor fauna is significantly different from the three modern

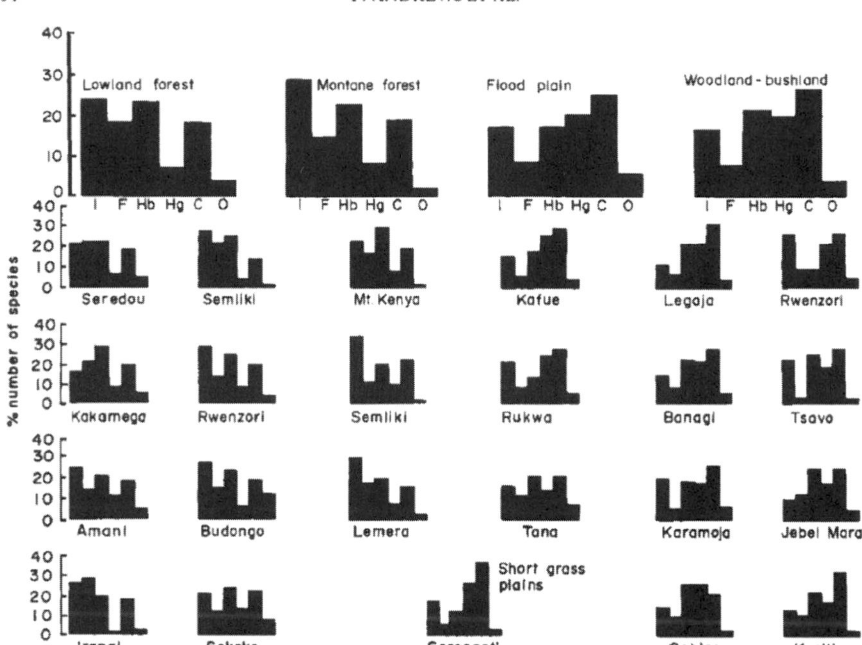

Figure 6. Ecological diversity by feeding adaptation for 23 present-day localities in Africa. The vertical scale gives the percentage numbers of species in the categories of feeding adaptation, which are divided as follows: I, insectivores; F, frugivores; Hb, herbivorous browsers; Hg, herbivorous grazers; C, carnivores; O, omnivores. The localities are listed in Table 2.

non-forest community types in all four categories, the level of significance varying from 0.05 to 0.001, but it is not significantly different from either of the modern forest communities; the Rusinga and Karungu faunas do not show any clear trends, although in terms of general similarity of ecological diversity patterns the Rusinga fauna resembles present-day forest faunas and the Karungu fauna resembles non-forest ones; the Fort Ternan fauna is significantly different from present-day forest faunas in three of the four categories and most closely resembles the woodland–bushland faunas in two categories and flood plain faunas in two categories; and finally the Olduvai fauna is also significantly different from the forest communities in all four categories at high levels of probability (0.01 to 0.001) and resembles woodland–bushland faunas in two categories and grassland faunas in two categories.

The calculation of correlation coefficients in general corroborated the contingency tests, but few of the correlations were significant, even at the 0.05 level of probability. Only in the locomotor analysis were the results consistently significant at this level. It can be observed, however, that in terms of the highest correlation coefficient values, the Songhor fauna correlated best with lowland forest on three categories out of four, the other being for montane forest; the correlations for the Rusinga fauna were similar although the values were generally lower, only two being significant; the Karunga fauna had a wide range of correlations; three out of four of the highest correlations for Fort Ternan were with the flood plain communities, one of these significant, and one was for montane forest; and the Olduvai fauna scored two with the flood plain community mean, one of which was significant, and one each for grassland (significant) and montane forest (not significant).

Ecological diversity indices

The ranges of ecological diversity indices for the 23 modern and five fossil localities are shown in Fig. 7. The composite indices fall into three groups. The forest communities are on the right side of the figure with high index values and therefore high diversity (see Figs 4, 5) in the size and locomotor zonal categories, but they have relatively low food habit diversity. The second group consisting of the woodland and bushland communities have higher food habit diversity but lower size and locomotor diversity. The grassland community of Serengeti, linked somewhat surprisingly with the wooded grasslands of the Kagera lake flats (Table 2) make up the third group with lower diversity in all three categories. The low size diversity of the three flood plain communities is interesting, but it is not clear whether this has any significance.

Forest, woodland–bushland and grassland communities are thus distinguished by their different values of the ecological diversity index. Two of the fossil faunas are comparable with them, namely Songhor with the forest localities and Fort Ternan with the woodland–bushland communities. From this it appears that

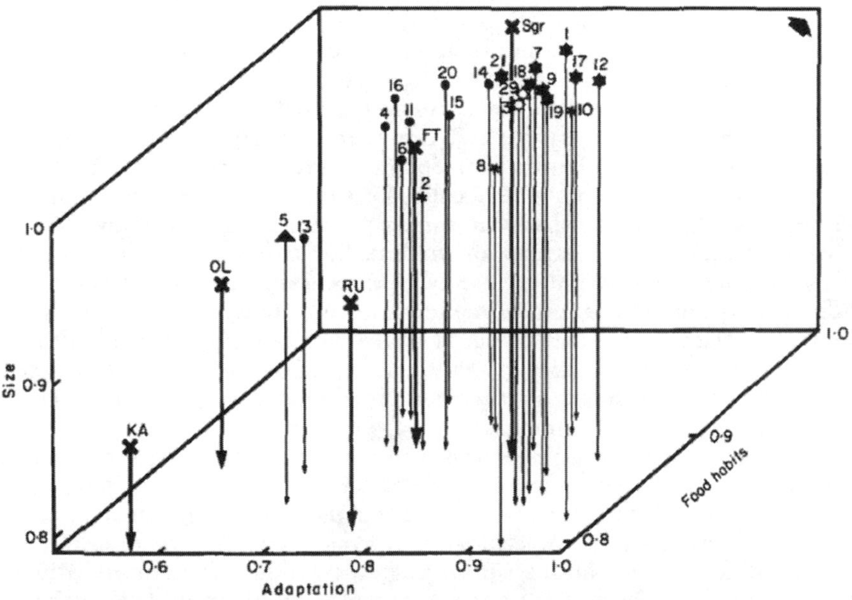

Figure 7. Ecological diversity of 22 modern and five fossil African mammal communities showing diversity indices for locomotor adaptation, food habits and body size. The arrow indicates the point of maximum diversity. Localities are as follows: *lowland forest:* 1, Irangi; 7, Usambaras; 9, Kakamega; 12, Guinea; 17, Sokoke; 18, Budongo; 19, Rwenzori; 21, Semliki; *montane forest:* 3, Mt Kenya; 22, Semliki; *woodland and bushland:* 4, Lake Legaja; 6, Banagi; 11, Karamoja; 13, Gabiro; 14, Kapiti; 15, Jebel Mara; 16, Tsavo; 20, Rwenzori; *grassland:* 5, Serengeti; *flood plain:* 2, Kafue; 8, Rukwa; 10, Tana. *Fossil localities:* Sgr Songhor, Ru Rusinga Island, KA Karungu, FT Fort Ternan, OL Olduvai Gorge.

these two fossil faunas had comparable levels of diversity with living faunas despite the fact that the sample sizes (Fig. 2) were relatively low. The position of Olduvai close to the Serengeti short grass plains fauna is mainly due to the low diversity of both faunas in locomotor adaptations.

DISCUSSION

The present use of ecological diversity in mammalian communities has been developed as a way of analysing the palaeoecology of fossil faunas without depending on single species as environmental indicators. It is quite common to find in the literature statements that a certain environment was present at a fossil locality because a particular species or group of species is present in the fauna, but not only does this exclude evidence from the rest of the fauna, but it also carries the assumption that somehow or other the authors know the habitat preferences of the extinct animal. Our own first approach to the subject was along these lines (Andrews & Van Couvering, 1975), although it was also backed up by geomorphological and floral evidence. Later Van Couvering (in press) with Solounias & Evans (in preparation) developed a method of scoring the inferred ecological preferences of extinct species by their degree of relationship to living species of known ecology, and this will be published separately. In this paper we have been concerned in trying out this new approach which makes no assumptions at all about the habitat preferences of any extinct animal, namely comparisons of ecological diversity of living and fossil faunas.

When the fossil faunas are compared with present-day communities, it can be seen that there are certain similarities between their patterns of ecological diversity and also certain differences. For example the Songhor early Miocene fauna is similar to the lowland forest communities of the present day except for the low numbers of artiodactyls and carnivores and higher numbers of small mammal species. Several faunas from Songhor were analysed, including the bed 5 fauna that formed part of the 1972 collection (Andrews & Pickford, 1973) which is illustrated here, and they had very similar patterns of ecological diversity. The differences between the patterns of the Songhor fauna and those of modern forest communities were not great enough to be statistically significant, but the differences from non-forest communities were nearly always statistically significant. The values of the ecological diversity index fell into the middle of the ranges for forest communities so that the fauna can be seen to have retained a comparable level of diversity with present-day communities despite its bias in one direction, towards small mammals.

It is not possible to say at present what sort of forest habitat the Songhor fauna indicates. Four different types of lowland forest, ranging from evergreen to deciduous associations and with correspondingly differing mammalian faunas, had very similar patterns of ecological diversity. The lack of discrimination between different associations within single community formations is both a strength and a weakness of the analysis of ecological diversity. On one hand it shows the existence of common patterns within the spectrum of variation based on reasonable sample sizes: thus the lowland forest and woodland–bushland patterns are each based on eight communities, the montane forest and flood plain patterns on three each, and only grassland is based on a single community. Faunas with unknown ecological affinities can then be compared with the five groups and their degrees of resemblance tested statistically. On the other hand it is not possible with the present data to determine the community type with any great degree of precision. A fauna can be said to characterize lowland forest conditions, but not what sort of forest; or to characterize woodland–bushland conditions, but not whether it is actually woodland, wooded grassland, bushland or bushed grassland (Pratt, Greenway & Gwynne, 1966). The differentiation between faunas from such habitats could be improved by including the numbers of individuals of each species, thereby giving the complete numerical composition of the mammalian fauna. This would immediately remove the problem of placing too much emphasis on species represented by few individuals, and it would reveal differences between habitats which share the same species but in different proportions. Unfortunately it has not been possible to do this because of lack of data.

The Rusinga early Miocene fauna appears intermediate in nature and shows no clear affinities in its pattern of ecological diversity. It could perhaps be said that it shows a greater degree of resemblance to forest communities than to non-forest, but the conclusion that the fauna represents a forested environment would be unjustified. Similarly the ecological diversity index is low and suggests either a mixed or impoverished fauna. The Rusinga fauna used in this study was in fact drawn from two faunal associations at the same level in the Kaswanga Point series (Andrews & Van Couvering, 1975). The two excavations (KG and KF) were determined on the basis of indicator species to have forest and waterside faunas respectively (Andrews & Van Couvering, 1975; Fig. 3), and this accounts for the ecological diversity results. It would be useful to analyse the two faunas separately, but unfortunately this has not been possible because the sample sizes are too small. An alternative might be to analyse various combinations of two or more present-day fauna communities to try and match the Rusinga pattern.

The Karungu early Miocene fauna is even harder to interpret than that from Rusinga. The fauna comes from one fossiliferous level, the central section of bed 16 (Oswald, 1914; Andrews, 1974) and although there were clearly two depositional events (Andrews, 1974) the species present in both cases were the same. The fauna is dominated by two species of ochotonids and elephant shrews, which make up 74% of the numbers of individuals (Andrews, 1974). These animals have been presumed to have been grazers on account of their hypsodont dentitions; and the emphasis placed by our method on the high proportion of browsing species, therefore, is probably misleading, for they represent a much smaller proportion of the whole community than the grazers. The locomotor diversity pattern of the fauna is indicative of a bushland community but in the other categories the patterns are different from any of the recent communities. The ecological diversity index is also anomalous and indicates perhaps that either the fauna is impoverished or the sample is too small. The sample size was less than half that of any of the other fossil localities, but on the other hand the bed 16 level at Karungu was very thoroughly excavated and the animal remains excavated must be considered representative of the level. Despite the small number of species, therefore, the possibility must be considered that the fauna is impoverished, either because of bias during fossilization, because the animal community from which it was derived was impoverished, or because the community was unlike any of the recent ones that we have examined.

The Fort Ternan middle Miocene fauna is similar to the flood plain and bushland–woodland communities of the present day in its patterns of ecological diversity. There is no significant difference between them, although the Fort Ternan fauna differs significantly in the ecological categories from the present-day forest faunas. It has low proportions of carnivores, in common with all the fossil faunas in this study, but more unusually it also has low proportions of small mammals; this is illustrated by the proportions of rodents and insectivores in the taxonomic analysis, of AB class mammals in the size analysis, of SGM class mammals in the locomotor analysis, and of insectivorous and frugivorous mammals in the feeding analysis. These differences could either be due to a size bias in the course of fossilization or they could indicate flood plain conditions in the area of deposition. The fauna is most highly correlated with present-day flood plain communities, and there is independent geological evidence for the presence of a drainage channel which runs through the Fort Ternan deposits and in which the fossils are concentrated (Walker, in press), but on the other hand the presence of deep palaeosols in the deposits (Bishop & Whyte, 1962) suggests that flooding could not have been extensive. It is quite likely that the small mammal populations were affected by local flooding at Fort Ternan, although the possibility that a size bias has been introduced through differential bone transport cannot be ruled out. The presence of a bias would normally reduce the diversity of the fauna below that of living communities, as suggested in the case of the Karungu fauna, but at Fort Ternan the apparent disproportion has not reduced the faunal diversity. In this event greater reliance can be put on the similarity in ecological diversity patterns of the Fort Ternan fauna with present-day flood plain and woodland–bushland communities.

The early Pleistocene level at Olduvai Gorge (FLK N1 L/3 Leakey, 1971) has a fauna that is very similar to present-day woodland–bushland to grassland communities in its patterns of ecological diversity. Once again carnivores are low in numbers, and small mammals are in slightly higher proportions than would be expected, but the tests of association showed no differences from woodland–bushland and grassland communities and highly significant differences from forest communities. The ecological diversity index suggests greater similarity with grassland communities. It seems possible, therefore, that the Olduvai fauna is derived very largely from grassland habitats and that the woodland contribution was limited.

A number of issues emerge from the foregoing discussion on the five fossil faunas. Analysis of ecological diversity combines the data on all the animals of specific faunas, with reduced emphasis on single indicator species, but the results from the two approaches are not always incompatible. For instance the Songhor fauna has both the diversity patterns and indicator species of faunas from forest habitats, and the Olduvai fauna has similarly consistent results indicating a mosaic of woodland and grassland. In mixed faunas, for example at Fort Ternan, the diversity patterns which link it with the woodland–bushland communities provide a more useful model than the few indicator species which suggest the presence of forest. Where there is good cause to believe that indicator species are associated with certain types of animal community and environment and not with others, and provided all indicator species point the same way, it does seem reasonable to use them as a preliminary guide to the palaeoecology of fossil faunas, but where there is any doubt, if for instance the fauna is from more than one source, analysis of ecological diversity becomes much more important.

It is very likely that all fossil faunas are, at least to a certain extent, derived from more than one source. The sequence of events leading from living animal communities to death assemblages and then to fossil ones is most complex and is open to a bewildering variety of forces. Death assemblages of animals killed by any extraordinary event, such as major floods or droughts, may be derived from a number of living communities, but since the particular event is unknown, or even if it occurred at all, it is hard to make adjustments for the resulting bias. In the second stage of the process, converting the dead animals into fossils, a whole set of completely different factors may be operational. For example the bones may be scattered or brought together by carnivores or scavengers. Some bones may be destroyed by weathering or decay while more resistant ones remain; fragile bones from animals such as birds, amphibians and bats may be totally destroyed, leaving no evidence for their presence in the fauna. Bones may be scattered, sorted, collected or eroded by water action; the sorting action may select for different body parts, depending on the density of the bone, or for different species, according to their size, and partially articulated skeletons may be affected differently from isolated bones and isolated bones differently from broken fragments of bone. It is only too common, for instance, to find small mammal horizons, where large mammals are rare or absent, or conversely large mammal horizons where small mammals are missing. Finally the sediments in which the bones are first deposited and possibly partially fossilized may be reworked and redeposited elsewhere under completely different conditions.

All the fossil faunas analysed in this study had some kind of bias. None of them contained any fragile-boned species, so that not only was the analysis restricted to mammals, but within these the bats had to be omitted so that the recent faunal communities would be comparable to the biased fossil ones. Carnivores were also found to be relatively uncommon in the fossil faunas compared with the recent ones; this could be a genuine difference, particularly as regards the small carnivores, such as viverrids and mustelids, which have many fewer species in the African Miocene faunas than in the Recent, but it would not be true of the much more recent Olduvai fauna, which is comparatively modern in most respects. It is interesting that investigations into bone proportions present in modern African environments have shown that proportions of carnivore bones are lower than predicted on the basis of censuses of the large mammal populations (Behrensmeyer, personal communication), and part of the explanation for this must probably be sought in the position carnivorous animals occupy in the ecological food chain, which results in their being less common within an ecosystem than their prey, so that they stand less chance of being fossilized. Because of their central ecological role in living mammal communities, carnivores have been included in the analyses, and their low proportion in the fossil faunas must be kept in mind when interpreting the diversity patterns of these faunas.

There is some evidence that collecting agencies operated in the accumulation of all the fossil faunas analysed in this study. The Olduvai fauna may well have been accumulated on a living floor, brought together by early man, although the wide range of animal sizes in the fauna indicates that this was not the only agency. The Songhor and Karungu faunas both have low proportions of large animals, which suggests that they were eliminated by a size-sorting process such as water action or scavenger activity, while the Fort Ternan fauna by contrast has low proportions of small mammals. Despite these potential sources of bias, however, the diversity patterns of the Olduvai, Songhor and Fort Ternan faunas closely resemble the patterns of living communities, and so it can be concluded that the biases are less significant than might have been expected.

The Rusinga fauna is a special case because it is known to have been derived from more than one faunal association (Andrews & Van Couvering, 1975). One of these was a waterside community containing many small and large mammals, and as the sedimentological evidence shows that the sediments and bones accumulated in flood plain conditions, it is likely that this fauna was autochthonous. The second faunal association included the forest indicator species and consisted entirely of small mammals, and so it is likely that it was both allochthonous and size sorted when it was transported to the flood plain environment. The analysis of ecological diversity of this combined fauna showed that it was mixed, and this could be used as a safeguard against making ecological prediction on faunas which have less detailed excavation records and which could also be from more than one level.

One last point to be considered concerns the possibility of change in the ecosystems between the Miocene and present. Such changes have been shown to have occurred since the Mesozoic (Olson, 1966), and one possibility of change since the Miocene is in the low proportions of small carnivores already mentioned. Another possibility is seen at Karungu, the fauna of which has no definitely assigned indicator species and has patterns of ecological diversity that are different from any of the modern communities. It may be that this lack of definition is due entirely to the low number of species in the fauna, so that it formed part of one or more larger communities, but it is also possible that the fauna represents a habitat type that is no longer represented in Africa today: the dominant faunal elements are ochotonids, that do not survive today in Africa, and myohyracine elephant shrews which have no living descendants and which have ungulate-like hypsodont teeth, unique to this group. These two species dominate the Karungu fauna at two stratigraphical levels, making up over 74% of known individuals, so that not only does the fauna differ strikingly from any living African fauna, but also the differences would not be detected by the method used here which is based on species numbers. Comparative data on relative abundance of all species in living mammal communities are not available at present, so it is difficult to assess how unusual the Karungu fauna is, but it must be concluded that the ecological diversity analyses do suggest a fundamental difference from living communities.

The other four fossil faunas have diversity patterns that are remarkably similar in most respects to living communities. Geomorphological evidence suggests that the early Miocene localities in western Kenya were geographically and climatically integrated with the central African equatorial forest belt (Andrews & Van Couvering, 1975), and the fossil flora from Rusinga, which is known in some detail (Chesters, 1957), is largely made up of extant genera of equatorial forest trees and shrubs. The distribution of the forest ecosystems was therefore different in the Miocene, but since the trees were very similar to modern species, it seems likely that the ecological niche structure was very similar to that of the present day. Similarly, the ecological adaptations of the faunas, in terms of feeding, locomotor or size categories, seems to be fully modern, so that the faunas are directly comparable with present-day ones.

CONCLUSIONS

The community structure of mammalian faunas from five fossil localities in East Africa has been analysed using their species diversity expressed in four taxonomic and ecological categories. The ecological diversity patterns were found to be significantly different for the five fossil faunas.

As a basis for comparison with the fossil faunas, the community structure of mammalian faunas from 23 present-day communities in Africa was analysed in the same way. Five types of community were distinguished by this method, and they could be related to five vegetation or habitat types: lowland forest, montane forest, woodland–bushland, grassland and flood plain. The ecological diversity patterns were consistent, with only minor variations, within each habitat type, so that they can be given a predictive value and can be used to identify the habitats of other faunas.

The fauna from the early Miocene locality of Songhor, Kenya, showed very close similarities to the present-day lowland forest communities both in terms of overall diversity and in terms of the patterns of diversity in the taxonomic, locomotor and feeding categories. It is concluded therefore that the Songhor fauna was probably derived from a lowland forest habitat.

The fauna from the other two Kenyan early Miocene localities, Rusinga Island and Karungu, both had lower diversity levels than any present-day community and were clearly biased in some way. The faunas may have been derived from more than one source, they may have been greatly impoverished by selective factors, or, in the case of Karungu, may be representative of a community type no longer extant in Africa.

The fauna from the middle Miocene locality of Fort Ternan, Kenya, showed closest affinities with flood plain and woodland–bushland communities of the present day. The similarities with flood plain communities might be an artifact, arising from a taphonomical bias against small mammals, but both the overall diversity and the diversity patterns are very similar to those of unbiased woodland–bushland communities. On the other hand, the presence of species which are regarded as forest indicators in the fauna suggests that there may have been some forest in the area. It can be concluded therefore that the Fort Ternan fauna was derived mainly from a woodland–bushland habitat, perhaps subject to local flooding related to the channel running through the deposits, and probably with a forested area somewhere not too far away.

Finally, the Pleistocene deposits from bed 1 in Olduvai Gorge, Tanzania, have yielded a fauna which also resembles the woodland–bushland communities but which has strong similarities also with the single grassland community available to us. The low overall diversity of the Olduvai fauna is unlikely to have been due to bias, because it was the largest of the fossil faunas analysed here, and it may be concluded that it does suggest that the fauna was for the most part derived from a grassland habitat.

ACKNOWLEDGEMENTS

We should like to thank Mr J. Kingdon for his advice and help with the faunal lists. Drs M. Coe, R. Barnes, K. Joysey and J. H. Van Couvering kindly read and commented on various drafts of this paper and we should like to thank them for this. We should also like to thank Dr M. Taieb and Dr D. Chivers for the opportunity of putting forward these ideas at seminars they organized.

REFERENCES

ALLEN, G. M. & LOVERIDGE, A., 1927. Mammals from the Uluguru and Usambara Mountains, Tanganyika Territory. *Proceedings of the Boston Society of Natural History, 38:* 413–441.

ANDERSON, G. D. & TALBOT, L. M., 1965. Soil factors affecting the distribution of the grassland types and their utilisation by wild animals of the Serengeti plains, Tanganyika. *Journal of Ecology, 53:* 33–56.

ANDREWS, P., 1973. *Miocene Primates (Pongidae, Hylobatidae) of East Africa.* Ph.D. thesis, University of Cambridge.

ANDREWS, P. & PICKFORD, M., 1973. Report on the 1972 Songhor field season. Kenya National Museum, unpublished.

ANDREWS, P., 1974. Report on the Karungu 1973 Expedition. Unpublished manuscript, Kenya National Museum, Nairobi.

ANDREWS, P., 1974a. New species of *Dryopithecus* from Kenya. *Nature, 249:* 188–190, 680.

ANDREWS, P., GROVES, C. P. & HORNE, J. F. M., 1975. Ecology of the Lower Tana River Flood Plain (Kenya). *Journal of the East Africa Natural History Society, 151:* 1–31.

ANDREWS, P. & VAN COUVERING, J. H., 1975. Palaeoenvironments in the East African Miocene. In F. S. Szalay (Ed.), *Approaches to Primate Paleobiology:* 62–103. Basel: Karger.

ANDREWS, P. & WALKER, A., 1976. The primate and other fauna from Fort Ternan, Kenya. In G. Isaac & E. R. McCown (Eds), *Human Origins:* 279–304. Menlo Park, California: Benjamin.

BEHRENSMEYER, A. K., 1975. The Taphonomy and Paleoecology of Plio-Pleistocene Vertebrate Assemblages East of Lake Rudolf. *Bulletin of the Museum of Comparative Zoology at Harvard College, 146:* 473–578.

BISHOP, W. W. & WHYTE, F., 1962. Tertiary mammalian faunas and sediments in Karamoja and Kavirondo, East Africa. *Nature, 196:* 1283–1287.

BUTLER, P. M., 1956. Erinaceidae from the Miocene of East Africa. *Fossil Mammals of Africa, 11:* 1–75. London: British Museum (Natural History).

BUTLER, P. M., 1965. East African Miocene and Pleistocene Chalicotheres. Fossil Mammals of Africa No. 18. *Bulletin of the British Museum (Natural History) (Geology),* 10: 163–237.

BUTLER, P. M., 1969. Insectivores and bats from the Miocene of East Africa: new material. In L. S. B. Leakey (Ed.), *Fossil Vertebrates of Africa, 1:* 1–38. London: Academic Press.

BUTLER, P. M. & HOPWOOD, A. T., 1957. Insectivora and Chiroptera from the Miocene rocks of Kenya Colony. *Fossil Mammals of Africa, 13:* 1–35. London: British Museum (Natural History).

CHESTERS, K. I. M., 1957. The Miocene flora of Rusinga Island, Lake Victoria, Kenya. *Palaeontographica, 101(B):* 30–67.

CHURCHER, C. S., 1970. Two new upper Miocene Giraffids from Fort Ternan, Kenya, East Africa: *Palaeotragus primaevus* n. sp. and *Samotherium africanum* n. sp. In L. S. B. Leakey & R. J. G. Savage (Eds), *Fossil Vertebrates of Africa, 2:* 1–105. London: Academic Press.

COE, M. J., 1967. *The Ecology of the Alpine Zone of Mount Kenya.* The Hague: Junk.

COE, M. J. & FOSTER, J. B., 1972. The mammals of the northern slopes of Mt. Kenya. *Journal of the East Africa Natural History Society, 131:* 1–18.

CURRY-LINDAHL, K., 1956. Ecological studies on mammals, birds, reptiles and amphibians in the Eastern Belgian Congo. *Annales du Musée r. du Congo Belge, 42:* 1–78.

DELANY, M. J., 1964a. A study of the ecology and breeding of small mammals in Uganda. *Proceedings of the Zoological Society of London: 142:* 347–370.

DELANY, M. J., 1964b. An ecological study of the small mammals in Queen Elizabeth Park, Uganda. *Revue de Zoologie et de Botanique Africaines, 70:* 129–147.

DELANY, M. J., 1975. *The Rodents of Uganda.* London: British Museum (Natural History).

DOLLMAN, G., 1914. Notes on mammals collected by Dr Christy in the Congo. *Revue de Zoologie et de Botanique Africaines, 4:* 75.

DORST, J. & DANDELOT, P., 1970. *A Field Guide to the Larger Mammals of Africa.* London: Collins.

DUNCAN, P. & WRANGHAM, R. W., 1971. On the ecology and distribution of subterranean insectivores in Kenya. *Journal of Zoology, 164:* 149–163.

EFREMOV, J. A., 1940. Taphonomy: A new branch of Paleontology. *Pan-American Geologist, 74:* 81–93.

FIELD, C. R. & LAWS, R. M., 1970. The distribution of large herbivores in the Queen Elizabeth National Park, Uganda. *Journal of Applied Ecology, 7:* 273–294.

FLEMING, T. H., 1973. Numbers of mammal species in north and central American forest communities. *Ecology, 54:* 555–563.

FOSTER, J. B. & COE, M. J., 1968. The biomass of game animals in Nairobi National Park. *Journal of Zoology, 155:* 413–425.

GENTRY, A. W., 1970. The Bovidae (Mammalia) of the Fort Ternan Fossil Fauna. In L. S. B. Leakey & R. J. G. Savage (Eds), *Fossil Vertebrates of Africa, 2:* 243–323. London: Academic Press.

GRZIMEK, M. & GRZIMEK, B., 1960. Census of plains animals in the Serengeti National Park, Tanganyika. *Journal of Wildlife Management, 24:* 27–37.

HANNEY, P., 1965. The Muridae of Malawi. *Journal of Zoology, 146:* 577–633.

HAPPOLD, D. C. D., 1966. The mammals of Jubel Marra, Sudan. *Journal of Zoology, 149:* 126–136.

HARRINGTON, G. N. & ROSS, I. C., 1974. The savanna ecology of Kidepo Valley National Park. *East African Wildlife Journal, 12:* 93–105.

HARRISON, J. L., 1962. The distribution of feeding habits among animals in a tropical rain forest. *Journal of Animal Ecology, 31:* 53–64.

HARTMAN, D. A., 1967. A June–July census of small mammals on the Athi Plains, Kenya. *Journal of the East Africa Natural History Society, 26:* 1–4.

HOOIJER, D. A., 1966. Miocene rhinoceroses of East Africa. Fossil Mammals of Africa No. 21. *Bulletin of the British Museum (Natural History), 12:* 120–190.

HOOIJER, D. A., 1968. A rhinoceros from the late Miocene of Fort Ternan, Kenya. *Zoölogische Mededeelingen, 43:* 77–92.

KARR, J. R., 1971. Structure of avian communities in selected Panama and Illinois habitats. *Ecological Monographs, 41:* 207–233.

KAY, R. F., 1977. Diet of early Miocene African hominoids. *Nature, 268:* 628–630.

KINGDON, J., 1971. *East African Mammals, I.* London: Academic Press.

KINGDON, J., 1974. *East African Mammals, II.* London: Academic Press.

KRUUK, H., 1972. *The Spotted Hyena.* Chicago: University of Chicago Press.

LAURIE, W. A., 1971. The food of the barn owl in the Serengeti National Park, Tanzania. *Journal of the East Africa Natural History Society, 28:* 1–4.

LAVOCAT, R., 1964. Fossil Rodents from Fort Ternan, Kenya. *Nature, 202:* 1131.

LAVOCAT, R., 1973. Les rongeurs du Miocene d'Afrique orientale. *Memoires et Travaux de l'Institut de Montpellier, 00:* 1–284.

LEAKEY, L. S. B., 1943. A Miocene anthropoid mandible from Rusinga, Kenya. *Nature, 152:* 319–320.

LEAKEY, L. S. B., 1967. Notes on the Mammalian Faunas from the Miocene and Pleistocene of East Africa. In W. W. Bishop & J. D. Clark (Eds), *Background to Evolution in Africa:* 7–29. Chicago: University of Chicago Press.

LEAKEY, M. D., 1971. *Excavations in Beds I & II, 1960–1963. Olduvai Gorge, III.* Cambridge: University Press.

LE GROS CLARK, W. E., 1952. Report on fossil hominoid material collected by British–Kenya Miocene expedition. *Proceedings of the Zoological Society of London, 122:* 273–286.

LE GROS CLARK, W. E. & LEAKEY, L. S. B., 1951. The Miocene Hominoidea of East Africa. *Fossil Mammals of Africa, 1:* 1–117. London: British Museum (Natural History).

LE GROS CLARK, W. E. & THOMAS, D. P., 1951. Associated jaws and limb bones of *Limnopithecus macinnesi. Fossil Mammals of Africa, 3:* 1–27. London: British Museum (Natural History).

MACIN, D. G., 1942. Miocene and post-Miocene Proboscidea from East Africa. *Transaction of the Zoological Society, London, 25:* 33–106.

MACINNES, D. G., 1936. A new genus of fossil dear from the Miocene of Africa. *Zoological Journal of the Linnean Society, 39:* 521–530.

MACINNES, D. G., 1943. Notes on the East African Miocene Primates. *Journal of the East Africa Natural History Society, 17:* 141–181.

MACINNES, D. H., 1951. Miocene Anthracotheridae from East Africa. *Fossil Mammals of Africa, 4:* 1–24. London: British Museum (Natural History).

MACINNES, D. G., 1953. The Miocene and Pleistocene Lagomorpha of East Africa. *Fossil Mammals of Africa, 6:* 1–30. London: British Museum (Natural History).

MACINNES, D. G., 1956. Fossil Tubulidentata from East Africa. *Fossil Mammals of Africa, 10:* 1–38. London: British Museum (Natural History).

MACINNES, D. G., 1957. A new Miocene rodent from East Africa. *Fossil Mammals of Africa, 12:* 1–35. London: British Museum (Natural History).

MEESTER, J. & SETZER, H. W., 1971. *The Mammals of Africa. An Identification Manual.* Washington: Smithsonian Institution.

MICHELMORE, A. P. G., 1939. Observations on Tropical African Grasslands. *Journal of Ecology, 27:* 282–312.

MISONNE, X., 1965. *Rongeurs. Exploration du Parc National de la Kagera, 1* (2): 77–118.

MISONNE, X. & VERSCHUREN, J., 1966. Les Rongeurs et Lagomorphes de la region du Parc National du Serengeti (Tanzanie). *Mammalia, 30:* 517–537.

MOREAU, R. E., 1935. A synecological study of Usambara, Tanganyika Territory. *Journal of Ecology, 23:* 1–43.

MOREAU, R. E. 1944. Kilimanjaro and Mt. Kenya: some comparisons, with special reference to the Mammals and Birds; and with a note on Mt. Meru. *Tanganyika Notes and Records, 18:* 28–68.

MOREAU, R. E., 1944a. A contribution to the Biology and Bibliography of Mt. Kenya. *Journal of the East Africa Natural History Society, 18:* 61–92.

NAPIER, J. R. & DAVIS, P. R., 1959. The fore-limb skeleton and associated remains of *Proconsul africanus. Fossil Mammals of Africa, 16:* 1–69. London: British Museum (Natural History).

NEAL, B. R., 1967. *The Ecology of Small Rodents in the Grassland Community of the Queen Elizabeth National Park, Uganda.* Ph.D. thesis, Southampton University.

OLSON, E. C., 1966. Community evolution and the origin of mammals. *Ecology, 47:* 291–302.

OSWALD, F., 1914. The Miocene beds of the Victoria Nyanza, and the geology of the country between the Lake and the Kisii Highlands. *Quarterly Journal of the Geological Society of London, 70:* 128–162.

PATTERSON, B., 1965. The Fossil Elephant Shrews (Family Macroscelididae). *Bulletin of the Museum of Comparative Zoology at Harvard College, 133:* 295–335.

PICKFORD, M., 1975. New fossil Orycteropodidae (Mammalia, Tubulidentata) from East Africa. *Netherlands Journal of Zoology, 25:* 57–88.

PICKFORD, M., 1975a. Late Miocene sediments and fossils from the Northern Kenya Rift Valley. *Nature, 256:* 279–284.

PICKFORD, M. & WILKINSON, A., 1975. Stratigraphic and phylogenetic implications of new Listriodontinae from Kenya. *Netherlands Journal of Zoology, 25:* 132–141.

PRATT, D. J., GREENWAY, P. J. & GWYNNE, M. D., 1966. A Classification of East African Rangeland with an appendix on terminology. *Journal of Applied Ecology, 3:* 369–382.

RAHM, U., 1965. Distribution et écologie de quelques mammifères de l'est du Congo. *Zoologica Africa, Cape Town, 1:* 149–164.

RAHM, U., 1966. Les mammifères de la forêt équatoriale de l'est du Congo. *Annales de Musée r. de l'Afrique Central, 00:* 000–000.

RAHM, U., 1967. Les Muridés des environs du Lac Kivu et les régions voisines (Afrique Centrale) et leur écologie. *Revue Suisse Zoologie, 74:* 439–519.

RAHM, U. & CHRISTIAENSEN, A., 1963. Les mammifères de la région du Lac Kivu. *Annales de Musée r. de l'Afrique Central, 118:* 1–83.

RICHARDS, P. W., 1952. *The Tropical Rain Forest. An Ecological Study.* Cambridge: Cambridge University Press.

RICHARDS, P. W., 1973. The Tropical Rain Forest. *Scientific American, 229:* 58–67.

ROCHE, J., 1971. Recherches mammalogiques en Guinée forestière. *Bulletin du Muséum National d'Histoire Naturelle, 16:* 737–781.

ROSS, I. C., 1968. The practical aspects of implementing a controlled burning scheme in the Kidepo Valley National Park. *East African Wildlife Journal, 6:* 101–105.

SACHS, R., 1967. Liveweights and body measurements of Serengeti game animals. *East African Wildlife Journal, 5:* 24–36.

SAVAGE, R. J. G., 1965. The Miocene carnivora of East Africa. Fossil Mammals of Africa No. 19. *Bulletin of the British Museum (Natural History) (Geology), 10:* 239–316.

SAVAGE, R. J. G., 1973. *Megistotherium,* gigantic hyaenodont from Miocene of Gebel Zelton, Libya. *Bulletin of the British Museum (Natural History) (Geology), 22:* 485–512.

SCHALLER, G. B., 1972. *The Serengeti Lion.* Chicago: University Chicago Press.

SHACKLETON, R. M., 1951. A contribution to the geology of the Kavirondo rift valley. *Quarterly Journal of the Geological Society of London, 106:* 345–383.

SHEPPE, W. & OSBORNE, T., 1971. Patterns of use of a flood plain by Zambian mammals. *Ecological Monographs, 41:* 179–205.

SIMPSON, G. G., 1964. Species diversity of North American recent mammals. *Systematic Zoology, 13:* 57–73.

SIMPSON, G. G., ROE, A. & LEWONTIN, R. C., 1960. *Quantitative Zoology.* New York: Harcourt Brace.

STEWART, D. R. M. & STEWART, J., 1963. The distribution of some large mammals in Kenya. *Journal of the East Africa Natural History Society, 24:* 1–52.

SWYNNERTON, G., 1958. Fauna of the Serengeti National Park. *Mammalia, 22:* 435–450.

SWYNNERTON, G. H. & HAYMAN, R. W., 1950. A checklist of the land mammals of the Tanganyika Territory and Zanzibar Protectorate. *Journal of the East Africa Natural History Society, 20:* 274–392.

TALBOT, L. M. & TALBOT, M. H., 1963. The high biomass of wild ungulates on East African savanna. *Transactions of the North American Wildlife and Natural Resources Conference, 28:* 465–476.

VAN COUVERING, J. H., in press. Community evolution and succession in East Africa during the late Cenozoic.

VAN COUVERING, J. A. & MILLER, J. A., 1969. Miocene stratigraphy and age determinations, Rusinga Island, Kenya. *Nature, 221:* 628–632.

VERSCHUREN, J., 1965. Contribution à l'étude des chéiroptères du Parc National de Serengeti (Tanzanie). *Revue de Zoologie et de Botanique Africaines, 71:* 371–375.

VESEY-FITZGERALD, D. F., 1963. Central African grasslands. *Journal of Ecology, 51:* 243–273.

VESEY-FITZGERALD, D. F., 1964. Mammals of the Rukwa Valley. *Tanganyika Notes and Records, 62:* 61–72.

WALKER, E. P., 1964. *Mammals of the World.* Baltimore: John Hopkins.

WATSON, J. M., 1948. The wild mammals of Teso and Karamoja. I. *Uganda Journal, 12:* 200–229.

WATSON, J. M., 1949. The wild mammals of Teso and Karamoja. II. *Uganda Journal, 13:* 39–60.

WATSON, J. M., 1949a. The wild mammals of Teso and Karamoja. III. *Uganda Journal, 13:* 182–201.

WATSON, J. M., 1950. The wild mammals of Teso and Karamoja. IV. *Uganda Journal, 14:* 53–84.

WATSON, J. M., 1950a. The wild mammals of Teso and Karamoja. V. *Uganda Journal, 14:* 163–203.

WATSON, J. M., 1951. The wild mammals of Teso and Karamoja. VI. *Uganda Journal, 15:* 92–106.

WATSON, J. M., 1951a. The wild mammals of Teso and Karamoja. VII. *Uganda Journal, 15:* 193–202.

WATSON, J. M., 1952. The wild mammals of Teso and Karamoja. VIII. *Uganda Journal, 16:* 89–93.

WHITWORTH, T., 1954. The Miocene Hyracoids of East Africa. *Fossil Mammals of Africa, 7:* 1–58. London: British Museum (Natural History).

WHITWORTH, T., 1958. Miocene ruminants of East Africa. *Fossil Mammals of Africa, 15*: 1–50. London: British Museum (Natural History).

WILKINSON, A. F., 1976. The Lower Miocene Suidae of Africa. In R. J. G. Savage & S. C. Coryndon (Eds), *Fossil Vertebrates of Africa, 4*: 173–282. London: Academic Press.

WILLIAMS, J. G., 1967. *A Field Guide to the National Parks of East Africa*: 352 pages. London: Collins.

PART

4 Paleoenvironmental Reconstruction

Edited by Anna K. Behrensmeyer and Caroline A. E. Strömberg

Ecologists typically work within a known environment and focus on finding evidence for how life is adapted to this environment. In contrast, a time-honored goal of paleoecology has been to work in the opposite direction, using fossil organisms and their inferred ecological requirements to reconstruct past environments. Modern ecologists can readily see that their target organisms inhabit ponds, river floodplains, coral reefs, and so on, and they can measure physical and chemical characteristics of these environments. Reconstructing environmental context is less straightforward for assemblages of fossils preserved in rocks. Even the best geological analysis of fossil-bearing sediments may provide only general information about the depositional setting where the plants and animals lived and died. Paleoecologists usually want more than this if they can get it—the same kinds of information ecologists measure in modern environments, including physiographic setting, temperature, rainfall, water chemistry and depth, vegetation structure, and many other variables.

In most of the rock record, ecological information has been altered by taphonomic filters and other geological processes that control how the organic record transitions from living ecosystems to dead remains and ultimately to preserved fossils. We have learned that the available subsamples of past life often are not always faithful recorders of ecological variables, even when these samples appear to be exceptionally preserved "snapshots" of an ancient community. Most paleontologists are appropriately cautious in making assumptions about how the dead may or may not represent the once living, but taphonomic research has also resulted in increased confidence that the dead do, in fact, accurately record many types of ecological information. Powerful methods have been developed for distilling correct and detailed information about past environments from fossils. The beginnings of a number of these methods are represented in the foundational papers in this section.

The authors of these papers all took new and creative approaches to reconstructing critical information about ancient environments using fossils and, in some cases, modern analogs for these environments. They had to figure out how to reliably distill such information from different kinds of fossils, and their approaches to this challenge fall under several major themes.

1. *Morphology as an Environmental Indicator*. Organisms are shaped primarily by their evolutionary history, but they also display phenotypic plasticity in response to environmental parameters during their lifetimes. Paleoenvironmental reconstruction for a particular time and place depends on finding morphological characteristics that are reliable indicators of particular environmental conditions, in spite of underlying phylogenetic constraints. Jack Wolfe used years of empirical research on modern and fossil leaf assemblages to establish that leaf margins are reliable indicators of regional temperature, and this work proved foundational in making paleobotany a major player in paleoclimate research (e.g., Wilf et al. 2003). Although he overplayed the role of climate versus phylogenetic controls in shaping leaf margin patterns (Little et al. 2010), Wolfe's work opened the door to documenting global and regional climate trends over large portions of the Phanerozoic land record. Ongoing work to make leaf shape and venation methods increasingly objective and precise build on these fundamental insights for improved climate inferences in deep time (e.g., Peppe et al. 2011).

   The last few decades have witnessed an explosion of novel, organism-based environmental proxies. The correlation between density of leaf pores (stomata) and atmospheric CO_2 level (McElwain and Chaloner 1995) has allowed reconstruction of Phanerozoic CO_2 fluctuations (Beerling and Royer 2011). Various features of mammalian faunas have been found to correlate with habitat type and climate (mainly rainfall), such as diversity and distribution of body sizes, tooth shape, and limb bone shape, and these have been used to infer local environment in deep time (Andrews and Nesbit Evans 1979; Legendre 1986; Fortelius et al. 2002; van Dam 2006).
2. *Biofacies in Time and Space*. Since William Smith and the birth of stratigraphy, geologists and paleontologists have used fossils to correlate similar-aged strata. Although Smith and many others realized that there was spatial variation in contemporaneous fossil assemblages, the default assumption has long been that equivalent faunas or floras mean age concordance as well. Manley Natland countered this in the early twentieth century by showing that the composition of modern assemblages of foraminifera off the California coast was strongly and predictably depth dependent. Comparisons with nearby Pliocene assemblages revealed that different-aged faunas could be similar because of environment—that is, water depth. His careful spatial documentation of modern foraminifera assemblages also broke new ground as an analogue study for interpreting the past. Maxim Elias also understood marine biofacies as depth related, but he saw these as communities of organisms that moved back and forth in "bands" with cyclic sea-level changes in the Carboniferous mid-continental seas of North America. He used the characteristic associations of marine taxa as indicators for particular depths and positions relative to the ancient shorelines, thus establishing an approach to paleoenvironmental interpretation that continues as a vibrant component of modern paleobiology (e.g., Hendy and Kamp 2004). In spite of general acceptance of the time-transgressive nature of biofacies throughout the geological record, however, the original default assumption about time equivalence still figures in debates about environmental versus temporal controls on fossil assemblage composition. For example, the exact temporal relationships of faunal and floral assemblages near the Cretaceous-Paleogene boundary—vital for understanding the timing and nature of dinosaur extinction and biotic recovery—continues to be debated (e.g., Lofgren et al. 2004 vs. Fox and Scott 2011).
3. *Biogeochemical Evidence*. We take stable isotopes for granted today as essential tools for paleoenvironmental analysis, but the origin and gestation of this method in the 1960s and 1970s required visionary scientists willing to cross discipline boundaries and do the

hard work of establishing the credibility of their methods. Deriving original biochemical signals from fossils went against a long-held assumption that the fossilization process involved serious "alteration" by a host of diagenetic processes. It took decades of research to establish when isotopic evidence from fossils was reliable and when it was not (e.g., Schoeninger and DeNiro 1982; Kohn and Cerling 2002). The importance of calibrating fractionation factors for different organisms, tissue types, and dietary pathways was recognized early on and continues today (DeNiro and Epstein 1978; Sponheimer et al. 2003; Clementz et al. 2009). John Vogel and Nikolas Van der Merwe published the first use of stable carbon isotopes as an indicator of paleodiet in 1977, building on earlier work on C_3 and C_4 metabolic pathways in plant biochemistry (Bender 1971; Smith and Brown 1973) and applying this to $\delta^{13}C$ ratios in collagen extracted from the bones of native North Americans. Their paper the following year (Van de Merwe and Vogel 1978) provided broader interpretation of the results, but the 1977 paper established the methodology and is recognized as the birth of stable isotope proxies for ancient diet and paleoclimate.

Given the lack of collagen in the more ancient fossil record, researchers explored the use of bone and tooth minerals (Sullivan and Krueger 1981), eventually concluding that enamel provides the most diagenesis-resistant recorder of original isotopic ratios (Ayliffe et al. 1992). The next major innovation occurred when Thure Cerling and colleagues used tooth enamel and pedogenic carbonates to track a major late Cenozoic transition from C_3 to C_4 vegetation in southern Asia and North America. Their paper demonstrated how isotopic signals from a succession of land vertebrate fossils can provide information on global-scale environmental change, although this may not have been be linked, as Cerling and his colleagues proposed, to lowered global CO_2 levels. This research helped to generate a continuing wave of related studies showing that the late Cenozoic C_3 to C_4 isotopic transition occurred at different times on different continents (MacFadden et al. 1996; Fox and Koch 2003; Levin et al. 2004; review in Edwards et al. 2010) and also (surprisingly) that the rise of hypsodonty in mammalian grazers occurred well before the later expansion of C_4 grasslands (Wang et al. 1994; review in Strömberg 2011). Increasingly, materials other than enamel are being used for isotopic analysis, such as pollen (Nelson et al. 2008), plant cuticle (Arens et al. 2000), and organic molecules (Feakins et al. 2005). Isotopes beyond carbon and oxygen also are being explored, including hydrogen (Huang et al. 2007) and strontium (Hoppe et al. 1999).

In the marine realm, research on stable isotopes in foraminferal tests was well established by the 1980s as a "tracer" for oceanographic conditions and long-term trends, particularly with respect to water temperature. James Zachos, Michael Arthur, and Walter Dean were the first, however, to use carbon and oxygen isotope ratios from foraminifera as a proxy for global biological productivity—thereby opening up a new direction in paleoenvironmental investigation. Their innovations included comparing carbon isotopic signals in benthic versus planktonic foraminifera to calibrate the recovery time of the ocean ecosystems following the Cretaceous-Paleogene bolide impact. This work provided a new way to measure changes in oceanic productivity over time that profoundly influenced our ability to understand the resiliencies and vulnerabilities of global-scale environmental states. For example, isotopic data used to track changes in oceanic productivity associated with the Eocene-Oligocene global cooling event (e.g., Diester-Haass and Zahn 2001) have provided valuable clues to the links between climate and the carbon cycle.

The three major areas of foundational research in part 4 of this book—morphology, biostratigraphy/biogeography, and biogeochemistry—show how fossils can be used to understand paleoenvironments. Biogeochemistry is currently the most prominent and widely used of these approaches, and these methods are increasingly combined with other lines of evidence (e.g., morphology, biogeography) to answer questions about large-scale patterns in paleoecology and evolution. Advances in laser-ablation sampling of thin dental enamel have allowed Kimura (Kimura et al. 2013) to use stable isotopes to examine early niche partitioning in the ancestors of modern rats and mice. Stable isotope analyses of mammalian enamel and pedogenic carbonates in the Miocene Siwalik sequence of Pakistan have documented ecological change through time as well as vegetation gradients across the depositional basin (Badgley et al. 2008; Morgan et al. 2009). A combination of stable isotopes and morphology has provided unique insights into the evolution of feeding strategies of Oligocene whales (Clementz et al. 2014). Pairing stable isotopes with tooth-wear studies made it possible to test what the first horses with grazer-type teeth in Europe really ate (Tütken et al. 2013).

The scientific power of all of these methods is not only because they allow us to use fossil organisms to accurately reconstruct past environments but also because they can track environmental parameters across long spans of time. This contributes to understanding how ecological processes operate over geological timescales, which is becoming increasingly essential as evidence for how the Earth-life system will function in the future.

Literature Cited

Andrews, P., and E. Nesbit Evans. 1979. The environment of *Ramapithecus* in Africa. *Paleobiology* 5 (1): 22–30.

Arens, N. C., A. H. Jahren, and R. G. Amundson. 2000. Can C_3 plants faithfully record the carbon isotopic composition of atmospheric carbon dioxide? *Paleobiology* 26:137–64.

Ayliffe, L. K., A. M. Lister, and A. R. Chivas. 1992. The preservation of glacial-interglacial climatic signatures in the oxygen isotopes of elephant skeletal phosphate. *Palaeogeography Palaeoclimatology Palaeoecology* 99:179–91.

Badgley, C., J. C. Barry, M. E. Morgan, S. V. Nelson, A. K. Behrensmeyer, T. E. Cerling, and D. Pilbeam. 2008. Ecological changes in Miocene mammalian record show impact of prolonged climatic forcing. *PNAS* 105: 12145–49.

Beerling, D. J., and D. L. Royer. 2011. Convergent Cenozoic CO_2 history. *Nature Geoscience* 4:418–20.

Bender, M. M. 1971. Variation in the 13C/12C ratios of plants in relation to the pathway of photosynthetic carbon dioxide fixation. *Phytochemistry* 10:1339–44.

Clementz, M. T., R. E. Fordyce, S. L. Peek, and D. L. Fox. 2014. Ancient marine isoscapes and isotopic evidence of bulk-feeding by Oligocene cetaceans. *Palaeogeography Palaeoclimatology Palaeoecology* 400:28–40.

Clementz, M. T., K. Fox-Dobbs, P. V. Wheatley, P. L. Koch, and D. F. Doak. 2009. Revisiting old bones: coupled carbon isotope analysis of bioapatite and collagen as an ecological and palaeoecological tool. *Geological Journal* 44:605–20.

DeNiro, M. J., and S. Epstein. 1978. Influence of diet on the distribution of carbon isotopes in animals. *Geochimica Cosmochimica Acta* 42:495–506.

Diester-Haass, L., and R. Zahn. 2001. Paleoproductivity increase at the Eocene-Oligocene climatic transition: ODP/DSDP sites 763 and 592. *Palaeogeography Palaeoclimatology Palaeoecology* 172:153–70.

Edwards, E. J., C. P. Osborne, C. A. E. Strömberg, and S. A. Smith. 2010. The origins of C_4 grasslands: integrating evolutionary and ecosystem science. *Science* 328: 587–91.

Feakins, S. J., P. B. deMenocal, and T. I. Eglinton. 2005. Biomarker records of late Neogene changes in northeast African vegetation. *Geology* 33:977–80.

Fortelius, M., J. T. Eronen, J. Jernvall, L. Liu, D. Pushkina, J. Rinne, A. Tesakov, et al. 2002. Fossil mammals resolve regional patterns of Eurasian climate change during 20 million years. *Evolutionary Ecology Research* 4:1005–16.

Fox, D. L., and P. L. Koch. 2003. Tertiary history of C_4 biomass in the Great Plains, USA. *Geology* 31:809–12.

Fox, R. C., and C. S. Scott. 2011. A new, early Puercan (earliest Paleocene) species of *Purgatorius* (Plesiadapiformes, Primates) from Saskatchewan, Canada. *Journal of Paleontology* 85:537–48.

Hendy, A. J. W., and P. J. J. Kamp. 2004. Late Miocene to early Pliocene biofacies of Wanganui and Tarankaki Basins, New Zealand: applications to paleoenvironmental and sequence stratigraphic analysis. *New Zealand Journal of Geology and Geophysics* 47:769–85.

Hoppe, K. A., P. L. Koch, R. W. Carlson, and S. D. Webb. 1999. Tracking mammoths and mastodons: reconstruction of migratory behavior using strontium isotope ratios. *Geology* 27:439–42.

Huang, Y., S. C. Clemens, Y. Wang, and W. L. Prell. 2007. Large-scale hydrological change drove the late Miocene C_4 plant expansion in the Himalayan foreland and Arabian Peninsula. *Geology* 35:531–34.

Kimura, Y., L. L. Jacobs, T. E. Cerling, K. T. Uno, K. M. Ferguson, L. J. Flynn, and R. Patnaik. 2013. Fossil mice and rats show isotopic evidence of niche partitioning and change in dental ecomorphology related to dietary shift in late Miocene of Pakistan. *PLoS One* 8: e69308.

Kohn, M. J., and T. E. Cerling. 2002. Stable isotope compositions of biological apatite, In *Phosphates: Geochemical, Geobiological, and Materials Importance*, edited by M. J. Kohn, J. Rakovan, and J. M. Hughes, 455–88. Reviews in Mineralogy and Geochemistry, vol. 48. Washington, DC: Mineralogical Society of America.

Legendre, S. 1986. Analysis of mammalian communities from the Late Eocene and Oligocene of southern France. *Palaeovertebrata* 16:191–212.

Levin, N. E., J. Quade, S. W. Simpson, S. Semaw, and M. Rogers. 2004. Isotopic evidence for Plio-Pleistocene environmental change at Gona, Ethiopia. *Earth and Planetary Science Letters* 219:93–110.

Little, S. A., S. W. Kembel, and P. Wilf. 2010. Paleotemperature proxies from leaf fossils reinterpreted in light of evolutionary history. *PLoS One* 5:e15161.

Lofgren, D. L., J. A. Lillegraven, W. A. Clemens, and P. D. Gingerich. 2004. Paleocene biochronology; the Puercan through Clarkforkian Land Mammal Ages. In *Late Cretaceous and Cenozoic Mammals of North America: Biostratigraphy and Geochronology*, edited by M. O. Woodburne, 43–105. New York: Columbia University Press.

MacFadden, B. J., T. E. Cerling, and J. Prado. 1996. Cenozoic terrestrial ecosystem evolution in Argentina: evidence from carbon isotopes of fossil mammal teeth. *Palaios* 11:319–27.

McElwain, J. C., and W. G. Chaloner. 1995. Stomatal density and index of fossil plants track atmospheric carbon dioxide in the Palaeozoic. *Annals of Botany* 76: 389–95.

Morgan, M. E., A. K. Behrensmeyer, C. Badgley, J. C. Barry, S. Nelson, and D. Pilbeam. 2009. Lateral trends in carbon isotope ratios reveal a Miocene vegetation gradient in the Siwaliks of Pakistan. *Geology* 37: 103–6.

Nelson, D. M., F. S. Hu, D. R. Scholes, N. Joshi, and A. Pearson. 2008. Using SPIRAL (Single Pollen Isotope Ratio AnaLysis) to estimate C_3- and C_4-grass abundance in the paleorecord. *Earth and Planetary Science Letters*, 269:11–16.

Peppe, D. J., D. L. Royer, B. Cariglino, S. Y. Oliver, S. Newman, E. Leight, G. Enikolopov, et al. 2011. Sensitivity of leaf size and shape to climate: global patterns and paleoclimatic applications. *New Phytologist* 190: 724–39.

Schoeninger, M. J., and M. J. DeNiro. 1982. Carbon isotope ratios of apatite from fossil bone cannot be used to reconstruct diets of animals. *Nature* 297:577–78.

Smith, B. N., and W. V. Brown. 1973. The Kranz syndrome in the Gramineae as indicated by carbon isotope ratios. *American Journal of Botany* 60:505–13.

Sponheimer, M., T. Robinson, L. Ayliffe, B. Passey, B. Roeder, L. Shipley, E. Lopez, et al. 2003. An experimental study of carbon-isotope fractionation between diet, hair, and feces of mammalian herbivores. *Canadian Journal of Zoology* 81:871–76.

Strömberg, C. A. E. 2011. Evolution of grasses and grassland ecosystems. *Annual Review of Earth and Planetary Sciences* 39:517–44.

Sullivan, C. H., and H. W. Krueger. 1981. Carbon isotope analysis of separate chemical phases in modern and fossil bone. *Nature* 292:333–35.

Tütken, T., T. M. Kaiser, T. Venneman, and G. Merceron. 2013. Opportunistic feeding strategy for the earliest old world hypsodont equids: evidence from stable isotope and dental wear proxies. *PLoS One* 8:e74463.

van Dam, J. A. 2006. Geographic and temporal patterns in the late Neogene (12–3 Ma) aridification of Europe: the use of small mammals paleoprecipitation proxies. *Palaeogeography, Palaeoclimatology, Palaeoecology* 238:190–218.

Van de Merwe, N. J., and J. C. Vogel. 1978. ^{13}C content of human collagen as a measure of prehistoric diet in woodland North America. *Nature* 276:815–16.

Wang, Y., T. E. Cerling, B. J. MacFadden, and J. D. Bryant. 1994. Fossil horses and carbon isotopes: new evidence for Cenozoic dietary, habitat, and ecosystem changes in North America. *Palaeogeography, Palaeoclimatology, Palaeoecology* 107:269–80.

Wilf, P., K. R. Johnson, and B. T. Huber. 2003. Correlated terrestrial and marine evidence for global climate changes before mass extinction at the Cretaceous-Paleogene boundary. *PNAS* 100:599–604.

PAPER 23

The Temperature- and Depth-Distribution of Some Recent and Fossil Foraminifera in the Southern California Region (1933)

M. L. Natland

Commentary

MARTY BUZAS

Natland's early paper on paleoenvironmental reconstruction is unique because within the contribution he presents and interprets both modern and fossil data. He shows unequivocally that correlation based on the similarity of biotas may not necessarily result because they are contemporaneous but, instead, because of the similarity of the past environments.

The recent data consists of the abundance of foraminiferal species from 165 localities along a traverse from Long Beach, California, to Santa Catalina Island, California. The depths range from less than 1 m to 2,500 m. The distribution of the foraminifera was arranged into five depth zones (biofacies or communities). For each zone, the temperature range was recorded, and, of course, these data indicate a decrease in range as well as cooler temperatures with increasing depth.

In nearby Hall Canyon, California, a section beginning in the lower Pliocene was sampled. Most of the fossil species are still extant and live in the nearby sediments off Santa Catalina. As we proceed up the section, a progression from the deepest zone to shallower zones is evident. Because he believed the observed zonation in the modern fauna was due mainly to temperature, Natland interpreted the change as a gradual warming of the water with time, which may have been the result of shallowing.

Since this pioneering study, numerous contributions from all over the world have shown a strong depth correlation of modern foraminiferal species, and these depth- related biofacies are easily recognized in the fossil record (Phleger 1960; Walton 1964; Culver 1988; Murray 2006). The exact role of temperature is hotly debated and a host of other environmental variables are also considered as important. Reconstruction of ancient depth-related biofacies has become routine. Probably unknown to Natland, his choice of a modern and nearby fossil section was ideal for paleoenvironmental reconstruction because subsequent work has shown that some species of foraminifera change their depth distributions on a regional basis.

Literature Cited

Culver, S. J. 1988. New foraminiferal zonation of the northwestern Gulf of Mexico. *Palaios* 3:69–85.

Murray, J. W. 2006. *Ecology and Applications of Benthic Foraminifera*. Cambridge: Cambridge University Press.

Phleger, F. B 1960. *Ecology and Distribution of Recent Foraminifera*. Baltimore: Johns Hopkins Press.

Walton, W. R. 1964. Recent foraminiferal ecology and paleoecology. In *Approaches to Paleoecology*, edited by J. Imbrie and N. D. Newell, 151–237. New York: Wiley.

From *Bulletin of the Scripps Institution of Oceanography*, Technical Series 3:225–30. A larger reproduction of the chart within this article can be found at www.press.uchicago.edu/sites/lyons/.

THE TEMPERATURE- AND DEPTH-DISTRIBUTION OF SOME RECENT AND FOSSIL FORAMINIFERA IN THE SOUTHERN CALIFORNIA REGION

BY

MANLEY L. NATLAND

THIS PAPER OFFERS an explanation of the accompanying tabular presentation of the bathymetric and temperature ranges of some Recent species of foraminifera off Long Beach, California, and the stratigraphic ranges of these species in a section of marine sediments exposed in Hall Canyon, about one mile northeast of Ventura, California. Many of the species found in the sedimentary section are now living in the near by sea; knowledge of the environmental conditions of the living representatives of these species, therefore, should make it possible to deduce the conditions under which the older sediments were deposited. The present paper is preliminary to a more comprehensive contribution which the writer hopes to complete in the near future.

Work on this project was begun in July, 1927. After that it was continued intermittently until June, 1931, when financial assistance obtained from the National Research Council and the Marsh Fund of the National Academy of Sciences made it possible to carry the work uninterruptedly to its present state. Valuable help and suggestions were received from Dr. T. Wayland Vaughan, Director of the Scripps Institution of Oceanography, La Jolla; Dr. A. O. Woodford, of Pomona College; Dr. R. R. Morse, of Los Angeles; and Mr. Guy E. Miller, of Long Beach, California. The excellent workmanship on the accompanying table is to be credited to Mr. W. L. Hildebrand, of the drafting department of the Shell Oil Company of California. For both the financial assistance and the other help which I received, I express my sincere appreciation.

From July, 1931, to November, 1932, Mr. Alexander Clark and I collected material from 153 localities in the channel between Long Beach and Santa Catalina Island, California. In order to extend the section studied into deeper water, several samples were kindly offered me by the Scripps Institution of Oceanography. Nine samples were from localities of Dr. Parker D. Trask, in the deeps between Santa Catalina Island and Santa Cruz Island. Three were from localities at which bottom samples were taken by the Coast and Geodetic Survey ship "Guide." The following data were recorded at these stations:

	Latitude N	Longitude W	Depth, fathoms	Bottom temperature
"Guide" Station.... 2	31° 55′	119° 37′	1,370	2.40° C
"Guide" Station.... 3	32 02	119 43	1,390	3.00
"Guide" Station.... 17	33 30	118 17	553	6.00

J. A. Cushman reported on the foraminiferal content of these samples in his publication on the "Recent foraminifera from off the west coast of North America."[1]

The sample from the littoral zone at Crescent City Beach, in the northern part of California, was collected by A. R. May, of Bakersfield, California. The temperature of the water at this locality was 16.1° C.

Six hundred samples were collected from the section exposed in Hall Canyon in order to compare them with this Recent material. So far as possible the samples were collected at ten-foot stratigraphic intervals. The sands and conglomerates were not sampled, because in this region, as a rule, they are barren of foraminifera. Since the same fauna was usually found in a number of consecutive samples the collections were combined into 89 groups. Sample 90 is from the lowest beds exposed at the axis of the Ventura anticline along the west bank of the Ventura River, about two miles west of the Hall Canyon section. These beds are several hundred feet lower, stratigraphically, than the lowest beds exposed at the axis in Hall Canyon. The list of species under Zone V is composite, including the foraminifera which occur in the "Lower Pico" or "Repetto" beds of the Ventura and Los Angeles basins. As is indicated, the top of this zone is encountered at a depth of about six thousand feet in wells at the axis of the Ventura anticline in Hall Canyon. The sediments studied from the shallow well in Ventura County are thought to be stratigraphically higher than the highest foraminiferal sediments outcropping in the canyon.

On the left half of the tabular presentation are plotted the data regarding the Recent material, and on the right half are recorded the data regarding the fossil species collected in Hall Canyon. In the upper left-hand corner is an enlarged tracing of a portion of Coast and Geodetic Survey Chart no. 5102, on which the localities of the samples from the channel are plotted. The locality of sample 79 is the lagoon near Sunset Beach.

The upper curve below the map is a depth profile of the channel. The vertical scale is approximately ten times the horizontal scale. Without exaggeration of this magnitude it would be difficult to show graphically the shallow depths off Long Beach. The lower curve indicates the bottom temperatures encountered at each locality. From an examination of this profile it should be noticed that from the surface to a depth of 1000 feet the bottom temperature decreases 16 degrees C, while from 1000 feet down to 8340 feet the de-

[1] Bull. Scripps Inst. Oceanog., Tech. Ser., 1:119–188, pls. 1–6, 1927.

crease in temperature is only 5.5 degrees C. The temperature gradient is steepest from the surface down to 1000 feet, but below 1000 feet the gradient is very gentle. In the shallower depths where the temperature declines rapidly there is a corresponding rapid change in the foraminiferal fauna, while in the deeper part of the channel where the temperature changes are very slight, even with an increase in depth of several thousand feet the changes in the foraminiferal fauna are much slower. Thus it would seem that temperature has a far greater influence than depth on some foraminifera.

The difference between the temperature at locality 1, near Long Beach, and that at 69, in Avalon Bay, both stations having about the same depth, arises from the fact that the temperature at locality 1 was taken in July, while that at locality 69 was taken in November.

The depth and temperature, together with the species of foraminifera found at each locality, are plotted on the same vertical line, except in a few places where projection was necessary. The estimated relative abundance of each species is indicated by symbols explained in the legend. The species opposite whose names a black square appears are pelagic. These are the only species found in abundance all the way across the channel.

The list of species and its alphabetical index at the left of the tabular presentation apply to both Recent and fossil localities and species.

The foraminifera, according to their distribution, have been divided into five faunal-, or life-zones, characterized as follows:

Zone I. Shallow, brackish-water lagoon
Depth at low tide, 1 foot; at high tide, 4 to 7 feet
Temperature of water, August, 1931, at low tide, 24.44° C
Temperature of air at the same time, 3:00 P.M., 28.33° C
Temperature of water, January, 1933, at low tide, 7:30 A.M., 6:54° C
Temperature of air at the same time, 5.66° C
Abundant: *Rotalia beccarii* (Linnaeus)

Zone II. Bottom temperature range, 21.43° C–13.20° C
Depth range, 14 feet–125 feet (open ocean)
Abundant: *Nonion scapha* (Fichtel and Moll)
Elphidium articulatum (d'Orbigny) ?
hannai Cushman and Grant
hughesi Cushman and Grant
spinatum Cushman and Valentine
Buliminella elegantissima (d'Orbigny)
Eponides ornata (d'Orbigny)

Zone III. Bottom temperature, range 13.20° C–8.50° C
Depth range, 125 feet–900 feet
Abundant: *Cassidulina californica* Cushman and Hughes
limbata Cushman and Hughes
tortuosa Cushman and Hughes
Eponides repanda (Fichtel and Moll)
Polymorphina charlottensis Cushman
Quinqueloculina akneriana d'Orbigny
Robertina charlottensis (Cushman)
Sigmomorphina frondiculariformis (Galloway and Wissler)
Triloculina trigonula (Lamarck)

Zone IV. Bottom temperature range, 8.50° C–± 4.° C
Depth range, 900 feet–± 6,500 feet

Abundant:

Bolivina subadvena **v.** ***spissa*** **Cushman**
argentea **Cushman**
Cassidulina cushmani **Stewart and Stewart**
Globobulimina pacifica **Cushman**
Pulvinulinella pacifica **Cushman**
Uvigerina peregrina **Cushman**

Zone V. Botton temperature range, ± 4.° C–± 2.40° C
Depth range, ± 6,500 feet–± 8,340 feet

Abundant:

Bulimina rostrata **H. B. Brady**
Pullenia bulloides **(d'Orbigny)**
Nonion umbilicatula **(Montagu)**
v. ***pacifica*** **Cushman**

This list also appears at the left of the depth and temperature curves.

On the Long Beach side of the channel, sediments outcrop which are rich in the same species of foraminifera as are found in the Recent section near by, therefore the possibility of the redeposition of the older forms within the Recent fauna cannot be ignored. Fortunately, however, the southern portion of Santa Catalina Island is composed of igneous and metamorphic rock from which the great abundance of foraminifera found in Zone III on the Catalina side of the channel could not have been derived. Zone III on the Long Beach side has the same depth range as it has on the Catalina side, and none of the species characterizing this zone were found in the middle of the channel. It is therefore difficult to see how the foraminifera of Zone III off Catalina could have been washed from outcrops near Long Beach without depositing some of the material in the middle of the channel. There are, however, exceptional occurrences of reworking. *Pullenia bulloides* (d'Orbigny), *Bulimina rostrata* H. B. Brady, and *Plectofrondicularia californica* Cushman and Stewart were found on the ridge at localities 6, 7, and 9. These species were probably washed from outcrops on a submarine ridge parallel to the shore line and about six miles south of Long Beach. For this reason they were not plotted.

On the steep slope off Catalina, *Cassidulina tortuosa* Cushman and Hughes and many other species characteristic of Zone III extended into deeper water than on the other side of the channel. Since no living representatives of these species were seen below 920 feet on either side of the channel, it is believed that the large number of worn specimens below this depth near Catalina were in some manner transported down the steep slope. Because it is easier for material to be carried down than up a slope, the shallowest occurrence of a species should be more reliable as a zone limit than the lowest range of a species.

The depth and temperature limits of Zones IV and V cannot be definitely determined from the data at hand.

As indicated by the abundance symbols in the tabular presentation, there was a far greater number of foraminifera per cubic centimeter of sediment off Catalina than there was off Long Beach. Off Catalina, approximately 90 per cent of the sediment from Zone III was composed of foraminifera and shell fragments and about 10 per cent was composed of silt. In Zone III off Long Beach the converse was true. Most of the foraminifera in Zone III off Long Beach were alive or had not been lying on the bottom long, judging from the state of their preservation when dredged. Evidently they were being covered with sediment as fast as they were accumulating. But off Catalina, where the specimens are very abundant, only 5 per cent or 10 per cent were living, while the other tests were badly worn and broken, indicating that they had been lying uncovered on the bottom for a long time. Thus it is suggested that the rate of sedimentation is much slower off Catalina than off Long Beach, which is to be expected, considering the relative sizes of the rivers and the areas contributing sediments to the two sides of the channel. From such facts it follows that, other factors being equal, the rate of sedimentation largely controls the abundance of foraminifera.

On the map in the upper right-hand corner are plotted the locations of the eighty-nine groups of samples from outcrops in Hall Canyon. The geologic section below the map illustrates roughly the lithologic character and structural features of the sediments there exposed. The species of foraminifera found at each locality are plotted vertically below each locality. By this arrangement the stratigraphic ranges of the species are apparent.

The strata from which samples 1 to 10 were taken are composed mostly of sand and conglomerate, containing an abundance of reworked species which have not been recorded on the chart. Although some reworking is prevalent throughout the entire section, most of the species below sample 10 were considered to be autochthonous.

The species characterizing the five Recent zones have limited ranges in the Pliocene and Pleistocene sediments of southern California. *Rotalia beccarii* (Linnaeus), diagnostic of Recent Zone I, was found in the sediments from the shallow well indicated on the left of the table for Hall Canyon. As *Rotalia beccarii* (Linnaeus) prefers to live in lagoons and is seldom found in the open ocean, it seems likely that the sediments containing this species from the shallow well were of lagoonal origin. The fauna of Zone I was not present on the Catalina side because on that side there were no lagoons. Zone II, marked by an abundance of specimens of *Elphidium,* extends from sample 11 to 26. It is evident from an examination of the table that the fauna from Zone II in Hall Canyon is more comparable with the assemblage of Zone II off Long Beach than with the assemblage of Zone II off Catalina. It is therefore suggested that the Hall Canyon sediments in Zone II were laid down in a shallow open sea and not in the shelter of some land mass such as Catalina. Zone II in Hall Canyon contains an abundance of *Elphidium hannai* Cushman and Grant, which was not seen in any available material from off southern California.

This species, however, is abundant off the northern part of California. For this reason the sample from the littoral zone at Crescent City Beach, California, was included.

Zone III, characterized by an abundance of *Cassidulina limbata* Cushman and Hughes, extends from sample 27 to sample 37, and includes sediments commonly called "the mud pit shale," which contains an abundance of *Pecten caurinus* Gould, a cool-water species which does not now range south of Oregon. The fauna of Zone III in Hall Canyon differs from the fauna in this horizon in other out-crop sections in that it contains many species indicative of cooler water. It is therefore surmised that the sediments of Zone III in Hall Canyon were deposited in deeper water, and that Zone III in Hall Canyon compares more closely with Zone III off Long Beach than it does with Zone III off Catalina. These facts indicate that Zone III in Hall Canyon was laid down in an open ocean.

Zone IV, carrying abundant *Uvigerina peregrina* Cushman, includes samples from 37 to 90, and extends down to the top of the Lower Pico or Repetto beds, which are encountered in wells at the axis of the Ventura anticline at a depth of about six thousand feet. The species listed under Zone V are found in the sediments below six thousand feet and above the Miocene in the wells at the axis. Of the species which are still living, *Pullenia bulloides* (d'Orbigny) and *Nonion umbilicatula* (Montagu) are most clearly diagnostic of Zone V. *Plectofrondicularia californica* Cushman and Stewart is also diagnostic of Zone V, but to date it has not been found living.

The changes in the foraminiferal fauna show that the lowest beds examined, considered to be of lower Pliocene age, were deposited in the coldest water and that progressively toward the topmost beds of the section a gradual warming of the water took place. Whether this warming resulted entirely from shallowing cannot be determined with the information at present available.

The work here detailed has shown that on the present sea floor, which may some day be a fossil horizon in a sedimentary section, like that exposed in Hall Canyon, there are five distinct faunal assemblages in close proximity. Therefore dissimilar faunas may be contemporaneous and, conversely, the correlation of two widely separated outcrops based on similarity of their foraminiferal assemblages alone is apt to be erroneous. The resemblance of two assemblages indicates similar environments but not necessarily contemporaneity.

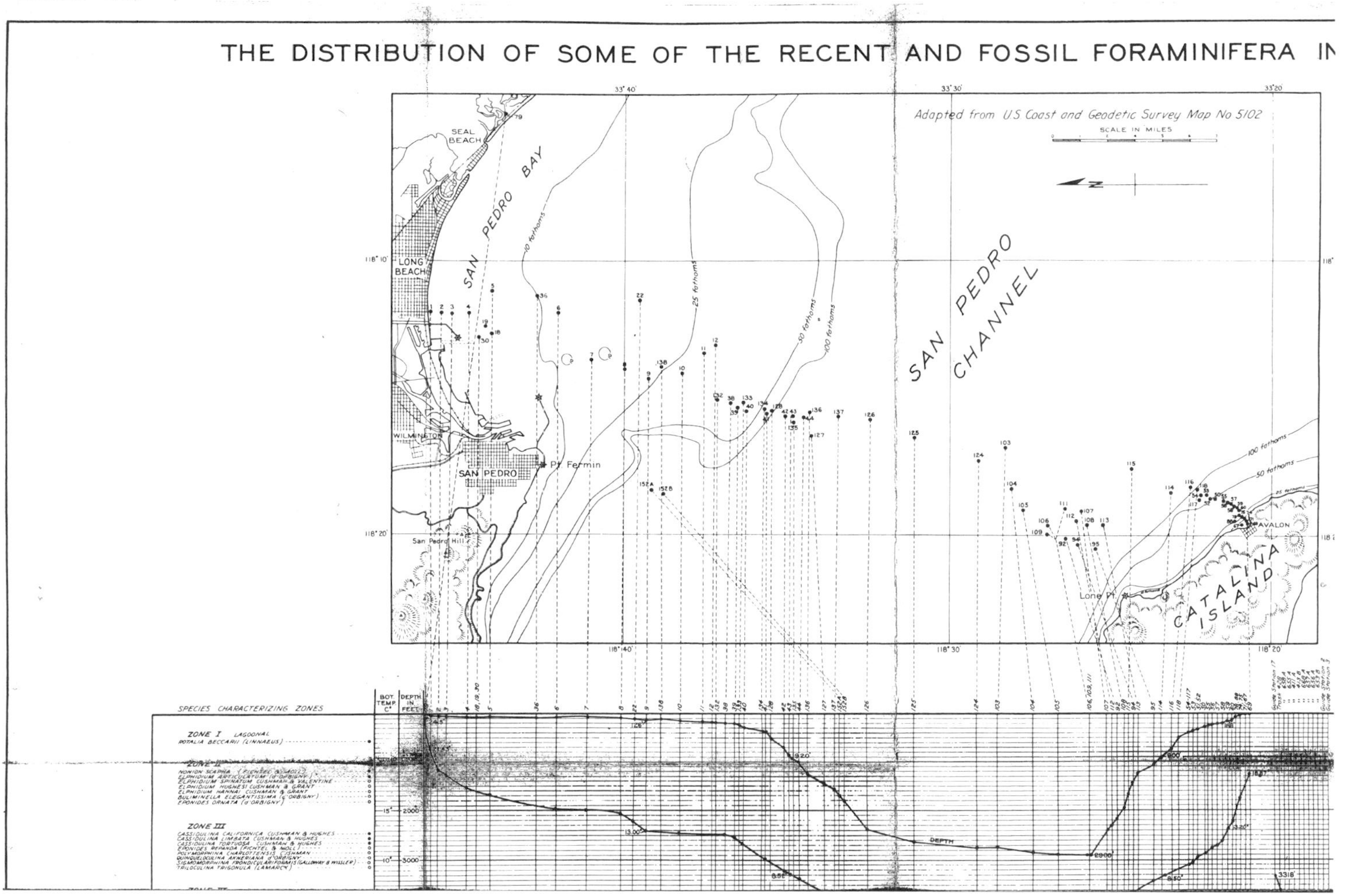
THE DISTRIBUTION OF SOME OF THE RECENT AND FOSSIL FORAMINIFERA IN
Adapted from US Coast and Geodetic Survey Map No 5102
SCALE IN MILES
SAN PEDRO CHANNEL
SAN PEDRO BAY
SEAL BEACH
LONG BEACH
WILMINGTON
SAN PEDRO
Pt Fermin
San Pedro Hill
CATALINA ISLAND
AVALON
Lone Pt
100 fathoms
50 fathoms
25 fathoms
10 fathoms
33°20'
33°30'
33°40'
118°
118°10'
118°20'
118°30'
118°40'
DEPTH
SPECIES CHARACTERIZING ZONES
BOT. TEMP. C°
DEPTH IN FEET
ZONE I LAGOONAL
ROTALIA BECCARII (LINNAEUS)
ZONE II
NONION SCAPHA (FICHTEL & MOLL)
ELPHIDIUM ARTICULATUM (d'ORBIGNY)
ELPHIDIUM SPINATUM CUSHMAN & VALENTINE
ELPHIDIUM HUGHESI CUSHMAN & GRANT
ELPHIDIUM HANNAI CUSHMAN & GRANT
BULIMINELLA ELEGANTISSIMA (d'ORBIGNY)
EPONIDES ORNATA (d'ORBIGNY)
ZONE III
CASSIDULINA CALIFORNICA CUSHMAN & HUGHES
CASSIDULINA LIMBATA CUSHMAN & HUGHES
CASSIDULINA TORTUOSA CUSHMAN & HUGHES
EPONIDES REPANDA (FICHTEL & MOLL)
POLYMORPHINA CHARLOTTENSIS CUSHMAN
QUINQUELOCULINA AKNERIANA d'ORBIGNY
SIGMOMORPHINA FRONDICULARIFORMIS (GALLOWAY & WISSLER)
TRILOCULINA TRIGONULA (LAMARCK)
15°
2000
10°
3000
2900
920
126
1300

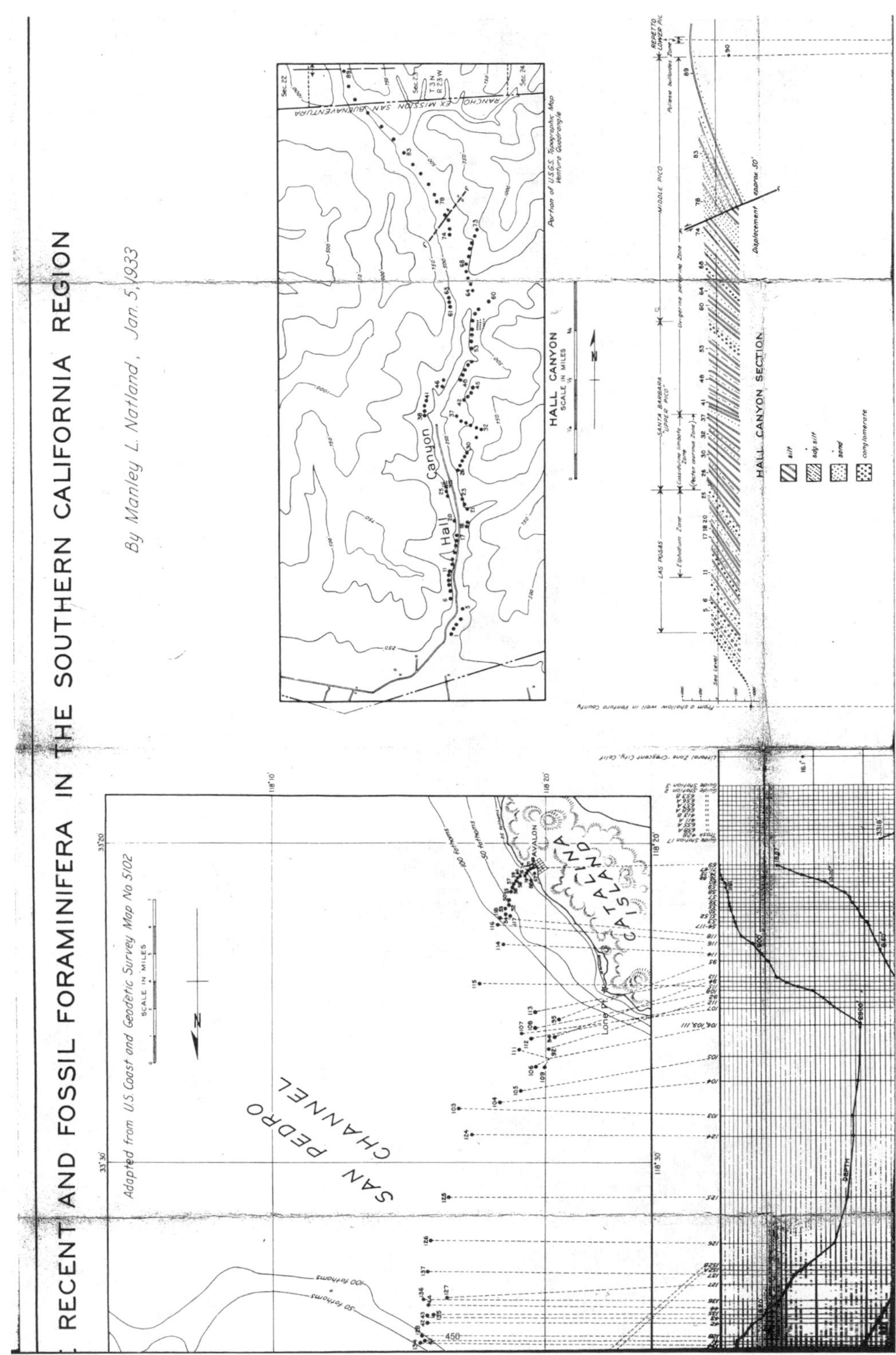
RECENT AND FOSSIL FORAMINIFERA IN THE SOUTHERN CALIFORNIA REGION
By Manley L. Natland, Jan. 5, 1933
Portion of U.S.G.S. Topographic Map
Ventura Quadrangle
HALL CANYON
SCALE IN MILES
Hall Canyon
HALL CANYON SECTION
LAS POSAS
SANTA BARBARA
MIDDLE PICO
silt
sand
conglomerate
Adapted from U.S. Coast and Geodetic Survey Map No 5102
SCALE IN MILES
SAN PEDRO CHANNEL
CATALINA ISLAND
AVALON
100 fathoms
50 fathoms
DEPTH

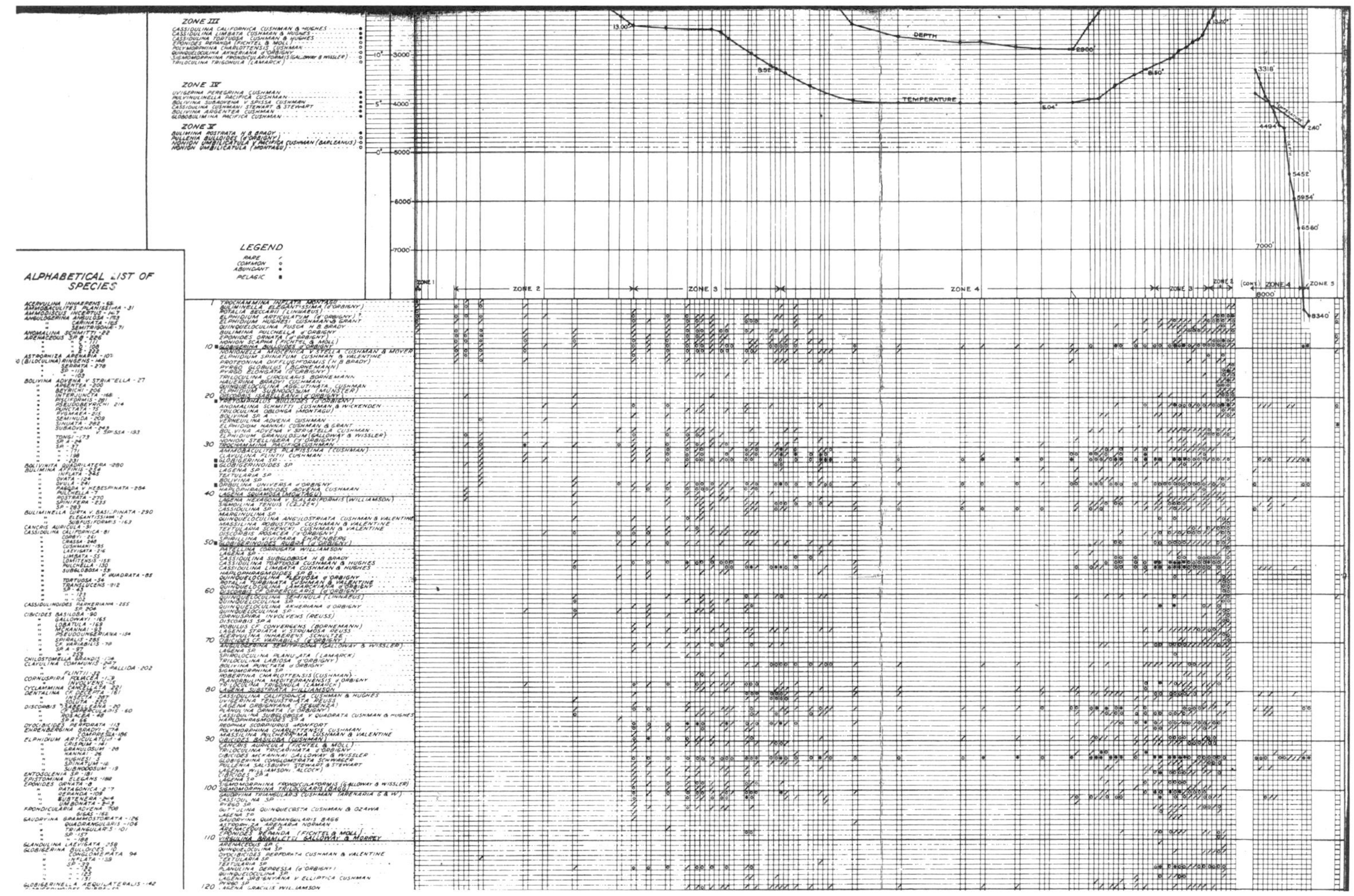
ZONE III
ZONE IV
ZONE V
LEGEND
RARE
COMMON
ABUNDANT
PELAGIC
ALPHABETICAL LIST OF SPECIES
DEPTH
TEMPERATURE
ZONE 1
ZONE 2
ZONE 3
ZONE 4
ZONE 5

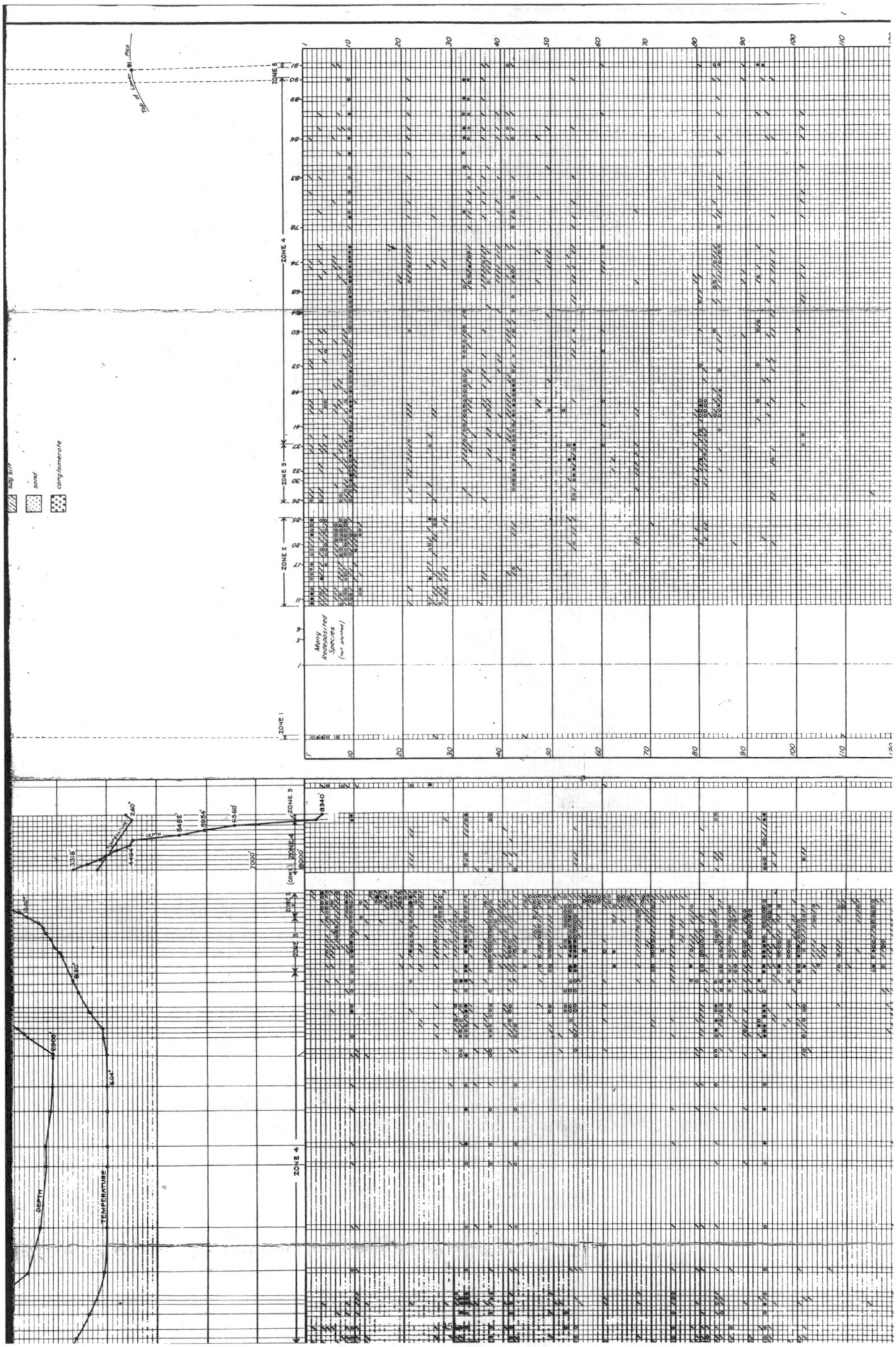

ZONE 1
ZONE 2
ZONE 3
ZONE 4
ZONE 5

Many Redeposited Species (not plotted)

sand
conglomerate

DEPTH
TEMPERATURE

" GIGAS -162
GAUDRYINA GRAMMOSTOMATA -126
" QUADRANGULARIS -106
" TRIANGULARIS -101
" SP -157
" " -188
GLANDULINA LAEVIGATA -258
GLOBIGERINA BULLOIDES -10
" CONGLOMERATA -94
" INFLATA -139
" SP -33
" " -122
" " -125
" " -131
GLOBIGERINELLA AEQUILATERALIS -142
GLOBIGERINOIDES RUBRA -56
" SP 34
GLOBOBULIMINA PACIFICA -194
GLOBOROTALIA CANARIENSIS -143
" TRUNCATULINOIDES -172
GLOBULINA SP -137
GUTTULINA QUINQUECOSTA -104
" SP -177
GYROIDINA SOLDANII -273
" " V ALTIFORMIS -218
" " V ROTUNDIMARGO -252
HAPLOPHRAGMOIDES ADVENA -39
" LATIDORSATUM -219
" SCITULUM -197
" SP B -56
" " A -86
" " -121
HAUERINA BRADYI -17
HYPERAMMINA SP -234
JACULELLA ACUTA -257
LAGENA ELONGATA -180
" GRACILLIMA -203
" GRACILIS -120
" HEXAGONA -192
" " V SCALARIFORMIS -41
" HISPIDA -152
" ORBIGNYANA -83
" " V ELLIPTICA -118
" SQUAMOSA -40
" STRIATA V STRUMOSA -68
" SUBSTRIATA -80
" WILLIAMSONI -96
" SP -35
" " -52
" " -72
" " -98
" " -105
" " -127
" " -132
" " -140
" " -146
" " -187
" " -213
" " -237
" " -238
" " -239 & 240
LAGENAMMINA SP -230
LATICARININA PAUPERATA -191
MARGINULINA SP -44
MASSILINA PULCHERRIMA -88
" ROBUSTIOR -46
NODOGENERINA CF LEPIDULA -268
" SP -288
NODOSARIA ARUNDINEA -263
" CF SCALARIS -126
" SUBSTRIATULA -151
" TOSTA -260
" SP -178
" " -223
" " -250
" " -262
" " -267
" " -269
NONION SCAPHA -9
" STELLIGERA -29
" CF TURGIDA -205
" UMBILICATULA -265
" " V PACIFICA -253
NONIONELLA MIOCENICA V STELLA -11
NUMMULOCULINA SP -190
ORBULINA UNIVERSA -38
PATELLINA CORRUGATA -31
PELLATISPIRA PAUPERATA -191
PLACOPSILINA CENOMANA -225
PLANORBULINA MEDITERRANENSIS -78
PLANULINA DEPRESSA -116
" ORNATA -84
PLECTOFRONDICULARIA CALIFORNICA -286
PLEUROSTOMELLA SP -266
POLYMORPHINA CHARLOTTENSIS -88
PROTEONINA DIFFLUGIFORMIS -13
PSAMMOSPHAERA FUSCA -224
PSEUDOPOLYMORPHINA DOANEI -158
PULLENIA BULLOIDES -272
" QUINQUELOBA -210
" SALISBURYI -95
PULVINULINELLA BRADYANA -196
" CULTER -276
" PACIFIC -176
PYRGO ELONGATA -15 } (NOTE: SEE BILOCULINA)
" GLOBULUS -14 }
QUINQUELOCULINA AGGLUTINATA -18
" AKNERIANA -63
" ANGULOSTRIATA -45
" FLEXUOSA -57
" FUSCA -5
" LAMARCKIANA -59
" SEMINULA -61
" SP -62
" " -64
" " -112
" " -117
" " -166
" " -179
" " -279
REOPHAX DENTALINIFORMIS -184
" EXCENTRICUS -34
" SCORPIURUS -87
" SP -233
RHABDAMMINA ABYSSORUM -135
ROBERTINA CHARLOTTENSIS -77
ROBULUS CF CONVERGENS -67
" CUSHMANI -149
" STRONGI -167
ROTALIA BECCARII -3
" TURBINATA -58
" SP -209
RUPERTIA SP -256
SARACENARIA ITALICA -175
SIGMOILINA TENUIS -42
SIGMOMORPHINA FRONDICULARIFORMIS -99
" TRILOCULARIS -100
" SP 76
" SP -160
SPHAEROIDINA BULLOIDES -264
SPHAEROIDINELLA DEHISCENS -271
SPIRILLINA VIVIPARA -49
SPIROLOCULINA DEPRESSA -129
" PLANULATA -73
TEXTULARIA GRAMEN -138
" SCHENCKI -27
" SP -36
" " -114
" " -115
" " -133
" " -164
" " -183
" " -275
CF TEXTULARIA SP -128
TRETOMPHALUS BULLOIDES -21
TRILOCULINA CIRCULARIS -16
" INSIGNIS -170
" LABIOSA -14
" OBLONGA -23
" SUBORBICULARIS -156
" TRICARINATA -92
" TRIGONULA -79
TROCHAMMINA INFLATA -1
" PACIFICA -30
" SP A -144
" " B -145
UVIGERINA AMPULLACEA -228
" CANARIENSIS -251
" FARINOSA -207
" PEREGRINA -217
" CF PYGMAEA -242
" SENTICOSA -189
" TENUISTRIATA -82
" SP -227
VAGINULINA LEGUMEN -150
VALVULINERIA ARAUCANA -211
" INAEQUALIS -199
VERNEUILINA ADVENA -25
" SP -229
VIRGULINA BRAMLETTI -110

LAGENA SP
GAUDRYINA QUADRANGULARIS BAGG
ASTRORHIZA ARENARIA NORMAN
ARENACEOUS SP D
EPONIDES REPANDA (FICHTEL & MOLL)
110 VIRGULINA BRAMLETTI GALLOWAY & MORREY
ARENACEOUS SP C
QUINQUELOCULINA SP
DYOCIBICIDES PERFORATA CUSHMAN & VALENTINE
TEXTULARIA SP
TEXTULARIA SP
PLANULINA DEPRESSA (d'ORBIGNY)
QUINQUELOCULINA SP
LAGENA ORBIGNYANA V ELLIPTICA CUSHMAN
PYRGO SP
120 LAGENA GRACILIS WILLIAMSON
HAPLOPHRAGMOIDES SP
■ GLOBIGERINA SP
CASSIDULINA SP
BULIMINA OVATA d'ORBIGNY
■ GLOBIGERINA SP
GAUDRYINA GRAMMOSTOMATA GALLOWAY & WISSLER
LAGENA SP
CF TEXTULARIA (TRUNCATA)
SPIROLOCULINA DEPRESSA d'ORBIGNY
130 CASSIDULINA PULCHELLA d'ORBIGNY
■ GLOBIGERINA SP
LAGENA SP
TEXTULARIA SP
REOPHAX EXCENTRICUS CUSHMAN
RHABDAMMINA ABYSSORUM H B CARPENTER
NODOSARIA CF SCALARIS BATSCH
GLOBULINA SP
TEXTULARIA GRAMEN d'ORBIGNY
GLOBIGERINA INFLATA d'ORBIGNY
140 LAGENA SP
ELPHIDIUM CRISPUM (LINNAEUS)
■ GLOBIGERINELLA AEQUILATERALIS (H B BRADY)
■ GLOBOROTALIA CANARIENSIS (d'ORBIGNY)
TROCHAMMINA SP A
TROCHAMMINA SP B
LAGENA SP
AMMODISCUS INCERTUS (d'ORBIGNY)
PYRGO RINGENS (LAMARCK)
ROBULUS CUSHMANI GALLOWAY & WISSLER
150 VAGINULINA LEGUMEN LINNAEUS
NODOSARIA SUBSTRIATULA CUSHMAN
LAGENA HISPIDA REUSS
ANGULOGERINA ANGULOSA (WILLIAMSON)
CIBICIDES PSEUDOUNGERIANA (CUSHMAN)
CASSIDULINA LOMITENSIS GALLOWAY & WISSLER
TRILOCULINA SUBORBICULARIS d'ORBIGNY
GAUDRYINA SP
PSEUDOPOLYMORPHINA DOANEI (GALLOWAY & WISSLER)
CORNUSPIRA FOLIACEA (PHILIPPI)
160 SIGMOMORPHINA SP
DENTALINA CF DECEPTA (BAGG)
FRONDICULARIA GIGAS CHURCH
BULIMINELLA SUBFUSIFORMIS CUSHMAN
TEXTULARIA SP A
CIBICIDES GALLOWAYI CUSHMAN & VALENTINE
QUINQUELOCULINA SP
ROBULUS STRONGI CHURCH
BOLIVINA INTERJUNCTA CUSHMAN
CIBICIDES LOBATULA (WALKER & JACOB)
170 TRILOCULINA INSIGNIS (H B BRADY)
BOLIVINA SP
■ GLOBOROTALIA TRUNCATULINOIDES (d'ORBIGNY)
BOLIVINA TONGI CUSHMAN
CHILOSTOMELLA GRANDIS CUSHMAN
SARACENARIA ITALICA DEFRANCE
PULVINULINELLA PACIFIC CUSHMAN
GUTTULINA SP
NODOSARIA SP
QUINQUELOCULINA SP C
180 LAGENA ELONGATA (EHRENBERG)
ENTOSOLENIA SP
EPISTOMINA ELEGANS (d'ORBIGNY)
TEXTULARIA SP B
REOPHAX DENTALINIFORMIS H B BRADY
ANGULOGERINA CARINATA CUSHMAN
EHRENBERGINA COMPRESSA CUSHMAN
LAGENA SP
GAUDRYINA SP
UVIGERINA SENTICOSA CUSHMAN
190 NUMMULOCULINA SP
LATICARININA PAUPERATA (PARKER & JONES)
LAGENA HEXAGONA (WILLIAMSON)
BOLIVINA SUBADVENA V SPISSA CUSHMAN
GLOBOBULIMINA PACIFICA CUSHMAN
CASSIDULINA CUSHMANI STEWART & STEWART
PULVINULINELLA BRADYANA CUSHMAN
HAPLOPHRAGMOIDES SCITULUM (H B BRADY)
BOLIVINA SP
VALVULINERIA INAEQUALIS (d'ORBIGNY)
200 BOLIVINA ARGENTEA CUSHMAN
VIRGULINA SP
CLAVULINA COMMUNIS V PALLIDA CUSHMAN
LAGENA GRACILLIMA (SEQUENZA)
CASSIDULINOIDES SP
NONION CF TURGIDA (WILLIAMSON)
BOLIVINA BEYRICHI REUSS
UVIGERINA FARINOSA HANTKEN
FRONDICULARIA ADVENA CUSHMAN
BOLIVINA SEMINUDA CUSHMAN
210 PULLENIA QUINQUELOBA (REUSS)
VALVULINERIA ARAUCANA (d'ORBIGNY)
CASSIDULINA TRANSLUCENS CUSHMAN & HUGHES
LAGENA SP
BOLIVINA PSEUDOBEYRICHI CUSHMAN
BOLIVINA PYGMAEA H B BRADY
CASSIDULINA LAEVIGATA d'ORBIGNY
UVIGERINA PEREGRINA CUSHMAN
GYROIDINA SOLDANII V ALTIFORMIS STEWART & STEWART
HAPLOPHRAGMOIDES LATIDORSATUM (BORNEMANN)
220 DENTALINA SOLUTA (REUSS)
CYCLAMMINA CANCELLATA H B BRADY
ARENACEOUS SP E
CF NODOSARIA SP
PSAMMOSPHAERA FUSCA SCHULZE
PLACOPSILINA CENOMANA d'ORBIGNY
ARENACEOUS SP
UVIGERINA SP
UVIGERINA AMPULLACEA H B BRADY
VERNEUILINA SP
230 LAGENAMMINA SP ?
BOLIVINA SP
VIRGULINA SCHREIBERSIANA CZJZEK
REOPHAX SP
HYPERAMMINA SP
BULIMINA SPINIFERA CUSHMAN
VIRGULINA CORNUTA CUSHMAN
LAGENA SP
LAGENA SP
LAGENA SP
240 LAGENA SP
BULIMINA OVULA d'ORBIGNY
UVIGERINA CF PYGMAEA d'ORBIGNY
EPONIDES UMBONATA (REUSS)
EPONIDES SUBTENERA (GALLOWAY & WISSLER)
BULIMINA INFLATA SEQUENZA
VIRGULINA SP
CLAVULINA COMMUNIS d'ORBIGNY
CASSIDULINA CRASSA d'ORBIGNY
BOLIVINA SUBADVENA CUSHMAN
250 NODOSARIA SP
UVIGERINA CANARIENSIS d'ORBIGNY
GYROIDINA SOLDANII V ROTUNDIMARGO STEWART & STEWART
NONION (BARLEANUS) UMBILICATULA V PACIFICA CUSHMAN
BULIMINA AFFINIS d'ORBIGNY
CASSIDULINOIDES PARKERIANA (H B BRADY)
RUPERTIA SP
JACULELLA ACUTA H B BRADY
GLANDULINA LAEVIGATA d'ORBIGNY
CIBICIDES SP
260 NODOSARIA TOSTA SCHWAGER
CASSIDULINA CORBYI CUSHMAN & HUGHES
NODOSARIA SP
NODOSARIA ARUNDINEA SCHWAGER
SPHAEROIDINA BULLOIDES d'ORBIGNY
NONION UMBILICATULA (MONTAGU)
PLEUROSTOMELLA SP
NODOSARIA SP
NODOGENERINA CF LEPIDULA (SCHWAGER)
NODOSARIA SP
270 BULIMINA ROSTRATA H B BRADY
SPHAEROIDINELLA DEHISCENS (PARKER & JONES)
PULLENIA BULLOIDES (d'ORBIGNY)
GYROIDINA SOLDANII d'ORBIGNY
EHRENBERGINA BRADYI CUSHMAN
TEXTULARIA SP
PULVINULINELLA CULTER (PARKER & JONES)
EPONIDES PATAGONICA (d'ORBIGNY)
PYRGO SERRATA (L W BAILEY)
QUINQUELOCULINA SP
280 BOLIVINITA QUADRILATERA (SCHWAGER)
BOLIVINA CF PISCIFORMIS GALLOWAY & MORREY (AENARIENSIS V SPINULOSA)
BOLIVINA SINUATA GALLOWAY & WISSLER
BULIMINA SP (INFLATA V APICULATA)
BULIMINA PAGODA V HEBESPINATA STEWART & STEWART

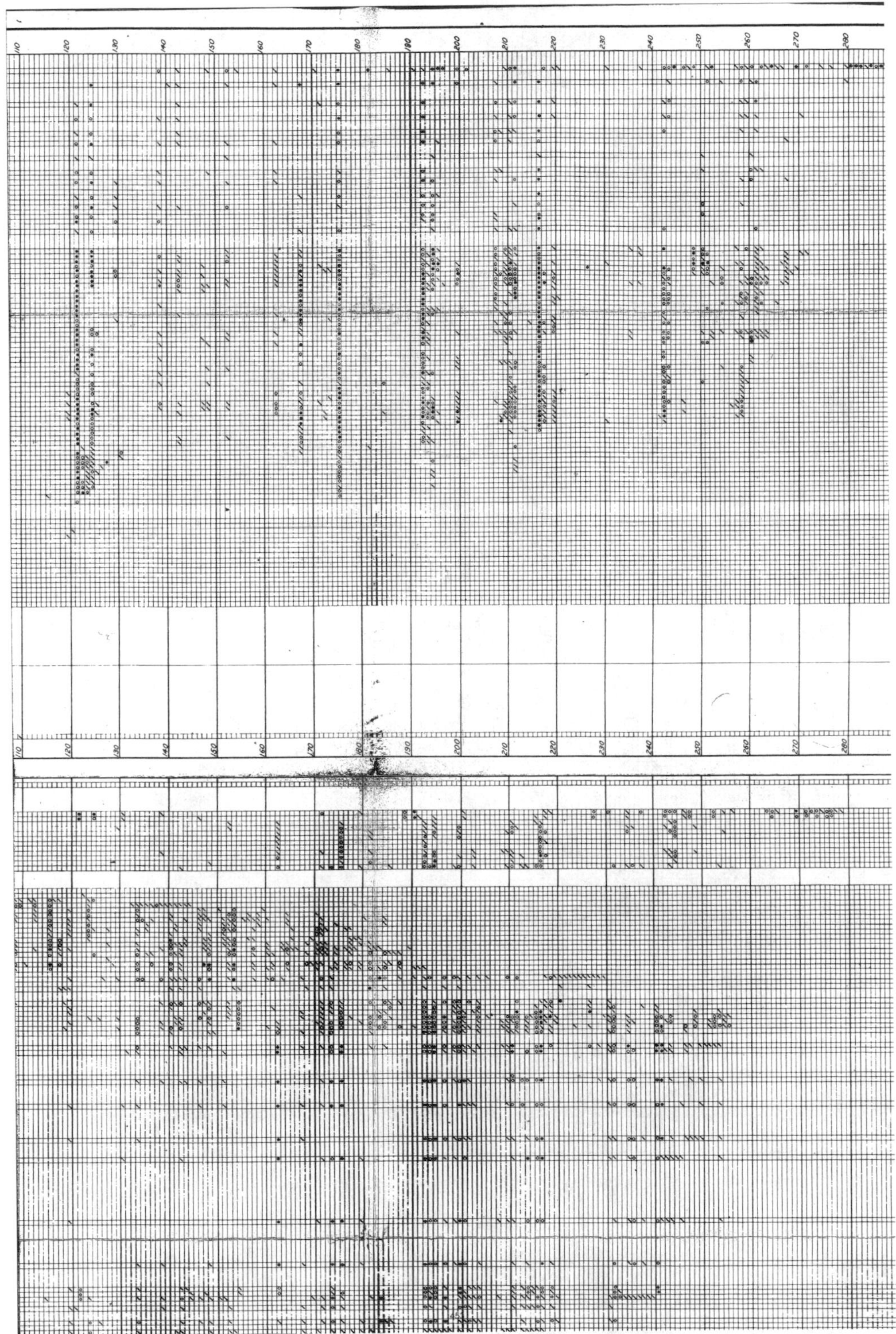

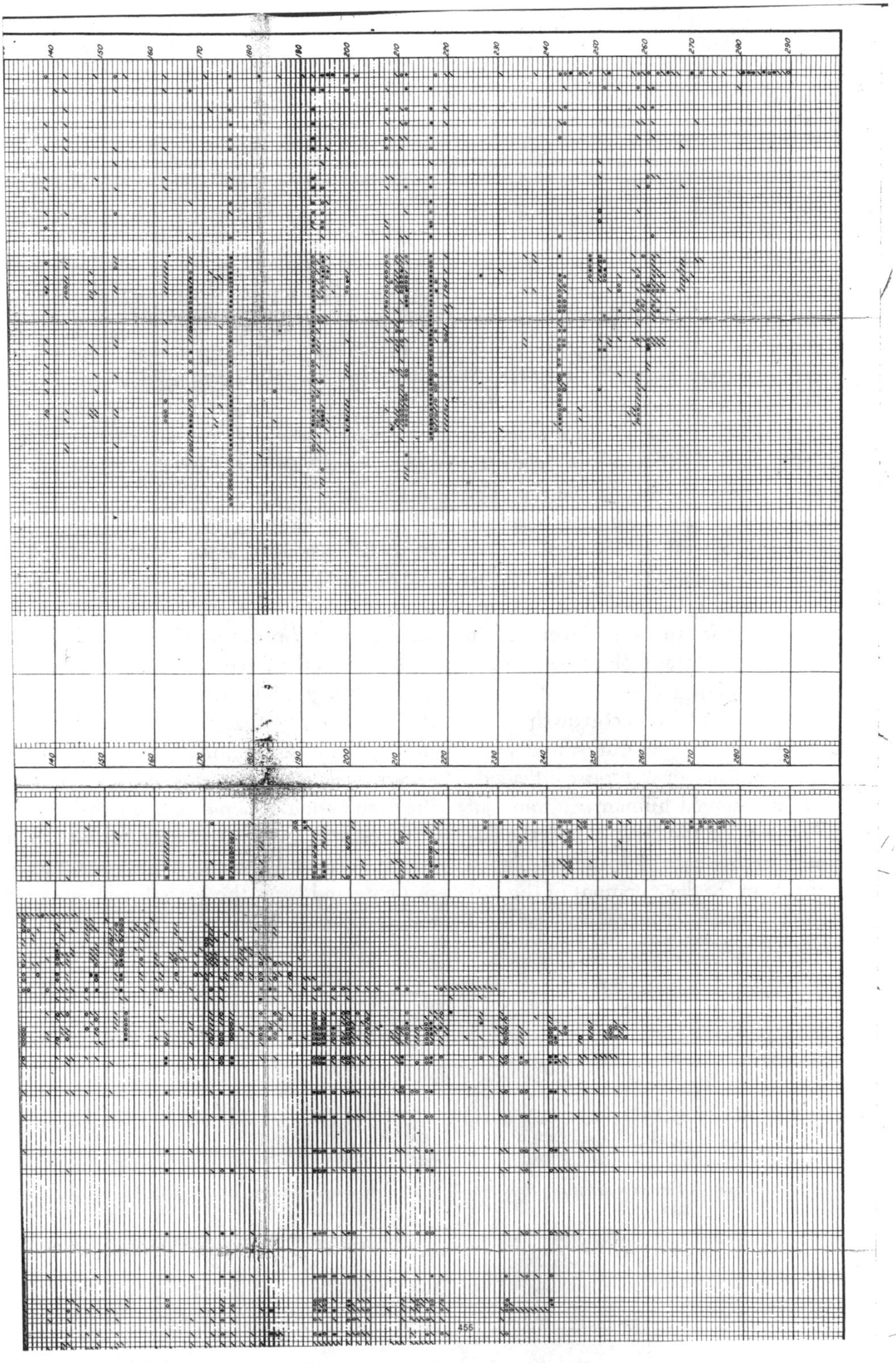

140 150 160 170 180 190 200 210 220 230 240 250 260 270 280 290

140 150 160 170 180 190 200 210 220 230 240 250 260 270 280 290

455

PAPER 24

Isotopic Evidence for Early Maize Cultivation in New York State (1977)

J. C. Vogel and N. J. Van der Merwe

Commentary

NOREEN TUROSS

Almost sixty years ago, Harmon Craig (1953) documented the existence of grasses that were highly enriched in the heavy isotope of carbon and isotopically distinct in $\delta^{13}C$ from the majority of plants that were measured. The biochemical explanation of Craig's observation is found in the work of Marshall Hatch and Charles Slack (1966). It is interesting to note that the original work on what became known as the C4 pathway was developed in an economically important plant, sugarcane (Kortschak et al. 1965).

The combination of higher growth rates, tolerance for higher temperatures, and well-developed mechanisms for water use efficiency in C4 plants has served human programs of domestication well. In the New World, the domestication of maize (corn), a C4 plant, played a dominant role in the development of the human subsistence base over many thousands of years (Piperno and Flannery 2001). The process of maize domestication and the spread of this important staple through space and time is an active and ongoing area of research in archaeology and plant biology. Today, the isotopic enrichment found in maize is transported in the grain we eat, the animals that are fed this staple, and the ethanol used to power some of our vehicles.

Vogel and van der Merwe utilized the isotopic differences in maize and the majority of the dietary plant base (C3) to examine the introduction of maize into the North American diet of ancient peoples. By carefully choosing populations that lived in C3-dominated ecosystems, Vogel and van der Merwe documented the addition of maize to the North American Indian diet by approximately 1000 AD. The issues raised in this important paper, including the preparation of bone collagen, the historical changes in maize isotopic composition, and the isotopic translation from diet to consumer, still resonate and form the basis for many studies today.

Literature Cited

Craig, H. 1953. The geochemistry of the stable carbon isotopes. *Geochimica et Cosmochimica Acta* 3:53–92.

Hatch, M. D., and C. R. Slack. 1966. A new carboxylation reaction and the pathway of sugar formation. *Biochemistry Journal* 101:112–11.

Kortschak, H. P., C. E. Hartt, and G. O. Burr. 1965 Carbon dioxide fixation in sugarcane leaves. *Plant Physiology* 40:209–13

Piperno, D. R., and K. V. Flannery. 2001 The earliest archaeological maize (*Zea mays* L.) from highland Mexico: new accelerator mass spectrometry dates and their implications. *PNAS* 98 (4): 2101–3.

From *American Antiquity* 42:238–42.

FIELD 3 SEGMENT
01 LT LATERAL SEGMENT OF PORTION
02 MID MEDIAL SECTION OF PORTION
03 AN ANTERIOR SEGMENT OF PORTION
04 PO POSTERIOR SEGMENT OF PORTION
05 AL ANTEROLATERAL SEGMENT OF PORTION
06 PL POSTEROLATERAL SEGMENT OF PORTION
07 AM ANTEROMEDIAL SEGMENT OF PORTION
08 PO POSTEROMEDIAL SEGMENT OF PORTION
09 HF HALF SEGMENT LATERAL MEDIAL ANTERIOR POSTERIOR INDET
10 FR UNSPECIFIED FRAGMENT OF PORTION
11 CS COMPLETE SHAFT TUBULAR

Bökönyi, S.
1972 Zoological evidence for seasonal or permanent occupation of prehistoric settlements. In Man, settlement and urbanism, edited by P. J. Ucko, R. Tringham, and G. W. Dimbleby, pp. 121-26. Gerald Duckworth, London.

Bouchud, Jean
1966 Essai sur le renne et la climatologie du paléolithique moyen et supérieur. Centre National de la Recherche Scientifique. Magne, Périgueux.

Casteel, Richard W.
1972 Some archaeological uses of fish remains. American Antiquity 37:404-19.

Control Data Corporation
1976 Sort/merge reference manual, Version 4, Revision J. Control Data Corporation, Minneapolis.

Crader, Diana C.
1974 The effects of scavengers on bone material from a large mammal: an experiment conducted among the Bisa of the Luangwa Valley, Zambia. In Ethnoarchaeology, edited by C. B. Donnan and C. W. Clewlow, Jr. Institute of Archaeology, University of California, Los Angeles, Monograph 4:161-73.

Creighton, R., P. Packard, and H. Lien
1972 Selgem retrieval: a general description. Smithsonian Institution Procedures in Computer Sciences, Vol. 1, No. 1.

Daly, Patricia
1969 Approaches to faunal analysis in archaeology. American Antiquity 34:146-53.

Higgs, E. S.
1967 Early domesticated animals in Libya. In Background to evolution in Africa, edited by W. W. Bishop and J. D. Clark, pp. 165-73. University of Chicago Press, Chicago.

Jarman, M. R.
1972 European deer economies and the advent of the Neolithic. In Papers in economic prehistory, edited by E. S. Higgs, pp. 125-47. Cambridge University Press, Cambridge, England.

Klein, Richard G.
1974 Environment and subsistence of prehistoric man in the Southern Cape Province, South Africa. World Archaeology 5:249-84.

McArdle, John
1976 A numerical (computerized) method for quantifying zooarchaeological comparisons. Paléorient.

Nie, N., C. Hull, J. Jenkins, K. Steinbrenner, and D. Bent
1970 Statistical package for the social sciences, 2nd edition, McGraw-Hill, New York.

Payne, S.
1972 On the interpretation of bone samples from archaeological sites. In Papers in economic prehistory, edited by E. S. Higgs, pp. 65-81. Cambridge University Press, Cambridge, England.

ISOTOPIC EVIDENCE FOR EARLY MAIZE CULTIVATION IN NEW YORK STATE

J. C. VOGEL
NIKOLAAS J. VAN DER MERWE

Plants metabolize carbon dioxide photosynthetically either through a 3-carbon (Calvin) or 4-carbon pathway. Most plants are of the C-3 type; C-4 plants are primarily grasses adapted to hot, arid environments. Since C-4 plants have a higher $^{13}C/^{12}C$ ratio than C-3 plants, animals and humans with a significant C-4 plant food-intake will have higher $^{13}C/^{12}C$ ratios as well. Maize is a C-4 plant, hence maize cultivators living in predominantly C-3 plant environments should show significant isotopic differences from local hunter-gatherers in their skeletal remains; the importance of maize in their diet should also be measurable. The practicability of this method is demonstrated for New York State archaeological materials and wider implications are mentioned.

Maize has long been known to have a higher $^{13}C/^{12}C$ ratio than most terrestrial plants (Vogel 1959; Münnich and Vogel 1958; Bender 1968). Since its $^{14}C/^{12}C$ ratio is also higher, radiocarbon dates of maize remains are known to be consistently too young if no correction is made for the fractionation effect. This is due to the fact that *Zea mays*, together with many other plants (mainly grasses) from semi-arid and arid regions utilize the recently discovered 4-carbon or Hatch-Slack photosynthetic pathway and not the more widespread 3-carbon or Calvin pathway for carbon-dioxide fixation (Kortschak, Hartt, and Burr 1965; Hatch and Slack 1966; Hatch, Slack, and Johnson 1967). Such C-4 species are not common in temperate regions; the syndrome generally constitutes an adaptation to a hot and dry environment.

The ^{13}C content of carbon-dioxide in the free atmosphere is 7 per mil (per thousand) lower than in average marine limestone used as a reference standard (Craig 1953). The latter is known as the PDB standard; the deviation from the standard is expressed as $\delta^{13}C = -7‰$. The 3-carbon photosynthetic pathway depletes the ^{13}C abundance, on the average, by a further 19 per mil. C-3 plants thus have an average ^{13}C content of −26 per mil, with the range extending from −22 to −34 per mil. In contrast, the isotopic abundance in C-4 plants ranges between −9 and −16 per mil, with a modal value of about −12.5 per mil (Troughton 1971; Vogel n.d.). Plants utilizing the Crassulacean Acid Metabolism (CAM) for carbon-dioxide fixation have an isotopic content similar to C-4 plants; CAM plants, however, are succulents adapted to xerophytic conditions (Neales, Patterson, and Hartley 1968) and need not be considered in the present context.

Since both animals and humans ultimately derive their carbon from plants, the carbon isotope ratio can be used to determine the relative intake of C-3 and C-4 plants at the beginning of their food chain. Where humans living in a predominantly C-3 plant environment have access to a C-4 cultigen which forms an important dietary staple, the relative importance of such a cultigen in the diet should be measurable through an isotopic study of skeletal remains. More specifically, it should be possible to detect the presence of maize in the diet of prehistoric peoples in many regions of North America, thus distinguishing them from peoples of the same regions who subsisted on hunting and the gathering of indigenous C-3 plants (van der Merwe 1973).

To demonstrate the practicability of this idea, human skeletal material from four archaeological sites in New York State was investigated. Two of the sites date from pre-horticultural times, while the other two post-date the introduction of maize horticulture to the state. This event, which is generally dated to about the eleventh century A.D., followed a period of adaptation in North America during which the Northern Flint variety of maize was developed from precursors derived ultimately from Central America (Winter 1971). In carrying out this analysis it was assumed that the few C-4 plants indigenous to New York (e.g., *Panicum virgatum*) would not have had an appreciable isotopic impact on humans, either pre- or post-maize. It can be predicted, in any case, that the isotopic effect of maize would be of sufficient magnitude to mask other minor contributions.

Samples of human bone (one rib each) were obtained from the following sites:

Frontenac Island: Archaic period, suggested date 2500-2000 B.C. Archaeological evidence indicates heavy emphasis on fishing (lacustrine), with hunting secondary. Collecting of wild plant foods inferred; no evidence of horticulture. Excavated in the 1940s and early 1950s. (See Ritchie 1965 for this and the following two sites.)

Vine Valley: Early Woodland period, suggested date 400-100 B.C. Hunting-collecting subsistence pattern; no evidence of horticulture. Excavated in 1937.

Snell: Late Woodland period, suggested date A.D. 1000-1300. Hunting-collecting subsistence pattern, with some evidence for horticulture at other sites of the same period. No direct evidence for maize cultivation was found at the Snell site. Excavated in the 1940s and 1950s.

Engelbert: Historic period, presumably Susquehannock. Associated radiocarbon date A.D. 1450±100 (Y-2617); European trade goods date the burials to the 1500s. Considerable evidence for horticulture and hunting-collecting in this period. Excavated 1967-68 (Lipe and Elliott 1970).

The carbon isotope ratio in purified collagen

from samples of bone belonging to seven different individuals was measured by means of a mass spectrometer (one measured twice). For interpreting the results it is necessary to know that a slight degree of ^{13}C enrichment occurs in the process of collagen formation in the body (Figs. 1 and 2). The magnitude of this effect is about 6 per mil (Vogel and Waterbolk 1967, 1972; Vogel, unpublished). Thus the relative ^{13}C content of bone collagen in individuals with a diet consisting exclusively of C-3 plants is about $-26 + 6 = -20$ per mil, while that of, for instance, grazing antelopes living entirely on C-4 grasses is expected to be about $-13 + 6 = -7$ per mil. The latter holds true for grazers living in the savannahs of southern Africa.

The results obtained for the human bone samples are given in Table 1. The difference between the pre-horticultural hunter-gatherers and the horticulturalists is evident. While the former show the same isotopic ratio as do animals and humans from Europe, the values for the latter suggest that C-4 plants played an important part in their diet. A relative ^{13}C content of −13 per mil requires that about 50% of the carbon in the sample be derived from C-4 plants—in this case presumably maize. No appreciable admixture of C-4 plants in the diet can be detected before the introduction of maize; considerable reliance on this new food source provides the only reasonable explanation for the observed isotopic changes.

These results suffice to demonstrate that carbon isotope abundance measurements can be used, given the right environmental circumstances, to measure the impact of the introduction of maize on prehistoric communities in many regions of North America. The extent of maize utilization at the Snell site is surprisingly high in this respect, suggesting that horticulture was adopted with remarkable rapidity and efficiency. Further applications of this method in North America and other regions readily suggest themselves, provided that sufficient background information on the edible plant population of a given area exists. Distribution maps of C-3 and C-4 plants are not as readily available as one could wish for, but are rapidly being compiled.

Elsewhere in the world, the introduction of millet and sorghum cultigens (both C-4 plants) to temperate or tropical regions should be traceable in the same manner. Wheat, barley, and rice are C-3 plants and, therefore, not detectable.

In a similar vein, the dependence of coastal hunting-fishing communities on marine life

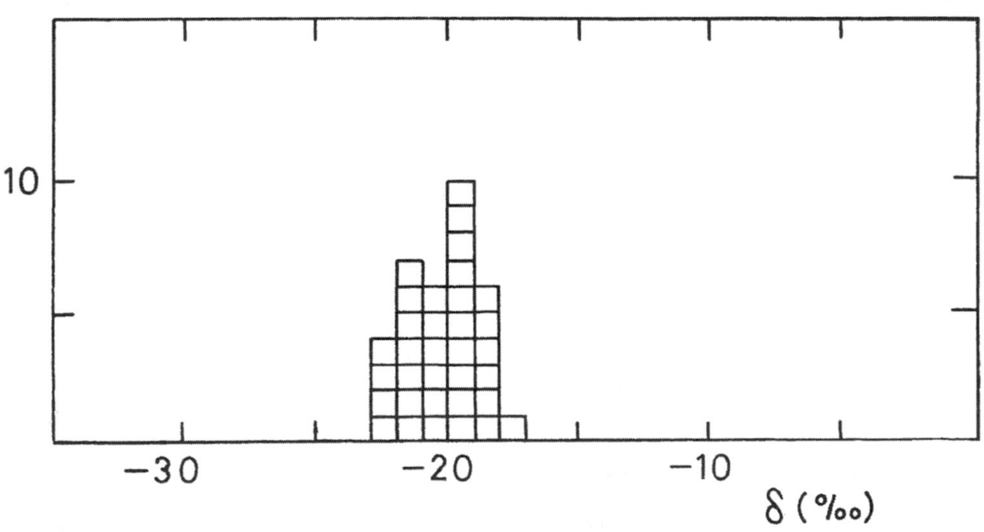

Fig 1. Histogram showing distribution of relative ^{13}C-contents of collagen from bone samples collected in Europe (4). The average is $\delta = -20.2$ per mil indicating that the ^{13}C is enriched by 6 per mil as compared with average plant foliage ($\delta = -26$ per mil).

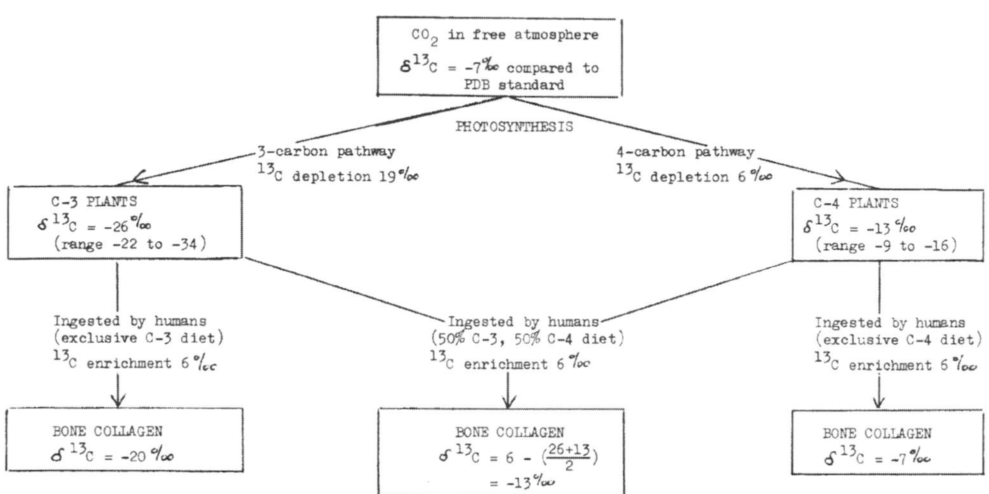

Fig. 2. Flow diagram of carbon fixation in the air-plant-human chain, showing the isotopic results of three hypothetical diets. ^{13}C abundance in bone changes differentially as a result of two metabolic pathways in plants and dietary selection of humans.

should be discernible by the same method, since marine organisms have a ^{13}C abundance similar to that of C-4 plants. In a hypothetical situation where a community moved between C-4, C-3, and coastal environments on a seasonal basis the problem could become rather complex, but such an example would be an extreme.

Table 1. Relative Carbon-13 Content of Collagen in Human Skeletons from Four Archaeological Sites in New York State.

Anal. No. MC-	*Site and Sample*	*Archaeological Period*	*Date*	*^{13}C-content $\delta^{13}C$(‰)*	*C-4 proportion*(%)*
	Pre-maize				
822	Frontenac 41270	Archaic	2500-2000 B.C.	−21.3	0
823	Vine Valley 38447	Early Woodland	400-100 B.C.	−18.9	0
830	Vine Valley 38447			−18.9	
831	Vine Valley 38465	Early Woodland	400-100 B.C.	−19.8	0
	Maize horticulture				
824	Snell 4	Late Woodland	A.D. 1000-1300	−14.0	43
828	Snell 5	Late Woodland	A.D. 1000-1300	−16.6	24
821	Engelbert B54A	Susquehannock	A.D. 1450±100 (Y-2617)	−13.6	46
827	Engelbert B81A	Susquehannock	A.D. 1450±100 (Y-2617)	−13.5	47

*Percentage contribution of carbon from C-4 plants (presumably maize) in the bone collagen of the individuals. The uncertainty is ±15%.

Other applications of the difference in isotopic composition of C-3 and C-4 plants include the determination of browsing/grazing ratios for certain antelopes in areas with an abundance of C-4 grasses and C-3 shrubs and trees (e.g., African savannah regions); changes in vegetation cover due to climatic change in subtropical countries, measured in the skeletal remains of certain animal species; and possibly determination of origin in forensic investigations.

Acknowledgments. We are indebted to Charles Gillette, New York State Museum, and the Anthropology Department, State University of New York at Binghamton, for making bone samples available to us. Thanks are due to Miss E. Lursen for performing the laboratory analyses.

Bender, M. M.
1968 Mass spectrometric studies of carbon-13 variations in corn and other grasses. *Radiocarbon* 10:468-72.

Craig, H.
1953 The geochemistry of the stable carbon isotopes. *Geochimica et Cosmochimica, Acta* 3:53-92.

Hatch, M. D., and C. R. Slack
1966 Photosynthesis by sugarcane leaves. A new carboxylation reaction and the pathway of sugar formation. *Biochemical Journal* 101: 103-11.

Hatch, M. D., C. R. Slack, and H. S. Johnson
1967 Further studies on a new pathway of photosynthetic carbon dioxide fixation in sugar-cane and its occurrence in other plant species. *Biochemistry Journal* 102:417-22.

Kortschak, H. P., C. E. Hartt, and G. O. Burr
1965 Carbon-dioxide fixation in sugarcane leaves. *Plant Physiology* 40:209-13.

Lipe, W. D., and D. N. Elliott
1970 *The Engelbert Site Project*. Binghamton, N.Y.

Münnich, K. O., and J. C. Vogel
1958 Durch atomexplosionen erzengter radiokohlenstoff in der atmosphäre. *Naturwissenschaften* 45:327-29.

Neales, T. F., A. A. Patterson, and V. J. Hartley
1968 Physiological adaptation to drought in the carbon assimilation and water loss of xerophytes. *Nature* 219:469-72.

Ritchie, W.
1965 *The archaeology of New York State*. New York: Natural History Press.

Troughton, J. H.
1971 Aspects of the evolution of the photosynthetic carboxylation reaction in plants. In *Photosynthesis and photorespiration*, edited by M. D. Hatch, C. B. Osmond, and R. O. Slater. New York: Wiley-Interscience.

van der Merwe, N. J.
1973 New wrinkles in radiocarbon. Paper delivered at Plains Anthropological Conference, Columbia, Mo.

Vogel, J. C.
1959 Isotopentrennfaktoren des Kohlen stoffs im Gleichgewichtssystem Kohlendioxyd–Bikarbonat–karbonat. Ph.D. thesis, Heidelberg.
n.d. Fractionation of the carbon isotopes during photosynthesis. *Plant Physiology*, in press.

Vogel, J. C., and H. T. Waterbolk
1967 Groningen radiocarbon dates VII. *Radiocarbon* 9:107-55.
1972 Groningen radiocarbon dates X. *Radiocarbon* 14:6-110.

Winter, J.
1971 A summary of Owasco and Iroquois maize remains. *Pennsylvania Archaeologist* 41.

RIDGE-BACK TOOLS OF THE COLORADO DESERT

W. MORLIN CHILDERS

Knapping experiments and examination of flaked stone objects distributed throughout portions of southwestern Imperial Valley and northern Baja California indicate that criteria other than presence or absence of percussion bulbs can be used to support the artifactual nature of certain fractured stones.

ARCHAEOLOGICAL CONTEXT

The ridge-back industry was described briefly by Childers (1974:2) as a uniface tradition in which trimming blows produced low-angle edges, with the scars characteristically meeting at a point or ridge along the dorsal side. The pieces are easily distinguishable from the uniface flat-top cores and scraper planes noted for this region (Rogers 1966; Carter 1957). The ridge-backs are characterized by the lower edge angle, longer flakes, and immense size of the largest pieces, and the presence of many pointed forms (Figs. 1a, b, and 4). The raw material used for ridge-backs crossed all mineral lines, indicating that these attributes resulted from a chipping technique and are not a natural

PAPER 25

Depth of Deposition of the Big Blue (Late Paleozoic) Sediments in Kansas (1937)

M. K. Elias

Commentary

THOMAS D. OLSZEWSKI

Almost as soon as fossils were recognized as the remains of once-living organisms, they were used as the basis for inferring ancient environmental conditions (Ladd and Gunter 1957); nineteenth-century scientists studying the distribution of living marine invertebrates explicitly understood the relevance of their work for the interpretation of sedimentary rocks (e.g., Forbes 1844). Maxim Elias's great contribution to paleoecology was to take the concept of living benthic associations developed by Forbes, Petersen (1924), and others and use it to infer ancient environmental gradients in the fossil record.

Although contemporaries like Raymond Moore and Norman Newell knew and worked with Elias (Merriam 2000), it took almost 30 years for his pioneering approach to become more widely adopted, and even then, like many great ideas, it may have arisen as an intellectual homoplasy. Alfred Ziegler, in his classic 1965 study using fossil communities in Wales to map benthic zones on a Silurian seafloor, did not cite Elias but essentially used the same approach and even cited inspiration from the same reports of modern benthic zonation. Peter Bretsky (1969), however, enthused that Elias's 1937 paper "continues to stand out as one the most substantial contributions to the study of Paleozoic fossil association."

Although the actual use of fossil associations to interpret sedimentary rocks was a genuinely original contribution, some of Elias's interpretations reflect the limited understanding of depositional environments, particularly carbonates, of his time. In particular, he interpreted fusulinid packstones as the deepest-water environment of the Big Blue Series, and when John Imbrie, Léo Laporte, and Daniel Merriam (1964) suggested they were much shallower (essentially a modern interpretation), Elias (1964) staunchly defended his original zonation. Despite their disagreement on the details, Imbrie et al. (1964) acknowledged Elias, who "through his classic paper on 'Big Blue' cyclothems (1937), provided much background and inspiration for the present study." Both Laporte and Imbrie incorporated the concept of biotic associations into some of their most significant subsequent work (e.g., Walker and Laporte 1970; Imbrie and Kipp 1971).

Reconstructing ancient depositional environments using fossil associations in the way that Elias did continues to be a basic tool in both academic and industrial contexts. At the same time, fossil associations are critical to current research synthesizing ecological theory, sequence stratigraphy, and quantitative methods to understand the dynamics and evolution of ecological communities. Elias's analysis of the Big Blue Series using fossil associations truly is foundational for paleoecology not just for the durability of the idea at its core but also for the research directions that the idea has subsequently nourished.

From *Geological Society of America Bulletin* 48:403–32.

Literature Cited

Bretsky, P. W., Jr. 1969. Evolution of Paleozoic benthic marine invertebrate communities. *Palaeogeography, Palaeoclimatology, Palaeoecology* 6:45–59.

Elias, M. K. 1964. Depth of late Paleozoic sea in Kansas and its megacyclic sedimentation. In *Symposium on Cyclic Sedimentation*, edited by D. F. Merriam, 87–106. Kansas Geological Survey Bulletin, vol. 169. [Lawrence]: [State Geological Survey of Kansas, University of Kansas].

Forbes, E. 1844. Report on the Mollusca and Radiata of the Aegean Sea, and on their distribution, considered as bearing on geology. In *Report of the 13th Meeting of the British Association for the Advancement of Science*, 130–93. London: John Murray.

Imbrie, J., and N. G. Kipp. 1971. A new micropaleontological method for paleoclimatology: application to a Late Pleistocene Caribbean core. In *The Late Cenozoic Glacial Ages*, edited by R. F. Flint and K. K Turekian, 71–181. New Haven, CT: Yale University Press.

Imbrie, J., L. F. Laporte, and D. F. Merriam. 1964. Beattie limestone facies (Lower Permian) of the northern midcontinent. In *Symposium on Cyclic Sedimentation*, edited by D. F. Merriam, 219–38. Kansas Geological Survey Bulletin, vol. 169. [Lawrence]: [State Geological Survey of Kansas, University of Kansas].

Ladd, H. S., and G. Gunter. 1957. Development of marine paleoecology. In *Treatise on Marine Ecology and Paleoecology*. Vol. 2, *Paleoecology*, edited by H. S. Ladd, 67–74. Geological Society of American Memoir, vol. 67. [New York:] National Research Council (U.S.), Committee on a Treatise on Marine Ecology and Paleoecology.

Merriam, D. F. 2000. Memorial to Maxim Konrad Elias, 1889–1982. *Geological Society of America Memorials* 31:73–75.

Petersen, C. G. J. 1924. A brief survey of the animal communities in Danish waters, based upon quantitative samples taken with the bottom sampler. *American Journal of Science* 7:343–54.

Walker, K. R., and L. F. Laporte. 1970. Congruent fossil communities from Ordovician and Devonian carbonates of New York. *Journal of Paleontology* 44:928–44.

Ziegler, A. M. 1965. Silurian marine communities and their environmental significance. *Science* 207:270–72.

PL. 1

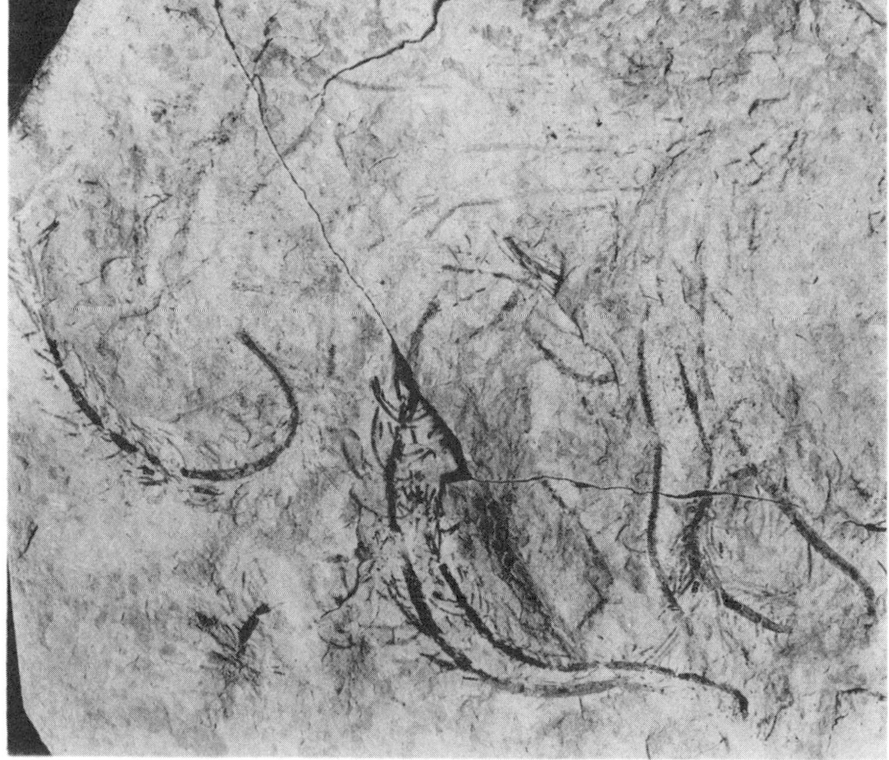

LEATHERY SEA-WEEDS

Remains probably related to living Rhodochroaceae or red algae. From the Luta limestone, Big Blue series, Dickinson County, Kansas.

BULLETIN OF THE GEOLOGICAL SOCIETY OF AMERICA
VOL. 48, PP. 403-432, 1 PL., 4 FIGS. MARCH 1, 1937

DEPTH OF DEPOSITION OF THE BIG BLUE (LATE PALEOZOIC) SEDIMENTS IN KANSAS

BY MAXIM K. ELIAS

CONTENTS

ILLUSTRATIONS

HISTORY AND ACKNOWLEDGMENTS

This paper is based on observations, made chiefly during field research, and detailed mapping of the Big Blue series in the north-central part of Kansas in Marshall, Washington, Clay, and Dickinson counties in 1932 for the State Geological Survey of Kansas. Various points of cyclic analysis and problems of environments of Late Paleozoic sedimentation in the northern Mid-Continent have been discussed with Dr. R. C. Moore and Dr. N. D. Newell, of the University of Kansas. The manuscript has been read by Dr. Moore and Dr. W. H. Twenhofel, whose criticism and suggestions are gratefully acknowledged.

The application of the theory of cyclic sedimentation to the Big Blue series was reported to the Paleontological Society in 1933,[1] and its significance in stratigraphic classification of Late Paleozoic rocks in Kansas was discussed jointly with R. C. Moore at that time.[2] An attempt to estimate the depth of deposition of the Big Blue sediments in the light of cyclic analysis, and by comparison with the distribution of benthonic life in modern seas, was reported to the Society of Economic Paleontologists and Mineralogists at Wichita in 1935 and to the Paleontological Society at New York in 1935.[3]

Some evidence of cyclic repetition in the Big Blue series in northern Kansas, in which only lithologic changes were considered, has been independently discovered and discussed by Professor J. M. Jewett, of Wichita University.[4]

STRATIGRAPHIC POSITION AND CHARACTER OF THE BIG BLUE SERIES

The Big Blue series is the uppermost of the predominantly marine Late Paleozoic rocks of Kansas, above which are the largely continental red beds of the Permian Cimarron series (Fig. 1-E). The Big Blue corresponds to the Wolfcamp beds of Texas, to the Dunkard group of the Appalachian region, to the Autunian of France, the Lower Rotliegende of Germany, and the Upper Uralian of Russia.[5]

These Late Paleozoic rocks present an alternation of innumerable limestones, shales, sands, and coals. The separate formations and beds are famous for their lateral persistency. Many of the least conspicuous

---

[1] M. K. Elias: *Cycles of sedimentation in the Big Blue series of Kansas* (abstract), Geol. Soc. Am., Pr. 1933 (1934) p. 366.

[2] R. C. Moore and M. K. Elias: *Sedimentation cycles as a guide to stratigraphic classification of the Kansas Lower Permian* (abstract), Geol. Soc. Am., Pr. 1933 (1934) p. 100.

[3] M. K. Elias: *Depth of deposition of the Big Blue Sediments* (abstract), Geol. Soc. Am., Pr. 1935 (1936) p. 375.

[4] J. M. Jewett: *Evidence of cyclic sedimentation in Kansas during the Permian period,* Kans. Acad. Sci., Tr., vol. 36 (1933) p. 137-140.

[5] The Big Blue series is classed as Lower Permian, but it is recognized that this age assignment is dependent on definition of the basal Permian boundary in Russia. If the base of the type Russian Permian is placed at the bottom of the Artinsk beds, then most of the Big Blue series should be classed as Upper Carboniferous.

limestones have been traced across Kansas and adjacent parts of Nebraska and Oklahoma, and some zones, only a few inches thick, marked by distinctive lithology and fossil content, occupy the same relative position within the limestone throughout their outcrop. An equal persistency of succession, modified to some extent by gradual changes in prominence of calcareous sediments, which are apparently due to deposition in slightly deeper portions of the epeiric sea, has been discovered in deep wells drilled in western Kansas, where the Paleozoic rocks are covered by Cretaceous and Tertiary beds. The alternation of marine sediments, mainly limestones and calcareous shales, with non-marine sediment, sands, red shales, and coals, is explained as a result of repeated invasions and retreats of oceanic waters.

The regular, predominantly limestone-and-shale, alternation is broken by moderately strong disconformities, which signify erosional intervals that were followed by deposition of prominent sands and sandy shales in local erosion channels. The youngest of the locally recognized Late Paleozoic major unconformities[6] marks the beginning of the Big Blue series, which differs from the preceding, Late Pennsylvanian, series by (1) still greater persistency and uniformity of its limestones and shales, (2) the nearly total disappearance of sands and conglomerates, (3) the disappearance of coals (except a few thin, locally developed beds), (4) the prominent development of red and green shales, and (5) the introduction of some gypsum and salt, especially in the upper part of the series.

Clearly, after the last slight rejuvenation of topography at the beginning of Big Blue time, which resulted in some deposition of sands at and near the base of the series, the subsequent invasions and retreats of the sea resulted in deposition of only silts and clays intercalated between the marine limestones. Undoubtedly, the sea waters made still more level and featureless the platform, which, even before Big Blue time, was an almost perfectly flat plain.

The leveling process was accompanied by simplification of the marine life. The diversified fauna of the preceding Pennsylvanian epochs disappeared rapidly. For example, of 40 to 45 species of brachiopods known from the lower part (Foraker limestone and below) of the Big Blue beds, not more than 10 survive above the Fort Riley limestone of the middle of the series. A considerable decrease in number of species is also noticeable in other groups of marine invertebrates. At the same time, the few surviving species become extremely numerous, at places excluding other organic remains. In the last invasions of the Big Blue sea, some marine invertebrates disappear completely in the following order:

[6] R. C. Moore and R. G. Moss: *Permian-Pennsylvanian boundary in the northern Mid-Continent area* (abstract), Geol. Soc. Am., Pr. 1933 (1934) p. 100.

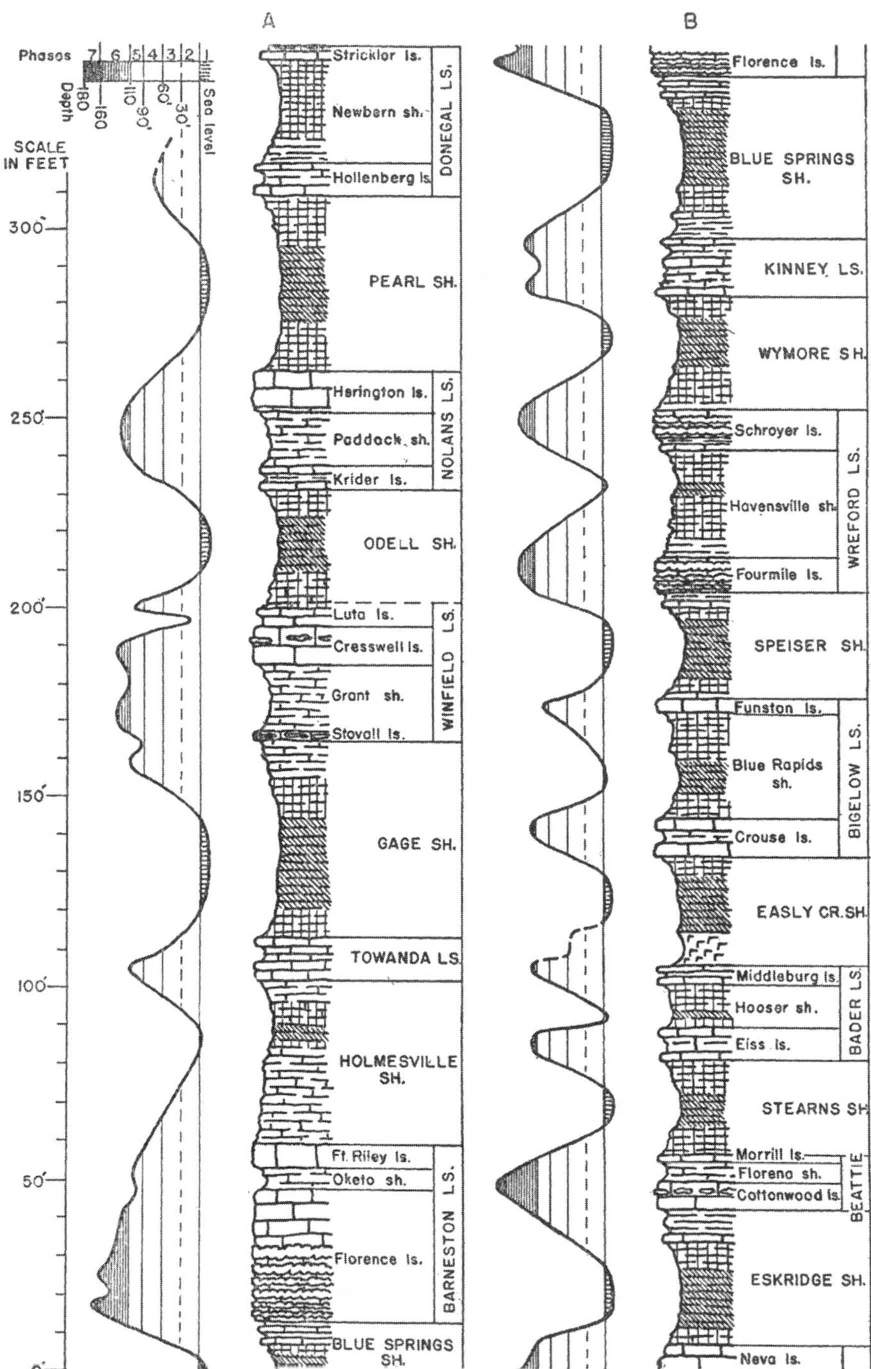

FIGURE 1.—*Cycles in the Big Blue series*

A, B, and C.—Composite geologic section of the part of the Big Blue series in northern Kansas, interpreted in terms of cyclic sedimentation and depth of deposition.

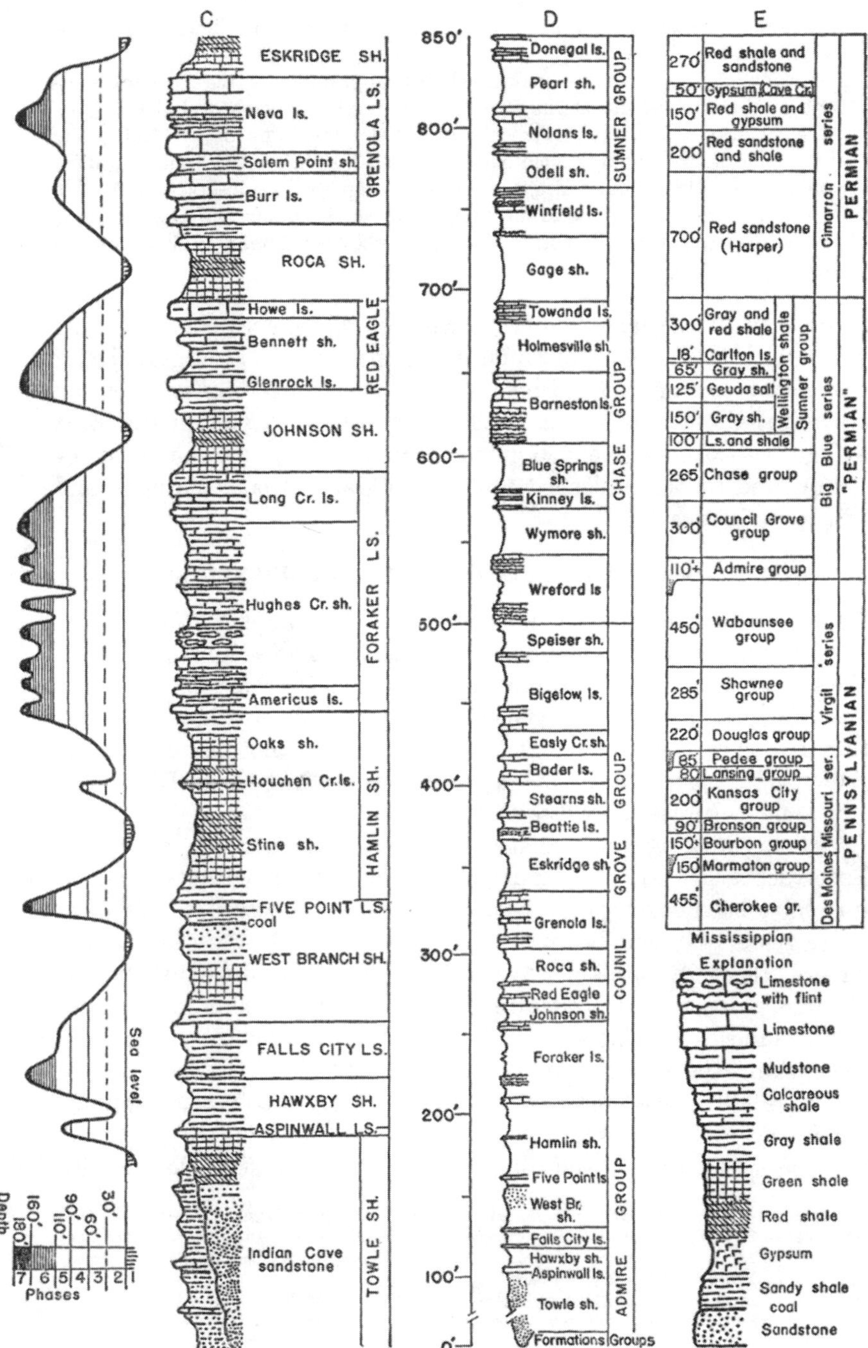

FIGURE 1.—*Cycles in the Big Blue series—continued*

D.—Key section for A, B, and C, showing formations and groups of the Big Blue series.
E.—Key section, showing position of the Big Blue series among the Late Paleozoic rocks of Kansas.

1. Fusulinids are observed for the last time in the Florence limestone of the Barneston cyclothem.
2. Brachiopods and bryozoans make their last appearance in the Paddock shale and at the base of the Herington limestone of the Nolans cyclothem.
3. The larger mollusks appear practically for the last time in the Hollenberg limestone of the Donegal cyclothem, in which they are rare. Small *Myalina* and a few other small pelecypods and gastropods are observed in a thin unnamed limestone in the lower Wellington shale and in the Carlton limestone. A few pelecypods and gastropods are known from the Cimarron red beds.

In the light of the next chapter discussion, it becomes clear that the order in which these groups of animals disappear in the stratigraphically highest cyclothems of the Big Blue indicates general decrease in extension or gradual shallowing of the last marine invasions. Later in Big Blue time, lagoons became numerous, in which rock salt and gypsum accumulated, followed by the prolonged continental stage of later Permian and Early Mesozoic time.

CYCLIC SEDIMENTATION OF THE BIG BLUE SERIES

The cyclic nature of the succession of sediments and faunas was reported to the Paleontological Society in 1933.[7] The sedimentary cycle involves both marine and continental deposits. They are intermediate between the great cycles of the geologic periods [8] and the minute seasonal rhythm of varved clays. They are comparable in magnitude to the sedimentary cycles observed in Kansas and elsewhere in beds of Dinantian, Pennsylvanian, Jurassic, and Eocene age.

The guiding lithologic units that determine the limits of the succeeding cycles of deposition are the persistent unfossiliferous red shales, which seem to have been developed in times of prolonged emergence. Between these are green, gray, and brown marine shales and light-gray limestones, which are, generally speaking, very fossiliferous. As indicated by the fossils, these beds were deposited in shallow waters. The orderly occurrence of mollusks, brachiopods, and fusulinids was found most helpful in establishing the cyclic character of sedimentation. In successively higher levels of the fossiliferous strata, the most common marine animals appear in the following order: (1) corneous brachiopods,[9] (2) mollusks (chiefly pelecypods, less plentifully gastropods, and still more rarely

---

[7] M. K. Elias: *Cycles of sedimentation in the Big Blue series of Kansas* (abstract), Geol. Soc. Am., Pr. 1933 (1934) p. 366. The cyclic succession of some lithologic units in the Big Blue series has been noticed also by J. M. Jewett: *Evidence of cyclic sedimentation in Kansas during the Permian period*, Kans. Acad. Sci., Tr., vol. 36 (1933) p. 137-140.

[8] J. W. Dawson: *Acadian Geology* (Geologic cycles) 2d ed. (1868) p. 135-138.

J. S. Newberry: *Circles of deposition in American sedimentary rocks*, Am. Assoc. Adv. Sci., Pr., vol. 22 (1874) p. 185-196.

[9] This phase of the progressive hemicycle has been actually observed only in some Pennsylvanian cycles below the Big Blue series, but it was observed in some regressive hemicycles of the latter series.

nautiloids), (3) calcareous brachiopods, and (4) fusulinids. Above the zone of fusulinids, these marine forms appear in exactly opposite order.

Although the general order of appearance of these faunal groups is always the same, they only rarely make clean-cut faunal zones, except the fusulinids, which generally are alone. Ordinarily, even as in modern seas, the faunal groups overlap each other to greater or lesser extent. It is noticed that the stoutest or most greatly inflated among the observed pelecypods[10] (*Allorisma* and *Pinna*, Fig. 2) seem to prefer the society of calcareous brachiopods and of fusulinids, whereas the flat-shelled pelecypods alone (*Aviculopecten, Pleurophorus*) were found in a few beds with the corneous brachiopods (*Lingula, Orbiculoidea*). Among the calcareous brachiopods, *Juresania* is most constantly associated with molluscan faunas.

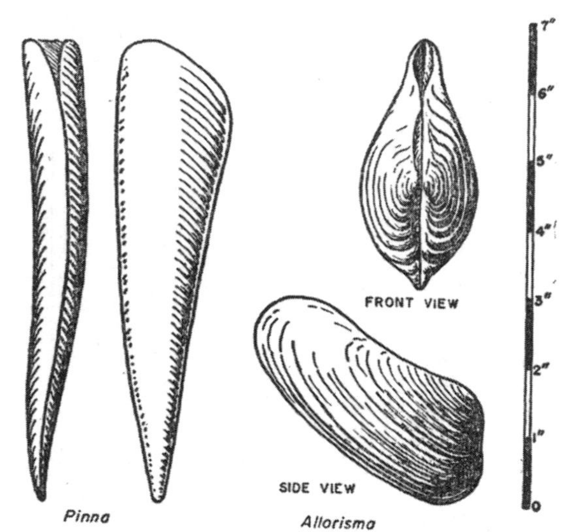

FIGURE 2.—*Burrowing clams*

Most common burrowing marine clams from the Late Paleozoic rocks of Kansas, shown in their natural position, which they acquired when they buried themselves in originally unconsolidated marine sediments of the Late Paleozoic.

In addition to these fossils, which are most useful in establishing the cyclic succession of the faunal zones, less common ones have also a fairly definite position within the cycles.

The few corals, which seem to belong to *Lophophyllum*, are found exclusively in the brachiopod and mixed-fauna phases. The much more common Bryozoa, mostly represented by fenestrate forms, *Rhombopora, Meekopora,* and others, are found in brachiopod(6) and mixed fauna(5) phases and also extend into the adjoining part of the molluscan(4) phase. The gastropods are found in the molluscan(4) and mixed fauna(5) phases, where also belong the few collected nautiloids and ammonoids. Ostracodes are numerous in the molluscan phases(4) of some cyclothems, but they may be present also in the deeper water phases. Rare fish remains (chiefly scales) are observed in the molluscan(4) and mixed(5) phases.

[10] These were undoubtedly burrowing forms.

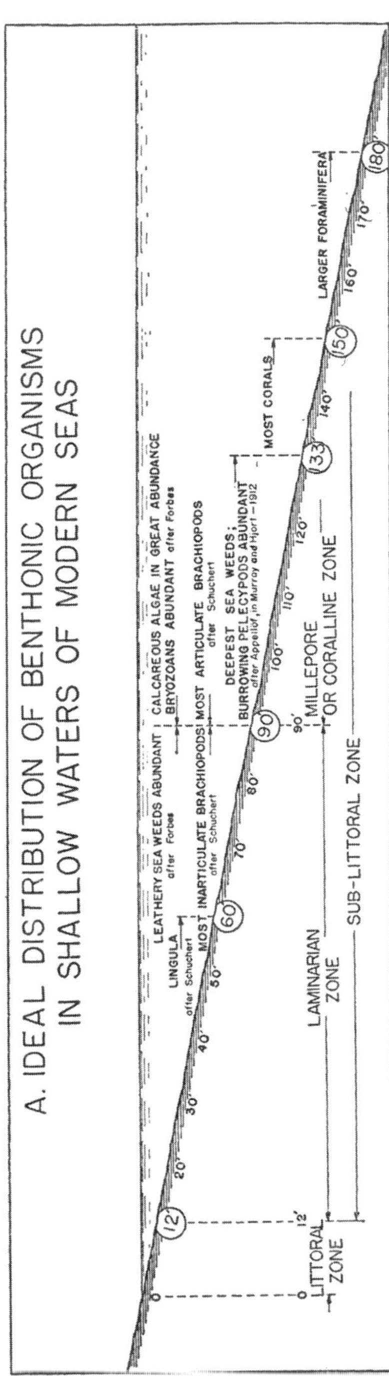

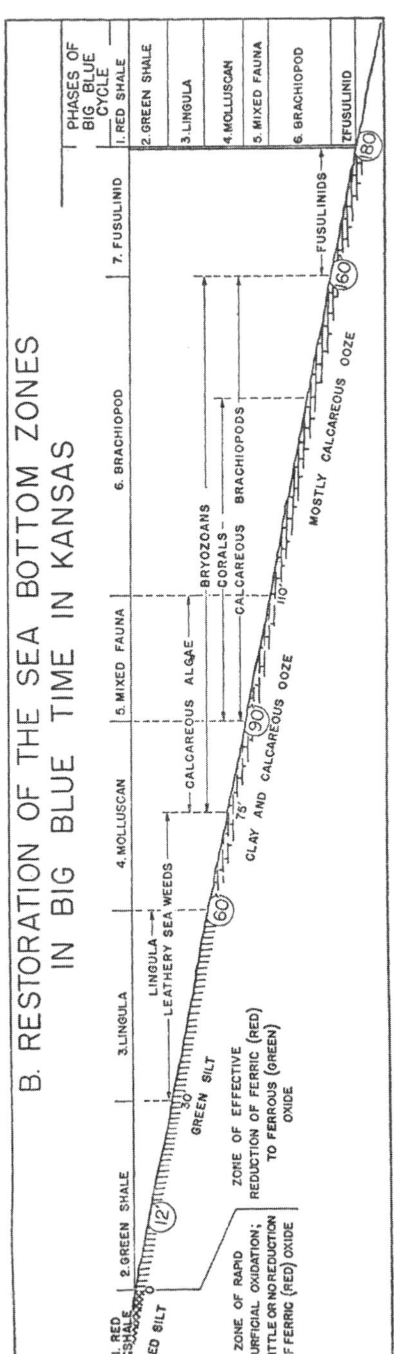

FIGURE 3.—*Sea-bottom zones in modern and in Big Blue seas*

A.—Ideal distribution of benthonic organisms in shallow waters of modern seas.
B.—Restoration of the sea bottom zones in Big Blue time in Kansas.

The few remains of terrestrial plants have no particular environmental significance, and they are naturally found in greatest quantities in the deposits that were formed nearest to the shore—that is, in the molluscan (4) phase, *Lingula* (3) phase, and the green shale beds (2). Much

TABLE 1.—*Idealized Big Blue cycle of deposition in north-central Kansas*

	No.	Phases established chiefly on paleontologic evidence	Corresponding typical lithology
Regressive hemicycle	1.	Red shale................	Clayey to fine sandy shale, rarely consolidated.
	2r.	Green shale................	
	3r.	Lingula phase............	Sandy, often varved (?), rarely clayey shale.
	4r.	Molluscan phase..........	Clayey shale, mudstone to bedded limestone.
	5r.	Mixed phase..............	Massive mudstone, shaly limestone.
	6r.	Brachiopod phase.........	
	7.	Fusulinid phase...........	Limestone, flint, calcareous shale.
Progressive hemicycle	6p.	Brachiopod phase.........	
	5p.	Mixed phase..............	Massive mudstone, shaly limestone.
	4p.	Molluscan phase..........	Clayey shale, mudstone to bedded limestone.
	3p.	Lingula phase	Sandy, often varved (?), rarely clayey shale.
	2p	Green shale................	Clayey to fine sandy shale, rarely consolidated.
	1.	Red shale................	

more important ecologically are marine algae. These include (1) the impressions and pseudomorphic replacements of thalloid-like branchlets, which apparently belong to Rhodochroaceae or red algae (Pl. 1), and (2) the massive calcareous encrustations and branching colonies, which were described from the local Pennsylvanian under the names *Cryptozoon, Osagia, Ottonosia,* and *Somphospongia* (which is not a sponge). The thalloid-like impressions are found exclusively in green shales (2) and in impure limestones and dolomites of the molluscan phase (4), but the calcareous encrusting algae may be collected from the molluscan (4) and mixed fauna (5) phases. In the Pennsylvanian cycles, they are also in the brachiopod (6) and rarely in fusulinid (7) phases.

No single cycle of the Big Blue rocks shows all phases of the ideal cycle, but the missed phases in one cycle appear in proper position in neighboring cycles above and below. The nearest approach to the ideal is reached in the Beattie and Barneston cycles. Excepting phase 3p, these cycles are practically complete.

The lithologic expression of the faunal phases is somewhat variable.

Some phases of different cycles, however, are expressed by characteristic and persistent lithologic units. For instance, the progressive mixed fauna phases (5p) of the Wreford and Florence cycles form persistent ledges of mudstone, separated by calcareous brachiopod-bearing shale(6p) from the main solid limestones. Although a general increase in content of calcium carbonate is found in the strata in passing from the red shale(1) to the fusulinid(7) phase, the beds of thickest and hardest limestone do not everywhere correspond exactly to the fusulinid(7) and brachiopod(6) phases of a cycle. In some cycles, especially toward the top of the Big Blue series, the mixed (5) and the pelecypod(4) phases are found in the thickest limestones of the cycle, whereas the brachiopod(6) phase is expressed partly or entirely in the form of calcareous shale. Although flint nodules and bands appear generally within phases 6 and 7, some thin and local bands of flint are observed rarely in phases 4 and 5.

ZONAL DISTRIBUTION OF BENTHONIC LIFE IN SHALLOW DEPTHS OF MODERN SEAS

GENERAL STATEMENT

The distribution of organisms on the bottom of modern seas is greatly affected not only by the depth but also by various other factors, such as temperature, degree of clearness and salinity of the water, presence of currents, and character of the bottom. Some generalizations in regard to the relative position of benthonic zones, largely controlled by their depths, have been reached. These are summarized and the main conclusions are presented in graphic form (Fig. 3-A).

OBSERVATIONS BY FORBES

The extensive study in the Aegean sea, and around the coasts of Great Britain, made by Forbes [11] is still one of the main sources of information on the distribution of benthonic organisms, and his principal shallow-bottom zones are generally accepted.

In 1844, Forbes [12] defined the littoral zone (first region) as the zone of tides, extending down to a depth of 2 fathoms, in which *Pinna squamosa* is abundant. The deeper portions of the sea were divided into regions as follows: Second region, depths ranging from 2 to 10 fathoms, burrowing conchifera abundant; third region, depths ranging from 10 to 20 fathoms, corallines more plentiful.

---

[11] Edward Forbes: *Report on the Mollusks and Radiata of the Aegean Sea, and on their distribution as bearing on geology,* Brit. Assoc. Adv. Sci., 13th Rept. (1844) p. 130-193; *On the connection between the distribution of the existing fauna and flora of the British Isles, and the geological changes which have affected their area . . . ,* Geol. Surv. Great Britain, Mem., vol. 1 (1846) p. 336-432; *Report on the investigations of British marine zoology by means of the dredge,* pt. 1, Brit. Assoc. Adv. Sci., 20th Rept. (1851) p. 192-263.

[12] Edward Forbes: *Report on the Mollusks and Radiata of the Aegean Sea, and on their distribution as bearing on geology,* Brit. Assoc. Adv. Sci., 13th Rept. (1844) p. 154-161.

Subsequently, Forbes[13] defined the zones included between depths of 15 and 20 fathoms as the region of *Nullipora* and remarks that below 20 fathoms "no decided algae were met with." This statement, however, refers primarily to leathery sea-weeds (= decided algae), because, at that time most calcareous algae were believed to be related to corals. He introduced the terms "Laminarian region" for low-water mark, down to 15 fathoms, and "Coralline region" for the interval between 15 and 40 fathoms. The latter region is subdivided into an upper part, from 15 to 25 fathoms, and middle and lower parts, from 25 to 40 fathoms.[14] Forbes speaks in this paper also of the shallow waters below the zone of tides as the "sublittoral" part of the sea bottom.

Summarizing, the following scheme, modified after Forbes,[15] is derived. The slope of shallow seas is divided into three principal zones: (1) Littoral zone, between highest and lowest tide marks; (2) laminarian zone, from low-tide mark to a depth of 15 fathoms, characterized by leathery sea-weeds, which afford shelter to many mollusks; (3) coralline zone, from 15 to about 40 fathoms, characterized by abundance of calcareous algae, leathery sea-weeds being only sparingly present in the shallow end of this zone. According to Appellöf[16] and Foslie,[17] however, lime-secreting algae are abundant not only in the coralline zone, but also in the laminarian zone and in the zone of tides. It is the *abundance* of leathery sea-weeds in the laminarian zone, therefore, and their disappearance at greater depths, that defines the limit between the laminarian and the coralline zones. The most typical lime-secreting plants of the coralline zone are the red algae, encrusting *Lithothamnium,* and branching *Corallina.* Bryozoa are abundant in this zone.

LITTORAL AND SUB-LITTORAL ZONES

Some modern writers use the terms "littoral" and "sublittoral" in a different sense than that originally defined by Forbes. For instance, Grabau[18] designated as littoral "all that part of the sea above the deep-sea portion, *i.e.*, approximately above the hundred-fathom line"; the original littoral zone of Forbes, he calls "littoral in restricted sense" or "shore zone." Peterson[19] uses the term littoral in the sense of Grabau. On the

[13] Edward Forbes: *Report on the investigation of British marine zoology by means of the dredge,* pt. 1, Brit. Assoc. Adv. Sci., 20th Rept. (1851) p. 246, 247.

[14] *Op. cit.,* p. 252.

[15] Edward Forbes: *On the connection between the distribution of the existing fauna and flora of the British Isles and the geological changes which have affected their area . . . ,* Geol. Surv. Great Britain, Mem., vol. 1 (1846) p. 375.

[16] A. Appellöf: *Invertebrate bottom fauna* in J. Murray and Johan Hjort: *Depths of the ocean* (1912) p. 459-526.

[17] M. Foslie: *Norwegian forms of Lithothamnion* (1894) Trondhjem, Norway; *Lithothamnia of the Maldives and Laccadives,* in Stanley Gardiner: *Fauna and geography of the Maldive and Laccadive Archipelagoes,* vol. 1 (1903) p. 460-471. Cambridge Univ. Press.

[18] A. W. Grabau: *Principles of stratigraphy* (1913) p. 470.

[19] C. G. J. Peterson: *On the animal communities of the sea-bottom in the Skagerak, the Christiania Fjord and the Danish waters,* Danish Biol. Stat., Copenhagen, Rept. XXIII (1915).

other hand, Appellöf [20] applies "littoral" to the portion of the sea bottom from shore line to a depth of 30 to 40 meters, and Neaversen [21] uses this term for the zone between the shore and a depth of 15 to 20 fathoms. These authors apply the term "sub-littoral" to the region below littoral down to 150 meters (Appellöf) or to 80 fathoms (Neaversen).

In view of the confusion arising from these varying definitions, Allee [22] suggests that "littoral" be used only in the original sense of Forbes. This conforms with a general tendency of modern science to follow rules of priority in use of terms. As the lower limit of the "sub-littoral" zone was not defined by Forbes, it is permissible to designate this according to Appellöf—that is, a depth of 30 to 40 meters, which is "almost as far down as there are sea-weeds." [23] The upper limit of the "sub-littoral" zone, however, remains as indicated by Forbes—that is, at low tide (12 feet).

OBSERVATIONS BY APPELLÖF

Important additional data on the shallow zones as observed in the Norwegian sea and adjacent part of the Atlantic are given by Appellöf. A zone extending downward to a depth of 30 or 40 meters is characterized by "periodic changes in temperature and salinity, strong light, and a great variety in the materials at the bottom, such as loose stones, solid rock, sand with or without coarse or fine fragments of different kinds of shells, mud and 'mixed mud'—that is to say, sand, mud, and stones are mixed together. Here we find the whole vegetation collected, consisting of fucoids, green and red algae, *Laminaria,* and *Zostera* [eelgrass], all of which, as a rule, form big independent communities that are very often arranged in belts." [24] The next deeper zone, extending down to 150 meters is marked by absence of vegetation, greater uniformity in the nature of bottom deposits, higher and more constant salinity and less pronounced differences in temperature.[25] Below this is the "continental deep-sea zone, ranging from 150 to 1,000 meters or more."

The following typical organisms were observed by Appellöf in the zone of tides, only those belonging to classes that are found preserved as fossils being here mentioned. (1) On rocky bottom: Barnacles (*Balanus balanoides*), mussels (*Mytilus edulus*), gastropods (cap-shaped limpet, *Patella vulgata;* periwinkles, *Littorina littorea, L. rudis;* purple snail, *Purpula lapillis*). Except the limpets, these forms live also on the algae. Algae, however, and *Fucus* in particular usually have their special fauna

---

[20] A. Appellöf: *op. cit.,* p. 459.

[21] Ernest Neaversen: *Stratigraphical paleontology* (1928) p. 87. Macmillan and Co., London.

[22] W. C. Allee: *Studies in marine ecology,* Biol. Bull., vol. 44 (1923) p. 169.

[23] A. Appellöf: *op. cit.,* p. 459. The sea-weeds, however, extend downward to about 200 feet.

[24] *Ibid.*

[25] *Op. cit.,* p. 459-460.

consisting chiefly of attached forms. Bryozoa (*Alcyonidium hirsutum, Flustrella hispida, Bowerbankia imbricata*) are common and fucoids are often densely thronged by small white spiral-shaped tube-worms (*Spirorbis*). (2) On sandy bottoms: Burrowing forms predominate. Pelecypods (sandgaper, *Mya arenaria;* cockle, *Cardium edule*) are common.

Appellöf does not subdivide the belt between low-tide mark and 40 meters depth. He mentions the sea-weeds, *Laminaria hyperborea, L. digitata* and *L. saccharina,* growth of which begins immediately below the fucoids of the zone of tides. Although *Laminaria* grows in exposed situations where there are waves and strong currents, there is a hard bottom that serves as a place of attachment. The eelgrass (*Zostera marina*) grows at the same depths on soft muddy bottoms of sheltered pools and estuaries. On blades of Laminaria are found the encrusting bryozoan *Membranipora membranacea,* calcareous sponges, and a few siliceous sponges. The importance of the 40-meter mark is emphasized, because, among the forms characteristic of the "littoral" zone, "there are very few that do not occur in all its depth"; these few, such as *Mya arenaria* and *Cardium edule,* are restricted to the "tidal area."[26]

Appellöf notes that "the limits between a littoral and non-littoral zone seem to be less clearly defined in the Arctic than in the Boreal region. The reason for this is obvious enough if we remember that temperature largely controls distribution."[27]

INFLUENCE OF COMPOSITION OF SEA-BOTTOM

The nature of the sea-bottom is important in the distribution of benthonic animals, and, in some cases, this overbalances depth. For instance, Herdman[28] states that one can find within a single zoological province a similar fauna on muddy bottoms in depths of from 10 to 50 fathoms. Sumner,[29] likewise, states that the character of the bottom, considered chiefly in relation to texture, is "foremost among the conditions determining the distribution of the bottom dwelling organisms." It is clear, however, that, wherever there is little difference in the physical character of the sediments, depth is the main control in distribution of the bottom-dwelling organisms that arrange themselves in belts more or less parallel to the shore line.

OBSERVATIONS BY PETERSON

Peterson's detailed studies in the Danish waters are particularly instructive. He found that certain species of pelecypods, echinoids, and

[26] *Op. cit.,* p. 479.

[27] *Op. cit.,* p. 526.

[28] W. A. Herdman: *Marine zoology of the Irish sea,* Brit. Assoc. Adv. Sci., Rept. (1894) p. 318-334.

[29] F. B. Sumner: *An intensive study of the fauna and flora of restricted area of sea bottom,* U. S. Bur. Fish., Bull., vol. 28 (1908) p. 1229.

starfishes are narrowly restricted to particular depths. His classification follows:

(1) Depth about 5 meters, *Macoma balthica* (pelecypod)
(2) Depth about 10 meters, *Venus gallica, V. ovata* (pelecypods)
(3) Depth about 15 meters, *Echinocardium, Venus* (pelecypods)
(4) Depth about 20 to 30 meters, *Echinocardium* (pelecypod), *Amphiura filiformis* (starfish)
(5) Depth about 40 to 80 meters, *Brissopsis lyrifera* (echinoid), *Amphiura chiajei* (starfish)
(6) Depth about 100 to 150 meters, *Brissopsis lyrifera* (echinoid), *Ophyoglypha sarsii* (starfish)
(7) Depth down to 670 meters, *Amphilepis norvegica* (starfish), *Pecten vitrens* (pelecypod).

The first four zones make up the Laminarian zone of Forbes. Peterson's fifth zone approximately corresponds to the Coralline zone of Forbes. Zones 5 and 6 comprise the sub-littoral zone of Appellöf.

OBSERVATIONS BY FISHER

Fisher [30] lists gastropods that are restricted in bathymetric range.

(1) Littoral zone: *Littorina, Hydrobia, Assiminea, Rissoa, Truncatella, Cerithium, Natica, Pyramidella, Nassa, Purpura, Murex, Conus.*

(2) Laminarian zone (from low tide to depth of 90 feet): *Phasianella, Xenophora, Triforis, Rissoa, Aclis, Daphnella, Lacuna, Terebellum, Pterocera, Marginella, Mitra, Nassa, Phos, Drillia, Pleurotoma.* According to Woodward,[31] the species of these genera are mostly phytophagous, feeding on algae, but here are also found carnivorous species of *Buccinum, Nassa,* and *Natica.*

(3) Zone from 90 to 300 feet: mainly carnivores of the genera *Bela, Buccinum, Cassis, Cassidaria, Chenopus, Eulima, Fossarus, Fusus, Nassa, Natica, Pleurotoma, Trichotropis, Tritonium, Trophon,* and *Velutina.* According to Woodward,[32] the vegetable-eaters are represented by *Fissurella, Emarginula, Pileopsis, Eulima,* and *Chemnitzia.*

The conclusions derived from the extensive observations of Forbes, Appellöf, Peterson, and others, indicate that "one of the most important controlling factors is depth, and we find, running roughly parallel to the coasts, a series of bathymetrical zones, each characterized by a community, or by a series of communities determined by other controlling factors such as mineral composition of the sea-bottom." [33]

An example of differentiation of communities determined by character

[30] Paul Fisher: *Manual de Conchyliologie* (1887). Paris.

[31] S. P. Woodward: *Manual of the Mollusca,* 4th ed. (1880) or reprint (1890) p .151. Lockwood and Son, London.

[32] *Op. cit.,* p. 152.

[33] A. M. Davies: *Sequence of Tertiary faunas,* in *Tertiary faunas,* vol. 2 (1934) p. 53-55 [Cycles of sedimentation]. Thomas Murby and Co., London.

of sea bottom within a single bathymetrical zone has been given by Forbes for his littoral zone.

"Throughout Europe, where it consists of *rock,* it is characterized zoologically by species of *Littorina,* botanically by *Corallina;* where *sandy,* by the presence of certain species of *Cardium, Tellina* and *Solen;* where *gravelly* by *Mytilus;* where *muddy* by *Latraria* and *Pullastra.*" [34]

BRACHIOPODS

In 1911, Schuchert [35] concluded that "between low-water and above 90 feet of depth the great majority of inarticulate or corneous brachiopods live, or 21 species of a total of 29 (*Lingula,* 11; *Glottidia,* 4; *Discina,* 1; *Discinisca,* 4; and *Crania,* 1). Of the articulate brachiopods, 23 out of the 129 live in these shallow waters." The remaining seven species of living inarticulate (corneous) brachiopods (*Discinisca,* 1; *Crania,* 6) appear from 90 feet down to 600 feet. Schuchert adds that the most conspicuous of the "shallow-water forms are of *Lingula* and *Discina,* genera that are restricted" to the zone "from the strand-line to a depth of probably not much more than 60 feet." [36] This places these living shells within the Laminaria zone of Forbes. The position of this zone on chart (Fig. 3A) corresponds closely to that of the *Lingula* phase in chart (Fig. 3B).

It is interesting to add that "many of the species of lingulids occur in bays and estuaries, indicating that they prefer a habitat more or less freshened by land waters." [37]

Schuchert points out that "of the 129 articulate species about 19 percent (25 species) live in less than 90 feet of water. Their real habitat, however, is in the deeper water between 90 and 600 feet, where nearly 46 percent (59 species) live. Down to 600 feet occur 84 articulate and 28 inarticulate forms, or, in other words, more than 70 percent of brachiopods are at home in these shallower waters. We may, therefore, conclude that the greatest abundance of living brachiopods is in the stormless waters between 90 and about 500 feet depth." [38]

The 90-foot mark, which Schuchert selects as the important lower boundary for the zone of abundant inarticulates, corresponds satisfactorily to the boundary on the restored Big Blue sea floor between the zone of pelecypods, containing some corneous brachiopods, and the next underlying zone, in which calcareous brachiopods begin to predominate.

[34] Edward Forbes: *On the connection between the distribution of the existing fauna and flora of the British Isles, and the geological changes which have affected their area . . . ,* Geol. Surv. Great Britain, Mem., vol. 1 (1846) p. 371.

[35] Charles Schuchert: *Paleogeographic and geologic significance of recent Brachiopoda,* Geol. Soc. Am., Bull., vol. 22 (1911) p. 260.

[36] *Op. cit.,* p. 262.

[37] *Ibid.*

[38] *Op. cit.,* p. 265.

CORALS

The reef-forming corals do not flourish below 25 fathoms,[39] as estimated originally by Darwin and confirmed by extensive recent investigations. Few reef-builders are found at greater depths (*Dendrophyllia* to 50 fathoms) in tropical and temperate waters, together with a few non-colonial corals. The most common solitary corals (*Balanophyllum, Coryophyllia*) in northern European seas are found in the zone of low tides and are widely distributed in shallow waters of the Mediterranean (*Caryophyllia*) and Indian Ocean (*Paracyathus*).[40]

FORAMINIFERA

The small planktonic Foraminifera are known to accumulate at depths of 3,000 to 5,000 meters, but the bathymetric position of the larger Foraminifera is of interest in considering probable depths indicated by fusulinids. According to Gálloway,[41] "the fusulinids lived in shallow water and must have been benthonic, for such large heavy tests could scarcely have floated. Beede has expressed the opinion that *Schwagerina* may have been pelagic, but its wide geographic range could be explained by its dispersal in the nepionic stage." Thus, it seems probable that the fusulinids lived at depths comparable to those inhabited by modern large benthonic Foraminifera. The modern larger Foraminifera of the families Alveolinellidae, Camerinidae, and others, are almost exclusively tropical, in waters less than 30 fathoms deep.[42]

BRYOZOA

The relationship of certain types of bryozoan zoaria to habitat is discussed by Stach,[43] on the basis of observations, chiefly along the coasts of Australia. He distinguishes nine principal zoarial types for living and Cenozoic cheilostome bryozoans. Most of these can be recognized also among Paleozoic Gymnolaemata. The encrusting forms of *Fistulipora, Tabulipora,* and other "stony" bryozoans, for instance, belong to Stach's *membraniporiform* type of zoarium. *Meekopora* and other bilaminate forms are eschariform; *Fenestrellina* ("*Fenestella*" of authors) and *Polypora* are reteporiform; and *Rhombopora* and other rod-shaped rigid branching types with solid core possess vinculariform zoaria. Among these four types only the encrusting membraniporiform zoaria are found in both littoral and sub-littoral zones. These extend also "to deeper waters but are there numerically unimportant." [44] The bilaminate eschariform zoa-

[39] T. W. Vaughan: *Physical conditions under which Paleozoic coral reefs were formed,* Geol. Soc. Am., Bull., vol. 22 (1911) p. 248.

[40] S. J. Hickson: *Introduction to the study of recent corals,* Univ. Manchester Publ., Biol. ser., no. 4 (1924) p. 37-38.

[41] J. J. Galloway: *Manual of Foraminifera* (1933) p. 391. The Principia Press, Indiana.

[42] J. A. Cushman: *Foraminifera* (1928, *also* 1933) p. 39.

[43] Leo W. Stach: *Correlation of zoarial form with habitat,* Jour. Geol., vol. 44 (1936) p. 60.

[44] *Op. cit.,* p. 61.

rium of modern Bryozoa is "adapted for life in sublittoral zones at depths of at least 10 fathoms" (60 feet). "It may extend to deeper water, but not to the littoral zone"—that apparently means above 60 feet. The reteporiform or fenestrate type of zoarium "is adapted for life in regions where wave action and currents are strong" and "is most prolific in sublittoral regions." The branching vinculariform types "are adapted for life in deep or sheltered waters where wave action is absent and currents scarcely active." Only the fenestrate types of Bryozoa, *Fenestrellina* (*Fenestella* of authors) and *Polypora,* have been observed by the writer in the molluscan (4) and mixed (5) phases of the Big Blue cycles, but all four indicated types of zoaria are represented in the brachiopod phase. These observations provide further support for the conclusion that the phase of calcareous brachiopods belongs to bottoms below 60 feet.

MOLLUSKS

The position of pelecypods and other mollusks in the Big Blue cycle is fairly conformable to that in modern seas. According to Forbes,[45] the mollusks are more or less evenly distributed from the shore line down to a depth of about 500 feet, but at greater depths their number diminishes rapidly. Within the region down to 500 feet, gastropods are somewhat more conspicuous in the littoral zone (down to 12 feet) and pelecypods at depths of 120 to 210 feet.[46] To this must be added, however, that the great banks of oysters in the north "are usually in 4 or 5 fathoms water; the scallop (*Pecten*) banks at 20 fathoms,"[47] and that pearl fisheries[48] of the south also belong chiefly to the same zone, rarely extending down to 120 feet.

Allee[49] reports that under favorable circumstances the pelecypod *Modiola modiolus* may be so abundant in the littoral zone as to cover the whole bottom with a layer of living mollusks. It appears, therefore, that gregarious pelecypods prefer waters down to a depth of 120 feet.

Appellöf cites abundant burrowing pelecypods in the clayey and sandy bottoms of the area between the shore and the 130-foot mark.

The quoted data are not at variance with the general position of the gastropods and pelecypods, including especially the burrowing pelecypods (*Allorisma, Pinna*), in the Big Blue cycles.

ALGAE

The general conclusion of Forbes concerning predominance of leathery sea-weeds to a depth of 90 feet, and the common occurrence of calcareous

---

[45] Edward Forbes: *Report on the mollusks and Radiata of the Aegean Sea, and on their distribution as bearing on geology,* Brit. Assoc. Adv. Sci., 13th Rept. (1844) p. 130-193.

[46] *Op. cit.,* p. 169-171.

[47] S. P. Woodward: *Manual of the Mollusca* (1858) p. 8.

[48] G. W. Tryon: *Structural and systematic conchology* (1882) p. 176.

[49] W. C. Allee: *Organization of marine coastal communities,* Univ. Chic., Ecol., Mon., vol. 4 (1934) p. 547.

sea-weeds (encrusting *Lithothamnium* and branching *Corallina*) below this depth, have already been mentioned. To this may be added the fact that the vertical distribution of the lime-secreting algae is known to extend to somewhat greater depths than that of the reef-building corals, but ordinarily it is limited to about 200 feet.[50]

The encrusting structures suggestive of lime-secreting red algae of the *Lithothamnium* type are occasionally found in the Big Blue series and are common in the underlying Pennsylvanian rocks of Kansas. They are found in association with pelecypods, brachiopods, and some even with fusulinids. The impressions of leathery sea-weeds have been observed only in the pelecypod(4) and *Lingula*(3) phases of the Big Blue cycles. This occurrence of leathery and calcareous algae in the Big Blue sediments compares well with the general distribution of this marine vegetation in modern seas.

SIGNIFICANCE OF CYCLIC SEDIMENTATION

GENERAL FACTORS

Although character of sea bottom, temperature, salinity, food supply, and other factors of environment undoubtedly influenced the distribution of marine organisms during Big Blue time, periodic change of sea level was probably the main factor producing the cyclic recurrence of stratigraphic units. The gentle sea-level oscillations that accompanied epeirogenic pulsations of the continents are quite sufficient to produce the observed faunal and lithologic changes. Gradual subsidence of the whole Mid-Continent platform under weight of accumulating sediments must have been taking place according to laws of isostasy, but, as this could counterbalance only a fraction of the thickness of cyclic sediments, the combination of accumulation and isostatic subsidence evidently cannot alone explain the symmetrical sedimentary cycles in the Big Blue series. They should have contributed, however, to fluctuations of the sea. Greater sinking than that possibly due to compensating isostatic movements could have been developed through general nonisostatic epeirogenic subsidence of local portions of the crust. However, mere accumulation of sediments could not have produced complete emergence between two cyclic marine transgressions, because the thickness of sediments of a single marine cycle averages 50 feet and, therefore, is not sufficient to fill depths of 160 to 180 feet, which depths are indicated by the biota of the deepest phases of the cycles. Comparison of the zonal distribution of living marine organisms with the cyclic distribution of the fossils in the sediments of the Permian rocks of Kansas indicates that the maxi-

[50] W. H. Twenhofel: *Treatise on sedimentation*, 2d ed. (1932) p. 307.

mum cyclic subsidence of the Late Paleozoic sea of Kansas amounted to about 180 feet.

If it were possible that the maximum depth of deposition in Big Blue time in Kansas was one-half or one-third as shallow as 180 feet, the depth at which fusulinids and other invertebrates of the deepest phases lived must have been 90 feet or less, which is much shallower than the zone of most of the recent articulate brachiopods, as summarized by Schuchert.[51] Environmental factors other than depth of water, unless these are definitely related to depth, produce effects other than those observed in the most typical and complete sedimentary cycles under discussion.

INFLUENCE OF SALINITY

Considerable changes in salinity, such as are observed in estuaries where water is greatly freshened by rivers, or in lagoons where great salinity is due to excessive evaporation, produce noticeable changes in the bottom faunas. These changes are recognized, for example, in Carboniferous and Permian sediments in which the cycles of deposition under discussion are observed. The so-called non-marine pelecypods, which are always found in closed communities, and occasionally in association with other organisms (*Lingula* and others) that indicate unusual degree of freshness of water, or, on the contrary, of great salinity, exemplify the influence of character of water. Moderate changes in salinity are associated with other factors such as depth, clearness of water, and penetration of light, to produce the environments that form successive zones paralleling the shore.

Observations by Appellöf, Forbes, and others clearly indicate that some other factors have greater effect upon the bottom dwellers of the shallow belts than moderate changes in salinity have. Although changes in salinity are more pronounced toward the shore line, here, other influences, such as character of bottom and currents, also become stronger.

The great variation of all these factors is characteristic of the near-shore environments. It results partly in separation of different organisms into communities that are adapted to different portions of the near-shore belts, and, in part, it results in adaptation of all organisms of a belt to those changes that are periodic and affect equally the whole belt. Most pronounced is the rise and fall of water in the tidal zone. According to Appellöf, changes in salinity as well as temperature are periodic at depths down to 30 or 40 meters. Insofar as analysis of the discussed cycles is concerned, it is noteworthy that each recognized phase of the sedimentary cycles in Kansas is remarkably uniform both in

[51] Charles Schuchert: *Paleogeographic and geologic significance of recent Brachiopoda*, Geol. Soc. Am., Bull., vol. 22 (1911) p. 258-275.

organic and in lithologic content. This indicates that there was little or no differentiation into facies within the zones or belts defined by the sediments that make up these phases.

INFLUENCE OF TEMPERATURE AND LIGHT

The influence of temperature and light are well known and need not be here discussed. It suffices to say that changes in both temperature and light are intimately connected with change of depth.

PHYSICAL CHARACTER OF BOTTOM

The shallow-bottom life in modern seas is arranged in belts more or less parallel to the shore line, but differences in the physical character of the bottom result in separation of communities within each belt, particularly in the shallowest ones. Only insignificant, if any, changes of that sort can be observed along the outcrops of the Big Blue rocks, probably because no clastic sediments other than fine muds, silts, and some fine sand were deposited on the floor of the Big Blue sea in Kansas.

It is interesting to observe that a change from clayey to calcareous bottom in the advancing sea and a reverse change from calcareous to clayey bottom in the regressing sea apparently resulted in slight differences in faunal content in the corresponding phases of the progressive and regressive hemicycles, creating a certain asymmetry of the otherwise symmetrical cycles of the Big Blue. For instance, the observed abundance of gastropods other than bellerophontids in the molluscan phase of regressive hemicycles and the rarity or absence in the corresponding phase of progressive hemicycles may be due to the difference in character of the bottom in the progressive and regressive hemicycles.

CORRESPONDENCE OF BIG BLUE PHASES TO ZONES IN SHALLOW MODERN SEAS

FOSSIL-BEARING PHASES

Comparison of the general distribution of benthonic organisms in modern seas with the succession of faunal phases in the Big Blue cycles suggests clearly the possibility of interpreting the latter in terms of depth of deposition. This interpretation is based on two assumptions:

1. The classes of marine animals and plants of Late Paleozoic time lived largely in the same environments as the corresponding classes of organisms today. The order of succession of these organisms in the Big Blue sedimentary cycles, as well as in other Late Paleozoic cycles elsewhere in the world, strongly supports this conclusion.

2. The distribution of benthonic organisms in the Big Blue sea was controlled chiefly by changes in relative depth of sea level.

The first of these two assumptions may be criticised on the ground that there is lack of absolute proof that extinct animals and plants lived in about the same environments as the corresponding modern classes of organisms. Most naturalists agree, however, that unless the principle that "the present is the key to the past" is accepted, there is little on which to base an approach to the problem of environments of organisms of the past. Checking the probable environments in which certain extinct organisms lived by evidence supplied by associated organic remains (and by lithology) is recommended by many prominent paleontologists (Stanton,[52] Neaversen,[53] and others). Such check for the case here discussed is provided by abundant marine animal and plant remains of varied sorts in the Big Blue series.

The chief evidence in support of the second assumption is the remarkable similarity between the order of succession of faunal phases in each sedimentary cycle of the Big Blue series and the order of succession of marine life zones defined by depth of water in modern seas.

The influence of physical character of the bottom, which considerably affects distribution of organisms in modern seas, apparently can be neglected in considering the Big Blue sedimentary cycles, because the bottom of the Big Blue seas must have been very uniform in physical character within each zone of depth. Undoubtedly, these uniform conditions must have tended to minimize, if not to eliminate, possible fluctuations of temperature (other than daily or seasonal changes which affected the whole region), salinity, and clearness of water, within each belt of depth that produced a distinguishable phase in the Big Blue cycles.

In 1911, on the base of various data, E. O. Ulrich concluded that "the average depth of the Paleozoic continental seas [of North America] were less than 200 feet, and that none attained depths exceeding 100 fathoms." [54]

In 1929, when discussing the environment of Pennsylvanian life in North America and particularly in the northern Mid-Continent, R. C. Moore concluded that the depths of the Pennsylvanian epi-continental seas "were possibly, in fact probably, almost nowhere so much as 100 fathoms," that the sea "may be conceived to have been always shallow," and that they "were not only very shallow but they were excessively fluctuating." [55] It now appears that the sea that covered Kansas and adjacent territory in Late Paleozoic time was not deeper than about 200 feet.

---

[52] T. W. Stanton: *The significance of ammonites in interpreting depth and temperature of the waters in which they lived,* unpublished manuscript quoted by W. H. Twenhofel: *Treatise on sedimentation,* 2d ed. (1932) p. 179.

[53] Ernest Neaversen: *Stratigraphical paleontology* (1928) p. 89, 90. Macmillan and Co., London.

[54] E. O. Ulrich: *Revision of the Paleozoic systems,* Geol. Soc. Am., Bull., vol. 22 (1911) p. 362.

[55] R. C. Moore: *Environment of Pennsylvanian life in North America,* Am. Assoc. Petr. Geol., Bull., vol. 13 (1929) p. 459-487.

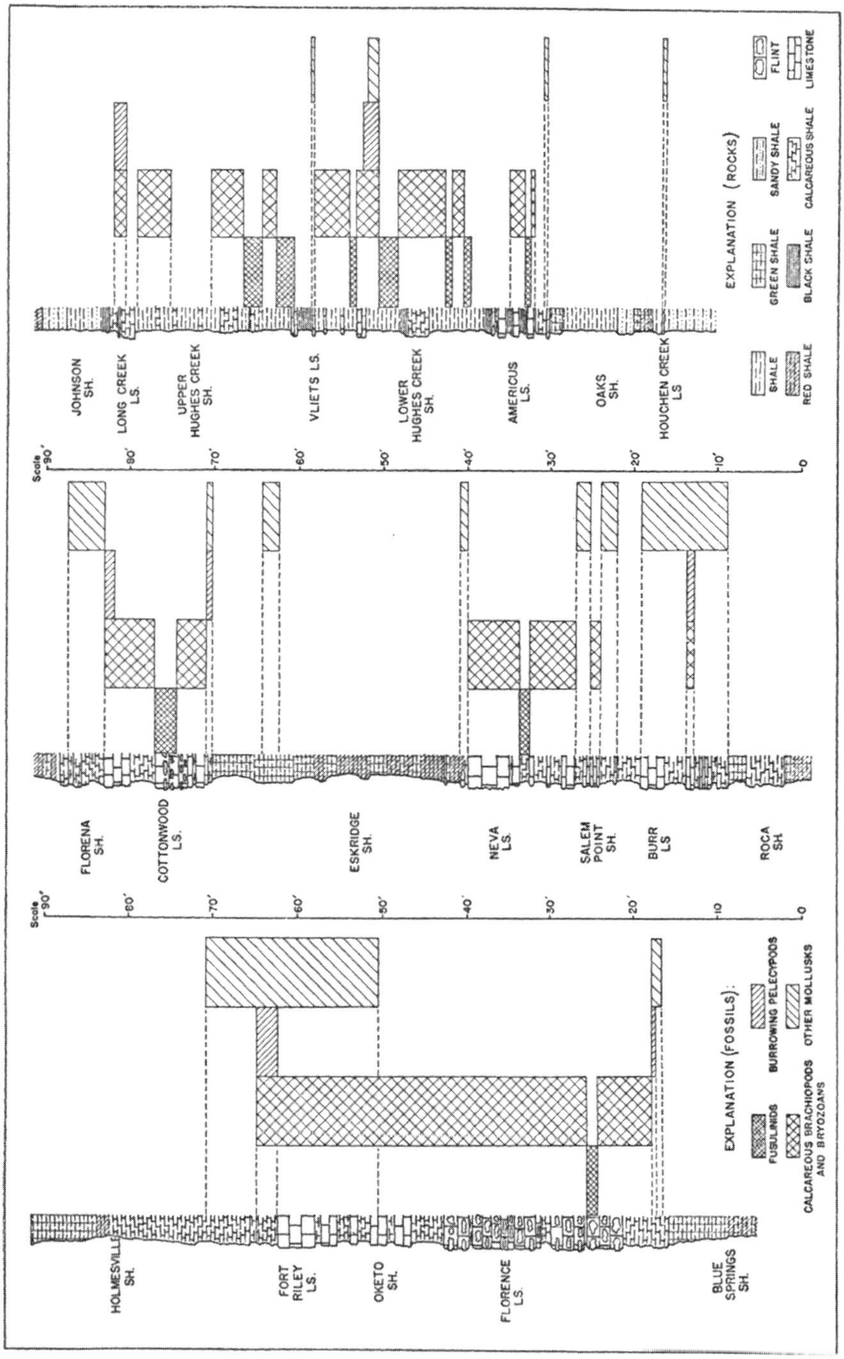

FIGURE 4.—*Detailed sections and phase thicknesses*

Illustrating the most complete sedimentary cycles of the Big Blue series and showing comparative thickness of the phases differentiated by their fossil content.

The main evidence for this conclusion is the position of fusulinids in cycles of the Big Blue, which indicates that they belong to the deepest bottoms. Inasmuch as the modern larger benthonic Foraminifera live in waters under 180 feet, this depth is selected as probably the deepest at which the Big Blue fusulinids lived (Fig. 3 B). This selection seems to be supported by the following observations. In some fusulinid facies from the Virgil series, which underlies the Big Blue beds, encrusting calcareous structures referable to *Osagia* have been observed (in the upper fusulinid phase of the lower Oread, in the fusulinid phase of the lower Deer Creek, in the Tarkio fusulinid limestone). Although these presumably algal incrustations are rather rare in the listed phases, their appearance within the fusulinid zones suggests a depth of 180 feet or shallower, because, below this depth, photosynthetic processes essential for chlorophyll-containing algae are considered practically impossible, even in the clearest waters. The very existence of the modern larger fusulinids is thought to be dependent on the particular commensal algae, and, thus, the distribution of these algae also delimits the depth at which these Foraminifera live.

Another basis for selection of the 180-foot depth is provided in the fact that next to the fusulinid (7) phase in the cycles of the Big Blue is the prominent phase (6) of calcareous brachiopods, bryozoans, and corals, a combination suggestive of depths between 90 and 150 feet. Furthermore, in the Americus, lower Oread, and other cycles, burrowing pelecypods have been observed in the fusulinid phase.

Intimately interbedded phases 7 and 6, as in the Foraker cycle of the Big Blue series or in the upper Oread cycle (Plattsmouth cycle of the Oread megacycle) of the Virgil series, suggest the proximity of the fusulinid and brachiopod zones in the Late Paleozoic seas, because the repeated change from one phase to another within a single cycle is much more easily explained as due to slight changes of ocean water level (or some other subtle changes of environment) than as being a result of rapidly repeated considerable changes of sea level, which shifted the depth from a much shallower brachiopod-bryozoan-coral environment to a much deeper fusulinid environment several times within what seems to be a single marine invasion (Fig. 1C).

The critical 90-foot depth, below which the most modern articulate or calcareous brachiopods live, is inferred to correspond to the line of division between the mixed fauna (5) phase of the Big Blue, with its abundant calcareous brachiopods, and the molluscan (4) phase, in which calcareous brachiopods are rare. The 60-foot depth is selected to mark the contact of the *Lingula* (3) and the molluscan (4) phases in the cycle of the Big

Blue, because this corresponds to the deepest occurrence of modern *Lingula*. The division of the deeper waters of the Big Blue sea (from 90 to 180 feet) between the 7 and 6 + 5 phases is made on the ground of the relative thickness of these phases. In so doing the following assumptions are made:

1. The advance and retreat of the sea, which resulted in deposition of the observed sedimentary cycles, were, on the average, of the same rate throughout the cycles.

2. The rate of deposition of the largely calcareous sediments of the brachiopod and fusulinid phases was generally the same.

Comparison of the thickness of 7 and 6 + 5 phases in the four most complete cycles of the Big Blue series (Foraker, Grenola, Beattie, and Barneston) permits the conclusion that phases 6 + 5, on the average, are about two to four times as large as phase 7 (Fig. 4). Inasmuch as the thin fusulinid (7) phases of the Grenola and the Barneston cycles may, and probably do, represent only a fraction of a complete deep phase, which here did not develop in full or to a depth of 180 feet, the larger proportion of phase 7 to phases 6 + 5, or a proportion of 1 to 3½, is selected to represent the more likely proportional subdivision of 90 to 180 feet depth between the corresponding zones. Thus, the brachiopod (6) and the mixed (5) zones are thought to stretch from a depth of 90 to 160 feet and the fusulinid (7) zone from 160 to 180 feet. It is interesting to note that King reports that calcareous brachiopods with thinner shell, indicative of deeper water, are found together with radiolarians in some Late Paleozoic sediments of Texas (Leonard formation), and these apparently belong to a deeper phase than that indicated by fusulinids. The depth of 150 feet corresponds, also, as has already been pointed out, to the depth down to which most modern corals are distributed, which thus corresponds approximately to the occurrence of corals in the Big Blue (as well as in the earlier Pennsylvanian) cycles.

RED AND GREEN BEDS

The change from green to red silt, at the shallow end of the Big Blue cycle, is interpreted to indicate the upper limit of the zone of tides. The development of 30 or 40 feet of red shale was probably subaerial and most probably represents combined wind and water transfer of the red soil developed on somewhat higher land. In the Big Blue cycles there is, as a rule, a green shale phase, which is about as thick as the red shale phase and in places contains *Lingula* in the part adjacent to pelecypod phase. The change from red to green shale is marked in many places by interbedding, by a zone of mottled red and green shale, or by beds that change from duller red, maroon to purple, to dark blue-green. The more earthy colors of these shales seem to be due not to fine-grained glauconite, but to the reducing action of organic matter upon iron oxides of the original red silt. The common observation of a greenish halo around organic remains embedded in reddish sediments is the best illustration of this process. The alternate submergence and emergence of the zone of tides seems to be favorable for rapid oxidation of the dead organic remains, but their well-known great accumulation in this zone seems to be responsible for reducing action in the zone, as pointed out by Twenhofel.[56] Therefore, the zone of tides is included in the area in which reduction of the red to green silt has taken place. The change from predominantly red to largely green is thus selected as indicating the upper limits of deposition in the zone of tides. The change from the molluscan (4) to *Lingula* (3) phase is chosen as corresponding to 60-foot depth, down to which lives, according to Schuchert, modern *Lingula*. The boundary between the *Lingula* (3) phase and the unfossiliferous green shale (2) is arbitrarily concluded to indicate deposition at a depth of about 30 feet.

CYCLES AND DEPTH OF DEPOSITION IN THE PENNSYLVANIAN SEA

A cyclic succession of faunal zones similar to that in the Big Blue series is observed throughout the Pennsylvanian rocks of Kansas and adjacent territory. In these cycles the fusulinid phase also is in a central position and is flanked above and below by the brachiopod-bryozoan-coral phase. Most of the sedimentary cycles in these rocks are less regular than in the Big Blue series, which may have been due to less even bottom conditions, greater influence of current and wave action, and other possible factors that interfered with the control provided by depth of deposition. Owing to influence of one or more of these factors at any particular time and place, the normal position in a cycle of various benthonic organisms (as detected in the Big Blue series) may be somewhat shifted up or down outside of the depths which they normally occupied. In view of all this, it seems reasonable to conclude that the maximum depth of the Pennsylvanian sea in eastern Kansas, also determined by the presence of the deepest fusulinid phase, was about 200 feet. In the western part of the State the depth was probably slightly greater, as this area was farther away from the shore line.

CONCLUSIONS

In the Big Blue series of Kansas the following cyclic succession is observed: (1) red shale, (2) green shale, (3) *Lingula* phase, (4) molluscan phase, (5) mixed fauna (molluscan and brachiopod) phase, (6) brachiopod phase, (7) fusulinid phase; after which the same phases appear in

[56] Letter of January 7, 1936.

reverse order until the red shale phase is again reached. Bryozoans and corals are restricted to phases 5 and 6; burrowing pelecypods are most prominent in phase 5 but appear also in phases 6 and 7. Leathery seaweeds have been found in phases 3 and 4 and calcareous algal incrustations in phases 4 and 5 (also in 6 and 7 in the Pennsylvanian rocks below the Big Blue). Comparison of distribution of organisms in the cycles of the Big Blue with the benthonic zones of modern seas suggests the following:

1. The various classes of marine organisms in Late Paleozoic time were adapted to about the same environments as they are today.
2. Observed cyclic repetition of sedimentary rocks and the enclosed organic remains indicate advance and retreat of sea, each cycle representing one major marine invasion.
3. Red shales that separate marine phases of two neighboring cycles are continental deposits and indicate emergence.
4. Maximum depth of the Big Blue sea did not exceed 180 feet, and the depth of the earlier Pennsylvanian sea in Kansas probably did not exceed 200 feet.
5. The Late Paleozoic benthonic organisms that characterize the phases of the Big Blue cycles were adapted to the following approximate depths: fusulinids, from 160 to 180 feet; calcareous brachiopods, from 90 to 160 feet; corals, from 90 to 140 feet; bryozoans, from 75 to 160 feet; burrowing pelecypods, from 90 to 180 feet; other pelecypods, from 60 to 110 feet; and leathery sea-weeds, from 30 (or probably from shore line) to 75 feet.

Although the suggested ranges of depths which the different Late Paleozoic benthonic organisms occupied are selected in a somewhat arbitrary way, it appears that these depths, expressed in round figures, approximate the actual depths of the benthonic zones of Late Paleozoic seas.

BIBLIOGRAPHY OF MARINE FACIES AND SEDIMENTARY CYCLES

Abrard, R.: *Faciès et associations paléontologiques,* Arch. Mus. (Paris), 6th ser., vol. II (1927) p. 81-109.

Allee, E. J.: *On the fauna and bottom-deposits near the thick fathom line from the Eddystone Grounds to Start Point,* Marine Biol. Assoc. United Kingdom, Jour., vol. 5 (1897-1899).

Allee, W. C.: *Studies in marine ecology,* Biol. Bull., vol. 44 (1923) p. 167-191.

———: *Organization of marine coastal communities,* Univ. Chicago, Ecol. Mon., vol. 4 (1934) p. 541-554.

Andrée, K.: *Über stetige und unterbrochene Meeres-sedimentation, ihre Ursachen, so wie über deren Bedeutung für die Stratigraphie,* N. Jahrb. f. Mineral., Geol. und Pal., Beil. Bd. 25 (1908) p. 366-421.

Antevs, Ernst: *Retreat of the last ice sheet in eastern Canada,* Geol. Surv. Canada, Mem. 146 (1925) 142 pages.

———: *Quaternary marine terraces in nonglaciated regions and changes of level of sea and land,* Am. Jour. Sci., 5th ser., vol. 17 (1929) p. 35-49.

———: *A geological chronometer; the varved glacial clays give an accurate measure of the ages,* Canad. Min. Jour., vol. 51 (1930) p. 388-390.

Antevs, Ernst: *Varved sediments; conditions of formation of the varved glacial clay,* Rept. Com. Sed., no. 92 (1930) p. 61-65; Nat. Res. Council, no. 98, 1929-1930 (1931) p. 51-53; no. 89, 1930-1932 (1932) p. 89, 90.

Appellöf, A.: *Invertebrate bottom fauna,* in J. Murray and Johan Hjort: *Depths of the Ocean* (1912) p. 459-526.

Arkell, M. A.: *Review on Brinkmann's Statistische-Biostratigraphische Huntersuchungen . . . ,* Geol. Mag., vol. 68 (1931) p. 373-376.

———: *Jurassic system in Great Britain* [Epeirogenic oscillations and subsidence of the troughs; cyclic sedimentation] (1933) p. 51-59. Clarendon Press, Oxford.

Ashley, George H.: *Pennsylvanian cycles in Pennsylvania,* Ill. Geol. Surv., vol. 60 (1931) p. 241-245.

Barrell, J.: *Rhythms and the measurements of geologic time,* Geol. Soc. Am., Bull., vol. 28 (1918) p. 745-904.

Boswell, P. G. H.: *Sedimentation, environment and evolution of past ages,* Liverpool Biol. Soc., Tr., vol. 35 (1921) p. 5-28.

Bradley, W. H.: *Shore phases of the Green River formation in northern Sweetwater County, Wyoming,* U. S. Geol. Surv., Prof. Pap. 140 (1926) p. 121-131.

———: *Varves and climate of the Green River epoch,* U. S. Geol. Surv., Prof. Pap. 158-E (1929) p. 87-110.

———: *Origin and microfossils of the oil shale of the Green River formation of Colorado and Utah,* U. S. Geol. Surv., Prof. Pap. 168 (1931) 58 pages.

Brinkmann, Roland: *Statistische-Biostratigraphische Untersuchungen an Mitteljurassischen Ammonitea über Artbegriff und Stammesenwicklung,* Abh. Ges. Wiss. zu Göttingen, Math. Phys. Kl., n. f., 3 (1929).

Broeck, van den, E.: *Note sur un nouveau mode de classification et de notation graphique des dépôts géologiques,* Mus. Roy. Nat. Hist., Bruxelles, Bull., vol. 2 (1883) p. 341-360.

Brooks, C. E. P.: *Cycles in natural phenomena,* Nature, vol. 125 (1930) p. 18, 19.

Brough, J.: *On rhythmic deposition in the Yoredale series,* Univ. Durham, Philos. Soc., Pr., vol. 8 (1929) p. 116-126.

Cushman, J. A.: *Foraminifera, their classification and economic use,* 1st ed. (1928) p. 1-401; 2d ed. (1933) p. 1-349. Sharon, Mass.

Davies, A. M.: *Sequence of Tertiary faunas* in *Tertiary faunas,* vol. 2 (1934) p. 53-55 [Cycles of sedimentation]. Thomas Murby and Co., London.

Dawson, J. W.: *Acadian geology* [Geologic cycles] 2d ed. (1868) p. 135-138.

Dorsey, G. E.: *Origin of the color of red beds,* Jour. Geol., vol. 34 (1926) p. 131-141.

Eaton, J. E.: *By-passing and discontinuous deposition of sedimentary materials,* Am. Assoc. Petr. Geol., Bull., vol. 13 (1929) p. 713.

Efimov, I. N.: *Distribution of organic remains in the roofing of Coal Measures of the Donetz Basin,* Acad. Sci., U. S. S. R., C. R., vol. 2 (1934) p. 376-381.

———: *Modifications of faunal conditions along the strike of the roofing of Coal Measures in the Donetz Basin,* Acad. Sci., U. S. S. R., C. R., vol. 2 (1935) p. 492-498.

Elias, M. K.: *Cycles of sedimentation in the Big Blue series of Kansas* (abstract), Geol. Soc. Am., Pr. 1933 (1934) p. 366.

——— *with* Moore, R. C.: *Sedimentation cycles as a guide to stratigraphic classification of the Kansas Lower Permian* (abstract), Geol. Soc. Am., Pr. 1933 (1934) p. 100.

Fenton, C. L., and Fenton, M. A.: *Ecologic interpretations of some biostratigraphic terms, I. Faunule and Zonule,* Am. Mid. Nat., vol. 11 (1928) p. 1-23.

Fenton, C. L., and Fenton, M. A.: *Ecologic interpretation of some biostratigraphic terms, II. Zone, subzone, facies, phase,* Am. Mid. Nat., vol. 12 (1930) p. 145-153.

Fisher, Paul: *Manual de Conchyliologie* (1887). Paris.

Forbes, Edward: *Report on the mollusks and Radiata of the Aegean Sea, and on their distribution as bearing on geology,* Brit. Assoc. Adv. Sci., 13th Rept., 1843 (1844) p. 130-193.

———: *On the connection between the distribution of the existing fauna and flora of the British Isles, and the geological changes which have affected their area,* Geol. Surv. Great Britain, Mem., vol. 1 (1846) p. 336-432.

———: *Report on the investigation of British marine zoology by means of the dredge,* Pt. 1, Brit. Assoc. Adv. Sci., 20th Rept., 1850 (1851) p. 192-263.

Ford, E.: *Animal communities of the level sea-bottom in the waters adjacent to Plymouth,* Marine Biol. Assoc. United Kingdom, Jour., vol. 13 (1923) p. 164-224.

Foslie, M.: *Norwegian forms of Lithothamnion* (1895). Trondhjem, Norway.

———: *New or critical calcareous algae* (1900). Trondhjem, Norway.

———: *Lithothamnia of the Maldives and Laccadives, in* Stanley Gardiner: *Fauna and geography of the Maldive and Laccadive Archipelagoes,* vol. 1 (1903) p. 460-471. Cambridge Univ. Press.

Fraser, H. J.: *Experimental study of varve deposition,* Royal Soc. Canada, Tr., vol. 23, sec. 4 (1929) p. 49-60.

Frebold, H.: *Ammonitenzonen und sedimentation-zyklen und ihrer Beziehung zueinander,* Centr. f. Mineral., Geol., und Pal. (1924) p. 313-320.

———: *Über cyclische Meeressedimentation* (1925). Leipzig.

———: *Die Paläogeographische analyse der epirogenen Bewegungen und ihre Bedeutung für die Stratigraphie,* Geol. Archiv., vol. 4 (1927) p. 223-240.

Galloway, J. J.: *Manual of Foraminifera* (1933) 483 pages. The Principia Press, Indiana.

Gardiner, J. Stanley, *et al.: Fauna and geography of the Maldive and Laccadive Archipelagoes,* vol. 1 (1903) p. 1-472; vol. 2 (1906) p. 473-1079.

Grabau, A. W.: *Principles of Stratigraphy* (1913) particularly p. 470.

Herdman, W. A.: *Marine zoology of the Irish Sea,* Brit. Assoc. Adv. Sci., Rept. (1894) p. 318-334.

Hickson, Sydney J.: *Introduction to the study of recent corals,* Univ. Manchester Publ., Biol. ser., no. 4 (1924) p. 1-257.

Hind, Wheelton: *Carboniferous rocks of the Pennine system,* Yorkshire Geol. Soc., Pr., n. s., vol. 14, pt. 3 (1902) p. 449.

Hudson, R. G.: *On the rhythmic succession of the Yoredale series in Wensleydale,* Yorkshire Geol. Soc., Pr., n. s., vol. 20 (1924) p. 125.

Hull, E.: *On iso-diametric lines as a means of representing the distribution of sedimentary clay and sandy strata,* Geol. Soc. London, Quart. Jour., vol. 18 (1862) p. 127-146.

Jewett, J. M.: *Evidence of cyclic sedimentation in Kansas during the Permian period,* Kans. Acad. Sci., Tr., vol. 36 (1933) p. 137-140.

Johnston, W. A.: *Sedimentation in Lake Louise, Alberta, Canada,* Am. Jour. Sci., 5th ser., vol. 4 (1922) p. 376-386.

Kendall, P. F.: *Upper Carboniferous, (B), Coal Measures,* in *Handbook of the Geology of Great Britain* (1929) p. 259-260.

Kindle, E. M.: *Sedimentation in a glacial lake,* Jour. Geol., vol. 38 (1930) p. 81-87.

King, Philip B.: *Limestone reefs in the Leonard and Hess formations of Trans-Pecos Texas,* Am. Jour. Sci., 5th ser., vol. 24 (1932) p. 337-354.

Klüpfel, W.: *Über die Sedimente der Flachsee im Lothringer Jura,* Geol. Rundschau, vol. 7 (1916) p. 97-109.

Krogh, A.: *Conditions of life in the ocean,* Univ. Chicago, Ecol. Mon., vol. 4 (1934) p. 421-429.

Merriam, J. C., *et al.: Symposium on climatic cycles,* Nat. Acad. Sci., Pr., vol. 19 (1933) p. 349-388.

Moore, R. C.: *Environment of Pennsylvanian life in North America,* Am. Assoc. Petr. Geol., Bull., vol. 13 (1929) p. 459-487.

———: *Sedimentation cycles in the Pennsylvanian of the northern Mid-Continent* (abstract), Geol. Soc. Am., Bull., vol. 41 (1930) p. 51, 52.

———: *Pennsylvanian cycles in the northern Mid-Continent region,* Ill. Geol. Surv., Bull. 60 (1931) p. 247-257.

Murray, J., and Hjort, Johan: *Depths of the ocean* (1912) 821 pages. Macmillan and Co., London.

Neaversen, Ernest: *Stratigraphical paleontology* [Faunas in relation to habitat] (1928) p. 87-107. Macmillan and Co., London.

Newberry, J. S.: *Circles of deposition in American sedimentary rocks,* Am. Assoc. Adv. Sci., Pr., vol. 22 (1874) p. 185-196. Portland, Maine.

Peterson, C. G. J.: *On the animal communities of the sea-bottom in the Skagerak, the Christiania Fjord and the Danish waters,* Danish Biol. Stat. (Copenhagen), Rept. 23 (1915).

Phillips, J.: *Geology of Oxford* (1871) p. 393-394.

Plummer, F. B.: *Pennsylvanian sedimentation in Texas,* Ill. Geol. Surv., Bull. 60 (1931) p. 259-269.

Pruvost, Pierre: *Sédimentation et subsidence,* Soc. Géol. France, Livre Jubilaire, vol. 2 (1930) p. 545-564.

Reeds, C. A.: *Weather and glaciation,* Geol. Soc. Am., Bull., vol. 40 (1929) p. 597-629; Science, vol. 70 (1929) p. 587.

Reger, David B.:*Pennsylvanian cycles in West Virginia,* Ill. Geol. Surv., Bull. 60 (1931) p. 217-239.

Robertson, T.: *Mertyr Tydfil* (2d ed.), Part 5 of the *Geology of South Wales coal fields,* Geol. Surv. England and Wales, Mem. (1932) p. 92, 93.

Rutot, A.: *Les phénomènes de la sédimentation marine étudiés dans leurs rapports avec la stratigraphie régionale,* Mus. Roy. Hist. Nat., Bruxelles, Bull., vol. 2 (1883) p. 41-83.

Sayles, R. W.: *New interpretation of the Permo-Carboniferous varves at Squantum,* Geol. Soc. Am., Bull., vol. 40 (1929) p. 541-546.

Schuchert, Charles: *Paleogeographic and geologic significance of recent Brachiopoda,* Geol. Soc. Am., Bull., vol. 22 (1911) p. 258-275.

Stach, Leo W.:*Correlation of zoarial form with habitat,* Jour. Geol., vol. 44 (1936) p. 60-65.

Stamp, L. D.: *On cycles of sedimentation in the Eocene strata of the Anglo-Franco-Belgian Basin,* Geol. Mag., vol. 58 (1921) p. 108-114, 146-157, 194-200.

———: *Introduction to stratigraphy* [Phenomena of sedimentation] (1923, also 1934, 2d ed.) p. 12-19 (1st and 2d ed.).

Stout, Wilber: *Pennsylvanian cycles in Ohio,* Ill. State Geol. Surv., Bull. 60 (1931) p. 195-216.

Suess, Edward: *Face of the earth* [Paleozoic seas, cycles] (English translation), vol. 2, chap. 5 (1906) p. 217-219. Oxford.

Sumner, F. B.: *An intensive study of the fauna and flora of restricted area of sea bottom,* U. S. Bur. Fish., Bull., vol. 28, pt. 2 (1908, 1910) p. 1227-1263.

Tchernyshew, B.: *Carbonicola, Anthracomya, and Najadites of the Donetz Basin,* Geol. Prosp. Service, U. S. S. R., Tr., fasc. 72 (1931) 125 pages.

Tryon, G. W.: *Structural and systematic conchology* (1882).

Twenhofel, W. H.: *Treatise on sedimentation,* 2d ed. (1932) p. 124, 612.

Udden, J. A.: *Geology and mineral resources of the Peoria quadrangle, Illinois,* U. S. Geol. Surv., Bull. 506 (1912) 103 pages.

Ulrich, E. O.: *Revision of the Paleozoic systems,* Geol. Soc. Am., Bull., vol. 22 (1911) p. 281-680; index in vol. 22 (1913) p. 625-668.

Vaughan, T. W.: *Physical conditions under which Paleozoic coral reefs were formed,* Geol. Soc. Am., Bull., vol. 22 (1911) p. 238-252.

Wanless, H. R.: *Pennsylvanian section in western Illinois,* Geol. Soc. Am., Bull., vol. 42 (1931) p. 801.

——— and Weller, J. M.: *Correlation and extent of Pennsylvanian cyclothems,* Geol. Soc. Am., Bull., vol. 43 (1932) p. 1003-1016.

——— and Shepard, F. P.: *Sea level and climatic changes related to Late Paleozoic cycles of sedimentation,* Geol. Soc. Am., Bull., vol. 47 (1936) p. 1177-1206.

Watts, W. W.: *Geology as geographical evolution,* Geol. Soc. London, Quart. Jour., vol. 67 (1911) p. 42-93.

Weller, J. M.: *Cyclic sedimentation of the Pennsylvanian period and its significance,* Jour. Geol., vol. 38 (1930) p. 97.

———: *Conception of cyclic sedimentation during the Pennsylvanian period,* Ill. State Geol. Surv., vol. 60 (1931) p. 163-193.

———: *Sedimentary cycles in the Pennsylvanian strata* [a reply], Am. Jour. Sci., 5th ser., vol. 21 (1931) p. 311-329.

White, David: *Notes on the fossil floras of the Pennsylvanian in Missouri,* in H. Hinds and F. C. Greene: *Stratigraphy of the Pennsylvanian series in Missouri,* Mo. Bur. Geol. Mines, 2d ser., vol. 13 (1915) p. 256-262 (*see* p. 256).

Willis, Bailey: *Conditions of sedimentary deposition,* Jour. Geol., vol. 1 (1893) p. 180.

Wilson, A. W.: *Theory of the formation of sedimentary deposits,* Can. Rec. Sci., vol. 9 (1903) p. 112-132.

Woodward, S. P.: *Manual of the Mollusca,* 4th ed. (1880) or reprint (1890) 542 pages and 85 pages appendix (4th ed.). Lockwood and Son, London.

Wright, W. B., *et al.: Geology of the Rossendale anticline,* Geol. Surv. England, Mem. (1927) p. 8-34. London.

———: *Upper Carboniferous (A), Millstone grit,* in *Handbook of the Geology of Great Britain* (1929) p. 253-254.

State Geological Survey, University of Kansas, Lawrence, Kansas.
Manuscript received by the Secretary of the Society, September 17, 1936.
Read before the Paleontological Society, December 27, 1935.
Published with the permission of the State Geologist.

PAPER 26

Geochemical Evidence for Suppression of Pelagic Marine Productivity at the Cretaceous/Tertiary Boundary (1989)

J. C. Zachos, M. A. Arthur, and W. E. Dean

Commentary

STEVEN D'HONDT

This study exemplifies the use of geochemical data to advance understanding of ancient ecological systems at the broadest possible scale. It demonstrated that the large end-Cretaceous impact changed the marine biological cycling of carbon for more than half a million years.

In 1980, Luis Alvarez and his colleagues proposed that a large asteroid or comet struck Earth at the end of the Cretaceous. Shortly after Alvarez et al.'s paper, Wally Broecker and Tsung-Hung Peng (1982) suggested that the global darkness that might result from such an impact could have caused a "Strangelove ocean," an ocean in which biological production of organic matter essentially stopped. In 1985, Ken Hsü and Judy McKenzie interpreted stable carbon isotope ratios in the skeletons of benthic foraminifera (which inhabited the seafloor) and mixed nannoplankton carbonate (which was mostly precipitated in the surface ocean) as direct evidence that the end-Cretaceous impact caused a Strangelove ocean. They showed that these carbon isotope ratios converged at the impact horizon. Their interpretation relied on the recognition that carbon 12 is preferentially fixed as organic matter (relative to carbon 13, which has an additional neutron) and that sinking of organic matter from the surface ocean to the deep sea and consequent oxidation of the sinking organic matter enriches deep water in carbon 12 relative to carbon 13.

Jim Zachos, Mike Arthur, and Walt Dean synthesized a broader range of evidence to show that the drastic change in marine carbon cycling at the end of the Cretaceous lingered for many hundreds of thousands of years. For example, they showed that carbon isotope differences between the skeletons of planktonic foraminifera (which inhabited the near-surface ocean) and the skeletons of benthic foraminifera did not recover for hundreds of thousands of years and rates of calcium carbonate (planktonic skeletal) accumulation did not locally recover for more than two million years after the end of the Cretaceous. Like Hsü and McKenzie, Zachos and colleagues interpreted the decrease in planktonic-to-benthic carbon isotope ratios as evidence of decreased primary productivity.

Subsequent studies expanded the findings of Zachos and colleagues to other regions of the ocean (e.g., Stott and Kennett 1989) and showed that carbon isotopic differences between planktonic and benthic organisms did not recover for more than three million years after the end-Cretaceous impact (D'Hondt et al. 1998). Many more recent studies have focused on relationships between the carbon-system recovery, accumulation rates of marine microfossils, and the recovery of open-ocean biological diversity following the mass extinction.

Present interpretations of these geochemical patterns are generally more nuanced than the low-productivity interpretations of Zachos and colleagues and of other early studies. Marine phytoplankton typically double their abundance on timescales of hours to days. Once sunlight recovered following the impact, how could marine productivity be suppressed? Given this issue, D'Hondt and colleagues (1998) proposed that rates of marine biological productivity may have recovered rapidly but that surface-ocean ecosystem structure was greatly altered and the fraction of organic production that sank to deep water was reduced until the time of marine carbon-system recovery. Whatever the ultimate interpretation of its cause, the central geochemical point of Zachos and colleagues remains clear—marine biological cycling of carbon was drastically altered for a very long time following the end-Cretaceous mass extinction.

Literature Cited

Alvarez, L. W., W. Alvarez, F. Asaro, and H. V. Michel. 1980. Extraterrestrial cause for the Cretaceous-Tertiary extinction. *Science* 208:1095–108.

Broecker, W. S., and T.-H. Peng. 1982. *Tracers in the Sea.* Palisades, NY: Lamont-Doherty Geological Observatory, Columbia University.

D'Hondt, S., P. Donaghay, J. C. Zachos, D. Luttenberg, and M. Lindinger. 1998. Organic carbon fluxes and ecological recovery from the Cretaceous-Tertiary mass extinction. *Science* 282:276–79.

Hsü, K. J., and J. McKenzie. 1985. A "Strangelove" ocean in the earliest Tertiary. In *The Carbon Cycle and Atmospheric* CO_2*: Natural Variations Archean to Present,* edited by W. S. Broecker and E. T. Sundquist, 487–92. Geophysical Monograph 32. Washington, DC: American Geophysical Union.

Stott, L. D., and J. P. Kennett. 1989. New constraints on early Tertiary palaeoproductivity from carbon isotopes in foraminifera. *Nature* 342 (6249): 526–29.

NATURE VOL. 337 5 JANUARY 1989 LETTERS TO NATURE 61

Geochemical evidence for suppression of pelagic marine productivity at the Cretaceous/Tertiary boundary

James C. Zachos*, Michael A. Arthur* & Walter E. Dean†

* Graduate School of Oceanography, University of Rhode Island, Narragansett, Rhode Island 02882, USA
† US Geological Survey, PO Box 25046, Denver, Colorado 80225, USA

The normal, biologically productive ocean is characterized by a gradient of the $^{13}C/^{12}C$ ratio from surface to deep waters. Here we present stable isotope data from planktonic and benthic microfossils across the Cretaceous/Tertiary boundary in the North pacific, which reveal a rapid and complete breakdown in this biologically mediated gradient. The fluxes of barium (a proxy for organic carbon) and $CaCO_3$ also decrease significantly at the time of the major marine plankton extinctions. The implied substantial reduction in oceanic primary productivity persisted for ~0.5 Myr before the carbon isotope gradient was gradually re-established. In addition, the stable isotope and preservational data indicate that environmental change, including cooling, began at least 200 kyr before the Cretaceous/Tertiary boundary, and a peak warming of ~3 °C occurred 600 kyr after the boundary event.

The section at DSDP Site 577 on the Shatsky Rise (32°26′ N, 157°43′ E) was collected by hydraulic piston core and yielded the first continuous, undisturbed sequence of pure nannofossil ooze (>90% $CaCO_3$) across the Cretaceous/Tertiary (K/T) boundary. Because of the relatively shallow burial depth of the K/T boundary sequence (<110 m below the sea floor (m.b.s.f.)[1]), the calcareous microfossils from Site 577 have not undergone significant recrystallization related to burial[2-4]. The position of the K/T boundary is placed at the last appearance of planktonic foraminifera of the *Abathomphalus mayaroensis* zone[4] and first appearance of calcareous nannofossils of the *Marcalius inversus* zone[2], which also coincides with peak abundance of magnetite-glauconite spherules at 109.09 m.b.s.f. (refs 4, 5). A peak irridium anomaly of 61 ng cm^{-2} also occurs at the boundary, as defined by nannofossils, in adjacent Hole 577B (ref. 6). The palaeodepth of Site 577 is estimated as having been 2,400 m at the time of the K/T boundary[3]. The record at DSDP Site 577 is adequate to resolve the nature of environmental changes in pelagic settings that occurred in conjunction with the K/T boundary event, and provides important constraints on hypotheses for extinction.

To reconstruct changes in the surface-to-deep-water carbon isotope gradient of total dissolved carbon (TDC), stable-isotope analyses were conducted on fine fraction (<63-μm) carbonate and several species of planktonic and benthic foraminifera in closely spaced samples over a 10-m interval spanning the K/T boundary[3,4,7] (Fig. 1). The $\delta^{13}C$ of fine-fraction carbonate (primarily calcareous nannoplankton) and of planktonic foraminifera from this site decreases by about 1.5‰ at the K/T boundary relative to values just below the boundary. The $\delta^{13}C$ values of biserial benthic foraminifera *Aragonia* and of trochospiral benthic foraminifera *Nuttallides*, *Gavelinella* and *Gyroidinoides* show an enrichment across the boundary converging with $\delta^{13}C$ values of both planktonic foraminifera and fine-fraction carbonate just above the boundary (Fig. 2).

Convergence of $\delta^{13}C$ values of planktonic foraminifera, nannoplankton (fine-fraction) and benthic foraminifera indicates that the biologically mediated gradient in carbon isotopes between the surface and deep waters, which is characteristic of productive oceans, effectively disappeared at the K/T boundary. Such conditions have been termed a 'Strangelove' ocean on the basis of predictions for events following a mass extinction[8,9]. The loss of a surface-to-deep-water carbon isotope gradient could only be caused by a major reduction in primary production or by extremely rapid rates of oceanic turnover. At Site 577 the lack of a carbon isotope gradient suggests that a poorly productive ocean persisted for at least 0.5×10^6 yr following the K/T boundary event (Fig. 2). The gradual development of a surface-to-deep-water $\delta^{13}C$ gradient in the period following this interval probably represents a progressive increase in primary production.

The changing offset of $\delta^{13}C$ between biserial and trochospiral

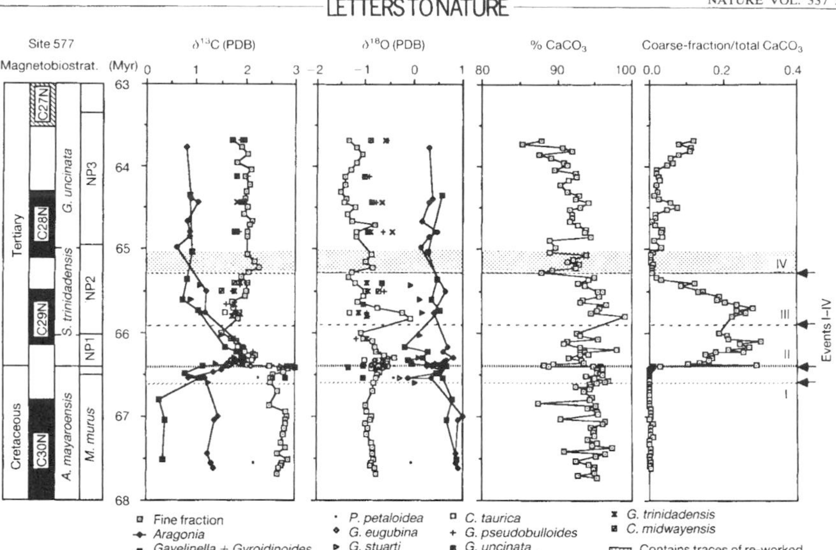

Fig. 1 Age plots of carbon and oxygen stable isotope compositions of fine-fraction carbonate (<63 μm), benthic foraminifer *Aragonia*, trochospiral genera *Gavelinella + Gyroidinoides*, and *Nuttallides*, various species of planktonic foraminifera, and carbonate content of whole rock samples. All stable isotope data are expressed in the delta notation, δ, in ‰ relative to the PDB marine carbonate standard[27]. The final column shows the weight ratio (%) of coarse fraction (>63 μm) to total carbonate. Age determinations were made using available magnetostratigraphy[28] and nannofossil stratigraphy[13] (presented on the far left) and the most recently published timescale[29]. Four separate events are recognized in the evolution of the stable isotope and carbonate content records and are denoted by the four dashed lines. The shaded area represents a portion of the sequence which contains re-worked Cretaceous fossils. The gap in sampling from 65.8 to 65.95 Myr encompasses the interval of an organic geochemistry core removed on board ship.

benthic foraminiferal genera (Fig. 1) is particularly interesting because of the observation that many species of modern benthic foraminifera do not reside directly on the sea floor but instead within the upper few centimetres of sediment, and therefore secrete their tests in carbon isotope equilibrium with the surrounding pore waters rather than with bottom waters[10]. *In situ* decay of organic matter in the upper few centimetres of sediment produces a pore-water $\delta^{13}C$ gradient with a magnitude which is, in part, a function of the accumulation rate of organic carbon (C_{org})[11]. Convergence of $\delta^{13}C$ values of *Aragonia* and that of *Gavelinella + Gyroidinoides* across the K/T boundary may indicate that the pore-water $\delta^{13}C$ gradient is eliminated as a result of the reduction in one C_{org} flux to sediments. The absence of benthic foraminiferal carbon isotope differences during the recovery period suggests that new C_{org} flux levels were lower than that of the latest Cretaceous.

Other evidence which suggest earliest Palaeocene reduced primary productivity involves carbonate and barium mass-accumulation-rate data. In the absence of changes in the rate of dissolution, the rate of accumulation of biogenic carbonate should be proportional to production in surface water. Carbonate accumulation rates decline on average by a factor of four across the K/T boundary in all pelagic marine sequences examined[3]. In general, preservation of calcareous microfossils improves across the K/T boundary at most localities[4,12]. Therefore, the decrease in carbonate accumulation rates from Maestrichtian to the early Palaeocene cannot be easily attributed to increased rates of dissolution and must have resulted from a decrease in the rate of $CaCO_3$ production in surface waters.

The distribution of dissolved barium in the modern oceans is nutrient-like and appears to be mediated by biological activity[13,14]. In the modern Pacific Ocean, Ba and organic-carbon sediment-trap fluxes are commonly proportional[15]. Because sediment pore waters are near saturation with respect to barite[16], barite is typically preserved in sediments whereas C_{org} is oxidized. Thus, Ba accumulation rates can be used as a proxy for C_{org} flux to the sediments[17]. At Site 577, Ba accumulation rates decrease from >0.2 to <0.1 mg cm^{-2} per 10^3 yr across the K/T boundary (Fig. 3), reflecting the decreased rain rate of C_{org}.

We have recognized four discrete episodes in the evolution of the stable-isotope and carbonate records across the K/T boundary and in the early Palaeocene at Site 577. The first of these events begins at least 10^5 yr before the main plankton extinctions and the last occurs nearly 1.5×10^6 yr later. We believe that these events represent significant changes in oceanic fertility, circulation and climate, which are accompanied by consequent effects on biotic evolution.

Event I (66.6 Myr), which precedes the K/T boundary event, is marked by enrichment in both planktonic and benthic microfossil carbon isotope records, accompanied by a slight increase in carbonate content and an improvement in microfossil preservation. In addition, Ba accumulation rates decrease. The increase in $\delta^{13}C$ of both the planktonic and benthic records probably indicates a change in $\delta^{13}C$ of the oceanic TDC reservoir. This could have resulted from increased organic-carbon burial on continental margins (or decreased carbon fluxes from weathering) that are related to the latest Cretaceous transgression, which began at about 66.7 Myr (ref. 18). In addition, there may have been changes in deep-water sources, increased oxygenation of deep waters and/or lower organic-carbon flux at Site 577 at this time. Enriched fine-fraction $\delta^{18}O$ values also indicate cooling of surface waters, this suggests that the rate of oceanic turnover and ventilation of the deep reservoir increased before the K/T boundary as the result of global cooling and intensification of thermohaline-driven circulation. The increase in relative proportions of the temperate planktonic foraminiferal

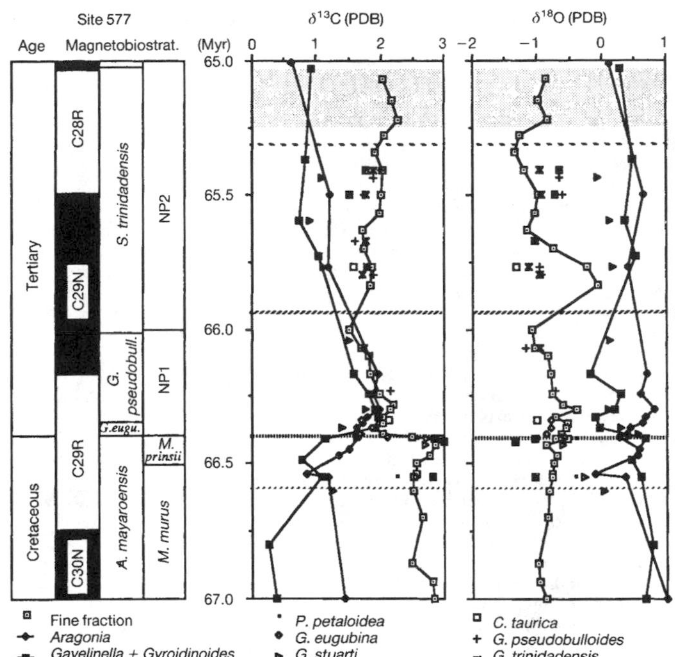

Fig. 2 Expanded 2-Myr record across the K/T boundary of the stable carbon and oxygen isotope data from Fig. 1. Shaded area represents a zone containing traces of re-worked Cretaceous microfossils.

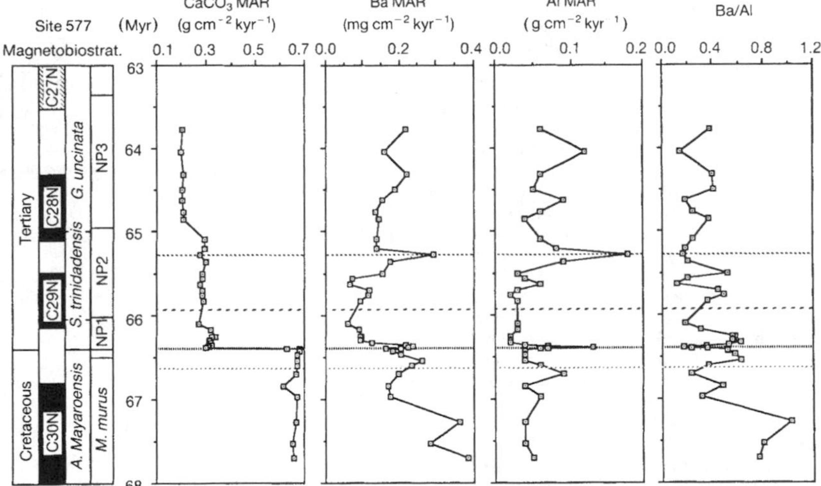

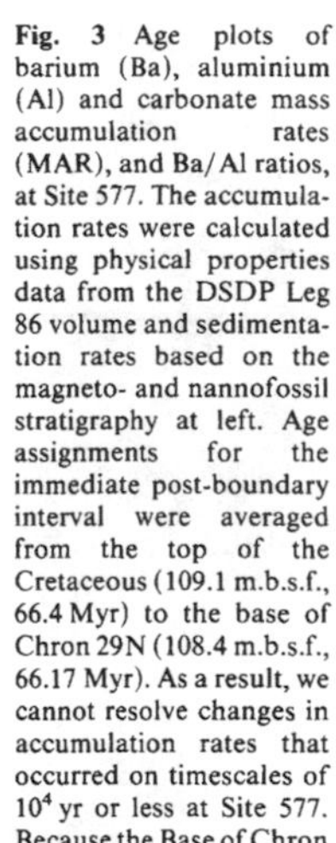

Fig. 3 Age plots of barium (Ba), aluminium (Al) and carbonate mass accumulation rates (MAR), and Ba/Al ratios, at Site 577. The accumulation rates were calculated using physical properties data from the DSDP Leg 86 volume and sedimentation rates based on the magneto- and nannofossil stratigraphy at left. Age assignments for the immediate post-boundary interval were averaged from the top of the Cretaceous (109.1 m.b.s.f., 66.4 Myr) to the base of Chron 29N (108.4 m.b.s.f., 66.17 Myr). As a result, we cannot resolve changes in accumulation rates that occurred on timescales of 10^4 yr or less at Site 577. Because the Base of Chron 30N (C30N) or the *M. murus* zone were not reached at Hole 577, the thickness of the *M. murus* zone (8.48 m) was extrapolated from adjacent Hole 577A. The Ba/Al ratio represents weight ratio of Ba to Al_2O_3.

genus *Heterohelicidae* at this low latitude provides additional support for cooling before the K/T boundary event[4].

Event II: K/T boundary (66.4 Myr), the main plankton extinction and productivity crisis, features the rapid collapse of the surface-to-deep-water carbon isotope gradient and decreases in carbonate and Ba accumulation rates described above. A puzzling aspect of this event is the 100-fold increase recorded in the concentration of foraminifera relative to total carbonate (Fig. 1). There are three possible explanations for this phenomenon: (1) intensified deep circulation resulting in winnowing of the fine fraction; (2) less fragmentation and improved preservation of the more dissolution-susceptible planktonic foraminifera as a result of deepening of the carbonate compensation depth (CCD) and/or lowered rates of *in situ* dissolution brought about by decreased C_{org} decay within sediment pore waters; and (3) an ecological trophic-structure response to the extinctions and environmental change in which calcareous nannoplankton are replaced by siliceous or organic-walled phytoplankton. We favour preservation changes (2) as the possible origin of the increase in foraminifera concentrations. It is generally accepted that coccoliths tend to be more dissolution-resistant than foraminifera and that calcite dissolution above the calcite saturation horizon is driven mainly by titration by metabolic CO_2 derived from C_{org} decay at or near the sediment-water interface[19]. Dissolution at the shallow palaeodepth of the Shatsky Rise near site 577 would have been controlled to a large extent by metabolic CO_2 production. Decay of C_{org} at the sediment-water interface would decline as a result of reduced C_{org} flux during the 'Strangelove' ocean event making bottom and pore water less corrosive to calcite. Furthermore, taking into account the large initial decrease in $CaCO_3$ accumulation that occurred as a result of the productivity crisis, and assuming that riverine input of inorganic C and alkalinity remained constant, one would expect a rapid deepening of the CCD and much improved preservation of foraminifera across the K/T boundary. Thus we explain the increased foraminiferal flux in the Palaeocene, implied by the higher relative proportion of foraminifers, by enhanced preservation.

Event III (65.9 Myr) is characterized by the gradual return of the surface-to-deep-water carbon isotope gradient, a slight increase in Ba accumulation rates and a gradual decrease in the planktonic foraminifera concentrations. These changes indicate that surface-water productivity began to increase ~0.5 Myr after the K/T boundary event, albeit at relatively low levels. Climate fluctuations of several degrees are indicated by variations in nannofossil and foraminiferal $\delta^{18}O$. Rates of evolution of planktonic forams remained high[20], suggesting continuing environmental instability.

Event IV (65.3 Myr) signals the full expression of the surface-to-deep-water carbon isotope gradient and a decrease in planktonic foraminiferal concentrations to normal values. Aluminium accumulation rates also increased at this time. Because Al accumulation rates are probably a reflection of flux of aeolian detritus to this site[21], the increased Al accumulation rates may represent climate change at this time and possibly an increase in wind strength. Environmental stability is indicated by a nearly constant $\delta^{13}C$ gradient, relatively stable $\delta^{18}O$ values and low rates of change of planktonic foraminiferal faunas in the *G. uncinata* zone[20].

The extinction of phyto- and zooplankton and disappearance of the carbon isotope gradient is abrupt and occurs over a period of less than 10^4 yr. It is unlikely that a gradual environmental change occurring over a longer period of time (that is, 10^5–10^6 yr) could have resulted in such a rapid and extreme collapse of the marine ecosystem unless some as yet unknown environmental threshold was exceeded[22]. The instantaneous environmental stress that would be brought about by a bolide impact provides a possible explanation for the collapse of primary productivity and its coincidence with an abundance of spherules and platinum-group metals at the K/T boundary, as exhibited at Site 577. It fails to explain, however, why primary productivity remained low for such a long period after the event.

It is possible that the 0.5-Myr 'Strangelove' ocean may represent the normal recovery time for the biosphere following a sudden mass extinction event. An alternative explanation is that the slow recovery of primary productivity was the result of climate and oceanic circulation instability. The rate of speciation was extremely high for the first several 10^6 years above the K/T boundary at Site 577, with one assemblage of planktonic foraminifera quickly replacing another[20]. Because the distributions of planktonic foraminifera are limited to particular water masses, the high turnover rate of assemblages suggests that surface-water conditions changed rapidly during the Danian. As a consequence, an efficient, highly structured ecosystem may have been incapable of developing. The origin of this climatic instability is uncertain but may prove to be related to feedback mechanisms involving the global carbon cycle[9,23-25]. Models predict that, following a rapid extinction event, atmospheric p_{CO_2} would have increased by as much as a factor of two or three on a timescale of several oceanic mixing cycles[26], as the result of nearly complete cessation of oceanic productivity. Such increases in p_{CO_2} would be expected to result in global warming. The nannofossil data, which represent the most complete time series, suggest a gradual 3 °C warming of surface waters during the initial 0.5 Myr of the Palaeocene.

We thank D. McLean for comments on this manuscript, and M. Bender, K. Caldeira, B. Corliss, D. Fastovsky, A. Mackensen, D. McCorkle, K. Miller, C. Officer, M. Rampino and R. Thunell for discussions. Special thanks are extended to J. Gerstel for assistance during sample preparation. This research was supported by the NSF.

Received 11 April; accepted 9 November 1988.

1. Heath, G. R. *et al. Init. Rep. DSDP* **86** (1985).
2. Monechi, S. *Init. Rep. DSDP* **86,** 301–306 (1985).
3. Zachos, J. C. & Arthur, M. A. *Paleoceanography* **1,** 5–26 (1986).
4. Gerstel, J., Thunell, R. C., Zachos, J. C. & Arthur, M. A. *Paleoceanography* **1,** 97–117 (1986).
5. Smit, J. & Romein, A. J. T. *Earth planet. Sci. Lett.* **74,** 155–170 (1985).
6. Michel, H. V., Asaro, F., Alvarez, W. & Alvarez, L. W. *Init. Rep. DSDP* **86,** 533–538 (1985).
7. Zachos, J. C., Arthur, M. A., Thunell, R. C., Williams, D. F. & Tappa, E. J. *Init. Rep. DSDP* **86,** 513–532 (1985).
8. Broecker, W. S. & Peng, T. H. *Tracers in the Sea* (ELDIGIO, New York, 1982).
9. Hsu, K. J. & McKenzie, J. A. in *The Carbon Cycle & Atmospheric CO_2: Natural Variations Archean to Present* (eds Sundquist, E. T. & Broecker, W. S.) 487–492 (Am. Geophys. Un., Washington DC, 1985).
10. Corliss, B. H. *Nature* **314,** 435–438 (1985).
11. McCorkle, D. C., Emerson, S. R. & Quay, P. D. *Earth planet. Sci. Lett.* **74,** 13–26 (1985).
12. Thierstein, H. R. *Soc. Econ. Paleontol. Miner. spec. Publn* **32,** 355–394 (1981).
13. Dehairs, F., Lambert, C. E., Chesselet, R. & Risler, N. *Biogeochemistry* **4,** 119–139 (1987).
14. Chan, L. H., Drummond, D., Edmond, J. M. & Grant, B. *Deep Sea Res.* **24,** 613–649 (1977).
15. Dymond, J. *Am. Geophys. Un. Ocean Sci Mtg* (1986).
16. Church, T. M. in *Reviews in Mineralogy*, Vol. 6 (ed. Burns, R. G.) 175–209 (Miner. Soc. Am., Washington DC, 1976).
17. Schmitz, B. *Paleoceanography* **2,** 63–78 (1987).
18. Haq, B. U., Hardenbol, J. & Vail, P. R. *Science* **235,** 1156–1167 (1987).
19. Emerson, S. E., Bender, M. A. *J. Mar. Res.* **39,** 139–162 (1981).
20. Gerstel, J., Thunell, R. & Ehrlich, R. *Geology* **15,** 665–668 (1987).
21. Doh, S-J., King, J. W. & Leinen, M. *Paleoceanography* **3,** 89–112 (1988).
22. Crowley, T. S. & North, G. R. *Science* **240,** 996–1002 (1988).
23. McLean, D. M. *Cret. Res.* **6,** 235–259 (1985).
24. Kasting, J. F., Richardson, S. M., Pollack, J. B. & Toon, O. B. *Am. J. Sci.* **286,** 361–389 (1986).
25. Rampino, M. & Volk, T. *Nature* **332,** 63–65 (1988).
26. Caldeira, K. G., Rampino, M. R. & Zachos, J. C. *Eos* **69,** 377 (1988).
27. Craig, H. *Geochim. cosmochim. Acta.* **12,** 133–149 (1955).
28. Bleil, U. *Init. Rep. DSDP* **86,** 441–458 (1985).
29. Berggren, W. A., Kent, D. V., Flynn, J. J. & Van Couvering, J. A. *Bull. geol. Soc. Am.* **96,** 1407–1418 (1985).

PAPER 27

Expansion of C_4 Ecosystems as an Indicator of Global Ecological Change in the Late Miocene (1993)

T. E. Cerling, Y. Wang, and J. Quade

Commentary

DAVID L. FOX

Unraveling the origins of terrestrial ecosystems dominated by C_4 grasses occupies the attentions of a diverse community of Earth and life scientists. The brief paper by Cerling et al. included in this volume acts as something of a synecdoche for the large body of research by Cerling and many of his students, including both junior authors on this paper, into the geological history of C_4 grasses using the stable carbon isotope composition of fossil mammal teeth and paleosols. The approach, pioneered by Cerling for paleosols, takes advantage of differences in carbon isotopic fractionation during photosynthesis by C_3 and C_4 plants, which are recorded in the tissues of primary consumers and the organic matter and authigenic minerals of soils. This particular paper is important in the development of both paleoecology and research on C_4 grasslands in several regards. First, it linked geographically distinct but coeval regional records (North America, South Asia), demonstrating for the first time that C_4 grasses increased in ecological importance globally only during the late Neogene. Second, it proposed an elegant hypothesis—decreased atmospheric CO_2—to account for the apparently synchronous increase in C_4 biomass beginning around 8 Ma, linking local paleoecological records to global change. A later paper expanded this hypothesis into an even more elegant Earth-systems concept that called on chemical weathering in the recently uplifted Himalayas to decrease atmospheric CO_2 and drive the expansion of C_4 grasslands. Third, and probably most importantly, this paper, along with others from the Cerling research group, inspired both additional investigations into the geological history of C_4 grasslands and integration of paleoecology into the broad, multidisciplinary study of the evolution of C_4 grasses and grasslands. These more recent studies reveal a more complex story, with no simple, synchronous increase in C_4 biomass in all regions, multiple evolutionary origins of C_4 photosynthesis among grass lineages long before any came to dominate specific grasslands, and atmospheric CO_2 levels that appear to have been close to preindustrial concentrations since the early Miocene. While no longer correct in every detail, this paper beautifully demonstrates how local paleoecological data can be connected to global phenomena.

Reprinted by permission from Macmillan Publishers Ltd: *Nature* 361:344–45, copyright 1993. https://doi.org/10.1038/361344a0.

LETTERS TO NATURE

Expansion of C4 ecosystems as an indicator of global ecological change in the late Miocene

Thure E. Cerling*, Yang Wang* & Jay Quade†

* Department of Geology and Geophysics, University of Utah, Salt Lake City, Utah 84112, USA
† Department of Geosciences, University of Arizona, Tucson, Arizona 85705, USA

THE most common and the most primitive pathway of the three different photosynthetic pathways used by plants is the C3 pathway, or Calvin cycle, which is characterized by an initial CO_2 carboxylation to form phosphoglyceric acid, a 3-carbon acid. The carbon isotope composition ($\delta^{13}C$) of C3 plants varies from about −23 to −35‰[1-3] and averages about −26‰. Virtually all trees, most shrubs, herbs and forbs, and cool-season grasses and sedges use the C3 pathway. In the C4 pathway (Hatch–Slack cycle), CO_2 initially combines with phosphoenol pyruvate to form the 4-carbon acids malate or aspartic acid, which are translocated to bundle sheath cells where CO_2 is released and used in Calvin cycle reactions[1-4]. The carbon isotope composition of C4 plants ranges from about −10 to −14‰, averaging about −13‰ for modern plants[1-3]. Warm-season grasses and sedges are the most abundant C4 plants, although C4 photosynthesis is found in about twenty families[5]. The third photosynthetic pathway, CAM, combines features of both C3 and C4 pathways. CAM plants, which include many succulents, have intermediate carbon isotope compositions and are also adapted to conditions of water and CO_2 stress. The modern global ecosystem has a significant component of C4 plants, primarily in tropical savannas, temperate grasslands and semi-desert scrublands. Studies of palaeovegetation from palaeosols and palaeodiet from fossil tooth enamel indicate a rapid expansion of C4 biomass in both the Old World and the New World starting 7 to 5 million years ago. We propose that the global expansion of C4 biomass may be related to lower atmospheric carbon dioxide levels because C4 photosynthesis is favoured over C3 photosynthesis when there are low concentrations of carbon dioxide in the atmosphere.

Fossil soils and fossil tooth enamel are important indicators of the presence of C4 biomass in local ecosystems. The carbon isotope composition of soil carbonate is related to the isotope composition of the local biomass, being enriched in ^{13}C by about 14 to 17‰ compared to the local biomass[6,7]; the carbon isotope composition of tooth enamel is enriched in ^{13}C by about 12‰ compared to the diet[8,9]. Palaeosol (fossil soil) carbonate and fossil tooth enamel are robust indicators of palaeoenvironment[10-13] and palaeodiet[14-16], respectively, and have been shown to retain their isotope composition through diagenesis[17,18]. The large range in the $\delta^{13}C$ values for C3 plants (where $\delta^{13}C = [(^{13}C/^{12}C)\ \text{sample}/(^{13}C/^{12}C)_{PDB} - 1] \times 1{,}000$ and PDB is the stable isotope reference standard) results in some ambiguity in interpretation of the $\delta^{13}C$ values for pedogenic carbonate and tooth enamel. Under conditions of moisture stress, the isotope composition of C3 plants can be several ‰ enriched in ^{13}C (refs 3, 19, 20), so that ecosystems with $\delta^{13}C$ values as high as −23‰ could be essentially pure C3 ecosystems, with resultant $\delta^{13}C$ of soil carbonate of −8‰ and tooth enamel of −10‰. As many animals are highly selective in their dietary intake, $\delta^{13}C$ values as high as −8 or −9‰ could result from a diet of C3 plants in semi-arid regions that are isotopically enriched in ^{13}C compared to more mesic ecosystems. $\delta^{13}C$ values of −20‰ have been reported for C3 plants growing in arid environments where stomatal conductance is limited[21].

Palaeosols from Pakistan show a very well defined change in the isotope composition of palaeosol carbonate from about −9 to −12‰ beginning at about 7.4 Myr, to −2 to +2‰ by about 5 Myr[12]. (Figure 1 shows data from ref. 12 as a five-point running average, with a sampling density of about 1 palaeosol per 130,000 years over 17 Myr.) The palaeosol data from Pakistan indicate a virtually pure C3 ecosystem before 7 Myr, which was replaced by an ecosystem dominated by C4 biomass by 5 Myr. It is unlikely that more than a small fraction of the biomass could have been using the C4 or CAM photosynthetic pathways before 7 Myr. Because of the possibility of C3 plants being isotopically enriched in semi-arid environments, which are those in which pedogenic carbonate formation is favoured[22,23], even the most positive $\delta^{13}C$ values for Pakistan palaeosols before 7.5 Myr ago represent essentially pure C3 ecosystems.

The isotope composition of fossil tooth enamel from mammals from the Siwaliks in Pakistan and from horses in western North America indicate that C4 plants were an important part of the diet starting between 7 and 6 Myr ago (Fig. 1), which is the same time that C4 biomass became important in Pakistan. Major faunal turnover, including the local extinction of hominoids, occurred in the Siwaliks between 7 and 5 Myr ago[24-27]. The shift in the $\delta^{13}C$ of enamel from North American horse teeth does not correspond to the noted change in hypsodonty that occurred in the early to middle Miocene[28]. This observation will require a revision in the interpretation of mammalian dentition and its relationship to the spread of savanna ecosystems in the Miocene. Preliminary studies of palaeosol carbonate from East Africa suggest that C4 biomass was of minor importance in the Miocene, but expanded significantly in the Pliocene and Pleistocene[13].

The palaeosol carbonate and fossil tooth enamel data suggest that the expansion of C4 ecosystems took place rapidly between 7 and 5 Myr ago in both the New World and the Old World. Synchronous expansion of C4 ecosystems in both the New World and the Old World suggests a change in global conditions, rather than local development and gradual expansion around the world. We suggest that the change may have been due to a decrease in atmospheric CO_2 levels. The C4 pathway is an adaptation to reduce photorespiration that occurs in C3 plants. Under conditions of increased CO_2, C3 photosynthesis is more efficient than C4 or CAM photosynthesis[4]. But when the CO_2 concentration in intercellular gases in C3 plants falls below about 400 p.p.m.v., the CO_2/O_2 ratio decreases sufficiently to increase photorespiration rates, reducing the inherent advantage of C3 plants over C4 plants in carbon fixation[4]. At sufficiently low concentrations of atmospheric CO_2, combined with high temperatures and/or moisture stress, C4 plants are more efficient than C3 plants[4]. Therefore a rapid global expansion of C4 ecosystems may have been a result of atmospheric CO_2 falling below a critical threshold for efficient C3 photosynthesis, probably around 400 to 500 p.p.m.v. Previous studies on the palaeosols indicates that the atmospheric CO_2 during the Miocene and Pliocene was less than 800 p.p.m.v. (ref. 29). The midpoint of this rapid change in global terrestrial ecosystems, about 6.3 Myr, is also a period when the carbon isotope composition of the oceans changed significantly[30-32]. An increase the proportion of C4 biomass or a decrease in the total terrestrial biomass would result in an isotope shift in the direction observed in the oceans. The near-temporal coincidence in the changes in the isotope composition of the terrestrial and marine carbon pools may be further evidence that a critical threshold was passed that affected the carbon budget of the biosphere.

Previously we attributed the dramatic shift in $\delta^{13}C$ of palaeosol carbonate to an intensification of the Asian monsoon[12], based in part on the less dramatic $\delta^{18}O$ shift in the isotope composition of palaeosol carbonate that slightly preceeded the $\delta^{13}C$ shift. It now appears that the changes observed in the sediments of the Siwaliks are related to global changes in both the carbon budget and the meteoric water cycle, which do not preclude a link to the Asian monsoon system.

The modern global ecosystem includes tropical savannas, temperate grasslands, and semi-desert shrublands that have abundant C4 biomass. Our results indicate that C4 plants underwent significant expansion starting about 7 Myr ago in both the Old World and the New world. Although our results do not preclude C3 grasslands before 7 Myr ago, modern C3 grasslands are restricted to regions with cool growing seasons, such as montane or boreal conditions, or when the growing season is in the spring or winter (as in the Mediterranean climate). It is possible that C3 temperate and tropical grasslands could have existed under CO_2 conditions significantly higher than modern values. However, modern ecosystems with significant C4 biomass seem to be a feature of the late Neogene to present. □

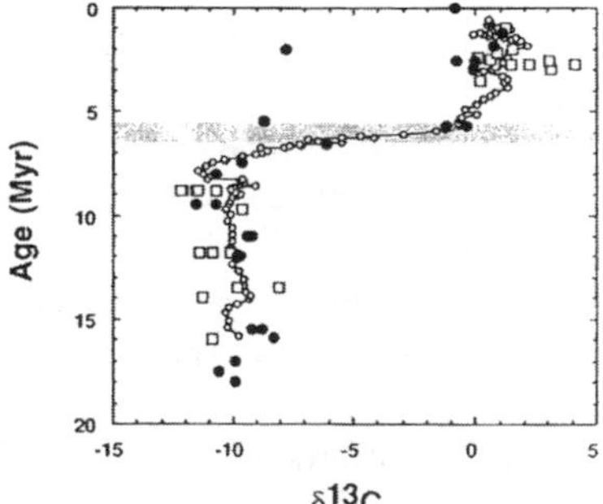

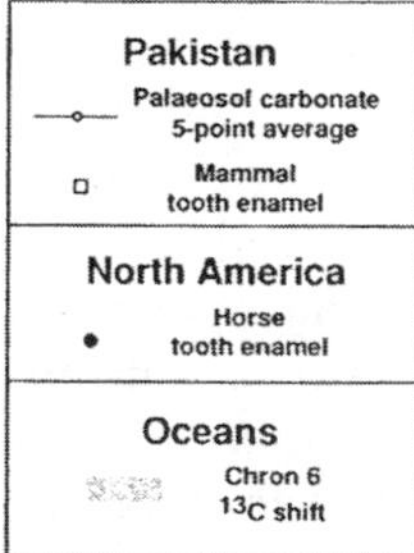

FIG. 1 Isotope transitions in North America and in Pakistan and the late Miocene carbon isotope shift in the oceans. Palaeosols from Pakistan (5-point running average shown as connected circles)[12], tooth enamel from Pakistan (squares; ref. 16, and our unpublished results) and North America (filled circles). Each shows a 10-12‰ shift between 7 and 5 Myr. The age of the marine carbon shift[30-32] is shown as a dotted area.

Received 1 September; accepted 30 November 1992.

1. Deines, P. in *Handbook of Environmental Isotope Geochemistry. 1. The Terrestrial Environment* (eds Fontes, J. C. & Fritz, P.) 329-406 (Elsevier, Amsterdam, 1980).
2. O'Leary, M. H. *Bioscience* **38**, 325-326 (1988).
3. Farquhar, G. D., Ehleringer, J. R. & Hubik, K. T. *An. Rev. Plant Physiol. Plant molec. Biol.* **40**, 503-537 (1989).
4. Ehleringer, J. R., Sage, R. F., Flanagan, L. B. & Pearcy, R. W. *Trends Ecol. Evol.* **6**, 95-97 (1991).
5. Smith, B. N. in *CRC Handbook of Biosolar Resources* (ed. Zaborsky, O. R.) 99-118 (CRC Press, Baton Rouge, 1982).
6. Cerling, T. E., Quade, J. & Bowman, J. R. *Nature* **341**, 138-139 (1989).
7. Quade, J., Cerling T. E. & Bowman, J. R. *Geol. Soc. Am. Bull.* **101**, 464-475 (1989).
8. Sullivan, C. H. & Krueger, H. W. *Nature* **292**, 333-335 (1981).
9. Lee-Thorp, J. A. & van der Merwe, N. J. *South Afr. Jour. Sci.* **83**, 712-713 (1987).
10. Cerling, T. E. & Hay, R. L. *Quat. Res.* **25**, 63-78 (1986).
11. Cerling, T. E., Bowman, J. R. & O'Neil, J. R. *Palaeogeogr. Palaeoclimat. Palaeoecol.* **63**, 335-356 (1988).
12. Quade, J., Cerling, T. E. & Bowman, J. R. *Nature* **342**, 163-165 (1989).
13. Cerling, T. E. *Palaeogeogr. Palaeoclimat. Palaeoecol.* **97**, 241-247 (1992).
14. Lee-Thorp, J. A., van der Merwe, N. J. & Brain, C. K. *J. hum. Evol.* **18**, 183-190 (1989).
15. Thackerey, J. F. *et al. Nature* **347**, 751-753 (1990).
16. Quade, J. *et al. Chem. Geol.* **94**, 183-194 (1992).
17. Koch, P. L., Zachos, J. C. & Gingerich, P. D. *Nature* **358**, 319-322 (1992).
18. Cerling, T. E. *Am. J. Sci.* **291**, 377-400 (1991).
19. Ehleringer, J. R., Field, C. B., Lin, Z. F. & Kuo, C. Y. *Oecologia* **70**, 520-526 (1986).
20. Ehleringer, J. R. & Cooper, T. A. *Oecologia* **76**, 562-566 (1988).
21. Delucia, E. H., Schlesinger, W. H. & Billings, W. D. *Ecology* **69**, 303-311 (1988).
22. Birkeland, P. W. *Soils and Geomorphology* (Oxford Univ. Press, New York, 1984).
23. Jenny, H. *The Soil Resource* (Springer, Berlin, 1980).
24. Barry, J. C., Lindsay, E. H. & Jacobs, L. L. *Palaeogeogr. Palaeoclimat. Palaeoecol.* **37**, 95-130 (1982).
25. Flynn, L. J. & Jacobs, L. L. *Palaeogeogr. Palaeoclimat. Palaeoecol.* **33**, 129-138 (1982).
26. Barry, J. C., Johnson, N. M., Raza, S. M. & Jacobs, L. L. *Geology* **13**, 637-640 (1985).
27. Barry, J. C. & Flynn, L. J. in *European Neogene Mammalian Chronology* (ed. Lindsay, E. H.) 557-571 (Plenum, New York, 1990).
28. MacFadden, B. J. *Fossil Horses: Systematics, Paleobiology and Evolution of the Family Equidae* (Cambridge Univ. Press, New York, 1992).
29. Cerling, T. E. *Global Biogeochem.* **6**, 307-314 (1992).
30. Keigwin, L. D. *Earth planet. Sci. Lett.* **45**, 361-382 (1979).
31. Hodell, D. A. & Kennett, J. P. *Paleoceanography* **1**, 285-311 (1986).
32. Hodell, D. A., Benson, R. H., Kennett, J. P. & Bied, K. R. *Paleoceanography* **4**, 467-482 (1989).

ACKNOWLEDGEMENTS. We thank T. M. Bown, D. J. Bryant, E. H. Lindsay, B. J. MacFadden and D. Winkler for assistance in obtaining specimens from North American fossil horses, J. C. Barry, A. K. Behrensmeyer and the Geological Survey of Pakistan for logistical support in Pakistan, and J. R. Ehleringer for discussions and use of laboratory facilities. Work on palaeosols was supported by the National Science Foundation. Field work in Pakistan was supported by the Smithsonian Institution Foreign Currency Program.

PAPER 28

A Paleobotanical Interpretation of Tertiary Climates in the Northern Hemisphere (1978)

J. A. Wolfe

Commentary

PETER WILF

Jack Wolfe was one of the most innovative paleobotanists, and his influence extended far outside his field. He was a highly regarded expert on Cretaceous and Cenozoic floras and paleoclimates as well as living vegetational types and biomes. Wolfe played a starring role in the great modernization of angiosperm paleobotany during the 1960s and 1970s. His major focus was the vast scientific potential of fossil angiosperm leaves, which he advanced in two major ways. First, Wolfe exhaustively studied how leaf shape and venation vary among living angiosperms and applied this knowledge to the description and analysis of fossils. Second, highlighted here, he devoted much of his career to a massive, quantitative exploration of how leaf-shape traits vary with climate and how this variation can be applied to infer paleoclimates (a core topic for paleoecology). There are many examples of this variation, the best known being that proportionally more species in a flora have toothed (jagged-edged) leaves in cooler climates than in warmer climates. By expanding the modern reference data and developing the statistical toolkit over several decades, Wolfe took leaf paleoclimatology from qualitative inference to quantitative science, and numerous other workers followed his lead. Through dozens of papers based on hundreds of fossil floras, Wolfe and colleagues made pioneering studies of climate change and biotic response, climate change during critical intervals, paleolatitudinal temperature gradients, marine-terrestrial climate correlations, paleoelevation history, and much more (Wolfe 1995). Thus, Wolfe made paleobotany into an integral part of mainstream paleoclimatology.

This paper is a snapshot of Wolfe's paleoclimate research during the late 1970s. It remains heavily cited because it contains famous graphs showing some of the first quantitative estimates of Cenozoic climate change on land, including regional variation within North America, all presented with lively and detailed discussion. The inferences have been refined over the years but remain fundamentally accepted. Wolfe rightly highlights his major role in recognizing many important trends, especially the rapidity of global cooling and concurrent vegetational change during the Eocene-Oligocene transition. Prior to the widespread recognition of greenhouse-gas drivers, Wolfe speculates here about other forcing mechanisms.

The core assumption of Wolfe's paleoclimate work, reiterated here, is that climate is much more important than phylogenetic history in molding leaf shape, thus allowing quantitative paleoclimate estimates based on modern calibration data to be reliably compared across great reaches of time and space. It turns out that leaf traits incorporate significant components of evolutionary history as well as climatic adaptation, and thus the core assumption has not proven correct (Little et al. 2010). Nevertheless, under a fairly broad set of conditions, qualitative comparisons based on leaf

From *American Scientist* 66:694–703.

traits remain valid, such as the inference of regional paleoclimate trends, as shown here. How to improve quantitative paleoclimatology from leaf traits by accommodating phylogenetic signal is probably the next great question arising from Jack Wolfe's legacy.

Literature Cited

Little, S. A., S. W. Kembel, and P. Wilf. 2010. Paleotemperature proxies from leaf fossils reinterpreted in light of evolutionary history. *PLoS One* 5:e15161

Wolfe, J. A. 1995. Paleoclimatic estimates from Tertiary leaf assemblages. *Annual Review of Earth and Planetary Sciences* 23:119–42.

Jack A. Wolfe

A Paleobotanical Interpretation of Tertiary Climates in the Northern Hemisphere

Data from fossil plants make it possible to reconstruct Tertiary climatic changes, which may be correlated with changes in the inclination of the earth's rotational axis

Anyone who has even a slight acquaintance with paleoclimatic literature is well aware that the last 1 to 1.5 million years of the Quaternary have been characterized by major episodic glaciations of the continents of the Northern Hemisphere, and hence the period is atypical of much of geologic time. A commonly accepted thesis on climates preceding Quaternary glaciation is that, from some time in the Late Cretaceous or early Tertiary (some 40–80 m.y. ago), when the earth's climate was characterized by generally higher temperatures and higher equability of temperature than now, both overall temperature and equability have gradually decreased, culminating in Quaternary glaciation. Further, some researchers have maintained that even as long ago as the early Tertiary, temperatures were only moderately higher than now, even at high latitudes. (See Table 1 for the geologic time span dealt with in this article.)

An increasing accumulation of data from a multitude of sources has, however, largely negated such once commonly accepted theses. A significant warm episode during the Miocene (see Fig. 1) was first documented in Europe by Mai (1964) and has subsequently been substantiated in other regions such as Japan (Tanai and Huzioka 1967), western North America (Wolfe and Hopkins 1967; Addicott 1969), and New Zealand (Devereux 1967). Alpine glaciation is known to have begun in Alaska during the Miocene (Denton and Armstrong 1969; Plafker and Addicott 1976), when at least part of the Antarctic ice sheet was also present (Kennett 1977).

The most dramatic climatic event, however, occurred during the middle of the Tertiary. MacGinitie (1953) recognized that, if certain floras in Oregon were as close in time as some stratigraphic evidence indicated, a rapid and major climatic change must have occurred, a decrease in temperature that was considered significant but gradual by Nemjč (1964) and Zhilin (1966). Utilizing newly available radiometric ages, Wolfe and Hopkins (1967) demonstrated that this major climatic deterioration had occurred within 1 or 2 m.y.

This temperature decrease has subsequently been recognized in many regions. I had previously (1971) termed it the "Oligocene deterioration," but since recent work in relating the marine and nonmarine chronologies indicates that, in the widely accepted chronology based on marine plankton, the event occurred at the end of the Eocene, I will refer to it as the "terminal Eocene event." In the Southern Hemisphere, the terminal Eocene event is closely associated with the initiation of cold bottom water in the oceans (Kennett 1977), while on the continents of the Northern Hemisphere the event is emphasized by a major decrease in equability of temperature (Wolfe 1971).

Foliar physiognomy

A thorough review of all pertinent paleoclimatic data for the Tertiary would be a lengthy and prodigious task. In this paper I will largely limit the discussion to the paleoclimatic data based on fossil plants from middle to high latitudes (<30°) of the Northern Hemisphere. For the Paleocene and Eocene, the North American data, which are based on leaf remains, are the most relevant. In Europe, the major Eocene floral sequence is based on fruit and seed floras. In eastern Asia, some Oligocene floras have been described (Tanai 1970), but most of the Paleocene and Eocene assemblages remain undescribed and unanalyzed (Tanai 1967).

There are several advantages in basing paleoclimatic interpretations on

Table 1. The subdivisions of the Cenozoic Era

		Million years ago
Quarternary		
Holocene		.012
Pleistocene		1.5
Tertiary		
Pliocene	Neogene	5
Miocene	Neogene	23
Oligocene	Paleogene	33
Eocene	Paleogene	53
Paleocene	Paleogene	65

In 1957, Erling Dorf delivered the Ermine Cowles Case Memorial Lecture at the University of Michigan, a lecture sponsored by Sigma Xi. His lecture, on Tertiary climates from a paleobotanist's viewpoint, which was later published in American Scientist, *shows some parallels in conclusions with the present article, but much of the basic information Dorf accepted has undergone major revision by subsequent paleobotanical and stratigraphic work. A version of the present paper was also delivered as a Case Lecture in October 1975. Jack A. Wolfe was educated at Harvard and Berkeley and is a geologist at the U.S. Geological Survey. His interests are in Cenozoic floras of western North America and systematics and phylogeny of angiosperms. Address: Paleontology and Stratigraphy Branch, U.S. Geological Survey, 345 Middlefield Road, Menlo Park, CA 94025.*

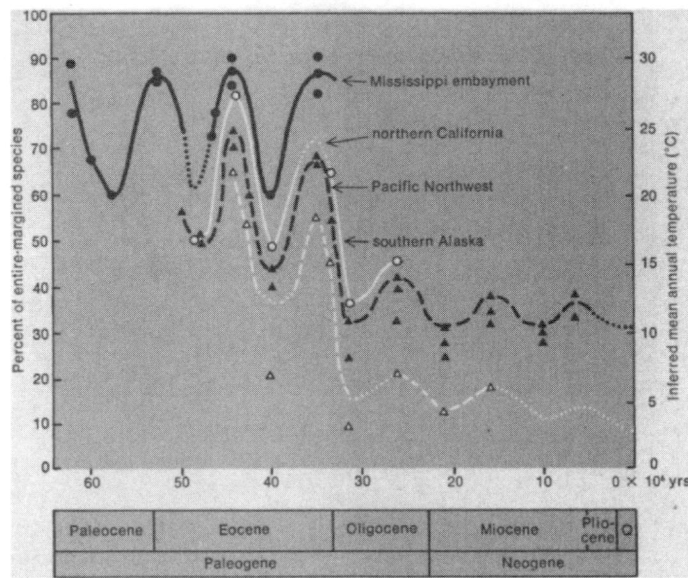

Figure 1. From the estimated percentages of species with entire-margined leaves in four locations in North America, it is inferred that a sharp drop in mean annual temperature took place in the early Oligocene and has continued—at least at high latitudes—to the present day. At middle latitudes mean annual temperature has, overall, not changed since the Oligocene. Dotted intervals indicate that leaf-margin data are either lacking or not considered reliable.

the physical aspects (or physiognomy) of fossil leaf assemblages. The physical characteristics of vegetation occupying similar climates in widely separated regions are highly similar, although the regions may have only a few taxa in common. Among the most conspicuous of such characteristics are the similarities in the appearance of foliage. On the other hand, vegetation occupying dissimilar climates in one region typically has different physical characteristics, although many taxa may occur throughout the region.

Thus the environment tends to select plants that have certain physical aspects for a given climatic type, whether this climatic type is separated by oceans or, presumably, by major periods of time. Just as we can expect the Tropical Rain forest in Africa to have the physical characteristics of Tropical Rain forest in other regions of the world, so we can also expect the present Tropical Rain forest to have the physical characteristics of the Tropical Rain forest of the Eocene.

On the other hand, the Tropical Rain forest of Indonesia has a floristic composition markedly different from the composition of the Tropical Rain forest of Brazil. Such differences have resulted from a variety of historical factors, both geographic and evolutionary. These historical factors will also result in floristic differences between the Tropical Rain forest of the Eocene in a given region and any part of the modern Tropical Rain forest. Considering that the floristic composition of any vegetational type is continually undergoing change, then the determination of the vegetational type (and hence climatic type) represented by a fossil assemblage is best accomplished by analyzing the physical characteristics of the assemblage rather than its floristic composition. And, the further back in time, the more dissimilar the floristic associations are to present associations and the more problematic become climatic inferences.

Among the most useful physiognomic characters of broad-leaved foliage are: type of margin, size, texture, type of apex, and type of base and petiole (see Fig. 2). In areas of high mean annual temperature and precipitation, for example, leaves typically have "entire"margins (i.e. lacking lobes or teeth), are large, are coriaceous ("leatherlike"—an indication of an evergreen habit), and have a high proportion of attenuated apices (i.e. "drip-tips," particularly common on lower-story plants); and a moderate number have cordate (heart-shaped) bases associated with palmate venation and joints ("pulvini") in the petiole—a combination of characters typically associated with the vine, or liana, habit.

The general correlation between type of leaf margin and climate was first

Figure 2. Physical characteristics of leaves largely represent adaptations to the environment and are thus good indicators of climate. The small (microphyllous) leaf (*top*), an alder leaf from southern Alaska, has an incised margin and is characteristic of cool climates. The vine leaf (*bottom*) from the Philippines has a swollen and jointed petiole, palmate venation, and a cordate (heart-shaped) base, as do the leaves of most vines. The attenuated tip (drip-tip) indicates a humid habitat, and large (mesophyllous) size and entire (smooth) margin are characteristic of most tropical plants. White bars represent 30 mm.

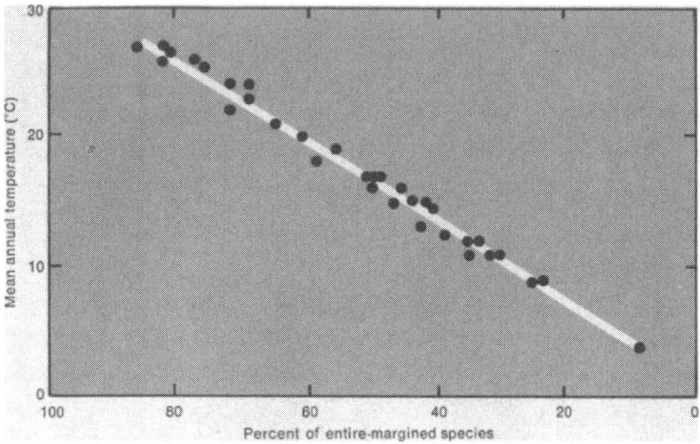

Figure 3. The percentage of species with entire-margined leaves in the humid to mesic broad-leaved forests of eastern Asia increases in direct proportion to the mean annual temperature of the particular forest.

documented by Bailey and Sinnott (1915) and has since been sporadically applied to interpretations of fossil assemblages. A recent compilation of analyses of woody vegetation in the humid to mesic (moderately humid) forests of eastern Asia has shown a strong correlation between the percentage of species with entire leaf margins and mean annual temperature (Fig. 3). Compilations of leaf-margin data of secondary vegetation—vegetation on disturbed sites that has not reached a climax stage—and of the broad-leaved element in coniferous forests do not display such a correlation.

Although leaf size is an important criterion in studies of extant vegetation, the application of this parameter to fossil assemblages is highly problematic, because leaves of different sizes may be differentially selected in the process of transport and preservation (Spicer 1975). Further, leaf-size changes can be related to precipitation and soils as well as to temperature. In the following discussion, I have used a generalized and modified version of the Raunkiaer (1934) system of leaf sizes: *mesophyll* for the larger mesophyll and larger classes, *notophyll* for the smaller mesophyll class (Webb 1959), and *microphyll* for the smaller classes.

The significance of physiognomy to the paleobotanist attempting paleoclimatic reconstructions is that the major physiognomic subdivisions of vegetation (which are partly based on foliar characters) have been found to correspond closely with certain major temperature parameters (Wolfe, in press). Figure 4 shows that mean annual temperature (an approximation of heat accumulation) is of major significance in determining what type of vegetation prevails, as are warm-month means. Only two cold-month means are of major significance. The 1°C mean separates dominantly broad-leaved evergreen (above 1°C) from broad-leaved deciduous (below 1°C) forests; in the areas that have cold-month means between 1°C and −2°C, notophyllous broad-leaved evergreens occur as an understory element, and in regions of even greater winter cold, notophyllous broad-leaved evergreens are lacking. The 18°C cold-month mean—a commonly accepted boundary between "tropical" and "subtropical"—has no relevance to the distribution of vegetation.

Estimates of mean annual temperature can be based on the percentage of entire-margined species in a given fossil assemblage. More difficult to infer is the mean annual *range* of temperature, which, in some cases, can be estimated only within broad parameters. In other cases, however, mean annual range of temperature can be accurately inferred. For example, if two succeeding assemblages have the same leaf-margin percentage of 50% (mean annual temperature ~17°C), and if the younger assemblage is dominantly microphyllous and the older assemblage is dominantly notophyllous, then reference to the framework of Figure 4 indicates that a mean annual range of temperature of 6°C was reached some time between the two assemblages. A second example is that of the Miocene Seldovia Point flora, which represents vegetation slightly inland from the coast of southern Alaska. Foliar physiognomic (as well as floristic) criteria indicate a mean annual temperature of 6–7°C (Wolfe and Tanai, in press). Other paleobotanical data indicate that the broad-leaved deciduous forest represented by the Seldovia Point flora merged with coniferous forest toward the coast. Again, reference to Figure 4 indicates a mean annual range of temperature of about 26–27°C.

Two major problems that have hampered many climatic inferences from paleobotanical data have been the lack of floras in even moderately close stratigraphic successions and the total misinterpretations of the age and climatic significance of high-latitude Tertiary floras. These misinterpretations arose from acceptance of the undocumented concept of an "Arcto-Tertiary Geoflora"—that the Eocene vegetation in Alaska represented temperate broad-leaved deciduous forest that, unchanged, gradually migrated southward to middle latitudes. In North America, both problems have, to a high degree, been overcome. In Alaska there are stratigraphic sequences of floras—many independently dated—that represent most of the Tertiary. In the Pacific Northwest, numerous floras—again, many in stratigraphic succession and/or independently dated—occur in early Eocene and younger rocks. In the Mississippi embayment region, an almost complete sequence of floras represents most of Paleocene and Eocene time. Almost all these floras represent coastal plain vegetation, and thus one major variable in interpreting the significance of paleoclimatic inferences—altitude—is held approximately constant. Certain floras from interior areas add other dimensions to paleoclimatic models, but the altitu-

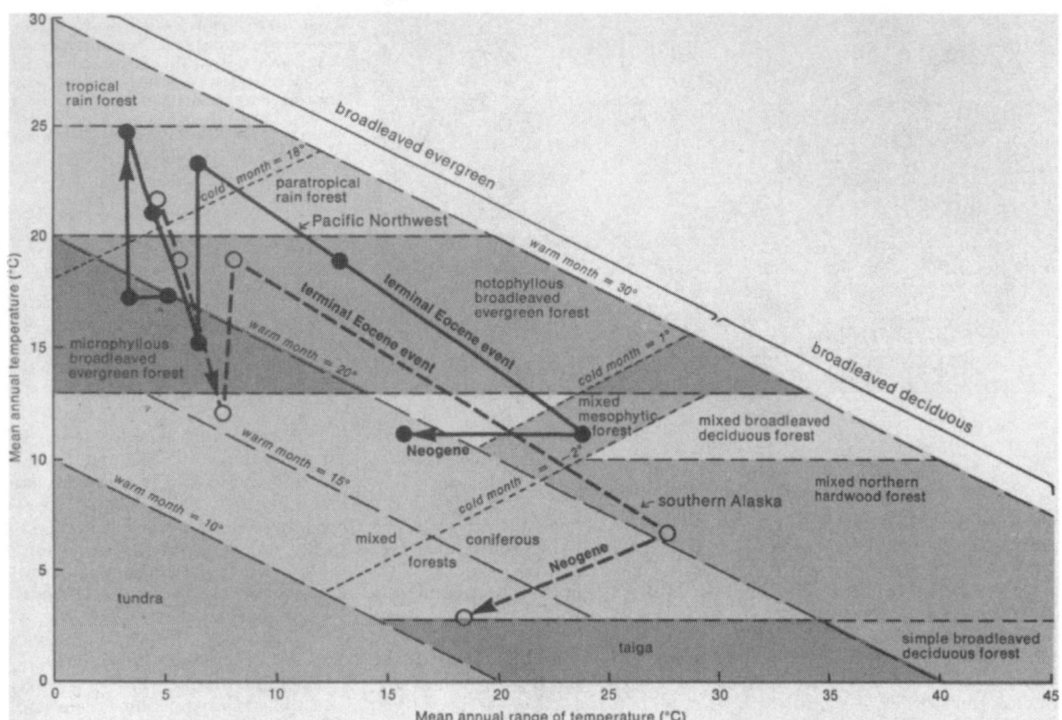

Figure 4. The humid to mesic forests of the Northern Hemisphere can be approximately circumscribed by various major temperature parameters. By comparing leaf assemblages in southern Alaska and the Pacific Northwest to the modern vegetation, we can infer the mean annual temperature and mean annual range of temperature for the assemblages. Major changes in temperature parameters are indicated for the time span between the middle Eocene and the Quaternary—showing a dramatic increase in mean annual temperature and an increase in mean annual range of temperature during the terminal Eocene event.

dinal factor introduces a problematic variable.

Paleocene and Eocene climates

The most complete sequence of Paleogene leaf floras in a small area is that of the Puget Group in western Washington (Wolfe 1968). The Puget assemblages extend from an estimated 50 m.y. ago (late early Eocene) up to about 34 m.y. ago (latest Eocene). In this sequence, the floras all contain numerous leaf species that have drip-tips and/or probable liana leaf physiognomy, and coriaceous (i.e. ≅ evergreen) texture dominates. Thus, all the assemblages represent vegetation that apparently grew under abundant year-round precipitation and would be classed as broad-leaved evergreen rain forests. Major changes in margin and size of the leaf assemblages occurred, however, during deposition of the Puget Group (Fig. 5). The leaf-margin data alone indicate major (perhaps 7–9°C) fluctuations in mean annual temperature, and, in a general manner, the leaf-size data also indicate climatic fluctuations. Sequences of floras in western Oregon, eastern Oregon, and northeastern California parallel the Puget sequence (Wolfe 1971).

The Puget leaf-size data are, however, possibly significant in a context other than mean annual temperature. In the lower part of the sequence, although the leaf-margin data do not significantly change, there is a pronounced movement from a notophyllous to a microphyllous forest. If mean annual temperature was approximately constant during this interval, then mean annual range of temperature must have decreased (i.e. equability of temperature increased). Indeed, the combination of physiognomic data indicates a mean annual range of temperature about half that at present in coastal western Washington, which is highly equable now in comparison to most other mid-latitude areas of the Northern Hemisphere. Although other workers have on questionable floristic interpretations (e.g. Berry 1914, p. 66–67) suggested that the Eocene was characterized by high equability, the Puget physiognomic data provide strong evidence that the Eocene was in fact highly equable.

One of the major corollaries of high equability during the Eocene has been generally overlooked, particularly by the proponents of the concept of an "Arcto-Tertiary Geoflora," who long argued that the Eocene vegeta-

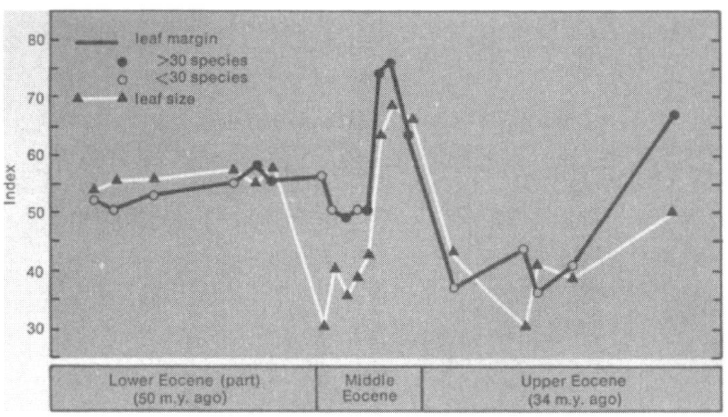

Figure 5. Foliar physiognomic data for Eocene assemblages in the Puget Group of western Washington show that both leaf margin and size underwent great changes between the early and late Eocene, indicating major fluctuations in mean annual temperature. The leaf size index is % microphyllous species + % notophyllous species × 2 + % mesophyllous species × 3 − 100 × 0.5.

tion in regions such as Alaska represented temperate broad-leaved deciduous forest. As in the now highly equable areas of the Southern Hemisphere or on tropical mountains, temperate and mesic broad-leaved deciduous forests could *not* have existed in the Northern Hemisphere during the Eocene. That is, the latitudinal temperature gradient would fall far to the left side of Figure 4—far from temperatures that would support temperate broad-leaved deciduous forest.

Notable also in the Puget analyses is that, during the late middle Eocene (ca. 45 m.y.), the vegetation was marginally Tropical Rain forest. A contemporaneous assemblage in northern California—the Susanville flora (lat 40°)—has a leaf-margin percentage of 82, concomitant with a leaf-size index of 79. The other physiognomic data are consistent with inferring the Susanville flora to be Tropical Rain forest. This indicates that Tropical Rain forest (and the 25°C isotherm) occurred at least 20° and possibly 30° poleward of the present northern limit.

Many assemblages that those who argued for an "Arcto-Tertiary Geoflora" interpreted as temperate broad-leaved deciduous forest are indeed that type; however, these assemblages are typically of Neogene age (cf. the radiometric data of Triplehorn et al. 1977), rather than Eocene, as they had supposed. In fact, only two small Eocene assemblages were known from Alaska until the last decade.

Collections from the Eocene at 60–61° latitude in the Gulf of Alaska region (Wolfe 1977) represent the latter half of the epoch. The late middle Eocene floras—correlative with the Susanville flora—represent Paratropical Rain forest and indicate the warmest climate (ca. 22°C mean annual temperature) of the Tertiary in Alaska (Wolfe 1972). The warmth indicated by the foliar physiognomic data is fully substantiated by the floristic evidence: included are feather and fan palms, mangroves, and members of other families now dominantly or entirely tropical (Wolfe 1977).

Recent geologic data (e.g. Jones et al. 1977) indicate that parts of southern Alaska were once at low latitudes and have drifted northward. The drift and accretion of these plate fragments to Alaska were, however, accomplished by the beginning of the Tertiary. The various major models of plate tectonics are unanimous in suggesting that, in general, western North America rotated southward during the Tertiary. That is, the paleolatitudes of these western North American floras were probably higher than the present latitudes of the fossil localities.

The latitudinal temperature gradient along the Pacific Coast of North America is today very moderate—about 0.5°C/1° latitude. During the late middle Eocene, however, the gradient was even lower. The Susanville (ca. 27°C mean annual temperature at 40° N.), the Puget (ca. 25°C mean annual temperature at 48° N.), and the Gulf of Alaska (ca. 22°C mean annual temperature at 60° N.) data indicate a temperature gradient of about 0.25°C/1° latitude.

The Alaskan Paleocene assemblages are exceedingly difficult to interpret climatically. In southeastern Alaska on Kupreanof Island (lat 57°) there are large assemblages that in all aspects of physiognomy represent Notophyllous Broad-leaved Evergreen forest (mean annual temperature approximately 18°C). Yet, only 2–4° latitude northward, the bulk of the broad-leaved evergreen element is unrepresented in the even larger assemblages of the Chickaloon and West Foreland formations. In features of foliar physiognomy such as margin and size, the Chickaloon assemblages would appear to correspond to Notophyllous Broad-leaved Evergreen forest (Wolfe 1972), yet the Chickaloon assemblages are dominantly broad-leaved deciduous, and even the broad-leaved deciduous element is not as diverse as in present temperate broad-leaved deciduous forests. The minor broad-leaved evergreen element includes palms and certain dicotyledonous families that are not to be expected in temperate broad-leaved deciduous forests. Thus, although almost all physiognomic characters and limited floristic data point to temperatures that should support dominantly broad-leaved evergreen vegetation, the vegetation was dominantly deciduous. Such assemblages occur throughout much of Alaska and Siberia north of latitude 60° during the Paleocene.

In the southeastern United States a large number of Paleogene leaf assemblages that represent coastal plain vegetation occur. Detailed work by many stratigraphers allows an accurate placement of these assemblages in stratigraphic sequence (Fig. 6). The oldest assemblages—early Paleocene (Midway Group)—are those from Naborton and Mansfield, Louisiana. The physiognomy of these assemblages is clearly indicative of Tropical Rain forest (mean annual temperature about 27°C). The succeeding late Paleocene (lower part of Wilcox Group) assemblages represent Paratropical Rain forest—an indicated cooling consonant with data from the continental interior (Wolfe and Hopkins 1967). In the earliest Eocene assemblages (upper part of Wilcox Group), however, leaf size is reduced, the probable liana type of leaf is not as common as earlier, and drip-tips are uncommon; at the same time, the leaf-margin data suggest a warm interval.

How much of a hiatus exists between the Wilcox and Claiborne assemblages is uncertain, but I am assuming that most of the late early and early middle Eocene is missing, at least in the floral sequence. The large assemblages from Puryear, Tennessee, and Granada, Mississippi (the bulk of the "Wilcox flora" of various authors, but actually Claiborne in age; cf. Dilcher 1973a), are of late middle Eocene age. Dilcher (1973b) considered the climatic inferences based on such assemblages to be puzzling; however, the scarcity of probable lianas, the scarcity of drip-tips, and the small leaf size concomitant with a high leaf-margin percentage are characteristic of dry tropical vegetation (cf. Rzedowski and McVaugh 1966).

It is perhaps also significant that these Claiborne assemblages contain a diversity of Leguminosae, a family common in dry tropical vegetation today. Apparently a cooling occurred near the Claiborne-Jackson boundary, but the one cool assemblage (interestingly, once interpreted by Berry, 1916, to be of Pleistocene age) is unfortunately small. In any case, the Mississippi embayment sequence indicates a pronounced drying trend from the Paleocene into at least the middle Eocene.

In the continental interior, the Paleocene floral sequence also shows a definite cooling from the early into the late part of the epoch (Wolfe and Hopkins 1967; Wolfe, in press). Hickey (1977) suggests a renewed warming trend near the Paleocene-Eocene boundary, which would parallel the Mississippi embayment trend. As in that area, the interior

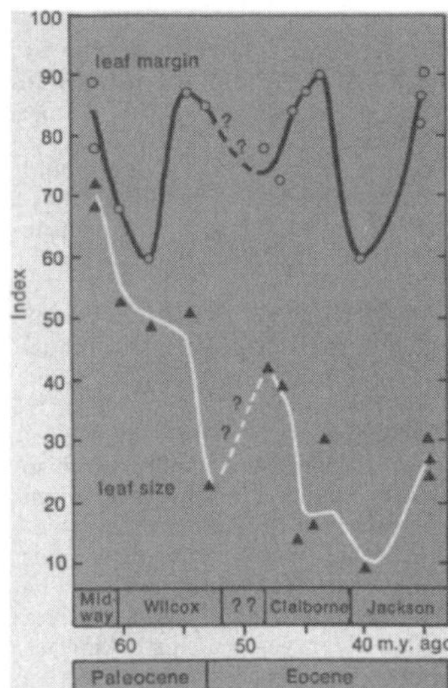

Figure 6. Indexes from foliar physiognomic data for Paleocene and Eocene assemblages in the Mississippi embayment region indicate a pronounced drying trend from the Paleocene into at least the middle Eocene.

Paleocene assemblages all indicate humid to mesic vegetation.

In the Eocene, however, the pattern in the interior becomes greatly complicated. The late early and early middle Eocene assemblages (the earliest Eocene assemblages are unstudied) from central and northern Wyoming represent definite mesic conditions, but the assemblages from southern Wyoming and adjacent Colorado and Utah represent pronounced dry conditions (MacGinitie 1969, 1974). How this situation is related to the presence of mountains and consequent rain shadows is uncertain. Today, the predominant sources of moisture for this region are southerly; one would expect the more southern area (southern Wyoming) to be moister if the Eocene circulation pattern were similar. Later Eocene leaf assemblages from this region are poorly known, except for the latest Eocene Florissant flora of central Colorado (MacGinitie 1953). The climatic significance of this flora is problematic because the altitude at which the Florissant beds were deposited is unknown.

To the west, a number of floras are known from an ancient uplifted area that stretched from Nevada north into British Columbia. The known assemblages represent mesic coniferous forest. Two—the Princeton, British Columbia (Arnold 1955), and the Republic, Washington (Berry 1929)—are of early middle Eocene age and represent the same cool interval as documented in western Washington. The Copper Basin and Bull Run floras from northern Nevada (Axelrod 1966) are correlative with the late Eocene cool interval.

The Paleocene and Eocene floras from North America thus provide the basis for a number of climatic inferences. (1) An overall gradual warming took place from the Paleocene into the middle Eocene, with gradual cooling until the terminal Eocene event. (2) Cool intervals occurred during the late Paleocene, the late early to early middle Eocene, and the early late Eocene. The difference between the intervening warm intervals was, in mean annual temperature, about 7°C. (3) The cool intervals were about 4 to 5°C (mean annual temperature) warmer than the present. (4) Mean annual range of temperature during the middle Eocene was about half that of the present. (5) Mean annual range of temperature decreased from the early into the middle Eocene and possibly increased slightly until the end of the Eocene. (6) The latitudinal temperature gradient during the middle Eocene along the west coast of North America was about half that of the present. (7) The west coast of North America received abundant precipitation during that period. (8) The southeastern United States experienced a pronounced drying trend from the Paleocene into at least the middle Eocene.

Oligocene and Neogene climates

The most profound climatic event of the Tertiary took place at the end of the Eocene. In middle to high latitudes of the Northern Hemisphere, the vegetation changed drastically. Within a geologically short period of time, areas that had been occupied by broad-leaved evergreen forest became occupied by temperate broad-leaved deciduous forest. A major decline in mean annual temperature occurred—about 12–13°C at latitude 60° in Alaska and about 10–11°C at latitude 45° in the Pacific Northwest. Just as profound, however, was the shift in temperature equability: in the Pacific Northwest, for example, mean annual range of temperature, which had been at least as low as 3–5°C in the middle Eocene, must have been at least 21°C and probably as high as 25°C in the Oligocene (Fig. 4; Wolfe 1971).

One of the major aspects of early Oligocene floras at middle to high latitudes is their lack of diversity, which was followed by enrichment during the remainder of the Oligocene (Wolfe 1972, 1977). The lack of diversity would be expected following a major and rapid climatic change such as the one that characterized the terminal Eocene event—that is, few lineages were preadapted or could rapidly adapt to the new temperature extremes.

Although the late early to early middle Miocene warming has been recognized throughout the world, some evidence indicates a warm interval during the late Oligocene (see references cited by Wolfe 1971) and perhaps, to a lesser extent, a warming during the latest Miocene (Wolfe 1969; Barron 1973). These warm intervals, however, were not as warm in comparison to adjoining cool intervals as were the Paleocene-Eocene warm intervals.

The climatic trends following the terminal Eocene event, aside from the minor fluctuations, are of great significance. One trend that can be demonstrated in areas north of latitude 30° is an increase in equability, a trend that runs counter to putative models of Neogene climatic change (cf. Axelrod and Bailey 1969).

Mean annual range of temperature was, during the Oligocene in western Oregon, as great as 21–25°C, but the present value is 12–16°C. At latitude 60° in Alaska, the mean annual range of temperature during the Miocene warm interval was at least as high as 26–27° C, in contrast to the current value of 18°C in the same area (Fig. 4). Similar declines in mean annual range of temperature can be demonstrated in other areas of the Northern Hemisphere, for example, in eastern Asia (Wolfe and Tanai, in press).

Overall trends in mean annual temperature since the terminal Eocene event are dependent on latitude. In southern Alaska (lat 60°), a decline of about 4°C can be documented since the early to middle Miocene. The salient feature of this high-latitude trend is that almost all the change appears to be the result of a decline in summer temperature, which would greatly enhance the "over-summering" of snow fields and, in turn, the initiation of widespread glaciation.

In the Pacific Northwest (lat 42–46° N.), no overall change in mean annual temperature appears to have occurred since the terminal Eocene event. In California and Nevada, climatic inferences from Neogene floras are so greatly complicated by altitudinal and rain shadow factors that extension of these inferences to other areas would at present be unjustified. The few Neogene floras based on leaf remains from eastern North America are too small to be of value in this context.

In Europe, the Neogene floras are found at about the same or higher latitudes as those in the Pacific Northwest; correcting for plate tectonic movements, the Pacific Northwest and European Miocene floras would have been at about equivalent latitudes. The European Neogene sequences typically display—as in the Pacific Northwest—an overall change from broad-leaved deciduous (with a broad-leaved evergreen element, particularly in the Miocene warm interval) to coniferous forest. This implies predominantly a decrease in mean annual range of temperature, possibly along with some decline in warm-month and consequently in mean annual temperatures.

In eastern Asia, the assemblages from Sakhalin (lat 50°) and Kamchatka (lat 55°) show much the same temperature trend as those in Alaska, whereas in Hokkaido (lat 42–45°) only a decrease in mean annual range of temperature occurred (Wolfe and Tanai, in press). South of Hokkaido the floras of Oligocene and Neogene age apparently indicate a contradictory trend—at least in part. In the early Miocene, for example, broad-leaved deciduous forest occupied lowland Kyushu (lat 32°) and southern Honshu (lat 35°; Tanai 1961)—areas now occupied by broad-leaved evergreen forest. Although it can be inferred that mean annual range of temperature has decreased by about 2–4°C, the major point is that mean annual temperature has increased by 3–4°C (Fig. 7). In the middle Miocene of northern Taiwan (lat 25°)—an area now occupied by Paratropical Rain forest—the lowland vegetation was Notophyllous Broad-leaved Evergreen forest (Chaney and Chuang 1968), indicating an increase of at least 2°C in mean annual temperature.

It is significant in this context that Muller (1966) has recorded pollen of elements such as alder and spruce from Borneo (lat 5°), while Graham and Jarzen (1969) have recorded similar cool-climate indicators from the Oligocene to the Miocene in Puerto Rico (lat 18°). In these instances the authors explained the presence of the cool element by suggesting the existence of mountains even higher than those now in the respective areas—although there are no geologic data to support such inferences. Graham (1976), however, explained the presence of cool-climate indicators in the Miocene of Veracruz (lat 19°) by suggesting that temperatures were cooler than now. I suggest that the presence of cool-climate indicators is consistent with the data from Taiwan and Kyushu and implies that, following the terminal Eocene event, low latitudes were cooler than at present.

It is noteworthy that the amount of change in mean annual range of temperature increases as latitude gets higher. At lower latitudes, the major change was apparently an increase in winter temperature that resulted in an overall increase in mean annual temperature and a slight decrease in mean annual range. At about 45° latitude, winter temperature increased by about the same amount as summer temperature decreased, with the result that mean annual temperature remained constant while mean annual range decreased moderately. At high latitudes, summer temperature decreased significantly, leading to a moderate decrease in mean annual range.

One of the obvious consequences of the above trends is an increase in the latitudinal temperature gradient. Such an increase would necessarily increase the intensity of the subtropical high-pressure cells (Willett and Sanders 1959), which, in turn, would bring increasing drought—particularly in summers—to the west coasts of the continents. Such an increase in

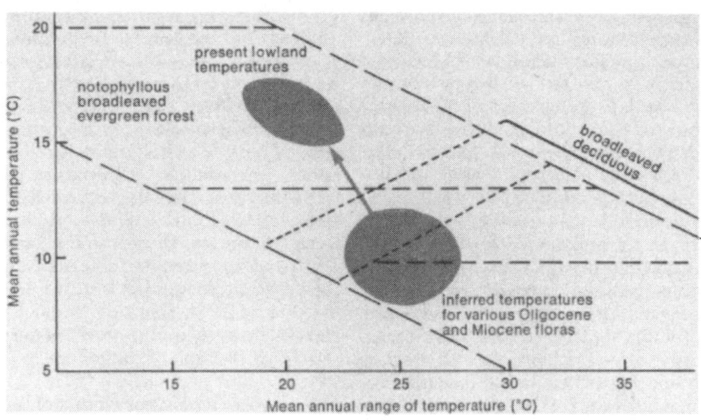

Figure 7. It can be inferred from the leaf assemblages that, in southern Japan, broad-leaved deciduous forest in the early Miocene gave way to broad-leaved evergreen forest and that mean annual temperature increased by 3–4°C.

summer drought during the Neogene has been well documented in western North America (Chaney 1944).

Causes of major climatic changes

Milankovitch (1938) proposed that the episodic glaciations of the Quaternary resulted from changes in the inclination of the earth's rotational axis. The changes Milankovitch considered were those due to perturbations, precession, and other known phenomena that result in minor changes in the inclination. In turn, inclinational changes would cause changes in insolation (the amount of radiation received from the sun) that would vary according to latitude. These astronomical explanations of Quaternary climatic change have received considerable support from many researchers, who consider that the paleoclimatic data fit well the timing based on calculations of minor inclinational changes.

Is it entirely coincidental that the divergent latitudinal patterns following the terminal Eocene event fit very well a model resulting from a significant decrease in the inclination of the earth's rotational axis? According to Milankovitch's hypothesis, under conditions of decreasing inclination (and assuming an atmospheric circulation pattern similar to that of the present), the pattern of climatic change would be (1) an increase in winter temperatures at lower latitudes (resulting in an increase in mean annual temperature), (2) an increase in winter temperatures about equal to a decrease in summer temperatures at latitude 43° (resulting in no increase in mean annual temperature), (3) a decrease in summer temperatures at higher latitudes (resulting in a decrease in mean annual temperature), and (4) a decrease in mean annual range of temperature proportional to the latitudes.

These are precisely the changes that are inferred from paleobotanical data for the Oligocene and Neogene and would indicate that a significant decrease in the earth's inclination has occurred during the last 30 million years.

Conditions during the Paleocene-Eocene were strikingly different from those during the Oligocene and younger epochs. During the thermal and equability maximum of the middle Eocene, the west coast of North America was wet and the southeastern United States was comparatively dry. The presence of humid broad-leaved evergreen forests in Alaska would, concomitant with the other data, argue for a circulation that involved the poleward flow of warm, moist air along the west coast; as this flow returned equatorward over the eastern part of the continent, the heating air would, of course, become drier. That is, the Eocene pattern may have been dominated by north-south (meridional) flow rather than being dominated by regional cells, as at present. What factor(s) could bring about such a pattern?

Readers who are plant physiologists may have been startled to learn that during the Eocene, broad-leaved evergreen forest extended north of latitude 60°. The principles of plant physiology would argue against such vegetation under the prolonged dark winters at such latitudes (Mason 1947; van Steenis 1962). The present distribution of broad-leaved evergreens is consistent with such principles—that is, notophyllous evergreens occur on mountains in California that have the same fundamental temperature parameters as lowland areas at more northern latitudes, where notophyllous broad-leaved evergreens are absent (Wolfe, in press). Notophyllous broad-leaved evergreens—except for a very few conspicuous taxa—today do not occur north of 50° latitude, and most are equatorward of 40–45°. Such considerations provide strong evidence that the middle Eocene Alaskan assemblages, with their diverse and dominant notophyllous to mesophyllous broad-leaved evergreen element, could not have existed under the present light conditions at the latitude of the fossil localities. Light was, in fact, considered a negative factor in the possibility of explaining trans-Pacific disjunctions of tropical broad-leaved evergreen groups via the land bridge that is now the Bering Straits (e.g. van Steenis 1962), and yet such groups are now known to have occurred in the Beringian region (van Buesekom 1971; Wolfe 1972).

Under conditions of high temperatures and prolonged winter darkness, the predicted vegetation would be broad-leaved deciduous with a minor broad-leaved evergreen element (van Steenis 1962), composed of those evergreens tolerant of low light levels (as a limited number of broad-leaved evergreen taxa are today). This is apparently the type of vegetation that existed in Alaska north of latitude 60°, during the Paleocene.

I am suggesting that the data thus far indicate that, during the middle Eocene, the light conditions at latitude 60° were more favorable to the growth of broad-leaved evergreens than at present. During the Paleocene, in fact, latitude 57° (but not 60°) supported a diverse broad-leaved evergreen forest. The only factor that could produce more light at these northern latitudes is a significantly smaller (at least 15° less during the middle Eocene than now) inclination of the earth's rotational axis. A smaller inclination would certainly be consistent with the low mean annual range of temperature during the Paleocene and Eocene: today this temperature parameter is primarily (although not entirely) a function of latitudinal position because of the inclination.

A suggestion that the inclination was, in comparison to today, considerably smaller is, if the obvious warmth of the Alaskan middle Eocene is accepted, contradicted by the Milankovitch calculations, i.e. a smaller inclination would yield lower insolation at high latitudes and hence lower temperatures. These calculations were, however, based on the assumption that the insolational values could be directly translated into temperature values—i.e. that the present atmospheric circulation pattern was, in general, constant throughout geologic time. But we have seen that the Paleocene and Eocene circulation pattern could not have been like that of the present. Could an increased insolational gradient under a low inclination have been the major driving force for a dominantly meridional circulation, a circulation that would have more than compensated for decreased annual insolational values at high latitudes? Perhaps at some critical value of inclination, the atmospheric circulation changes from one that is dominantly cellular (as it is today and was during the Oligocene and Neogene) to one that is dominantly meridional.

If the major climatic trends during the Tertiary were largely the result of inclinational changes, then from the Paleocene to the middle Eocene, inclination decreased gradually from a value of perhaps 10° to a value approaching 5°. The inclination then began to increase slightly until the end of the Eocene, when the inclination increased rapidly to 25–30°. Since then, the inclination has gradually decreased to the present average value of 23.5°.

The drastic change in inclination suggested as the cause of the terminal Eocene event would have had a profound effect on the earth's crust. It is significant that a number of researchers have suggested major tectonic changes at the end of the Eocene. For example, Molnar et al. (1975) suggest that the tectonic patterns of the South Pacific were different in the pre-Oligocene than now and that the current patterns were achieved at the end of the Eocene. Menard (1978) similarly indicates major changes in the northeastern Pacific at the end of the Eocene.

Yet, even assuming the validity of this model of inclinational change, the several fluctuations in mean annual temperature are not explained. From available radiometric data, the fluctuations appear to represent a cycle about 9.5 m.y. in duration (Wolfe 1971). Presumably such regular fluctuations would result from fluctuations in the amount of solar radiation reaching the earth; certainly no model of plate tectonic movements could explain such fluctuations.

Much additional information is needed, particularly from the continental interiors, to develop accurate models of temperature and precipitation distribution during the Paleocene and Eocene. Low-latitude leaf floras of Oligocene and Neogene age are needed to determine whether the low-latitude data thus far accumulated are anomalous or typical. More studies of modern depositional environments are needed to understand fully the significance of what is actually found in fossil assemblages. More information from rigidly controlled experiments is needed to determine the physiological response of broad-leaved woody plants to low light levels.

This review has shown the type of data that can be obtained from fossil plants. Brooks (1949) noted that the evidence available to him could not be explained solely by geographic factors, and even attempts to explain Paleocene and Eocene climates by the changing positions of the continental plates are only partially satisfactory (Frakes and Kemp 1973). The evidence now available indicates even more radical differences between present climates and those of the past than were recognized by Brooks. Attempting to fit such data into a "steady-state" hypothesis would be doing an injustice to the data similar to that done to geologic data prior to the general acceptance of plate tectonics.

References

Addicott, W. O. 1969. Tertiary climatic change in the marginal northeast Pacific Ocean. *Science* 165:583–86.

Arnold, C. A. 1955. Tertiary conifers from the Princeton coal field of British Columbia. *Michigan Univ. Mus. Paleontol. Contr.* 12:245–58.

Axelrod, D. I. 1966. The Eocene Copper Basin flora of northeastern Nevada. *Calif. Univ. Pubs. Geol. Sci.* 59:1–125.

Axelrod, D. I., and H. P. Bailey. 1969. Paleotemperature analysis of Tertiary floras. *Palaeogeography, Palaeoclimatology, Palaeoecology* 6:163–95.

Bailey, I. W., and E. W. Sinnott. 1915. A botanical index of Cretaceous and Tertiary climates. *Science* 41:831–34.

Barron, J: A. 1973. Late Miocene-early Pliocene paleotemperatures for California from marine diatom evidence. *Palaeogeography, Palaeoclimatology, Palaeoecology* 14: 277–91.

Berry, E. W. 1914. The Upper Cretaceous and Eocene floras of South Carolina and Georgia. U.S.G.S. Prof. Paper 84.

———. 1916. The Mississippi River bluffs at Columbus and Hickman, Kentucky, and their fossil flora. *U.S. Natl. Mus. Proc.* 48: 293–303.

———. 1929. A revision of the flora of the Latah formation. U.S.G.S. Prof. Paper 154-H, pp. 225–65.

Brooks, C. E. P. 1949. *Climate through the Ages*, 2nd ed. London: Benn.

Chaney, R. W. 1944. Summary and conclusions. In *Pliocene Floras of California and Oregon*, ed. R. W. Chaney, pp. 353–83. Carnegie Inst. Washington publ. 553.

Chaney, R. W., and G. C. Chuang. 1968. An oak-laurel forest in the Miocene of Taiwan (Part 1). *Geol. Soc. China* 11:3–18.

Denton, G., and R. L. Armstrong. 1969. Miocene-Pliocene glaciations in southern Alaska. *Am. J. Sci.* 267:1121–42.

Devereux, I. 1967. Oxygen isotope paleotemperature measurements on New Zealand Tertiary fossils. *New Zealand J. Sci.* 10: 988–1011.

Dilcher, D. L. 1973a. Revision of the Eocene flora of southeastern North America. *Palaeobotanist* 20:7–18.

———. 1973b. A paleoclimatic interpretation of the Eocene floras of southeastern North America. In *Vegetation and Vegetational History of Northern Latin America*, ed. A. Graham, pp. 39–59. Amsterdam: Elsevier.

Frakes, L. A., and E. M. Kemp. 1973. Paleogene continental positions and evolution of climate. In *Implications of Continental Drift to the Earth Sciences*, ed. D. H. Tarling and S. K. Runcorn, pp. 535–58. Academic Press.

Graham, A. 1976. Studies in neotropical paleobotany. II: The Miocene communities of Veracruz, Mexico. *Missouri Bot. Garden Annals* 63:787–842.

Graham, A., and D. M. Jarzen. 1969. Studies in neotropical paleobotany. I: The Oligocene communities of Puerto Rico. *Missouri Bot. Garden Annals* 56:308–57.

Hickey, L. J. 1977. Stratigraphy and paleobotany of the Golden Valley Formation (early Tertiary) of western North Dakota. *Geol. Soc. Am. Mem.* 150.

Jones, D. L., N. J. Silberling, and J. Hillhouse. 1977. Wrangellia—a displaced terrane in northwestern North America. *Can. J. Earth Sci.* 14:2565–77.

Kennett, J. P. 1977. Cenozoic evolution of Antarctic glaciation, the circum-Antarctic ocean, and their impact on global paleoceanography. *J. Geophys. Res.* 82:3843–60.

MacGinitie, H. D. 1953. *Fossil Plants of the Florissant Beds, Colorado.* Carnegie Inst. Washington publ. 599.

———. 1969. The Eocene Green River flora of northwestern Colorado and northeastern Utah. *Calif. Univ. Pubs. Geol. Sci.* 83:1–140.

———. 1974. An early middle Eocene flora from the Yellowstone–Absaroka volcanic province, northwestern Wind River basin, Wyoming. *Calif. Univ. Pubs. Geol. Sci.* 108:1–103.

Mai, D. H. 1964. Die Maxtixioideen-Floren im Tertiär der Oberlausitz. *Palaontolog. Abh.*, Abt. B, 2:1–92.

Mason, H. L. 1947. Evolution of certain floristic associations in western North America. *Ecol. Monographs* 17:201–10.

Menard, H. W. 1978. Fragmentation of the Farallon plate by pivoting subduction. *J. Geol.* 86:99–110.

Milankovitch, M. 1938. Astronomische Mittel zur Erforschung der erdgeschichtlichen Klimate. *Handbuch der Geophysik* 9: 593–698.

Molnar, P., T. Atwater, J. Mammerick, and S. M. Smith. 1975. Magnetic anomalies, bathymetry, and the tectonic evolution of the South Pacific since the late Cretaceous. *Royal Astron. Soc. Geophys. J.* 40:383–420.

Muller, J. 1966. Montane pollen from the Tertiary of northwestern Borneo. *Blumea* 14:231–35.

Nemjč, F. 1964. Biostratigraphic sequence of floras in the Tertiary of Czechoslovakia. *Časopis pro mineralogii a geologii, Rocnik* 9:107–9.

Plafker, G., and W. O. Addicott. 1976. Glaciomarine deposits of Miocene through Holocene age in the Yakataga Formation along the Gulf of Alaska margin, Alaska. In *Symposium on Recent and Ancient Sedimentary Environments in Alaska*, ed. T. P. Miller, pp. Q1–Q23. Alaska Geological Society.

Raunkiaer, C. 1934. *The Life Forms of Plants and Statistical Plant Geography.* Oxford: Clarendon Press.

Rzedowski, J., and R. McVaugh. 1966. La vegetacion de Nueva Galicia. *Michigan Univ. Herbarium Contr.* 9:1–123.

Spicer, R. A. The sorting of plant remains in a Recent depositional environment. 1975 diss., London Univ.

Tanai, T. 1961. Neogene floral change in Japan. *Hokkaido Univ. Fac. Sci. J.*, ser. 4, 11: 119–298.

———. 1967. On the Hamamelidaceae from the Paleogene of Hokkaido, Japan. *Palaeont. Soc. Japan Trans. Proc.*, N.S., 66:56–62.

———. 1970. The Oligocene floras from the Kushiro coal field, Hokkaido, Japan. *Hokkaido Univ. Fac. Sci. J.*, ser. 4, 14:383–514.

Tanai, T., and K. Huzioka. 1967. Climatic implications of Tertiary floras in Japan. In *Tertiary Correlation and Climatic Changes in the Pacific*, ed. K. Hatai, pp. 89–94. Sendai: Sasaki Printing and Publishing Co.

Triplehorn, D. M., D. L. Turner, and C. W. Naeser. 1977. K-Ar and fission-track dating of ash partings in coal beds from the Kenai Peninsula, Alaska: A revised age for the Homerian Stage-Clamgulchian Stage boundary. *Geol. Soc. Am. Bull.* 88:1156–60.

van Buesekom, C. F. 1971. Revision of *Meliosma* (Sabiaceae) section, *Lorenzanea* excepted, living and fossil, geography and phylogeny. *Blumea* 19:355–529.

van Steenis, C. G. G. J. 1962. The land-bridge theory in botany. *Blumea* 11:235–372.

Webb, L. J. 1959. Physiognomic classification of Australian rain forests. *J. Ecol.* 47:551–70.

Willett, H. C., and F. Sanders. 1959. *Descriptive Meteorology.* Academic Press.

Wolfe, J. A. 1968. Paleogene biostratigraphy of nonmarine rocks in King County, Washington. U.S.G.S. Prof. Paper 571.

———. 1969. Neogene floristic and vegetational history of the Pacific Northwest. *Madroño* 20:83–110.

———. 1971. Tertiary climatic fluctuations and methods of analysis of Tertiary floras. *Palaeogeography, Palaeoclimatology, Palaeoecology* 9:27–57.

———. 1972. An interpretation of Alaskan Tertiary floras. In *Floristics and Paleofloristics of Asia and Eastern North America*, ed. A. Graham, pp. 201–33. Amsterdam: Elsevier.

———. 1977. Paleogene floras from the Gulf of Alaska region. U.S.G.S. Prof. Paper 997.

———. In press. Temperature parameters of humid to mesic forests of eastern Asia and relation to forests of other regions of the Northern Hemisphere and Australasia. U.S.G.S. Prof. Paper 1106.

Wolfe, J. A., and D. M. Hopkins. 1967. Climatic changes recorded by Tertiary land floras in northwestern North America. In *Tertiary Correlation and Climatic Changes in the Pacific*, ed. K. Hatai, pp. 67–76. Pacific Sci. Cong., 11th, Tokyo, Aug. 1966, Symp. 25.

Wolfe, J. A., and T. Tanai. In press. The Miocene Seldovia Point flora from the Kenai Group, Alaska. U.S.G.S. Prof. Paper 1105.

Zhilin, S. G. 1966. A new species of *Carya* from the late Oligocene. *Paleont. Zhur.* 4:104–8 (in Russian). English translation available from Telberg Book Co., New York.

"I don't know what it measured. The Richter scale is down there."

PART

5 Species Interaction

Edited by Conrad C. Labandeira and Hans-Dieter Sues

Species interactions formed much of the basis of ecological research even before Darwin's *Origin of Species* was first published more than 150 years ago (Darwin 1859). However, soon after Darwin's work (1859) British naturalist explorers (Wallace 1862; Bates 1863; Belt 1874) started a tradition of detailing the ecological interactions among various species of animals and plants in tropical terrestrial ecosystems. In Germany, Möbius introduced a major new concept, the living community, and developed quantitative approaches for the study of interactions among marine organisms (Möbius 1877). Later research aimed at understanding the structure of ecological communities at higher organizational levels, a trend that eventually eclipsed studies involving the natural history of species interactions. Exemplary of this shift in ecology were the debates centered on contending hypotheses of plant-community structure. One such debate centered on whether plants were organized as ecologically linked species groups forming successional states (Clements 1936) or as unlinked individualistic units constituting random associations (Gleason 1926). In the second half of the twentieth century, the equilibrium theory of island biogeography (MacArthur and Wilson 1967), experimental manipulations of ecosystems (Simberloff and Wilson 1969), and food-web theory (Williams and Martinez 2000) emerged as major advances.

Partly because of the nature of fossil evidence, which revealed few interactions based on difficult-to-ascertain evidence, paleoecology was late in coming toward an understanding of the role of species interactions in ecology, particularly compared to earlier advancements in ecology. For paleoecology, these trends were developed during the late 1960s and the 1970s primarily based on data originating from Paleozoic marine invertebrates. The first trend, more of a consequence of the compass of the discipline rather than a directed research program, was a scale of analysis that ranged from the global (Bambach 1977) to broadly regional (Bretsky 1969) in extent. A second, related aspect was the long temporal durations that were involved, often comprising hundreds of thousands to millions of years (De Keyser 1977). Third was the ecological level at which observations were made, which emphasized spatiotemporally recurring and community-wide assemblages (Brett and Liddell 1978) rather than detailing the intricacies of interspecies ecological relationships that appeared later within the discipline (Baumiller and Macurda 1995; Gahn and Baumiller 2005). Fourth was the importance of the environment and other abiotic factors, focusing on controls such as regional tectonics, sediment type, depositional environment, and long-term climatic trends, often related to habitat type or community development, in the regulation of ecosystem properties (Bretsky 1969; Anderson 1971). However, most important of all, is the fifth trend, involving the explanations offered for understanding ecological structure, again historically related to the nature of paleoecological data. Explanations were largely based on the dual focus of abiotic environmental factors as controls (Bretsky 1969; Anderson 1971) and antagonistic, mostly predator-driven interactions as explanations (Sheehan and Lesperance 1978). Our selection of seven papers is emblematic of this focus on environmental control and predation-based species interactions.

This long-standing effort for understanding the ecology of Paleozoic marine communities provides a context for our selection of seven articles that represent a sample of the salient issues that have animated paleoecology during the last third of the twentieth century. Some papers, focusing on marine ecosystems, were major advances in the earlier tradition of Paleozoic marine ecology, such as Thayer's (1979) demonstration of the replacement of sedentary suspension feeders by mobile deposit feeders throughout the Phanerozoic. Vermeij (1977, 1987) explored the ramifications of the Mesozoic Marine Revolution, which represented a global episode of escalation between predator efficiency and prey defense in a variety of marine habitats. However, other advances in paleoecology involved continental ecosystems as well, emphasizing the advent of herbivory in vertebrate communities, beginning in the 1960s with Olson's (1966) account on successive replacements of differently structured communities of continental vertebrates during the Permian and Triassic. Olson documented pulsed transitions from Early Permian food chains linked to aquatic habitats composed of multiple faunivores and ending in Late Permian and Triassic fully terrestrial food chains with a major herbivore component. Soon thereafter, Martin (1973) provided the novel and highly controversial "overkill hypothesis" regarding the ecological impact of the peopling of the Americas. Assuming a Beringian source of a limited number of colonists, the southward colonization of such a massive land mass would have produced a demographic explosion within 1,000 years to the time in which southern South America was colonized, and during the course of which the New World mammalian megafauna became extirpated through human hunting. An ecological consequence of this catastrophic demise of large mammalian herbivores was Janzen and Martin's (1982) observation that seed dispersal syndromes and geographical distributions of Neotropical plants with large, fleshy fruits consumed by vertebrate herbivores were dramatically altered and the plants underwent spatial constriction of ranges. Accordingly, the later European introduction of horses and other livestock may have expanded the ranges of the same tree species to European precontact population densities. In a very different vein, but also emphasizing the theme of herbivory in terrestrial ecosystems, Wilf and Labandeira (1999) applied to the fossil record the modern observation that insect herbivory increases in diversity and intensity along a pole-to-equator latitudinal and increased thermal gradient, such that the highest consumption values occur in tropical ecosystems. This modern pattern was borne out when records from the late Paleocene to early Eocene interval in Wyoming at constant paleolatitude but increasing temperature revealed elevated attack frequencies and per-host increases in herbivore damage.

Van Valen's (1973) paper certainly belongs in its own, separate category, not only for its provocative title but also, more importantly, for providing the quantitative taxonomic backdrop for what paleobiologists informally had described in a more ecological context (Vermeij 1977). The Red Queen hypothesis suggested that all major groups of animals go extinct at a stochastically constant rate, deemed a "Law of Extinction" (Van Valen 1973). A deep-time ecological description of the pattern would posit that predators and prey escalate in predation pressure and prey counter defense, such that long-term fitness of both trophic roles would increase. This theoretical construct likely underpins the explanation for many or possibly all of the studies presented in this section. Examples would include the turnover of sedentary suspension feeders by mobile deposit feeders (Thayer 1979), the evolution of synapsid-dominated communities of continental vertebrates during the late Paleozoic (Olson 1966), and the demise of the New World mammalian megafauna soon after the arrival of human hunter-gatherers (Martin 1973).

What is notable for all of the articles in this section, as well for paleoecology in general, is the focus on predation and other forms of antagonistic species interactions over commensalistic or mutualistic associations in the fossil record. Perhaps as a consequence of Thompson's (2005) influential book on the role of mutualisms forming geographic mosaic in modern community-level processes, there now may be a greater incentive to ferret out data on mutualisms and coevolutionary associations in the fossil record. However, such data are more difficult to obtain than the comparatively more abundant evidence for antagonisms.

Literature Cited

Anderson, E. J. 1971. Environmental models for Paleozoic communities. *Lethaia* 4:287–302.

Bambach, R. K. 1977. Species richness in marine benthic habitats through the Phanerozoic. *Paleobiology* 3: 152–67.

Bates, H. W. 1863. *The Naturalist on the River Amazons: A Record of the Adventures, Habits of Animals, Sketches of Brazilian and Indian Life, and Aspects of Nature under the Equator, during Eleven Years of Travel.* 2 vols. London: John Murray.

Baumiller, T. K., and D. B. Macurda Jr. 1995. Borings in Devonian and Mississippian blastoids (Echinodermata). *Journal of Paleontology* 69:1084–89.

Belt, T. 1874. *The Naturalist in Nicaragua: A Narrative of a Residence at the Gold Mines of Chontales; Journeys in the Savannahs and Forests; with Observations on Animals and Plants in Reference to the Theory of Evolution of Living Forms.* London: John Murray.

Bretsky, P. W. 1969. Evolution of Paleozoic benthic marine invertebrate communities. *Palaeogeography, Palaeoclimatology, Palaeoecology* 6:45–59.

Brett, C. E., and D. Lidell. 1978. Preservation and paleoecology of a Middle Ordovician hardground community. *Paleobiology* 4:329–48.

Clements, F. E. 1936. Nature and structure of the climax. *Journal of Ecology* 24:252–84.

Darwin, C. 1859. *On the Origin of Species by Means of Natural Selection; or, The Preservation of Favoured Races in the Struggle for Life.* London: John Murray.

De Keyser, T. L. 1977. Late Devonian (Frasnian) brachiopod community patterns in western Canada and Iowa. *Journal of Paleontology* 51:181–96.

Gahn, F. J., and T. K. Baumiller. 2005. Arm regeneration in Mississippian crinoids: evidence of intense predation pressure in the Paleozoic. *Paleobiology* 31:151–64.

Gleason. H. A. 1926. The individualistic concept of plant association. *Bulletin of the Torrey Botanical Club* 53: 7–26.

Janzen, D. H., and P. S. Martin. 1982. Neotropical anachronisms: the fruits the gomphotheres ate. *Science* 215: 19–27.

MacArthur, R. H., and E. O. Wilson. 1967. *The Theory of Island Biogeography*. Princeton, NJ: Princeton University Press.

Martin, P. S. 1973. The discovery of America. *Science* 179: 969–74.

Möbius, K. A. 1877. *Die Auster und die Austernwirtschaft.* Berlin: Wiegand, Hempel and Parey.

Olson, E. C. 1966. Community evolution and the origin of mammals. *Ecology* 47:291–302.

Sheehan, P. M., and P. J. Lesperance. 1978. Effect of predation on the population dynamics of a Devonian brachiopod. *Journal of Paleontology* 52:812–17.

Simberloff, D. A., and E. O. Wilson. 1969. Experimental zoogeography of islands: the colonization of empty islands. *Ecology* 50:278–96.

Thayer, C. W. 1979. Biological bulldozers and the evolution of marine benthic communities. *Science* 203: 458–61.

Thompson, J. N. 2005. *The Geographic Mosaic of Coevolution*. Chicago: University of Chicago Press.

Van Valen, L. 1973. A new evolutionary law. *Evolutionary Theory* 1:1–30.

Vermeij, G. J. 1977. The Mesozoic marine revolution: evidence from snails, predators, and grazers. *Paleobiology* 3:245–58.

———. 1987. *Evolution and Escalation: An Ecological History of Life*. Princeton, NJ: Princeton University Press.

Wallace, A. R. 1862. *The Malay Archipelago: The Land of the Orang-Utan and the Bird of Paradise*. 2 vols. London: Macmillan.

Wilf, P., and C. C. Labandeira. 1999. Response of plant-insect associations to Paleocene-Eocene warming. *Science* 284:2153–56.

Williams, R. J., and N. D. Martinez. 2000. Simple rules yield complex food webs. *Nature* 404:180–83.

PAPER 29

Neotropical Anachronisms: The Fruits the Gomphotheres Ate (1982)

D. H. Janzen and P. S. Martin

Commentary

JESSICA THEODOR

Janzen and Martin was a unique contribution in paleoecology: a revelation to ecologists who seemed unaware that the past might be important to modern ecosystem function and a reminder to paleoecologists of the time frame of recovery from extinction. It revealed, more than any other paper on Pleistocene extinction, the ghost gomphothere in the room: that profound changes to terrestrial ecosystems had taken place over the last 12,000 years, and that those ecosystems were still in the process of responding to the loss of the megafauna. The paper showed that a number of these now geographically restricted plants produced more seeds than could be dispersed by the animals that currently distribute them and that these plants had undoubtedly been adapted for dispersal by the now extinct megafauna.

This idea that living plants might be anachronistic survivors of extinction, coevolved with extinct megafauna and struggling to survive, has been widely cited. It highlighted the Red Queen's role in mid-action: with the extinction of the gomphotheres, were these plants also doomed to extinction, or could they attract and adapt to new seed dispersers? This paper reminded the research community that living ecosystems are not static and have a historic dimension to them, that Eurasia, the Americas, Australia, and New Zealand had a depauperate terrestrial fauna, and that those extinctions mattered not only to the paleontologists who studied them but also to the ecosystems they lived in, and to our understanding of how the systems are still changing today.

From *Science* 215:19–27. Reprinted with permission from AAAS.

comparing results obtained simultaneously in the field with a sampling system having a 2-second residence time. The correction factors are in reasonable agreement with coagulation calculated theoretically for particles in Brownian motion.
28. J. Ogren, *J. Aerosol Sci.* 11, 427 (1980).
29. S. Twomey, *J. Comput. Phys.* 18, 188 (1975).
30. Thermosystems, Inc., Saint Paul, Minnesota.
31. R. Flagan and S. Friedlander, in *Recent Developments in Aerosol Science*, D. Shaw, Ed. (Wiley-Interscience, New York, 1978), pp. 25-59.
32. D. Taylor and R. Flagan, paper presented at the Western States Section spring meeting of the Combustion Institute, Irvine, Calif. (1980).
33. A. Sarofim, J. Howard, A. Padia, *Combust. Sci. Technol.* 16, 187 (1977).
34. M. McElroy and R. Carr, *EPRI Rep. WS-97-220* (1981), vol. 1.
35. A. Sarofim and R. Flagan, *J. Prog. Energy Combust. Sci.* 2, 1 (1976).
36. G. Fisher and D. Natusch, *Analytical Methods for Coal and Coal Products* (Academic Press, New York, 1979), pp. 489-541; D. F. S. Natusch, J. R. Wallace, C. A. Evans, Jr., *Science* 183, 202 (1974); J. Ondov, R. Ragaini, A. Biermann, *Environ. Sci. Technol.* 13, 946 (1979).
37. J. Gooch and G. Marchant, *EPRI Rep. FP-792* (1978), vol. 3.
38. B. Piper, S. Hersh, D. Mormile, *EPRI Rep. CS-1995* (1981).
39. R. Carr and J. Ebrey, *Environ. Prot. Agency (U.S.) Rep. EPA-600/9-80-039a* (1980).
40. W. Smith, R. Carr, K. Cushing, G. Gilbert, *Fifth International Fabric Filter Forum* (American Air Filter Co., Louisville, Ky., in press).
41. D. Giovanni, Ed., *Arapahoe Update* (Electric Power Research Institute, Palo Alto, Calif., 1980-1981), Nos. 1-4.
42. We acknowledge the contribution of A. Shendrikar, who was instrumental in gathering the trace element data. We also acknowledge the efforts of the staff of Meteorology Research, Inc., who performed the field measurements. Special thanks are expressed to the various electric utility companies that participated in the field test programs.

Neotropical Anachronisms: The Fruits the Gomphotheres Ate

Daniel H. Janzen and Paul S. Martin

Summary. Frugivory by extinct horses, gomphotheres, ground sloths, and other Pleistocene megafauna offers a key to understanding certain plant reproductive traits in Central American lowland forests. When over 15 genera of Central American large herbivores became extinct roughly 10,000 years ago, seed dispersal and subsequent distributions of many plant species were altered. Introduction of horses and cattle may have in part restored the local ranges of such trees as jicaro (*Crescentia alata*) and guanacaste (*Enterolobium cyclocarpum*) that had large mammals as dispersal agents. Plant distributions in neotropical forest and grassland mixes that are moderately and patchily browsed by free-ranging livestock may be more like those before megafaunal extinction than were those present at the time of Spanish conquest.

New World terrestrial biotas have long contained a rich fauna of large herbivores. During the Pleistocene, until around 10,000 years ago, the North American mammalian megafauna was comparable in its number of genera of large mammals (those exceeding 40 kilograms in adult body weight) to that of Africa in historical time (*1*). Although quantitative estimates of prehistoric biomass cannot be obtained directly from the fossil record, the high carrying capacity for domestic mammals of New World ranges—a capacity similar to that of African game parks—indicates that the Pleistocene biomass of native New World large herbivores was high. Martin (*2*) estimated an average preextinction biomass for unglaciated North America north of Mexico at 21 animal units per square kilometer or 28.2×10^6 metric tons on 7.8×10^6 square kilometers (1 unit = 1 cow plus a calf or 1 horse = 449 kilograms). While patchily distributed, the megafaunal biomass of lowland Central America must have been comparable, exceeding 50 animal units per square kilometer on favorable sites.

The number of species of large Central American Pleistocene herbivores in Neogene deposits of the last 10 million years greatly exceeds the number present in the past 10,000 years. Tapir, deer, peccaries, monkeys, and capybara occur as Pleistocene fossils, but the remains of gomphotheres (mastodon-like proboscidians), ground sloths, glyptodonts, extinct equids, *Mixotoxodon*, *Toxodon*, and other extinct large herbivorous animals (Table 1) are more common. If Neotropical ecologists and evolutionary biologists wish to determine who eats fruit, who carries sticky seeds, and who browses, grazes, tramples, and voids that segment of the habitat that would have been within reach of a variety of megafaunal trunks, tusks, snouts, tongues, and teeth, the missing megafauna must be considered.

There are prominent members of the lowland forest flora of Costa Rica whose fruit and seed traits can best be explained by viewing them as anachronisms. These traits were molded through evolutionary interactions with the Pleistocene megafauna (and earlier animals) but have not yet extensively responded to its absence. We first examine this evolutionary and ecological hypothesis by reconstructing the interaction between an extant palm and its Pleistocene megafauna. Without concerning ourselves with what caused the Pleistocene megafaunal extinctions (*3*), we are considering a portion of what happened when roughly three-quarters of all the species and individuals of large mammals were suddenly removed from a dry tropical region and its adjacent rain forests. The present-day analogy is a tropical, forested African habitat stripped of its elephants, rhinoceroses, zebras, elands, bush pigs, and other large herbivores and left alone for 10,000 years.

We focus on the trees that did not go extinct when their dispersal agents were removed. We do this because (i) tree-disperser interactions are not so tightly coevolved that a reasonable natural history consequence is extinction of one immediately following extinction of the other; (ii) if there is a large extinct Pleistocene megaflora in tropical America, it has so far escaped detection by paleobotanists; (iii) the plants that did go extinct cannot be directly studied; and (iv) we are confronted with a number of puzzling fruit and seed traits whose mystery disappears when interpreted in the light of the extinct Pleistocene megafauna. Although megafaunal extinction resulted in

Daniel H. Janzen is a professor in the Department of Biology, University of Pennsylvania, Philadelphia 19104, and Paul S. Martin is a professor in the Department of Geosciences, University of Arizona, Tucson 85721.

 0036-8075/82/0101-0019$01.00/0

major changes in intrahabitat plant species composition and population traits, 10,000 years is too short a time to expect all the surviving trees to have come to a new evolutionary equilibrium with the surviving animals and other plants.

A Reconstruction of the Fruiting of *Scheelea* 12,000 Years Ago

We shall reconstruct an event from the Costa Rican lowlands about the time a portion of the megafauna vanished. Toward the end of the dry season in the Pacific coastal plain, at a time when nutritious forage is scarce, there is the major peak in ripe fruit fall from the large forest palm *Scheelea rostrata*. In the dense riparian palm groves and upland mixed forest, the yellow egg-sized drupes fall by the thousands. The fruit fall attracts a herd of five gomphotheres (*Cuvieronius*), members of the family Gomphotheriidae and more closely related to the extinct North American mastodonts (*Mammut*) than to mammoths (*Mammuthus*) (*4*). They forage here daily and consume about 5000 *Scheelea* fruits per day. The hard nut wall (bony fruit endocarp) protects the large soft seeds from the gomphotheres' massive molars and most of the nuts are defecated intact. Below most palms, the ground is picked clean of the fallen fruit. The palm groves and individual palms are connected by well-traveled trails along which small piles of defecated *Scheelea* nuts are common. Such piles of nuts are also scattered about in other areas where the gomphotheres browse, such as in tree-falls, along river banks, and at forest edges.

Nut-rich dung is frequented by agoutis (*Dasyprocta punctata*) and other small rodents that remove the nuts. They gnaw some open and bury others, which are disinterred when food is scarce. Occasionally, when an agouti finds an intact *Scheelea* fruit, it eats the oily sweet pulp and discards the nut. The palm fruits that escape the gomphotheres and agoutis are eaten by tapirs (*Tapirus bairdii*) and collared peccaries (*Tayasu tajacu*). These animals chew off the pulp and spit out the hard nuts. Some *Scheelea* fruits and nuts are taken by squirrels (*Sciurus variegatoides*) which prey on the seeds.

Insect seed predators (adult bruchid beetles) oviposit on exposed nuts in the gomphothere dung. The larvae destroy virtually all the seeds in the nuts left on the ground surface. By ovipositing on nuts before the rodents get them, these insects even kill many of the seeds in the nuts buried by rodents.

The palm population occurs in riparian vegetation, dry hillsides, and wooded patches in grassland and is largely main-

Fig. 1. Fruits (all to the same scale) in Santa Rosa National Park, Guanacaste Province, Costa Rica, that were probably eaten by Pleistocene megafauna: (A) *Crescentia alata* (Bignoniaceae), (B) *Enterolobium cyclocarpum* (Leguminosae), (C) *Sapranthus palanga* (Annonaceae), (D) *Annona purpurea* (Annonaceae), and (E) *Acrocomia vinifera* (Palmae) (*19*). The white portion of the rule in (B) is 15 centimeters long.

tained by the seed input from the gomphothere dung. A seedling commonly appears many kilometers from its parent yet in the vicinity of conspecific adults. There are even adults in habitats where seedlings have extremely low survival probabilities because the gomphotheres generate repeated palm recruitment attempts in them. Many seeds are killed by seed predators, and most seedlings grow from seeds that were missed by both bruchids and agoutis because they were deeply buried in dung or were carried far from the concentrations of seed predators near the parent trees. Also, the rodents fail to retrieve some of the nuts they bury. The fruit phenology (that is, the timing of fruit fall within the day and season), fruit nutrient content, nut shape and hardness, seed crop size, germination timing, and other reproductive traits are molded and maintained by complex interactions in which the gomphothere, with its huge stomach, massive molars, and peripatetic behavior, plays a central role.

Then the gomphotheres are gone. The palm fruits fall as usual; in a month as many as 5000 accumulate below each fruit-bearing *Scheelea* palm. The first fruits to fall are picked up by agoutis, peccaries, and other animals that are soon satiated. As the pulp rots off fallen fruits beneath the parent palm, the bruchids oviposit on virtually all of the exposed nuts. The bulk of the seeds perish directly below the parent. Even if they escape the predators, the seedlings from the undispersed seeds are overshadowed by an adult conspecific, one of the strongest competitors in the habitat. In the next century the distribution of *Scheelea* begins to shrink. In several thousand years the local distribution of *Scheelea* has reached a new equilibrium

Table 1. Missing large herbivores of Central America.

Scientific name	Common name	Size in animal units (1 = 440 kilograms)	Habitat	Food	Origin of fossil record
Edentata					
Megatheridae					
Eremotherium (including *Megatherium*)	giant ground sloth	8	Lowland tropical forest, savanna	Leafy browse (*39*)	Guatemala (*40*), Panama (*41*)
Mylodontidae	mylodont ground sloth	2 to 4	Savanna	Grass (*42*), browse (*43, 44*)	Guatemala (*40*), Venezuela (*45*)
Megalonychidae	megalonychid ground sloth	1 to 2	Forest	Browse	Nicaragua (*46*)
Dasypodidae					
Pampatherium	giant armadillo	1 to 2	Savanna	Omnivore (terrestrial)	Venezuela (*45*)
Chlamytherium					Guatemala (*40*)
Glyptodontidae					
Glyptodon	glyptodont	1 to 2	Arid lowland tropics, warm temperate	Grass (*23*), fruit, carrion	Venezuela (*45*), Guatemala (*40*)
Rodentia					
Hydrochoeridae					
Neochoerus	giant capybara	0.3	Riparian forest	Riparian and aquatic plants	Venezuela (*45*)
Carnivora					
Ursidae					
Arctodus *Tremarctos*	extinct bear	1 to 1.5	Forest, savanna	Meat, fruits, foliage	Venezuela (*45*)
Notoungulata					
Toxodontidae					
Toxodon	toxodon	3	Savanna	Grass, low browse	El Salvador (*47*), Nicaragua (*46, 48*)
Liptoterna					
Macraucheniidae					
Macrauchenia	macraucheniops	2	Savanna	High browse	Venezuela (*45*)
Proboscideae					
Gomphotheriidae					
Haplomastodon	gomphothere	5 to 8	Tropical forest, savanna	Fruits, browse	El Salvador (*47*), Brazil (*49*)
Cuvieronius					Costa Rica (*50*), Venezuela (*51*)
Elephantidae					
Mammuthus	mammoth	10 to 15	Forest, savanna	Grass, browse	El Salvador (*47*)
Perissodactyla					
Equidae					
Equus (*Amerhippus*)	native horse	1	Savanna, forest edge	Grass, browse, fruits	Central America (*23*), Guatemala (*40*), Nicaragua (*48*), Costa Rica (*19a*), Venezuela (*45*)
Artidactyla					
Tayassuidae					
Platygonus	flat-headed peccary	0.3	Savanna, forest edge	Grass, browse, fruits	Mexico (*52*)
Camelidae					
Paleolama	extinct llama	0.7	Savanna	Grass, low browse	Venezuela (*45*)
Bovidae					
Bison	extinct bison	1	Savanna, forest edge	Grass, low browse	Guatemala (*40*), Nicaragua (*48*)

pattern that involves fewer habitat types and a lower density of adult trees. The palm grows only in those microhabitats so favorable that recruitment occurs with minimal seed disperal and escape from seed predators.

Now enter the biologists, assuming that they are studying a coevolved system that approximates an evolutionary equilibrium. They search the morphological and behavioral features of the existing biota for adaptive meanings. They study *Scheelea* nut wall thickness and hardness (*5*), size of fruits and dispersal agents (*6, 7*), the ratio of one- to two- to three-seeded nuts (*6, 8*), the spatial pattern of seed predation (*9*), fruiting phenology (*5, 9*), seed predator satiation (*5, 6*), and the balance between the fruit pulp reward and the seed content reward to the foraging rodent (*10*). These investigators notice the huge surplus of fallen nuts that remain directly below the parent *Scheelea* and attribute it to contemporary removal of dispersers by hunters or simply poor adjustment of seed crop size to the disperser guild. If they were working in Africa, however, they would notice the *Scheelea*-elephant interaction; in Central America they do not consider the former *Scheelea*-gomphothere interaction. The investigators attend only to the living fauna, although they take care to study native, not introduced, animals in a seemingly natural habitat.

Researchers have regarded nut wall thickness as an evolutionary adaptive response by *Scheelea* to the drilling abilities of bruchid larvae and the gnawing abilities of rodents. The main selective pressure determining nut wall thickness, however, could well have been the crushing force of a gomphothere's molars, and bruchids and rodents might simply have evolved to where they could penetrate this defense. The researchers assumed that the reward of fruit pulp should exceed the work expended by a rodent to get at the edible seed minus the value of that seed; throughout most of the evolutionary history of *Scheelea*, however, terrestrial rodents may have gotten fruit pulp only rarely. Coevolution of rodents and *Scheelea* fruits was assumed; the alternative hypothesis was not considered; the rodent is simply making use of a food source that was suddenly plentiful because of Pleistocene megafaunal extinction. Biologists did not suspect that flowering schedules, plant heights, leaf replacement rates, fruit crop size and phenology, or even the genetic structure of a palm population could now be seriously anachronistic if it was evolved to match the habitats occupied and type of population distribution pattern that is generated by dispersal through an extinct wide-ranging large mammal. If the fruiting traits of *S. rostrata* are now in major part anachronistic, as we suggest, then much of its interaction with present-day animals may hardly be evolved, to say nothing of coevolved (*11*).

Table 2. Native trees and large shrubs of lowland Pacific coastal deciduous forests in or near Santa Rosa National Park, Guanacaste Province, Costa Rica (*19*), whose seeds were probably dispersed by extinct megafauna.

Family	Scientific name	Common name
Anacardiaceae	*Spondias mombin*	jobo
	Spondias purpurea	jobo
	Spondias radlkoferi	jobo
Annonaceae	*Annona purpurea*	soncoya
	Annona holosericea	soncoya
	Annona reticulata	anona
	Sapranthus palanga	palanco
Bignoniaceae	*Crescentia alata*	jicaro
Bromeliaceae	*Bromelia karatas*	piñuela
	Bromelia penguin	piñuela
Ebenaceae	*Diospyros nicaraguensis*	persimmon
Euphorbiaceae	*Hippomane mancinella*	manzanillo
Leguminosae	*Acacia farnesiana*	huisache
	Andira inermis	almendro del monte
	Caesalpinia coriaria	divi divi
	Dioclea megacarpa	ojo de buey
	Enterolobium cyclocarpum	guanacaste
	Hymenaea courbaril	guapinol
	Pithecellobium mangense	
	Pithecellobium saman	cenizero
	Prosopis juliflora	mesquite
Malpighiaceae	*Bunchosia biocellata*	cerezo
	Byrsonima crassifolia	nance
Moraceae	*Brosimum alicastrum*	ramon
	Chlorophora tinctoria	mora
	Ficus spp.	higo, fig
Palmae	*Acrocomia vinifera*	coyol
	Bactris guinensis	biscoyol
	Bactris major	biscoyol
Rhamnaceae	*Zizyphus guatemalensis*	naranjillo
Rubiaceae	*Alibertia edulis*	trompillo
	Genipa americana	guaitil blanco
	Guettarda macrosperma	mosqueta
	Randia echinocarpa	
Sapotaceae	*Manilkara zapota*	nispero
	Mastichodendron capiri	tempisque
Tiliaceae	*Apeiba tibourbou*	peine de mico

The Megafaunal Dispersal Syndrome

In the lowland deciduous forest of Guanacaste Province, Costa Rica, there are at least 39 species of trees or large shrubs (Table 2) that are reasonable candidates for a reconstruction such as that envisioned for *Scheelea* palms and gomphotheres. These trees and shrubs display a set of fruit and seed traits in common—traits that are puzzling if examined only in the context of the potential native dispersal agents. We view these traits as part of the following megafaunal dispersal syndrome.

1) The fruits are large and indehiscent (Fig. 1) and contain sugar-, oil-, or nitrogen-rich pulp. The seeds they contain are obviously not dispersed abiotically as are the seeds in the large explosive schizocarp of *Hura crepitans* (Euphorbiaceae) or the large samara-filled dehiscent fruit of *Swietenia macrophylla* (Meliaceae).

2) The fruits look, feel, and taste like those eaten by large seed-dispersing mammals in Africa and have seeds and nuts of similar size, hardness, and shape to those in African fruits that are eaten by large mammals.

3) The large nuts or seeds (Fig. 2) are usually protected by a thick, tough or hard endocarp or seed coat that usually allows them to pass intact by the molars and through the digestive tract when eaten by introduced large mammals such as horses, cows, and pigs. Seed scarification in the animal digestive tract sometimes occurs during dispersal, and some scarified seeds are digested.

4) If the seeds are soft or weak, they are very small (as in figs) or imbedded in a hard core or nut like those in *Spondias*, *Scheelea*, and *Hippomane*. Fruits with soft seeds may also contain seed-free hard sections in the pulp or core that

block occlusion of the molar mill, as in the sweet and woody fruit of *Guazuma ulmifolia*.

5) Different species bear ripe fruits at different times of the year in a given habitat.

6) Many of the fruits fall off the tree upon ripening or even well before they ripen; this is best described as behavioral presentation of fruits to earth-bound dispersal agents.

7) The fruits usually attract few or no arboreal or winged dispersal agents such as bats, guans, or spider monkeys. If these animals are attracted, as they are to figs or *Spondias* fruits, there is usually a much larger fruit crop than they can eat.

8) In present-day forests, a high proportion of a tree's fruit crop rots in the tree or on the ground beneath it without being tasted by any potential dispersal agent. This is true even in those national parks where sizable wild vertebrate populations may equal or exceed their pre-Columbian densities.

9) Peccaries, tapirs, agoutis, and small rodents usually act as seed predators and dispersers of these trees; these animals do not act purely as dispersal agents, but at present they are often the only ones.

10) The fallen fruits are avidly eaten by introduced horses, pigs, or cattle (or by more than one). Free-ranging populations of these animals at carrying capacity normally consume all of the fallen fruit in most trees' crops. At least some of the seeds pass through the digestive tract of these animals and eventually germinate. The introduced large herbivores may re-enact many portions of the interaction the trees had with the extinct megafauna.

11) The natural habitats (such as alluvial bottoms or gentle slopes) of these trees are on the edges of grassland and in adjacent forest that are likely to be attractive to herbivorous megafauna and usually not on steep rocky outcrops and precipitous slopes.

As we come to know more of the natural history of the Costa Rican trees, more species will undoubtedly be added to the list in Table 2. For example, in southwestern Costa Rica in the lowland evergreen rain forest of Corcovado National Park, at least the following have most or all of the traits listed above: *Achmaea magdalenae, Astrocaryum standleyanum, Calophyllum macrophyllum, Dusia macrophylata, Enallagma latifolia, Elais melanococa, Hymenaea courbaril, Parkia pendula, Pouteria* spp., *Raphia taedigera, Scheelea rostrata, Simaba cedron, Terminalia catappa, Theobroma* sp., *Zamia* spp. *Coumarouna panamensis* nuts come from a tree common in many Panamanian and Costa Rican rain forests; the nuts are dispersed by contemporary mammals (*12*) and were probably dispersed by gomphotheres as well.

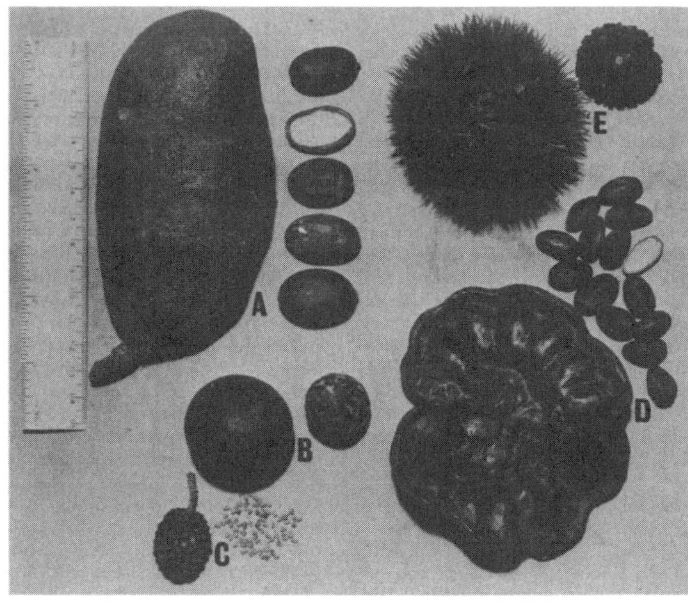

Fig. 2. Fruits and their seeds from Santa Rosa National Park that were probably dispersed by Pleistocene megafauna. The seeds to the right of each fruit represent a normal quantity of seeds found in each fruit. (A) *Hymenaea courbaril* (Leguminosae). (B) *Acrocomia vinifera* (Palmae). (C) *Guazuma ulmifolia* (Sterculiaceae). (D) *Enterolobium cyclocarpum* (Leguminosae). (E) *Apeiba tibourbou* (Tiliaceae) (*19*).

Certain species listed in Table 2 have instructive exceptions to the traits listed above. Although *Acacia farnesiana* has no sweet flavor or other attractant easily perceptible to humans in the mesocarp of its dry, pulpy, and indehiscent fruit, cattle and horses seek out and eat the fruits (*13*), just as do African big game animals with African *Acacia* (*14*). *Prosopis juliflora* (mesquite) is especially interesting in this context. In the arid southwestern United States, horses and cattle are known to have aided in the dispersal of mesquite seeds and the ripe pods of various *Prosopis* species are often sweet and pleasant tasting to people. In Guanacaste, the ripe pods of *P. juliflora* are only slightly sweet and somewhat astringent. Horses and cattle in Guanacaste eat the pods but not as eagerly as do these animals in northern Mexico, Texas, and southern Arizona. Because of the very local and patchy distribution of *P. juliflora* in Guanacaste (landward margins of mangrove swamps and high beach dunes), it has had minimal contact with livestock.

The relation between habitat and palatable fruit production is important. In Guanacaste, the species in Table 2 occur on relatively flat ground on terrain suitable for large mammal movement. On steep rocky slopes in the dry tropical forest (short-tree forest) of southern Sonora, terrain unsuitable for foraging of large mammals, Gentry (*15*) listed 32 prominent woody species, none of which have fruits or seeds adapted for large mammal transport. These include *Ceiba acuminata, Bursera simaruba, Willardia mexicana, Conzattia sericea, Caesalpinia platyloba, C. standleyi, Cassia emarginata, Lysiloma divaricata, L. watsoni, Tabebuia palmeri, T. chrysantha, Haematoxylon brasiletto, Jatropha platanifolia, J. cordata*, and *Ipomoea arborescens*. On the adjacent floodplains and arroyo bottoms there are species that have fruits adapted for megafaunal dispersal: *Sassafridium macrophyllum, Vitex mollis, Guazuma ulmifolia, Pithecellobium dulce, P. mexicanum, P. undulatum, Prosopis chilensis*, and *Randia echinocarpa*. Thus, in southern Sonora, where deciduous tropical forest reaches its northern limit, at about 28°N, the trees with hard seeds and sweet fruits that are palatable to large mammals, including humans, are found in canyon bottom habitats that would have been the natural corridor for movement of the extinct megafauna, just as they are for introduced livestock.

The diets of the extinct neotropical herbivorous megafauna. Many large mammals (Table 1), including edentates, gomphotheres, notoungulates, and at least some equids, were in contact with neotropical and subtropical floras for tens of millions of years, an ample period for the evolution of a plant-megafauna dispersal syndrome. On the basis of field studies (*13, 14, 16, 17*), we assume that, just as contemporary large grazing and browsing mammals and some large carnivores readily consume wild fruits and defecate the seeds alive, the extinct ones did as well.

Hypotheses and Tests

Our evolutionary hypothesis can be tested by comparing the array of fruits eaten and seeds dispersed by large mammals in Africa and Asia with the fruits of tropical America on the one hand and with the fruits of New Guinea or tropical Australia on the other; the latter two tropical land masses have never had a mammalian fauna that would select for a well-developed megafaunal dispersal syndrome. We can also test our hypothesis by reintroducing Pleistocene mammals such as horses (*18*) to the neotropics and observing their response to the fruits and the response of the plant populations to the mammals. Since the experiment has been running for 400 years, a number of the relevant tree populations may have already regained population structures that are more similar to those of the Pleistocene than they are to those of recent pre-Columbian times. Nevertheless, on a very local scale the opportunity exists for experimentation with tree population structures by the introduction of horses, as does the opportunity to study horse responses to detailed fruit and seed traits.

The interaction between Costa Rican range horses and jicaro trees (*Crescentia alata*) is an example. In Santa Rosa National Park (*19*), a horse population that is usually on an unsupplemented diet ranges freely through a portion of the mixed deciduous forest and grassland where there are Pleistocene fossil horse remains (*19a*). The contemporary horses in Santa Rosa eat substantial amounts of fallen fruit of jicaro as well as fruits of many other trees listed in Table 2.

In this park, as elsewhere in Mexico and northern Central America (*20*), jicaro grows above small patches of grass in diffuse, nearly monospecific stands. Scattered individuals also occur in the adjacent forest. Reaching a height of 3 to 4 meters, jicaro has the spreading and shrubby shape of a savanna tree (Fig. 3). It would not look out of place in Nairobi National Park in Kenya.

The spherical fruits of jicaro (6 to 15 centimeters in diameter) contain 200 to 800 seeds that are similar in size and shape to broadened cantaloupe seeds and are embedded in a slippery, fibrous pulp. Although the seeds are stiff and solid, they are more rubbery than hard. Toward the end of the dry season (March to May), and again in mid-rainy season (August to September), the still-green hard fruits fall from the tree. After a month or more the fruit turns brown and is ripe. There is a very thin layer of sugar on its outer surface at this time. During ripening, the inner light-colored pulp changes from one with a flat and slightly astringent flavor to a slimy black mass that is quite sweet. Despite a penetrating fetid odor, the pulp is quite palatable to humans (*21*). In horse-free habitats the indehiscent fruits lie on the ground and rot in the rainy season, and fermentation of the fruit pulp kills the seeds. A falling fruit occasionally cracks open on impact, but one of us (D.H.J.) has not found seedlings to be produced as a result. When the jicaro tree is in or near forest, an occasional fruit is chewed open by squirrels. These rodents remove the seeds from the fruit pulp and chew them up. This seed predation results in occasional seed dispersal, since the animal may carry the fruit to a site better protected from predators and drop some seeds along the way or leave some inside the fruit. The vast majority of jicaro fruits are not subject to this treatment.

Fig. 3. Adult *Crescentia alata* with full-sized immature fruit during the dry season. Naturally fallen ripening fruits are visible on the ground to the left of the tree (*19*).

When range horses are free to forage below the trees, they quickly consume the crop of jicaro fruits. The hard fruits are broken between the incisors (Fig. 4), an act that requires a pressure of about 200 kilograms (*22*). The gooey pulp is scooped out with the tongue and incisors and swallowed with little chewing. For more than ten consecutive days, three captive and well-fed range horses ate the fruit pulp of 10 to 15 fruits in each of two meals a day, one in the morning and one in the evening (*22*). A herd of 17 range horses broke and consumed 666 jicaro fruits in one 24-hour period (*22*). The percentage of seeds that survive passage through the gut of a horse is not known, but the dung becomes filled with viable jicaro seeds on the second day after the horse starts to eat the fruits. About 97 percent of these filled seeds germinate after they are washed out of the horse dung and placed on moist soil or paper. Seeds washed out of the pulp and placed on moist paper also show 97 percent germination. Sapling jicaro trees are commonplace in horse pasturing areas inside and outside of Santa Rosa National Park, provided that the habitats are not burned annually. Seedling and sapling jicaro trees are extremely rare in those areas of the park where horses do not have access, even in grass and forest habitats that have dense stands of adults and are rarely burned.

These observations indicate that Pleistocene horses were an important part of the disperser coterie of *Crescentia alata*. Since the Pleistocene horse evolved in the New World (*23*), there might even be elements of coevolution in the interaction of horses and jicaro fruits.

Today, jicaro and its congener (*Crescentia cujete*) are widespread in the drier parts of Central America (*20*). This distribution is probably the result of both the immediate pre-Columbian distribution and the post-Columbian spread of *Crescentia* by introduced horses. In addition, the hard fruits are used by humans as household tools such as bowls, ladles, and rattles, and the trees are therefore dispersed in this way too (*21*). At the time that the domestic horse was introduced, *C. alata* was very likely a relatively rare tree, occurring in small patches in relatively open vegetation such as marsh edges, along topographic breaks, and on floodplains, just as it is now in lowland Costa Rican habitats free of horses. With essentially no seed dispersal, the trees were limited to those sites where populations could survive with minimal seedling recruitment. The return of horses after 10,000 years resulted in intensified seed dispersal and has undoubtedly resulted in the appearance of more adult jicaro trees in many more kinds of habitats.

The postulated constriction of the range of *C. alata* after the extinction of the Pleistocene horse may affect other animals in the habitat. For example, nectarivorous bats would be affected by a reduction in jicaro density. The flowers of *C. alata* are nocturnal, abundant, and heavily visited by four species of nectarivorous bats in Guanacaste deciduous forests (*24*), and are the only common nectar source available to bats in the park forests during several months of the rainy season. The decline of the jicaro population would have strongly affected the population dynamics and structure of the many other plant species that are pollinated or dispersed by bats in the Central American deciduous forest lowlands.

Fig. 4. Range horse breaking a ripe fruit of *Crescentia alata* between its incisors (*19*).

Jicaro fruits are not the only fruits readily eaten by introduced horses. A similar interaction takes place between the fruits of *Enterolobium cyclocarpum* (guanacaste) (*25*), *Guazuma ulmifolia* (guacimo), and *Pithecellobium saman* (cenizero) and horses and cattle.

Additional Considerations

Partial loss of dispersal agents. Although some frugivores may be little more than fruit thieves (*26*) or deposit the seeds in lethal sites, a tropical tree usually has a complex seed shadow produced by several quite different types of animals (*12, 27*). Extinction of the Pleistocene megafauna would eliminate some of a tree's disperser coterie and thereby excise part of the tree's seed shadow. For example, two bat-generated seed shadows (*28*) of *Andira inermis* (Table 2) contained many fruits that fell below the parent tree and were passed over by pigs, cattle, and horses, perhaps because of antibiotic compounds in the fruit pulp. The seeds in such fallen fruits are killed by the larvae of weevils (*28*), and the fallen and wasted seeds were viewed by biologists as a cost of having a sloppy seed disperser and perhaps as due to the tree's being in an area where the human-disturbed bat populations are lower than those to which the fruiting behavior of the tree is genetically adjusted. However, we suspect that during the Pleistocene the fallen fruits would have been picked up by foraging gomphotheres, toxodons, and other animals that dispersed the nut-encased, soft seeds more effectively, and perhaps to quite different places.

Bats and other aerial or arboreal vertebrates would generally have taken their share of a fruit crop before it was available to the terrestrial megafauna, and therefore megafaunal extinction should have had little direct effect on them or the seed shadows that they generate. However, monkeys, squirrels, guans, and curassows, animals that forage for fruit both on the ground and in the tree crown, would have had more opportunity to harvest fruits after the megafauna extinction. Some increased seed dispersal by these groups could be expected and this might have compensated in part for the loss of the larger dispersers.

Response by seed predators. Vertebrate seed predators such as agoutis, peccaries, and small rodents experienced a substantial increase in their food supply after the megafaunal extinction. As food availability increased, so should their populations, habitat coverage, and species density.

Arthropod fruit eaters and seed predators were also affected by megafaunal extinction. Three species of *Cleogonus* weevils feed on the ripening fruit of *Andira inermis*, and their larvae develop in the fruit pulp and seeds of fallen fruits (*28*). If fruits were removed from below *Andira* trees by large vertebrates, there would not be the sizable weevil populations that there are at present. The density of *Zabrotes interstitialis* bruchids, and thus their intensity of seed predation on seeds of *Cassia grandis*, is greatly increased when the fruits are left on the trees until they rot (*29*). When a *Pithecellobium saman* fruit crop falls, its primary insect seed predator, *Merobruchus columbinus*, has just left the fruits (*30*); we suspect that the risk of being eaten by a large mammal (now extinct) accounts for the insects' rapid exits. Ripe fruits are rotted by their occupant microbes as a way of defending this resource against large herbivores (*31*); a major selective pressure for such microbial behavior disappeared when the Pleistocene Neotropical megafauna disappeared. Likewise, other associates of large mammals, such as dung beatles (Scarabaeidae), ticks, horse flies (Tabanidae), cowbirds, and vampire bats, must have been depleted by the loss of the Pleistocene megafauna.

Vegetative defenses against an extinct megafauna. The extinct tropical Pleistocene herbivores consumed substantial

Fig. 5 (left). Spines, 7 to 11 centimeters long, on the underside of the petiole of the leaf of sapling *Acrocomia vinifera* (*19*). Fig. 6 (right). *Desmodium* (Leguminosae) beggar's-ticks stuck to the forelegs of a free-ranging horse on the edge of the Costa Rican rain forest (*38*).

amounts of browse as well as fruits and seeds. We expect that some "functionless" but potentially defensive vegetative traits exhibited by trees in modern habitats are Pleistocene anachronisms. Spininess of African plants developed as a defense against large herbivores (*32*). There are numerous New World spiny plants in habitats where causal herbivores are missing. Spines on palm trunks are probably important in keeping climbing rodents from getting at developing fruits (for example of *Bactris* spp. and *Astrocaryum* spp.), but the long spines on leaves of *Bactris* and *Acrocomia* (Fig. 5) cannot be explained this way. An attempt to explain the spines without visualizing large browsing mammals as part of the interaction has led to construction of a model in search of a realistic selective pressure (*33*). In Santa Rosa National Park and elsewhere in Central America, prominent spines on the trunks and sometimes leaves of *Hura crepitans*, *Ceiba pentandra* (saplings only), *Ceiba aesculifolia*, *Acrocomia vinifera*, *Bombacopsis quinatum*, *Xanthoxylum setulosum*, and *Chlorophora tinctoria* (saplings only) are defenses of trees, especially young trees, against a browsing megafauna. Although such mechanical defenses may be diminishing because of the relaxation of selection for them, they have not yet disappeared. The recurved thorns on the twigs and leaves of *Mimosa guanacastensis*, *Pithecellobium platylobum*, *Acacia riparia*, *A. tenuifolia*, and *Mimosa eurycarpa* could easily have deterred ground sloths or gomphotheres. The same applies to the needle-sharp tips of the leaves of the understory shrub *Jacquinia pungens*, which is leafy in Costa Rica only during the dry season (*34*). On well-armed deciduous forest trees, the spines are commonly best developed within 4 to 6 meters of the ground in the neotropics just as they are on African trees. In open vegetation in southern Sonora, we observed that the shrubby cymbal-spine acacia, *Acacia cochliacantha*, is extremely thorny. Nearby taller conspecific trees growing in regenerated low forest are almost entirely unarmed.

External seed dispersal. Contemporary beggar's-ticks (*Desmodium* spp.) stick tightly to the hair of domestic horses (Fig. 6). Although they failed to adhere to the sleek coat of an adult captive tapir, or to that of a paca, collared peccary, and white-lipped peccary, experiments and observations by D.H.J. in Santa Rosa National Park show that the bur fruits of *Pisonia macranthocarpa*, *Desmodium* spp., *Krameria cuspidata*, *Triumfetta lappula*, *Aeschynomene* sp., *Petiveria alliacea*, and *Bidens riparia* stick tightly to the denser coats of horses and cattle. Except for *Pisonia* and *Krameria*, these plants are herbaceous; they depend on early colonization of open or nearly open ground for survival. With the loss of a megafauna we suspect that many of these plants declined severely in density and some even suffered local extirpation, as the once open and well-trampled habitats were reforested and as seeds were no longer dispersed by large shaggy beasts such as gomphotheres, toxodons, and ground sloths.

Discussion

In this addition to current evolutionary thought about the equilibrium state of contemporary neotropical habitats, we propose an answer to the riddle of why certain trees produce far more edible fruits than their current dispersal agents will remove, produce fruits that are not eaten by contemporary dispersal agents, bear fruits that resemble those eaten by African megafauna, and bear fruits avidly eaten by introduced livestock. These are traits of a megafaunal dispersal syndrome that has not been evolutionarily eradicated after the extinction of the dispersal agents 10,000 years ago. An alternative hypothesis is that these trees are not closely coevolved with particular frugivores and that the system is just very inefficient, as has been suggested for a Panamanian rain forest tree (*35*).

The fate of fruit crops in African game preserves is instructive in considering these two hypotheses. Observations by D.H.J. in Uganda and Cameroon forests suggest that it is indeed a rare event when the intact animal fauna does not consume all of the fallen fruit crop. For example, in a portion of Kibale Forest near Fort Portal, Uganda, where all the elephants had been killed, the fruits of *Balanites wilsoniana* (100 to 150 grams and 10 to 15 centimeters long) were abundant and rotting on the ground below parent trees. The fruits of *B. wilsoniana* contain a 40-gram nut and are about the same size and flavor as sapotaceous fruits of the Costa Rican rain forest which often lie rotting in large numbers below parent trees. *Balanites* fruits are swallowed by elephants (*36, 37*) and in the portions of Kibale Forest where elephants were numerous, all the fallen *Balanites* had been immediately and thoroughly removed by them. In this portion of Kibale, there are germinating *B. wilsoniana* seeds in elephant dung along forest trails.

Even if our hypothesis were to be rejected because it could be shown that in certain truly pristine neotropical habitats the extant animals can fully process the annual fruit fall, the intriguing matter of the fate of those seed species that were dispersed by Pleistocene mammals is not explained. Even if most population structures are now adjusted to the loss of the dispersal megafauna, we do not think that this is likely to be the case with evolutionary or coevolutionary equilibria. We doubt that those trees with lifespans of 100 to 500 years have experienced sufficient generations since the Pleistocene to replace the syndrome that is no longer highly functional. Let us assume that the agouti was once a trivial dispersal agent and figured primarily as a seed predator. With the removal of the Pleistocene megafauna, the agouti suddenly has the opportunity for a variety of evolved and coevolved interactions. However, it may well not have yet exploited the opportunity (*11*). It may shift its day-to-day activities in ways that serendipitously serve the dispersal needs of certain species of tree moderately well, even though no evolution has taken place in plant or animal.

Our discussion has focused on neotropical plants and animals, but it can be generalized to the sweet-fleshed large fruits of the Kentucky coffee bean *Gymnocladus dioica* and honey locust *Gleditsia triacanthos* (Leguminosae), osage orange *Maclura* (Moraceae), pawpaw *Asimina* (Annonaceae), and persimmon *Diospyros* (Ebenaceae). When there was a megafauna available to disperse their seeds, such genera may have been denser and had much wider ranges. The extreme spininess of various New World extra-tropical shrubs that are found in moist as well as arid regions has not been well explained. The vesicatory ripe fruits and weak-walled nuts of *Gingko biloba* might even have been evolved in association with a tough-mouthed herbivorous dinosaur that did not chew its food well.

References and Notes

1. P. S. Martin, *Nature (London)* **212**, 339 (1966).
2. ———, *Science* **179**, 969 (1973).
3. L. Van Valen, *Proc. N. Am. Paleo. Conf. Congr. E.* (1969), p. 469; L. Medway, *Asian Perspect.* **20**, 51 (1979).
4. B. Kurten and E. Anderson, *Pleistocene Mammals of North America* (Columbia Univ. Press, New York, 1980).
5. D. H. Janzen, *Principes* **15**, 89 (1971); R. A. Kiltie, thesis, Princeton University (1980); T. T. Struhsaker and L. Leland, *Biotropica* **9**, 124 (1977).
6. L. R. Heaney and R. W. Thorington, *J. Mammal.* **59**, 846 (1978).
7. N. Smythe, *Am. Nat.* **104**, 25 (1970).
8. D. F. Bradford and C. C. Smith, *Ecology* **58**, 667 (1977).
9. D. E. Wilson and D. H. Janzen, *ibid.* **53**, 954 (1973).
10. C. C. Smith, in *Coevolution of Animals and Plants*, L. E. Gilbert and P. H. Raven, Eds. (Univ. of Texas Press, Austin, 1975), pp. 53–77.
11. D. H. Janzen, *Evolution* **34**, 611 (1980).
12. F. J. Bonaccorso, W. E. Glanz, C. M. Sandford, *Rev. Biol. Trop.* **28**, 61 (1980).
13. M. J. C. Basáñez, *Biotica* **2**, 1 (1977); D. H. Janzen, unpublished observations.
14. M. D. Gwynne, *East Afr. Wild. J.* **7**, 176 (1969).
15. H. S. Gentry, *Carnegie Inst. Washington Publ. 527* (1942).
16. D. Y. Alexandre, *Terre Vie* **32**, 47 (1978); B. D. Burtt, *J. Ecol.* **17**, 351 (1929); R. M. Laws, *Oikos* **21**, 1 (1970); D. H. Janzen, *Ecology*, in press.
17. D. H. Janzen, *Oikos*, in press.
18. H. Ryden, *America's Lost Wild Horses* (Dutton, New York, 1978).
19. Santa Rosa National Park, on the Pacific coastal plain in extreme northwestern Costa Rica, in Guanacaste Province, has a mix of deciduous forest, old abandoned cattle pastures, and evergreen riparian forest (0 to 350 m in elevation). The photographs in Figs. 1 to 5 were taken there.

19a. L. D. Gomez (personal communication) reports that Pleistocene horse remains are common in the Costa Rican lowlands, particularly in Guanacaste Province.

20. J. Rzedowski and R. McVaugh, *Contrib. Univ. Michigan Herb.* **9**, 1 (1966).
21. I. C. Mercado, master's thesis, University of San Carlos, Guatemala (1975); J. F. Mortion, *Econ. Bot.* **22**, 273 (1968).
22. D. H. Janzen, unpublished data.
23. B. Patterson and R. Pascual, *Q. Rev. Biol.* **43**, 409 (1968).
24. D. J. Howell and D. Burch, *Rev. Biol. Trop.* **21**, 281 (1974); E. R. Heithaus, T. H. Fleming, P. A. Opler, *Ecology* **56**, 841 (1975).
25. D. H. Janzen, *Ecology* **62**, 587 and 593 (1981); *Biotropica* **13** (Suppl.), 59 (1981).
26. H. F. Howe and G. A. Vande Kerckhove, *Ecology* **60**, 180 (1979).
27. D. H. Janzen, *Annu. Rev. Ecol. Syst.* **10**, 13 (1979).
28. D. H. Janzen *et al.*, *Ecology* **56**, 1068 (1976).
29. D. H. Janzen, *ibid.* **52**, 964 (1971).
30. ———, *Trop. Ecol.* **18**, 162 (1977).
31. ———, *Am. Nat.* **111**, 691 (1977).
32. A. S. Boughey, *Ohio J. Sci.* **63**, 193 (1963); W. L. Brown, *Ecology* **41**, 587 (1960); A. I. Dagg and J. B. Foster, *The Giraffe: Its Biology, Behavior and Ecology* (Van Nostrand Reinhold, New York, 1976).
33. G. B. Williamson, *Principes* **20**, 116 (1976).
34. D. H. Janzen, *Biotropica* **2**, 112 (1970).
35. H. F. Howe, *Ecology* **61**, 944 (1980).
36. B. D. Burtt, *J. Ecol.* **17**, 351 (1929).
37. D. H. Janzen, in *Ecology of Arboreal Folivores*, G. G. Montgomery, Ed. (Smithsonian Institution Press, Washington, D.C., 1979), pp. 73–84.
38. The photograph was taken in Corcovado National Park, a lowland rain forest on the Osa Peninsula, southwestern Costa Rica.
39. C. S. Churcher, *Can. J. Zool.* **44**, 985 (1966).
40. M. O. Woodburne, *J. Mammal.* **50**, 121 (1969).
41. C. L. Gazin, *Smithson. Inst. Annu. Rep. 4272* (1956), p. 341.
42. M. Salmi, *Acta Geog.* **14**, 314 (1955).
43. J. Semper and H. Lagiglia, *Rev. Cient. Invest. Mus. Hist. Nat. San Rafael (Mendoza)* **1**, 89 (1968).
44. R. M. Hansen, *Paleobiology* **4**, 302 (1978).
45. J. Royo y Gomas, *International Geologic Congress Report, 21st session* (1960), part 4, p. 154.
46. W. D. Page, in *Early Man in America*, A. L. Brian, Ed. (Occasional Paper 1, Department of Anthropology, University of Alberta, 1968), pp. 231–262.
47. R. A. Stirton and W. K. Gealey, *Geol. Soc. Am. Bull.* **60**, 1731 (1949).
48. T. R. Howell, *Geol. Soc. Am. Spec. Pap.* **121**, 143 (1969).
49. G. G. Simpson and C. P. Coute, *Bull. Am. Mus. Nat. Hist.* **112**, 129 (1957).
50. M. J. Snarskis, H. Gamboa, O. Fonseca, *Vinculos* **3**, 1 (1977): L. D. Gomez (personal communication) reports mastodon remains from the provinces of San Jose, Alajuela, and Guanacaste.
51. A. L. Bryan, R. M. Casamiquela, J. M. Cruxent, R. Gruhn, C. Ochsenius, *Science* **200**, 1275 (1978).
52. A. Mones, *An. Inst. Nac. Antropol. Hist.* **7**, 119 (1970–1971).

53. This study was supported by NSF grants DEB 77-4889 and 80-11558 to D.H.J. and NSF grants to P.S.M. Servicio de Parques Nacionales de Costa Rica provided habitats, field support, and experimental animals and plants. W. Freeland and D. McKey aided in making African observations. R. M. Wetzel and I. Douglas-Hamilton aided in locating references. The manuscript was constructively criticized by J. Brown, W. Hallwachs, D. J. Howell, and G. G. Simpson. It is dedicated to W. A. Haber who first told D.H.J. that horses eat *Crescentia* fruits and to G. Vega who first made D.H.J. aware of the same fact.

Reporting of Faculty Time: An Accounting Perspective

Arthur L. Thomas

The Office of Management and Budget's requirements for 100 percent reporting of faculty time on federally sponsored research, set out in circular A-21 (*1*), are provoking controversy (*2*, *3*). A much broader reporting dilemma, of which 100 percent reporting is a special case, is rooted in a theoretical difficulty so radical that there is nothing comparable to it in our ordinary experience. Yet, accountants in the business world must prepare similar reports under like difficulties and have learned to live with such requirements. Their experience, and comprehension of the real nature of the theoretical difficulty, may assist government agencies, universities, and investigators. First, I will briefly describe the federal reporting requirements, their context, the uses made of the figures, and reasons advanced for objecting to such reports.

Terminology, Requirements, and Controversies

For many years, our federal government has supported basic research in the sciences through grants and contracts. To simplify a bit, these have reimbursed universities for two things: "direct costs" and, symmetrically, "indirect costs." The distinction between the two may be explained by an example. Let us suppose that the government is sponsor-

The author is Arthur Young Distinguished Professor of Business, School of Business, University of Kansas, Lawrence 66045. This article is adapted from the 4 February 1981 inaugural address for the professorship.

0036-8075/82/0101-0027$01.00/0

PAPER 30

Biological Bulldozers and the Evolution of Marine Benthic Communities (1979)

C. W. Thayer

Commentary

MARY DROSER

With the publication of his 1979 *Science* paper on biological bulldozers, Charlie Thayer was a pioneer of a revolution in paleoecology that signaled the beginning of *evolutionary* paleoecology: using the fossil record to understand ecological processes and patterns that are not recognizable on a modern timescale. Thayer extended the concept of ecological feedback to the evolutionary timescale by suggesting that biological disturbance (bioturbation) increased markedly through the Phanerozoic and had significant impact on the fate of soft substrate suspension feeders as well as the structure and composition of whole marine benthos. The "fits and starts" of increasing bioturbation, the impact of extinction, sediment mixing, and morphological adaptations are all part of his sweeping hypothesis, which was based on an anecdotal review of the available literature. The power and novelty of his surmises is evinced by the fact that nearly 35 years later, paleontologists and ecologists are still refining the patterns and process he outlined, rather than changing the paradigm. Furthermore, several lines of research build on concepts presented in this seminal paper. For example, recognition of changing substrates from mat grounds to soft grounds and the adaptations of marine benthos—although not explicitly discussed—are implied in his paper. These topics are prevalent across the paleontological literature today.

Thayer's insights are even more impressive when one considers that, in the late 1970s, ichnology, the study of trace fossils and the record of bioturbation, was still in its infancy, and thus Thayer necessarily tackled the issue of large-scale patterns in biological disturbance by examining the fossil record of bioturbators, since there was little data on the direct record of bioturbation. Thayer's use of the fossil record to trace the evolution of bioturbation elegantly demonstrated that the fossil record had unique information to provide on evolutionary ecology, a message that remains worth emphasizing today.

From *Science* 203:458–61. Reprinted with permission from AAAS.

β angles can be seen in some crystals in this [010] direction (11).

Any non-[010] pattern that contains (101)* or (10$\bar{1}$)*: gives an image consisting of fringes of roughly 7-Å spacing. One such direction is shown for Swedish hollandite in Fig. 2b. There is no disordering evident; all the electron diffraction patterns and images of Stuor Njouskes hollandite are consistent with its sharp, readily interpreted x-ray pattern.

Viewing along the crystal lengths in the Priceless sample (Fig. 2c), we can see two types of tunnels-square and rectangular, corresponding, respectively, to hollandite and romanechite. Apparently the similarity of the double chains in both hollandite and romanechite permits coherent intergrowths of the two minerals as schematically illustrated in Fig. 1. No evidence of ordering of the intergrowths has been found; the streaking in the diffraction pattern reflects the degree of disorder. The same sorts of features occur in samples from the Rattlesnake mine; intergrowths of romanechite and hollandite are shown along the tunnel lengths in the image (Fig. 2d) of the Rattlesnake sample.

We can obtain some insight into the romanechite structure from the b axis images and diffraction patterns. An ambiguity in the literature on the crystal system of romanechite has led to its being reported as both monoclinic (12) and orthorhombic (13). All the romanechite in the romanechite-hollandite mixtures we have thus far observed appears to be monoclinic. Further HRTEM study of romanechite samples is necessary to confirm this deduction as it is possible that the hollandite influences the intergrown romanechite material.

Recently, silicate chains with widths greater than triple have been imaged by HRTEM (14). This is also the case with the octahedral chains of the Mn oxides. We interpret Fig. 3a as containing a quadruple chain associated with hollandite and romanechite, whereas Fig. 3b contains a septuple chain. The structures with greater width have thus far only been seen as isolated chains; corresponding minerals are as yet unknown. However, the crystal structure of todorokite, a Mn oxide found as a major constituent of deep-sea Mn nodules, has not yet been solved. Its structure is thought to be closely related to the hollandite and romanechite structures (4), and so it could possibly be similar to one of the greater-width structures reported here. HRTEM should be of aid in giving a general idea of the todorokite structure.

The known complexity of the Mn oxides derives from the small crystal size-in some cases samples are x-ray amorphous-and the potential for chemical solid solution and structural intergrowths. The present study has demonstrated that intergrowths are common between the hollandite and romanechite structures. Furthermore, HRTEM has been used to show that even wider tunnels can occur as coherent intergrowths. In light of these observations, it is not surprising that the mineralogy of the Mn oxides is complex and has been confusing. It may be expected that HRTEM will be a powerful technique for studying and helping to unravel the structural and chemical complexities of the Mn oxides.

SHIRLEY TURNER

Department of Geology, Arizona State University, Tempe 85281

PETER R. BUSECK

Departments of Geology and Chemistry, Arizona State University

References and Notes

1. Psilomelane has also been used as a mineral name, but the Commission of New Mineral Names voted on 25 August 1969 to make romanechite the name for the specific mineral and psilomelane the general term (M. Fleischer, personal communication).
2. The exact formulas for both hollandite and romanechite are a matter of dispute in the literature. Two suggested formulas are: for hollandite, (Ba,K,Pb,Na)$_x$(Mn,Fe,Al)$_8$(O,OH)$_{16}$ (3, 7); for romanechite, (Ba,H$_2$O)$_4$Mn$_{10}$O$_{20}$ (12).
3. A. E. Ringwood, A. F. Reid, A. D. Wadsley, *Acta Crystallogr.* 23, 1093 (1967); A. F. Reid and A. E. Ringwood, *J. Solid State Chem.* 1, 6 (1969).
4. M. M. Mouat, *Am. Mineral.* 47, 744 (1962).
5. R. G. Burns and V. M. Burns, in *Proceedings of an International Symposium on Manganese*, A. Kozawa and R. J. Brodd, Eds. (Electrochemical Society, Princeton, N.J., 1975), pp. 306-327.
6. This cation variability results in a family of minerals with the hollandite structure: Pb-coronadite, K-cryptomelane, and Na-manjiroite.
7. D. E. Appleman and H. T. Evans, *U.S. Geological Survey Computer Contribution No. 20* (National Technical Information Service, Springfield, Va., 1973).
8. A. Byström and A. M. Byström, *Acta Crystallogr.* 3, 146 (1950).
9. P. R. Buseck and S. Iijima, *Am. Mineral.* 59, 1 (1974).
10. L. A. Bursill and A. R. Wilson, *Acta Crystallogr. Sect. A* 33, 672 (1977).
11. S. Turner and P. R. Buseck, in preparation.
12. A. D. Wadsley, *Acta Crystallogr.* 6, 433 (1953).
13. B. Mukherjee, *Mineral. Mag.* 35, 643 (1965).
14. J. L. Hutchison, D. A. Jefferson, L. G. Mallinson, J. M. Thomas, *Mater. Res. Bull.* 11, 1557 (1976); D. R. Veblen, P. R. Buseck, C. W. Burnham, *Science* 198, 359 (1977); D. A. Jefferson, L. G. Mallinson, J. L. Hutchison, J. M. Thomas, *Contrib. Mineral. Petrol.* 66, 1 (1978); M. Czank and P. R. Buseck, *Z. Krist.*, in press.
15. We thank J. M. Cowley, I. Mackinnon, and D. Veblen for valuable discussion; J. Hunt for microprobe work; and J. Wheatley for assistance in the electron microscope laboratory in the Center for Solid State Science at Arizona State University. We thank D. J. Fischer for recommending and the Rocky Mountain Mineralogical Society for awarding S.T. a scholarship. The research was supported in part by grant EAR 77-00128 from the National Science Foundation Earth Sciences Division.

2 August 1978; revised 2 October 1978

Biological Bulldozers and the Evolution of Marine Benthic Communities

Abstract. *During the Phanerozoic, the diversity of immobile suspension feeders living on the surface of soft substrata (ISOSS) declined significantly. Immobile taxa on hard surfaces and mobile taxa diversified. Extinction rates of ISOSS were significantly greater than in other benthos. These changes in the structure of benthic communities are attributed to increased biological disturbance of the sediment (bioturbation) by diversifying deposit feeders.*

Most marine fossils are preserved in fine-grained, originally soft sediments. In the Paleozoic these habitats were dominated by immobile suspension feeders (*1, 2*): articulate brachiopods, dendroid graptolites, tabulate and rugose corals, bryozoa, cystoids, blastoids, and Archeocyatha (the only known extinct phylum). Physically equivalent environments are now occupied primarily by deposit feeders (*3, 4*), for example, protobranch bivalves, irregular echinoids, and certain crustacea, holothurians, and annelids. I suggest that newly evolved deposit feeders and other sediment-disturbing taxa displaced the immobile suspension feeders on soft substrata (ISOSS).

Deposit feeders "mine" organic particles from mud or sand (*5, 6*), whereas suspension feeders filter food particles from the water column. In Recent marine communities, deposit feeders may exclude suspension feeders by bioturbation (*7, 8*). This includes suspending sediment or feces that foul biological filters, fluidizing mud substrata (*9*), accidental ingestion (*10*), as well as bulldozing: overturning or burial of ISOSS. The effects of bulldozing will be greatest when the deposit feeders are much larger than the ISOSS; juvenile ISOSS will be especially susceptible (*3, 6*). If the ISOSS cannot complete their life cycles before the sediment is disturbed, they are unlikely to persist. Scavengers and predators may also act as bulldozers (*3, 11, 12*).

Modern bulldozers—holothurians, irregular echinoids (*13*), malacostracan

0036-8075/79/0202-0458$00.50/0

crustacea—diversified after the Silurian (Table 1). They rework sediment an order of magnitude faster than do living taxa that were present in the early Paleozoic (annelids, protobranch bivalves) (*14-16*). Among the latter, ophiuroids have blind guts (*17*) and are virtually sedentary (*18*), while scaphopods are selective carnivores (*19*) and, by analogy with ecologically similar polychaetes (*20*), produce minimal bioturbation. Trilobites had small mouths, suggesting selective ingestion of small volumes. Their tracks indicate surface feeding (*21*) and minimal disturbance of sediment (*6, 22*). The feeding of *Limulus*, a modern trilobite analog, reinforces this inference (*12, 23*).

Thus rates of bioturbation probably increased several orders of magnitude since the Silurian, and the record of trace fossils supports this conclusion. Other factors being equal, the rate of bioturbation is proportional to the cross-sectional area of the burrower (*12, 24*). Throughout the Phanerozoic, the width (diameter) of trace fossils from shelf sediments increased, and systematic foraging became more efficient (*25*). Most important, bioturbation seems to have increased with time (*26*). Although the critical rate of bioturbation cannot be determined without knowing deposition rate, this uncertainty can be diminished by comparing sediments which accumulated in the same kind of environment. My observations of Devonian and Cretaceous deltaic deposits are consistent with the trend.

There are several possible explanations for the continued diversification of bulldozers. (i) Deposit feeders exploit a seasonally stable food source (*27, 28*). (ii) This resource probably increased through geologic time. Land plants now dominate global primary productivity (*29*) and contribute detritus that is an important food for marine deposit feeders (*30*). The diversification of both land plants and holothurians in the Devonian may be coupled. (iii) Many deposit feeders occupied a relatively stable environment within the sediment (*31*). None of the infaunal deposit feeding taxa in Table 1 have become extinct. (iv) Diversification of predators in the Mesozoic placed a premium on escape by "infaunalization" (*2, 32*). (v) Predators (such as brachyurans and naticids) "followed" prey into the sediment.

There has been a major reorganization of marine benthic communities over the span of the Phanerozoic (Fig. 1). All benthic groups show a statistically significant ($P < .05$) reduction in ISOSS since the Devonian and a corresponding increase in mobile taxa. Hard substrata are not bulldozed, and the occupants of this adaptive zone increased significantly. My conservative definition of ISOSS consigned many probable ISOSS to an intermediate adaptive zone (immobile on hard or soft substrata), which also experienced significant decline. When analysis is restricted to soft substrata, both brachiopods and bivalves as well as other benthos show a significant ($P < .001$) decline of ISOSS.

Articulate brachiopods are the dominant preserved macrobenthos of most marine communities in the Paleozoic, but during the late Paleozoic and early Mesozoic their abundance and diversity declined to present minor levels. All articulates were immobile suspension feeders (*33*) and many were ISOSS. The obvious early success of ISOSS brachio-

Table 1. First appearances of bioturbation agents (bulldozers). Only trilobites are extinct. Chronostratigraphic data from (*49*). Malacostracan crustacea italicized.

Age	Deposit feeders	Nondeposit feeders
Tertiary (T)	*Hippidea*	
	Amphipoda	
Cretaceous (K)	Tellinacean bivalves	
Jurassic (J)	Irregular urchins	*Brachyura*
	Thalassinidae	
Triassic (Tr)	*Mysidacea*	Burrowing suspension feeding bivalves
Permian (P)	*Leptostraca*	
	Isopoda	
	Cumacea	
	Tanaida	
Carboniferous (C)		
Devonian (D)	Holothuroids (*50*)	
Silurian (S)		Scaphopods
Ordovician (O)	Protobranch bivalves (*51*)	
	Ophiuroids	
Cambrian (Є)	Agnath fish (*52*)	
	Trilobites	
Precambrian	Annelids	

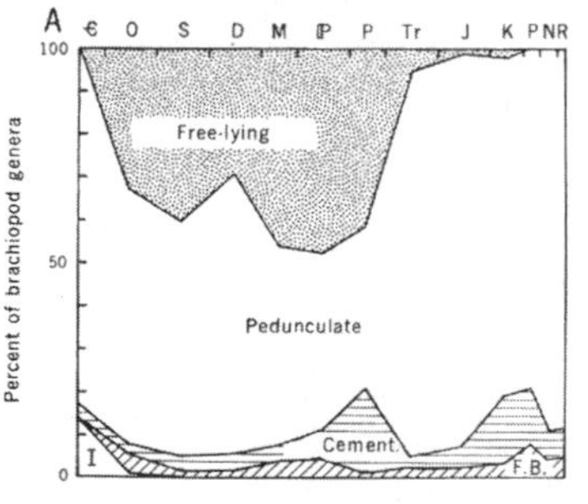

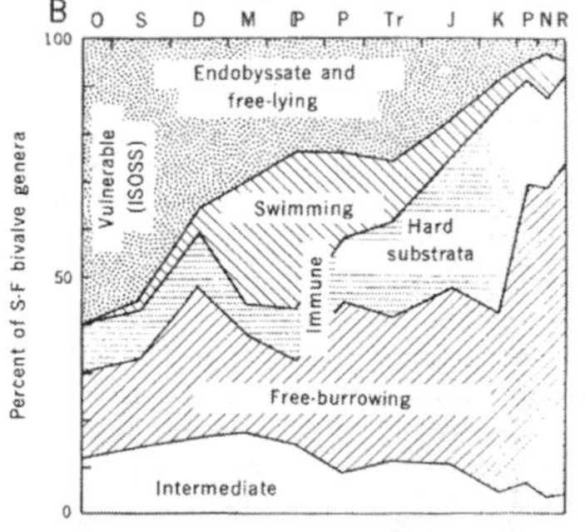

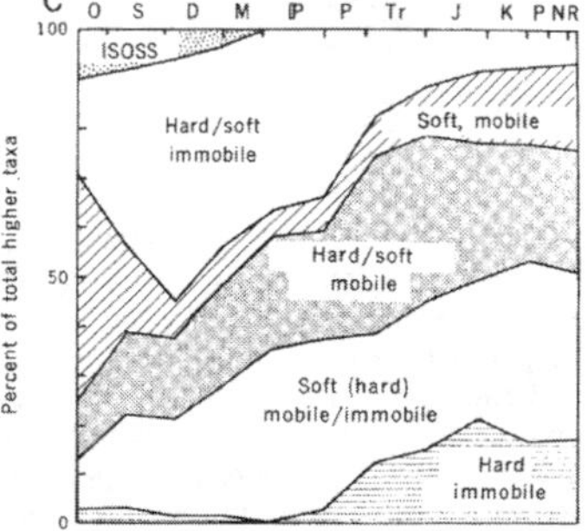

Fig. 1. Percentage of taxa in major adaptive zones as a function of geologic time. Brachiopods and bivalve mollusks are analyzed in detail because preserved, hard parts are reliable indicators of life habit. (A) Brachiopod genera. *Free-lying*, ISOSS; *I*, intermediate; *F.B.*, free-burrowing; *Cement*, and *Pedunculate*, animals attached to hard substrata. Chronostratigraphic data from (*53*). Life habits after (*53, 54*). (B) Genera of suspension feeding bivalves (*55*). Chronostratigraphic data from (*53*). Life habits after (*53, 56*). *Endobyssate*, byssally attached and partially buried in sediment. (C) Higher taxa (usually families) of well-skeletonized marine invertebrates (brachiopods and bivalves excluded). Chronostratigraphic data from (*49*). Twofold labels record hardness of substrate and mobility of organisms. Slashes represent intermediates: "hard/soft" inhabited soft bottoms but some attached to hard surfaces (such as dead shells). Often these objects were too small to stabilize the adults, which thus became functional ISOSS (for example, corals, bryozoa). "Soft (hard)" was predominantly soft.

pods posed an enigma that was compounded by the suggestion (*34*) that brachiopods are exceptionally tolerant of turbidity. If so, why do they no longer inhabit mud? Increased "bulldozing" may be responsible.

Stanley (*35, 36*) has described important evolutionary changes among the bivalves: a Mesozoic diversification of burrowing suspension feeders and a Paleozoic transition from partial burial and byssal attachment (ISOSS) to byssal attachment on hard substrata. Stanley attributed both radiations to the selective pressure of physically unstable sediments. But why did these changes occur relatively late in geologic time? In the early Paleozoic, bioturbation had not yet placed a premium on mobility or forced the ISOSS onto hard substrata.

The extinction at the end of the Permian was "one of the most striking events in the history of life"; a "crisis in earth history" (*37*), and the diversification of bulldozers probably contributed to it. In the brachiopods, bivalves, and other benthos (Fig. 1), the ISOSS suffered significantly ($P < .2$) greater extinction rates than non-ISOSS. For inhabitants of soft sediments only, P is $<.01$. Of course, bulldozing cannot explain the Permian extinction of pelagic taxa.

Living descendants of ISOSS occupy four types of refugia. (i) Articulate brachiopods are confined to hard substrates. Semi-infaunal byssate bivalves are abundant only in sediment which has been physically or biologically stabilized (*2*). Unlike Paleozoic corals, Recent scleractinians are more abundant and diverse on reefs as compared to soft bottoms. Perhaps their reef-forming ability (Triassic-Recent) was selected by the concurrent diversification of bulldozers. (ii) Mobile taxa such as gastropods, malacostraca, and regular echinoids were little affected by the Permo-Triassic "crisis." Originally immobile taxa have evolved mobility since the Paleozoic: comatulid crinoids (*38*), some ahermatypic corals (*39*), and lunulitiform bryozoa (*40*). Significantly, the most persistent "living fossils," *Lingula*, *Limulus*, and Monoplacophora, are mobile. (iii) The stalked crinoids typical of Paleozoic shelf seas are now concentrated in the deep sea (*38*) where bioturbation is extremely slow (*41*). Stromatolites survive where physical stress excludes burrowers and herbivores (*42*). (iv) Two bivalves, *Pinna* and *Placuna* persist as ISOSS on unstabilized sediment (*43*) and apparently grow rapidly to a "refuge-in-size."

Early deposit feeders (such as polychaetes and protobranchs) probably selected food particles from organic-rich sediments. After that adaptive zone had filled, a major resource remained unexploited: sediment with a low concentration of organic matter. Deposit feeders could not tap this food source until they evolved the capacity to process large volumes of sediment rapidly. This adaptive threshold delayed the development of holothurians, urchins, and thalassinids.

The diversification of these deposit feeders also may have been linked to the relatively late enrichment of marine sediments by land plants (*29, 30*). Infaunal suspension feeders did not diversify until bioturbation intensified. I attribute this to the energetics of infaunal as compared to epifaunal suspension feeders (ISOSS). Infaunal suspension feeders construct and maintain a conduit through which they pump large volumes of water. Furthermore, most are mobile. In contrast, ISOSS could deploy extensive passive filtration nets directly in the water column (for example, crinoids).

The Phanerozoic proliferation of bulldozers has other implications. The intensification of this relatively predictable disturbance (*44*) may have aided the increase in global diversity (taxon richness) of marine invertebrates (*26*). Regeneration (turnover rate) of nutrients from the sediment of the continental shelves was probably increased (*15, 29, 30, 45*), perhaps contributing to the Mesozoic diversification of phytoplankton (coccoliths, diatoms, dinoflagellates) and, via trophic linkage, zooplankton (radiolaria, foraminifera). Because of increased bioturbation, the temporal resolution of the stratigraphic record has probably decreased during the Phanerozoic. Biological mixing of strata obscures or obliterates the details of geologic history.

Bioturbation cannot be a panacea for extinction problems in general because there are as many possible extinction hypotheses as there are factors controlling the distributions of organisms. Complementary hypotheses invoke marine regression, plate tectonics, trophic resource regime (*46*), and increased predation (*2, 32*). However, modern predators appeared too late to contribute to the Permo-Triassic "crisis" and its long prelude (*47*). Nor does predation explain the differential success of immobile epifauna on hard versus soft substrata.

Unlike many extinction hypotheses, biological bulldozing offers testable corollaries. (i) Diversity and abundance of deposit feeders and other sediment reworkers increased; (ii) ISOSS decreased; (iii) ISOSS had greater extinction rates than non-ISOSS; (iv) among the immobile epifauna, diversity increased on hard substrata but decreased on soft; (v) rate of bioturbation increased with time; (vi) local extinctions of a given taxon were not synchronous (*37*); and (vii) relict ISOSS (such as endobyssate bivalves and stalked crinoids) occur in relatively nonbioturbated sediment.

Stochastic explanations have been offered for evolutionary patterns (*48*), but why be defeatist when there are credible deterministic alternatives? I have recorded some major evolutionary patterns and have suggested a specific cause: disturbance by deposit feeders and other bulldozers. This hypothesis is particulary appealing because it integrates a wide variety of data.

CHARLES W. THAYER
Department of Geology,
University of Pennsylvania,
Philadephia 19174

References and Notes

1. In terms of both abundance and diversity. For example, C. W. Thayer, *Lethaia* **7**, 121 (1974) and W. I. Ausich, *Geological Society of America Annual Meeting Abstracts* (1977), p. 884.
2. For example, S. M. Stanley, in *Patterns of Evolution*, A. Hallam, Ed. (Elsevier, New York, 1977), p. 209 and references therein.
3. S. A. Woodin, *J. Mar. Res.* **34**, 25 (1976).
4. R. G. Johnson, in *Approaches to Paleoecology*, J. Imbrie and N. D. Newell, Eds. (Wiley, New York, 1964), p. 107; E. G. Purdy, in *ibid.*, p. 238.
5. For example, A. C. Meyers, *J. Mar. Res.* **35**, 633 (1977).
6. For example, R. C. Aller and R. E. Dodge, *ibid.* **32**, 209 (1974).
7. D. C. Rhoads and D. K. Young, *ibid.* **28**, 150 (1970).
8. C. H. Peterson, *Mar. Biol.* **43**, 343 (1977).
9. D. C. Rhoads, in *Trace Fossils*, T. P. Crimes and J. C. Harper, Eds. (Seel House Press, Liverpool, 1970), p. 391.
10. G. Thorson, *Neth. J. Sea Res.* **3**, 267 (1966).
11. A. D. McIntyre, *J. Zool. (London)* **156**, 377 (1968); S. A. Woodin, *Ecol. Monogr.* **44**, 171 (1974).
12. D. C. Rhoads, *J. Geol.* **75**, 461 (1967).
13. Dendrochirotid holothuroids are suspension feeders [D. L. Pawson, in *Physiology of Echinodermata*, R. A. Boolootian, Ed. (Interscience, New York, 1966), p. 63] and some sand dollars (Astriclypeidae, Rotulidae, Scutellidae) may be [for *Dendraster*, see P. L. Timko, *Biol. Bull. (Woods Hole, Mass.)* **151**, 247 (1976)].
14. G. C. Cadée, *Neth. J. Sea Res.* **10**, 440 (1976); E. N. Powell, *Int. Rev. Gesamte Hydrobiol.* **62**, 385 (1977); R. H. Chesher, *Bull. Mar. Sci.* **19**, 72 (1969).
15. D. C. Rhoads, *Oceanogr. Mar. Biol. Ann. Rev.* **12**, 263 (1974).
16. C. W. Thayer, in preparation.
17. H. B. Fell, in *Physiology of Echinodermata*, R. A. Boolootian, Ed. (Interscience, New York, 1966), p. 129.
18. J. D. Woodley, *J. Exp. Mar. Biol. Ecol.* **18**, 29 (1975). Although some are carnivores or suspension feeders, most are deposit feeders (*17*).
19. J. E. Morton, *J. Mar. Biol. Assoc. U.K.* **38**, 225 (1959).
20. D. C. Gordon, *Limnol. Oceanogr.* **11**, 327 (1966).
21. R. A. Osgood, in *Study of Trace Fossils*, R. W. Frey, Ed. (Springer-Verlag, New York, 1975), p. 87.
22. A. Rowe, in *Deep-Sea Sediments*, A. Inderbitzen, Ed. (Plenum, New York, 1974), p. 381; D. Cullen, *Nature (London)* **242**, 323 (1973).
23. D. Schneider, *Nature (London)* **271**, 353 (1978).
24. D. J. W. Piper and N. F. Marshall, *J. Sed. Petrol.* **39**, 601 (1969); D. C. Rhoads (*12*) found insignificant disturbance by small polychaetes.

25. A. Seilacher, in *Patterns of Evolution*, A. Hallam, Ed. (Elsevier, New York, 1977), p. 359.
26. R. K. Bambach, *Paleobiology* **3**, 152 (1977).
27. M. A. Rex, *Deep-Sea Res.* **23**, 975 (1976); J. S. Levinton, *Am. Nat.* **106**, 472 (1972).
28. J. S. Levinton, *Palaeontology* **17**, 579 (1974).
29. D. M. McLean, *Science* **200**, 1060 (1978).
30. K. R. Tenore, in *Ecology of Marine Benthos*, B. C. Coull, Ed. (Univ. of South Carolina Press, Columbia, 1977), p. 37; K. L. Smith, Jr., *Mar. Biol.* **47**, 337 (1978).
31. C. W. Thayer, *Science* **186**, 828 (1974).
32. G. J. Vermeij, *Paleobiology* **3**, 245 (1977).
33. Chonetids may have been mobile; M. J. S. Rudwick, *Living and Fossil Brachiopods* (Hutchinson, London, 1970).
34. H. M. Steele-Petrović, *J. Paleontol.* **49**, 552 (1975).
35. S. M. Stanley, *ibid.* **42**, 214 (1968).
36. ——, *ibid.* **46**, 165 (1972).
37. F. H. T. Rhodes, in *The Fossil Record*, W. B. Harland *et al.*, Eds. (Geological Society, London, 1967), p. 57.
38. D. L. Meyer and D. B. Macurda, Jr., *Paleobiology* **3**, 74 (1977).
39. G. A. Gill and A. G. Coates, *Lethaia* **10**, 119 (1977).
40. R. Greely, *Geol. Soc. Am. Bull.* **78**, 1179 (1967); P. L. Cook, *Cah. Biol. Mar.* **4**, 407 (1963).
41. G. P. Glasby, *N. Z. J. Sci.* **20**, 187 (1977); B. C. Heezen and C. D. Hollister, *The Face of the Deep* (Oxford Univ. Press, New York, 1971); D. R. Schink and N. L. Guinasso, Jr., *Mar. Geol.* **23**, 133 (1977).
42. P. Garrett, *Science* **169**, 171 (1970).
43. Personal observation, at Discovery Bay, Jamaica, and J. Hornell [Report to the Government of Baroda on the marine zoology of Okhamandel in Kattiawar (1909)].
44. Compare P. Dayton and R. Hessler, *Deep-Sea Res.* **19**, 199 (1972).
45. Compare G. T. Rowe and K. L. Smith, Jr., in *Ecology of Marine Benthos*, B. C. Coull, Ed. (Univ. of South Carolina Press, Columbia, 1977), p. 55; D. C. Rhoads, R. C. Aller, M. B. Goldhaber, in *ibid.*, p. 113; D. C. Rhoads, K. Tenore, M. Brown, *Estuarine Res.* **1**, 563 (1975); G. R. Lopez, J. S. Levinton, L. B. Slobodkin, *Oecologica* **30**, 111 (1977).
46. N. D. Newell, *Geol. Soc. Am. Spec. Pap.* **89**, 63 (1967); J. W. Valentine, *Evolutionary Paleoecology of the Marine Biosphere* (Prentice-Hall, Englewood Cliffs, N.J., 1973).
47. Most extinctions were preceded by a long diversity decline beginning as early as the Devonian [data from (*49*)].
48. D. M. Raup, S. J. Gould, T. J. M. Schopf, D. S. Simberloff, *J. Geol.* **81**, 525 (1973); S. J. Gould, D. M. Raup, J. J. Sepkoski, Jr., T. J. M. Schopf, D. S. Simberloff, *Paleobiology* **3**, 23 (1977); C. A. F. Smith, III, *ibid.*, p. 41; R. W. Osman and R. B. Whitlach, *ibid.* **4**, 41 (1978).
49. W. B. Harland *et al.*, Eds., *The Fossil Record* (Geological Society of London, 1967).
50. Appeared in Ordovician [C. R. C. Paul, in *Patterns of Evolution*, A. Hallam, Ed. (Elsevier, New York, 1977), p. 123] but was minor until Devonian.
51. May have originated in the Cambrian [J. Pojeta and B. Runnegar, *Am. Sci.* **62**, 706 (1974)], but these early bivalves may have been epifaunal suspension feeders [M. J. S. Tevez and P. L. McCall, *Paleobiology* **2**, 183 (1976)]; J. A. Allen and H. L. Sanders, *Malacologia* **7**, 381 (1969).
52. J. E. Repetski, *Science* **200**, 529 (1978).
53. R. C. Moore (ed. 1) and C. Teichert (ed. 2), Eds., *Treatise on Invertebrate Paleontology* (Univ. of Kansas Press, Lawrence, 1953–1970).
54. R. R. Alexander, *Paleogeog. Paleoclimat. Paleoecol.* **21**, 209 (1977); C. W. Thayer and H. M. Steele-Petrović, *Lethaia* **8**, 209 (1975).
55. Deposit feeders may be less susceptible to extinction than suspension feeders (*28*) and are excluded. This reduces the increase of burrowers.
56. S. M. Stanley, *Geol. Soc. Am. Mem.* **125**, 1 (1970).
57. I thank E. Bird, L. Hammond, J. B. C. Jackson, H. Faul, R. McHorney, C. Soukup, G. Vermeij, S. A. Woodin, Friday Harbor Laboratories, and Discovery Bay Marine Laboratory for their contributions. Supported by NSF grant OCE76-04387.

12 May 1978; revised 26 September 1978

Chemotactic Factor-Induced Release of Membrane Calcium in Rabbit Neutrophils

Abstract. *The interaction of chemotactic factors (fMet-Leu-Phe and C5a) with rabbit neutrophils leads to rapid and specific release of membrane calcium, as evidenced by changes in the fluorescence of cell-associated chlorotetracycline. These two structurally different stimuli appear to interact with the same pool of membrane calcium.*

Chemotactic and secretory stimuli alter the ionic permeability of the plasma membrane and the intracellular concentrations of exchangeable calcium in neutrophils (*1*, *2*). The molecular mechanisms that cause these effects are of interest for an understanding of neutrophil physiology in particular and the contractile activities of nonmuscle cells in general. We reported recently that the interaction of chemotactic factors with neutrophil membranes leads to release of calcium from previously bound stores, and we postulated that the plasma membrane or other membranous cell components may act as such calcium stores (*2*). This tentative conclusion was based on indirect evidence derived from studies dealing with the effect of chemotactic factors on the movement of ^{45}Ca across rabbit neutrophil membranes. We report here the results of experiments that implicate membrane-associated calcium, and its release by chemotactic factors, in the initial events involved in neutrophil activation.

The fluorescence characteristics of the chelate probe chlorotetracycline are extremely sensitive to the concentrations of divalent cations within the hydrophobic environment into which it is preferentially partitioned. Chlorotetracycline fluoresces more intensely when complexed to divalent cations, and its emission and excitation spectra can be used to differentiate between its Ca^{2+} and Mg^{2+} chelates (*3*).

We took advantage of these fluorescence characteristics of chlorotetracycline, particularly the dependence on Ca^{2+}, to investigate the effects of chemotactic factors on the fluorescence of chlorotetracycline-loaded neutrophils. Rabbit peritoneal polymorphonuclear leukocytes (neutrophils) were incubated for 45 to 60 minutes at 37°C in the presence of 100 μM chlorotetracycline (Sigma Chemical Co., St. Louis, Mo.) in Hanks balanced salt solution containing 10 mM Hepes [4-(2-hydroxyethyl)-1-piperazineethanesulfonic acid] (pH, 7.3), 1.7 mM Ca^{2+}, and no Mg^{2+}, to which glucose and bovine serum albumin were added at 1 mg/ml each. The desired number of cells were then washed once, resuspended in Hanks balanced salt solution without Mg^{2+} and bovine serum albumin and with or without 0.5 mM Ca^{2+}, and transferred to a Perkin-Elmer MPF 2A fluorescence spectrophotometer equipped with a temperature control cuvette holder and stirrer. All the experiments were performed at 37°C with a cell density of 2.5×10^6 cells per milliliter. The viability of the cells, as measured by lactate dehydrogenase release, and their functional responsiveness, as indicated by their ability to release lysosomal enzymes in the presence of cytochalasin B and fMet-Leu-Phe (*4*), were found not to be affected by these experimental manipulations. Excitation and emission wavelengths were 390 and 520 nm, respectively. The excitation and emission slits were adjusted to maximize the signal-to-noise ratio. These experimental conditions are essentially the same as those described by other workers who used chlorotetracycline as a calcium probe (*3*). The synthetic chemotactic factor fMet-Leu-Phe was obtained as previously described (*5*). Its competitive antagonist Boc-Phe-Leu-Phe-Leu-Phe (*4*) was provided by R. J. Freer, Medical College of Virginia, Richmond. Partially purified C5a, the low-molecular-weight chemotactic fragment of the fifth component of complement, was generated by trypsin treatment of C5 as described by Cochrane and Müller-Eberhard (*6*). These preparations of C5a exhibited maximal biological activity (lysosomal enzyme release) at a dilution of 1 to 1000.

Immediately after the addition of either fMet-Leu-Phe or C5a, a rapid decrease in the fluorescence of chlorotetracycline-loaded neutrophils was observed (Fig. 1). Comparison of Fig. 1, A to C, shows that essentially similar results were obtained whether or not extracellular calcium was present at the time of stimulation. This indicates that the fluorescence signal reflects one of the initial molecular events that follows the binding of the chemotactic factor to its receptor rather than the net influx of calcium that occurs when chemotactic factors are added in the presence, but not in the absence, of extracellular Ca^{2+} (*1*, *2*). The dependence of the fluorescence changes on the concentration of the chemotactic factors is illustrated in Fig. 1B. Fluores-

PAPER 31

The Mesozoic Marine Revolution: Evidence from Snails, Predators, and Grazers (1977)

G. J. Vermeij

Commentary

PATRICIA H. KELLEY

A marked restructuring of shallow marine benthic communities occurred during the middle and late Mesozoic, spurred by an intensification of grazing and the diversification of durophagous (shell-breaking) predators. Bioturbation and infaunal life modes became more common, attached epifauna disappeared from shallow-water habitats, and prey shells became sturdier. Although previous authors such as Steven Stanley had noted some of these changes, Vermeij applied the memorable term "Mesozoic marine revolution" to this reorganization of communities, stimulating a wealth of research.

This paper documents qualitatively the evolution of gastropod shell architecture, including a post-Jurassic increase in the ability to remodel the shell interior that permitted antipredatory morphologies such as narrow apertures, apertural teeth, and pronounced ornamentation. The paper compares the timing of these architectural changes with that of diversification of key durophages and grazers (teleost fish, decapod crustaceans, carnivorous gastropods, and echinoids). Vermeij suggests the increase in durophagy and grazing produced not only changes in shell architecture but also sweeping reorganization of the benthos. Infaunal taxa (siphonate bivalves) largely replaced epifauna; mobile epifauna exhibited changes in shell structure; and attached epifauna (articulate brachiopods and crinoids) retreated to cryptic or deep-water habitats. The paper speculates about the processes (coevolution, species selection) and the environmental context (warm climate, continental breakup) that produced the Mesozoic marine revolution.

The paper fueled interest in evolutionary paleoecology, which recognizes causal links between ecological and macroevolutionary processes, by arguing convincingly that ecological processes such as predation affect biotic organization and diversity profoundly. It introduced concepts articulated more fully a decade later as Vermeij's escalation hypothesis (1987), which claims that biological hazards such as predation, and adaptations to those hazards, have increased during the Phanerozoic. Escalation, driven by adaptation to enemies, remains a fruitful though highly debated hypothesis in macroevolution. The Mesozoic marine revolution and other intervals of change potentially driven by predation (Cambrian and mid-Paleozoic) are not yet fully understood (e.g., regarding timing, causality, and links between benthic and planktic communities). However, the research spurred by Vermeij's paper has brought us closer to answering important questions such as: Do ecological interactions matter over evolutionary time? And what role do biotic interactions play in evolution?

Literature Cited

Vermeij, G. J. 1987. *Evolution and Escalation: An Ecological History of Life*. Princeton, NJ: Princeton University Press.

Paleobiology. 1977. vol. 3, pp. 245–258.

The Mesozoic marine revolution: evidence from snails, predators and grazers

Geerat J. Vermeij

Abstract.—Tertiary and Recent marine gastropods include in their ranks a complement of mechanically sturdy forms unknown in earlier epochs. Open coiling, planispiral coiling, and umbilici detract from shell sturdiness, and were commoner among Paleozoic and Early Mesozoic gastropods than among younger forms. Strong external sculpture, narrow elongate apertures, and apertural dentition promote resistance to crushing predation and are primarily associated with post-Jurassic mesogastropods, neogastropods, and neritaceans. The ability to remodel the interior of the shell, developed primarily in gastropods with a non-nacreous shell structure, has contributed greatly to the acquisition of these antipredatory features.

The substantial increase of snail-shell sturdiness beginning in the Early Cretaceous has accompanied, and was perhaps in response to, the evolution of powerful, relatively small, shell-destroying predators such as teleosts, stomatopods, and decapod crustaceans. A simultaneous intensification of grazing, also involving skeletal destruction, brought with it other fundamental changes in benthic community structure in the Late Mesozoic, including a trend toward infaunalization and the disappearance or environmental restriction of sessile animals which cannot reattach once they are dislodged. The rise and diversification of angiosperms and the animals dependent on them for food coincides with these and other Mesozoic events in the marine benthos and plankton.

The new predators and prey which evolved in conjunction with the Mesozoic reorganization persisted through episodes of extinction and biological crisis. Possibly, continental breakup and the wide extent of climatic belts during the Late Mesozoic contributed to the conditions favorable to the evolution of skeleton-destroying consumers. This tendency may have been exaggerated by an increase in shelled food supply resulting from the occupation of new adaptive zones by infaunal bivalves and by shell-inhabiting hermit crabs.

Marine communities have not remained in equilibrium over their entire geological history. Biotic revolutions made certain modes of life obsolete and resulted in other adaptive zones becoming newly occupied.

Geerat J. Vermeij. Department of Zoology. University of Maryland, College Park, Maryland 20742

Accepted: February 21, 1977

Introduction

When we view the history of life from its beginnings some three billion years ago, we seem to witness a pattern of comparatively sudden revolutions interspersed with long intervals of relative quiescence. The most notable revolutions, the development of hard parts in the Early Paleozoic and the marine events of the Late Mesozoic, are not the instantaneous take-overs or inventions with which we identify revolutions in human history, but lasted tens of millions of years; yet, these episodes are short relative to the hundreds of millions of years when comparatively little fundamental change took place in community organization.

In the Early Cambrian, diverse marine groups developed skeletons of calcium phosphate, calcium carbonate, and silica. Some groups were earlier in this development than others (for a review see Stanley 1976a), but by the Middle Cambrian most groups which were to play an important role in the fossil record had become skeletonized. By early Ordovician time, marine communities had acquired fundamental characteristics which they would retain for more than three hundred million years during the remainder of the Paleozoic and Early Mesozoic. During this period, reefs came and went (for reviews see Newell 1971; and Copper 1974), and groups diversified and declined; but most marine benthic communities were dominated by epifaunal

US ISSN 0094–8373

(surface-dwelling) or semi-infaunal elements (Stanley 1972, 1977).

Then, inexplicably, things began to change. Beginning in the Jurassic and continuing at an accelerated pace in the Cretaceous, changes in life habits took place which fundamentally altered relationships among organisms in shallow-water marine communities. These currents of change were unaffected by, or only temporarily delayed, by episodes of extinction, including the crisis at the end of the Cretaceous (Maastrichtian-Danian transition). Stanley (1977) and Meyer and McCurda (1977) have already treated some aspects of this reorganization, and I have speculated on some curious changes in shell geometry which affected gastropods in the Mesozoic (Vermeij 1975). The consensus reached from these investigations is that the intensity of predation has increased substantially since the Middle Mesozoic.

In this paper, I shall expand upon this theme. From further evidence of skeletal geometry and other data, I shall argue that predation as well as grazing has intensified and become more destructive to skeletons; and I shall ask, mostly in vain, what triggered this Mesozoic reorganization.

Trends in Gastropod Morphology

Tertiary and Recent marine gastropods differ from Mesozoic and especially from Paleozoic snails in many architectural ways and include in their diverse ranks a complement of mechanically sturdy forms quite unknown in earlier times. First, the incidence of shells possessing an umbilicus (cavity in the base of the cell created by the incomplete overlap of adjacent whorls) has significantly decreased in the course of geological time (Vermeij 1975). Paleozoic gastropod faunas were heavily dominated by archaeogastropods and by bilaterally symmetrical bellerophontaceans. (For the purposes of this paper, I shall here regard Bellerophontacea as belonging to the class Gastropoda, and follow for convenience the classificational scheme of Taylor and Sohl (1962). Runnegar and Pojeta (1974) have presented arguments in favor of transferring the bellerophonts to the Monoplacophora; see also Golikov and Starobogatov (1975) for further discussion.) I have estimated that 74% of the genera in the largely Paleozoic and Early Mesozoic superfamily Pleurotomariacea had umbilicate shells, while only 58 to 60% of the genera in the Mesozoic and Cenozoic Trochacea possess umbilici (Vermeij 1975). Non-umbilicate shells with often relatively high spires, placed by Knight et al. (1960) in the Murchisoniacea, Subulitacea, and Loxonematacea, were present from Late Cambrian or Early Ordovician time onward; but the great expansion of the largely non-umbilicate mesogastropods and neogastropods did not take place until the latter half of the Mesozoic. Thus, archaeogastropods still predominated in Triassic faunas, but Jurassic and Cretaceous assemblages already contained a large diversity of mesogastropods. Siphonate neogastropods and higher mesogastropods constitute the majority in some Late Cretaceous and nearly all Cenozoic gastropod assemblages (Sohl 1964).

The trend in the reduction in frequency of umbilicate shells seems to reflect an overall increase in the mechanical resistance of shells to crushing (Vermeij 1976). The large umbilicus and high whorl expansion rate (W) render the West Indian trochid *Cittarium pica* vulnerable to predation by crabs at larger shell diameters than the taxonomically and ecologically related western Pacific *Trochus niloticus*, which has a lower W and a smaller umbilicus (Vermeij 1976).

Several additional features of gastropod shell architecture support the thesis that predation intensity due to crushing enemies has increased since the Middle Mesozoic. The diversity, and especially the size, of bilaterally symmetrical (usually umbilicate) coiled shells among the gastropods has dramatically decreased since the Late Cretaceous (Vermeij 1975). Today, such planispiral shells are found principally in fresh water and on land, where they may still attain diameters as much as 5 cm. In the sea, however, all planispirally coiled snails are small (usually less than 0.5 cm in diameter), and most are found among algae, under stones, or in the plankton. Predation by shell crushing seems to be unimportant for very small snails (see e.g. Hobson 1974; Stein et al. 1975) and is likely to be of greater consequence for snails living on exposed rock surfaces than for those living among algae or under rocks (Vermeij 1974). In the latter two habitats, constraints on rapid locomotion and visual detection limit the activities of such potential crushing predators as fishes and crabs. Planispiral coiling among planktonic gastropods may be related to their swimming mode of life and seems to be associated with rapid locomotion. Most planktonic predators, moreover, do not appear to use crushing as a means of prey capture.

Pre-Cretaceous seas, in contrast, often supported large planispiral or near-planispiral snails (Macluritacea, Euomphalacea, Bellerophontacea, Pleurotomariacea, etc.), most of which were probably members of the benthic epifauna (Vermeij 1975). At least in some macluritaceans and euomphalaceans, the planispiral geometry was not associated with rapid locomotion but rather with a sedentary mode of life on quiet bottoms (Yochelson 1971).

Rex and Boss (1976) have summarized the taxonomic distribution of the 15 Recent species of gastropods exhibiting open coiling. Open coiling is defined by Rex and Boss (1976) as regular coiling based on the logarithmic spiral, in which adjacent whorls do not touch one another; thus, such groups as *Vermicularia*, *Siliquaria*, and Vermetidae, whose coiling is irregular and non-logarithmic, are excluded. Of the 15 Recent species, only 8 are marine, 3 are found in fresh water, and 4 are terrestrial. From estimates of Recent molluscan species numbers given by Boss (1971), I calculate that open coiling is found in about 1 out of every 2,200 marine gastropod species, 1 out of every 1,000 fresh-water species, and 1 out of every 3,750 terrestrial species of shelled gastropod. The fresh-water forms with open coiling all occur in cold lakes, where crushing predation is rare (Vermeij and Covich 1977). Most of the marine open-coiled snails are found either in very deep water (*Lyocyclus*, *Epitonium*) or in sand associated with cnidarians (*Spirolaxis*, *Epitonium*). The habits of *Extractrix* and *Callostracum* are not known but are probably sandy or muddy, judging from the distribution of other members of their respective families (Cancellariidae and Turritellidae). Predation through crushing is probably relatively uncommon in most of these environments. Bright (1970), for instance, found few shelled invertebrates in the diet of 81 deep-sea fish individuals he examined from the Gulf of Mexico, and these were ingested whole. Cnidarians often give associated animals a measure of protection from potential predators (see for example McLean and Mariscal 1973).

Open coiling was apparently much commoner in the Paleozoic and Early Mesozoic than now. From various sources (Knight et al. 1960; Wanberg-Eriksson 1964; Yochelson 1971), I have counted some 18 to 20 genera of Paleozoic, and 2 genera of Triassic gastropods with open coiling. Some of these, such as the Paleozoic *Macluritella* and various euomphalids, may have rested on soft muds (Yochelson 1971), while others (e.g. the Platyceratidae and the Late Devonian and Triassic Tubinidae) may have been sedentary reef-associated forms (Bowsher 1955; Knight et al. 1960).

Paleozoic nautiloids and Mesozoic ammonoids also included in their ranks many shells with open and sometimes irregular coiling, which could not have been well suited for rapid swimming (Kummel and Lloyd 1955). The only Recent externally shelled cephalopod (genus *Nautilus*) has rather tight coiling, and the only living cephalopod with open coiling (*Spirula*) lives at moderate depths and has an essentially internal shell. It is very likely that the fragile open coiling of fossil cephalopods would, if these animals were alive today, restrict them to great depths or to other marginal habitats where shell-destructive predation is rare.

The ability to remodel the interior of the shell has contributed greatly to the increased sturdiness of Late Mesozoic and especially Cenozoic gastropods (Mesogastropoda, Neogastropoda, and Neritacea). Carriker (1972) has suggested that the external spines of many muricid shells are resorbed by the left edge of the mantle as the exterior of the shell is obliterated from view by the encroaching inner lip during spiral growth. Spines of this type serve to increase the effective size of the shell, and often strengthen it against crushing or other destruction (Vermeij 1974 and references therein). Without resorption or remodeling, these spines and other collabral sculptural elements (those parallel to the outer lip) would protrude into the aperture and, if large enough, might interfere with retraction of the body into the shell. Although such protrusions are actually known in some fresh-water shells (e.g. *Biomphalaria*) and land snails (*Helicodiscus* and other genera), they are apparently exceedingly rare among marine gastropods. In the archaeogastropods (excluding Neritacea), where resorption appears to be poorly developed or altogether absent, this problem can be avoided by either (1) covering the sculpture with a thick inductural glaze, or (2) producing shells with reduced whorl overlap so that interference with previously built sculpture is minimal (see also Vermeij 1973). The first solution demands a relatively broad aperture and would restrict the sculptured surface to the body whorl. Examples of the second solution, which implies a tall spire that can often be cracked by predators, are provided by the sometimes open-coiled, strongly ribbed mesogastropod *Epitonium* and by the trochid archaeogastropod *Tectus*, which is sculptured with one or more spiral rows of knobs on the upper face of the whorls and is further characterized by a spire unusually high for an archaeogastropod (apical half-angle of Palauan *T. triseriata* is 26.3°).

In view of these and other limitations imposed by general shell geometry on sculpture (see also Vermeij 1971, 1973), it is perhaps not surprising that few Paleozoic snails had shells in which nodes, spines, or strong collabral ridges were the dominant sculptural elements. Examples of periodically produced collabral lamellae are known in some bellerophonts (the Middle to Upper Ordovician *Phragmolites*, the Middle Ordovician to Silurian *Temnodiscus*, and the nearly open-coiled Middle Silurian *Pharetrolites*). Collabral threads often associated with nodes or even spines were developed in the Lower to Middle Permian portlockiellid *Tapinotomaria* and by various Upper Devonian open-coiled Tubinidae. The high-spired Pseudozygopleuridae of the Carboniferous and Permian were usually characterized by evenly spaced collabral ridges.

In the pre-Cretaceous Mesozoic we see a larger number of gastropods with collabral sculpture: some Triassic and Jurassic Helicotomidae (Macluritacea); Triassic Porcelliidae, Zygitidae, and Jurassic Pleurotomariidae (Pleurotomariacea); open-coiled Triassic Tubinidae (Oreostomatacea); some Turbinidae (Trochacea), Middle Triassic and younger Cirridae and Amberleyidae (Amberleyacea); high-spired Triassic and Jurassic Zygopleuridae (Loxonematacea), Middle Jurassic and younger Neritidae (Neritacea); and Jura-Cretaceous high-spired nerineaceans. Yet, it is primarily the Cretaceous and Cenozoic mesogastropods and neogastropods in which nodes, spines, and collabral ribs have been most elaborately developed. Some contemporaneous trochid, turbinid, and cyclostrematid archaeogastropods have also produced such sculpture (recall the trochid *Tectus*, for example), but their numbers are less impressive.

Similarly, extensive development of protective dentition on the inner face of the outer lip is predominantly a post-Jurassic innovation which, in such groups as Neritidae, Muricidae, and many other mesogastropods and neogastropods would be impractical if it could not be secondarily resorbed as the position of the growing edge moves in a spiral direction. Without remodeling, teeth restricting accessibility into the aperture could be developed only in the adult stage of species with determinate growth, when spiral growth has ceased. Indeed, such teeth are commonly seen in cowries (Cypraeidae), helmet shells (Cassidae), and many tritons (Cymatiidae), as well as in many Paleozoic nautiloids and Mesozoic ammonoids. These molluscs forfeited the luxury of apertural defense during the active spiral phase of growth, when most gastropods are particularly vulnerable to predation.

The first snail with outer-lip dentition seems to be the Middle Permian subulitid *Labridens* (Knight et al. 1960). The only other gastropod with labial teeth to have evolved before the Late Cretaceous is the Upper Jurassic to Upper Cretaceous trochid *Chilodonta*. Within the internally resorbing Neritidae (Woodward 1892), which arose during the Triassic, the first genera to show dentition on the outer lip (*Myagrostoma*, *Velates*, and *Nerita*) date from sediments of Late Cretaceous age, when most of the higher mesogastropods and neogastropods also differentiated or expanded.

In *Conus*, *Olivella*, *Bullia*, Ellobiidae, and possibly other snails with markedly narrow and elongate apertures, extensive internal resorption of whorl partitions (Fischer 1881; Crosse and Fischer 1882; Morton 1955) secondarily widens the shell cavity to accommodate the visceral mass while at the same time effectively restricting entry into the aperture. Before the Cretaceous, relatively long, narrow apertures are known in but a few isolated genera: the Lower Carboniferous opisthobranch *Acteonina*, the Lower Carboniferous to Lower Permian *Soleniscus* and Middle Permian *Labridens* (Subulitiidae), and the Lower Jurassic to Upper Cretaceous neritid *Pileolus* (see Knight et al. 1960). The greatest diversity of snails with slit-like apertures is found in the post-Jurassic cypraeids, vexillids, conids, mitrids, olivids, columbellids, marginellids, and even strombids.

It may thus be concluded that Neritacea, mesogastropods, and especially neogastropods have, by evolving and taking advantage of the capacity to remodel the shell interior, significantly broadened the range of shell types that were available to earlier archaeogastropods. In particular, remodeling has enabled these snails to acquire combinations of short spires, narrow apertures, strong external sculpture, and apertural dentition which render the shell relatively impervious to predation by crushing or other shell-destructive modes of attack. In most archaeogastropods, possession of one architectural antipredatory attribute compromises another, so that the shell is rarely as well defended (in a geometrical sense) as in many post-Jurassic siphonate snails.

In passing, it is worth noting that all the gastropods in which internal remodeling has been employed extensively have a non-nacreous shell structure. Currey and Taylor (1974) have shown that nacre is in many ways mechanically tougher and stronger than other types of shell microarchitecture, yet nacre appears to be the primitive condition in most molluscan classes (Taylor 1973). The intriguing possibility that the advantages of remodeling in non-nacreous shells could have contributed to the replacement through time in many lineages of the nacreous type of structure with other forms of microarchitecture deserves further investigation.

Portmann (1967) and many others have pointed out that the external shell has been internalized or entirely lost in a large number of gastropod and cephalopod lineages. This loss is interpreted by these authors as a de-emphasis of the shell as a protective device. It is, of course, impossible to know how widespread this loss was in the course of Paleozoic and Mesozoic history; but it does seem that shell loss in Recent molluscs is associated either with great speed (cephalopods, heteropod gastropods) or with toxicity or unpalatability (many nudibranchs, onchidiids, perhaps some fissurellids?). Even in cowries (Cypraeidae), in which the well-developed shell is often covered during life with the mantle, Thompson (1969) demonstrated the existence of acid-secreting cells which may discourage predation by fishes. (See also Thompson 1960). Thus, internalization and loss of the shell may have been alternative solutions to the problem of increased predation intensity on molluscs.

Gastropod Predators

There is good reason to believe that the increase in snail-shell sturdiness beginning in the Cretaceous and continuing through the Tertiary has accompanied, and was perhaps in response to, the evolution of powerful, relatively small, shell-destroying (durophagous) predators. Table 1 summarizes the geological origins of the major groups of present-day molluscivores, and shows that most of these are in the Mesozoic era. Within the majority of durophagous groups (for example, in the teleosts, brachyurans, birds, and drilling gastropods), the ability to destroy or penetrate shells was not perfected until Late Cretaceous time or even later (Sohl 1969; Vermeij 1977), long after these groups first arose. The Asteroidea (sea stars) are the only group of important Recent molluscivores with a known pre-Mesozoic history, and they do not employ crushing or other shell-destructive methods to obtain their molluscan prey. Sea stars interpreted to have employed extraoral digestion co-occur with intraorally digesting sea stars in Upper Ordovician rocks (see Carter 1968).

There were, of course, durophagous predators which affected shelled molluscs in Paleozoic and Mesozoic communities but are no longer with us today. Several Late Devonian arthrodires and arthrodire relatives (e.g. *Mylostoma* and the ptyctodonts) had jaw dentition in the form of a crushing pavement, as did a number of Late Devonian to Permian hybodont sharks. Some of the latter were apparently quite small fishes (e.g. *Chondrenchelys* and *Helodus*) (Romer 1966). Hybodont sharks may have been responsible for the disarticulation and shell breakage observed

TABLE 1. Origins and modes of predation used by recent molluscivores.

Asteroidea; extraoral and intraoral digestion; arose and became molluscivores Late Ordovician; Carter (1968)
Dipnoi (fresh-water lungfishes); crushing; arose and became durophagous in Devonian; Thomson (1969)
Heterodontidae (sharks); crushing; arose and became durophagous in Jurassic; Romer (1966)
Batoidea (rays); crushing; arose Jurassic, became durophagous in Cretaceous; Bigelow and Schroeder (1953)
Stomatopoda (crustaceans); hammering and spearing; arose Jurassic, may have achieved durophagy by late Cretaceous; Holthuis and Manning (1969)
Palinuridae (spiny lobsters) and Nephropidae (lobsters); crushing; arose Jurassic. date of acquisition of durophagy uncertain; George and Main (1968)
Brachyura (crabs); crushing, apertural extraction; arose Jurassic, achieved durophagy by Latest Cretaceous or Paleocene; Glaessner (1969), Stevčić (1971), Vermeij (1977)
Aves (birds); crushing, swallowing whole, wrenching; arose Late Jurassic, were molluscivorous by Late Eocene; Cracraft (1973)
Muricacea (gastropods); drilling, apertural extraction; arose Albian (Early Cretaceous), drilling by Campanian (Late Cretaceous); Sohl (1969)
Naticacea (gastropods); drilling; arose Triassic, drilling by Campanian (Late Cretaceous); Sohl (1969)
Other Neogastropoda; apertural extraction; arose Early Cretaceous, molluscivory certainly established by Late Cretaceous; Sohl (1964, 1969)
Cynatiidae (Mesogastropoda); apertural extraction; arose and became molluscivorous Late Cretaceous; Sohl (1969)
Actinaria (sea anemones); swallowing whole; nothing known of fossil record
Teleostei (bony fishes); crushing, wrenching, swallowing whole; arose Triassic, durophagy by Late Cretaceous or earliest Tertiary; Schaeffer and Rosen (1961)

among some thick-shelled trigoniacean, heterodont, and taxodont-hinged bivalves in the Park City Formation of the Lower Permian of Wyoming (Boyd and Newell 1972). Earlier in the Paleozoic, cephalopods may have been important molluscivores, but no record of their activities remains.

Probable Mesozoic durophages include the remarkable Triassic marine placodont reptiles (von Huene 1956), the Triassic ichthyosaur *Omphalosaurus*; the Late Cretaceous mososaurid lizards, especially the genus *Globidens* (see Kauffman and Kesling 1960); and the Cretaceous ptychodontoid sharks (Romer 1966). Kauffman (1972) believes ptychodonts to have been important predators on inoceramid bivalves (see also Speden 1971) and has described some suggestive tooth-marks on these large shells. Tooth-marks on the rostra of Jurassic belemnites from southwest Germany are also thought to have been made by various fishes (Riegraf 1973).

It is, of course, difficult to assess the impact of extinct predators; yet, the large number of taxa which have perfected durophagy in the Cenozoic suggests that shell-crushing has become a relatively more important form of predation on molluscs over the course of time. Westermann (1971) noted that the frequency of repaired shell injuries among Mesozoic ammonoids was less than the 19% frequency reported by Eichler and Ristedt (1966) for juvenile Recent *Nautilus pompilius*. Preliminary results from studies on repaired injuries in high-spired Cretaceous and Cenozoic gastropod shells (Vermeij, Zipser, and Dudley, in preparation) point to the same conclusion.

Moreover, drilling appears to be a relatively new form of predation invented by gastropods and octopods in the Late Cretaceous (Sohl 1969). Problematical perforations in Paleozoic shells, especially brachiopods, seem in most cases to have been made after death of the affected animal and therefore did not constitute predation (Carriker and Yochelson 1968; Sohl 1969). Even if the boreholes observed by Rohr (1976) in Silurian brachiopods of the genus *Dicaelosia* were made during the life of the brachiopod, they occur in less than 3% of the population, and were thus extremely rare compared to the holes made in Tertiary shells.

Infaunalization and Grazing

Stanley (1977) has summarized additional changes in the composition and structure of marine benthic communities through geological time. Foremost among the long-term trends recognized by him is what may be labeled an infaunalization of soft-bottom benthic animals, which becomes particularly noticeable during the Late Cretaceous. In the Paleozoic, many brachiopods and stalked crinoids lived on the surface of, or were partially buried in, subtidal muds. In the Mesozoic, the epifaunal reclining habit is commonly seen in inoceramid bivalves. From the Late Cretaceous onward, however, these epifaunal and semi-infaunal elements have been largely replaced by infaunal forms (see also Rhoads 1970). Among bivalves, for example, there has been a relative decrease in the diversity of endobyssate (bysally attached semi-infaunal) forms since the Ordovician, and a corresponding dramatic increase in infaunal siphonate bivalves since the Paleozoic and especially the Late Mesozoic (Stanley 1968, 1970, 1972, 1977). Gastropods and echinoids exhibited a similar, disproportionate infaunal diversification in the Jurassic and Cretaceous (Kier 1974; Stanley 1977). All these changes may reflect intensification of predation or other sources of mortality at and above the sediment-water interface.

A further indication of this comes from architectural changes in the epifauna itself. Waller (1972) noted that most Paleozoic pectinacean bivalves (scallops) were characterized by prismatic shell structure; sculpture consisted of radial thickenings (costae). During the Mesozoic, shallow-water pectinaceans leading to the present-day Pectinidae became increasingly characterized by a foliated calcitic structure, which permitted the development of radial crinkles or folds (plicae). These, in turn, permit the necessarily thin shell of these swimming bivalves to be buttressed against what Waller (1972) believed to be an ever increasing number of shell-destroying predators.

Brachiopods today are found mostly in cryptic temperate and deep-sea environments but in the Paleozoic and pre-Cretaceous Mesozoic were dominant elements of hard- as well as soft-bottom epifaunal assemblages. C. W. Thayer has pointed out to me that, when a brachipod which normally lives attached by a peduncle becomes dislodged by some external agent, it cannot reattach itself and is therefore far more likely to die than is a byssally attached bivalve, which can in principle re-establish itself after being torn loose. It is these byssally attached bivalves, together with the post-Permian oysters and other bivalves cemented with one valve to the underlying substratum, that have come to predominate in Mesozoic and Cenozoic epifaunal assemblages. In the same way, stalked crinoids are today restricted to the deep sea but were important constituents of Paleozoic shallow-water bottom communities. They probably cannot reattach once dislodged and gave way to other organisms in the Mesozoic. The only surviving crinoids in shallow water are post-Permian comatulids which can change their position at will and are able to attach themselves facultatively by means of the cirri (see also Meyer and McCurda 1976).

Increased epifaunal predation may well be partly responsible for these far-reaching changes in benthic community structure (Stanley 1977; Meyer and McCurda 1977), but I am inclined to think that heavier grazing pressures have had a more profound influence (see also Garrett 1970). Two groups which today contain quantitatively important grazers, the teleostean fishes and echinoid echinoderms, have apparently been significant forces in shallow-water community structure only since the Late Mesozoic. Teleosts arose in the Triassic, but the critical morphological breakthroughs permitting them to browse algae, nip coral polyps, and scrape hard surfaces were not perfected until the Cretaceous (Schaeffer and Rosen 1961). The pycnodonts, an early (Jura-Cretaceous) holostean experiment in grazing and browsing, had teeth perhaps adapted for breaking off protruding parts of invertebrates (Romer 1966). (In a reinterpretation of pycnodonts, Thurmond (1974) suggests that the Jura-Cretaceous *Gyrodus* group was characterized by a shearing dentition, while the European Tertiary *Pycnodus* group had strictly crushing teeth adapted for feeding on small shelled invertebrates.) However, the scarids, acanthurids, kyphosids, and other typical grazers and browsers of tropical reefs are apparently of latest Cretaceous or, more probably, of earliest Tertiary extraction (see Hiatt and Strassburg 1960; Romer 1966; Hobson 1974). These fishes have a profound impact not only upon the algae which they eat, but equally on other epifaunal organisms which are scraped or sheared off the rock along with the algae (see, for example, Stephenson and Searles 1960).

Sea urchins (Echinoidea) arose no later than the Late Ordovician but were not common in intertidal or shallow-water communities until the latter half of the Mesozoic (Kier 1974; Bromley 1975). For example, they were not the important destroyers of reef limestone in the Paleozoic that they are in reefs of the present day (Newell 1957; Copper 1974). Like grazing fishes, echinoids scrape rocks bare of macroscopic algae and epifaunal animals and are known to have a profound influence on the composition and architecture of tropical as well as temperate hard-bottom communities (see e.g. Sammarco et al. 1974; Dayton 1975). The Aristotle's Lantern, with which hard surfaces are scraped by the urchin, began its development as the perignathic girdle in the Permian but was not perfected until shortly before the Maastrichtian in the Late Cretaceous, at which time the pentaradial traces characteristic of modern echinoid grazing first became common in shallow-water communities (Bromley 1975).

Grazing chitons and gastropods, which today have less impact on the epifauna than do sea urchins and teleosts (Stephenson and Searles 1960; Dayton 1975), have probably been present in shallow-water and intertidal communities since the Early Cambrian. Although certain deeply scouring limpets (post-Permian Acmaeidae and Patellidae) may be able to graze the algal turf to a lower height than can other gastropods, I doubt that grazing gastropods have contributed materially to the demise of the brachiopods and stalked crinoids, or to their restriction to cryptic or deep-water environments. The relative impact of other, post-Paleozoic and usually post-Cretaceous, grazers among crabs, mammals, birds, and reptiles has not been investigated thoroughly even in Recent communities and cannot be evaluated for ancient assemblages.

Other Mesozoic Changes

Reefs have developed at various times in the Phanerozoic, but post-Jurassic reefs differ in many important ways from all earlier versions of this community. Not only were rock-destroying fishes and echinoids absent or unimportant in Paleozoic reefs, but the structure of modern reefs is far more cavernous than that of Paleozoic and perhaps even Late Cretaceous rudist reefs (Jackson et al. 1971; Kauffman and Sohl 1974). Copper (1974) has noted that such fast-growing palmate or highly branched reef builders as the present-day acroporid and pociloporid corals were unknown in Paleozoic reefs. Skeletons of this type probably contribute heavily to the creation of caves and other interstices and did not arise among corals until later in the Cretaceous. Possibly they reflect more intense competition for light than was typical of reef builders in earlier epochs.

Two ecologically important communities, the sea-grass bed and the mangrove swamp or mangal, were not present before Late Cretaceous time (see Brasier 1975). The flowering plants which give these communities their names possess roots which stabilize and baffle the sediment, thus providing a habitat for a great diversity of marine organisms.

In benthic communities as a whole, many groups (echinoids, gastropods, bivalves, crustaceans, fishes) seem to have diversified dramatically beginning in the Middle Jurassic and continuing through the Cretaceous and Tertiary periods. This diversification is so widespread and pervasive that I am inclined to concur with Valentine (1969) that it is a real phenomenon and not merely the consequence of preservational biases (see Raup 1972 for a careful discussion on this point).

The alterations which affected marine communities during the Mesozoic were not limited to the benthos. Beginning in the Jurassic, there was a marked expansion in the diversity of marine microplankton, including the appearance of the first planktonic Foraminfera and Coccolithophorida; planktonic diatoms joined the assemblage in the Cretaceous (for a review see Lipps 1970).

Events on land are more spread out over geological time than were those in the sea. With the evolution of endothermy among therapsid reptiles in the Permian and probably dinosaurs in the Triassic came important alterations in trophic energy transfer which were to persist, with a short exceptional period in the Paleocene, to the present day (Bakker 1972, 1975). Dinosaurs evolved cursorial (fast-running) habits by Middle or Late Triassic time, and large cursorial vertebrates were important constituents of terrestrial communities throughout the rest of the Mesozoic and Cenozoic, again excepting the Paleocene (Bakker 1975). The only events on land which coincide in time with the Late Mesozoic marine reorganization are the evolution in the Early Cretaceous of angiosperms, their rapid rise to dominance in the Late Cretaceous and Cenozoic (Muller 1970), and the unprecedented radiation of insects which pollinated their flowers and ate their leaves (Ehrlich and Raven 1964). The diversification of birds may also be related to the rise of angiosperms and insects (Morse 1975).

Not all groups were, however, affected by

the Mesozoic reorganization. S. A. Woodin has pointed out to me that, judging from the meager fossil record, most polychaete families were established in the Early Paleozoic and that no significant or detectable morphological changes have taken place among the worms since then. Nuculacean bivalves, inarticulate brachiopods, and other groups characteristic of physiologically stressed or marginal environments seem also to have remained much the same morphologically through time. Although further, more detailed studies are required, it seems that the most profound changes affected communities in physiologically favorable environments.

Discussion

Throughout geological time, there has been a tendency in many skeletonized groups for morphologically simple, unadorned forms to give rise to, or to be replaced by, types with more elaborate skeletons. These elaborate forms are then largely obliterated in the major episodes of extinction, and the process begins anew. Thus, Valentine (1973) has commented that the skeletons of Lower Cambrian brachiopods, trilobites, and molluscs are often simple ("grubby," in his terminology), while Permian representatives often have elaborately sculptured or otherwise highly "specialized" skeletons. After the Permo-Triassic extinctions, Early Triassic fossils are once again simple, while Cretaceous skeletons are often bizarre and, in the case of ammonoids with complex sutures or irregular coiling, even a bit rococo. The frequent extinctions which have ravaged the planktonic world have affected primarily the elaborate and the adorned forms, and it is generally the morphologically conservative types which survive and rediversify after an episode of extinction (Lipps 1970; Cifelli 1976; Fischer and Arthur 1977).

Is this gradual process of skeletal elaboration and specialization comparable to the revolutionary events in the Early Paleozoic and Late Mesozoic? I believe not. The long-term skeletal specialization may be one expression of a continuing process of coevolution or coadaptation, in which predators increase their prey-capturing efficiency, while the prey are continually seeking ways to evade, repel, or otherwise thwart their enemies. Species less well adapted to the constantly changing biotic scene are more likely to become extinct or to be pushed into marginal habitats than are those which can successfully coexist with predators and competitors. In other words, the process which Stanley (1975, 1977) has called species selection may well account for the long-term trend toward greater predator-prey and competitor-competitor coadaptation. If this same species selection is to account for the Late Mesozoic community reorganization (Stanley 1977), then it is surprising that the various seemingly independent events contributing to it did not take place at widely different times but rather took place more or less in concert; and we must wonder why the critical adaptive breakthroughs either came so late in time or lay dormant for tens to hundreds of millions of years before spawning a major diversification. In other words, the Mesozoic revolution, recognized by the preferential survival of powerful consumers and well-armed prey over species less well adapted to predation, reflects unusually intense species selection; it upset the status quo which had been established since the Early Paleozoic. The species selection which prevailed during this long interval of "equilibrium" produced more powerful or more elaborate variations on a few adapted themes but produced very few new themes. Thus, I envision one of a few physical or geographical events as having permitted a degree of speciation and coevolution that would not have been possible under earlier conditions.

In the Early Paleozoic, we can point to several events which could have triggered, or at least greatly speeded up, the diversification of the Metazoa and the development of hard parts in many of their lineages. Skeletonization in the Cambrian is broadly correlated with the demise of stromatolites and with the greater activity of grazers, small burrowers, and perhaps predators (Garrett 1970; Awramik 1971; Stanley 1973). Rhoads and Morse (1971) have conjectured that these developments were made possible when the dissolved oxygen concentration exceeded about 1 ml O_2 per liter of water, since calcification at lower dissolved oxygen concentrations would have unacceptably compromised other competing functions involving calcium (see also Towe 1970). Stanley (1976b) has further suggested that the invention of sex late in the Precambrian promoted both speciation and adaptive radiation, diversifying what by modern standards must have been a monotonous biota.

What features made the Middle Mesozoic earth so special that skeletal destruction developed or became more important among both predators and grazers in shallow benthic environments? In attempting to account for the concomitant burst of biotic diversification in the Middle and Late Mesozoic, Valentine (1969) and Valentine and Moores (1970) pointed to the breakup of continents beginning in the Jurassic and to the gradual steepening of the latitudinal climatic gradient during the Tertiary. These tectonic and climatic events created biogeographical provinces where only one or a few had existed before and thus promoted geographical isolation and subsequent speciation. Moreover, the resulting constant raising and lowering of geographical barriers would have led to even more isolation, differentiation, and eventual contact with species which had previously evolved under different biological regimes. This continual biological rearrangement and opportunity for speciation could promote coevolution and coadaptation and speed up the spread of a major adaptive breakthrough.

Speciation would have been even more favored if ocean circulation (and thus planktonic larval dispersal) were reduced during periods of high diversity (polytaxic episodes), as has been suggested by Fischer and Arthur (1977). Slow circulation and high diversity, moreover, were generally associated with high stands of sea level and, therefore, with large areas available for occupation by benthic and planktonic marine life (Fischer and Arthur 1977; Valentine and Moores 1970). It is important to remember, however, that episodes of provinciality had occurred in earlier epochs —the Late Ordovician and Permian, for example—without resulting in the kind of reorganization of communities witnessed in the Mesozoic.

The events which shaped the Mesozoic revolution probably took place in a tropical setting, since some of the most characteristic symptoms (intense grazing by teleosts, predation by crushing) are today predominantly warm-water phenomena (Earle 1972; Vermeij 1977). Hence, one could argue that the perhaps unusually benign climate during much of the Mesozoic (at least in polytaxic times) created conditions favorable to the revolution (for a review of climate see Cracraft 1973). Again, however, it must be recalled that mild climatic episodes may have characterized the Devonian and perhaps other Paleozoic epochs without initiating major upheavals in marine community organization. Furthermore, intense grazing and predation by crushing are typical of Recent tropical marine assemblages despite a steep latitudinal gradient in climate and the large extent of temperate and polar oceanic waters.

In short, continental breakup and long-lasting favorable climate may have created conditions favorable to the evolution of skeleton-destroying predators and well-armored prey and had far-reaching consequences for benthic marine communities as a whole. Whether the Mesozoic era differed fundamentally from the Paleozoic in these respects remains to be determined, and the problem of dormant adaptive breakthroughs is not yet satisfactorily solved.

Lipps and Mitchell (1976) and Fischer and Arthur (1977) have independently pointed to the diversification of very large open-ocean predators during Mesozoic and Tertiary polytaxic episodes. During these times of high diversity and reduced oceanic circulation, the prey populations upon which the very large predators depended must have been large enough and stable enough to permit exploitation by a specialized animal. Many of the gigantic consumers became extinct during oligotaxic episodes (periods of low diversity and more rapid oceanic circulation) and were therefore only ephemeral elements in the community. In this way, they differ importantly from the benthic products of the Mesozoic revolution, since these animals persisted in spite of the periodic crises that beset the biotic world as a whole. The resources upon which the new predators depended were therefore more permanent and more reliable through time.

The gradual occupation of new adaptive zones may have contributed to the evolution of new and more specialized predators. Stanley (1968) has documented the invasion of deep sediments by shelled bivalves. Though this invasion began in the Early Paleozoic, it was primarily a Mesozoic phenomenon associated with the evolution of siphons through mantle-lobe fusion. This ecological expansion may have led to an overall increase in the availability and abundance of prey protected by a hard shell. Whether such an expansion was independent of, or correlated with, the appearance of more specialized predators remains an open question.

Another example of a possible increase in shelled food supply, and of the occupation of a new adaptive zone, is the evolution and diversification of hermit crabs (Paguridea). These crustaceans, which are known from the Jurassic onward (Glaessner 1969), normally live in the discarded shells of snails but do not usually feed on the previous adult inhabitants of their domiciles. As a result, hermit crabs could have enormously extended the ecological life-span of the average snail shell, thereby dramatically increasing the number and abundance of shelled prey. This effect, which would be particularly strong in the tropics (where most hermit crabs live today), could permit predators to specialize on shelled prey where such specialization would have been trophically unfeasible before the hermit crabs arose. The propensity of shell-crushing fishes, crabs, and other predators (which take both hermit crabs and snails) to attack living gastropod prey too large for them to crush, a common phenomenon today (Vermeij 1976; Miller 1975), would be encouraged if attacks on hermited shells of equal size are sometimes successful. In such cases, the hermited shell may have been weakened by boring organisms, or the crab may be somewhat too large for its shell, thus being more vulnerable to successful attack than is the living snail which built the shell. Indeed, Rossi and Parisi (1973) have shown that hermited shells are more vulnerable to crushing by the crab *Eriphia verrucosa* than are conspecific shells inhabited by living snails. Shell characteristics promoting resistance to predators might become more strongly selected for as more predators assault the living shelled snail.

Just how much hermit crabs have added to the ecological life-span of a given gastropod shell may be difficult to assess in the fossil record, but it may be possible to show whether a given fossil shell was inhabited by a hermit crab before being preserved in the sediment. If, in any assemblage, most fossils bear evidence of having been carried by a hermit crab, it may be concluded that the ecological life-span of the shell is relatively long. If reliable criteria for hermit-crab occupation would be established, then the proportion of hermited shells in a given fossil assemblage may give some relative indication of the extent to which the supply of shelled prey is constituted by hermit crabs.

The evolution of hermit crabs would not explain why drilling was developed among predatory gastropods and octopods, since at least gastropods do not normally drill hermited shells. However, the adaptive expansion of the bivalves may possibly be connected with the development of this new method of predation.

Finally, the consequences of the Mesozoic community reorganization underscore the fact that what ecologists perceive as long-term equilibria in communities may from time to time be fundamentally shaken by revolutionary events. Not only is the functional architecture of component species in a community affected by these events but so are such community properties as species diversity and trophic structure. Certain modes of life become obsolete or restricted to a small number of marginal environments, while other adaptive zones are newly occupied and previously untapped resources are newly exploited. From present evidence, it seems that periods of equilibrium have been interrupted by destabilizing revolutions and that these revolutions are not strictly connected with the major biotic crises.

Acknowledgments

I am deeply indebted to Robert Bakker, Alfred Fischer, Jeremy Jackson, Douglas Morse, Norman Sohl, Steven Stanley, Thomas Waller, Ellis Yochelson, and Edith Zipser for helpful discussions and criticisms. The empirical data underlying many of the conclusions of this paper were obtained with the assistance of Grants FA-31777 and DES74-22780 of the Oceanography Section, National Science Foundation; and a fellowship from the John Simon Guggenheim Memorial Foundation.

Literature Cited

Awramik, S. M. 1971. Precambian columnar stromatolite diversity: reflection of metazoan appearance. Science. *174*:825–827.

Bakker, R. T. 1972. Anatomical and ecological

evidence of endothermy in dinosaurs. Nature. *238*: 81–85.

BAKKER, R. T. 1975. Experimental and fossil evidence for the evolution of tetrapod bioenergetics. pp. 365–399. In: Gates, D. and R. Schmerl, eds. Perspectives of Biophysical Ecology. Springer-Verlag; New York.

BIGELOW, H. B. AND W. C. SCHROEDER. 1953. Fishes of the Western North Atlantic. Part II, Sawfishes, guitarfishes, skates and rays. Mem. Sears Found. Mar. Res. *1*:1–585.

BOSS, K. J. 1971. Critical estimate of the number of Recent Mollusca. Occas. Pap. on Mollusks, Mus. Comp. Zool. *3*:81–135.

BOUCOT, A. J. 1975. Evolution and Extinction Rate Controls. 427 pp. Elsevier Sci. Publ. Co.; Amsterdam, Netherlands.

BOWSHER, A. L. 1955. Origin and adaptation of platyceratid gastropods. Kansas Univ. Paleontol. Contrib. *17* (Mollusca, Article 5):1–11.

BOYD, D. W. AND N. D. NEWELL. 1972. Taphonomy and diagenesis of a Permian fossil assemblage from Wyoming. J. Paleontol. *46*:1–14.

BRASIER, M. D. 1975. An outline history of seagrass communities. Palaeontology. *18*:681–702.

BRIGHT, T. J. 1970. Food of deep-sea bottom fishes. Texas A&M Univ. Oceanogr. Studies. *1*: 245–252.

BROMLEY, R. G. 1975. Comparative analysis of fossil and Recent echinoid bioerosion. Palaeontology. *18*:725–739.

CARRIKER, M. R. 1972. Observations on the removal of spines by muricid gastropods during shell growth. Veliger. *15*:69–74.

CARRIKER, M. R. AND E. L. YOCHELSON. 1968. Recent gastropod boreholes and Ordovician cylindrical borings. U.S. Geol. Surv. Prof. Pap. *593B*: B1–B23.

CARTER, R. M. 1968. On the biology and palaeontology of some predators of bivalved Mollusca. Paleogeogr., Paleoclimatol., Paleoecol. *4*:29–65.

CIFELLI, R. 1976. Evolution of ocean climate and the record of planktonic Foraminifera. Nature. *264*:431–432.

COPPER, P. 1974. Structure and development of Early Paleozoic reefs. Second Int. Coral Reef Symp. *1*:365–386.

CRACRAFT, J. 1973. Continental drift, paleoclimatology, and the evolution and biogeography of birds. J. Zool. London. *169*:455–545.

CROSSE, H. AND P. FISCHER. 1882. Note complémentaire sur la resorption des parois internes du teste chèz *Olivella*. J. Conchyliol. *30*:181–183.

CURREY, J. D. AND J. D. TAYLOR. 1974. The mechanical behaviour of some molluscan hard tissues. J. Zool. London. *173*:395–406.

DAYTON, P. K. 1975. Experimental evaluation of ecological dominance in a rocky intertidal algal community. Ecol. Monogr. *45*:137–159.

EARLE, S. A. 1972. The influence of herbivores on the marine plants of Great Lameshur Bay, with an annotated list of plants. Nat. Hist. Mus. Los Angeles County Sci. Bull. *14*:17–44.

EHRLICH, P. R. AND P. H. RAVEN. 1964. Butterflies and plants: a study in coevolution. Evolution. *18*:586–608.

EICHLER, R. AND H. RISTEDT. 1966. Untersuchungen zur Fruhontogenie von *Nautilus pompilius* (Linné). Palaontol. Z. *40*:173–191.

FISCHER, A. G. AND M. A. ARTHUR. 1977. Secular variation in pelagic realms. In: Enos, P. and H. Cook, eds., Soc. Econ. Petr. Mineral. Special Publ. *25*: In press.

FISCHER, P. 1881. Note sur le genre *Olivella*. J. Conchyliol. *29*:31–35.

GARRETT, P. 1970. Phanerozoic stromatolites: noncompetitive ecological restriction by grazing and burrowing animals. Science. *169*:171–173.

GEORGE, R. W. AND A. R. MAIN. 1968. The evolution of spiny lobsters (Palinuridae): a study of evolution in the marine environment. Evolution. *22*:803–820.

GLAESSNER, M. F. 1969. Decapoda. pp. R399–R533. In: Moore, R. C., ed. Treatise on Invertebrate Paleontology. Part R, Arthropoda 4 (2). Univ. Kansas Press; Lawrence, Kansas.

GOLIKOV, A. M. AND Y. I. STAROBOGATOV. 1975. Systematics of prosobranch gastropods. Malacologia. *15*:105–232.

HIATT, R. W. AND D. W. STRASSBURG. 1960. Ecological relationships of the fish fauna on coral reefs of the Marshall Islands. Ecol. Monogr. *30*:65–127.

HOBSON, E. S. 1974. Feeding relationships of teleostean fishes on coral reefs in Kona, Hawaii. Fish. Bull. *72*:915–1031.

HOLTHUIS, L. B. AND R. B. MANNING. 1969. Stomatopoda. Pp. R535–R552. In: Moore, R. C., ed. Treatise on Invertebrate Paleontology. Part R, Arthropoda 4 (2). Univ. Kansas Press; Lawrence, Kansas.

VON HUENE, F. R. 1956. Paläontologie und Phylogenie der niederen Tetrapoden. 716 pp. Gustav Fischer; Jena, Deutschland.

JACKSON, J. B. C., T. F. GOREAU, AND W. D. HARTMAN. 1971. Recent brachiopod-coralline sponge communities and their paleoecological significance. Science. *173*:623–625.

JELETZKY, J. A. 1965. Taxonomy and phylogeny of fossil Coleoidea (= dibranchiata). Geol. Surv. Pap. Can. *65-2*:72–76.

KAUFFMAN, E. G. 1972. *Ptychodus* predation upon a Cretaceous *Inoceramus*. Palaeontology. *15*:439–444.

KAUFFMAN, E. G. AND R. V. KESLING. 1960. An Upper Cretaceous ammonite bitten by a mososaur. Contrib. Mus. Paleontol. Univ. Mich. *15*:193–248.

KAUFFMAN, E. G. AND N. F. SOHL. 1974. Structure and evolution of Antillean Cretaceous rudist frameworks. Verhandl. Naturf. Ges. Basel. *84*: 399–467.

KIER, P. M. 1974. Evolutionary trends and their functional significance in the post-Paleozoic echinoids. J. Paleontol., Paleontol. Soc. Mem. 5, 48, Part II of II: 1–95.

LIPPS, J. H. AND E. MITCHELL. 1976. Trophic model for the adaptive radiations and extinctions of pelagic marine mammals. Paleobiology *2*:147–155.

McLEAN, R. B. AND R. N. MARISCAL. 1973. Protection of a hermit crab by its symbiotic sea anemone *Calliactis tricolor*. Experientia. *29*:128–130.

MEYER, D. L. AND B. MCCURDA. 1977. Adaptive radiation of the comatulid crinoids. Paleobiology. *3*:74–82.

MILLER, B. A. 1975. The biology of *Terebra gouldi* Deshayes, 1859, and a discussion of life history similarities among other terebrids of similar proboscis type. Pacific Sci. *29*:227–241.

MORSE, D. H. 1975. Ecological aspects of adaptive radiation in birds. Biol. Rev. *50*:167–214.

MORTON, J. E. 1955. The evolution of the Ellobiidae with a discussion on the origin of the Pulmonata. Proc. Zool. Soc. London. *125*:127–168.

MÜLLER, J. 1970. Palynological evidence of early differentiation of angiosperms. Biol. Rev. *45*:417–450.

NEWELL, N. D. 1957. Paleoecology of Permian reefs in the Guadalupe Mountains area. Geol. Soc. Am. Mem. *67*:407–436.

NEWELL, N. D. 1971. An outline history of tropical organic reefs. Am. Mus. Novit. *2465*:1–37.

PORTMANN, A. 1967. Animal forms and patterns: a study of the appearance of animals. 254 pp. Schocken Books, New York.

RAUP, D. M. 1972. Taxonomic diversity during the Phanerozoic. Science. *177*:1065–1071.

REX, M. A. AND K. J. BOSS. 1976. Open coiling in Recent gastropods. Malacologia. *15*:289–297.

RHOADS, D. C. 1970. Mass properties, stability, and ecology of marine muds related to burrowing activity. In: Grimes, T. P. and J. C. Harper, eds. Trace Fossils. Geol. J. Special Issue. *3*:391–406.

RHOADS, D. C. AND J. W. MORSE. 1971. Evolutionary and ecologic significance of oxygen-deficient marine basins. Lethaia. *4*:413–428.

RIEGRAF, W. 1973. Biszspuren auf Jurassischen Belemniten-rostren. N. Jb. Geol. Palaontol. M.H. *8*:494–500.

ROHR, D. M. 1976. Silurian predator borings in the brachiopod *Dicaelosia* from the Canadian Arctic. J. Paleontol. *50*:1175–1179.

ROMER, A. S. 1966. Vertebrate Paleontology. 468 pp. Univ. Chicago Press; Chicago, Ill.

ROSSI, A. C. AND V. PARISI. 1973. Experimental studies of predation by the crab *Eriphia verrucosa* on both snail and hermit crab occupants of conspecific gastropod shells. Boll. Zool. *40*:117–135.

RUNNEGAR, B. AND J. P. POJETA, JR. 1974. Molluscan phylogeny: the paleontological viewpoint. Science. *186*:311–317.

SAMMARCO, P. W., J. S. LEVINGTON, AND J. C. OGDEN. 1974. Grazing and control of coral reef community structure by *Diadema antillarum* Philippi (Echinodermata: Echinoidea): a preliminary study. J. Mar. Res. *32*:47–53.

SCHAEFFER, B. AND D. E. ROSEN. 1961. Major adaptive levels in the evolution of the actinopterygian feeding mechanism. Am. Zool. *1*:187–204.

SOHL, N. F. 1964. Neogastropoda, Opisthobranchia and Basommatophora from the Ripley, Owl Creek, and Prairie Bluff Formations. U.S. Geol. Surv. Prof. Pap. *331*-B:153B–344B.

SOHL, N. F. 1969. The fossil record of shell boring by snails. Am. Zool. 9:725–734.

SPEDEN, I. G. 1971. Notes on New Zealand fossil Mollusca—2. Predation on New Zealand Cretaceous species of *Inoceramus*. N.Z. J. Geol. Geophys. *14*:56–70.

STANLEY, S. M. 1968. Post-Paleozoic adaptive radiation of infaunal bivalve molluscs—a consequence of mantle fusion and siphon formation. J. Paleontol. *42*:214–229.

STANLEY, S. M. 1972. Functional morphology and evolution of byssally attached bivalve molluscs. J. Paleontol. 46:165–212.

STANLEY, S. M. 1973. An ecological theory for the sudden origin of multicellular life in the late Precambrian. Proc. Nat. Acad. Sci. U.S.A. *70*:1486–1489.

STANLEY, S. M. 1975. A theory of evolution above the species level. Proc. Nat. Acad. Sci. U.S.A. *72*: 646–650.

STANLEY, S. M. 1976a. Fossil data and the Precambrian-Cambrian evolutionary transition. Am. J. Sci. *276*:56–76.

STANLEY, S. M. 1976b. Ideas on the timing of metazoan diversification. Paleobiology. *2*:209–219.

STANLEY, S. M. 1977. Trends, rates, and patterns of evolution in the Bivalvia. In: Hallam, A., ed. Patterns of Evolution. Elsevier; Amsterdam. In press.

STEIN, R. A., J. F. KITCHELL, AND B. KNEZEVIC. 1975. Selective predation by carp (*Cyprinus carpio* L.) on benthic molluscs in Skadar Lake, Yugoslavia. J. Fish. Biol. 7:391–399.

STEVČIĆ, Z. 1971. The main features of brachyuran evolution. Syst. Zool. *20*:331–340.

STEPHENSON, W. AND R. B. SEARLES. 1960. Experimental studies on the ecology of intertidal environments at Heron Island. I. Exclusion of fish from beach rock. Aust. J. Mar. Fresh-Water Res. *11*: 241–267.

TAYLOR, D. W. AND N. F. SOHL. 1962. An outline of gastropod classification. Malacologia. *1*:7–32.

TAYLOR, J. D. 1973. The structural evolution of the bivalve shell. Palaeontology. *16*:519–534.

THOMPSON, T. E. 1960. Defensive adaptations in opisthobranchs. J. Mar. Biol. Assoc. U.K. *39*:123–134.

THOMPSON, T. E. 1969. Acid secretion in Pacific Ocean gastropods. Aust. J. Zool. *17*:755–764.

THOMSON, K. S. 1969. The biology of the lobe-finned fishes. Biol. Rev. *44*:91–154.

THURMOND, J. T. 1974. Lower vertebrate faunas of the Trinity division in north-central Texas. Geoscience and Man. *8*:103–129.

TOWE, K. N. 1970. Oxygen-collagen priority and the early metazoan fossil record. Proc. Nat. Acad. Sci. U.S.A. *65*:781–788.

VALENTINE, J. W. 1969. Patterns of taxonomic and ecological structure of the shelf benthos during Phanerozoic time. Palaeontology. *12*:684–709.

VALENTINE, J. W. 1973. Evolutionary paleoecology of the Marine Biosphere. 511 pp. Prentice Hall Inc.; Englewood Cliffs, New Jersey.

VALENTINE, J. W. AND E. M. MOORES. 1970. Plate-tectonic regulation of faunal diversity and sea level: a model. Nature. *228*:657–659.

VERMEIJ, G. J. 1971. The geometry of shell sculpture. Forma et Functio. *4*:319–325.

VERMEIJ, G. J. 1973. Adaptation, versatility, and evolution. Syst. Zool. *22*:466–477.

VERMEIJ, G. J. 1974. Marine faunal dominance and molluscan shell form. Evolution. *28*:656–664.

VERMEIJ, G. J. 1975. Evolution and distribution of left-handed and planispiral coiling in snails. Nature. *254*:419–420.

VERMEIJ, G. J. 1976. Interoceanic differences in vulnerability of shelled prey to crab predation. Nature. *260*:135–136.

VERMEIJ, G. J. 1977. Patterns in crab claw size: the geography of crushing. Syst. Zool. *26*: In press.

WALLER, T. R. 1972. The functional significance of some shell microstructures in the Pectinacea (Mollusca: Bivalvia). Pp. 48–56. Int. Geol. Congr. 24th Session, Montreal, Can. Sect. 7, Paleontol.

WÄNBERG-ERIKSSON, K. 1964. *Isospira reticulata* n.sp. from the Upper Ordovician Boda Limestone, Sweden. Geol. Forenings I Stockholm Förhandl. *86*:229–237.

WESTERMANN, G. E. G. 1971. Form, structure and function of shell and siphuncle in coiled Mesozoic ammonoids. Life Sci. Contrib. R. Ontario Mus. *78*:1–39.

WOODWARD, B. B. 1892. On the mode of growth and the structure of the shell in *Velates conoideus*, Lamk., and other Neritidae. Proc. Zool. Soc. London. 528–540.

YOCHELSON, E. L. 1971. A new Late Devonian gastropod and its bearing on problems of open coiling and septation. Smithson. Contrib. Paleobiol. *3*:231–241.

PAPER 32

Response of Plant-Insect Associations to Paleocene-Eocene Warming (1999)

P. Wilf and C. C. Labandeira

Commentary

ELLEN D. CURRANO

In the 117 years between the earliest descriptions of fossil leaf mines (Fritsch 1882; Hagen 1882) and the publication of Wilf and Labandeira, quantitative studies of fossil insect folivory were rarely performed. This may have been due to a lack of interest among paleobotanists or because percentage of leaf area damaged, the metric most often used by neoecologists studying plant-insect interactions, is difficult and time consuming to measure on fossil leaves. Wilf and Labandeira demonstrate the importance of studying plant-insect interactions in the fossil record and present a new method of doing so that can be quickly, easily, and reproducibly applied to any well-preserved fossil flora.

Despite being just over three pages long, Wilf and Labandeira's manuscript touches on many critical issues in ecology and paleoecology. Herbivore damage on fossil floras from the late Paleocene and early middle Eocene is compared to provide insight on the latitudinal gradient in insect herbivory, the response of plant-insect interactions to global warming, and the relationship between host abundance and insect attack rate. Data were collected using unbiased insect damage censuses, in which every identifiable leaf is scored for the presence/absence of clearly defined insect damage morphotypes. Wilf and Labandeira describe 41 damage types; today, over 250 damage types have been described, and this number steadily increases as more floras are studied. The damage-type system has been used to study mass extinctions (e.g., Labandeira et al. 2002; Wilf et al. 2006), abrupt global warming (Currano et al. 2008), past latitudinal herbivory gradients (e.g., Sunderlin et al. 2010; Currano et al. 2010; Wappler and Denk 2011), the evolution of terrestrial food webs (Labandeira 2006), and much more. Furthermore, paleobotanical collections targeting damaged leaves have proved useful in studying the evolutionary history of insects (e.g., Winkler et al. 2009).

Beauty is in the eye of the beholder. One paleobotanist's trash is another paleobotanist's treasure. The systematic paleobotanist searches for pristine leaf fossils that perfectly preserve every detail. Wilf and Labandeira, in contrast, illustrate the importance of the chewed, mined, galled, and otherwise damaged fossil leaves, the ones that were all too often carelessly tossed in the slag heap.

Literature Cited

Currano, E. D., P. Wilf, S. L. Wing, C. C. Labandeira, E. C. Lovelock, and D. L. Royer. 2008. Sharply increased insect herbivory during the Paleocene-Eocene Thermal Maximum. *PNAS* 105:1960–64.

Currano, E. D., C. C. Labandeira, and P. Wilf. 2010. Fossil insect folivory tracks paleotemperature for six million years. *Ecological Monographs* 80:547–67.

Fritsch, A. 1882. Fossile Arthropoden aus der Steinkohlen

From *Science* 284:2153–56. Reprinted with permission from AAAS.

und Kreideformation Bohemens. *Beiträge zur Paläontologie und Geologie Österreich-Ungarns und des Orients* 2:1–7.

Hagen, H. A. 1882. Fossil insects of the Dakota Group. *Nature* 25:265–66.

Labandeira, C. C. 2006. The four phases of plant-arthropod associations in deep time. *Geologica Acta* 4: 409–38.

Labandeira, C. C., K. R. Johnson, and P. Wilf. 2002. Impact of the terminal Cretaceous event on plant-insect associations. *PNAS* 99:2061–66.

Wappler, T., and T. Denk. 2011. Herbivory in early Tertiary Arctic forests. *Palaeogoegraphy, Palaeoclimatology, Palaeoecology* 310:283–95.

Wilf, P., C. C. Labandeira, K. R. Johnson, and B. Ellis. 2006. Decoupled plant and insect diversity after the end-Cretaceous extinction. *Science* 313:1112–15

Winkler, I. S., C. Mitter, and S. J. Scheffer. 2009. Repeated climate-linked host shifts have promoted diversification in a temperate clade of leaf-mining flies. *PNAS* 106:18103–8.

REPORTS

Response of Plant-Insect Associations to Paleocene-Eocene Warming

Peter Wilf[1]* and Conrad C. Labandeira[1,2]

The diversity of modern herbivorous insects and their pressure on plant hosts generally increase with decreasing latitude. These observations imply that the diversity and intensity of herbivory should increase with rising temperatures at constant latitude. Insect damage on fossil leaves found in southwestern Wyoming, from the late Paleocene–early Eocene global warming interval, demonstrates this prediction. Early Eocene plants had more types of insect damage per host species and higher attack frequencies than late Paleocene plants. Herbivory was most elevated on the most abundant group, the birch family (Betulaceae). Change in the composition of the herbivore fauna during the Paleocene-Eocene interval is also indicated.

Terrestrial plants and insects today make up most of Earth's biodiversity (*1*), and almost half of insect species are herbivores (*2*). Consequently, understanding how plant-insect associations respond to warming events is a vital component of global change studies (*3*). The fossil record offers a unique opportunity to examine plant-insect response to climate change over long time intervals through analysis of insect damage on fossil plants (*4*, *5*).

In modern insect faunas, decreasing latitude is associated with increased diversity of insect herbivores per host plant and greater herbivore pressure; the latter is expressed as higher attack frequency (*6*, *7*). For this study, we used insect damage on fossil plants to test for these trends at constant latitude, in the context of the global warming interval that began in the late Paleocene and reached maximum Cenozoic temperatures by the middle early Eocene, about 53 million years ago (*8*). We also examined whether the diversity of herbivory and increase in attack rates was highest on the most abundant hosts and addressed whether a compositional change in the Paleocene-Eocene herbivore fauna occurred.

The Great Divide, Green River, and Washakie basins of southwestern Wyoming, U.S.A. (Fig. 1), bear diverse and abundant floral assemblages containing well-preserved insect damage (Fig. 2 and Table 1) (*9*). We compared two floral samples from this region, from the latest Paleocene and middle early Eocene (*10*). Both samples were originally deposited in fine-grained sediments on humid, swampy floodplains (*9*), which allowed us to use an isotaphonomic (*11*) approach that helps to factor out biases such as depositional regime, paleotopography, and past moisture levels. Previous analysis of these samples (*9*, *12*) showed that, from the latest Paleocene to the middle early Eocene, (i) mean annual temperatures rose from an estimated 14.4° ± 2.5°C to 21.3° ± 2.2°C, (ii) plant species turnover exceeded 80%, (iii) all dominant plant species were replaced, and (iv) plant diversity increased significantly.

We identified 41 types of insect damage (Table 1 and Fig. 2) on 39 Paleocene and 49 Eocene species of terrestrial flowering plants at 49 Paleocene and 31 Eocene localities (Fig. 1) (*9*, *10*, *13*). A database was constructed in which the presence or absence of each damage type was scored for each species in each sample (Table 1). We also quantitatively took field censuses of the four plant localities with highest diversity and best preservation (two Paleocene and two Eocene) for insect damage on dicot leaves (*14*).

Census data were analyzed for all leaves and separately for Betulaceae and all non-betulaceous taxa. A single species of Betulaceae was a dominant component of the vegetation in both the Paleocene (*Corylites* sp.) and

[1]Department of Paleobiology, National Museum of Natural History, Smithsonian Institution, Washington, DC 20560–0121, USA. [2]Department of Entomology, University of Maryland, College Park, MD 20742–4454, USA.

*To whom correspondence should be addressed (after August 1999) at the Museum of Paleontology, University of Michigan, Ann Arbor, MI 48109–1079, USA. E-mail: pwilf@umich.edu

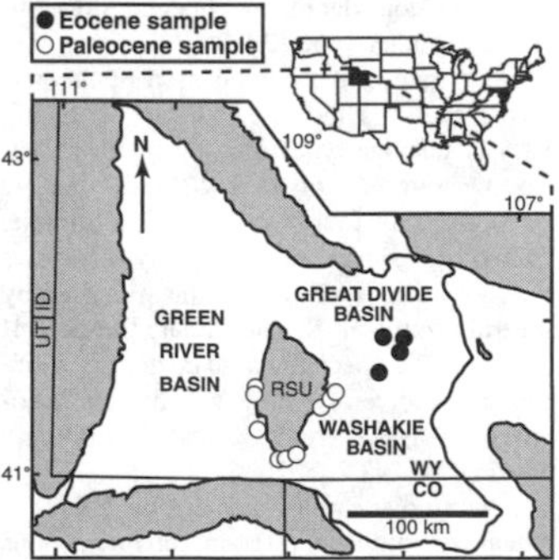

Fig. 1. Sampling areas. The most northeastern circle for each set includes the localities for insect damage censusing (Figs. 3 and 4). Gray areas are uplifts. RSU, Rock Springs Uplift. Redrawn after (9).

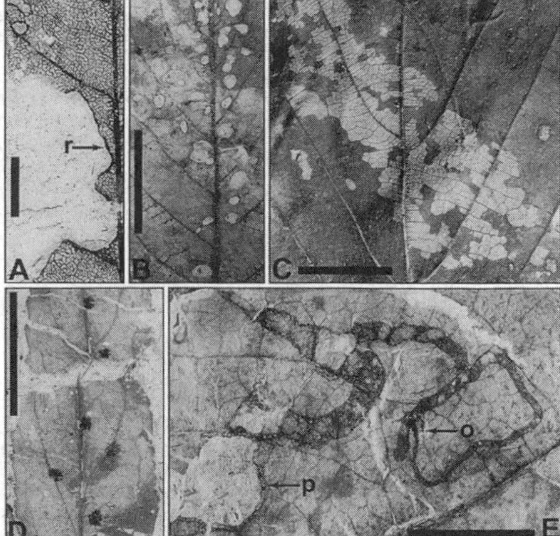

Fig. 2. Examples of Paleocene-Eocene insect damage. Panels (A) and (C) are Paleocene and (B), (D), and (E) are Eocene. All scale bars equal 1 cm. **(A)** Margin feeding to primary vein on *Persites argutus* Hickey (Lauraceae), USNM 498036, USNM locality (loc.) 41292. Note thick reaction tissue (r). **(B)** Polymorphic, elliptical hole feeding on *Alnus* sp. (Betulaceae), USNM 498177, USNM loc. 41339. Note reaction tissue bordering holes. **(C)** Broad, rectangular skeletonization of *Corylites* sp. (Betulaceae), USNM 498176, USNM loc. 41270. Note fine detail of exposed venation. **(D)** Galls on primary and secondary veins of *Stillingia casca* Hickey (Euphorbiaceae), USNM 498175, USNM loc. 41341. **(E)** Serpentine mine (type E) on new dicot sp. RR37, USNM 498091, USNM loc. 41353. The mine crosses tertiary and higher order veins. The oviposition site (o) and the site of the pupation chamber (p) are both preserved.

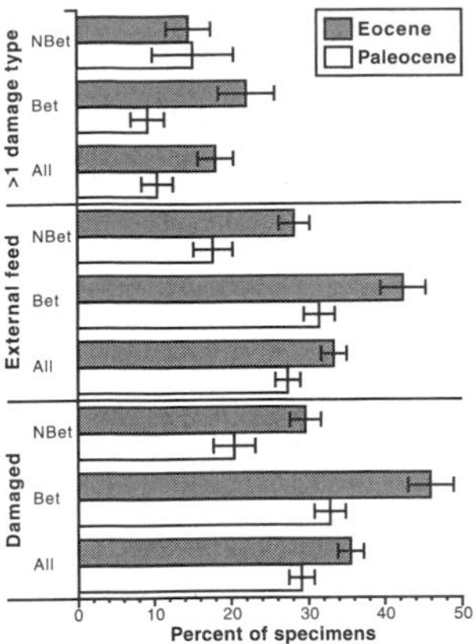

Fig. 3. Damage census data. From bottom to top: leaves with any insect damage, leaves externally fed, and the percentage of damaged leaves bearing more than one damage type (Table 1). These categories are each analyzed separately for all leaves (All), Betulaceae only (Bet), and non-Betulaceae only (NBet). Error bars are one standard deviation of binomial sampling error (*27*). Sample sizes for Paleocene and Eocene, respectively: All (749, 791); Bet (524, 285); and NBet (225, 506). Total leaf area examined in censuses, derived from Webb leaf-area categories (*28*): 2.26 m^2 (Paleocene) and 2.12 m^2 (Eocene). Paleocene = USNM locs. 41270 and 41300 combined; Eocene = USNM locs. 41342 and 41352 combined.

Table 1. Insect damage types. The presence (+) or absence (−) of each type in the Paleocene (Pal) and Eocene (Eoc) samples is indicated and their relative degree of specialization (Spec): 1 = most generalized, 3 = most specialized. Terminology modified from (*26*). Genus names or morphotype numbers of host plant species are listed for the most specialized damage types and those that exhibit turnover (*9*).

Damage type	Pal	Eoc	Spec
External feeding			
Constant width, elongate, branching	+	+	2
Strip-feeding between secondary veins (*Zingiberopsis*)	−	+	3
Window feeding, generalized	+	+	1
Hole feeding			
Generalized, unpatterned	+	+	1
Bud feeding (*Alnus, Hovenia, Schoepfia*)	−	+	2
Curvilinear	+	+	2
Elliptical	+	+	1
Elongated slot	+	+	2
Large, ovoidal or circular	+	+	1
Large, polylobate	+	+	1
Exceptionally thick necrotic tissue	+	+	1
Polymorphic, generally elliptical	+	+	2
Ring (aff. *Ocotea*)	+	−	1
Small, ovoidal or circular	+	+	1
Small, polylobate	+	+	1
Margin feeding			
Generalized, usually cuspate	+	+	1
Apex feeding	+	+	1
Free feeding (*Platycarya, Populus*)	−	+	2
To primary vein	+	+	1
Trenched (deeply incised)	+	+	2
Skeletonization			
General, reaction rim weak	+	+	1
General, reaction rim well developed	+	+	1
Broad, with rectangular pattern (*Corylites*)	+	−	2
Curvilinear (*Persites*)	+	−	2
Highest order venation removed (*Platycarya*)	−	+	2
Linear pattern (*Alnus*)	−	+	2
Skeletonization (cont.)			
Ovoidal, adjacent to midvein	+	+	2
Multiple, subparallel, curvilinear tracks (*Corylites*, new dicot sp. RR31)	+	+	3
Mining			
Blotch, central chamber (*Persites*, Magnoliales sp., aff. *Sloanea*)	+	+	3
Blotch, large (>2 cm diam.), no central chamber ("*Ampelopsis*")	+	−	3
Circular, with case (*Corylites*)	+	−	3
Serpentine A: long, undulatory; frass particulate (*Corylites, Alnus, Cinnamomophyllum*, new dicot sp. RR20)	+	+	3
Serpentine B: length medium, width rapidly increasing, margin irregular (*Corylites*)	+	−	3
Serpentine C: length short, frass trail solid ("*Dombeya*", cf. Magnoliales sp. RR12, *Alnus*)	−	+	3
Serpentine D: long, frass tightly sinusoidal, frass trail narrow (*Cinnamomophyllum*)	−	+	3
Serpentine E: length medium, margin irregular, oviposition site and terminus well defined (new dicot sp. RR37)	−	+	3
Galling			
On blade, other than major veins	+	+	2
On primary vein(s) only	+	+	2
On secondary veins only	+	+	2
Piercing and sucking			
Scale or puncture, circular depression (Magnoliaceae sp. FW07, palm leaf, new dicot sp. RR48)	+	+	3
Scale or puncture, elliptical depression (palm leaf)	+	−	3

the Eocene (*Alnus* sp.) (*15*). These two species fit the traditional model of "apparent" plants in that they were abundant, conspicuous hosts that formed significant ecological islands (*16*). Like all modern Betulaceae, whose leaves are heavily consumed by insects (*17, 18*), *Corylites* and *Alnus* (alder) were thin-leaved and deciduous, adding to their presumed palatability (*7, 19*). We hypothesized that these taxa were frequently consumed by a high diversity of herbivores.

The census data show that, overall, damage frequency is significantly higher in the Eocene sample, indicating elevated levels of herbivory (Fig. 3) (*20*). Betulaceous leaves were attacked significantly more often than nonbetulaceous leaves within both sampling levels, and their damage frequency (Fig. 3), multiple damage frequency (Fig. 3), and damage diversity (Figs. 4 and 5) increased markedly from the Paleocene to the Eocene (*21*). *Alnus* palatability was probably enhanced by elevated leaf nitrogen content resulting from an actinorhizal association with nitrogen-fixing symbionts, as in all modern *Alnus* (*18, 22*).

Bootstrap curves derived from the census data (Fig. 4) show increased minimum and maximum damage diversity at a local scale during the Eocene. All of the Paleocene taxa except one (aff. *Ocotea*) have nearly identical bootstrap curves. Four Eocene species have bootstrapped values higher than all of the Paleocene taxa (*Alnus, Cinnamomophyllum*, "*Dombeya*", and *Populus*). Three other Eocene species have bootstrap values that are lower than the Paleocene mode represented by *Corylites* (*Allophylus*, Apocynaceae sp., and aff. *Sloanea*) but still higher than the Paleocene minimum (aff. *Ocotea*).

The diversity of insect damage per host species increases with the percentage of localities where a given host occurred because increased sampling raises the probability of discovering damage types (Fig. 5). However, when comparison is made at equal frequency of occurrence, greater herbivore diversity per host plant is again found in the Eocene than in the Paleocene. The Eocene slope in Fig. 5 is higher, even though 37% fewer localities are in the Eocene sample and less geologic time is represented (*10*). Also, the five largest positive residuals are all Eocene species. Finally, the single abundant monocot (Eocene *Zingiberopsis*) has a large effect. If dicots alone are considered, the Eocene slope increases another 15% (*23*).

A change in the composition of the herbivore fauna is indicated (Table 1). In all, 17% of damage types only occur in the Paleocene sample, whereas 20% of damage types are only found in the Eocene sample. Each of the generalized damage types (scores of 1 in Table 1) may have been caused by several groups of distantly related insects. If only the 27 specialized damage types are counted (scores of 2 or 3 in Table 1), Paleocene-only types are 22% and Eocene-only types 30% (*24*).

This study demonstrates that the effects of global warming on plant-insect interactions are detectable in the fossil record. Climate change also provides a largely unexplored context for related areas of inquiry, such as the histories of plant-pollinator relations and insect diversification.

REPORTS

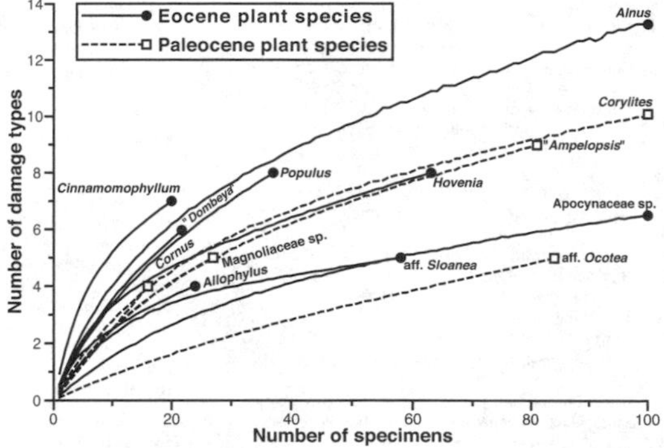

Fig. 4. Bootstrapped damage diversity, derived from the census data, for species with >15 specimens in total census counts. For each positive integer n along the horizontal axis up to the total number of specimens for a species (N), 5000 subsamples of n specimens were taken at random and the mean number of damage types calculated (vertical axis). The line graphs connect the N mean values for each species. Shown only to $n \leq 100$ for greater detail. Maximum $\sigma = 1.8$ (for *Alnus*, $n = 80$). Family or generic names only are shown; see (9) for complete nomenclature. "aff." = morphological affinity to indicated genus, a qualified identification.

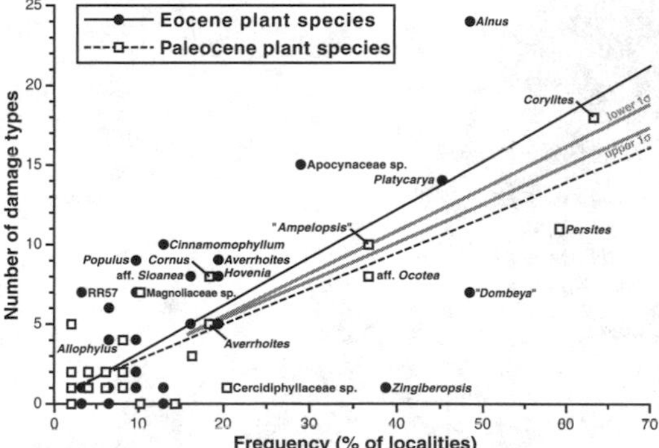

Fig. 5. Diversity of insect damage per plant host species (vertical axis), plotted against the percentage of localities (49 Paleocene, 31 Eocene) at which the species occurs. Each data point is one species; many data points overlap at the lower left; survivors are plotted twice. Gray lines show divergence of 1σ (68%) confidence intervals for the two regressions. Paleocene regression: $y = 22.3x + 0.545$, $r^2 = 0.775$, $P < 10^{-12}$ (r^2 is the coefficient of determination). Eocene regression: $y = 30.1x + 0.117$, $r^2 = 0.538$, $P < 10^{-8}$. Family or generic names are shown for plant species that are abundant, plot with large residuals, or appear in Fig. 4.

References and Notes

1. N. E. Stork, *Biol. J. Linn. Soc.* **35**, 321 (1988).
2. L. M. Schoonhoven, T. Jermy, J. J. A. van Loon, *Insect-Plant Biology* (Chapman & Hall, London, 1997).
3. E. D. Fajer, M. D. Bowers, F. A. Bazzaz, *Science* **243**, 1198 (1989); R. L. Lindroth, K. K. Kinney, C. L. Platz, *Ecology* **74**, 763 (1993); J. A. Arnone, J. G. Zaller, C. Ziegler, H. Zandt, C. Körner, *Oecologia* **104**, 72 (1995); R. A. Fleming, *Silva Fenn.* **30**, 281 (1996); C. S. Awmack, R. Harrington, S. R. Leather, *Global Change Biol.* **3**, 545 (1997); P. D. Coley, *Clim. Change* **39**, 455 (1998); S. J. Dury, J. G. Good, C. M. Perrins, A. Buse, T. Kaye, *Global Change Biol.* **4**, 55 (1998); J. B. Whittaker and N. P. Tribe, *J. Anim. Ecol.* **67**, 987 (1998).
4. C. C. Labandeira, D. L. Dilcher, D. R. Davis, D. L. Wagner, *Proc. Natl. Acad. Sci. U.S.A.* **91**, 12278 (1994); C. C. Labandeira, *Annu. Rev. Earth Planet. Sci.* **26**, 329 (1998).
5. A. L. Beck and C. C. Labandeira, *Palaeogeogr. Palaeoclimatol. Palaeoecol.* **142**, 139 (1998).
6. T. L. Erwin, *Coleopt. Bull.* **36**, 74 (1982); P. D. Coley and T. M. Aide, in *Plant Animal Interactions: Evolutionary Ecology in Tropical and Temperate Regions*, P. W. Price, T. M. Lewinsohn, G. W. Fernandes, B. B. Benson, Eds. (Wiley, New York, 1991), pp. 25–49; P. D. Coley and J. A. Barone, *Annu. Rev. Ecol. Syst.* **27**, 305 (1996); M. G. Wright and M. J. Samways, *Oecologia* **115**, 427 (1998).
7. Y. Basset, *Acta Oecol.* **15**, 181 (1994).
8. M.-P. Aubry, S. G. Lucas, W. A. Berggren, Eds., *Late Paleocene-Early Eocene Climatic Events in the Marine and Terrestrial Records* (Columbia Univ. Press, New York, 1998). Mean annual temperatures increased 7° to 9°C in the Rocky Mountain region from the latest Paleocene to the middle early Eocene (9, 25), and sea surface temperatures in southern high latitudes rose from between 10°–12° to 14°–16°C, the latter 13° to 15°C warmer than today [J. C. Zachos, L. D. Stott, K. C. Lohmann, *Paleoceanography* **9**, 353 (1994)]. A significant atmospheric increase in the partial pressure of CO_2 has been postulated as a cause of Paleocene-Eocene warming [D. K. Rea, J. C. Zachos, R. M. Owen, P. D. Gingerich, *Palaeogeogr. Palaeoclimatol. Palaeoecol.* **79**, 117 (1990)], but this hypothesis is not yet supported by proxy and model data [E. Thomas, in *Late Paleocene–Early Eocene Climatic and Biotic Events in the Marine and Terrestrial Records*, M.-P. Aubry, S. G. Lucas, W. A. Berggren, Eds. (Columbia Univ. Press, New York, 1998), pp. 214–235].
9. P. Wilf, *Geol. Soc. Am. Bull.*, in press.
10. The late Paleocene sample is sample 2 of (9), a lumped Clarkforkian assemblage. The Eocene sample, from the Cenozoic thermal maximum, is the middle Wasatchian Sourdough flora, Great Divide Basin, sample 5 of (9). The Paleocene sample is time-averaged over ~0.5 million years, during which some temperature increase occurred (9, 12, 25), whereas the more diverse Eocene sample is not significantly time-averaged and is derived from fewer localities. A total leaf count was not made because not all identifiable specimens were collected from every locality. An unbiased measure is that 7511 leaves have been identified from the 15 localities that were censused on the outcrop (9); the total number of leaves examined from all 80 localities was far greater. Although floras from intervening stratigraphic intervals are known (9), the two samples used here are among the best preserved and are derived from many more localities. Deciduous taxa were more abundant than evergreens in both samples, although evergreens were more diverse in the Eocene sample (9). "Species" is used here to indicate both described species and undescribed forms considered as operational species in (9). Voucher collections, from the 1994 to 1996 and 1998 field seasons, are housed at the U.S. National Museum of Natural History (USNM), with supplementary material at the Denver Museum of Natural History and the Florida Museum of Natural History (9).
11. A. K. Behrensmeyer and R. W. Hook, in *Terrestrial Ecosystems Through Time*, A. K. Behrensmeyer *et al.*, Eds. (Univ. of Chicago Press, Chicago, IL, 1992), pp. 15–136.
12. P. Wilf, K. C. Beard, K. S. Davies-Vollum, J. W. Norejko, *Palaios* **13**, 514 (1998).
13. A damage type (Table 1) was assigned if a distinctive insect feeding mode, or ecotype, was represented. No damage was found on conifers, cycads, or aquatic angiosperms, and damage on ferns was rare.
14. The primary goal of initial collecting (1994 to 1996 field seasons) was to reconstruct floral diversity and leaf morphology, with some resulting collection bias against common species and consequently against their insect damage. The field censusing of insect damage (in 1998) allowed unbiased sampling of herbivory on all species to complement that known from collections and also permitted the observed damage frequencies to be related as directly as possible to the relative abundances of host plants in the source forests [R. J. Burnham, S. L. Wing, G. G. Parker, *Paleobiology* **18**, 30 (1992)].
15. The *Corylites* sp., from the Paleocene sample only, was described as the presumed foliage of *Palaeocarpinus aspinosa* Manchester and Chen [S. R. Manchester and Z. Chen, *Int. J. Plant Sci.* **157**, 644 (1996)]. The *Alnus* sp., undescribed, is known from leaves, female cones, and staminate inflorescences in early Eocene deposits of southern and northern Wyoming (9, 25). These taxa occurred at the largest number of localities (Fig. 5) and also were most frequently the dominant species at individual localities (9).
16. D. H. Janzen, *Am. Nat.* **102**, 592 (1968); P. A. Opler, *Am. Sci.* **62**, 67 (1974); P. P. Feeny, in *Biochemical Interaction Between Plants and Insects*, J. Wallace and R. L. Mansell, Eds. (Plenum, New York, 1976), pp. 1–40.
17. J. Reichholf, *Waldhygiene* **10**, 247 (1974); J. M. Cobos-Suarez, *Bol. Sanid. Veg.* **14**, 1 (1988); B. Gharadjedaghi, *Anz. Schaedlingskd. Pflanz. Umweltschutz* **70**, 145 (1997); B. Gharadjedaghi, *Forstwiss. Centralbl.* **116**, 158 (1997); J. Oleksyn *et al.*, *New Phytol.* **140**, 239 (1998).
18. O. Q. Hendrickson, W. H. Fogal, D. Burgess, *Can. J. Bot.* **69**, 1919 (1991).
19. P. D. Coley, *Oecologia* **74**, 531 (1988).
20. Damage frequency in fossil floras has been significantly lower than modern values in several studies (5), as we find here. Although insect damage may well have increased through time, it is likely that several factors would make damage appear less prevalent in fossil assemblages, including taphonomic bias against damaged leaves, the rarity of complete fossil leaves, the inability to observe completely consumed leaves, and the low probability of preservation for minute damage types, such as piercing and sucking. We consider good preservation of leaves (highest order venation visible on the majority of specimens, more than half of the original leaf usually present) and a fine-grained matrix, as in this study, to be prerequisites for censusing of insect damage. Insect damage by its nature reduces the preservability of leaves by

creating tear points, although this bias needs to be quantified in actualistic studies.
21. *Corylites* and *Alnus* fit well with the resource availability hypothesis [P. D. Coley, J. P. Bryant, F. S. Chapin III, *Science* **230**, 895 (1985)] in which high herbivory rates are correlated with short leaf lifespan (*Corylites* and *Alnus* were deciduous), high growth rates and relatively early successional status (*Corylites* and *Alnus* had tiny, wind- or water-dispersed fruits and colonized disturbed environments on floodplains), and low concentrations of defensive compounds (implied).
22. G. Bond, in *Symbiotic Nitrogen Fixation in Plants*, P. S. Nutman, Ed. (Cambridge Univ. Press, Cambridge, 1976), pp. 443–474; C. P. Onuf, J. M. Teal, I. Valiela, *Ecology* **58**, 514 (1977); W. J. Mattson Jr., *Annu. Rev. Ecol. Syst.* **11**, 119 (1980); R. E. Ricklefs and K. K. Matthew, *Can. J. Bot.* **60**, 2037 (1982); J. J. Furlow, in *Magnoliophyta: Magnoliidae and Hamamelidae*, vol. 3 of *Flora of North America*, Flora of North America Editorial Committee, Ed. (Oxford Univ. Press, New York, 1997), pp. 507–538. Phylogenetic analysis of DNA sequences from the *rbcL* gene places all actinorhizal plants within a single clade, indicating that actinorhizal association is ancient [D. E. Soltis *et al.*, *Proc. Natl. Acad. Sci. U.S.A.* **92**, 2647 (1995)].
23. Insect damage on the single dicot species that was abundant in both samples (*Averrhoites affinis*) increased from five types in the Paleocene to nine in the Eocene (Fig. 5).
24. Although the most specialized damage types are rare, sampling was intensive (*10, 14*), which supports our view that the inferred turnover of herbivores is not a sampling artifact. The percentages listed should be regarded as minima given the difficulty of evaluating the more generalized feeding groups.
25. S. L. Wing, H. Bao, P. L. Koch, in *Warm Climates in Earth History*, B. T. Huber, K. MacLeod, S. L. Wing, Eds. (Cambridge Univ. Press, Cambridge, 1999), pp. 197–237.
26. R. N. Coulson and J. A. Witter, *Forest Entomology: Ecology and Management* (Wiley, New York, 1984).
27. P. Wilf, *Paleobiology* **23**, 373 (1997).
28. L. J. Webb, *J. Ecol.* **47**, 551 (1959). The logarithmic mean area was used for each Webb category to estimate total leaf area [P. Wilf, S. L. Wing, D. R. Greenwood, C. L. Greenwood, *Geology* **26**, 203 (1998)].
29. We thank A. Ash, R. Schrott, K. Werth, and others for field and laboratory assistance, Western Wyoming Community College for logistical support, and W. DiMichele, P. Dodson, R. Horwitt, B. Huber, S. Wing, and two anonymous reviewers for helpful comments on the manuscript. P.W. was supported by Smithsonian Institution predoctoral and postdoctoral fellowships, the Smithsonian's Evolution of Terrestrial Ecosystems Program (ETE), a University of Pennsylvania Dissertation Fellowship, the Geological Society of America, Sigma Xi, and the Paleontological Society. C.C.L. was supported by the Walcott Fund of the National Museum of Natural History. This is ETE contribution number 68.

19 March 1999; accepted 5 May 1999

On the Weakening Relationship Between the Indian Monsoon and ENSO

K. Krishna Kumar,[1*†] Balaji Rajagopalan,[2] Mark A. Cane[2]

Analysis of the 140-year historical record suggests that the inverse relationship between the El Niño–Southern Oscillation (ENSO) and the Indian summer monsoon (weak monsoon arising from warm ENSO event) has broken down in recent decades. Two possible reasons emerge from the analyses. A southeastward shift in the Walker circulation anomalies associated with ENSO events may lead to a reduced subsidence over the Indian region, thus favoring normal monsoon conditions. Additionally, increased surface temperatures over Eurasia in winter and spring, which are a part of the midlatitude continental warming trend, may favor the enhanced land-ocean thermal gradient conducive to a strong monsoon. These observations raise the possibility that the Eurasian warming in recent decades helps to sustain the monsoon rainfall at a normal level despite strong ENSO events.

Most parts of India receive a major proportion of their annual rainfall during the summer (June to September) monsoon season. Extreme departures from normal seasonal rainfall, such as large-scale droughts and floods, seriously affect agricultural output and regional economies. By the early 1900s, investigators had identified the two large-scale forcings still thought to be most important for predicting monsoon anomalies: Himalayan/Eurasian snow extent (*1*) and the ENSO cycle (*2*). The former is generally believed to provide an indication of the premonsoon thermal condition over the Asian land mass. Warmer conditions are thought to aid the buildup of a strong land-sea thermal gradient during the summer (*3, 4*). ENSO, the largest known climatic forcing of interannual monsoon variability, acts through the east-west displacement of large-scale heat sources in the tropics (*5*). Numerous studies (*6*) have shown a significant simultaneous association between the monsoon rainfall over India and ENSO indices. However, secular variations in the relationships between monsoon rainfall and its predictors have also been noted (*7*). These variations have been found to be linked to changes in ENSO characteristics such as amplitude and period (*8*).

Almost all the statistical seasonal prediction schemes of monsoon rainfall rely heavily on the change in magnitude in various ENSO indices (*8, 9*) from winter [December to February (DJF)] to spring [March to May (MAM)]. Numerical general circulation models (GCMs) are also used for seasonal rainfall prediction. The monsoon simulated in these GCMs is more sensitive to the sea surface temperatures (SSTs) specified in the Pacific (*10*) than to other external boundary forcings. Hence, the success of seasonal forecasts of monsoon rainfall depends on the stationarity of the monsoon-ENSO relationship.

We used data on Indian monsoon rainfall, SST, velocity potential fields, and global surface temperatures (*11*) to examine the simultaneous relationship between the monsoon and ENSO during the last 142 years, and to explore possible roles for other climatic forcings.

Low-frequency variations in the monsoon rainfall and a widely used measure of ENSO, the NINO3 index (*11*), show a clear resemblance until the late 1970s (Fig. 1A), but diverge thereafter. This change reflects the recent modest increase in the monsoon rainfall despite an increase in the magnitude and frequency of ENSO warm events. Sliding correlations on a 21-year moving window between monsoon rainfall and the NINO3 index are strong during the entire data period with the exception of the recent two to three decades (Fig. 1B), not withstanding the considerable impact of the 1982 and 1987–88 ENSO events on the monsoon. The drop in correlations in the recent decades is found to be significant on the basis of bootstrap confidence limits (*12*). The loss of monsoon-ENSO correlation is not particular to NINO3, but appears with any ENSO index. Correlation patterns of monsoon rainfall with SSTs in the Pacific show a coherent region of strong correlations in the central and eastern equatorial Pacific before 1980 and no region with statistically significant correlations thereafter.

The conventional description of the ENSO-induced teleconnection response in the monsoon is through the large-scale east-west shifts in the tropical Walker circulation. During an El Niño event, the tropical convection and the associated rising limb of the Walker circulation normally located in the western Pacific shift toward the anomalously

[1]International Research Institute (IRI) for Climate Prediction, Lamont-Doherty Earth Observatory (LDEO) of Columbia University, Post Office Box 1000, Route 9W, Palisades, NY 10964–8000, USA. [2]LDEO of Columbia University, Post Office Box 1000, Route 9W, Palisades, NY 10964–8000, USA.

*Permanent address: Indian Institute of Tropical Meteorology, Dr. Homi Bhabha Road, Pashan, Pune 411008, India.

†To whom correspondence should be addressed. E-mail: krishna@iri.ldeo.columbia.edu

PAPER 33

Community Evolution and the Origin of Mammals (1966)

E. C. Olson

Commentary

KENNETH D. ANGIELCZYK

"Community Evolution and the Origin of Mammals" is a foundational paper in paleoecology, but not for its insight into mammal origins. Mammals, nonmammalian synapsids, and the evolution of mammalian characters receive little attention. Instead, Olson paints a compelling picture of changes in community organization, particularly trophic relationships, that occurred as terrestrial communities evolved. Although Olson presented these ideas in earlier and later works, this paper's narrative style and clear graphics make it an approachable entry point into this aspect of his research.

Olson posits that terrestrial communities have three basic trophic structures. Type I communities remain dependent on aquatic environments: their main primary producers and primary consumers are aquatic, and terrestrial animals feed on aquatic food sources and/or each other. Type II and type III communities are fully terrestrial, with land plants as primary producers. Abundant, diverse terrestrial vertebrate or invertebrate herbivores, respectively, link the plants to terrestrial secondary consumers. The types also show temporal succession: Early Permian and older communities are type I, but in the Middle Permian, type II communities appear, perhaps because of increased terrestrial adaptation among their members. By the Late Permian, type II communities are dominant and maintain their hegemony to today (Olson considered type III communities to be a rare but important step in the evolution of certain type II communities). Olson's theory that terrestrial communities of modern structure first appeared in the Permian, coincident with the radiation of tetrapod herbivores, remains the foundation for our understanding of this transition. It also spurred interest in the evolution and diversification of tetrapod herbivores and in when and where the first herbivore-dominated communities appeared.

Although Olson never explicitly mentions it, the paper raises an intriguing question: Do the community types' emergent properties influence their persistence over geological timescales? Recent modeling has shown that the trophic networks of ancient communities vary in their ability to inhibit the spread of disturbances that can cause the extinction of member species. If type II trophic networks have properties such as these that increase their members' fitnesses, it would provide an alternate explanation for their rapid rise and subsequent dominance.

From *Ecology* 47:291–302.

Horton, J. S., and C. J. Kraebel. 1955. Development of vegetation after fire. Ecology 36: 244-260.
Howard, W. E., R. L. Fenner, and H. E. Childs, Jr. 1959. Wildlife survival on brush burns. J. Range Mgmt. 12: 230-234.
Hutchinson, C. B., and E. I. Kotok. 1942. The San Joaquin experimental range. Agric. Exp. St. Bull. 663: 3-145.
Love, R. M., and B. J. Jones. 1952. Improving California brush ranges. Calif. Agric. Exp. Sta. Circ. 371: 4-38.
Marle, M. 1963. Homing and orientation of deermice. J. Mamm. 44: 338-349.
Neiland, B. J. 1958. Forest and adjacent burn in the Tillamook burn area of northeastern Oregon. Ecology 39: 660-671.
Pearson, O. P. 1959. A traffic survey of Microtus-Reithrodontomys runways. J. Mammal. 40: 169-180.
Pruitt, W. O., Jr. 1953. An analysis of some physical factors affecting the local distribution of the shorttail shrew Blarina brevicauda in the northern part of the lower peninsula of Michigan. Univ. Mich. Misc. Publ. Zool. No. 79, 39 p.
Quast, J. C. 1954. Rodent habitat preferences on foothill pastures in California. J. Mamm. 35: 515-521.
Sampson, A. W. 1944. Plant succession on burned chaparral lands. Univ. Calif. Agric. Exper. Sta. Bull. No. 685: 1-144.
Scheffer, T. H. 1931. Habits and economic status of pocket gopher. U.S.D.A. Tech. Bull. No. 224: 1-9.
Sweeney, J. R. 1956. Responses of vegetation to fire; a study of the herbaceous vegetation following chaparral fires. Univ. Calif. Publ. Bot. 28: 143-250.
Tevis, L., Jr. 1956. Effect of slash burn on forest mice. J. Wildl. Mgmt. 20: 405-409.
Thornthwaite, C. W. 1940. Atmospheric moisture in relation to ecological problems. Ecology 21: 17-28.
Wieslander, E. E., and Clark H. Gleason. 1954. Major brushland areas of the coast ranges and Sierra Cascade foothills in California. Misc. paper No. 15. Southeast Forest and Range Exp. Sta., Berkeley, California.
Williams, O. 1955. Distribution of mice and shrews in a Colorado montane forest. J. Mammal. 36: 221-231.

COMMUNITY EVOLUTION AND THE ORIGIN OF MAMMALS[1]

Everett C. Olson
Department of Geophysical Sciences, University of Chicago, Chicago, Illinois

Abstract. The evolutionary course from primitive pelycosaurian reptiles through therapsids to mammals can be profitably studied in relationship to modifications of the structure of the communities in which these reptiles existed. For this purpose the community is defined in very broad terms. Three types of communities are recognized upon the basis of the nature of the food chain. Each has an important tetrapod component.

Early phases of the evolution that culminated in mammals took place in communities that were strongly tied to water by the structure of the food chain. The physiological bases of the development of mammals appear to have been related to this environmental restriction. In successive pulses, however, the pelycosaur–therapsid communities developed terrestrial reptilian herbivores and thereby broke with the water-based food chain. More strictly terrestrial communities developed concurrently, with the insects, which were a food source for the reptiles, as the principal herbivores. From this sort of community came the terrestrial lepidosaurian–archosaurian reptilian radiation.

The terrestrial communities so developed came into competition. In this competition the therapsid lines were temporarily unsuccessful, leaving only small, but very mammal-like, representatives in the late Triassic. After a long period with relatively little adaptive radiation, these remnants provided the basis for the radiations of mammals that led to the great successes of the Cenozoic era.

Introductory Explanations

Studies of vertebrate evolution centered around the concept of faunal modifications have constituted one of the major fields of interest of the writer over the last decade and a half. A number of publications which have resulted from this interest, as cited specifically in the following text, have stressed the changes of communities with the passage of geological time. The present paper represents a continuation and extension of this kind of work. Most of the data upon which it is based have been included in the earlier studies, but the synthesis is somewhat more general than any attempted previously, and the interpretations are more directly ecological.

Studies, such as this one, which involve broad areas of paleoecology are necessarily cast at rather different levels from those of most neoecological investigations. Naturally, as well, a strong element of speculation must enter in, for assumptions of a rather sweeping nature are necessary and conclusions often must be based on complexly interwoven threads ot evidence. Yet the resulting insights into the relationships of ecology and evolutionary processes are such that, even though crude, the interpretations are stimulating in them-

[1] The research leading to this paper was supported by NSF grants 19093 and 2543.

selves and frequently direct investigations into new paths.

The kinds of data that enter into these studies defy a short and concise description. They have been presented in some detail, along with consideration of the problems that they pose in two earlier papers (Olson 1958, 1962). The most reliable information on fossils, sediments and associations of the materials comes from day to day field studies made in the course of assembling collections of fossils, with ecological analyses as one of the principal goals.

Sites containing fossil vertebrates may yield materials ranging from fragments of bones and single bones, through all stages to the complete skeleton. Specimens may occur in assemblages where many individuals are associated in various states of preservation. All types of occurrences have meaning for ecological studies. They assume greater significance when their materials are related to the types of sediments in which they occur and the ways that the specimens are preserved in those sediments. Time relationships, and geographical distribution are equally significant.

A brief consideration of the nature of the record used in the present study may aid in clarifying the nature of the evidence. Many of the sites that have yielded vertebrates from the Permian and Triassic contain sediments that were deposited under conditions that can be broadly defined as deltaic. On the basis of sediment characteristics such as composition and size distributions of particles, the gross shape of a deposit and relationships of a bed to those adjacent to it, the characteristics of the environment of deposition can be reasonably well determined. In deltaic deposits sediments formed in stream channels, in lakes and ponds, and on flood plains can generally be recognized without great difficulty (see Olson 1962, for a full discussion of this matter). Other kinds of depositional sites may be determined in many cases, but generally with less ease and less certainty.

The animals themselves, of course, are good indicators of conditions of their own existence, but raise problems with respect to the relationships of the death assemblage and the living populations. In an extensive study of the related problems of the death and accumulation of animals and the nature of deposition of the containing sediments, Efremov (1950) treated in detail a process which he called taphonomy, which concerns the whole spectrum of events during the passage of animals from living populations to death assemblages. Many fossil associations are what he terms taphonomic assemblages, those in which the animals and plants which were available to the processes of erosion and transportation that produced a particular deposit were not all members of the same life assemblages. Such associations are characteristic of accumulations produced by stream deposits, which may tap a variety of life zones along the course of the stream and its tributaries. Such deposits are of great use in determining the general nature of the life of a given time, but are not as useful in paleoecological studies.

Assemblages that do reflect the associations that existed during life are especially instructive, although frequently difficult to recognize. Pond deposits may be of this type. Assemblages formed in standing water in the Clear Fork deposits of Texas, for example, show a remarkable consistency. It has been possible, on the basis of considerable field experience in the area, to specify with about 75 to 80% accuracy the generic content of particular examples of such deposits in the field by observation of the beds alone, before examination of their fossil content.

Various assemblages also tell a good deal about the life habits of the organisms. A small, somewhat worm-like amphibian of the Clear Fork Permian of Texas, *Lysorophus* occurs characteristically in high concentration, with a hundred or more individuals more or less evenly distributed over roughly circular areas with diameters generally from 10 to 20 ft. There is little question that these represent an aestivating phase of the life of the animal. Furthermore, study of size distributions shows that each of these assemblages tends to have individuals of very limited size range. Each aestivating swarm appears to represent an age group, probably an annual group. *Gnathorhiza,* an extinct lung fish, similarly was an aestivator, spending part of its life in a cylindrical burrow (Romer and Olson 1954). These, and various other examples, are taken up in reports on the details of studies which have contributed to this synthesis (Olson 1958, and a series of 13 preliminary papers cited in that reference).

To the extent that collectors have documented their materials with regard to conditions of preservation, associations, sedimentation, stratigraphy and so on, general museum collections may be used for such interpretations and may be of considerable importance. For the current study, the collections of fossils and field notes made by the writer and his associates over the last 25 years in the course of work in the Permian of Texas and Oklahoma have been the most useful. Much weight has also been given to collections from older Permian beds made by those collectors who kept records of the physical and biological associations of their materials. Very useful, as well, have been the collections from the Kazanian of the Russian Permian housed in the Museum of Paleontology in Moscow, U.S.S.R. These in many instances reflect the faunal point of view of I. A. Efremov and his colleagues and predecessors, in particular Sushkin and Bystrow.

Potentially most promising but unfortunately of less than full value are collections from the Permian and Triassic series of South Africa. Until rather recently less than sufficient heed has been given to associations of the fossil organisms, to the physical conditions of their occurrence, and to stratigraphy. The lack of published data on these critical matters, and my own lack of first hand field study and of study of some of the larger collections are reflected in the fact that the weakest part of the analysis that follows is in those sections dealing with these materials.

A first study along the lines followed in the present paper dealt with the lower Permian Clear Fork beds of Texas and their faunas (Olson 1952). This was elaborated and somewhat modified in a later summary work cited earlier (Olson 1958). More recently other publications have broadened the temporal and evolutionary scope of the investigations and have attempted to delineate some of the evolutionary implications of community structures (Olson 1961, 1962). Some matters taken up in the following pages were first developed in the last two cited references.

The nature of the communities

One of the problems that inevitably arises is the coordination of the meaning of terms that are convenient and useful in paleoecology with those used in neoecological studies. At best only an approximation of common meaning is possible. The term community as used in this paper represents such a case. As employed here the term covers a very extensive unit of biological association, and lies at one limit of the spectrum of range of usages. Furthermore, as used here the term is basically conceptual but in some instances, as is clear in context, it refers to actual systems that existed at given times over finite geographic areas.

References are made later to a community type, which is conceptual, or to an example of a type which consists of a community that existed at a given time and place, for instance the Clear Fork community of the early Permian of Texas.

It is common practice among paleontologists to give a taxonomic designation to an assemblage of animals that is encountered repeatedly in the record. Such assemblages are often called complexes and the designation is generally taken from a common element of the complex. These assemblages are usually considered to include only the animals, or only the invertebrate or the vertebrate animals, and thus are less comprehensive in scope than the community as used in this paper. They do, however, provide convenient nomenclatural units to express some temporal or taxonomic stage of a community type. Efremov, whose work has been used extensively in later parts of this paper, has employed phrases such as deinocephalian complex, based on a taxon of infraordinal rank, and pareiasaur complex, where the term of designation is generic. In the discussion of the sequence of communities in evolution, particular units are specified by these and similar terms.

There is an additional complicating factor, one not present in the same sense in the communities of neoecology. The actual communities of paleoecology had duration in time (Olson 1952). The property of temporal duration through some appreciable segment of geological time means, of course, that the existence of a community is not dependent upon persistence of its individual biological components, the animals and plants of a particular time, for these vanish to be replaced either by their own kind or by ecological equivalents. The key rather is the pattern of interaction expressed in the structure of the community. It is the persistence of this pattern that provides the continuity of existence. The biological, geographical and temporal limits are thus necessarily somewhat indefinite and are in no sense independent variables that can be treated in separate contexts. Conceptually, however, the units here designated as communities are, in spite of the variations with time, roughly determinable. They form a suitable framework for the type of work presented in this paper. The breadth of this framework is essential at the level at which such studies must be conducted.

There undoubtedly existed a large number of community types of this general sort within the major habitats available during the Permian and Triassic Periods. Presumably the community units had various and temporally varying limits of discreteness, some with little dependence upon other communities and others with a large measure of dependence. Our primary concern will be with those systems that had a terrestrial vertebrate component, and as a result the classification necessary in this study is quite limited. Among such communities three types have been distinguished, and representatives of each type seem to have existed as relatively independent entities over appreciable spans of geological time.

There are, of course, many different ways in which the community types might be designated and different arrangements emerge from application of different criteria. With interest focused upon the vertebrates, the following two aspects of the system appear to be the most important:

1. The contribution by its vertebrates, considered as ecological types, to the composition of the community. Taxonomic composition is secondary, but to the extent that the taxonomic position of a particular vertebrate relates directly to its contribution to the community pattern, taxonomy may be considered as important.

2. The paths of flow of food-derived energy through the community. The focus is upon the structure of the food chain, especially as it relates to sources of energy of the vertebrates.

Consideration of the many problems associated with determination of diets and feeding patterns among extinct organisms, pertinent to the second point is much too complex for treatment here. Some points, however, are particularly important. At best, interpretations tend to stretch the capacities of rational inference to the limit. The primary data, of course, are those of morphology. The dangers of using such data alone for determination of diets are quite evident among living animals. Yet some broad limits can be drawn.

Another source of information is in the analysis of the potentialities and limitations of feeding interactions imposed by temporary, environmental and geographic distributions. The presence of organisms may indicate that they were a potential food source for an appropriate consumer, but the absence of some types from the record does not necessarily mean that they were not living in the community from which the sample of the organisms was drawn. The inferred role of insects in the Permian Clear Fork community is a good case in point. No insects have been found in association with the plants and animals that have been collected. In adjacent areas, from sites of about the same age, well-preserved insects have been found. These sites contain no vertebrate remains. The differences between sites are related, without much question, to conditions of preservation. Thus I freely infer the existence of insects as components of the Clear Fork community, and call upon them as a food source of the vertebrates.

Coprolites, or fossil feces, are occasionally useful sources of information on diet. Very often plant and animal remains are preserved in coprolites and in some cases identifications of the contained organic materials can be made. Some kinds of rather general information can be had. It is notable, for example, that in the Clear Fork almost all coprolites contain a wealth of debris from animals but very little from plants. When more specific information is sought, the difficulty of association of the coprolite with its donor assumes importance. In the Clear Fork this has been possible, with some assurance, for only two genera, *Dimetrodon,* a large predaceous pelycosaurian reptile, and *Xenacanthus,* a fresh water shark. From Isheevo, a Permian site in the U.S.S.R., comes a series of coprolites that seem to pertain to a large, predaceous amphibian known in the same deposits. Their yield of remains of crossopterygians, lung fish and predaceous reptiles, as well as of their own kind, gives considerable insight into the feeding habits of the presumed donor.

In the analysis of diet, as in all such analyses, it is the bits of evidence from many sources that finally may be woven into a reasonably satisfactory picture.

Community evolution

The meaning of the phrase community evolution is a complex and somewhat nebulous matter. The way that it is viewed in the present context will help to set the stage for analysis of the relationships of the three community types considered later. Community evolution is thought of as crudely analogous to organic evolution. If not carried to an extreme, particularly with respect to mechanism, the analogy presents a convenient and comprehensible model for analysis. Stability of structure over a long period of time, splitting, convergence and threshold phenomena are all evident. Communities also may merge and this can, by some stretch of the imagination, be analogized to hybridization. At this point, however, the model of a population is unsuitable and such a concept clouds rather than clarifies the situation.

Community types, again like populations of organisms, are limited in their possible spectra of development by the various conditions that are functions of the time in which they exist. These are in large part products of the generally emergent nature of evolutionary processes and are thus both a product and function of the very evolution that is being studied. Any particular time provides a limited suite of settings in which both physical and biological restrictions limit the types of communities that exist. Each community is itself a part of the existing array, supplying one element of the self-limiting complex. The evolution of a complex of communities is related to changes of each of its components, as their modifications alter living conditions of communities with which they have some sort of association.

During the Permian period communities that included terrestrial vertebrates suffered rather severe restrictions, partly as a result of the relatively recent emergence of vertebrates onto land, partly as a function of the type of floral covering of the land, and partly in relationship to the level of biological organization of the vertebrates at the stage of evolution reached at that time. The relaxation of these restrictions is one of the important events that we trace as we follow the semiterrestrial and terrestrial complexes through the Permian and Triassic periods.

Community types of the Permian and Triassic

Three distinct community types which include terrestrial vertebrates can be identified in the known Permian deposits. Each is reasonably discrete, but geographic and ecological overlaps do occur. Two of these types appear to persist into the Triassic, although evidence for one of them is not too satisfactory. The three types, which are designated as I, II, and III, are as follows:

Type I (Figs. 1, 2): This type of community has its food chain based primarily in aquatic plants.

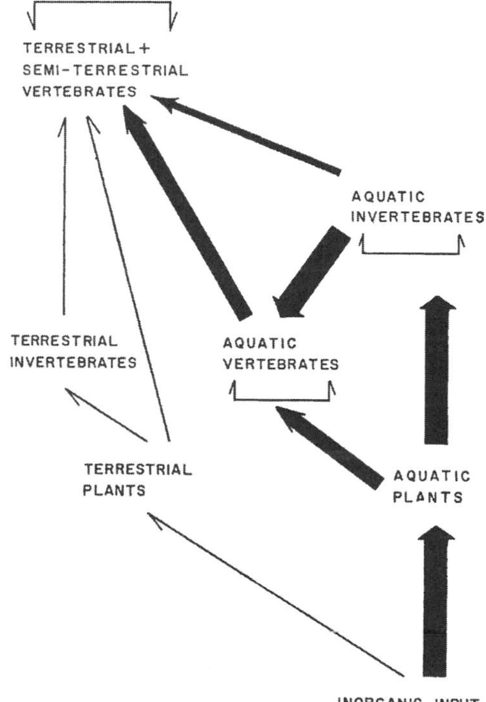

FIG. 1. Type I community with the food chain based primarily in aquatic plants, consumed by invertebrates and vertebrates, the food sources for both aquatic and terrestrial vertebrates. The direction of flow (by feeding) is indicated by the arrows. Main avenues, heavy arrows; subsidiary avenues, light arrows. Arrows on brackets leading to the same group indicate that some members of the particular group fed upon others of the same group.

These provide the primary source of food energy, supplied to the community by synthesis of organic matter from inorganic. Terrestrial and semiterrestrial invertebrates are included, but are of only minor importance. Aquatic and semiaquatic plants occupy a dominant position, with terrestrial plants of much less significance.

Following the primary conversion of inorganic substances to organic, chiefly by aquatic plants, energy follows a course that leads through small aquatic invertebrates and vertebrates to aquatic, semiaquatic and terrestrial vertebrates. Once the community has been established, organic debris, supplied from many sources, presumably became an important food source. The course of energy flow is complex, with multiple channels and alternative routes. An attempt to map out some of these is shown in Figure 2, the case for the Permian complex in the lower part of the Clear Fork (Arroyo formation). The arrows indicate the direction of flow of energy based upon presumed feeding relationships. A complicated structure, such as this, tends to make for high stability, not only in the persistence of the basic ecological types, but also in temporally extended occupancy of primary roles by particular genera and species. *Dimetrodon* is a pertinent example, as illustrated in the upper right hand corner of Figure 2. This large pelycosaurian predator persisted in its role as principal predator with no evident morphological change in a slowly modifying community for a period of several million years. Its taxonomic position and those of other genera as well as of higher categories, are summarized in the appendix.

The community type I was derived from a more ancient, essentially aquatic type of community, by expansion into the terrestrial realm without important modification of the preexisting structure. It can be recognized from the base of the Permian, with some Pennsylvanian (Carboniferous) traces that suggest its earlier existence, through to the end of the Kazanian (or possibly early Tatarian) of the upper Permian of the Soviet Union. The latest clear-cut example is from the vicinity of Isheevo, Tatar, U.S.S.R. No clearly identifiable examples of this type are discernable in the Permo-Triassic Beaufort series of South Africa, although some of the assemblages are suggestive.

Type I communities appear to have provided the framework within which the members of the lines of synapsid reptiles that eventually lead to mammals existed during the early and middle stages of this evolutionary history. Ophiacodonts, sphenacodonts, phthinosuchids, gorgonopsians and possibly therocephalians are known examples of

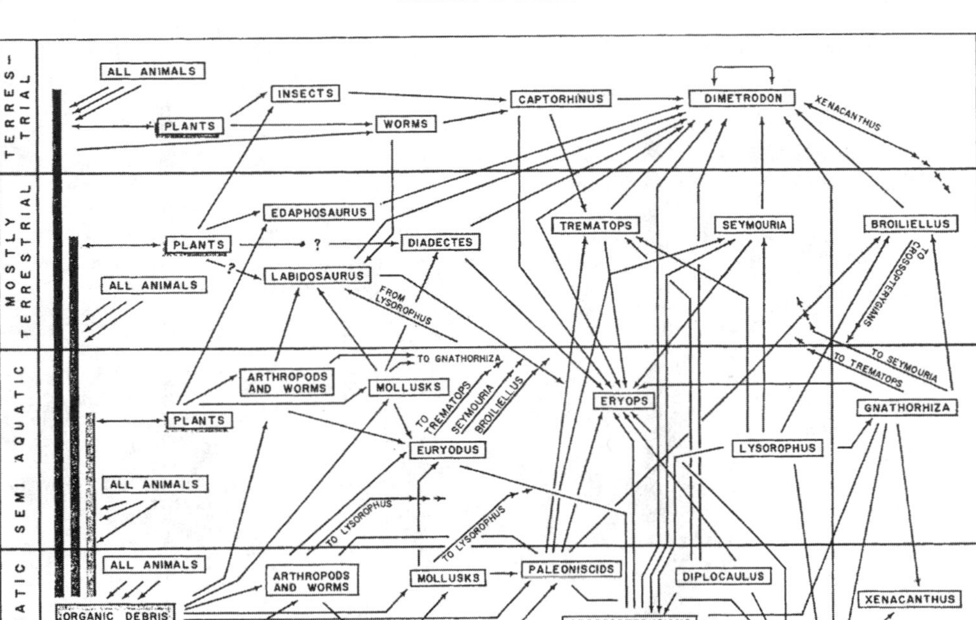

FIG. 2. An example of the complex patterns of flow of food-derived energy in a particular community of type I. The diagram is based primarily upon the well-known vertebrates of the Arroyo (Clear Fork) early Permian community in northern Texas. It is, of course, overly simple as compared to conditions that must have actually existed. Fish: *Xenacanthus* (predaceous shark), *Gnathorhiza* (lung fish), crossopterygians (lobe fins), paleoniscoids (primitive actinopterygians). Amphibians: *Lysorophus, Diplocaulus, Trimerorhachis, Euryodus, Trematops, Broiliellus, Seymouria, Diadectes* (see Romer 1964) Reptiles: *Edaphosaurus, Labidosaurus, Dimetrodon, Captorhinus.*

such lines. Many features of mammals, both morphological and physiological, seem to find reasonable explanations in the long duration of aquatic ties of this community type. Tatarinov (1959) in particular has emphasized the role of aquatic ancestors in the development of mammalian structure. Many of the morphological and physiological features that are considered reptilian because they are present in modern reptiles may never have existed in the reptilian stocks which gave rise to mammals, but themselves became extinct late in the Triassic period.

Evolving within community type I, but playing a relatively minor role, were certain elements that several times broke with their aquatic heritage. Community types II and III, considered in the following paragraphs, probably had their origin within these minor components. Final dissolution of type I appears to have taken place in the very late Permian, as the bonds to water were progressively broken and durable representatives of other community types emerged. Later, during the Triassic, new versions of type I appeared once more. With the development of birds and mammals, the chances of such communities were reduced. Type I, however, is approximated in some of the largely fish-based communities that exist in subtropical swamp regions. The assemblages of the Florida Everglades, in their unmodified form, depart only moderately from this type of community.

Type II: This community type (Fig. 3) is dominated by terrestrial vertebrates, invertebrates and plants. The primary sources of food energy of the vertebrates are terrestrial plants. These are consumed directly by terrestrial herbivorous vertebrates. The invertebrate component, which may be highly significant numerically and in terms of biomass, plays a secondary role in the vertebrate food chain, although it is, of course, important in the totality of the functions of the community type.

The structure of this type of community can be extremely complex, as witnessed by representative cases among some of the mammal-dominated communities of type II in existence today. Among modern communities, of course, extensive subdivision is possible. Such opportunities are more limited in older communities, both because of a

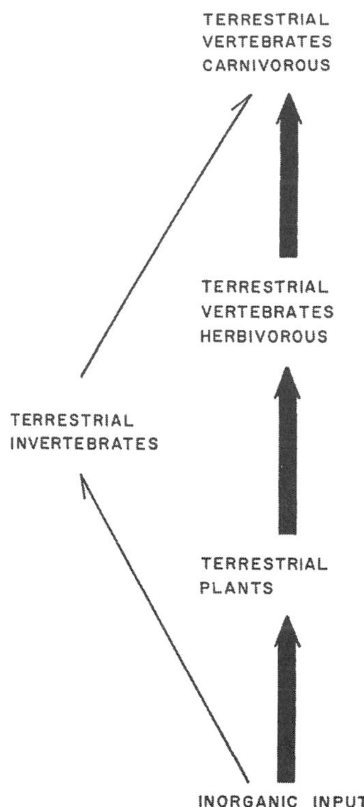

FIG. 3. Type II community. Dominantly terrestrial with food sources in terrestrial plants consumed directly by terrestrial, vertebrate herbivores. Symbols as for Figure 1.

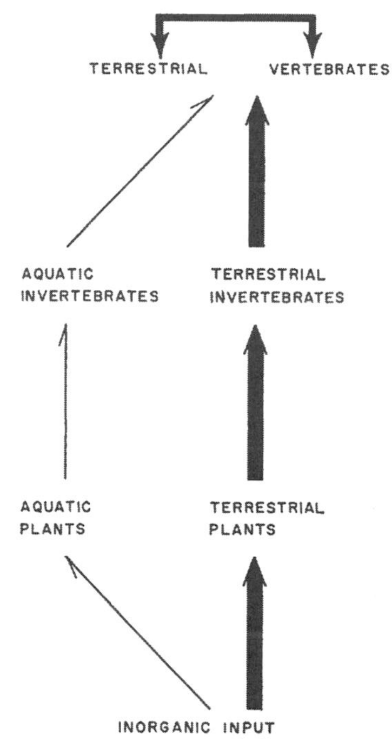

FIG. 4. Type III community. Dominantly terrestrial food sources in terrestrial plants consumed primarily by terrestrial invertebrates which in turn are principal food source of terrestrial vertebrates. The important differences from type II is the near absence of terrestrial vertebrate herbivores, and the primary role of invertebrates as plant feeders. Symbols as in Figure 1.

presumably greater simplicity, especially before the Cenozoic, and because of the nature of the evidence available for their interpretation.

The first definitely recognized occurrence of a community of type II is in the very early part of the late Permian of North America, in the San Angelo-Flowerpot formation complex. This seems to be a continuation of an earlier stage, from the Hennessey formation of Oklahoma, but the fauna of the Hennessey is not well enough known for a definite conclusion.

Communities of type II have risen many times and from many sources. The late Permian and early Triassic communities of the Beaufort Series of South Africa seem to be mostly of this type, derived in part from type I and in part from type III. Prior to the Triassic, South African assemblages show some features reminiscent of type I communities; and the degree of independence from aquatic ties, in particular from aquatic food ties, needs further investigation.

Type III: This (Fig. 4) is a type of community that consists predominantly of terrestrial organisms. The primary conversion of inorganic to organic matter is accomplished by terrestrial plants and the principal consumers of plant food are terrestrial invertebrates, primarily insects. Vertebrates are small to moderate-sized insect eaters and predators. The difference in the paths which plant food follows in providing food energy to the animal components of the community constitutes the principal distinction between types II and III. In the former, the primary herbivores are vertebrates; in the latter, they are invertebrates. Type II communities may develop either from type I or type III, but the resultant structure will initially, at least, be distinctly different.

Type III is a restricted community which could exist today only under special conditions of isolation from the dominant type II communities. Type III has been specifically identified only in the Kazanian of the Russian Permian in this study. The small number of deposits preserving this type of community results largely from the fact that uplands were the common abode. Almost all existing deposits of the Permian and early Triassic periods formed either under marine circumstances or in lowland areas marginal to the seas. There are essentially no truly upland deposits known. Occurrence of assemblages of strictly terrestrial organisms are unlikely. Members of upland communities that were carried to lower levels lost their life associations and these can be reconstructed only with serious reservations.

At the time that upland deposits first appear in the record, later in the Triassic, some of them, specifically those characterized by lepidosaurs and archosaurs, show that there had been a fairly long history of strictly terrestrial evolution before their formation. The composition of the vertebrate assemblages that they contain suggests that these communities originated from type III communities.

The role of type III communities in tetrapod evolution of the Permian and Triassic, as far as direct and unequivocal evidence is concerned, remains somewhat problematical until more adequate data are at hand. It seems very likely, however, that the lepidosaur-archosaur radiations of the Mesozoic era had their origins in stocks that gained freedom from aquatic ties by the development of such a community structure. The known examples of type III, from the Russian Kazanian, have taxonomic compositions only partially appropriate to the occupants of this ancestral position. From one site, the Mezen locality in northern European Russia, there is evidence of an eosuchian. But the others lack this element.

The fact that less than enough is known of the early stages of the lepidosaur-archosaur radiation has led to strong differences of opinion concerning the ultimate source of the stocks. Some elements that seem to have been involved, milleretids and younginids (eosuchians), occur associated with early therapsid, anomodont-theriodont communities of type II. The time of occurrence, the ecological associations, and the lack of any marked trend among these eosuchians toward archosaurian diapsids all seem to suggest that the sources of the lepidosaur-archosaurian radiations did not lie within the anomodont-theriodont complexes, but were developing elsewhere under other circumstances.

It is possible, of course, that these occurrences represent a stage of sorting out, in which the eosuchians were developing an independence from the early therapsid complexes that led to type III communities. Or more probably, as the timing suggests, such ancestral communities were already well established, and the eosuchians in the early therapsid complexes were immigrants from more stable type III communities, which also included the initial stages of the archosaurian radiation. It cannot be determined from the published information just how intimate the actual associations of the eosuchians and anomodont-theriodont therapsid complexes were.

We shall assume in the synthesis in the next section that a community of type III provided the base of the lepidosaur-archosaur radiation. We shall assume also, for want of evidence of a better solution, that this base had its source in the terrestrial, presumably insect-feeding captorhinomorphs of the type I communities of the early Permian. Whether these assumptions are precisely true or not, the critical point is that nearly complete freedom from aquatic environments must have been attained by a potential lepidosaur-archosaurian assemblage or by two assemblages independently, earlier and in a manner different from the emancipation of the therapsid communities. The reptilian characters of morphology and physiology in the radiation are considered to reflect the results of an early and complete abandonment of aquatic ties. Later successes are believed to have resulted from a consequent high level of adaptation to land.

Evolution of communities leading to mammals

The essence of the scheme of faunal evolution suggested here is shown in Figure 5. The situation is complex and the fact that the phylogeny involves communities rather than the more usual systematic units may add to the problems of following it through. The names used for communities are of three types, to conform with designations commonly used in the literature on fossil vertebrates. For the North American Permian, where the communities are known over rather wide areas, stratigraphical terms have been used, for example Wichita and Clear Fork. Some of the Russian sites are local, and place names, such as Yeshovo and Shikhovo-Chirki, are entered. However, Bashkirian and Kargalian, have a stratigraphic connotation, once again because of the extensive area over which the collections have been taken. The names of certain animal complexes are entered in Figure 5, covering several sites and stratigraphic terms in some instances, to indicate the relationship of the community types to such designated units. In the upper Permian and Triassic and in entries for more recent times, taxonomic designations are used, since these communities cannot be readily designated either in units of space or of time. The taxonomic terms are given in systematic form in the appendix, but

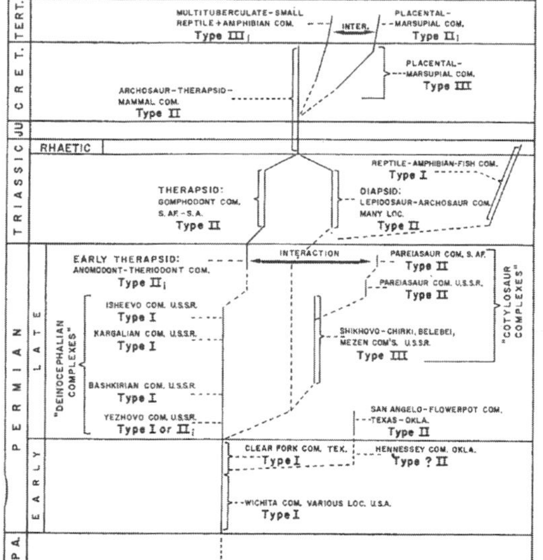

FIG. 5. A diagram of the scheme of interrelationships of community types as represented by particular communities in the series of events leading to the origin and early development of mammals. Essentially this is a phylogeny of communities. It is described in the text in more detail. The various representatives of community types are indicated by locality names, stratigraphic names and taxonomic names as appropriate to accord with the circumstances of preservation and common usage in the literature concerning each. The solid lines indicate presumed continuous spans of duration of a particular community type, as indicated. In some instances an incipient community of another type is indicated by the subscript i. The specifically noted instances of community types, for example Clear Fork or Shikhovo-Chirki, in essence represent samples taken from the geographic and temporal ranges of the community type which they represent. Complexes, such as the deinocephalian complex, may have considerable duration, as indicated by the extent of the brackets. The dashed lines show the presumed line of descent between the community types, portraying the phylogeny of the community systems. Thus the Wichita community of type I led to the Clear Fork community of type I and also, provisionally, to the community type II of the Hennessey formation. This in turn is the probable progenitor of the San Angelo-Flowerpot type II community. The type I is continued through the U.S.S.R. samples represented by sites and stratigraphically named assemblages. These are broadly grouped under a series of deinocephalian complexes, as indicated. This general pattern of interpretation carries through the rest of the phylogeny, with supplemental explanation in the text.

for convenience are also entered in the caption for Figure 5.

The early Permian communities that contained terrestrial vertebrate elements were all of type I. This was an extremely persistent type which probably began in the Pennsylvanian. There is considerable evidence of it in the later parts of this period, but none of the assemblages is complete enough for clear interpretation.

Within the early Permian communities there can be identified small segments that had capacities for breaking from the parent type. The edaphosaurids and incipient caseids, primitive herbivorous pelycosaurian reptiles, provided potential terrestrial herbivores. Neither appears to have been very abundant in the known assemblages. Captorhinomorphs, rather lizard-like anapsid reptiles, probably were the main feeders on invertebrates. Of interest in this regard is a recent paper by Eaton (1964) in which he has presented evidence, based on what appears to be an actual case of predation caught in the record, that captorhinomorphs were to some extent at least predators on small vertebrates. It seems very probable that these reptiles, like modern ones, ate in large part what was available to them, within limits of their capabilities of ingesting and utilizing the food. Thus it may be supposed that captorhinomorphs fed on insects, small vertebrates, and any other creatures that they encountered and could handle.

In the latest Clear Fork there is a suggestion of a type II community in the Hennessey formation of Oklahoma, but it is in the beds composing the San Angelo and Flowerpot formations of Texas and Oklahoma that a complex example of this type can first be authenticated. It is based on plant feeding terrestrial vertebrates, caseid in large part, and carnivores, mainly advanced sphenacodonts and phthinosuchids. Large captorhinomorphs played a numerically important role. Caseids, especially the large genus *Cotylorhynchus*, outnumber all other members of the fauna by a ratio of about 10 : 1 in the collections. They were very massive and up to 20 ft long. In mass they probably ran as high as 20 or 30 times the rest of the fauna. There is some evidence of the existence of such a community type in the upper Kazanian of Russia, but associations are poor and the record is far from unequivocal.

The main line of post Clear Fork evolution is found in the continuation of the type I community, as seen in a succession of sites in the Kazanian (late Permian) of Russia (Yeshovo, Bashkirian, Kargalian and Isheevo). These are the deinocephalian complexes of Efremov, so-named because they usually include specimens of these large, primitive therapsids. Even the youngest of these occurrences, that at Isheevo, maintains a full array of the properties manifest in the early Permian.

From several places in the Soviet Union have come assemblages that give a tangible basis for recognition of communities of type III; for example, Mezen, Shikhovo-Chirki and Belebei. These represent the cotylosaur complex of Efremov (see Efremov and Vjushkov 1954; Olson 1958, 1962) and include primitive procolophons (the cotylosaurs), carnivorous phthinosuchids and, at Mezen, an eosuchian. Amphibians occur in the same beds, but the association is to be considered taphonomic rather than ecological; that is, the death assemblage has sampled different life zones along the stream course and has brought together animals that did not form a life assemblage. It is significant that the cotylosaur complexes of type III, and the deinocephalian complexes, which are of type I, are contemporary but geographically separated. Members of the two are not mixed in the record.

From this type of community, as represented in the materials from Belebei and Shikhovo-Chirki, emerged the Russian pareiasaur complex, classified as type II. This shows strong ecological resemblances in its major constituents to the earlier San Angelo-Flowerpot community of North America, characterized by an animal complex including *Cotylorhynchus*, sphenacodonts and phthinosuchids. The taxonomic position of the predominant herbivores, however, was very different. Presumably the Russian and also the South African pareiasaur complexes had a common origin, but the latter clearly includes important increments from a type I community.

By gradual relinquishment of aquatic ties, type I communities, such as those seen in the Russian Kazanian, gave rise to type II communities, as seen in the upper Permian of South Africa. For a time these emerging communities were in close association and perhaps merged with pareiasaurian communities.

From this gradual shift emerged the early therapsid anomodont-theriodont complex of the uppermost Permian of South Africa. Such communities surely were developing elsewhere in the world, but there is no other good record of them at this time in earth history. In South Africa, during the late Permian, there appears to have existed a complex community, dependent primarily upon food energy synthesized from inorganic sources by terrestrial plants. Early in the Triassic period internal evolutionary advances, it would seem, resulted in transference of the primary herbivorous role from the anomodonts to gomphodont cynodonts, and similar creatures. This represented a change in constitution but one that did not result in fundamental alteration of the type II community. It is striking that strictly terrestrial diapsids are not a component of this community type, although they surely were in existence and undergoing a strong evolutionary development. Some diapsids, such as rhynchosaurids, specialized rhynchocephalians, contribute to the community, but these somewhat aquatic creatures are not part of the main stream of diapsid evolution.

Small, primitive diapsids, as has been noted, were constituents of the anomodont-theriodont complex of the later Permian. If they came from the millerettids, as is generally believed, and if the millerettids came from captorhinomorphs, as Parrington (1958) among others including the writer have argued, then they may be thought of as direct descendants and continuations of the insectivorous increment represented earlier by the captorhinomorphs. This stock, so interpreted, forms the base of the terrestrial radiation of diapsids. Initially its terrestrial role was acquired as a progenitor of a type III community while technically a part of a very early (Pennsylvanian) community of type I, the same type as that in which the synapsids began their early radiation. The best-known genus of the captorhinomorphs (*Captorhinus*) did not itself occupy this position, for it is too late and structurally too specialized. Rather less specialized genera from the very early Permian or Pennsylvanian approach the actual ancestry more closely.

Whatever the final solution of this problem may be, it is evident that at present we lack much-needed information about the origin and early history of the diapsids and the places of their evolution. It is significant, as noted earlier, that our records tap only relatively lowland areas, close to the sea, prior to the middle or late part of the Triassic. The important fact concerning the diapsids, even emphasized by the poor record, is that their development took place independently of that of the communities that contain the lines of therapsid evolution.

Through much of the Triassic, the highly terrestrial diapsid (lepidosaur-archosaur) communities of type II continued independently of the contemporary therapsid communities. In view of their apparently long history of independence from aquatic environments, it may be supposed that the diapsid communities had become more highly adapted to terrestrial existence than were the contemporary therapsid complexes. As the latter pushed into more strictly terrestrial environments, during the middle and late Triassic, and the lepidosaur-archosaur lines continued to deploy in these environments and increasingly to exploit less strictly terrestrial, semiaquatic zones, the two great community arrays, each of which had attained a type II organization independently of the other, came into close contact and undoubtedly into competition. For the first time the two are found intimately associated in the deposits; this occurs more or less simultaneously in several widely separated regions.

Rapid impoverishment of the therapsid stocks followed, and most of the large, prominent types became extinct. Some of its elements, however, were incorporated into the more successful lepidosaur-archosaur complexes. These were for the most part small insectivorous and herbivorous remnants, those that we consider to have been extremely advanced mammal-like reptiles, such as the trityolodonts, and primitive mammals, the triconodonts, symmetrodonts, and morganucodontids. In beds of Rhaetic age, uppermost Triassic, this type of community, type II, is found in many places in the world. Perhaps the finest known examples are in the Lu Feng deposits in Yunnan, China.

Too little is known to permit speculation upon the exact role of the Mesozoic mammals in the communities of the Jurassic and Cretaceous. At length the "insectivorous mammals," incipient carnivores, broke from the dominant communities, very probably to form a community type resembling what has been defined earlier as type III. Eventually there emerged from this a primitive mammalian community of type II, the ancestral type from which the familiar mammalian assemblages characteristic of the Cenozoic arose. In this was reconstituted much of the structure of the type II therapsid community of earlier times, but it is clear that the mammalian community structure, while similar, did not derive directly from the very mammal-like reptilian community of the Triassic.

Conclusion

The framework that has been established and in which the evolution of mammals has been cast is obviously based on less-than-sufficient evidence at many levels. On the other hand, there is good evidence for part of it and more detailed studies may add strength elsewhere. Some parts, such as the evolution of initial stages of the lepidosaur-archosaur lines may never be known. Much of what has been reconstructed is an extrapolation from the relatively extensive information about parts of the Permian. The rationale of evolution of the mammals that emerges fits the facts as I now understand them. The emergence of mammals is related to the persistence of aquatic habits which are thought to be reflected in many of the mammalian features of structure, physiology and behavior, in particular the premium on activity. The long, slow maturing of mammalian features during the interim period from the beginning Jurassic to late Cretaceous, during which they play an inconspicuous role in ecology, probably gave full measure of gestation to the incipient mammalian characters. The stage was thus set for the rapid and vast radiation that commenced in the latest Cretaceous and reached full swing early in the Cenozoic.

Literature Cited

Eaton, T. H. Jr. 1964. A captorhinomorph predator and its prey (Cotylosauria). Amer. Mus. Nat. Hist. Novitates, no. 2169, 3 p.

Efremov, I. A. 1950. Taphonomy and the geological record. Tr. Paleon. Inst. Acad. Sci. U.S.S.R., **24**: 3-177. (In Russian)

Efremov, I. A., and Vjushkov, B. P. 1955. Catalogue of Permian and Triassic terrestrial vertebrates in the territories of the U.S.S.R. Tr. Paleon. Inst. Acad. Sci. U.S.S.R., **46**: 1-185. (In Russian)

Olson, E. C. 1952. The evolution of a Permian vertebrate chronofauna. Evolution **6**: 181-196.

———. 1957. Catalogue of localities of Permian and Triassic terrestrial vertebrates of the territories of the U.S.S.R. J. Geol., **65**: 196-226.

———. 1958. Fauna of the Vale and Choza: 14. Summary, review and integration of the geology and the faunas. Fieldiana: Geology **10**: 397-448.

———. 1961. The food chain and the origin of mammals., Internat. Colloq. on Evol. Mammals. Kon. Vlaamse Acad. Wetensch. Lett. Sch. Kunsten Belgie, Brussels, 1961, pt. I, p. 97-116.

———. 1962. Late Permian terrestrial vertebrates, U.S.A., and U.S.S.R. Trans. Amer. Philos. Soc., n.s., 52 pt. 2, p 224

Parrington, F. R. 1958. The problem of the classification of reptiles. J. Linn. Soc. London, **44**: 99-115.

Romer, A. S. 1964. *Diadectes* an amphibian? Copeia **1964**: 718-719.

Romer, A. S. and E. C. Olson. 1954. Aestivation in a Permian lungfish. Mus. Comp. Zool., Harvard, Breviora, no. 30. 8 p.

Tatarinov, L. P. 1959. Origin of reptiles and some principles of their classification. Paleon. J., Acad. Sci., U.S.S.R., no. 4. p. 66-84. (In Russian)

Appendix

Class Reptilia

- Subclass Anapsida
  - Order Captorhinomorpha
    - *Captorhinus*
    - *Labidosaurus*
  - Order Procolophonia
  - Order Pareiasauria
- Subclass Diapsida
  - Infraclass Lepidosauria
    - Order Eosuchia
      - Family Millerettidae
      - Family Younginidae
    - Order Rhynocephalia
      - Family Rhynchosaurida
  - Infraclass Archosauria
    - Order Thecodontia
- Subclass Synapsida
  - Order Pelycosauria
    - Suborder Ophiacodontia
      - *Ophiacodon*

Suborder Sphenacodontia
Dimetrodon
Suborder Edaphosauria
Edaphosaurus
Suborder Caseosauria
Family Caseidae
Cotylorhynchus

Order Theriodonta
Suborder Theriodonta
Infraorder Gorgonopsia
Family Phthinosuchidae
Family Gorgonopsidae
Infraorder Therocephalia
Infraorder Cynodontia
Family Gomphodontidae

Suborder Anomodontia
Suborder Tritylodontia

Class Mammalia
Subclass Allotheria
Order Multituberculata

Subclass(es) Uncertain
Order Triconodonta
Order Symmetrodonta
Order Docodonta
Family Morganucodontidae

Subclass Theria
Order Pantotheria
Order Marsupialia
Order Placentalia

PAPER 34

The Discovery of America (1973)

P. S. Martin

Commentary

ANTHONY D. BARNOSKY

By 1973, Paul Martin's idea that aboriginal people had hunted the Pleistocene megafauna to extinction had been in the mainstream scientific literature for seven years. He had been dealing with two major criticisms. How could just a few people with stone-age weapons wipe out so many different kinds of animals? And where were the bodies with spear points in them? His paper "The Discovery of America" was his answer. It was destined to become a classic from the start, both because he was so innovative in how he approached the problem and because it dealt with something we still grapple with today: the impact of humans on other species. His paper treated *Homo sapiens* as an invasive species interacting in the global ecosystem. Among his clever approaches were integrating equations of population growth, estimations of megafaunal biomass, information about caloric needs, and so on to come up with a number for just how few humans would be required to hunt large-bodied animals to extinction. It was very few indeed—he concluded that the American extinctions could have resulted within 1,000 years after 100 prehistoric hunters began a southward journey near Edmonton, Canada. For Martin, that also explained the lack of kill sites: it all happened so fast, the preservation probability was vanishingly low. Neither argument did much to convince his opponents—the debate raged throughout the twentieth and into the twenty-first century. Newer information ended up rejecting Martin's model of a Clovis-first entry and a simple north-to-south migration of *Homo sapiens*. Also failing the test of further data was extinction solely due to hunting and all within 1,000 years. And the idea of absence of evidence being evidence has not found many supporters. Nevertheless, the fact remains that early on, working at the frontiers of knowledge in his field in his day, Paul Martin got it basically right. By assembling paleoecological data and using his remarkably creative mind to interpret it, he brought to the forefront the idea that people could indeed wipe out many species. His central argument—that there are no continents or islands where major Quaternary extinction pulses preceded arrival by humans—has only become better supported with the research his pioneering ideas stimulated.

From *Science* 179:969–74. Reprinted with permission from AAAS.

The Discovery of America

The first Americans may have swept the Western Hemisphere and decimated its fauna within 1000 years.

Paul S. Martin

America was the largest landmass undiscovered by hominids before the time of *Homo sapiens*. The Paleolithic pioneers that crossed the Bering Bridge out of Asia took a giant step. They found a productive and unexploited ecosystem of over 10^7 square miles (2.6×10^7 square kilometers). As Bordes has said (*1*), "There can be no repetition of this until man lands on a [habitable] planet belonging to another star."

At some time toward the end of the last ice age, big game hunters in Siberia approached the Arctic Circle, moved eastward across the Bering platform into Alaska, and threaded a narrow passage between the stagnant Cordilleran and Laurentian ice sheets. I propose that they spread southward explosively, briefly attaining a density sufficiently large to overkill much of their prey.

Overkill without Kill Sites

Pleistocene biologists wish to determine to within 1000 years at most the time of the last occurrence of the dominant Late Pleistocene extinct mammals. If one recognizes certain hazards of "push-button" radiocarbon dating (*2*), especially dates on bone itself, it appears that the disappearance of native American mammoths, mastodons, ground sloths, horses, and camels coincided very closely with the first appearance of Stone Age hunters around 11,200 years ago (*3*).

Not all investigators accept this circumstance as decisive or even as adequately established. No predator-prey model like Budyko's (*4*) on mammoth extinction has been developed to show how the American megafauna might have been removed by hunters (*5*). Above all, prehistorians have been troubled by the following paradox.

In temperate parts of Eurasia, large numbers of Paleolithic artifacts have been found in many associations with bones of large mammals. Although the evidence associating Stone Age hunters and their prey is overwhelming, not much extinction occurred there. Only four late-glacial genera of large animals were lost, namely, the mammoth (*Mammuthus*), woolly rhinoceros (*Coelodonta*), giant deer (*Megaloceros*), and musk-ox (*Ovibos*).

In contrast, the megafauna of the New World, very rarely found associated with human artifacts in kill or camp sites (*6*), was decimated. Of the 31 genera of large mammals (*7*) that disappeared in North America at the end of the last ice age, only the mammoth (*Mammuthus*) is found in unmistakable kill sites. The seven kill sites listed by Haynes (*8*) lack the wealth of cultural material, including art objects, associated with the Old World mammoth in eastern Europe and the Ukraine. It is not surprising that some investigators discount overkill as a major cause of the extinctions in America.

But if the new human predators found inexperienced prey, the scarcity of kill sites may be explained. A rapid rate of killing would wipe out the more vulnerable prey before there was time for the animals to learn defensive behavior, and thus the hunters would not have needed to plan elaborate cliff drives or to build clever traps. Extinction would have occurred before there was opportunity for the burial of much evidence by normal geological processes. Poor paleontological visibility would be inevitable. In these terms, the scarcity of kill sites on a landmass which suffered major megafaunal losses becomes a predictable condition of the special circumstances which distinguish a sudden invasion from more gradual prehistoric cultural changes in situ. Perhaps the only remarkable aspect of New World archeology is that *any* kill sites have been found (*9*).

Megafaunal Biomass

Bordes (*1*) and Haynes (*8*) believe that the Stone Age hunters found abundant game in America. Although the fauna was diverse (*7*), no estimates of the size of the Late Pleistocene game herd have been attempted. I propose two crude but independent methods of estimating the biomass of the native megafauna, both of which utilize present range-carrying capacity. In the first method one projects estimates of the biomass of large mammals in African game parks to areas of comparable range productivity in the New World. The other method is based on the assumption that present managed livestock plus game populations in the Americas would equal, and probably exceed, the maximum herd size of the Late Pleistocene.

Estimates of biomass in various African parks are shown in Table 1. The drier parks such as Tarangire Game Reserve, Kafue, Kagera, and others not included in Table 1 such as Kruger National Park, South Africa, and Tsavo National Park, Kenya, support 10 to 20 animal units (*10*) per section (1.8 to 3.5 metric tons per square kilometer). In the Americas during the Pleistocene, similar values might be expected on drier ranges (mean annual precipitation, 400 to 600 millimeters) dominated by mammoth, horse, and camel. The carrying capacity would have been much less in the driest regions (annual precipitation less than 200 millimeters).

African game parks supporting the highest biomass, over 100 animal units per section (over 18 metric tons per square kilometer), occur in tall grass savannas along the margin of wet tropical forest. The dominant species are elephant, buffalo, and hippo. In the tropical American savannas, along coasts, and on floodplains of the temperate regions, one might expect a similar biomass in the Pleistocene. The dominant species were mastodons and large edentates.

For the 3×10^6 sections of land

The author is professor of geosciences, University of Arizona, Tucson 85721.

Table 1. Large mammal biomass in some African parks and game reserves [from Bourliere and Hadley (*32*)]. [Courtesy of Annual Reviews, Inc., Palo Alto, Calif.]

Location	Habitat	Number of species	Biomass (metric tons per square kilometer)	Biomass (animal units per square mile)
Tarangire Game Reserve, Tanzania	Open *Acacia* savanna	14	1.1	6
Kafue National Park, Zambia	Tree savanna	19	1.3	7
Kagera National Park, eastern Rwanda	*Acacia* savanna	12	3.3	18
Nairobi National Park, Kenya	Open savanna	17	5.7	32
Serengeti National Park, Tanzania	Open and *Acacia* savannas	15	6.3	36
Queen Elizabeth National Park, western Uganda	Open savanna and thickets	11	12	68
Queen Elizabeth National Park, western Uganda	Same habitat, overgrazed	11	27.8–31.5	158–179
Albert National Park, northern Kivu	Open savanna and thickets, overgrazed	11	23.6–24.8	134–141

(7.8×10^6 square kilometers) in the unglaciated United States, I propose the following average stocking capacities, each covering roughly 10^6 sections: (i) savannas, forest openings, floodplains, and other highly to moderately productive habitats, 50 animal units per section, or 22.7×10^6 metric tons; (ii) arctic, boreal, semiarid, short grass ranges, and other low to moderately productive habitats, ten animal units per section, or 4.6×10^6 metric tons; (iii) closed canopy forest, extreme desert, barren rock, and other habitats unproductive for large herbivores, two animal units per section, or 0.9×10^6 metric tons. The total for North America north of Mexico is 62×10^6 animal units or 28.2×10^6 metric tons.

Turning from the African analogy to estimates based on current livestock plus game populations, one obtains a higher biomass. The United States alone supported 1.20×10^8 animal units (all types of livestock) in 1900 and 1.48×10^8 in 1945. Adding wild game, I project these values for the Western Hemisphere south of Canada to a total of 5.00×10^8 animal units, or 2.30×10^8 metric tons (*11*). Presumably this value, based on managed herds, exceeds the natural Pleistocene biomass.

A herd of 2.50×10^8 animal units during the Late Pleistocene would seem more realistic in terms of the African values. A hemispheric estimate of 10^8 animal units should be far too low for the primary plant productivity available to the native herbivores but would still be a sizable resource for the first Paleolithic hunters.

The alternate view, that the American large mammal biomass was in eclipse during the late glacial (*12*), cannot be tested quantitatively on the basis of fossils alone. Bones do not provide reliable estimates of past biomass (*13*). But the great numbers of mastodon, mammoth, extinct horse, camel, and bovid bones found in late-glacial sediments hardly suggest scarcity. Evidence that the Late Pleistocene megafauna was declining in numbers and diversity before 12,000 years ago, as Kurtén found in Late Paleolithic sites of the Old Wo:d (*14*), is lacking in the New World (*11*).

America's First Population Explosion

The *minimum* growth rate required to attain the estimated (A.D. 1500) population of the New World is negligible, 0.1 percent annually. Slow, imperceptible growth is what demographers are prone to project into the Paleolithic (*15*). They have no choice. Neither bones nor artifacts will reveal instantaneous rates of change. A century of maximum growth, followed by a year of massive mortality, would escape detection by archeologists.

It seems likely that, when entering a new and favorable habitat, any human population, whatever its economic base, would unavoidably explode, in the sense of Deevey [see (*15*)], with a force that exceeded ordinary restraint. The environment of the New World should have been particularly favorable. The hunters who conquered the frozen tundra of eastern Siberia and western Alaska must have been delighted when they first detected milder climates as their route turned southward. Predation loss seems improbable (*11*). More important, the major hominid diseases endemic to the Old World tropics were unknown in the New World (*16*). Hunting accidents undoubtedly occurred, but presumably less often when New World elephants were at bay than in the case of the more wary and experienced mammoths of Eurasia.

When they reached the American heartland, the Stone Age hunters may have multiplied as rapidly as 3.4 percent annually, the rate Birdsell (*17*) reported for the settlement of Pitcairn Island and elsewhere. Anthropologists regard one person per square mile (0.4 person per square kilometer) as maximum for a preagricultural economy on its best hunting grounds. Had such a population density been attained throughout the Americas, it would represent a total population of 10^7 in the 10^7 square miles (2.6×10^7 square kilometers) outside of Canada and other glaciated regions. At a rate of population growth of 3.4 percent annually, or a doubling every 20 years, 340 (17 generations) would be the minimum time needed for a band of 100 invaders to saturate the hemisphere. Even at a rate of 1.4 percent annually, or a doubling every 50 years, saturation would require only 800 years. Presumably a population crash would soon follow the extinction of the megafauna (*4*).

One need not demand that a maximum growth rate was maintained for long, or that the New World Paleo-Indian population ever totaled 10^7 at any one time. Animal invaders expand along an advancing front (*18*). I propose that the human invasion of the Americas proceeded in the manner Caughley [see (*18*)] has reported for exotic mammals spreading through New Zealand (see Fig. 1). A high population density was concentrated only along the periphery. The advance of the hunters was determined partly by the abundance of fresh game within the front and partly by cultural limits to the rate of human migration. In a decade or less, the population of vulnerable large animals on the front would have been severely reduced or entirely obliterated. As the fauna vanished, the front swept on, while any remaining human population would have been driven to seeking new resources.

For the North American midcontinent, I assume the arrival near Edmonton of a band of 100 hunters (Fig. 2). If the average southward movement did not exceed 16 kilometers per year (*19*), 184 years would have been required to develop a population of 61,000, large enough to continue to expand southward at the required rate while maintaining the required frontal density of 0.4 person per square kilometer across an arc 160 kilometers deep. By then the front would have advanced southward 640 kilometers beyond Edmonton (Fig. 2).

Further expansion would be limited not by the maximum rate of population growth but by the assumed cultural limits to migration. A maximum population for North America would be about 600,000, with half that number on the front when it reached the Gulf of Mexico, 3300 kilometers south of Edmonton. The concordant radiocarbon ages Haynes finds among midcontinent man-mammoth sites (*8*) are conformable with the proposed rapid sweep of the hunters. Alternate solutions based on computer simulation are shown in Table 2.

Under the conditions of the model, the front reached Panama at 10,930 years ago. At this point a second slight lag ensued, imposed by the need to develop a broad front into South America after passage of the Panamanian bottleneck (Fig. 3). In this case, a larger initial population seems likely. Within about 130 years, a population growth rate of 3.4 percent annually would again begin to be limited by cultural restraints. By 10,500 years ago, 1000 years after the arrival of the hunters at Edmonton (1200 years after arrival in Alaska), Tierra del Fuego would be within view (Fig. 3).

Table 2. Simulated values for New World population growth.* In example 1, 104 people reached Edmonton 11,500 years ago. They double in numbers every 20 years until limited by their southward migration rate, 16 kilometers per year. Population growth fills a sector of 90° and is concentrated along the arc ("front") through a depth of 160 kilometers. The density on the front is 0.4 person per square kilometer and behind the front is 0.04 person per square kilometer. Example 2 is the same as example 1 except that the population doubles every 30 years. Example 3 is the same as example 1 except that the migration rate is 8 kilometers per year. Example 4 is the same as example 1 except that the migration rate is 25 kilometers per year. Example 5 is the same as example 1 except that the front is 80 kilometers deep. Example 6 is the same as example 1 except that the front is 240 kilometers deep. Example 7 is the same as example 1 except that the front is 0.02 person per square kilometer.

Example	Filling the front		Point at which frontal advance is limited by migration rate			Front reaches the Gulf of Mexico	
	Time (years)	Population	Time (years)	Distance (km)	Population	Time (years)	Population
1	125	8,000	184	393	61,000	345	590,000
2	188	8,000	299	583	102,000	440	590,000
3	125	8,000	159	207	26,000	519	590,000
4	125	8,000	199	583	102,000	294	590,000
5	86	2,800	172	457	40,000	326	435,000
6	149	18,000	193	349	81,000	358	750,000
7	125	8,000	180	368	53,000	344	450,000

* Computer simulations programmed by D. P. Adam.

Modeling Overkill

The impact of the hunters is best visualized if one considers a representative area on their front. If a sizable biomass, say, 50 animal units per square mile, were exposed for 10 years to hunters whose density is one person per square mile on the average, what removal rates would be necessary to reduce the fauna? The fraction of the standing crop of moose available annually to wolves on Isle Royale is 18 percent (*20*); mainly older and young animals are taken. For animals larger than moose, an annual removal rate of 20 percent of the biomass attributable to all predators would very likely be lethal in a few years. Smaller animals (between 50 and 400 kilograms in adult body weight) reproduce at higher rates, but their vulnerability would increase if the hunters preyed less selectively than wolves, taking a higher percentage of adult females.

An annual removal of 30 percent of their biomass should exceed normal replacement by reproduction for all the mammals lost in the Late Pleistocene. If one person in four did all of the hunting, destroying one animal unit (450 kilograms) per week from an animal population on the front averaging 50 animal units per section, he would eliminate 26 percent of the biomass in 1 year. Regions with a higher biomass (more animals), resembling the richest African game parks of today, would not have escaped if the density of hunters rose accordingly.

Provided that each carcass was carefully butchered and dried and all edible portions were ultimately consumed, the minimum caloric requirements for one person per section could have been met by the annual removal of only 5 percent of the assumed 50 animal units per section. But much more wasteful consumption is to be expected, especially if tempting, new prey were easily accessible (*11*). Wheat [see (*3*)] has

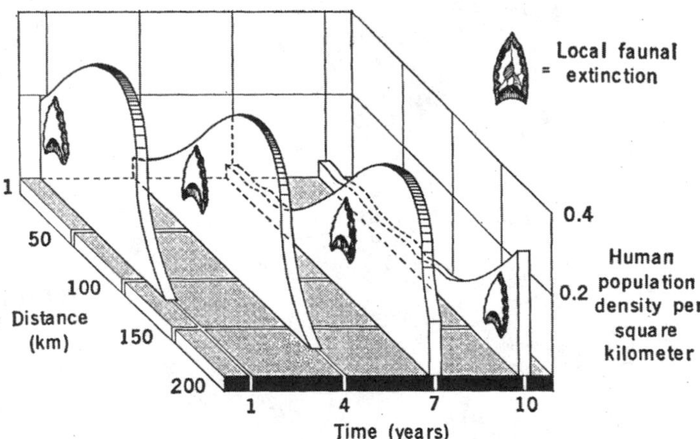

Fig. 1. Passage of the front showing theoretical changes in human population density. At any one point, the big game hunters and the extinct animals coexisted for no more than 10 years. Poor paleontological visibility of kill sites is thus inevitable.

reviewed the historic records of extraordinary meat consumption and occasional extreme waste among the Plains Indians.

Unless one insists on believing that Paleolithic invaders lost enthusiasm for the hunt and rapidly became vegetarians by choice as they moved south from Beringia, or that they knew and practiced a sophisticated, sustained yield harvest of their prey, one would have no difficulty in predicting the swift extermination of the more conspicuous native American large mammals. I do not discount the possibility of disruptive side effects, perhaps caused by the introduction of dogs and the destruction of habitat by man-made fires. But a very large biomass, even the 2.3×10^8 metric tons of domestic animals now ranging the continent, could be overkilled within 1000 years by a human population never exceeding 10^6. We need only assume that a relatively innocent prey was suddenly exposed to a new and thoroughly superior predator, a hunter who preferred killing and persisted in killing animals as long as they were available (*21*).

With the extinction of all but the smaller, solitary, and cryptic species, such as most cervids, it seems likely that a more normal predator-prey relationship would be established. Major cultural changes would begin. Not until the prey populations were extinct would the hunters be forced, by necessity, to learn more botany. Not until then would they need to readapt to the distribution of biomes in America in the manner Fitting (*22*) has proposed.

An explosive model will account for the scarcity of extinct animals associated with Paleo-Indian artifacts in obvious kill sites. The big game hunters achieved high population density only during those few years when their prey was abundant. Elaborate drives or traps were unnecessary.

Sudden overkill may explain the absence of cave paintings of extinct animals in the New World and the lack of ivory carvings such as those found in the mammoth hunter camps of the Don Basin. The big game was wiped out before there was an opportunity to portray the extinct species.

Finally, the model overcomes any objections that acceptable radiocarbon dates of around 10,500 years ago on artifacts from the southern tip of South America (*23*) require a crossing of the Bering platform thousands of years earlier (*24*).

As Birdsell (*17*) found in the case of Australia, it appears that prehistorians have overlooked the potential for a population explosion in what ecologists must regard as a uniquely favorable environment—the New World when first discovered. An outstanding difficulty remains, the question of "pre-Paleo-Indians" or "early-early man."

The Hunt for Early-Early Man

The population and overkill model I have proposed predicts that the chronology of extinction is as effective a guide to the timing of human invasion as the oldest artifacts themselves. According to Haynes (*8*), well-dated New World mammoth kill sites cluster tightly around 11,200 years ago. The population growth model presented here requires that the time of human entry into Alaska need be no older than 11,700 years ago to bring the hunters to Arizona by 11,200 years ago and to

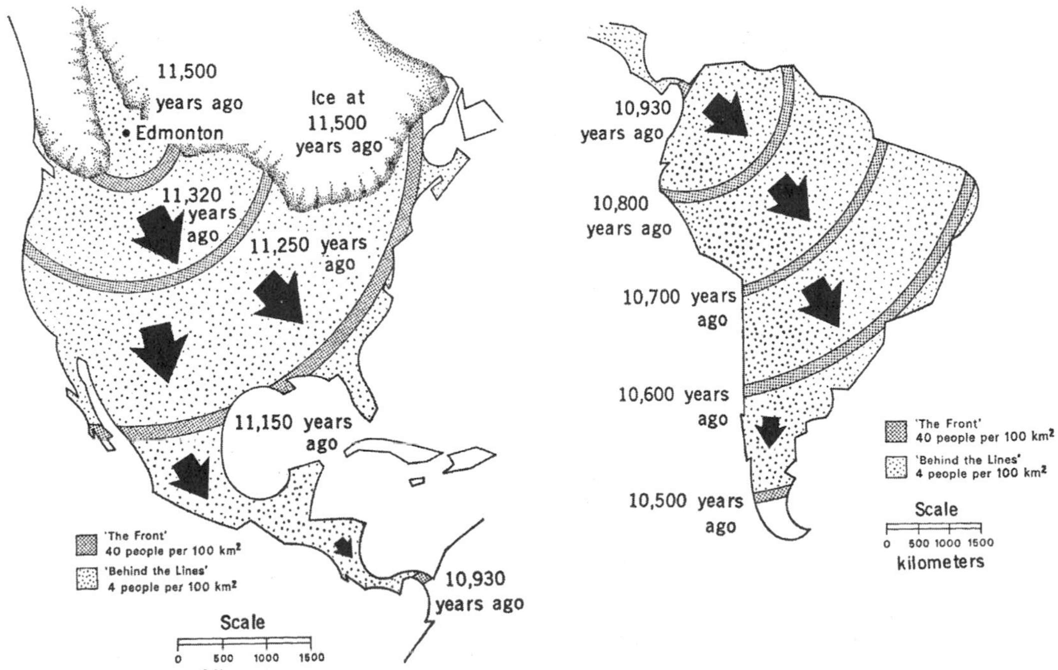

Fig. 2 (left). Sweep of the front through North America. As local extinction occurs, the hunter moves on. Fig. 3 (right). Sweep of the front through South America. Local extinction accompanies passage of the front. (Figures 2 and 3 are not drawn to scale.)

the tip of South America by 10,500 years ago.

A growing number of claims and reviews of sites considered to be at least 13,000 years old or older, including some proposed to be over 20,000 years old, have appeared recently (*25*). The presence of people in the New World long before the big game hunters of 11,200 years ago seems all but conclusively established. Most prehistorians assume that the Americas were occupied by 15,000 years ago (*26*). However, questions of evidence loom.

An ephemeral or scarcely detectable invasion by or before 15,000 years ago implies slow population growth and a low population density. Few would claim that the putative early-early Americans were numerous, and Irwin-Williams [see (*25*)] concluded that they were scarce. A sizable hunt for new evidence of early-early Americans is under way. The more spectacular the claim, the more interest is generated in the announcement (*27*).

The nature of death assemblages, the subtleties of rebedding and redeposition, the uncertainty in diagnosing artifacts, and, especially, the limitations of various dating methods under ordinary field conditions are certain to generate difficulties even for the most careful investigator. Although replication or the critical verification of an original excavation assumes major significance, it is not often attempted.

In a notable exception, the reexcavation of Tule Springs, Nevada, a well-funded team of geologists, ecologists, and archeologists failed to verify the impressive claim of a 23,000-year-old human occupation (*28*). The oldest evidence of occupation that could be verified at Tule Springs occurred in depositional units considered to be between 11,000 and 13,000 years old (*29*).

Their research material has made behavioral scientists especially sensitive to interpreter and experimenter effects. According to Rosenthal, "Perhaps the greatest contribution of the skeptic, the disbeliever, in any given scientific observation is the likelihood that his anticipation, psychological climate, and even instrumentation may differ enough so that his observation will be a more independent one" (*30*). Site replication established the contemporaneity of ancient man and extinct fauna in the New World (*31*).

Replication is now needed if the early-early man sites are to be regarded seriously. The enthusiastic search and the growing number of claims can be viewed as destructive, not supportive, of the early-early man theory. At this point, the more unreplicated claims that are filed, the more likely that their authors may be victims of an experimenter, or, in this case, excavator, effect.

Begging each claim is an ecological paradox: If *Homo sapiens* was clever enough to master a technology that allowed him to penetrate the Arctic or the marine barriers standing in the way of discovery of the New World, why did he fail to exploit the highly productive ecosystem he found in warmer parts of this hemisphere? Why did he fail to leave a trail of evidence at least as obvious as the Mousterian, Gravettian, Solutrean, and other Middle and Upper Paleolithic cultures so abundant in Europe? My questions simply rephrase an objection voiced long ago by Hrdlička and revised by Graham and Heizer (*31*).

For the present, American archeologists can rest assured that ecological principles are not violated by evidence at the known validated sites. A brief moment of big game hunting, not only of mammoth but also of many other species, could have led to megafaunal extinction around 11,000 years ago and to major cultural readaptation in most of the hemisphere afterward. It is not necessary to postulate human invasion by or before 15,000 years ago.

Invasion by a slowly growing and chronically sparse population is not impossible. But it requires major ecological constraints that have yet to be identified in the American environment. Given the biology of the species, I can envision only one circumstance under which an ephemeral discovery of America might have occurred. It is that sometime before 12,000 years ago, the earliest early man came over the Bering Straits without early woman.

Summary

I propose a new scenario for the discovery of America. By analogy with other successful animal invasions, one may assume that the discovery of the New World triggered a human population explosion. The invading hunters attained their highest population density along a front that swept from Canada to the Gulf of Mexico in 350 years, and on to the tip of South America in roughly 1000 years. A sharp drop in human population soon followed as major prey animals declined to extinction.

Possible values for the model include an average frontal depth of 160 kilometers, an average population density of 0.4 person per square kilometer on the front and of 0.04 person per square kilometer behind the front, and an average rate of frontal advance of 16 kilometers per year. For the first two centuries the maximum rate of growth may have equaled the historic maximum of 3.4 percent annually. During the episode of faunal extinctions, the population of North America need not have exceeded 600,000 people at any one time.

The model generates a population sufficiently large to overkill a biomass of Pleistocene large animals averaging 9 metric tons per square kilometer (50 animal units per section) or 2.3×10^8 metric tons in the hemisphere. It requires that on the front one person in four destroy one animal unit (450 kilograms) per week, or 26 percent of the biomass of an average section in 1 year in any one region. Extinction would occur within a decade. There was insufficient time for the fauna to learn defensive behaviors, or for more than a few kill sites to be buried and preserved for the archeologist. Should the model survive future findings, it will mean that the extinction chronology of the Pleistocene megafauna can be used to map the spread of *Homo sapiens* throughout the New World.

References and Notes

1. F. Bordes, *The Old Stone Age* (McGraw-Hill, New York, 1968).
2. P. S. Martin, in *Pleistocene Extinctions, the Search for a Cause*, P. S. Martin and H. E. Wright, Jr., Eds. (Yale Univ. Press, New Haven, Conn., 1967), pp. 87–89.
3. Over the past two decades radiocarbon dates have been published which, if taken at face value, appear to show that mammoths, mastodons, ground sloths, and other common members of the extinct American megafauna lasted into the postglacial. Since my 1967 review (*2*), the following dates, all younger than 10,000 years old, have appeared: UCLA-1325 (fossil wood below Pleistocene mammal bones); ISGS-17 A,B,C (mastodon ivory and bone); OWU-224 A,B (gyttja with mastodon); A-806 A,C,D; A-874 C; A-876 C; A-195; A-584; A-619; A-536; I-2244; M-1764; M-1765 (bone apatite, acid-soluble organic matter, enamel, and other materials from mammoth bones); UGa-79 (sloth bone). For a complete description of field and laboratory treatment of the samples and for laboratory designations, see *Radiocarbon* **9–13** (1967–1971). In those cases in which stratigraphically associated charcoal dates were available, the bone dates were all significantly younger. They may be suspected of contamination by younger carbon. In another set of especially interesting cases, much younger organic material was found associated with mammoth bones (M-2361, 3310 ± 160 years, conifer cones), a mastodon rib (M-2436, 4470 ± 160 years, conifer log), and sloth dung (UCLA-1069, 2400 ± 60 years; UCLA-1223, 2900 ± 80 years on artifacts). On the basis of other information, the collectors could discount the associations as probably secondary. As long as sample selec-

tion is not foolproof and because bone contamination is not always avoidable, we may expect a steady increase in postglacial dates of the sort which misled me years ago [P. S. Martin, in *Zoogeography*, C. Hubbs, Ed. (Publication No. 51, AAAS, Washington, D.C., 1958), p. 397]. Admittedly, there is no theoretical reason why a herd of mastodons, horses, or ground sloths could not have survived in some small refuge until 8000 or even 4000 years ago. But in the past two decades, concordant stratigraphic, palynological, archeological, and radiocarbon evidence to demonstrate beyond doubt the postglacial survival of an extinct large mammal has been confined to extinct species of *Bison* [see S. T. Shay, *Publ. Minn. Hist. Soc.* (1971); J. B. Wheat, *Sci. Amer.* **216**, 44 (January 1967); *Amer. Antiquity* **37** (part 2) (No. 1) (1972); D. S. Dibble and D. Lorrain, *Tex. Mem. Mus. Misc. Pap. No. 1* (1968)]. No evidence of similar quality has been mustered to show that mammoths, mastodons, or any of the other 29 genera of extinct large mammals of North America were alive 10,000 years ago. The coincidence in time between massive extinction and the first arrival of big game hunters cannot be ignored.

4. M. I. Budyko, *Sov. Geogr. Rev. Transl.* **8** (No. 10), 783 (1967).
5. According to R. F. Flint [*Glacial and Quaternary Geology* (Wiley, New York, 1971), p. 778], "The argument most frequently advanced against the hypothesis of human agency is that in no territory was man sufficiently numerous to destroy the large numbers of animals that became extinct."
6. Apart from postglacial records of extinct species of *Bison*, very few kill sites have been discovered. J. J. Hester [in *Pleistocene Extinctions, the Search for a Cause*, P. S. Martin and H. E. Wright, Jr., Eds. (Yale Univ. Press, New Haven, Conn., 1967), p. 169], A. J. Jellinek (*ibid.*, p. 193), and G. S. Krantz [*Amer. Sci.* **58**, 164 (1970)] have all raised this point as a counterargument to overkill.
7. The North American megafauna that I believe disappeared at the time of the hunters includes the following genera: *Nothrotherium*, *Megalonyx*, *Eremotherium*, and *Paramylodon* (ground sloths); *Brachyostracon* and *Boreostracon* (glyptodonts); *Castoroides* (giant beaver); *Hydrochoerus* and *Neochoerus* (extinct capybaras); *Arctodus* and *Tremarctos* (bears); *Smilodon* and *Dinobastis* (saber-tooth cats); *Mammut* (mastodon); *Mammuthus* (mammoth); *Equus* (horse); *Tapirus* (tapir); *Platygonus* and *Mylohyus* (peccaries); *Camelops* and *Tanupolama* (camelids); *Cervalces* and *Sangamona* (cervids); *Capromeryx* and *Tetrameryx* (extinct pronghorns); *Bos* and *Saiga* (Asian antelope); and *Bootherium*, *Symbos*, *Euceratherium*, and *Preptoceras* (bovids).
8. C. V. Haynes, in *Pleistocene and Recent Environments of the Central Great Plains*, W. Dort, Jr., and J. K. Jones, Jr., Eds. (University of Kansas Department of Geology Special Publication No. 3, Lawrence, 1971), p. 77.
9. A. Dreimanis [*Ohio J. Sci.* **68**, 257 (1968)] estimates that there are more than 600 mastodon occurrences in northeastern North America. If we suppose that 500 of these were of late-glacial age and assign an equal probability of death, burial, and discovery to each in the time span from 10,500 to 15,500 years ago, then an average of ten may be expected for any given century and one for any decade. I assume that in any one region local extinction was swift. The elephants and their hunters were associated for no more than a decade. Even if the temporal overlap between the elephants and their hunters were as much as 100 years, and if half (five) of the finds represent animals killed by hunters, it is clear that the probability of the field evidence actually being detected and appreciated by the discoverers of the bones is small. Had the hypothetical mastodon kill sites been located on the uplands rather than in bogs or on lake shores, the probability of discovery becomes smaller still. My pessimistic appraisal should not deter those engaged in the search for more kill sites. It should refute the view that extinction by overkill would yield abundant fossil evidence.
10. One animal unit can be used as a standard for paleoecological comparison in the sense range managers have used it for comparing stocking rates under common use. One animal unit equals 1000 pounds (450 kilograms), or approximately the adult weight of one steer, one horse, one cow, four hogs, five sheep, or five deer. Some possible Pleistocene equivalents would be 0.2 mammoth (*Mammuthus columbi*), 0.3 mastodon (*Mastodon americanus*), 0.6 large camel (*Camelops*), one large horse (*Equus occidentalis*), one woodland musk-ox (*Symbos*), three woodland peccaries (*Mylohyus*), or ten cervicaprids (*Tetrameryx*).
11. P. S. Martin, in *Arctic and Alpine Research*, J. Ives and R. Barry, Eds. (Methuen, London, in press).
12. P. F. Wilkinson, in *Models in Archaeology*, D. L. Clarke, Ed. (Methuen, London, 1972).
13. R. D. Guthrie [*Amer. Midl. Natur.* **79**, 346 (1968)] concluded that mammoths constituted 3 to 11 percent of the megafauna in the rich Late Pleistocene deposits near Fairbanks, Alaska. Their size meant that they comprised 20 to 50 percent of the *relative* biomass. But fossil mammal deposits are obviously not randomly distributed. Within a deposit, the rates of bone deposition are unknown and apparently unknowable. There is no prospect of estimating accurately the size of a past population from its fossil bones.
14. B. Kurtén, *Acta Zool. Fenn.* **107**, 1 (1965). The late Würm fossil carnivores of Levant Caves reveal a shrinkage in range, decline in numbers, and reduction in body size.
15. F. Lorimer, in *The Determinants and Consequences of Population Trends* (Population Studies No. 17, United Nations, New York, 1953), pp. 5–20; A. Desmond, *Population Bull.* **18** (No. 1), 1 (February 1962); E. S. Deevey, Jr., *Sci. Amer.* **203**, 195 (September 1960). Deevey proposed an increase of 1.4 times in a 28-year generation, or 1.3 percent annually, as prehistoric man's best effort. C. V. Haynes, [*Sci. Amer.* **214**, 104 (June 1966)] used this value to estimate the population growth of mammoth hunters. Much more rapid growth rates can be assumed.
16. Disease can be discounted. Microbiologists generally regard the Paleolithic as a healthy episode [R. Hare, in *Diseases in Antiquity*, D. Brothwell and A. T. Sandison, Eds. (Thomas, Springfield, Ill., 1967), pp. 115–131]. Their reasons are based less on detailed knowledge of skeletal pathologies or the scarcity of major parasites in prehistoric feces [G. F. Fry and J. G. Moore, *Science* **166**, 1620 (1969); R. F. Heizer and L. K. Napton, *ibid.* **165**, 563 (1969)] than on biological inference. Lacking closely related hosts, the New World held no major reservoir of hominid diseases. Cholera and African sleeping sickness never became established. American Indians suffered catastrophic losses in historic time, presumably through lack of prior exposure to Old World diseases such as smallpox and tuberculosis [H. F. Dobyns, *Curr. Anthropol.* **7**, 395 (1966)].
17. J. B. Birdsell, *Cold Spring Harbor Symp. Quant. Biol.* **22**, 47 (1957).
18. C. S. Elton, *The Ecology of Invasions by Animals and Plants* (Methuen, London, 1958). Introduced populations of the giant African snail, *Achatina fulica*, attain highest values at the time of establishment, declining rapidly after initial introduction [A. R. Mead, *The Giant African Snail* (Univ. of Chicago Press, Chicago, 1961)]. Exotic large mammals spreading through New Zealand attain peak population densities and maximum reproduction rates at the margin of their range [T. Riney, *Int. Union Conserv. Nature Publ. (n.s.) No. 4* (1964), p. 261; G. Caughley, *Ecology* **51**, 53 (1970)].
19. The proposed migration rate is well within the distance covered by groups of Zulus known to have moved from Natal to Lake Victoria (3000 kilometers) and halfway back in half a century [J. D. Clark, *The Prehistory of Southern Africa* (Penguin Books, Harmondsworth, Middlesex, England, 1959), p. 168].
20. P. A. Jordan, D. B. Botkin, M. L. Wolfe, *Ecology* **52**, 147 (1970). G. B. Schaller [*Natur. Hist.* **81**, 40 (1971)] says Serengeti predators remove roughly 10 percent of the prey biomass.
21. Even when most of their calories come from plants [see R. B. Lee, in *Man the Hunter*, R. B. Lee and I. Devore, Eds. (Aldine, Chicago, 1969)], men of modern nonagricultural tribes devote much time to the hunt. The arctic invaders of America had come through a region notably deficient in edible plants. As long as large mammals were flourishing, there was no need to devise new techniques of harvesting, storing, and preparing less familiar food. None of their artifacts suggests that the first American hunters also stalked the wild herbs, and none of their midden refuse suggests that the succeeding gatherers knew the extinct mammals.
22. J. E. Fitting, *Amer. Antiquity* **33**, 441 (1968).
23. J. B. Bird, *ibid.* **35**, 205 (1970).
24. M. Bates, *Where Winter Never Comes* (Scribner, New York, 1952); R. F. Black, *Arctic Anthropol.* **3**, 7 (1966); H. T. Irwin and H. M. Wormington, *Amer. Antiquity* **34**, 24 (1969).
25. A. L. Bryan, *Curr. Anthropol.* **10**, 339 (1969); unpublished data; C. S. Chard, *Man in Prehistory* (McGraw-Hill, New York, 1969); W. N. Irving, *Arctic Anthropol.* **8**, 68 (1971); C. Irwin-Williams, paper presented at the Conference on Pleistocene Man in Latin America, San Pedro de Atacoma, Chile, 1969; E. Lanning, *World Archaeol.* **2**, 90 (1970); T. F. Lynch, *Occas. Pap. Idaho Univ. State Mus. No. 21* (1967); ——— and K. A. R. Kennedy, *Science* **169**, 1307 (1970); R. S. MacNeish, R. Berger, R. Protsch, *ibid.* **168**, 975 (1970); R. S. MacNeish, *Sci. Amer.* **224**, 36 (February 1971); H. Müller-Beck, *Science* **152**, 1191 (1966); P. C. Orr, *Prehistory of Santa Rosa Island* (Santa Barbara Museum of Natural History, Santa Barbara, Calif., 1968); B. E. Raemsch, *Yager Mus. Publ. Anthropol. Bull. No. 1* (1968); A. Stalker, *Amer. Antiquity* **34**, 428 (1969).
26. K. W. Butzer, *Environment and Archaeology: An Ecological Approach to Prehistory* (Aldine-Atherton, Chicago, ed. 2, 1971); J. M. Cruxent, in *Biomedical Challenges Presented by the American Indian* (Pan-American Health Organization Science Publication 165, Washington, D.C., 1968), pp. 11–16; J. E. Fitting, *The Paleo-Indian Occupation of the Holcombe Beach* (Michigan University Museum of Anthropology Anthropological Paper No. 27, Ann Arbor, 1966).
27. One of the boldest claims of great antiquity is that of L. S. B. Leakey, R. De E. Simpson, and T. Clements [*Science* **160**, 1022 (1968)] near Calico Hills, California. Flaked cherts have been reported in fan deposits first considered to be at least 40,000 years old, and later judged to be much older. A study based on a visitation to the quarry in October 1970 by 60 leading American geologists and archeologists sustained the view that the deposits are of great age. It failed to satisfy skeptics that the alleged artifacts were definitely man-made and in a cultural context [C. Behrens, *Sci. News* **99**, 98 (1971)].
28. M. R. Harrington and R. De E. Simpson, *Tule Springs, Nevada, with Other Evidences of Pleistocene Man in North America* (Southwest Museum Paper No. 18, Los Angeles, 1961).
29. C. V. Haynes, in *Pleistocene Studies in Southern Nevada*, H. M. Wormington and D. Ellis, Eds. (Nevada State Museum of Anthropology Paper No. 13, Reno, 1967); R. Shutler, Jr., *Current Anthropol.* **6**, 110 (1965).
30. R. Rosenthal, *Experimenter Effects in Behavioral Research* (Appleton-Century-Crofts, New York, 1966).
31. A. Hrdlička, "Early man in South America" [*Bur. Amer. Ethnol.* **52**, 4 (1912)]; J. A. Graham and R. F. Heizer, *Quaternaria* **9**, 225 (1968).
32. F. Bourliere and M. Hadley, *Annu. Rev. Ecol. Syst.* **1**, 138 (1970).

33. For comments and counterarguments I am grateful to D. P. Adam, J. B. Birdsell, A. L. Bryan, J. E. Guilday, E. W. Haury, C. V. Haynes, Jr., J. J. Hester, A. J. Jelinek, R. G. Klein, G. S. Krantz, L. S. Lieberman, E. H. Lindsay, Jr., D. I. Livingstone, A. Long, R. H. MacArthur, M. Martin, J. H. McAndrews, P. Miles, J. E. Mosimann, C. W. Ogston, B. Rippeteau, J. J. Saunders, G. G. Simpson, W. W. Taylor, N. T. Tessman, and most especially, R. F. Heizer. I thank D. P. Adam for programming the computer simulations in Table 2. Special thanks for editorial aid are due B. Fink. This study was supported in part by NSF grant 27406. Contribution No. 56, Department of Geosciences, University of Arizona, Tucson.

PAPER 35

A New Evolutionary Law (1973)

L. Van Valen

Commentary

ANDY PURVIS

What a title! And what a paper! Leigh Van Valen's "new evolutionary law" needed him to set up a new evolutionary journal in order to publish it. The first thirty typewritten pages of *Evolutionary Theory* ("dedicated to the primacy of content over display"), their Greek letters inked in by hand, documented a new general pattern in the history of life and proposed a new worldview—Red Queen evolution—to explain it.

Let's begin with the pattern. Van Valen collated stratigraphic ranges of many sets of closely related subtaxa, from species of planktonic foraminifera to orders of mammals, and constructed survivorship curves. To his initial surprise, survivorship was log-linear: within each taxon, subtaxa appeared to have a constant extinction hazard. How could this be?

To explain the pattern, Van Valen posited that the "effective environment of the members of any homogeneous group of organisms deteriorates at a stochastically constant rate" (16) and pinned the blame on natural enemies. In this worldview, species within an adaptive zone compete—directly or indirectly—in a zero-sum evolutionary game: gains in the fitness of one are at the expense of the others. Species are therefore continually having to adapt in order to maintain their abundance—having to run to keep in the same place. Van Valen read more widely than many whole faculties; I don't suppose it was long before the metaphor came to him.

It's a beautiful metaphor, one that still drives much macroevolutionary research (Benton 2009), as well as having entered the microevolutionary mainstream. Not all aspects of the paper have fared so well. Raup (1972) showed analytically that the "law" of constant extinction can't hold simultaneously both for species and for higher taxa; McCune (1981) showed that log-linear survivorship curves imply only that extinction hazard is independent of age—not that it has been constant; and adaptive zones fell from favor in the 1980s, casualties of the cladomasochism that left Van Valen unimpressed. Now, however, growing evidence of diversity dependence in clade dynamics (e.g., Alroy 2010) and of stable adaptive peaks in morphospace (e.g., Hunt 2007) suggest that adaptive zones are ripe for rehabilitation.

Literature Cited

Alroy, J. 2010. The shifting balance of diversity among major marine animal groups. *Science* 329:1191–94.

Benton, M. J. 2009. The Red Queen and the Court Jester: species diversity and the role of biotic and abiotic factors through time. *Science* 323:728–32.

Hunt, G. 2007. The relative importance of directional change, random walks, and stasis in the evolution of fossil lineages. *PNAS* 104:18404–8.

McCune, A. R. 1981. On the fallacy of constant extinction rates. *Paleobiology* 6:610–14.

Raup, D. M. 1972. Taxonomic diversity during the Phanerozoic. *Science* 177:1065–71.

From *Evolutionary Theory* 1:1–30.

A NEW EVOLUTIONARY LAW

Leigh Van Valen
Department of Biology
The University of Chicago
Chicago, Illinois 60637

ABSTRACT:

All groups for which data exist go extinct at a rate that is constant for a given group. When this is recast in ecological form (the effective environment of any homogeneous group of organisms deteriorates at a stochastically constant rate), no definite exceptions exist although a few are possible. Extinction rates are similar within some very broad categories and vary regularly with size of area inhabited. A new unit of rates for discrete phenomena, the macarthur, is introduced. Laws are appropriate in evolutionary biology. Truth needs more than correct predictions. The Law of Extinction is evidence for ecological significance and comparability of taxa. A non-Markovian hypothesis to explain the law invokes mutually incompatible optima within an adaptive zone. A self-perpetuating fluctuation results which can be stated in terms of an unstudied aspect of zero-sum game theory. The hypothesis can be derived from a view that momentary fitness is the amount of control of resources, which remain constant in total amount. The hypothesis implies that long-term fitness has only two components and that events of mutualism are rare. The hypothesis largely explains the observed pattern of molecular evolution.

* * *

Introduction

During a study (Van Valen, submitted) on the effects of extinction I wanted to show that a model I was using was oversimplified. It assumed no correlation of probability of extinction with age of the group, and I thought that generally more vulnerable groups should die out first. A test using data from Simpson (1953) showed to my astonishment that the assumption was reasonably correct in these cases. I did not believe it could be generally true and so tested these and other cases in more detail. The assumption proved to be consistent with all available data. Others (unpublished results) have now confirmed this finding for individual taxa. I will present a more extended treatment elsewhere; the present paper is condensed.

The Evidence

The method is an application of the survivorship curve of population ecology (including demography). It is a simple plot of the proportion of the original sample that survive for various intervals. In this case the sample is the set of all known subgroups of some larger group, no matter when in absolute time each subgroup originated. A logarithmic ordinate, standard in ecology, gives the property that the slope of the curve at any age is proportional to the probability of extinction at that age. Simpson (1944, 1953) compiled two well-known taxonomic survivorship curves but used an arithmetic ordinate (1-4).

The results (Figs. 1-5) for over 25,000 subtaxa show almost uniform linearity for extinct taxa except for effects attributable to sampling error (5,6). Sampling error is most noticeable at the bottom of the graphs, where

Evol. Theory 1:1-30 (July 1973)

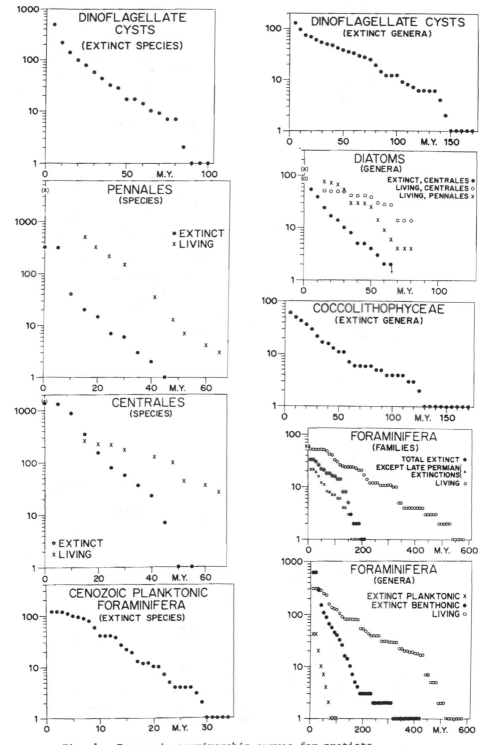

DINOFLAGELLATE CYSTS (EXTINCT SPECIES)

DINOFLAGELLATE CYSTS (EXTINCT GENERA)

PENNALES (SPECIES)
• EXTINCT
x LIVING

DIATOMS (GENERA)
EXTINCT, CENTRALES •
LIVING, CENTRALES o
LIVING, PENNALES x

COCCOLITHOPHYCEAE (EXTINCT GENERA)

CENTRALES (SPECIES)
• EXTINCT
x LIVING

FORAMINIFERA (FAMILIES)
TOTAL EXTINCT •
EXCEPT LATE PERMIAN EXTINCTIONS x
LIVING o

CENOZOIC PLANKTONIC FORAMINIFERA (EXTINCT SPECIES)

FORAMINIFERA (GENERA)
EXTINCT PLANKTONIC x
EXTINCT BENTHONIC •
LIVING o

Fig. 1. Taxonomic survivorship curves for protists.

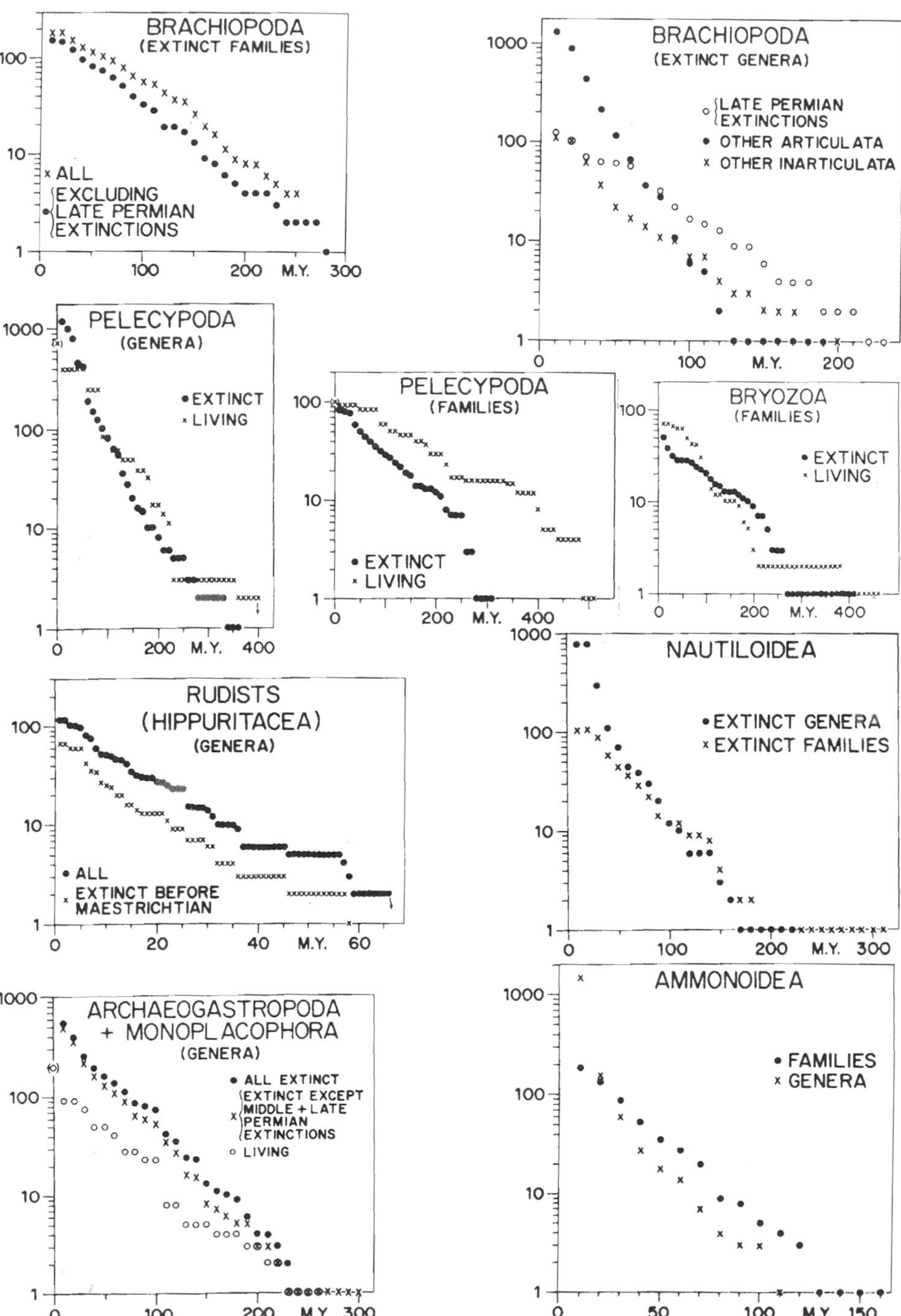

Fig. 2. Taxonomic survivorship curves for Mollusca and Brachiopoda.

4

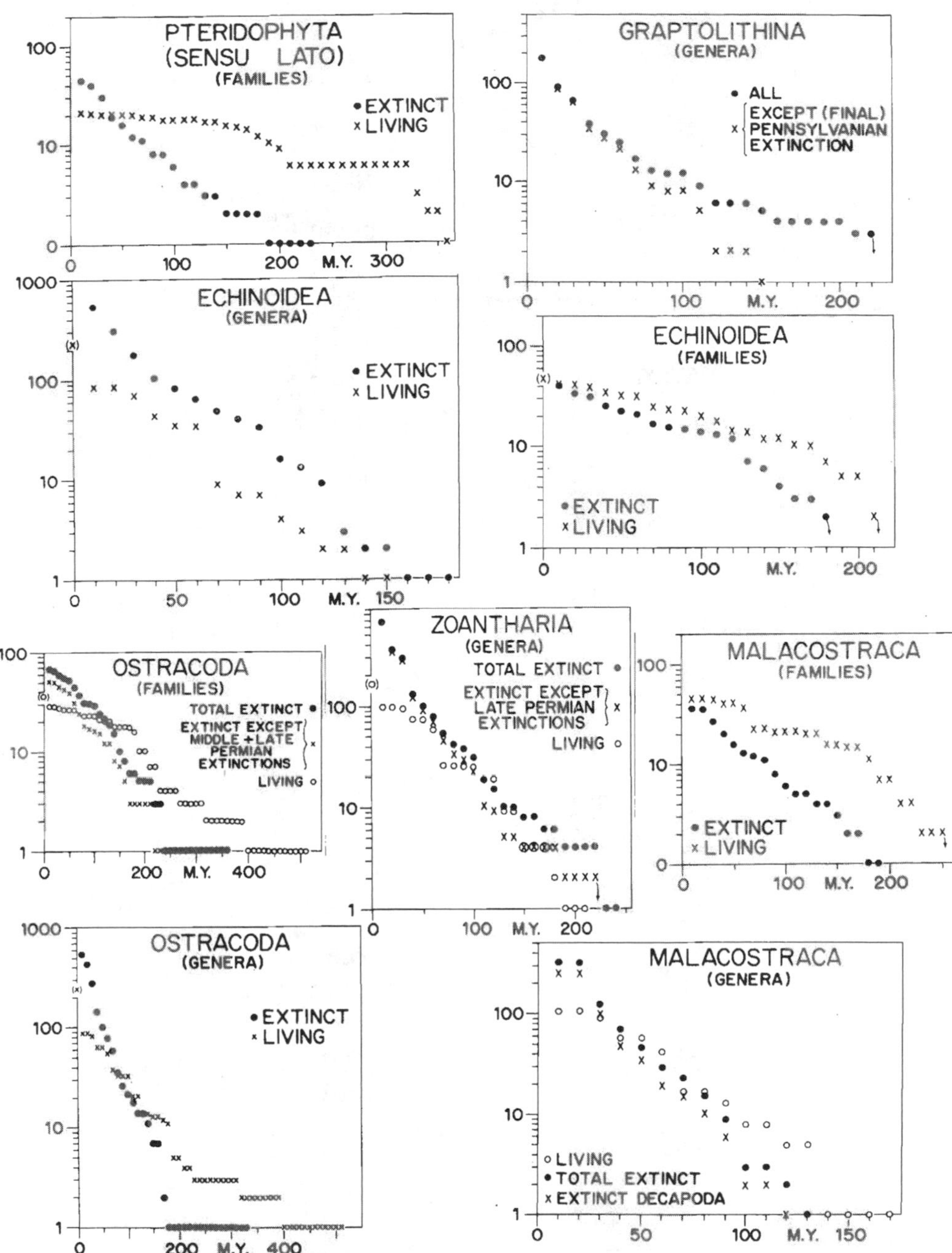

Fig. 3. Taxonomic survivorship curves for plants and invertebrates.

5

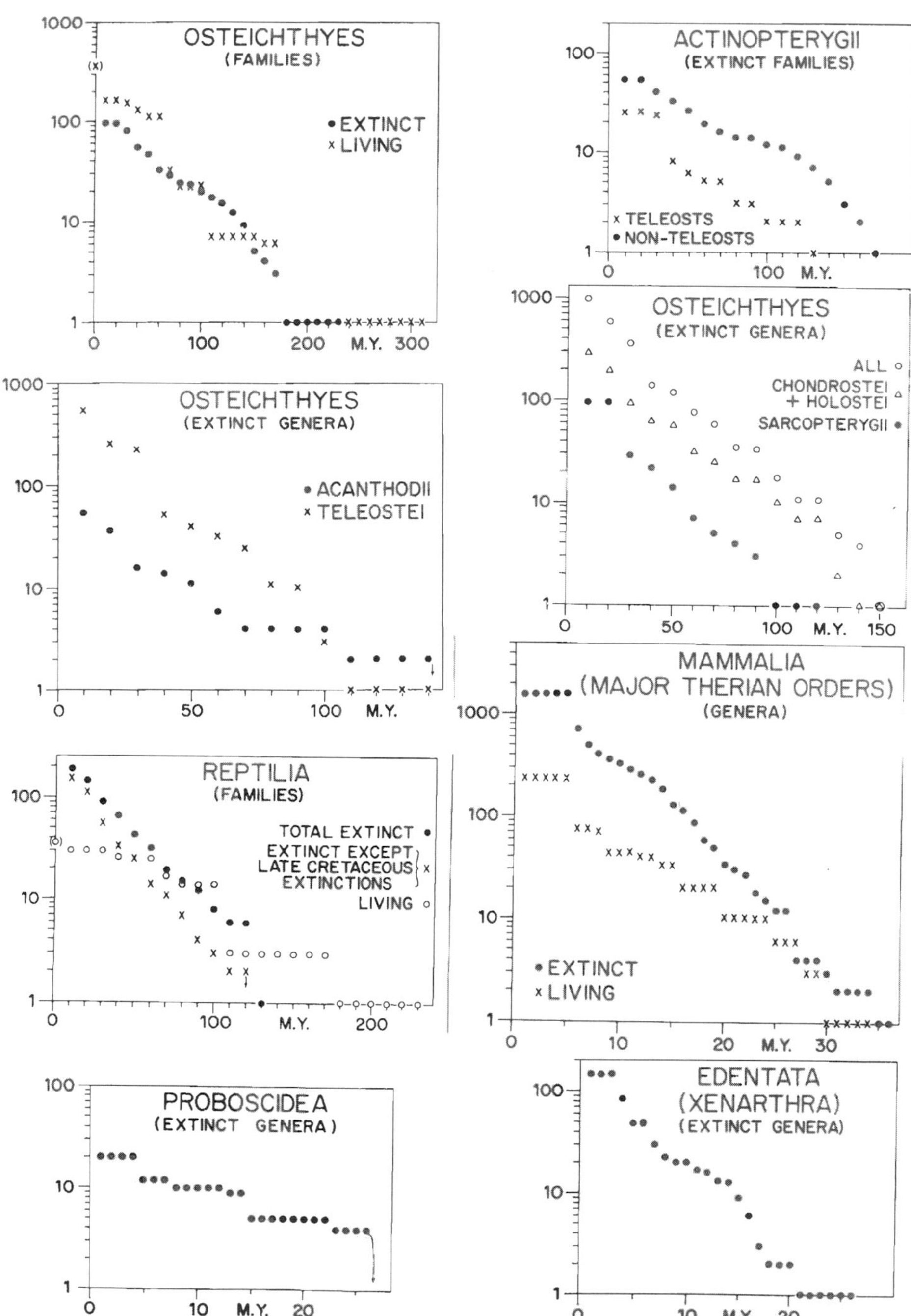

Fig. 4. Taxonomic survivorship curves for vertebrates. "Major therian orders" are those with individual plots.

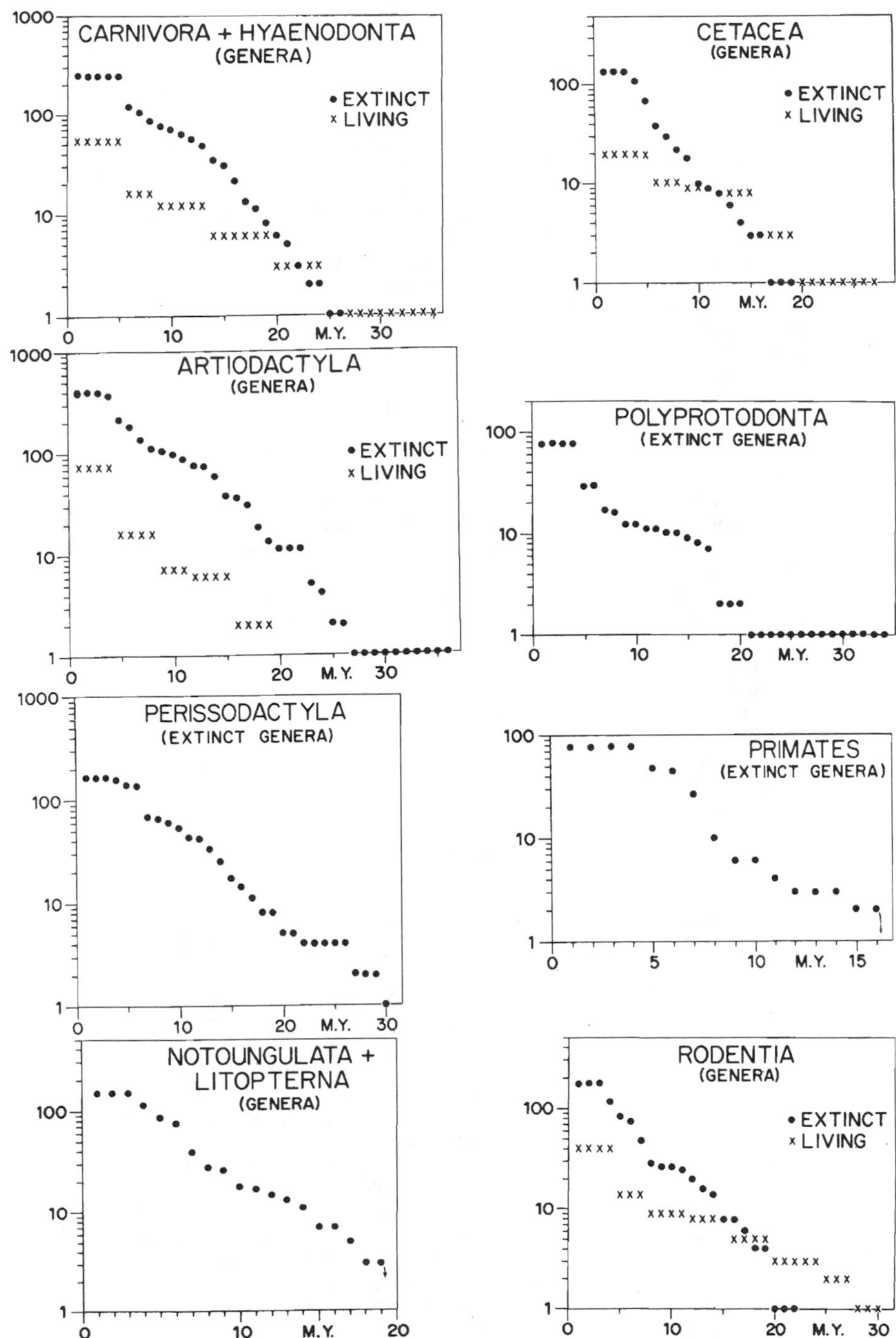

Fig. 5. Taxonomic survivorship curves for mammals. For Primates, Madagascar genera are omitted because the island lacks pre-Pleistocene fossils. Polyprotodonta includes Caenolestoidea.

a few taxa have disproportionate weight, and at the left side, where inaccuracies in dating are most important (7). Usually, the more taxa included the closer is the approach to linearity.

For living taxa linearity of the distribution requires both constant extinction and constant origination. It further requires that both be more nearly constant over absolute time than does a distribution for extinct taxa, which normally spans many overlapping half-lives of taxa, and that probability of discovery not decrease appreciably with age of strata. It is therefore surprising that many even of these distributions are linear (8).

Linearity is not an artifact of the method. Most survivorship curves for individuals are either markedly concave or markedly convex (9). Most small passerine birds are exceptional in having linear survivorship curves. The very wide diversity (biologically and stratigraphically) of groups plotted here argues against any kind of special artifact.

The sources of data usually do not distinguish between real extinction of a lineage and pseudo-extinction by evolution of one taxon into a successor taxon (Simpson, 1953). The latter proves to be negligible. Most taxa which give rise to successor taxa continue in their original form for an appreciable period after the branching. For families of mammals, for which I am familiar with the phylogeny, a maximum (and surely inflated) estimate is 20 per cent pseudo-extinction; a more likely estimate is 5 per cent. Ammonites (1) give 6 per cent (12 of 188 taxonomic extinctions for families). All other available phylogenies give similar results. The pattern of constant extinction remains unchanged whether pseudo-extinctions are included or excluded. Additionally, it is plausible that even pseudo-extinction usually implies the end of an adaptive mode and so would fit into the hypothesis given below. Pseudo-extinctions are probably more common at lower taxonomic levels, despite the counter-claim made or implied by Ruzhentsov (1963), MacGillavry (1968), and Eldredge (1971; Eldredge and Gould, 1972) that they do not occur for species. Any example of an ancestral taxon co-existing with a descendant taxon violates a basic premise of cladistic systematics (Hennig, 1966). As noted, such examples not merely exist but greatly predominate.

Apparent Exceptions

There are some real and some spurious exceptions to constant extinction rates. As noted above, the ages of living taxa are not directly comparable to those of extinct taxa. Sampling error also causes deviations from linearity.

Some short intervals of geologic time have had massive extinctions of some kinds of organisms (Newell, 1963, 1967, 1971). All graptolites became extinct in the Pennsylvanian and almost all stony corals (Zoantharia) did so in the late Permian. This is clearly a different sort of event from the usual process of extinction and does not fit the general explanation given below. In the late Permian, almost everything in the adaptive zone of stony corals was eliminated (10). The adaptive zone itself was demolished for a while. If we eliminate such extinctions as being different from events within an adaptive zone (done supplementarily in parts of Figs. 1-4), linearity is never reduced and sometimes, as with corals, increased (11). When, as with brachiopod genera, extinctions at such a time are sufficiently numerous to be plotted separately, the slope is less than for others, as for the ages of genera living today and for the same trivial reason.

There is no accepted classification of corals at the family level. Two more or less orthogonal classifications are plotted in Figure 6. One is linear, the other convex. This gives some evidence that the Treatise

8

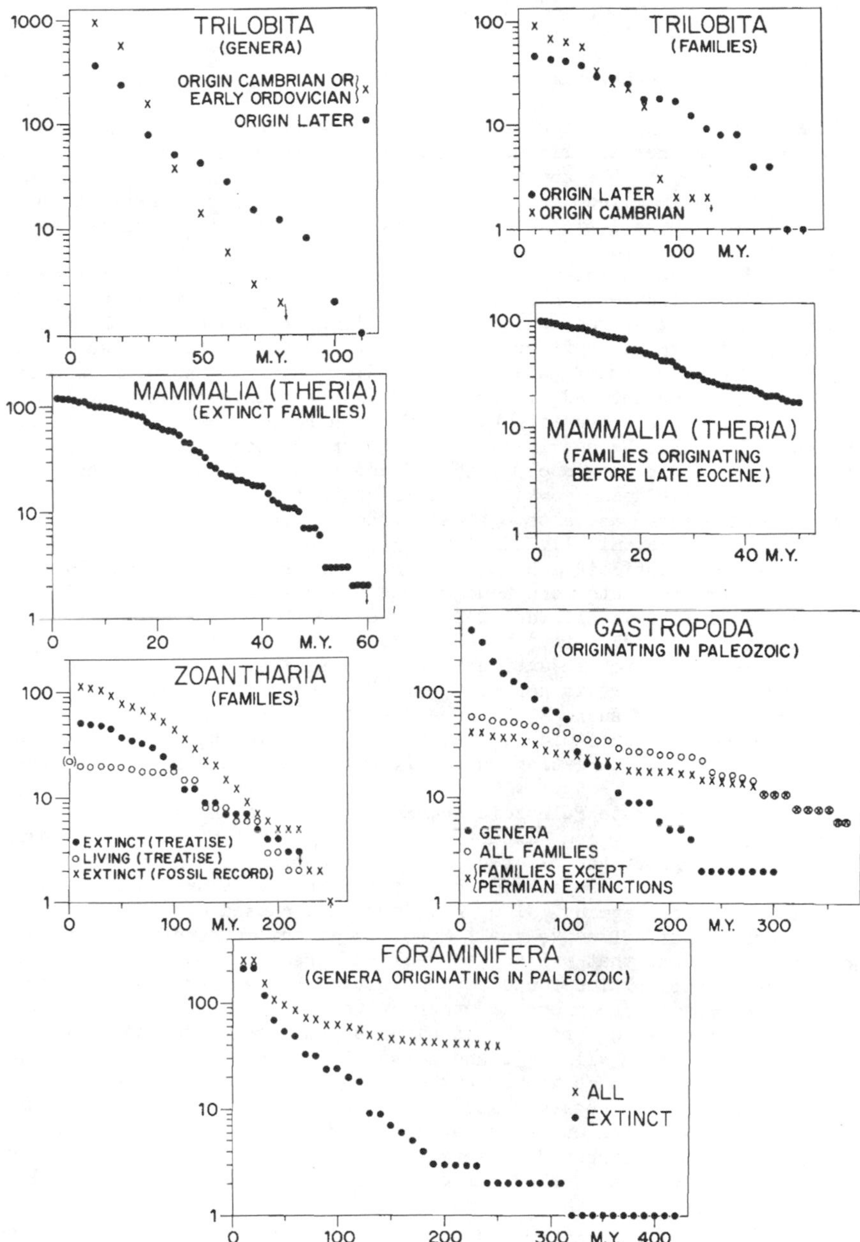

Fig. 6. Taxonomic survivorship curves for apparent exceptions to linearity. See text. Zoantharia families are from very different classifications.

classification is ecologically more realistic, but a decision must ultimately come from phylogeny.

Mammalian families also give a convex curve (Fig. 6). As shown by plotting only families (extinct and living) originating before the late Eocene, however, this is an artifact of the short duration of the Cenozoic in relation to the durations of many families. Living families seem similar to extinct ones, because they jointly determine a single linear curve. This is also true for gastropods but not for foraminiferans.

This difference in foraminiferans of some living taxa from most extinct ones also occurs elsewhere, notably in the Pteridophyta. It is indeterminable from these data whether real exceptions are involved or whether the deviant taxa occupy sufficiently different adaptive zones as not to interact appreciably with the majority of related taxa. Simpson (1944) noted this phenomenon for pelecypods and made it the basis of his bradytelic evolution.

Deviant adaptive zones are obvious for genera of most of the different mammalian orders (Van Valen, 1971) and also for rudists, coral-like pelecypods which formed reefs in the late Cretaceous and seem to have caused the extinction of many corals while doing so (12). Their extinction rate is greater than that of normal pelecypods, although also greater than that of genera of corals, which may not be taxonomically similar (13).

Ammonite and nautiloid genera, but not families, give ostensibly concave curves. When grouped into more homogeneous classes (Fig. 7), approximate linearity results. This also applies to the lengths of the terminal branches (after each final branch-point) of the phylogeny of ammonite families (14). Removal of the late Devonian extinctions has a similar effect for genera as for terminal branches but is not plotted. I did not try to subdivide the nautiloids. I further suspect that pseudo-extinction is more common for ammonite and perhaps nautiloid genera than is true elsewhere, which would increase concavity by increasing short-lived taxa at the expense of longer-lived lineages.

The need to separate Paleozoic and Mesozoic ammonites shows that there is at least a descriptively real exception here. The extinction rate is definitely not constant throughout the existence of the group. The same is true for trilobites (Fig. 6), the separation here being in the Ordovician. Linearity again holds for each segment. I do not know what caused either change, but it may be relevant that effectively all Mesozoic ammonites descended from one Paleozoic lineage and that trilobites declined greatly in the Ordovician. In each case the division time found for extinction rates corresponds to the greatest separation in the group on other criteria (15).

Moreover, because of preservational bias and incomplete collecting and study, short-lived taxa will be found less often than long-lived taxa. However, for equally frequent groups the effect of this bias will be a reduction in observed longevity by a constant absolute amount. This will leave linearity and even the slope of the curve unchanged. Rarer groups will have a greater expected reduction in observed longevity than commoner groups. Any effect of this property depends on whether rarity is correlated with longevity, and I know of no relevant data.

Any combination of subgroups with unequal constant extinction rates produces a concave resultant curve. The amount of inequality can be estimated by this concavity, but unfortunately any concavity manifests itself most in the regions of greatest sampling error. Linearity is nevertheless sufficient that exceptions must be rare or slight. Abrupt ends to distributions, as with echinoid families, occur when (as we have seen with mammal families) the stratigraphic range of the group impinges on the possibility of long-lived taxa having already become extinct. Linearity is unaffected by multiplication or addition with a logarithmic (ordinate) or arithmetic (abscissa) constant.

10

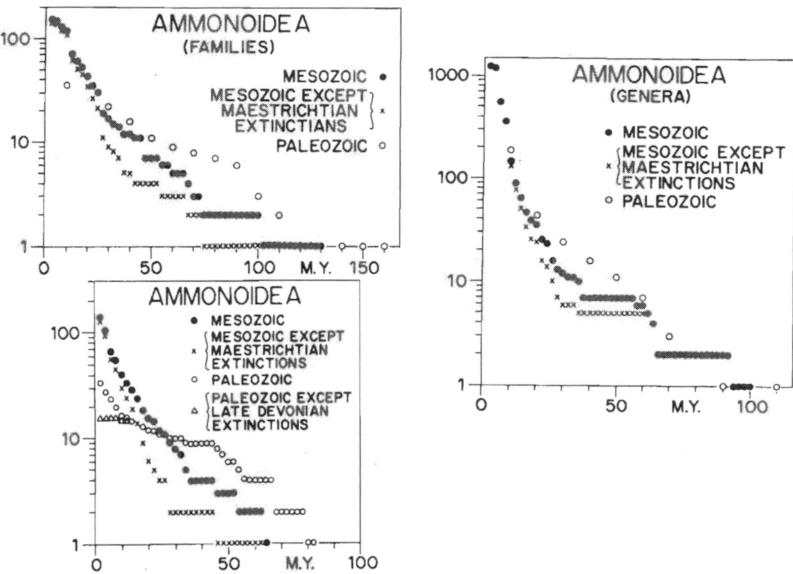

Fig. 7. Taxonomic survivorship curves for ammonites. See text. The graph on the lower left is for terminal twigs of a phylogeny of families.

* * *

We see that the exceptions are either spurious, rare, slight, doubtful, or exceptions only in part. The pattern is therefore sufficiently general that the minor exceptions are best explained by unusual circumstances peculiar to each case. There is a strong first-order effect of linearity, and it is this, rather than its perturbations by special and diverse circumstances, that deserves primary attention.

Related Phenomena

The constancy of extinction in terms of survivorship does not imply constancy over geological time, and vice versa. Any form of survivorship curve has a mean, as does any pattern of extinction rates over absolute time, and this is the only formal connection between the two phenomena. There may nevertheless be a deeper causal connection.

I give some extinction and origination curves over geological time to illustrate their variability in the measurement framework proposed here (Fig. 8). There is an extraordinary exponential decrease in origination rate of mammalian families, by two orders of magnitude, from the beginning of the Cenozoic. An inverse phenomenon occurs in diatoms, where the species of Pennales (a largely benthonic groups, unlike the Centrales) have increased exponentially, as in the log phase of bacterial culture, in the same interval (16; Fig. 1). We can look at Figure 8F inversely: there is a nearly linear increase in the proportion of families that had originated by a given time and that are now extinct.

Large foraminiferans (17) have originated about 41 times independently from smaller foraminiferans, almost always from much smaller ones. About 33 of these clades have existed since the end of the early Cretaceous, when an approximate equilibrium was established. Figure 9 shows that the numbers of clades and genera present simultaneously have followed more or less log-normal

11

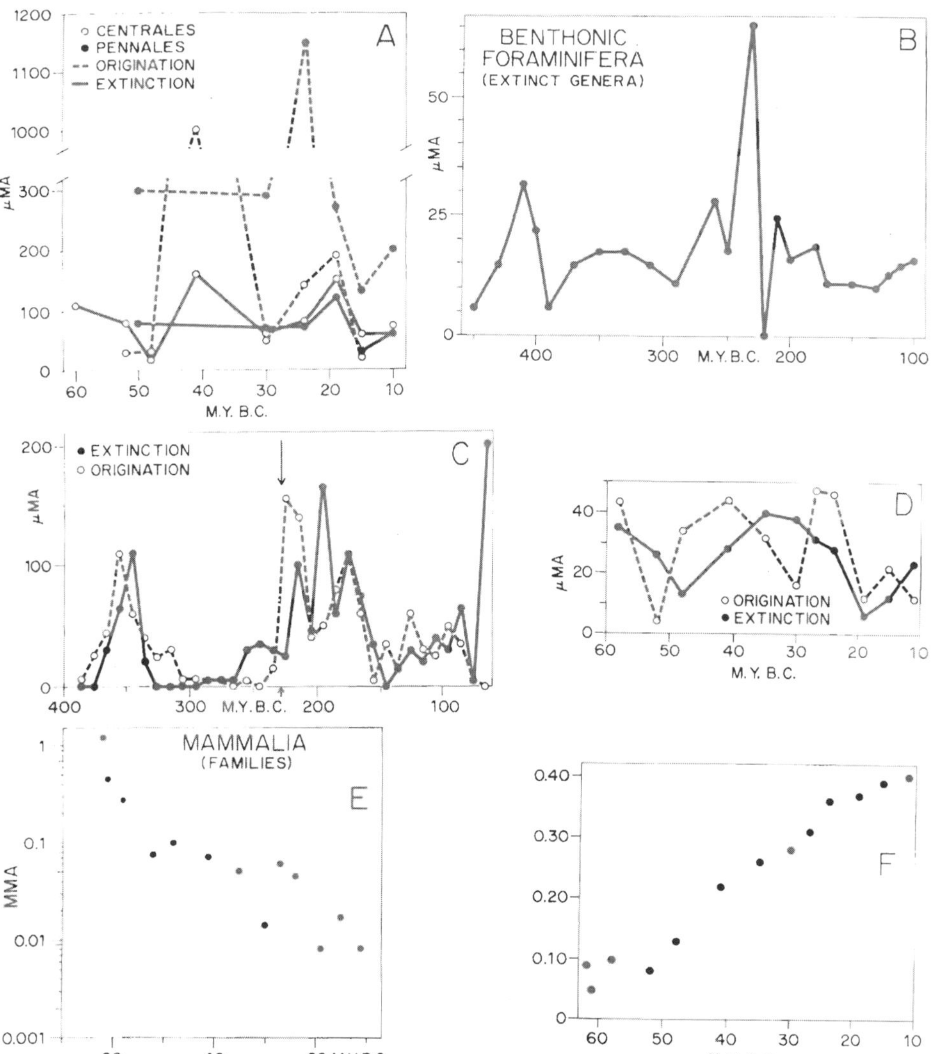

Fig. 8. A: Origination and extinction rates for species of diatoms. B: Extinction rates for genera of benthonic foraminiferans. C: Origination and extinction rates for families of ammonites. High origination rates precede high extinction rates except for the Permian and latest Cretaceous. D: Extinction rates of mammalian family lineages, and origination rates for mammalian families now surviving. E: Origination rates for all mammalian families (new families per family in previous age per unit time). F: For mammalian families, (cumulative recent families originated)/(cumulative total families originated).

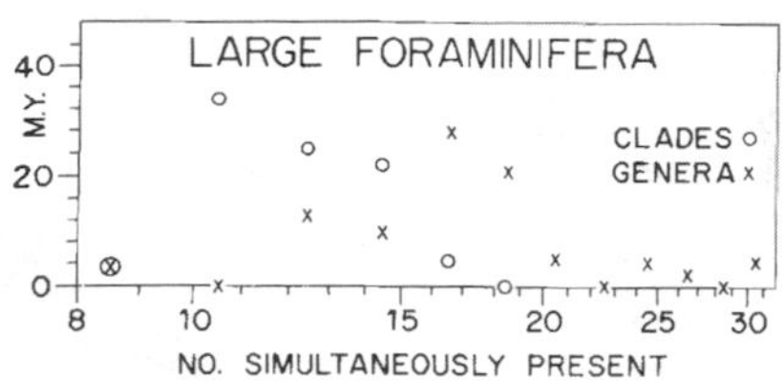

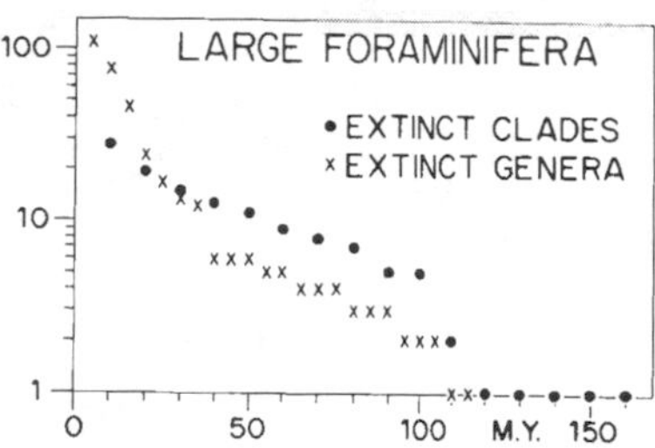

Fig. 9. Survivorship curves for genera and clades of large foraminiferans, and distribution of number of genera and clades simultaneously present after the early Cretaceous (Albian).

* * *

distributions since that time. Survivorship curves for clades and genera of this taxonomically very heterogeneous, but ecologically rather homogeneous, group are linear within sampling error and differ from the curve for genera of all benthonic foraminiferans.

It would be useful to know whether the extinction of entire lineages also occurs at a constant rate, or if some taxa can avoid it by evolving successor taxa. This appears to represent an unsolved problem in graph theory (18).

Measurement of Rates

I propose a general unit of rates for phenomena which can be treated as discrete.

A macarthur (ma) is the rate at which the probability of an event per 500 years is 0.5. Robert H. MacArthur showed the importance of extinction in ecology (19).

Let P be the probability per t thousand years.

$$ma = -\log_2(1-P^{2t}) \qquad (1)$$

With respect to extinction, one macarthur is the rate of extinction (Ω) giving a half-life of 500 years. With respect to origination, one macarthur is the occurrence of one origin per thousand years per potential ancestor. With respect to molecular evolution, one millimacarthur (mma) is the rate giving one substitution per million years (20).

Apparent equilibrium extinction rates of bird species on islands that have been studied (21) are 0.5 to 10 ma. Mammal species in the late Pleistocene of Florida (Martin and Webb, in press) had a rate of regional turnover (time from immigration to local extinction) of about 7 mma (Fig. 10).

Table 1 gives estimated extinction rates for the taxa studied. The sedentary marine benthos is remarkably homogeneous, and motile marine organisms have somewhat higher rates less similar among themselves. Mammalian genera (and Mesozoic ammonites) have the highest rates, but rates for reptilian families are as high as those for mammalian families. For everything except mammals the extinction rate for families is about half that for genera. Groups in relatively new adaptive zones (at least to them) usually have higher rates than long-established groups. Not surprisingly, ecology is a better predictor of evolutionary rate than is amount of information-bearing DNA (22).

If genera of a family go extinct independently, the extinction rate of families will depend directly on the number of contemporaneous genera per family

TABLE 1: Extinction Rates (in μma).
Abbreviations: fam., family; gen., genus; sp., species

GROUP	RATE		
	fam.	gen.	sp.
Pteridophyta	20		
diatoms		50	90
Coccolithophyceae		25	
Dinoflagellata		20	55
Foraminifera	10		
benthonic		20	
planktonic		30	100
large		50	
Ostracoda	10	25	
Graptolithina		30	
Bryozoa	7		
Brachiopoda	15		
Articulata		45	
Inarticulata		25	
Malacostraca	12	30	
Trilobita			
early	20	80	
late	10	35	
Echinoidea	10	25	
Zoantharia	10	25	
Ammonoidea			
Paleozoic	20	35	
Mesozoic	75	150	
Nautiloidea	15	30	
Archaeogastropoda + Monoplacophora		20	
Gastropoda			
Paleozoic	4	20	
Pelecypoda	8	20	
rudists		50	
Osteichthyes	15	30	
Teleostei	20	35	
Holostei + Chondrostei	12	25	
Sarcopterygii		30	
Acanthodii		30	
Reptilia	30		
Mammalia (Theria)	30	150	
Polyprotodonta		150	
Primates		220	
Rodentia		160	
Carnivora + Hyaenodonta		120	
Edentata		180	
Proboscidea		60	
Notoungulata + Litopterna		170	
Perissodactyla		120	
Artiodactyla		120	
Cetacea		200	

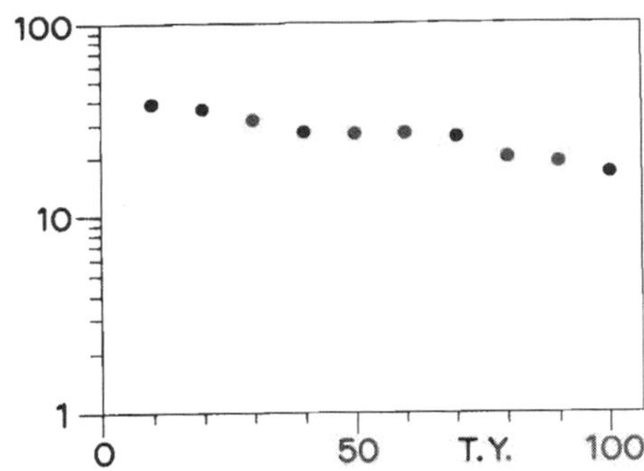

Fig. 10. Regional survivorship curve for mammalian species in Florida during the late Pleistocene.

* * *

if there is no origination of new genera in the family. Rough estimates, weighted by the number of taxa in each interval, give 2.9 contemporaneous genera per family of Ostracoda, 2.8 for Echinoidea, and 5.1 and 5.8 species per genus respectively for pennate and centric diatoms. With 2.9 genera per family, independent extinction will give an extinction rate per genus 2.6 times that per family. This ratio would be 2.5 for echinoids and 3.5 for average diatom species. I take the expected time of family extinction to be when the expected number of genera per family reaches 0.5. With branching (origination of new genera), the families will of course have longer expected longevity and so the expected ratio will increase. If the rate of branching equals that of extinction, simulation in the above range indicates roughly a doubling of the expected ratio, although a high variance among families in genera per family probably increases it more.

Table 1 shows that genera of echinoids and ostracodes have about 2.5 times the extinction rate of families, a result indistinguishable from that on the assumptions of independence and no branching. Because branching exists, some degree of correlated extinction of genera within families seems probable. The observed rate for species of diatoms is only twice that for genera, much less than the expected value of 3.5, so there is a strong correlation in extinction here even without considering branching.

Contemporaneous Subgroups

Extinction rate obviously depends on the area considered as well as on the inhabitants. Fig. 11 gives a relation between area and extinction rate, using all available data (21). Approximate linearity holds over 11 orders of magnitude on a log scale. With the extinction rate in $\log_{10}$ macarthurs and $\underline{A}$ the area in $\log_{10}$ sq. km, the regression is $\Omega = -0.66\underline{A} + 1.53$ when calculated from the birds and mammals and excluding the world fauna. A regression from complete data would presumably differ somewhat. As expected, the arthropods on Simberloffia seem to require less area for the same extinction rate than do vertebrates (23). A continent seems to be the largest area in which organisms can interact more or less as they do in any smaller area that is large enough for a population.

Therefore species and higher taxa occupying smaller areas, as will often be the case when they originate, have an expected rate of extinction greater than that of more widespread taxa. Similarly, as Small (1946, 1948b) noted in

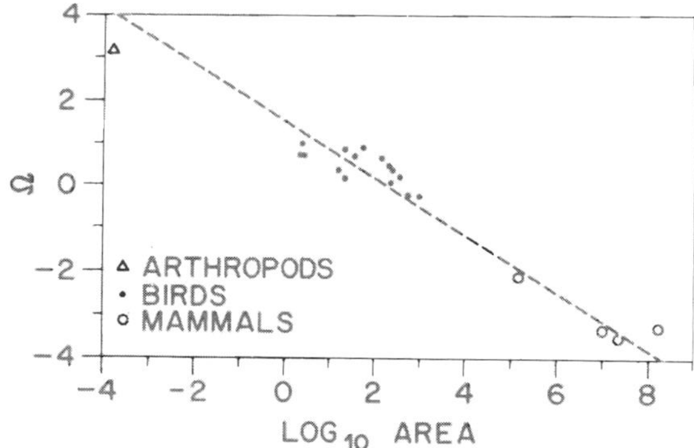

Fig. 11. Extinction rate (Ω,in $\log_{10}$ ma) of species as a function of area (sq. km.).

* * *

a different context for diatoms, genera with fewer species have a higher probability of extinction than do genera with more species. This is of course because branching at the specific level increases the probability of survival of the entire sub-tree, since the species will usually have different ecologies.

For these and other, less well-documented, reasons, there are definable and large subgroups that have a higher extinction rate than others, and moreover these subgroups should be relatively most frequent among younger taxa. Therefore a concave survivorship curve should result. How can it be that the curves are nevertheless linear?

There are two possibilities: incorrect assumptions and compensatory feedback. It may be that, by the time taxa have an appreciable probability of appearing in the fossil record, they have the same mean area occupied as older taxa. This sort of rapid increase may also be true, although it doesn't seem as likely, for number of species in a genus. Enough data are probably available to resolve the latter point. Compilations by Sloss (1950) and Simpson (1953), Small's studies (e.g. 1946) on diatoms, and data by Berggren (1969) and others on planktonic foraminiferans, suggest that there is indeed usually an appreciable lag before the number of subordinate taxa approaches its maximum value. The kinds of evidence in these papers are by no means conclusive on the present point, however. Furthermore, there may well be a threshold (like those found by MacArthur [1972] and others for population size in simplified models of colonization) above which survival is virtually assured and below which rapid extinction is likely. Other possible inadequacies of the assumptions, that the effects are too small to be detected or that they affect only a small proportion of taxa, are clearly unrealistic.

Even if an initial concavity in survivorship curves is avoided by the rapid changes invoked above, there still remains a major heterogeneity among the surviving taxa with respect to area, number of species or genera, and other factors relevant to survival. There must then be some kind of compensatory feedback to give the observed degree of linearity. Two kinds of feedback seem possible. There might be an interaction between age and area (or number of species) such that older and younger taxa have different effects of area on probability of extinction. No simple model fits this biologically implausible alternative. However, if for new taxa small area and low number of species were not harmful, or as harmful as later, the expected initial concavity would be reduced or eliminated.

The other kind of feedback is by the continual appearance at all ages of taxa of low area and other characters giving greater susceptibility to extinction. In fact this could be regarded as an aspect of the extinction process: taxa would become more susceptible at a constant average rate per taxon per time. This transformation need not be by feedback per se, but it does require that the production of greater susceptibility be part of the same process as the extinction. There is no requirement that each subordinate taxon be equally likely at any given time to become more susceptible, but the overall control must compensate for survival by greater susceptibility at some other time or for some other species. This second alternative is plausible (given linear survivorship) but largely untested. It predicts that, at least after a short initial period, criteria of susceptibility will have a steady-state distribution in survivorship time independent of the age of the taxon. Small (1946, 1947, 1948a,c) found such a steady-state distribution in real time for generic sizes of diatoms through the Cenozoic.

An Evolutionary Law

The effective environment (24) of the members of any homogeneous group of organisms deteriorates at a stochastically constant rate.

This law goes a bit beyond the observations in postulating the cause to be extrinsic rather than intrinsic, like nuclear decay. There is of course much other evidence (Simpson, 1953) for such a step. The law is also stated in terms of real time rather than survivorship time, although it applies to both and ties them together. A more neutral statement is that extinction in any adaptive zone occurs at a stochastically constant rate.

"Homogeneous" is a necessarily ambiguous word, and its meaning in a particular case must depend on the particular circumstances and on the degree of precision desired. Paleozoic and Mesozoic ammonites may differ sufficiently to be nonhomogeneous, but the entire marine benthos could be treated together almost as well as subdividing it. This does not lead to circularity; no subdivision of a group with racial senescence (25) would give a set of linear survivorship curves. The homogeneity required (and so its verification) is entirely ecological and is in terms of ultimate regulatory factors (Van Valen, 1973) of population density. Mice and fruit-eating flies might be homogeneous with each other but not with blood-sucking flies, although a scaling problem would remain. We could say that there always exists a degree of homogeneity at which the law is true, and try to measure it independently. Homogeneity does not imply equality of ability to respond to the deterioration; but counterexamples to a constant distribution of such abilities would disprove the law in its first form as a universal phenomenon (26).

Like any law (27), the effects of the law of extinction are observable only under appropriate conditions (here, persistence of the width of the adaptive zone). Unlike many laws (28), it may not be universal. If it is not universal, it should be possible to limit its domain in an objective manner. Such limitation should derive from an understanding of the causal basis for the law.

It is not fashionable to speak of laws in evolutionary biology or for historical processes generally. I think this is based on both a misunderstanding of the regularity of actual processes (29) and on an over-reaction to poorly formulated laws of earlier workers (30). Laws are propositions that specify sufficient conditions for a result; given the conditions, the result will occur, although some of the conditions (the bounds of the domain) may be implicit. The degree of confirmation of a law is of course a different matter from whether a proposition is (or represents) a law in this sense. Any general statement of the nature of a causal process states a law (31).

The Red Queen's Hypothesis (32)

The probability of extinction of a taxon is then effectively independent of its age. This suggests a randomly acting process. But the probability is strongly related to adaptive zones. This shows that a randomly acting process cannot be operating uniformly. How can it be that extinction occurs randomly with respect to age but nonrandomly with respect to ecology?

We can consider the situation in terms of an ensemble of mutually incompatible optima. It is selectively advantageous for a prey or host (including plants) to decrease its probability of being eaten or parasitized. It is often selectively advantageous for a predator or parasite species, and much more often for a predator or parasite individual, to increase its expected rate of capture of food. It is selectively advantageous for a competitor for resources in short supply (food and space in the broadest senses, and sometimes also externally supplied adjuncts to reproduction or dispersal) both to increase its own effect on its competitors and to decrease the effect of its competitors on itself (33).

Every species does the best it can in the face of these pressures. Probably all species are affected importantly by them at least over intervals of a few generations. Response to one kind of pressure may well decrease resistance to some other, at that time weaker, kind.

The various species in an adaptive zone (Van Valen, 1971), whether or not this zone is sharply delimited from others, can be considered together. We can assume as an approximation that a proportional amount w of successful response by one species produces a total negative effect of $v = w$ on other species jointly, usually less than v for any one of these species. For n species, the mean decrement of fitness per species is v/n. For m successful responses simultaneously (in an interval τ), the mean total decrement per species is mv/n. Over some interval t (in units of τ), the decrement is then mvt/n. To maintain itself as before, the species must increase its fitness by an amount mwt/n. Since each decrement generates a response, $m = n$. Most species will be able to recoup their loss, more or less while it occurs, but in doing so they jointly produce another disadvantage of v for the average species. This process of successive overlapping decrements of v may continue indefinitely. Species at least locally new may replace those for which the rate of environmental deterioration has been too great.

For the momentary fitness F of a random species,

$$\phi \equiv \frac{dF}{dt} = \frac{m(w-v)}{n} . \qquad (2)$$

For a given species, ϕ will of course vary from time to time as its specific decrements and possible responses covary. However, since $w = v$ on the average, $E(\phi) = 0$, i.e. the mean fitness in the adaptive zone is constant in the long run. If we take a standard average fitness F_1, the real average fitness will be $F_0 = F_1 - v$, because any response of one species above F_0 will bring a corresponding decrement from the counter-response of other species. v is the environmental load. Thus, even for a single species, the total selection pressure is constant over intervals that are long enough to average out irregularities.

ϕ is of course itself a variable among the species at any time, and its variation (or that of F) will determine the extinction rate in the adaptive zone. Species may (and do, but to an unknown extent) differ in the spectral threshold T (Reyment and Van Valen, 1969) of ϕ or its components to which they can no longer respond successfully and at which they therefore become extinct. Which species have which values of ϕ and T will change markedly with the nature of the stresses at a given time. The observed pattern of extinction seems to imply that the stresses are sufficiently diverse that they affect most species similarly over a very long interval. Assume, as justified in part by the Central Limit Theorem, that both ϕ and T are normally distributed, with means μ_1 and μ_2 respectively. Then, since both distributions range over the same set of species and so have the same area, the rate of extinction is the overlap of the distributions. In the case of equal variances for ϕ and T,

$$\Omega \equiv \frac{dE}{dt} = \frac{1}{\sqrt{2\pi}} e^{-\frac{(\mu_1 - \mu_2)^2}{8}} . \tag{3}$$

Uncompensated departures from normality and from equal variances in the distribution will of course affect the accuracy of this expression but not the dependence of Ω on the real area of overlap.

On this self-contained system we must now impose the physical environment and irregular biotic perturbations. Almost all changes in the physical environment of an adaptive zone will be deleterious to its inhabitants, either directly or by permitting the establishment of other species which can then survive there. Therefore regular changes in the physical environment can be treated as constant factors of ϕ. The observed constancy of Ω is evidence that such a treatment is permissible.

Important perturbations do, however, occur, from both biotic and physical causes. They can be treated together. The probability of extinction is not constant over geological time, as Fig. 8 and the extinctions at the end of the Cretaceous suffice to remind us. When major perturbations occur well within the geological range of a group, as for brachiopods in the late Permian, Paleozoic ammonites at the end of the Devonian, or coccoliths at the end of the Cretaceous, it happens that if we ignore the subgroups that became extinct then, the remainder show a constant rate of extinction (34). There is no important lasting effect of the unique event, unless total extinction occurred, nor do its effects extend detectably beyond the subgroups prematurely eliminated. Newly appearing subgroups after the event continue in the same way as those that lived through it. Smaller perturbations occur much more frequently, however (Fig. 8), and it is the very resultant of these perturbations that determines the long-term constancy.

Therefore the effects of these minor perturbations are not independent of each other over time. If they were, they should not be distributed about a constant mean. The rate of deterioration of the effective environment has what we can descriptively call the property of homeostasis. A large change in one interval is compensated for by a small one later, on the average, and vice versa; the mean itself does not undergo a random walk. It is the organisms in the adaptive zone that can link the perturbations across time in that the effects of one perturbation may depend on the effects of those before it. Species easily removed by one kind of perturbation are not there if it comes again soon, and may accumulate if it does not come for a long time. Meanwhile other kinds of perturbations are taking their own toll, more or less independently of the first kind (35).

We see here a major difference from the usual theory of genic selection. The latter depends only on the current distribution of frequencies of alleles and their interactions with each other and the environment. It does not depend at all on the process by which the current distribution was obtained. In formal language, it is a Markov process. But any process can be made Markovian by choice of a suitable level of analysis. With extinction we can see that a non-Markovian analysis of selection is appropriate. This is probably true for the general case also (36).

The Red Queen does not need changes in the physical environment, although she can accommodate them. Biotic forces provide the basis for a self-driving (at this level) perpetual motion of the effective environment and so of the evolution of the species affected by it (37).

The Red Queen's Hypothesis is a sufficient explanation of the law of extinction but not one that is yet derivable with confidence from lower-level knowledge of the causes of individual extinctions and the nature of species interactions. There may be other sufficient explanations that I have not been imaginative enough to see, and predictions by the Red Queen may be duplicated by such undiscovered explanations. Disproof of the Red Queen's Hypothesis is possible in several ways, but fully adequate confirmation must await derivation of it from what we can reasonably regard as facts.

We can go a step further by thinking of an adaptive landscape in a resource space (Van Valen, 1971). The amount of resources is fixed and can be thought of as an incompressible gel neutrally stable in configuration, supporting the peaks and ridges. If one peak is diminished there must be an equal total increase elsewhere, in one related peak or more uniformly. Similarly, increase in a peak results in an equal decrease elsewhere.

Species occupy this landscape and can be thought of as trying to maximize their share of whatever resource is scarcest relative to its use and availability. This resource will take the role of the gel, and the momentary fitness of a species will be proportional to the amount of gel under its area (the amount of the limiting resource it controls). To a sufficiently close approximation this momentary fitness seems to be what natural selection maximizes.

The landscape is changing continuously, at three levels. Species displace each other from areas of the adaptive surface. This can be by resistance to predation as well as other means, and some parts of the landscape may in this way or others (e.g. severe weather) be under-occupied. Incompatible optima maintain diversity. Secondly, the distribution of gel within the landscape changes, as with climatic change or (for herbivores) when the flora changes. Finally, the total amount of gel (or the amount in fact used) can change. This can occur by another necessary resource becoming limiting, by constraints on occupation of the adaptive zone, or by a change in the amount of the same limiting resource. The entire landscape can in this way disappear.

The first two kinds of change in the landscape lead to the the Red Queen's Hypothesis, while the third is, without compensatory changes, inconsistent with it. Empirically extinction is less independent of age when extinctions caused by major constrictions of adaptive zones are included. The observed degree of independence suggests that the third kind of change is relatively unimportant on the evolutionary time scale. We can hardly regard such a conclusion as established, but it does show that the question of variation in the size of major adaptive zones is fundamental.

Molecular Evolution

The Red Queen's Hypothesis provides reason to expect a long-term approximate constancy in the rate of evolution of individual proteins within any single adaptive zone (38). It does so without ad hoc assumptions, although such assumptions could be invoked to explain away exceptions just as they have been invoked (Clarke, 1970) to explain away constancy by the usual selective framework. Constancy of the rates of protein evolution is often regarded as the most important evidence for what King and Jukes (1969) called non-Darwinian evolution, despite serious misinterpretations (39).

A further prediction is that rates of protein evolution will be related monotonically to rates of taxonomic evolution, to the extent that the subtaxa in different groups are comparable (13,40). Some evidence (41) is available that rates are not always constant, and preliminary results I obtained long before discovering constant extinction suggest that the mean rate of protein evolution during the relatively short time while the orders of placental mammals diverged from each other was greater than that later (42). *Lingula*, a brachiopod genus present since the Ordovician or Silurian, and *Triops cancriformis*, a branchiopod (*not* brachiopod) species not detectably changed since at least the early Triassic (1), would be expected to have proteins more similar to those of the common ancestor than would their more divergent relatives. The evidence for approximately constant evolutionary rates of proteins comes from organisms (mainly vertebrates) which have all evolved appreciably since their separation from each other.

The Red Queen in her simplest gown also predicts a perhaps real phenomenon, which Ohta and Kimura (1971) noted as mysterious: that irregularities in the rate of molecular evolution seem to be more or less cancelled out over long intervals by a seemingly negative autocorrelation.

Implications of the Red Queen

Thoday (1953) proposed a concept of fitness as the probability of survival of a lineage to some very distant time in the future. More generally, we can take

$$F_t = \int_0^{\infty} w(t)P(t)\,dt \qquad (4)$$

where $P(t)$ is the probability (43) at time 0 of survival to time t, and $w(t)$ is a weighting function (the same for all lineages and perhaps integrating to 1 for scale) for which I would choose exponential decay at a low rate (44). Time 0 is the variable present.

The Red Queen proposes that this fitness has only two components for almost any real lineage: which adaptive zone is occupied and what the probability distribution is of new sublineages occurring by branching. Artiodactyls have largely replaced perissodactyls in the same adaptive zone, yet their extinction rates are identical (Tables 1,2). It is the rates of branching that are decisive. Monotremes still linger on in two isolated subzones, one of which has already been invaded by marsupials. The gradual extinction of multituberculates (Van Valen and Sloan, 1966) occurred by herbivorous placentals diversifying into parts of the joint adaptive zone formerly held by multituberculates. One multituberculate survived 15 or 20 million years after the rest of the extinction was completed (45); the latter had taken only about 10 million years. The Red Queen's proposal implies that extinction of lineages (subtrees) will prove not to be constant when it can be measured.

For the Red Queen, curiously, does not deny progress in evolution. By group selection such as this, as well as by individual selection, properties of communities can change in a directional manner. It may well be that the average Cenozoic species could outcompete the average Cambrian one; some information on this can probably be derived from functional analysis. It is

TABLE 2: Extinction and origination rates (in ma) for genera of two competing orders of mammals.
E: early; M: middle; L: late

Epoch	EXTINCTION Perissodactyla	Artiodactyla	ORIGINATION Perissodactyla	Artiodactyla
E. Eocene	120	160	1300	500
M. Eocene	80	90	340	920
L. Eocene	120	90	290	260
E. Oligocene	120	100	80	150
M. Oligocene	100	160	40	130
L. Oligocene	170	130	130	80
E. Miocene	90	130	130	260
M. Miocene	30	90	90	90
L. Miocene	40	60	60	100
E. Pliocene	90	130	90	320
L. Pliocene	90	80	20	70

* * *

well known (cf. Romer [1966]) that such functional progress has occurred in vertebrate evolution. The Red Queen measures environmental deterioration on a scale that is determined by the resistance of contempory species to it, so the scale and real deterioration themselves may well change without a change in the measured deterioration. Darwin's example of wolf and deer exemplifies this.

The Red Queen proposes that events of mutualism, at least on the same trophic level, are of little importance in evolution in comparison to negative interactions, although she does not consider other cases where mutualism is so great that the mutualists function as an evolutionary unit, as with lichens and perhaps chloroplasts. She considers the usual contrary view to be a result of wishful thinking, the imposition of human values on nonhuman processes.

The existence of the law of extinction is evidence for ecological significance and ecological comparability of taxa from species to family, within any adaptive zone.

We can think of the Red Queen's Hypothesis in terms of an unorthodox game theory (46). To a good approximation, each species is part of a zero-sum game against other species. Which adversary is most important for a species may vary from time to time, and for some or even most species no one adversary may ever be paramount. Furthermore, no species can ever win, and new adversaries grinningly replace the losers. This is a direction of generalization of game theory which I think has not been explored.

From this overlook we see dynamic equilibria on an immense scale, determining much of the course of evolution by their self-perpetuating fluctuations. This is a novel way of looking at the world, one with which I am not yet comfortable. But I have not yet found evidence against it, and it does make visible new paths and it may even approach reality.

Acknowledgments

I thank the National Science Foundation for regularly rejecting my (honest) grant applications for work on real organisms (cf. Szent-Györgyi, 1972), thus forcing me into theoretical work. This paper has been circulating in *samizdat* since December, 1972, and I have given talks based on it before and after then. I thank Dr. R. A. Martin for unpublished data and Drs. P. Billingsley, J. Cracraft,

J. F. Crow, D. H. Janzen, T. H. Jukes, S. A. Kauffman, H. W. Kerster, M. Kimura, E. G. Leigh, J. S. Levinton, R. C. Lewontin, J. F. Lynch, V. C. Maiorana, J. Maynard Smith, P. Meier, D. M. Raup, G. A. Sacher, T. J. M. Schopf, and E. O. Wilson for discussion. The Louis Block Fund of the University of Chicago paid for the preparation of the figures.

Notes

(1). The taxonomic data I used are from the following sources: For vertebrates, Romer (1966), with a few modifications from various later work. For invertebrates and genera and families of Foraminifera, R.C. Moore (ed.), Treatise on Invertebrate Paleontology (Boulder, Colorado: Geological Society of America and Univ. Kansas Press), 1953-present, including revisions (C. Teichert, ed.), with supplementation or (for Bryozoa) substitution from The Fossil Record (W. B. Harland et al., eds.; London: Geological Society of London [1967]). I deliberately ignored later pertinent work (e.g. Yochelson [1969]) on invertebrates. Pteridophytes and coccoliths are also from The Fossil Record. For species of Foraminifera, Berggren (1969). For Dinoflagellata, Sarjeant (1967). For diatoms, Small (1945,1946).

(2). The time scale I used is from the following sources: Berggren (1972); Anonymous (1964); Everndon, Savage, Curtis, and James (1964); Everndon and Curtis (1965); Kauffman (1970); and Van Valen and Sloan (1966). Experiments show that the results are robust to reasonable changes in the time scale.

(3). Various conventions are necessary in any such compilation. Some general ones I used are the following: I took the duration of a taxon as from the middle of the epoch (or other shortest interval) before the first record, to the middle of the epoch of the last record. I ignored questioned records and unrecognizable taxa. For data accurate only to period I plotted the range as ending in the middle unit of the period. I used all data regardless of degree of precision (unless imprecise beyond a period) but used them to the precision allowable. Also, summaries of ranges (even of individual taxa) are too often inaccurate, as shown by unquestioned records of subordinate taxa beyond the stated limits, so whenever possible I compiled data at the level of genus.

(4). I knew of this deficiency for many years but saw no reason to pursue it, as I expected the shape of the curves to remain concave. In hindsight one could also expect that the group as such might have a progressively higher probability of extinction as the biota around it evolves while it does not. This would give convex survivorship curves.

(5). In addition to the data plotted, I made several dozen plots of subsets of the same data, using such criteria as exclusion of a major extinction or subtaxon, or restriction of the time interval used. All important deviations from the total distribution are included in the figures given here. There is a bias in using all extinct subtaxa of a living group in that the longer-lived of subtaxa originating recently are still alive. Tests show that this effect is negligible for long-ranging groups (not, e.g., for families of mammals or echinoids) and obviously all living and extinct subtaxa cannot be combined into one useful curve (Simpson, 1953).

(6). I used all groups for which adequate data were available. Omissions are due to poverty of the fossil record or its study (e.g. Insectivora), small number of extinct taxa, error of dating being a substantial part of estimated durations (e.g. Archaeocyatha), or lack of adequate compilation (e.g. Gymnospermae).

(<u>7</u>). This is the reason for the apparently flat tops of some distributions.

(<u>8</u>). The initial figure in parentheses in distributions of most living taxa represents the total now alive, including those not known as fossils. Many living taxa in some groups are not easily fossilizable, and our knowledge of the present fauna is better than that at any other time even aside from this.

(<u>9</u>). Examples can be found in the following: Pearl (1928); Allee, Emerson, Park, Park, and Schmidt (1949); Kurten (1953); Odum (1971).

(<u>10</u>). Coral-like brachiopods of the Permian also became extinct, as did many other organisms superficially less similar to corals.

(<u>11</u>). Tests are necessarily weak, but data from several groups give no indication that a major extinction affects taxa of different ages to a different extent.

(<u>12</u>). A third (17 of 51) of the families of stony corals present in the late Cretaceous became extinct then, but only one of these did so at the major crisis at the end of the Cretaceous. <u>The Fossil Record</u> families are the only ones with adequate data on this point. Rudists diversified throughout the late Cretaceous and abruptly disappeared at its close.

(<u>13</u>). I give several ways of evaluating comparability of taxa elsewhere (Van Valen, in press).

(<u>14</u>). The phylogeny is from the <u>Treatise</u> (<u>1</u>) and excludes 5 families of unknown ancestry. Schindewolf (1961-1968) has given a rather different phylogeny, but parts of it are not detailed enough for this application.

(<u>15</u>). A brief discussion of possible factors in the crisis for trilobites can be found in <u>The Fossil Record</u> (<u>1</u>), p. 54. Some other curves also probably exhibit real deviations from linearity, but the ones discussed are the largest.

(<u>16</u>). Small (1946,1952) noticed this exponential increase for some individual genera, as Small (1950) and Tappan and Loeblich (1971) did for the entire group.

(<u>17</u>). I arbitrarily define these as having a maximum diameter of at least 5 mm. Most are highly complex internally, and most are discoidal, fusiform, or low conical.

(<u>18</u>). What is necessary is the expected distribution of the durations of all sub-trees, including parts of larger sub-trees, given a constant extinction rate and a branching rate that varies from time to time. Harris (1963, p. 32) considered the problem for a constant branching rate and found it intractable. It would further be useful if some sub-trees could be ignored as effectively infinite (extending an unknown amount beyond some absolute date such as the present). The point is directly resolvable by simulation, but this requires more money for computer use than I have.

(<u>19</u>). MacArthur died (at the age of 42) two weeks after the discovery of constant extinction.

(<u>20</u>). 1 ma is the reciprocal of a half-life of 500 years; in general, ma = (half-life in units of 500 years)$^{-1}$. Macarthurs apply also to phenomena (such as origination rates) for which half-lives are inappropriate. Applications in the text are to phenomena variously with and without replacement of the item sampled; the derivations therefore differ in detail. Some computational aids: $ma = (\log_{10}2)^{-1}(-\log_{10}(1 - p^{2\underline{t}}))$; $P_{0.5}$(1 day) = 183 kilomacarthurs (kma); $P_{0.5}$(1 minute) = 263 megamacarthurs (Mma). Given $P(\underline{t})$ of an event per interval $\underline{t}$, $P(\underline{kt}) = [P(\underline{t})]^{\underline{k}}$. The rate for interval <u>kt</u> is $ma(\underline{kt}) = \underline{k}^{-1}ma(\underline{t})$, given the same probability for both intervals. Haldane (1949) proposed an analogous measure for rates of continuous phenomena: a <u>darwin</u> is the rate giving a change by a factor of e per million years. Kimura (1969) defined a unit of rates for molecular evolution: a pauling is, in this context, the same as a millimacarthur.

(21). For birds, the data are for islands and of variable quality: MacArthur and Wilson (1967); Lack (1942); Diamond, (1969,1971). The islands are, in order of size, Los Coronados, Santa Barbara, Anacapa, St. Kilda, the Scillies, Krakatoa, San Miguel, San Nicolas, San Clemente, Santa Catalina, Santa Rosa, Santa Cruz, Karkar, Man, and the Orkneys. For arthropods, Simberloff and Wilson (1969) have approximate data from mangrove islets. The inaccuracy of the estimated extinction rates for birds and arthropods, in the present context, is about an order of magnitude. Estimates for both groups are probably too high, but I use available values. Mammal data are from Kurten (1968) for Europe, Martin and Webb (in press) for Florida, Webb (1969) for North America, and the data of Table 1 for the world. There are now about 4 species per genus of mammals in both North America and the world; if they go extinct independently, the extinction rate of species is three times that of genera. This inaccurate assumption, which ignores the correlation between number of species present and number of new species produced, is reasonable at the scale used. The grouping of birds and mammals into one equation is less defensible but is supported by the Florida data. The same is true for grouping data from islands, a peninsula, and two partly isolated continents.

(22). Treatments of evolutionary rates in paleontology not in other notes can be found in, e.g., the books by Schindewolf (1950), Zeuner (1958), and Kurten (1968); a symposium (J. Paleont. 26: 297-394 [1952]); and papers by Williams (1957), Kurten (1960), Bone (1963), Lerman (1965), House (1967), Newell (1967), Valentine (1969), Lipps (1970), Kurten (1971), Olsson (1972), and Cooke and Maglio (1972). DNA estimates are readily accessible in the Atlas (Dayhoff, 1972); various guesses exist on the proportions that are informational.

(23). A more extreme divergence from the regression would be expected for smaller organisms. Cairns, Dahlberg, Dickson, Smith, and Waller (1969) in fact give data for protozoans on blocks of an artificial substrate in a lake. An extinction rate of about 12 kma [P(extinction) = 0.043 per species per day] can be derived from their data. However, the substrate was a foam and so the effective area is unknown; the area of the top was 5×10^{-9} sq. km. Furthermore, the extinction rate may be higher than that in naturally occurring isolated substrates of the same effective area, and the experiment lasted only about 40 days. The glass slides that Patrick (1967) used as islands for diatoms would seem an excellent model system for such estimations, especially because the effect of area itself can be isolated from that of spatial heterogeneity and both studied together. The estimated extinction rate on Simberloffia may also be higher than a rate comparable to that for vertebrates; Simberloff and Wilson (1969) say that most "extinctions" seem to have been of species that couldn't colonize the islets under any circumstances, so no real populations of them existed to become extinct. This is a serious bias even if one difficult to overcome, and illustrates the danger of letting what we can easily measure determine what we think we want to measure, the tyranny of epistemology on ontology.

(24). The effective environment of any organism is its adaptive zone (Van Valen, 1971) plus the effects of any other organisms within that adaptive zone.

(25). A hypothesis recently revived in terms of DNA by Bachmann, Goin and Goin (1972).

(26). If we look at too narrow a part of the zone, with only a few taxa, discrete single events will be individually noticeable, as with any random process in the real world. In a causal universe a claim of randomness is a badge of ignorance. With evolutionary diversification the causes of seemingly random patterns may well be important and discoverable (Van Valen and Sloan [1972] give an example). The law of extinction is on the next level of abstraction from such causes.

(27). Gravitation does not cause an object resting on the floor to fall. Lakatos' critique (1963-1964) of mathematical proof is based on the difficulty of delimiting domains objectively.

(28). But like, e.g., Mendel's Laws or the gas laws.

(29). I have treated this subject elsewhere (Van Valen, 1972).

(30). For instance, E.D. Cope proposed a famous law in the nineteenth century that primitive taxa have a greater expected longevity than their descendants. This has never been adequately tested and should be re-formulated in terms of degree of primitiveness (assuming a threshold is absent) and a definition of primitiveness by entrance to an adaptive zone.

(31). A law need not be quantitative (although the law of extinction is). The contrary tradition is a myth derived, as Egbert Leigh has said, from physics envy.

(32). "Now here, you see, it takes all the running you can do, to keep in the same place." (L. Carroll, *Through the Looking Glass*.)

(33). Fisher (1930) and others, including Darwin and especially Lyell (1832), foreshadowed the Red Queen's Hypothesis but had no reason to impose the crucial constraint of constancy, and did not do so. I regard interference competition as causally a mechanism of resource competition, a proximal rather than ultimate regulator.

(34). Whether the total group does also depends on the distribution of longevities of the subgroups omitted.

(35). On one level, the probability of extinction of a group is related to its own properties because different groups go extinct for different reasons and so at different times. But on the next level, the Red Queen says that having one set of properties is not appreciably better than having another because the expected time to extinction is the same.

(36). Levine and Van Valen (1964) showed experimentally for *Drosophila* that natural selection has rather non-Markovian aspects. Lewontin (1966) later elaborated the point theoretically but without specific results.

(37). Origination-extinction equilibria are implicit in Simpson's work (1944,1953), and in Lyell's (1832). I realize that the Red Queen's Hypothesis is at least a simplification of reality. It is directly analogous to Newton's third law of motion.

(38). That the Red Queen in her simplest gown implies long-term constancy in total evolutionary rate is obvious. For any single protein we must invoke an analogue of the Central Limit Theorem: pervasive pleiotropy makes the rate for one protein roughly proportional to that for all, or linkage effects have a similar result. For instance, many proteins are to some extent attached, and the other components of the attachment may change for extraneous reasons, making the previous structure nonoptimal. Dickerson (1971) and others have made a similar point from the other end of the microscope.

(39). Stebbins and Lewontin (1972) actually think that "the entire argument is based upon a confusion between an average and a constant." What is remarkable is, however, precisely that the average rate (over shorter segments of a phylogeny) is so nearly constant (among these segments) for a given protein, rather than reflecting a branching random walk or some other process.

(40). More precisely, the prediction is of a monotone relationship of the average rate of protein evolution with the average rate of change among phenotypic characters (including the origin of new characters).

(41). Horne (1967), Kohne (1970), Ohta and Kimura (1971), Uzzell and Pilbeam, (1971), Jukes and Holmquist (1972). Also, constancy predicts the same expected number of changes in each lineage after the latest common ancestor. The data for hemoglobins in the Atlas (Dayhoff, 1972) seem inconsistent with this expectation. The assumption of total constancy leads to the expectation, presented seriously by D. Boulter (1972) and Ramshaw et al. (1972) that angiosperms originated in the early or middle Paleozoic.

(42). The approach used the protein sequence data of the Atlas, 1969 edition, and probability estimates of various alternative placental phylogenies as determined by myself.

(43). Probability in the sense of a propensity, a property of any single lineage.

(44). The rank-order of different groups with respect to F_t can depend on the choice of $w(t)$. It is an almost universal mistake to think that evolution locally maximizes fitness. Evolutionary fitness is F_t except to some population geneticists, but evolution doesn't maximize it. Selection at any level locally maximizes momentary fitness for that level, but the optima of different levels need not coincide. This is obvious between prezygotic and individual selection but is equally true for higher levels. Individual selection, the most important evolutionary force, can decrease F_t until extinction by, e.g., forcing the occupation of only a temporary niche.

(45). The latest record has now been found by J.F. Sutton, University of Kansas (talks at 1971 and 1972 meetings of the Society of Vertebrate Paleontology).

(46). Lewontin (1961), Warburton (1967), and Maynard Smith (1972) have made applications of game theory to evolution within the usual evolutionary framework.

Literature Cited

Allee, W.C., A.E. Emerson, O. Park, T. Park, and K.P. Schmidt. 1949. Principles of Animal Ecology. Philadelphia: Saunders. 837 pp.

Anonymous. 1964. Geological Society Phanerozoic time-scale 1964. Quart. Jour. Geol. Soc. London 120 S: 260-262.

Bachmann, K., O.B. Goin, and C.J. Goin. 1972. Nuclear DNA amounts in vertebrates. In evolution of Genetic Systems (H.H. Smith, ed.), pp. 419-450. New York: Gordon and Breach.

Berggren, W.A. 1969. Rates of evolution in some Cenozoic planktonic foraminifera. Micropaleontology 15: 351-365.

_____. 1972. A Cenozoic time-scale--some implications for regional geology and paleobiogeography. Lethaia 5: 195-215.

Bone, E.L. 1963. Paleontological species and human speciation. South African Jour. Sci. 59: 273-277.

Boulter, D. 1972. Protein structure in relationship to the evolution of higher plants. Sci. Prog. 60: 217-229.

Cairns, J., M.L. Dahlberg, K.L. Dickson, N. Smith, and W.T. Waller. 1969. The relationship of fresh-water protozoan communities to the MacArthur-Wilson equilibrium model. Amer. Nat. 103: 439-454.

Clarke, B. 1970. Darwinian evolution of proteins. Science 168: 1009-1011.

Cooke, H.B.S., and V.J. Maglio. 1972. Plio-Pleistocene stratigraphy in East Africa in relation to proboscidean and suid evolution. In Calibration of Hominoid Evolution (W.W. Bishop and J.A. Miller, eds.), pp. 303-329. Scottish Academic Press.

Dayhoff, M.O. 1972. Atlas of Protein Sequence and Structure 1972. Washington: National Biomedical Research Foundation. 124+382 pp.

Diamond, E.M. 1969. Avifaunal equilibria and species turnover rates on the Channel Islands of California. Proc. Natl. Acad. Sci. U.S.A. 64: 57-63.

__________. 1971. Comparison of faunal equilibrium turnover rates on a tropical island and a temperate island. Proc. Natl. Acad. Sci. U.S.A. 68: 2742-2745.

Dickerson, R.E. 1971. The structure of cytochrome c and the rates of molecular evolution. Jour. Molec. Evol. 1:26-45.

Eldredge, N. 1971. The allopatric model and phylogeny in Paleozoic invertebrates. Evolution 25: 156-167.

__________ and S.J. Gould. 1972. Punctuated equilibria: an alternative to phyletic gradualism. In Models in Paleobiology (T.J.M. Schopf, ed.), pp. 82-115. San Francisco: Freeman, Cooper.

Everndon, J.F., and G.H. Curtis. 1965. The potassium-argon dating of late Cenozoic rocks in East Africa and Italy. Cur. Anth. 6: 343-385.

Everndon, J.F., D.E. Savage, G.H. Curtis, and G.T. James. 1964. Potassium-argon dates and the Cenozoic mammalian chronology of North America. Amer. Jour. Sci. 262: 145-198.

Fisher, R.A. 1930. The Genetical Theory of Natural Selection. Oxford: Clarendon Press. 272 pp.

Haldane, J.B.S. 1949. Suggestions as to the quantitative measurement of rates of evolution. Evolution 3: 51-56.

Harris, T.E. 1963. The Theory of Branching Processes. Berlin: Springer-Verlag. 230 pp.

Hennig, W. 1966. Phylogenetic Systematics. Urbana: Univ. Illinois Press. 263 pp.

Horne, S.L. 1967. Comparisons of primate catalase tryptic peptides and implications for the study of molecular evolution. Evolution 21: 771-786.

House, M.R. 1967. Fluctuations in the evolution of Paleozoic invertebrates. In The Fossil Record (W.B. Harland et al., eds.), pp. 41-54. London: Geological Society of London.

Jukes, T.H., and R. Holmquist. 1972. Evolutionary clock: nonconstancy of rate in different species. Science 177: 530-532.

Kauffman, E.G. 1970. Population systematics, radiometrics and zonation--a new biostratigraphy. Proc. North American Paleont. Conv. (F): 612-666. Lawrence: Allen Press.

Kimura, M. 1969. The rate of molecular evolution considered from the standpoint of population genetics. Proc. Natl. Acad. Sci. U.S.A. 63: 1181-1188.

King, J.L., and T. H. Jukes. 1969. Non-Darwinian evolution. Science 164: 788-798.

Kohne, D.E. 1970. Evolution of higher-organism DNA. Quart. Rev. Biophys. 3: 327-375.

Kurtén, B. 1953. On the variation and population dynamics of fossil and recent mammal populations. Acta Zool. Fennica 76: 1-122.

________. 1960. Rates of evolution in fossil mammals. Cold Spring Harbor Symp. Quant. Biol. 24 (for 1959): 205-215.

________. 1968. Pleistocene Mammals of Europe. London: Weidenfeld and Nicholson. 317 pp.

________. 1971. Time and hominid brain size. Comment Biol. Soc. Sci. Fennica 36: 1-8.

Lack, D. 1942. Ecological features of the bird faunas of British small islands. Jour. Anim. Ecol. 11: 9-36.

Lakatos, I. 1963-1964. Proofs and refutations. Brit. Jour. Philos. Sci. 14: 1-25, 120-139, 221-245, 296-342.

Lerman, A. 1965. On rates of evolution of unit characters and character complexes. Evolution 19: 16-25.
Levine, L., and L. Van Valen. 1964. Genetic response to the sequence of two environments. Heredity 19: 734-736.
Lewontin, R.C. 1961. Evolution and the theory of games. Jour. Theor. Biol. 1: 382-403.
__________. 1966. Is nature probable or capricious? BioScience 16: 25-27.
Lipps, J.H. 1970. Plankton evolution. Evolution 24: 1-21.
Lyell, C. 1832. Principles of Geology. 1st ed., vol. 2. London: J. Murray. 330 pp.
MacArthur, R.H. 1972. Geographical Ecology. New York: Harper and Row. 269 pp.
__________ and E.O. Wilson. 1967. The Theory of Island Biogeography. Princeton: Princeton Univ. Press. 203 pp.
MacGillavry, H.J. 1968. Modes of evolution mainly among marine invertebrates. Bijdr. Dierk. 38: 69-74.
Martin, R.A., and S.D. Webb. In press. Late Pleistocene mammals from Devil's Den, Levy County. In Pleistocene Mammals of Florida (S.D. Webb, ed.) Gainesville: Univ. Florida Press.
Maynard Smith, J. 1972. On Evolution. Edinburgh: Edinburgh Univ. Press. 125 pp.
Newell, N.D. 1963. Crises in the history of life. Sci. Amer. 28(2): 76-92.
__________. 1967. Revolutions in the history of life. Geol. Soc. Amer. Spec. Pap. 89: 63-91.
__________. 1971. An outline history of tropical organic reefs. Amer. Mus. Novit. 2465: 1-37.
Odum, E.P. 1971. Fundamentals of Ecology. 3rd. Ed. Philadelphia: Saunders. 574 pp.
Ohta, T., and Kimura, M. 1971. On the constancy of the evolutionary rate of cistrons. Jour. Molec. Evol. 1: 18-25.
Olsson, R.K. 1972. Growth changes in the Globorotalia fohsi lineage. Eclogae Geol. Helvetiae 65: 165-184.
Patrick, R. 1967. The effect of invasion rate, species pool, and size of area on the structure of the diatom community. Proc. Natl. Acad. Sci. U.S.A. 58: 1335-1342.
Pearl, R. 1928. The Rate of Living. New York: Knopf. 185 pp.
Ramshaw, J.A.M., D.L. Richardson, B.T. Meatyard, R.H. Brown, M. Richardson, E.W. Thompson, and D. Boulter. 1972. The time of origin of the flowering plants determined by using amino acid sequence of cytochrome c. New Phytol. 71: 773-779.
Reyment, R., and L. Van Valen. 1969. Buntonia olokundudui sp. nov. (Ostracoda, Crustacea): a study of meristic variation in Paleocene and Recent ostracods. Bull. Geol. Inst. Univ. Uppsala (N.S.) 1: 83-94.
Romer, A.S. 1966. Vertebrate Paleontology. 3rd ed. Chicago: Univ. Chicago Press. 468 pp.
Ruzhentsov, V. Ye. 1963. The problem of transition in paleontology. Paleont. Zhur. 1963(2): 3-16. (Translated 1964, Int. Geol. Rev. 6: 2204-2213).
Sarjeant, W.A.S. 1967. The stratigraphical distribution of fossil dinoflagellates. Rev. Palaeobot. Palynol. 1: 323-343.
Schindewolf, O.H. 1950. Der Zeitfaktor in Geologie und Paläontologie. Stuttgart: Schweizerbart. 114 pp.
__________. 1961-1968. Studien zur Stammesgeschichte der Ammoniten. Akad. Wiss. Lit. Mainz, Abhandl. Math.-Naturw. Kl. 1960: 635-744; 1962: 425-572; 1963: 285-432; 1965: 137-238; 1966: 323-454, 719-808; 1968: 39-209.
Simberloff, D., and E.O. Wilson. 1969. Experimental zoogeography of islands: the colonization of empty islands. Ecology 50: 278-296.
Simpson, G.G. 1944. Tempo and Mode in Evolution. New York: Columbia Univ. Press. 237 pp.

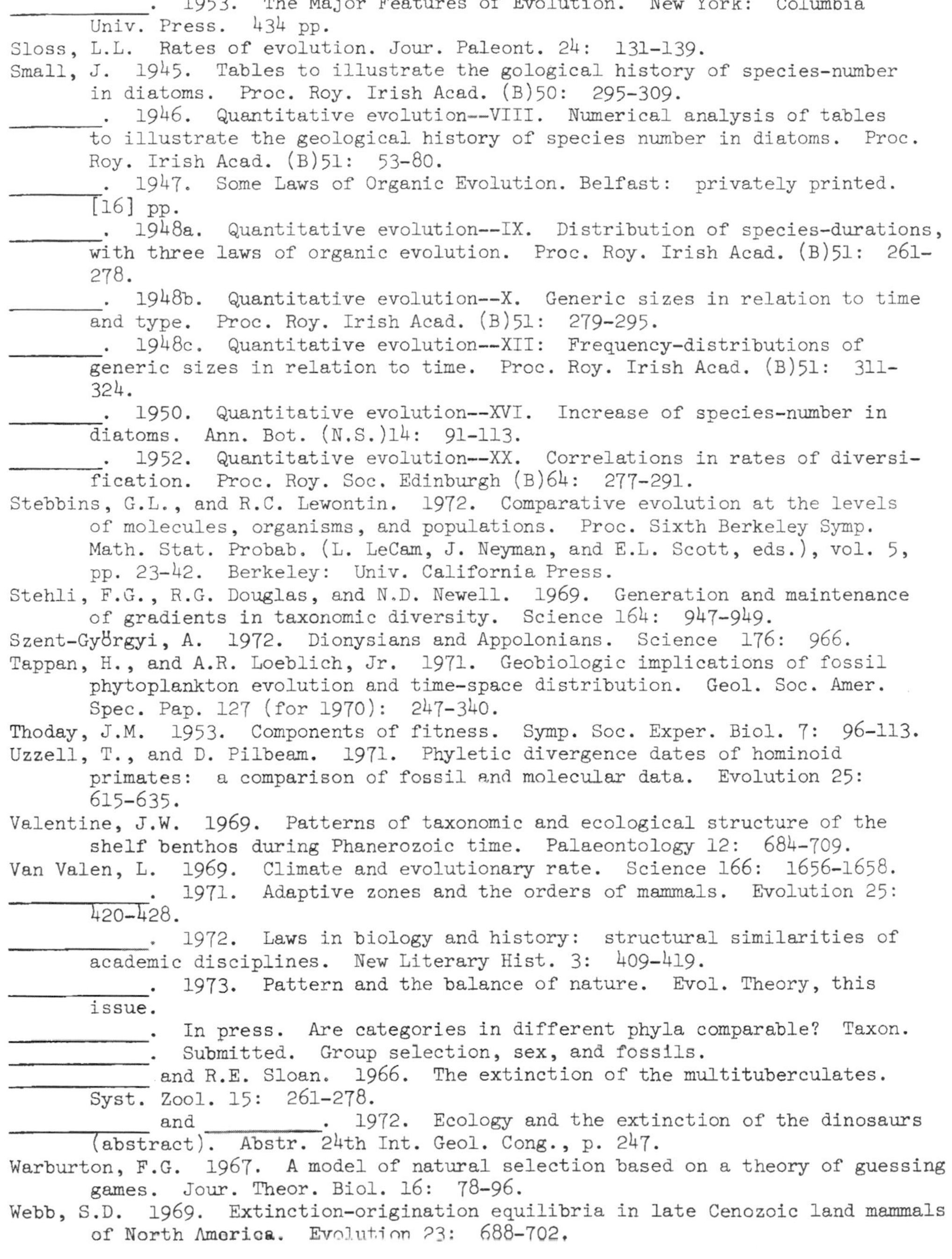

________. 1953. The Major Features of Evolution. New York: Columbia Univ. Press. 434 pp.

Sloss, L.L. Rates of evolution. Jour. Paleont. 24: 131-139.

Small, J. 1945. Tables to illustrate the gological history of species-number in diatoms. Proc. Roy. Irish Acad. (B)50: 295-309.

________. 1946. Quantitative evolution--VIII. Numerical analysis of tables to illustrate the geological history of species number in diatoms. Proc. Roy. Irish Acad. (B)51: 53-80.

________. 1947. Some Laws of Organic Evolution. Belfast: privately printed. [16] pp.

________. 1948a. Quantitative evolution--IX. Distribution of species-durations, with three laws of organic evolution. Proc. Roy. Irish Acad. (B)51: 261-278.

________. 1948b. Quantitative evolution--X. Generic sizes in relation to time and type. Proc. Roy. Irish Acad. (B)51: 279-295.

________. 1948c. Quantitative evolution--XII: Frequency-distributions of generic sizes in relation to time. Proc. Roy. Irish Acad. (B)51: 311-324.

________. 1950. Quantitative evolution--XVI. Increase of species-number in diatoms. Ann. Bot. (N.S.)14: 91-113.

________. 1952. Quantitative evolution--XX. Correlations in rates of diversification. Proc. Roy. Soc. Edinburgh (B)64: 277-291.

Stebbins, G.L., and R.C. Lewontin. 1972. Comparative evolution at the levels of molecules, organisms, and populations. Proc. Sixth Berkeley Symp. Math. Stat. Probab. (L. LeCam, J. Neyman, and E.L. Scott, eds.), vol. 5, pp. 23-42. Berkeley: Univ. California Press.

Stehli, F.G., R.G. Douglas, and N.D. Newell. 1969. Generation and maintenance of gradients in taxonomic diversity. Science 164: 947-949.

Szent-Györgyi, A. 1972. Dionysians and Appolonians. Science 176: 966.

Tappan, H., and A.R. Loeblich, Jr. 1971. Geobiologic implications of fossil phytoplankton evolution and time-space distribution. Geol. Soc. Amer. Spec. Pap. 127 (for 1970): 247-340.

Thoday, J.M. 1953. Components of fitness. Symp. Soc. Exper. Biol. 7: 96-113.

Uzzell, T., and D. Pilbeam. 1971. Phyletic divergence dates of hominoid primates: a comparison of fossil and molecular data. Evolution 25: 615-635.

Valentine, J.W. 1969. Patterns of taxonomic and ecological structure of the shelf benthos during Phanerozoic time. Palaeontology 12: 684-709.

Van Valen, L. 1969. Climate and evolutionary rate. Science 166: 1656-1658.

________. 1971. Adaptive zones and the orders of mammals. Evolution 25: 420-428.

________. 1972. Laws in biology and history: structural similarities of academic disciplines. New Literary Hist. 3: 409-419.

________. 1973. Pattern and the balance of nature. Evol. Theory, this issue.

________. In press. Are categories in different phyla comparable? Taxon.

________. Submitted. Group selection, sex, and fossils.

________ and R.E. Sloan. 1966. The extinction of the multituberculates. Syst. Zool. 15: 261-278.

________ and ________. 1972. Ecology and the extinction of the dinosaurs (abstract). Abstr. 24th Int. Geol. Cong., p. 247.

Warburton, F.G. 1967. A model of natural selection based on a theory of guessing games. Jour. Theor. Biol. 16: 78-96.

Webb, S.D. 1969. Extinction-origination equilibria in late Cenozoic land mammals of North America. Evolution 23: 688-702.

Williams, A. 1957. Evolutionary rates of brachiopods. Geol. Mag. 94: 201-211.
Yochelson, E.L. 1969. Stenothecoida, a proposed new class of Cambrian Mollusca. Lethaia 2: 49-62.
Zeuner, F.E. 1958. Dating the Past. 4th ed. London: Methuen. 516 pp.

PART

6 Taphonomy

Edited by Nicholas D. Pyenson

The study of the dead—taphonomy—must overcome a fundamental fact about the fossil record: the near entirety of life that has ever existed on Earth has left no trace. Through pathways of dismemberment and decay, biogeochemical recycling processes prevent most organic remains from being incorporated into the geologic record. It is striking, thus, that we know anything at all about life in the past. Yet, paleontologists do know a lot about the fossil record, and while some periods of time remain woefully incomplete, other intervals are represented by dense collections of organic remains. Taphonomy is an integrative approach that provides a means of understanding what these differential snapshots of life in the past represent, in terms of their fidelity to the original biota. Studying how organisms become fossils, and escape the great recycling, involves different flavors of investigations, from experimental studies in the laboratory (Briggs 2003), to comparative surveys across different depositional environments (Kidwell 2002), and to long-term studies on the fate of organic remains in modern environments (Behrensmeyer et al. 2000). For paleoecologists, these approaches ground basic questions about the diversity, organization, and evolution of ecosystems through time.

The contributions in this section span the entirety of this still-growing discipline within the broader landscape of paleobiology. Taphonomy has its origins in the early to mid-twentieth century with several workers, including Johannes Weigelt (1890–1948) and Ivan Efremov (1908–72) and Wilhelm Schäfer (1912–81), whose innovative, actualistic research programs were only appreciated decades after their inception (Cadée 1991). Despite the delay for Weigelt, Efremov, and Schäfer's works (in original or translated forms) to penetrate into the libraries of their English-speaking counterparts, taphonomic research expanded during the late 1960s and the 1970s, as students of paleoecology came to realize that direct readings of the fossil record may be misleading, biased, and/or distorted. One of the more telling characteristics of research at this time was the effort to plumb modern analogues with tools that allowed these settings to be compared to fossil assemblages (e.g., Johnson 1965). Disparate research lines among sedimentologists, paleobiologists, and stratigraphers were catalyzed during the 1980s, punctuated by Behrensmeyer and Kidwell (1985)'s seminal paper in *Paleobiology*, which served as a clarion call for emergent and synthetic taphonomic research programs, no matter the depositional setting, temporal span, or taxonomic concentration.

In recent decades, paleoecology has benefited from taphonomy's flourishing, which has generated clear methodologies and spurred new modes of thinking about the fossil record. Foremostly, taphonomic approaches have clarified the scope of long-standing problems in paleoecology, including the generation and maintenance of biodiversity, the structure of ancient communities, and the function of ecosystems at different times in the geologic past (National Research Council 2005). The influence of taphonomy on modern paleoecological research covers many different themes. For example, the live-dead studies that initiated taphonomic thinking among paleobiologists in the 1960s have recently undergone a renaissance, reiterating the widespread fidelity of death records to source communities, for a variety of taxa in both terrestrial and marine environments (Olszewski and Kidwell 2007; Terry 2010; Miller 2011; Pyenson 2011). The high fidelity of death records appears especially enhanced by protracted time- and spatial-averaging, underscoring the value of such biases formerly considered as "overprints" that needed to be eliminated. Fidelity studies have also quickly become key pieces of evidence for the nascent discipline of conservation paleobiology, which seeks to expand the historical baselines of populations beyond the timespan of human anecdotes and written records (Kidwell 2007; Dietl and Flessa 2011). In this regard, taphonomic insights have fueled the expansion of paleoecology's purview to encompass those traditionally ascribed to historical ecology. A taphonomic lens on paleoecology has also yielded novel research programs at the largest temporal scale, geologic time. Taphonomy plays a major role in reconstructing the history of sedimentary systems (Holland 2000), and it has also spurred the search for so-called megabiases—distortions in the quality of the fossil record at >10 million–year timescales (Kowalewski and Flessa 1996; Kosnik et al. 2011).

Looking ahead, taphonomy's contributions to paleoecology remain mostly undiscovered future generations of investigators. Open questions remain, for example, about the universality of major modes of fossilization across different sedimentary systems, across periods of geologic time, and among different phyla and taxa. Experimental decay studies have only scratched the surface of understanding the basic biochemical pathways that prevent tissues from completely decaying, which in turn constrains the differential preservation of some taxa rather than others. Lastly, still more modern studies, in the same vein as Efremov, Weigelt and Schäfer's *aktuopalaeontologie*, are sorely needed to resolve the modern analogues for taphonomic patterns in the fossil record that remain enigmatic. Although modern investigators have many more tools at their disposal than taphonomy's founders (e.g., digital cameras, chemical assays, and radiographic imaging), the contributions in this section all can be viewed as a part of a larger conversation about the value of asking the right questions about the dead to understand the ancient ecology of the very dead.

Literature Cited

Behrensmeyer, A. K., and S. M. Kidwell. 1985. Taphonomy's contributions to paleobiology. *Paleobiology* 11:105–19.

Behrensmeyer, A. K., S. M. Kidwell, and R. A. Gastaldo. 2000. Taphonomy and Paleobiology. In *Deep Time: Paleobiology's Perspective*, edited by D. H. Erwin and S. L. Wing, 103–44. Lawrence, KS: Paleontological Society.

Briggs, D. E. G. 2003. The role of decay and mineralization in the preservation of soft-bodied fossils. *Annual Review of Earth and Planetary Sciences* 31: 275–301.

Cadée, G. C. 1991. The history of taphonomy. In *The Processes of Fossilization*, edited by S. K. Donovan, 3–21. New York: Columbia University Press.

Dietl, G. P., and K. W. Flessa. 2011. Conservation paleo-

biology: putting the dead to work. *Trends in Ecology and Evolution* 26:30–37.
Holland, S. M. 2000. The quality of the fossil record: a sequence stratigraphic perspective. In *Deep Time: Paleobiology's Perspective*, edited by D. H. Erwin and S. L. Wing, 148–68. Lawrence, KS: Paleontological Society.
Kidwell, S. M. 2002. Time-averaged molluscan death assemblages: palimpsests of richness, snapshots of abundance. *Geology* 30:803–6.
———. 2007. Discordance between living and death assemblages as evidence for anthropogenic ecological change. *PNAS* 104:17701–6.
Kosnik, M. A., J. Alroy, A. K. Behrensmeyer, F. T. Fürsich, R. A. Gastaldo, S. M. Kidwell, M. Kowalewski, R. E. Plotnick, R. R. Rogers, and P. J. Wagner. 2011. Changes in shell durability of common marine taxa through the Phanerozoic: evidence for biological rather than taphonomic drivers. *Paleobiology* 37:303–31
Kowalewski, M., and K. Flessa. 1996. Improving with age: the fossil record of lingulid brachiopods and the nature of taphonomic megabiases. *Geology* 24:977–80.
Johnson, R. G. 1965. Pelecypod death assemblages in Tomales Bay, California. *Journal of Paleontology* 39: 80–85
Miller, J. H. 2011. Ghosts of Yellowstone: multi-decadal histories of wildlife populations captured by bones on a modern landscape. *PLoS One* 6:e18057.
National Research Council, Committee on the Geologic Record of Biosphere Dynamics 2005. *Geological Record of Ecological Dynamics*. Washington, DC: National Academies Press.
Olszewski, T. D., and S. M. Kidwell. 2007. The preservational fidelity of evenness in molluscan death assemblages. *Paleobiology* 33:1–23
Pyenson, N. D. 2011. The high fidelity of the cetacean stranding record: insights into measuring diversity by integrating taphonomy and macroecology. *Proceedings of the Royal Society B: Biological Sciences* 278: 3608–16.
Terry, R. C. 2010. The dead do not lie: using skeletal remains for rapid assessment of historical small-mammal community baselines. *Proceedings of the Royal Society B: Biological Sciences* 277 (1685): 1193–1201.

PAPER 36

An Approach to the Paleoecology of Mammals (1955)

J. A. Shotwell

Commentary

PATRICIA A. HOLROYD AND SUSUMU TOMIYA

Much of current paleobiological research on diversity still deals primarily with taxonomic richness. By contrast, ecologists embrace taxonomic abundance as a crucial dimension of diversity. It is commonplace today to count and report total numbers of identified specimens and estimate minimum numbers of individuals in fossil assemblages, but that practice was unusual in the early part of the twentieth century. Although the best methods for estimating the number of individuals and interpreting the processes that generate differences in their abundance are still debated, Shotwell's paper in *Ecology* began that conversation in paleobiology. Written at a time when ecologists were also trying to find general laws for how abundances in living communities are structured (e.g., Preston 1948), Shotwell presented an explicit way to measure the relative taxonomic abundance in fossil assemblages that are composed of parts of individuals and suggested an interpretive framework for understanding community structure, mediated by taphonomic processes.

Today, most references to Shotwell disagree with his quantitative method of community reconstruction or simply cite it in a list of references for minimum numbers of individuals and numbers of identified specimens. What is lost in such typically ahistorical treatments is how this early paper highlighted the importance of taxonomic abundance in paleoecology and the taphonomic distortion of this information in allochthonous fossil assemblages. Shotwell recognized that a method for the precise (and hopefully accurate) estimation of abundance in the original, spatially coherent live assemblages based on fossil counts was needed if paleontology was to contribute to the broader study of ecology; and he explicated one such method. His seminal work prompted subsequent researchers to collect and document assemblages in more systematic ways. At the same time, it also inspired neotaphonomic studies to better understand skeletal transport mechanisms that affect fossil preservation potential (e.g., Voorhies 1969).

Literature Cited

Preston, F. W. 1948. The commonness, and rarity, of species. *Ecology* 29:254–83.

Voorhies, M. R. 1969. *Taphonomy and Population Dynamics of an Early Pliocene Vertebrate Fauna, Knox County, Nebraska*. Contributions to Geology, Special Paper no. 1. Laramie: University of Wyoming.

From *Ecology* 36:327–37.

Summary

In Rondeau Bay, perch, *Perca flavescens* (Mitchill), were the most abundant fish large enough to be caught by gill-nets. In summer the two chief groups of perch were (a) two- and three-year-old migratory perch which were concentrated at the end of the bay nearer Lake Erie and migrated daily into the lake and (b) two- and three-year-old non-migratory perch remaining in the interior of the bay.

The activity patterns, as measured by gill-net catches, indicated that all the perch were very inactive during the hours of darkness. The catches of migratory perch showed daily sunrise and sunset peaks. The low catch during midday was due to decreased local abundance. The non-migratory perch became active at sunrise, showed a peak of activity at about 1700 hours, and became inactive after sunset.

An approximate route of the daily migration of perch between Rondeau Bay and Lake Erie is presented. The migration was detected chiefly by tabulating the direction in which the perch were headed when caught in the net, by a chronological succession of peaks along the migration route and by analyses of the stomach contents.

Light appeared to play an important role in the timing of the beginning and ending of the day's activity. The time of the migration changed with the change in the time of sunrise and sunset during the season.

References

Calhoun, J. B. 1945. Twenty-four hour periodicities in the animal kingdom. Part II, The Vertebrates. Jour. Tenn. Acad. Sci., 20: 228-232.

Carlander, K. D., and R. E. Cleary. 1949. The daily activity patterns of some freshwater fishes. Amer. Midl. Nat., 41: 447-452.

Hasler, A. D., and J. E. Bardach. 1949. Daily migrations of perch in Lake Mendota, Wisconsin. Jour. Wildl. Mgt., 13: 40-51.

Pearse, A. S., and Henrietta Achtenberg. 1920. Habits of yellow perch in Wisconsin Lakes. Bull. U.S. Bur. Fish., 36: 294-366.

Sieh, J. G., and John Parsons. 1950. Activity patterns of some Clear Lake, Iowa, fishes. Iowa Acad. Sci., 57: 511-518.

Welsh, J. H. 1938. Diurnal rhythms. Quart. Rev. Biol., 13: 123-139.

AN APPROACH TO THE PALEOECOLOGY OF MAMMALS

J. Arnold Shotwell

Museum of Natural History, University of Oregon, Eugene, Oregon

Introduction

The paleoecology of Tertiary invertebrates and plants has been widely and effectively studied by applying the uniformitarian concept in comparison of living and fossil species. This approach makes use of the occurrence in Tertiary deposits of species which are essentially the ecological equivalents of those now living, as judged from their close morphological similarity. Paleobotanists have applied this most successfully to community groupings (the work of R. W. Chaney and students for example), whereas the invertebrate paleontologist has more often used one or more stenotopic forms to indicate environmental conditions (Smith 1919; Durham 1950). This method of reference to living environments has provided data concerning changes in distribution of faunas and floras which are interpreted as reflecting climatic changes. The fact that invertebrates and plants had established their adaptive radiations by or soon after the beginning of the Tertiary makes such an approach possible.

Mammals, however, had only begun their adaptive radiation by the beginning of the Tertiary. Most living families of mammals are not known before the Oligocene and few living genera are known before the Pliocene. It is obvious then, that the approach of the invertebrate paleontologist and the paleobotanist cannot be applied in the determination of the paleoecology of mammals. These fields of study, especially floristic paleobotany, however, offer much information to the student of fossil mammals regarding climate and, to a lesser extent, the topographic diversity during Tertiary time.

The rapid evolution and adaptive radiation of mammals in the Tertiary has presented an enigma to the mammalian paleontologist as to criteria for an approach to the paleoecology of mammals. Some use has been made of morphology alone as suggesting environmental situations. Community structure, though, cannot be approached in this way without a background of criteria based on quantitative analysis of quarry samples.

Acknowledgments

The author is indebted to a number of colleagues who have read the manuscript at various stages and offered suggestions and criticisms. They are R. W. Chaney, D. E. Savage, Frank Pitelka, Richard Tedford and Malcolm McKenna. The author, however, must be held responsible for the approach as it appears here. Appreciation is also expressed to the staff of the University of California, Museum of Paleontology, for use of the Coffee Ranch Hemphill collection in their care.

This study was in part supported by funds of the Museum of Natural History of the University of Oregon and the General Research Committee of the same institution.

Illustrations are by the author.

The Quarry as a Source of Quantitative Data

Localized concentrations of fossil bones often occur in the sediments. These concentrations make excavation on a large scale practical, and are referred to as quarries. They vary greatly in the density of materials and in total volume. Such concentrations often furnish a large number of specimens which in varying degree represent the fauna once living locally. The nature of this representation must first be determined in order to make use of the quanitative data available.

Many quarry samples obviously do not present a true picture of the fauna they represent. This may be indicated by the presence of only one species in the sample. The *"Stenomylus"* and *"Coryphodon"* quarries are examples of this. A sample may also be useless in which only one ontogenetic age group is present. Another questionable type of quarry sample, for our purposes here, is that in which there is a strict selection of size of individuals included. These are sometimes explained as carnivore lairs. Other quarry samples may be more diverse in forms and age groups. The nature of their representation is more subtle.

In living mammalian faunas a general relationship is evident between the number of carnivores (individuals) and herbivores present in a given area. This relationship is often referred to as the "Eltonian Pyramid." The herbivorous mammals form the base of the pyramid with the lesser number of carnivores making up the upper part of the pyramid. This represents a generalized food-energy cycle controlling the balance in a fauna. Presumably these restrictions in food-energy cycles are transferable to fossil faunas. We can use this as a rough indicator of balance in a fauna. If a quarry sample, supposedly a large enough one, does not indicate the gross balance in the fauna it cannot be considered as a true representation of the fauna. On the other hand if the sample appears to show the balance of numbers of herbivores and carnivores which was present in the original living fauna then it presents the most desirable quantitative data for a study of the paleoecology it represents. As an example the relative number of individuals of herbivores and carnivores for the two faunas employed in the approach outlined here, McKay Reservoir and Hemphill, and those of Rancho LaBrea are illustrated in Figure 1. It is readily seen that the Rancho LaBrea figure does not indicate the necessary balance of the fauna in the sample whereas those from the other two faunas do reflect this balance. They appear, at least in a broad way, to provide a more probably true quantitative representation of the living fauna they were originally derived from.

Problems Involved in the Quarry Sample

Even after the quarry sample is determined to be a likely true quantitative representation of the once living fauna, other problems arise as to its possible errors in representation.

Forms which are nearly always rare or missing in fossil mammalian faunas, irrespective of their probable abundance in the area, are those with volant or arboreal habits. This characteristic has hindered the study of such groups as bats, primates, and flying squirrels. Their usual small size and fragility does not seem to be the important factor since insectivores and small rodents are not uncommon in quarries. Their habits which keep them away from sites of deposition more than other mammals must be responsible. Other forms may not be found in the quarry sample due to their low density in the area supplying material to the deposit. A paleontologist can never expect to have a complete faunal list.

It may be of interest at this point to make an estimate of the completeness of our faunal lists. The faunas used in the present study are assigned to the Hemphillian Provincial Age (about mid-Pliocene.) Paleobotanical evidence indicates that our present diversity of environments was becoming established by Hemphillian time. The diversity of Hemphillian mammals as shown in numbers of genera might be expected to be roughly equal to the number of living genera. This allows for additions due to increasing diversity and subtractions brought about by migration and extinction. It is necessary in such an estimate to rule out forms which we know are usually poorly represented, arboreal and volant, in order to obtain a comparative estimate. The fossil evidence of Hemphillian mammals for North America is almost completely restricted to the United States so that the numbers of living genera used in the

FIG. 1. Relative numbers of individuals of carnivores to herbivores in the McKay Reservoir local fauna, the Coffee Ranch quarry of the Hemphill local fauna, and the Rancho LaBrea local fauna. Data for Rancho LaBrea from Stock 1939.

estimate are those found in the United States. Table I lists the number of genera of living and Hemphillian mammals by orders. The classification used is that of Simpson (1945).

Ninety-one living genera of mammals are listed. This includes thirteen genera of bats. When these are removed there remain seventy-eight genera.

TABLE I. *Numbers of Recent and Hemphillian Genera of Mammals*

Order	Recent	Hemphillian
Marsupialia	1	0
Insectivora	10	2
Chiroptera	13	Unassignable material
Edentata	1	1
Lagomorpha	4	4
Rodentia	34	21
Carnivora	17	19
Proboscidia	0	3
Perissodactyla	0	6
Artiodactyla	7	11
Total	91	67

The table shows a total of sixty-seven genera of Hemphillian mammals.

If the number of living genera of mammals is an indication of the number of fossil mammals to be expected to have lived in the same general area and in a similar diversity of environments, then the recovery of fossil mammals, Hemphillian age at least, is more complete than is usually assumed (about 85% in this example.) This, of course, means completeness of a faunal list, not quantitative completeness. Following the same comparative method used for the entire U. S. Hemphillian fauna a rough estimate of the same type may be made for local faunas. The data for the number of living genera is derived from descriptions of local faunas of various parts of the United States. These are usually lists of the mammals of some small county, a game reserve or research reserve. The number of genera averages about 30-35. The number of genera appearing in the two faunas which are used in this study are 22 and 28. This represents about the same degree of completeness as was evident when considering the entire United States fauna. Admittedly this is a rough estimate but if the method used in obtaining it is reasonable, then the results are of interest, much more so than the usual subjective remarks offered.

Another problem common to many quarries centers around the fact that quarries frequently contain representatives of more than one community. These community mixtures are so common that a special term has been used for them, the "Thanatocoenosis," suggesting that only in death are these organisms found together. In any approach to the paleoecology of any group of animals or plants there must be a separation of these communities.

Chaney (1924), working with leaves, has compared quarry accumulations at the Bridge Creek flora localities and those accumulating at present in a number of small depositional basins in Muir

Woods, California. The plant association at Muir Woods compares closely with that of the Bridge Creek flora. His results indicate that there is a high correlation between the number of plants locally present of a given species, and the number of leaves of that species present in the basin. A high incidence of leaves of a species indicates then, proximity of that species to the site of deposition. The relative abundance of leaves of various species in the accumulating deposits at Muir Woods is similar to those from the quarry at the Bridge Creek locality. Although the actual mechanics of transportation of leaves to the depositional site and their preservation are not the same as those for mammals, it is probable that there is a corresponding relationship between the materials preserved and the once-living associations. The leaves of some species of woody plants do not occur in deposits except in rare instances because of the structure of the leaf. This selection, due to structure, seems not to be evident in mammals.

It can be seen that even in quarry samples which by broad standards appear to represent well the fauna they are derived from there are a number of inherent misrepresentations. These may be listed as:

1. A quarry sample does not produce a complete faunal list of any of the communities represented.
2. Often more than one community is represented in a quarry.
3. Materials are perserved in the quarry in relation to the habits of the species, density of the species in the area and proximity of the the species habitat to the site of deposition.

The present method, in part, makes use of some of these "misrepresentations" to derive an interpretation of the quantitative data available. For instance, it was suggested above that some separation of communities may be possible by determining proximity of the habitat of a mammal to the site of deposition. It will be seen below that the mechanics for doing this are largely based on the apparent quantitative misrepresentations in the sample.

Method of Collecting the Sample

The method used in obtaining the quarry sample is the most important part of any quantitative analysis. The sample must be assumed to have been taken at random. The collection of material then must be made so that every specimen in the volume of sediment worked is retained. This includes every scrap, not just complete or nearly complete material. Collecting only the anatomically complete material from quarries of course makes results useless for an ecological quantitative analysis and certainly cannot be considered valid for quantitative morphological analysis. In fact, this practice has already destroyed the possibility for any type of quantitative analysis in many North American Tertiary faunas. The sample collected from McKay Reservoir, Oregon, the principal fauna used in this approach, was made by working small blocks of wet sediment by hand. The light-colored specimens were easily distinguished from the dark sediments which made possible a high percentage of recovery. The Coffee Ranch quarry sample (Univ. Calif. Coll.) of the Hemphill Fauna (Texas), used in the contrast of faunas below, was collected with care to recover all material present. The McKay Reservoir sample includes 751 specimens while the Hemphill sample includes 3,259 specimens.

Approach to Proximity and Density Problems

If a species of mammal is a member of a community in close proximity to the site of deposition, individuals may be expected to have died more often near the site of deposition and thus become a part of the accumulating deposit than those of a species which lived in a distant community. Species are excluded which, owing to their habits, are not commonly preserved (volant or arboreal forms). Mammals from distant communities appear in the deposit as materials introduced by the activities of carnivores, by stream action carrying them in, or by some other method of chance occurrence. They may be expected to be represented by fewer bones of their skeleton per individual present simply because they have less opportunity of being completely preserved. As a corollary, it is assumed that the completeness of representation indicates the proximity of the habitat of the species to the site of deposition.

When there are few or no articulated specimens in a sample it is not possible to determine exactly the number of individuals of a species represented. A useful indication can be found by determining the smallest number of individuals which could produce the material present. This *minimum number* is arrived at by examining all the material referable to each species, fragmental and complete, noting the element present most frequently. Thus fragments or complete specimens of five femurs of a species, three lefts and two right, demand three individuals as the minimum number represented. In some instances the most frequent element may not be the deciding one, for instance in the example above, if there had been four tibia, all lefts, it would have been necessary to allow four individuals as the minimum number. The element used may be any recognizable part of the skeleton, teeth, limb elements, or skull fragments.

Another quantity that can be useful from the data of each species in the deposit is the *number of specimens* which represent it in the sample. This should not be confused with the minimum number. The *number of specimens* is simply the number of recognizable bones and teeth which are present in the sample. For instance, in Table III it is indicated that the number of specimens of *Osteoborus* in the Hemphill sample is 362. However it is only necessary to propose 15 individuals (the minimum number) to account for these specimens. A ratio of the two numbers

$$\left(\frac{\textit{number of specimens}}{\textit{minimum number}}\right)$$

gives the number of specimens per individual and thus an indication of the relative completeness of preservation when applied to a number of species.

A difficulty in this approach lies in the variation of the number of skeletal elements to be found in different genera of mammals. For instance, a rhino may be expected to contribute more recognizable elements to a deposit than a horse (that is, later Tertiary horses) owing to the larger number of foot elements in the rhino. It may also be expected to produce more elements than a small rodent. The rodent has a similar number of elements, but rodent foot bones are difficult to assign. There is therefore a variation in the *number of specimens* due to the difference in the number of skeletal elements various mammals can contribute and also to the difference in the number of skeletal elements which the worker can readily assign. In order to eliminate such variation, the data should be corrected so that the *corrected number of specimens* represents the number which would be expected if all genera contributed the same number (the *standard number of elements*) of recognizable elements. This is done by setting up the following proportion:

TABLE II. *McKay Reservoir Quarry Sample*

Genus	Number of individuals	Relative abundance	Number of specimens	Estimated no. elements	Corr. no. specimens	Specimens per individual
Scapanus	1	1.1	2	23	5.0	5.0
Desman	2	2.3	7	23	17.3	8.7
Chiropterid	1	1.1	2	15	7.6	7.6
Ochotona	11	12.5	17	7	138.2	12.6
Hypolagus	11	12.5	203	45	257.1	23.4
Liodontia	1	1.1	1	41	1.4	1.4
Marmota	2	2.3	3	41	4.2	2.1
Citellus (O)	5	5.7	25	41	34.8	7.0
Citellus (C)	1	1.1	1	41	1.4	1.4
Pliosaccomys	2	2.3	4	23	9.9	4.9
Perognathus	1	1.1	1	23	2.5	2.5
Geomyid	1	1.1	1	23	2.5	2.5
Dipoides	14	15.9	219	41	304.4	21.6
Microtid	1	1.1	4	23	9.9	9.9
Pliozapus	2	2.3	2	23	4.9	2.5
Plesiogulo	1	1.1	2	75	1.5	1.5
Felis	1	1.1	5	75	3.8	3.8
Machairodus	1	1.1	5	75	3.8	3.8
Osteoborus	1	1.1	7	75	5.3	5.3
Canis	4	4.5	24	75	25.8	6.4
Martes	1	1.1	1	75	0.8	0.8
Hipparion	3	3.4	10	81	7.0	2.3
Neohipparion	1	1.1	1	81	0.7	0.7
Teleoceras	4	4.5	108	126	49.3	12.3
Paracamelus / *Procamelus*	3	3.4	16	57	16.0	5.3
Prosthennops	11	12.5	69	79	49.8	4.5
Proboscidian	1	1.1	1	49	1.1	1.1

elements for the pika (*Ochotona*) of the McKay Reservoir, Table II, appears to be a special case. Its *minimum number* is 11, the same as for *Hypolagus*. However, the number of specimens representing *Ochotona* is only 17 while that for *Hypolagus* is 203. Examination of the pika material shows only elements of the head to be present in the sample. A careful search of all the small mammal material reveals no elements of any other part of the skeleton to be present. It seems, at least in this case, that there is reason to believe that the rest of the skeleton was destroyed at the time of death. Perhaps the pika's predators, owls, canids or some other carnivore which takes in the whole body found the head of the pika obstructive or for some other reason disgorged it when the

$$\textit{Corrected number of specimens} = \frac{\textit{number of specimens} \times \textit{standard number of elements}}{\textit{estimated number of elements}}$$

The *estimated number of elements* for a genus is arrived at by determining how many elements the worker could distinguish for each genus involved if all its different elements were available. In the present approach ribs and vertebrae, which do not show much variation in numbers between groups of mammals, were not considered or counted.

The determination of the *estimated number of*

pika was killed. Whatever the cause it is too frequent an occurrence in this case to be considered as a matter of chance. For this reason the *estimated number of elements* of the pika is much lower than for any of the other mammals listed.

The *standard number of elements* is completely arbitrary since any number should give the same results. In this example fifty-seven was used because it was about median in the range of the *estimated numbers of elements* for the mammals in the sample. The use of a median number in the range of the estimated numbers of elements as a *standard number of elements* has the advantage of keeping the amount of correction to a minimum, some genera requiring no correction at all (those with an *estimated number of elements* of 57). The *estimated number of elements* will vary with different authors. This difference however will probably not present any appreciable source of error in the approach here presented.

An illustration may help to clarify the operation described above. *Hypolagus,* in the sample from the McKay Reservior fauna (Table II) is represented by 203 specimens (*number of specimens*). The *estimated number of elements* for the genus is determined as 45. Using 57 as the *standard number of elements* the

$$\textit{Corrected number of specimens} = \frac{203 \times 57}{45} = 257.1$$

The results of the determination of the *corrected number of specimens* for each of the mammals found in the samples from McKay Reservoir and the Hemphill are included in Tables II and III. Also included are the *numbers of specimens per individual* for each genus. This figure is the result of the division of the *corrected number of specimens* by the *minimum number* (number of individuals.) It is an expression of the completeness of preservation of each species in the sample. It

Table III. *Hemphill Fauna Quarry Sample*

Genus	Number of individuals	Relative abundance	Number of specimens	Estimated no. elements	Corr. no. specimens	Specimens per individual
Sloth	2	1.3	5	87	3.3	1.5
Hypolagus	3	2.0	24	45	30.4	10.1
Mylagaulus	3	2.0	16	29	31.5	10.5
Canis	1	0.65	10	75	7.6	7.6
Osteoborus	15	9.9	362	75	274.5	18.3
Plesiogulo	1	0.65	2	75	1.5	1.5
Vulpes	2	1.3	18	75	13.7	6.8
Agriotherium	1	0.65	11	75	8.4	8.4
Machairodus	3	2.0	108	75	82.1	27.3
Aphelops	9	5.9	374	126	169.0	18.8
Teleoceras	1	0.7	1	126	0.5	0.5
Nannippus	5	3.3	191	81	134.0	26.9
Neohipparion	3	2.0	126	81	88.7	29.4
P. Astrohippus	7	4.6	224	81	157.4	22.5
Pliohippus	63	41.5	1157	81	815.0	12.4
Prosthennops	5	3.2	65	79	46.9	9.4
Pediomeryx	3	2.0	40	57	40.0	20.0
Capromeryx	2	1.3	42	57	42.0	21.0
Pliauchenia	21	13.8	416	57	416.0	19.8
Paracamelus	3	2.0	37	57	37.0	12.3
Alticamelus	3	2.0	22	57	22.0	7.3
Proboscidian	2	1.3	8	49	9.3	4.7

should be remembered that relative completeness of preservation is assumed, in this approach, to be an indication of the proximity of the habitat of a mammal to the site of deposition.

The data in Tables II and III which are to be used in the remainder of the discussion are the *relative abundance* (based on *minimum number*) and the *number of specimens per individual.* These data are presented graphically in what are called here *faunal analysis diagrams.* (Figs. 2, 3.) These diagrams consist of a circle graph divided to represent the relative abundance in the sample of the mammal indicated. These divisions are blacked in with a radial bar graph which shows the number of *specimens per individual* for the same mammal. The members of the fauna are arranged counterclockwise on the diagram in the decreasing order of their *number of specimens per individual,* or in other words, as assumed in this approach, their order of proximity to the site of deposition.

In order to designate the community of genera which are not common to both faunas it is necessary to delimit the proximal community on the faunal analysis diagram. The point at which the average *number of specimens per individual* occurs on the diagram is used here as the limit of the proximal community. It is noted on the diagrams by a heavy concentric line which includes all of its members appearing in the sample. Earlier it was assumed that completeness of preservation, measured by the number of specimens per individual, is an indication of proximity of the habitat of a mammal to the site of deposition. Mammals must then have a higher-than-average *number of specimens per individual* to be considered as members of a proximal community. Those below average belong to intermediate and distant communities.

In faunas representing different environments those mammals in common to both faunas, which appear as members of the proximal community on the faunal analysis diagram of one of the faunas cannot in most cases be expected to appear in the proximal community of the other fauna. This characteristic presents a method of evaluating some of the assumptions upon which the interpretation of the faunal analysis diagram is based. If the interpretation of the diagrams of faunas of different environments based on the assumptions presented earlier bears out the necessary mutually exclusive nature of most of the members of the association of the proximal community then it may be said to form a reasonable working hypothesis for an approach as presented here. A test is made below using the McKay Reservoir, Oregon and

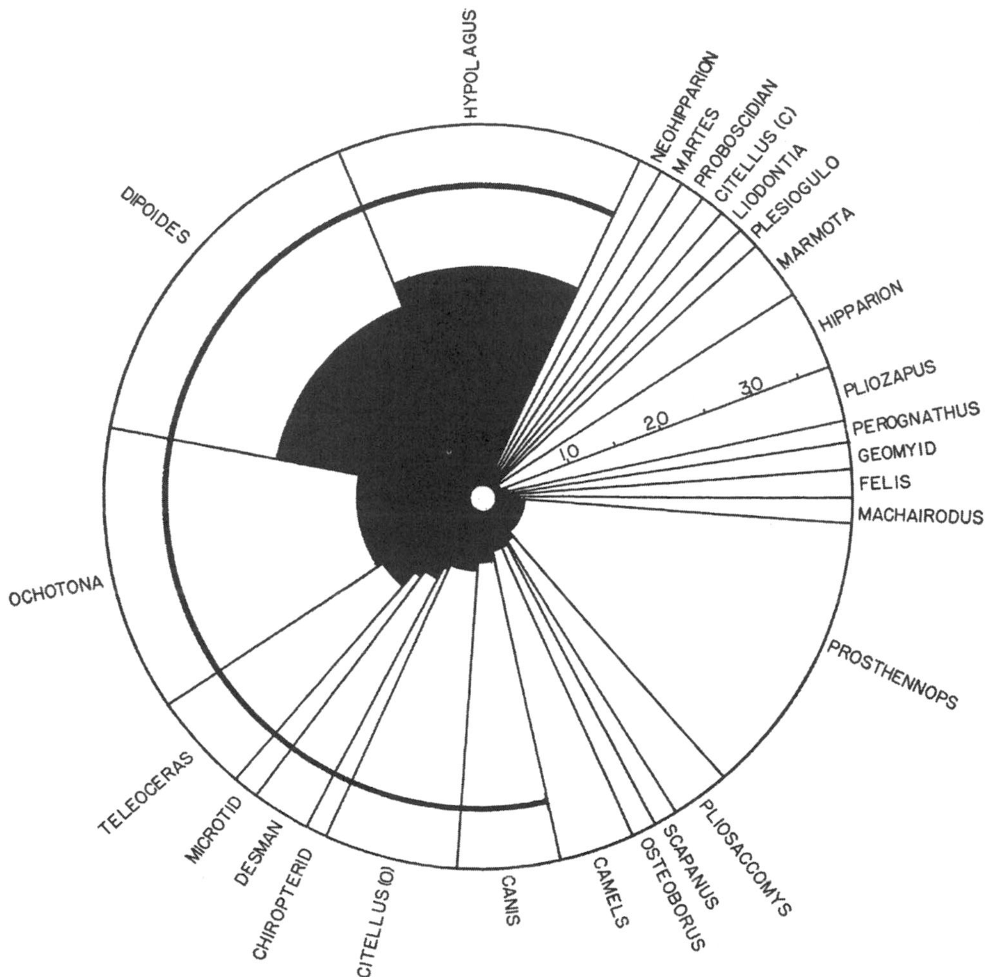

FIG. 2. Faunal Analysis Diagram of the quarry sample from McKay Reservoir, Oregon. The divisions of the circle graph represent relative abundance of the mammals indicated. The radial bar graph represents the number of specimens per individual. The heavy concentric line includes genera assigned to the proximal community.

Hemphill (Coffee Ranch) Texas quarry samples. These two faunas are contemporary by present standards of correlation. They both contain a number of short-lived genera (in geologic time). Although there are a number of mammals common to both faunas, as known from the quarry samples, the faunal lists of the two are otherwise quite different. Since these are essentially contemporary faunas the difference expressed is likely due to environmental differences at the two sites of deposition at the time these animals lived. Other evidence which might be obtained from plants in association with the mammals in the sediments is not available. Such associations occur only rarely. However, Chaney and Elias (1936) have shown that the Great Plains environment was well established by the time of the Hemphill Fauna. Chaney (1944) has shown also that at the same time the environments of the Pacific Coast were becoming more diverse. It seems probable then that the gross differences in the McKay Reservior and Hemphill (Coffee Ranch) faunas are due to their existence under different local environmental conditions.

Using the McKay Reservoir and Hemphill faunas illustrated in the diagrams (Figs. 2, 3) we note that nine genera of mammals are found in both faunas. Six of these appear on the diagrams as members of proximal communities. These are *Neohipparion* (horse), *Machairodus* (saber-

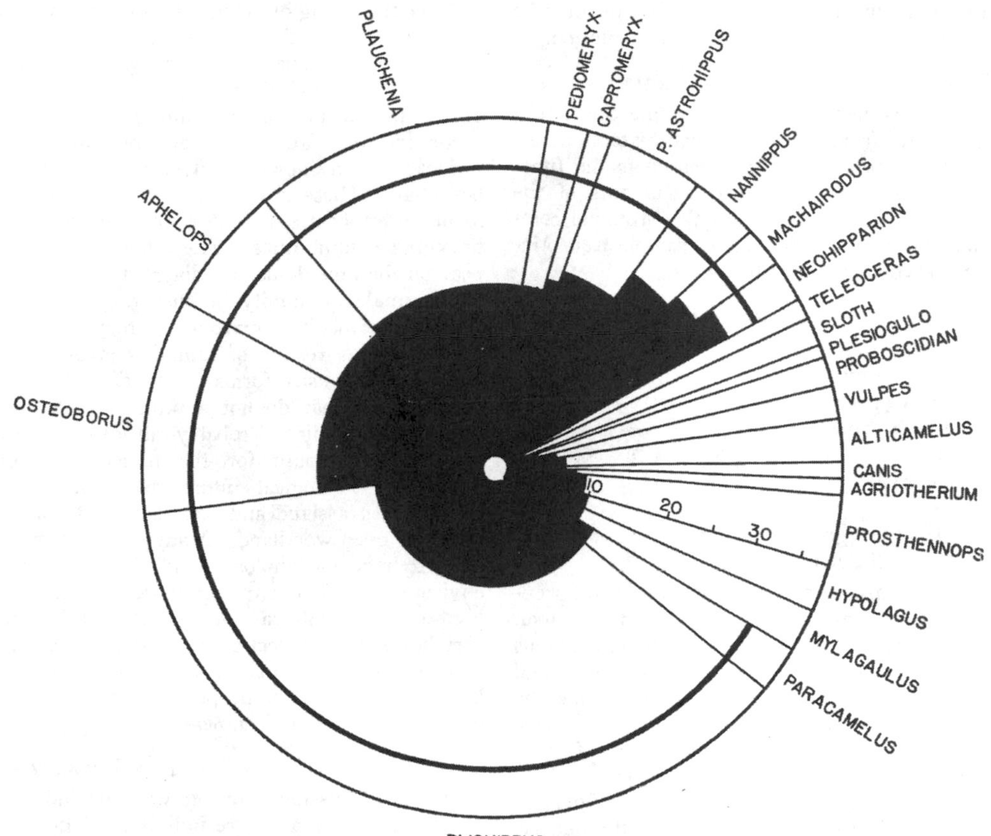

FIG. 3. Faunal Analysis Diagram of sample from the Coffee Ranch quarry, Hemphill local fauna, Hemphill County, Texas.

toothed cat), *Osteoborus* (hyaenoid dog), *Paracamelus* (camel), *Hypolagus* (rabbit) and *Teleoceras* (rhino). *Neohipparion, Machairodus, Osteoborus* and possibly *Paracamelus* are representatives of the proximal community in the Hemphill (Fig. 3) but these mammals represent distant communities in the McKay Reservoir fauna (Fig. 2.) *Hypolagus* and *Teleoceras* are members of a proximal community at McKay Reservoir (Fig. 2) but are in more distant communities at Hemphill. These are the results which are to be expected if the assumptions upon which the interpretation of the faunal analysis diagram is based are valid. Apparently they are, at least as far as this evaluation can test them.

The relative abundance of the various members of the proximal community as they appear on the faunal analysis diagram are probably indicative of their relative density in the once living fauna. These relative abundances are shown on the diagrams (Figs. 2, 3) by the size of the division of the circle. The data for mammals of distant communities, of course, have little relationship to their relative densities in their home community since they appear in the quarry sample as only chance occurrences. It is interesting to note the implications of the relative densities of various forms in the proximal communities. In the Hemphill (Fig. 3) for instance, *Machairodus* (saber-toothed cat) has a very low density, while *Osteoborus* (hyaenoid dog) on the other hand is a much more common form. Possibly *Machairodus* was a solitary animal or worked in pairs, whereas *Osteoborus* traveled in packs. The hipparion horses (*Nannippus* and *Neohipparion,* three-toed) and the smaller pliohippine horse (*P. Astrohippus,* single-toed) are of low density while the large pliohippine horse (*Pliohippus,* single-toed) is the most common form in the community. Apparently the smaller Pliocene horse did not occur in great herds here but the larger ones did. A realization of the relative densities of various mammals as expressed in the faunal analysis diagram may be helpful in studies of their taxonomic evolution.

Habitats of the Proximal Communities

When it is necessary to determine the nature of the habitat from the mammalian material alone, the best source of such information is the functional morphology of all the inhabitants of the community. The members of the proximal community appearing in the quarry sample from McKay Reservoir are listed below.

McKay Reservoir

Hypolagus (rabbit)
Dipoides (beaver)
Ochotona (pika)
Teleoceras (rhino)
microtid
desman mole
bat
Citellus (O) (ground squirrel)
Canis (small dog)

Among these are several mammals which are probably aquatic or amphibious. They are the desman mole, the beaver (*Dipoides*), and the amphibious rhino (*Teleoceras.*) Some type of large pond or small lake must be suggested to provide the habitat for these forms. The presence of ducks, crayfish and amphibians in the fauna further emphasizes the presence of a body of water. Presumably this body of water was the site of deposition of the materials which make up the quarry. The other members of the proximal community consist of rodent and lagomorph grazers. The habitat of the proximal community then appears to have been a pond-bank association.

The Hemphill proximal community is as follows:

Hemphill

Machairodus (saber-toothed cat)
Neohipparion (three-toed horse)
Nannippus (small three-toed horse)
P. Astrohippus (small single-toed horse)
Capromeryx (antelope)
Aphelops (rhino)
Osteoborus (hyaenoid dog)
Pliohippus (large single-toed horse)
Pediomeryx (large deer-like mammal)
Paracamelus (large camel)
Pliauchenia (small camel)

The Hemphill proximal community consists of a number of large grazing herbivores that are characterized by high crowned teeth and lengthened metapodials. The other members of the community are their large predators. This group of mammals represents a grassland habitat.

After the listing of the members of the two proximal communities there still remain eighteen mammals in the two quarry samples which do not appear as members in either proximal community on the faunal analysis diagrams. Three of these occur in both faunas. They are *Prosthennops* (peccary), *Plesiogulo* (wolverine) and the proboscidian. These three may represent an environment intermediate to those represented by the proximal communities. The others may not appear on the faunal analysis diagram as members of a proximal community because they also represent an intermediate environment, they may appear as they do as results of sampling error, or they may be low-density forms of one of the proximal communities that do not appear as such due to random sampling. Probably all three of these suggestions account for the fifteen unassigned mammals. A logical intermediate environment between a grassland and pond-bank association is brush or open woodland. Many of the forms considered have morphologies indicative of such an environment. These are omnivores, browsing herbivores and small carnivores. It will be necessary however to collect a quarry sample which has a proximal community habitat of an open woodland to determine the proper assignment of mammals to such an environment.

Distribution of Habitats Locally

In the discussion above it was concluded that three major habitats were indicated. These were pond-bank association, grassland, and an intermediate open woodland. At both quarries all three habitats were recognized as either a proximal, distant or intermediate community in relation to the site of deposition. These quarry samples from contemporary faunas can then be considered as ecofacies of the total fauna of the time. Figure 4 is an illustration of a possible distribution of the habitats and sites of deposition which would produce the results described.

Summary

The approach to the paleoecology of mammals presented here is based on the following assumptions:

1. Reasonably large collections from quarry accumulations are a random sample of what is present in the quarry providing all specimens are retained from the volume of sediment worked.
2. An indication of density of mammals of the proximal community may be obtained from use of the *minimum number*.
3. Mammals whose community in life was close to the site of deposition will be more com-

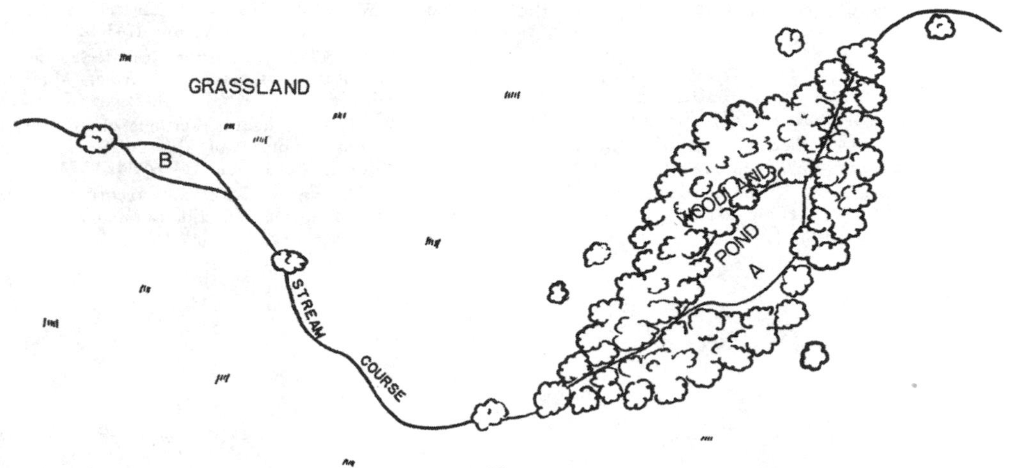

FIG. 4. Geographic relationship of community habitats. (a) Site of deposition of the ecofacies type which appears as the proximal community at McKay Reservoir. (b) Site of deposition of the ecofacies type which appears as the proximal community of Hemphill.

pletely represented than will those whose community was farther away.

4. If a community other than the proximal one is represented in a quarry sample, the habitat of that community must be present in the region contributing specimens to the quarry.

These assumptions form the basis for the interpretation of the quantitative data presented by the quarry sample. They are in part tested by applying the interpretation to the data of two faunas of different environments. The result demonstrates the necessary difference in membership of the proximal communities. It is felt therefore that the interpretation and the assumptions upon which it is based form a reasonable working hypothesis for studying the paleoecology of mammals. The advantage of this approach is that it not only indicates the habitat various fossil mammals lived in but determines their relative density in their community, lists other members of the community, separates the several communities represented in a quarry and suggests the geographic relationships of the habitats in which the communities are found.

Climatic conditions are not found by this approach except in so far as the limitations of the general habitat type demand. Mammalian morphology is of little help in the determination of temperature ranges, rainfall, topographic diversity and other pertinent facts necessary to the synthesis of a climate picture. Climate, however, is a regional characteristic, and consequently the climatic conditions implied by the occurrence of a certain habitat type may be confirmed if there are other occurrences of the same type and of equivalent age within the region. Terrestrial climatic conditions are best determined from the evidence of fossil plants. Such evidence only rarely occurs in association with mammals. Thus it is necessary to transpose information from fossil plant occurrences to those of fossil mammals. This demands strict stratigraphic control. A synthesis of data readily available from paleobotany with that derived from the application of the present approach can give a picture of fossil mammal environments, both organic and physical.

If this type of approach, an approximation at best, is applied to superimposed successions of faunas, data may be obtained to illustrate the dynamics of community change over long periods of time. Such information is rare at present. It can give the ecologist the historical background in community dynamics necessary to answer many of the questions in living biota.

The approach outlined and illustrated in this study is a preliminary suggestion. It will undoubtedly be improved by analysis of more quarry samples and constructive criticism. However, if this study encourages other workers to collect with similar approaches in mind so that they do not discard or destroy valuable fragmentary material then it will have served a good purpose. Vertebrate paleontology should present the historical background for the understanding of biogeography, evolution and ecology. It should not be the mere mechanical determination of phylogenies and stratigraphic relations, techniques which are tools of vertebrate paleontology, not the subject itself.

REFERENCES

Chaney, R. W. 1924. Quantitative studies of the Bridge Creek flora. Amer. Jour. Sci., (5) **8**: 127-144.

Chaney, R. W. 1944. Pliocene floras of California and Oregon. Carnegie Inst. Washington, Pub. 553.

Chaney, R. W., and M. K. Elias. 1936. Late Tertiary floras from the High Plains. Carnegie Inst. Washington, Pub. 476, + 72 pp.

Durham, J. W. 1950. Cenozoic marine climates of the Pacific Coast. Bull. G.S.A., **61**: 1243-1264.

Simpson, G. G. 1945. The principles of classification and a classification of mammals. Amer. Mus. Nat. Hist., Bull. 185.

Smith, J. P. 1919. Climatic relations of the Tertiary and Quaternary faunas of the California region. Proc. California Acad. Sci., (4) **60**: 123-173.

Stock, C. 1949. Rancho LaBrea, a record of Pleistocene life in California. 4th Ed. Los Angeles County Museum, Science Ser. 3, Paleontology 8.

PAPER 37

Live and Dead Molluscs in a Coastal Lagoon (1969)

J. E. Warme

Commentary

SUSAN KIDWELL

Are natural accumulations of skeletal remains—death assemblages—strongly biased by postmortem transportation, or do they provide reliable pictures of the original spatial distribution of living communities? Although John Warme's classic 1969 paper on Mugu Lagoon, California, was not the first live-dead molluscan analysis, its focus on postmortem transportation as a (potential) biasing force is very much of its time. However, transportation is only a framing device for one of the most thoughtful and well-written theoretical discussions of time-averaging in the paleontological literature to this day.

Warme's basic findings—that the magnitude of out-of-habitat transport is insignificant for most paleoecological purposes and that most live-dead differences simply reflect the time-averaging of local populations—have proven to be the rule rather than exception for mollusks in diverse settings. Warme argued that, even in the absence of between-habitat transportation of specimens, death assemblages are likely to include many "dead only" occurrences of species owing to (1) patchiness in the spacing of individuals within living populations, (2) the rarity of individuals of some species, and (3) natural shifts in habitats within the time frame of death-assemblage formation. Across the intertidal and shallow subtidal landscape of Mugu Lagoon, he found that, although birds and currents could be observed moving individual shells out of their life habitat, genuinely allochthonous specimens and species in fact constituted only a small fraction of any local death assemblage. At the level of key individual taxa (those with strong habitat preferences), whole communities (cluster analysis), and relative abundances within communities (agreement among top-ranked species), live-dead differences were best explained by the simple time-averaging of local skeletal production. These were encouraging words for paleoecological analysis.

Today, postmortem transportation remains the first concern of most newcomers to paleontology and death assemblages and, like most neuroses, can be hard to shake—there will always be extreme examples of long-distance and/or wholesale postmortem transport of shells to cite (e.g., table 9 in Kidwell and Bosence 1991). However, meta-analyses of dozens of molluscan live-dead studies conducted in intertidal to deep-shelf settings reveal that, for diverse ecological measures and analytic approaches, Warme's basic findings are the norm. Live-dead agreement is remarkably high and differences are attributable largely—and in many instances entirely—to the limited frame of the live data rather than to taphonomic bias per se (Kidwell and Bosence 1991; Kidwell 2001), with the strongest live-dead discrepancies limited to areas with strong recent anthropogenic change (the dead remember the pre-impact living community; Kidwell 2007). Reef-dwelling corals and mollusks and land mammals, although less studied, are revealing similar patterns and modeling is becoming a standard component of analysis (e.g., using spatially replicate samples of the living to simulate temporal variability; Edinger et al. 2001; Zuschin and Oliver 2003; Western and Behrensmeyer 2009; Terry 2010; Miller 2011; Tomašových and Kidwell 2011). Warme's 1969 paper nonetheless remains a touchstone of conceptual and analytic clarity, and one of the founding papers of paleoecology and taphonomy.

Literature Cited

Edinger, E. N., J. M. Pandolfi, and R. A. Kelley. 2001. Community structure of Quaternary coral reefs compared with recent life and death assemblages. *Paleobiology* 27:669–94.

Kidwell, S. M. 2001. Preservation of species abundance in marine death assemblages. *Science* 294:1091–94.

———. 2007. Discordance between living and death assemblages as evidence for anthropogenic ecological change. *PNAS* 104 (45): 17701–6.

Kidwell, S. M., and D. W. J. Bosence. 1991. Taphonomy and time-averaging of marine shelly faunas. In *Taphonomy, Releasing the Data Locked in the Fossil Record,* edited by P. A. Allison and D. E. G. Briggs, 115–209. New York: Plenum Press.

Miller, J. H. 2011. Ghosts of Yellowstone: multi-decadal histories of wildlife populations captured by bones on a modern landscape. *PLoS One* 6 (3): e18057. doi:10.1371/journal.pone.0018057

Terry, R. C. 2010. The dead do not lie: using skeletal remains for rapid assessment of historical small-mammal community baselines. *Proceedings of the Royal Society of London B* 277 (1685): 1193–1201.

Tomašových, A., and S. M. Kidwell. 2011. Accounting for the effects of biological variability and temporal autocorrelation in assessing the preservation of species abundance. *Paleobiology* 37:332–54.

Western, D., and A. K. Behrensmeyer. 2009. Bone assemblages track animal community structure over 40 years in an African savanna ecosystem. *Science* 324 (5930): 1061–64.

Zuschin, M., and P. G. Oliver. 2003. Fidelity of molluscan life and death assemblages on sublittoral hard substrata around granitic islands of the Seychelles. *Lethaia* 36:133–50.

JOURNAL OF PALEONTOLOGY, V. 43, NO. 1, P. 141–150, 2 TEXT-FIGS., JANUARY 1969

LIVE AND DEAD MOLLUSCS IN A COASTAL LAGOON

JOHN E. WARME
Rice University, Houston, Texas

ABSTRACT—Living animals and empty shells have been collected in 55 samples from Mugu Lagoon, coastal southern California. The abundances and distributions of 73 molluscan species from these samples were studied in order to evaluate post-mortem movement of shells. Although transportation is common over short distances within the lagoon, several lines of evidence indicate that most empty shells were buried about where they lived. Taxa collected alive are adequately reflected by the empty shells accumulating in the lagoon, whether live and dead are compared on the basis of individual taxa, whole communities, or relative abundances within communities. Post-mortem transportation within this environment is insignificant for most paleoecological purposes.

INTRODUCTION

A PERSISTENT problem is the difficulty of determining whether fossils were buried exactly where they lived, were moved only short distances, or were transported beyond their biotopes before final burial. It is widely held that most shells have been repeatedly disturbed or transported before incorporation into the geological record (Ager, 1963, p. 199–200; Hallam, 1965, p. 148; Craig, 1966, p. 131; Wilson, 1967, p. 369). In this paper the results of quantitative sampling of live and dead molluscs in a modern coastal lagoon are related to the question of post-mortem transportation of invertebrate skeletal remains.

Several criteria have been used to determine post-mortem transportation in fossils. Johnson (1962, p. 125–127) and Warme (1967a) list most of these criteria and comment on some of their inadequacies. In many cases it is impractical to recover large numbers of fossils, properly oriented, and inspect them for signs of transportation. This is especially true if specimens are poorly preserved or occur in highly indurated or fractured rocks. Yet, large numbers of fossils are usually needed to study the dynamics of fossil populations or to reconstruct validly fossil communities. It is important to know how much and what kind of post-mortem transportation can be expected in natural depositional environments.

One approach to this problem has been to collect samples from modern environments and to compare the distributions of animals that occur alive with those represented only by skeletal remains. Few investigations of this kind have been published. Miyadi & Habe (1947, p. 110) evaluated the live and dead species in samples from several Japanese bays; they state that transportation of molluscan shells was not important, and that most of the species were "autochthonous." Valentine (1961, p. 324) compared live and dead molluscs from Santa Monica Bay, California, and judged that "all or nearly all of the dead shells represent animals that formerly lived about where they were found." Johnson (1964, p. 121) analyzed 72 samples from Tomales Bay, northern California, and concluded that dead pelecypod assemblages from many of these samples did not reflect well the live taxa collected with them. Upon reevaluation of these and additional samples, Johnson (1965, p. 85) regarded most of the dead shells as having been buried about where they lived. He also pointed out many of the difficulties of interpreting such samples.

SAMPLES FROM MUGU LAGOON

Fifty-five samples have been collected from Mugu Lagoon, near Point Mugu, southern California (long. 119°05′ W.; lat. 34°06′ N.; text-fig. 1). Mugu Lagoon lies in a region of Mediterranean-type climate and fringes the eastern end of the Oxnard Plain, a subaerial deltaic complex. The lagoon is about 6 km long, and extends up to 2 km in width behind well-defined barrier spits. It is shallow, being no more than 3 meters deep at low tide except where artificially dredged. Hydrography is tidally controlled, and the maximum tidal range within the lagoon is about 1.4 meters. No perennial streams enter the lagoon. Salinity is usually that of the open coast (approx. 34‰) except after infrequent local rainstorms.

The faunal samples (UCLA fossil loc. nos. 5329–5384) were collected from several environments within Mugu Lagoon (tidal flats, tidal channels, eelgrass ponds, etc.) each containing somewhat different substrate types, hydrographic regimes, floral cover, and animal assemblages. All samples are from the eastern arm of the lagoon. This region is in a relatively natural condition; it is under the supervision of the U.S. Navy, and public entry ordinarily is prohibited.

Each sample was isolated in situ by using a

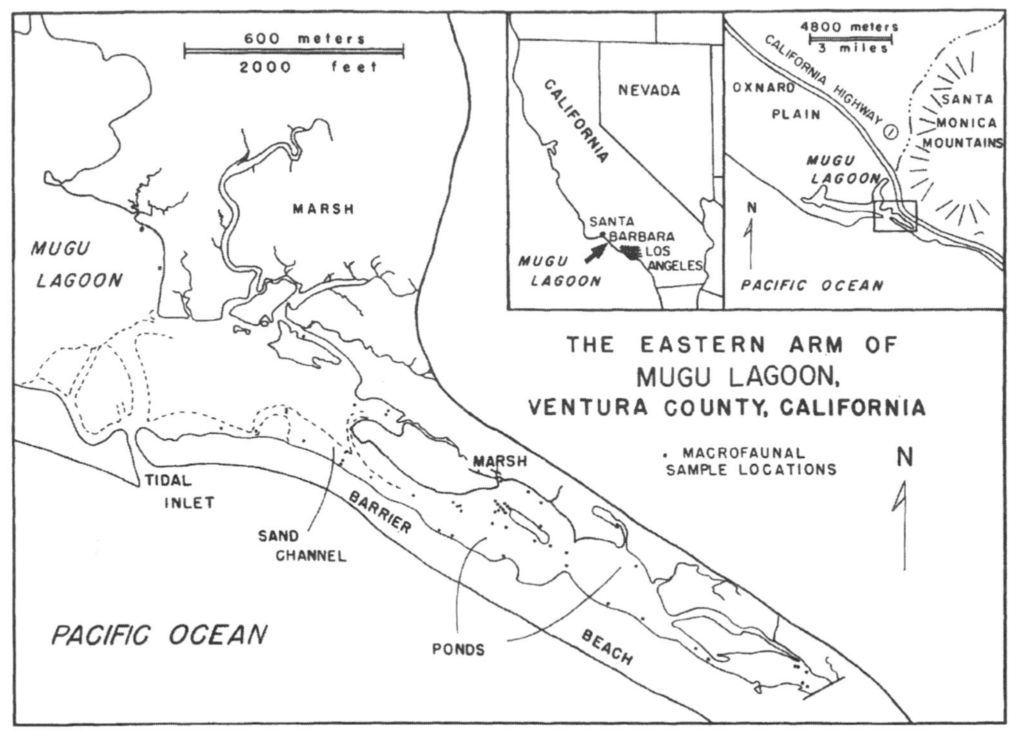

TEXT-FIG. *1*—Geographic location and major topographic features in the region of the eastern arm of Mugu Lagoon, California. Macrofaunal sample locations are indicated by dots.

special sampling cylinder of ¼square meter cross-sectional area and 75 cm height. It was then collected by shovel or with an underwater vacuum modified after apparatus described by Brett (1964). Depth of sampling ranged from 0.3 to 0.8 meters below the water-sediment interface. Sampling was continued until it was judged that all live macroscopic animals had been collected.

Specimens from each sample were sorted, identified and counted. Several phyla were collected, but only shelled molluscs (gastropods and pelecypods) are considered here. The samples contained 33 species of gastropods, represented by 511 live and 10,313 dead individuals, and 40 species of bivalves represented by 3,237 live and 15,856 dead individuals (two valves of the same species being counted as one dead individual). There were from 0 to 13 live molluscan species represented per sample, and from 7 to 40 dead species per sample. Abundances of individuals ranged from 0 to 262 live and from 13 to 1,627 dead per sample. For details of sampling apparatus, procedure, and taxa collected see Warme (1966, Ph.D. Thesis, Dept. Geology, Univ. California, Los Angeles).

PRELIMINARY CONSIDERATIONS

Some basic ecological concepts should be considered before evaluation is made of the correspondence between the live and dead animals in samples from modern environments. Several important factors make it unlikely that any sample will contain a complete species-for-species correspondence between its live and dead components. This would be true even if no post-mortem transportation had taken place. Three of these factors are: (1) crowding or patchiness in the spacing of individuals within live populations, (2) rarity of individuals belonging to some species within natural communities, and (3) shifting of habitats within natural depositional environments. The first two of these factors are almost universal characteristics of live populations and communities. The third factor commonly influences the composition of dead assemblages.

Irregular spacing of live individuals.—Individuals belonging to most living species populations are neither regularly nor randomly spaced, even under what might be considered "ideal" or uniform environmental conditions. Instead, individuals are distributed in patches or clumps.

This spacing is variable in scale and plan, varies among species and among localities, and has many causes. Ecologists have generally found it difficult to estimate average densities and sizes of populations because of this irregular spacing. Census efficiency depends on the size and nature of the patches, and the size, number, and distribution of the samples (see discussions in Hairston, 1959; Macfadyen, 1963, p. 101–106; MacArthur & Connell, 1966, p. 41–51; Lloyd, 1967). Regardless of the geometry of spacing, it is unlikely that any sample of reasonable size will include all indigenous species living in a given habitat.

If the location of patches of a species does not reflect some permanent difference in microhabitats, as is generally accepted, but instead reflects chance processes and perhaps historical accidents in the depositional environment, then the patches can be expected to shift throughout the habitat with time. This may occur from one generation to the next, with the change in seasons, or in a more random fashion sometimes attributable to food-searching or breeding activities. MacGinitie (1939, p. 43) described the migration of large patches of the echinoid *Lytechinus anamesus* in Newport Bay, California:

> Sometimes it requires a week of dredging to find a colony of these urchins, but when one does locate them a short dredge haul will yield several bucketfuls. In returning for a new supply within a week, one can be fairly sure of finding them at approximately the same place; but if a month elapses several dredge hauls will be necessary before they can be relocated, and they may have migrated in any direction. So far, however, there is evidence that they are found only from one quarter to one mile off shore.

Numerous similar cases have been documented, especially from investigations of commercially valuable shellfish populations.

Population movement of this kind results in a widespread distribution of skeletal remains as individuals within migrating patches are subject to normal mortality. This distribution will be more uniform than the distribution of the live animals at any given moment. Population migration may help explain the wider distributions of dead shells compared to their living representatives, as described by Miyadi & Habe (1947, p. 110) and Buchanan (1958), and as was true in parts of Mugu Lagoon.

Rarity of some species within communities.—Natural habitats commonly are occupied by communities in which a few species are represented by a large number of individuals, and many more species by smaller numbers of individuals. The relative abundances of different species within a community usually approaches a logarithmic series (Williams, 1964), a lognormal distribution (Hairston, 1959, p. 405–406), or a similar, generally logarithmic pattern (Odum, Cantlon, & Kornicker, 1960; MacArthur, 1965). The reasons for these patterns in relative abundances are not yet clear. However, almost all communities that have been studied exhibit a few abundant species and many rare ones.

TABLE 1—Bivalves, gastropods and echinoids collected in 9 samples from the main sand channel, Mugu Lagoon. Indicated are the number of live individuals (column A) and the number of dead individuals (column B), and the number of samples from which they came (in parentheses). The species list was derived from cluster analysis based on the live fauna. (Two valves were regarded as one bivalve; half numbers were rounded to the next higher whole number.)

Species	A Live	B Dead
Sanguinolaria nuttallii	676 (9)	454 (7)
Cryptomya californica	203 (6)	294 (8)
***Dendraster excentricus*	42 (7)	54 (9)
Diplodonta orbella	15 (5)	5 (4)
**Olivella biplicata*	3 (1)	16 (8)
Chione californiensis	2 (2)	6 (4)
Spisula dolabriformis	1 (1)	2 (1)
**Nassarius fossatus*	1 (1)	1 (1)
**Lunatia lewisi*	1 (1)	1 (1)
**Polinices reclusianus*	1 (1)	1 (1)

*gastropod
**echinoid

Species collected in the samples from Mugu Lagoon have been assembled into recurring groups or "communities" using a computerized cluster analysis. The most clearly-defined community in the lagoon is confined to clean sand (>95% sand) substrates. This community is present in channels near the tidal inlet or on washover fans behind the barrier beach. Live species of the sand channel community are listed in table 1, with the total number of live specimens collected and the number of samples in which each species occurred (column A).

It is unrealistic to expect that the rarer species of table 1 be collected alive in every sample from their natural habitat, even though they represent normal components of the assemblage living in that habitat. Underwater observations reveal that the four gastropod species of table 1 are common members of the sand channel community. They are relatively uncommon, however, compared to the numerically more "dominant" species collected there. These gastropods belong to higher trophic levels than the more abundant detritus-eaters. Food-web relationships suggest that they should be less abundant (Elton, 1927, p. 55–70; Lindeman, 1942).

The clumping of live populations and the rarity of some species in any natural assemblage are real obstacles to sampling all live species in a given habitat. Only a sample that is very large relative to the size of the habitat would be expected to contain all resident species. Furthermore, it may be increasingly difficult to include patches of species as they become rarer; an inverse relationship has been demonstrated between patchiness and abundance in some environments (Hairston, 1959, p. 415).

Shifting habitats.—Additional problems arise when comparing the live and dead fauna in modern samples because the dead assemblages usually have accumulated over a long period of time. The history of faunal change at any sample site is represented in the dead assemblage. This will ordinarily increase the number of both dead species and dead individuals relative to the live ones.

Faunal responses to change in ecological factors in shallow marine waters have been described in numerous cases. Samples of the benthos commonly contain empty shells that lived under former conditions at the sample site, in addition to live and dead animals from the habitat being sampled. For this reason there can be a substantial discrepancy between taxa present in the live and dead assemblages even though no post-mortem transportation occurred.

Sampling in Mugu Lagoon has revealed a habitat shift that affected a whole assemblage and that took place probably within the last 100 years. Low water-current velocities and a muddy substrate now characterize the extreme eastern end of the lagoon. The live fauna recovered in samples from this part of the lagoon is composed of only a few mud-tolerant species (*Cerithidea californica, Tagelus californianus, Protothaca staminea*) occurring in low densities. However, the dead assemblage from these samples also contains a number of species that now live only in the clean sand substrates of the lagoon (e.g. *Sanguinolaria nuttallii, Cryptomya californica, Dendraster excentricus*). These species numerically dominate the community in that habitat (table 1). Part of the dead assemblage thus represents former conditions in the eastern end of the lagoon, when the tidal inlet was closer than at present (U.S.C.G.S. Topo. Surv. No. T893, 1857; Inman, 1950, p. 37) and when the substrate was clean sand. A radiocarbon date was obtained on shells recovered at a depth of 75 cm beneath the tidal flat in the eastern arm of the lagoon. The shells dated represent a species indicative of freely-circulating lagoonal waters (Warme, 1967b, p. 545), and were too young to date accurately, being probably less than 100 years old (Rainer Berger, 1964, personal communication). This suggests a definite shift of habitat in the eastern arm of the lagoon within the last century. A muddy veneer has accumulated since the inlet migrated westward. Samples collected in the eastern extremity of the lagoon represent two or more habitats, recorded by changes with depth in both substrate type and skeletal remains.

With time there should be a general enrichment of species in the accumulating dead assemblage relative to the live assemblage in any environment. It is not expected that every species collected dead in a sample would be matched by live representatives. "Dead only" occurrences cannot, without further evidence, be considered solely as transported or "exotic" shells. However, a species represented by live individuals in a sample from a stable environment should normally also be represented by some dead individuals if there is good distributional correspondence between live and dead.

COMPARISON OF LIVE AND DEAD MOLLUSCS

The following comparisons have been made in order to evaluate post-mortem transportation in Mugu Lagoon: (1) species-by-species comparison of live and dead molluscs in each sample, (2) comparison of communities constructed by numerical analysis and based on the separate live and dead assemblages in all of the samples, (3) comparison of the relative abundances of species occuring dead in one clearly-defined community, and (4) comparison of the live and dead distributions of members of species that are environmentally restricted within the lagoon during life.

Species-by-species comparisons.—Each species collected in each sample falls into one of four categories: it may occur as (1) dead only, (2) as both dead and live, (3) as live only, or (4) it may be absent. In a matrix of 4,015 possible occurrences (the 55 samples from Mugu Lagoon times the 73 molluscan species collected in the lagoon), there were 1,383 actual occurrences of a species in a sample (as live or dead or both; see table 2). Of these, 1,043 (75.4%) are occurrences in which only empty shell(s) of a species were present (table 2, Group I). By themselves, these can be regarded as evidence neither for nor against significant post-mortem transportation, because of the factors of spacing, rarity and habitat shifts discussed above.

Group II (table 2) includes 297 (21.5%) occurrences in which both live and dead individuals of a species were associated in a sample. This group obviously indicates good live-dead correspondence.

TABLE 2—Occurrences of 73 species of molluscs in 55 samples (¼ sq. m.) from Mugu Lagoon, and the numbers of individuals in each category. An *occurrence* is defined as the presence of any one species in any one sample.

Type of occurrence	Number of occurrences	Number of specimens in occurrences	
Group I (dead only)	1043 (75.4%)	14,772 (49.4%)	
Group II (live and dead)	297 (21.5%)	live with dead 3,687 (12.3%) dead with live 11,397 (38.1%)	15,084 (50.4%)
Group III (live only)	43 (3.1%)	61 (0.2%)	
Totals	1383 (100%)	29,917 (100%)	

Group III consists of 43 (3.1%) occurrences of individuals whose species were collected live in a sample without dead shell remains. This group is significant because its members lack the live-dead correspondence of group II forms. Group III may be represented by populations whose shells are transported away from their habitat as they die, or even by individuals that have been moved from their normal habitat while still alive. There are, however, other explanations.

"Live only" occurrences may be due to simple expansions of existing populations by larval settlement or postlarval colonization of newly-deposited sediment, or by colonization of a new area. In these cases a dead assemblage has not yet developed exactly where the samples are collected. Of the 43 occurrences in Group III, 22 include only 4 taxa (39 specimens); these may be expanding populations in Mugu Lagoon. Fluctuations in population size, whether cyclic or erratic, are common in natural environments (e.g., Elton, 1927, p. 127; Moore, 1933; Slobodkin, 1962, p. 82). Newly deposited sediment is quickly colonized in most parts of Mugu Lagoon. Seventeen of these 43 occurrences are from samples collected in newly-deposited channel sand.

When the total number of individuals appearing in Group III (table 2) is considered (rather than occurrences of species on a presence-or-absence basis as above), the animals live and belonging to species not represented in the dead assemblage of their respective samples comprise less than 2 percent of the total number of live individuals from all of the samples. The 43 occurrences of Group III involve only 61 of the 29,917 molluscs collected. However, species occurring both live and dead (Group II) in a given sample are represented by many individuals (15,084 individuals or 50.4%). The remaining 14,772 specimens (49.4%) are "dead only" occurrences; as indicated above, interpretation of these cases is equivocal when no other criteria are being employed.

These data are regarded as indicating good correspondence between the place of life and the place of burial of molluscan populations in Mugu Lagoon. Taken alone, however, they could be interpreted differently. If the dead shells of all species were transported and deposited throughout the lagoon, almost every species collected live should also be represented by dead shells in the same sample. Such a condition is unlikely as indicated by the fact that only about one third of the possible occurrences were realized (1,383 out of 4,015). An analysis of complete assemblages argues further against such an interpretation.

Assemblage comparisons. — "Communities" have been constructed from the samples collected in Mugu Lagoon using a weighted pair-group method of cluster analysis. The analysis was performed first on the basis of the live and then on the basis of the dead assemblages recovered. This numerical technique is designed to cluster in both the Q- and R-modes. The Q-mode program clusters samples on the basis of their faunal lists; the R-mode program clusters species on the basis of their joint occurrences with other species. Jaccard's coefficient of mutual occurrences (Fager, 1963, p. 420) has been used. For details of cluster analysis see Sokal & Sneath (1963) and Valentine & Peddicord (1967). One R-mode cluster ("community") is presented in table 1. The composition, distribution, and ecology of the communities delimited in this fashion will be treated elsewhere.

The clustering of samples, based on their species lists (Q-mode), is shown in the dendrograms of text-figure 2. Dendrogram A represents the clusters of samples developed on the

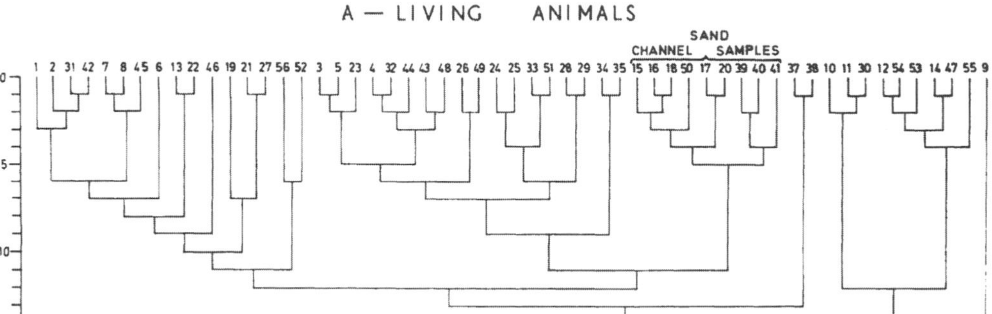

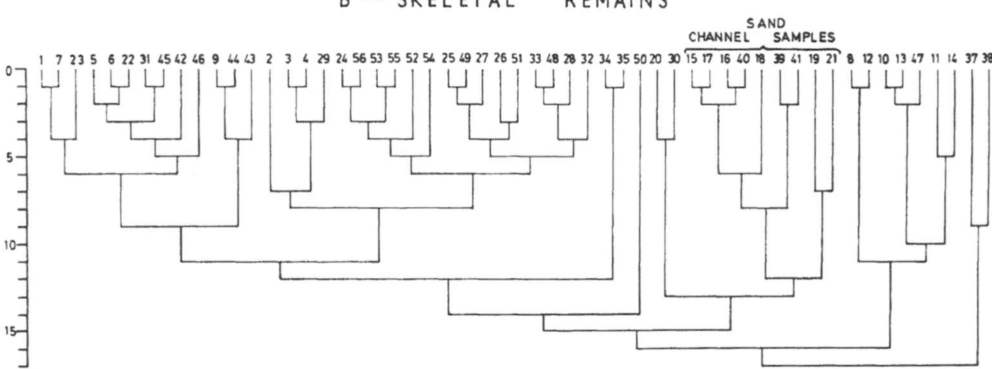

TEXT-FIG. 2—Dendrogram constructed by cluster analyses. Numbers across the top represent individual samples; the vertical scale represents the number of times that the matrix was clustered and then recalculated, becoming less significant downward. Dendrogram *A* depicts the affinities and sequence of clustering of samples based on species lists of live shelled invertebrates which they contained. Dendrogram *B* was constructed similarly from the same samples but based on empty shells.

basis of the live taxa collected. Dendrogram B represents clusters of samples similarly developed but on the basis of the dead taxa. The affinity of the samples and the clusters of samples become less significant downward.

Comparison of dendrograms in text-figure 2 reveals that clusters based on the dead species are generally less compact than those based on the live species. This is because of the greater diversity of dead species in every sample collected.

Entire assemblages of live and dead species can be directly compared in one subenvironment in Mugu Lagoon, without the possibility of the number of dead species having been increased by a history of habitat change at the sample site. The sand channel immediately inside the tidal inlet is underlain by more than one meter of clean sand in most places. Only one of the samples, taken at the end of the channel, penetrated the older lagoonal mud that is known to lie beneath parts of the sand. With the exception of shells that may have washed into the channel from other environments, both the live and dead assemblages in samples from within this channel represent species from only one habitat. These samples all cluster together on the basis of both their live and dead assemblages ("sand channel," text-fig. 2). Samples 20 and 50 are also included with the sand channel cluster based on the live fauna. These samples were collected on washover fans behind the barrier beach, and represent the same surface substrate (clean sand). Samples 19 and 21 are included with the analogous "sand channel" cluster based on dead occurrences; they contain shell remains from the complete clean sand fauna as well as the underlying muddy substrate fauna, and are thus separated from the main cluster, joining at a lower level of significance.

Samples from other identifiable habitats in the lagoon (eelgrass, tidal flat, shifting tidal delta, etc.) also tend to form clusters based on the live assemblages that match those based on the dead assemblages, though not as clearly as in the case of the sand channel. These similar clusters developed regardless of the fact that cluster analysis as performed was based on species presence-or-absence. The introduction of one shell representing a species displaced from another environment counts as heavily as many shells of that species. A small amount of post-mortem mixing can drastically change the composition of the clusters.

Distinctive pairs of samples (34–35, 37–38) are matched in the dendrograms of text-figure 2. Furthermore, the two largest clusters in dendrogram A (samples 1 to 52 and 3 to 35 from left to right) are similar to the equally large clusters in dendrogram B (samples 1 to 43 and 2 to 32), having 57 percent and 75 percent samples in common, respectively. This is the case even though it is known that some dead assemblages represent more than one community, and it may be explained by the sequence of sediments filling the lagoon. Depositional environments migrate in a predictable manner as sedimentation proceeds. All of the samples collected in a given environment tend to reflect similar depositional histories and to contain similar sequences of dead assemblages.

Relative abundances.—Listed in order of relative abundances, species collected dead rank very close to those collected live. The five most abundant pelecypod species collected live are among the six most abundant dead species. The four most abundant species of gastropods taken live are among the six most abundant dead species. Some taxa were abundant as dead yet were never observed or collected live; apparently they were not living in the lagoon during the time it was sampled.

The ranked relative abundances of species comprising the sand channel fauna are listed in table 1. With one exception, the rank for each species is the same whether based on their live or dead numbers. On the basis of these samples, the dead assemblages of shelled molluscs accumulating in this channel accurately portray the relative abundances of the species living there.

The low number of dead shells compared to the live animals listed in table 1 is probably a reflection of the high rate of sedimentation measured in this channel. Although ratios of dead to live fluctuate greatly between samples, in total numbers the dead outnumber the live by about 10 to 1.

Spatial relationships.—Another method of detecting the post-mortem transportation is to determine the distribution of empty shells in environments where they could not have lived. *Macoma nasuta* is ideal for this purpose because it is the most abundant pelecypod present in the 55 samples from the lagoon, yet it is restricted to muddy substrates. It is represented by 884 live specimens and 9,297 separate empty valves. No live specimens are present in collections from the sand channel near the tidal inlet. Empty shells of *M. nasuta,* however, were found in the channel. In two samples collected on the tidal delta, immediately inside the inlet (text-fig. 1), valves of this species were absent. They were present in samples collected progressively eastward in the sand channel in numbers of 3, 3, 6, 2, 3, 11, 22, and 38. In the last sample, at the eastern end of the channel, 10 specimens were also found live, buried in mud beneath newly-deposited channel sand. Of the 9,297 empty valves of this species, only 50 (0.54%) occurred in the 9 samples (16.4% of total samples) from the sand channel. Many of these were identified from hinge fragments or were juveniles. In contrast, up to 970 empty *M. nasuta* valves have been counted in one sample and several hundred valves in other samples in which this species was also collected live (live densities ranged from 1 to 104 per ¼ square meter). In samples containing both live and dead *M. nasuta,* most empty valves are complete, and matching pairs of valves are common.

The distribution of *Leptopecten latiauritus* provides another example. This pelecypod commonly lives attached to eelgrass by byssal threads. It is most abundant in eelgrass at the end of the sand channel where circulation of tidal waters is good. Of 1,869 empty *L. latiauritus* valves present in all samples, only 10 valves were present in samples from the channel.

DISCUSSION

Although it is not difficult to observe shells that are undergoing transportation in modern environments, even in regions of "low energy," it is difficult to ascertain quantitatively the importance of such transportation.

Most observations regarding shell movement have been confined to beaches, tidal flats, or similar areas near the strand line, a vertical zone where physical energy is concentrated by tidal and other currents, waves, and wind. Shell movement is expected in these places. As an extreme example, Kristensen (1957) stated that over a period of several years almost all dead cockles (*Cardium edule*) were removed from some carefully studied tidal flats in the Wadden

Sea, yet a thriving population of live animals remained. Wilson (1967, p. 369) has shown that live and dead populations differed in several important respects on tidal flats in a Scottish estuary.

There is undoubtedly some mixing of shells in Mugu Lagoon. Shells indigenous to the lagoon are swept out of the tidal inlet and deposited on the beach. Others are scattered across the surface of tidal flats and salt marsh by birds, tidal waters, and even the wind. Many shells float after drying at low tide.

Shells of species that live on the open beach are swept into Mugu Lagoon by waves washing across the barrier spit. Abraded and broken fragments of the large *Tivela stultorum* and complete valves of the smaller *Donax gouldii* are abundant in the sands of the main channel and of washover fans along the barrier side of the lagoon. A mixing of these open-beach species with more quiet water forms is also reported in collections from the California Pleistocene, and could have been accomplished in the same manner (Valentine & Mallory, 1965, p. 696). Nevertheless, analyses of the live and dead animals collected in samples from Mugu Lagoon indicate that mixing of shells between habitats is not typical within this environment.

Physical energy and lagoon transportation.—The sand channel leading into the eastern arm of Mugu Lagoon has been used to illustrate several points for two reasons: (1) it contains the most distinctive and easily identified community in the subtidal portion of the lagoon, and (2) it is floored by a thick and uniform substrate of clean channel sand, largely discounting the possibility of more than one habitat being represented in the dead assemblage of samples collected there.

This channel contains the coarsest sediments and greatest tidal-current velocities of any environment in Mugu Lagoon. It is almost constant in width and funnels water to and from the entire eastern arm of the lagoon with changing of the tides. Current velocities of two knots have been measured during neap tides; spring tidal-current velocities are five knots and more. From the various analyses presented above, however, it is evident that the assemblage of invertebrates with hard parts that inhabits this channel is faithfully reflected in the sediments accumulating there.

It is difficult to show any simple relationship between water-current velocities and shell transportation in the natural environments of Mugu Lagoon. The experiments of several investigators indicate that much has yet to be learned about the interaction of shell size and shape, current velocity, and substrate characteristics as they relate to the problem of shell transportation and burial (Lever, 1958; Lever et al., 1961; Kelling & Williams, 1967). For instance, owing to current eddies and subsequent scouring of loose substrate around shells, increased current velocities may result in shell burial rather than transportation (Menard & Boucot, 1951; Johnson, 1957).

Time represented in Recent samples.—Every sample from Mugu Lagoon contained more dead than live taxa. This is explained by the fact that most dead assemblages represented more time than live assemblages. The effects of irregular spacing of live individuals, rarity of some species, and the many arbitrary events that can influence the composition of the biota at any one moment, all tend to become blurred and to have less significance the longer that the environment is occupied. Organisms capable of fossilization and inhabiting a given depositional environment should become increasingly better represented with time as their remains or traces accumulate in that environment. Samples from modern environments are expected to contain an increasing number of dead indigenous taxa as all forms collect throughout the environment. In addition, the longer the habitat is occupied, the greater will be the proportion of dead over live individuals of all taxa.

SUMMARY

Almost 30,000 live and dead molluscan specimens are present in 55 quantitative samples from Mugu Lagoon. The distribution of live and dead shells has been analyzed in order to evaluate post-mortem transportation.

In a comparison of live and dead taxa in any sample, the live individuals obviously belong where they were collected. The majority of the dead individuals recovered from samples that also contain live representatives of their species probably also lived, died, and were buried about where they were collected. Those taxa represented only by dead specimens, however, may represent: (1) individuals whose species live nearby in the same habitat, but are clumped in their distribution or rare in that habitat, and simply missed in the sample, (2) individuals that were buried where they lived, but whose species can no longer live at the sample site owing to habitat changes, or (3) individuals displaced from other habitats and transported to the site after death.

Half of the individual molluscan specimens collected in Mugu Lagoon, whether dead or live, are from samples in which both live and dead members of their species are present. Almost all other specimens are from samples in which their species occurred only as empty shells. Although the presence of a species as "dead only" in a sample does not necessarily indicate post-mortem transportation, neither does it discount significant mixing of skeletal remains.

The distributions of recurrent assemblages ("communities") indicate that widespread mixing has not occurred. Cluster analysis has been used to compare the similarity of samples based on their entire faunal lists. This computerized numerical technique was applied to the live species collected and separately to the dead species collected in all of the samples. The analyses resulted in clusters that were somewhat similar to nearly identical. As an example, all seven samples collected from the main sand channel in Mugu Lagoon formed a significant cluster, whether based on the live or on the dead components in the samples. It is very unlikely that this should happen unless (1) transportation of dead shells into the channel was negligible, or (2) transportation was systematic, effecting all samples equally. Although an example of the latter circumstance has been documented, all other lines of evidence strongly suggest that shells in Mugu Lagoon are not transported away from their proper habitats after death.

The sand channel is a "high energy" environment compared to other parts of the lagoon, yet the dead shell assemblages recovered there adequately reflect the community inhabiting the channel. Many environments within the lagoon have more restricted circulation; it might be expected that shell transportation is relatively minor in these places. Nevertheless, other investigations in more open environments (Miyadi & Habe, 1947; Valentine, 1961; Johnson, 1965) have also yielded the conclusion that post-mortem transportation may not be important for many paleontological purposes.

This evaluation of post-mortem shell transportation does not consider local shifting and minor movements of shells within their habitats; such movement certainly occurs owing to both physical and biological processes. However, the good correspondence between place of life and place of burial of the shelled invertebrate populations studied should be encouraging to those attempting to investigate paleo-population dynamics or to reconstruct natural assemblages of fossils from similar depositional environments in the geologic record. Other environments should be investigated. Detailed information is lacking particularly on the subtidal regions of shelf and inland seas, depositional environments which are probably represented in the bulk of the presently exposed fossiliferous marine rocks.

ACKNOWLEDGMENTS

Sincere gratitude is expressed to the following for helpful criticism of the manuscript: Drs. J. T. Enright, Scripps Institution of Oceanography, University of California, San Diego; C. A. Hall, University of California, Los Angeles; J. W. Valentine, University of California, Davis; A. Hallam, Oxford University, England; Prof. G. Y. Craig, University of Edinburgh, Scotland; and Mr. J. Gillespie, University of Texas, Austin. J. W. Valentine generously provided the cluster analyses and arranged for computer time. The United States Navy has provided continued access to Mugu Lagoon.

REFERENCES

AGER, D. V., 1963, Principles of paleoecology: New York, McGraw-Hill, 371 p.

BRETT, C. E., 1964, A portable hydraulic diver-operated dredge-sieve for sampling subtidal macrofauna: Jour. Marine Res., v. 22, p. 205–209.

BUCHANAN, J. B., 1958, The bottom fauna communities across the continental shelf off Accra, Ghana (Gold Coast): Proc. Zool. Soc. London, v. 130, p. 1–56.

CRAIG, G. W., 1966, Concepts in palaeoecology: Earth-Science Rev., v. 2, p. 127–155.

ELTON, CHARLES, 1927, Animal ecology; London, Sedgwick and Jackson, 209 p.

FAGER, E. W., 1963, Communities of organisms, *in* The sea, v. 2, Hill, M. N., editor: New York, Interscience Pub., p. 415–437.

HAIRSTON, N. G., 1959, Species abundance and community organization: Ecology, v. 40, p. 404–416.

HALLAM, A., 1965, Environmental causes of stunting in living and fossil marine benthonic invertebrates: Palaeontology, v. 8, p. 132–155.

INMAN, D. L., 1950, Report on beach study in the vicinity of Mugu Lagoon, California: U.S. Dept. Army, Corps of Engineers, Beach Erosion Board Tech. Memo. no. 14, 47 p.

JOHNSON, R. G., 1957, Experiments on the burial of shells: Jour. Geology, v. 65, p. 527–535.

——, 1962, Mode of formation of marine fossil assemblages of the Pleistocene Millerton Formation of California: Geol. Soc. America Bull., v. 73, p. 113–130.

——, 1964, The community approach to paleoecology, *in* Approaches to paleoecology, Imbrie, J., and Newell, N.D., editors: New York, John Wiley, p. 107–134.

——, 1965, Pelecypod death assemblages in Tomales Bay, California: Jour. Paleontology, v. 39, p. 80–85.

KELLING, GILBERT, & WILLIAMS, P. F., 1967, Flume studies of the reorientation of shells: Jour. Geology, v. 75, p. 243–267.

KRISTENSEN, I., 1957, Differences in density and growth in a cockle population in the Dutch Wadden Sea: Arch. Néerl. Zool., v. 12, p. 350–453.

LEVER, J., 1958, Quantitative beach research. I. The "left-right phenomenon": sorting of lamellibranch valves on sandy beaches: Basteria, v. 22, p. 21–51.

——, KESSLER, A., VAN OVERBEEKE, A. P., &

Thijssen, R., 1961, Quantitative beach research. II. The "hole effect": Neth. Jour. Sea Res., v. 1, p. 339–358.

Lindeman, R. L., 1942, The trophic-dynamic aspect of ecology: Ecology, v. 23, p. 399–418.

Lloyd, Monte, 1967, "Mean crowding": Jour. Animal Ecology, v. 36, p. 1–30.

MacArthur, R. H., 1965, Patterns of species diversity: Biol. Rev., v. 40, p. 510–533.

——, & Connell, J. H., 1966, The biology of populations: New York, John Wiley, 200 p.

Macfadyen, A., 1963, Animal ecology. Aims and methods: London, Pitman, 344 p.

MacGinitie, G. E., 1939, Littoral marine communities: Am. Midland Naturalist, v. 21, p. 28–55.

Menard, H. W., & Boucot, A. J., 1951, Experiments on the movement of shells by water: Am. Jour. Sci., v. 249, p. 131–151.

Miyadi, D., & Habe, T., 1947, On thanatocoenoses of bays (in Japanese with English abstract): Physiology and Ecology, Tokyo, v. 1, p. 110–124.

Moore, H. B., 1933, A comparison of the sand fauna of Port Erin Bay in 1900 and 1933: Proc. Malacol. Soc. London, v. 20, p. 285–294.

Odum, H. T., Cantlon, J. E., & Kornicker, L. S., 1960, An organizational hierarchy postulate for the interpretation of species-individual distributions, species entrophy, ecosystem evolution, and the meaning of a species-diversity index: Ecology, v. 41, p. 395–399.

Slobodkin, L. B., 1962, Growth and regulation of animal populations: New York, Holt, Rinehart and Winston, 184 p.

Sokal, R. R., & Sneath, P. H. A., 1963, Principles of numerical taxonomy: San Francisco, Freeman and Sons, 359 p.

Valentine, J. W., 1961, Paleoecologic molluscan geography of the California Pleistocene: Univ. California Pub. Geol. Sci., v. 34, p. 309–442.

——, & Peddicord, R. G., 1967, Evaluation of fossil assemblages by cluster analysis: Jour. Paleontology, v. 41, p. 502–507.

——, & Mallory, Bob, 1965, Recurrent groups of bonded species in mixed death assemblages: Jour. Geology, v. 73, p. 683–701.

Warme, J. E., 1967a, Comparisons of living and dead mollusks in quantitative samples from a coastal lagoon: Internat. Sed. Congr., Reprints of Papers, Reading and Edinburgh, August 1967.

——, 1967b, Graded bedding in the Recent sediments of Mugu Lagoon, California: Jour. Sed. Petrology, v. 37, p. 540–547.

Williams, C. B., 1964, Patterns in the balance of nature: New York, Academic Press, 324 p.

Wilson, J. B., 1967, Paleoecological studies on shell beds and associated sediments in the Solway Firth: Scottish Jour. Geology, v. 3, p. 329–371.

Manuscript received February 5, 1968

PAPER 38

Taphonomy: New Branch of Paleontology (1940)

I. A. Efremov

Commentary

ANNA K. BEHRENSMEYER

Paleontologists and other natural historians, including Charles Darwin, were thinking about links between modern organisms and the fossil record long before the Soviet scientist Ivan A. Efremov wrote his thoughtful—and as it turned out—seminal proposal to make this a new field of science. As a scientist trained in vertebrate paleontology, Efremov was exposed to popular concepts of his time relating to fossilization—biostratinomy and actuopaleontology. Biostratinomy relates to processes of burial and fossilization inferred from examining rock strata, while actuopaleontology investigates processes of death, destruction, and burial that can be observed in the modern world. Efremov saw these as different approaches to the same problem—how organic remains become fossils. He thus proposed a more comprehensive field with a simpler name to focus attention on intersections of the realms of the living and the dead. His short paper defines "taphonomy" as "the science of the laws of embedding" and provides multiple reasons why it is important to study "the transition (in all its details) of animal remains from the biosphere to the lithosphere" in order to make sense of what we find in the fossil record.

It is not often that a new field of science appears on the landscape, and the success of Efremov's proposal owes much to his well-organized rationale as well as his strong conviction that paleontologists need taphonomy in order to realize the potential of the fossil record. His ideas might not have made an impact outside of the Soviet Union, however, without the enthusiastic support of his American friend and colleague, Everett C. Olson. "Ole" quickly adopted the term and its underlying concepts in his studies of vertebrate paleoecology, thus spreading it to vertebrate paleontologists in North America and Europe. In the 1950s and 1960s, taphonomy became a focus for invertebrate paleontologists and paleobotanists reconstructing ancient animal and plant communities. Today, paleoanthropologists and archeologists in Africa and North America use taphonomic methods to help distinguish evidence for early human versus nonhuman processes in Plio-Pleistocene bone assemblages. Taphonomy continues to evolve, with recent interest in links between the living and the dead in modern ecosystems. In the 60 years since its birth, taphonomy has helped to stimulate research across the boundaries of biology, paleobiology, and geology through its focus on information flow between different Earth systems—the living, the dead, and the fossil record.

Note: This article has also been referenced as: Efremov, I. A., 1940. Taphonomy, a new branch of paleontology: Akad. Nauk. S.S.S.R. Biul., Biol. Ser., no. 3, p. 161–68.

From *Pan-American Geologist* 74:81–93.

PAN-
AMERICAN GEOLOGIST

Vol. LXXIV SEPTEMBER, 1940 No. 2

TAPHONOMY: NEW BRANCH OF PALEONTOLOGY

By Prof. J. A. Efremov
Soviet Academy of Sciences, Moscow

At the present time paleontology has already passed through the stage during which primary actual data are gathered. The former iconographic papers are replaced by a growing number of articles developing more or less the evolutionary theory based on paleontological data. The phylogeny of different groups of organisms, the interpretation of the biology of forms now extinct, the influence of outer surroundings on the organism in process of time, paleozoogeography, all these problems are commonly at least touched upon in every considerable paleontological research. For the better development of this "theoretic" part of paleontology, it is subdivided into different subparts: Biostratigraphy, paleo-ecology (paleobiology), paleopathology, and others; each subpart unites a series of different problems, not a single one of which can be easily solved.

The greatest difficulty in all paleontological reconstructions is the desultory and incomplete character of the material and casuality of its preservation in the rock. Therefore, Darwin's theory of the incompleteness of the geological chronicle is of special importance in paleontology. And this is still so, in spite of the voluminous data which have been gathered since Darwin's time.

It is to be exceedingly regretted that paleontologists do not take into consideration this incompleteness of the geological chronicle; that there have hardly been any attempts to correlate paleontological data with the animal world actually existing in the past. Many phylogenetic theories and general laws have been formed on the basis of data obtained on fossils as if these latter presented a true and complete image of animal life during past geological eras.

All that has just been said is of special importance for the ancient faunas. The older the age of the geological formation is, the scarcer is the actual material, and there are more and more ways in which it can be interpreted. All the general conceptions of ancient faunas and ancient paleozoogeography are very primitive and not convincing. Our knowledge is still more vague about such groups of forms, "faunas," as those which have been found in conditions, showing that they were embedded in alien surroundings and in tanatoceonoses.

The study of sediments which contain some type of fossilized biocœnoses of stationary marine fauna, together with the study of the distribution of different forms in different facies, and in space, such research brings good results for estimating the outer conditions (physico-geographical and biotic ones), and their influence on the fauna as a whole, and on separate forms.

Such methods of paleoecological analysis become hardly possible when a fauna is found in casual groups, tanatoceonoses, and not at the place where it existed when living. Still, this problem is a much more simple one for marine organisms, because the transference of the remains along the bottom of the sea is generally insignificant. Moreover, the areas where a fauna is found when alive are nearly always the areas of sedimentation, and the surrounding conditions are simply variations of one large bionomical factor. Thus, the study of sediments containing abundant faunas, and the study of the variations in the sediments along large areas and during great intervals of time leads to perceptible results.

But when we turn to the terrestrial faunas we come upon very different conditions. All the localities where these faunas are found are tanatocoenoses which have been formed in alien vital surroundings. The areas occupied by these faunas when alive are generally regions of wash-out. The fossilized remains of land-organisms are found on very small disconnected areas, so that the incompleteness of the geological chronicle for the land-faunas is much greater than for that of the sea.

Meanwhile, the terrestrial organisms, both insects and vertebrates, are highly organized, having passed through a long period of evolutional development. Because of this land-organisms are of special interest as material for establishing evolutionary laws; but it is just for this group that the precision of theoretical deduction is not great, because the actual material is very incomplete.

The chief and most precise method of paleontology, detailed morphological study followed by comparative analysis, can be of use only for objects which have been well preserved. But it cannot be of any use at all for reconstructing that part of the fauna, which has not reached us even partly. The paleobiological study of ancient land-forms also does not give any precise notion of the life of the fauna; the remains are preserved in tanatocœnoses which bear no part in the life-surroundings of the fauna; so the analysis of individual adaptive deviation is not sufficiently authentic.

All the other known methods of paleoecological research are also, for the same reason, quite helpless in the study of terrestrial forms. Taking all this into consideration, it will be easy to surmise why it chanced that the first paleontologists who found it indispensable to work from a new point of view, were those who studied the fossilized remains of the vertebrates. This new outlook created a basis for a critical examination of the place which paleontological data must occupy in the tangible animal-life of the past.

It is evident that, apart from the study of fossilized objects in itself, there is another way to the knowledge of the animal world of the past eras, *i.e.*, the study of the conditions in which paleontological records have been preserved, a comparative study of the localities where the remains have been found.

Many of the works published recently contain chapters devoted to the analysis of the conditions under which remains of the terrestrial animals are found in the rocks, fossilization and biostratonomy (Walther, Deecke, Abel, Weigelt, Wepfer, etc.), *i.e.*, an analysis of the processes of embedding. To be able to study comparatively the gathered data it became necessary to study the contemporary processes of embedding. After a series of short articles by different authors (Besser, Braun, Zelizko, Hoernes, Moos, Richter, and others) devoted to separate observations on the destruction of recent animals and the formation of accumulations of their remains. In 1927 there was published Weigelt's book devoted specially to the problem of the finding of remains of vertebrates in natural surroundings and their paleobiological significance.

The author has gathered together a great deal of data. But the main purpose of this work is the explanation of the causes of death of animals, based on the position of their remains at the time when they were embedded; so, for all that the author has disclosed many general laws, the general problems of embedding of animal remains have not been formulated. The data were discussed only from the point of view of how such study could help forward paleobiology.

Already in the following year (1928) Richter separated the study of contemporary embedding from paleobiology and named this new branch of science "actuopaleontology." The author formulated it thus: "Actuopaleontology is the science of the way paleontological documents are at present formed to be afterwards preserved as fossils."

In his work Richter gave a detailed analysis of the problems and methods of paleobiology and the new actuopaleontology. The author considers that the following three groups of problems are embraced by actuopaleontology:

(1) The science of life-marks (Lebensspuren of Abel), which can be named ichnology;

(2) The science of the destruction of animals and of the embedding of their remains. It can be subdivided into, (*a*) tanatology, the causes of death and its immediate results, (*b*) comidology, the transportation of animal remains, (*c*) biostratonomy (Weigelt), the science of embedding and (*d*) necrology, the decay of animal remains down to diagenesis.

(3) The science of biofacies. This comprises such different parts of ecology (in the strict sense of the word) as bionomy (of Walther) and morphonomy; they are analyzed from the point of view so important for the paleontologist, the areal differences of life as a reaction to the outer surroundings.

For all that actuopaleontology is subdivided into many branches and embraces many different parts of zoology (botany), it studies only the contemporary conditions of embedding of organic remains. Actuopaleontology on the one hand studies the problem of embedding only from the narrow point of view of explaining individual laws, but on the other hand, the science of biofacies has to combat very wide problems which by far exceed, both as to methods and theme, the actual possibilities of actuopaleontology.

Actuopaleontology by no means exhausts in full the problems of the processes of embedding, as it does not study this problem comparatively during the geological history, but separates the conditions of the present from those of the past. Besides this, the name "actuopaleontology" is badly chosen, being cumbrous and without any meaning when translated from Greek.

In 1934 the author of this article published a short analysis of the reasons why transitional forms are not included into the geological chronicle; and in 1936—a short outline of the problems that must be studied regarding the process of embedding of ancient terrestrial vertebrates. The new branch of paleontology is outlined by joining together all the different attempts of analyzing the processes of embedding. The development of this new science has become imperative.

The chief problem of this branch of science is the study of the transition (in all its details) of animal remains from the biosphere into the lithosphere, *i.e.*, the study of a process in the upshot of of which the organisms pass out of the different parts of the biosphere and, being fossilized, become part of the lithosphere.

The passage from the biosphere into the lithosphere occurs as a result of many interlaced geological and biological phenomena. That is why, when this process is analyzed, the geological phenomena must be studied in the same measure as the biological ones.

In the indissoluble unity of geological-biological analysis lies the key to the following most important problems of paleontology, which cannot be determined by the usual methods.

(1) As we go down into the ages we find that the number of forms that are preserved out of the general mass gets less and less. The cause thereof is the general denudation of continents, the sinking of great areas of the Earth's crust, below the level of the sea, or under masses of less ancient sediments, the metamorphism of the rocks, and, lastly, the small degree in which large areas of the globe are yet investigated. In what measure that part of the fossil animal world that is known to us for each given space of time tallies with the real living fauna of that period, to what extent is

our notion of the geographical localization, centers of development, and expansion of the ancient faunas, casual, is our knowledge of the climate which is characteristic for definite groups of forms in the area within which they lived, correct, these are the main problems which must be solved in the general part of the work.

(2) Each complex of land-forms found in one locality, and called by us a "fauna," is in truth but an accidental accumulation of animal remains. The formation of each locality depends on many causes, and, firstly, on the coincidence in a given place of a concentration of animal remains with geological conditions favorable to the conservation of these remains.

The concentration of animal remains depends on the number of species of that, or other forms, the rapidity with which they perished, or the length of time during which the carcasses of the animals were concentrating in a given place. Because of this it is the more numerous species which are more often embedded in the site, where the conditions are favorable for the conservation and fossilization of the remains. So, nearly every bed of fossilized remains is only a selection (and an accidental one) which does not reflect the real composition of the fauna at the place of its formation.

The processes which form the sediments are also favorable to selection in embedding. The animal remains are located mechanically in accordance with the strength of the flow. The smaller remains are carried along farther than the large ones; so, in accordance with the place which the given locality has in the flow, we can meet one containing only large or only small forms.

Why different forms are embedded in this or that locality depends on the kind of their food, the character of their adaptation, and its relation to the general physico-geographical features of the region where the conditions of sedimentation are favorable for embedding.

In this short outline it is, of course, impossible to enumerate all the causes of "selection," and incompleteness in embedding. For example, the classical localities of Late Carbonic amphibia, Stegocephalians, in Bohemia, Saxony, Ireland, and Canada can be indicated. Here are found in great numbers in argillaceous layers in the coal series small Stegocephalians, chiefly larval specimens. For a long time this fauna was taken for the true terrestrial Carbonic fauna, showing a definite phase in the development of the land vertebrates. Therefore, the subsequent findings in older rocks of large and highly organized Stegocephalia (lower Coal Measures, England), or of a diverse and abundant reptilian fauna in rocks but in a slight degree younger (U. S. A.), were very amazing.

Meanwhile, from the point of view of the tenet of embedding, all the localities of Late Carbonic Stegocephalians just described represented pools or small lakes located in marshy forests. These pools swarmed with larvæ, or young Stegocephalians, so that later, when the pools were turned into beds of fossilized remains, they were found to be filled with a singular fauna where young species predominated and which contained some small predacious forms which fed upon the youngest larvæ. These latter, having no bony skeletons, have been but rarely preserved.

Another example can be found in the well-known localities of the Permian fauna of Pareiasaurians on the Little North Dvina River. Near the largest lens, "Sokolky," is found the lens "Zavrajie." Here have been found the remains of Pareiasaurians (Scutosaurus) with greatly thickened bones. In the chief lens, "Sokolky," is found a large accumulation of Pareiasaurians (Scutosaurus) with normal skeletons, whereas in the lens "Zavrajie" we have undoubtedly a selected embedding of specimens with thickened bones. If the lens "Zavrajie" had been known before the lens "Sokolky" was discovered, and not the other way, as actually happened, the exceptional fauna found there should have been considered typical, and our idea of the Pareiasaurian fauna of the U. S. S. R. should have been initially erroneous.

A great number of such examples of the findings of terrestrial vertebrates are known. They all show to what poor extent the forms found in a given locality reflect the fauna which actually existed there.

(3) In a great number of localities containing land vertebrates, we are met by an apparently unexpected appearance of great masses of bony remains. The roof and the bed-rock of the layer containing animal remains are formed by rocks similar to the layer, but of much greater thickness. These latter rocks generally do not contain any organic remains.

All the numerous so-called "faunas" of ancient vertebrates appear, it would seem quite suddenly, without any "roots" in the underlying rocks. The existence of such localities is generally explained by the swift destruction of large masses of animals, by their sudden migration into the region or by the influence of catastrophic changings in the outer surroundings. Nevertheless, we do not generally meet in dumb series with conditions suitable for embedding, although the fauna has lived for a long period in this region. Much more rarely are found real "death-fields" caused by anastrophical changings in the outer surroundings of the animals.

The solution of the problem of the character of physicogeographical alterations and of the reasons why large accumulations of animal remains are formed, is of great importance for understanding the processes of embedding.

(4) The fragmentary character of the findings of accidental animal complexes explains the loss of prospect into time. Synchronous complexes of animal forms found in different localities and, because of strong differences in embedding conditions, different one from another, have often been taken for different faunas. Sometimes the same form is found in different localities; in one case with a background of primitive forms, and in the other, of forms of a later appearance. This gives occasion to call such a form a "stable" one in the run of time.

On the other hand, forms of different age, grooped into large systematical complexes with a common age definition (for example "Permian") on a morphological basis, seem to shorten the period of evolution, to augment the number of forms which have reached us from a given space of geological time. In reality, if we distribute, for example, all the Permian forms of terrestrial faunas known to us along the run of time represented by the 40 million years of the Permian era, the seeming diversity and numerousness of the forms known to us shall appear essentially other than now. So we see that the problem of dated detailed coordination of the faunas and the gathering of exact data the whole world over during short periods of time is also one of the problems solved by studying the processes of embedding.

The work of solving the above mentioned problems can be divided into two parts.

The first part, the study of geological processes of the transition of animal remains from the biosphere into the lithosphere, and in the first place, the detailed study of the localities where a terrestrial fauna is found. Up to the present these problems have not been studied, notwithstanding their significance.

The second part, the study of the laws by which are governed contemporary embedding processes of animal remains (in part Richter's actuopaleontology).

The chief work of the first part is as follows:

1. Biostratonomy of the localities. The study of the spatial distribution of animal remains and their distribution relatively to the planes of stratification, denudation, cross-bedding, relative to shingle and different mineralogical fractions. The construction of biostratonomical diagrams.

2. The study of the lithology of sedimentary rocks, both of those that contain animal remains and those that surround the given locality. These researches conducted reciprocally make it possible to define the character of animal remains, the cause of their disposition in a given manner, and the possibility that part of the fauna has been destroyed during the formation of the locality. Besides this, it is possible to ascertain the direction and hydrodynamical regime of the streams that precipitated the sediments, the distribution of the regions of denudation, the speed of sedimentation, embedding, and lithification, the degree of inspissation and deformation of the sediments, and the fossils contained in them, and the retracing of the physico-geographical conditions in the regions of embedding. In the end all these researches will show to what degree the given locality is a "selected" one, and will make it possible indirectly to approach the solving of the problem of correlation of the fauna embedded with that which actually existed in the region.

3. The study of the processes of fossilization of animal remains. Microscopical and chemical analyses of fossilized remains, conducted together with experimental work of artificial fossilization, and with observations of the destruction of the surfaces of organic remains in different surroundings. The study of the processes of rounding off of the bones during transportation and the calculation for each locality of the relations between the rounded off bones and the object as a whole. Here also must be studied the conservation of organic matter in different rocks, which is of great importance for the definition both of the facial conditions of the region where the locality is situated, the fossilization, and the relative periods of

time. To these problems can yet be added the study of pathological changings and damages in skeleton parts and the search for tokens of great episootical illnesses; here can be found an answer to the separate causes of the destruction of animals and the formation of accumulations of remains appertaining to one species. The investigations indicated above make it possible to ascertain the condition of animal remains before fossilization, *i.e.*, to explain the primary stages of the process of embedding and to investigate the process of destroying of animal remains before fossilization, and while it lasted.

4. Estimating the reserves of fossils. The outlining and the calculation of the mass of the layer containing animal remains for each locality. The calculation of the relations of the areas of the localities to the general area of a given layer of continental sediments. The figuring out of the average quantity of animal remains in different parts of the locality and the probable average quantity of remains in the whole locality. With the help of this latter the number of species and, sometimes, the number of forms found in the locality can be calculated.

The above mentioned research paves the way to:

5. The general reckoning up for definite lengths of geological time of the areas where the continental series are developed and the bulk of these sediments, and within them the districts containing localities of embedded land-fauna; this solves the question of the relation between the areas of sedimentation and embedding. The summing up of the results of the above mentioned work for all the continents and then the definition of the relation between the survived areas of sedimentation and the whole area of the continents of the given period, the era where the terrestrial fauna existed. In the end we can reckon up the per cent of embedded fauna.

The general comparative facial analysis of the conditions which form thick series of rocks, on this basis, the explanation of the general laws of embedding of the fauna which are typical for a given length of geological time (historical faciology of the localities). All the data obtained during the first part of the work is verified during the second:

(1) The laws of disposition and conservation of contemporary animal remains in sediments just formed or being formed now.

It is quite evident that this work is carried on in unison with the study of the origin of contemporary continental sediments. This comprises the study of biostratonomical laws in the contemporary sediments, the laws of the concentration and distribution of animal remains in the process of embedding, the laws of the concentration and scattering of animal remains in the process of contemporary sedimentation, etc.

(2) The establishing of general laws in the relations between the areas where a land fauna exists, the regions of continental sedimentation, and inside these latter, the districts where localities containing remains of terrestrial fauna are formed. The establishing of the subsequent wash-out of the sediments formed already and of the localities of sub-fossil fauna, the establishing of the comparative longevity for different types of continental sediments. This latter has a great significance for the study of facies, but rarely preserved in fossilized condition, for instance, the sediments of inclines, mountain streams, river terraces, small bogs and lakes, caves, etc. The calculation of the comparative value of the contemporary continental sediments and the areas of wash-out of the continents which will give the clue for establishing comparative data about the areas of wash-out of ancient continents.

(3) Biological observations of the number of terrestrial fauna, the dynamics of the population as a whole, and particularly, in separate biocœnoses. The establishing of the comparative quantity of destroyed species, the explanation of the causes of destruction *en masse,* and of the concentration of the remains of animals in habitual surroundings. The establishing of the laws of the dispersion and destruction of animal remains on the sub-aëral surface, and the calculation of the relation between the quantity of destroyed species and forms in general and the quantity of those which fall into such conditions that they are preserved. The establishing of the stability in process of time of separate biocœnoses, the establishing of the relations (quantitative and qualitative) between the population of areas of denudation and sedimentation, the establishing of the fossilized and contemporary terrestrial faunas and a synthesis the first and second areas; these are the chief directions of the work of this (biological) group.

All the works outlined above of both parts give sufficient data for a comparative analysis of the embedding of animal remains both of the fossilized and contemporary terrestrial faunas and a synthesis of the general laws of this process.

Moreover, by comparing the Quaternic fauna of land vertebrates with that of the present time, one can already come near to calculating the approximate numbers of destroyed representatives of terrestrial fauna before it was embedded.

Inasmuch as the Quaternic fauna of terrestrial vertebrates is comparatively well known and in quality (and, one must think, in quantity) greatly resembles contemporary fauna, the calculation of the general number of species and forms found in the Quaternic localities, when compared with the number of species and forms of contemporary faunas, can give an idea of the part of animal life which has been destroyed.

With due accuracy of calculations, bringing in corrections on the unexplored regions, the difference in time and series, and taking into account the changing of the mass of generations, we can obtain a sufficiently probable figure.

Such comparative figures can be obtained by comparing the areas of habitation, continental sedimentation, and of embedding, on contemporary and ancient material for separate geological units. These figures show the limits of accuracy of all our evolutionary and faunal generalities. All that has been said outlines sufficiently distinctly the problems, and methods of solving them, which are studied in the processes of embedding of terrestrial faunas.

In great part neither the problems nor methods are new. Research in the directions outlined above is going on already, being done not only by separate geologists and paleontologists, but also by the whole scientific institutions, national parks and reservations. These latter register the contemporary terrestrial fauna and study its dynamics; the work is of great importance for the comparative analysis of the processes of embedding. However, all such research is deficient from the point of view of paleontology, in the absence of a general plan or uniformity in methods; besides, it all bears an episodic character. It is self-evident that the study of the processes of embedding, as it has been said earlier, is also significant for floras and marine faunas. In this article I have outlined the field of activity for studying the processes of embedding on the material of terrestrial fauna because for these such processes have a greater significance. Because of this they have been better studied and the methods of work are more advanced. It is probable that in the future the study of the processes of embedding will develop on material of the same kind. For the marine faunas one can go along another and more easy line of research, the reconstruction of a whole fauna by the biocœnoses which have been better preserved; this is quite impossible for the terrestrial faunas.

It is necessary to unite all the separate "inclines" and directions in the study of the processes of embedding into a new branch of paleontology, the science of embedding. Earlier I have tried to show the significance of the problems solved by this branch of science, and the complexity of the methods and of the different branches of knowledge embraced by it. All this shows that the new branch of paleontology is evidently a separate and an important one.

I propose for this part of paleontology the name of "TAPHONOMY," the science of the laws of embedding. I find that this name will best reflect the chief direction of work in this new branch of paleontology. Taphonomy is certainly not a separate science. It stands on the border of paleontology, uniting it both with geology and biology into one general geo-biological historical method of study. From this point of view it is not necessary to subdivide it into taphonomy of contemporary fauna, actuotaphonomy, and of fossilized faunas, paleotaphonomy.

Taphonomical research allows us to glance into the depth of ages from another point of view than that which is in general use in paleontology. Therefore, these researches are also of importance for geology, paleogeography, faciology and sedimentology.

As regards paleontology, it is well to remember Darwin's significant thesis: "The life of each species is in greater dependence on other organic forms already established than on climate." Meanwhile, in the estimation of the general balance of fauna, paleontology is mostly helpless, so that a most important criterion of the knowledge of the processes of evolution is absent. Taphonomy produces a method for controlling the reconstructions on an evolutionary basis, and so aids these reconstructions to acquire the preciseness of which they are greatly in want.

PAPER 39

Taphonomic and Ecologic Information from Bone Weathering (1978)

A. K. Behrensmeyer

Commentary

RAYMOND R. ROGERS

Dry old bones rotting on the ground in East Africa—to most people nothing more than the macabre residue of once interesting animals now left for the recyclers. Then along came Kay Behrensmeyer. Kay methodically scoured the vast plains, woodlands, bush, and swamps of the Amboseli Basin for any trace of mortality, and meticulously collected postmortem details for thousands of specimens. She recovered volumes of data from the dead and found a way to coax valuable ecological and paleobiological insights from skeletons on the ground. One of the many outcomes of this ambitious (and on-going) endeavor was a concise paper published in *Paleobiology* in 1978 that treated bone weathering.

The fundamental contribution of "Taphonomic and Ecologic Information from Bone Weathering" is the enduring classification scheme that describes and categorizes the weathering stages of bone. Kay recognized six weathering stages and considered them in relation to weathering processes and to time since death. She provided exacting instructions for evaluating bone weathering and highlighted the types of questions that can be addressed with the resultant data. With Kay's paper in hand, any diligent researcher, be they in the field or in a museum, could generate a data set and begin to explore larger questions that relate to modes of accumulation (attritional vs. catastrophic mortality), time-averaging, and other potential biases (e.g., underrepresentation of the small and the young in vertebrate collections). The paper provided clear direction and described clear outcomes at a time when vertebrate taphonomy was just beginning to gain momentum, and anyone dealing with modern or ancient bone took note.

I first encountered "Taphonomic and Ecologic Information from Bone Weathering" as a fledgling student of vertebrate taphonomy working on Montana's dinosaur bonebeds. It was definitely one of the first papers that I read as I began to digest the taphonomic literature, and it was the only paper in my nascent library that yielded almost instant results (all of my bones had a weathering history). And I was not alone. A quick Google Scholar search indicates that "Taphonomic and Ecologic Information from Bone Weathering" has been highly and widely cited in diverse venues catering to ecologists, paleontologists, conservation biologists, anthropologists, and archaeologists! The rate of citation has been steadily increasing—a testament to the enduring importance and relevance of this paper. "Taphonomic and Ecologic Information from Bone Weathering" truly stands as one of the seminal contributions to vertebrate taphonomy and paleoecology, and from all indications it has weathered well (stage 0—still fresh).

Paleobiology, 4(2), 1978, pp. 150–162

Taphonomic and ecologic information from bone weathering

Anna K. Behrensmeyer

Abstract.—Bones of recent mammals in the Amboseli Basin, southern Kenya, exhibit distinctive weathering characteristics that can be related to the time since death and to the local conditions of temperature, humidity and soil chemistry. A categorization of weathering characteristics into six stages, recognizable on descriptive criteria, provides a basis for investigation of weathering rates and processes. The time necessary to achieve each successive weathering stage has been calibrated using known-age carcasses. Most bones decompose beyond recognition in 10 to 15 yr. Bones of animals under 100 kg and juveniles appear to weather more rapidly than bones of large animals or adults. Small-scale rather than widespread environmental factors seem to have greatest influence on weathering characteristics and rates. Bone weathering is potentially valuable as evidence for the period of time represented in recent or fossil bone assemblages, including those on archeological sites, and may also be an important tool in censusing populations of animals in modern ecosystems.

Anna K. Behrensmeyer. Peabody Museum, Yale University; New Haven, Connecticut 06520

Accepted: January 6, 1978

Introduction

Vertebrate bones decompose on subaerial surfaces, and their destruction can be viewed as part of the normal process of nutrient recycling in and on soils. Whether or not a bone survives to become fossilized depends on the intensity and rate of various destructive processes and the chance for permanent burial prior to total destruction. Surprisingly little is known about how these processes affect bones, and how they may consequently bias the vertebrate fossil record.

Bones freed of covering tissue and exposed on the ground surface usually undergo rather rapid changes in appearance. Various researchers interested in bone decomposition have conducted small-scale experiments or have collected observations on recent bones which indicate that weathering may follow broadly similar patterns in different environments (Voorhies 1969; Isaac 1967; Brain 1967; Hill 1975, pers. comm.; D. Gifford, pers. comm.; Tappen and Peske 1970). Ecologists have recently become interested in aging vertebrate carcasses using weathering characteristics (D. Western, pers. comm.). However, little systematic work has been done to define types of bone weathering and relate these to specific processes acting over known periods of time.

This paper will present a descriptive categorization of weathering based on systematic observations of recent bones in Amboseli Park, Kenya. Weathering categories or "stages" are set up to provide a basis for descriptive comparison with bones from other contexts, both fossil and recent. Within the Amboseli ecosystem, it is possible to show how the weathering stages relate to rates of decomposition and to surface processes characteristic of different macro- and micro-habitats. Such information provides a basis for assessing the potential value of bone weathering as a source of ecologic and taphonomic data, both within the Amboseli Basin and in a more general context.

Background

The study of bone weathering grew out of a broader taphonomic sampling program initiated in 1975. The results of the overall study of the recent bone assemblage in the Amboseli Basin will be reported elsewhere (Behrensmeyer and Dechant, in prep.).

The Amboseli Basin lies at the northern edge of Mt. Kilimanjaro in southern Kenya (Fig. 1)

0094–8873/78/0402–0004/$1.00

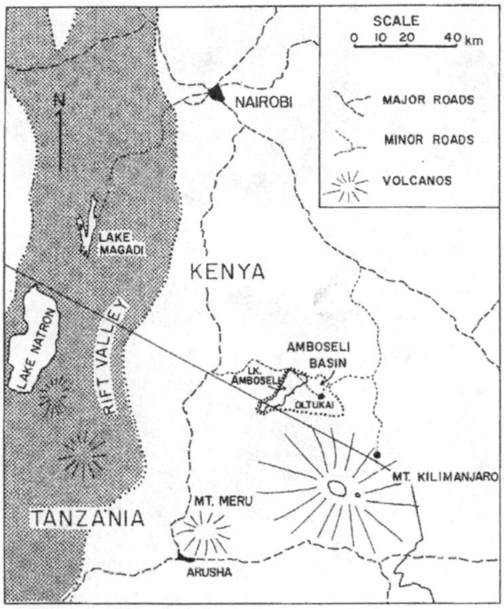

FIGURE 1. Map showing general location of the Amboseli Basin in southern Kenya. Lake Amboseli is dry except during periods of heavy rainfall.

and includes 600 km^2 of flat-lying terrain covered by a mosaic of vegetation zones. The basin is internally drained, and soils are generally alkaline. Amboseli National Park covers a part of the basin area which is used by a diverse fauna of wild herbivores and carnivores, as well as Maasai people, domestic stock, and tourists. Numbers of animals in the park fluctuate seasonally, and highest concentrations occur during dry periods, when the only available water flows from the base of Kilimanjaro (Western and Van Praet 1973).

Sampling of the bone assemblage in the central basin was done using linear transects covering parts of six major habitats. These habitats included swamp, dense woodland, open woodland, plains, bush and lakebed. The lakebed is dry most of the year, with sporadic inundation for 1 to 2 months following periods of heavy rain. Weathering stages defined by easily observable criteria were established early in the sampling program and used throughout the study. Reference bones and photographs were used as standards of comparison. Considerable effort has gone into making the descriptions given below simple and straightforward so that they can be readily learned and applied in the field.

Bone Weathering Stages

Stage 0.—Bone surface shows no sign of cracking or flaking due to weathering. Usually bone is still greasy, marrow cavities contain tissue, skin and muscle/ligament may cover part or all of the bone surface.

Stage 1.—Bone shows cracking, normally parallel to the fiber structure (e.g., longitudinal in long bones). Articular surfaces may show mosaic cracking of covering tissue as well as in the bone itself. Fat, skin and other tissue may or may not be present. (Fig. 2a)

Stage 2.—Outermost concentric thin layers of bone show flaking, usually associated with cracks, in that the bone edges along the cracks tend to separate and flake first. Long thin flakes, with one or more sides still attached to the bone, are common in the initial part of Stage 2. Deeper and more extensive flaking follows, until most of the outermost bone is gone. Crack edges are usually angular in cross-section. Remnants of ligaments, cartilage, and skin may be present. (Fig. 2b)

Stage 3.—Bone surface is characterized by patches of rough, homogeneously weathered compact bone, resulting in a fibrous texture. In these patches, all the external, concentrically layered bone has been removed. Gradually the patches extend to cover the entire bone surface. Weathering does not penetrate deeper than 1.0–1.5 mm at this stage, and bone fibers are still firmly attached to each other. Crack edges usually are rounded in cross-section. Tissue rarely present at this stage. (Fig. 2c)

Stage 4.—The bone surface is coarsely fibrous and rough in texture; large and small splinters occur and may be loose enough to fall away from the bone when it is moved. Weathering penetrates into inner cavities. Cracks are open and have splintered or rounded edges. (Fig. 2d)

Stage 5.—Bone is falling apart in situ, with large splinters lying around what remains of the whole, which is fragile and easily broken by moving. Original bone shape may be difficult to determine. Cancellous bone usually exposed, when present, and may outlast all traces of the former more compact, outer parts of the bones. (Fig. 2e)

In deciding what stage to record for a particular bone, the following guidelines are used;

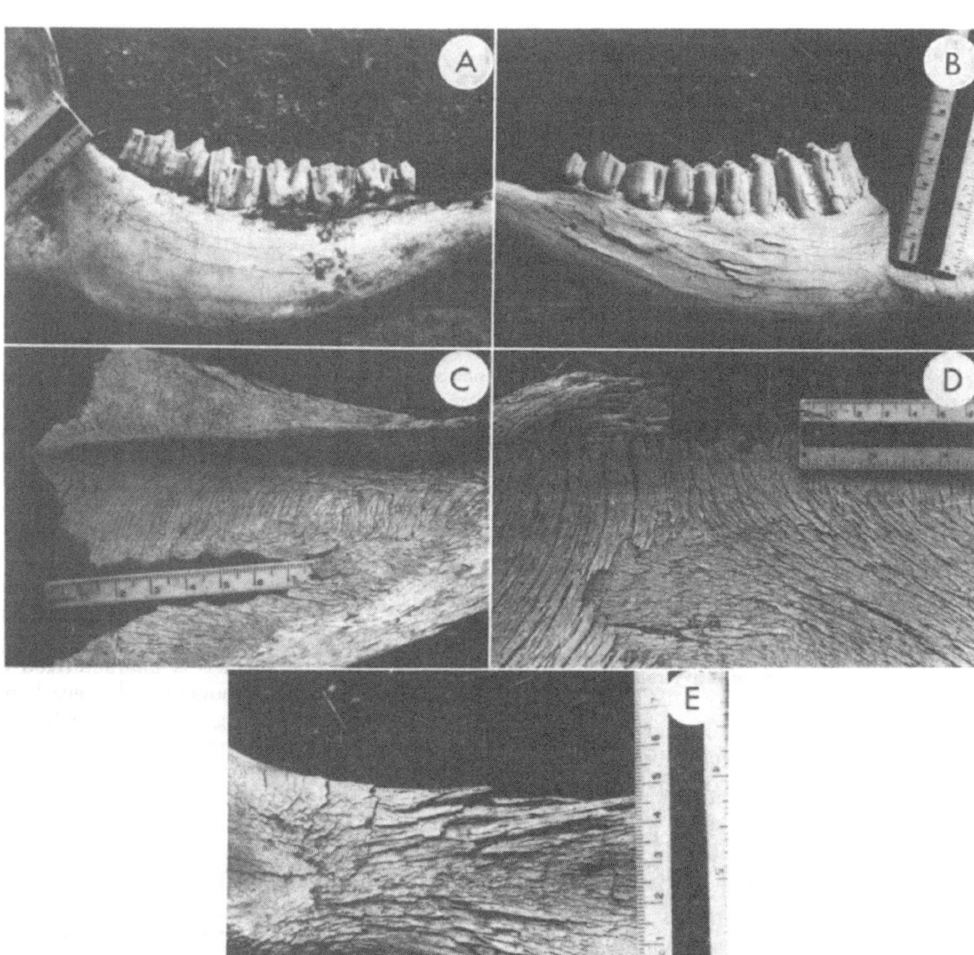

FIGURE 2. Weathering stages for the Amboseli recent bone assemblage. a: *Stage 1*: cow mandible showing initial cracking parallel to bone fiber structure; b: *Stage 2*: opposite side of same cow mandible showing flaking of outer bone layers; c: *Stage 3*: bovid scapula showing fibrous, rough texture and remnants of surface bone near lower right border; d: *Stage 4*: part of scapula showing deep cracking, coarse, layered fiber structure; *e*: *Stage 5*: scapula blade showing final stages of deep cracking and splitting. (15 cm scale in all photographs)

1) The most advanced stage which covers patches larger than 1 cm^2 of the bone's surface is recorded.
2) Whenever possible shafts of limb bones, flat surfaces of jaws, pelves, vertebrae, or ribs are used, not edges of bones or areas where there is evidence of physical damage (e.g., gnawing).
3) All observers must agree concerning the stage before it is recorded.

The larger skeletal parts are easiest to categorize. Small, compact bones such as podials and phalanges weather more slowly than other elements of the same skeleton and do not exhibit all the diagnostic characteristics of the weathering stages. In deciding the stage for a carcass with many parts, rather than for an isolated bone, it is advisable to examine several different bones (e.g., a limb bone, rib, ilium, vertebral spine) to assess the most advanced weathering characteristics.

The six weathering stages impose arbitrary

divisions upon what was observed to be a continuous spectrum, and this created some problems of categorization. For instance, many bones were equally covered by Stage 2 and Stage 3 surfaces, and proved difficult to place in one or the other category. However, by employing the criterion of "most advanced stage covering more than 1 cm^2," most of these problems were resolved. The success of the scheme has been indicated by the ease with which new observers recognize the stages, and there may be some "natural" component in the classification. This could reflect the fact that bones spend relatively longer periods within each stage than between them. However, at this point the stages should be treated as provisional descriptive categories only.

In general the six weathering stages are only applicable to mammals larger than 5 kg body weight, and a weathering classification has not been attempted for smaller animals. Bones of mammals such as the African hare (*Lepus capensis*) clearly differ in weathering characteristics from the larger species in Amboseli and seem more subject to cracking and splintering rather than flaking. Bones of birds, reptiles and fish also differ from mammals in weathering features, and each group will eventually require separate study.

In Amboseli the weathering characteristics of teeth could not be related in any consistent way to the bone stages. There are mandibles in Stage 3–4 with uncracked teeth, and Stage 1–2 mandibles with severely fractured teeth. Teeth seem to split easily when subjected to surface desiccation, but the lack of any clear pattern of weathering suggests that the individual characteristics of each tooth, including stage of eruption, wear, ratio of enamel to dentine and overall morphology, may be most important in determining characteristics and rates of weathering. More rapid fragmentation of teeth relative to bone cannot be related simply to arid climate, as suggested by Toots (1965).

Not all observed weathering characteristics in Amboseli fall into the six stages given above. A few bones exhibited a striking pattern of mosaic surface flaking (Fig. 3) and others showed extensive flaking without significant cracking. The causes for the mosaic flaking and variability in cracking are not yet understood. Another type of unusual weathering was observed in an assemblage accumulated by hyenas, within a shallow depression on a caliche "hard pan" surface. Many of the bones were chalky and had irregular small-scale flaking without extensive cracking. This weathering was not seen in the overall surface assemblage.

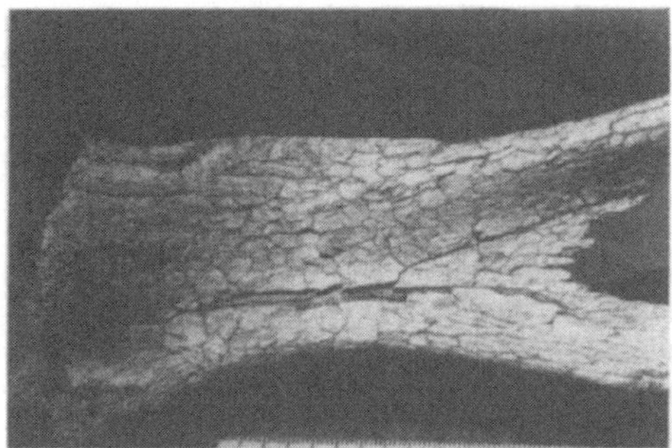

FIGURE 3. "Mosaic" pattern of bone surface cracking and flaking which is rare in the Amboseli assemblage compared to textures shown in Fig. 2. Cause of this weathering pattern is not yet understood.

Processes of Weathering

Weathering is defined for the purposes of this paper as the process by which the original microscopic organic and inorganic components of a bone are separated from each other and destroyed by physical and chemical agents operating on the bone in situ, either on the surface or within the soil zone. Physical damage caused by carnivore mastication, trampling, fluvial transport, and geochemical changes which take place diagenetically during fossilization are excluded from consideration here, although such processes are closely related to weathering in its broader context.

Evidence for the important weathering processes in Amboseli is derived from observed variation in weathering characteristics in different macro- and micro-environmental contexts. At present, hypotheses concerning weathering processes are based on comparative data and inferences which have defined the potentially important variables but await empirical confirmation.

Variation in weathering stages on a single bone provides important evidence concerning process. In particular, bones in Amboseli are usually weathered more on upper (exposed) than lower (ground contact) surfaces. Bones with Stage 0 on the lower surface and Stage

2 on the upper are not uncommon. Bones or parts of bones that project more than 10 cm above the ground surface are often less weathered than those close to the ground. Thus it seems that greatest bone weathering occurs in the zone immediately above the soil.

Recent work by Hare (1974; pers. comm.) indicates that organic breakdown in bones is enhanced by fluctuating temperature and humidity. Soil surfaces in open habitats can be subject to wide diurnal fluctuations in temperature and in moisture evaporating through the soil (Coe, pers. comm.; Ricklefs 1973). It is relatively easy to observe that such fluctuations occur in Amboseli; one can feel that bone surfaces are drier and hotter during midday than during morning or evening. The lower surfaces of bones are less subject to extremes in temperature and moisture than upper surfaces, and they can also be observed to weather relatively more slowly. Buried bones frequently show no sign of weathering even when exposed parts are in Stage 4 or 5. Thus, it seems reasonable to associate temperature and moisture fluctuations at the soil surface with bone weathering in Amboseli. Preliminary analysis of Amboseli bones of different weathering stages which indicates a correlation between the stages and progressive amino acid racemization in the collagen supports this interpretation (Hare, pers. comm.).

It seems probable that the physical stresses of repeated heating and cooling, and wetting and drying, contribute to the weathering characteristics in addition to microscopic changes in organic components. Large longitudinal cracks often appear on limb bones only a few days after the death of the animal. The relative importance of such physical stress may be affected by the original bone thickness and density and hence by the physical condition of the animal before death. This factor is of unknown significance in Amboseli, but it has been recently studied by Binford and Bertram (1977) in relation to bone fracturing and seems potentially important.

In some cases bones are more weathered on lower than upper surfaces. This usually occurs on highly alkaline soils where salts (Na_2CO_3, NaCl) crystallize on the bone surfaces (Fig. 4). The lower parts of the bones often are encrusted with salt crystals and show flaking and splitting caused by the force of crystallization. In situations where lower surfaces are not in contact with salt-precipitating conditions, bacterial activity, roots or other organisms may be responsible for distinctive patterns of weathering. Dendritic patterns of shallow grooving observed on some Amboseli bones are interpreted as the result of dissolution by acids associated with the growth and decay of roots or fungus in direct contact with bone surfaces. Roots may also cause splitting and fragmentation of buried bones (Fig. 5). Termites have been found in bones obviously chewed by a small organism, and in some areas (particularly on red, non-alkaline soils north of the basin) presumed termite damage to bones is not uncommon (Fig. 6). Distinctive grooving known to be caused by larvae of the moth *Tinea deperdella*, which feed on the organic components of horns (Hill 1975; Behrensmeyer 1975), has been observed on many of the bovid horn cores in Amboseli (Fig. 7).

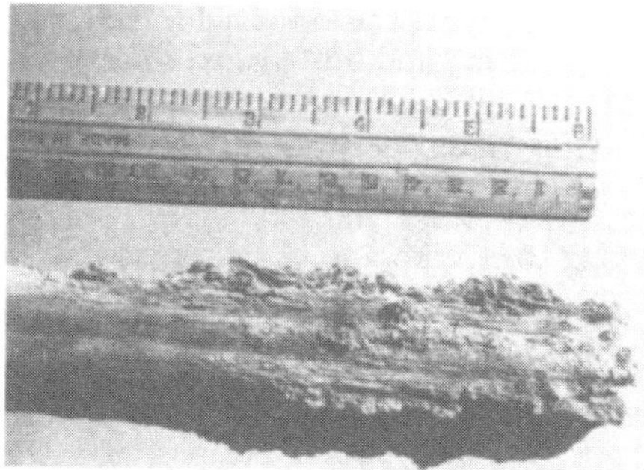

FIGURE 4. Bone surface cracking, splitting and fragmentation due to the precipitation of salts (Na_2CO_3 and NaCl) in pore spaces. Salts present in the soils and ground water precipitate as surface crusts during periods of net evaporation and may preferentially destroy those portions of a bone in direct contact with the ground.

Another weathering process can produce distinctive polish on bone surfaces. Bone fragments swallowed by hyenas and either regurgitated or passed as fecal matter have a characteristic polished and "dissolved" appearance, often with sharply pointed ends. Bones can also acquire polish and edge rounding through continued trampling, as demonstrated by Brain (1967), and more work needs to be done in order to establish how to distinguish these two processes.

Overall, the major weathering effects in Amboseli appear to be caused by a combina-

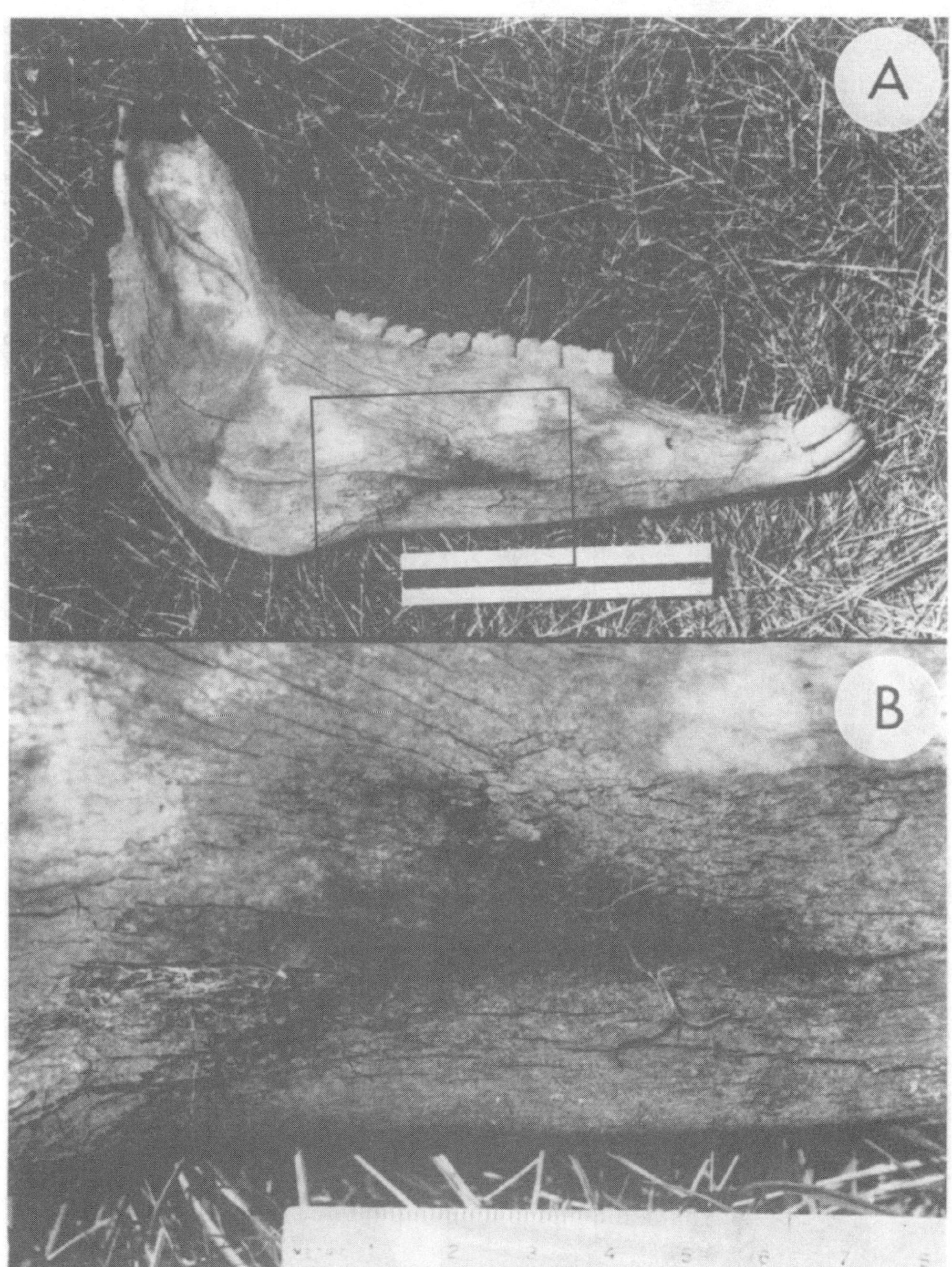

FIGURE 5. Cracking and expansion of a zebra (*Equus burchelli*) mandible due to root growth within the body of the ramus. Enlarged area in 5b is designated in 5a.

FIGURE 6. Cranium of cow showing perforations attributed to termites or possibly other unidentified insects. The bone was in a fragile state in 1975 and had fallen apart when relocated a year later.

tion of temperature and moisture fluctuations leading to bone decomposition at the ground surface. The weathering stages apparently are a result both of the intensity and duration of this process, and much local variation in weathering can be expected due to microclimatic conditions. In general, the less equable the immediate environment of the bone, the faster it should weather. Local variations in the patterns of weathering may occur due to soil chemistry, insects, and the effects of bone-collecting by hyenas. The relative importance of bacteria, root growth, and soil acids associated with plant decomposition cannot yet be assessed, except to note that the good condition of most buried bones argues against these as significant factors in bone destruction in Amboseli.

Rates of Bone Weathering

Weathering rates for Amboseli bones have been determined using 35 carcasses with known dates of death and one year or two successive years of observation. Results are pre-

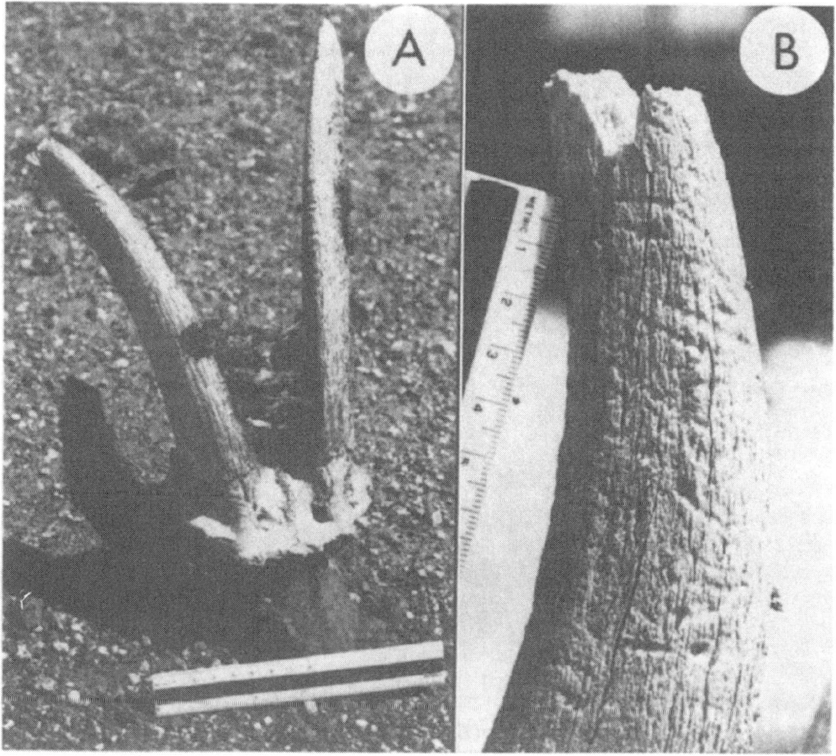

FIGURE 7. Grooving on horn cores of a male Grant's gazelle (*Gazella granti*) caused by larvae of the moth *Tinea deperdella*. Arrow in 7a indicates enlarged area in 7b. Bone fiber structure is vertical with grooving almost perpendicular to it.

TABLE 1. Numbers of carcasses of known age since death that can be assigned to each weathering stage category. The total sample consists of 35 carcasses, 17 of which have been observed 2 yr in succession for a total of 52 observations.

Years since death	Weathering stage						Total number observations
	0	1	2	3	4	5	
<1	6	1	–	–	–	–	7
1–2	–	7	–	–	–	–	7
2–3	–	3	3	–	–	–	6
3–4	–	1	5	–	–	–	6
4–5	–	–	–	3	–	–	3
5–6	–	–	–	3	–	–	3
6–7	–	–	1	4	1	1	7
7–8	–	–	–	3	3	–	6
8–9	–	–	–	–	2	–	2
9–10	–	–	–	1	1	–	2
10–15	–	–	–	1	1	1	3
							52

sented in Tables 1 and 2 and Fig. 8. Information on dates of death has been provided by D. Western. Weathering stage determinations were done in 1975 and 1976.

Weathering stages appear to be predictably linked with time since death as shown in Fig. 8, in spite of the variation in habitat and micro-environments represented by the carcass sample. Most bones falling into Stages 0, 1 and 2 have lain exposed for three years or less. Carcasses exposed longer than three years show a broad range of weathering stages, but none fall into Stage 1 or 0. Thus it is possible to distinguish carcasses exposed for less than 3 yr with fair certainty using weathering stages. Otherwise, the initial sample indicates that weathering stages are most useful in providing an estimate of minimum number of years since death (Table 2). With additional sampling and monitoring of the known weath-

TABLE 2. Weathering stages related to number of years since death for known-age carcasses in Amboseli.

Weathering stage	Possible range in years since death
0	0–1
1	0–3
2	2–6
3	4–15+
4	6–15+
5	6–15+

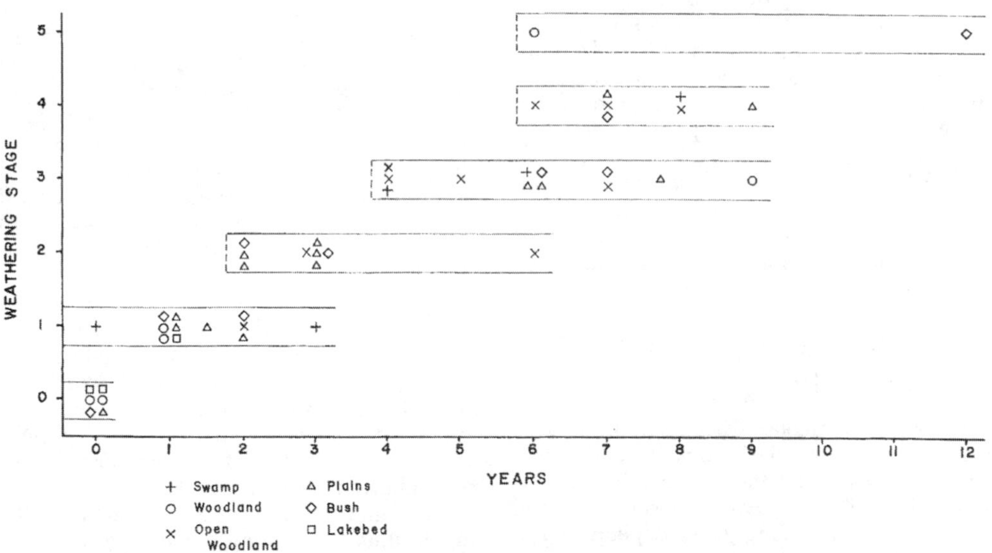

FIGURE 8. Weathering stage versus number of years since death for known-age carcasses in Amboseli. Each point represents a single weathering stage observation; in some cases the same carcass was observed in 2 successive years and thus appears twice on the graph. The total sample consists of 35 carcasses. Open-ended bars indicate that the maximum ages for the weathering stages are not yet known; minimum ages are more certain based on the present sample.

ering-age of carcasses over several years, it should be possible to judge accurately the most probable number of years of exposure of any bone of a given weathering stage. This will be further refined by increased understanding of how weathering rates vary in different habitats and micro-habitats.

There is no consistent association between habitat and weathering stage revealed by the data presented in Fig. 8. This implies that localized conditions (e.g., vegetation, shade, moisture) are more important to bone weathering than overall characteristics of the habitats. However, the samples from each habitat are relatively small and may not reveal important trends. In the total bone sample, Stages 0–1 are more common in the swamp habitat, and weathering in general seems to be slower where bones are kept moist and protected by swamp vegetation. As evidence of this, a rhinoceros (*Diceros bicornis*) killed in the swamp in 1961 was Stage 3 weathering in 1975, while a number of carcasses half its age were Stages 4 or 5 in other habitats.

In the total bone sample from Amboseli, 81% of the recorded carcasses were Stage 0–3. Based on the information available at present, it seems safe to say that nearly all of these carcasses were exposed for 10 yr or less. Sixty-two per cent of the total sample is Stage 0–2, and this, for the most part, should represent the last five years of accumulation. Bones in Stages 3–5 may have been exposed for as much as 15 yr but most are probably less than this. Thus, in Amboseli, bones exposed continuously on the surface show significant weathering by the time they are 3–5 yr old, and most disintegrate in less than 15 yr, although some may last longer under favorable micro-environmental conditions.

Weathering Stages in an Attritional Bone Assemblage

Overall the Amboseli bone sample is attritional, i.e., composed of carcasses added continually through death by predation, starvation, disease, etc. High mortality during drought periods among water dependent species adds a seasonal "mass death" component to the attritional assemblage which does affect its faunal composition (Behrensmeyer and Dechant, in prep.). Bones of animals dying during dry periods are subjected initially to intense desiccation unless protected within a mummified carcass. The potential effects of these initial conditions on subsequent weathering may be important but are unknown at present.

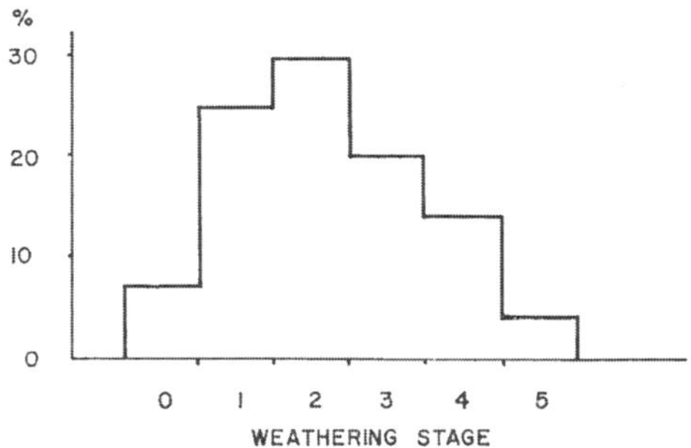

FIGURE 9. Percentages of carcasses (minimum numbers of individuals) in each weathering stage for the total mammal sample (1534 carcasses). The peak in Stage 2 may be due to heavy drought mortality in 1972–1973, but it could also mean that bones remain in this stage for a longer period than in other stages.

The relative numbers of carcasses in the six weathering stages are given in Fig. 9. Thirty-eight percent of the sample is "significantly weathered" (Stage 3–5), 55% is "slightly weathered" (Stage 1–2) and only 7% is "fresh" (Stage 0). Although how the Amboseli sample compares with other attritional surface assemblages is not yet known, it seems highly probable that these will all include some bones that are fresh, some that are slightly weathered and some that are weathered. This is significant in terms of the fossil record, since it provides a means of identifying an attritional assemblage.

Weathering has often been noted by paleontologists, usually as the cause of destruction of a valuable piece of morphological information on a specimen. However, in its weathering characteristics, the same specimen may hold important paleoecological data. Shipman (1977) has applied the weathering stage scheme given above to an excavated Miocene fauna from Ft. Ternan, Kenya. She notes that fossilization may obscure or change bone surface features, leading to difficulties in stage assignment.

In applying weathering stages to fossils, the worker must be sure that surface textures on bones are primary, and this limits the usable

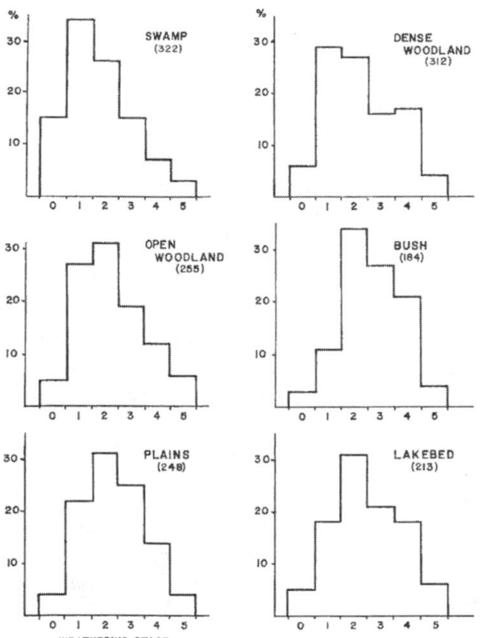

FIGURE 10. Percentages of carcasses in the six weathering stages for the major habitats in Amboseli. Numbers in parentheses are the total mammal sample for each habitat. Two × two χ^2 test on the numbers of carcasses in Stages 1 and 2 show that there is no significant difference ($P = .5–.75$) between swamp and dense woodland but that for swamp vs. open woodland, $P = .90–.95$ and for swamp vs. bush, $P \gg .995$.

sample to carefully excavated bones or "float" specimens with matrix covering bone surfaces which have not been changed during recent erosion. Three broad categories: fresh (Stage 0), slightly weathered (Stage 1–2) and weathered (Stage 3–5) are most likely to be recognizable on fossils. Simple ratios of fresh to weathered bones may also provide important taphonomic evidence concerning the attritional or non-attritional nature of the original assemblage.

Weathering Stages and Habitats

The six major habitats in Amboseli: swamp, dense woodland, open woodland, plains, lakebed and bush, proved variable in the relative numbers of carcasses showing the different weathering stages (Fig. 10). This variation suggests what might be expected in an attritional fossil assemblage derived from different habitats. Two important factors may be responsible: 1) differential rates of weathering 2) changes in patterns of habitat utilization and/or mortality among the major herbivores over the last 10 yr.

Censuses of the living herbivores in the various habitats have been done by Western (1973 and pers. comm.) since 1967. During 1967–1969, the open woodland, plains and lakebed habitats were utilized heavily by the major wild herbivores, zebra and wildebeest. In 1973 these species increased their utilization of the dense woodland and swamp habitats during a drought period (D. Western pers. comm.). Bones from animals that died in 1967–69 should be Stage 2 and over, those from deaths in 1973 and later should be Stage 0–2. Assuming that weathering rates do not vary significantly from habitat to habitat, the areas used more heavily in 1967–69 should show a greater abundance of carcasses of Stage 2 and over. The data presented in Fig. 10 for lakebed and bush, in contrast to swamp, agree with this hypothesis. However, open woodland, plains and dense woodland do not fit as well, and it is apparent that other factors must be considered.

The more equable environments with respect to weathering processes are the swamp and dense woodland, where moisture and shade tend to moderate the diurnal and seasonal fluctuations in surface temperature and humidity. Slower weathering in these habitats could lead to relatively greater numbers of bones in Stage 0–1 than in other habitats, all other factors being equal. Such a weathering effect could be added to the changes in habitat utilization to give modes in Stage 1 for the swamp and dense woodland, but a Stage 2 mode in the others.

The biological and the taphonomic factors in the patterns of weathering stage distribution cannot be separated at present. However, habitat effects on bone weathering rates are being monitored, and once this factor is known it should be possible to isolate the ecological component in the weathering stage distributions. Surface bone assemblages could then become a source of information for the history of habitat utilization in modern vertebrate communities.

Weathering and Body Size

There is evidence that bones of relatively small animals (≤100 kg) weather more rapidly than those of large, thereby introducing a size-related bias into the Amboseli bone assemblage. Size-biasing is also caused by a variety of other processes which are discussed by Behrensmeyer and Dechant (in prep.). This paper deals with evidence relating weathering as previously defined to the preferential destruction of bones of small animals, including both small adults and juveniles of larger species.

Among the major herbivores, there are 15% fewer carcasses in Stage 3–5 for species smaller than 100 kg, compared with larger forms. Populations have remained relatively stable in the basin, and the carcass input per year from all species has been more or less constant. Therefore all species should have proportional numbers of carcasses in each weathering stage unless there is variability in weathering rates. Effects of habitat should not be important because the smaller species are distributed over much the same areas as the large. It is possible that smaller bones do not always exhibit the typical characteristics of Stages 3–5, or that they achieve them more slowly during weathering. The rate of decomposition may also be greater for smaller bones once they reach Stage 3, resulting in their more rapid elimination from the surface assemblage.

Bones of juvenile animals can be classified using the weathering stages, although it is clear that the immature bone varies from adult in weathering characteristics and in rates of decomposition. Bone of very young or foetal animals associated with bones of the mother usually were more weathered by one or two stages. Numbers of juvenile carcasses also vary with weathering stage. The ratio of juvenile to adult wildebeest (*Connochaetes taurinus*) is about .70 for the live population (Andere 1975). There is high juvenile mortality, but many carcasses are completely destroyed by carnivores. Those remaining should represent a constant proportion at each weathering stage unless: 1) weathering rates are different from those of adult bones, 2) mortality of juveniles versus adults has changed during the sample period, 3) carnivore consumption of juvenile versus adult carcasses has varied during the sample period. The ratio of juvenile to adult carcasses in Stage 0 is .50, Stage 1: 3.7, Stage 2: 1.0, Stage 3: .20. There were no juvenile wildebeest bones recorded in Stages 4 or 5. Variations in ratios between Stages 0 and 3 may be caused by any one of the three factors given above, but it is difficult to explain the total absence of juvenile bones in Stages 4 and 5 unless weathering has decomposed them beyond recognition. The evidence thus indicates that immature bones weather more rapidly than adult. The record of juvenile mortality is lost for the bone sample older than 6–8 yr in Amboseli.

It is not yet possible to separate the effects of the two variables, small size and immature bone structure, on weathering rates. Overall the latter appears to be more significant because there is a fairly large proportion of Stages 4 and 5 in the adult bone sample for species under 100 kg, while these stages are almost unrepresented for juveniles. Monitoring of marked carcasses will eventually provide calibration for the size-related weathering effects.

Weathering in Areas Other Than Amboseli

A preliminary survey of known-age carcasses in Nairobi National Park suggests that the Amboseli weathering stages may be generally indicative of the number of years since death, even under different climatic and soil conditions. Seven carcasses of large mammals were all in Stage 2 after 2.5–3.0 yr of weathering. This agrees well with the Amboseli data. A single carcass 4.5 yr old was Stage 3. Nairobi Park has a higher rainfall than Amboseli, cooler average temperatures due to its higher elevation and generally more vegetation. If weathering rates (at least through Stage 2) are partly independent of overall climatic regime, then weathering stages may be broadly applicable as an ecological tool. However, further sampling over a broad range of climatic regimes will be essential to establish this, and in temperate climates, the added effects of freezing and thawing will need to be tested.

A weathering rate experiment conducted by M. Posnansky and G. Isaac (Isaac 1967) at Olorgesailie, Kenya showed that after 7 yr bones of a cow and juvenile goat had disintegrated to a cracked and friable state probably corresponding to Stage 3 or 5. Experiments and observations east of Lake Turkana, Kenya, indicate that bones go through stages similar to those in Amboseli, and that weathering rates are highly dependent on local conditions of moisture and vegetation cover (Hill 1975; Gifford, pers. comm.). Bones of a young topi (*Damaliscus korrigum*) placed on a roof several meters above the ground and fully exposed to sun and rain at Lake Turkana weathered to Stages 2 and 3 after 3 yr. This implies that conditions near the ground are not required to produce the recognizable weathering textures.

Observation of surface bone weathering in a variety of environments on several continents has shown that textural characteristics of the different stages are generally recognizable. This implies that structural features of the bones themselves have a major influence in the weathering characteristics regardless of external conditions. Temperature and humidity may exert strong control of the rate of this weathering, however.

Potential Uses of Weathering in the Fossil Record

Weathering features on fossils can provide evidence of taphonomic processes, and if primary weathering can be distinguished from transport abrasion and diagenetic effects, then primary weathering can give specific information concerning surface exposure of a bone prior to burial and the time period over which bones accumulated. Voorhies (1969; p. 31) has used the lack of variation in weathering as evidence that animals preserved in the Pliocene Verdigre Quarry died and were buried in a relative short period of time. If bones exhibit only a single weathering stage, this may indicate catastrophic death, but it could also mean that local conditions (e.g., moisture, rapid burial) inhibited weathering of gradually accumulating skeletal remains. An assemblage with bones in all stages of weathering may be attritional, representing long term accumulation over periods of years or tens of years. However, it could also reflect highly variable micro-environmental conditions in which some bones of the same carcass could remain in Stage 1 while others weathered more rapidly, even to the point of total distintegration. The relative importance of micro-environmental factors versus the length of time of accumulation can be sorted out using: 1) the spatial distributions of bones of different weathering stages, 2) the variation in weathering on bones of a single animal, 3) the relationship of weathering stages to different sedimentary environments. If bones in all weathering stages are homogeneously mixed in a single deposit, then it is highly likely that they represent attritional accumulation. On the other hand, if bones of the same weathering stage are clustered together, then they may reflect local variation in weathering conditions and may or may not be attritional. Careful study of fossil remains in situ will obviously be important in establishing the history of accumulation and surface exposure of any particular vertebrate assemblage, and further study of modern land surfaces may reveal other helpful lines of evidence.

In archeological sites, weathering could provide important evidence for relative duration of occupation, recurring occupations, or the presence of a "background" of skeletal material that was not related to site formation. Recent ethnoarcheological investigation of modern campsites has shown that much bone refuse may be buried by trampling and remain in Stage 0–1 (D. Gifford, pers. comm.; Gifford and Behrensmeyer, 1977). If bones were present on the site prior to occupation, they could be similarly buried but should represent a wide range of weathering categories, unlike the cultural refuse of a short-term occupation. The potential use of weathering in an archeological context seems promising, but systematic investigations using information from recent bones have yet to be done.

Conclusion

Bone weathering is a potentially useful tool in paleoecology, archeology and recent ecology. The presence of a wide range of weathering stages in bones from a single deposit may indicate that it is an attritional assemblage. This can strongly affect paleoecologic interpretations concerning the faunal composition, relative abundances of taxa and age-structure of the preserved populations because it indicates that taphonomic biases inherent in an attritional bone assemblage must be taken into consideration. The presence of a single weathering stage or a mixture of various stages on an archeological site can indicate important differences in duration of occupation, local conditions of burial and/or the presence of non-cultural bone material. Bone weathering rates, when known for a variety of environmental conditions, should provide ecologists with a means of censusing animals and surveying habitat utilization over periods of perhaps 1–15+ yr.

Work on bone weathering has only begun, and readers are advised to treat statements in this paper as hypotheses which need testing through additional research on both recent and fossil bones. The major point of emphasis is that weathering characteristics do record meaningful information which deserves to be recognized and investigated further. The weathering stages given in this paper may well need elaboration and revision as more work is done. They should be considered provisional, but hopefully will help to establish a useful basis of communication and promote interest in weathering rates and processes.

Acknowledgments

Information and assistance of major importance to the study of bone weathering in Amboseli was generously given by Dorothy E. Dechant, Elizabeth Oswald and David Western. Permission to work in Amboseli National Park was granted by Dr. Perez Olindo, then Director of National Parks, and Mr. E. K. Ruchiami of the Office of the President of Kenya. Their cooperation, and also that of Mr. Joe Kioko, Park Warden, was much appreciated. The staff of the National Museum of Kenya were very helpful in providing osteological material for the study. Chris and Alison Hillman contributed information and help in aging carcasses in Nairobi National Park. I also thank Diane Gifford, I. Findlater, E. P. Hare, Andrew Hill, G. Isaac, and Judith Van Couvering for their contributions, ideas and stimulating skepticism regarding bone weathering.

The study of recent bones in Amboseli Park was supported primarily by the National Geographic Society (Grant #1508), with additional funding provided by National Science Foundation Grant #GS 268607A-1.

Literature Cited

Behrensmeyer, A. K. 1975. The Taphonomy and paleoecology of Plio-Pleistocene vertebrate assemblages of Lake Rudolf, Kenya. Bull. Mus. Comp. Zool. *146*:473–578.

Behrensmeyer, A. K. and D. E. Dechant. In prep. The recent bones of Amboseli Park, Kenya in relation to East African Paleoecology. To be in: Behrensmeyer, A. K. and A. Hill, eds. Fossils in the Making. In Prep. (Wenner-Gren Foundation for Anthropological research symposium 69.)

Binford, L. R. and J. B. Bertram. 1977. Bone frequencies—and attritional processes. Pp. 77–153. In: Binford, L. R., ed. For Theory Building in Archeology. Academic Press; New York.

Brain, C. K. 1967. Bone weathering and the problem of bone pseudotools. South Afr. J. Sci. *63*(3): 97–99.

Gifford, D. P. and A. K. Behrensmeyer. 1977. Observed formation and burial of a recent human occupation site in Kenya. Quaternary Res. *8*:245–266.

Hare, E. P. 1974. Amino acid dating of bone—the influence of water. Carnegie Inst. of Washington Year Book. *73*:576–581.

Hill, A. 1975. Taphonomy of contemporary and late Cenozoic East African vertebrates. 331 pp. Unpubl. Ph.D. thesis; Univ. of London, London.

Isaac, G. I. 1967. Towards the interpretation of occupation debris: some experiments and observations. The Kroeber Anthropol. Soc. Pap. *37*: 31–57.

Rickleffs, R. E. 1973. Ecology 861 pp. Chiron Press; Portland, Oregon.

Shipman, P. 1977. Paleoecology, taphonomic history, and population dynamics of the vertebrate fossil assemblage from middle Miocene deposits exposed at Fort Ternan, Kenya. Unpubl. Ph.D. thesis; New York Univ. (Anthropology), New York.

Tappen, N. C. and G. R. Peske. 1970. Weathering cracks and split-line patterns in archeological bone. Am. Antiq. *35*:383–386.

Toots, H. 1965. Sequence of disarticulation in mammalian skeletons. Contrib. Geol. *4*:37–39.

Voorhies, M. 1969. Taphonomy and population dynamics of an early Pliocene vertebrate fauna, Knox County, Nebraska. 69 pp. Contrib. Geol., Spec. Pap. No. 1. Univ. Wyo. Press; Laramie, Wyoming.

Western, D. 1973. The structure, dynamics and changes of the Amboseli ecosystem. 385 pp. Unpubl. Ph.D. thesis; Univ. of Nairobi (Zoology), Nairobi.

Western, D. and C. van Praet. 1973. Cyclical changes in the habitat and climate of an East African ecosystem. Nature. *241*:104–106.

PAPER 40

Conceptual Framework for the Analysis and Classification of Fossil Concentrations (1986)

S. M. Kidwell, F. T. Fürsich, and T. Aigner

Commentary

ADAM TOMAŠOVÝCH

Fossil concentrations are, foremostly, rich in fossils, which provides a very good sample size for estimates of various aspects of paleoecological variability. However, they also allow us to look at the dynamic of past biological productivity under a magnifying glass because, in order to form fossil concentrations, either some part of the input in fossil supply has to be maximized and/or some part of the output has to be minimized. Both sides of this dynamic—the input and output of preservable organic materials—are determined by the ecology of skeletal producers and destroyers. In this way, patterns of this dynamic capture information about variation in productivity, which is a fundamental component of ecosystem dynamic (Bambach 1993). Building on empirical examples of gradients in shell concentrations (e.g., the Middle Triassic of southern Germany, the Upper Jurassic of France and England, and the Miocene of Maryland), Kidwell et al. introduced a methodological protocol that allows quantifying such variations in space and time, effectively introducing a new approach for looking at paleoecological dynamics (Kidwell and Brenchley 1994). Of course, fossil concentrations still represent one end-member of the gradient ranging from fossil-poor to fossil-rich assemblages, but Kidwell et al. developed an approach with which a discrete and relatively simple categorization of fossil concentrations can be easily quantified and traced through time and space in the fossil record.

Kidwell et al. presented criteria for the (1) descriptive and (2) genetic classification of fossil concentrations. Their descriptive criteria are based on a simple but robust methodological protocol that was later used for standardized descriptions of fossil concentrations directly in the field in many paleoecological and taphonomic studies. Rather than just emphasizing one aspect, the genetic classification looked at fossil concentrations from three fundamental angles, including biologic, sedimentologic and diagenetic modes of fossil-concentration origin. However, Kidwell et al. also stressed that many concentrations will have composite origins and stressed especially that many sedimentologic concentrations are conditioned by biogenic concentrations. This argument also influenced sedimentologists and sequence stratigraphers because Kidwell et al. predicted that contributions of three genetic types will vary along paleoenvironmental gradients and thus will be diagnostic for specific environments and temporal segments of basinal history. Although the variation in composition and preservation of fossil concentrations can be immense, Kidwell et al.'s concept was successful because it looked at shared similarities among concentrations rather than trying to account for all of the possible attributes of concentrations, thereby maximizing the trade-off between being explaining a sufficiently complex phenomenon while still presenting tractable analyses.

From *Palaios* 1:228–38.

Literature Cited

Bambach, R. K. 1993. Seafood through time: changes in biomass, energetics and productivity in the marine ecosystem. *Paleobiology* 19:372–97.

Kidwell, S. M., and P. J. Brenchley. 1994. Patterns of bioclastic accumulation through the Phanerozoic: changes in input or in destruction? *Geology* 22:1139–43.

Conceptual Framework for the Analysis and Classification of Fossil Concentrations

SUSAN M. KIDWELL[1]

Department of Geosciences, University of Arizona, Tucson, Arizona 85721 U.S.A.

FRANZ T. FÜRSICH

Institut für Paläontologie und historische Geologie, Universität München, Richard-Wagner-Strasse 10, 8000 München 2, West Germany

THOMAS AIGNER[2]

Geologisches Institut, Universität Tübingen, Sigwartstrasse 10, 7400 Tübingen, West Germany

PALAIOS, 1986, V. 1, p. 228–238

Densely fossiliferous deposits are receiving increasing attention for their yield of paleobiologic data and their usefulness in sedimentology and stratigraphy. This trend has created a pressing need for standardized descriptive terminology and a genetic classification based on a coherent conceptual framework. The descriptive procedure outlined here for skeletal concentrations stresses four features—taxonomic composition, bioclastic fabric, geometry, and internal structure—that can be described readily in the field by nonspecialists. The genetic classification scheme is based on three end members, representing biologic, sedimentologic, and diagenetic factors in skeletal concentration. Concentrations created through the simultaneous or sequential action of two or more factors are classified as mixed types. As a conceptual framework for comparative biostratinomic analysis, the broad categories of this ternary classification scheme should facilitate recognition of large-scale temporal and spatial patterns in skeletal accumulation. The usefulness of this approach is suggested by the good agreement between biostratinomic patterns observed in ancient onshore-offshore facies tracts and those predicted across paleobathymetric transects based on modern processes of skeletal concentration.

INTRODUCTION

Concentrations of biologic hardparts are common and conspicuous features of the stratigraphic record. They have come under increasing scrutiny in recent years, both by paleontologists concerned with post-mortem bias of fossil assemblages and by geologists concerned with the determination of paleohydraulic regimes, facies analysis, and marker-bed correlation (e.g., Baird and Brett, 1983; Kreisa, 1981; Fürsich, 1978, 1982; Futterer, 1982; Hagdorn, 1982; Kidwell and Jablonski, 1983; Aigner, 1985; Kidwell, 1985; Seilacher, 1985). Nonetheless, densely fossiliferous horizons are often not fully exploited for the information they record, and there have been few attempts to relate them to one another in terms of genetic classifications or models.

Schäfer (1962, transl. in 1972) provided one of the earliest biostratinomic classifications, subdividing all marine biofacies—hardpart-rich and hardpart-poor—into five types based on water oxygenation and energy. This scheme was broadened and recast by Rhoads (1975), who used trace and body fossil abundances to classify facies. The "Fossil-Lagerstätten" scheme of Seilacher and Westphal (1971; Seilacher et al., 1985) focused on deposits which are unusually rich in paleontologic information, irrespective of whether the deposit is densely or sparsely fossiliferous. Such beds are classified according to process of concentration (condensation, placer, passive fissure trap) or mode of preservation (stagnation, rapid burial, conservation trap).

To facilitate biostratinomic analysis and comparison, this paper summarizes our joint efforts 1) to standardize nomenclature for the field description of skeletal concentrations, using both new and existing terms, and 2) to devise a conceptual framework or genetic classification scheme for skeletal concentrations. These range from localized fecal pods and channel lags to reefs and shelf-wide bioclastic sands, so a major challenge is to identify significant similarities among the many possible kinds of concentrations rather than to differentiate further among them. We have aimed for a system that is simple and yet as inclusive and general as possible.

Our descriptive and genetic schemes for fossil concentrations address only a subset of all fossil assemblages, of course, since fossils are not always found in great abundance. The schemes are thus intended to supplement rather than to replace more general models for the formation of fossil assemblages (e.g., Fagerstrom, 1964; Johnson, 1960), which stress the biasing effects of hardpart destruction, transport, and time-averaging on paleoecological data.

Fossil Concentrations and Fossil Assemblages

We define a skeletal (or fossil) concentration as *any relatively dense accumulation of biologic hardparts, irrespective of taxo-*

[1]Department of Geophysical Sciences, University of Chicago, 5734 South Ellis Avenue, Chicago, IL 60637, USA
[2]Shell Research (KSEPL), Volmerlan 6, 2288GD Rijswijk, The Netherlands

0883-1351/86/0002-0228/$03.00

nomic composition, state of preservation, or degree of post-mortem modification. Concentrations of steinkerns are included in this category. Although macroinvertebrate and vertebrate hardparts are more conspicuous, ostracods and algae can constitute concentrations as do even smaller elements such as sponge spicules, radiolarians, and coccoliths. Concentrations are not necessarily restricted in physical scale, and they can reflect fossil accumulation over very brief or very prolonged periods of time.

Fagerstrom (1964) has defined a fossil assemblage as "*any group of fossils from a suitably restricted stratigraphic interval and geographic locality*" (his italics). Most authors discriminate among types of assemblages by the extent and nature of preburial alteration of the remains of the original living community (Boucot, 1953; Johnson, 1960; Craig and Hallam, 1963; Fagerstrom, 1964; Hallam, 1967; Lawrence, 1968). Scott (1970) employed a comparative procedure based on a series of questions. Are the fossils in their life positions or disturbed? If disturbed, do they occur in the original or in some foreign substratum; that is, were they transported out of the original life habitat or not? Is the assemblage ecologically homogeneous or heterogeneous, and, if heterogeneous, were specimens mixed by biological or physical processes? In such an analysis, the abundance of fossils in the stratigraphic interval is not a factor. A fossil assemblage may occur in a bed containing abundant and densely packed fossils or in one where only widely dispersed specimens are present (Johnson, 1960).

A skeletal concentration may consist of a single, homogeneous assemblage, or it may be heterogeneous, consisting of several subsidiary fossil assemblages. We characterize these assemblages in conventional European terms that have come into common use in North America and are broadly synonymous with categories established by Scott (1970).

Autochthonous assemblages are composed of specimens derived from the local community and preserved in life positions. This category is largely synonymous with Scott's "in-place assemblage," Johnson's (1960) Model I assemblage, and Fagerstrom's (1964) end-member "fossil community." Many autochthonous assemblages are ecologically homogeneous and are undisturbed records of mass kills. Autochthonous assemblages can, however, be ecologically heterogeneous owing to time-averaging (Peterson 1977) or faunal condensation (Fürsich 1978) of successive, ecologically dissimilar species, which colonize the substratum in response to fluctuations in water salinity or oxygenation, progressive change in substratum mass properties (dewatering of mud, accumulation of bioclastic debris), or autogenic, biotically-driven succession.

Parautochthonous assemblages are composed of autochthonous specimens that have been reworked to some degree but not transported out of the original life habitat. Specimens can be reoriented, disarticulated, and concentrated by biologic agents (bioturbators, predators, scavengers) and by physical processes. This category is thus synonymous with Scott's (1970) "disturbed-neighborhood assemblage" and includes Johnson's (1960) Model II assemblage and Fagerstrom's (1964) "residual fossil assemblage," in that some original elements of the life assemblage can be missing due to selective destruction or transport out of the habitat.

Allochthonous assemblages, composed of specimens transported out of their life habitats and occurring in a foreign substratum, are equivalent to the "transported assemblages" of Johnson (1960: Model III), Fagerstrom (1964), and Scott (1970). Assemblages composed of specimens of different origins are referred to as *mixed autochthonous-parautochthonous, parautochthonous-allochthonous,* or *autochthonous-allochthonous.* These types subdivide the "mixed assemblage" category of Fagerstrom (1964) and Scott (1970).

All of these terms refer to skeletal material that is roughly contemporaneous in age with the embedding sediment. Following Craig and Hallam (1963), we use *remanie* to refer to significantly older material that is reworked into the assemblages, and we use *leaked* to indicate younger material piped down into the assemblage through burrows or fissures.

We emphasize that there is no necessary one-to-one correspondence or synonymy between types of fossil assemblage and types of fossil concentration. Fossil concentrations are formed by sedimentologic as well as biologic processes, so their analysis and classification complements paleoecological terminology and the relation of fossil assemblages to the once-living communities from which they have been derived.

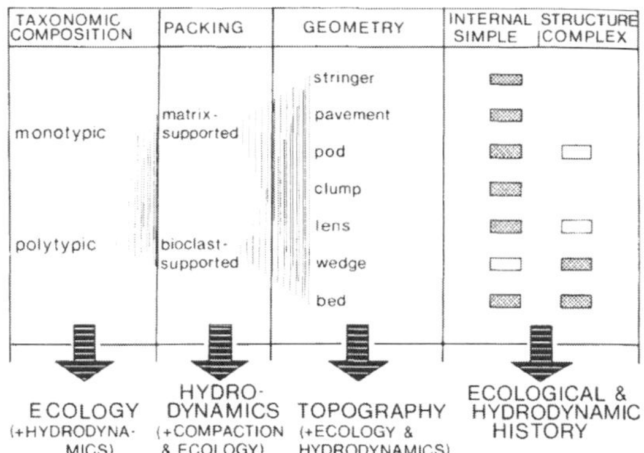

FIGURE 1—Procedure for description of skeletal concentrations, with standardized terms for major features and their genetic significance. Not all combinations of features are likely. For example, stringers and pavements are almost always simple in internal structure (shaded box), whereas among the thicker beds, both simple and complex internal structures are common.

DESCRIPTIVE NOMENCLATURE FOR SKELETAL CONCENTRATIONS

Of the many possible field observations of skeletal concentrations (Ager, 1963), our descriptive scheme stresses only four: 1) taxonomic composition, 2) bioclastic fabric (specifically close-packing), 3) geometry, and 4) internal structure of the deposit (Fig. 1). Most of these characters show a continuous range of states, so our terminology incorporates arbitrary subdivisions. The descriptive categories could be quantified, but here we stress visual definitions in order to maximize practicality for field studies.

Although reefs and other biologic buildups are certainly

skeletal concentrations, they are not included in this discussion. Nomenclature for the description of boundstones is quite distinctive and usually inappropriate for level-bottom marine and terrestrial fossil concentrations (Heckel, 1974; James and MacIntyre, 1985).

Taxonomic Composition

This criterion depends upon the biologic structure of living communities of skeletonized organisms—the source of hardparts—and also upon the hydrodynamics of their accumulation, since differential fragmentation and transport of skeletal elements affect the final taxonomic composition of the deposit (Hollmann, 1968; Force, 1969; Cadée, 1968; Aigner and Reineck, 1982). Taxonomic composition is also a function of differential preservation of skeletal elements in consequence of post-mortem exposure on the sea floor and early diagenesis (Driscoll, 1970; Peterson, 1976), reworking or stratigraphic leakage of hardparts, and time-averaging or faunal condensation of successive ecologic associations (Fürsich, 1978; Peterson, 1977; Kidwell and Aigner, 1985).

Concentrations are characterized as *monotypic* or *polytypic* by whether they are composed of a single or several types of skeletons. We broaden Fagerstrom's (1964) original definition of monotypy to apply to any taxonomic level appropriate to the study, not to species alone. A monotypic accumulation can be composed solely of pelecypods, of ostreids, or of *Crassostrea virginica*. The lower the taxonomic level, of course, the greater the ecological or hydrodynamic significance of monotypy. Monospecific and polyspecific are useful variants, and the term paucispecific is valuable for reference to an assemblage of only a few species, especially if it is strongly dominated by one species.

Biofabric

Biofabric refers to the three-dimensional arrangement of skeletal elements in the matrix. It includes skeletal orientation, close-packing, and sorting by size and shape. The biofabric depends primarily on the hydrodynamics of hardpart concentration, but it may also reflect rotation and disarticulation of elements during compaction; preferential destruction of the matrix by pressure solution; the ecology and necrology of the organisms, including their life positions; and the ecology of other organisms that modified the skeletal remains (predators, scavengers, bioturbators).

In the plane of bedding surfaces, orientation is usually described by rose diagrams (Figure 2). It has been used by many workers as a paleocurrent indicator (review, Potter and Pettijohn, 1963; Nagle, 1967; Brenner, 1976; Futterer, 1982). In assemblages comprising more than one modal size or multiple shapes, it is commonly necessary to evaluate the orientation of each mode or shape separately in order to detect the unimodal or bimodal alignment of elongate elements (Urlichs, 1971).

For orientations observed in cross section, no consistent terminology has been proposed. We suggest *concordant* to describe the parallel to subparallel alignment with bedding of long axes of hardparts or flat surfaces of platey elements; *perpendicular* to describe elements arranged largely at right angles to bedding; and *oblique* for elements exhibiting intermediate positions (Fig. 2). We recommend concordant as a less ambiguous term than parallel or subparallel, which can equally well refer to the orientation of hardparts relative to one another. Special terms are reserved for fabrics that are characteristically ordered in cross section; these include *imbrication, edgewise, stacking,* and *nesting. Telescoping* is an unusual but highly diagnostic fabric in which conical elements interpenetrate. It has been recognized in nautiloid concentrations, where the apex of one shell pierces the septa of another (Tasch, 1955), and in scaphopod and hyolithid concentrations (Spath, 1936; Yochelson and Fraser, 1973; James and Klappa, 1983).

Close-packing in fossiliferous deposits ranges from highly dispersed fabrics to densely packed concentrations containing 70% or more hardparts by volume. Becaue the visual estimation of close-packing (e.g., charts of Schäfer, 1969) is a function of the shape and orientation of hardparts as well as their volumetric abundance, we adapt Dunham's (1962) petrographic fabric terms to describe skeletal accumulations. *Matrix-supported* refers to concentrations in which the hardparts "float" in the matrix, and *bioclast-supported* indicates that hardparts are in physical contact with each other, supporting the bed structurally or as a house-of-cards. These biostratinomic categories are roughly comparable to Dunham's mudstone/wackestone and packstone/grainstone categories.

The hydrodynamic and ecological significance of hardpart size-frequency distributions is poorly understood owing to the effects of shape and density on hydraulic equivalence (Maiklem, 1968; Braithwaite, 1973) and the complexities of population dynamics, ecology, compaction, and differential preservation potential according to size (Shimoyama, 1985; Green et al., 1984; Mancini, 1978; Price, 1982). We describe the size ranges of species that predominate in numbers of individuals or skeletal volume and record whether the distribution is unimodal or polymodal, well sorted or poorly sorted. The frequency distribution of hardpart shapes is difficult to quantify, but it is implied by taxonomic composition and by the extent of fragmentation and disarticulation, which are described as aspects of shape and size sorting.

Geometry

The geometry of a fossil concentration depends upon a number of biologic and physical factors. These include 1) the inherited topography of the depositional surface (e.g., crevice and burrow fills); 2) the mode of life of the hardpart producer (e.g., epibenthic clumps of gregarious taxa such as oysters and mussels); 3) activities of other living organisms (e.g., fecal pellets, gastric residues, shell-lined burrows, and shell lags produced through selective deposit feeding); and 4) physical processes of hardpart concentration that produce syngenetic topography (e.g., lags produced by migrating ripples, channels, and shoals).

Innumerable terms have been used to describe the geometries of fossil concentrations. Some of the most common ones are illustrated in Figure 3. Thin, virtually two-dimensional bedding-plane concentrations are referred to as *pavements* and *stringers*, depending on whether the concentration is laterally continuous or only patchily developed at the scale of a single outcrop. A stringer indicates a very localized and usually elongate concentration. Where thickness exceeds that of one shell

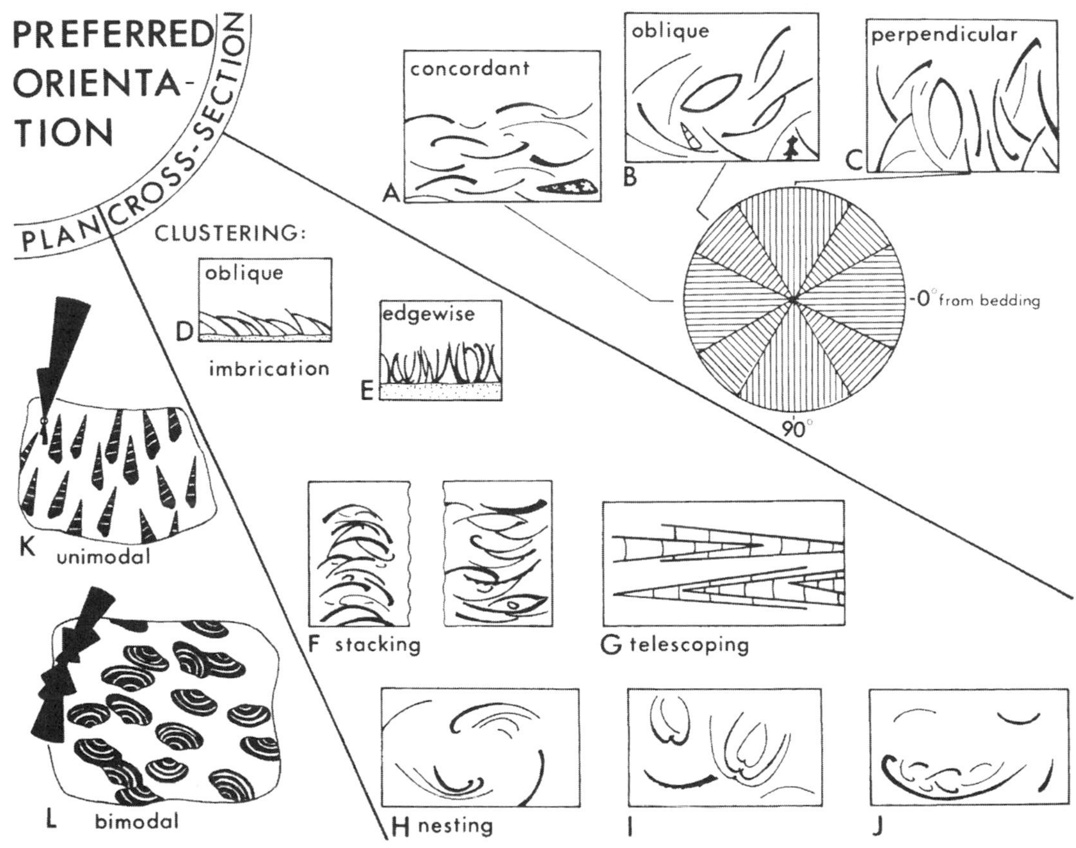

FIGURE 2—Terminology for hardpart orientation on bedding planes and in cross section of bed.

or other skeletal element, the terms *clump, pod, lens, wedge,* and *bed* are applied. A pod denotes a small-scale, irregular concentration with well-defined edges (e.g., isolated burrow fill, some fecal masses), and a clump denotes a concentration with poorly defined margins (e.g., clusters of shells in life position, or small concentrations that have been disturbed by sediment bioturbation). Small or large lenses have a more regular geometry with tapered lateral terminations (pinched out or erosionally beveled). They include ray-pit fills, channel and scour fills, isolated shoals, and biohermal structures. Wedges taper laterally in a more complex manner, in only one direction, whereas beds are continuous tabular or sheet-like accumulations. Beds can exhibit considerable variation in thickness depending on the topography of their upper and lower boundaries, and are distinguished from horizons of physically discrete lenses or pods.

Internal Structure

Lateral and vertical variation in the taxonomic composition, biofabric, and matrix of fossil deposits provides important evidence of complex histories of hardpart accumulation.

Simple skeletal concentrations are internally homogeneous (invariable) or exhibit at most some monotonic trend in features, such as lateral or upward fining in the grain size of matrix or bioclasts. Shelly turbidites and offshore tempestites (Aigner, 1982) are examples of skeletal concentrations showing simple internal variation.

Complex concentrations exhibit more complicated patterns of variation in one or more features, such as alternating horizons of articulated and disarticulated, reoriented hardparts. Concentrations produced through lateral or vertical amalgamation of smaller-scale concentrations (Aigner, 1982; Fürsich and Oschmann, in press; Kidwell and Aigner, 1985) are examples of complex internal structures.

Simple and complex are used here in a strictly descriptive sense: they do not necessarily indicate simple or complex origins for the concentration, nor do they translate directly into the single- and multiple-event genetic categories of Aigner et

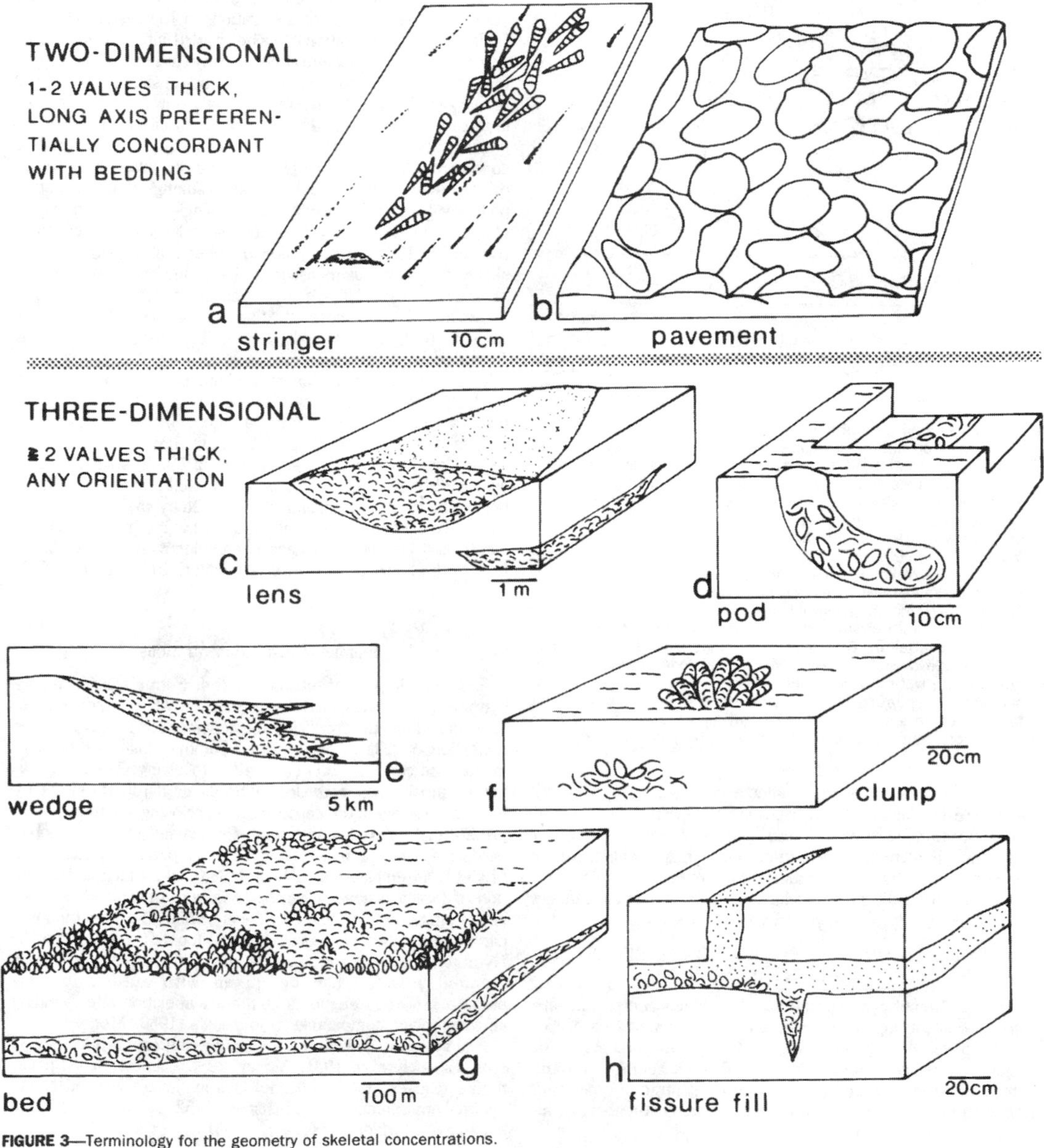

FIGURE 3—Terminology for the geometry of skeletal concentrations.

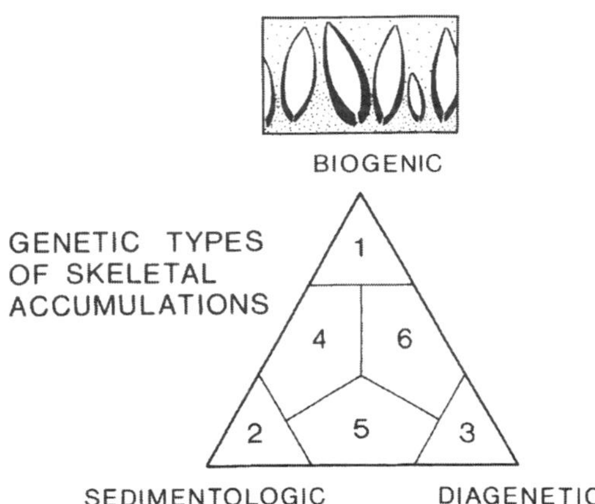

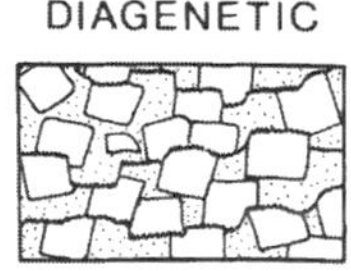

FIGURE 4—Conceptual framework for genesis of skeletal concentrations based on three end-member sets of concentrating processes. Biogenic concentrations (area 1) are produced by the gregarious behavior of skeletonized organisms (*intrinsic* biogenic) or by the actions of other organisms (*extrinsic* biogenic). Sedimentologic concentrations (area 2) form through hydraulic reworking of hardparts as particles and/or through nondeposition or selective removal of sedimentary matrix. Diagenetic concentrations (area 3) include residues of concentrated skeletal material along pressure solution seams and compaction-enhanced fossil horizons. Mixed origin concentrations (areas 4–6) reflect the strong influence of two or more different kinds of processes, for example, hydraulic overprinting of a biogenic precursor concentration. End-member concentrations (areas 1–3) can record single or multiple events of skeletal concentration; beds of mixed origin will most commonly reflect more than one episode of skeletal concentration. Concentrations of any of the six types can form rapidly (a few hours) or very slowly (hundreds to thousands of years); long-term concentrations will typically be mixed in origin.

al. (1978). Many internally simple concentrations probably originate in single events that bring hardparts together, but concentrations with complicated, multiple-event histories may be simple in structure because of thorough admixing of earlier and later skeletal concentrations. Moreover, a laterally complex concentration can be produced by a single event if it affects a geomorphically or bathymetrically variable area.

BIOSTRATINOMIC CLASSIFICATION OF SKELETAL CONCENTRATIONS

In our genetic classification, skeletal concentrations are grouped according to the inferred relative importance of biologic, physical sedimentologic, and diagenetic processes (or agents) of concentration (Figure 4). Fossil assemblages of the three end-member and three mixed concentration types recognized here can be autochthonous, parautochthonous, or allochthonous in origin.

Biogenic Concentrations

Although individual hardparts are produced by biologic processes, not every fossil concentration owes its high density to biologic processes alone. Those that do are categorized here as biogenic, including the ecologic shell beds of Aigner et al. (1978) and the biologic concentrations of Kidwell (1982a).

Biogenic concentrations can be divided into two subtypes. *Intrinsic biogenic* concentrations are created by the organisms that produce the hardparts: concentration results from intrinsic gregarious behavior—in life or death—of the skeletonized organisms themselves. The component fossil assemblages are typically autochthonous or parautochthonous. Intrinsic biogenic concentrations can record 1) preferential colonization by larvae of sites with abundant adults, as seen among vermetid gastropods, oysters, and some scallops; 2) single colonization events of opportunistic species characterized by large population sizes (Levinton, 1970); and 3) dense ephemeral aggregations of skeletonized organisms associated with feeding, spawning, or moulting (Waage, 1964; Speyer and Brett, 1985).

Extrinsic biogenic concentrations are produced by other organisms that interact with skeletonized organisms or their discarded hardparts. Component fossil assemblages are typically parautochthonous or allochthonous. Examples include hardpart-rich fecal masses (Mellett, 1974; Korth, 1979; Freeman, 1979), subsurface lags produced by "conveyor belt" deposit feeders (Rhoads and Stanley, 1965; van Straaten, 1952; Cadée, 1976; Trewin and Welsh, 1976), shell-filled pits produced by bottom-feeding predators and scavengers (Gregory et al., 1979), accumulations produced by shell-transporting birds (Teichert and Serventy, 1947; Lindberg and Kellogg, 1982), and burrows that have been selectively lined or backfilled with shells (e.g., *Diopatra*; Schäfer, 1972; Kern, 1978).

Sedimentologic Concentrations

Sedimentologic concentrations result from physical, usually hydraulic, processes of concentration, in which hardparts behave as sedimentary particles and nonbioclastic matrix is either reworked or fails to accumulate. Common histories of sedimentologic concentration include the following: 1) Concentration of initially dispersed, locally produced hardparts either by hydraulic sorting of the hardparts and enclosing sediment, or by preferential removal of fine sediment, leaving a lag of immobile hardparts; such a fossil assemblage is parautochthonous, a Model II assemblage of Johnson (1960). 2) Gradual accumulation of locally produced hardparts during a period of low net sedimentation; this yields a concentration of autochthonous-parautochthonous hardparts, also a Model II assemblage. 3) Hydraulic transport and selective redeposition of allochthonous elements, which can be mixed with autochthonous-parautochthonous elements at the accumulation site, yielding an assemblage comparable to Johnson's (1960) Model III.

Examples of sedimentologic concentrations include shelly storm lags (Kreisa, 1981; Aigner, 1982), aeolian beach pavements (Carter, 1976), channel lags in fluvial, intertidal, and subtidal environments (van Straaten, 1952; Schäfer, 1972), and shell-paved turbidites (Tucker, 1969).

Diagenetic Concentrations

These are skeletal concentrations created or significantly enhanced by physical and chemical processes acting after burial. Most fossil assemblages are altered in some way by diagenesis, but only where fossil density is significantly *increased* are these classified as diagenetic concentrations. The most important diagenetic processes of hardpart concentration are probably 1) compaction, which can increase the close-packing of shells, particularly in a fine-grained matrix (Fürsich and Kauffman, 1984), and 2) selective pressure solution of matrix in bioclastic limestones, which leaves enriched fossil horizons along stylolitic seams (Wanless, 1979; Eller, 1981). Relative concentrations of hardparts can also result from the diagenetic destruction of hardparts in adjacent beds (Fürsich, 1982; Haszeldine, 1984).

Concentrations of Mixed Origin

Concentrations inferred to have formed by the interplay of two or more kinds of processes, or by the strong overprinting of a precursor concentration of one type by later processes of a different kind, are mixed in origin and plot in one of the intermediate fields of our ternary diagram (areas 4–6, Fig. 4). Biostromes built up by alternating gregarious settlement and hydraulic reworking are common examples of deposits of mixed origin, formed by multiple events of hardpart concentration.

The mixed biogenic-sedimentologic category of Figure 4 (area 4) includes skeletal accumulations with two common types of history. One is the enrichment of an initial biogenic concentration through a later episode or episodes of physical reworking, which yields a largely parautochthonous fossil assemblage. Examples from the Maryland Miocene include localized pavements of flat-lying, articulated *Pinna* which have been eroded out of their upright, semi-infaunal life positions; beds of disarticulated, convex-up valves of the gregarious infaunal bivalve *Glossus*; and pavements of wave-oriented shells of the gregarious infaunal gastropod *Turritella* (Kidwell, 1982a). Because hydraulic reworking can obscure or even erase evidence of the original biogenic nature of a concentration, many concentrations classified as purely sedimentologic in origin may in fact be strongly overprinted biogenic deposits. Such overprinting can usually be inferred for small-scale pavements and lenses of hydraulically-oriented specimens whose taxonomic composition is similar to that of biogenic concentrations in the same or adjacent facies. An interpretation of mixed origin is strengthened by other evidence for gregarious behavior, as where closely related or homomorphic living taxa create intrinsically biogenic concentrations. If reworked skeletal elements are judged to be allochthonous, the concentration is classed as sedimentologic rather than overprinted biogenic.

Sedimentologic concentrations that have been recolonized by bottom-dwelling organisms represent a second major mode of formation of mixed biogenic-sedimentologic concentrations. Because the sedimentologic concentration of hardparts provides a hard or coarse-textured substratum in a setting that was previously characterized by soft, fine-grained sediments, the new colonists, and hence their skeletal remains, can be quite different from those of the initial concentration. The resulting fossil assemblage is either parautochthonous (but ecologically mixed) or mixed parautochthonous-allochthonous, depending on the source of hardparts in the sedimentologic phase of hardpart concentration (Kidwell and Jablonski, 1983; Kidwell and Aigner, 1985).

Early cementation of current- and wave-generated shell pavements (Sepkoski, 1978; McCarthy, 1977) results in the preferential preservation of hardparts and illustrates diagenetic enhancement of sedimentologic concentrations (area 5 of ternary diagram in Figure 4). Early concretionary cementation of the products of mass mortality enhances biogenic concentrations in a similar way. In both cases, initial concentrations are enriched relative to the failure of dispersed hardparts outside the concentration to survive early diagenesis. Diagenesis can also secondarily enhance biogenic concentrations through compaction of their muddy matrix (Fürsich and Kauffman, 1984).

PALEOECOLOGICAL AND SEDIMENTOLOGICAL APPLICATIONS

This classification is more of a conceptual framework than a precise model for biostratinomic analysis. The ternary diagram (Fig. 4) is not a graph in the strict, quantitative sense: the axes are unscaled and the boundaries of the fields representing six broad genetic categories are arbitrary. This schematic diagram draws attention to possible interactions among three distinctive sets of processes or agents that are involved in the formation of skeletal concentrations. As in the case of the Konstruktions-Morphologie scheme of Seilacher (1970), the relative roles of end-member agents or processes are difficult to quantify. They refer to variables that are incommensurate, and so the scheme cannot be based on any single data set.

Like all classification systems, our ternary scheme groups phenomena at the expense of loss of detail. For example, hydraulically reworked biogenic concentrations are lumped together with recolonized sedimentologic concentrations in the mixed biogenic-sedimentologic category. The processes involved in the formation of these concentrations act in opposite sequence: the first starts in area 1 and moves down into area 4 of the diagram (Fig. 4), the other begins in area 2 and moves up into area 4. The cost of losing such detailed historical information is outweighed, we feel, by the scheme's ability to accommodate concentrations whose origins are more complex and less clear-cut, such as concentrations formed by repeated cycles of both hydraulic overprinting and recolonization. Because the scheme can include such a variety of concentrations in its six categories, it should facilitate the identification of broad patterns in the fossil record.

Environmental Gradients

Comparative analysis using the ternary scheme is especially worthwhile along environmental gradients. Figure 5 depicts relative abundances of hardpart concentrations expected across an onshore-offshore transect in a terrigenous, nondeltaic depositional setting. This represents our present best estimate of an actualistic biostratinomic facies model, and is derived from qualitative patterns reported in the published literature and personal observations of modern environments.

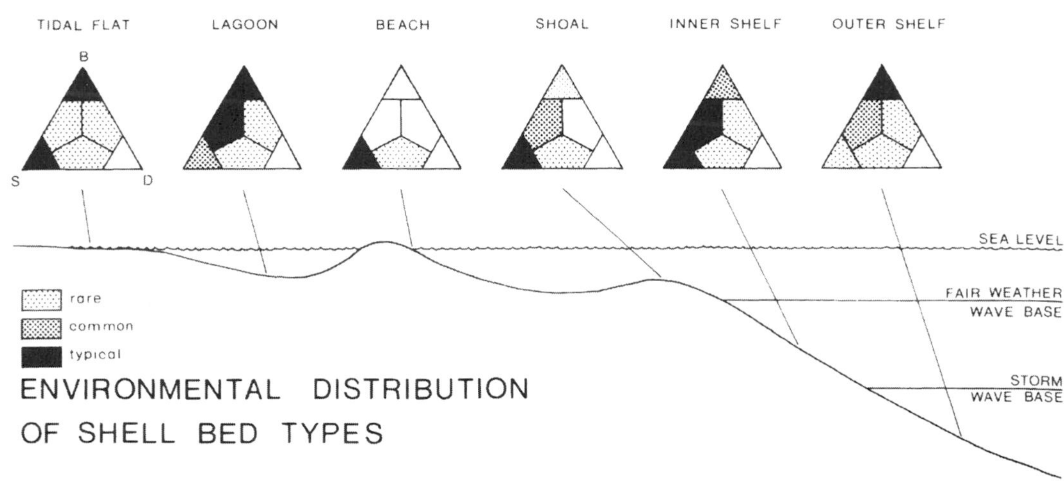

FIGURE 5—Expected relative abundances of skeletal concentrations along an onshore-offshore transect in a marine setting dominated by terrigenous sedimentation. Rates of sediment accumulation are assumed to be constant across the transect. In this preliminary actualistic model, sedimentologic concentrations decrease in abundance from beach to outer shelf because of diminishing water energy at the sea floor, and biogenic concentrations increase in relative abundance.

We suggest a diverse assortment of both biogenic and sedimentologic concentrations in intertidal and backbarrier environments, and an offshore trend of increasing biogenic and decreasing sedimentologic concentrations seaward from the coast. This reflects diminishing energy of physical reworking in deeper, more offshore waters and assumes equivalent rates of sediment accumulation in onshore and offshore settings.

Intertidal and Supratidal Flats

Biogenic concentrations include channel-margin oyster bars, mussel clumps, beds of deep-burrowing bivalves in life position, winnowed lags of shallow-burrowing opportunists, *Arenicola*-graded shell beds, ray pits, and bird nests. Sedimentologic concentrations include those produced by the lateral migration of channels and a variety of shell pavements and spits produced by storm surges.

Lagoons and Bays

Biogenic concentrations include those produced by cohorts of opportunistic species; mass mortalities related to fluctuations in salinity, temperature, or oxygen, as well as to stranding; and colonization of exposed shell by encrusters and other sessile epifauna. Sedimentologic concentrations here are typified by storm lags, flood deposits (plants, freshwater macroinvertebrates), and distal washover deposits associated with barrier beaches.

Beaches

Few species colonize sandy beaches owing to their instability, so biogenic concentrations and hydraulically overprinted biogenic concentrations are usually absent. Bedding plane lags of allochthonous and parautochthonous shells, shell lags in scour structures, aeolian shell pavements, and shell beds formed through lateral migration of tidal inlets and storm-surge channels are most typical.

Subtidal Shoals

Owing to the relatively high energy of the environment, virtually all biogenic concentrations are reworked to some degree. In addition, many sedimentologic concentrations produced through storm reworking and fair-weather winnowing provide substrata for colonization by benthos, which increases the relative frequency of mixed biogenic-sedimentologic shell beds in this facies.

Open Shelves

Above storm-wave base, sedimentologic concentrations dominate. These are typified by individual and amalgamated shelly storm lags, but also include reworked biogenic concentrations and recolonized sedimentologic concentrations. Further offshore, biologic processes are responsible for the formation of most concentrations (e.g., biostromes, bioherms, clumps, fecal concentrations). Episodes of sea-floor reworking or winnowing are too infrequent in these settings or too low in energy to generate significant numbers of sedimentologic and overprinted-biogenic shell beds. Moreover, bioturbation is likely to remove evidence of physical reworking. Sedimentologic concentrations in this deepest and most distal part of the shelf are largely limited to those formed during very rare storm events and under regimes of reduced net sedimentation. The latter provide excellent opportunities for colonization by shell gravel taxa, as seen in the colonization of relict shell gravels on modern continental shelves. In tide-dominated systems, sedimentologic concentrations will be more common on the outer shelf.

Preliminary Results and Discussion

Our generalizations and expected trends are largely corroborated by the fossil record, where we have independent sedimentologic and paleoecologic evidence for paleobathymetric interpretations. For example, within regressive sequences of the Maryland Miocene (Kidwell, 1984), nearshore shallow-water facies are consistently dominated by a variety of sedimentologic and mixed sedimentologic-biogenic concentrations, whereas offshore facies are characterized by biogenic and mixed sedimentologic-biogenic deposits (Table 1). Each of the four regressive sequences includes benthic assemblages dominated by different infaunal genera, so the observed biostratinomic trends are not linked to specific organisms or life habits.

A similar, very strong paleobathymetric trend in skeletal concentrations has been recognized in the Pliocene Purisima Formation of central California by Norris (1986). His quantitative assessment indicates that skeletal concentrations from nearshore facies in this high energy setting are primarily sedimentologic in origin, whereas biogenic and mixed biogenic-sedimentologic concentrations predominate in middle shelf facies. The outer shelf is marked by bone beds formed under conditions of sediment starvation. As in the Purisima, the deepest water facies of the Maryland Miocene also contains a starved, condensed bone accumulation (Myrick, 1979; Kidwell, 1984).

The stratigraphic frequency of skeletal concentrations of a single type can also be used as a paleobathymetric criterion within basins. Aigner (1982, 1985) has demonstrated this by contouring the frequency of storm-generated shell beds in the Triassic Muschelkalk of southern Germany. Tempestite frequency decreases offshore owing to the increasing rarity of storm currents with sufficient energy to rework deeper water sediments. The degree of amalgamation of individual concentrations also decreases offshore in this basin. Similarly, Norris (1986) has documented an offshore decrease in the frequency and extent of amalgamation of biogenic as well as sedimentologic shell beds in the Purisima Formation, a pattern that is also apparent, but less clearly developed, in the lower energy, shallow-marine record of the Maryland Miocene (Table 1).

These trends in the relative abundance of types of concentration should be reasonably consistent along paleobathymetric transects, although factors such as coastal energy and seafloor slope will shift the absolute depths and relative widths of the biostratinomic facies belts. The thickness of individual fossil concentrations and their stratigraphic spacing within facies must also vary significantly with background rates of sedimentation and hardpart supply, both as a function of basin history (e.g., Kidwell and Jablonski, 1983) and climatic regime. Onshore-offshore biostratinomic trends are expected to depart from the actualistic model further back in the Phanerozoic. The environmental expansion, diversification, and intensification of bioturbation over time (Thayer, 1983) have surely altered the balance of factors involved in skeletal concentration, because bioturbation disperses sedimentologic concentrations and accelerates dissolution of calcareous hardparts (Aller, 1982). The decrease in frequency of flat-pebble conglomerates and other storm lags in shallow subtidal sediments after the early Paleozoic (Sepkoski, 1982) is probably one expression of long-term change in biostratinomic patterns that has resulted from biologic evolution.

TABLE 1—Biostratinomic trends along onshore-offshore gradients in regressive facies tracts of the Maryland Miocene (Calvert and Choptank Formations), showing decreased relative abundance of sedimentologic concentrations in offshore facies. Onshore sand facies indicate intertidal to very shallow subtidal conditions based on sedimentary structures and paleoecology; offshore silty sand facies record subtidal environments below fairweather wave base. PP = Plum Point Member, Calvert Formation. CT = Choptank Formation. Further documentation of these data is in preparation (S.M. Kidwell).

Stratigraphic Interval	Onshore Sand Facies	Offshore Silty Sand Facies	Relative Abundance of Genetic Types: Sedimentologic	Mixed Origin	Biogenic
	Mytilus		++	+/−	++
CT-0		*Turritella-Glossus*	−	++	+
		Tellina	−	−	+
	Pandora		+	++	−
PP-3		*Isognomon*	−	++	+
		Turritella	−	++	++
		Glossus	−	++	+/−
	Chione		+	+	+
PP-2		*Glossus*	−	++	+/−
	Corbula		++	+	+
PP-0		*Glossus-Tellina*	+/−	++	+

++ = frequent + = occasional +/− = rare − = absent

CONCLUSIONS

1. The straightforward descriptive procedure developed here should be applicable in the field by geologists as well as paleontologists. We have attempted to standardize existing nomenclature for four major features of hardpart concentrations—taxonomic composition, biofabric, geometry, and internal structure—which have genetic significance but are themselves noninferential. This procedure should facilitate systematic characterization of local sections in terms of their skeletal concentrations, which are at present underexploited in the differentiation and mapping of sedimentary facies.
2. We have identified a minimum number of genetic types of skeletal concentration based on three end-member sets of processes and their interactions. This scheme is intended to provide a conceptual framework for biostratinomic analysis. Genetic classification can be made more precise by the addition of modifiers (for example, single- versus multiple-event biogenic concentrations) and by subdivision (for example, reworked biogenic versus recolonized sedimentologic concentrations within the mixed biogenic-sedimentologic category).

3. The genetic scheme does not specify taxonomic composition, physical scale, or the time scale of formation of skeletal concentrations. It should thus facilitate the recognition of broad biostratinomic trends in the stratigraphic record. Stratigraphic data as well as observations of modern processes indicate that skeletal concentrations are not distributed randomly along environmental gradients. They reflect differences in hydraulic energy, the ecology of skeletonized organisms, rates and styles of bioturbation, rates of sediment accumulation, and diagenetic regimes, among other variables.
4. Expected onshore-offshore trends exhibited by marine skeletal concentrations imply different kinds of post-mortem preservational bias among sedimentary facies and different opportunities for biotic recolonization of skeletal material. These factors further alter the composition of prolific fossil assemblages. Onshore-offshore biostratinomic trends also provide a potentially useful criterion for paleobathymetric interpretation.

ACKNOWLEDGMENTS

This collaboration was made possible by a Heisenberg Fellowship (FTF) and by grants from the Petroleum Research Fund of the American Chemical Society (SMK) and the West German Sondersforschungbereich Project 53 (TA and SMK), for which we are grateful. We thank D. Jablonski, A. Seilacher, D.J. Bottjer, R.D. Norris, C.E. Brett, and anonymous reviewers for helpful comments and criticism. R.D.K. Thomas provided excellent criticisms and editorial suggestions.

REFERENCES

AGER, D. V., 1963, Principles of Paleoecology: New York, McGraw-Hill, 371 p.

AIGNER, T., 1982, Calcareous tempestites: storm-dominated stratification in Upper Muschelkalk limestones (Triassic, SW-Germany), *in* Einsele, G., and Seilacher, A., eds., Cyclic and Event Stratification: Berlin, Springer-Verlag, p. 180–198.

AIGNER, T., 1985, Storm Depositional Systems. Dynamic Stratigraphy in Modern and Ancient Shallow-Marine Sequences: Berlin, Springer-Verlag, 174 p.

AIGNER, T., and REINECK, H.-E., 1982, Proximality trends in modern storm sands from Helgoland Bight (North Sea) and their implications for basin analysis: Senck. Marit., v. 14, p. 183–215.

AIGNER, T., HAGDORN, H., and MUNDLOS, R., 1978, Biohermal, biostromal, and storm-generated coquinas in the Upper Muschelkalk: N. Jb. Geol. Paläont. Abh., v. 157, p. 42–52.

ALLER, R. C., 1982, Carbonate dissolution in nearshore terrigenous muds: the role of physical and biological reworking: Jour. Geol., v. 90, p. 79–95.

ARNTZ, W. E., BRUNWIG, D., and SARNTHEIN, M., 1976, Zonierung von Mollusken und Schill im Rinnensystem der Kieler Bucht (Westliche Ostsee): Senck. Marit., v. 8, p. 189–269.

BAIRD, G. C., and BRETT, C. E., 1983, Regional variation and paleontology of two coral beds in the Middle Devonian Hamilton Group of western New York: Jour. Paleont., v. 57, p. 417–446.

BOUCOT, A. J., 1953, Life and death assemblages among fossils: Amer. Jour. Sci., v. 251, p. 25–40.

BRAITHWAITE, C. J. R., 1973, Settling behaviour related to sieve analysis of skeletal sands: Sedimentology, v. 20, p. 251–262.

BRENNER, K., 1976, Ammoniten-Gehause als Anzeiger für Paläo-Strömungen: N. Jb. Geol. Paläont. Abh., v. 151, p. 101–118.

BRETT, C. E., and BROOKFIELD, M. E., 1984, Morphology, faunas, and genesis of Ordovician hardgrounds from southern Ontario, Canada: Palaeogeogr. Palaeoclim. Palaeoecol., v. 46, p. 233–290.

CADÉE, G.C., 1968, Molluscan biocoenoses and thanatocoenoses in the Ria de Arosa, Galicia, Spain: Rijksmuseum Nat. Hist. Leiden, Zool. Verhandl., v. 95, p. 1–121.

CADÉE, G. C., 1976, Sediment reworking by *Arenicola marina* on tidal flats in the Dutch Wadden Sea: Neth. Jour. Sea Research, v. 10, p. 440–460.

CARTER, R. W. G., 1976, Formation, maintenance, and geomorphological significance of an aeolian shell pavement: Jour. Sed. Petrology, v. 46, p. 418–429.

CRAIG, G. Y., and HALLAM, A., 1963, Size-frequency and growth-ring analyses of *Mytilus edulis* and *Cardium edule*, and their palaeoecological significance: Palaeontology, v. 6, p. 731–750.

DRISCOLL, E. G., 1970, Selective bivalve shell destruction in marine environments, a field study: Jour. Sed. Petrology, v. 40, p. 898–905.

DUNHAM, R. J., 1962, Classification of carbonate rocks according to depositional texture: Amer. Assoc. Petrol. Geol. Mem., v. 1, p. 108–121.

ELLER, M. G., 1981, The red chalk of eastern England: a Cretaceous analogue of Rosso Ammonitico: Rome, Edizioni Technoscienza, Rosso Ammonitico Symp. Proc., p. 207–231.

FAGERSTROM, J. A., 1964, Fossil communities in paleoecology: their recognition and significance: Geol. Soc. Amer. Bull., v. 75, p. 1197–1216.

FORCE, L. M., 1969, Calcium carbonate size distribution on the West Florida shelf and experimental studies on the microarchitectural control of skeletal breakdown: Jour. Sed. Petrology, v. 39, p. 902–934.

FREEMAN, E. F., 1979, A Middle Jurassic mammal bed from Oxfordshire: Palaeontology, v. 22, p. 135–166.

FÜRSICH, F. T., 1978, The influence of faunal condensation and mixing on the preservation of fossil benthic communities: Lethaia, v. 11, p. 243–250.

FÜRSICH, F. T., 1979, Genesis, environments, and ecology of Jurassic hardgrounds: N. Jb. Geol. Paläont. Abh., v. 158, p. 1–63.

FÜRSICH, F. T., 1982, Rhythmic bedding and shell bed formation in the Upper Jurassic of East Greenland, *in* Einsele, G., and Seilacher, A., eds., Cyclic and Event Stratification: Berlin, Springer-Verlag, p. 209–222.

FÜRSICH, F. T., and KAUFFMAN, E. G., 1984, Palaeoecology of marginally marine sedimentary cycles in the Albian Bear River Formation of southwestern Wyoming (USA): Palaeontology, v. 27, p. 501–536.

FÜRSICH, F. T., and OSCHMANN, W., Storm shell beds of *Nanogyra virgula* in the Upper Jurassic of France: N. Jb. Geol. Paläont. Abh. (in press).

FUTTERER, E., 1982, Experiments on the distinction of wave and current influenced shell accumulations, *in* Einsele, G., and Seilacher, A., eds., Cyclic and Event Stratification: Berlin, Springer-Verlag, p. 175–179.

GREEN, R. H., MCCUAIG, J., and HICKS, B., 1984, Testing a paleoecology matrix model: estimates for a *Sphaerium* population in equilibrium with its death assemblage: Ecology, v. 65, p. 1201–1205.

GREGORY, M. R., BALLANCE, P. F., GIBSON, G. W., and AYLING, A. M., 1979, On how some rays (Elasmobranchia) excavate feeding depressions by jetting water: Jour. Sed. Petrology, v. 49, p. 1125–1130.

HAGDORN, H., 1982, The "Bank der kleinen Terebrateln" (Upper Muschelkalk, Triassic) near Schwäbisch Hall (SW-Germany)—a tempestite condensation horizon, *in* Einsele, G., and Seilacher, A., eds., Cyclic and Event stratification: Berlin, Springer-Verlag, p. 263–285.

HALLAM, A., 1967, The interpretation of size-frequency distributions in molluscan death assemblages: Palaeontology, v. 10, p. 25–42.

HASZELDINE, R. S., 1984, Muddy deltas in freshwater lakes, and tectonism in the Upper Carboniferous Coalfield in NE England: Sedimentology, v. 31, p. 811–822.

HECKEL, P. H., 1974, Carbonate buildups in the geologic record: a review, *in* Laporte, L.F., ed., Reefs in Time and Space: Soc. Econ. Paleontologists Mineralogists Spec. Publ., v. 18, p. 90–155.

HOLLMAN, R., 1968, Zur Morphologie rezenter Mollusken-Bruchschille: Paläont. Zeitschr., v. 42, p. 217–235.

JAMES, N. P., and KLAPPA, C. F., 1983, Petrogenesis of early Cambrian reef limestones, Labrador, Canada: Jour. Sed. Petrology, v. 53, p. 1051–1096.

JAMES, N. P. and MACINTYRE, I. G., 1985, Carbonate depositional environments, modern and ancient. Part I: Reefs: zonation, depositional facies, diagenesis: Colo. School Mines Quart., v. 80, p. 1–70.

JOHNSON, R. G., 1960, Models and methods for analysis of the mode of formation of fossil assemblages: Geol. Soc. Amer. Bull., v. 71, p. 1075–1086.

KERN, J. P., 1978, Paleoenvironment of new trace fossils from the Eocene Mission Valley Formation, California: Jour. Paleont., v. 52, p. 186–194.

KIDWELL, S. M., 1982a, Time scales of fossil accumulation: patterns from Miocene benthic assemblages: Third North Amer. Paleont. Conv. Proc., Montreal, v. 1, p. 295–300.

KIDWELL, S. M., 1982b, Stratigraphy, invertebrate taphonomy and depositional history of the Miocene Calvert and Choptank Formations, Atlantic Coastal Plain [unpubl. Ph.D. dissert.]: New Haven, CT, Yale Univ., 514 p.

KIDWELL, S. M., 1984, Outcrop features and origin of basin margin unconformities in the Lower Chesapeake Group (Miocene), Atlantic Coastal Plain: Amer. Assoc. Petrol. Geol. Mem., v. 36, p. 37–58.

KIDWELL, S. M., 1985, Palaeobiological and sedimentological implications of fossil concentrations: Nature, v. 318, p. 457–460.

KIDWELL, S. M. and AIGNER, T., 1985, Sedimentary dynamics of complex shell beds: implications for ecologic and evolutionary patterns, *in* Bayer, U., and Seilacher, A., eds., Sedimentary and Evolutionary Cycles: Berlin, Springer-Verlag, p. 382–395.

KIDWELL, S. M. and JABLONSKI, D., 1983, Taphonomic feedback: ecological consequences of shell accumulation, *in* Tevesz, M.J.S., and McCall, P.L., eds., Biotic Interactions in Recent and Fossil Benthic Communities: New York, Plenum Press, p. 195–248.

KORTH, W. A., 1979, Taphonomy of microvertebrate fossil assemblages: Annals Carnegie Museum, v. 48, p. 235–285.

KREISA, R. D., 1981, Storm-generated sedimentary structures in subtidal marine facies with examples from the Middle and Upper Ordovician of southwestern Virginia: Jour. Sed. Petrology, v. 51, p. 823–848.

LAWRENCE, D. R., 1968, Taphonomy and information losses in fossil communities: Geol. Soc. Amer. Bull., v. 79, p. 1315–1330.

LEVINTON, J. S., 1970, The paleoecological significance of opportunistic species: Lethaia, v. 3, p. 69–78.

LINDBERG, D. R., and KELLOGG, M. G., 1982, Bathymetric anomalies in the Neogene fossil record: the role of diving marine birds: Paleobiology, v. 8, p. 402–407.

MAIKLEM, W. R., 1968, Some hydraulic properties of bioclastic carbonate grains: Sedimentology, v. 10, p. 101–109.

MANCINI, E. A., 1978, Origin of micromorph faunas in the geologic record: Jour. Paleont., v. 52, p. 321–333.

MCCARTHY, B., 1977, Selective preservation of mollusc shells in a Permian beach environment, Sydney Basin, Australia: N. Jb. Geol. Paläont. Mh., 1977, p. 466–474.

MELLETT, J. S., 1974, Scatological origins of microvertebrate fossil accumulations: Science, v. 185, p. 349–350.

MYRICK, A. C., 1979, Variation, taphonomy, and adaptation of the Rhabdosteidae (Eurhinodelphidae) (Odontoceti, Mammalia) from the Calvert Formation of Maryland and Virginia [unpubl. Ph.D. dissert.]: Los Angeles, Univ. California, 411 p.

NAGLE, J. S., 1967, Wave and current orientation of shells: Jour. Sed. Petrology, v. 37, p. 1124–1138.

NORRIS, R. D., 1986, Taphonomic gradients in shelf fossil assemblages: Pliocene Purisima Formation, California: Palaios, v. 1, p. 252–266.

PETERSON, C. H., 1976, Relative abundances of living and dead molluscs in two California lagoons: Lethaia, v. 9, p. 137–148.

PETERSON, C. H., 1977, The paleoecological significance of undetected short-term temporal variability: Jour. Paleont., v. 51, p. 976–981.

POTTER, P. E., and PETTIJOHN, F. J., 1963, Paleocurrents and Basin Analysis: New York, Academic Press, 296 p.

PRICE, A. R. G., 1982, Western Arabian Gulf echinoderms in high salinity waters and the occurrence of dwarfism: Jour. Nat. Hist., v. 16, p. 519–527.

RHOADS, D. C., 1975, The paleoecological and environmental significance of trace fossils, *in* Frey, R. W., ed., The Study of Trace Fossils: New York, Springer-Verlag, p. 147–160.

RHOADS, D. C., and STANLEY, D. J., 1965, Biogenic graded bedding: Jour. Sed. Petrology, v. 35, p. 956–963.

SCHÄFER, K. A., 1969, Vergleichs-Schaubilder zur Bestimmung des Allochemgehaltes bioklastischer Karbonatgesteine: N. Jb. Geol. Paläont. Mh. 1969, p. 173–184.

SCHÄFER, W., 1972, Ecology and Palaeoecology of Marine Environments [transl. I. Oertel]: Chicago, University of Chicago Press, 568 p.

SCOTT, R. W., 1970, Paleoecology and paleontology of the Lower Cretaceous Kiowa Formation, Kansas: Univ. Kansas Paleont. Contrib. Art., 52 (Cretaceous 1), 94 p.

SEILACHER, A., 1970, Arbeitskonzept zur Konstruktions-Morphologie: Lethaia, v. 8, p. 393–396.

SEILACHER, A., 1985, The Jeram model: event condensation in a modern intertidal environment, *in* Bayer, U., and Seilacher, A., eds., Sedimentary and Evolutionary Cycles: Berlin, Springer-Verlag, p. 336–341.

SEILACHER, A. and WESTPHAL, F., 1971, Fossil-Lagerstätten, *in* Miller, G., ed., Sedimentology in parts of Central Europe; Guidebook to Excursions held during the VII Int. Sed. Congress: Frankfurt am Main, W. Kramer, p. 327–335.

SEILACHER, A., REIF, W. E., and WESTPHAL, F., 1985, Sedimentological, ecological and temporal patterns of fossil Lagerstätten: Phil. Trans. Roy. Soc. Lond. B, v. 311, p. 5–23.

SEPKOSKI, J. J., 1978, Taphonomic factors influencing the lithologic occurrence of fossils in Dresbachian (Upper Cambrian) shaley facies [abstr.]: Geol. Soc. Amer. Abstr. W. Progr., v. 10, p. 490.

SEPKOSKI, J. J., 1982, Flat pebble conglomerates, storm deposits, and the Cambrian bottom fauna, *in* Einsele, G., and Seilacher, A., eds., Cyclic and Event Stratification: Berlin, Springer-Verlag, p. 371–385.

SHIMOYAMA, S., 1985, Size-frequency distribution of living populations and dead shell assemblages in a marine intertidal land snail, *Umbonium (Suchium) moniliferum* (Lamarck), and their palaeoecological significance: Palaeogeogr. Palaeoclim. Palaeoecol., v. 49, p. 327–353.

SPATH, L. F., 1936, So-called *Salterella* from the Cambrian of Australia: Geol. Mag., v. 73, p. 433–440.

SPEYER, S. E., and BRETT, C. E., 1985, Clustered trilobite assemblages in the Middle Devonian Hamilton Group: Lethaia, v. 18, p. 85–103.

TASCH, P., 1955, Paleoecologic observations on the orthoceratid coquina beds of the Maquoketa at Graf, Iowa: Jour. Paleont., v. 29, p. 510–518.

TEICHERT, C., and SERVENTY, D. L., 1947, Deposits of shells transported by birds: Amer. Jour. Sci., v. 245, p. 322–328.

THAYER, C. W., 1983, Sediment-mediated biological disturbance and the evolution of marine benthos, *in* Tevesz, M.J.S., and McCall, P.L., eds., Biotic Interactions in Recent and Fossil Benthic Communities: New York, Plenum Press, p. 479–625.

TREWIN, N. H., and WELSH, W., 1976, Formation and composition of a graded estuarine shell bed: Palaeogeogr. Palaeoclim. Palaeoecol., v. 19, p. 219–230.

TUCKER, M. E., 1969, Crinoidal turbidites from the Devonian of Cornwall and their paleogeographic significance: Sedimentology, v. 13, p. 281–290.

URLICHS, M., 1971, Alter und Genese des Belemnitenschlachtfeldes im Toarcium von Franken: Geol. Bl. Nordost-Bayern, v. 21, p. 65–83.

VAN STRAATEN, L. M. J. U., 1952, Biogenic textures and formation of shell beds in the Dutch Wadden Sea: Proc. K. ned. Akad. Wet. (B), v. 55, p. 500–516.

WAAGE, K. M., 1964, Origin of repeated fossiliferous concretion layers in the Fox Hills Formation: Kansas Geol. Surv. Bull., v. 169, p. 541–563.

WANLESS, H. R., 1979, Limestone response to stress: pressure solution and dolomitization: Jour. Sed. Petrology, v. 59, p. 437–462.

YOCHELSON, E. L., and FRASER, G.D., 1973, Interpretation of depositional environment of the Plympton Formation (Permian), southern Pequop Mountains, Nevada, from physical stratigraphy and a faunule: U.S. Geol. Surv. Jour. Res., v. 1, p. 19–32.

PAPER 41

Quantitative Studies of the Bridge Creek Flora (1924)

R. W. Chaney

Commentary

LEO J. HICKEY

Two papers representing fundamentally different approaches to the study of fossil plants appear in the issue of the *American Journal of Science* for August 1924. The first, by the then-dean of American paleobotany, Edward W. Berry, describes a purported fossil cashew nut from South America and follows the prevailing interest of the field up until then with the identity and temporal ranges of plant fossils. The second, a report titled, "Quantitative Studies on the Bridge Creek Flora," by the neophyte Ralph W. Chaney, is a radical departure for the field and marks the beginning of efforts to rigorously reconstruct plant paleoecology and paleofloral dynamics. In the words of Daniel Axelrod (1971), previous paleobotanists "were concerned chiefly with the 'what' and the 'when,' of fossil plants," while his teacher was concerned with the why.

Educated as both a biologist and a geologist at the University of Chicago, Chaney came under the influence of one of the founders of the science of plant ecology, Frederic E. Clements, on a visit to Washington, DC, in 1916 and was to spend much of the remainder of his career applying Clements's concepts of vegetation and dynamical changes in floras to the fossil record. In Clementsian terms, a flora represents a static list of the species present, while vegetation represents the dynamic interactions of its components. This made it essential to establish the quantitative aspects of a flora, such as percentage representation and cover of its species.

This paper—the product of one herculean week when Chaney and a single assistant excavated and identified 20,000 plant fossils from the type locality of the Bridge Creek Flora in Oregon—represents the first effort to derive such information from the fossil record by a field census and by quantitative comparison of a fossil and an analogous modern vegetational assemblage. It also shows his surprisingly modern understanding of the taphonomic factors that limit paleovegetational reconstructions. Publication of the Bridge Creek study marks the foundation of a major school of paleoecologically oriented paleobotany that is still in vigorous flower through Ralph Chaney's students and their intellectual descendants.

Literature Cited

Axelrod, D. I. 1971. Ralph Works Chaney. In *Yearbook of the American Philosophical Society 1971*, 115–20. Philadelphia: American Philosophical Society.

From *American Journal of Science*, 5th ser., 8:127–44.

Art. VIII.—*Quantitative Studies of the Bridge Creek Flora;* by Ralph W. Chaney. (With Plates V, VI.)

In the study of the plant life of the past, exact data concerning the relative abundance of the species making up a flora are as important to a paleobotanist as are similar data to a student of modern plants. Both the botanical and geological aspects of the study make it essential to determine which are the dominant and which are the accidental species in a fossil flora. To the geologist, the dominant species are of the greater value in correlating fossil-bearing deposits of two regions; it is these species which he most commonly finds, and which have the greatest weight when, as is not infrequently the case, only the most fragmentary material is available. The rare species are of much less interest to him, but it is essential that they should be recognized as rare, so that their general absence from the record can be properly interpreted. The dominant species give to the botanist a clew as to the general character of the vegetation which enables him to reconstruct it in terms of modern plant associations with similar dominants. The rare species are likewise of interest to him, indicating the border forms which are less readily made a part of the fossil record, or in some cases forms which are just beginning to become established or just being eliminated. To the combined viewpoint of geologist and botanist, in the interpretation of the fossil plant record in terms of earth history,—climate, topography and conditions of sedimentation, a knowledge of the dominant or accidental character of the various species is also essential.

Such descriptive phrases as "a large number," "many hundreds," "most abundant," and "quite rare," which are examples of the sort most commonly used in papers on fossil botany, give an imperfect idea as to the numerical representation of the various fossil species in the flora. Further, they may be based not on the actual numbers of individuals present in the rocks at a given locality but on the numbers of individual specimens represented in collections from that locality. The writer recently had occasion to comment on the representation of a species of fossil sequoia, *Sequoia langsdorfii,* in one of the Tertiary floras of the West. Judging from the number of specimens in his collections he would have decided that it was a rare species; but an actual count at the fossil locality showed that it made up over fifteen percent of the total of a large number of individual leaves examined. In this case, its rarity in the collections was misleading, and was due in the main to the fact that few specimens had been taken, since sequoia fossils show little variation.

In the course of a recent study of the Bridge Creek flora the need for exact quantitative data regarding the fossil species has become apparent. This flora is found in volcanic shales referred to the lower part of the John Day series, which is commonly considered to be of upper Oligocene age, and is rather widespread in central Oregon. The composition of the fossil flora suggests its relationship to the modern redwood forest, as has been pointed out.[1] Table I shows the Bridge Creek species which are related to species living in or on the borders of the redwood forest.

In addition there is a fossil species each of *Pteris* and *Equisetum* which are much like species of these genera now living on the stream borders in the redwood forest.

There are also two elements in the Bridge Creek flora which do not have any corresponding species in the forest of the Redwood Belt. One of these is made up of such genera as *Juglans, Platanus,* and *Celtis,* of which there are species now living in the West, but not associated with the redwood. The other includes *Carpinus, Tilia, Ulmus,* and *Fagus,* genera which do not reach the Pacific Coast to-day although there is abundant evidence that they have formerly been widespread over the whole of the northern hemisphere.

This paper will discuss the methods used in determining the relative abundance of the species making up the Bridge Creek flora, and will make suggestions as to the significance of the figures. The type locality on Bridge Creek, nine miles northwest of Mitchell, Oregon, was selected as the most suitable place for making a count of fossil specimens because of the abundance there

[1] Chaney, R. W., A Comparative Study of the Bridge Creek Flora and the Modern Redwood Forest, Carnegie Inst. Wash. Pub. 337, Contributions to Paleontology, vol. 1.

TABLE I

Bridge Creek	Redwood Forest
Pinus knowltoni Chan.	Pinus ponderosa Laws.—Western yellow pine
Tsuga sp.[2]	Tsuga heterophylla Sarg.—Western hemlock
Sequoia langsdorfii Heer.	Sequoia sempervirens Endl.—coast redwood
Torreya sp.[2]	Torreya californica Torr.—California nutmeg
Myrica sp.	Myrica californica Cham.—California waxberry
Corylus macquarrii Heer.	Corylus rostrata Ait. var. californica A.DC.—California hazel
Alnus carpinoides Lesq.	Alnus rubra Bong.—red alder
Quercus consimilis Newb.	Quercus[3] densiflora Hook. & Arn.—tanbark oak
Berberis simplex Newb.	Berberis nervosa Pursh.—Oregon grape
Umbellularia sp.	Umbellularia californica Nutt.—California laurel
Philadelphus sp.	Philadelphus lewisii Pursh.—syringa
Crataegus newberryi Cock.	Crataegus rivularis Nutt.—hawthorne
Rosa hilliae Lesq.	Rosa nutkana Presl.—wild rose
Franxinus sp.	Franxinus oregona Nutt.—black ash
Acer osmonti Kn.	Acer macrophyllum Pursh.—broadleaf maple
Rhamnus sp.	Rhamnus purshiana C.—Cascara Sagrada
Cornus sp.	Cornus nutallii Aud.—Western dogwood

of leaves sufficiently well preserved to be readily identified. This locality is a hill about thirty-five feet high, made up of alternating layers of reddish-gray shale containing leaf impressions and layers of yellowish clay. From the top, referred to as Pit I, the loose soil and rock fragments were cleared away from an area approximating fifteen feet in length and averaging four feet wide. Excavation was then made into a layer of leaf-bearing shale a foot in thickness, giving a total volume of fossil-bearing material of sixty cubic feet. The shale was then split by the writer and Mr. Frederick H. Frost and the fossil impressions counted and recorded. Fig. 1 (Pl. V) shows the general character of the country at Bridge Creek, and fig. 2 (Pl. V) the fossil-bearing slabs of shale excavated at Pit I. During the first day's work we averaged 114 specimens apiece an hour, but after a

[2] Forms referred to in this way are new species which have not yet been described or species whose relation to previously described fossil forms is not fully understood at present.

[3] The generic name Lithocarpus is often used to emphasize the intermediate position of this form between the oaks and chestnuts. But the name Quercus is applied here to correspond with the generic reference of its fossil equivalent.

Am. Jour. Sci., Vol. VIII, August, 1924. Plate V.

FIG. 1.—The type locality of the Bridge Creek flora in the John Day Basin of Oregon. The dominant plant is *Artemisia tridentata,* and the only trees present are scattered junipers, *Juniperus occidentalis.*

FIG. 2.—Pit I at the top of the hill shown in the preceding figure. The slabs piled around the pit show the character of the leaf-bearing shale as it was excavated.

few days this average was more than doubled. A total of 15,653 specimens were counted and recorded from Pit I during the nearly forty hours (eighty man-hours) we worked there. This gave an average of 261 specimens to a cubic foot; and when it is realized that, in spite of the fineness of our splitting, we were unable to expose and count more than perhaps 75% of the total of the leaves enclosed in the shale slabs, the abundance of the fossils is even more striking.

The shale at Pit II, eight feet below the top of the hill, did not contain as well-preserved material as at Pit I, and only about 8 cubic feet were excavated here. From this a total of 1723 specimens were counted and recorded, giving an average of 215 specimens to the cubic foot. From Pit III, near the base of the hill and 25 feet below Pit I, 30 cubic feet of shale were removed. This was not as finely bedded as that from the other pits and could not be so finely split; it yielded only 3234 specimens, or an average of 108 to the cubic foot. From all three pits there was excavated 98 cubic feet of fossiliferous shale which yielded a total of 20,611 specimens. These include 31 different forms, whose numbers for each pit and for the total are shown, together with their percentages, in Table II.[3a] Unless otherwise stated the figures represent numbers of leaves except in the case of *Sequoia langsdorfii*, where twigs bearing a score or more needles are the deciduous units. The figures for *Juglans* sp., *Fraxinus* sp., *Rhus* sp., and *Rosa hilliæ*, which have compound leaves, refer to the numbers of leaflets.

It will be noted that of the species listed in Table I *Pinus knowltoni, Tsuga* sp., *Torreya* sp., *Myrica* sp., *Berberis simplex,* and *Rhamnus* sp. have not been recorded in the count at the type locality, although they have been found previously at this or other localities of the Bridge Creek flora in Oregon. None of these species is at all common, half a dozen specimens or less representing the total of each which have ever been found. It is therefore not surprising that they were not recorded during the count at the type locality.

There will be no attempt in this paper to make a taxo-

[3a]. There is a slight variation in the relative numbers of the species at the three pits, but this difference is of no great significance to the discussion of this paper.

nomic revision, but the writer wishes to point out the doubtful status of three of the species listed in Table II. The leaves of *Quercus clarnoensis* differ from those of *Quercus consimilis* only in being more slender; there appears, however, to be a gradation in shape between these two forms, and it is possible that they should be

TABLE II.

	PIT I		PIT II		PIT III		TOTAL	
	No.	%	No.	%	No.	%	No.	%
Alnus carpinoides	7827	50.00	951	55.19	2268	70.13	11046	53.59
Sequoia langsdorfii	2327	14.86	334	19.33	445	13.76	3106	15.07
Quercus consimilis	1594	10.18	119	6.91	134	4.14	1847	8.96
Umbellularia sp.	1667	10.65	113	8.56	48	1.48	1828	8.82
Quercus clarnoensis	726	4.63	62	3.60	36	1.11	824	4.00
Ulmus speciosa	458	2.92	53	3.08	123	3.80	634	3.08
Platanus aspera	178	1.13	10	.58	48	1.48	236	1.15
Platanus condoni	159	1.01	5	.29	24	.74	188	.91
Grewia crenata	156	.99	7	.41	10	.31	173	.84
Acer seed	121	.77	8	.46	12	.37	141	.68
Ulmus newberryi	86	.55	10	.58	7	.22	103	.50
Crataegus flavescens	68	.46	4	.23	2	.06	74	.36
Tilia sp.	72	.43	1	.06			73	.35
Quercus seed	62	.39	9	.52	1	.03	72	.34
Philadelphus sp.	46	.29	9	.52	5	.15	60	.29
Juglans sp.	6	.04	6	.35	42	1.29	54	.26
Alnus female ament	19	.12	8	.46	1	.03	28	.13
Fraxinus sp.	12	.07	6	.35	7	.22	25	.12
Sequoia cone	10	.06	3	.17	6	.18	19	.09
Potomogeton (?)	11	.07	1	.06	3	.09	15	.07
Rhus sp.	14	.08					14	.07
Rosa hilliae	7	.04	2	.12	1	.03	10	.05
Cornus sp.	4	.03			6	.18	10	.05
Carpinus grandis	5	.03	1	.06			6	.03
Celtis sp.	1	.006			5	.15	6	.03
Alnus male ament	6	.04					6	.03
Carpinus seed	5	.03					5	.02
Hydrangea bendirei (flower)	3	.02	1	.06			4	.02
Corylus macquarryi	2	.01					2	.01
Equisetum sp.(stem)	1	.006					1	.005
Leguminous pod	1	.006					1	.005
	15654	99.92	1723	99.86	3234	99.93	20611	99.93

NOTE.—Line 4, column 4 read 6.56, not 8.56.

considered as a single species. In this case, the percentages of *Quercus consimilis* in the table would be notably increased. The presence of a species of *Grewia* in a flora made up of genera which are temperate in distribution to-day seems unlikely, as has been pointed out.[4] Except for the bluntly serrate margins of the leaves of *Grewia crenata*, it could better be referred to *Cercis;* the

[4] Op. cit. p. 2.

leguminous pod listed closely resembles the pods of Cercis, a genus now living on the Pacific Coast. There is no certainty that the flower listed as *Hydrangea bendirei* is referred to the proper genus, but it and the other two doubtful forms are listed here under their old names.

It seemed probable that the fruiting parts are related to the leaf species of corresponding genera. Thus the alder aments are referable to *Alnus carpinoides,* the acorns to *Quercus consimilis,* the sequoia cones to *Sequoia langsdorfii,* and the *Carpinus* seed to *Carpinus grandis.* There were no leaves of *Acer* recorded in this count at the type locality, but a few have been found there previously and elsewhere at the same horizon, and referred to *Acer osmonti.* It has already been suggested that the leguminous pod may be related to the leaves of *Grewia crenata* and that both are referable to Cercis. The reference of the fruits of *Alnus, Sequoia, Quercus,* and *Carpinus* to the related leaf species would reduce the total number of forms to twenty-six, and increase slightly the percentages of these leaf species.

Referring again to Table II, the question may be raised as to the actual significance of the figures for the various species. Can it be assumed that the numbers of leaves and fruiting parts are an accurate indication of the relative abundance of the species they represent in the Bridge Creek forest? The general aspects of this matter have been discussed by the writer,[5] and five factors have been named which may condition the abundance of the several species of leaf fossils in the record. However, it has been difficult to consider the significance of these factors in more than a theoretical way with no other data than the fossil record. And it has been wholly impossible to determine the numbers of trees in the Bridge Creek forest, since the record is made up only of disconnected tree fragments,—leaves and fruits, and pieces of wood. A study of conditions in the most nearly related living forest has seemed to present the

[5] Op. cit. p. 2. The factors are: (1) the distance of a species from the cite of deposition; (2) the original thickness of the leaf, determining its ability to be transported without destruction; (3) the size and shape of the leaf as related to its transportation in air and water; (4) the habits of the tree with regard to shedding its leaves; and (5) the height of the stem of the plant, involving its arborescent, shrubby, or herbaceous habit.

most promising method of supplying the needed information regarding the Tertiary forest. The results of this study are here presented with the full realization that they are in no sense complete or final, and that this attempt to compare living and past conditions is handicapped by the many factors involved, some of which may not have been properly evaluated or may have been wholly disregarded.

The Bridge Creek flora has been shown to have many elements in common with the redwood forest now living in the Coast Ranges of California, and to resemble it more closely than any other living forest.[6] Measurements have been made of the redwood forest in and adjacent to the Muir Woods National Monument, which is located in Marin County, California,[7] and which occupies a canyon on the south side of Mt. Tamalpais. This canyon is traversed by Redwood Creek, a stream only a few feet wide and with a volume so limited during much of the year that most of its water is collected in basins at the bends. It is in these basins that sediments and leaves are accumulating and they are taken to represent situations similar to the sites of deposition of the Bridge Creek shales. It should be pointed out that the Bridge Creek deposits appear to have been laid down in basins of deposition of a stream of considerably greater size than Redwood Creek, a fact which introduces certain possible errors which will be mentioned below.

Measurements were made along the course of Redwood Creek throughout the main portion of the redwood forest, beginning at a point 800 feet downstream from the lower boundary of Muir Woods and extending about 500 feet upstream from its upper boundary, a distance along the stream of about seven-eighths of a mile. In each basin a count was made of the leaves on one square foot of the bottom, and in most cases the downstream end of the basin was selected because the leaves were more abundant there. At the suggestion of Dr. Frederic E. Clements, a stiff wire one foot square was used, and this was ordinarily tossed into the basin to avoid unconscious selection of the spot to be measured. Fig. 1 (Pl. VI) shows the general aspect of the forest at Muir

[6] Op. cit. p. 2.

[7] See U. S. Geol. Surv., Tamalpais Sheet, central and east central rectangles.

Am. Jour. Sci., Vol. VIII, August, 1924. Plate VI.

Fig. 1.—The redwood forest at Muir Woods, California. The four dominant species, *Sequoia sempervirens, Alnus rubra, Quercus densiflora,* and *Umbellularia californica,* are present here, together with *Rhododendron occidentale* and others.

Fig. 2.—Station 28 in the bed of Redwood Creek at Muir Woods, showing leaves of *Sequoia sempervirens, Alnus rubra,* and *Quercus densiflora* in the square foot counted.

Woods, and fig. 2 (Pl. VI) the locus of one of the stations in the stream bed of Redwood Creek. Following the recording of the leaves, a count was made of all of the plants within a radius of fifty feet, and the distance noted from the nearest branch of each species to the point where the leaf count was made. Several basins were not measured because of their small size (less than fifteen feet in length). In counting the surrounding plants, of which many of the common tree species have the habit of growing in more or less united groups, the practice was followed of recording as separate trees all stems which were units to the ground, regardless of the fact that they were in some cases connected at or below the surface. In the case of such shrubs as *Corylus, Rubus,* and *Rhododendron,* the large numbers of associated stems were not counted, but referred to as thickets which were recorded as one individual. No attempt was made to count the actual number of ferns or herbs, which are referred to in the tables in such general terms as common and abundant, since none of these are present in the fossil record with which the Muir Woods record is to be compared. Altogether forty-two basins were measured, yielding a total of 8422 leaves and fruits, and the data secured are summarized in Table III. Unless otherwise stated, the figures represent numbers of leaves, except in the case of *Sequoia sempervirens,* where twigs bearing a score or more of needles are the deciduous units. In the cases of the species having compound leaves, the figures refer to the numbers of leaflets.

This table will be briefly compared with Table II, which shows the occurrence of the fossil species at Bridge Creek. Twenty-seven species are present in the part of the modern forest which was studied, of which the leaves of nineteen have been noted in the deposits along the stream; six of the species represented by leaves are also represented by fruits. Twenty-five, or possibly twenty-six species, are known to have been present in the fossil forest, of which 21 have entered the fossil record as leaves; four, or possibly five, of these latter are also represented by fruits. Of the 19 species represented in the deposits in the living forest, nine have equivalents in the fossil record, and at least four other species have been recorded from the Bridge Creek flora previous to the count at the type locality. The four

TABLE III.

Summary of Records at 42 Stations Along Redwood Creek.

	Total No. of specimens	% of total	Total No. plants	% of total	Distance Av.	Distance Max.	Min
Sequoia sempervirens Endl.	3314	39.35	281	19.84	10	50	0
Alnus rubra Bong.	2304	27.36	209	14.76	3	100	0
Umbellularia californica Nutt	1118	13.27	231	16.24	4	50	0
Quercus densiflora Hook. & Arn.	460	5.46	240	16.95	4	35	0
Rhododendron occidentale Gray.	321	3.81	138	9.74	30	125	0
Acer macrophyllum Pursh	241	2.86	26	1.84	21	120	0
Corylus rostrata Ait.	116	1.38	142	10.03	21	100	0
Salix lasiolepis Benth.	20	.24	4	.28	29	100	0
Vaccinium ovatum Pursh.	7	.08	34	2.40	17	30	0
Rubus parviflorus Nutt.	5	.06	65	4.59	15	40	5
Aesculus californica Nutt	5	.06	11	.78	12	25	3
Rubus vitifolius C. & S.	4	.05	Fairly Common		8	10	4
Cornus nuttallii Aud.	3	.03	4	.28	97	100	95
Myrica californica Cham.	2	.02	2	.14			
Quercus agrifolia Nee.	2	.02	0		375	500	250
Aspidium munitum Kaulf.	2	.02	Very Common		16	20	12
Pteris aquilina pubescens Underw	1	.01	Common		8	8	8
Asarum caudatum Lindl.	1	.01	Fairly Common		20	20	20
Rosa nutkana Presl.	1	.01	8	.56	8	8	8
Sequoia sempervirens cone	241	2.86			8	45	0
Alnus rubra - female ament	124	1.47			2	18	0
Alnus rubra - male ament	1	.01			25	25	0
Umbellularia seed	87	1.03			3	15	0
Quercus seed	38	.45			6	15	0
Acer seed	4	.05			6	15	0
Asplenium filix-femina Bernh.	0		A few				
Woodwardia chamissoi Brack.	0		A few				
Equisetum telmateia Ehrh.	0		2	.14			
Pseudotsuga taxifolia Britt.	0		2	.14			
Berberis nervosa Pursh.	0		Fairly Common				
Ribes sp.	0		2	.14			
Sambucus sp.	0		7	.49			
Arbutus menziesii Pursh.	0		1	.07			
Euonynus occidentalis Nutt.	0		6	.42			
Dirca occidentalis Gray.	0		1	.07			
	8422	99.97	1416	99.90			

most abundant species, making up a total of 85.44% in the modern forest, are the equivalents of the four most abundant fossil species, which make up 86.44% of the total in the count at Bridge Creek. Of the living species lacking representatives in the fossil flora, only one, *Rhododendron occidentale,* comprises more than one per-cent of the total. The fossil species lacking representatives in the modern forest are for the most part members of genera which are not associated with the redwood to-day. Clearly the most characteristic species of the two floras show a relationship which serves further to emphasize the similarity between the Bridge Creek fossil flora and the flora of the living redwood forest. In view of this relationship a somewhat detailed analysis will be made covering the data secured in the modern forest, since it may be expected to throw light on the conditions which prevailed during the Bridge Creek epoch.

It seems desirable at the outset to determine whether there is a well-established relationship between numbers of leaves and numbers of adjacent trees, or in other words to establish the degree of correlation between the numbers of leaves of each species found at 42 stations and the numbers of individual plants of these species in their immediate vicinity, within a radius of 50 feet of each station. The writer wishes to acknowledge at this point the assistance given him by Dr. R. H. Franzen, of the University of California, in computing and interpreting the statistical data of this paper. Table IV shows the record for *Alnus rubra.*

TABLE IV.

Record of Leaf and Plant Frequency for *Alnus rubra.*

Station	No. of leaves in one square foot X	No. of plants within a 50-foot radius Y
1	186	18
2	61	18
3	121	5
4	40	6
5	85	25
6	97	23
7	68	6
8	69	6
9	123	28
10	56	25
11	105	5
12	112	4

Station	No. of leaves in one square foot X	No. of plants within a 50-foot radius Y
13	72	8
14	15	0
15	33	0
16	142	1
17	100	7
18	1	0
19	13	2
20	103	1
21	9	2
22	29	2
23	90	2
24	0	1
25	0	0
26	88	2
27	56	2
28	108	2
29	63	2
30	75	3
31	79	2
32	0	0
33	5	0
34	69	1
35	1	0
36	0	0
37	0	0
38	0	0
39	0	0
40	0	0
41	0	0
42	0	0
	2304	209

By means of a correlation formula,[8] a correlation value of .486 is secured which indicates that there is a distinct tendency for the number of Alnus leaves to be large at stations about which Alnus trees are most numerous, and to be small where the trees are few in number or

[8]
$$r_{xy} = \frac{\frac{\Sigma XY}{n} - M_x M_y}{\sqrt{\frac{\Sigma X^2}{n} - M_x^2}\sqrt{\frac{\Sigma Y^2}{n} - M_y^2}}$$

In this formula X is used as the value of leaf number, and Y as the value of plant number. The correlation between leaf number and plant number, r_{XY}, is equal to the mean of the product of X and Y, $\frac{\Sigma xy}{n}$, minus lacking. Correlations were made in a similar way for the other species which have Bridge Creek equivalents,[9] with the following results:

TABLE V.

Correlation Values of Deposited Leaves with Numbers of Plants in Muir Woods.

	Leaf
Sequoia sempervirens[10]	.493
Alnus rubra	.486
Umbellularia californica	.649
Quercus densiflora	.569
Acer macrophyllum[11]	.339
Corylus rostrata	.134
Cornus pubescens	.477
Rosa nutkana	.364

The correlation values for the leaves of all but *Corylus rostrata* and *Rosa nutkana* are sufficiently high to indicate a relationship between the numbers of leaves in the stream deposits and the numbers of adjacent trees. These correlations make possible a series of predictions as to the numbers of trees of the various species which were present in the Bridge Creek forest.[12] Without going into the details of the mathematical procedure, it may be stated that the predictions are comparatively accurate in the case of all of the species which have the higher correlation values, and that if the number of leaves were determined in many other cases the probable error (P. E. of the table) would represent the range of

the product of the mean of X and the mean of Y, $M_X M_Y$, divided by the product of the sigmas of X and Y. The sigmas are the standard deviations of X and Y from their means.

[9] Excepting Equisetum, which is so rare in the redwood forest as to give insufficient distributional data.

[10] The deciduous units are not single leaves in this case, but twigs bearing a score or more leaves (needles).

[11] In this case the correlation is made between numbers of seeds and trees. The leaf correlation was much higher, but the seed correlation is more significant in this case because only seeds are included in the fossil count.

[12] The prediction formula is $Y = r_{xy} \frac{\sigma y}{\sigma x} (X - M_x) + M_y$, in which is the predicted number of trees, which is equal to the product of the correlation value of X and Y, r_{xy}, the sigma of Y divided by the sigma of X, and of the deviation of X from the average plus the mean Y, $(X - M_x) + M_y$.

prediction of the numbers of trees in 50% of the cases. In making these predictions it has been assumed that a cubic foot of shale represents the equivalent of a square foot of area in the Bridge Creek site of deposition. The actual thickness of the organic and inorganic deposits in the square foot units in Muir Woods is much less than one foot; but deposition of sediments here is relatively slow as compared with the accumulation of volcanic material during the Bridge Creek epoch. Since the average number of leaves in a square foot at Muir Woods is 200, and since the average number of fossil leaves in a cubic foot of Bridge Creek shale is 210, it appears reasonable to suppose that a cubic foot of the shale represents essentially the same value as a square foot of area in the modern site of deposition. Even if this supposition is wholly incorrect, making the actual numbers of the trees of each species incorrect in the predictions, the relative numbers of each species will still be correct. And it is these relative numbers which are most important in the reconstruction of the characters of the fossil forest.

TABLE VI.

Predictions of number of plants of different species, within a radius of 50 feet of the average site of deposition in the Bridge Creek forest, and the probable errors of these predictions.

	Number of Plants	P. E.
Alnus carpinoides	9.373	± 4.555
Quercus consimilis	7.327	± 3.853
Sequoia langsdorfii	5.416	± 3.153
Umbellularia sp.	4.791	± 2.416
Corylus macquarrii	3.219	± 1.862
Acer osmonti	1.014	± .553
Rosa hilliae	.195	± .337
Cornus sp.	.136	± .232

The predictions for *Alnus carpinoides, Quercus consimilis, Sequoia langsdorfii,* and *Umbellularia* sp. may be considered to be relatively more accurate, since they are based upon correlations of .49 or higher. The other four predictions cannot be given equal weight, since they are based on lower correlations; they indicate, however, that *Corylus macquarrii, Acer osmonti, Rosa hilliae,* and

Cornus sp. were represented by relatively few individuals in the Bridge Creek forest.

Since the four species, *Alnus carpinoides, Quercus consimilis, Sequoia langsdorfii,* and *Umbellularia* sp. make up 86.44% of the fossil record, the predictions regarding them give a fairly close idea as to the content of the Bridge Creek forest. The total number of alder trees which were in a position to contribute leaves to the part of the sedimentary record which was measured was 922,[13] the oak trees numbered 718, the sequoias 531, and the laurels 470. But in view of the present distribution of the related species in Muir Woods, it does not seem reasonable to suppose that their distribution was uniform throughout the Bridge Creek Valley. At Muir Woods the alder is the most abundant tree in the lower portion of the valley and is rare or entirely lacking in its upper portion; the occurrence of the oak is exactly reversed, since its greatest abundance is upstream and it is absent in the lower stretches. Moreover, leaves of the oak are absent in the sites of deposition downstream, due to the fact that Redwood Creek is so small that it does not transport them for any great distance down the valley. The high prediction of numbers of oak trees in the Bridge Creek forest can therefore be interpreted in two ways:—first, that the distribution of oak and alder was different during the Bridge Creek epoch than it is at present, involving their association in the same places and in approximately equal numbers along the stream course, and second, that the stream which deposited the Bridge Creek shales was sufficiently large to transport the oak leaves from the upper stretches where oak trees were common and deposit them in the lower part of the stream course where alder trees were common and where alder leaves entered into the sedimentary record. If it could be assumed that the distribution of oak and alder trees in the Tertiary forest was so strikingly different from their present distribution, the value of all studies in paleoecology would be seriously brought to question. But if their distribution has always been controlled by the same physical factors that are in operation to-day, the second alternative seems to be established. It is significant that a stream of considerable size is also

[13] A total of 98 cubic feet, or 98 times the prediction as listed in Table 6.

indicated by the size and distribution of the shale deposits bearing leaves. If it is assumed that the Bridge Creek leaf-bearing shales were laid down in the lower course of the stream, as indicated by the abundance of alder leaves in the fossil record, and by the abundance of the leaves of *Sequoia langsdorfii* and *Umbellularia* sp., both of which species are also common there, it seems necessary also to assume that the stream was of sufficient size[14] to transport leaves of *Quercus consimilis* from its upstream habitat down to the point where alder trees were common, and where their leaves entered the stream deposits in large numbers.

The assumption that the Bridge Creek stream was larger than Redwood Creek and that it was a more important agent in transporting leaves, carries with it the inference that a more significant comparison could be made between the fossil leaf record and the leaf record along a modern stream of larger size. It seems desirable in any case to compare the leaf deposits of two streams which have essentially the same transporting power. Further it seems desirable that the counts of modern leaves be made in basins of deposition in the lower portions of the stream course, to correspond with the suggested site of deposition of the Bridge Creek leaves. The writer is planning to carry on further studies of leaf and tree relationships along the major stream courses in the extensive redwood forests in Humboldt County, California.

At the time of beginning the quantitative studies in the modern forest, it was hoped that the results would indicate the relative importance of all of the factors which determine the abundance of leaves in sedimentary deposits. These factors include (1) the distance of the plant from the basin of deposition where the leaf lodges and enters the sedimentary record; (2) the thickness of the leaf; (3) the size and shape of the leaf; (4) the habits of the plant in shedding its leaves; and (5) the length of the plant stem, involving its habit as an herb, a shrub,

[14] The term "size" as applied to the stream has particular reference to its depth: a shallow stream does not permit extensive transportation of leaves since they lodge against obstructions. The term also has reference to the volume of the stream; a stream with a large volume would be more likely to transport leaves long distances before they became water-soaked and sank.

or a tree. Table III shows preliminary data on the distance factor, but no generalizations can be made upon these in view of the probable combination with them of certain of the other factors which have not been isolated. It may be pointed out, however, that of the four most abundant species, all are most commonly found near the stream edge or the site of deposition. Two of the four, the laurel and oak, have thick leaves, which facilitate their transportation without destruction. The alder, laurel and oak all have broad leaves of medium size, which are well suited to being carried by air currents and floated on the surface of the stream. Only one of the four, the alder, is strictly deciduous; the others shed their leaves irregularly, the average leaf remaining on the tree for from two to four years. This is a disadvantage to them, since it reduces the annual number of leaves which fall and may enter the stream deposits. However, all four have an advantage in this respect over plants of the herbaceous type, whose leaves are never definitely shed, but dry up and gradually disintegrate. All four of the abundant species are trees, which means that their leaves become loosened at a higher elevation than those of shrubs and herbs, and that they have a better opportunity of being widely borne by air currents and of getting into the basins of deposition.

Observations on the difference between the abundance of ferns in the forest at Muir Woods and their representation in the stream deposits of Redwood Creek bear out the idea, previously expressed by the writer,[15] that ferns and herbs may have been much more common in the Tertiary than much of the fossil record would indicate. *Aspidium munitum* is very common on the flats and the valley slopes, in some cases growing on the very borders of the pools where leaf counts were made, but it is represented in the leaf counts by only a single fragment. *Pteris aquilina* is common, especially in the more open stretches of the valley, but its representation in the leaf counts is also limited to a single fragment. Two other species of ferns, *Asplenium felix-femina* and *Woodwardia chamissoi*, occur on the valley flats, but they have no representation in the stream deposits.

[15] The Flora of the Eagle Creek Formation, Cont. Walker Museum, vol. 2, no. 5, pp. 130-31, 1920.

Clearly a plant whose leaves are not definitely shed, but which dry up and disintegrate while still attached, is not likely to be represented in a fossil flora. Even where a leaf fragment is loosened from the plant, it is so near the ground that it is not commonly carried far by the wind. Half a dozen specimens of Pteris have been found in the Crooked River locality of the Bridge Creek flora, but the other fern genera have no known representation in it.[16] In the case of *Asarum caudatum,* which is perhaps the most abundant herb in Muir Woods, there has been only one leaf recorded in the stream deposits, for this species suffers the same disadvantages as do the ferns. It is of interest to note that a single specimen of what appears almost certainly to be a fossil species of *Asarum* has been found in the Bridge Creek flora at the Crooked River locality. Clearly the small numbers of ferns and other herbs in the Bridge Creek flora cannot be taken as an indication of their scarcity in the Bridge Creek forest.

Summary.

The abundance of fossil leaves of *Sequoia langsdorfii, Alnus carpinoides, Quercus consimilis* and *Umbellularia* sp., which make up 86.44% of the total of 20,611 specimens counted at the type locality of the Bridge Creek flora, suggests their probable dominance in the Bridge Creek forest. A count of 8422 leaves and fruits in basins of deposition in the living redwood forest at Muir Woods, Marin County, California, and of the trees surrounding these basins of deposition, indicates that the modern equivalents of these four species, *Sequoia sempervirens, Alnus rubra, Quercus densiflora* and *Umbellularia californica,* are the dominants in this forest, and that their leaves are entering the stream deposits in essentially the same relative numbers, making up 85.44% of the total of plant remains now accumulating. Correlations between leaf and tree numbers in Muir Woods have been made, by which it is possible to predict the approximate numbers of trees of the four dominant species in the Bridge Creek forest, and the exact relative

[16] A possible explanation of the abundance of fossil ferns in certain other floras is that whole plants were buried by sediments laid down on their stream or lake-border habitats during abnormally high water.

numbers of these species. The inferences drawn from these predictions are in accord with the idea based upon other evidence, that the stream which deposited the leaf-bearing Bridge Creek shales was of sufficient size to transport leaves for considerable distances. Since the stream in Muir Woods is not large enough to be an important factor in leaf transportation, it seems desirable to make further studies of the relative numbers of leaves accumulating in modern valleys of larger size, and to compare these with the record of the fossil flora.

Of the remaining 13.46% of the total of the fossil count, there are small numbers of leaves and other remains of eight species whose modern equivalents are present in relatively small numbers in the living redwood forest, and in addition there are ten forms which do not have generic representation in this forest at the present time including species of such eastern genera as *Carpinus, Tilia, Ulmus* and *Fagus*. These indicate that the redwood forest of the Tertiary was somewhat more diversified than is the redwood forest of to-day. But the dominants are closely similar, both in kind and in numbers, and it seems reasonable to assume that the general aspect of both forests was essentially the same. A comparison of the character of the vegetation now living near the Bridge Creek locality, as shown in fig. 1, with the vegetation of the past, whose modern equivalent is shown in fig. 1 (Pl. VI), suggests the extent of the climatic change which has taken place in central Oregon since the Tertiary.

Carnegie Institution of Washington,
Berkeley, Cal.

PAPER 42

Taphonomy and Information Losses in Fossil Communities (1968)

D. R. Lawrence

Commentary

CARLTON E. BRETT

In his seminal paper of 1968, David Lawrence, then professor at the University of South Carolina, presented a terminology for taphonomy, a cautionary note, and an example of how preservation may bias the fossil record. His oft-reprinted figure 1, showing the relationship of taphonomy to paleoecology, introduced English-speaking paleontologists to the dual concepts of biostratinomy—processes affecting organic remains between the death of organisms and their final burial—and fossil diagenesis—geochemical and physical factors acting on those remains largely after burial. Lawrence also admonished paleoecologists that they must strip away the "taphonomic overprint" before proceeding to make inferences about ancient communities and illustrated how that might be done, using the example of modern and ancient oyster reefs. He paid particular attention to "redundant information" on soft-bodied organisms, in the form of borings and malformations in calcitic oyster shells. Yet, even with some salvaging of information on these bases, he concluded that perhaps as little as 22% of the original taxonomic diversity was recorded. Lawrence's mental experiments on taphonomic degradation presage in situ experimental approaches to taphonomy (e.g., Parsons-Hubbard et al. 2011).

This paper was a key contribution in a long line of "live-dead" studies that compared the record of skeletal parts accumulating in sediments with living communities (e.g., Johnson 1964, 1965), and the outlook was bleak. Decay and disintegration processes in marine environments not only remove the obvious, soft tissue portions of ancient communities but also bias the record of skeletal hardparts, both by differential breakdown of lightly skeletonized and aragonitic skeletons and by transport of remains from living communities to their final burial sites. More than four decades of study since the time of Lawrence's paper have brightened the picture somewhat; syntheses of numerous live-dead studies (Kidwell 2002; Kidwell and Holland 2002) have shown not only that out-of-habitat transport is rare in offshore marine environments but also that the accumulating remains in these settings, though time-averaged, nonetheless may show a high fidelity of taxonomic composition and even relative abundance with original communities. Lawrence anticipated neither the ways in which taphonomy can actually lead to information *gain* through time-averaging nor that preservational aspects of fossils can provide valuable data for reconstructing paleoenvironments ("taphofacies"; Brett and Baird 1986) or for the assessment of anthropogenic change ("conservation paleobiology"; Kidwell 2007; Dietl and Flessa 2011). Instead, during the heyday of community paleoecology studies, he offered a sobering, yet critically important caveat: "Ignore taphonomy at your peril."

From *Geological Society of America Bulletin* 79:1315–30.

Literature Cited

Brett, C. E., and G. C. Baird. 1986. Comparative taphonomy: a key to paleoenvironmental interpretation using fossil preservation. *Palaios* 1:207-227.

Dietl, G. P., and K. W. Flessa. 2011. Conservation paleobiology: putting the dead to work. *Trends in Ecology and Evolution* 26 (1): 30–37

Johnson, R. G. 1964. The community approach to paleoecology. In *Approaches to Paleoecology*, edited by J. Imbrie and N. D. Newell, 107–34. New York: Wiley.

———. 1965. Pelecypod death assemblages in Tomales Bay, California. *Journal of Paleontology* 39:80–85.

Kidwell, S. M. 2002. Time-averaged molluscan death assemblages: palimpsests of richness, snapshots of abundance. *Geology* 30 (9): 803–6

———. 2007. Discordance between living and death assemblages as evidence for anthropogenic ecological change. *PNAS* 104 (45): 17701–6.

Kidwell, S. M., and S. M. Holland. 2002. Quality of the fossil record: implications for evolutionary biology. *Annual Review of Ecology and Systematics* 33: 561–88.

Parsons-Hubbard, K. M., S. E. Walker, and C. E. Brett, eds. 2011. The Shelf and Slope Experimental Taphonomy Initiative (SSETI): thirteen years of taphonomic observations on carbonate and wood in the Bahamas and Gulf of Mexico. Special Issue, *Palaeogeography, Palaeoclimatology, Palaeoecology* 312:195–208.

DAVID R. LAWRENCE *Department of Geology, University of South Carolina, Columbia, South Carolina 29208*

Taphonomy and Information Losses in Fossil Communities

Abstract: Taphonomy explores post-mortem relations between organic remains and their external environment. Paleoecologists must first be taphonomists, because studies of life-environmental histories require prior knowledge of post-mortem events. Two types of post-mortem events can seriously hamper the recognition and analysis of fossil communities: (1) information losses through nonpreservation, and (2) losses through transport away from the life setting. Although many organisms are capable of leaving multiple or redundant evidence of their presence, redundancy cannot fully counterbalance the many processes leading to information losses.

At the present stage of our knowledge, *in situ* preservation is the only valid criterion we possess for recognizing fossil communities. Information losses through transport must be minimal in these cases; major losses occur through nonpreservation. Data on *potential* information losses through nonpreservation are not numerous. The available analyses of Recent marine communities indicate that from 7 to 67 percent of these communities' species are soft-bodied and have little potential for preservation. Fewer data exist concerning *actual* information losses for fossil communities of any type.

Actual losses can be estimated for oyster communities from the Atlantic Coastal Plain in North Carolina. An Oligocene *Crassostrea gigantissima* (Finch) community at Belgrade contains about 18 preserved megascopic species. Documented processes contributing to information loss included dissipation of organic soft parts and dissolution of chitinous, siliceous, and aragonitic skeletons. A Recent minimal *Crassostrea virginica* (Gmelin) community from the Beaufort area includes 80 megascopic animal species. This Recent community, if subjected to the same history as the fossil one, would yield similar types and amounts of preserved information. The comparative analysis suggests, therefore, that the Oligocene community contained the same major taxa as the Recent community; over 75 percent of the species in the original Oligocene community have not been preserved.

CONTENTS

Geological Society of America Bulletin, v. 79, p. 1315–1330, 4 figs., October 1968

INTRODUCTION

Organisms and their fossil remains undergo long and complex environmental histories. European workers have long recognized four important events in the environmental history of a given fossil: birth, death, final burial, and discovery by a scientist (Fig. 1). Studies of environmental relations of fossil organisms between their birth and death form the basis of *paleoecology;* paleoecologists treat fossils as once-living organisms. The discipline of *biostratinomy* (formerly biostratonomy; etymology required change, A. Seilacher, 1964, personal commun.) explores pre- and syn-burial interrelations between dead organisms and their external environment (Müller, 1951; 1963, p. 17). *Diagenetic studies* ("Fossildiagenese," Müller, 1963, p. 2) unravel the post-entombment histories of organic remains. Biostratinomic and diagenetic studies are combined in the field of *taphonomy* (Efremov, 1940, *see* Fig. 1). Paleoecologists must first be taphonomists, since once-living organisms can only be studied and reconstructed through adequate knowledge of their post-mortem history. Success in paleoecology depends largely upon the worker's ability to strip away the taphonomic overprint.

In the study of well-preserved fossil communities, these time intervals and disciplines can only be separated by careful analysis. Understanding the post-mortem history of one fossil group often requires knowledge of the life histories of associated and interacting organisms. Other complexities are often present. Yet responses of organisms and their remains show distinctive changes in the defined time intervals. During their lifetime organisms can react to external stimuli· after death their remains are passive elements subject to decay, degradation, and transport. These processes and their products change throughout the total environmental history of the organisms and their remains, and these changes, in turn, demand the use of separate terms to indicate the time intervals and their respective environmental disciplines. In addition, widespread and logical use of the terms "taphonomy," "biostratinomy," and "diagenetic studies" should help clear up confusion about the basic nature and the structure of paleoecology.

It is obvious that many organic remains have not survived, and do not now survive the post-mortem period. An analysis of post-mortem losses of information should help us understand what information on life habits and life habitats can rightly be expected in the fossil record.

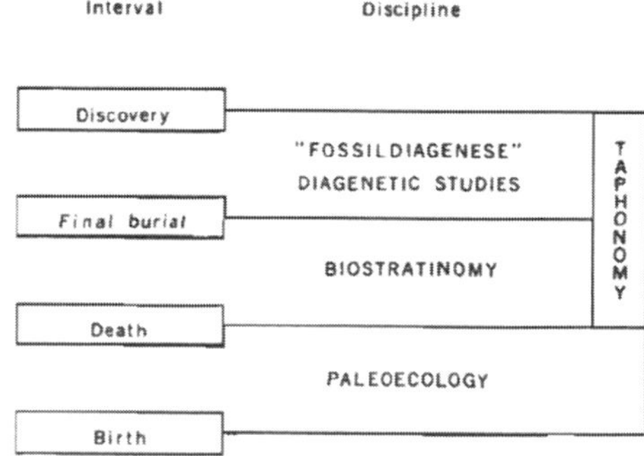

Figure 1. The environmental aspects of paleontology. Disciplines based upon time interval in history of organism or organisms being studied.

Theoretical discussion of these losses is followed by comparative analysis of Recent and Oligocene oyster communities from the Atlantic Coastal Plain. This analysis documents the types of information losses in the Oligocene community and shows that an "average" Recent community, if subjected to the same post-mortem happenings, would yield similar types and amounts of information if studied in the future. This analysis has evolutionary implications, since it suggests that the gross biotic composition of communities in the *Crassostrea soleniscus* lineage (Sohl and Kauffman, 1964) has not changed since Oligocene time.

ACKNOWLEDGMENTS

This paper is adapted from a doctoral dissertation submitted to the Department of Geology of Princeton University in 1966; Erling Dorf and Alfred G. Fischer were instrumental in the development of this section of the dissertation. Martin A. Buzas, Derriel B. Cato, A. F. Chestnut, A. G. Fischer, Thomas G. Gibson, Barbara Jean Moore, Hugh Porter, Gayle M. Robinson, and David M. Raup gave help during the preparation of this paper. Special thanks are due Ashley C. Lawrence, who prepared various typescripts and drafted the illustrations. However, the author assumes full responsibility for ideas and data presented herein. Field and initial laboratory work were supported by the Department of Geology of Princeton University and by the National Science Foundation through its

Graduate Fellowships Program. The final stages of this work were supported by a grant-in-aid from the University of South Carolina.

TAPHONOMIC LOSSES OF INFORMATION

Two general types of taphonomic events commonly make paleoecologic work difficult: (1) loss of information by nonpreservation, and (2) loss of information by transport of organisms from the site of the original community.

Losses Through Nonpreservation

Many organisms in a typical marine community are soft-bodied and hence subject to dissipation after death. Scavengers, decomposer organisms, and various physical forces all contribute to these data losses. Possession of resistant hard parts is one obvious aid to preservation.

Modern marine benthic communities show a wide range in percentages of animals with resistant hard parts. R. G. Johnson (1964) collated and analyzed data from nine recent surveys, comprising 534 samples, of marine level-bottom communities. The percentage of species with resistant hard parts ranged from 7 to 67 percent, and averaged about 30 percent in the various communities sampled. Individuals with preservable hard parts ranged from 1 to 87 percent and also averaged about 30 percent in the analyzed samples. More recently, Craig and Jones (1966) analyzed present-day infaunal and epifaunal species in the Irish Sea. Here about 33 percent of the more abundant epifauna species had hard parts and thus stood a chance of being preserved in the fossil record (Craig and Jones, 1966, p. 35). Thus, a minimum of about 40 to 70 percent of a marine community's species can be expected to leave no trace of their existence.

Yet possession of hard parts is no guarantee of preservation. Before and after burial, physical, chemical and biological agents or processes can destroy hard parts. Current or wave action may abrade and disintegrate resistant parts of marine organisms; selective chemical solution may obliterate many remains, or chemical diagenesis may alter mineral form; boring and burrowing organisms may likewise remove evidence of species and of past organic habits and habitats. These various aspects of taphonomy have recently been reviewed by Chave (1964).

Neither does lack or loss of hard parts necessarily imply nonpreservation. Evidence is often duplicated, and parts of the original message may be decipherable. Tasch (1965, p. 323) has applied communications theory to this concept and has used the term "redundancy" for repetition of the same message about a fossil organism.

Analysis of redundancy depends strictly upon the concept of the message. If the message is defined as the presence of a given organism in an ancient community, then body parts are one obvious transmission. The message may be duplicated by means of tracks, borings, coprolites, casts and molds, and other means. Message duplication may originate during the organism's lifetime or may arise from postmortem happenings. Some or all of the redundant transmissions may subsequently be lost and thus be unavailable to the decoding paleontologist. Nonetheless, when the presence of a fossil community member is based upon evidence other than original body parts or upon repeated and preserved evidence, redundancy certainly occurred during some time interval of the message transmission. Although a more detailed discussion of information theory is beyond the general scope of this paper, future application of this theory to paleontology should crystallize many ideas and pinpoint many difficulties in paleoecologic and taphonomic work.

Polydorid bristleworms provide an excellent example of information redundancy due to life activity of soft-bodied organisms. Some species of polydorids are shelter commensals within or upon oyster shells. The two groups—oysters and bristleworms—interact in peculiar ways. Distinctive records of this interaction, internal U-shaped tubes or surficial openings and blisters, are left in the oyster valves (Whitelegge, 1890; Lunz, 1941; Loosanoff and Engle, 1943). This evidence can be used to reconstruct the living worms even though all evidence of their soft bodies is gone. Similarly, the postmortem formation of casts or molds may allow analysis of shelled community members which would otherwise have been obliterated. Redundancy tends to counteract information losses through nonpreservation. Yet redundancy cannot normally counteract the major losses of information through transport of organisms.

Losses Through Transport

During their lifetime, organic communities occupy a discrete unit of space. Any disruption of these spatial relations constitutes a major loss of information. Shelled infaunal organisms may be relocated by burrowers, and they, as well as epifaunal elements, may be uprooted or moved by waves or currents, or both uprooting and transportation may occur. Important and critical studies of post-mortem transport by waves and current have been made in recent years (Lever, 1958; Lever and others, 1961, 1964; Nagle, 1964a, 1964b). Differentiation of wave and current distribution patterns (Nagle, 1964a) holds great promise for studies of taphonomy and basin analysis, but additional work is needed. Many problems are yet unsolved. All of these information losses make original communities and environments hard to decipher.

RECOGNITION OF FOSSIL COMMUNITIES

Paleoecologists have developed and used three criteria for recognizing fossil communities: (1) evidence of *in situ* burial, (2) taxonomic analogy with modern communities, and (3) recognition of recurrent organic groupings which are analogous to those observed in modern communities (Johnson, 1964, p. 109). Fagerstrom (1964) has summarized the many criteria used in the second and third approaches and Johnson (1964) has also discussed the practical limitations and pitfalls of these approaches. Recently Craig and Oertel (1966) have pointed out complexities in the analysis of size-frequency distributions; these distributions have often been cited as an important clue in community recognition (Fagerstrom, 1964, p. 1200–1202 and references therein).

Quite sophisticated mathematical techniques have been developed which have the potential to reduce and analyze data on recurrent death assemblages; cluster analysis is but one recent example (Valentine and Peddicord, 1967). Widespread use of such techniques in community recognition, however, seems premature. Recurrent death groupings may also represent organic remains, originally drawn from many settings, which are redeposited in response to similar, widespread, and recurrent hydrodynamic regimes. They may reflect no pattern of life association unless it can be shown that the recurrent groupings are similar to nearby, *in situ* communities (as Johnson, 1965, p. 84, has done in Tomales Bay, California). Until we better understand the mechanics of transport of organisms and their remains, recurrent death assemblages must remain a dubious criterion for recognizing fossil communities.

Yet both the recurrent grouping and taxonomic analogy approaches may also be attacked on more theoretical grounds. Both these approaches rely heavily or entirely upon the use of neontological information in community recognition. Recent organic distributions are applied statically to the geologic past. This use of Recent data requires a rigid and unjustified view of uniformitarianism. Indeed, Geikie's dictum—The Present is a Key to the Past—may often be an oversimplification. The assumption of a static natural history in any group of organisms (substantive uniformitarianism; Gould, 1965) is certainly contrary to our knowledge of the development of the organic world through time. Even though the *law* of substantive uniformitarianism should be abandoned (Gould, 1965, p. 226), the *concept* of substantive uniformity must not be lost. Substantive uniformity takes on a new focus if we remember that both static and dynamic natural histories must be established, not assumed. The nature of this establishment becomes quite important.

The inductive basis of paleontology certainly hampers any attempt to establish static natural histories through time. The very recognition of fossil oysters, for example, would be impossible if we did not extend our knowledge of Recent oysters to the material shells we find in ancient strata. Induction of life habits and life habitats of fossil organisms is obviously hampered by the fact that the organisms cannot recur and allow their life histories to be observed directly. Given this unavoidable framework, we can recognize three levels of proof concerning the natural histories of fossil organisms and their remains: (1) reconstruction based entirely upon direct observation on ancient strata, (2) reconstruction based upon direct observations in the fossil record, supplemented by evidence from the Recent, and (3) reconstruction based solely upon the transfer of Recent data to the ancient setting. This third level of proof has been widely used in paleoecology and normally results from the transferred ecology approach described by Craig (1964, p. 589); fossils present in an area are catalogued and then the environmental tolerances of Recent counterparts are used to reconstruct the ancient physical setting. Yet this third level of proof is no proof at all; working methods presuppose static natural histories through time. The transferred ecology approach *may* be helpful in some paleoenvironmental reconstructions if the interpretations drawn from this approach are sufficiently broad (marine versus non-marine, for example).

However, to increase the validity and detail of their reconstructions, paleoecologists must focus upon those aspects of their science amenable to direct observations in ancient settings. For example, orientation and spatial relations of organisms of *in situ* fossil communities can be directly measured at outcrops (Lawrence, unpublished doctoral dissertation, Princeton University, 1966). Analyses of orientation patterns result in first-order proofs of randomness or nonrandomness. Comparison of orientation patterns in Recent and fossil communities can lead to first-order proofs of community statics or community dynamics, with respect to orientation patterns. These and similar aspects of paleoecology deserve increased emphasis.

The reconstruction of an Oligocene oyster community, given later in this paper, is a prerequisite to traditional paleoecologic work upon the community. The reconstruction also provides an example of a second-order proof in paleoecology. Preserved members of the *in situ* community are identified, and known processes leading to information losses are documented through observations on the outcropping strata. A Recent oyster community is then "subjected" to the same processes; the effects upon preservation in both Recent and fossil communities are compared. These working methods are in direct opposition to the working methods of transferred ecology. In transferred ecology, Recent data and processes are imposed upon the fossil record. In the second-order proof just outlined, documented geologic processes are "imposed" upon Recent organisms and the results compared with observations upon fossil organisms. The interpretations drawn from this comparative analysis are more firmly based than any that could have been made through transferred ecology.

Thus the three types of proof have differing degrees of validity. Most paleoecologic reconstructions involve complex chains of data analysis and synthesis. Any reconstruction is only as valid as the weakest link in the complex pattern of reasoning. The framework presented herein may provide a means for determining the validity of reconstructions at all levels, from single species to community complexes.

PRESERVATION OF FOSSIL COMMUNITIES

Data on information losses may be recast in terms of these latter theoretical considerations. In the present state of our knowledge, *in situ* preservation is our principal guide to fossil communities. Under these circumstances, information loss by transport is minimized. Major losses occur through nonpreservation.

The cited works of Johnson (1964) and Craig and Jones (1966) provide data on the *potential* losses of information due to lack of hard parts and consequent nonpreservation. Yet few or no data are available on *actual* losses within any given community type. Studies of actual losses must analyze information redundancy and the possible loss of skeletal remains during the post-mortem history, as well as initial preservability. Quantitative effects of losses may be estimated also. These latter analyses have four requisites: (1) a well-preserved fossil community must be amenable to analysis, (2) Recent counterpart communities must be well-documented, (3) a comparative analysis of the Recent and fossil communities must be possible, and (4) temporal changes in community composition must be open to analysis. These conditions are met by fossil and Recent oyster communities from the Coastal Plain of North Carolina. Each of the requisites will be examined, in turn.

THE FOSSIL OYSTER COMMUNITY

Geologic Occurrence and Setting

In situ deposits dominated by the oyster *Crassostrea gigantissima* (Finch) are well-known from Mid-Tertiary strata of the southern Atlantic Coastal Plain (Miller, 1912; Loughlin and others, 1921; Sandy and others, 1966). One exceptionally well-preserved deposit occurs at Belgrade, North Carolina. Belgrade is located in east-central North Carolina, about 30 km from the Atlantic Ocean (Fig. 2). During World War II, the Superior Stone Company began quarrying operations near the banks of the Whiteoak River in Belgrade. Numerous pits were excavated and then abandoned. Strata used commercially by the Superior Stone Company include over six of calcareous beds, characterized by species of *Anomia* and *Mercenaria*. Correlations, based mainly upon ostracods, date these basal calcareous strata and the overlying oyster deposit as Late Oligocene (Philip M. Brown, 1966, personal commun.).

From the mid-fifties to 1964, the Company worked a pit in the northwest corner of the quarry. The oyster locality occurs along the access road to this pit (Fig. 2). In this area, a channel was cut into the basal calcareous

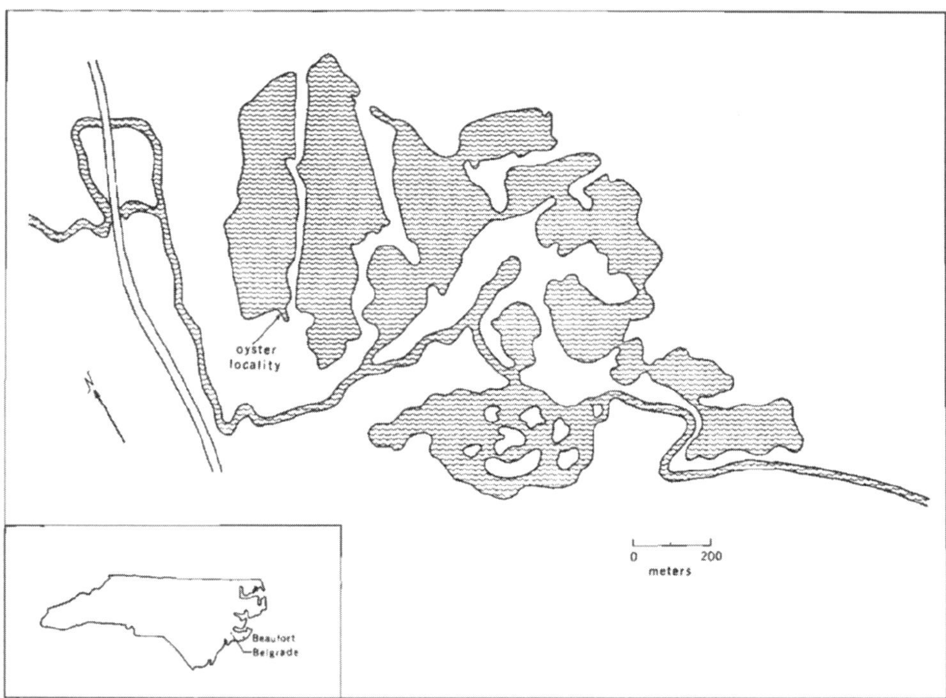

Figure 2. Belgrade, North Carolina, quarry of the Superior Stone Company. Sketch map shows abandoned pits, which are now water-filled lakes, and location of oyster deposit described in text. Quarry is shown as it appeared in June, 1964. Map drawn from aerial photographs supplied by Superior Stone Company.

strata. The oyster community settled and lived in this channel.

Much of the channel's history can be read from a complete cross section mapped and shown in Figure 3. In overview, the oyster community first settled in the deepest and middle portion of the channel. It migrated laterally, up the right or south side of the channel, while the channel was being filled with sandy sediments. Regional geologic evidence indicates that a marine regression occurred during Late Oligocene time in the east-central North Carolina near (Brown, 1963, p. 12–13; 1966, personal commun.). Evidence from the channel complements this interpretation. Details of the channel's history are reported by Lawrence (Ph.D. dissert. Princeton University, 1966, 187 p.); they will be published separately.

Composition

Large and massive specimens of *Crassostrea gigantissima* (Finch) are the most distinctive fossils in the channel. Most commonly, they are densely packed and occur in their upright life positions. Oysters are associated with foraminiferans, clionid sponges, encrusting bryozoans, boring bryozoans, polydorid bristle-worms, a spirorbid worm, boring pelecypods, a low-spired gastropod, a barnacle, and echinoid remains.

The sponges, bryozoans, worms, molluscs, and barnacles occur within or upon oyster shells. In all but one case their life position can be ascertained. Two species of clionid sponges bored labyrinthine galleries within the oyster shells. A form analogous to *Cliona celata* Grant has large external perforations and relatively smooth-sided galleries, while another form, with smaller perforations and more delicate galleries, is analogous to the *Cliona vastifica* species complex (for descriptions of the Recent species *see* Old, 1941; Hartman, 1958). No siliceous spicules are preserved within the oyster shells.

Bryozoan remains are not widespread and have not yet been studied in detail. By inspection, encrusting cheilostomatous bryozoans are represented by six or seven species. Other shell markings are apparently due to boring bryozoans of the Suborder Stolonifera. Impressed pits connected by linear furrows are quite similar to markings made by the linear stolons and zooids of boring bryozoans (Bassler, 1953, p. G35–G37). Two marking patterns occur on the shell surface of Belgrade oysters. One is simple and linear; the other, arborescent. These patterns probably represent two species.

Polydorid bristleworms are recorded by oyster shell decalcification patterns or by blisters formed from secretions of the oyster in response to polydorid activity. Two behavioral types of polydorids occur in the deposit. One lived within the shell, in the marginal blisters made by the oysters. The other lived near or underneath projecting lamellae on shell exteriors. These two behavioral types represent either one or two species (Korringa, 1951, p. 91–101).

Many oyster shells are penetrated by larger tubular cavities. Cavities are subcircular in cross section and may extend over 40 mm into the shell. One tube contained a small fragment of a thin-shelled, antero-posteriorly elongate pelecypod. Cavities are grossly similar, and a single species of boring pelecypod may have been responsible for all of them.

Epibiont barnacles are widespread in the deposit. All specimens represent a single species (Arnold Ross, 1966, personal commun.). Spirorbid worms and gastropods are each represented by a single occurrence. The original calcareous tubes of the worm are present. The low-spired gastropod is preserved as an impressed mark or "ghost" on the attachment area of one oyster shell.

In summary, the preserved megascopic elements which doubtless belong to this Oligocene oyster community fall into five animal phyla: Porifera, Bryozoa, Annelida, Mollusca, and Arthropoda. About 16 to 18 species are preserved. Nine or ten are represented by calcareous skeletons; seven or eight, largely or completely by redundant information concerning their presence. Evidence from the literature on Recent oyster communities, when compared with field observations, suggests that these elements are the preserved remnant of an original, large, and diverse biota.

RECENT OYSTER COMMUNITIES

Diversity of Forms

The diversity of organic forms in Recent oyster beds has been noted for many years. Möbius (1877) stated that oyster beds are richer in all kinds of animal life than any other portion of the sea bottom. In his classic study of the sea-flats of Schleswig-Holstein, Möbius listed 24 species of animals and 4 plant taxa associated with the German oyster beds.

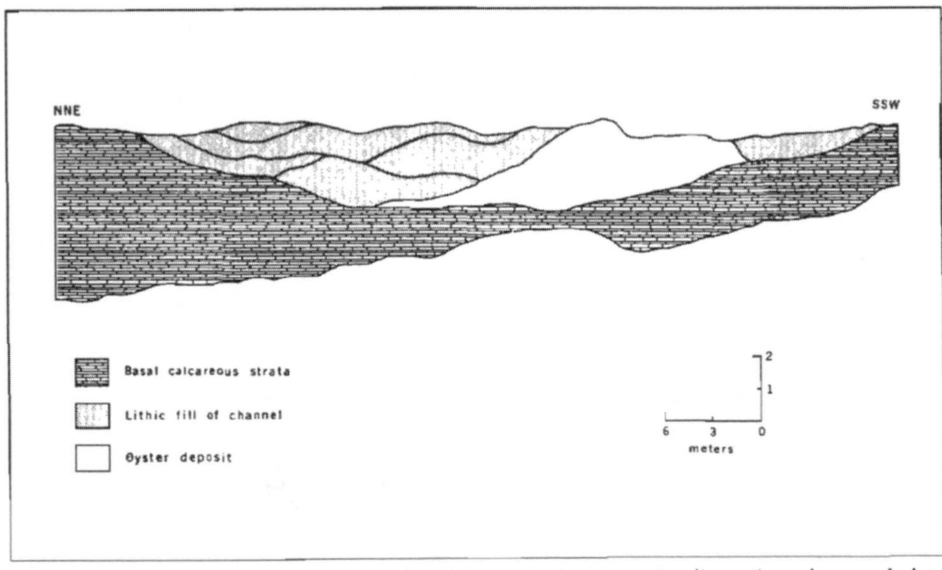

Figure 3. Cross section of the channel deposit at Belgrade, North Carolina. Channel exposed along east side of access road to northwesternmost quarry pit.

Additional and more recent work has increased the numbers, until Wells (1961) listed 303 animal species, representing at least 10 phyla, associated with beds of *Crassostrea virginica* (Gmelin) near Beaufort, North Carolina. Table 1 shows some of the taxa known from association with Recent oysters, and gives one example from each group. The list is collated from the works of Möbius (1877), Miyazaki (1938), Pearse and Wharton (1938), Lund (1945), Mattox (1949), Korringa (1951), and Wells (1961). It is not meant to be a complete listing; it does, nonetheless, show the breadth and diversity of forms associated with oysters. Classification follows the categories of Simpson and Beck (1965, p. 837–842).

Composition of a Single Community

To help fully explain and complement the direct observations on Belgrade community preservation, it is essential to know what members of the total Recent biota would be present in a *single* modern oyster community. A detailed look at Wells' 1961 North Carolina study should help solve this problem.

Wells sampled *Crassostrea virginica* (Gmelin) communities from nine localities in the Beaufort Inlet and Newport River area near Beaufort, North Carolina. The nine locales covered a range of eight nautical miles in the inlet-river complex. He recognized a total fauna of 303 species from these localities, including 2 protozoans, 12 sponges, 14 coelenterates, 8 flatworms, 4 nemerteans, 99 molluscs, 42 annelids, 2 sipunculids, 76 arthropods, 20 bryozoans, 5 echinoderms, and 19 chordates. Wells made no effort to identify the plants, many of the protozoans, and the copepods, ostracods, or nematodes. Despite these limitations, the fauna is extremely well documented and diverse.

Wells sampled the nine locales over an 18-month period, and he gathered a gallon of material at each sampling. This sample included live oysters, epibiont-containing shells of dead oysters, and associated organisms. Mean number of species per sample collection ranged from 16 to 67, while total number of species per locality ranged from 56 to 220. Both of these numbers decreased inshore. Thus, numbers of species were influenced by salinity and other environmental factors.

Wells found 80 megascopic species associated with 20 percent or more of the collections. These 80 most common and prevalent organisms should approximate a minimal oyster community in the Beaufort area. The organisms fall into 9 animal phyla and represent 68 genera. Table 2 presents this data and shows the still-present diversity of forms.

PRESERVATION OF OYSTER COMMUNITIES

General Statement

Oyster communities may be subject to all

TABLE 1. SOME TAXA ASSOCIATED WITH RECENT OYSTER COMMUNITIES*

Phylum Schizomycetes—bacilli
Phylum Mastigophora—dinoflagellates
Phylum Sarcodina—foraminiferans
Phylum Sporozoa—haplosporidians
Phylum Ciliophora—heterotrichids
Phylum Chlorophyta—grass-green algae
Phylum Chrysophyta—diatoms
Phylum Phaeophyta—brown algae
Phylum Rhodophyta—red algae
Phylum Porifera
  Class Demospongiae—clionids
  Class Calcispongiae—grantiids
Phylum Coelenterata
  Class Hydrozoa—hydroids
  Class Scyphozoa—jellyfishes
  Class Anthozoa—anemones
Phylum Ctenophora—comb jellies
Phylum Platyhelminthes—turbellarians
Phylum Nemertea—ribbon worms
Phylum Aschelminthes—rotifers
Phylum Bryozoa
  Class Endoprocta—pedicellinids
  Class Ectoprocta—bugulids
Phylum Annelida
  Class Polychaeta—polydorids
  Class Oligochaeta—peloscolexids
  Class Gephyra—sipunculids
Phylum Mollusca
  Class Amphineura—chitons
  Class Gastropoda—muricids
  Class Pelecypoda—mytilids
Phylum Arthropoda
  Class Crustacea—balanoid cirripeds
  Class Arachnida—acarids
  Class Insecta—dipterans
Phylum Echinodermata
  Class Asteroidea—asteriids
  Class Ophiuroidea—brittle stars
  Class Echinoidea—sea urchins
  Class Holothuroidea—sea cucumbers
Phylum Chordata
  Subphylum Tunicata—ascidians
  Subphylum Vertebrata—fishes

* (with a description or example of each group; sources of data given in text)

TABLE 2. COMPOSITION AND CHARACTERISTICS OF THE BEAUFORT, NORTH CAROLINA RECENT MINIMAL OYSTER COMMUNITY*

Taxa	Total Species	Soft-Bodied	With preservable hard parts Ca	Ch	Si	Ph	Possible Redundancy
Porifera	5	—	—	—	5	—	3
Coelenterata	6	5	1	—	—	—	—
Platyhelminthes	1	1	—	—	—	—	—
Nemertea	2	2	—	—	—	—	—
Bryozoa							
Ectoprocta	7	4	3	—	—	—	—
Annelida							
Polychaeta	13	13	—	—	—	—	4
Mollusca							
Gastropoda	9	—	9	—	—	—	1
Pelecypoda	13	—	13	—	—	—	2
Arthropoda							
Crustacea	19	10	4	5	—	—	5
Arachnida (?)	1	1	—	—	—	—	—
Insecta	1	1	—	—	—	—	—
Chordata							
Tunicata	2	2	—	—	—	—	—
Vertebrata	1	—	—	—	—	1	1
Totals	80	39	30	5	5	1	15
Percentages of total community	100	49	38	6	6	1	19

* (data from Wells, 1961) Among the arthropods, only decapod crabs with relatively well-calcified and/or well-tanned exoskeletons have been included with the organisms with hard parts.
Ca = calcareous
Ch = chitinous
Si = siliceous
Ph = phosphatic

the taphonomic events described in early sections of this paper. Scavengers, decomposer organisms, and chemical diagenetic processes are important agents in the loss of information about these *in situ* communities. In fact, analysis of the literature on Recent oyster biotas suggests that evidences of only four types of organisms are likely to be initially preserved in the fossil record. First, oysters and organisms juxtaposed with them may possess hard parts conducive to preservation; second, associates may demineralize hard skeletons in particular patterns; third, other organisms may cause external aberrations of oyster form; and fourth, associates may modify the internal oyster shell surface in some peculiar way. These last three categories give rise to redundant bits of information concerning the presence of individual community members.

Diverse oyster community members belong in these various categories. Prominent groups with hard parts include the protozoan foraminifers, molluscs, echinoderms, and arthropods. In his study of the Beaufort oyster beds, Wells found hard-shelled molluscs living in crevices between oyster shells, living in the substrate between or under oysters, and actively burrowing into the oyster shells (Wells, 1961, p. 248).

Molluscs also occur in the second group—those which demineralize hard parts. Boring naticid, muricid, and thaidid gastropods drill holes in shells and feed upon soft parts of oysters, while boring pelecypods may penetrate shells to gain shelter (Carriker, 1955, 1961; Yonge, 1963). Clionid sponges and some polydorid worms also decalcify shells to obtain shelter (Korringa, 1951, p. 52–54, 91–101); xanthid crabs chip margins while preying on oysters (McDermott, 1960); and a flatworm (*Pseudostylochus*) apparently bores through shells and feeds on the oysters (Woelke, 1956). Decalcifying groups are diverse.

Numerous groups have the potentiality to

cause external modifications of oyster form. The attachment area of oyster left valves faithfully replicates details of the substrate. Right valves also conform to patterns established by left valves (Gregg, 1948). Thus any adjacent and sedentary organism possessing hard parts may have its details imprinted on oyster shells.

A much smaller group of organisms can cause misshapen internal cavities of oyster shells. Prominent members come from two genera—the turbellarian flatworm *Stylochus* and the annelid *Nereis.* Both of these groups enter between the valves and attempt to prey on the oysters. Oysters react by building internal partitions to seal off these organisms, and these partitions are distinctive features (Pearse and Wharton, 1938, p. 633). Internal blisters, formed from secretions of oysters in response to irritating polydorids, also fall in this category. The Belgrade and Beaufort communities include representatives of all four types of preservable organisms.

Chemical diagenesis may further affect the preservation of information. Organic skeletons are composed of a wide range of mineral species (Lowenstam, 1963, Fig. 2); these minerals have differing resistances to the effects of chemical diagenetic processes. For example, the stability of low-magnesium calcite skeletons and instability of aragonite and high-magnesium calcite skeletons in the zone of circulating fresh ground water has been well documented (Chave, 1964, p. 382 and references). Redistribution of carbonate may often lead to the formation of casts and molds.

The following section documents the processes leading to information losses in the Belgrade setting and shows the effects the same processes would have on preservation of the Beaufort community.

Comparison of Belgrade and Beaufort Communities

Diversity of forms. Five phyla and at least 16 species occur in the fossil setting; the 80 megascopic species of the Beaufort minimal community represent 9 animal phyla. The coelenterates, flatworms, and nemerteans present at Beaufort are missing in the Belgrade channel setting. Chordates are present in both the Recent and ancient settings. However, it is impossible to demonstrate that bone fragments in the Belgrade channel represent members of the original Oligocene oyster community. Boring bryozoans are present at Belgrade, but were not recorded in the Recent minimal community; this is the only major taxonomic addition in the fossil setting. Tables 2 to 4 present and expand upon the data presented in this and following sections of the comparison.

Soft-bodied forms. Polydorid bristleworms and boring bryozoans are the only soft-bodied organisms recorded in the fossil deposit. Evidence of these three or four species is based upon redundant message transmissions—their distinctive oyster shell markings.

At Beaufort, 39 species (over 48 percent) are soft-bodied. Only four soft-bodied forms, a polydorid bristleworm and three tube-building worms, can initially leave a record of their presence. But one species, the polydorid, would remain if the tubes were subsequently lost; the bristleworm would be recorded through redundancy. Fossil preservation and Recent preservability are analogous.

Forms with hard parts. Only nine or ten calcitic species are taxonomically identifiable in the Belgrade setting. However, evidence from the quarry indicates that aragonitic, siliceous, and chitinous skeletons were present and were subsequently lost during the post-mortem history of the oyster deposit. The main evidence for these losses comes from decodable information redundancy.

Several lines of evidence point to large-scale losses of aragonitic mollusc skeletons. The most compelling evidence comes from boring pelecypod remains. Cavities attributable to boring pelecypods occur on 34 oysters (24 percent of the sample) in the deposit, but only one minute aragonitic fragment of the borer remains. The impressed ghost of the low-spired gastropod, preserved on an oyster shell, may well represent another aragonitic mollusc. Yet other aragonitic remains may have been lost without having left any trace of their prior existence. Thus it is impossible to tell whether or not such common oyster associates as *Crepidula* were present in the original community.

Observations on the oysters themselves complement the idea of aragonite loss. The principal skeletal parts of Recent crassostreids are calcitic, while pads for muscle attachment are composed of aragonite (Stenzel, 1963). This mineral fabric apparently also occurred in the Oligocene *Crassostrea gigantissima* (Finch), since the area of the adductor pads is now a voided region, while the rest of the shell is well preserved. There is no evidence of mixed aragonitic-calcitic shells (with appreciable aragonite) preserved in the fossil deposit.

TABLE 3. EFFECT OF ARAGONITE LOSS UPON PRESERVATION OF THE RECENT MINIMAL OYSTER COMMUNITY

Taxa	Number of species	A	C	A + C
Coelenterata	1	—	1	—
Bryozoa	3	—	3	—
Mollusca				
Gastropoda	9	9	—	—
Pelecypoda	13	4	3	6
Arthropoda				
Crustacea	4	—	4	—
Totals	30	13	11	6
Percentage of total community	38	16	14	8

A = completely aragonitic skeletons
C = entirely calcitic skeletons; calcitic skeletons with very minor aragonite, as in crassostreids
A + C = appreciable aragonite plus calcite in skeletons
data sources given in text

Evidence for the loss of chitinous exoskeletons comes from the broken oyster shell margins which are reminiscent of valve chippings done by predatory crabs (McDermott, 1960). The perforations and galleries of clionid sponges are recorded on oyster shells, but no siliceous spicules remain. Thus evidence indicates the preservation of nine or ten calcitic skeletons and the loss of siliceous, aragonitic, and chitinous hard parts. How would a similar post-mortem history affect the Recent minimal community?

Of the Recent species, 41 (over 51 percent of the community) possess hard parts. These include forms with aragonitic, calcitic, mixed aragonitic-calcitic, siliceous, chitinous, and phosphatic skeletons. Calcareous barnacles, bryozoans, and coelenterates are present and 22 (about 26 percent) are calcareous molluscs.

Table 3 summarizes the possible effects of aragonite loss upon the preservation of the Recent community members with calcareous skeletons. Data on skeletal mineralogy are taken from the available reviews of Boggild (1930), Chave (1954, 1964), and Lowenstam (1954, 1963), and from X-ray analyses of conspecific or congeneric forms from the Beaufort area. The distinction between low-Mg and high-Mg calcite was not included in this analysis for three reasons. First, readily available and complete data are lacking in the literature. Second, Chave (1962) demonstrated that a continuous spectrum of Mg-calcite exists in nature; this spectrum is present even within individual skeletons (Chave and Wheeler, 1965). Third, the destruction of high-Mg calcite normally proceeds not by wholesale solution but by an as yet not clearly understood "replacement" by low-Mg calcite, with skeletal preservation.

The possibilities for information losses due to disappearance of aragonitic and mixed aragonitic-calcitic skeletons are striking. Of the 30 calcareous forms, only 11 would remain. This loss of 19 species represents removal of over 23 percent of the original community.

Disappearance of the five siliceous forms would not constitute a major loss of information, since three species of boring clionid sponges would leave other evidence of their existence. Loss of chitinous skeletons would probably obliterate the record of the five decapods. Although different crabs produce distinctive chipping patterns (McDermott, 1960), these nuances may not be readily decodable in the fossil record.

As more skeletons are lost, a larger percentage of our knowledge about the community would be based upon redundant message transmissions. Table 4 expands the previous discussion by analyzing preservability in three cases: (1) entire community preserved, (2) soft bodies lost, all forms with hard parts preserved, and all redundant transmissions preserved, and (3) soft-bodied, aragonitic, aragonitic-calcitic, siliceous, and chitinous species lost; tube-building worms and valve chippers not taxonomically identifiable. Of the 80 species in the Recent minimal community, only 18 (22 percent) would remain in the third case.

TABLE 4. PRESERVABILITY OF RECENT MINIMAL OYSTER COMMUNITY, AND COMPARISON WITH BELGRADE OYSTER COMMUNITY

	Recent A	Minimal B	Community C	Belgrade Fossil Community D
Phyla represented	9	7	7	5
Species present	80	45	18	16–18
Percentage of total community preserved	100	56	23	—
Percentage of information through redundant transmissions—preservation of non-body parts	0	7	33	~ 44

Column A: original community; all organisms preserved
Column B: all hard parts and all redundant information preserved
Column C: aragonitic, mixed aragonitic-calcitic, chitinous, and siliceous skeletons lost; redundant information preserved but oyster valve chippers and mud-tube-building worms not taxonomically identifiable
Column D: Belgrade community, for comparison with C

Comparison of columns C and D in Table 4 shows that the minimal Beaufort community, if subjected to the same history as the Oligocene community, would yield similar types and amounts of preserved information. The original Oligocene community must have approximated Recent communities in diversity and number of individual species present. To estimate numbers and diversity more exactly, paleontologists need data on the preservability of Recent oyster communities showing a range in organic diversity. The gathering and analysis of this data is now underway. This latter work may help paleontologists to analyze spatial and temporal aspects of organic diversity in oyster communities. Yet even the present, broad conclusion has added significance since both *Crassostrea gigantissima* and *C. virginica* belong to the *C. soleniscus* lineage as defined by Sohl and Kauffman (1964). There is no reason to suspect drastic changes in the major taxa of communities, within the lineage, since Oligocene time.

Contribution of negative evidence. Negative evidence (lack of aragonite within the oyster shells) has helped substantiate the idea of mineral losses. Yet negative evidence also suggests that a few individual species preservable in the Recent minimal community could not have been present in the original fossil community.

Absence of some distinctive oyster form modifications indicates that some common oyster associates were not present in the Oligocene community. The distinctive shell borings of predatory gastropods do not occur on Belgrade oysters, even on thin and susceptible valve portions. Thus predatory muricid gastropods were apparently absent. Other evidence indicates that calcitic skeletons should be preserved in the deposit. Lack of calcitic non-ostreid pelecypods suggests that oyster associates like *Anomia* were missing from the original community or were transported away from the site of the community. Negative evidence can eliminate some possible individual members of the fossil community.

This comparative analysis, the various information losses, and the contributions of negative evidence may be summarized by reconstructing the original Oligocene community.

THE ORIGINAL OLIGOCENE COMMUNITY

The original Belgrade oyster community was quite diverse. Plants were perhaps well represented and megascopic animal species numbered in the high tens or low hundreds; these species came from about nine animal phyla. Post-mortem events reduced this number to about 16 to 18 preserved species, representing five phyla. Over 75 percent of the species have not withstood post-mortem environmental events. Knowledge of seven or eight of these species is based upon redundant pieces of information.

Nearly 50 percent of the original community was composed of soft-bodied organisms, mainly forms like bryozoans, coelenterates, various worms, and arthropods with fragile and unmineralized exoskeletons. The bodies of these creatures were completely dissipated after death. Bryozoan borings and polydorid bristle-worm markings on oyster shells are the only remnants of this diverse group; about three or four species are preserved.

An additional five to seven percent of the fauna consisted of arthropods with mineralized exoskeletons. These forms were also completely lost after death of the organisms. Probable oyster valve chippings, due to predatory crabs, are the only evidence of any arthropods. Less than five percent of the fauna consisted of sponges. Although skeletal remains are lacking, the original biota contained at least two species of boring clionids.

The remainder of the fauna consisted of forms with calcareous skeletons, and included bryozoans, a spirorbid worm, a cirriped crustacean, and molluscs. The first three taxa persisted in the fossil record. However, molluscs did not fare so well. Aragonitic and mixed aragonitic-calcitic shells of dead molluscs were almost completely dissolved and lost. The original diversity of mollusc forms is impossible to determine. Only the oysters themselves, cavities from aragonitic boring pelecypods, and one imprint of a low-spired gastropod remain. Predatory muricid gastropods were not present, and calcitic non-ostreid molluscs were probably absent from the biota.

Three species of calcareous foraminifers and echinoid spines also occur in the oyster deposit, but these remains are not in obvious life juxtaposition with other community members. Environmental evidence, summarized by Lawrence (Ph.D. dissert., Princeton University, 1966), suggests that the foraminifers were endemic. The echinoid spines may represent an introduced and foreign element. Vertebrates were doubtless present in the general areas, but only minute fragments were found in the quarry sample of the channel.

Over half of the preserved species represent "the 'insoluble residue' of a much larger, diverse fauna and flora, most of which has left no record" (Chave, 1964, p. 384). Information from this chemically stable residue is supplemented by information from originally redundant evidences. Recent oyster communities, if subjected to similar taphonomic histories, would yield similar types and amounts of information. There are no valid reasons for postulating major changes in the taxa present, in communities of the *Crassostrea soleniscus* lineage, since Oligocene times. This analysis of post-mortem history and composition of the original community is a necessary background for paleoecologic work. Community maturity, dominance within the community, spatial patterns of organic diversity, life habits of individual community members, and all other

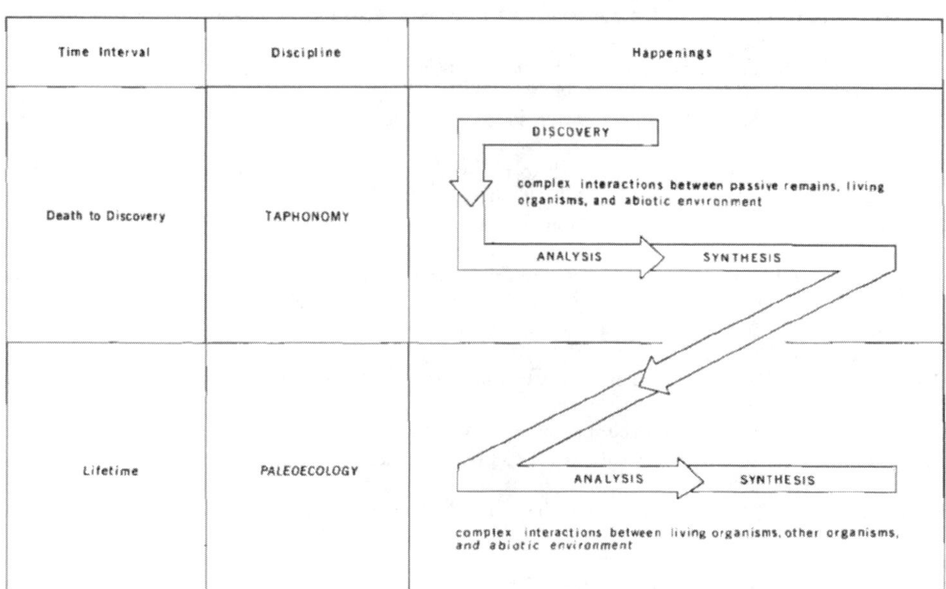

Figure 4. The reconstruction pathway in paleoecology.

aspects of the once-living organisms must be evaluated with reference to this prior analysis.

SUMMARY

Paleoecologists, like other historical scientists, must be detectives. Fossil organisms interact with other living organisms and the abiotic environment in a complex manner. After death of the organisms in question, these complex interactions continue; the passive remains are acted upon and degraded by successive living organisms and physical or chemical environmental factors. Common post-mortem events include the nonpreservation of soft or hard parts and the transport of organic remains away from their life habitat. Nonpreservation of soft-bodied organisms may eliminate from 40 to 70 percent or more of a given community; subsequent elimination of unstable hard parts may further increase information losses. The processes and results of transport need further study by environment-oriented paleontologists.

These post-mortem events commonly result in the preservation of only a fragmentary, blurred picture of the original community. These events must be understood before paleoecologic work may be done; in analysis of a fossil community, taphonomy must precede paleoecology. Thus the ultimate reconstruction pathway in paleoecology is a complex one (Fig. 4). Only by careful analysis and synthesis at each successive stage can we reconstruct the once-living organisms. The taphonomic overprint must be considered in any future studies of working methods in paleoecology.

REFERENCES CITED

Bassler, R. S., 1953, Bryozoa, Part G, Treatise on invertebrate paleontology: Lawrence, University of Kansas Press, 253 p.

Boggild, O. B., 1930, Shell structure of mollusks: Danske Vidensk. Selsk. Skrifter, v. 9, pt. 2, p. 235–326.

Brown, P. M., 1963, The geology of northeastern North Carolina: North Carolina Dept. Conserv. and Development, Spec. Publ., 44 p.

Carriker, M. R., 1955, Critical review of biology and control of oyster drills *Urosalpinx* and *Eupleura:* [U. S.] Fish and Wildlife Service, Spec. Sci. Rept.—Fisheries No. 148, 150 p.

—— 1961, Comparative functional morphology of boring mechanisms in gastropods: Am. Zoologist, v. 1, p. 263–266.

Chave, K. E., 1954, Aspects of the biogeochemistry of magnesium, Pt. 1: Calcareous marine organisms: Jour. Geology, v. 62, p. 266–283.

—— 1962, Factors influencing the mineralogy of carbonate sediments: Limnology and Oceanography, v. 7, p. 218–233.

—— 1964, Skeletal durability and preservation, p. 377–387 *in* Imbrie, J., and Newell, N., *Editors,* Approaches to paleoecology: New York, John Wiley, 432 p.

Chave, K. E., and Wheeler, B. D., Jr., 1965, Mineralogical changes during growth in the red alga, *Clathromorphum compactum:* Science, v. 147, p. 621.

Craig, G. Y., 1964, An ecological approach to the study of fossil marine invertebrates, p. 583–590, *in* Nairn, A. E. M., *Editor,* Problems in Palaeoclimatology: New York, John Wiley, 705 p.

Craig, G. Y., and Jones, N. S., 1966, Marine benthos, substrate and palaeoecology: Palaeontology, v. 9, p. 30–39.

Craig, G. Y., and Oertel, G., 1966, Deterministic models of living and fossil populations of animals: Geol. Soc. London Quart. Jour., v. 122, p. 315–355.

Efremov, I. A., 1940, Taphonomy, a new branch of paleontology: Akad. Nauk. S.S.S.R. Biul., Biol. Ser., no. 3, p. 405–413.

Fagerstrom, J. A., 1964, Fossil communities in paleoecology: Their recognition and significance: Geol. Soc. America Bull., v. 75, p. 1197–1216.

Gould, S. J., 1965, Is uniformitarianism necessary?: Am. Jour. Sci., v. 263, p. 223–228.

Gregg, J. H., 1948, Replication of substrate detail by barnacles and some other marine organisms: Biol. Bull., v. 94, p. 161–168.

Hartman, W. D., 1958, Natural history of the marine sponges of southern New England: Peabody Mus. Yale Univ., Bull. 12, 155 p.

Johnson, R. G., 1964, The community approach to paleoecology, p. 107–134 *in* Imbrie, J., and Newell, N., *Editors,* Approaches to paleoecology: New York, John Wiley, 432 p.

—— 1965, Pelecypod death assemblages in Tomales Bay, California: Jour. Paleontology, v. 39, p. 80–85.

Korringa, P., 1951, The shell of *Ostrea edulis* as a habitat: Archives Neerlandaises Zoologie, v. 10, p. 32–152.

Lever, J., 1958, Quantitative beach research. I. The "left-right phenomenon": sorting of lamellibranch values on sandy beaches: Basteria, v. 22, p. 21–51.

Lever, J., Kessler, A., Van Overbeeks, A. P., and Thijssen, R., 1961, Quantitative beach research. II. The "hole effect": a second mode of sorting of lamellibranch valves on sandy beaches: Netherlands Jour. Sea Research, v. 1, p. 339–358.

Lever, J., van den Bosch, M., Cook, H., van Dijk, T., Thiadens, A. J. H., and Thijssen, R., 1964, Quantitative beach research. III. An experiment with artificial valves of *Donax vittatus:* Netherlands Jour. Sea Research, v. 2, p. 458–492.

Loosanoff, V. L., and Engle, J. B., 1943, *Polydora* in oysters suspended in the water: Biol. Bull, v. 85, p. 69–78.

Loughlin, G. F., Berry, E. W., and Cushman, J. A., 1921, Limestones and marls of North Carolina: N. C. Geol. and Econ. Survey, Bull, 28, 211 p.

Lowenstam, H. A., 1954, Factors affecting aragonite-calcite ratios in carbonate secreting marine organisms: Jour. Geology, v. 62, p. 284–322.

—— 1963, Biological problems relating to the composition and diagenesis of sediments, p. 137–195 *in* Donnelly, T. W., *Editor,* The earth sciences: Chicago, University of Chicago Press, 195 p.

Lund, S., 1945, On *Colpomenia peregrina* Sauv. and its occurrence in Danish waters: Danish Biol. Station Rept., v. 47, p. 3–16.

Lunz, G. R., Jr., 1941, *Polydora,* a pest in South Carolina oysters: Elisha Mitchell Sci. Soc. Jour., v. 57, p. 273–283.

McDermott, J. J., 1960, The predation of oysters and barnacles by crabs of the family Xanthidae: Pa. Acad. Sci. Proc., v. 34, p. 199–211.

Mattox, N. T., 1949, Studies on the biology of the edible oyster, *Ostrea rhizophorae* Guilding, in Puerto Rico: Ecol. Mon., v. 19, p. 339–356.

Miller, B. L., 1912, The Tertiary formations, p. 171–266 *in* Clark, W. B., Miller, B. L., Stephenson, L. W., Johnson, B. L., and Parker, H. N., The coastal plain of North Carolina: North Carolina Geol. and Econ. Survey, v. 3, 552 p.

Miyazaki, I., 1938, On fouling organisms in the oyster farm: Japanese Soc. Sci. Fish. Bull., v. 6, p. 223–232.

Möbius, Karl, 1877, Die Auster und die Austernwirthschaft: Verlag von Wiegandt, Hempel, und Parey, 126 p.

Müller, A. H., 1951, Grundlagen der Biostratonomie: Deutsch. Akad. Wiss. Berlin Abh., Jahrg. 1950, no. 3, 147 p.

—— 1963, Lehrbuch der Paläozoologie Band 1. Allgemeine Grundlagen: Gustav Fischer Verlag Jena, 387 p.

Nagle, J. S., 1964a, Wave and current orientation of shells (abstract): Geol. Soc. America, Program 1964 Ann. Mtgs., p. 138.

—— 1964b, Differential sorting of shells in the swash zone (abstract): Biol. Bull., v. 127, p. 353.

Old, M. C., 1941, The taxonomy and distribution of the boring sponges (Clionidae) along the Atlantic Coast of North America: Chesapeake Biol. Lab. Contr., Publ. 44, 30 p.

Pearse, A. S., and Wharton, G. W., 1938, The oyster "leech" *Stylochus inimicus* Palombi associated with oysters on the coasts of Florida: Ecol. Mon., v. 8, p. 605–655.

Sandy, J., Carver, R. E., and Crawford, T. J., 1966, Stratigraphy and economic geology of the coastal plain of the Central Savannah River area, Georgia: Geol. Soc. America, Southeastern Sect., Guidebook, 1966 Field Trip No. 3, 30 p.

Simpson, G. G., and Beck, W. S., 1965, Life: An introduction to biology, 2d edition: New York, Harcourt, Brace, and World, 869 p.

Sohl, N. F., and Kauffman, E. G., 1964, Giant Upper Cretaceous oysters from the Gulf Coast and Caribbean: U. S. Geol. Survey Prof. Paper 483–H, 22 p.

Stenzel, H. B., 1963, Aragonite and calcite as constituents of adult oyster shells: Science, v. 142, p. 232–233.

Tasch, P., 1965, Communications theory and the fossil record of invertebrates: Kansas Acad. Sci. Trans., v. 68, p. 322–329.

Valentine, J. W., and Peddicord, R. G., 1967, Evaluation of fossil assemblages by cluster analysis: Jour. Paleontology, v. 41, p. 502–507.

Wells, H. W., 1961, The fauna of oyster beds, with special reference to the salinity factor: Ecol. Mon., v. 31, p. 239–266.

Whitelegge, T., 1890, Report on the worm disease affecting the oysters on the coast of New South Wales: Australian Mus. Recs., v. 1, p. 41–54.

Woelke, C. E., 1956, The flatworm *Pseudostylochus ostreophagus* Hyman, a predator of oysters: Natl. Shellfish. Assoc. Proc., v. 47, p. 62–67.

Yonge, C. M., 1963, Rock-boring organisms, p. 1–24 *in* Sognnaes, R. F., *Editor*, Mechanisms of hard tissue destruction, A.A.A.S. Publ. 75, 776 p.

Manuscript Received by the Society August 4, 1967
Revised Manuscript Received February 21, 1968

PAPER 43

Models and Methods for Analysis of the Mode of Formation of Fossil Assemblages (1960)

R. G. Johnson

Commentary

RICHARD K. BAMBACH

This paper by Ralph Gordon Johnson established the discipline of taphonomy in the United States. It touches all the bases, from the range of potential fates of individuals after death to the ways in which exposure and transportation can modify the accumulation of a fossil assemblage to sampling methodology. Although every paleontology textbook began with a chapter on preservation, serious evaluation of how postmortem processes influenced the preserved fossil record was not systematically investigated in the United States until Schäfer's classic book was translated from the original German and published in English (Schäfer 1972). Before that, Johnson's paper stood almost as a lone beacon lighting the scope of issues that need attention if we are to reliably interpret the fossil record. Johnson is very clear that the severity of the spectrum of potential biases he is describing was simply unknown at the time he wrote. He even began his concluding paragraph by commenting, "Until supplementary knowledge concerning the sources of bias is available, analysis of the mode of formation and preservation of fossil assemblages must be based upon theoretical foundations."

Johnson expressed deep concern that interpretation of the record was not meaningful unless one knew whether the actual biological signal had been preserved. Fortunately, his worst fears about loss of data generally are not realized. Johnson himself documented that transport was not a severe problem for most reasonably well-preserved marine assemblages (Johnson 1965). Driscoll (1970) showed that shells were simply destroyed if they remained unburied for long, and Clifton (1971) discovered that burial from bioturbation alone was remarkably rapid (commonly within weeks), thus explaining why disturbance and transportation were not crippling biases for interpreting the ecologic setting of many assemblages. Susan Kidwell, a professor in Johnson's old department at the University of Chicago, along with various colleagues, has produced a series of studies over the past quarter-century that have put the flesh of observation on the framework of theory of Johnson's seminal paper (e.g., Kidwell and Bosence 1991; and Kidwell and Rothfus 2010, among many others).

Literature Cited

Clifton, H. E. 1971. Orientation of empty pelecypod shells and shell fragments in quiet water. *Journal of Sedimentary Petrology* 41:671–82.

Driscoll, E. G. 1970. Selective bivalve shell destruction in marine environments, a field study. *Journal of Sedimentary Petrology* 40:898–905.

Johnson, R. G. 1965. Pelecypod death assemblages in Tomales Bay, California. *Journal of Paleontology* 39, pp. 80–85.

Kidwell, S. M., and D. W. J. Bosence. 1991. Taphonomy and time-averaging of marine shelly faunas. In *Taphonomy: Releasing the Data Locked in the Fossil*

From *Geological Society of America Bulletin* 71:1075–86.

Record, edited by P. A. Allison and D. E. G. Briggs, 115–209. Topics in Geobiology, vol. 9. New York: Plenum Press.

Kidwell, S. M., and T. A. Rothfus. 2010. The living, the dead, and the expected dead: variation in life span yields little bias of proportional abundances in bivalve death assemblages. *Paleobiology* 36:615–40.

Schäfer, W. 1972. *Ecology and Palaeoecology of Marine Environments*. Translated by I. Oertel. Chicago: University of Chicago Press.

BULLETIN OF THE GEOLOGICAL SOCIETY OF AMERICA
VOL. 71, PP. 1075-1086, 2 FIGS. JULY 1960

MODELS AND METHODS FOR ANALYSIS OF THE MODE OF FORMATION OF FOSSIL ASSEMBLAGES

By Ralph Gordon Johnson

Abstract

Theoretical aspects of the formation of fossil assemblages are explored for the purpose of obtaining criteria and methods for the reconstruction of circumstances of preservation of shallow-water marine organisms. Models are developed which represent: (1) a death assemblage preserved under conditions of rapid burial; (2) an assemblage preserved *in situ* under conditions of gradual accumulation; and (3) an assemblage composed almost entirely of remains transported to the site of burial.

The histories represented by the models influence the following features of fossil assemblages: faunal composition, morphologic composition, density, disassociation of hard parts, fragmentation, surface condition of fossils, chemical and mineralogical composition of fossils, orientation, dispersion, and the texture and structure of the sedimentary aggregate. The expressions of these features indicate that biological criteria are more indicative of the mode of accumulation than physical criteria.

The stretched-line method of sampling provides a means of obtaining objective and repeatable measures of features of the fossil assemblages in place. It is restricted to sediments in which fossils can be recovered readily. A rank-correlation analysis of 11 samples from the Pleistocene Millerton formation of Tomales Bay, California, is given as an example of a means of evaluating the interrelations of variables measured by the line technique.

CONTENTS

TEXT

ILLUSTRATIONS

TABLES

Introduction

At present we do not know enough about the processes that have acted upon dead organisms to retrace in detail the history of a particular fossil assemblage. Many of these processes (*e. g.*, diagenesis) are not amenable to ordinary methods of study. In the absence of direct knowledge of these processes and their results, reconstructions of the circumstances of preservation have been based upon general assumptions. Several attempts have been made to strengthen the theoretical framework of paleoecology by redesigning and refining these assumptions (recently Boucot, 1953; Miller and Olson, 1955; Olson, 1957). This paper explores the circumstances leading to the preservation of shallow-water marine organisms. Model fossil assemblages are used to examine criteria and methods for reconstruction of the mode of formation of actual fossil assemblages.

In the summer of 1956 the author began a general investigation of the preservability of marine communities. The assemblages of the Pleistocene Millerton formation of Tomales

Bay, California, offer a unique opportunity for evaluation of paleoecological theory and technique. The fauna comprises extant species of which at least 36 per cent are now living in Tomales Bay. The sediments associated with the Millerton fauna resemble closely the sediments of the bay. These circumstances permit a detailed reconstruction of the environment of deposition of the Millerton and the relationships of its fossils to living communities. The reconstructions may then be compared to the conclusions reached by paleontological means. Thus, the theory and methods of paleoecological analysis may be evaluated under as controlled conditions as can normally be achieved with the basic materials of paleontology. The present paper is concerned with the theoretical aspects of this project and is the first in a series dealing with the preservation of the Millerton fauna and the formation of death assemblages in the modern environments of Tomales Bay. While much of what follows is commonly implied in the literature, the detail sought in modern paleoecological studies requires the explicit statement of basic hypotheses so that these may be examined and made amenable to quantification.

Acknowledgments

Financial aid furnished by The Penrose Bequest of The Geological Society of America made possible the field study of the Millerton deposits. Mr. Robert Hessler assisted greatly in the field. Profs. J. Marvin Weller, Robert L. Miller, and Everett C. Olson have consulted with the author throughout this work. Dr. Joel W. Hedgpeth, Director of the Pacific Marine Station, has aided immeasurably in the field both as a stimulating consultant and jovial companion.

Formation of Fossil Assemblages

Possible Fates of Elements of Life and Death Assemblages

Figure 1 shows the possible fates of elements of life and death assemblages pertaining to the formation and preservation of fossil assemblages. The ten cases shown for individuals of a life assemblage[1] at a particular site are variations of four kinds of sequences. An organism may emigrate in the normal course of its existence as a predator or as a planktonic form (case 1). Emigration may result from an unfavorable change in the environment such as a sudden local increase in the amount of sediment deposited, The ability of active animals to detect and avoid unfavorable changes favors the preservation of sedentary forms. If an organism is removed by a predator or currents following death, it cannot contribute to the local death assemblage (case 2). The remaining possibilities for members of a life assemblage result in contributions to a death assemblage but differ in the ultimate disposition of these contributions. Cases 7, 8, and 9 represent the fates of infaunal organisms[2] that die within the sediment. Dying individuals of some infaunal species may emerge from the sediments to perish on the bottom (MacGinitie, 1935). Case 10 represents nonliving parts, such as molts, discarded during a life cycle. Such parts are exposed to the possibilities shown as 3, 4, 5, and 6. Excluding the variations introduced by an infauna, at least four kinds of histories are possible for elements of the exposed death assemblage. Probably the most common history in shallow seas is the destruction of remains by decomposition, mechanical and chemical action, and the action of scavengers; or removal by currents (case 3). The history of particular interest herein is that of an organism which, upon death, enters the exposed death assemblage and persists in some form so that it enters the local buried death assemblage (cases 4, 5, and 6). The period between death and burial is critical as the remains are exposed to a variety of destructive forces. The alteration of an element of a death assemblage depends partially on the length of time it remains exposed and the rates of the biological and physical processes acting upon it. Two dissimilar histories may produce the same net effect upon the same kinds of materials if different periods of exposure are involved.

Remains from other death assemblages may be introduced into the local death assemblage. An element of a buried death assemblage may be reintroduced into the exposed death assemblage by erosion (case 15). Foreign remains may be removed subsequently (case 11) or persist and be included in the local buried

[1] "Life assemblage" as used herein means an assemblage of living organisms, "death assemblage", an assemblage of dead organisms, parts of dead organisms and/or the discarded parts of living organisms. The term "exposed death assemblage" is used for a death assemblage not buried in sediments.

[2] Thorson (1957) has defined the infauna as ". . . comprising all animals inhabiting the sandy or muddy surface layers of the sea . . . *i.e.*, living buried or digging in a substratum."

assemblage (cases 12, 13, 14). Thus, two or more distinctive communities which were separated in life by space and/or time may be mixed after death.

sediments deposited in the area of life and remains introduced into the site of deposition from elsewhere. Additional criteria are required for the assessment of the alteration of a fossil

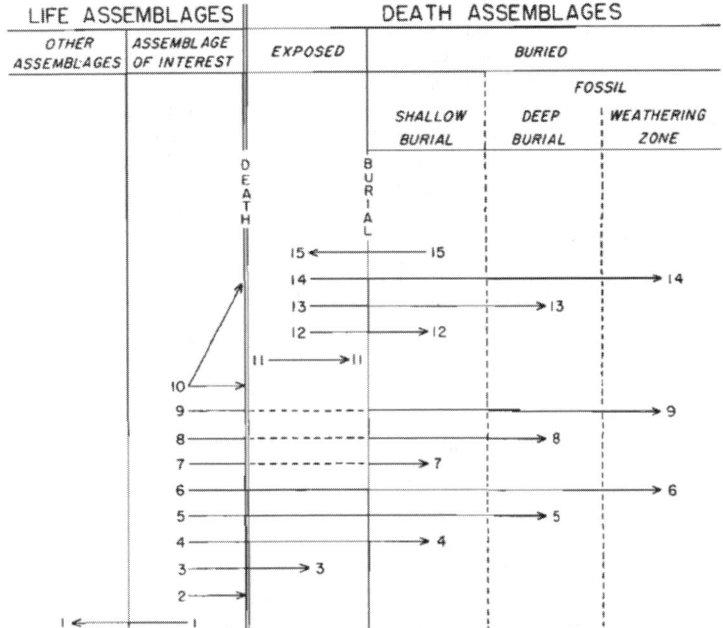

FIGURE 1.—POSSIBLE FATES OF ELEMENTS OF LIFE AND DEATH ASSEMBLAGES

Case 1—emigration; 2—death and immediate removal; 3—remains enter exposed death assemblage but are destroyed or removed; 4, 5, 6—remains enter exposed death assemblage and are buried; 7, 8, 9—remains of infaunal organisms *in situ*; 10—nonliving parts discarded during lifetime; 11—remains from other death assemblages are introduced but destroyed or removed subsequently; 12, 13, 14—remains from other death assemblages are introduced and persist to be buried; 15—buried remains are reintroduced to exposed death assemblage by erosion.

Cases 4, 7, 12—remains are destroyed or removed while buried at a shallow depth; 5, 8, 13—remains are destroyed or removed by diagenetic processes; 6, 9, 14—remains persist and eventually are exposed to weathering.

The remains of organisms that become buried in the sediment may be destroyed or removed while buried shallowly and still under the influence of the local ecosystem (cases 4, 7, 12). They may persist and be more deeply buried only to be destroyed by the physical processes operating within the body of rock (cases 5, 8, 13). Or they may survive and become available for sampling (cases 6, 9, 14). As the death assemblage is brought into the weathering zone, the rate of alteration of the fossils may increase rapidly. This "last-moment" alteration is one of the factors that limits geochemical investigations of primary composition.

The immediate concern here is to develop criteria that will permit distinction between the remains of organisms that are buried in

assemblage that has taken place between the time of death and recovery. For studies of evolution, ecology, and geology, the elements not preserved may be as important as or even more important than those that are preserved. Some knowledge of the missing components may be inferred by analogy and the application of principles of ecological integration or by recognition of the physical and chemical effects of organisms upon one another and upon the sediment. The uncertainty inherent in this kind of inferential knowledge constitutes a fundamental limitation of paleontology.

Models of Death Assemblages

A fossil assemblage can contain any proportion of fossils representing different histories.

Each history can produce a wide variety of exposure effects depending upon the rate, effectiveness and period of operation of the processes involved. Thus a broad spectrum of fossil assemblages is possible with regard to mode of accumulation and alteration. The continuum of variation represented by fossil assemblages cannot be expressed by a system of classification recognizing discrete variants. We can obtain a basis for their comparison by constructing special models which feature the environmental conditions that result in the common and extreme combinations of modes of accumulation and exposure effects. These might serve as reference types, similar in function to the type specimen of taxonomy. Three such models are proposed here. While the conditions described are hypothetical, they approximate actual conditions in certain modern environments. The models represent (1) a death assemblage buried at the site of life with a minimum of disturbance, (2) an assemblage intermediate with regard to exposure effects and the introduction of foreign elements; and (3) a death assemblage composed almost entirely of transported remains.

MODEL I: A life assemblage lives in a restricted area of the sea bottom below the low-tide mark but above the maximum wave base. The water mass moves over the bottom at a low velocity which is occasionally raised to erosional competency. The substrate is composed of clastic sediments. Intermittently, small amounts of similar sediments are introduced into the area together with minor amounts of the durable remains of organisms. The abundant and diverse life assemblage consists of soft-bodied organisms and organisms bearing hard parts. Continually and under the conditions of normal mortality, elements of this life assemblage enter the local death assemblage. These decompose; some are buried after varying periods of exposure. Then, suddenly, the entire life assemblage is buried and killed by the rapid introduction of a large amount of sediment not unlike the material of the former substrate. The sedimentary body containing the death assemblage is gradually compacted as a result of further accumulation of sediment at the site.

MODEL II: A life assemblage lives in an environment similar to that described for Model I. Continually and under the conditions of normal mortality, elements of the life assemblage enter the local death assemblage. These decompose; some of the remains are carried away, whereas others of durable composition are buried after varying periods of exposure. In time, the local environment changes and the life assemblage is eventually replaced by another of different composition. The death assemblage of interest becomes more deeply buried, and the sedimentary body becomes gradually compacted as a result of further accumulation of sediment at the site.

MODEL III: A life assemblage lives in a restricted area of the sea bottom below the low-tide mark but above the maximum wave base. The water mass moves at a moderate velocity over a bottom of clastic sediments consisting, in large part, of the durable remains of organisms. Frequently the velocity of the water mass is high enough to move sediment and organic detritus through the area. The hydrodynamic circumstances favor the accumulation of organic remains at the site. The sparse fauna consists of a few epifaunal species of scavengers, boring and encrusting organisms. Continually and under the conditions of normal mortality, elements of this life assemblage enter the local death assemblage. These decompose; some of the remains are carried away, whereas others of durable composition are buried after varying periods of exposure. Eventually, the rate of accumulation of organic remains at the site decreases as changes occur in the source areas of the sediment and debris. In time, the zone containing the high concentration of durable remains is buried and gradually compacted as a result of further accumulation of sediment at the site.

The majority of the elements of a fossil assemblage formed under the conditions of Models I and II will have had a history shown as cases 6 and 9 in Figure 1. The majority of remains brought together as in Model III will have had a history shown as case 14.

The unique feature of Model I is the rapid burial of a life assemblage. Such a history is probably rare in the fossil record. Submarine slumping, storms, or rapid deposition off the mouth of a flooding river might achieve this result. Assemblages resembling those to be expected under the conditions of Model II have been encountered in the superficial sediments of Tomales Bay. At most sites sampled, the death assemblage consisted chiefly of representatives of the local life assemblage. A small fraction of each of the samples, however, consisted of the remains of animals now living elsewhere. The conditions described in Model III probably are more common in shallow water than those of Model I but less common than those of Model II. Environments in which durable hard parts accumulate under some of the circumstances described in Model III have been encountered in Tomales Bay in channels bordering tidal flats, off wave-eroded shores,

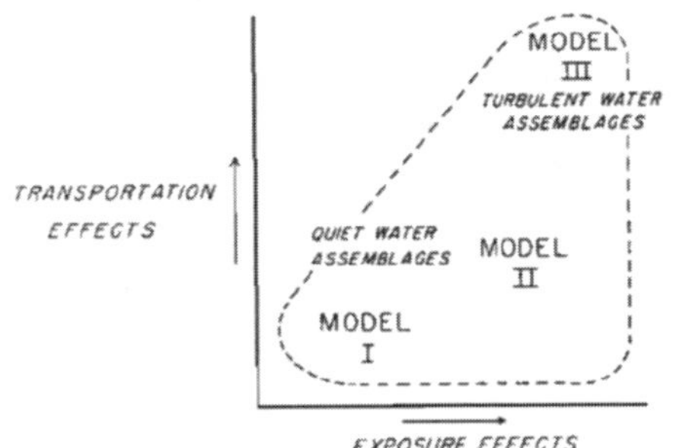

FIGURE 2.—INFERRED RELATIONSHIPS OF THE MODELS

Hypothetical diagram representing the relative alteration of the fossil assemblages resulting from three modes of formation. Model I represents the sudden burial of a life assemblage; Model II, the gradual accumulation and burial of the remains of organisms living at the site of deposition; and Model III, an assemblage comprised almost entirely of transported remains. Exposure effects include degree of abrasion, solution, fragmentation, decomposition, encrustation and boring. Transportation effects include spurious association of species and size sorting. Most fossil assemblages probably occur within the area of the dotted line. Assemblages formed in quiet waters should cluster near Models I and II while those of turbulent water should resemble Model III.

and in very shallow water adjacent to certain beaches. The death assemblages at these sites consist of shells in all states of disintegration. Many complete single valves are present; nearly all show abrasion, boring, and encrustation. Most of the organic sediment, however, consists of riddled and stained angular fragments of shells. Commonly several life assemblages are represented. The inorganic component varies from place to place in composition, size of particles, and proportion to the total sediment. In one area, wave action is eroding fossiliferous Millerton deposits. Off this shore, the fossils of locally extinct Pleistocene species are mingled with the remains of modern forms.

Figure 2 shows the inferred relationships between the model assemblages with regard to exposure and transportation effects. The results of exposure and transportation cannot be sharply distinguished in real circumstances. The introduction or subtraction of elements implies the action of moving water, a principal agent of degradation during local exposure. Introduced elements will contribute previously developed exposure effects to the assemblage. Elements can be subtracted by local destruction as well as by transportation. An assemblage composed largely of exotic elements but exhibiting a miniumum of exposure effects could result from the mass transport of bottom forms and sediments under unusual wave or current action such as might be attained during severe storms. Such assemblages probably are rarely formed in shallow waters. On the other hand, indigenous assemblages can exhibit the entire conceivable range of exposure effects, depending upon the forces in action and the period of time involved. In spite of these complications, Figure 2 provides a medium for viewing a variety of fossil assemblages and a departure point for finding distinguishing features.

The environmental settings and the modes of death and burial specified in the models result in assemblages differing in their relationships to associated sediments and in the degree of exposure effects they exhibit. They should, in theory, influence the following features of the resultant fossil assemblages:

(1) faunal composition
(2) morphologic composition
(3) density of fossils
(4) size-frequency distribution of fossils
(5) disassociation of hard parts
(6) fragmentation of remains
(7) surface condition of fossils
(8) chemical and mineralogical composition of hard parts
(9) orientation of fossils
(10) dispersion of fossils
(11) texture and structure of the sedimentary aggregate.

Table 1 shows the relative expression of these features expected for fossil assemblages formed as in the models. Only the most distinctive or most probable developments of these features are shown.

FAUNAL COMPOSITION: It is expected that many of the species preserved under the conditions of Model I would be associated in death with the species and sediments with which they were associated in life. This is probably also true for Model II assemblages but not for Model III assemblages. By analogy, a faunal list may be translated into collections of modes of life and environmental tolerances. These provide a basis for evaluating the consistency of a particular combination of species with regard to their ecological connotations. For the

Table 1.—Relative Expressions of Features Developed under the Conditions of the Models

Only the most distinctive or most probable development is shown.

Feature	Expression		
	Model I	Model II	Model III
Faunal composition	Ecologically coherent assemblage of species	As in Model I	Not necessarily as in Models I and II
Morphologic composition	Delicate structures and heterogeneous shapes and sizes may be preserved. Suites of shapes represented are ecologically consistent.	As in Model I	May consist only of the durable parts of the species present; may be homogeneous in shape and sizes
Density of fossils	Wide range of densities possible	As in Model I	High
Size-frequency distribution	Many species exhibit a size-frequency distribution conforming to an ideal distribution for an indigenous population (Olson, 1957).	Some species as in Model I	Not as in Models I and II
Disassociation	High proportion of articulated remains; disassociated parts represented in appropriate relative numbers for a species	Moderate proportion of articulated remains; disassociated parts as in Model I	Not as in Models I and II
Fragmentation	Low proportion of remains are fragments.	Moderate proportion of remains are fragments.	High proportion of remains are fragments.
Surface condition of fossils	Surfaces of preserved structures as in life.	Various states of wear represented	As in Model II
Chemical and mineralogical composition	No general expectations are warranted by present knowledge in this area.	As in Model I	As in Model I
Orientation	Some species may retain orientation as at time of death.	Majority of fossils oriented with long axis parallel to bedding plane (for exceptions see text)	As in Model II (for exceptions see text)
Dispersion	Articulated remains of some species may retain pattern of dispersion as in life.	Not as in life	As in Model II
Sediment structure and texture	Consistent with inferred tolerances of the fauna and with relatively quiet waters	As in Model I	Not necessarily consistent with inferred tolerances of the fauna; consistent with relatively turbulent waters

ancient fossil record only very broad ecological inferences can be made concerning particular species; other lines of evidence may be necessary in evaluating the ecological homogeneity of ancient assemblages of species. Evidence concerning the composition and sedimentary associations of broadly contemporaneous fossil assemblages can be used to strengthen the analogy.

MORPHOLOGIC COMPOSITION: An assemblage exhibiting delicate structures and heterogeneous shapes and sizes may be preserved under the conditions of Models I and II. Model III assemblages, however, may consist only of the most durable structures of the species represented and under extreme conditions may consist of elements of similar shape and size. In addition to reflecting exposure effects, the morphologic composition of the assemblage may provide evidence as to the ecological consistency. Shape, growth habit, and shell thickness can serve in this manner for some groups of animals.

DENSITY OF FOSSILS: If assemblages of colonial organisms are excluded, extremely high fossil densities are more probable under the conditions of Model III than for those of Models I and II. Fossil density, as expressed by some convenient measure of the relative proportions of organic remains and sediment, is a function of source (biomass), exposure, rate of deposition, and diagenesis (chiefly compaction). Other sources of evidence are therefore required in order to disentangle the roles played by these factors in any real fossil assemblage.

SIZE-FREQUENCY DISTRIBUTION OF FOSSILS: Under the conditions of Model I, some species should show a size-frequency distribution approximating that of a living population. This result is not probable under the conditions of Models II and III. Further, assemblages of Models I and II should contain forms whose size-frequency distributions conform to those of the complete death assemblage of an indigenous population. This result is not probable under the conditions of Model III. Olson (1957) shows that very precise biological information is required before the size-frequency distributions observed among fossils can be interpreted in these terms.

DISASSOCIATION OF HARD PARTS: Assemblages resulting from the events of Model I would be expected to include a relatively large number of articulated remains. Disassociated skeletal parts would be expected to be represented in appropriate relative numbers in the assemblages resembling Model I. Thus the right and left valves of a species of pelecypod should be present in approximately equal numbers. No such result could be deduced for assemblages resembling Model III. Model II represents assemblages intermediate in these regards. The differences in the strength and means of articulation in different parts of the skeleton and in different organisms complicate attempts to estimate the degree of exposure from the amount of disassociation.

FRAGMENTATION OF REMAINS: The history described for Model III should result in an assemblage of fragments of larger skeletal units. This is certainly the case in the Tomales Bay death assemblages that correspond to Model III. In contrast, a large proportion of the elements of assemblages formed under the conditions of Model I should be unbroken. Model II should result in an assemblage consisting of both broken and unbroken elements. The relative durability of the resistant parts of organisms represented must be considered when assemblages are compared as to fragmentation.

SURFACE CONDITION OF FOSSILS: The surface condition of most of the elements brought together as in Model I should be similar to that in life. The elements of a Model III assemblage, however, might show considerable surface damage as a consequence of wear, collisions in transportation, solution, or the drillings and encrustations of other organisms. Assemblages formed as in Model II would be expected to be intermediate in this regard. Diagenetic processes may also contribute to surface markings or polishing (Pettijohn, 1957, p. 68–72).

CHEMICAL AND MINERALOGICAL COMPOSITION OF HARD PARTS: If primary compositions and diagenetic histories are similar (an unwarranted assumption in most real cases) the chemical and mineralogical composition of elements of a Model III assemblage are likely to be more altered than those of Model I and Model II. At present, our knowledge of the multiple factors involved is insufficient to permit prediction of the geochemical consequences of a specific sequence of events.

ORIENTATION OF FOSSILS: Most of the elements of a Model II or III assemblage would be expected to be oriented with the long axes parallel to the bedding plane where associated with well-sorted sediments of sand size or smaller where the density of the fossils is not high. Under these circumstances, elements may also show a preferred orientation in the plane of bedding in response to a prevailing current direction. A significant proportion of single valves would be expected to be oriented with the concave surfaces downward (Johnson, 1957). When these conditions of sediment and fossil density are not met, interference between shells or between shells and large particles may result in the orientation of the elements at various attitudes. Some of the forms living at the time of rapid burial under the circumstances described for Model I would be expected to retain their primary orientation. Compaction probably would cause some reorientation, but it is not possible to predict the extent without specifying original position, specific features of the sedimentary aggregate, and the amount of compaction.

DISPERSION OF FOSSILS: Under the conditions described for Model I, the horizontal patterns of dispersion in life may be retained among several species of the death assemblage. No such result could be expected for Models II and III. Life patterns of dispersion are known for only a few marine invertebrates (Connell, 1956; Johnson, 1959). Observation shows that within their general area of habitation, sedentary species are seldom evenly or randomly dispersed. Within the aggregation of individuals, however, random, even, and clustered patterns of distribution have been encountered in different species. The vertical aspect of a life pattern of dispersion in most fossil assemblages is expected to be readily obscured by succession of generations and compaction.

TEXTURE AND STRUCTURE OF THE SEDIMENTARY AGGREGATE: Assemblages of the Model I type would be expected to be associated with sedimentary features consistent with the inferred tolerances of the fauna and with relatively quiet waters. Assemblages arising under the conditions of Model III could be associated with sediments inconsistent with the fauna (*e.g.*, rock-bottom forms associated with sands), and they may exhibit the effects of strong waves or currents. Thus sedimentary features may indicate the ecological consistency of the fauna with respect to the inferred substrate and water movements. Particle size, sorting, and wave and current structures may imply current systems competent enough to effect the transportation of elements to and from the site.

For each of the features of a fossil assemblage it is possible to conceive of circumstances in which a similar expression results from dissimilar histories. Thus compound evidence should be used to determine mode of formation. One inherent weakness of this principle is the tacit assumption that each line of evidence can contribute equally to the strength of a working hypothesis.

The models considered here indicate that biological criteria, such as ecological coherency, are less ambiguous than physical characters. Ecological coherency and size-frequency distribution reflect transportation more directly than exposure effects, which can be developed without transportation. The biological criteria, however, depend on fundamental knowledge which usually is not available. It appears that knowledge of the organization and dynamics of populations of marine organisms (as well as of the physical processes operating on death assemblages) is critical to understanding the nature of fossil assemblages.

In their most distinctive development and taken all together, the features considered here discriminate between assemblages of the Model I and III types. Model II gains its distinctive character, under these conditions, by compromising the differences between the others. As noted previously, real fossil assemblages vary as a continuous series with regard to their suite of characters, so that no set of criteria can delimit all possible variations. It seems possible, theoretically at least, to compare actual fossil assemblages by reference to the extreme development of such features as described here. As knowledge of marine biology increases and the formation of modern death assemblages is studied, the criteria almost certainly will be refined and the number of criteria increased.

METHODS OF STUDY

The principal objective in most paleontological field work is the collection of fossils. Observations upon mode of occurrence of fossils, if made at all, generally are recorded in qualitative terms as supplementary information. For these purposes the field observations of a careful worker may suffice. As attention is focused upon paleoecology, however, systematic and objective observations are required. Fossil assemblages must be compared in the context of their mode of formation. Measures which can be used to reconstruct the history of a fossil assemblage must be developed in connection with studies of modern death assemblages. Only in this way can observational data be substituted for many of the speculations now relied upon to relate measures to events. Many measures appropriate to the features discussed in the preceding section have been in use for some time (*e.g.*, Imbrie, 1955).

Many of the features of interest herein can be observed only in the undisturbed assemblage at an outcrop. Taking samples to the laboratory and breaking down the matrix destroys some of the most important relationships. Careful dissection of block samples is so time consuming as to severely restrict the number of samples that can be used in a particular study. The writer in an investigation of the Pleistocene Millerton formation of Tomales Bay, California, adapted the "stretched-line" method used by plant ecologists to the sampling of outcrops. In this method a line is stretched across the face of an outcrop, parallel to the planes of bedding. The distance between each fossil encountered along the line is measured. The orientation and state of articulation and preservation are recorded for the entire structure or debris pile represented in part on the sampling horizon. Each specimen is collected, measured, examined for surface features, and identified. From these data a scaled, strip diagram can be constructed showing the distribution and properties of the fossils at the outcrop as sampled along the line. Such a diagram can be used to study the linear pattern of dispersion.

The universe that is sampled in this manner is the face of the outcrop. The number of lines and length of each of them are determined by the objectives of the study and the size and form of the outcrop. In the Millerton study, the zone of interest seldom exceeded 1 m vertically and 3–4 m horizontally. Each line extended the length of the zone, and no more than three lines, randomly place one above the other, were required. As a check on the sampling and the criteria used to select the zone of interest, one narrow zone was sampled throughout its length at four horizons. The data show remarkably good agreement between lines.

The advantages of this sampling procedure are: it provides a means of observing and retaining for study the features of the assemblage as observed in place; it provides a means of treating each outcrop identically; it establishes a context for the precise definition of measures; a relatively large number of features can be measured in a short period of time; and finally, it reduces bias in the collection of fossils.

The disadvantages of this sampling method are, in general, the same as those for the other methods now in use. It is almost entirely restricted to sediments in which fossils can be recovered readily. Care must be taken to sample relatively fresh surfaces in order to avoid, so far as possible, the mechanical effects of weathering. Another disadvantage of the method, as used in the Millerton study, is that the horizontal aspects of the assemblage are overly emphasized. This objection might be overcome by using both vertical and horizontal lines. Many problems also arise in using such samples for statistical inference of the properties of the fossil assemblage in three dimensions. This last weakness certainly is not peculiar to the method of sampling proposed here. The progress of other sciences concerned with somewhat similar problems leads the writer to believe that a satisfactory solution can be achieved.

Relationships Between Measures

If compound evidence is to be used to select the most probable history for a fossil assemblage, it is necessary to know something of the relationships of the several kinds of evidence relied upon. One way of gaining such information from the data provided by line samples can be illustrated as follows. The relationships between 9 variables were studied in 11 line samples of the Millerton formation, involving the occurrence of 607 fossils. These variables were taken as representing some of the features of fossil assemblages that can be influenced by the mode of formation as discussed earlier. The variables studied were line density, percentage of fragments, percentage of articulated remains (pelecypods), percentage of pelecypod valves oriented with concave surfaces downward, median size of sediment, coefficient of sediment sorting, median size of fossils, and coefficient of fossil sorting. The data were analyzed by computing a rank-correlation coefficient for all possible and meaningful combinations (Dixon and Massey, 1951). Thirty-three such combinations were chosen. Table 2 shows that the hypothesis of independence was rejected in only three instances. These were percentage of fragments and shell sorting, percentage of articulated remains and median shell size, and percentage of articulated remains and shell sorting. In the last case a negative relationship is indicated by the test. These results appear reasonable, but more diversified samples are needed before general conclusions can be drawn concerning the interrelationships of the features represented by the measures. An expanded study of this kind is essential to reduce circularity and to evaluate the relative contributions of such evidence in development of hypotheses on mode of formation.

Discussion

A fossil assemblage can have little paleoecological significance if it is not possible to reconstruct, at a high level of confidence, the sequence of events leading to its formation. This contention is supported by the number and nature of the alternative histories possible for elements of life and death assemblages. The geological history of the remains of any organism is a record of continuous alteration leading to its ultimate destruction as a biological entity. Samples are drawn from assemblages in varying stages of degradation. If these materials operating upon modern death assemblages should provide a basis for more refined interpretations of the discrepancies of the fossil record. In addition, studies on the ecological

TABLE 2.—RANK-CORRELATION COEFFICIENTS FOR MEASURES TAKEN IN 11 LINE SAMPLES OF THE MILLERTON FORMATION OF TOMALES BAY, CALIFORNIA

	%f	%a	%I	%-	mss	css	msf	csf
ld	−0.10	−0.26	−0.55	0.51	0.07	0.31	0.00	0.48
%f	..	†	0.30	−0.16	0.12	0.05	−0.53	0.61*
%a	..	..	0.08	0.24	−0.18	−0.30	0.60*	−0.79*
%I	..	..	..	0.21	0.08	−0.38	−0.18	0.01
%-	..	..	..	..	0.40	0.10	0.31	−0.18
mss	..	..	..	..	..	†	0.30	0.00
css	..	..	..	..	..	..	−0.10	0.28
msf	..	..	..	..	..	..	..	†

* hypothesis of independence rejected at the 5 per cent level of significance

† considered not meaningful

ld—line density; %f—percentage of fragments; %a—percentage of articulated remains; %I—percentage of elements oriented with the long axis parallel to the bedding plane; %-—percentage of pelecypod valves oriented with concave surfaces downward; mss—median size of sediment; css—coefficient of sediment sorting; msf—median size of fossils; csf—coefficient of fossil sorting

are to be used to obtain new knowledge, we must be able to eliminate effects introduced by the mode of formation of fossil assemblages and their subsequent geologic history. This might be accomplished in several ways. We could frame our study so as to employ only those features of a fossil assemblage that can be assumed to have been the least modified (*See* Miller and Olson, 1955 for a more complete discussion of this argument). If we do so, we must concentrate on a very few aspects of the assemblage and have a strong belief in our assumptions. Another approach is to attempt to understand the operation and results of the processes affecting death and fossil assemblages. With such knowledge it may be possible to reduce the inherent bias by recognizing and correcting for its effects in our analysis of data. This approach is potentially the most efficient means of utilizing the fossil record but requires fundamental knowledge in biology and geology which, for the most part, is not now available.

There appear to be three principal sources of bias in the record of an ancient community. These are (1) the selectivity of the destructive processes operating upon death assemblages, (2) the mixing of indigenous and exotic elements in a death assemblage, and (3) the physical and chemical alteration of fossils and enclosing sediment. Direct study of the selective processes structure of marine communities are needed for inference of some, at least, of the ecologic roles not represented by fossils. However, since organisms have evolved and physical environments have changed in time, high levels of confidence cannot be placed in explicit inferences concerning those portions of the community that have not been preserved.

The mixing of indigenous and exotic elements can affect both the physical and biological nature of the resultant death assemblages. The implications developed from the models discussed herein are that biological criteria are more indicative of transportation than the physical features. Use of biological criteria requires knowledge of morphology and adaptation, patterns of spatial distribution, size-frequency distributions in stable and unstable populations, and the environmental tolerances and preferences of kinds of marine organisms. The application of such knowledge to ancient communities involves a high degree of uncertainty but the information available from the composition and associations of broadly contemporaneous fossil assemblages can be used to strengthen the chain of analogy.

The bias that results from the physical and chemical alteration of fossils and enclosing sediments can be recognized and corrected only with the aid of supplementary data not now available. Field and laboratory studies of diagentic processes are greatly needed.

Until supplementary knowledge concerning the sources of bias is available, analysis of the mode of formation and preservation of fossil assemblages must be based upon theoretical foundations. The models proposed here are suggested as devices for comparing actual fossil assemblages in these regards. Their use permits us to relate fossil assemblages as to the general degree of disturbance without specifying, in detail, the causes of disturbance. New methods of study must be developed in the context of both theory and the practical limitations of the outcrop. The "stretched-line" method for obtaining field data is suggested as one effective means of obtaining pertinent data.

Selected Bibliography

Boucot, A. J., 1953, Life and death assemblages among fossils: Am. Jour. Sci., v. 251, p. 25–40

Boucot, A. J., Brace, W., DeMar, R., 1958, Distribution of brachiopod and pelecypod shells by currents: Jour. Sed. Petrology, v. 28, p. 321–332

Connell, J. H., 1956, Spatial distribution of two species of clams *Mya arenaria* L. and *Petricola pholadiformis* Lamarck in an intertidal area: Investigations Shellfisheries Mass., rept. no. 8, p. 15–28

Dixon, W. J., and Massey, F. J., 1951, Introduction to statistical analysis: N. Y., McGraw-Hill Book Co., 370 p.

Imbrie, J., 1955, Biofacies analysis, p. 449–464 *in* Poldervaart, Arie, *Editor*, Crust of the earth: Geol. Soc. America Special Paper 62, 762 p.

Johnson, R. G., 1957, Experiments on the burial of shells: Jour. Geology, v. 65, p. 527–535

——1959, Spatial distribution of *Phoronopsis viridis* Hilton: Science, v. 129, p. 1221

Ladd, H. S., 1957, Paleoecological evidence, p. 31–66 *in* Ladd, H. S., *Editor*, Treatise on marine ecology and paleoecology, v. 2, Paleoecology: Geol. Soc. America Mem. 67, 1077 p.

MacGinitie, G. E., 1935, Ecological aspects of a California marine estuary: Am. Midland Naturalist, v. 16, p. 629–765

Menard, H. W., and Boucout, A. J., 1951, Experiments on the movement of shells by water: Am. Jour. Sci., v. 249, p. 131–151

Miller, R. L., 1956, Speculation on water currents in a black shale environment (Abstract): Am. Geophys. Union Trans., v. 37, p. 354–355

Miller, R. L., and Olson, E. C., 1955, The statistical stability of quantitative properties as a fundamental criterion for the study of environment: Jour. Geology, v. 63, p. 376–387

Olson, E. C., 1957, Size-frequency distribution in samples of extinct organisms: Jour. Geology, v. 65, p. 309–333

Payne, T. G., 1942, Stratigraphical analysis and environmental reconstruction: Am. Assoc. Petroleum Geologists Bull., v. 26, p. 1697–1770

Pettijohn, F. J., 1957, Sedimentary rocks: 2d ed., N. Y., Harper Bros., 718 p.

Thorson, G., 1957, Bottom communities (sublittoral or shallow shelf), p. 461–534 *in* Hedgpeth, J. W., *Editor*, Treatise on marine ecology and paleocology, v. 1, Eoclogy: Geol. Soc. America Mem. 67, 1296 p.

Department of Geology, University of Chicago, Chicago, Illinois

Manuscript Received by the Secretary of the Society, December 9, 1958

PAPER 44

Sedimentological, Ecological, and Temporal Patterns of Fossil Lagerstätten (1985)

A. Seilacher, W.-E. Reif, and F. Westphal

Commentary

DEREK E. G. BRIGGS

This paper was delivered by Seilacher at a meeting titled Extraordinary Fossil Biotas: Their Ecological and Evolutionary Significance at the Royal Society in London in February 1985. Seilacher first applied the term "Lagerstätten" to deposits that preserve an unusual amount of paleontological information in a 1970 paper in German in *Neues Jahrbuch*. He identified two types: Konservat-Lagerstätten or conservation deposits, which preserve skeletons without disarticulation and sometimes soft tissues, versus Konzentrat-Lagerstätten or concentration deposits, which are characterized by an unusual abundance of specimens. Fossil Lagerstätten were the subject of a major 1970s Tübingen research project on paleoecology (Sonderforschungsbereich 53) directed by Seilacher. Although results from the study had been published, mainly in *Neues Jahrbuch*, it was not until the 1985 meeting that the concept of fossil Lagerstätten was propelled into the international limelight.

Soft-bodied fossil deposits provide a more complete picture of diversity than the shelly fossil record, but Seilacher noted in his 1970 paper that Lagerstätten "convey no adequate picture of contemporary life" (34) because they are only preserved in unusual environmental conditions. By 1985, however, it was clear that Konservat-Lagerstätten, in particular, yield paleoecological data of considerable value, provided that their taphonomy is understood. Stagnant bottom conditions, such as those represented by the Jurassic Posidonia shales, may eliminate life on the seabed but still preserve a remarkable record of animals that lived in the water column. Rapid burial, as in the Cambrian Burgess Shale and Devonian Hunsrück Slate, can capture a census of benthic animals but fewer actively swimming forms. The faunas preserved in lithographic limestones, like that of Jurassic age at Solnhofen, reflect the influence of both taphonomic factors. Figure 1 of the 1985 paper illustrates how such Lagerstätten can be classified in terms of environment.

Konservat-Lagerstätten, particularly those preserving soft tissues, are fundamental to our understanding of the phylogeny and history of almost all groups of organisms. But they are also critical to paleoecological studies, particularly of the Ediacaran and early Paleozoic. Seilacher and his colleagues emphasized the potential of "a comparative analysis of fossil Lagerstätten . . . over the whole fossil record" (21). One of the major challenges that remains is integrating results based on extraordinary preservations with those from the more continuous record of time-averaged shelly fossil assemblages.

Literature Cited

Seilacher, A. 1970. Begriff und Bedeutung der Fossil-Lagerstätten. *Neues Jahrbuch für Geologie und Paläontologie, Monatshefte*1970:34–39.

From *Philosophical Transactions of the Royal Society of London B* 311:5–24, by permission of the Royal Society.

Phil. Trans. R. Soc. Lond. B **311**, 5–23 (1985) [5]
Printed in Great Britain

Sedimentological, ecological and temporal patterns of fossil Lagerstätten

By A. Seilacher, W.-E. Reif and F. Westphal
Institut und Museum für Geologie und Paläontologie, Sigwartstrasse 10, *D*-7400 *Tübingen, F.R.G.*

[Plate 1]

Preservation of non-mineralized structures (including plants) and of articulated skeletons results from extraordinary hydrographic, sedimentational and early diagenetic conditions. The corresponding chief causative effects (stagnation, obrution and bacterial sealing) define a conceptual continuum into which individual occurrences may be mapped. A more pragmatic, typological classification of conservation deposits, using a standard questionnaire, reveals ecological replacements, as well as trends related to the evolution of the biosphere, through geological time.

The theme of this symposium, 'extraordinary fossil biotas: their evolutionary and ecological significance', recalls a project carried out some time ago in the Tübingen special research division 53 on palaeoecology. Our term 'fossil Lagerstätten' (alternative spelling: Lagerstaetten) is difficult to translate into English. Corresponding to economic Lagerstätten of minerals and ores, fossil Lagerstätten were defined as rock bodies unusually rich in palaeontological information, either in a quantitative or qualitative sense. This means that the term embraces not only strata with an unusual preservation, but also less spectacular deposits such as shell beds, bone beds and crinoidal limestones. The concept also implies that there is no sharp boundary with 'normal' fossiliferous rocks. Rather, the preservation of any fossil is to be considered as an unusual accident that deserves attention and questioning.

The rationale behind such a broad approach is a sedimentological one. Fossil Lagerstätten are considered as end members of ordinary sedimentary facies, in which the unusual amount and quality of the palaeontological test bodies allows us to identify better the factors responsible for such a facies at all levels of its genesis: biotope conditions (palaeobiology), fate of the soft parts and organic skeletal material (necrolysis), sedimentary transport and burial (biostratinomy) and fate of the mineralized skeletons (fossil diagenesis).

On the other hand this approach is based on the contention that in spite of all the possible combinations and variations of biotic and abiotic factors involved there is a finite number of situations that lead to the formation of fossil Lagerstätten and that prospecting for them is therefore a realistic objective. A first step in this direction was a genetic classification (figure 1), the broad outlines of which are still valid.

In the meantime, our understanding of *concentration deposits* has been greatly improved by the concepts of dynamic and event stratigraphy with far reaching applications in basin analysis (see Einsele & Seilacher (1982) and Bayer & Seilacher (1985) for examples and references).

With respect to *conservation deposits*, on which this symposium is focused, many new examples have been discovered and old ones analysed in more detail, including case histories presented in this symposium. Each of them has, of course, its merit as a peephole in the screen of imperfect

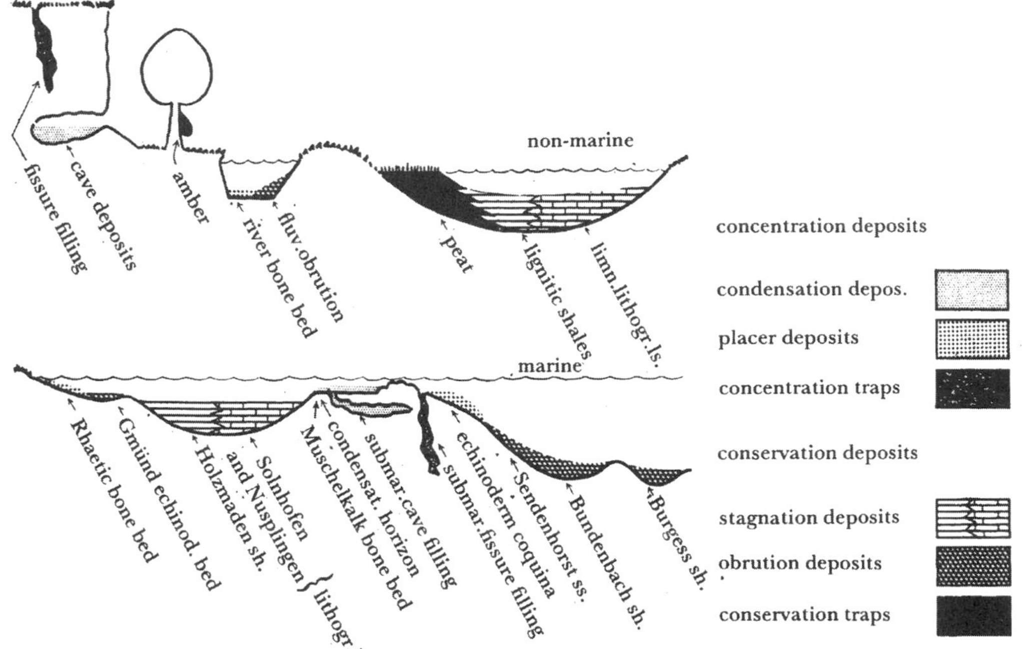

FIGURE 1. Synopsis and classification of fossil Lagerstätten as given by Seilacher & Westphal (1971).

preservation, through which we can see unknown anatomical details and new, commonly problematic taxa. But each case is also important as an additional test for environmental and diagenetic boundary conditions in our classification.

Instead of adding new examples, we shall here choose three well studied and representative examples from the Jurassic of southern Germany as a reference, and try to derive from them a questionnaire that can be applied to other conservation deposits.

In doing so, we shall emphasize the invertebrate rather than the vertebrate, microfossil and plant records, because invertebrates cover a wider spectrum with regard to geological time, biomaterials and environments.

1. The echinoderm Lagerstätte of Gmünd (L. Sinemurian): an obrution deposit

Being only a few centimetres thick and possibly no more than a few metres in lateral extent, this deposit can not compete with those of Holzmaden and Solnhofen. But its fossil content, the stratigraphic context and low level of genetic complexity as well as a careful analysis by Rosenkranz (1971) make it a very suitable prototype of what we call obrution deposits.

(a) *Stratigraphic situation*

Stratigraphically, the fossil-bearing lens of black shale is separated from contact with the underlying terrestrial Keuper marls by a conglomeratic shell bed 0.5–0.6 cm in thickness and overlain by a similar conglomeratic limestone. Palaeontological evidence, however, shows that this basal conglomerate of the Jurassic transgression represents in fact a rather complex environmental history.

The erosional nature of the Triassic–Jurassic contact is shown first by the absence of the Rhaetic, which in this area is usually represented by coal-bearing swamp deposits which pass into marine sands and clays farther to the west. Second, the conglomerate consists largely of reworked calcareous concretions which are characteristic of the underlying Keuper marls ('Knollen-Mergel'). Third, the sole of the conglomeratic bed is covered by casts of vertical rhizocoralliid spreite burrows, whose perfectly preserved scratch marks suggest that the Keuper marl, into which they were dug, was already well compacted and rather stiff. Such burrows are characteristic of the *Glossifungites* ichnofacies of shallow marine firmgrounds. While other species of the ichnogenus *Glossifungites* are oblique and in the size-class of the related *Rhizocorallium*, the smaller ones at the base of the Jurassic can be compared to the vertical burrows of the modern amphipod *Corophium* and may well be the works of related intertidal crustaceans. In any case these burrows show that the conglomerate is not simply the depositional phase of the event that made the erosion. Rather, the two processes were separated by a quiescent firmground period lasting long enough to allow the establishment of the *Glossifungites* community. Moreover, deposition of the conglomerate bed at this place was not preceded by major erosion, which would have destroyed the burrows.

Nor does the conglomerate bed itself represent a simple sedimentational process. Besides lithoclasts it contains many shells, predominantly of byssate and cemented epifaunal bivalves (*Plagiostoma*, *Lima*, *Entolium*, *Inoceramus*, *Liostrea*). But poorly preserved specimens of *Cardinia* indicate that mud-burrowers were also present originally, but were largely eliminated by diagenetic solution of their aragonitic shells. Accordingly, we interpret this bed as a composite record of muddy periods interrupted by major storm events. These winnowed the mud away, but their erosive effect was stopped by the growing lag of shells and pebbles, which later allowed the colonization by an epifaunal post-event community of epibyssate and cemented bivalves before the mud took over again.

This interplay of sedimentation, shell production and storm reworking, comparable to situations known from modern littoral environments (Seilacher 1985), was periodically interrupted by the deposition of black mud containing the echinoderms, which, unlike the previous and following mud-layers, locally escaped subsequent storm erosion.

The previous style of storm-winnowing, plus the lateral introduction of reworked Keuper nodules, continued in the thicker conglomeratic bed overlying the echinoderm horizon. But its top is modelled by large oscillation ripples, suggesting that the eventual shift to the muddy sedimentation of the Psilonotus Clays happened between the sand and the mud phase of a single tempestite.

(*b*) *Faunal spectrum and preservation*

The well preserved echinoderm fauna (30 asteroid, 15 ophiuroid, 15 crinoid, 40 echinoid specimens; Rosenkranz (1971)) is embedded in the lower 1–2 cm of the clay lens (figure 2, 1–5). It represents an ordinary shelly bottom community with respect to functional adaptation, trophic diversity and age structure. Its only unusual feature is the well articulated preservation in a muddy sediment not corresponding to the substrate on which such forms would normally live. It is surprising, however, that groups other than echinoderms are not equally well represented.

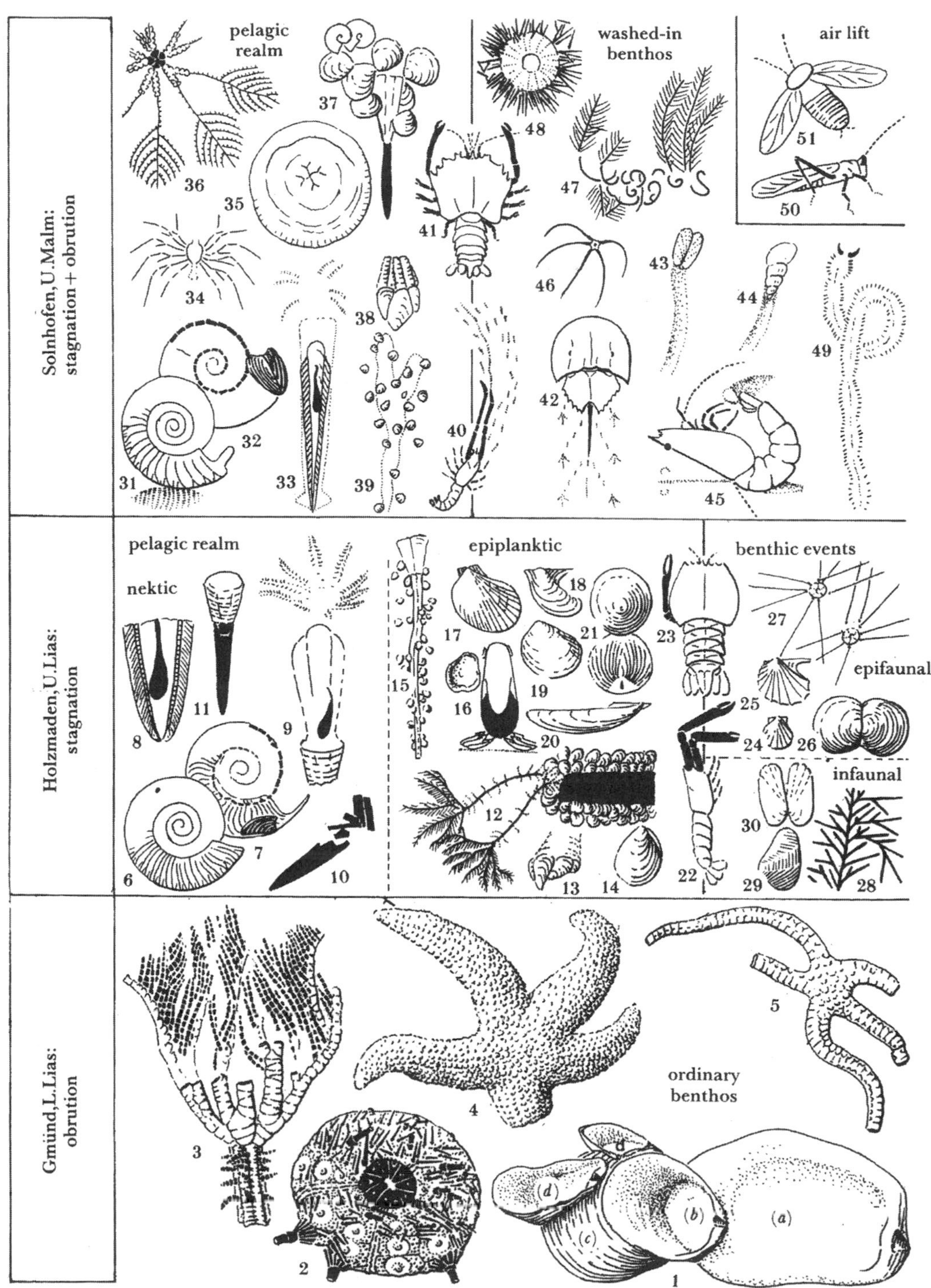

FIGURE 2. For description see opposite.

The only common non-echinoderm associate is a small oyster (*Liostrea irregularis*) which encrusted pebbles as well as shells of its own kind. Such oyster clusters (figure 2, 1) may comprise several generations, the changing orientations of which attest to repeated overturn during an extended exposure at the sediment surface. However, the generations could also represent different shelly phases separated by mud-covered periods. Nevertheless, Rosenkranz found that about 50 % of the oysters on top of the basal shell bed were double-valved and probably became victims of the same event that killed the echinoderms.

In contrast, epibyssal bivalves, being represented within the basal shell bed, are conspicuously lacking on its echinoderm-bearing top.

Rosenkranz's explanation is convincing: the shelly bottom epifauna became buried alive by rapid mud sedimentation. Among the bivalves, this was fatal for the oysters, while pectinids and limids could escape by swimming up. For the echinoderms, however, the fine sediment also had a smothering effect because it blocked their ambulacral systems. It is this 'Achilles heel' that accounts for the high-level taxonomic selectivity of their preservation.

(*c*) *Other examples*

As Rosenkranz (1971) has shown, many, if not the majority of articulated echinoderms in the fossil record are found in a corresponding situation: at the tops of condensed tempestitic

---

FIGURE 2. Ecological spectra of representative Jurassic fossil Lagerstätten in southern Germany.

1, Reclining clusters of oysters show sequences of generations separated by overturning (and burial?) events.

2–5, The exclusive, articulated preservation of ecologically diverse echinoderms suggests that they fell victims to a mud-smothering event.

6–11, Among the vertebrates as well as the invertebrates the Holzmaden fauna is dominated by pelagic organisms. Their perfect preservation (associated aptychi, periostracal film and zig-zag siphuncle in ammonites; ink sacks preserved in jet and other soft parts in coleoids 8, 9) suggests rapid settling after death and the absence of benthic scavengers. In the belemnites, only carcasses sunk by predator bites show soft parts (including the non-calcified proostracum, 9), which were otherwise lost during the necroplanktonic drifting stage (11).

12–21, Normally benthic groups are mainly represented by forms attached to floating objects such as drift wood (12–24) and belemnite species whose non-calcified rostrum permitted an extended necroplanktonic drift (15), or to live ammonites (16–21). For details see Seilacher (1982).

22–23, Crustaceans are represented only by very rare specimens of apparently benthic forms. The fact that corresponding species are also found in deposits of the Solnhofen type (40–41; Osteno; Lebanon) suggests, however, that they were not normal benthos.

24–30, Truly benthic organisms (including foraminifers and ostracods) are characterized by small size and monotypic occurrence on particular bedding planes. This suggests that they colonized the mud during benthic events and then became smothered in an articulated fashion. Main representatives are epibyssate bivalve recliners (24–26) and small, long-spined echinoids (27). Endobenthos is represented by tiered *Chondrites* horizons (28) and very rare burrowing bivalves (29, 30 from Riegraf (1977)).

31–36, In Solnhofen the pelagic guild is similar to Holzmaden, but contains additional elements such as crustacean larvae (34), jellyfish (35) and the microcrinoid *Saccocoma* (36, reconstruction), which in places is the most common fossil.

37–39, Epiplankton is less diverse (rarity of ammonites, absence of driftwood may be a reason); but it includes floats assembled by byssate bivalves from necroplanktonic shells of hibolitid belemnites and small ammonites (37). Overgrown seaweeds (39) were probably washed in from the shore.

40–41, Among the diverse crustacean fauna, the common occurrence of *Mecochirus* (40, with trace of death march) and of *Eryon* (41) is reminiscent of Holzmaden.

42–49, In Solnhofen truly benthic organisms do not occur as autochthonous benthic horizons, but as lateral import by turbidity currents. Some of them were still alive enough to leave a short track (42–44), others dead, showing belly-up landing marks (5; figure 5), current orientation (46) or *post mortem* contortion possibly indicative of hypersaline bottom waters. Immobile and burrowing forms (except a rare *Solemya*, 43) were not imported.

50–51, Rare, but extremely diverse insects (most commonly dragonflies), like the flying reptiles and *Archaeopteryx*, represent import by air or surface currents.

shell beds. It could be added that this situation is most likely to occur in the transgressive phases of larger or smaller cycles.

In all cases, smothering by finer sediment is the dominating process, while oxygen deficiency in the pore water (gyttja condition) was either absent or helpful only in the sense that it inhibited infaunal scavengers or bulldozers that would have secondarily disarticulated the buried skeletons.

2. The bituminous Posidonia shales of Holzmaden (L. Toarcian): a stagnation deposit

Since the famous fossil Lagerstätte of Holzmaden and its fossil content have been adequately described (see Kauffman 1981; Riegraf 1984), we can focus here on the points that are important for a comparison with the other examples.

(i) While the Gmünd deposit is at the very base of the Jurassic transgression, the Toarcian marks its peak. This situation favours the establishment of anoxic conditions in many epicontinental basins, whether this is due to changes in the hydrographic regime or to the mobilization of brines from underlying salt deposits (Jordan 1974).

(ii) In contrast to the Gmünd echinoderm bed, here we are dealing with an enormous rock body that spreads over large parts of central Europe with a thickness of tens of metres and representing a time span of the order of millions of years. This difference is important to keep in mind in view of the neverending discussions as to whether we are dealing with a sapropel underneath a stratified, oxygen deficient water body or with a gyttja, in which oxygen deficiency was essentially restricted to the pore water. In such spatial and temporal dimensions, both conditions must be expected at different times and places, as they are in modern stagnant basins (Savrda *et al.* 1984). The question is simply, which situation dominated in space and time and was responsible for the unusual preservation (including soft parts), with which we are concerned here.

(iii) The ambiguity between the two alternative, or rather alternating, models is expressed in contrasting sedimentological as well as ecological evidence. Sedimentologically, the presence of a millimetric lamination that can be correlated over tens of kilometres (H. Roscher, personal communication) as well as landed ammonites that dropped their overgrowth before they eventually tilted over without changing place or position (Seilacher 1982) indicate very quiet water. On the other hand, the orientation and linear accumulation of ammonite shells and other fossils at many levels (Brenner 1976; Seilacher 1982) record uniform currents of up to 20 cm s^{-1}.

(iv) In the ecological spectrum (figure 2, 6–30) we observe the general absence of benthic organisms. Encrusting or reclining brachiopods, bivalves, serpulids and crinoids, otherwise standard elements of Liassic soft bottoms, are represented only as epiplanktonic float on ammonite shells (figure 2, 16–20), belemnite shells with a non-calcified rostrum (figure 2, 15) or drift wood (figure 2, 12–14, Seilacher 1982), while nektonic forms (ammonites, coleoids, fish, ichthyosaurs and plesiosaurs) are the dominant fossils. On the other hand there are individual horizons covered by monotypic small epibenthics, such as diademoid echinoids or byssate bivalves (*Posidonia*, figure 2, 24–26), whose articulated state of preservation would be in conflict with large-scale lateral import. More common are horizons with abundant benthic microfauna (foraminifers, ostracodes; Riegraf 1985), while tiered bioturbation horizons, dominated by the low-oxygen trace fossil *Chondrites* (figure 2, 28; Bromley & Ekdale 1984) are restricted to a few levels, which become more numerous towards the margins of the basin.

This picture is in accord with observations in modern stagnant basins, where benthic faunas decrease in body size and diversity with decreasing oxygen levels and advance towards the centre of the basin during periods of reduced stagnation (Savrda *et al.* 1984). In the shallower Posidonia shale basin, such benthic events were probably brought about by extreme storms, but they allowed only a few generations to flourish before abiotic conditions took over again.

3. The Solnhofen lithographic limestones (Tithonian): an obrutionary stagnation deposit

(*a*) *Stratigraphic and environmental setting*

This fossil Lagerstätte, most famous for the preservation of feathered *Archaeopteryx* and medusae, was formed just before the regressive end of the Jurassic cycle. With a thickness of up to 90 m it exceeds the south German Posidonia Shale, but both its lateral extent (basins a few kilometres in diameter in a buried sponge–reef topography) as well as its time equivalent (0.5 Ma) are considerably smaller. Also different is the scarcity of fossils: had there been no quarries, the unit would have possibly been mapped as 'non-fossiliferous'.

After all the criteria for emergence (*Limulus* tracks interpreted as bird tracks, desiccation cracks, rain drop impressions, etc.) have been discredited, the old lagoonal model is no longer relevant. Most authors now agree that we are dealing with permanently submerged restricted basins, in which storm wave action was confined to a few metres of depth and that there was some kind of stratification in the water body excluding macrobenthos. This stratification had probably the form of a halocline (Keupp 1977) and it is a secondary question, whether or not the resulting stagnation was also expressed by an oxycline. In any case the bitumen content, even if secondarily lost by weathering, was originally lower than in the Posidonia Shales, owing to the higher rate of carbonate mud sedimentation and lower productivity in the surface water.

(*b*) *Ecological spectrum*

As in the Posidonia Shales, the fauna (figure 2, 31–51) is dominated by pelagic organisms. But instead of ammonites, the most common fossil is the minute stemless crinoid *Saccocoma* (figures 2, 36 and 3*b*), whose anatomy and preservation (see below) suggests a medusoid-like mode of life. There is also a large variety of fish and aquatic reptiles, but among these, land-related forms are more common than the open-ocean ichthyosaurs.

Epiplanktonic fauna is also present, but on a broader spectrum of floats. Driftwood is hardly ever found, probably because in such small water bodies it would always be washed ashore. Since ammonites are rarer as a whole, the small percentage of overgrown shells amounts only to a few specimens. Some of them carry oysters, other lepadomorph barnacles (figure 2, 38), which in Holzmaden occur only on wood (figure 2, 13). Instead we find fouled belemnites, not in the form of an encrustation that occurs when the rostra lie at the bottom, but as a loose attachment of byssal bivalves that prefer the conotheca rather than the rostrum (figure 2, 37). Obviously, the now dominant hibolitid belemnites had a longer necroplanktonic drifting time, which made them more attractive for hitchhikers. The common association of several belemnites, or of belemnites and ammonites, in such clusters suggests that the bivalves could actively enlarge their float by byssal assemblage. Algal fronds are yet another kind of float, but in this case we deal with seaweeds that were torn off their rocky littoral habitat by storms and washed into the basin.

Benthic elements are by no means missing in the Solnhofen limestones, but unlike the

Holzmaden situation they never represent autochthonous burrow or other benthic horizons (except perhaps of Foraminifera; Groiss 1967). Instead we are dealing with inhabitants of the shallow margins that became accidentally washed-in. This view is supported by several lines of evidence. Most spectacular are the death marches recorded by tracks behind carcasses in an environment in which other tracks are never found. The length of these tracks relative to the animal is largest in the low-oxygen bivalve *Solemya* (figure 2, 43) and decreases (with decreasing hardiness in *Limulus* (figure 2, 42), *Mecochirus* (figure 2, 40) and *Eryon* (figure 2, 41). Other crustaceans are never found with a track, but the marks left by the dorsal side of *Penaeus* (figure 2, 45 and figure 5, plate 1), or by the dorsal fins of various fish, show that these swimmers were already dead when they landed at the bottom (Mayr 1967).

A second indication is the selectivity of benthic import. With the exception of *Solemya* (figure 2, 43), no endobenthos (including burrows) is represented, neither is sessile epibenthos. Mobile epibenthic species are found, particularly ones that, on perturbation, would be able to swim up, such as *Antedon* (figure 2, 47), *Limulus* (figure 2, 42) and a variety of crustaceans. It is also noteworthy that almost all imported limulids are juveniles not adults, whose much larger sizes can be reconstructed from tracks (not death marches) found in marginal areas of Solnhofen basins (Painten). This could be because juvenile limulids more readily swim up on disturbance and are therefore more prone to be washed away than adults.

(*c*) *Necrolytic features*

A dorsal bend of the vertebral column, comparable to *post mortem* deformations of drying modern carcasses (Schäfer 1955) is a familiar feature in Solnhofen *Archaeopteryx*, *Pterodactylus* and *Compsognathus*, but occurs also in fishes. In the latter case it cannot be a desiccation feature, because associated belly-up landing marks and tail fins disrupted during the bending of the vertebral column (figure 7; Mayr 1967) clearly show that the deformation took place only after the carcass had sunk to the bottom. Therefore it may be assumed that the bottom water was hypersaline to effect the necessary dehydration.

Analogous necrolytic deformations may be observed in articulated skeletons of invertebrates. In a carcass of *Penaeus* (figures 2 and 5) with a belly-up landing mark, the ventral flexure of the tail scraped off sediment after the body had tilted over. A possibly related phenomenon is the contortion of Solnhofen crinoids. In *Saccocoma*, the feather-like form of the arm tips (figure 2, 36) is invariably obscured by intense inward coiling (figure 3*b*). *Antedon*, in contrast, has the proximal parts of the arms intensely coiled (figure 3*c*) and sometimes disrupted (figure 2, 47), while the tips are conspicuously straight. From such deformations we may derive flexibility patterns, and hence the mode of swimming in different species (figure 3*d*, *e*), but we may also take them as environmental signals that seem to be lacking in Holzmaden-type deposits.

(*d*) *Biostratinomic features*

The scarcity of fossils in Solnhofen precludes the statistical measurement of preferred azimuth orientations. Strong currents are indicated, however, by tool and roll marks left by ammonite shells (Seilacher 1963) and other objects. Their association with incipient flute casts further suggests that these currents were sediment-loaded. This is in accord with the preferred position of the Solnhofen fossils at the bases of beds and with the occurrence of grading in other Upper Jurassic occurrences (Nusplingen; Temmler 1964). Turbidity currents are also indicated by

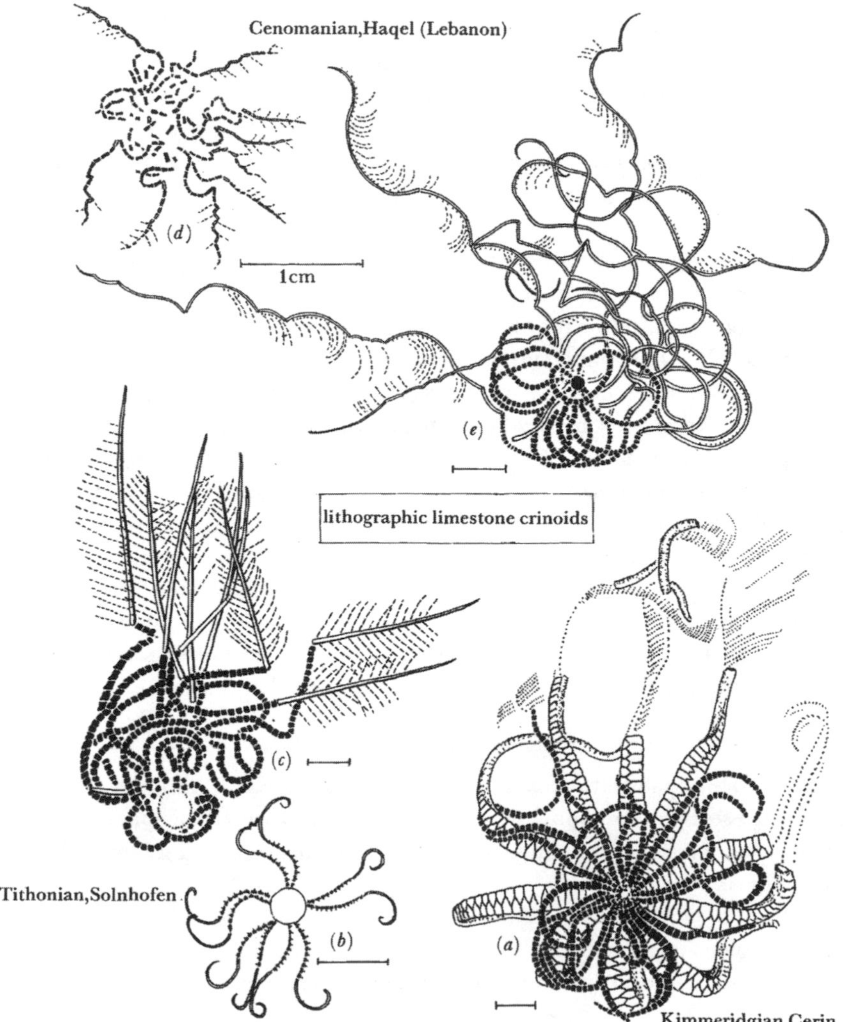

FIGURE 3. *Post mortem* contortion of echinoderm carcasses in lithographic limestones reveals flexibility patterns along the arms as well as the dehydrating effect of hypersaline bottom waters. (*a*) *Antedon thiollieri*, U. Jurassic, Cerin. Drag marks (dotted) suggest secondary contortion of the arm tips in a current-oriented carcass. (*b*) In Solnhofen *Saccocoma* (GPIT 1630/1) maximum contortion is in the pinnulated arm tips (see figure 2, 36) suggesting that these were the active parts in filter swimming, while the arm bases were held out as an umbrella. (*c*) Solnhofen '*Antedon*', in contrast had stiff arm tips moved as oars by the more flexible proximal parts. (*d*) This small comatulid has a style of contortion similar to (*c*) (GPIT 1630/2). (*e*) In the larger species, proximal arm sections are coiled like in (*c*), while the distal portion deformed in an arcuate fashion (GPIT 1630/3). Scale bars are 1 cm.

the radial current directions mapped from fossil orientations around the much more fossiliferous and better exposed lithographic limestone basins of the Lebanese Cretaceous (Hückel 1970, figure 13), which is in marked contrast to the uniform current directions of the Posidonia Shales (Brenner 1976).

(*e*) *Diagenetic features*

As to be expected in such different lithologies, certain biomaterials suffered different transformations in bituminous shales and lithographic limestones. In Solnhofen, for instance, both vertebrate coprolites and the ink sacks of coleoid cephalopods are phosphatic, while in Holzmaden only coprolites are phosphatized, while the ink is preserved as jet. In both cases, however, mineralization happened so early that these originally soft materials did not become deformed by compaction.

Aragonite solution seems to have been a similarly early process, as shown by the preservational history of ammonites (Seilacher *et al.* 1976). Their compactional deformation, varying with different shell geometries, shows that the aragonite was dissolved not at the surface but within the sediment. On the other hand the plastic lateral deformation of Solnhofen ammonites in slumped layers ('Krumme Lagen') indicates that the shells had already been reduced to periostracal films within a few metres below the sediment surface.

The real problem is not the difference, but the identity of diagenetic signatures in the Holzmaden and Solnhofen cases in spite of their lithological differences. The flattened periostracal foil preservation of empty ammonite phragmocones that the two localities have in common contrasts sharply with 'normal' situations, in which such phragmocones are either lined with pyrite (non-bituminous Liassic black shales; Hudson (1982)) or filled with blocky calcite (micritic limestones of the Upper Jurassic) and are therefore little compacted. This indicates that the special conditions responsible for the exceptional preservation of other fossils were not restricted to the open water, but had an effect also on the pore water system.

(*f*) *Prokaryotic scum: the neglected factor*

Diagenesis is traditionally viewed as the result of physicochemical processes. Within the upper centimetres of the sediment column, however, in which the observed soft-part permineralization and aragonite solution appears to have taken place, chemical conditions are mainly controlled by microbiological activity.

To reconstruct the microbiology of ancient mud bottoms is largely a geochemical task. Still there are a few morphological clues. Scanning electron microscope studies by Keupp (1977) have demonstrated that, apart from coccoliths (Hemleben 1977*b*), coccoid cyanobacteria are

---

Description of plate 1

Figures 4–9. Cyanobacterial mats, probably a most important factor in preservational processes, can be inferred from indirect evidence. In Solnhofen, their presence is indicated not only by lamination and near-absence of erosion, but also by surface features.

Figure 4. Halo of ripped-up scum around swaying fish tail (from Barthel (1978), plate 61, figure 1).

Figure 5. Preservation of landing marks (rostral carina, eye tips, abdomen) besides a fallen-over carcass of *Penaeus* and of scrape mark plus scum heap produced during *post mortem* inflection of tail end (GPIT 1630/4).

Figure 6. Ruffling around roll mark of an ammonite shell (GPIT 1630/5) and drag marks (Eichstätt Museum).

Figure 7. Cast of reticulate ridge pattern (bottom surface, GPIT 1630/6) around fish with *post mortem* contortion of vertebral column.

Figure 8. Radial arrangement of such ridges around an ammonite shell, whose 'pedestal' elevation is a later diagenetic feature. (From Janicke (1969), plate 6, figure 4.)

Figure 9. Corresponding pattern produced by mechanical creasing of a depressurized blister. (Courtesy of Professor Frei Otto, University of Stuttgart.)

Phil. Trans. R. Soc. Lond. B, volume 311

Seilacher, et al., plate 1

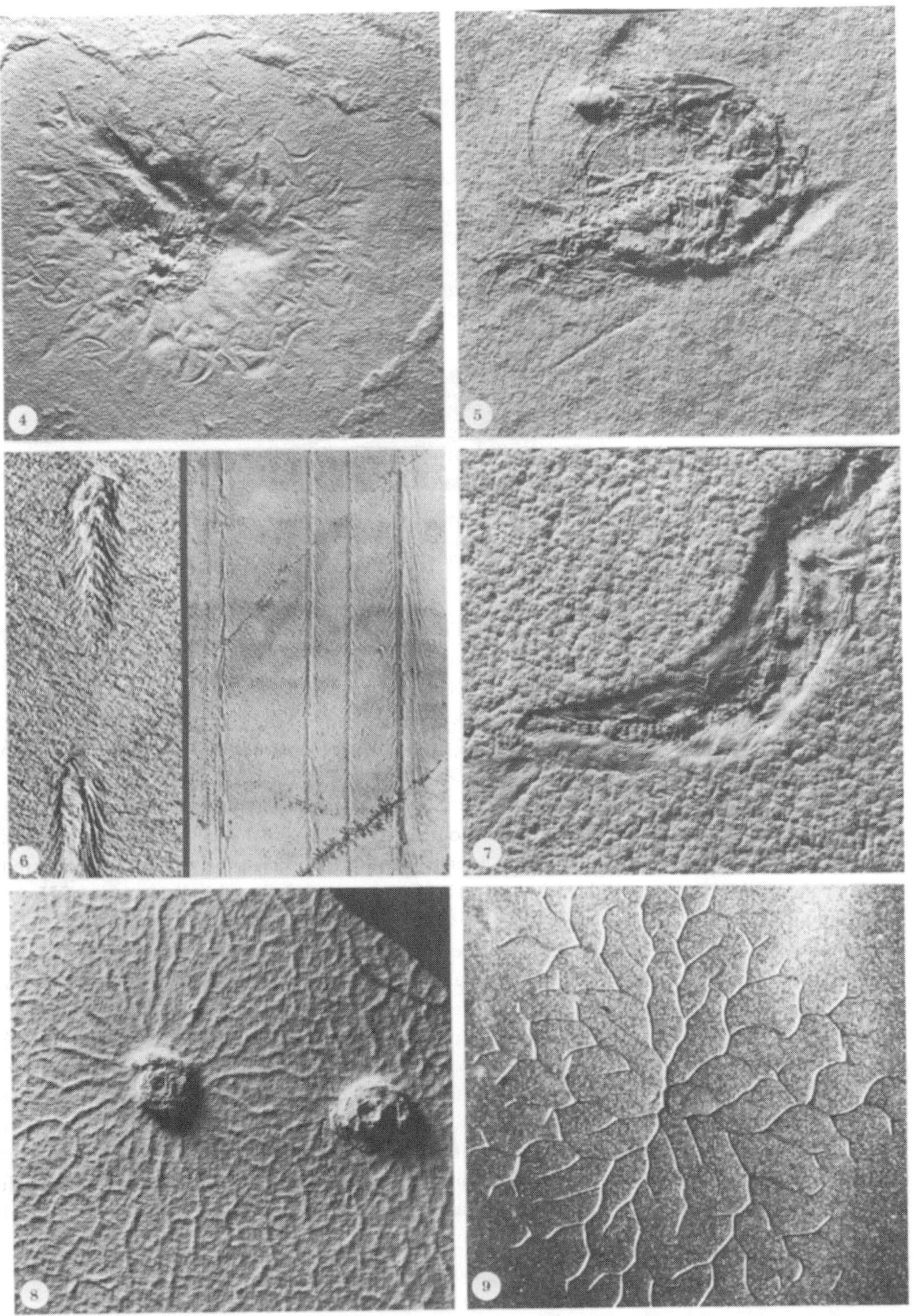

FIGURES 4–9. For description see opposite.

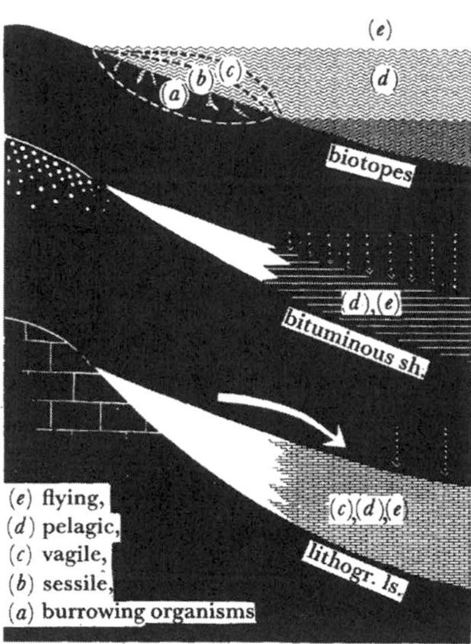

FIGURE 10. Stagnant basins of the Holzmaden type (mostly bituminous shale facies) and Solnhofen type (lithographic carbonates) both lack autochthonous benthos (except during short benthic events) and are dominated by drop fauna. The geometry of Solnhofen-type basins, however, allows also the lateral import, by turbidity currents, of littoral benthos, preferably of vagile forms.

a major constituent of the Solnhofen muds. Their presence is also expressed macroscopically by a scum on the bed surface. This scum allowed tracks and roll marks to be preserved, caused the ruffling of roll and drag marks (figure 6; Mayr 1967, plate 13, figures 1–3) and was visibly ripped off around a swaying fish body (figure 4). Its presence is also expressed by reticulate ridges on the tops of Solnhofen beds (figure 7), which have been variously explained as rain drop impressions (Mayr 1967), syneresis cracks (Janicke 1969) or load casting. Their radial arrangement around an ammonite (figure 8), however, suggests that we deal with the creasing of a film similar to the tepee structures of modern algal mats or the mechanical creasing of blister membranes (figure 9). From other evidence Hemleben & Freels (1977*a*) have considered such prokaryotic scum as the chief factor in the Cretaceous lithographic limestones of Hvar, Yugoslavia.

Today, cyanobacterial films of this kind are largely restricted to hypersaline environments. But, like cyanobacterial stromatolites, they could have had a much wider distribution in earlier times, particularly in the Precambrian. Their effect would be (Keupp 1977): (i) to protect soft sediments against erosion; (ii) to favour the preservation of tracks and other markings; (iii) to serve as food source during benthic events; (iv) to protect carcasses against decay; (v) to act as a 'carbonate pump' into the sediment (Walker & Diehl, this symposium); (vi) to seal the particular microenvironment responsible for the absence of bioturbation (Krumbein 1983) and for the unusual preservational histories of ordinary fossils such as ammonite shells.

As yet, this eludes macroscopic investigation, but it may become more transparent through ultramicroscopic studies in ancient rocks and particularly through microbiological and geochemical analyses in adequate modern environments.

4. Other examples

As epitomized by the faunal spectra and modes of preservation, the Holzmaden and the Solnhofen deposits have obviously much in common. Differences are minor and largely explained by the different palaeogeographic frames (figure 11). On the one hand we have a large stagnant basin with little influx from the margins, but with the possibility of storm-induced water-mixing and the short-term establishment of small-sized and monotypic benthic faunas. In the much smaller basins of the Solnhofen type, storms could affect only the shallow margins; but the steeper slopes favoured the episodic introduction, by slumping and turbidity currents, of fine-grained sediment and transportable benthic organisms from the more oxygenated nearshore environments. In such an event, the imported bodies (some still alive) would reach the bottom first, soon to be covered by the mud settling from suspension. Therefore, we would define the Solnhofen lithographic limestones primarily as a stagnation deposit, but one in which obrution was also an important factor.

There are many counterparts to these two types in the Phanerozoic record, each with minor but significant pecularities. Thus the bituminous shales of the Lower Lias of southern Germany (Ölschiefer, Sinemurian) resemble Holzmaden except that, due to the thickness of only a few decimetres and poor outcrops, their fauna is much less spectacular; but they do contain benthic horizons of small echinoids ('*Cidaris*' *olifex*) as well as tiered bioturbation horizons. Equally comparable are the Upper Triassic bituminous shales of southern Switzerland, but the shales themselves lack periostracal impressions of ammonite shells, which are only found in the hard dolomitic layers, where they were fixed by early diagenetic cementation (Rieber 1973).

A less stagnant modification of the Holzmaden type is found in marginal areas of the Toarcian basin, where benthic horizons are more common and the aragonitic ammonite shells are not dissolved. The same appears to be true for parts of the Oxford and Kimmeridge clays in Britain, where the benthic element includes reclining and bone-encrusting oysters (Martill, this symposium; Aigner 1980). Nevertheless, the shales are bituminous and vertebrate skeletons have remained articulated.

The sponge reef topography and the regression of the Upper Jurassic has led to the formation of lithographic limestones in other parts of central Europe also. In Nusplingen (south of Tübingen) fauna and preservations are similar to Solnhofen except that *Saccocoma* is lacking, some beds are clearly graded and the mud consists largely of sponge spicules, indicating that sponge reefs around the small basin were still growing (Temmler 1964, 1966). The faunal difference of the French locality of Cerin may be largely due to its slightly older age (Kimmeridgian, Bernier *et al.* 1983). In this case, however, the muds did at times emerge, as shown by horizons with dinosaur tracks.

The most similar extra-Jurassic representatives of the Solnhofen type are in the Cretaceous of Lebanon. These lithographic limestones are associated and interbedded with huge slump masses in what appear to be rhythmically subsiding pull-apart basins related to the nearby Jordan rift system (Hückel 1970; Hemleben 1977*a*). Fossils are very common and diversified, with washed-in benthos dominating over the pelagic rain. *Post mortem* contortions of fish and echinoderm skeletons (figure 3) suggest a halocline. In the mid-Triassic lithographic rocks of Alcover (Spain; Esteban Cerda *et al.* 1977; Via Boada 1977; Hemleben & Freels 1977*b*) the relation to an ancient reef topography in a regressing sea resembles Solnhofen. Minor differences in preservation are due to the dolomitic nature of the rock, which, together with gypsum crystals, also suggests increased salinity in this case.

Identification becomes more difficult with increasing temporal distance from the Jurassic-type examples. The lithographic limestones of Monte Bolca (Eocene, N Italy; Sorbini 1983), probably deposited in a volcanogenic topography by graded turbidite sedimentation, can still be compared to Solnhofen, although many index groups of organisms have fallen victims to the Cretaceous extinction. But what about the 'lithographic' Green River Shales (Eocene) of North America? Being deposited in a basin too large to produce turbidity currents, are they rather a stagnation deposit comparable to the Posidonia shales, but in a salinity régime that favoured bacteria-mediated carbonate precipitation? Similarly enigmatic is the case of the thin-bedded chalks of the Niobrara formation. In addition to the famed vertebrate skeletons and the floating crinoid *Uintacrinus* they contain *Inoceramus* shells that appear too large to have been epiplanktonic. However, the double-valved, closed preservation of these giant bivalves, equal encrustation on both valves and their lying always on the same valve does not either fit a reclining mode of life. Also reclining oysters, which are such a regular element in other chalks, are conspicuously absent. Could it be that the Niobrara chalks are the equivalent of a dark bituminous shale at a time when excessive planktonic carbonate production diluted the sapropel into a white mud without the necessity of lateral carbonate import?

On the other hand, the Lower Liassic deposits of Osteno (see Pinna, this symposium) are dark-coloured, finely laminated and bituminous. But they consist, like the lithographic limestones of Nusplingen, largely of sponge spicules. Also their faunal spectrum relates them to Solnhofen rather than to Holzmaden, with crustaceans, worms and other washed-in benthos dominating over the pelagic guild and with benthic events (including bioturbation horizons) being completely absent. The preservation of the rare ammonites also resembles Solnhofen (flattened, non-pyritized, zig-zag siphuncle).

What we can learn from the comparisons so far is that lithology, colour and bitumen content should not be taken as the primary criteria in a genetic classification of fossil Lagerstätten, because they may change with the hydrographic situation and with shifts in biological carbonate mud production. Nor should the salinity factor be rated too highly.

The Jurassic deposits of the Karatau (Kazakhstan), for example, are a valid counterpart of the Solnhofen situation although they were deposited as dolomitic muds in a tectonic, probably hypersaline lake, in which the 'washed-in benthos' is mainly replaced by river-imported insects.

Also deposited in a tectonic lake are the Eocene shales of Messel, Germany (Franzen, this symposium). Their ecological spectrum (fishes, washed-in land and flying vertebrates, insects) resembles Karatau, but this sediment is bituminous and non-calcareous without indications of increased salinity. Nevertheless the complete absence of autochthonous benthos suggests that the stagnant condition was permanent and not interrupted by either storms (small basin geometry) or floods (thermocline).

Environmental identifications become still more difficult if we deal with Palaeozoic or even Precambrian examples because modes of life are increasingly difficult to assess. There is no problem in tracing back the obrution deposits of the Gmünd type as long as we have echinoderms to go by. The Holzmaden type is also well represented by bituminous shales such as the Ohio shales of the Upper Devonian and Lower Carboniferous of the eastern United States (Barrow & Ettenson 1980). But we do have problems to identify the Solnhofen type.

The Silurian Mississinewa shales of Indiana, for instance, were deposited in inter-reef basins, in which a turbiditic mode of sedimentation is indicated by grading and position of the fossils at the bases of the silty dolomite beds (Erdtmann & Prezbindowski 1974). But the fauna is more diverse than in Solnhofen and includes such immobile benthic forms as sponges, brachiopods and a large variety of dendroid graptolites. Also the beds are commonly bioturbated from the top, indicating a rather diversified autochthonous benthos. But the dominant dendroid graptolite genus is *Dictyonema*, generally taken to have been pelagic. Also there are many kinds of cephalopods (little compacted, but not pyritized) and perfectly, three-dimensionally preserved crinoid crowns, whose arms and pinnules are integrated into an umbrella that would fit a pelagic mode of life well. Thus we are dealing with a similar situation to Solnhofen combined with a benthic element unfamiliar in Mesozoic lithographic limestones. There appears to be some similarity, however, to argillaceous dolomites of similar age in Wisconsin (Mikulic *et al.*, this symposium).

Also unfamiliar, from a Mesozoic point of view, are the Hunsrück Shales of the Lower Devonian of Germany (Kuhn 1961; Seilacher & Hemleben 1966), because they combine soft-part preservation (non-mineralized skeletons of arthropods; articulation) with pyritization and a diverse benthic fauna including a variety of burrows. Silty microturbidites are a common element that also allows the preservation of various arthropod tracks. Current action is also indicated by sedimentary structures and fossil orientations, while wave ripples are conspicuously absent. These features and the predominance of echinoderms suggest that obrution was the major factor; but oxygen deficiency must also have been involved, if only in the form of gyttja, to allow soft-part preservation. It is also noteworthy that the echinoderms lack *post mortem* contortions.

Still stranger forms of conservation deposits are represented in the Cambrian. Among these we note the combination of a trilobite shell hash suggestive of storm winnowing with a bituminous lithology and phosphatized arthropod 'soft parts' in the Upper Cambrian Alum Shales (Müller; this symposium) and, of course, the famous Burgess Shale (Collins *et al.* 1983; Conway Morris, this symposium). Here the problematic nature of many fossils and the absence of more familiar groups such as echinoderms and molluscs make it difficult to interpret the ecological spectrum. As a whole, the benthic element appears to dominate in the Burgess Shale, but in a washed-in rather than a benthic-event fashion. However, there are some assumed pelagics such as *Marrella*, the most common faunal element. This looks like the Solnhofen spectrum, though a non-carbonate version. The discovery of new localities in addition to the classical one suggests, however, that we deal not simply with a unique physical setting, but that a kind of time signature is also involved.

The most challenging problem in this respect is the classification of the Ediacara-type conservation deposits of Vendian age (Fedonkin, this symposium): a task that is largely independent of the taxonomic problems involved (see Seilacher (1984) and this symposium). There is no question of dealing with the impressions of soft-bodied organisms in a sandy facies that contrasts with the fine-grained sediments discussed so far.

It is true that we do know sandy obrution deposits also in the Phanerozoic. But they refer to vertebrate skeletons in continental régimes, to exceptional turbidite events (fish deposit of Sendenhorst; Siegfried 1954) or to sand-smothered echinoderms (Seilacher 1968) and not to soft-bodied organisms. The Vendian fossiliferous sandstones, in contrast, appear to be largely storm-generated (Goldring & Curnow 1967), that is, depositionally similar to many shallow marine sandstones in the Phanerozoic, in which such soft-body impressions never occur. To invoke for this unusual preservation a unique physiographic constellation is no more justified, given that Ediacara-type fossils have been found in more than 20 localities all over the world (Glaessner 1984, figure 1.8) and also in sandstones that are claimed to be deep-water turbidites

(Anderson & Conway Morris 1982). Clearly this preservation must be due to an 'anactualistic' factor (in a Phanerozoic view) of a more global nature.

It has been claimed that this factor was a lack of bioturbation. But burrows made by worm-like organisms and possibly coelenterates have been found in many Ediacaran fossil localities and with surprisingly advanced sediment feeding strategies. It should also be remembered that microbial degradation would destroy a shallowly buried soft-bodied carcass. A more relevant clue may come from sedimentological data (J. Gehling, unpublished, and personal observations in the field). They suggest the presence of a rather resistant cyanobacterial scum that protected these sand surfaces against storm erosion and amalgamation and allowed flexible sand-shards to be ripped off. This would go along with the general impression that cyanobacterial mats (with or without stromatolitic structures) were much more widespread in Precambrian than in later environments; but in exactly what way this non-uniformitarian feature should have influenced the preservation of soft-bodied organisms still remains to be explained.

5. Conceptual frame and classification

Our short review of conservation deposits is admittedly very incomplete. For instance it does not cover a group that we previously called 'conservation traps' (Seilacher & Westphal 1971), which is a holding bag for such different cases as mammoths in permafrost crevices, amber insects or bogs. Of significance in our present discussion are only the early diagenetic concretions in the form of coal balls (Scott & Rex, this symposium), 'orsten' (Müller, this symposium), Mazon Creek nodules (Baird *et al.*, this symposium) or the Lower Cretaceous fish concretions from Brazil (Müller, this symposium). Certainly, such concretions are important in that they facilitate fossil hunting and that they provide, through the right preparation techniques, relatively uncompacted fossils. But in a genetic sense, they are only a subset of stagnation (and obrution?) deposits.

In the Posidonia Shales, for instance, beautifully preserved reptiles, fish and coleoid cephalopods come from such concretions; but most of the nodules are barren and most of the fossils are found in layers in which no such concretions occur.

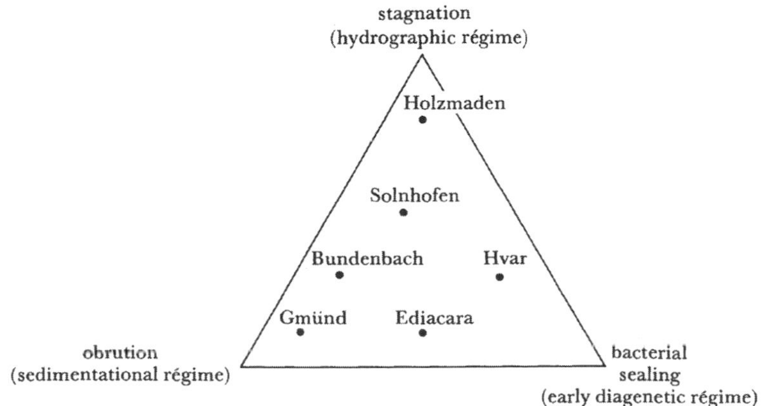

Figure 11. Among the many factors involved in the formation of conservation deposits, stagnation, obrution and cyanobacterial sealing are considered the most dominant. They define a conceptual continuum, into which particular examples may be 'mapped'.

TABLE 1. TENTATIVE QUESTIONNAIRE FOR CONSERVATION DEPOSITS

Tübingen questionnaire
Fossillagerstätten
Conservation deposits

locality:
age:

(1) basin situation
size (km): 10^{-1} 10^{0} 10^{1} 10^{2} 10^{3}
setting:
oceanic
epicontinental
terrestrial
origin:
tectonic
volcanic
astroblemic
subrosional
recifal
sedimentary
glacial
geographic frame:
limestones
clastics
crystalline

(2) stratigraphy
thickness (m): 0.01 0.1 1 10 100
duration in absolute time:
vertical context:
transgressive
peak transgression
regressive
fining-up cycles
coarsening-up cycles
lateral sequence:

(3) sedimentology
lithology:
biograins (coccoliths, forams, radiolarians, spicules, etc.):
sedimentary structures:
lamination (varves, algal, etc.)
slump horizons
graded horizons
current ripples
wave ripples
emersion marks (tidal channels, mud cracks, etc.)

(4) geochemistry
evaporitic precipitates (aragonite, calcite, dolomite, gypsum, halite):
pyrite concretions (globular, discoid):
isotopic deviations:
C_{org} (particles, kerogene, bitumen):

(5) taxonomic spectrum
dominated by (priority):
echinoderms
cephalopods
vertebrates
crustaceans
trilobites
others:

(6) ecological spectrum
burrows (episodic, continuous):
tracks (with or without bodies):
endobenthos:
hemisessile
vagile
epibenthos (episodic, continuous):
sessile
vagile
able to swim
pelagics:
nekton
floaters
epiplankton (on wood, cephalopod shells, live or dead)
terrestrial organisms:
tracks
skeletons
land plants (twigs, leaves, trunks)
flyers

(7) necrolytic criteria
death marches:
landing marks (live, dead):
soft parts (impressions, films):
organic cuticles:
articulation (vertebrates, echinoderms, arthropods, bivalves, aptychi):

(8) stratinomic criteria
life positions:
roll marks (of what?):
convex up or down:
current orientation (azimuth):
wave orientation (azimuth):

(9) diagenetic criteria
aragonite preservation (what?):
early aragonite solution (composite casts):
pyritic steinkerns:
concretionary cementation (nucleus, pressure shadow, buckle, pedestal):
compactional deformation
incoalation:
phosphatization (of what?):
replacement (shells, bones):

General conclusions:
stagnation (thermal or halocline):
obrution:
algal sealing:

In spite of its incompleteness this review has shown, however, that, apart from the two principal factors of stagnation and obrution and their various combinations, a large number of palaeogeographic, biological, sedimentological, diagenetic and time factors may contribute to the formation of conservation deposits. The result is that each case has a 'personality' that defies rigorous classification.

Of course, there is some virtue in typological groupings. Thus we may distinguish between marine, hypersaline, lacustrine and swamp deposits, or between vertebrate, echinoderm, arthropod and plant Lagerstätten. Or group into Solnhofen, Holzmaden, Bundenbach and Burgess types of deposits. Grouping according to age (Conway Morris, this symposium; Walker & Diehl, this symposium) may also be revealing. But while being useful to emphasize patterns, such classifications can never be binding, because unlike in taxonomy there will be no agreement as to which criteria should have primacy.

Nevertheless we should not be content with descriptive registrations. Instead of expecting the classification to provide a standard set of pigeon holes, we should better consider it as a conceptual framework for heuristic purposes. In figure 11 we propose to use the critical hydrographic, sedimentational and early diagenetic exponents to define a triangular space into which each particular case can be mapped. In addition, we present a tentative questionnaire (table 1) in order to standardize this mapping job, in which by necessity a large number of specialists must be involved.

What lies in front of us, is more than fossil hunting, the unravelling of taxonomic relationships and facies analysis. Expanded over the whole fossil record, a comparative analysis of fossil Lagerstätten could become a genuine contribution of palaeontologists to the integrated view of Earth history.

References

Aigner, T. 1980 Biofabrics and stratinomy of the Lower Kimmeridge Clay (U. Jurassic, Dorset, England). *N. Jb. Geol. Paläont. Abh.* **159**, 324–338.

Anderson, M. M. & Conway Morris, S. 1982 A review, with descriptions of four unusual forms, of the soft-bodied fauna of the Conception and St John's Groups (late Precambrian), Avalon Peninsula, Newfoundland. *Third N. Am. Paleont. Conv., Proc.* **1**, 1–8.

Barrow, L. S. & Ettensohn, F. R. 1980 *A bibliography of the paleontology and paleoecology of the Devonian–Missisippian black-shale sequence in North America.* 86 pages. Morgantown, West Virginia: U.S. Department Energy, Information Center.

Barthel, K. W. 1978 *Solnhofen. Ein Blick in die Erdgeschichte.* 393 pages. Thun: Ott-Verlag.

Bayer, U. & Seilacher, A. 1985 Sedimentary and evolutionary cycles. In Friedmann, G. M. (ed.) *Lecture notes in earth sciences*, vol. 1, 465 pages. Berlin, Heidelberg, New York, Tokyo: Springer.

Bernier, P., Barale, G. & Bourseau, J.-P. 1983 Le chantier de fouilles paleoecologiques de Kimmeridgien supérieur de Cerin (Jura, France). *Int. Congr. Paleoecol. Lyon, Guide Exc.* **1**, 1–19.

Brenner, K. 1976 Ammoniten-Gehäuse als Anzeiger von Paläo-Strömungen. *N. Jb. Geol. Paläont., Abh.* **151**, 101–118.

Bromley, R. G. & Ekdale, A. A. 1984 *Chondrites*: a trace fossil indicator of anoxia in sediments. *Science, Wash.* **224**, 872–873.

Collins, D., Briggs, D. & Conway Morris, S. 1983 New Burgess Shale fossil sites reveal Middle Cambrian faunal complex. *Science, Wash.* **222**, 163–167.

Einsele, G. & Seilacher, A. (ed.) 1982 *Cyclic and event stratification.* 536 pages. Berlin, Heidelberg, New York: Springer.

Erdtmann, B.-E. & Prezbindowski, D. 1974 Niagaran (Middle Silurian) interreef fossil burial environments in Indiana. *N. Jb. Geol. Paläont., Abh.* **144**, 342–372.

Esteban Cerda, M., Calzada, S. & Via Boada, L. 1977 Ambiente depositionalde los yacimentos fosiliferos del Muschelkalk superior de Alcover–Mont-Ral. *Cuadernos Geol. Iberica* **4**, 189–200.

Freels, D. 1975 Plattenkalk-Becken bei Pietraroia (Prov. Benevento, S-Italien) als Voraussetzung einer Fossillagerstättenbildung. *N. Jb. Geol. Paläont., Abh.* **148**, 320–352.

Glaessner, M. F. 1984 *The dawn of animal life. A biohistorical study.* 244 pages. Cambridge: University Press.

Goldring, R. & Curnow, C. N. 1967 The stratigraphy and facies of the late Precambrian at Ediacara, South Australia. *J. geol. Soc., Aust.* **14**, 195–214.

Groiss, J. T. 1967 Mikropaläontologische Untersuchungen der Solnhofener Schichten im Gebiet um Eichstätt (Südliche Frankenalb). *Erlanger geol. Abh.* **66**, 75–92.

Hemleben, C. 1977*a* Rote Tiden und die oberkretazischen Plattenkalke im Libanon. *N. Jb. Geol. Pal., Mh.*, pp. 239–255.

Hemleben, C. 1977*b* Autochthone und allochthone Sedimentanteile in den Solnhofener Plattenkalken. *N. Jb. Geol. Pal., Mh.*, pp. 257–271.

Hemleben, C. & Freels, D. 1977*a* Algen-laminierte und gradierte Plattenkalke in der Oberkreide Dalmatiens (Jugoslawien). *N. Jb. Geol. Paläont., Abh.* **154**, 61–93.

Hemleben, C. & Freels, D. 1977*b* Fossilführende dolomitisierte Plattenkalke aus dem 'Muschelkalk superior' bei Montral (Prov. Tarragona, Spanien). *N. Jb. Geol. Paläont., Abh.* **154**, 186–212.

Hudson, J. D. 1982 Pyrite in ammonite-bearing shales from the Jurassic of England and Germany. *Sedimentology* **29**, 639–667.

Hückel, U. 1970 Die Fischschiefer von Haqel and Hjoula in der Oberkreide des Libanon. *N. Jb. Geol. Paläont., Abh.* **135**, 113–149.

Janicke, V. 1969 Untersuchungen über den Biotop der Solnhofener Plattenkalke. *Mitt. bayer. Staatssamml. Paläont. hist. Geol.* **9**, 117–181.

Jordan, R. 1975 Salz- und Erdöl/Erdgas-Austritt als Fazies bestimmende Faktoren im Mesozoikum Nordwest-Deutschlands. *Geol. Jb., Reihe A*, Heft **13**, 64 pages.

Kauffman, E. G. 1981 Ecological reappraisal of the German Posidonienschiefer (Toarcian) and the stagnant basin model. In *Communities of the past* (ed. J. Gray, A. J. Boucot & W. B. N. Berry); pp. 311–381. Stroudsburg Pa: Dowden.

Keupp, H. 1977 Ultrafazies und Genese der Solnhofener Plattenkalke (Oberer Malm, südliche Frankenalb). *Abh. naturhistor. Ges. Nürnberg* **37**, 128 pages.

Krumbein, W. E. 1983 *Microbial chemistry.* 330 pages. Oxford: Blackwell.

Kuhn, O. 1961 Die Tierwelt der Bundenbacher Schiefer. *Neue Brehm-Bücherei* **274**, 48 pages. Wittenberg.

Mayr, F. X. 1967 Paläobiologie und Stratinomie der Plattenkalke der Altmühlalb. *Erlanger geol. Abh.* **67**, 40 pages.

Rieber, H. 1973 Ergebnisse paläontologisch-stratigraphischer Untersuchungen in der Grenzbitumenzone (Mittlere Trias) des Monte San Giorgio (Kanton Tessin, Schweiz). *Ecl. geol. Helvetiae* **66**, 667–685.

Riegraf, W. 1977 *Goniomya rhombifera* (Goldfuss) in the Posidonia Shales (Lias epsilon). *N. Jb. Geol. Paläont. Mh.*, pp. 446–448.

Riegraf, W. 1984 *Der Posidonienschiefer. Biostratigraphie, Fauna und Fazies des südwestdeutschen Untertoarciums.* 195 pages. Stuttgart: Enke.

Riegraf, W. 1985 Mikrofauna, Biostratigraphie und Fazies im unteren Toarcium Süddeutschlands und Vergleiche mit benachbarten Gebieten. *Tübinger Mikropal. Mitt.* **3**, 232 pages.

Rosenkranz, D. 1971 Zur Sedimentologie und Ökologie von Echinodermen-Lagerstätten. *N. Jb. Geol. Paläont., Abh.* **138**, 221–258.

Savrda, C. E., Bottjer, D. J. & Gorsline, D. S. 1984 Development of a comprehensive oxygen-deficient marine biofacies model: evidence from Santa Monica, San Pedro, and Santa Barbara Basins, California Continental Borderland. *Bull. Am. Ass. Petrol. Geol.* **68**, 1179–1192.

Seilacher, A. 1963 Umlagerung und Rolltransport von Cephalopoden-Gehäusen. *N. Jb. Geol. Paläont., Mh.*, pp. 593–619.

Seilacher, A. 1968 Origin and diagenesis of the Oriskany Sandstone (Lower Devonian, Appalachians) as reflected in its shell fossils. *Recent developments in carbonate sedimentology in Central Europe*, pp. 175–185. Berlin, Heidelberg, New York: Springer.

Seilacher, A. 1982 Posidonia shales (Toarcian, S. Germany) – stagnant basin model revalidated. In *Paleontology, essential of historical geology*, Internat. meeting Venice 1981 (ed. E. Montanaro-Gallitelli), pp. 25–55. Modena.

Seilacher, A. 1984 Late Precambrian and early Cambrian Metazoa: preservational or real extinction? In *Patterns of change in earth evolution*, Dahlem-Konferenzen 1984 (ed. H. D. Holland & A. F. Trendall), pp. 159–168. Berlin, Heidelberg, New York, Tokyo: Springer.

Seilacher, A. 1985 The Jeram model: event condensation in a modern intertidal environment. In *Sedimentary and evolutionary cycles*, Lecture Notes in Earth Sciences (ed. G. M. Friedmann), vol. 1, 336–342. Berlin, Heidelberg, New York, Tokyo: Springer.

Seilacher, A., Andalib, F., Dietl, G. & Gocht, H. 1976 Preservational history of compressed Jurassic ammonites from Southern Germany. *N. Jb. Geol. Paläont., Abh.* **152**, 307–356.

Seilacher, A. & Hemleben, C. 1966 Spurenfauna und Bildungstiefe der Hunsrückschiefer (Unterdevon). *Notizbl. hess. Landesamt Bodenforsch* **94**, 40–53.

Seilacher, A. & Westphal, F. 1971 Fossil-Lagerstätten. In *Sedimentology of parts of Central Europe*, Guidebook 8. Int. Sediment. Congr., pp. 327–335. Heidelberg.

Siegfried, P. 1954 Die Fisch-Fauna des westfälischen Ober-Senons. *Palaeontographica* A **106**, 1–36.
Sorbini, L. 1983 The fossil fish deposit of Bolca (Verona), Italy. *Int. Congr. Paleoecol., Lyon, Guide Exc.* 11A, 23–33.
Temmler, H. 1964 Über die Schiefer- und Plattenkalke des Weissen Jura der Schwäbischen Alb (Württemberg). *Arb. Geol.-Paläont. Inst. der TH*, N.F. **43**, 107 pages.
Temmler, H. 1966 Über die Nusplinger Fazies des Weissen Jura der Schwäbischen Alb (Württemberg). *Z. deutsch. geol. Ges.* **116**, 891–907.
Via Boada, L., Villalta, J. F. & Esteban Cerda, M. 1977 Paleontologia y paleoecologia de los yacimentos fosiliferos del Muschelkalk superior entre Alcover y Mont-Ral. *Cuadernos Geol. Iberica* **4**, 247–256.

Discussion

R. Riding (*Department of Geology, University College, Cardiff CF1 1XL, U.K.*). I wish to query use of the term obrution. Reference to stagnation and smothering (obrution) as alternative modes of formation of conservation Lagerstätten is slightly confusing because stagnation indicates the nature of the environment whereas smothering is a type of asphyxiation. It would be clearer in this case to refer to rapid burial rather than to smothering. Stagnation and rapid burial could then be regarded as different processes, either of which can result in the asphyxiation of organisms. Asphyxiation occurs above the substrate in the case of stagnation and below it in the case of smothering. This usage then clearly distinguishes environmental conditions (stagnation, rapid burial) from the actual mode of death which is, in both cases, asphyxiation.

E. N. K. Clarkson (*Grant Institute of Geology, University of Edinburgh, West Mains Road, Edinburgh EH9 3JW, U.K.*). The Lower Lias obrution deposit at Gmünd is, as Dr Seilacher has noted, dominated by echinoderms and he had suggested that in such a case the active elements in the benthos might have escaped, whereas the echinoderms could not.

Is it possible to distinguish such an obrution deposit from one in which the original fauna was dominated by echinoderms and little else? I ask this with particular reference to some horizons rich in intact echinoderms in the Scottish Silurian.

H. B. Whittington, F.R.S. (*Department of Earth Sciences, Sedgwick Museum, University of Cambridge*). The Burgess Shale fauna includes the hard parts of characteristic Middle Cambrian animals – trilobites, sponges, brachiopods, molluscs, hyolithids and echinoderms – as well as a remarkable soft-bodied fauna. I have argued in detail (Whittington 1971) that *Marrella* was a benthic animal, the thousands of specimens having been buried in varied orientations in the deposit resulting from a turbidity current. The highly fossiliferous layers in the Phyllopod bed of the Burgess Shale appear to originate from such a mode of transport and burial (Whittington 1980). Comparison with the Solnhofen deposit would thus involve consideration of a similar mechanism for its formation.

References

Whittington, H. B. 1971 Redescription of *Marrella splendens*, (Trilobitoidea) from the Burgess Shale, Middle Cambrian, British Columbia. *Geol. Surv. Canada Bull.* **209**, 19–20.
Whittington, H. B. 1980 The significance of the fauna of the Burgess Shale, Middle Cambrian, British Columbia. *Proc. Geol. Ass.* **91**, 129–132.

A. Seilacher. 'Obrution' means rapid burial, whether this process only preserved carcasses or also killed the organisms. But only in the latter case can we expect a selective preservation. In the meantime, the analysis of clustered trilobites that appear to have been selectively killed and then buried by the muddy clouds of storm events has added another beautiful example of such obrution deposits. Whether the striking absence or under-representation of non-trilobites or non-echinoderms in these cases be considered an original feature or the outcome of better resistivity to such accidents is a matter of ecological taste.

Reference

Speyer, S. E. & Brett, C. E. 1985 Clustered trilobite assemblages in the Middle Devonian Hamilton Group. *Lethaia* **18**, 85–103.

Contributors

Kenneth D. Angielczyk
Department of Geology
Field Museum of Natural History
1400 South Lake Shore Drive
Chicago, IL 60605

Richard B. Aronson
Department of Biological Sciences
Florida Institute of Technology
150 West University Boulevard
Melbourne, FL 32901

Catherine Badgley
Department of Ecology and Evolutionary Biology and Museum of Paleontology
University of Michigan
1109 Geddes Road
Ann Arbor, MI 48109

Richard K. Bambach
Department of Paleobiology
National Museum of Natural History
Smithsonian Institution [NHB, MRC 121]
P.O. Box 37012
Washington, DC 20013

Anthony D. Barnosky
Department of Integrative Biology and Museums of Paleontology and Vertebrate Zoology
University of California, Berkeley
Berkeley, CA 94305

Anna K. Behrensmeyer
Department of Paleobiology
National Museum of Natural History
Smithsonian Institution [NHB, MRC 121]
P.O. Box 37012
Washington, DC 20013

Claire M. Belcher
Department of Geography
College of Life and Environmental Sciences
University of Exeter
Hatherly Laboratories
Prince of Wales Road
Exeter, EX4 4PS, UK

Carlton E. Brett
Department of Geology
510 Geology-Physics Building
University of Cincinnati
Cincinnati, OH 45221

Derek E. G. Briggs
Department of Geology and Geophysics
Peabody Museum of Natural History
Yale University
P.O. Box 208118
New Haven, CT 06520

Andrew M. Bush
Department of Ecology and Evolutionary Biology and Center for Integrative Geosciences
University of Connecticut
75 North Eagleville Road
Storrs, CT 06269

Marty Buzas
Department of Paleobiology
National Museum of Natural History
Smithsonian Institution [NHB, MRC 121]
P.O. Box 37012
Washington, DC 20013

Ellen D. Currano
Departments of Botany and Geology and Geophysics
University of Wyoming
Laramie, WY 82071

Larisa R. G. DeSantis
Department of Earth and Environmental Sciences
5721 Science and Engineering Building
Vanderbilt University
Nashville, TN 37240

Steven D'Hondt
Graduate School of Oceanography
University of Rhode Island
Narragansett Bay Campus
South Ferry Road
Narragansett, RI 02882

Mary Droser
Department of Earth Sciences
Geology 1464
University of California, Riverside
Riverside, CA 92521

Seth Finnegan
Department of Integrative Biology
University of California Museum of Paleontology
University of California, Berkeley
Berkeley, CA 94305

Michael Foote
Department of the Geophysical Sciences
University of Chicago
5734 South Ellis Avenue
Chicago, IL 60637

David L. Fox
Department of Earth Sciences
University of Minnesota
310 Pillsbury Drive SE
Minneapolis, MN 55455

Ian Glasspool
Department of Geology
Field Museum of Natural History
1400 South Lake Shore Drive
Chicago, IL 60605

Eric C. Grimm
Illinois State Museum
Research and Collections Center
1011 East Ash Street
Springfield, IL 62703

Leo J. Hickey
Department of Geology and Geophysics
Yale University
210 Whitney Avenue
New Haven, CT 06511

Patricia A. Holroyd
University of California Museum of Paleontology
University of California, Berkeley
1101 Valley Life Sciences Building
Berkeley, CA 94720-4780

Gene Hunt
Department of Paleobiology
National Museum of Natural History
Smithsonian Institution [NHB, MRC 121]
P.O. Box 37012
Washington, DC 20013

Nathan Jud
Florida Museum of Natural History
University of Florida
3215 Hull Road
Gainesville, FL 32611

Patricia H. Kelley
Department of Earth Sciences
University of North Carolina at Wilmington
601 South College Road
Wilmington, NC 28403

Susan Kidwell
Department of the Geophysical Sciences
University of Chicago
5734 South Ellis Avenue
Chicago, IL 60637

Conrad C. Labandeira
Department of Paleobiology
National Museum of Natural History
Smithsonian Institution [NHB, MRC 121]
P.O. Box 37012
Washington, DC 20013

Cindy V. Looy
Department of Integrative Biology
University of California Museum of Paleontology
University of California, Berkeley
Berkeley, CA 94305

S. Kathleen Lyons
School of Biological Sciences
University of Nebraska–Lincoln
Lincoln, NE 68502

Arnold I. Miller
Department of Geology
University of Cincinnati
609 Geology-Physics Building
Cincinnati, OH, 45221

Thomas D. Olszewski
Department of Geology and Geophysics
Texas A&M University
Halbouty Room 163
MS 3115
College Station, TX 77843

Mark E. Patzkowsky
Department of Geosciences
539 Deike Building
Penn State University
University Park, PA 16802

Shanan E. Peters
Department of Geoscience
University of Wisconsin–Madison
1215 West Dayton Street
Madison, WI 53706

Matthew G. Powell
Department of Earth and Environmental Sciences
Juniata College
1700 Moore Street
Huntingdon, PA 16652

Surangi Punyasena
Department of Plant Biology
Program in Ecology, Evolution and Conservation
University of Illinois at Urbana-Champaign
139 Morrill Hall
505 South Goodwin Avenue
Urbana, IL 61801

Andy Purvis
Division of Biology
Silwood Park Campus
Imperial College London
Ascot, Berkshire, SL5 7PY, UK

Nicholas D. Pyenson
Department of Paleobiology
National Museum of Natural History
Smithsonian Institution [NHB, MRC 121]
P.O. Box 37012
Washington, DC 20013

Raymond R. Rogers
Geology Department
Macalester College
1600 Grand Avenue
St. Paul, MN 55105

Hallie J. Sims
1 Bella Vista Place
Iowa City, IA 52245

Dena M. Smith
Department of Geological Sciences
CU Museum of Natural History
265 UCB
University of Colorado
Boulder, CO 80309-0265

Caroline A. E. Strömberg
Department of Biology and Burke Museum of Natural History and Culture
University of Washington
24 Kincaid Hall, Box 351800
Seattle, WA 98195

Hans-Dieter Sues
Department of Paleobiology
National Museum of Natural History
Smithsonian Institution [NHB, MRC 121]
P.O. Box 37012
Washington, DC 20013

Shinya Sugita
Institute of Ecology
Tallinn University
Tallinn 10120, Estonia

Jessica Theodor
Department of Biological Sciences
University of Calgary
2500 University Drive NW
Calgary AB, Canada T2N 1N4

Ellen Thomas
Department of Geology and Geophysics
Yale University
P.O. Box 208109
New Haven CT 06520-8109

Adam Tomašových
Geological Institute
Slovak Academy of Sciences
Dubravska cesta 9
Bratislava, 84005, Slovakia

Susumu Tomiya
Department of Geology
Field Museum of Natural History
1400 South Lake Shore Drive
Chicago, IL 60605

Noreen Tuross
Department of Human Evolutionary Biology
Harvard University
11 Divinity Avenue
Cambridge, MA 02138

Peter J. Wagner
Department of Earth and Atmospheric Sciences
School of Biological Sciences
University of Nebraska–Lincoln
Lincoln NE 68502

Peter Wilf
Department of Geosciences
539 Deike Building
Penn State University
University Park, PA 16802

Scott Wing
Department of Paleobiology
National Museum of Natural History
Smithsonian Institution [NHB, MRC 121]
P.O. Box 37012
Washington, DC 20013

Author Index

Subject Index